Powers and Compensation in Circuits with Nonsinusoidal Current

Powers and Compensation in Circuits with Nonsinusoidal Current

Leszek S. Czarnecki

OXFORD UNIVERSITY PRESS

Great Clarendon Street, Oxford, OX2 6DP,
United Kingdom

Oxford University Press is a department of the University of Oxford.
It furthers the University's objective of excellence in research, scholarship,
and education by publishing worldwide. Oxford is a registered trade mark of
Oxford University Press in the UK and in certain other countries

Published in the United States of America by Oxford University Press
198 Madison Avenue, New York, NY 10016, United States of America

British Library Cataloguing in Publication Data

Data available

Library of Congress Control Number: 2023941304

ISBN 9780198879206
ISBN 9780198879213 (pbk.)

DOI: 10.1093/oso/9780198879206.001.0001

Printed and bound by
CPI Group (UK) Ltd, Croydon, CR0 4YY

The manufacturer's authorised representative in the EU for product safety is
Oxford University Press España S.A. of El Parque Empresarial San Fernando de Henares, Avenida
de Castilla, 2 – 28830 Madrid (www.oup.es/en or
product.safety@oup.com). OUP España S.A. also acts as importer into Spain
of products made by the manufacturer.

To my wife Maria, the "Driving Force" behind this book

Preface

Electrical circuits and systems provide the "blood", the electric energy for our homes, commercial buildings, and technological processes. This energy costs. Even if it comes from unlimited energy reservoirs, such as river flows, wind, sunlight, sea waves, or geothermal processes, its "harvesting" is not free. It has a price. This "harvesting" can deteriorate or even damage our natural environment. Moreover, the priceless resources of the Earth, such as coal, crude oil, or natural gas, can be wasted to produce electric energy. Therefore, with the increasing social concern about where our energy comes from and at what cost, we should use it as sparingly as possible. Also, the improvement in the effectiveness of electric energy transfer becomes more and more important.

We cannot deliver this energy to customers without some energy loss, thus much more energy has to be produced than is used. The difference between the produced and the consumed energy, meaning the energy wasted as losses, is the lowest when the voltages and currents are sinusoidal, without any phase shift. In three-phase circuits they should be, moreover, symmetrical. These situations offer the lowest demands as to the capability and the cost of energy-transmitting equipment.

Sinusoidal, symmetrical currents that are in phase with the supply voltage are drawn only by balanced purely resistive loads supplied with a symmetrical sinusoidal voltage, however. Only three-phase electrical heaters draw such currents. Electric energy is too expensive to be wasted on heating, however. Other common loads do not draw currents that are in phase with the supply voltage. These currents do not have to be, moreover, sinusoidal and symmetrical.

The phase shift between the supply voltage and the current can be eliminated and sinusoidal waveforms and current symmetry can be restored by compensators. The knowledge of the power properties of circuits with nonsinusoidal and/or asymmetrical currents is necessary in order to develop these kinds of compensators.

Students taking university courses on circuits are taught mainly about electric powers and compensation when voltages and currents are sinusoidal. At the same time, LED or fluorescent bulbs, computers, printers, TV and audio equipment, microwaves, or induction stoves draw currents that are very far from sinusoidal ones. Distribution systems supply very high-power rectifiers in chemical plants and power electronics-driven variable speed drives in manufacturing processes and transportation. From loads in our homes to ultra-high-power metallurgic plants with power electronics-driven crushers, mills, and arc furnaces, the power used is sometimes comparable to cities of 1 million people and these-electrical loads cause current distortion and asymmetry.

This book provides solid fundamentals of circuit analysis and the development of compensators that can operate at nonsinusoidal voltages and currents in the presence of load imbalance. Such an analysis provides reliable conclusions as to energy flow-related physical phenomena in such systems, which in turn create a well-founded basis for developing methods of compensation. Before this analysis, the credibility of various opinions on energy flow-related phenomena and compensation methods should be verified and misconceptions identified. Therefore, a substantial part of this book is dedicated to the critical analysis of existing interpretations of power-related phenomena and their conclusions upon compensation.

The book is dedicated to scientists, engineers, graduate and undergraduate students interested in energy flow in systems saturated with power electronics converters, renewable sources of energy, or those interested in microgrid development. Such systems are usually not very strong, and consequently, they are susceptible to voltage and current distortion and asymmetry. The assumption that voltages and currents in such systems are sinusoidal and symmetrical can result in substantial errors.

Investigations on power properties of circuits with nonsinusoidal voltages and currents have more than a century-long history, with hundreds of scientists involved and several "schools" that explain power phenomena and define power quantities in such circuits in different ways. Mutually conflicting conclusions regarding the possibility and methods of compensation have been drawn in these investigations and disseminated in the power engineering community. Different quantities have been suggested to be regarded as reactive, apparent, or other powers. These "schools" also differ concerning the physical interpretations of power-related phenomena in electrical circuits. Some of them and their effects are misinterpreted. For example, the opinion, which is very common in the electrical engineering community, that energy oscillations between energy sources and customer loads are responsible for the degradation of the effectiveness of energy transfer is one of the most remarkable misinterpretations. Even the presence of such oscillations is controversial. The book presents a detailed analysis and interpretation of the various, energy transfer-related phenomena and their effect on the methods of compensation in circuits with nonsinusoidal and asymmetrical voltages and currents.

Leszek S. Czarnecki

Biography

Leszek S. Czarnecki, IEEE Life Fellow, A.M. Lopez Distinguished Professor at Louisiana State University, Titled Professor of Technological Sciences, granted by the President of Poland (Wikipedia). He received an M.Sc., Ph.D., and D.Sc. degrees in electrical engineering from the Silesian University of Technology, Poland. For two years he was with the Power Engineering Section of the National Research Council (NRC) of Canada, as a Research Officer, and for two years with the Electrical Engineering Department at Zielona Gora University, Poland. In 1989 Dr. Czarnecki joined the Electrical and Computer Engineering Department of Louisiana State University (LSU), Baton Rouge. For developing power theory of three-phase systems with nonsinusoidal and asymmetrical voltages and currents and for methods of compensation of such systems he was elected to the grade of Fellow IEEE in 1996. The Currents' Physical Components (CPC)-based power theory, which explains all physical phenomena that specify power properties of electrical systems and creates fundamentals for compensation in circuits of any complexity, was Dr. Czarnecki's major contribution to electrical engineering, for which he was nominated to the IEEE Proteus Charles Steinmetz Award. Dr. Czarnecki was decorated by the President of Poland for his public activity in the United States, aimed at the acceptance of Poland into NATO, with the Knight Cross of the Medal of Merit of the Republic of Poland.

An avid mountaineer and underwater photographer, he has climbed, without oxygen support, Lhotse (No. 4 in the World) in the Himalayas (8350 m); the main ridge of the Ruwenzori in Africa (nineteen summits of the average high of 5000 m), Kilimanjaro, and Mt. Kenya; traversed on ski (500 km) Spitsbergen in the deep Arctic; climbed in the Alps and Andes; climbed McKinley, the highest mountain in North America, solo, and traveled to Antarctica. As a photographer Dr. Czarnecki dived at coral reefs of Red Sea, Great Reef Barrier, Philippines, Caribbean, Belize, Tahiti, Maui, Bonaire, Hawaii, New Zealand and at reefs of Mombasa.

Contents

List of Abbreviations

3p3w	Three-phase three-wire (circuit)
3p4w	Three-phase four-wire (circuit)
AC	Alternating current
ACN	Algebra of complex numbers
A/D	Analog to Digital (converter)
CPC	Currents' Physical Components
CPT	Conservative power theory
crms	Complex rms (value)
CTHD	Current total harmonic distortion
DA	Data acquisition (system)
DC	Direct current
DFT	Discrete Fourier Transform
DSP	Discrete signals processing (system)
ECP	Energy Conservation Principle
EE	Electrical engineering (community)
EMI	Electromagnetic interference
FBD	Fryze-Buchholz-Depenbrock (Method)
FFT	Fast Fourier Transform
FVL	Fast varying load
GA	Geometric algebra
HBC	Harmonics blocking compensator
HC	Hybrid compensator
HF	High frequency
HGL	Harmonics generating load
HV	High voltage
IEC	International Electrotechnical Commission
IGBT	Insulated Gate Bipolar Transistors
IRP	Instantaneous Reactive Power
ISC	Inverter Switches Control
KCL	Kirchhoff's current law
KVL	Kirchhoff's voltage law
K&M	Kusters and Moore's (power theory)
LC	Inductive-capacitive (circuit)
LF/HF	Low-frequency/high-frequency (compensator)
LQ	Loading quality
LTI	Linear, time-invariant (load)
mF	miliFarad
MOSFET	Metal Oxide Semiconductor Field Effect Transistors
MVA	Megavoltampers
NRC	National Research Council
nv&c	Nonsinusoidal voltages and currents

PBP	Power Balance Principle
PCC	point of common coupling
PE	Power electronics
PF	Power factor
PQ, SQ, LQ	Power, supply, and loading quality
PR	Positive Real (function)
PT	Power theory
PV	Poynting Vector
PWM	Pulse width modulated (inverter)
RBC	Reactance balancing compensator
RC	Resistive-capacitive (load)
RHF	Resonant harmonic filter
RL	Resistive-inductive (load)
RLC	Resistive-inductive-capacitive (load)
rms	Root mean square (value)
S&Z	Shepherd and Zakikhani's (power theory)
SC	Switching compensator
SQ	Supply quality
SSC	Shunt switching compensator
sv&c	Sinusoidal voltages and currents
TCS	Thyristor-controlled susceptance
TER	Two-element reactance (compensator)
THD	Total harmonic distortion
TRIAC	Bidirectional triode thyristor
TSI	Thyristor switched inductor
TT	Tellegen Theorem
VA	Volt-ampers
VPC	Voltage physical components
VSI	Voltage Source Inverter
VTHD	Voltage total harmonic distortion

List of Symbols

$t, \tau, t_k, \Delta\, t, T$ time, time delay, instant of time, time interval, period

f function, frequency

ω, ω_k variable radial frequency, fixed radial frequency

ω_1, ω_n fundamental radial frequency, radial frequency of the nth order harmonic

$i(t), i, j(t), j$ current instantaneous value

$u(t), u, e(t), e$ voltage instantaneous value

$\bar{u}, \dot{u}$ voltage integral and derivative

U, I, J rms value of sinusoidal voltage u and currents i, j

$||u||, ||i||, ||j||$ rms values of nonsinusoidal voltage u and currents i, j

$\boldsymbol{U}, \boldsymbol{I}, \boldsymbol{X}$ complex rms value of sinusoidal voltage, current, and quantity $x(t)$

R, S, T or a, b, c three-phase terminals

$i_{\mathrm{R}}, i_{\mathrm{S}}, i_{\mathrm{T}}$ line R, S, and T current instantaneous values

$u_{\mathrm{R}}, u_{\mathrm{S}}, u_{\mathrm{T}}$ line-to-neutral voltage instantaneous value of lines R, S, and T

$p(t), P, Q, S, \boldsymbol{C}$ instantaneous, active, reactive, apparent, and complex powers

S_{sc} short-circuit power

$D_{\mathrm{s}}, D_{\mathrm{u}}$ scattered and unbalanced powers

Q_{B}, D Budeanu's reactive and distortion powers

$i_{\mathrm{a}}, i_{\mathrm{s}}, i_{\mathrm{r}}$ active, scattered, and reactive currents

$\boldsymbol{i} \stackrel{\mathrm{df}}{=} \begin{bmatrix} i_{\mathrm{R}} \\ i_{\mathrm{S}} \\ i_{\mathrm{T}} \end{bmatrix}$ vector of three-phase line current instantaneous value

$\mathfrak{i} \stackrel{\mathrm{df}}{=} \begin{bmatrix} i_{\mathrm{R}} \\ i_{\mathrm{S}} \end{bmatrix}$ reduced vector of three-phase line current instantaneous values

$\boldsymbol{u} \stackrel{\mathrm{df}}{=} \begin{bmatrix} u_{\mathrm{R}} \\ u_{\mathrm{S}} \\ u_{\mathrm{T}} \end{bmatrix}$ vector of three-phase voltage instantaneous value

$\mathfrak{u} \stackrel{\mathrm{df}}{=} \begin{bmatrix} u_{\mathrm{R}} \\ u_{\mathrm{S}} \end{bmatrix}$ reduced vector of three-phase voltage instantaneous values

$\boldsymbol{U} \stackrel{\text{df}}{=} \begin{bmatrix} U_{\text{R}} \\ U_{\text{S}} \\ U_{\text{T}} \end{bmatrix}$ vector of crms values of three-phase voltages

$\boldsymbol{I} \stackrel{\text{df}}{=} \begin{bmatrix} I_{\text{R}} \\ I_{\text{S}} \\ I_{\text{T}} \end{bmatrix}$ vector of crms values of three-phase currents

$||\boldsymbol{u}||, ||\boldsymbol{i}||, ||\boldsymbol{j}||$ three-phase rms values of voltage vectors $\boldsymbol{u}$ and currents vectors $\boldsymbol{i}, \boldsymbol{j}$

(x, y) scalar product of $x(t)$ and $y(t)$ quantities

$(\boldsymbol{x}, \boldsymbol{y})$ scalar product of $\boldsymbol{x}(t)$ and $\boldsymbol{y}(t)$ three-phase vectors

n harmonic order

N_0 set of all harmonic orders n, $N_0 \stackrel{\text{df}}{=} \{0, 1, 2, 3...\}$

N set of all harmonic orders n without $n = 0$; $N \stackrel{\text{df}}{=} \{1, 2, 3...\}$

N_{d} set of all harmonic orders n without $n = 1$; $N_{\text{d}} \stackrel{\text{df}}{=} \{0, 2, 3...\}$

$\delta_{\text{u}}, \delta_{\text{i}}$ voltage and current Total Harmonic Distortion

U_n, I_n, X_n crms value of nth order harmonic of voltage, current, and quantity $x(t)$

$\boldsymbol{x}^{\text{p}}, \boldsymbol{x}^{\text{n}}, \boldsymbol{x}^{\text{z}}$ positive, negative, and zero sequence symmetrical components of $\boldsymbol{x}(t)$

$X^{\text{p}}, X^{\text{n}}, X^{\text{z}}$ the crms values of the positive, negative and zero sequence components

$\boldsymbol{X}^{\text{p}}, \boldsymbol{X}^{\text{n}}, \boldsymbol{X}^{\text{z}}$ three-phase vectors of crms values of symmetrical components of $\boldsymbol{x}(t)$

$X^{\text{s}} \stackrel{\text{df}}{=} \begin{bmatrix} X^{\text{p}} \\ X^{\text{n}} \end{bmatrix}$ reduced three-phase vectors of symmetrical components crms values

$\mathbf{1}^{\text{p}}, \mathbf{1}^{\text{n}}, \mathbf{1}^{\text{z}}$ unite three-phase vectors of positive, negative, and zero sequence

$\beta_n \stackrel{\text{df}}{=} 1e^{jn\frac{2\pi}{3}}$ complex rotation coefficient of the nth order harmonic

$\mathbf{1}_n \stackrel{\text{df}}{=} \begin{bmatrix} 1 \\ \beta_n^* \\ \beta_n \end{bmatrix} = \begin{cases} \mathbf{1}^{\text{p}}, & \text{for } n = 3k+1 \\ \mathbf{1}^{\text{n}}, & \text{for } n = 3k-1 \\ \mathbf{1}^{\text{z}}, & \text{for } n = 3k \end{cases}$ unite three-phase vector of the nth order harmonic

$S_{\text{A}}, S_{\text{G}}, S_{\text{B}}, S_{\text{C}}$ arithmetic, geometric, Buchholz's, and Czarnecki's apparent powers

$\lambda_{\text{A}}, \lambda_{\text{G}}, \lambda_{\text{B}}$ arithmetic, geometric, and Buchholz's power factors

$\lambda, \lambda_1, \lambda_{\text{d}}$ power factor, fundamental harmonic, and distortion power factor

$\boldsymbol{i}_{\text{a}}, \boldsymbol{i}_{\text{r}}, \boldsymbol{i}_{\text{s}}, \boldsymbol{i}_{\text{u}}$ vectors of the active, reactive, scattered, and unbalanced currents

$\boldsymbol{i}_{\text{C}}, \boldsymbol{i}_{\text{G}}$ vectors of supply source (C) and the load-originated (G) currents

$\boldsymbol{u}_{\text{C}}, \boldsymbol{u}_{\text{G}}$ vectors of supply source (C) and the load-originated (G) voltages

$P_{\text{C}}, P_{\text{G}}$ - supply source (C) and the load-originated (G) active powers

$Y_{\text{e}}, Y_{\text{b}}, Y_{\text{u}}$ equivalent, balanced, and unbalanced admittance

$Y_{\text{b}n}$ balanced admittance for the nth order harmonic

Y_{d} voltage asymmetry-dependent unbalanced admittance

$\boldsymbol{a} = ae^{j\psi}$ complex coefficient of the supply voltage asymmetry

$\mathbf{Y}_{\mathrm{u}}^{\mathrm{p}}, \mathbf{Y}_{\mathrm{u}}^{\mathrm{n}}, \mathbf{Y}_{\mathrm{u}}^{\mathrm{z}}$ unbalanced admittance of the positive, negative, and zero sequence

$\boldsymbol{i}_{\mathrm{u}}^{\mathrm{p}}, \boldsymbol{i}_{\mathrm{u}}^{\mathrm{n}}, \boldsymbol{i}_{\mathrm{u}}^{\mathrm{z}}$ unbalanced currents of the positive, negative, and zero sequence

$P^{\mathrm{p}}, P^{\mathrm{n}}, P^{\mathrm{z}}$ positive, negative, and zero sequence active power

$D_{\mathrm{u}}^{\mathrm{p}}, D_{\mathrm{u}}^{\mathrm{n}}, D_{\mathrm{u}}^{\mathrm{z}}$ positive, negative, and zero sequence unbalanced power

$P_{\mathrm{w}}, P_{\mathrm{r}}, P_{\mathrm{d}}$ working, reflected, and detrimental active power

$\boldsymbol{i}_{\mathrm{w}}, \boldsymbol{i}_{\mathrm{f}}, \boldsymbol{i}_{\mathrm{d}}$, vectors of working, reflected and detrimental current

$\vec{\boldsymbol{E}}, \vec{\boldsymbol{H}}$ vectors of electric and magnetic fields intensity

$\vec{\boldsymbol{D}}, \vec{\boldsymbol{B}}$ vectors of electric and magnetic induction

$\vec{\boldsymbol{P}}$ Poynting Vector

$u_{\alpha}, u_{\beta}, u_{0}$ three-phase voltages in Clarke's coordinates α, β,

$i_{\alpha}, i_{\beta}, i_{0}$ three-phase currents in Clarke's coordinates

$i_{\alpha\mathrm{p}}, i_{\beta\mathrm{p}}$ instantaneous active current in Clarke's coordinates

$i_{\alpha\mathrm{q}}, i_{\beta\mathrm{q}}$ instantaneous reactive current in Clarke's coordinates

p, q instantaneous active and reactive powers

$\mathbf{C} \overset{\mathrm{df}}{=} \sqrt{\frac{2}{3}} \begin{bmatrix} 1 & -\frac{1}{2} & -\frac{1}{2} \\ 0 & \frac{\sqrt{3}}{2} & -\frac{\sqrt{3}}{2} \\ \frac{1}{\sqrt{2}} & \frac{1}{\sqrt{2}} & \frac{1}{\sqrt{2}} \end{bmatrix}$ Clarke's Transform matrix

$\mathbf{D} \overset{\mathrm{df}}{=} \begin{bmatrix} \sqrt{2/3}, & -1/\sqrt{6} \\ 0, & 1/\sqrt{2} \end{bmatrix}$ line-to-line voltages to Clarke's coordinates conversion matrix

$\mathbb{C} \overset{\mathrm{df}}{=} \begin{bmatrix} \sqrt{3/2}, & 0 \\ 1/\sqrt{2}, & \sqrt{2} \end{bmatrix}$ Clarke's Transform reduced matrix

$\boldsymbol{u}_{\mathrm{C}} \overset{\mathrm{df}}{=} \begin{bmatrix} u_{\alpha} \\ u_{\beta} \\ u_{0} \end{bmatrix}$ vector of three-phase voltages in Clarke's coordinates

$\mathbb{i}_{\mathrm{C}} \overset{\mathrm{df}}{=} \begin{bmatrix} i_{\alpha} \\ i_{\beta} \end{bmatrix}$ reduced vector of three-phase currents in Clarke's coordinates

$\boldsymbol{i}_{\mathrm{C}} \overset{\mathrm{df}}{=} \begin{bmatrix} i_{\alpha} \\ i_{\beta} \\ i_{0} \end{bmatrix}$ vector of three-phase currents in Clarke's coordinates

${}^{\mathrm{R}}\boldsymbol{U} \overset{\mathrm{df}}{=} \begin{bmatrix} U_{\mathrm{R}} \\ U_{\mathrm{T}} \\ U_{\mathrm{S}} \end{bmatrix} = \boldsymbol{U}^{\#}$ vector of crms values of three-phase voltages with reference voltage u_{R}

$^{S}\boldsymbol{U} \overset{\text{df}}{=} \begin{bmatrix} U_S \\ U_R \\ U_T \end{bmatrix}$ vector of crms values of three-phase voltages with reference voltage u_S

$^{T}\boldsymbol{U} \overset{\text{df}}{=} \begin{bmatrix} U_T \\ U_S \\ U_R \end{bmatrix}$ vector of crms values of three-phase voltages with reference voltage u_T

α, μ firing angle, commutation angle

$x^{\approx}(\mathrm{t})$ periodic extension of $x(t)$ observed in T-long window

s0 state of arc furnace in balanced mode of operation

s1 state of arc furnace with one arc extinct

s2 state of arc furnace with one unidirectional arc

PART A

CIRCUITS WITH NONSINUSOIDAL CURRENTS AND VOLTAGES

Analysis
Currents' Physical Components
Powers

Introduction

Three powers are commonly used to describe the power properties of electrical circuits: the active power P, the reactive power Q, and the apparent power S. These powers were introduced into electrical engineering at the end of the 19th century. The energy supplier delivers energy to customers with the apparent power S, while the customer consumes this energy at a rate equal to the active power P. According to the knowledge at that time, energy transfer is accompanied by the reactive power Q, so that they satisfy the power equation

$$P^2 + Q^2 = S^2.$$

Back in 1892, Steinmetz noticed in an experiment with an arc bulb, that even if the reactive power Q was equal to zero, the apparent power S was greater than the active power P, thus, electrical powers have to satisfy the inequality:

$$S \geq P.$$

This inequality means that the current that delivers energy to a customer is higher than needed. It increases energy loss at its delivery to customers, so that, it increases the demand for the produced electric energy, thus it increases its cost for consumers.

With the increasing cost of energy, physical reasons for this inequality and possibilities of its reduction have growing importance. This inequality also increases the cost of the equipment needed by energy distributors. Therefore, it is of fundamental importance for the power systems economy.

Steinmetz's observation has initiated research focused on finding physical causes of this inequality and on methods of reducing the apparent power S to the active power P.

This issue, apparently simple, has occurred to be one of the most difficult issues in electrical engineering for the whole 20th century. Thousands of engineers and scientists have worked to explain this inequality. They have attempted to describe electrical circuits in terms of powers and develop methods of the apparent power S reduction to the active power value.

This book presents the currently most advanced approach to this problem, namely, the Currents' Physical Components (CPC)-based power theory (PT) of electrical circuits.

The CPC – based PT has revealed all physical reasons for which the apparent power S can be higher than the active power P. It also provides theoretical fundamentals for methods of reducing the apparent power S, meaning compensation, in single-phase and three-phase circuits with linear, nonlinear, and time-variant loads.

Chapter 1
Doubts and Questions

1.1 Steinmetz's Experiment

During experiments in 1892 with an arc bulb in a circuit shown in Fig. 1.1, Steinmetz [2] observed that the power factor λ of such a bulb was lower than 1, although such a bulb had zero reactive power Q.

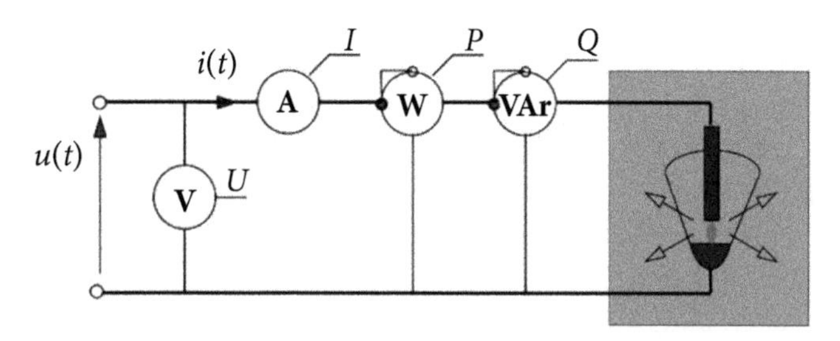

Figure 1.1 A circuit with an arc bulb.

He concluded that in such a circuit, despite the lack of inductance

$$S > P.$$

It seems that Steinmetz was confused with the results of his experiment, because in [2] he asked: "is there any phase shift between the current and the voltage" that could be blamed for the power factor decline, simultaneously observing that there is no reactive power Q at the load terminal. Thus, he did not associate the presence of the reactive power with the voltage and current phase-shift. We should be aware, however, that Steinmetz did not have any tools with which to observe the voltage and current waveform. In 1892 the oscilloscope was not yet invented. It was invented five years later.

Steinmetz's experiment has demonstrated that from the point of view of power properties, there is a major difference between linear and nonlinear loads. This difference stems from the fact that the current in the circuit with the arc bulb is not sinusoidal. Similar inequality occurs also in purely resistive three-phase circuits even with sinusoidal voltages and currents when the load currents are asymmetrical. How do we define powers in such a situation and what is the relation between them?

Steinmetz's observation has stirred up attempts aimed at the identification of physical phenomena that contribute to an increase of the apparent power, which means phenomena that contribute to the degradation of the effectiveness of the

Powers and Compensation in Circuits with Nonsinusoidal Current. Leszek S. Czarnecki, Oxford University Press.
 DOI: 10.1093/oso/9780198879206.003.0001

energy transfer. The knowledge of these phenomena is required for the development of compensators that would be capable of reducing the apparent power S.

Although the issue raised by Steinmetz appears, at a first glance, not to be very difficult to clarify, his observation initiated one of the toughest discussions in the electrical engineering community. It was also one of the longest discussions in the entire twentieth century, and even now it is not over. It was a major debate about how to explain the inequality of the active and the apparent powers, and consequently, a debate on power phenomena and definitions of power quantities in circuits with nonsinusoidal as well as asymmetrical voltages and currents.

In attempts aimed at an explanation of Steinmetz's observation, several concepts referred to as *power theories* were developed. These theories interpret the power properties of electrical circuits with nonsinusoidal voltages and currents, even in single-phase circuits, in diverse ways. The issue becomes even more controversial in three-phase circuits with asymmetrical voltages and currents. More than a century after Steinmetz's experiment, there is still substantial confusion in the electric engineering community in regard to powers and power phenomena in such circuits.

One should be aware, moreover, that the arc bulb in Steinmetz's experiment had the power of only a few hundred watts, thus his observation had an academic rather than practical importance. The present-time Steinmetz's experiment can be run with a high-power alternating current (AC) three-phase arc furnace, shown in Fig. 1.2.

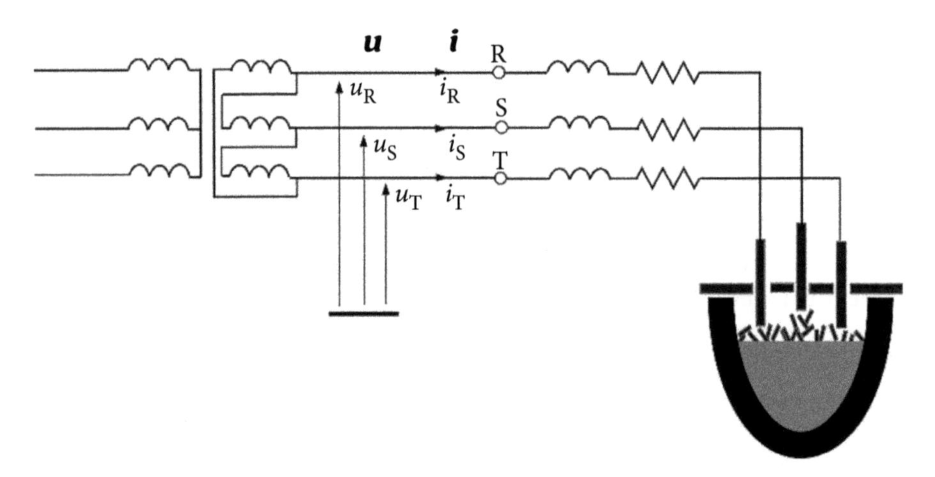

Figure 1.2 AC arc furnace and its supply structure.

Such an arc furnace can have power in the range of 750 MVA, which is equivalent to the power, approximately, of a million-population city. The annual bill for electric energy of such a furnace could be in the order of $500 million.

To stabilize the ignition of arcs, series inductors are inserted into the supply lines. Their inductance is selected such that their reactive power Q is comparable with the furnace active power P. It reduces the power factor for the fundamental frequency to the level of 0.7. When one of three arcs is not ignited, the furnace current is not only strongly distorted but also strongly asymmetrical and the power factor could be in the order of 0.4. Moreover, the power of the supply transformer can be comparable with the arc furnace power. To confine the supply voltage rms variation below a few percent, the supply transformers in distribution systems have usually a power rating at least twenty times higher than the load power. This is not possible, due to the required transformer power rating, in the case of the ultra-high-power arc furnaces. In effect, not only the line currents but also the arc furnace voltages are strongly distorted and asymmetrical. The improvement of the power factor by a compensator in such a situation is a real challenge. Considering the cost of electricity needed for operating the furnace, this challenge has to be faced.

As in the case of the original Steinmetz's experiment, the question: "of why the apparent power is higher than the active one?" remains a major question. The answer is to be found in a much more complex situation. Not only the furnace currents but also the furnace voltages are strongly distorted and asymmetrical.

1.2 Does the Reactive Power Occur Because of Energy Oscillations?

The author of this book had multiple opportunities to ask this question at seminars and in lectures for power system engineers, scientists, or graduate students. In the majority of cases the answer was confirmatory: the reactive power is caused by energy oscillations between the energy provider and the customers' load. The same answer can be found in the majority of textbooks for undergraduate students on electrical circuits. For example, in [177] one can find the following statement: "Reactive power represents an oscillatory energy exchange" or in [51] the following: "The reactive power Q is, by definition, equal to the peak value of the power component that travels back and forth on the line."

This interpretation of the reactive power is correct but only apparently. If the instantaneous power is decomposed, as shown in Fig. 1.3, as follows

$$p(t) = u(t)i(t) = 2\ UI \cos \omega t\ \cos(\omega t - \varphi) = p_{\mathrm{u}}(t) + p_{\mathrm{b}}(t),$$

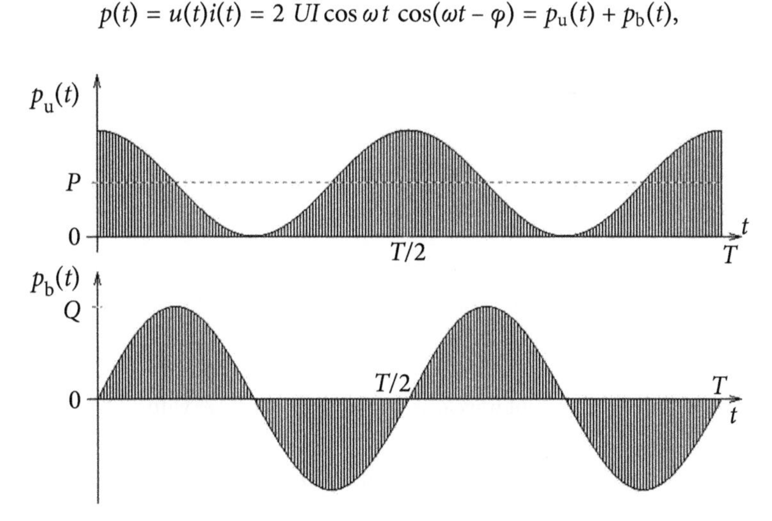

Figure 1.3 Components of the instantaneous power.

with the unidirectional component

$$p_{\mathrm{u}}(t) = P(1 + \cos 2\omega t),$$

then the bidirectional component is

$$p_{\mathrm{b}}(t) = Q \sin 2\omega t.$$

Thus indeed, the reactive power Q is the amplitude of the oscillating component of the instantaneous power. Let us consider, however, a purely resistive load with a bidirectional triode thyristor (TRIAC), shown in Fig. 1.4, supplied with a sinusoidal voltage.

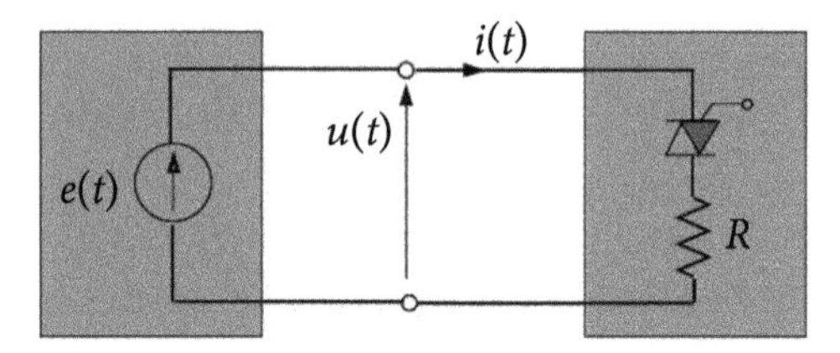

Figure 1.4 A resistive load with a TRIAC.

The waveform of the supply current at some value of the firing angle α of the TRIAC is shown in Fig. 1.5. The instantaneous power at the supply source terminals $p(t) = u(t)\, i(t)$ is always non-negative, thus the energy cannot flow to the supply source, so there are no energy oscillations in such a circuit. Despite that, assuming that the supply voltage is

$$u(t) = 220\sqrt{2} \sin \omega_1 t \text{ V},$$

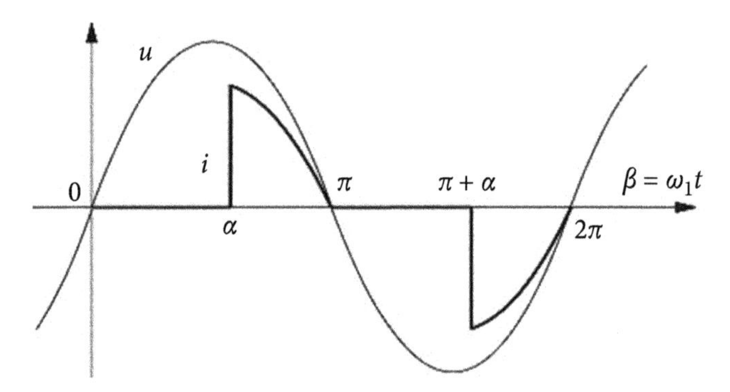

Figure 1.5 The current waveform in the resistive circuit with TRIAC.

the load resistance $R = 1\ \Omega$, and the TRIAC firing angle $\alpha = 135^{\circ}$, the meters provide the readings shown in Fig. 1.6; in particular, the measured reactive power $Q = 7.7$ kVAr.

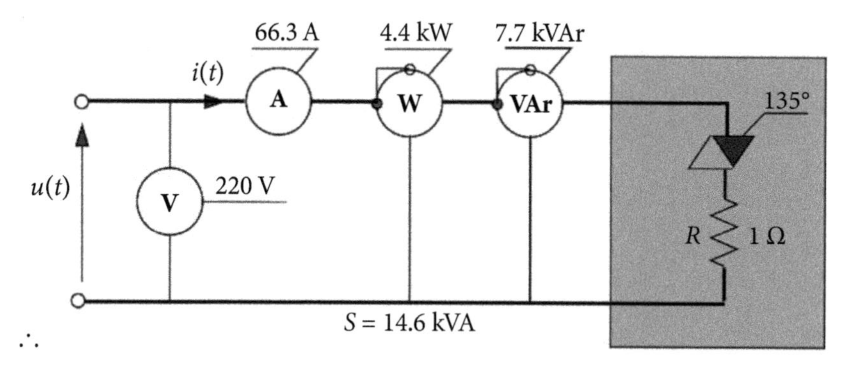

∴

Figure 1.6 Meter readings in a resistive circuit with TRIAC.

It means that the reactive power Q can exist in the circuit without any oscillations of energy. Its interpretation as the power that occurs because of energy oscillations is wrong. Moreover, it exists in circuits with purely resistive loads, without any capability of energy storage. Thus, the interpretation of the reactive power Q as the power that occurs because of energy storage in the load is also wrong. The meaning or interpretation of the reactive power Q should not depend on the load properties. It should be the same for linear or nonlinear loads. In particular, the reactive power measured in the circuit shown in Fig. 1.6 does not differ from the reactive power of linear loads, and even more so, this reactive power can be compensated by a capacitor connected as shown in Fig. 1.7.

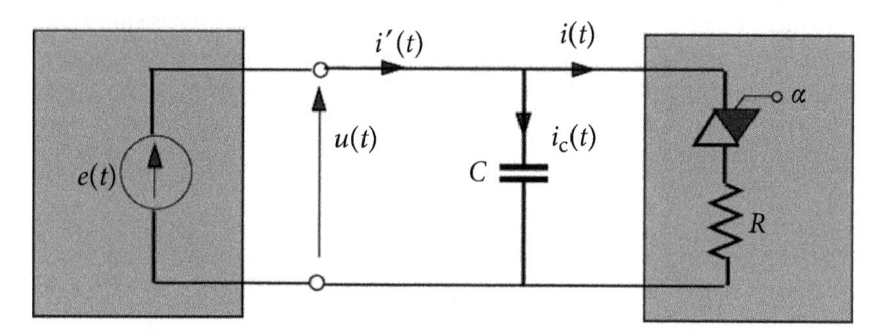

Figure 1.7 Capacitive compensation of the load with TRIAC.

Such a compensator can compensate the fundamental harmonic of the load current, equal to

$$i_1(t) = \sqrt{2}I_1 \sin(\omega_1 t - \varphi_1) = 40.3\sqrt{2} \sin(\omega_1 t - 60.3^\circ) \text{ A}$$

and shown in Fig. 1.8.

The phase shift of this fundamental harmonic versus the supply voltage explains the presence of the reactive power

$$Q = Q_1 = UI_1 \sin\varphi_1 = 220 \times 40.3 \sin(60.3^\circ) = 7.7 \text{ kVAr}.$$

A compensator of the capacitance

$$C = \frac{Q}{\omega_1 U^2}$$

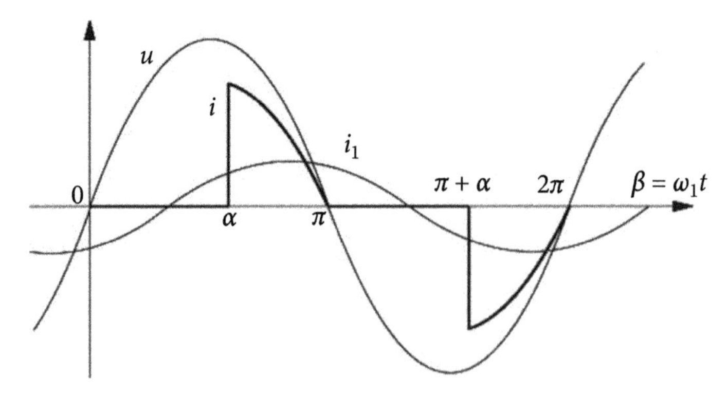

Figure 1.8 The TRIAC current and its fundamental harmonic.

can compensate this power, eliminate the phase-shift of the supply current fundamental harmonic, and improve the power factor.

1.3 Does Energy Oscillate in Three-Phase Supply Lines?

It is very common in power engineering community opinion that in the presence of the reactive power Q in single-phase circuits, the energy oscillates between the supply source and the load is, according to the author's observations, usually extrapolated to three-phase circuits. It is commonly thought that in three-phase lines energy oscillates forth and back with the double frequency of the distribution voltage.

This view is wrong, however. To show this, let us consider a three-phase circuit with resistive-inductive (RL) balanced load, shown in Fig. 1.9, supplied by a three-phase line with symmetrical and sinusoidal voltages. Let us assume that the supply voltage and current of the line R be equal to

$$u_{\mathrm{R}}(t) = \sqrt{2}\,U\cos\omega_1 t \quad i_{\mathrm{R}}(t) = \sqrt{2}\,I\cos(\omega_1 t - \varphi),$$

and let us assume that voltages be of a positive sequence.

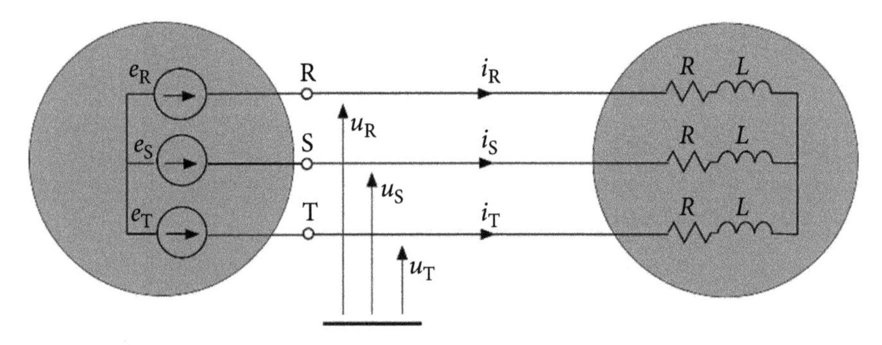

Figure 1.9 A three-phase circuit with a balanced RL load.

The load reactive power of the load is

$$Q = 3\ UI \sin\ \varphi.$$

The instantaneous power of the load, meaning the rate of energy flow in the three-phase supply line, is

$$\begin{aligned}
p(t) = \frac{dW}{dt} &= u_{\mathrm{R}}(t)\, i_{\mathrm{R}}(t) + u_{\mathrm{S}}(t)\, i_{\mathrm{S}}(t) + u_{\mathrm{T}}(t)\, i_{\mathrm{T}}(t) = \\
&= 2UI\Big[\cos \omega_1 t \cos(\omega_1 t - \varphi) + \\
&+ \cos\left(\omega_1 t - \frac{2\pi}{3}\right) \cos\left(\omega_1 t - \varphi - \frac{2\pi}{3}\right) + \\
&+ \cos\left(\omega_1 t + \frac{2\pi}{3}\right) \cos\left(\omega_1 t - \varphi + \frac{2\pi}{3}\right)\Big] = 3\ UI \cos \varphi = P;
\end{aligned}$$

thus it is constant, independently of the load-reactive power Q value. Thus, there are no energy oscillations between the load and the supply source. Such oscillations of energy in the three-phase supply line do not exist.

Some members of the power engineering community challenge this conclusion claiming that such oscillations do exist, but that they are confined to individual supply lines, R, S, and T, and cancel mutually. Indeed, the product $u_{\mathrm{R}}(t)\ i_{\mathrm{R}}(t)$ is equal to,

$$u_{\mathrm{R}}(t)\, i_{\mathrm{R}}(t) = \frac{P}{3}\ (1 + \cos 2\omega_1\, t) + \frac{Q}{3}\ \sin 2\omega_1\, t$$

thus it contains the oscillating component. Such components also exist in the remaining products of line voltages and currents. The question remains, however: does such a product at only one terminal of a three-phase line stand for the rate of energy flow? The adverse answer to the question is given in Section 24.1. Energy does not flow forth and back in three-phase lines. In balanced three-phase circuits with symmetrical and sinusoidal supply, voltage energy flows at a constant rate. This constant flow rate can only be disturbed by the voltage, and/or current asymmetry and/or waveform distortion.

1.4 Do Energy Oscillations Degrade Power Factor?

Energy oscillations between the energy provider and customers' loads are usually blamed for the power factor value drop from the unity value, thus for the reduction of the effectiveness of energy transfer from its provider to customers' loads. This opinion is usually supported by the observation that in single-phase linear circuits the reactive power Q is the amplitude of the bidirectional, or oscillating component of the instantaneous power. As it was shown in Section 1.3, the reactive power cannot be interpreted as the amplitude of energy oscillations, however, thus these oscillations cannot be regarded as the cause of the power factor drop.

Since in purely resistive circuits such as that shown in Fig. 1.4, the power factor is lower than 1, despite the lack of energy oscillations, thus there have to be other reasons for this. Further, when the reactive power in the circuit with TRIAC is compensated by a compensator, then the supply current has the waveform as shown in Fig. 1.10. Having the energy storage capability, such a capacitor enables energy oscillations between the supply source and the compensated load.

Observe that after compensation, there are intervals of time where the supply current has the opposite sign to the sign of the supply voltage. The instantaneous power $p(t) = u(t)\, i'(t)$ in such intervals is negative, thus energy flows back to the supply source.

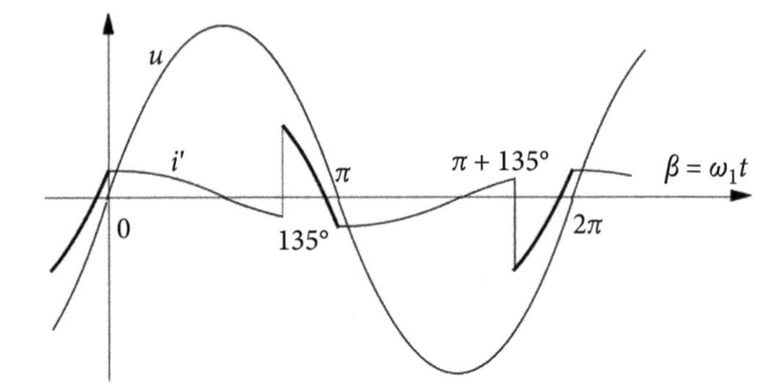

Figure 1.10 The supply current waveform after capacitive compensation.

Thus the improvement in the power factor value is accompanied by the emergence of energy oscillations. The power factor of the load is higher in the presence of energy oscillations than without them. To summarize, compensation of the reactive power to zero and the power factor improvement causes energy oscillations. It is just the opposite to what is commonly expected in power factor improvement.

1.5 What Are Harmonics and Their Complex Rms (crms) Value?

To describe mathematically nonsinusoidal but periodic quantities, such as voltages or currents, the Fourier series is commonly used. When a quantity has the angular frequency $\omega_1 = 2\pi f$, then a Fourier series of such a quantity is an infinite sum of sinusoidal components of the angular frequency $\omega_n = n\omega_1$, which is an integer multiplicity of the frequency ω_1. These sinusoidal components are referred to as *harmonics* and the integer multiplicity n of the frequency ω_1 is referred to as the *harmonic order*. Thus, any periodic quantity $x(t)$, it could be voltage $u(t)$ or current $i(t)$, can be presented in a form

$$x(t) = X_0 + \sqrt{2}\sum_{n=1}^{\infty} X_n \cos(n\omega_1 t + \alpha_n) = \sum_{n=0}^{\infty} x_n(t).$$

For example, a nonsinusoidal current composed of the fundamental harmonic, that is, $n = 1$, the second order harmonic, $n = 2$, and the thirteen order harmonic, $n = 13$, can have the form

$$i(t) = 100\sqrt{2}\ \cos\,(\omega_1 t - 90^\circ) + 30\sqrt{2}\ \cos\,(2\omega_1 t - 90^\circ) + 6\sqrt{2}\cos\,(13\,\omega_1 t - 90^\circ) =$$

$$= 100\sqrt{2}\ \sin\omega_1 t + 30\sqrt{2}\ \sin 2\,\omega_1 t + 6\sqrt{2}\ \sin 13\,\omega_1 t = i_1(t) + i_2(t) + i_{13}(t).$$

Its waveform and harmonics are shown in Fig. 1.11.

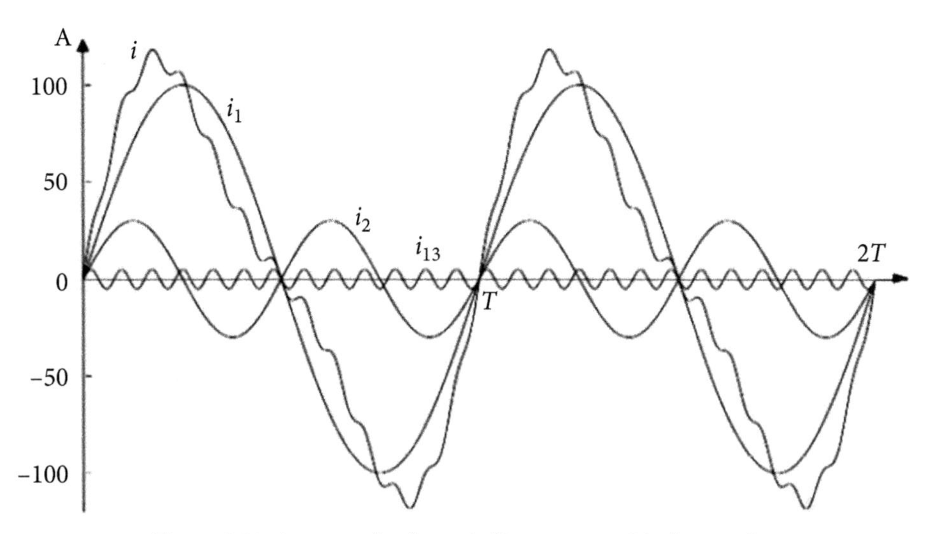

Figure 1.11 An example of a periodic current and its harmonics.

The rms value of such a periodic quantity is defined as

$$||\,x\,|| \stackrel{\mathrm{df}}{=} \sqrt{\frac{1}{T}\int_0^T x^2(t)\,dt}.$$

The Fourier Series as presented above has a form that is not convenient for circuit analysis. It will be used in this book in the form

$$x(t) = X_0 + \sqrt{2}\ \mathrm{Re}\sum_{n\in N} \boldsymbol{X}_n e^{jn\omega_1 t} = \sum_{n\in N_0} x_n(t)$$

where N is a set of harmonic orders n of harmonics $x_n(t)$ the quantity $x(t)$ is composed of, and

$$\boldsymbol{X}_n = X_n\, e^{j\alpha_n} = \frac{\sqrt{2}}{T}\int_0^T x(t)e^{-jn\omega_1 t}dt$$

is the *complex rms* (crms) value of the nth order harmonic. Its magnitude X_n specifies the rms value of the nth order harmonic and the argument α_n specifies its phase.

Harmonics were found very early to be responsible for the difference between the active and apparent powers in Steinmetz's experiment, and consequently, the harmonic approach has profoundly affected investigations of powers in circuits with nonsinusoidal voltages and currents.

1.6 Do Harmonics Exist as Physical Entities?

Phrases such as "*harmonics in power systems*" in article titles and conferences, as well as harmonic meters, analyzers, harmonic standards, or harmonics compensators, cause the voltage, and current harmonics are regarded in the electrical engineering community as physical entities.

The concept of harmonics is powerful. Without this concept, it would be difficult to express electrical quantities, often specified by graphs, tables, or sequences of instantaneous values, in an analytical form susceptible for using them in mathematical formulae.

A reader of this book should be aware that harmonics are only mathematical entities, however. A Fourier Series is only a kind of an intellectual tool that enables us to describe periodic quantities in mathematical terms. Consequently, harmonics do not exist as physical entities. To demonstrate this, let us consider the circuit shown in Fig. 1.12, composed of a direct current (DC) source with internal resistance equal to $R = 1\ \Omega$ and a periodic switch with the duty factor ON/OFF = ½.

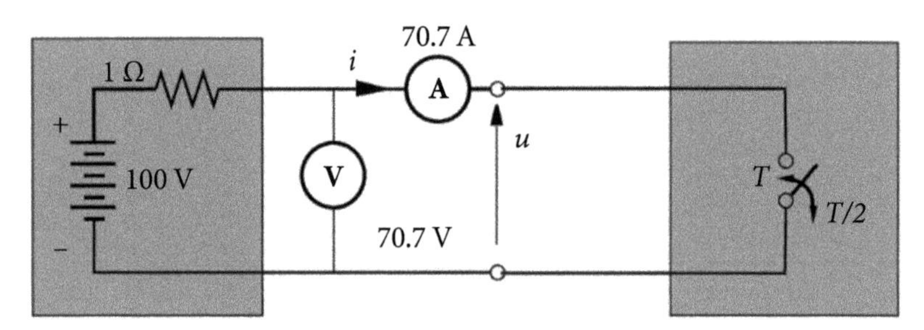

Figure 1.12 A circuit with a periodic switch.

The circuit is periodically open thus the supply current is zero, and short-circuited, thus the switch voltage is zero, while the current is limited only by the supply source resistance. Thus, the voltage and current at the switch terminal change as shown in Fig. 1.13.

The voltage and current at the switch terminals can be expressed using the Fourier Series as the sum of the voltage and current harmonics, namely

$$u(t) = U_0 + \sqrt{2}\sum_{n=1}^{\infty} U_n \cos(n\omega_1 t + \alpha_n) = \sum_{n=0}^{\infty} u_n$$

$$i(t) = I_0 + \sqrt{2}\sum_{n=1}^{\infty} I_n \cos(n\omega_1 t + \beta_n) = \sum_{n=0}^{\infty} i_n.$$

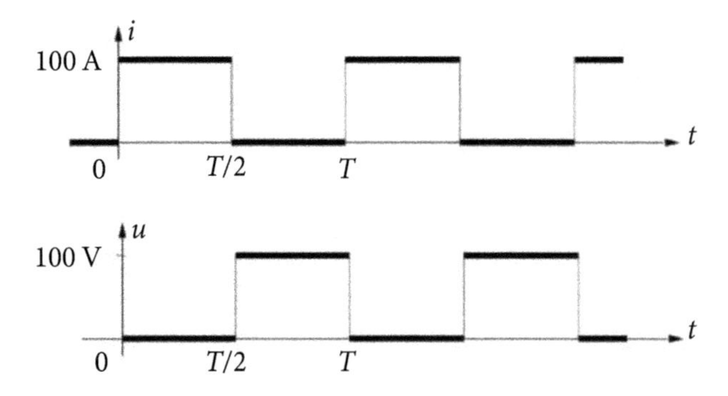

Figure 1.13 The voltage and current waveform at the switch terminals.

An individual voltage or current harmonic stands for a sinusoidal, continuous quantity. If this is a current harmonic, it cannot flow through the switch in the OFF state. If this is a voltage harmonic, it cannot occur on the switch in the ON state. It means any individual harmonic cannot exist as a physical entity. They are only mathematical entities that, in sum, approximate the voltage and current waveforms. Only their infinite sum in some interval of time approximates the voltage and current values.

1.7 How to Describe Single-Phase Circuits in Terms of Powers?

Steinmetz concluded [2] in 1892 that at nonsinusoidal current in a purely resistive, single-phase circuit, the apparent power S is higher than the active power, that is, $S > P$, so that how to describe such a circuit in terms of powers, especially that the supply voltage can be nonsinusoidal. The question applies to a simple circuit such as that shown in Fig. 1.14.

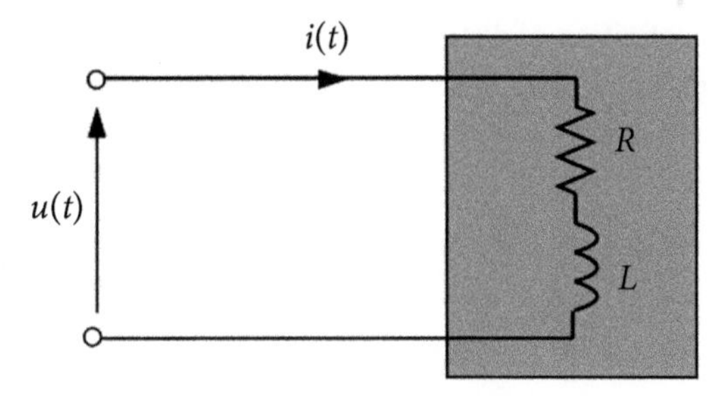

Figure 1.14 RL load.

Over the years, several different answers to this question have been suggested. The consensus was only regarding the active and apparent powers, namely they were defined as follows:

$$P \overset{\text{df}}{=} \frac{1}{T}\int_0^T u(t)i(t)dt$$

$$S \overset{\text{df}}{=} \sqrt{\frac{1}{T}\int_0^T u^2(t)dt}\sqrt{\frac{1}{T}\int_0^T i^2(t)dt} = ||\, u\,||\;||\, i\,||.$$

The main answers, focused on the difference between the apparent and active powers, in terms of their squares, are compiled below.

Budeanu [14] (1927). Assuming that

$$u(t) = U_0 + \sqrt{2}\sum_{n=1}^{\infty} U_n \cos(n\omega_1 t)$$

$$i(t) = I_0 + \sqrt{2}\sum_{n=1}^{\infty} I_n \cos(n\omega_1 t - \varphi_n),$$

then

$$S^2 - P^2 = Q_{\rm B}^2 + D_{\rm B}^2$$

with

$$Q_{\rm B} \overset{\text{df}}{=} \sum_{n=1}^{\infty} U_n I_n \sin\varphi_n$$

which is Budeanu's definition of the reactive power, and

$$D_{\rm B} \overset{\text{df}}{=} \sqrt{S^2 - (P^2 + Q_{\rm B}^2)}$$

which is Budeanu's definition of distortion power.

Fryze [17] (1931):

$$S^2 - P^2 = Q_{\rm F}^2$$

with

$$Q_{\rm F} \overset{\text{df}}{=} ||u||\;||i_{\rm rF}||$$

where$||i_{\rm rF}||$ is the rms value of the Fryze's reactive current defined as

$$i_{\rm rF}(t) \overset{\text{df}}{=} i(t) - \frac{P}{||u||^2}u(t).$$

Shepherd and Zakikhani [39] (1971):

$$S^2 = S_R^2 + Q_S^2.$$

The active power P in this power equation does not exist. Instead, there is a resistive apparent power

$$S_R \stackrel{\text{df}}{=} ||u||\ ||i_R||$$

where

$$i_R(t) \stackrel{\text{df}}{=} I_0 + \sqrt{2} \sum_{n=1}^{\infty} I_n \cos\varphi_n \cos(n\omega_1 t)$$

and the reactive power

$$Q_S \stackrel{\text{df}}{=} ||u||\ ||i_{rS}||$$

where

$$i_{rS}(t) \stackrel{\text{df}}{=} \sqrt{2} \sum_{n=1}^{\infty} I_n \sin\varphi_n \sin(n\omega_1 t).$$

Kusters and Moore [59] (1979):

$$S^2 - P^2 = Q_C^2 + Q_{Cr}^2$$

with

$$Q_C \stackrel{\text{df}}{=} ||u||\ ||i_{qC}||$$

referred to as a capacitive reactive power, where

$$i_{qC}(t) \stackrel{\text{df}}{=} \frac{(\dot{u}, i)}{||\dot{u}||^2}\dot{u}(t), \quad \dot{u} = du/dt, \quad (\dot{u}, i) \stackrel{\text{df}}{=} \frac{1}{T}\int_0^T \dot{u}(t)\, i(t) dt,$$

and

$$Q_{Cr} \stackrel{\text{df}}{=} ||u||\ ||i_{qCr}||$$

referred to as a residual capacitive reactive power

$$i_{Cr}(t) \stackrel{\text{df}}{=} i(t) - [i_a(t) + i_{qC}(t)].$$

Czarnecki [71] (1983):

$$S^2 - P^2 = D_s^2 + Q^2$$

with

$$D_s \stackrel{\text{df}}{=} ||u||\,||i_s||$$

referred to as a scattered power, where

$$i_s(t) \stackrel{\text{df}}{=} (G_0 - G_e)U_0 + \sqrt{2}\,\text{Re}\sum_{n=1}^{\infty}(G_n - G_e)\boldsymbol{U}_n\, e^{jn\omega_1 t}$$

and

$$G_e \stackrel{\text{df}}{=} \frac{P}{||u||^2}$$

is the equivalent conductance of the load, while G_n is its conductance for the nth order harmonic, and

$$Q \stackrel{\text{df}}{=} ||u||\,||i_r||$$

is the reactive power, where

$$i_r(t) \stackrel{\text{df}}{=} \sqrt{2}\,\text{Re}\sum_{n=1}^{\infty} jB_n\,\boldsymbol{U}_n\, e^{jn\omega_1 t}.$$

and B_n is the load susceptance for the nth order harmonic.

Tenti [191] (2003):

$$S^2 - P^2 = Q_T^2 + D_v^2$$

with

$$Q_T \stackrel{\text{df}}{=} ||u||\,||i_{rT}||$$

where

$$i_{rT}(t) \stackrel{\text{df}}{=} \frac{(\widehat{u}, i)}{||\widehat{u}||^2}\,\widehat{u}(t)$$

while $\widehat{u}(t)$ denotes the oscillating component of the supply voltage integral, and

$$(\widehat{u}, i) = \frac{1}{T}\int_0^T \widehat{u}(t)\, i(t)\, dt.$$

The second power in Tenti's equation is a void or distortion power, defined as

$$D_{\mathrm{v}} \stackrel{\mathrm{df}}{=} ||u||\ ||i_{\mathrm{v}}||$$

where

$$i_{\mathrm{v}}(t) \stackrel{\mathrm{df}}{=} i(t) - [i_{\mathrm{a}}(t) + i_{\mathrm{rT}}(t)].$$

Thus, eleven different powers can be used to describe the gap between the apparent and active power even for a simple load such as that shown in Fig. 1.14. It seems that the selection of the right power quantities even for single-phase circuits is beyond the capability of a common member of the power engineering community, especially though some of them were supported by national or international standards, which have recently been shown to be erroneous. They are discussed in Chapters 18–22.

1.8 How to Describe Harmonics Generating Loads in Terms of Powers?

Most of the present-day electrical loads in our homes, such as TV sets, computers and compute-like equipment, fluorescent lamps, and microwaves, draw a current which is distorted from a sinusoidal waveform even when supplied with a sinusoidal voltage. The same occurs in commercial buildings and manufacturing plants.

Since distorted currents can be described in terms of harmonics, such loads are classified as *Harmonics Generating Loads* (HGLs).

A strongly simplified model of such an HGL is shown in Fig. 1.15.

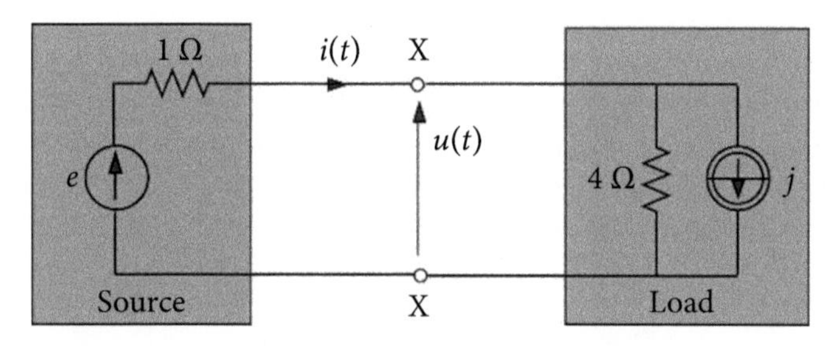

Figure 1.15 Circuit with Harmonics Generating Load (HGL).

Let us assume that the internal voltage of the supply source is

$$e(t) = 100\sqrt{2}\sin\omega_1 t\,\mathrm{V}$$

while the HGL generates only a third-order harmonic

$$j(t) = 50\sqrt{2}\sin 3\omega_1 t\,\mathrm{A}.$$

The voltage $u(t)$ in the cross-section x–x, calculated using the Superposition Principle, has the waveform

$$u(t) = 80\sqrt{2}\sin\omega_1 t - 40\sqrt{2}\sin 3\omega_1 t \text{ V}$$

and the current in that cross-section is

$$i(t) = 20\sqrt{2}\sin\omega_1 t + 40\sqrt{2}\sin 3\omega_1 t \text{ A.}$$

With such a voltage and current, the active power P in the cross-section x–x has zero value, because

$$P = \frac{1}{T}\int_0^T u(t)\, i(t)\, dt = 80 \times 20 - 40 \times 40 = 0.$$

Observe that there is no phase shift between the voltage and current harmonics however, thus the reactive power Q in the circuit considered is equal to zero. The voltage and current rms values in the cross-section x–x are equal to

$$||u|| = \sqrt{80^2 + 40^2} = 89.5 \text{ V}, \quad ||i|| = \sqrt{20^2 + 40^2} = 44.7 \text{ A},$$

and consequently, the apparent power S is

$$S = ||u||\,||i|| = 89.5 \times 44.7 = 4000 \text{ VA.}$$

Thus, at the selected structure and the circuit parameters the active power $P = 0$, and there is no reactive power Q, thus, how do we could write the power equation of such a load? On the left side of such an equation, we have the apparent power S, written usually in the square, but what is on its right side?

$$S^2 = ?$$

The answer to this question can be found in Chapter 6.

1.9 How to Calculate the Apparent Power in Three-Phase Circuit?

Most of the electric energy is distributed to customers by three-phase circuits and consequently, the power properties of three-phase circuits are of particular importance.

The apparent power S is a conventional quantity. According to the IEEE *Dictionary of Electrical and Electronics Terms* [154], the apparent power in circuits with sinusoidal voltages and currents can be defined as the *arithmetical apparent power*

$$S = S_A \stackrel{df}{=} U_R I_R + U_S I_S + U_T I_T$$

or as the *geometrical apparent power.*

$$S = S_G \stackrel{df}{=} \sqrt{P^2 + Q^2}.$$

There is also a definition of the apparent power suggested by Buchholz [11] in 1922:

$$S = S_B \stackrel{df}{=} \sqrt{U_R^2 + U_S^2 + U_T^2}\ \sqrt{I_R^2 + I_S^2 + I_T^2}.$$

The last definition, not being supported by the IEEE *Dictionary of Electrical and Electronics Terms*, is not commonly known. Nonetheless, it was used in the FDB (Fryze–Depenbrock–Buchholz) Method [90] by Depenbrock. Also, the apparent power in the frame of the Currents' Physical Components (CPC)-based power theory, developed by Czarnecki [93] for three-phase circuits, has the form of the Buchholz definition at sinusoidal voltages and currents.

As long as a three-phase load is balanced, the numerical value of the apparent power S, calculated according to these three different definitions, is the same. A difference in values occurs when the load is unbalanced, so that line currents are not symmetrical. Consequently, the power factor λ depends on the selection of the apparent power S definition. Such a situation is particularly visible in traction three-phase circuits since ac train engines are supplied from only one line. Other lines supply other sectors of the train system and consequently, from the supply source perspective, a train has to be regarded as an unbalanced load. It can be illustrated by the following situation.

A three-phase circuit, with an ideal transformer of turn ratio 1:1, is loaded, as shown in Fig. 1.16, by a purely resistive load with resistance $R = 2\ \Omega$. The rms values of the line currents are shown in this figure.

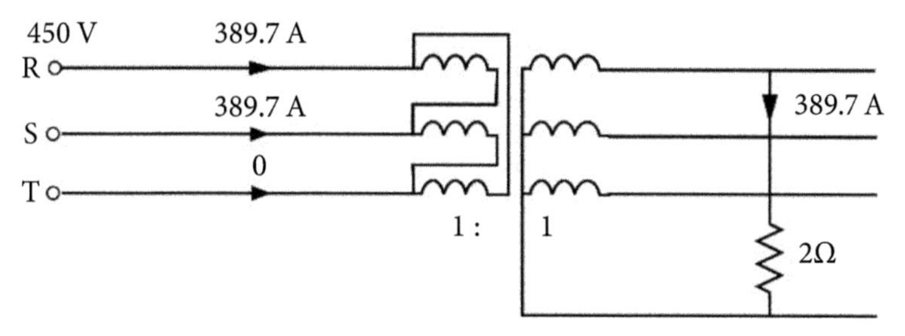

Figure 1.16 A three-phase unbalanced circuit.

The load active power is equal to $P = 304$ kW, the reactive power Q has zero value, because the load is purely resistive, while the apparent power S, depending on its definition, is equal to

$$S_A = 351 \text{ kVA}, \quad S_G = 304 \text{ kVA}, \quad S_B = 430 \text{ kVA}.$$

Consequently, the power factor depends on the selected apparent power S definition, namely

$$\lambda_A = \frac{P}{S_A} = 0.86, \qquad \lambda_G = \frac{P}{S_G} = 1, \qquad \lambda_B = \frac{P}{S_B} = 0.71.$$

The question of which of these three values is the true power factor value is not only academic. The cost of the energy delivery depends on the power factor and its value should affect billing for the energy delivered. The energy supplier might opt to increase its revenue, for λ_B value, while a customer, to reduce the energy bill, would prefer if the apparent power S be calculated using the geometrical definition of the apparent power and consequently be billed at the assumption that the power factor has λ_G value.

The effect of the power factor λ on tariffs for the electric energy delivery is commonly specified by local regulations or agreements between utilities and customers. Which apparent power, S_A, S_B, or S_G, specifies the loading of the supply source correctly is a key question for fair energy accounts, however.

This issue is analyzed in detail in Chapter 7. It is shown there that the power factor calculated using arithmetic or geometric definition of the apparent power has an erroneous value. It should be calculated using Buchholtz's definition of apparent power.

1.10 Is the Common Power Equation of Three-Phase Circuits Right?

Power properties of three-phase circuits with sinusoidal voltages and currents are commonly described using the power equation

$$S^2 = P^2 + Q^2.,$$

with the active and reactive powers equal to

$$P = \sum_{p\,=\,R,S,T} U_p I_p \cos \varphi_p$$

$$Q = \sum_{p\,=\,R,S,T} U_p I_p \sin \varphi_p.$$

This power equation is written on the condition that the apparent power S is defined as the geometrical apparent power $S = S_G$, however. The geometrical apparent power S_G in the presence of current asymmetry does not provide the right value for the power factor λ. The apparent power should be calculated using Buchholz's definition, meaning, $S = S_B$.

In a situation as illustrated in Fig. 1.16, the right value of the apparent power S, calculated according to Buchholz's definition, is

$$S = S_B \stackrel{\text{df}}{=} \sqrt{U_R^2 + U_S^2 + U_T^2}\sqrt{I_R^2 + I_S^2 + I_T^2} =$$

$$= \sqrt{450^2 + 450^2 + 450^2}\sqrt{389.7^2 + 389.7^2} = 450 \times 389.7 \times \sqrt{6} = 429.6 \text{ kVA},$$

while the active power

$$P = I^2 R = 389.7^2 \times 2 = 303.7 \text{ kW}.$$

Since the transformer was assumed to be ideal, that is, lossless, there is no other active power, while the load is purely resistive, so that $Q = 0$. With such powers, the power equation

$$S^2 = P^2 + Q^2$$

is not satisfied, however. This equation is wrong.

While not going into details discussed in Chapter 7, the power equation of three-phase, three-wire circuits with unbalanced LTI loads supplied with sinusoidal and symmetrical voltage has the form

$$S^2 = P^2 + Q^2 + D_u^2.$$

It should contain a new power quantity, introduced in the paper [93], called an *unbalanced power*, D_u.

1.11 Is the Reactive Power Caused by Energy Storage?

When a three-phase load has an inductive reactance as shown in Fig. 1.9, then some energy is stored in magnetic fields. Can this storage of the energy in the load be the cause of the reactive power Q?

Let us calculate the energy stored in the magnetic field of the load, assuming that it is a balanced load, and the supply voltage is sinusoidal, symmetrical, and of the positive sequence.

$$W(t) = W_R(t) + W_S(t) + W_T(t) = \frac{1}{2}L\left[i_R^2(t) + i_S^2(t) + i_T^2(t)\right] =$$
$$= \frac{1}{2}L2I^2\left[\cos^2(\omega t - \varphi) + \cos^2\left(\omega t - \varphi - \frac{2\pi}{3}\right) + \cos^2\left(\omega t - \varphi + \frac{2\pi}{3}\right)\right] =$$
$$= LI^2\left[\frac{1}{2} + \frac{1}{2}\cos 2(\omega t - \varphi) + \frac{1}{2} + \frac{1}{2}\cos 2\left(\omega t - \varphi - \frac{2\pi}{3}\right) + \right.$$
$$\left. + \frac{1}{2} + \frac{1}{2}\cos 2\left(\omega t - \varphi + \frac{2\pi}{3}\right)\right] =$$
$$= \frac{3}{2}LI^2 = \text{const.}$$

Thus, the energy stored in the magnetic fields of the load is constant. It is delivered to the load after the load is connected to the supply source, in other words in the transient state of the circuit, when line currents are non-periodic. It does not change in the steady-state when these currents are sinusoidal. Consequently, the presence of the reactive power Q cannot be explained by the energy flow to the magnetic field of the load.

With the change of the line currents, the energies stored in magnetic fields $W_R(t)$, $W_S(t)$, and $W_T(t)$, change as well. The line currents have to contain the reactive components, that is, components shifted by $\pi/2$ versus the active currents. Just these reactive components of the line currents contribute to the current rms value increase, thus to the power factor decline.

1.12 Why Can Capacitive Compensator Degrade Power Factor?

Capacitive compensation is the most common method of power factor improvement in distribution systems. Its fundamentals are explained in Section 9.7. Unfortunately, in the presence of the voltage or current distortion such compensation can sometimes be ineffective or even; instead of the power factor improvement such compensation degrades it.

It is illustrated in the following situation. An RL load, shown in Fig. 1.17, is supplied from a voltage source that has a short-circuit power S_{sc}, which is forty times higher than the load active power $P = 10$ kW. The reactance-to-resistance ratio of the source impedance is $X_s/R_s = 3$. At sinusoidal supply voltage, the power factor is equal to $\lambda = 0.5$.

The load is compensated by a capacitor of capacitance C that changes from zero to 2mF. The plot of the power factor at a sinusoidal supply voltage denoted as "Fund. harm. PF", is shown in Fig. 1.18. The power factor λ at a capacitance equal approximately to $C = 1.8$ mF reaches the maximum value of $\lambda = 1$.

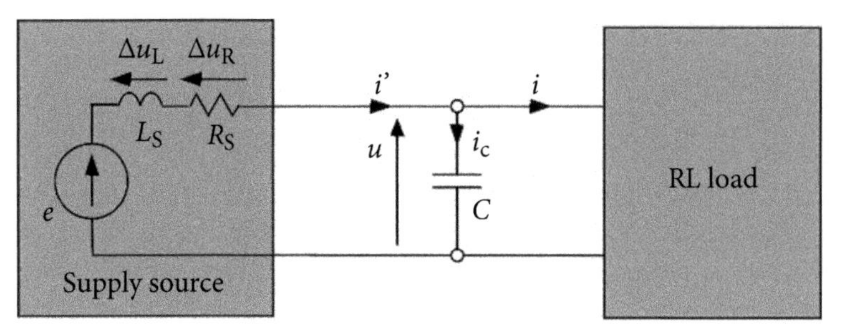

Figure 1.17 RL load with capacitive compensator.

It is enough that the supply voltage is distorted by 3% of the 5th order harmonic, 1.5% of the 7th order harmonic, and 0.5% of the 11th order harmonic, and the power factor changes with the capacitance change as shown in Fig. 1.18 by the bottom line, however. It cannot be higher than approximately $\lambda = 0.8$ and at the optimum capacitance at sinusoidal voltage $C = 1.8$ mF, it declines to the level of 0.2. The series resonance of the capacitor with the internal source inductance L_S is responsible for this decline in the power factor. Observe that this decline occurs even at a quite moderate distortion of the supply voltage.

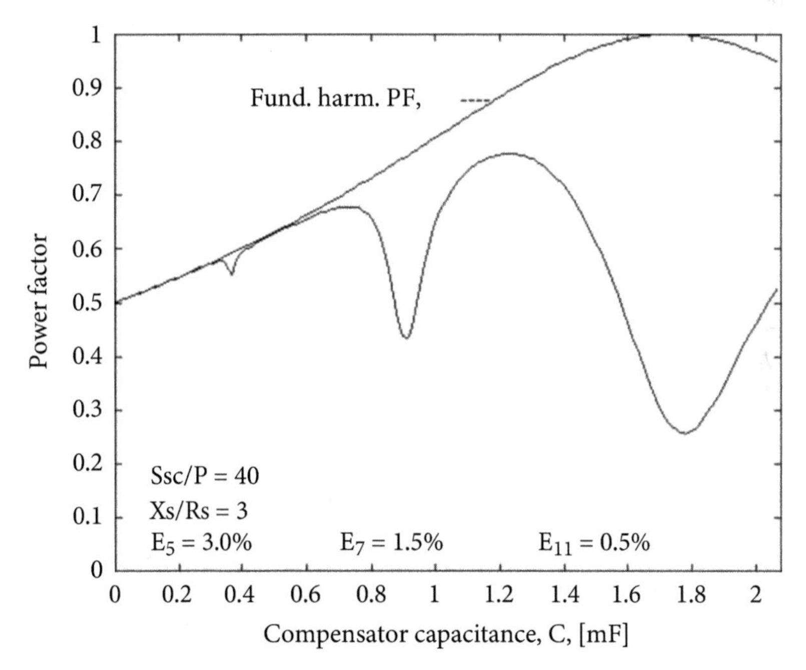

Figure 1.18 Plots of the power factor λ at a sinusoidal and distorted voltage versus compensator capacitance C.

1.13 Why the Term: "Power Quality" Could be Misleading?

Over the last few decades, the electrical engineering community has been increasingly concerned with the effects of the voltage and current distortion, short disturbances, asymmetry, a variation of the voltage and current rms value, or high-frequency noise on the power system and customer equipment. All these agents that would disturb the power system performance were regarded as components of "power quality". Meters and recorders of power quality were developed and a high number of papers on power quality were published, journals established, and conferences held.

Power quality is commonly evaluated based on the measurement of voltages and currents, performed at a point of common coupling (PCC). When it is known what kind of equipment is installed on the customer side and how the PCC is supplied, then such measurement can provide some information on the causes of power quality degradation. Without this knowledge, this information becomes ambiguous. When, for example, a voltage and current distortion is revealed by a power quality meter, it remains unclear what is the cause of this distortion. Is the current distorted because of the voltage distortion or the current is the cause of the voltage distortion? The same goes for the voltage and current asymmetry. If a power quality meter or recorder reveals asymmetry, it is not known what is the cause and what is the effect. The same applies to the voltage and current rms variations, short-time disturbances of the voltage and current waveform, and high-frequency noise.

These questions boil down to where the causes of the power quality degradation are located: on the customer or the energy provider side? Observe that without a correct answer to this question it is not possible to remove the cause, either by the system rearrangement, management, or compensation.

This also raises the question of whether the term "power quality" is appropriate. Power, or rather the energy provided by a supplier, has only one feature: its amount, but not quality. Terms like "high PQ", "low PQ", or good or bad kWhs, are a sort of a jargon. A concept of "*quality*" should be applied to the supply source and the load separately, rather than to the power. Instead of the term "power quality", it seems that more appropriate terms would be the "*loading quality*" LQ and "*supply quality*" SQ. The meanings of these two terms are illustrated in the diagram shown in Fig. 1.19.

A customer load has an ideal quality from the perspective of the energy provider when it is purely resistive and balanced it does not generate current harmonics and it is of the constant power P. Any deviation from such an ideal load, that is the presence of the reactive, unbalanced, or scattered currents and/or current harmonics, increases the energy loss at its delivery. Short transients or variations of the current rms value or a high-frequency noise might require that some expensive countermeasures be undertaken to keep the supply standards. Thus degraded LQ reduces the energy provider revenues.

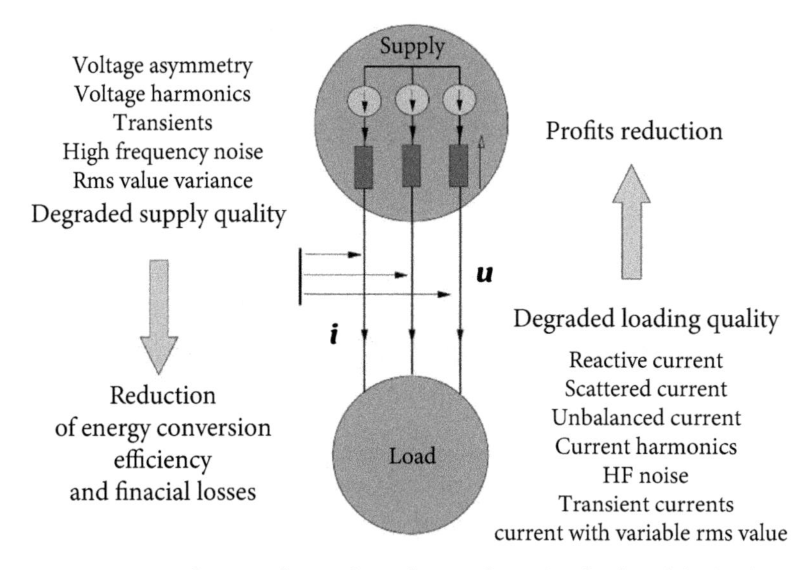

Figure 1.19 A diagram that explains the supply quality (SQ) and the loading quality (LQ) concepts.

A power supply has an ideal quality from a customer perspective when the supplied voltage is sinusoidal, symmetrical, not sensitive to the load power variation, and without any short-time transients, rms variation, or high-frequency noise. Any deviation from such an ideal supply, such as the presence of the voltage distortion, asymmetry, short transients, or variation, of the voltage rms value or high-frequency noise, can disturb the customer's equipment and increase the cost of use of the energy delivered. It can be regarded as degradation of the SQ. The concept of the loading quality and the supply quality is explained in the diagram shown in Fig. 1.19.

Summary

This chapter demonstrates that there are many confusing issues in electrical power systems, caused mainly by the voltage and current distortion or asymmetry. Most of them are related to the interpretation of power-related phenomena.

Chapter 2
Sources of Current and Voltage Distortion

2.1 Nonsinusoidal Voltages and Currents: General

Electric energy is provided to power systems mainly by high-power three-phase synchronous generators driven by steam or hydro turbines in power plants. Some amount of electric energy is produced in industrial plants that need hot steam for heating and various industrial processes. This energy to AC power systems can also be provided by high-power AC/DC power electronics converters supplied from high-voltage DC (HVDC) transmission lines which transmit energy from distant areas. It is also provided by renewable sources of energy, like wind turbines or solar panels. Such wind turbine-driven generators and solar panels are interfaced with the power systems by power electronics-based frequency inverters.

Transmission of electric energy in power systems and its utilization in customer loads are most effective when voltages and currents are sinusoidal. Moreover, in the case of three-phase systems, these quantities should be symmetrical. Therefore, there is considerable effort in the power engineering community aimed at supplying electric energy to power systems at sinusoidal and symmetrical voltage.

The assumption that voltages and currents, denoted generally as $x(t)$, change in electrical circuits as a sinusoidal quantity, namely as

$$x(t) = \sqrt{2}\, X \sin(\omega t - \phi),$$

provides a very useful mathematical model of these quantities. It is commonly used for the analysis, control, and development of power systems. Such an assumption that voltages and currents are sinusoidal can provide quite satisfactory results but sometimes they do not, due to distortion and asymmetry.

The distortion and asymmetry of the voltage produced at terminals of high-power synchronous generators are negligibly small. This voltage is usually considered as ideally sinusoidal and symmetrical. Unfortunately, the voltage as shown in Fig. 2.1 can be observed inside the distribution system.

Powers and Compensation in Circuits with Nonsinusoidal Current. Leszek S. Czarnecki, Oxford University Press.
 DOI: 10.1093/oso/9780198879206.003.0002

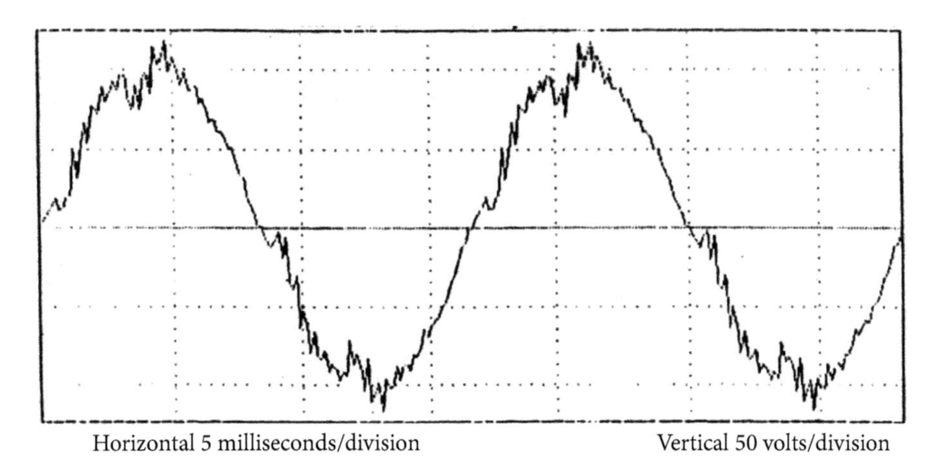

Figure 2.1 Example of a voltage waveform, recorded on 8 kV substation.

There are a few mechanisms that contribute to voltages and currents distortion in distribution systems:

1. The supply source may not deliver the sinusoidal voltage. This applies mainly to power electronics converters that interface HVDC lines, wind-driven generators or solar panels with the AC power system.
2. The nonlinearity of the magnetic core of distribution system transformers.
3. Nonlinearity and/or periodic switching in customer loads. Such loads are referred to as Harmonics Generating Loads.
4. Voltage and current resonances due to the presence in the distribution system and customer loads of both inductances and capacitances.

Concerning the current waveform, loads respond in two ways. Some loads, such as incandescent lamps or AC motors, at sinusoidal supply voltage, draw sinusoidal current. These are *linear time-invariant* (*LTI*) loads. Other loads, such as fluorescent lamps, rectifiers, or arc furnaces, even when supplied with sinusoidal voltage draw nonsinusoidal currents. These are nonlinear or periodically time-variant loads. Such loads that generate current harmonics are referred to as *harmonic generating loads (HGLs)*. The nonlinearity of the magnetic core of transformers and discharge from conductors of very high voltage transmission lines, known as *corona discharge*, contribute to voltage distortion because the discharge current from the line occurs when the voltage reaches its maximum value.

Electrical power system parameters are not constant, moreover. They can change because of switching operations or because of a variation in loads' and generators' power, and due to common disturbances of various kinds. In response to all these changes or transients, voltages and currents in electrical systems are *random quantities*. It would be difficult, however, to analyze the power system and design its components taking into account the randomness of voltages and currents.

All distortions can be categorized as *primary* or *secondary* distortions. The voltage distortion at terminals of the supply source is a primary distortion, or is the current distortion caused by the nonlinearity of magnetic cores of transformers, similarly to the current distortion caused by the load nonlinearity or its periodic switching. This distorted current causes distortion of the voltage drop Δu_s on the distribution system impedance, since, as shown in Fig. 2.2, the voltage at the load terminals is

$$u = e - \Delta u_S.$$

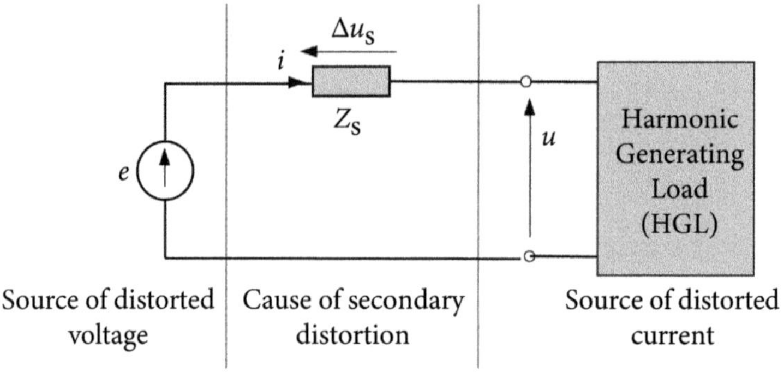

Figure 2.2 Illustration on how the secondary distortion occurs in electrical systems.

Such voltage distortion can be classified as secondary distortion. The same holds true with distortion caused by resonances in the distribution system.

Periodic electrical quantities distorted from a sinusoidal waveform can be presented as a sum of harmonics:

$$x_n = \sqrt{2} X_n \cos(n\omega_1 t + \alpha_n)$$

using the Fourier series. It is used in this book in the complex form

$$x(t) = \overline{x(t)} + \sqrt{2}\,\mathrm{Re}\sum_{n\in N} \boldsymbol{X}_n\, e^{jn\,\omega_1 t} = X_0 + \sum_{n\in N} x_n(t), \qquad \boldsymbol{X}_n = X_n\, e^{j\alpha_n}.$$

More details on Fourier series in the complex form are in Chapter 3. Symbol N in this formula denotes the set of harmonic orders n of dominating harmonics, including the fundamental, thus with $n = 1$. When the quantity $x(t)$ is known in an analytical form so that integration is possible, the Fourier series coefficients $\boldsymbol{X}_n$, referred to as the *complex rms (crms) value* of the nth order harmonic, can be calculated with the integral:

$$\boldsymbol{X}_n = \frac{\sqrt{2}}{T}\int_0^T x(t)\, e^{-jn\omega_1 t} dt. \tag{2.1}$$

Otherwise, having M samples x_m of $x(t)$, taken by an Analog to Digital (A/D) converter at instances:

$$t_m = \frac{T}{M}\, m$$

these coefficients can be calculated with the Discrete Fourier Transform (DFT)

$$\boldsymbol{X}_n = \frac{\sqrt{2}}{M} \sum_{m=0}^{m=M-1} x_m \, e^{-j\, n \frac{2\pi}{M} m}. \tag{2.2}$$

To avoid the spectrum aliasing, and consequently, an aliasing error, the number M of samples taken in one period T should satisfy the Nyquist Criterion, that is it should be at least twice much higher than the order n of the highest order harmonic of $x(t)$.

2.2 Distortion Measures

When the fundamental harmonic $u_1(t)$ is separated from the voltage $u(t)$, then the remaining part, composed of other harmonics, including the DC component U_0, stands for the *voltage distorting component*, $u_{\mathrm{h}}(\mathrm{t})$,

$$u_{\mathrm{h}}(t) = U_0 + \sum_{n\in N,\, n\neq 1} u_n(t) = \sum_{n\in N_{\mathrm{d}}} u_n(t)$$

thus

$$u(t) = u_1(t) + u_{\mathrm{h}}(t).$$

The voltage distortion can be specified quantitatively as the ratio of the rms value of the distorting voltage, $||u_{\mathrm{h}}||$, to the rms value of the voltage fundamental harmonic, $||u_1|| = U_1$. This ratio

$$\delta_{\mathrm{u}} \stackrel{\mathrm{df}}{=} \frac{||u_{\mathrm{h}}||}{U_1} = \frac{\sqrt{\sum_{n\in N_{\mathrm{d}}} U_n^2}}{U_1} \tag{2.3}$$

is referred to as the *Voltage Total Harmonic Distortion* and denoted with the acronym VTHD or simply, the voltage THD. Similarly, the current waveform $i(t)$ can be presented as the sum of its fundamental harmonic $i_1(t)$ and the current distorting component $i_{\mathrm{h}}(t)$, namely

$$i(t) = i_1(t) + i_{\mathrm{h}}(t)$$

with

$$i_{\mathrm{h}}(t) = I_0 + \sum_{n\in M,\ n\neq 1} i_n(t) = \sum_{n\in M_{\mathrm{d}}} i_n(t).$$

The set of orders M_{d} of harmonics that cause the current distortion could be different from the set of harmonics orders N_{d} of harmonics that cause the voltage distortion.

The current distortion can be specified quantitatively as the ratio of the rms value of the distorting current, $||i_h||$, to the rms value of the current fundamental harmonic, $||i_1|| = I_1$. This ratio

$$\delta_i \stackrel{df}{=} \frac{||i_h||}{I_1} = \frac{\sqrt{\sum_{n \in M_d} I_n^2}}{I_1} \tag{2.4}$$

is referred to as the *Current Total Harmonic Distortion* and denoted with the acronym CTHD or simply, the current THD.

Observe that THD values, δ_u, or δ_i, provide only limited information on the waveform distortion. Sometimes is important to know the content of individual harmonics. It could be not the same for some equipment whether the voltage is distorted by 5% of the third order harmonic or by 5% of the thirteen order.

It is recommended by IEEE Std 519 [110] that in the US power system the voltage THD, δ_u, is confined, depending on the voltage range, by the values compiled in Table 2.1.

Table 2.1 Recommended limits of the voltage THD.

Voltage rms range	δ_u
Above 161 kV	1.5 %
69 kV < U < 161 kV	2.5 %
Below 69 kV	5.0 %

These recommended limits are based on conclusions regarding the effects of distortion and a trade-off between energy providers, customers, and equipment manufacturers.

Distorted currents, injected into power systems by harmonic generating loads (HGLs), are dominating sources of waveform distortion. Such currents are injected by low-power, but numerous, single-phase HGLs, mainly in residential and commercial building distribution systems, and by high-power three-phase power electronics-driven loads in industrial distribution systems. These could be variable-speed drives of induction motors needed for manufacturing or transportation processes, in air conditioning systems of commercial buildings, or power electronics drives of pumps for crude oil pipelines.

The asymmetry of the voltage in power systems can be caused by their structural asymmetry and/or by asymmetrical currents. The structural asymmetry can be caused by a geometric asymmetry of three-phase transformers and supply lines. Transformers and transmission lines do not have geometrical symmetry, which depends on the line orientation to other objects, in particular towers and the ground. In effect, the resistance, inductance, and capacitance of individual conductors of the line can mutually differ.

2.3 Harmful Effects of Distortion

There is a great variety of harmful effects related to voltages and currents distortion. On the one hand, these effects can be as direct as a malfunction of equipment or a disturbance of industrial processes. On the other, they can be so distant as an increase in the difficulty and the cost of distribution systems analysis, design, and the prediction of systems performance.

Harmful effects of waveform distortion can be classified in a variety of ways. One of them is their predictability or possibility of evaluation of their effects. Namely

(1) *Easily predictable effects.* It is easy to predict, for example, the voltage distortion caused by the current distortion. It is enough to know the supply system impedance for harmonic frequencies. Distortion increases the current and voltage rms values which can be calculated easily if the harmonic contents are known. Similarly, the waveform distortion increases the active power loss. When the distribution system structure, parameters, and harmonic contents are known, it is not difficult to calculate the increase in the active power loss caused by the waveform distortion. It is similar to the reduction of the mechanical torque of rotating machines due to the supply voltage distortion. The supply voltage distortion also affects the performance of AC/DC converters and rectifiers. Both the output voltage and the supply current depend on the supply voltage harmonics, and these effects are relatively easy to evaluate.

(2) *Effects difficult to evaluate.* The active power loss due to waveform distortion in transformers, AC rotating machines, capacitor banks, or cables increases their temperature. The increased temperature may cause failures and speed up eldering processes and, consequently, may cause a reduction in the lifetime expectation of various equipment. It is very difficult, however, to predict to which extent this lifetime expectation is reduced. This depends on whether that equipment is loaded up to its operational capability or only partially loaded. Harmonics deteriorates the effectiveness of capacitor banks installed for improving the power factor. The resonance may occur between such a capacitor bank and the distribution system inductance in the presence of harmonics. The resonant frequency and its harmful effects are difficult to evaluate. Waveform distortion causes an increase in the measurement error of electric instruments. Its effect on energy meters is particularly important because bills for energy are dependent on this error. Calibration may be needed to conclude the measurement error value. This error changes, however, with distortion level. Also, it is not easy to evaluate how harmonic distortion affects the performance and reliability of the protective equipment or the level of harmonic interaction with telephone lines, which contributes to communication noise.

(3) *Unpredictable effects.* The harmful effects of voltage and current distortion classified in items (1) and (2) refer mainly to the traditional electrical equipment. The digital, computer-like equipment, based on digital signal processes for industrial control, information, science, or medicine, being developed over the last decades is subjected also to unpredictable effects. The operation of such digital equipment can be disturbed not only by harmonics of the supply voltage but also by harmonics in electric or magnetic fields. A random change of a single bit of information may have

drastic effects. Computer computations can be disrupted, important industrial processes can be stopped, and life-support systems can fail. The concern with waveform distortion is generated mainly in recent years by the interference of harmonics just with the digital equipment.

The harmful effects listed above contribute directly to an increase in the cost of electric energy utilization in the presence of voltage and current distortion. To reduce this cost, both technical and regulatory measures are adopted in distribution systems. First of all, the customer equipment should be built in such a way that it produces as few harmonics as possible, or at least that it satisfies regulatory requirements with the generated current harmonics. If this is not possible, filters are installed to reduce harmonics. On the other hand, electrical devices have to be built to be more immune to harmonics. All these countermeasures, as well as various studies on waveform distortion, also elevate the cost, though indirectly, of energy utilization in the presence of harmonics.

2.4 Distortion Caused by Ferromagnetic Core

Nonlinear ferromagnetic cores of power system transformers are one of the common causes of magnetizing current distortion. This can be explained by calculating the current of a coil with a ferromagnetic core, as shown in Fig. 2.3, assuming that coil voltage is sinusoidal.

The magnetic flux density B in the core is not proportional to the field intensity H.

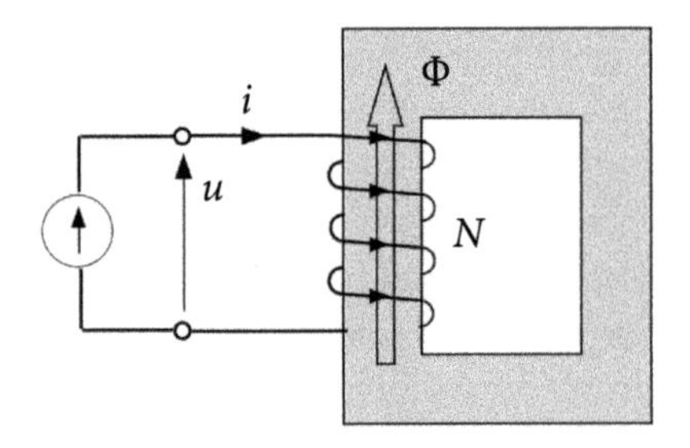

Figure 2.3 A coil with a ferromagnetic core.

Assuming that the hysteresis loop is very narrow, the relationship between the flux density B and the magnetic field intensity H, illustrated in Fig. 2.4, can be approximated by a polynomial:

$$H = a_0 B + b_0 B^3 + c_0 B^5 + d_0 B^7 + \dots \quad (2.5)$$

where a_0, b_0, c_0, and d_0 are coefficients dependent on the core material. The magnetic field intensity, H, in the core is proportional, according to Amper's Law, to the coil current i, namely

$$H = \frac{Ni}{l}$$

where N is the number of turns and l is the average length of the magnetic path.

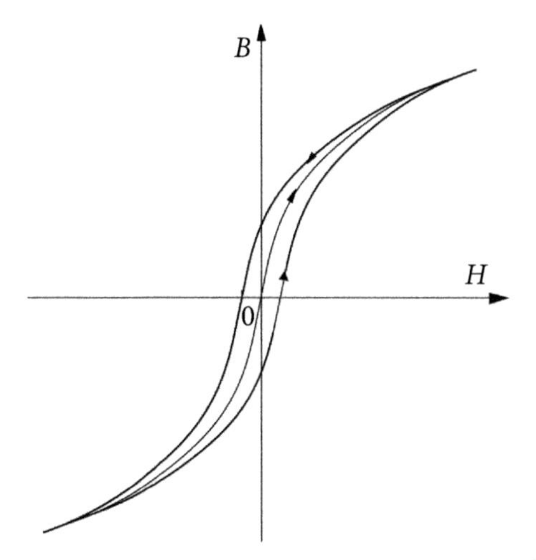

Figure 2.4 Hysteresis loop of a ferromagnetic core.

The voltage on the coil is equal to

$$u = N\frac{d\Phi}{dt} = NA\,\frac{dB}{dt}$$

where A is the core cross-section area. Assuming that the coil voltage is sinusoidal,

$$u = \sqrt{2}\,U\cos\omega_1 t$$

the magnetic flux density is equal to

$$B = \frac{1}{NA}\int u\,dt = \frac{\sqrt{2}\,U}{NA\omega_1}\,\sin\omega_1 t.$$

Taking into account the two first terms of the polynomial (2.3), we obtain

$$\frac{Ni}{l} = a_0\left(\frac{\sqrt{2}\,U}{NA\omega_1}\,\sin\omega_1 t\right) + b_0\left(\frac{\sqrt{2}\,U}{NA\omega_1}\,\sin\omega_1 t\right)^3.$$

Since $\sin^3\alpha = \frac{3}{4}\sin\alpha - \frac{1}{4}\sin 3\alpha$, the coil current can be expressed as

$$i = \sqrt{2}\,U(a + b\,U^2)\sin\omega_1 t - \sqrt{2}\,c\,U^3\sin 3\omega_1 t$$

where

$$a = \frac{a_0\,l}{N^2A\omega_1}\quad b = \frac{3}{2}\frac{b_0\,l}{N^4A^3\omega_1^3}\quad c = \frac{1}{2}\frac{b_0\,l}{N^4A^3\omega_1^3},\tag{2.6}$$

or as

$$i = \sqrt{2}\,I_1 \sin\omega_1 t + \sqrt{2}I_3 \sin(3\omega_1 t - \pi) = i_1 + i_3.$$

Thus, the coil current, meaning a magnetizing current of a transformer, is not sinusoidal but is distorted by the third order harmonic. Its waveform is shown in Fig. 2.5.

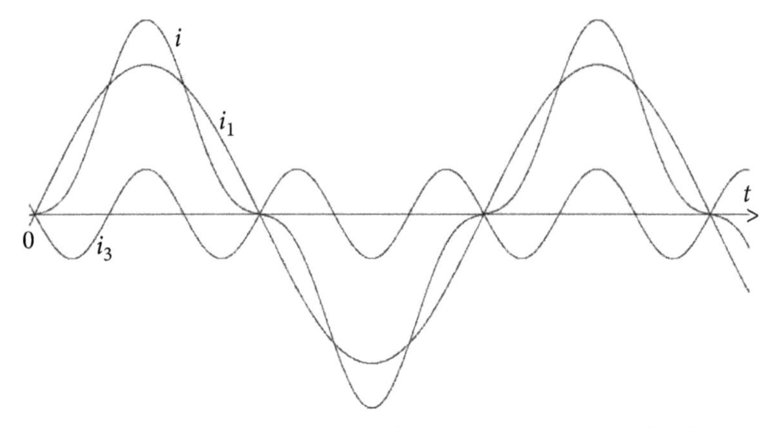

Figure 2.5 A waveform of the transformer's magnetizing current and its harmonics.

The coil current is composed of the fundamental and the third order harmonic. The content of the third order harmonic, at specified core dimension, A and l, number of turns N, and frequency ω_1, depends on the supply voltage rms value, U. Observe, that the third order harmonic increases with the third power of the supply voltage. Thus, the voltage strongly affects the distortion of inductors and transformer currents.

When more terms in the magnetizing curve approximation (2.5) are taken into account, meaning also terms with power 5, 7, and higher, then also harmonics of the order 5, 7, and higher will occur in the magnetizing current of the transformer. It may be easier, however, to find the contents of these harmonics by measurement rather than analytically. In particular, we find that mathematical models of a magnetic core, due to hysteresis phenomena, are usually not very accurate or coefficients a, b, and c in formula (2.6) are not available.

In effect of the primary distortion and distortion of the power system transformers magnetizing current is that voltages in the power systems are distorted. Distortion caused by HGLs is superimposed on the primary distortion and the distortion caused by transformers.

The voltage and current distortion have to be specified by some quantitative measures that would enable us to gain a rough comparison of the level of distortion. A concept of total harmonic distortion provides such a measure.

Current Distortion Caused by Loads with Odd *u*-*i* Relationship. Most single-phase nonlinear resistive loads have an odd voltage-current relationship (Fig. 2.6), meaning that with the change of the load voltage to a negative value, the current changes to a negative value as well.

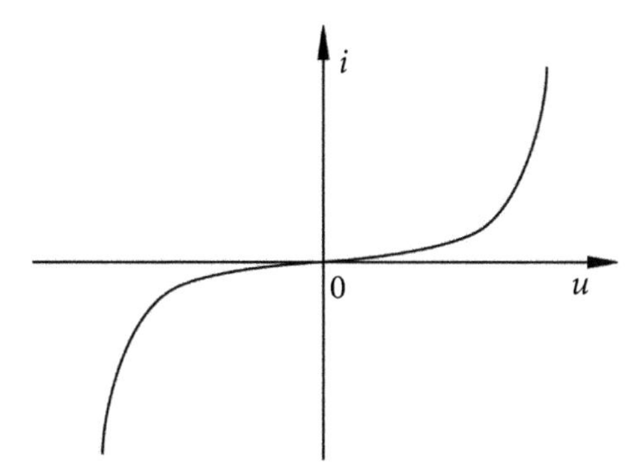

Figure 2.6 Example of the voltage-current odd relationship.

This relationship can be expressed by the polynomial

$$i = au + bu^3 + cu^5 + \dots$$

with non-negative coefficients, *a, b, c...* . Such a property has fluorescent bulbs and rectifiers in supply circuits of various video and computer-like equipment and DC power chargers. All of them have power ratings in the order of tens and usually no more than a couple of hundred volt-ampers (VA) but they are very numerous in every residential home and commercial building. An example of the current waveform is shown in Fig. 2.7.

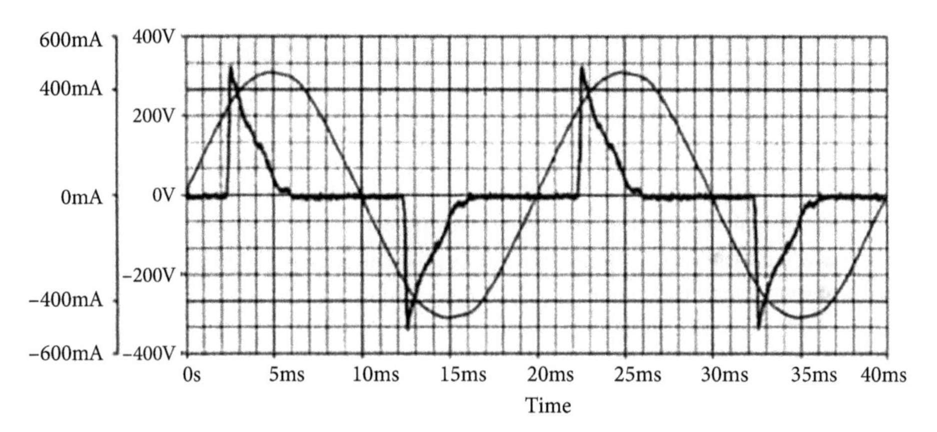

Figure 2.7 A waveform of a fluorescent bulb current.

Currents of such devices may have a variety of different waveforms but all of them have a *half-wave negative symmetry*, shown in Fig. 2.8, meaning that

$$i\left(t + \frac{T}{2}\right) = -i\,(t).$$

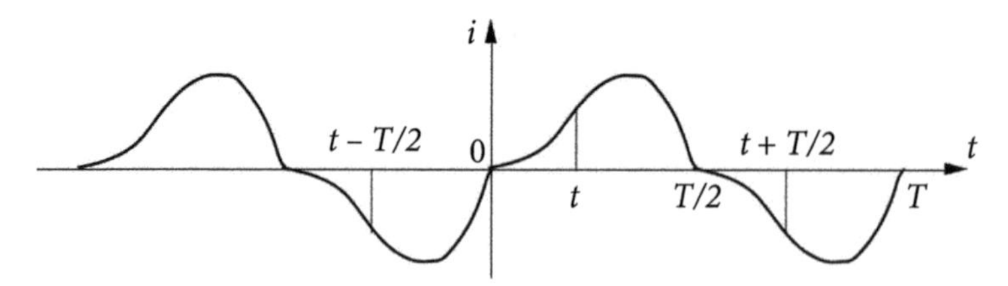

Figure 2.8 Current waveform with half-wave negative symmetry.

The same property has currents and voltages of distribution system transformers. Quantities with such a property do not have harmonics of the even order n. Thus, the currents in such systems can have only the form

$$i(t) = \sqrt{2} \sum_{n \in N} I_n \sin(n\omega_1 t - \alpha_n), \quad N \stackrel{\text{df}}{=} \{1, 3, 5, 7, ...\}.$$

The third order harmonic is usually the dominating current harmonic generated by single-phase nonlinear devices. Harmonics of the even order, $n = 0, 2, 4 \ldots 2k$, can occur only if the half-wave negative symmetry of the load current is not satisfied. This can happen only when some harmonics of the even order are present in the supply voltage and/or when a nonlinear load behaves at a negative voltage differently than at a positive one.

2.5 Current Distortion and the Power Factor

The supply voltage distortion is usually negligible compared to distortion of the load current so that the supply voltage rms value $||u|| = U_1$. In such a situation, only the current fundamental harmonic i_1 contributes to the energy transfer from the supply source to the load. When the load is purely resistive, then this fundamental harmonic is in phase with the supply voltage so that its phase shift $\varphi_1 = 0$, and hence the power factor of such a load is equal to

$$\lambda \stackrel{\text{df}}{=} \frac{P}{S} = \frac{U_1 I_1 \cos\varphi_1}{||u||\,||i||} = \frac{I_1}{||i||}. \tag{2.7}$$

It declines from the unity value not because of the current phase shift versus the voltage but only because of the current harmonic distortion. Therefore, it can be referred to as a *distortion power factor* λ_{d}.

$$\lambda \stackrel{\text{df}}{=} \frac{I_1}{||i||} = \lambda_{\text{d}}.$$

The THD of the current, $\delta_{\rm i}$, as defined by (2.3), can be expressed in the form

$$\delta_{\rm i} \stackrel{\rm df}{=} \frac{||i_{\rm h}||}{I_1} = \frac{\sqrt{||i||^2 - I_1^2}}{I_1} = \sqrt{\frac{1}{\lambda_{\rm d}^2} - 1}$$

meaning the distortion power factor $\lambda_{\rm d}$ specifies the current distortion, or the current distortion, $\delta_{\rm i}$, reduces the power factor $\lambda_{\rm d}$ to the value

$$\lambda_{\rm d} = \frac{1}{\sqrt{\delta_{\rm i}^2 + 1}}.$$

In the presence of reactive elements, the load current fundamental harmonic can be shifted against the supply voltage as shown in Fig. 2.9.

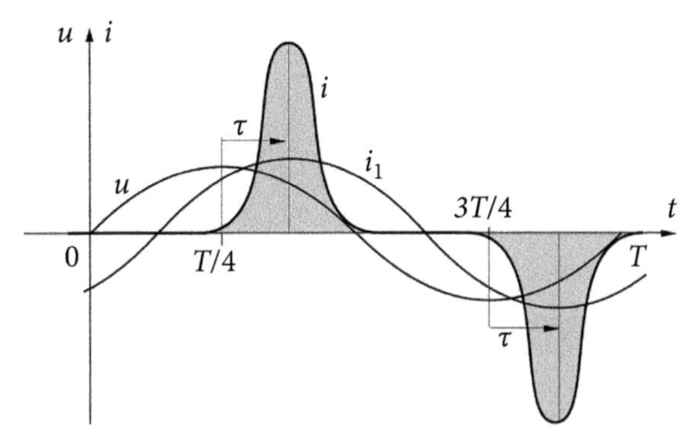

Figure 2.9 An example of the current and its fundamental harmonic of a HGL with reactive components.

Consequently, the power factor can be expressed in the form

$$\lambda \stackrel{\rm df}{=} \frac{P}{S} = \frac{I_1}{||i||} \cos\varphi_1 = \lambda_{\rm h} \times \lambda_1$$

where

$$\lambda_1 = \cos\varphi_1$$

is referred to as a *displacement power factor*, or the power factor of the load for the fundamental harmonic. This formula is important in situations when a compensator of the reactive power of the fundamental harmonic Q_1 for improving the power factor is installed on the system. A capacitor bank is usually used for that purpose. Indeed, such a compensator can improve the displacement power factor λ_1 to unity. Unfortunately, such a compensator in the presence of harmonics can increase the total harmonic distortion, $\delta_{\rm I}$, of the supply current. Consequently, such compensation may be not as effective as expected or can even worsen the power factor at the supply source terminals.

2.6 Lightning Systems as the Source of Distortion

In residential, commercial, and office distribution systems, lightning is the major source of the current distortion. Lightning is provided now mainly by gas-discharge bulbs and light-emitting diode (LED) bulbs. These are very low-power devices, usually below 10 watts, but they may be numerous. Only sodium or mercury gas-discharge lamps used for outdoor lighting systems have higher power, in the range of a few hundred watts.

All gas-discharge lamps operate with *a ballast,* which confines the lamp current. Usually an inductor, as shown in Fig. 2.10, serves as ballast, called a *magnetic ballast,* though an electronic converter can also be used for that purpose.

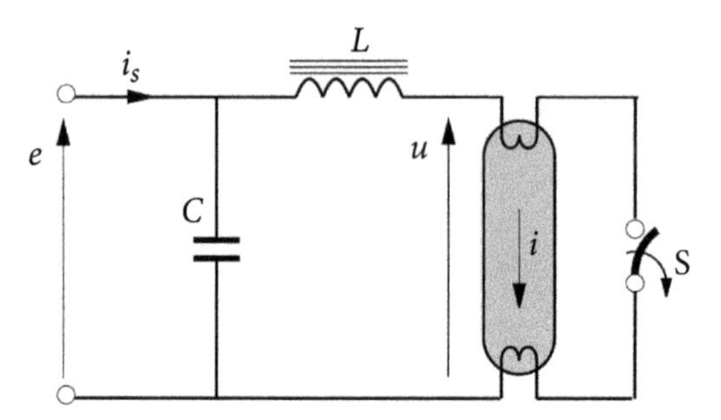

Figure 2.10 Fluorescent lamp with magnetic ballast.

A gas-discharge lamp is built with a tube filled with mercury vapor, argon or neon gas. The tube has a fluorescent phosphor coating and two pre-heated electrodes. The lamp operates in such a way that mercury vapor, argon or neon gas inside of the lamp produces, at some voltage, short-wave ultraviolet light. This light is converted by phosphor on the inner surface of the lamp into visible light. To enable the emission of electrons, the switch S is initially closed, and electrodes are heated. The bimetallic thermal switch, S, opens when the electrodes have sufficient temperature for electron emission. Its opening creates a high voltage pulse of a value higher than U_f, shown in the voltage–current relationship of a gas-discharge lamp in Fig. 2.11, which enables ionization of vapor in the lamp and strikes the arc between electrodes. The capacitor, C, reduces the reactive power Q of the magnetic ballast to improve its power factor.

A gas-discharge lamp starts to conduct when the voltage between electrodes reaches a firing U_f value, and consequently, the lamp current has a form of positive and negative short pulses. The current waveform of the whole device, including magnetic ballast, L, and the power factor correcting capacitor, C, strongly depends on the circuit parameters. Analysis of circuits with gas-discharge lamps is difficult because of the complexity of the mathematical model of physical processes inside the lamp. Consequently, a measurement is practically the only approach to the identification of properties of systems with such lamps.

Since LEDs need a DC supply, such diodes are integrated with rectifiers. Consequently, LED-based lighting systems are a sort of distributed rectifier circuit. The

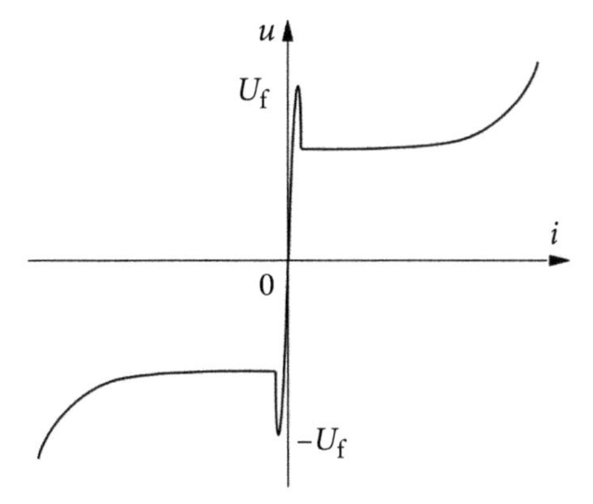

Figure 2.11 Voltage-current relationship of gas-discharge lamp.

same applies to video and computer-like residential and commercial devices. All of them are seen from the supply perspective as single-phase rectifiers.

2.7 Single-Phase Rectifiers

All video and computer-like home devices or battery chargers require a DC supply which is provided by single-phase rectifiers. Such rectifiers are built as full-wave rectifiers with a supply transformer or a voltage divider. Full-wave rectification eliminates a DC component from the supply current, while the supply transformer is necessary to fit the rectifier output voltage, for example, $U_d = 6$ V, to the fixed AC grid voltage, for example, with $U = 120$ V.

Single-phase rectifiers, depending on requirements as to the rectification efficiency, can be built with the output (DC) voltage filter in the form of a capacitor, an inductor, or without any filter. The most primitive rectifiers, such as car battery chargers, of the structure shown in Fig. 2.12, are built usually without any filter. It is important to observe that such a charger has an active load, meaning a load with an internal voltage E. The presence of a filter is of major importance for the supply current waveform.

Rectifier Without any Filter. In the case of rectifiers without any filter but with an active load, like that of the structure shown in Fig. 2.12, diodes D_1 and D_2 conduct on the condition that $u_a > E$, while diodes D_3 and D_4 conduct when $-u_a > E$. When $|u_a| < E$, the rectifier current is equal to zero.

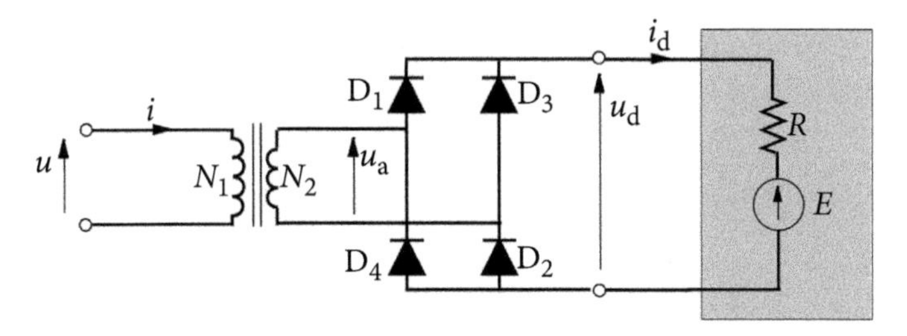

Figure 2.12 Single-phase, full-wave rectifier with active load.

Assuming that the supply voltage is

$$u = \sqrt{2}U \sin \omega_1 t$$

then the rectifier current can be approximated by the waveform specified as follows:

$$i = \begin{cases} \dfrac{N_2}{N_1}\dfrac{\sqrt{2}U_a \sin \omega_1 t - E}{R}, & \text{for} \quad u_a > E \\[2ex] 0, & \text{for} \quad |u_a| < E \\[2ex] \dfrac{N_2}{N_1}\dfrac{\sqrt{2}U_a \sin \omega_1 t + E}{R}, & \text{for} - u_a > E \end{cases} \tag{2.8}$$

The rectifier current waveform, approximated in such a way, is shown in Fig. 2.13. This approximation does not take diodes and transformer parameters into account. Diodes are regarded as ideal, meaning they are in the ON state at positive anode-cathode voltage, without any resistance. The transformer is also ideal, without stray inductance and any active power loss. Modeling is needed, however, to find the waveform of the rectifier current in a more realistic rectifier circuit.

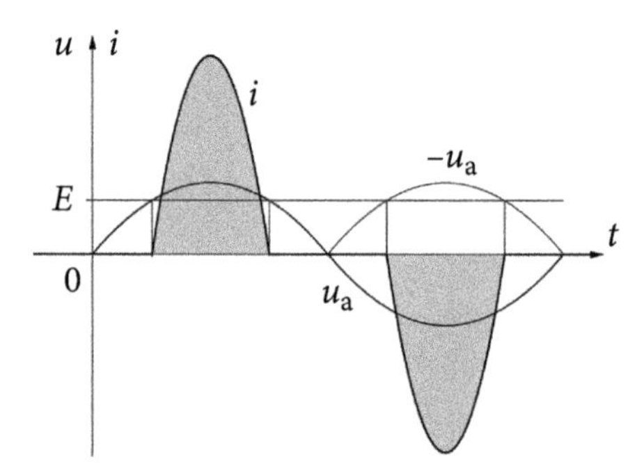

Figure 2.13 An approximation of the rectifier current.

An example of this current waveform obtained using PSpice modeling software is shown in Fig. 2.14. The current waveform obtained from modeling is closer, of course, to the real waveform than that obtained analytically in Fig. 2.13. Even if more accurate, modeling provides only a particular result. It does not provide general properties of the modeled object. Such properties can be found if a mathematical model, as specified by formula (2.8) is available. For example, we can find how the current THD, δ_i, or the power factor, λ, depends on the supply current pulse duration, α.

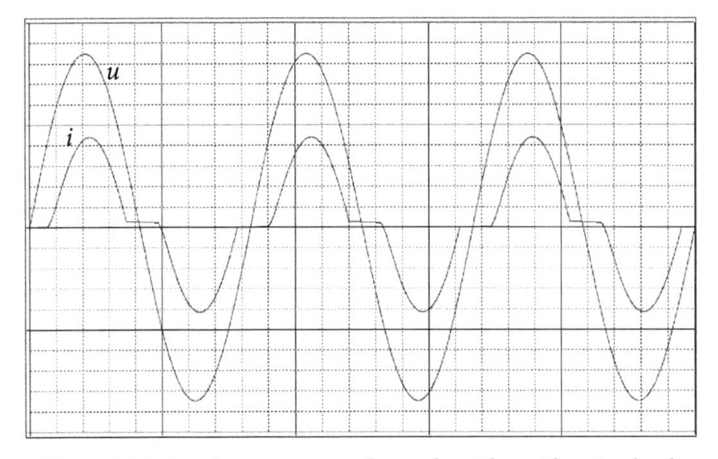

Figure 2.14 Supply current waveform of rectifier with active load.

Observe that the rectifier current can be decomposed into pulses shown in Fig. 2.15, specified in a half of a period as

$$x \stackrel{\text{df}}{=} \begin{cases} I_m\,[\cos(\omega_1 t) - \cos(\alpha/2)], & \text{for } \ |\omega_1 t\,| < \alpha/2 \\ 0, & \text{for } \ |\omega_1 t| \ > \alpha/2 \end{cases}.$$

The current maximum value I_m in Fig. 2.15 is equal to the current amplitude in the circuit shown in Fig. 2.12, when $E = 0$, meaning

$$I_m = \frac{N_2}{N_1}\frac{\sqrt{2}\,U_a}{R}$$

while the angle $\alpha/2$ has to fulfill the relation

$$\sqrt{2}\,U_a \cos(\alpha/2) - E = 0, \quad \text{thus} \quad \alpha = \cos^{-1}\left\{\frac{E}{\sqrt{2}\,U_a}\right\}$$

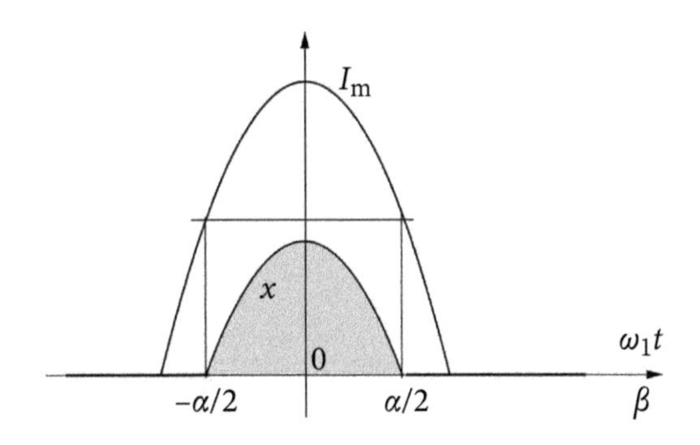

Figure 2.15 A current pulse waveform.

The rms value of the rectifier current can be calculated as

$$||i|| = \sqrt{2}\,||x|| = \sqrt{2}\sqrt{\frac{1}{2\pi}\int_{-\pi/2}^{\pi/2} x^2(\omega_1 t)\, d(\omega_1 t)} =$$

$$= \sqrt{2}\sqrt{\frac{1}{2\pi}\int_{-\alpha/2}^{\alpha/2} I_m^2\,[\cos(\omega_1 t) - \cos(\alpha/2)]^2\, d(\omega_1 t)} =$$

$$= I_m\sqrt{\frac{\alpha\,(2 + \cos\alpha) - 3\sin\alpha}{2\pi}}. \tag{2.9}$$

Assuming that the rectifier supply voltage is

$$u = \sqrt{2}\,U\cos\omega_1 t$$

then the supply current can be expressed as

$$i(t) = x(t) - x(t - T/2).$$

Therefore, (see Chapter 3) the crms value of the current fundamental harmonic is equal to

$$\boldsymbol{I}_1 = 2\boldsymbol{X}_1 = 2\frac{1}{\sqrt{2}\,\pi}\int_{-\pi/2}^{\pi/2} x(\omega_1 t)\, e^{-j\omega_1 t}\, d(\omega_1 t) =$$

$$= 2\frac{1}{\sqrt{2}\,\pi}\int_{-\alpha/2}^{\alpha/2} I_m\,[\cos(\omega_1 t) - \cos(\alpha/2)]\, e^{-j\omega_1 t}\, d(\omega_1 t) =$$

$$= \frac{I_m}{\sqrt{2}\,\pi}(\alpha - \sin\alpha). \tag{2.10}$$

Thus, the power factor of the rectifier λ changes with the pulse duration α as

$$\lambda \overset{\text{df}}{=} \frac{P}{S} = \frac{I_1}{||i||} = \frac{\alpha - \sin\alpha}{\sqrt{\pi}\,\sqrt{\alpha\,(2 + \cos\alpha) - 3\sin\alpha}}. \tag{2.11}$$

Illustration 2.1 *Let us calculate the power factor λ and the supply current total harmonic distortion, THD, of a rectifier with an active load, that operates at the DC voltage $E = U_a$.*

The duration of the supply current pulse at such voltage, measured in radians, is

$$\alpha = 2\cos^{-1}\left\{\frac{E}{\sqrt{2}U_a}\right\} = 2\cos^{-1}\left\{\frac{1}{\sqrt{2}}\right\} = \frac{\pi}{2}.$$

Thus, the power factor of the rectifier has the value

$$\lambda = \frac{\frac{\pi}{2} - \sin\frac{\pi}{2}}{\sqrt{\pi}\sqrt{\frac{\pi}{2}(2 + \cos\frac{\pi}{2}) - 3\sin\frac{\pi}{2}}} = \frac{\frac{\pi}{2} - 1}{\sqrt{\pi}\sqrt{\frac{\pi}{2}(2) - 3}} = 0.86.$$

The THD of the rectifier supply current is equal to

$$\delta_i = \sqrt{\frac{1}{\lambda^2} - 1} = \sqrt{\frac{1}{0.86^2} - 1} = 0.59$$

*****

Rectifier With a Capacitive Filter. Such a rectifier has the structure shown in Fig. 2.16. Ripples of the output DC voltage of a rectifier can disturb the operation of some DC loads. If needed, these ripples can be reduced by an inductive or a capacitive filter. Cheaper are usually capacitive filters and such filters are commonly used in low-power DC suppliers of video or computer-like home appliances.

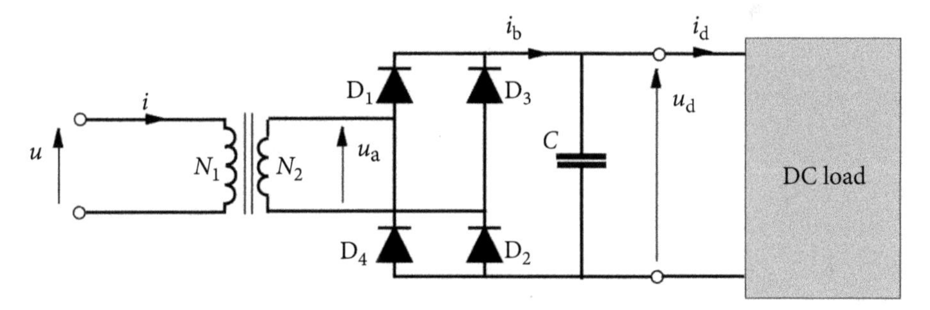

Figure 2.16 A rectifier with a capacitive filter structure.

The capacitor is charged when the module of the input voltage u_a is higher than capacitor voltage u_d, that is, when $|u_a| > u_d$, as shown in Fig 2.17, which presents the results of the circuit modeling with PSpice. The capacitor is discharged to a DC load in the remaining intervals of time and consequently the load voltage in these intervals declines exponentially. The output current of the diode bridge $i_b(t)$ has a form of a sequence of short pulses of the same positive polarity as shown in Fig 2.17. The supply current of the rectifier is composed of positive and negative charging pulses.

A PSpice model of the rectifier can be used for calculating not only currents and voltages waveform but also harmonics and THD of the rectifier current. Such a

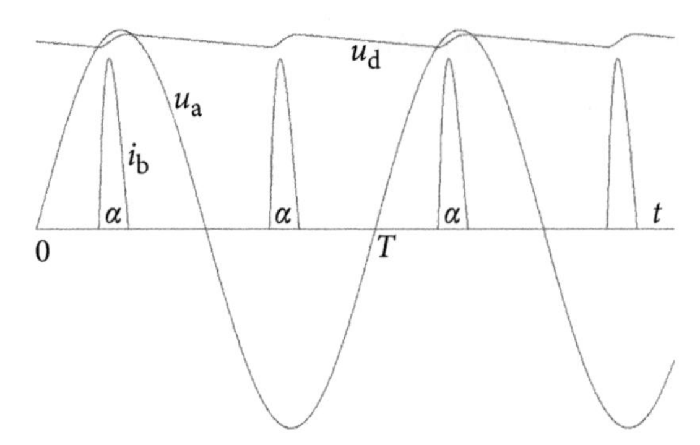

Figure 2.17 The charging current waveform of rectifier with capacitive filter.

model is not capable of showing the factors that affect δ_i value, however. Analytical results are needed for that. To calculate the THD of the supply rectifier current, some approximations have to be made.

The current distortion is not affected, of course, by the transformer turn ratio, N_1/N_2, so that we can assume that $N_1/N_2 = 1$. At a cost of some inaccuracy, we can assume that the transformer and rectifier diodes are ideal, so that $u_a = u$. Moreover, we can assume that filtering of the output voltage u_d ripples is ideal, meaning that $u_d = U_d = \text{const.} = \sqrt{2}U$.

Let us assume that the charging current pulse has average value I_b, as shown in Fig. 2.18, and duration, in radians, α, while the DC load current has mean value I_d. In the steady-state

$$u_d\,[(k+1)\,T] = u_d\,[kT]$$

on the condition that

$$I_b \times 2\alpha - I_d \times 2\pi = 0$$

thus

$$I_b = \frac{\pi}{\alpha} I_d \overset{\text{df}}{=} \frac{I_d}{d}.$$

The coefficient $d = \alpha/\pi$ is referred to as a *duty factor*.

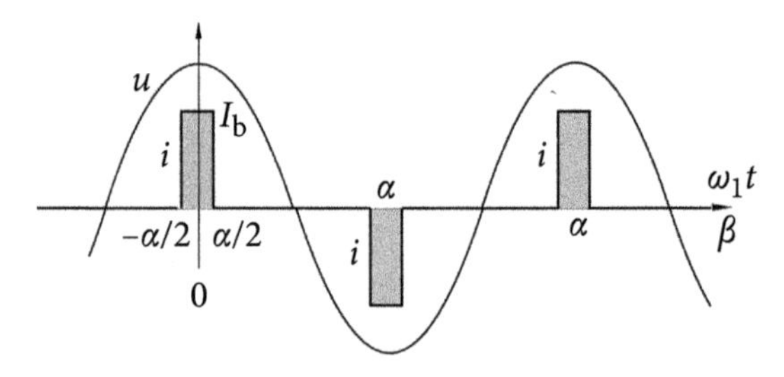

Figure 2.18 An approximation of the rectifier current000.

The rms value of the rectifier current $i(t)$ approximated by rectangular pulses is equal to

$$||i|| = \sqrt{2}\sqrt{\frac{1}{2\pi}\int_{-\alpha/2}^{\alpha/2} I_{\mathrm{b}}^2\, d(\omega_1 t)} = I_{\mathrm{b}}\sqrt{\frac{\alpha}{\pi}} = \frac{I_{\mathrm{d}}}{\sqrt{d}} \tag{2.12}$$

thus the current rms value increases with the duty factor, meaning the current pulse duration α declines. If, as assumed, transformer and rectifier diodes are ideal, then the active power of the rectifier at sinusoidal supply voltage

$$P = UI_1 \cos\varphi_1 \approx UI_1$$

is equal to the DC load power

$$P_{\mathrm{d}} = U_{\mathrm{d}} I_{\mathrm{d}} = I_{\mathrm{d}}\sqrt{2}\,U.$$

Thus, approximately

$$I_1 = \sqrt{2}\,I_{\mathrm{d}}.$$

Therefore, the power factor of the rectifier with a capacitive filter is approximately equal to

$$\lambda \stackrel{\mathrm{df}}{=} \frac{P}{S} = \frac{I_1}{||i||} = \sqrt{2d}$$

meaning it declines with the duty factor decline. Consequently, the THD value of the rectifier current is

$$\delta_{\mathrm{i}} = \sqrt{\frac{1}{\lambda^2} - 1} = \sqrt{\frac{1}{2d} - 1} = \sqrt{\frac{\pi}{2\alpha} - 1}. \tag{2.13}$$

For example, if the capacitor charging current has the duration α =15 deg, meaning

$$\alpha = \frac{15}{180}\pi = 0.083\,\pi\,[\mathrm{rad}],$$

then the THD of the rectifier current is approximately equal to

$$\delta_{\mathrm{i}} = \sqrt{\frac{\pi}{2\alpha} - 1} = \sqrt{\frac{\pi}{2 \times 0.083\,\pi} - 1} = 2.2$$

or δi = 220%. This means that the current of rectifiers with a capacitive filter can be strongly distorted. This distortion increases with the filter capacitance C, meaning that with the reduction of the rectifier output voltage u_{d} ripples. With their reduction, the charging current pulses become shorter, meaning their duration α declines.

Since the supply current of rectifiers with a capacitive filter has a form of short pulses with high THD, therefore, only at a very low power rating of the rectifier, such capacitive filtering is acceptable. At higher power ratings, inductive filters are preferable.

A Rectifier with an Inductive Filter has a structure as shown in Fig. 2.19. The inductance L of the filter is usually selected such that the inductor's reactance for the fundamental harmonic $\omega_1 L$ is much higher than the equivalent resistance of the DC load, R_d, meaning that $\omega_1 L \gg R_d$. When this condition is fulfilled, then the inductor current has an almost constant value, that is, $i_d \approx I_d$. Such a rectifier provides DC voltage u_d of the mean value

$$U_d = \frac{2\sqrt{2}}{\pi} U_a. \tag{2.14}$$

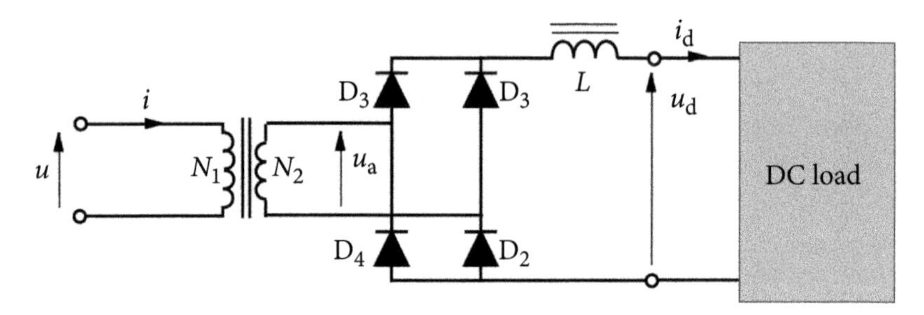

Figure 2.19 A rectifier with an inductive filter structure.

This almost constant i_d current is observed in the rectifier's supply lines approximately as a sequence of positive and negative pulses of the average amplitude equal to $N_2/N_1 \times I_d$, and half-period duration, $T/2$.

The supply current waveform, obtained from a PSpice model of a rectifier with an inductive filter, and $N_2/N_1 = 1$, is shown in Fig. 2.20.

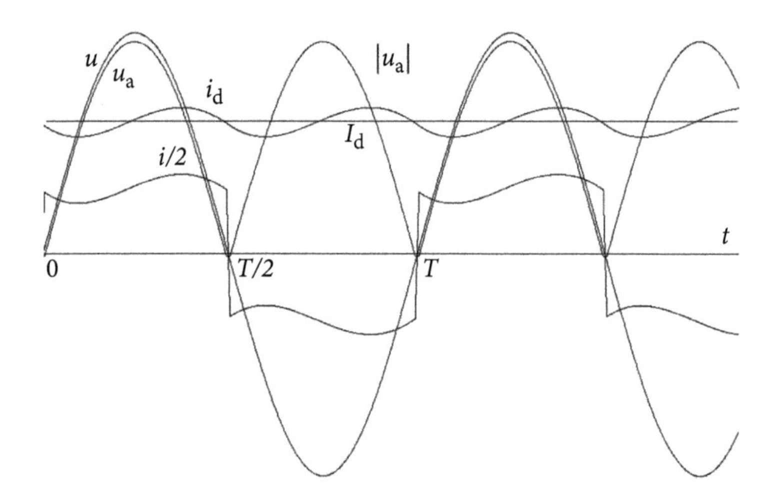

Figure 2.20 Voltages and currents waveforms of a rectifier with an inductive filter.

To differentiate the inductor and the supply current, the supply current is reduced by half in this figure. Let us calculate the THD of the supply current, assuming that the rectifier, transformer, and filtering inductor L are ideal, and the output voltage ripples are ideally filtered, meaning the rectifier output current is constant, that is:

$$i_{\mathrm{d}} = \text{const.} = I_{\mathrm{d}}.$$

Let us assume, moreover, that the turn ratio $N_2/N_1 = 1$. At such assumptions, the input current rms value $||i|| = I_{\mathrm{d}}$. Assuming that the rectifier, transformer, and filtering inductor are lossless, the active power at rectifier supply terminals is equal to the DC power at rectifier output, thus

$$UI_1 \cos \varphi_1 = U_{\mathrm{d}} I_{\mathrm{d}} = \frac{2\sqrt{2}}{\pi} UI_{\mathrm{d}}.$$

Since $\cos\varphi_1 \approx 1$, the rms value of the supply current fundamental harmonic is equal to

$$I_1 = \frac{2\sqrt{2}}{\pi} I_{\mathrm{d}}. \tag{2.15}$$

Therefore, the power factor of the rectifier with a capacitive filter is approximately equal to

$$\lambda \overset{\mathrm{df}}{=} \frac{P}{S} = \frac{I_1}{||i||} = \frac{2\sqrt{2}}{\pi} = 0.90.$$

Consequently, the THD value of the rectifier current is

$$\delta_{\mathrm{i}} = \sqrt{\frac{1}{\lambda^2} - 1} = \sqrt{\frac{1}{0.90^2} - 1} = 0.48.$$

This distortion of the supply current caused by real rectifiers, due mainly to the transformer stray inductance, is even lower. Because of this inductance, the transformer current, meaning the rectifier supply current, cannot change direction instantaneously. Time, referred to as *commutation time* is needed for that. This reduces high-order current harmonics and consequently the total harmonic distortion, δ_{i}.

2.8 Three-Phase Rectifiers

Single-phase HGLs, although numerous, have low power and consequently such loads are not usually the major source of harmonic distortion in power systems. Three-phase HGLs are mainly responsible for this distortion.

Although high-power rectifier banks in electrochemistry and, in particular, rectifiers for aluminum electrolysis, as well as high-power arc furnaces in metallurgy, are powerful sources of the current distortion in power systems, the role of power electronics in waveform distortion is continuously increasing. It is because the control of energy flow with power electronics (PEs) converters enables huge savings of energy in various industrial processes. In effect, more and more motors are supplied from such PEs converters. Current harmonics and distortion are by-products of these economical benefits.

Three-Phase Six-Pulse Rectifier is the most common three-phase HGL. Such rectifiers are used to provide DC supply voltage for high-power DC loads such as motors or electrochemical processes or as DC supply for PE converters. Such a rectifier, even without any filter, provides DC voltage with much lower ripples than those at single-phase rectification, and consequently, for some applications filters are not needed. A three-phase transformer, referred to as a *rectifier transformer*, is usually needed for fitting the rectifier, with the required DC voltage, to the existing grid voltage. The rectifier transformer can have Y/y, Y/d, Δ/y, or Δ/d configuration.

A structure of such a rectifier with Y/y transformer is shown in Fig. 2.21. The rectifier bridge, built of six diodes D1, D2, ...D6, operates as a selector of the highest line-to-line instantaneous voltage value from six voltages

$$u_{ab}, -u_{ab}, u_{bc}, -u_{bc}, u_{ca}, -u_{ca}.$$

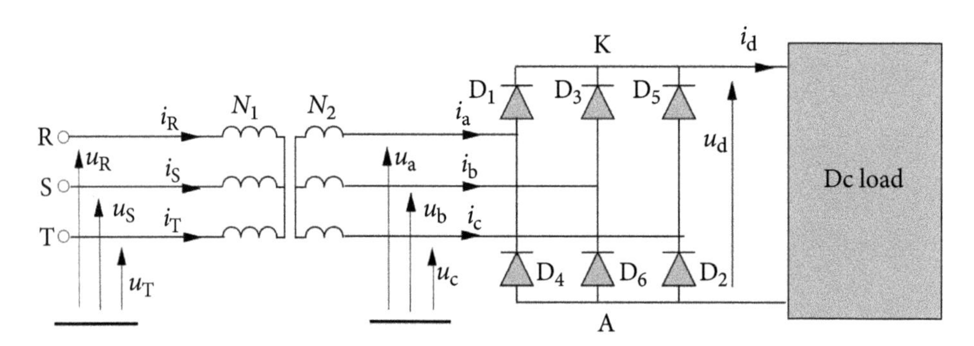

Figure 2.21 Three-phase six-pulse rectifier structure.

The highest of these voltages occurs at the rectifier output terminals, thus the output voltage u_d is composed of six pulses of $T/6$ duration, as shown in Fig. 2.22. Assuming that the stray inductance of the transformer is equal to zero, the line currents can change direction instantaneously, then at each instant of time only two out of three supply lines deliver the voltage and the current to the rectifier output.

Thus, each line is engaged in the voltage and current delivery to the DC load only over two-thirds of the period T. The supply voltage and current waveforms of line R of a rectifier without any filter, along with the output voltage u_d, for a situation where the rectifier has relatively low power as compared to the short-circuit power of the supply source, are shown in Fig. 2.22. The line supply current in such a situation changes almost instantaneously and the supply voltage remains almost unaffected.

The output voltage of the rectifier u_d is composed in one period T of six pulses of line-to-line voltages of 60° duration, shown in Fig. 2.23 so that its mean value is equal to

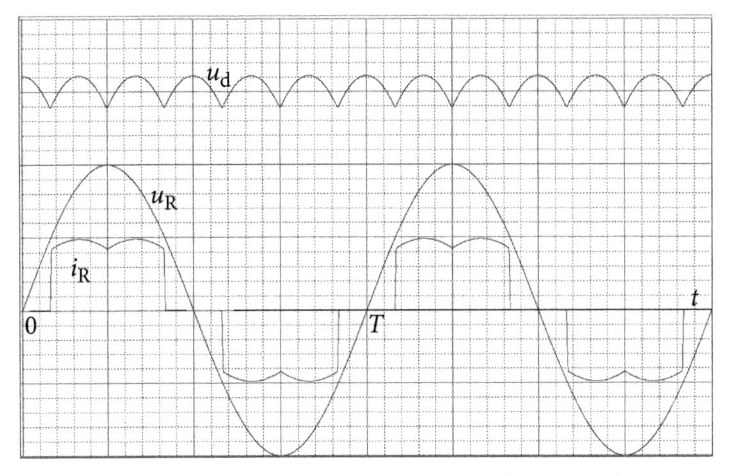

Figure 2.22 Waveforms of line R current and output voltage of a three-phase rectifier without a filter.

$$U_d = 6\,\frac{1}{2\pi}\int_{-30^0}^{30^0} \sqrt{3}\sqrt{2}\,U_a \cos\beta \,d\beta = \frac{3\sqrt{6}}{\pi}U_a. \tag{2.16}$$

To calculate the THD, δ_t, of the rectifier current, let us observe that the transformer turn ratio does not affect distortion. Let us assume that this turn ratio is equal to 1. Also assume that there is no energy loss in the rectifier and the transformer, thus

$$P = 3\,UI_1 \cos\varphi_1 \approx U_d\,I_d = \frac{3\sqrt{6}}{\pi}U\,I_d. \tag{2.17}$$

Since $\varphi_1 = 0$, the fundamental harmonic of the line current rms value is approximately equal to

$$I_1 \approx \frac{\sqrt{6}}{\pi}I_d. \tag{2.18}$$

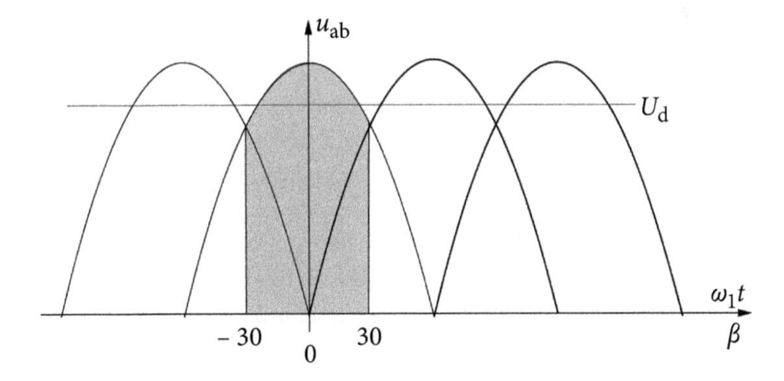

Figure 2.23 One pulse of the rectifier output voltage.

In one cycle, the supply current is composed of four (two positive and two negative) pulses $x(t)$, similar to the output voltage $u_{\rm d}$ pulses, shown in Fig. 2.23, of duration $T/6$ and approximately $I_{\rm d}$ value, thus its rms value is

$$||i|| = \sqrt{4\,||x||^2} = \sqrt{4\,\frac{1}{2\pi}\int_{-\pi/6}^{\pi/6} I_{\rm d}^2\, d\beta} = \sqrt{\frac{2}{3}}\, I_{\rm d}. \tag{2.19}$$

Thus, the power factor of three-phase rectifiers without a filter is equal to

$$\lambda \stackrel{\rm df}{=} \frac{P}{S} = \frac{I_1}{||i||} \approx \frac{\frac{\sqrt{6}}{\pi} I_{\rm d}}{\sqrt{\frac{2}{3}} I_{\rm d}} = \frac{3}{\pi} = 0.95$$

and the THD of the supply current is

$$\delta_{\rm i} = \sqrt{\frac{1}{\lambda^2} - 1} \approx \sqrt{\left(\frac{\pi}{3}\right)^2 - 1} = 0.31.$$

The rectifier supply currents in lines S and T are identical to the current in phase R, only they are shifted versus that current by one-third of the period T. Thus, at each instant of time each line current has to satisfy the relation

$$i(t) + i(t - T/3) + i(t + T/3) \equiv 0.$$

Therefore (see Chapter 3) the crms values $\boldsymbol{I}_n$ of the line current harmonics have to fulfill the relation

$$\boldsymbol{I}_n + \boldsymbol{I}_n\, e^{-jn\omega_1 \frac{T}{3}} + \boldsymbol{I}_n\, e^{jn\omega_1 \frac{T}{3}} \equiv 0$$

or

$$\boldsymbol{I}_n \left(1 + e^{-jn\frac{2\pi}{3}} + e^{jn\frac{2\pi}{3}}\right) \equiv 0.$$

This is possible only on the condition that for $n = 3k$, $k = 0, 1, 2...$, the crms value is $\boldsymbol{I}_n = 0$. Thus line currents of three-phase rectifiers cannot contain harmonics of the order that is a multiplicity of three, such as 3rd, 9th, 15th.... They do not have the zero, 6th, 12th, 18th... order harmonics as quantities with half-wave negative symmetry. Consequently, the rectifier current contains only harmonics of the order $n = 1, 5, 7, 11, 13, 17, 19... 6k \pm 1$. These harmonics are often referred to as *characteristic harmonics* of three-phase, six-pulse rectifiers.

Other harmonics can occur in three-phase system only when the line currents do not have three-phase symmetry. Unbalanced HGLs, in particular single-phase HGLs, can cause such line current asymmetry or an asymmetry in the supply voltage.

If the supply line current waveform, shown in Fig. 2.22, of three-phase rectifiers is approximated by rectangular, meaning pulses without ripples, then the rms values of the current harmonics change as

$$I_n = \frac{I_1}{n}, \quad n = 6k \pm 1, \quad k = 1,\ 2,\ 3...$$

meaning they decline hyperbolically with the harmonic order, n. Thus, as compared to the rms value of the fundamental harmonic, the supply current contains 20% of the fifth order harmonic, 14% of the seventh order, 9% of eleventh order, 7.7% of the thirteenth order, and so on. When measured on the rectifier terminals, these are usually the upper limits of these harmonics content. It is because the supply line currents of three-phase rectifiers do not change instantaneously between zero and, approximately, $\pm I_{\mathrm{d}}$ value. Some time is needed to change the current because of energy stored in the supply inductance.

As it can be seen on the plot of line current i_{R} in Fig. 2.22, there are four points of time in each period T when the line current changes between zero and, approximately, $\pm I_{\mathrm{d}}$ value. Because of the rectifier transformer's stray inductance this change cannot be instantaneous. Short intervals of time, referred to as *commutation intervals*, are needed for this. In each such interval, two diodes conduct, as shown in Fig. 2.24; namely the diode in the line where the current changes from zero to $\pm I_{\mathrm{d}}$ value and the diode in the line where the current changes from $\pm I_{\mathrm{d}}$ value to zero. This means that in the commutation intervals, the rectifier transformer is short-circuited by such two diodes. Despite this short-circuit, the line currents are not limited only by the distribution system impedance, as it is a common short-circuit state in power systems. The sum of currents of commutating lines is almost constant and equal to $\pm I_{\mathrm{d}}$.

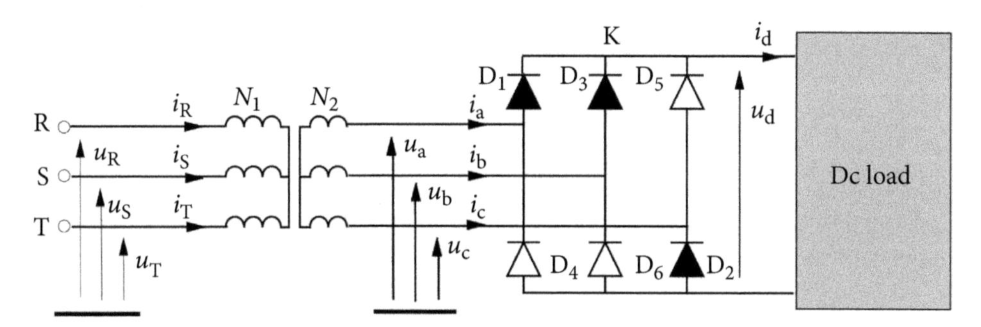

Figure 2.24 Conducting and not conducting diodes in a commutation interval of time.

Commutation affects the supply voltage: four *voltage notches* occur during the line current commutation in each period, as seen in Fig. 2.25.

These notches can disturb other loads, in particular those supplied from the bus at which the rectifier is installed. The effect of commutation increases with the increase in the rectifier power as compared to the short-circuit power, S_{SC}, at the point of supply.

Commutation will be analyzed in Chapter 3 where distortion caused by AC/DC converters will be discussed. From that analysis comes the result that the

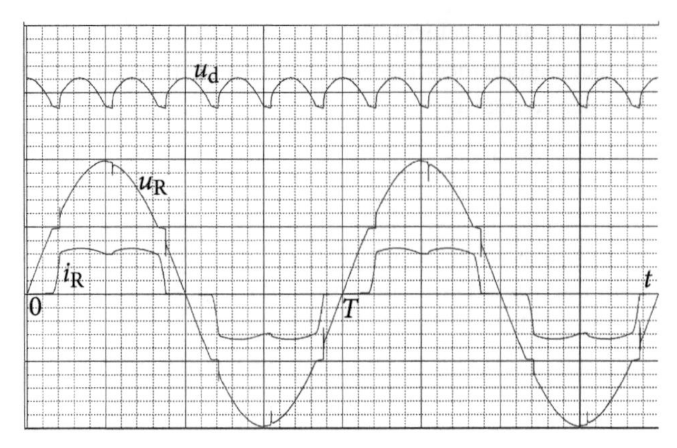

Figure 2.25 Supply current i_R of a rectifier waveform with commutation effects.

commutation angle, μ, meaning the commutation interval recalculated to an angle, for a three-phase rectifier is equal to

$$\mu = \cos^{-1}\left\{1 - \sqrt{\frac{2}{3}\frac{I_d}{I_{sc}}}\right\}. \tag{2.20}$$

Symbol I_{sc} denotes the short-circuit current rms value at the rectifier terminals, equal to

$$I_{sc} = \frac{S_{sc}}{3\,E} \approx \frac{S_{sc}}{3\,U},$$

and E is the rms value of the internal voltage of the distribution system. Taking into account that the rectifier active power is

$$P = \frac{3\sqrt{6}}{\pi} U I_d, \text{ and hence } I_d = \frac{\pi}{3\sqrt{6}}\frac{P}{U},$$

the commutation angle can be expressed in terms of the rectifier to short-circuit power ratio,

$$\mu = \cos^{-1}\left\{1 - \sqrt{\frac{2}{3}\frac{I_d}{I_{sc}}}\right\} = \cos^{-1}\left\{1 - \frac{\pi}{3}\frac{P}{S_{sc}}\right\}.$$

For example, if $S_{sc} = 50\,P$, then

$$\mu = \cos^{-1}\left\{1 - \frac{\pi}{3}\frac{P}{S_{sc}}\right\} = \cos^{-1}\left\{1 - \frac{\pi}{3}\frac{1}{50}\right\} = \cos^{-1}\{0.98\} = 11.7^0.$$

*****

Three-Phase Rectifier with Capacitive Filter. When a level of the rectifier output voltage ripples cannot be tolerated by a DC load, then, similarly as in single-phase systems, capacitive or inductive filters are used. Three-phase rectifier with a capacitive filter has a structure shown in Fig. 2.26. There are six current pulses in single period T charging the capacitor C to the input voltage maximum value. After it is charged, the capacitor discharges through the load. The output voltage ripples are effectively reduced. However, the supply current of the rectifier is strongly distorted.

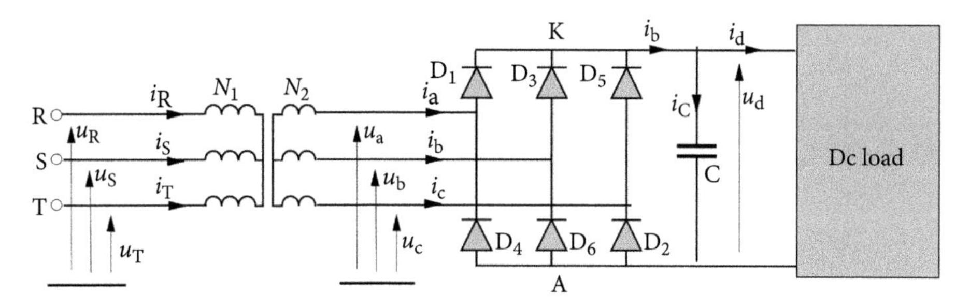

Figure 2.26 A three-phase six-pulse rectifier with a capacitive filter structure.

The supply current waveform of a rectifier with a capacitive filter, obtained with a PSpice model, is shown in Fig. 2.27. The current is composed of four positive and negative pulses and consequently, the supply current has usually a high THD coefficient.

To evaluate its value, let us approximate these pulses with rectangular pulses of I_b maximum value and τ duration. The capacitor is charged by six of such pulses, denoted as $x(t)$, and discharged with the DC load current of I_d mean value. The mean value of the capacitor current, i_C, is zero, if

$$6\, I_b\, \tau = I_d\, T, \text{ hence } I_b = \frac{T}{6\tau} I_d.$$

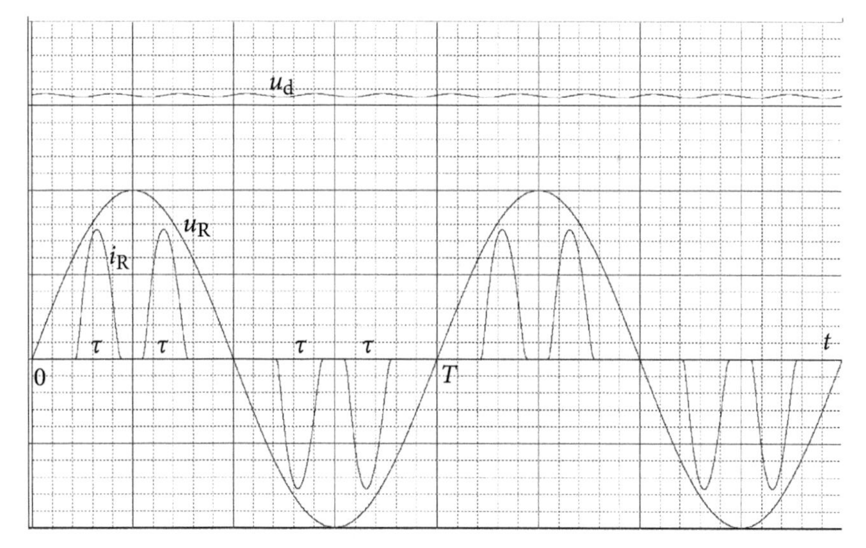

Figure 2.27 Waveforms of voltages and the supply current i_R of a three-phase rectifier with a capacitive filter.

Since the supply line current is composed of four, positive and negative, charging pulses $x(t)$, its rms value is equal to

$$||i|| = \sqrt{4\,||x||^2} = 2\sqrt{\frac{1}{T}\int_{-\tau/2}^{\tau/2} I_b^2\,dt} = 2\,I_b\sqrt{\frac{\tau}{T}} = \frac{1}{3}\sqrt{\frac{T}{\tau}}I_d. \tag{2.21}$$

Let us assume that the transformer turn ratio, which does not affect the current distortion, is 1 and both the transformer and diodes are ideal. Then the active power of the rectifier at sinusoidal supply voltage

$$P = 3\,UI_1\cos\varphi_1 \approx 3\,UI_1$$

is equal to the DC load power

$$P_d = U_d I_d = I_d\sqrt{2}\sqrt{3}\,U.$$

Thus, approximately

$$I_1 \approx \sqrt{\frac{2}{3}}\,I_d.$$

Therefore, the power factor of the rectifier with a capacitive filter is approximately equal to

$$\lambda \overset{\text{df}}{=} \frac{P}{S} = \frac{I_1}{||i||} \approx \sqrt{\frac{6\tau}{T}}.$$

Consequently, the THD value of the rectifier current is

$$\delta_i = \sqrt{\frac{1}{\lambda^2} - 1} \approx \sqrt{\frac{T}{6\,\tau} - 1}. \tag{2.22}$$

For example, the ratio of the supply current period T to the pulse duration τ for the current shown in Fig. 2.27, is approximately $T/\tau = 20/2.2$, thus the THD of the supply current is equal to

$$\delta_i \approx \sqrt{\frac{T}{6\,\tau} - 1} = \sqrt{\frac{20}{6\times 2.2} - 1} = 0.71.$$

*****

This distortion is substantially higher compared to its value for a rectifier without any filter. Therefore, even if this kind of filtering is acceptable in the case of low-power single-phase rectifiers, it is rather not tolerated in the case of high-power three-phase rectifiers.

Three-Phase Rectifier with Inductive Filter. The structure of a rectifier with an inductive filter is shown in Fig. 2.28. The inductance L is usually selected such that its reactance for the fundamental frequency is much higher than the load resistance. In effect, the output current i_d ripples are attenuated, and this current is almost constant, that is, $i_d \approx I_d$.

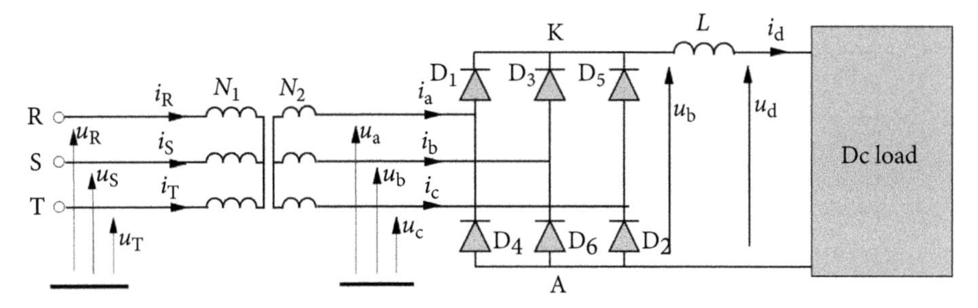

Figure 2.28 The structure of a three-phase rectifier with an inductive filter.

The rectifier bridge acts as a switch that commutates this constant current from line to line. Diodes with the highest line-to-line voltage are in the ON state, and the remaining diodes are in the OFF state. The supply current waveform in line R, obtained from a PSpice model, is shown in Fig. 2.29. It was assumed in the modeled rectifier that its power is substantially lower than the short-circuit power of the supply system and consequently, commutation intervals are not observed in Fig. 2.29.

The harmonic contents and THD of the supply current are practically the same as in the case of a rectifier without any filter. Consequently, they almost do not differ in their effect on the distribution system. Benefits are only on the rectifier DC side.

A three-phase rectifier can be supplied not only from a rectifier transformer in Y/y configuration but in any other possible configuration. This configuration may

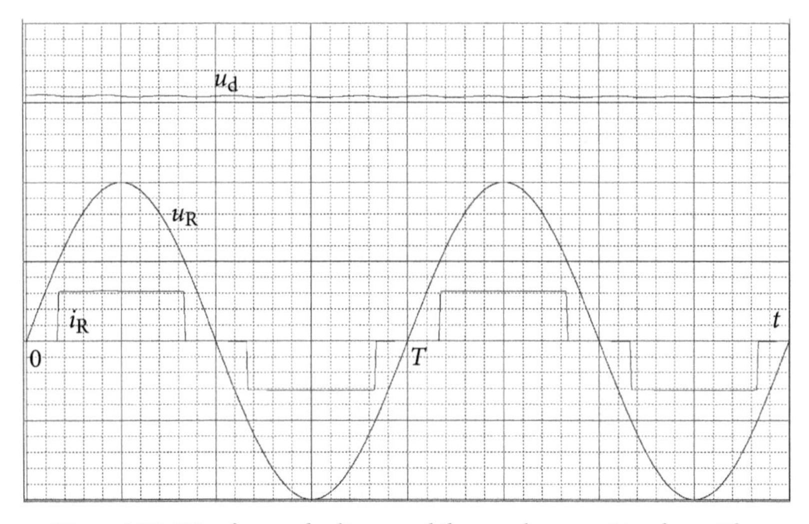

Figure 2.29 Waveforms of voltages and the supply current i_R of a rectifier with an inductive filter.

affect the supply current waveform. It is illustrated with a rectifier supplied from a transformer in Δ/y configuration, as shown in Fig. 2.30.

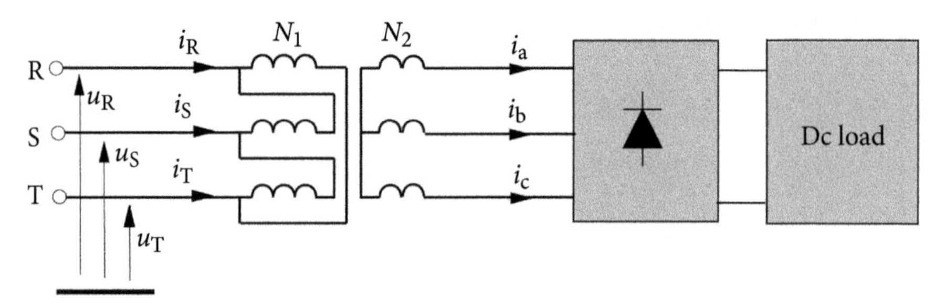

Figure 2.30 A rectifier with Δ/y transformer.

Observe that at the same turn ratio, N_1/N_2, reconfiguration of the primary side of the transformer from Y to Δ increases the line voltage rms value on the transformer secondary side by a root of three. It is because line-to-line voltages are on primary windings. Thus, such a rectifier is equivalent concerning the output voltage u_d, to a rectifier with Y/y transformer, only if the turn ratio, N_1/N_2, of the transformer in Δ/y configuration is reduced by a root of three.

The waveform of the supply current in line R, obtained from a PSpice model of a rectifier with Δ/y transformer and an inductive filter, is shown in Fig. 2.31.

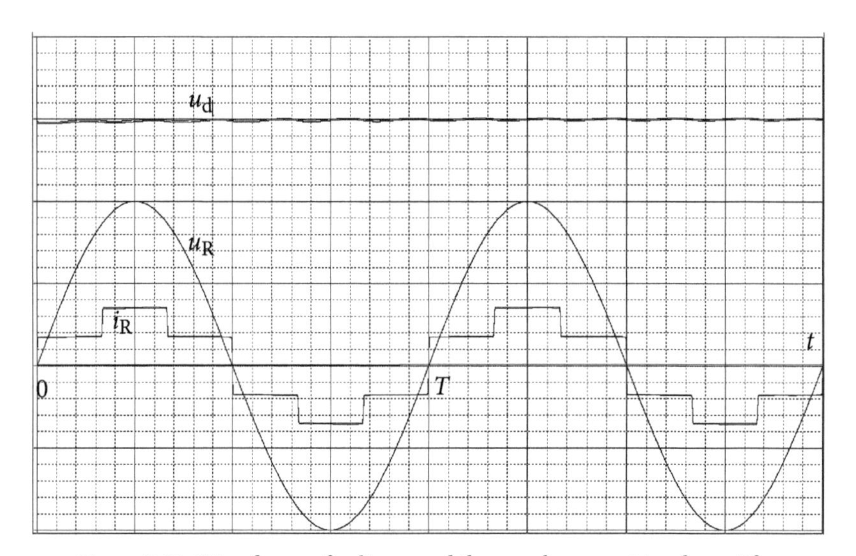

Figure 2.31 Waveforms of voltages and the supply current i_R of a rectifier with Δ/y transformer.

The current waveform might create an impression that it better approximates a sinusoidal current, so that it has lower THD than the supply current of the rectifier with Y/y transformer. Unfortunately, it is only apparent. The supply current on the primary side of the transformer, assuming, that the turn ratio $N_1/N_2 = 1$, is a difference of line currents on the secondary side. In particular,

$$i_{\mathrm{R}}(t) = i_{\mathrm{a}}(t) - i_{\mathrm{b}}(t) = i_{\mathrm{a}}(t) - i_{\mathrm{a}}(t - T/3),$$

therefore, crms values of the current harmonics have to satisfy (see Chapter 3) the relation

$$\boldsymbol{I}_{\mathrm{R}n} = \boldsymbol{I}_{\mathrm{a}n} - \boldsymbol{I}_{\mathrm{a}n}\, e^{-jn\,\omega_1 \frac{T}{3}} = \boldsymbol{I}_{\mathrm{a}n}\left(1 - e^{-jn\,\frac{2\pi}{3}}\right) = \begin{cases} \sqrt{3}\,\boldsymbol{I}_{\mathrm{a}n}\, e^{j30^\circ}, & n = 1, 7, \ldots n = 6k + 1 \\ \sqrt{3}\,\boldsymbol{I}_{\mathrm{a}n}\, e^{-j30^\circ}, & n = 5, 11, . \, n = 6k - 1 \end{cases}$$

which means, that the transformer does not change the current harmonic contents. Only the current waveform, but not its rms value and harmonic contents, is affected by the transformer's windings configuration.

Three-Phase Twelve-Pulse Rectifier. When rectifier power is very high, of the order of megawatts, then the supply current distortion caused by rectifiers discussed above and their impact on the distribution system are usually not acceptable. Twelve- or even twenty-four-pulse rectifiers are used instead.

A twelve-pulse rectifier is composed of two six-pulse rectifiers connected in parallel, but supplied with the three-phase voltage shifted mutually by 30°. This phase shift is provided by transformers that have different winding configurations; for example, a pair of transformers {Δ/Y and Δ/Δ}, or {Y/Y and Y/Δ}. It can be a single transformer with double secondary windings; for example, Y/(y+Δ), or Δ/(Δ+Y).

A structure of a three-phase twelve-pulse rectifier with {Y/Y and Y/Δ} transformer is shown in Fig. 2.32. To calculate the THD of the supply current, assuming that the turn ratio of the Y/Y transformer is one while for the Y/Δ transformer this ratio is $1/\sqrt{3}$. The supply voltage at the lower rectifier, at such assumption, is deleted by 30°, or $T/12$ versus the supply voltage of the upper rectifier. Thus,

$$u_{\mathrm{r}}(t) = u_{\mathrm{a}}\left(t - \frac{T}{12}\right)$$

and consequently, the line currents of the lower rectifier are deleted by $T/12$ versus the supply current of the upper rectifier.

The supply current in the line R of the rectifier is composed of i_{a}, i_{r}, and i_{s} currents and can be expressed as

$$\begin{aligned} i_{\mathrm{R}}(t) &= i_{\mathrm{a}}(t) + \sqrt{3}\left\{\frac{1}{3}[i_{\mathrm{r}}(t) - i_{\mathrm{s}}(t)]\right\} = \\ &= i_{\mathrm{a}}(t) + \frac{1}{\sqrt{3}}\left[i_{\mathrm{a}}\left(t - \frac{T}{12}\right) - i_{\mathrm{a}}\left(t - \frac{T}{12} - \frac{T}{3}\right)\right] = \\ &= i_{\mathrm{a}}(t) + \frac{1}{\sqrt{3}}\left[i_{\mathrm{a}}\left(t - \frac{T}{12}\right) - i_{\mathrm{a}}\left(t - \frac{5T}{12}\right)\right]. \end{aligned}$$

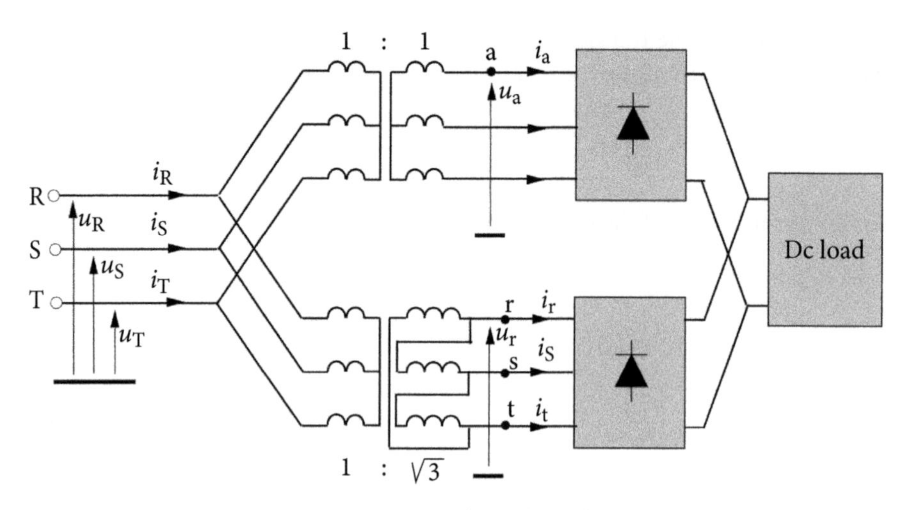

Figure 2.32 A structure of a twelve-pulse rectifier.

Because this current is a linear form of mutually shifted components, the crms value of this current harmonics are equal to

$$\boldsymbol{I}_{\mathrm{R}n} = \boldsymbol{I}_{\mathrm{a}n}\left[1 + \frac{1}{\sqrt{3}}\left(e^{-jn\,\omega_1 \frac{T}{12}} + e^{-jn\,\omega_1 \frac{5T}{12}}\right)\right].$$

The crms values $\boldsymbol{I}_{an}$ are non-zero only for $n = 6k \pm 1$, meaning for $n = 1, 5, 7, 11, 13, \ldots$ and for these harmonics

$$\boldsymbol{I}_{\mathrm{R}n} = \boldsymbol{I}_{\mathrm{a}n}\left[1 + \frac{1}{\sqrt{3}}\left(e^{-jn\frac{\pi}{6}} + e^{-jn\frac{5\pi}{6}}\right)\right] = \begin{cases} 0, & \text{for } n = 5,\ 7,\ 17,\ 19\ldots \\ 2\boldsymbol{I}_{\mathrm{a}n}, & \text{for } n = 1,\ 11,\ 13,\ 23,\ \ldots \end{cases}.$$

This means that the supply current, of the waveform in line R, shown in Fig. 2.33, apart from the fundamental harmonic, has only harmonics of the order

$$n = 12\,k \pm 1, \quad k = 1,\ 2\ldots$$

Their relative rms value, in the ideal case, when the commutation phenomenon is neglected, is equal to

$$I_n = \frac{I_1}{n},$$

thus,

$$I_{11} \approx 9.1\% \text{ of } I_1, \quad I_{13} \approx 7.7\% \text{ of } I_1, \quad I_{23} \approx 4.3\% \text{ of, } I_1, \quad I_{25} \approx 4.0\% \text{ of } I_1, \ \ldots$$

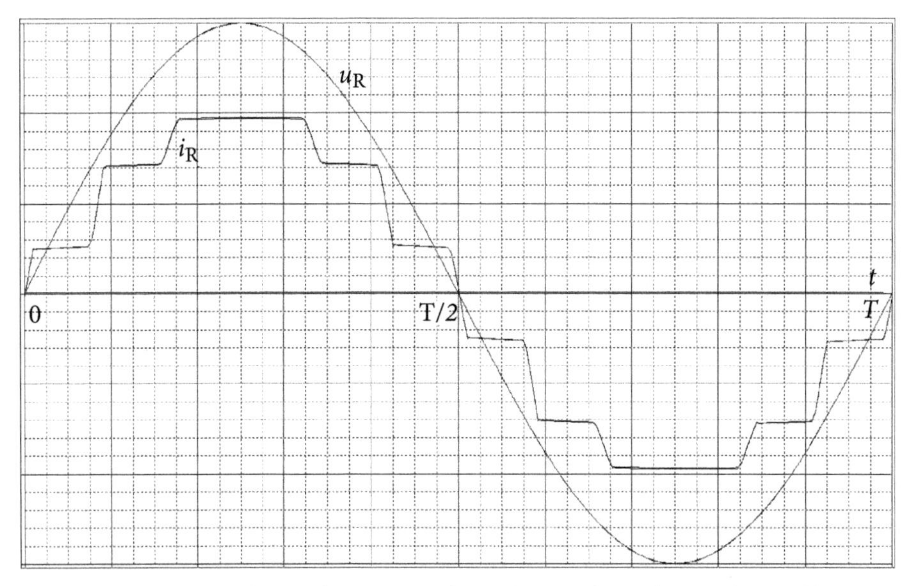

Figure 2.33 Waveforms of current i_R of 12-pulse rectifier with an inductive filter.

The THD of the supply current of three-phase twelve-pulse rectifiers is approximately equal to

$$\delta_i \stackrel{df}{=} \frac{||i_h||}{I_1} = \frac{\sqrt{\sum_{n \in N_d} I_n^2}}{I_1} \approx \sqrt{\frac{1}{11^2} + \frac{1}{13^2} + \frac{1}{23^2} + \frac{1}{25^2} + \frac{1}{35^2} + \ldots} = 0.16.$$

Practically, the supply current of a 12-pulse rectifier contains some amount of the fifth and the seventh order harmonics because conditions for their entire elimination can be fulfilled only approximately. In particular, the turn ratio of the Y/Δ transformer can only approximate the $1/\sqrt{3}$ value. Some asymmetries of the rectifier transformers can also contribute to the fifth and seventh order harmonic increase.

In situations when the level of the supply current total harmonic distortion δ_i provided by a 12-pulse rectifier is not acceptable, and especially in very high-power rectifiers, 24-pulse rectifiers can be built.

2.9 Three-Phase Six-Pulse AC/DC Converter

In situations when a controllable DC voltage is needed, and in particular in machine drive systems with high-power inertial loads, three-phase six-pulse AC/DC converters are used. Such converters enable the recovery of energy stored as the kinetic or potential energy of various mechanical loads such as elevators, centrifuges, or electric trains, the energy that otherwise at braking is lost as heat.

A three-phase six-pulse AC/DC converter has a structure identical to three-phase six-pulse rectifiers with an inductive filter, as shown in Fig. 2.34, only with diodes replaced by thyristors, T.

The supply current waveform of the AC/DC converter is similar to the supply current waveform of three-phase rectifiers with an inductive filter, with only one important difference, namely that this current can be shifted against the supply voltage by controlling thyristors' firing angle α. A trapezoidal approximation of the converter supply current in line R is shown in Fig. 2.35. Such a converter provides the voltage of the mean value, equal to

$$U_{\rm d} = \overline{u}_{\rm d} = \frac{3\sqrt{6}}{\pi} U_{\rm a} \cos\alpha$$

to the output.

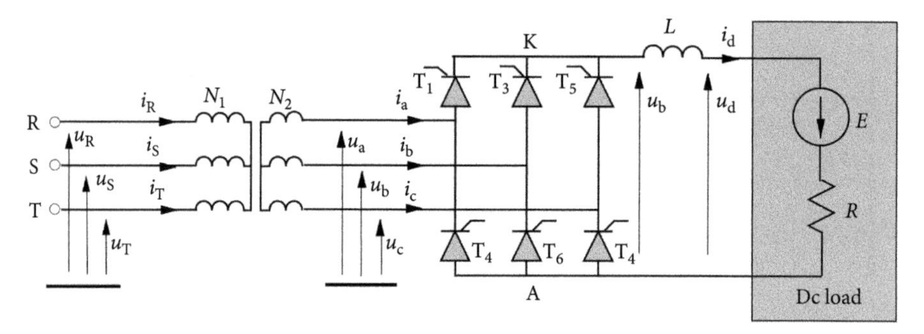

Figure 2.34 Structure of three-phase AC/DC converter.

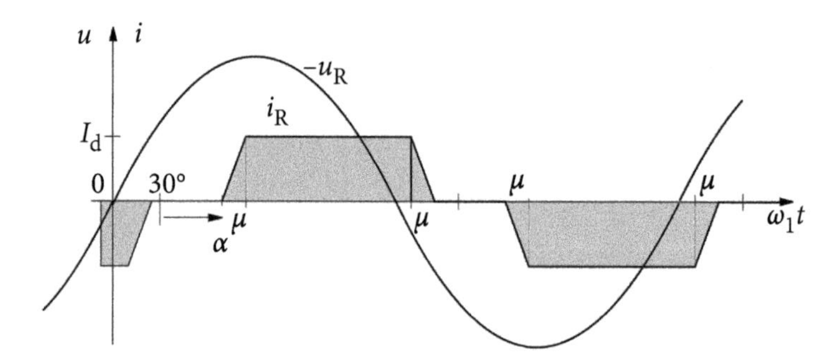

Figure 2.35 Trapezoidal approximation of the AC/DC converter supply current.

Observe, that for $\alpha > 90^\circ$, this voltage is negative. If the internal voltage of the DC load, E, is sufficiently high to keep the mean value of the output current, equal to

$$I_{\rm d} = \overline{i}_{\rm d} = \frac{U_{\rm d} + E}{R}$$

positive, despite negative voltage $U_{\rm d}$, then thyristor firing angle α can be changed from 0 to 180°. When the firing angle α is in the range of 0 to 90° the output voltage mean value U_d is positive so that the energy flows to the DC load. The converter operates in a *rectifier mode.*

When the firing angle α is in the range from 90° to 180° then the output voltage mean value U_d is negative so that the energy flows from the DC load to the supply AC system. The converter operates in an *inverter mode.* Thus, the converter operates in the inverter mode when the DC side of the converter is a source of energy. Such a situation occurs, for example, when an elevator, driven by a DC machine, is decelerated. The elevator at deceleration forms a *primary mover* for the DC machine which operates as a generator. Consequently, the declining kinetic energy of the slowing down elevator is conveyed through the AC/DC converter, operating in inverter mode, to the AC system.

The effect of AC/DC converters on the distribution system differs from such an effect of rectifiers in two ways. First, rectifiers behave as a nonlinear resistive load, meaning they do not practically load the supply with reactive power. AC/DC converters, due to the current phase shift, have to be considered as nonlinear RL loads with a non-zero reactive power, Q. Consequently, the converter can have a lower power factor, λ. The second difference is related to commutation. Commutation in converters can create voltage spikes and notches that are much more visible compared to those created by rectifiers.

Let us evaluate the power factor of AC/DC converter, assuming that the commutation is neglected, in other words that the commutation angle $\mu = 0$. At such an assumption, the phase shift of the current fundamental harmonic versus the supply voltage, φ_1, is equal to the firing angle α. Thus if the energy loss in the converter and the transformer is neglected, then the active and reactive powers at converter input terminals are

$$P = 3\,UI_1 \cos\alpha, \qquad Q = 3\,UI_1 \sin\alpha;$$

moreover,

$$P = 3\,UI_1 \cos\alpha \approx U_d I_d = \frac{3\sqrt{6}}{\pi} UI_d \cos\alpha$$

hence

$$I_1 \approx \frac{\sqrt{6}}{\pi} I_d.$$

The supply current rms value, similarly to a circuit with three-phase rectifiers, is equal to

$$||i|| = \sqrt{\frac{2}{3}}\, I_d \approx \frac{\pi}{3} I_1,$$

thus the power factor of AC/DC converters is equal to

$$\lambda \overset{\text{df}}{=} \frac{P}{S} = \frac{3\,UI_1 \cos\alpha}{3\,U||i||} \approx \frac{3}{\pi}\cos\alpha = \lambda_h\,\lambda_1$$

where $\lambda_h = 3/\pi$ is the harmonic power factor while $\lambda_1 = \cos\alpha$ is the displacement power factor. The total harmonic distortion of the AC/DC converters, because of the same supply current waveform as that of six-pulse rectifiers, is approximately equal to $\delta_i = 0.31$.

2.10 Commutation as the Source of Distortion

The supply current of AC/DC converters is the source of current distortion because this current has a trapezoidal waveform rather than a sinusoidal waveform. There is also a second mechanism of distortion, namely the commutation of conducting thyristors causing short-circuits.

As thyristors are sequentially fired with $T/6$ or 60° delay, the line currents have to be commutated from lines with the switched OFF thyristor to the lines with the switched ON thyristors. Thus, the converter transformer is six times short-circuited by two thyristors conducting at the same time. It creates notches and spikes in the thyristor bridge voltage waveform. Their rectangular approximation for the voltage u_a is shown in Fig. 2.36.

Each thyristor bridge voltage is affected by four commutations. For example, voltage u_a, shown in Fig. 2.36. is affected by the commutation of thyristors T1, T3, T4, and T6.

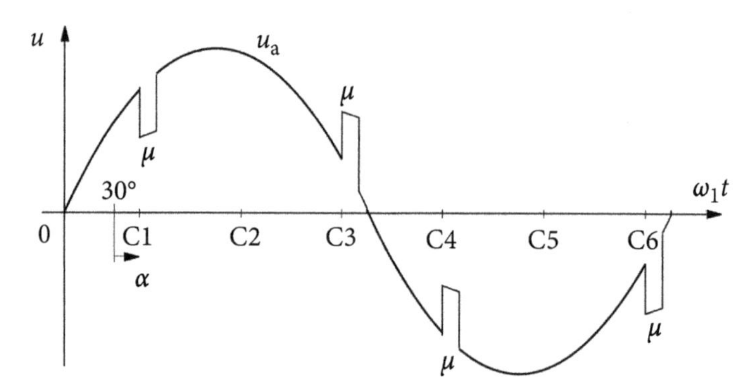

Figure 2.36 Voltage waveform on thyristor bridge.

Notches and spikes of the converter voltage penetrate the distribution system and can affect other equipment. Therefore, their duration μ, and magnitude, Δu, are the first important factors, that specify the effect of AC/DC converters upon the distribution voltage.

To evaluate these two parameters, let us analyze commutation C1, meaning thyristor T1 is switched ON and consequently, thyristor T5 is switched OFF. In effect of it the output current I_d, which has flown before commutation in lane "c", has to be switched to the line "a". This is illustrated on the equivalent circuit shown in Fig. 2.37.

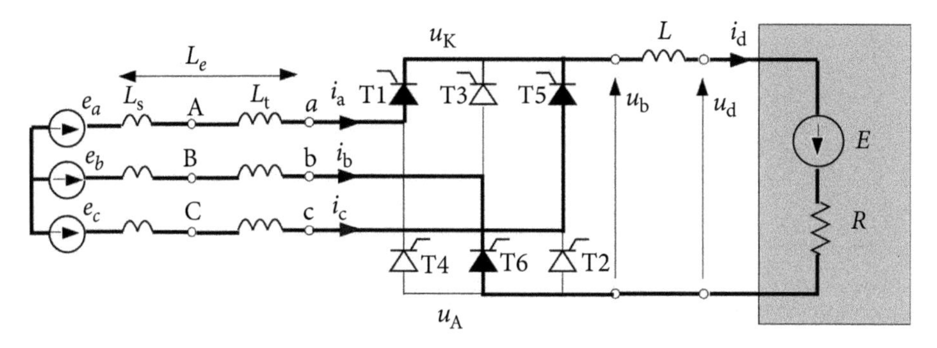

Figure 2.37 An equivalent circuit at commutation C1.

Let us assume that the output current is ideally filtered, so that $i_d = I_d =$ const. In such an assumption,

$$i_a + i_c = I_d, \qquad \text{so that} \qquad \frac{di_a}{dt} + \frac{di_c}{dt} = 0.$$

Neglecting the voltage drop on the conducting thyristor, usually below 1V, the cathode voltage u_K can be expressed as

$$u_K = e_a - L_e\frac{di_a}{dt}, \qquad u_K = e_c - L_e\frac{di_c}{dt},$$

and hence, by adding these two lines, we obtain

$$2u_K = e_a + e_c - L_e\left(\frac{di_a}{dt} + \frac{di_c}{dt}\right) = e_a + e_c.$$

Thus, the voltage on the cathode bar, which during commutation is observed at terminal "a", and "c", is equal to

$$u_K = u_a = u_c = \frac{e_a + e_c}{2} = \frac{e_a + e_c + e_b - e_b}{2} = -\frac{e_b}{2}$$

meaning it is equal to half of the negative value of the voltage at the line which is not involved in commutation. The voltage notch at commutation C1 is shown in Fig. 2.38.

The magnitude of the voltage u_a notch at firing angle α is equal to

$$\Delta u_a = e_a(\alpha) - u_K(\alpha) = e_a(\alpha) + \frac{e_b(\alpha)}{2} + = \sqrt{2}E\sin(30^0 + \alpha) +$$
$$+ \frac{\sqrt{2}E}{2}\sin(30^0 + \alpha - 120^0) = \sqrt{\frac{3}{2}}E\sin\alpha,$$

thus, it depends on the firing angle. The notch is deepest at $\alpha = 90^{\circ}$.

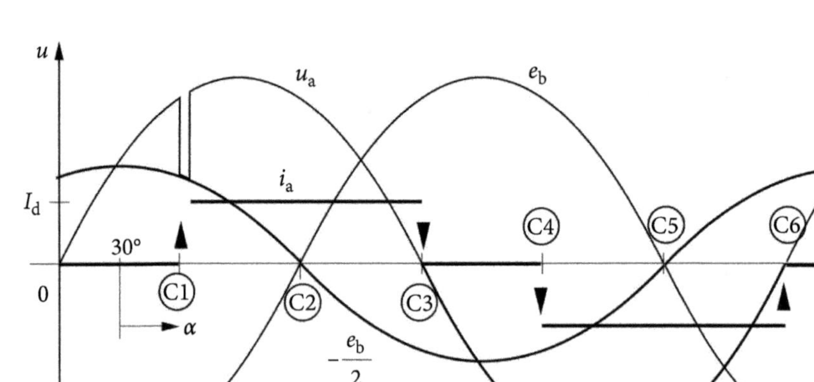

Figure 2.38 Waveforms of voltages at commutation C1.

The spike on the voltage u_a, shown in Fig. 2.36, is created during commutation C3, when thyristor T3 is fired so that thyristor T1 is switched OFF and consequently, the current I_d is commutated from the line "a" to line "b". Thus, at commutation C3,

$$i_a + i_b = I_d$$

thus,

$$\frac{di_a}{dt} + \frac{di_b}{dt} = 0.$$

The cathode voltage u_K can be expressed as

$$u_K = e_a - L_e\frac{di_a}{dt},\ u_K = e_b - L_e\frac{di_b}{dt}$$

and hence, the cathode voltage is

$$u_K = u_a = u_b = \frac{e_a + e_b}{2} = -\frac{e_c}{2}.$$

Since the voltage u_K, as shown in Fig. 2.39, is higher than e_a, thus commutation C3 creates a voltage spike. Its magnitude is equal to

$$\Delta u_a = u_K(\alpha + 120°) - e_a(\alpha + 120°) = -\sqrt{2}E\left[\frac{1}{2}\sin(\alpha + 270°) + \right.$$

$$\left. + \sin(30° + \alpha + 120°)\right] = \sqrt{\frac{3}{2}}E\sin\alpha.$$

Thus, notches and spikes have the same magnitude, with the highest value for $\alpha = 90^{\circ}$.

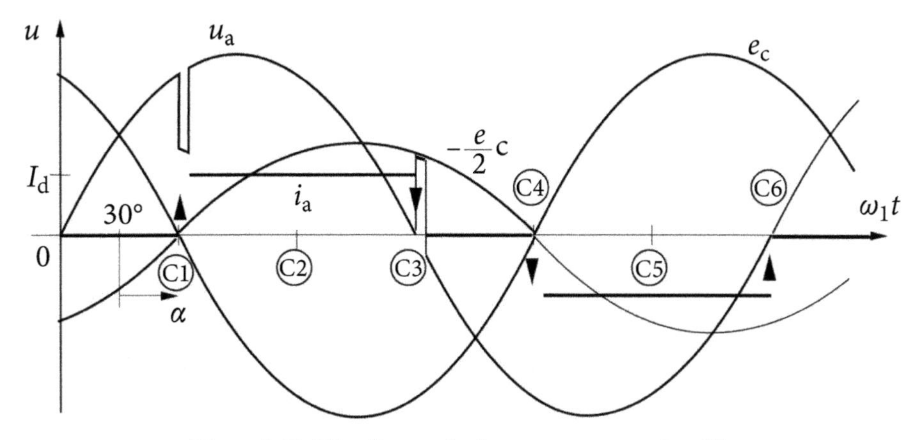

Figure 2.39 Waveforms of voltages at commutation C3.

After the notches or spikes magnitude, their duration in degrees, μ, is the next important parameter. To evaluate it, let us consider the commutation C1 in more detail. The equivalent circuit, with neglected resistance and the voltage drop on conducting thyristors, which are of secondary importance for the commutation, is shown for commutation C1 in Fig. 2.40.

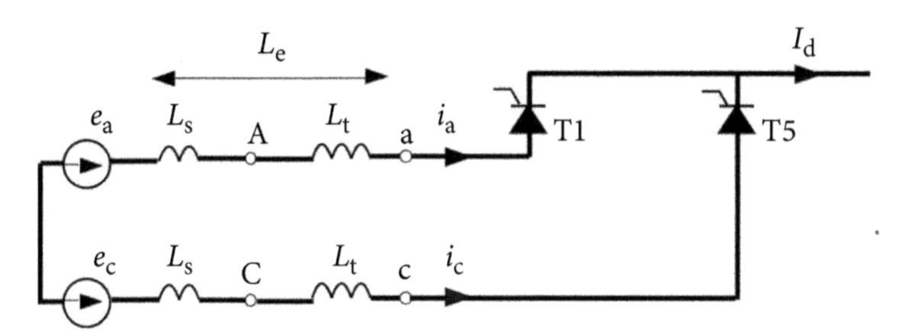

Figure 2.40 Equivalent circuit for commutation C1.

Observe that the line-to-line internal voltage, shown in Fig. 2.41, of the distribution system, e_{ac}, is the driving voltage for the current commutation from line "c" to "a". This voltage has to satisfy the relation

$$e_{ac} = 2L_e \frac{di_a(t)}{dt} = 2\omega_1 L_e \frac{di_a}{d(\omega_1 t)} = 2\omega_1 \, L_e \frac{di_a(\beta)}{d\beta}$$

where L_e is equivalent inductance as seen from thyristor's bridge, that is, the sum of the inductance of the converter transformer L_t and the distribution system, L_s. The line "a" current during commutation is equal to

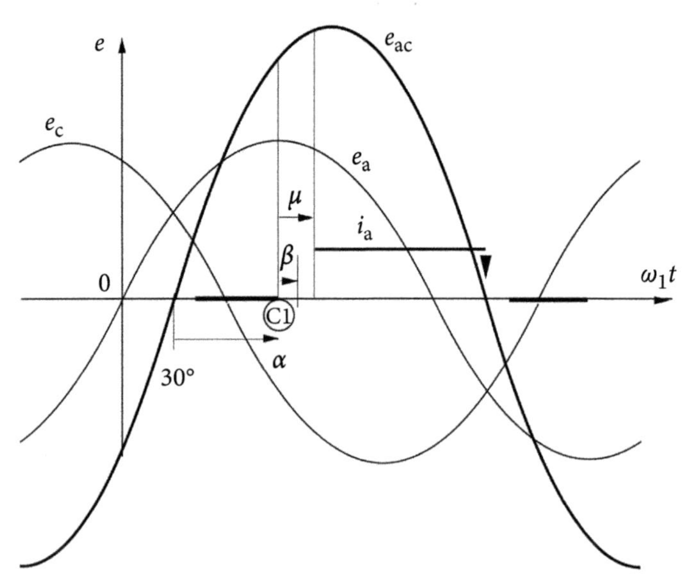

Figure 2.41 Voltage waveforms during commutation C1.

$$i_a(\beta) = \frac{1}{2\omega_1 L_e} \int_{\frac{\pi}{6}+\alpha}^{\frac{\pi}{6}+\alpha+\beta} e_{ac}\, d\beta \frac{1}{2\omega_1 L_e} \int_{\frac{\pi}{6}+\alpha}^{\frac{\pi}{6}+\alpha+\beta} = \sqrt{6}E \sin\left(\beta - \frac{\pi}{6}\right) d\beta =$$

$$= \sqrt{\frac{3}{2}} \frac{E}{\omega_1 L_e} [\cos\alpha - \cos(\alpha+\beta)].$$

Commutation terminates at $\beta = \mu$, when $i_a(\beta) = I_d$, thus

$$\sqrt{\frac{3}{2}} \frac{E}{\omega_1 L_e} [\cos\alpha - \cos(\alpha+\mu)] = I_d.$$

This equation can be solved versus the commutation angle, μ, resulting in

$$\mu = \cos^{-1}\left\{\cos\alpha - \sqrt{\frac{2}{3}} \frac{\omega_1 L_e}{E} I_d\right\} - \alpha. \tag{2.23}$$

Because

$$I_{sc} = \frac{E}{\omega_1 L_e}$$

is the rms value of a short-circuit current at the thyristor bridge terminals, this angle can be expressed in terms of output current and short-circuit current rms ratio

$$\mu = \cos^{-1}\left\{\cos\alpha - \sqrt{\frac{2}{3}\frac{I_d}{I_{sc}}}\right\} - \alpha. \tag{2.24}$$

For example, if the output current and short-circuit current rms ratio $I_{sc}/I_d = 40$, then at firing angle $\alpha = 45^{\circ}$, the commutation angle is equal to

$$\mu = \cos^{-1}\left\{\cos\alpha - \sqrt{\frac{2}{3}\frac{I_d}{I_{sc}}}\right\} - \alpha = \cos^{-1}\left\{\cos 45^{\circ} - \sqrt{\frac{2}{3}\frac{1}{20}}\right\} - 45^{\circ} = 3.2^{\circ}.$$

The commutation process is slowest, that is, $\mu = \mu_{max}$, at zero firing angle (or in rectifier circuit) since at such an angle, as can be seen in Fig. 2.41, the voltage e_{ac} has the lowest value. At the same time, the output current at the firing angle $\alpha = 0$ is maximum, $I_d = (I_d)_{max}$, thus

$$\mu = \mu_{max} = \cos^{-1}\left\{1 - \sqrt{\frac{2}{3}\frac{(I_d)_{max}}{I_{sc}}}\right\}.$$

This maximum commutation angle can be expressed in terms of the maximum active power of the converter and the short-circuit power at the thyristor bridge. Namely

$$(I_d)_{max} = \frac{(P_d)_{max}}{(U_d)_{max}} \approx \frac{\pi}{3\sqrt{6}}\frac{(P_d)_{max}}{E}.$$

Since the short-circuit power at the thyristor bridge terminals is equal to

$$S_{sc} = 3\,E\,I_{sc}, \qquad \text{thus} \qquad I_{sc} = \frac{S_{sc}}{3\,E}$$

and the maximum value of the commutation angle is

$$\mu_{max} = \cos^{-1}\left\{1 - \sqrt{\frac{2}{3}\frac{(I_d)_{max}}{I_{sc}}}\right\} = \cos^{-1}\left\{1 - \frac{\pi}{3}\frac{(P_d)_{max}}{S_{sc}}\right\}. \tag{2.25}$$

The commutation angle increases with the increase of the converter power to the short-circuit power ratio.

2.11 Arc Furnace

Arc furnaces, used for steel production, are the ones of the largest industrial loads, with power ratings from tens of MVA up to hundreds of MVA. They are built as DC arc furnaces, with one graphite electrode supplied from a three-phase rectifier or AC/DC converter, as shown in Fig. 2.42, or as AC arc furnaces with three graphite electrodes supplied directly from a three-phase furnace transformer. Due to the supply, DC arc furnaces cause the current and voltage distortion typical for three-phase rectifiers or AC/DC converters, as discussed in the previous sections of this chapter. The supply current distortion caused by AC arc furnaces is governed by physical phenomena inside the furnace cage.

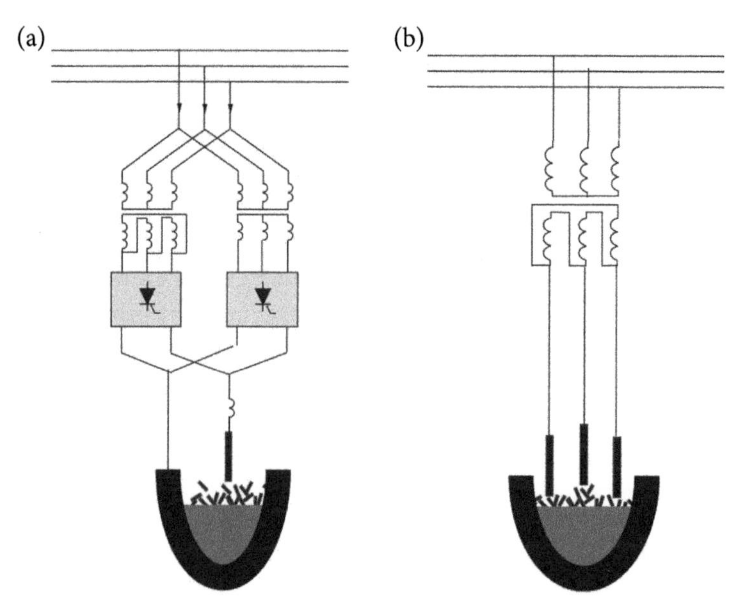

Figure 2.42 A supply structure of a DC (a) and AC (b) arc furnaces.

An electric arc is a nonlinear phenomenon thus the supply current of AC furnaces can be strongly distorted. Moreover, especially in the first interval of scrap melting, before the scrap is melted and a flat surface of liquid iron is formed, the electric arc starts randomly, and consequently, the arc current is not only distorted but also stochastic. Since arcs in particular lines are fired independently of each other, three-phase currents at the furnace supply can be strongly asymmetrical. Taking into account the very high power of arc furnaces, they can cause substantial distortion, asymmetry, and voltage rms value fluctuation in the distribution system. The power factor of arc furnaces can be of the order of $\lambda = 0.7$ and can fluctuate in a wide range in the melting process. Compensators of the reactive power are installed usually at the furnace terminals, for reducing voltage rms value fluctuation. Since the arc current is unidirectional, the even order harmonics, and in particular, the second order harmonic can occur in the supply current of arc furnaces. Harmonic filters are often installed for protecting the distribution system against current harmonics produced by AC arc furnaces.

The cage where steel melting takes place is lined inside with ceramic, and its coat is grounded. It keeps the cage coat voltage at zero level, but the ceramic resistance prevents the arc current from flowing to the ground. This current can return to the furnace transformer only by other electrodes with the arc in ON state. Therefore, despite grounding, the arc furnace has to be regarded as a three-phase load with a three-wire supply. The structure of the AC arc furnace along with the furnace transformer is shown in Fig. 2.43.

The graphite electrodes, mounted in a hydraulic assemble, can be moved down and up to initiate the electric arc and to control its current. Electrodes are supplied

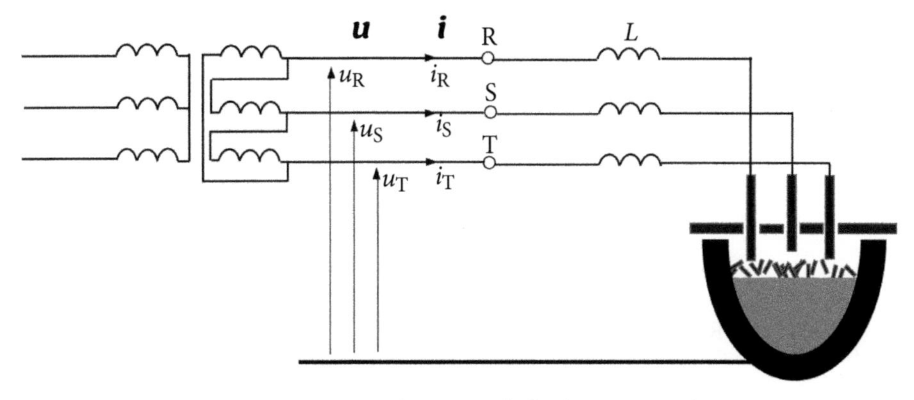

Figure 2.43 An AC arc furnace with the furnace transformer.

from three cables. To accommodate the free movement of the electrodes, the cables are flexible.

To stabilize the arc, the supply system has to have some inductance, thus the arc furnace burdens the supply source not only with the active power P but also with the reactive power Q. The inductance is provided by the furnace transformer and by cables that connect it with electrodes. Arc furnaces are operated usually in such a way that the reactive power Q is approximately equal to the active power P. The active power of the furnace changes in the melting process, therefore an inductor with controllable inductance is needed on the supply side of the furnace. It is connected usually to the primary side of the furnace transformer.

After a new metal scrap is loaded into the furnace cage, there are three stages in the furnace single cycle of operation usually of approximately one-hour duration. These are (i) the bore-down stage when the non-melted scrap changes the geometry of the arc, which affects its ignition causing randomness of the arc current. A hole in the middle of scrap is formed at this stage. When the hole reaches the bottom of the cage, the furnace proceeds to (ii) the melting stage, when there is still no melted scrap in the cage but the randomness of the arc current declines. Eventually, the furnace operation reaches (iii) the refining stage, when the cage is filled only with liquid steel and the arcs are the most stable.

The arc in the arc furnace is bidirectional, meaning the arc current can flow from the electrode to the steel in the cage or from the steel to the electrode. When the arc is ignited, then there is a direct voltage U_0 on the arc which does not depend on the arc current. It is equal to, approximately, $U_0 = 300\text{V}$. There is also a voltage component that changes with the arc current and depends on the resistance R_p of the arc plasma. The arc ignites when the magnitude of the voltage u between the electrode and the metal in the cage is higher than U_0. Therefore, the arc can be approximated with ideal diodes in series with DC voltage sources of the value U_0, as shown in Fig. 2.44b, and the arc furnace with the supply can be analyzed using the equivalent circuit shown in Fig. 2.45.

The resistance R in Fig. 2.45 denotes the combined resistance of the supply and the arc plasma. The equivalent circuit shown in Fig. 2.45 stands only for an

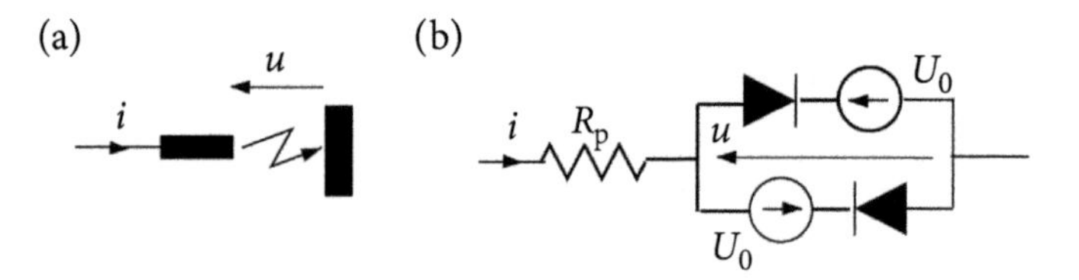

Figure 2.44 An equivalent circuit of the arc.

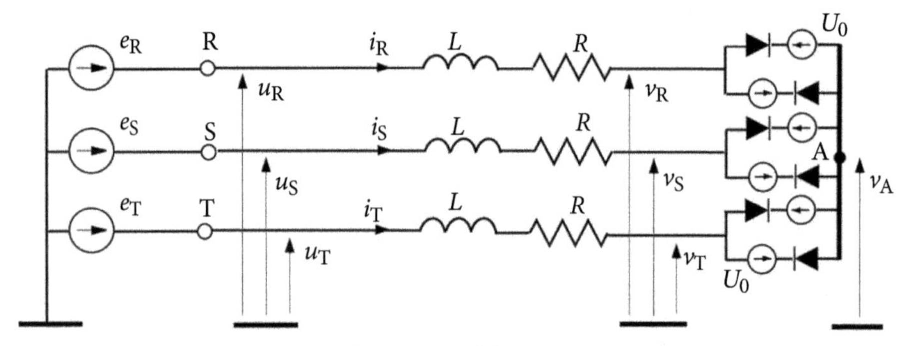

Figure 2.45 An equivalent circuit of the arc furnace and the supply.

approximation of the real arc furnace because several important characteristics of the furnace and its supply were simplified or ignored. Nonetheless, even such a simplified model can provide a lot of information about the furnace's performance.

First, the furnace arcs can be in different states, dependent on the value of the supply voltage and the electrodes' physical position versus the cage charge. In effect, all three or only two arcs are fully ignited, or some arcs are ignited only in one direction.

The arc ignition depends on the value of the supply voltage and the electrode distance to the cage charge. The conditions for ignition can be met for each electrode twice for each period T of the voltage variation. Assuming that arcs of electrodes R and T are a sort of reference with the longest ON state, the furnace arcs in a single period T can be in one of four states, s0, s1, s2, or s3, shown in Fig. 2.46.

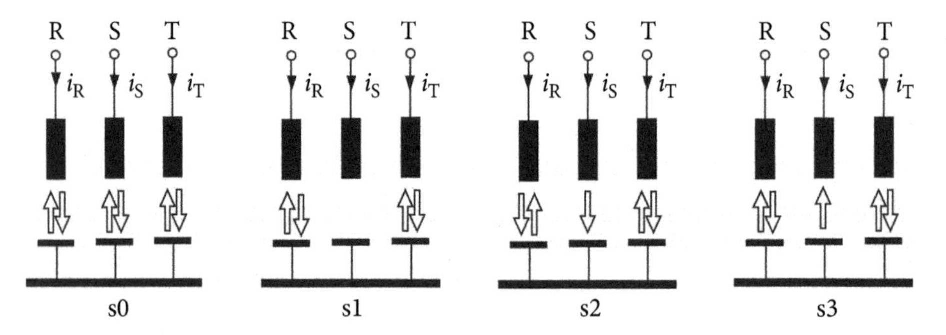

Figure 2.46 States of furnace arcs.

The specific state is determined randomly. The arc can extinct when its current reaches zero, therefore the specific furnace state will stay unchanged probably at least by one period T. The equivalent circuit of the arc, shown in Fig. 2.45, contains the arc plasma resistance R_p which depends on the plasma temperature and the arc geometry, thus its value changes randomly. In one period T time, neither the temperature nor the arc geometry changes sufficiently to affect the plasma resistance. It can be regarded as constant. Thus, we can assume that the randomness refers only to the furnace states but not to the period T-long intervals when the furnace state and its parameters are fixed.

When the arc furnace is in the refining stage (iii), the cage is filled with a homogenous liquid metal, the geometry of the arc plasma does not change, and the furnace operates in a quiet mode with all arcs ignited (s0). Some randomness in arc ignition can be visible in the melting stage (ii) when there is still some non-melted scrap in the cage. The highest randomness in arc ignition is at the beginning of the furnace operation, in the bore-down stage (i), when the charge of the cage is not yet melted. This stage can be specified only in statistical terms. It strongly depends on the sort of metal charge and how it is prepared for the melting process.

Let us analyze the furnace in the most common state, namely s0. At sufficiently high line inductances L all arcs are ignited, and the line currents are continuous. There is a DC voltage $a_x U_0$, in each line, with coefficient $a_x = \pm 1$ $(x = \mathrm{R, S, T})$, dependent on the current sign. Namely, if $i_x > 0$, then $a_x = 1$; if $i_x < 0$, then $a_x = -1$. If $G = 1/R$ denotes the line conductance, then the voltage v_{A0} of the joining point of arcs, that is, the melted steel, has to satisfy for DC voltage the nodal equation

$$v_{A0} = -\frac{G a_R U_0 + G a_S U_0 + G a_T U_0}{3G} = -\frac{1}{3}(a_R + a_S + a_T) U_0. \tag{2.26}$$

Its variation is shown in Fig. 2.47. In three-wire circuits

$$i_R(t) + i_S(t) + i_T(t) \equiv 0$$

thus at least one line current has to be negative, meaning at least one coefficient a_x is negative: $a_x = -1$. Since in one period T each of the line currents changes its sign twice, thus the voltage of the joining point of arcs

$$v_{A0} = \pm\frac{1}{3} U_0$$

and changes the sign six times in period T, as shown in Fig. 2.47.

Taking into account the variation of the voltage on the joining point of arcs, the voltage on the arc in line R is

$$v_R = a_R U_0 + v_{A0} = \frac{1}{3}(2a_R - a_S - a_T) U_0.$$

The variation of the voltage v_R is shown in Fig. 2.48.

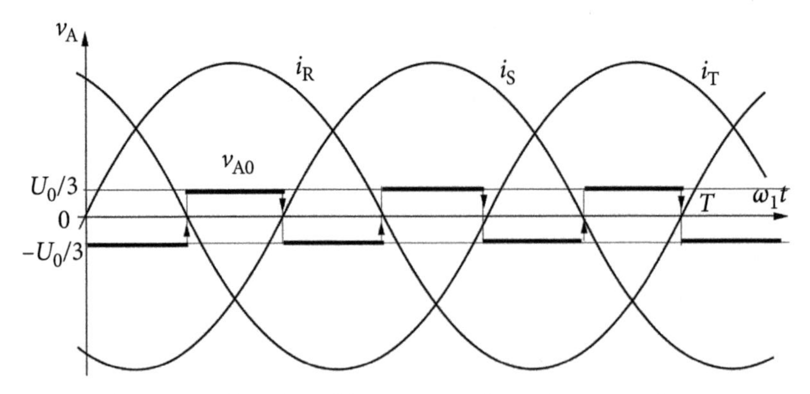

Figure 2.47 Variation of the voltage at the joining point of arcs.

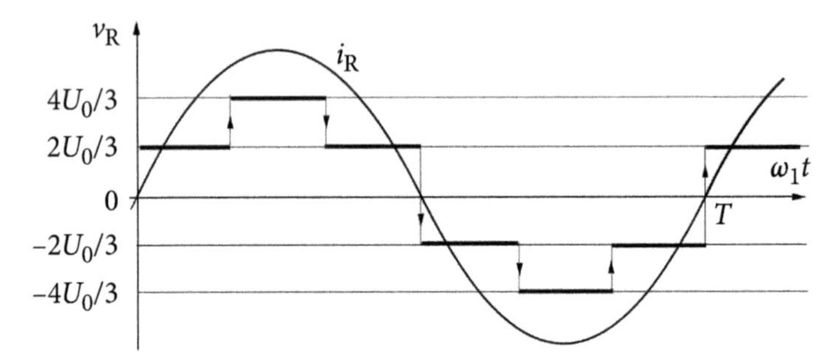

Figure 2.48 Variation of the voltage at the arc in line R.

One should observe that the voltages v_R, v_S, and v_T, shown in Fig. 2.48, are not voltages of the arc furnace electrodes, because the arc plasma resistance R_p is included in the line resistance R. Thus, these are only fictitious voltages. There are no physical points in the furnace that have such voltages. Nonetheless, such fictitious voltages enable us to separate two main features of arcs nonlinearity. These are: (i) a change of the arc plasma resistance with the current value, and (ii) a DC voltage on the arc. Consequently, there are two different mechanisms of heat release in the arc. The energy released in the furnace on this resistance is proportional to the active power.

Properties of arc furnaces will be illustrated in this book with results of modeling of a sort of a low-power furnace, with the parameters' proportions that are common for AC arc furnaces. It was assumed that the furnace has the line resistance equal to $R = 1\ \Omega$ and the reactance for the fundamental frequency equal to $X = R = 1\ \Omega$. The furnace transformer's secondary side voltage rms value can be between 400V and 1300V. The voltage $E = 700$V was selected for this illustration. The power ratings of the furnace transformer of ultra high-power furnaces, which now reaches 750 MVA, due to limited availability, are relatively low. It could be of the order of the furnace power or only a few times higher. Therefore, the line impedance of the transformer could be comparable to or only slightly lower than the furnace, including the supply cable impedance. The reactance-to-resistance ratio of the transformer was assumed to be $X_t/R_t = 5$. The electrical parameters of the modeled arc furnace and the transformer are shown in Fig. 2.49.

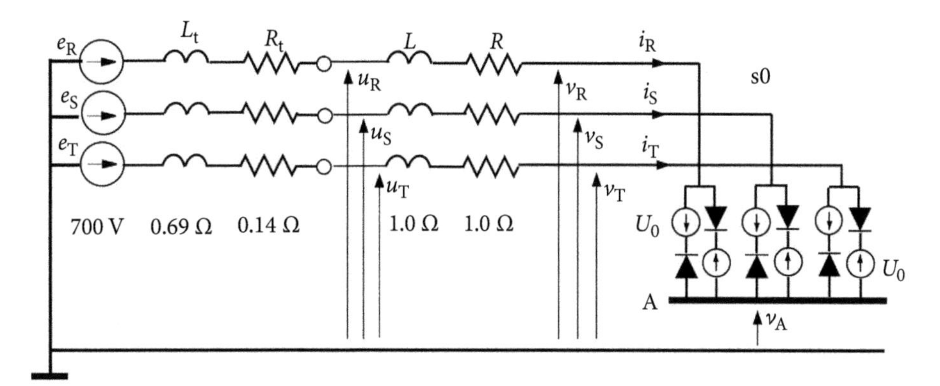

Figure 2.49 Arc furnace and transformer parameters in the illustration.

The waveforms of the internal voltage of the supply system, recalculated to the secondary side of the transformer, e_R, the voltage on the supply terminals of the furnace u_R, the fictitious voltage v_R, and the line current i_R, obtained from a computer modeling, are shown in Fig. 2.50.

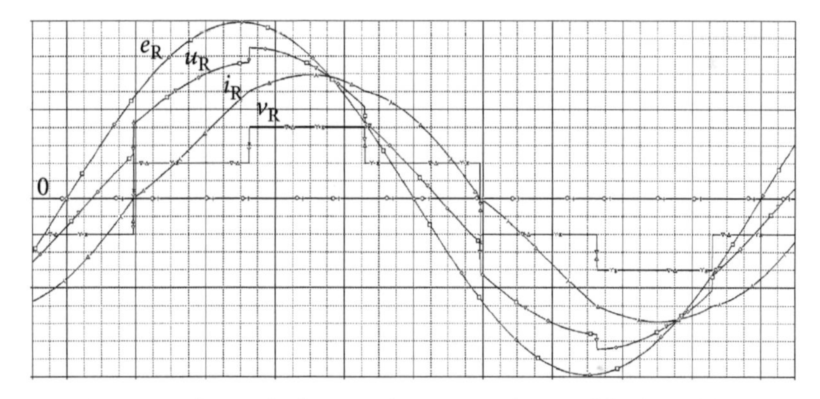

Figure 2.50 Waveforms of voltages and currents in line R of the furnace in state s0.

It can be observed from the plot in Fig. 2.50 that the furnace current in the state s0 is not strongly distorted. The harmonic content of the furnace current at state s1 is compiled in Table 2.2. When due to a movement of the metal charge, one arc, for example in line S extinguishes, as shown in Fig. 2.51, the furnace proceeds to the state s1.

Assuming that the line inductance is sufficiently high to preserve the current continuity, the line-to-line voltage v_{RT} is equal to

$$v_{RT} = 2a_R\, U_0 \tag{2.27}$$

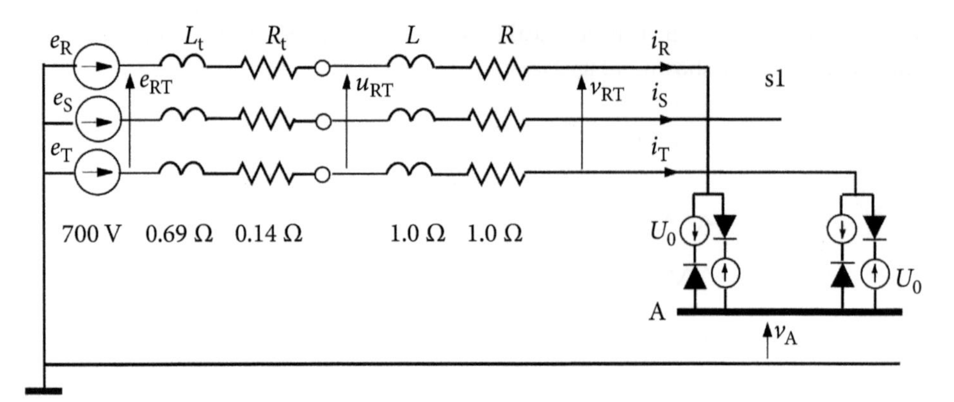

Figure 2.51 Arc furnace modeled in the illustration in state s1.

As it was in a balanced mode of operation, s0, with the voltage v_R, the voltage v_{RT} is only a fictitious voltage, because the arc plasma resistance R_p was moved to the line resistance R. The results of the modeling are shown in Fig. 2.52.

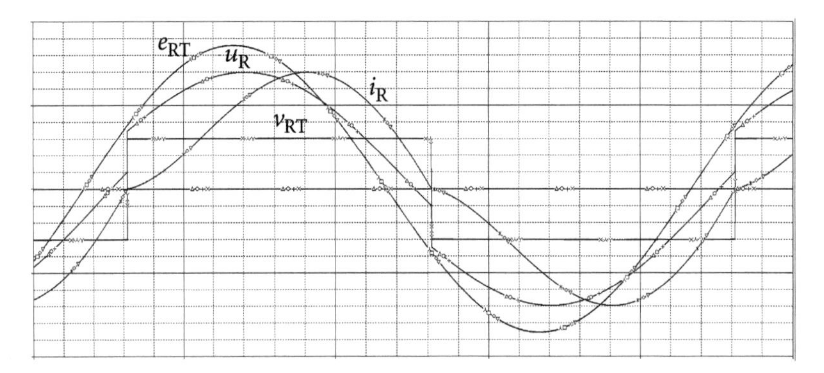

Figure 2.52 Waveforms of voltages and current in line R at two ignited arcs.

In arc furnaces during a quiet mode of furnace operation, arcs are bidirectional. During such a mode of operation, the line currents can be approximated with high accuracy by quantities with a negative symmetry versus the values shifted by a half of the period T,

$$x(t - T/2) \equiv -x(t) \tag{2.28}$$

Quantities with this property do not contain harmonics of the even order n. Moreover, due to current symmetry and three-wire without neutral supply lines, the third order harmonic cannot occur in the supply current. Consequently, the fifth order is the lowest order harmonic.

Published reports, based on measurements performed on arc furnaces [301], show that the arc current can contain harmonics of the even order, however. The second order current harmonic could even reach a level above 50% of the fundamental harmonic. This can be explained only if there are intervals of time when an arc in the furnace is unidirectional, meaning the identity (2.28) is no longer valid.

At assumptions as previously, the arc furnace with one, unidirectional arc, say in line S with a positive current, can be approximated as shown in Fig. 2.53.

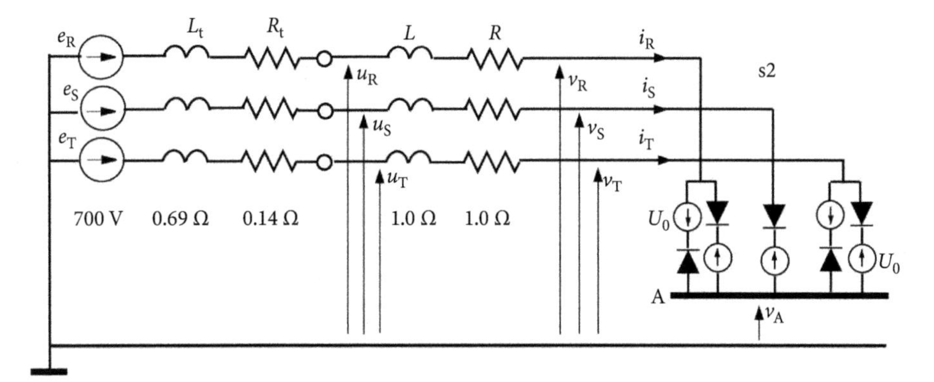

Figure 2.53 An equivalent circuit of an arc furnace with a unidirectional arc in line S.

When the arc in line S is ignited, then the furnace behaves as a balanced load, with the voltage v_A of the joining point of arcs as shown in Fig. 2.48.

When the arc in line S is not ignited, that is, when the current in line S reaches zero, then the DC voltage at the joining point of arcs is zero, because coefficients a_R and a_T have to be of the opposite sign, and hence

$$v_{A0} = -\frac{G a_R U_0 + G a_T U_0}{2G} = -\frac{1}{2}(a_R + a_T)U_0 = 0.$$

Only AC voltage occurs at point A in this state of the furnace operation. The crms value of this voltage fundamental harmonic is

$$\boldsymbol{V}_{A1} = \frac{Y_1 \boldsymbol{E}_R + Y_1 \boldsymbol{E}_T}{2Y_1} = \frac{1}{2}(\boldsymbol{E}_R + \boldsymbol{E}_T) = -\frac{1}{2}\boldsymbol{E}_S$$

meaning it changes as $e_S(t)/2$. The time interval with this voltage ends when the distribution system internal voltage $e_S(t)$, reaches the value $v_A + U_0$, and consequently, the arc in line S is ignited again. The time variance of voltages and current in the line R is shown in Fig. 2.54. The current does not have any negative symmetry with values shifted by half of the period T and consequently, it can have harmonics of the even order.

The results of computer modeling of the arc furnace under consideration in states s0, s1, and s2, with regard to the content of the high order harmonics in the line current i_R, are compiled in Table 2.2.

Observe, that the DC component of the current generated in the furnace in state s2 cannot be transformed to the primary side of the furnace transformer. However, it causes the transformer ferromagnetic core saturation, thus generation of the even order current harmonics on its primary side.

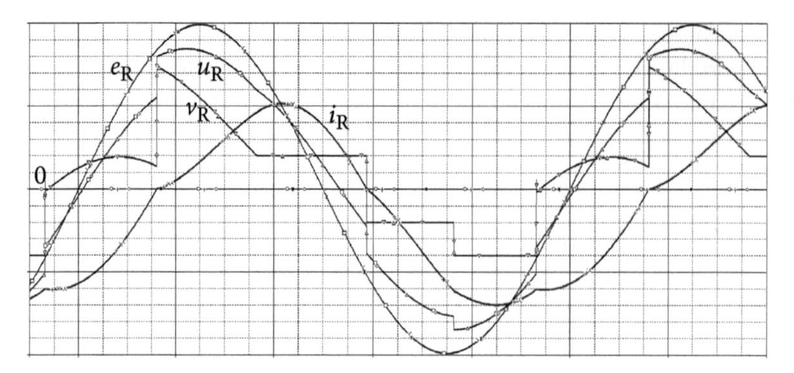

Figure 2.54 Waveforms of voltages and current in line R of the furnace with unidirectional arc, meaning in state s2.

Table 2.2 The rms value of the line current i_R harmonics.

—	Unit	s0	s1	s2
δ_i	%	2.2	7.5	40.7
I_0	A	0	0	56
I_1	A	331	260	266
I_2	A	0	0	44.9
I_3	A	0	17.7	5.8
I_4	A	0	0	10.0
I_5	A	6.4	6.4	4.8
I_6	A	0	0	2.5
I_7	A	3.2	3.2	2.9

2.12 Cycloconverter

Cycloconverters are power electronics devices used for a direct conversion of a fixed frequency voltage, usually 50 or 60 Hz, to a voltage of a lower and controlled frequency. AC/DC three-phase converters are used for that.

Such a converter operates as a cycloconverter if the thyristors' firing angle is not kept constant but increases linearly with time. If the firing angle of thyristors increases linearly in time as

$$\alpha = \Omega t + Const,$$

then the thyristor output voltage changes as

$$u = \frac{3\sqrt{6}}{\pi} U_R \cos(\Omega t + Const) = \sqrt{2}\, U \cos(\Omega t + \phi), \qquad \phi = Const$$

that is, with frequency Ω [rad/s]. The converter has to have the capability of providing a negative current, thus it has to be built as a *dual converter*, that is, of the structure shown in Fig. 2.55.

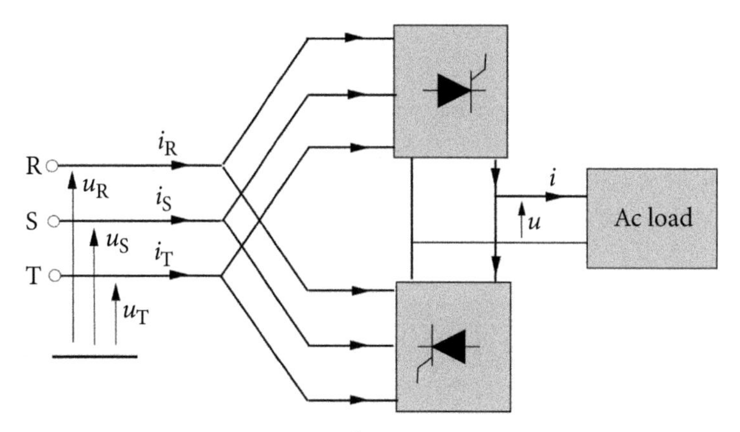

Figure 2.55 A cycloconverter structure.

The upper converter in this circuit operates at a positive output current i, while the lower converter operates at a negative current. Cycloconverters are built as variable speed drivers for induction motors, meaning for three-phase loads. Fig. 2.55 presents the structure of a cycloconverter for only one phase of a motor. Cycloconverters needed for the remaining phases supply differ only by φ value.

With the change of the firing angle, the supply current losses periodicity with the supply voltage, illustrated in Fig. 2.56, for the voltage and current at terminal R of the cycloconverter.

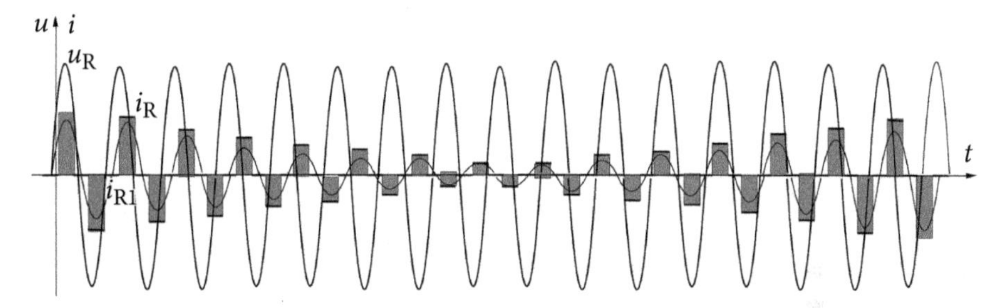

Figure 2.56 Voltage and current waveform at terminal R of a cycloconverter.

The supply current is composed of pulses, but their frequency is not equal to the frequency of the supply voltage, 50 or 60Hz, but to a frequency dependent on the coefficient Ω. which is a radial frequency of the output voltage of the converter. The fundamental component of these pulses i_{R1} does not have the frequency of the supply voltage. Consequently, neither the period of the supply voltage T nor the period of the supply current can be used for calculating the active power P at the cycloconverter's supply terminals. For calculating this power, a common period of the supply voltage and the supply current has to be found, and this is discussed in Section 3.1.

Summary

Over the past few decades, there has been a rapid increase in nonlinear loads and loads with switched parameters in electrical systems. These cause current and distribution voltage distortion. A description of the main sources of distortion was compiled in this chapter.

Chapter 3
Circuits with Nonsinusoidal Voltages and Currents Analysis

3.1 Periodic Quantities

Basic academic education on electrical circuit analysis is usually confined to single-phase circuits with sinusoidal voltages and currents. Sometimes this education is extended to three-phase circuits in balanced sinusoidal conditions, which can be analyzed "phase by phase" as single-phase circuits with sinusoidal voltages and currents. Such fundamentals are not sufficient or useful for the analysis of circuits in the presence of voltage and current distortion and/or three-phase circuit asymmetry.

Harmonics and three-phase asymmetry add up two levels of complexity to basic circuit analysis. Without proper mathematical tools and well-selected sets of symbols for nonsinusoidal and three-phase voltages and currents, analysis and its results could be confusing. Therefore, this chapter focuses on methods of describing single- and three-phase circuits in the simplest way possible, reducing the effect of the intrinsic complexity of such circuits upon the clarity of their description.

Quantity $x(t)$ is referred to as periodic if each of its values repeats sequentially, as shown in Fig. 3.1, after the same interval of time a, that is, for any integer n and interval of time a

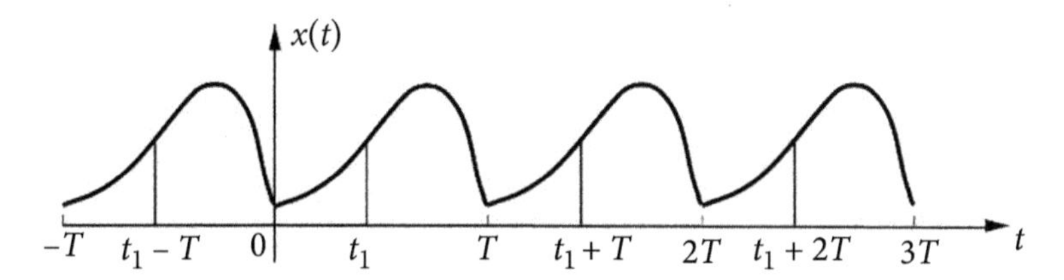

Figure 3.1 A periodic quantity.

$$x(t - na) \equiv x(t).$$

The shortest non-zero interval a is referred to in mathematics as a period and denoted by T. Thus, if the values of the quantity $x(t)$ are known in the interval of the period T duration, they are consequently known in the infinite interval of time, that is, in the interval $(-\infty, \infty)$.

The meaning of the period in electrical engineering could be different from that in mathematics, however. The smallest number T for which the quantity $x(t)$ satisfies the relation

Powers and Compensation in Circuits with Nonsinusoidal Current. Leszek S. Czarnecki, Oxford University Press.
 DOI: 10.1093/oso/9780198879206.003.0003

$$x(t - nT) \equiv x(t) \tag{3.1}$$

is the period of $x(t)$ in mathematical but not always in electrical engineering meaning.

Let us consider, for example, a three-phase, six-pulse rectifier. When such a device is supplied with a sinusoidal voltage of period T, then the output voltage is composed of pulses that repeat after each interval of time of $T/6$ duration. Thus, from the mathematical point of view, the output voltage is periodic with a period equal to $T/6$. However, such a voltage is considered in power engineering as periodic with the period T. It is because the period of the distribution voltage in the circuit is regarded usually as the period of all quantities in this circuit, even if their period is in the mathematical sense different.

Frequency of Periodic Quantities: The number of periods, T, of the quantity x in the time interval of one-second duration is referred to as the *frequency*, f, of this quantity. Sometimes, to distinguish it from harmonic frequencies, it is also referred to as the *fundamental frequency* and denoted by

$$f_1 = \frac{1}{T}.$$

The unit of frequency is Hz (hertz) and dimension [1/s].

The change, in radians, of the phase angle of a sinusoidal quantity with the period T in one second time is referred to as the *angular frequency*, or as the *fundamental angular frequency*,

$$\omega_1 = 2\pi f_1 = \frac{2\pi}{T}.$$

The angular frequency has no unit. Its dimension is [rad/s]. The adjective *angular* is commonly omitted, however, and the term "*frequency*" is applied both to the number of cycles per second, f, and to the change of the argument of the periodic quantity, in radians per second [rad/s], ω. We can infer which meaning one has in mind by looking at the frequency dimension, that is, is it specified in 1/s or rad/s?

Sum of Periodic Quantities: A very frequent question that refers to periodic quantities concerns their sum; namely, is the sum of periodic quantities a periodic quantity? To answer this question let us add up two sinusoidal quantities of frequency f_{a} and f_{b}

$$x = \sqrt{2}X_{\mathrm{a}} \sin(\omega_{\mathrm{a}} t - \alpha) + \sqrt{2}X_{\mathrm{b}} \sin(\omega_{\mathrm{b}} t - \beta).$$

Their sum is periodic if there is interval T such that for integers n and m,

$$\omega_{\mathrm{a}}(t - T) - \alpha = \omega_{\mathrm{a}} t - \alpha - n2\pi$$

$$\omega_{\mathrm{b}}(t - T) - \beta = \omega_{\mathrm{b}} t - \beta - m2\pi.$$

This condition is fulfilled only if the frequency of both components of the sum satisfies the condition

$$\frac{\omega_a}{\omega_b} = \frac{f_a}{f_b} = \frac{n}{m} \tag{3.2}$$

tha is, only if the ratio of these frequencies is a rational number. If the ratio of frequencies of the sum components is not a rational number, then such a sum is not a periodic quantity. For such a reason, the quantity

$$x = \sqrt{2}X_a \sin \omega t + \sqrt{2}X_b \sin \sqrt{2}\omega t$$

is not a periodic quantity.

Illustration 3.1 *Calculate the period T of a quantity composed of two components, one of the frequency* $f_a = 50$ Hz *and the second of the frequency* $f_b = 50.5$ Hz. *Since*

$$\frac{f_a}{f_b} = \frac{50}{50.5} = \frac{500}{505} = \frac{100}{101} = \frac{n}{m}$$

where $n = 100$ *and* $m = 101$ *are integer numbers, then their sum is a periodic quantity. Since*

$$\omega_a T = 2\pi f_a T = n 2\pi$$

hence the period is

$$T = \frac{n}{f_a} = \frac{100}{50} = 2\,\text{s}, \quad \text{thus,} \quad f = \frac{1}{T} = \frac{1}{2} = 0.5\,\text{Hz}.$$

Observe, that to calculate the period T, the ratio of frequencies has to be presented in a form of an irreducible fraction, n/m, that is, without any common divider other than 1.

****

A Shift of Integration Limits: When a periodic quantity $x(t)$ is integrated throughout the period T or its multiplicity, then the result of integration is independent of the integration limits, that is,

$$\int_0^T x(t)\,dt = \int_\tau^{T+\tau} x(t)\,dt$$

where τ is any real number. This property enables us to simplify integration by selecting limits of integration in such a way that the integrated quantity has the simplest mathematical form possible. It is also desirable that the integrated quantity has some symmetry against the upper and lower limit of integration. This simplifies its calculation.

Illustration 3.2 *A current of the waveform shown in Fig. 3.2 is to be integrated over the period.*

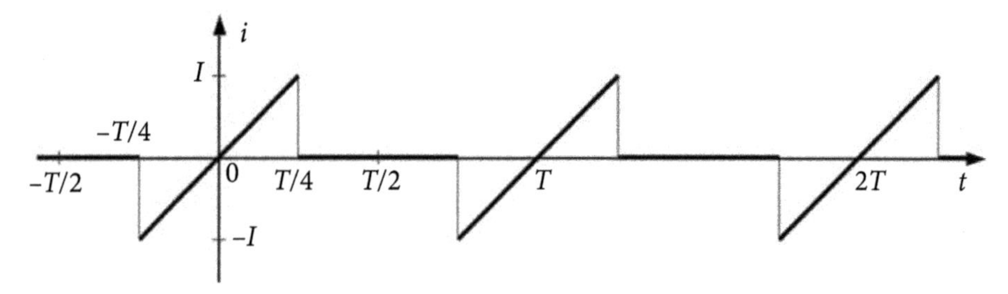

Figure 3.2 A periodic current waveform.

When the lower limit of integration is zero, then the following integral,

$$\int_0^T i(t)\,dt = \int_0^{T/4} \frac{4I}{T} t\,dt + \int_{3T/4}^{T} \frac{4I}{T}(t - T)\,dt$$

composed of two terms, has to be calculated. The calculation can be simplified by selecting value $\tau = -T/2$. Indeed,

$$\int_0^T i(t)\,dt = \int_{-T/2}^{T/2} i(t)\,dt = \int_{-T/4}^{T/4} \frac{4I}{T} t\,dt.$$

This illustration shows that a proper choice of integration limits may facilitate calculations. When the integrated quantity has symmetry as it does in this illustration, then selecting integration limits symmetrical against the point of symmetry simplifies integration.

*****

Quantities with Limited Power: When a periodic voltage $u(t)$ is applied to a resistor of conductance G, then the active power of such a resistor is equal to

$$P = \frac{1}{T}\int_0^T u(t)\,i(t)\,dt = G\frac{1}{T}\int_0^T u^2(t)\,dt.$$

The active power is limited on the condition that this integral is lower than infinity, and consequently, quantities that satisfy the condition

$$\frac{1}{T}\int_{0}^{T} x^2(t)\, dt < \infty \tag{3.3}$$

are referred to as quantities with *limited power.* Such quantities are also referred to as quantities *integrable with squares.*

Periodic quantities in physical circuits are always of limited power. Quantities without limited power may occur only in mathematical models of physical circuits, due to excessive simplifications. The voltage on an ideal inductor with a ramp-like current, shown in Fig. 3.3, is an example of such a situation.

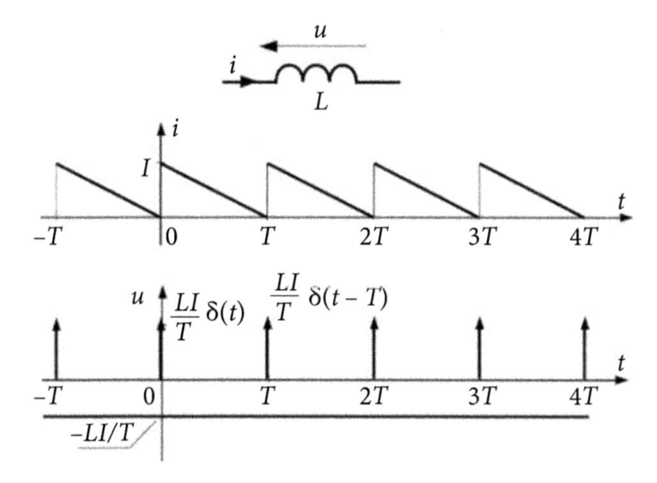

Figure 3.3 A voltage waveform of an ideal inductor with a ramp-like current.

If the inductor has no stray capacitance between conductors, then its current can have discontinuities as shown in the figure only if the voltage contains infinite pulses, $\delta(t - nT)$, referred to as Dirac pulses. Dirac pulse is defined as

$$\delta(t) \overset{\text{df}}{=} \begin{cases} 0, & \text{for } t \neq 0 \\ \infty, & \text{for } t = 0 \end{cases} .$$

$$\int_{-\infty}^{\infty} \delta(t)\, dt = 1.$$

In intervals where the current is continuous, it changes at the rate of I/T, thus the inductor voltage in such intervals is

$$u = L\frac{di}{dt} = -L\frac{I}{T}.$$

At points of discontinuity, pulses $A\delta(t-nT)$ occur in the voltage, where A is an unknown coefficient. The inductor current is periodic on the condition that the voltage mean value is equal to zero, that is,

$$\bar{u} = \frac{1}{T}\int_{-t}^{T-\tau} u dt = \frac{1}{T}\int_{-t}^{T-\tau}\left[A\delta(t) - L\frac{I}{T}\right] dt = \frac{1}{T}(A - LI) = 0.$$

Hence, the coefficient $A = LI$, and consequently, the inductor voltage can be presented in the form

$$u(t) = -L\frac{I}{T} + LI\sum_{-\infty}^{\infty}\delta(t-nT).$$

This is a periodic voltage. However,

$$\frac{1}{T}\int_{-T/2}^{T/2}\delta^2(t)\, dt = \infty$$

thus such a voltage is of infinite power. A source of infinite power is needed to obtain a discontinuous current in an ideal inductor, that is, to change energy stored in a magnetic field,

$$W = \frac{1}{2}Li^2$$

instantaneously. Thus, the model of an inductor with discontinuous current, shown in Fig. 3.3, is not accurate enough. Therefore, some results obtained from the analysis of models of real objects should be treated with limited credibility. They can result in unacceptable conclusions. To avoid it, models should be adequately enriched. In the situation analyzed above, to avoid Dirac pulses it should be assumed that a non-zero interval of time is needed for the inductor current to increase from zero to its maximum value, and it should be taken into account that a capacitance exists between the inductor's conductors.

Space $\boldsymbol{L}_\mathrm{T}^2$ of Periodic Quantities: All periodic quantities of limited power and the same period T form a linear space, $\boldsymbol{L}_\mathrm{T}^2$. Individual quantities $x(t) \in \boldsymbol{L}_\mathrm{T}^2$, can be referred to as elements or vectors of that space.

Since $\boldsymbol{L}_\mathrm{T}^2$ is a linear space, each linear form of vectors of that space belongs to $\boldsymbol{L}_\mathrm{T}^2$. Thus, if $x_\mathrm{a}(t) \in \boldsymbol{L}_\mathrm{T}^2$ and $x_\mathrm{b}(t) \in \boldsymbol{L}_\mathrm{T}^2$ then

$$x(t) = \alpha\, x_\mathrm{a}(t) + \beta\, x_\mathrm{b}(t) \in \boldsymbol{L}_\mathrm{T}^2 \tag{3.4}$$

for any real α, β.

Observe, however, that when the concept of a linear space is applied beyond mathematics, where quantities are dimensionless, in a field of science that involves physics, such as electrical engineering, where all quantities have physical dimensions, we have to remember that dimensional conflict cannot occur in linear forms. The space L_{T}^2 contains periodic, with period T, voltages, and currents of limited power. A linear form that would contain both of them, however, makes no sense.

Scalar Product: The functional on vectors $x(t) \in \boldsymbol{L}_{\mathrm{T}}^2$ and $y(t) \in \boldsymbol{L}_{\mathrm{T}}^2$

$$\frac{1}{T}\int_0^T x(t)\,y(t)\,dt \stackrel{\mathrm{df}}{=} (x, y) \tag{3.5}$$

defines a scalar product in $\boldsymbol{L}_{\mathrm{T}}^2$.

Observe that if $x(t)$ and $y(t)$ represent a load voltage and current, $u(t)$ and $i(t)$, such that $u(t) \in \boldsymbol{L}_{\mathrm{T}}^2$ and $i(t) \in \boldsymbol{L}_{\mathrm{T}}^2$, then their scalar product (u, i) is equal to the active power of the load, because

$$\frac{1}{T}\int_0^T u(t)\,i(t)\,dt = (u, i) = P. \tag{3.6}$$

Norm or Rms Value: The functional defined as

$$||x|| \stackrel{\mathrm{df}}{=} \sqrt{\frac{1}{T}\int_0^T x^2(t)\,dt} \tag{3.7}$$

in the space $\boldsymbol{L}_{\mathrm{T}}^2$ specifies the *norm* of the vector $x(t)$ or its *length*. In electrical engineering, this functional is called the *root mean square (rms) value* of $x(t)$. It specifies the heating effect of the resistor current $i(t)$ or resistor voltage $u(t)$. This is true, however, only if the resistance R does not depend on frequency. Meaning if $u(t) = R\,i(t)$, where $R = \text{const}$. In such a condition

$$P = \frac{1}{T}\int_0^T u(t)\,i(t)\,dt = (u, i) = \frac{1}{T}\int_0^T R\,i(t)\,i(t)\,dt = R\,||i||^2.$$

For example, in a presence of the *skin effect*, the resistance R of a conductor increases with the frequency. It is caused by an inductance inside a conductor which pushes the current from the conductor center towards its surface. In effect, the conductor resistance R increases. This effect increases with frequency and therefore there is no simple relationship between the active power released on the conductor and its current rms value. Similar to the skin effect is the *proximity effect*, when a magnetic field of one conductor changes the distribution of the current density in another one, causing its resistance to increase with frequency.

Therefore, the relation between the conductor voltage or current rms value and the active power, in the form

$$P = R\,||i||^2 = G\,||u||^2$$

is valid only on the condition that the resistance R or the conductance G do not change with frequency. Also, very common assumptions are that they cannot change in time and consequently, the rms value and its relationship with the active power on a conductor is commonly approximated with the above relation. Consequently, the concept of the rms value is fundamental for the analysis of the power properties of electric circuits.

Illustration 3.3 *The rms value of the current shown in Fig. 3.2 is equal to*

$$||i|| = \sqrt{\frac{1}{T}\int_{-T/2}^{T/2} i^2(t)\,dt} = \sqrt{\frac{1}{T}\int_{-T/4}^{T/4}\left(\frac{4\,I}{T}t\right)^2 dt} = \frac{1}{\sqrt{6}}.$$

******

Time Versus Angle: There are two different approaches to the description of time-varying quantities. In mathematics, such quantities are functions of the angle $\beta = \omega t$, specified in radians or degrees. The angle 2π in radians or 360° is a period in a mathematical sense. In electrical engineering or signal theory, quantities are specified as functions of time, t, meaning their plots are drawn as shown in Fig. 3.4, and the interval in which its value is repeated, is regarded as the period T.

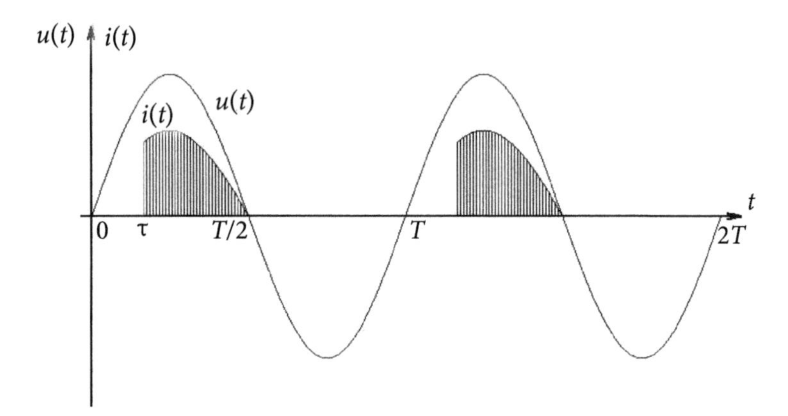

Figure 3.4 Thyristor current as a function of time.

In power electronics, however, it is easier to describe switching operations in terms of an angle, usually expressed in degrees, rather than in terms of time, usually milliseconds. Therefore, time-varying quantities in power electronics, unlike in electrical engineering, are described similarly to mathematics. Such quantities are specified in terms of angles. It is important to observe that according to the power electronics convention of presentation of time-varying quantities, their value does not change with frequency, for example like the firing angle α when expressed in degrees.

When the quantity is specified as a function of angle, then definitions of the scalar product and the rms value have to be adequately modified, namely

$$(x, y) = \frac{1}{2\pi} \int_0^{2\pi} x(\beta)\, y(\beta)\, d\beta \tag{3.8}$$

$$||x|| = \sqrt{\frac{1}{2\pi} \int_0^{2\pi} x^2(\beta)\, d\beta}. \tag{3.9}$$

Illustration 3.4 *The rms value of the thyristor current of the waveform, shown in Fig. 3.5, is equal to*

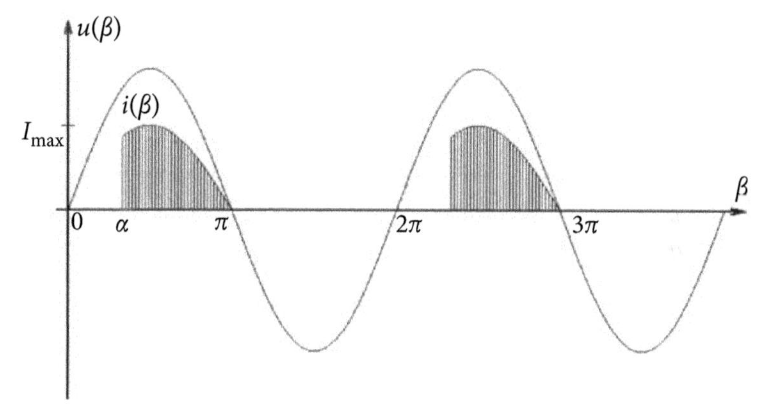

Figure 3.5 Thyristor current as a function of angle.

$$||i|| = \sqrt{\frac{1}{2\pi} \int_\alpha^\pi I_{\max}^2 \sin^2\beta d\beta} = I_{\max} \sqrt{\frac{1}{4\pi} \int_\alpha^\pi (1 - \cos 2\beta)\, d\beta} = \frac{I_{\max}}{2} \sqrt{1 + \frac{\sin 2\alpha - 2\alpha}{2\pi}}. \tag{3.10}$$

Assuming that $I_{\max} = 100$ A *and* $\alpha = 0$, *we obtain,*

$$||i|| = \frac{I_{\max}}{2} = 50\,\text{A}.$$

At firing angle increase to $\alpha = 90^\circ = \pi/2$, *the current rms value declines to*

$$||i|| = \frac{I_{\max}}{2} \sqrt{1 + \frac{\sin \pi - \pi}{2\pi}} = \frac{I_{\max}}{2\sqrt{2}} = 35.1\,\text{A.s.}$$

Observe that the conducting interval of the thyristor was reduced by half, but in effect the rms value declined only by $\sqrt{2}$.

*****

Mean Value: The functional on $x(t)$

$$\bar{x} = \frac{1}{T}\int_0^T x(t)\, dt = \frac{1}{2\pi}\int_0^{2\pi} x(\beta)\, d\beta \tag{3.11}$$

defines the *mean* or *average* value of $x(t)$ in the space $\boldsymbol{L}_T{}^2$. This value can be calculated, however, even if quantity $x(t)$ is not of limited power and does not belong to $x(t)$. Electrochemical effects, as well as the power of DC motors, are proportional to the current mean value.

Illustration 3.5 *The mean value of the current of the waveform shown in Fig. 3.6,*

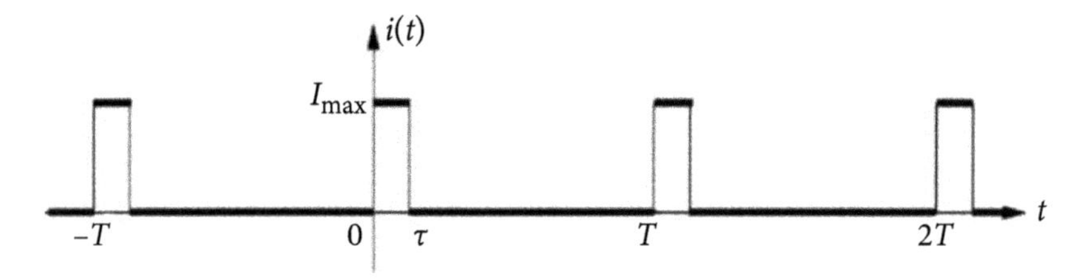

Figure 3.6 A current waveform.

assuming that $\tau/T = a$, *is equal to*

$$\bar{i} = \frac{1}{T}\int_0^{\tau} I_{\max}\, dt = \frac{\tau}{T} I_{\max} = a\, I_{\max}.$$

The ratio of the interval τ *when the current has a non-zero value to the period* T *is referred to as a* duty factor. *The current rms value is*

$$||i|| = \sqrt{\frac{1}{T}\int_0^{\tau} I^2_{\max} dt} = I_{\max}\sqrt{\frac{\tau}{T}} = \sqrt{a}\, I_{\max} \tag{3.12}$$

thus the mean value declines proportionally to the duty factor decline, while its rms value declines proportionally to the root of the duty factor, meaning the reduction in the pulse duration affects its mean value more than its rms value. For example, if the current maximum value is $I_{\max} = 100$ A, *and the duty factor* $a = 0.1$, *then we obtain*

$$\bar{i} = 0.1 \times 100 = 10\,\text{A}, \qquad ||i|| = \sqrt{0.1} \times 100 = 31.6\,\text{A}.$$

*****

A Linear Form of Mean Values: If quantities $x(t)$ and $y(t)$ are periodic with period T, and

$$z(t) = \alpha x(t) + \beta y(t)$$

where α and β are real numbers, then the mean value of $z(t)$ is

$$\overline{z} = \frac{1}{T}\int_0^T [\alpha x(t) + \beta y(t)]\, dt = \alpha\overline{x} + \beta\overline{y}.$$

The mean value of linear form is a linear form of mean values, with the same coefficients. It means that independently of their waveforms, mean values of periodic voltages and currents have to fulfill Kirchoff's current and voltage laws.

Illustration 3.6 *Currents in the node shown in Fig. 3.7 have to satisfy Kirchoff's current law, meaning*

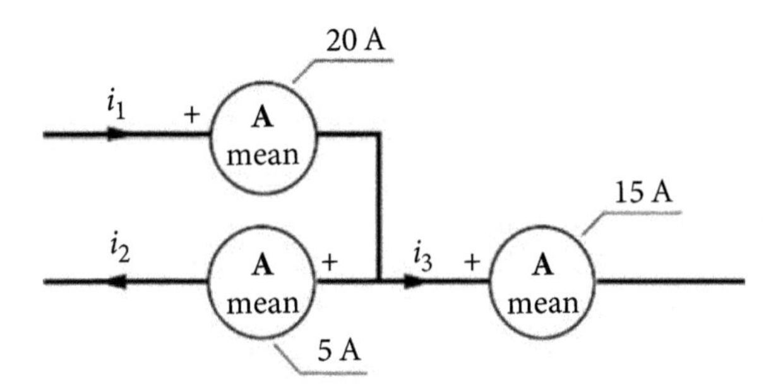

Figure 3.7 Node with meters of the current mean value.

$$i_3(t) = i_1(t) - i_2(t).$$

The mean values of currents i_1 and i_2 are measured with ammeters, which show

$$\overline{i_1} = 20\,\text{A}, \quad \overline{i_2} = 5\,\text{A}$$

thus, what is the mean value of the current i_3?

Since the mean value of the current i_3 is equal to the linear form of mean values of its components, the ammeter reading is

$$\overline{i_3} = \overline{i_1} - \overline{i_2} = 20 - 5 = 15\,\text{A}$$

independently on these currents' waveform.

*****

The rms Value of a Linear Form: Let us assume that quantities $x(t)$ and $y(t)$ are elements of the space L_T^2 and

$$z(t) = \alpha x(t) + \beta y(t)$$

is a linear form in that space. The rms value of $z(t)$ is

$$||z|| = \sqrt{\frac{1}{T}\int_0^T [\alpha x(t) + \beta y(t)]^2 dt} = \sqrt{\alpha^2||x||^2 + 2\alpha\beta(x,y) + \beta^2||y||^2} \qquad (3.13)$$

which is not a linear form of the rms values of $x(t)$ and $y(t)$ quantities. It means that the knowledge of the rms values of components of the sum does not allow us to calculate the rms value of this sum.

There is an exception from this general rule, however. If quantities are mutually proportional, meaning $y(t) = c\,x(t)$, $c \geq 0|$, thus $||y|| = c\,||x||$, then

$$||z|| = \sqrt{\frac{1}{T}\int_0^T [\alpha + \beta c]^2 x^2(t)\, dt} = |\alpha + \beta c|\ ||x||.$$

On the condition that coefficients α, β, and c are positive, then

$$||z|| = |\alpha + \beta c|\ ||x|| = \alpha\,||x|| + \beta\,||y||$$

meaning that the rms value of a linear form is the linear form of its components' rms values. Such situations often occur in electrical circuits, so this property is very useful for calculating the rms value of voltages and currents.

Illustration 3.7 *At supply voltages in the circuit shown in Fig. 3.8*

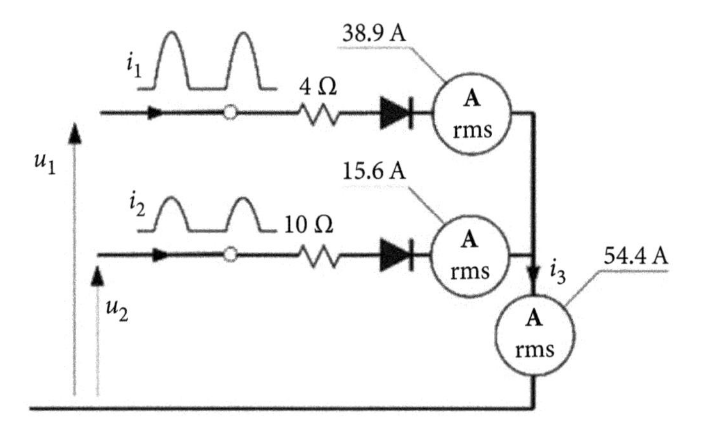

Figure 3.8 An example of a circuit.

equal to

$$u_1(t) = u_2(t) = 220\sqrt{2}\sin\omega t\,\mathrm{V}$$

while diode currents i_1 *and* i_2 *have the waveforms*

$$i_1(t) = 55\sqrt{2}\sin\omega t \times \mathrm{sgn}(\sin\omega t)\,\mathrm{A}$$

$$i_2(t) = 22\sqrt{2}\sin\omega t \times \mathrm{sgn}(\sin\omega t)\,\mathrm{A}$$

where the symbol sgn (x) *denotes the sign function, defined as*

$$\mathrm{sgn}(x) = \begin{cases} 1, & \text{if } \; x \geq 0 \\ -1, & \text{if } \; x < 0. \end{cases}$$

Diode currents have the rms values, respectively

$$||i_1|| = \frac{(I_1)_{\mathrm{max}}}{2} = \frac{55\sqrt{2}}{2} = 38.9\,\mathrm{A}, \quad ||i_2|| = \frac{(I_2)_{\mathrm{max}}}{2} = \frac{22\sqrt{2}}{2} = 15.6\,\mathrm{A}.$$

These two currents are mutually proportional, thus the rms value of their sum, the current i_3*, is equal to the sum of rms values of the diode currents* i_1 *and* i_3*, namely*

$$||i_3|| = ||i_1|| + ||i_2|| = 38.9 + 15.6 = 54.4\,\mathrm{A}.$$

In a situation where these currents are not mutually proportional, their scalar product (i_1, i_2) *has to be known for calculating the rms value of their sum,* i_3.

*****

Illustration 3.8 *If the voltage* u_2 *in the circuit shown in Fig. 3.8 is replaced by a* DC *voltage, for example* $u_2 = 156$ V*, as shown in Fig. 3.9,*

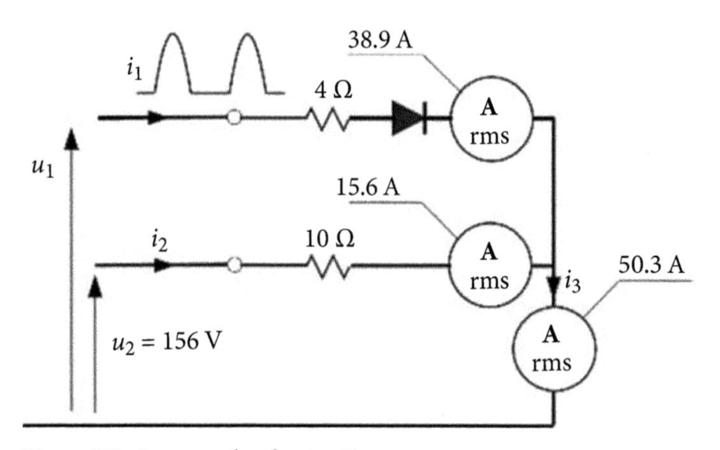

Figure 3.9 An example of a circuit.

then the rms values of currents i_1 *and* i_2 *remain unchanged. These currents are not mutually proportional, however. Their scalar product has to be calculated for calculating the rms value of their sum. It is equal to*

$$(i_1, i_2) = \frac{1}{T}\int_0^T i_1(t)\, i_2\, dt =$$

$$= \frac{1}{T}\int_0^{T/2} \left(55\sqrt{2}\sin \omega t\right) \times 15.6\, dt = 386.4\ \mathrm{A}^2.$$

Hence,

$$||i_3|| = \sqrt{||i_1||^2 + 2(i_1, i_2) + ||i_2||^2} = \sqrt{38.9^2 + 2 \times 386.4 + 15.6^2} = 50.3\ \mathrm{A}.$$

*****

3.2 Orthogonality

The norm of a linear form of vectors in space $\boldsymbol{L}_{\mathrm{T}}{}^2$, or simply the rms value of a linear form of periodic quantities that belong to that space can be calculated if the rms values of individual quantities are known only on the condition, excluding the case of their proportionality, that these quantities are orthogonal, meaning their scalar product

$$(x, y) = \frac{1}{T}\int_0^T x(t)\, y(t)\, dt$$

is equal to zero and then the rms value is equal to

$$||z|| = ||\alpha\, x(t) + \beta\, y(t)|| = \sqrt{\alpha^2||x||^2 + \beta^2||y||^2}.$$

Coefficient α and β are very often equal to ± 1 and then

$$||z||^2 = ||x + y||^2 = ||x||^2 + ||y||^2. \tag{3.14}$$

When triangle sides are interpreted as vectors, then this formula expresses the relationship between the side length of a right triangle.

The relation between the rms value of a quantity $x(t)$

$$x(t) = \sum_{\mathrm{c}} x_{\mathrm{c}}(t)$$

and rms values of its components $x_{\mathrm{c}}(t)$ is of utmost importance for interpreting the power properties of electrical circuits. It says that when these components are orthogonal, they affect the rms value of the quantity $x(t)$ independently of each other.

There are four general situations where two quantities $x(t)$ and $y(t)$ from the space L_T^2 are orthogonal:

(1) Quantities $x(t)$ and $y(t)$ have such waveforms that at each instant of time t when one of them is not equal to zero, the other one is equal to zero, meaning

$$x(t)\, y(t) \equiv 0.$$

(2) Quantities $x(t)$ and $y(t)$ are sinusoidal but mutually shifted by a quarter of the period,

$$x = \sqrt{2}X \sin(\omega t - \psi), \qquad y = \sqrt{2}Y \sin(\omega t - \psi \pm \pi/2).$$

(3) Quantities $x(t)$ and $y(t)$ are harmonics of a different order $r \neq s$.

$$x = \sqrt{2}X_r \sin(r\omega t - \psi), \quad y = \sqrt{2}Y_s \sin(s\omega t - \phi), \quad r \neq s.$$

(4) One of the quantities $x(t)$ and $y(t)$ is a derivative of the other one, for example

$$y(t) = \alpha \frac{d}{dt} x(t).$$

Situation (2) is a special case of situation (4) but is very common in form (2).

When components of a sum of two quantities have one of these four properties, then a calculation of the rms value of that sum can be substantially simplified. Also, when a quantity, which has a complex waveform, can be decomposed into components that have one of the properties (1)–(4), then such decomposition simplifies the rms value calculation.

Illustration 3.9 *The waveform of a current shown in Fig. 3.10a can be expressed as the difference of half-rectified currents, shown in Fig. 3.10b and 3.10c, of the rms value equal to $I_{max}/2$, thus* 50 A *and* 15 A, *respectively. The current can be expressed as the difference*

$$i(t) = i_a(t) - i_b(t).$$

Since their product is equal to zero, that is,

$$i_a(t) \cdot i_b(t) \equiv 0$$

these two currents satisfy condition (1), thus they are mutually orthogonal and consequently,

$$||i|| = \sqrt{||i_a||^2 + ||i_b||^2} = \sqrt{50^2 + 15^2} = 52.2\,\text{A}$$

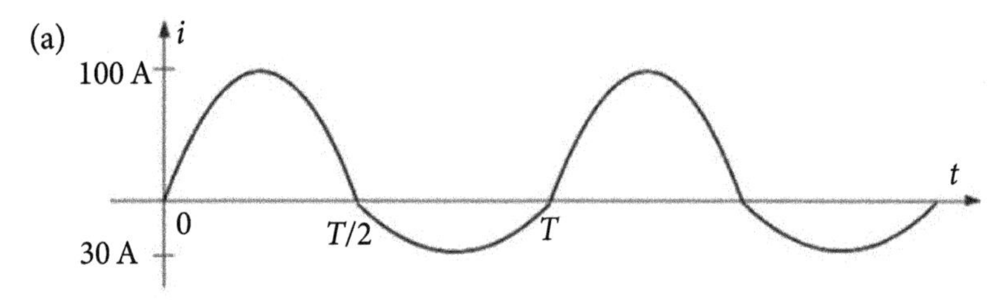

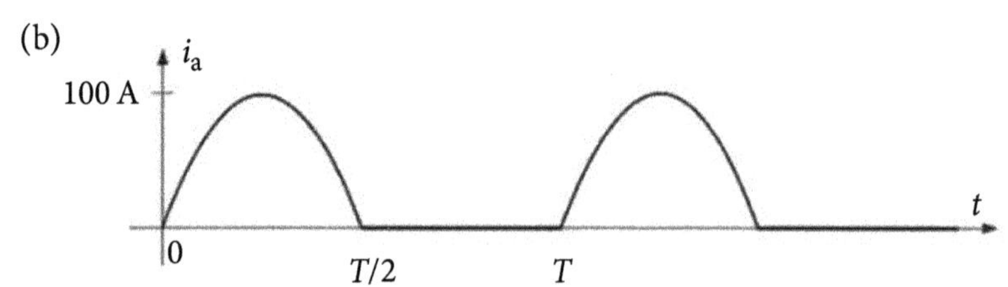

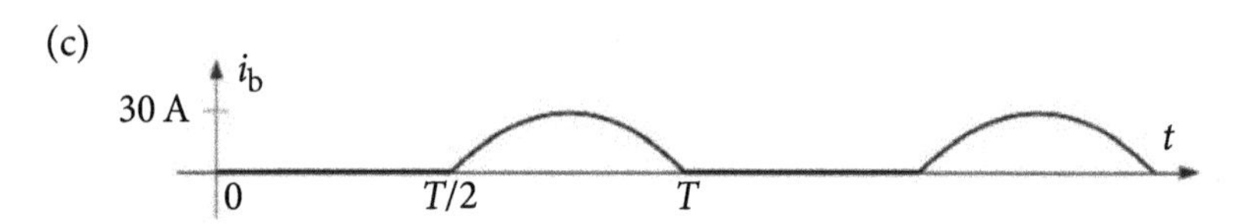

Figure 3.10 A current waveform, (a), and its components, (b) and (c).

*****

Illustration 3.10 *A periodic voltage shown in Fig. 3.11a, with unknown analytical waveform, is specified by a sequence of N = 9 values in equidistant instants, namely*

$$u_1 = 45\text{V}, u_2 = 90\text{V}, u_3 = 93\text{V}, u_4 = 70\text{V}, u_5 = 35\text{V},$$
$$u_6 = 3\text{V}, u_7 = -30\text{V}, u_8 = -45\text{V}, u_9 = -31\text{V}.$$

Having these values, the voltage can be approximated with a stair-like function, shown in Fig. 3.11b, and next, decomposed into N = 9 rectangular pulses, shown in Fig. 3.11c. These pulses are mutually orthogonal, since for r ≠ s,

$$u_r(t)u_s(t) \equiv 0.$$

The rms value of a single pulse is equal to

$$||u_n|| = \sqrt{\frac{1}{T}\int_0^{T/N} u_n^2\, dt} = \frac{u_n}{\sqrt{N}}.$$

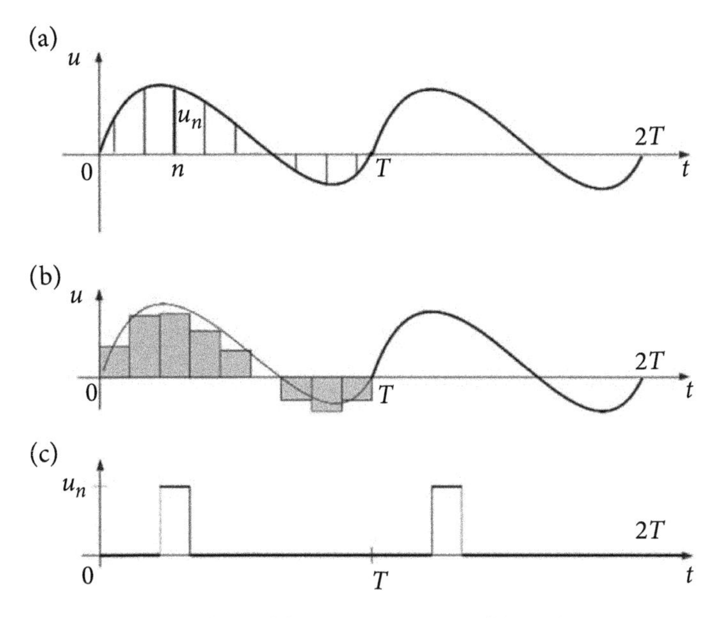

Figure 3.11 A voltage waveform, (a), its approximation (b), and pulse components (c).

Due to the orthogonality of these pulses, the voltage rms value can be approximated with

$$||u|| = \sqrt{\sum_{n=1}^{N} ||u_n||^2} = \sqrt{\frac{1}{N}\sum_{n=1}^{N} u_n^2}. \tag{3.15}$$

For data in this illustration,

$$||u|| = \sqrt{\frac{1}{N}\sum_{n=1}^{N} u_n^2} =$$

$$= \sqrt{\frac{1}{9}\left(45^2 + 90^2 + 93^2 + 70^2 + 35^2 + 3^2 + 30^2 + 45^2 + 31^2\right)} = 56.6\ \text{V}.$$

*****

Illustration 3.11 *The supply current of a three-phase rectifier with an inductive filter can be approximated with a trapezoidal waveform shown in Fig. 3.12a. Such approximated current can be decomposed into rectangular and saw-like pulses, shown in Fig. 3.12b and c. It contains two rectangular and four saw-like pulses. The rectangular pulse duration is $T/3$ and the duration of the saw-like pulse is τ.*

Assuming that $I_{max} = 100$ A and $\tau = T/24$, we obtain

$$||i_1|| = \sqrt{\frac{1}{T}\int_0^{T/3} (I_{max})^2 dt} = \frac{I_{max}}{\sqrt{3}} = 57.74 \text{ A}$$

$$||i_2|| = \sqrt{\frac{1}{T}\int_0^{\tau} \left(I_{max}\frac{t}{\tau}\right)^2 dt} = \frac{I_{max}}{\sqrt{3}}\sqrt{\frac{\tau}{T}} = \frac{100}{\sqrt{3}}\sqrt{\frac{1}{24}} = 11.79 \text{ A}.$$

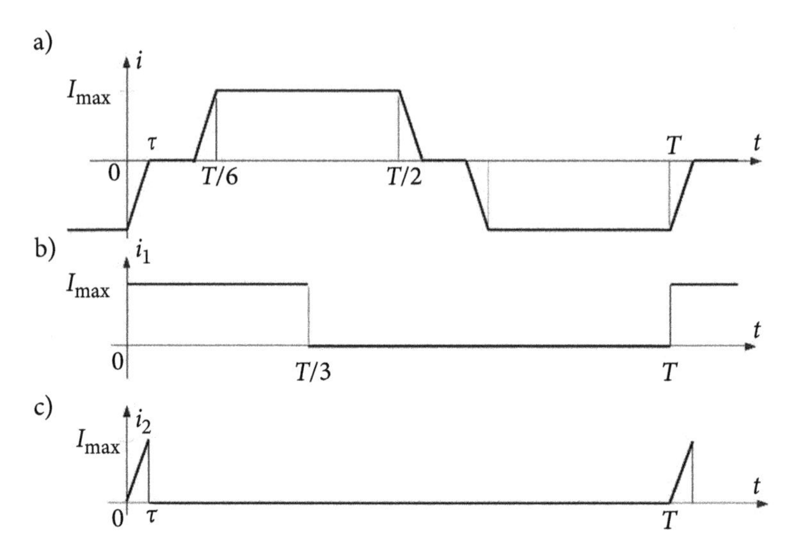

Figure 3.12 A trapezoidal approximation (a) of three-phase rectifier current and its orthogonal components (b and c).

All components are mutually orthogonal, thus the supply current rms value is

$$||i|| = \sqrt{2||i_1||^2 + 4||i_2||^2} = \sqrt{2 \times 57.74^2 + 4 \times 11.79^2} = 84.99 \text{ A}.$$

*****

Illustration 3.12 *Currents in a resistive and a capacitive branch of the load RC load shown in Fig. 3.13 have the following waveforms*

$$i_R = 30\sqrt{2}\sin(\omega_1 t + 20°) \text{ A}, \quad i_C = 40\sqrt{2}\cos(\omega_1 t + 20°) \text{ A}.$$

The branch currents are shifted mutually by a quarter of the period, thus they are orthogonal. Consequently, the supply current rms value is

$$||i|| = \sqrt{||i_R||^2 + ||i_C||^2} = \sqrt{40^2 + 30^2} = 50 \text{ A}.$$

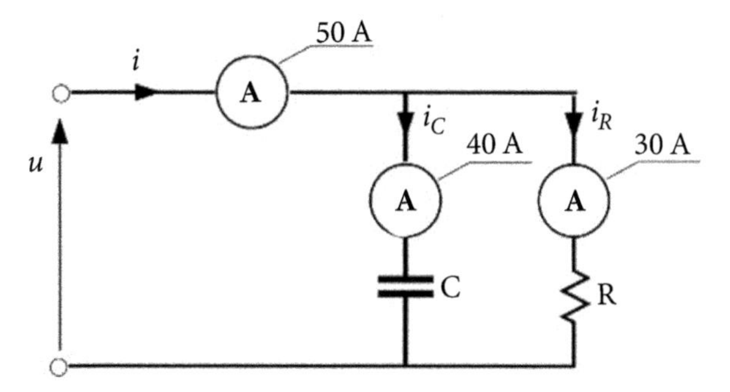

Figure 3.13 An *RC* load with ammeters of the rms values.

*****

This result can be obtained not only using condition (2) but also condition (4) for the orthogonality of two quantities. Namely, if $x(t) \in L_T^2$ and its derivative $dx(t)/dt \in L_T^2$, then, their scalar product is equal to

$$\left(x, \frac{dx}{dt}\right) = \frac{1}{T}\int_0^T x(t)\frac{dx(t)}{dt}dt = \frac{1}{T}\int_{x(0)}^{x(T)} x(t)\,dx(t) = 0 \tag{3.16}$$

since $x(T) - x(0)$. Thus, the derivative of a quantity is orthogonal to the quantity. In particular, for the load shown in Fig. 3.13, the scalar product of the resistive and capacitive branches current is equal to

$$(i_R, i_C) = \left(Gu, C\frac{du}{dt}\right) = GC\left(u, \frac{du}{dt}\right) = 0$$

independently on the supply voltage waveform. The relation

$$||i|| = \sqrt{||i_R||^2 + ||i_C||^2}$$

is satisfied not only for sinusoidal supply voltage but for voltage with any waveform that belongs to space L_T^2. A similar relation is satisfied for a load composed of a resistor and an ideal inductor connected in series, as shown in Fig. 3.14. Indeed,

$$(u_R, u_L) = \left(Ri, L\frac{dt}{dt}\right) = RL\left(i, \frac{di}{dt}\right) = 0$$

thus

$$||u|| = \sqrt{||u_R||^2 + ||u_L||^2}$$

independently of the supply voltage waveform.

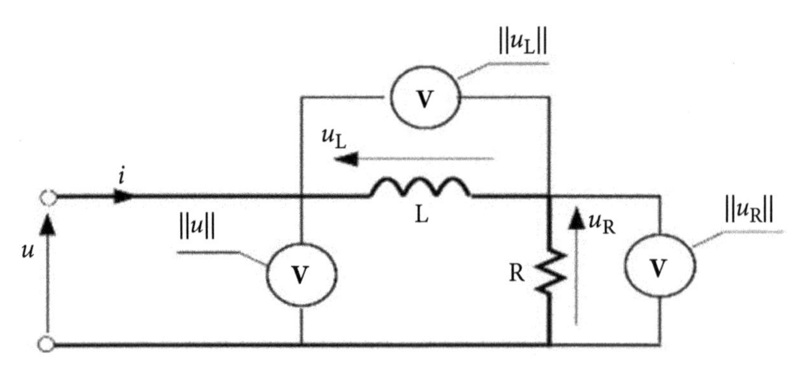

Figure 3.14 An *RL* circuit.

Illustration 3.13 *The supply voltage has the rms value* $||u|| = 220$ V, *but its waveform, although periodic, is not specified. Calculate the inductor voltage rms value, if the resistor voltage rms value is* $||u_R|| = 200$ V.

Assuming that the inductor is lossless, due to the resistor and the inductor voltage orthogonality, independently of the supply voltage waveform, the inductor voltage rms value is

$$||u_L|| = \sqrt{||u||^2 - ||u_R||^2} = \sqrt{220^2 - 200^2} = 91.6\,\text{V}.$$

This is, of course, only an approximate result, since the inductor always has some resistance r_L, *so that the voltage on a real inductor, including its resistance, is not orthogonal to the voltage on a series resistor R.*

*****

3.3 Fourier Series in a Complex Form

Elementary Vectors of Space L_T^2: A quantity that is a sum of two sinusoidal quantities of different periods T and T_r

$$x(t) = \sin\left(\frac{2\pi}{T}t\right) + \sin\left(\frac{2\pi}{T_r}t\right) + \cos\left(\frac{2\pi}{T_r}t\right)$$

has the period T only when there is a natural number r, such that $rT_r = T$, meaning

$$T_r = \frac{T}{r}.$$

Therefore

$$\sin\left(r\frac{2\pi}{T}t\right), \quad \cos\left(r\frac{2\pi}{T}t\right)$$

are also vectors in space $\boldsymbol{L}_{\mathrm{T}}^2$. Such vectors are referred to as *elementary vectors* of that space. A constant value $x(t) = 1$ is also an elementary vector. One should be aware however that although a linear form of any finite number of vectors from a linear space $\boldsymbol{L}_{\mathrm{T}}^2$ belongs to that space, an infinite number of such vectors may not belong to it. For example, if $t = nT$, then the series

$$\sum_{r=0}^{\infty} \cos\left(r\frac{2\pi}{T}t\right) - > \infty$$

is not limited. It is not integrable with square. Such a sum does not belong to space, $\boldsymbol{L}_{\mathrm{T}}^2$.

Fourier Series: Each quantity $x(t)$ that belongs to space $\boldsymbol{L}_{\mathrm{T}}^2$, has a Fourier series $x_{\mathrm{F}}(t)$, defined as

$$x_{\mathrm{F}}(t) \stackrel{\mathrm{df}}{=} a_0 + \sum_{n=1}^{\infty} a_n \cos n\omega_1 t + \sum_{n=1}^{\infty} b_n \sin n\omega_1 t$$

with

$$a_0 = \frac{1}{T}\int_0^T x(t)\, dt, \quad a_n = \frac{2}{T}\int_0^T x(t)\cos(n\omega_1 t)\, dt, \quad b_n = \frac{2}{T}\int_0^T x(t)\sin(n\omega_1 t)\, dt.$$

The angular frequency of quantities of the period T commonly is not denoted by ω, but by ω_1, namely

$$\omega_1 = \omega = \frac{1}{T}.$$

Approximation of nonsinusoidal but periodic quantities by an infinite series of sinusoidal components, referred to as *harmonics*, was invented by Jean Baptist Fourier [1] in 1807 in his research on heat transfer, published in 1822. There was a lot of controversy among mathematicians on conditions that a function has to fulfill to have a Fourier series. Fortunately, this controversy does not apply to physical quantities but only to their mathematical models.

The Fourier series $x_{\mathrm{F}}(t)$ is convergent to $x(t)$ at each point where $x(t)$ is continuous. Therefore, a continuous quantity $x(t)$ and its Fourier Series $x_{\mathrm{F}}(t)$ are identical.

When a quantity $x(t)$ has discontinuities, then its Fourier series $x_{\mathrm{F}}(t)$ is convergent to the mean value of the left and right limits of $x(t)$ at discontinuity points, namely, if $x(t)$ has a discontinuity at point $t = t_k$, then

$$x_{\mathrm{F}}(t_k) = \frac{1}{2}\left[\lim_{t\to t_k^+} x(t) + \lim_{t\to t_k^-} x(t)\right].$$

Since real voltages and currents cannot have discontinuities, there is no need to use different symbols for quantities and their Fourier series. Discontinuities can occur only in mathematical models of real quantities. However, even in such a case, the same symbols are used usually for such mathematical models of real quantities and their Fourier series.

The term

$$x_n(t) = a_n \cos n\omega_1 t + b_n \sin n\omega_1 t = c_n \cos(n\omega_1 t + \alpha_n) \tag{3.17}$$

with

$$c_n = \sqrt{a_n^2 + b_n^2}, \qquad \alpha_n = -\tan^{-1}\left\{\frac{b_n}{a_n}\right\} \tag{3.18}$$

is referred to as the nth *order harmonic* of $x(t)$. Harmonic of the first order, that is, for $n = 1$

$$x_1(t) = a_1 \cos \omega_1 t + b_1 \sin \omega_1 t = c_1 \cos(\omega_1 t + \alpha_1)$$

is referred to as the *fundamental harmonic*, sometimes simplified to only the *fundamental*. The mean value, a_0, is sometimes referred to as the *zero-order harmonic.*

The Fourier series in the form presented above will be referred to in this book as a Fourier series in a *traditional form*. Most textbooks on circuits and signals analysis as well as books on power electronics use the Fourier series in this traditional form.

Circuits with nonsinusoidal voltages and currents expressed in a traditional form of the Fourier series can be described in terms of sets of differential equations. Their analysis requires that the set of equations be integrated. Integration of such a set of differential equations usually is not easy, however.

The circuit analysis can be substantially simplified if such a set of differential equations is converted to an equivalent set of algebraic equations. To make such a conversion possible, the Fourier series in a traditional form has to be converted as follows. Since

$$e^{jn\omega_1 t} = \cos(n\omega_1 t) + j\sin(n\omega_1 t)$$

thus

$$\cos(n\omega_1 t) = \mathrm{Re}\{e^{jn\omega_1 t}\}, \quad \sin(n\omega_1 t) = \mathrm{Re}\{-je^{jn\omega_1 t}\}.$$

The Fourier series can be presented as

$$x(t) = a_0 + \sum_{n=1}^{\infty} a_n \cos n\omega_1 t + \sum_{n=1}^{\infty} b_n \sin n\omega_1 t = a_0 + \sum_{n=1}^{\infty} a_n \mathrm{Re}\{e^{jn\omega_1 t}\} + \sum_{n=1}^{\infty} b_n \mathrm{Re}\{-je^{jn\omega_1 t}\} =$$

$$= a_0 + \sqrt{2}\,\mathrm{Re}\sum_{n=1}^{\infty} \frac{a_n - jb_n}{\sqrt{2}} e^{jn\omega_1 t}.$$

If we denote

$$a_0 = \frac{1}{T}\int_0^T x(t)\,dt \overset{\text{df}}{=} X_0, \quad \boldsymbol{X}_n \overset{\text{df}}{=} \frac{a_n - jb_n}{\sqrt{2}} \tag{3.19}$$

the Fourier series can be expressed in the form

$$x(t) = \sum_{n=0}^{\infty} x_n(t) = X_0 + \sqrt{2}\,\text{Re}\sum_{n=1}^{\infty} \boldsymbol{X}_n e^{jn\omega_1 t} \tag{3.20}$$

referred to as a *complex form* of the Fourier series.

The complex coefficient

$$\boldsymbol{X}_n = \frac{a_n - jb_n}{\sqrt{2}} = \frac{1}{\sqrt{2}}\left[\frac{2}{T}\int_0^T x(t)\cos(n\omega_1 t)\,dt - j\frac{2}{T}\int_0^T x(t)\sin(n\omega_1 t)\,dt\right]$$

can be simplified to the form

$$\boldsymbol{X}_n = X_n e^{j\alpha_n} = \frac{\sqrt{2}}{T}\int_0^T x(t)e^{-jn\omega_1 t}dt \tag{3.21}$$

and it is referred to as a *complex rms (crms) value* of the nth order harmonic. The magnitude of the crms value $\boldsymbol{X}_n$ is equal to the harmonic rms value, X_n, while the argument of $\boldsymbol{X}_n$ is the harmonic phase angle, α_n.

In power electronics, where voltages and currents are specified usually as a function of angle, $\beta = \omega t$, the crms value of the nth order harmonic can be calculated from the formula

$$\boldsymbol{X}_n = \frac{\sqrt{2}}{\omega_1 T}\int_0^{\omega_1 T} x(\omega_1 t)e^{-jn(\omega_1 t)}d(\omega_1 t) = \frac{1}{\sqrt{2}\pi}\int_0^{2\pi} x(\beta)e^{-jn\beta}d\beta. \tag{3.22}$$

Nonsinusoidal voltages and currents can be expressed, according to Fourier decomposition, as a sum of an infinite number of harmonics. It is, of course, not possible in practical applications, therefore in such applications, the voltages and currents are only approximated by a sum of a finite number of harmonics. Such approximation reduces the Fourier series to a trigonometric polynomial of the form,

$$x(t) = X_0 + \sqrt{2}\,\text{Re}\sum_{n\in N} \boldsymbol{X}_n\, e^{jn\omega_1 t} = \sum_{n\in N_0} x_n(t)$$

where N_0 denotes the set of harmonic orders, n, along with $n = 0$, while N denotes the same set, but without $n = 0$. The name *trigonometric polynomial* is a mathematical term for such a sum, however it is not particularly used in electrical engineering, where it is usually still referred to as a Fourier series.

Illustration 3.14 *Fourier series of the current has the following traditional form*

$$i(t) = 10 + 100\cos(\omega_1 t) + 20\cos(3\omega_1 t) + 20\sin(3\omega_1 t) + 10\sin(7\omega_1 t)\ \text{A}$$

meaning that coefficients a_n and b_n for $N_0 = \{0,1,3,7\}$ are given. The complex form of the Fourier series is

$$i(t) = 10 + \sqrt{2}\,\text{Re}\left\{70,7e^{j\omega_1 t} + 20e^{-j45^\circ}e^{j3\omega_1 t} + 7,07e^{-j90^\circ}e^{j7\omega_1 t}\right\}\text{A.}$$

The crms values of harmonics in this illustration were calculated using the relationship between the crms values, $\boldsymbol{X}_n$, of harmonics and their a_n and b_n coefficients. When these coefficients are not known, the functional

$$\boldsymbol{X}_n = \frac{\sqrt{2}}{T}\int_0^T x(t)e^{-jn\omega_1 t}dt$$

has to be calculated. Observe, however, that calculation of this functional reduces the amount of calculation as compared to that needed for calculating a_n and b_n coefficients separately.

*****

Circuit modeling programs such as Pspice or MATLAB are equipped with programs for calculating the crms values of harmonics with a *Discrete Fourier Transform* (DFT), performed on a sequence of discrete values of the analyzed periodic quantity. These programs usually implement an algorithm that reduces the number of multiplications needed for the DFT calculation. This algorithm is known as the *Fast Fourier Transform* (FFT).

While these programs are convenient tools for calculating the crms values of harmonics, unfortunately they provide only specific but not general information on harmonics and their dependence on some features of the analyzed quantity. The complex rms values of harmonics in analytical form can be needed in studies on various aspects of electrical performance. The skill in their effective calculation can be beneficial for such studies. In particular, it is beneficial in studies on power phenomena in with nonsinusoidal voltages and currents and for compensation.

Illustration 3.15 *Let us calculate the crms values of harmonics of the sequence of rectangular pulses shown in Fig. 3.15, assuming that*

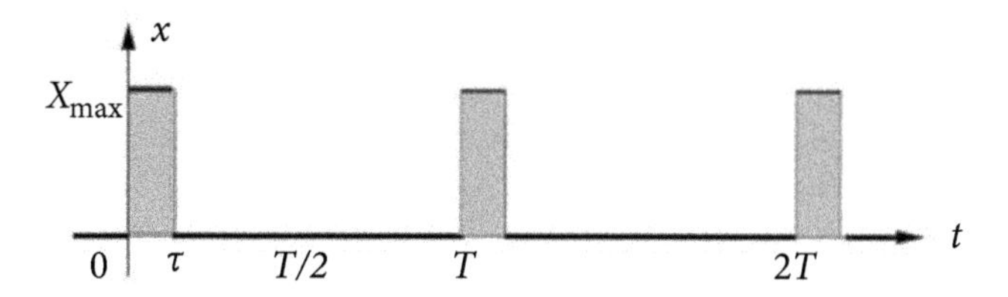

Figure 3.15 A periodic sequence of rectangular pulses.

$$X_{\max} = 100, \text{ and } \tau/T = a = 0.2.$$

The mean value of rectangular pulses is

$$X_0 = a\,X_{\max} = 0.2 \times 100 = 20.$$

The analytic formula for the crms values of harmonics has the form

$$\boldsymbol{X}_n = \frac{\sqrt{2}}{T}\int_0^T x(t)e^{-jn\omega_1 t}dt = \frac{\sqrt{2}}{T}\int_0^\tau X_{\max}e^{-jn\omega_1 t}dt = j\frac{X_{\max}}{\sqrt{2}\pi n}(e^{-jn2\pi\frac{\tau}{T}} - 1).$$

The magnitude and argument should be separated. Unfortunately, they are not separated in this formula. They can be separated as follows:

$$\boldsymbol{X}_n = j\frac{X_{\max}}{\sqrt{2}\pi\, n}(e^{-jn2\pi\frac{\tau}{T}} - 1) = -\sqrt{2}\frac{X_{\max}}{\pi n}\frac{(e^{-jn\pi\frac{\tau}{T}} - e^{jn\pi\frac{\tau}{T}})}{2j}e^{-jn\pi\frac{\tau}{T}} =$$

$$= \frac{\sqrt{2}}{\pi\, n}X_{\max}\sin\left(n\pi\frac{\tau}{T}\right)\, e^{-jn\pi\frac{\tau}{T}}.$$

For $X_{\max} = 100$ *oraz* $\tau/T = 0.2$, *numerical crms values* $\boldsymbol{X}_n$ *up to* $n = 10$ *are equal to*

$$\boldsymbol{X}_1 = 26.46e^{-j36°},\ \boldsymbol{X}_2 = 21.41e^{-j72°},\ \boldsymbol{X}_3 = 14.27e^{-j108°},\ \boldsymbol{X}_4 = 6.61e^{-j144°},\ \boldsymbol{X}_5 = 0,$$
$$\boldsymbol{X}_6 = 5.51e^{-j36°},\ \boldsymbol{X}_7 = 6.12e^{-j72°},\ \boldsymbol{X}_8 = 5.35e^{-j108°},\ \boldsymbol{X}_9 = 2.94e^{-j144°},\ \boldsymbol{X}_{10} = 0.$$

Thus, the Fourier series of these periodic rectangular pulses in the complex form is

$$x(t) = 20 + \sqrt{2}\,\text{Re}\,\left\{26.46e^{-j36°}e^{j\omega_1 t} + 21.41e^{-j72°}e^{j2\omega_1 t} + 21.41e^{-j108°}e^{j3\omega_1 t} + \ldots\right\}.$$

The sum of the first fifteen harmonics is shown in Fig. 3.16. Because rectangular pulses have discontinuities, the sum of harmonics converges at these discontinuity points to the mean value of 0 and $X_{\max}$, *meaning* $X_{\max}/2$.

*****

One can observe in Fig. 3.16 that as time approaches the points of discontinuities, some oscillations grow up in the sum of harmonics. These oscillations are known as a *Gibbs phenomenon.* Independently of how many harmonics are added, these oscillations do not disappear. Their frequency only increases with the number of added harmonics. It is not an error but a property of the Fourier series convergence at discontinuity points. Observe that when only harmonics and their sum are known, one can draw an incorrect conclusion of the quantity maximum value. Despite what can be seen in Fig. 3.16, the maximum value of the quantity $x(t)$ is not higher than $X_{\max}$.

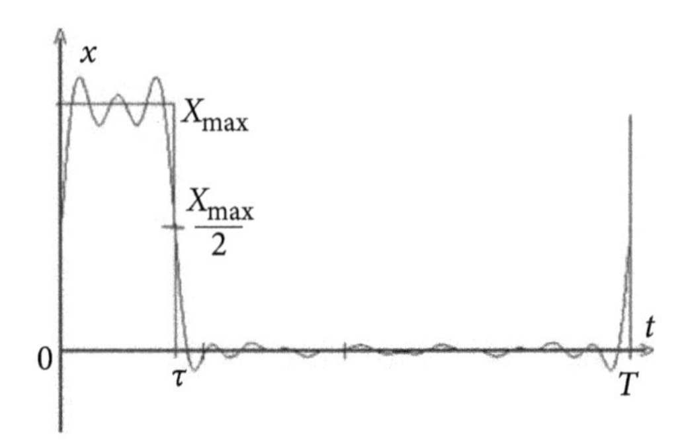

Figure 3.16 Sum of 15 harmonics of rectangular pulses.

When a quantity is specified analytically not as a function of time but as a function of angle, $\beta = \omega_1 t$, as is common in power electronics, the formula for the crms value X_n of harmonics has to be adequately modified as follows,

Illustration 3.16 *Let us develop the analytical formula for the crms value $\mathbf{X}_n$ of harmonics of thyristor current at sinusoidal supply voltage and resistive load and the firing angle α, that is, the current of the waveform shown in Fig. 3.17.*

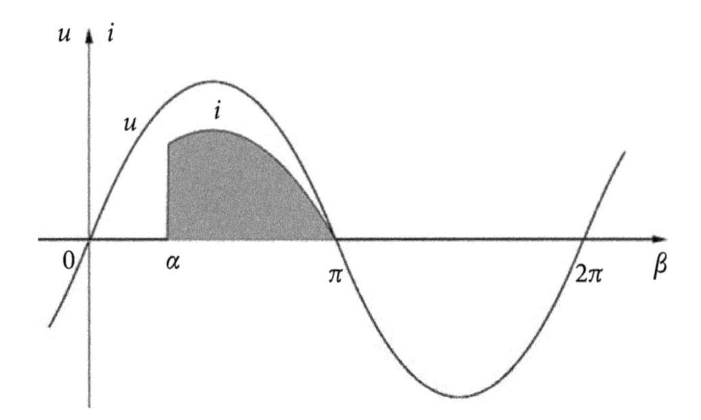

Figure 3.17 A thyristor's current waveform.

Calculate the crms values $\boldsymbol{X}_n$ of the current harmonics for $I_{\max} = 100$ A and $\alpha = 45^\circ$. The current mean value is

$$I_0 = \frac{1}{2\pi}\int_{\alpha}^{\pi} I_{\max}\sin\beta d\beta = \frac{I_{\max}}{2\pi}(1+\cos\alpha).$$

For the assumed numerical values

$$I_0 = \frac{100}{6.28}\left(1+\frac{\sqrt{2}}{2}\right) = 27.17\,\text{A}.$$

The crms values of the current harmonics

$$\boldsymbol{I}_n = \frac{1}{\sqrt{2\pi}}\int_{\alpha}^{\pi} I_{\max}\sin\beta\, e^{-jn\beta} d\beta = \frac{I_{\max}}{\sqrt{2\pi}}\int_{\alpha}^{\pi}\frac{e^{j\beta}-e^{-j\beta}}{2j}e^{-jn\beta}d\beta =$$

$$= \frac{I_{\max}}{2j\sqrt{2\pi}}\int_{\alpha}^{\pi}\left[e^{-j(n-1)\beta} - e^{-j(n+1)\beta}\right] d\beta.$$

For $n = 1,\ (n-1)\beta = 0;$ thus $e^{-j(n-1)\beta} = 1$, *hence*

$$\boldsymbol{I}_1 = \frac{I_{\max}}{2j\sqrt{2\pi}}\int_{\alpha}^{\pi}(1-e^{-j2\beta})\, d\beta = \frac{I_{\max}}{2j\sqrt{2\pi}}\left(\beta + \frac{e^{-j2\beta}}{2j}\right)\Bigg|_{\alpha}^{\pi} =$$

$$= \frac{I_{\max}}{2j\sqrt{2\pi}}\left[(\pi-\alpha) - j\frac{1}{2}(1-e^{-j2\alpha})\right] = \frac{I_{\max}}{4\sqrt{2\pi}}[(e^{-j2\alpha}-1) - j2(\pi-\alpha)] =$$

$$= \frac{I_{\max}}{4\sqrt{2\pi}}\left[(\cos 2\alpha - 1) - j2\left(\pi - \alpha + \frac{1}{2}\sin 2\alpha\right)\right].$$

For $\alpha = 45^\circ = \pi/4$ rad *and* $I_{\max} = 100$ *A, the crms value of the fundamental harmonic is*

$$\boldsymbol{I}_1 = \frac{100}{4\sqrt{2\pi}}\left[\left(\cos 2\frac{\pi}{4} - 1\right) - j2\left(\pi - \frac{\pi}{4} + \frac{1}{2}\sin 2\frac{\pi}{4}\right)\right] = -5.63 - j32.14 = 32.60\, e^{-j99.9^\circ}\ \text{A}.$$

For other current harmonics

$$\boldsymbol{I}_n = \frac{I_{\max}}{2j\sqrt{2\pi}}\left[\frac{e^{-j(n-1)\beta}}{-j(n-1)} - \frac{e^{-j(n+1)\beta}}{-j(n+1)}\right]\Bigg|_{\alpha}^{\pi} =$$

$$= \frac{I_{\max}}{2\sqrt{2\pi}}\left[\frac{e^{-j(n-1)\pi} - e^{-j(n-1)\alpha}}{n-1} - \frac{e^{-j(n+1)\pi} - e^{-j(n+1)\alpha}}{n+1}\right].$$

When n is odd, then $e^{-j(n-1)\pi} = e^{-j(n+1)\pi} = 1$, *and hence*

$$\boldsymbol{I}_n = \frac{I_{\max}}{2\sqrt{2}\pi}\left[\frac{1 - e^{-j(n-1)\alpha}}{n-1} - \frac{1 - e^{-j(n+1)\alpha}}{n+1}\right]$$

$$= \frac{I_{\max}}{2\sqrt{2}\pi(n^2-1)}[(n-1)e^{-j(n+1)\alpha} - (n+1)e^{-j(n-1)\alpha} + 2].$$

When n is even, then $e^{-j(n-1)\pi} = e^{-j(n+1)\pi} = -1$, *and hence*

$$\boldsymbol{I}_n = \frac{I_{\max}}{2\sqrt{2}\pi}\left[\frac{-1 - e^{-j(n-1)\alpha}}{n-1} - \frac{-1 - e^{-j(n+1)\alpha}}{n+1}\right]$$

$$= \frac{I_{\max}}{2\sqrt{2}\pi(n^2-1)}[(n-1)e^{-j(n+1)\alpha} - (n+1)e^{-j(n-1)\alpha} - 2].$$

Numerical values of $\boldsymbol{I}_n$, *up to* $n = 11$, *are*

$$\boldsymbol{I}_2 = 18.90e^{j164^\circ}\,\mathrm{A}, \boldsymbol{I}_3 = 5.63e^{j90^\circ}\,\mathrm{A}, \boldsymbol{I}_4 = 4.27e^{j95.5^\circ}\,\mathrm{A}, \boldsymbol{I}_5 = 4.19e^{j26.6^\circ}\,\mathrm{A},$$
$$\boldsymbol{I}_8 = 2.11e^{-j107^\circ}\,\mathrm{A}, \boldsymbol{I}_9 = 1.59e^{-j135^\circ}\,\mathrm{A}, \boldsymbol{I}_{10} = 1.84e^{j175^\circ}\,\mathrm{A}, \boldsymbol{I}_{11} = 1.35e^{j124^\circ}\,\mathrm{A}.$$

The sum of the first fifteen harmonics of the thyristor current is shown in Fig. 3.18.

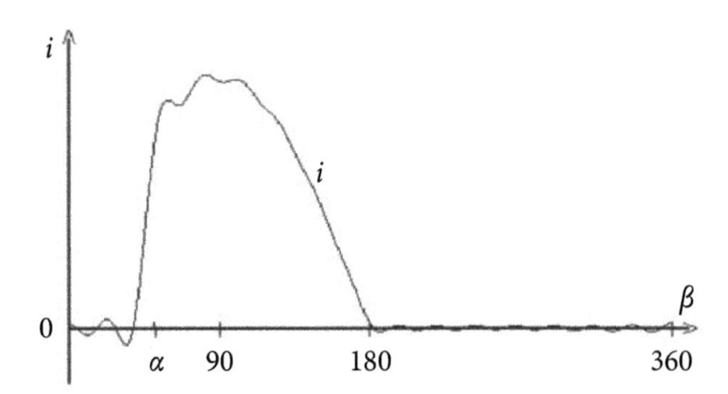

Figure 3.18 The sum of the first fifteen harmonics of thyristor current.

The Gibbs phenomenon can be observed in this sum. Also, observe that the sum approximates the original current much more accurately at points distant from the point of discontinuity than close to it.

*****

Illustration 3.17 *Let us develop the Fourier series in a complex form for the saw-like quantity shown in Fig. 3.19.*

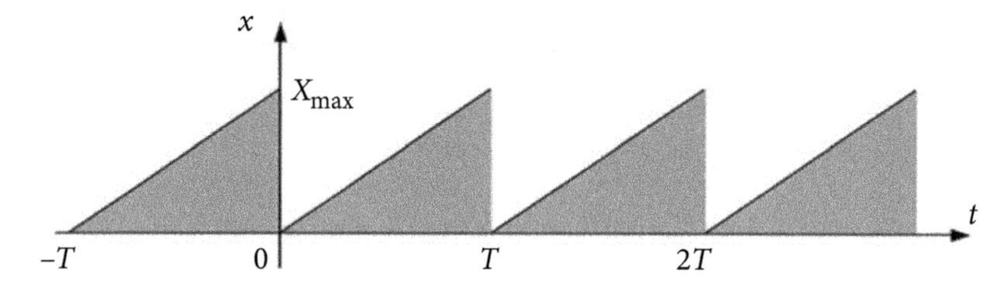

Figure 3.19 A saw-like waveform.

The crms values of harmonics are

$$\mathbf{X}_n = \frac{\sqrt{2}}{T}\int_0^T x(t)\, e^{-jn\omega_1 t}dt = \frac{\sqrt{2}}{T}\int_0^T \frac{X_{max}}{T}\, t\, e^{-jn\omega_1 t}dt =$$

$$= \frac{\sqrt{2}X_{max}}{T^2}\left(\frac{1}{-jn\omega_1}t\, e^{-jn\omega_1 t}\bigg|_0^T - \frac{1}{-jn\omega_1}\int_0^T e^{-jn\omega_1 t}dt\right).$$

The integral in the last term, as an integral of periodic quantity over the period, is equal to zero, thus

$$\mathbf{X}_n = \frac{\sqrt{2}X_{max}}{T^2}\left(\frac{1}{-jn\omega_1}t\, e^{-jn\omega_1 t}\bigg|_0^T\right) = \frac{X_{max}}{\sqrt{2}\pi}\frac{1}{n}e^{j90^\circ}.$$

Since $X_0 = X_{max}/2$, *the saw-like quantity has the Fourier series in the complex form,*

$$x = X_{max}\left(\frac{1}{2} + \sqrt{2}\,\mathrm{Re}\left\{\frac{1}{\sqrt{2}\pi}\sum_{n=1}^{\infty}\frac{1}{n}e^{j90^\circ}\, e^{jn\omega_1 t}\right\}\right).$$

*****

Illustration 3.18 *A Zener diode current at sinusoidal supply voltage and resistive load can be approximated by function x(t) shown in Fig. 3.20.*

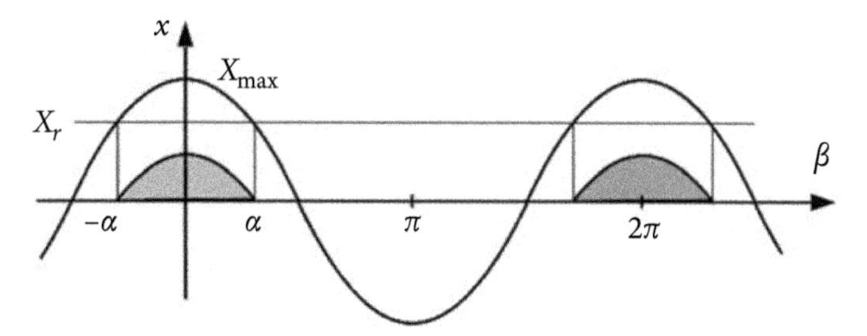

Figure 3.20 Approximation of Zener diode current.

Such a waveform has approximately the supply current of a half-wave rectifier with an active resistive load. It can be also regarded approximately as a component of the supply current of a single-phase rectifier with a capacitive filter.

Such a current can be described analytically as follows:

$$x(\beta)=\begin{cases} X_{\max}\cos\beta - X_{\mathrm{r}}, & \text{for } X_{\max}\cos\beta - X_{\mathrm{r}} > 0 \\ 0, & \text{for } X_{\max}\cos\beta - X_{\mathrm{r}} < 0 \end{cases}$$

The angle α specifies the instant of time when

$$X_{\max}\cos\alpha - X_{\mathrm{r}} = 0$$

meaning

$$\alpha = \arc\cos\left\{\frac{X_{\mathrm{r}}}{X_{\max}}\right\}.$$

The mean value of such a quantity is

$$X_0 = \frac{1}{2\pi}\int_{-\alpha}^{\alpha}(X_{\max}\cos\beta - X_{\mathrm{r}})d\beta = \frac{X_{\max}}{\pi}(\sin\alpha - \xi\alpha)$$

where $\xi = X_{\mathrm{r}}/X_{\max}$. The analytical formula for the crms values of the function harmonics is

$$\mathbf{X}_n = \frac{1}{\sqrt{2\pi}}\int_{-\alpha}^{\alpha}(X_{\max}\cos\beta - X_{\mathrm{r}})e^{-jn\beta}d\beta = \frac{X_{\max}}{\sqrt{2\pi}}\int_{-\alpha}^{\alpha}\left(\frac{e^{j\beta}+e^{-j\beta}}{2} - \frac{X_{\mathrm{r}}}{X_{\max}}\right)e^{-jn\beta}d\beta =$$

$$= \frac{X_{\max}}{2\sqrt{2\pi}}\int_{-\alpha}^{\alpha}\left(e^{-j(n-1)\beta} + e^{-j(n+1)\beta} - 2\xi e^{-jn\beta}\right)d\beta =$$

$$= \frac{X_{\max}}{2\sqrt{2\pi}}\left[\frac{e^{-j(n-1)\beta}}{-j(n-1)} + \frac{e^{-j(n+1)\beta}}{-j(n+1)} - \frac{2\xi}{-jn}e^{-jn\beta}\right]\Bigg|_{-\alpha}^{\alpha} =$$

$$= \frac{X_{\max}}{\sqrt{2\pi}}\left[\frac{\sin(n-1)\alpha}{(n-1)} + \frac{\sin(n+1)\alpha}{(n+1)} - \frac{2\xi}{n}\sin(n\alpha)\right]$$

on the condition, however, that $n \neq 1$. The crms value of the fundamental harmonic has to be calculated separately, namely, for $n = 1$,

$$\mathbf{X}_1 = \frac{X_{\max}}{2\sqrt{2\pi}}\int_{-\alpha}^{\alpha}(1 + e^{-j2\beta} - 2\xi e^{-j\beta})d\beta = \frac{X_{\max}}{2\sqrt{2\pi}}\left[\beta + \frac{e^{-j2\beta}}{-j2} - \frac{2\xi}{-j1}e^{-j\beta}\right]\Bigg|_{-\alpha}^{\alpha} =$$

$$= \frac{X_{\max}}{2\sqrt{2\pi}}[2\alpha + \sin(2\alpha) - 4\xi\sin(\alpha)].$$

Since $\xi = \cos(\alpha)$, the final formula for the crms value of the fundamental harmonic is

$$\boldsymbol{X}_1 = \frac{X_{\max}}{2\sqrt{2}\pi}[2\alpha - \sin(2\alpha)].$$

Assuming, for example, that, $X_{\mathrm{r}} = X_{\max}/\sqrt{2}$, thus $\alpha = \pi/4$, the numerical crms values of the quantity $x(t)$ harmonics are

$$X_0 = \frac{X_{\max}}{\pi}(\sin\alpha - \xi\alpha) = \frac{X_{\max}}{\pi}\left(\frac{1}{\sqrt{2}} - \frac{1}{\sqrt{2}}\frac{\pi}{4}\right) = 0.048\,X_{\max}$$

$$\boldsymbol{X}_1 = \frac{X_{\max}}{2\sqrt{2}\pi}[2\alpha - \sin(2\alpha)] = \frac{X_{\max}}{2\sqrt{2}\pi}\left[2\frac{\pi}{4} - \sin\left(2\frac{\pi}{4}\right)\right] = 0.064\,X_{\max}$$

$$\boldsymbol{X}_2 = \frac{X_{\max}}{\sqrt{2}\pi}\left[\sin\left(\frac{\pi}{4}\right) + \frac{1}{3}\sin\left(3\frac{\pi}{4}\right) - \frac{1}{\sqrt{2}}\sin\left(\frac{\pi}{2}\right)\right] = 0.053\,X_{\max}$$

$$\boldsymbol{X}_3 = \frac{X_{\max}}{\sqrt{2}\pi}\left[\frac{1}{2}\sin\left(\frac{\pi}{2}\right) + \frac{1}{4}\sin(\pi) - \frac{2}{3}\frac{1}{\sqrt{2}}\sin\left(3\frac{\pi}{4}\right)\right] = 0.037\,X_{\max}$$

*****

Harmonic Spectrum of Periodic Quantity: A set of crms values $\boldsymbol{X}_n$ of a quantity $x(t)$ that belongs to the space $\boldsymbol{L}_{\mathrm{T}}^2$ is referred to as a *harmonic spectrum* of $x(t)$, or simply, a *spectrum* of $x(t)$. Sometimes not the set of crms values $\boldsymbol{X}_n$ but a set of rms values X_n is regarded as a harmonic spectrum. Also, the term *phase spectrum*, meaning the set of phase angles of harmonics, is sometimes used. A harmonic spectrum, regarded as a set of rms values of harmonics, is usually visualized in a form of a discrete plot, with bars or lines length proportional to the rms value of particular harmonics.

The contents of particular harmonics could be at a level of a single percent or even a part of it, and consequently, such harmonics would be almost invisible on the spectrum plot. Such harmonics would be much more visible in the spectrum plot, if it is drawn in a logarithmic scale, meaning with the bar length proportional to the logarithm of the relative contents, X_n/X_1, of particular harmonics. An example of a harmonic spectrum, drawn on a logarithmic scale, is shown in Fig. 3.21.

A Time Domain and a Frequency Domain: Periodic quantities can be specified by a sequence of values in sequential instances of time, t, meaning $x(t)$. We say in such a case that the quantity is specified in a *time domain*. The waveform we see on a common oscilloscope is just specified in the time domain. The same periodic quantity can be specified by a sequence of the crms values of its harmonics, $\boldsymbol{X}_n$. We say in such a case that the quantity is specified in the *frequency domain*. A *spectrum analyzer* is a common device that specifies periodic quantities in the frequency domain.

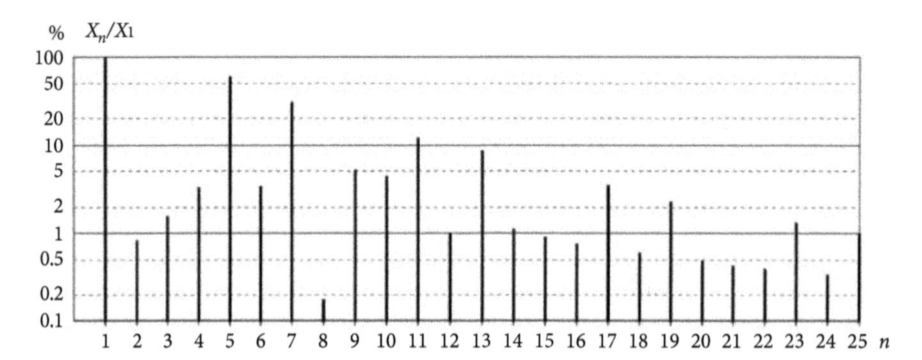

Figure 3.21 Example of harmonics spectrum in logarithmic scale.

A Hybrid Domain: Power plant generators produce almost sinusoidal voltage, while most distortion is caused by the load current. Thus all current harmonics, with the exception of the fundamental one, are orthogonal to the generator voltage. This means that only the fundamental harmonic of the generator voltage and current deliver energy to the power. Thus, the fundamental harmonic has particular importance from the point of view of energy flow. In several applications in particular, at compensation, only the fundamental harmonic is separated from voltages and current, meaning its crms value $\boldsymbol{X}_1$ is measured, while the remainder, $x(t)-x_1(t)$, is not decomposed into harmonics. In such a case the quantity is specified both in the frequency and in time or the *hybrid domain.*

Properties of electrical s and in particular, their power properties can be described in the time domain and the frequency domain. These two domains for quantities that belong to the space $\boldsymbol{L}_{\mathrm{T}}{}^2$ are entirely equivalent, with the only exception in points of discontinuity. This equivalence results from the properties of the Fourier Transform and the Inverse Fourier Transform. Consequently, results of analysis, at least for continuous quantities, obtained in these two domains, have to be mutually equivalent.

Despite their equivalency, the analysis and the mathematical form of results in these two domains are entirely different. Different mathematical methods are needed in the time domain and the frequency domain. Because of different mathematical forms, there can be differences in interpretations of these results. The technical implications of these two approaches can be different too, in particular those concerning instrumentation. There are also situations where the frequency domain approach to circuit analysis can lead to erroneous results.

The time domain approach to circuit analysis is fundamental and can be applied irrespective of the circuit properties. It is not the same with nonlinear circuits.

When the energy flows in circuits with nonsinusoidal voltages and currents then their waveforms can change. These changes in linear circuits affect individual harmonics of voltages and currents in such a way that their total effect describes the changes of voltages and currents properly. Therefore, such linear circuits can be analyzed using a harmonic-by-harmonic approach, meaning in the frequency domain. This approach involves the superposition principle for individual harmonics. This principle cannot be used for the analysis of nonlinear s and consequently, the harmonic-by-harmonic approach is not acceptable. Thus, linear s can be analyzed both in the time domain and in the frequency domain, while nonlinear s can be analyzed only in the time domain. In such a case, only the result of circuit analysis can be presented, if needed, in the frequency domain, meaning the crms values of harmonics can be calculated for voltages and currents, obtained however in a time domain analysis. Such voltages and currents in nonlinear s can be present in the form of the Fourier series but cannot be used for such analysis. Also, electrical powers, as some functionals of voltages and currents, can be specified in the frequency domain.

3.4 Scalar Product in the Frequency Domain

The scalar product and rms value of periodic quantities are defined in the time domain as functionals, meaning an operation that allocates a value to time-varying quantities. Their calculation often requires a toilsome integration of mathematically complex functions of time. When the crms values of harmonics of such quantities are known, meaning they are specified in the frequency domain, the scalar product and the rms value can be calculated in this domain. This is a result of the following reasoning.

Harmonics of different orders n are mutually orthogonal, thus

$$(x, y) = \frac{1}{T}\int_0^T x(t)\, y(t)\, dt = \frac{1}{T}\int_0^T \left[\sum_{n\in N_0} x_n(t)\right]^{\mathrm{T}} \left[\sum_{n\in N_0} y_n(t)\right] dt =$$

$$= \sum_{n\in N_0} \frac{1}{T}\int_0^T x_n(t)\, y_n(t)\, dt.$$

Thus the scalar product of a periodic quantity is equal to the sum of scalar products of individual harmonics,

$$(x, y) = \sum_{n\in N_0} (x_n, y_n).$$

The scalar product of the zero-order harmonic is $(x_0, y_0) = X_0 Y_0$, while the scalar product of the nth order harmonic can be expressed in the form

$$(x_n, y_n) = \frac{1}{T}\int_0^T x_n(t)\, y_n(t)\, dt = \frac{1}{T}\int_0^T \sqrt{2}\,\mathrm{Re}\{\boldsymbol{X}_n\, e^{jn\omega_1 t}\}\sqrt{2}\,\mathrm{Re}\{\boldsymbol{Y}_n\, e^{jn\omega_1 t}\}\, dt =$$

$$= 2\frac{1}{T}\int_0^T \left[\mathrm{Re}\boldsymbol{X}_n \cos(n\omega_1 t) - \mathrm{Im}\boldsymbol{X}_n \sin(n\omega_1 t)\right]\left[\mathrm{Re}\boldsymbol{Y}_n \cos(n\omega_1 t) - \mathrm{Im}\boldsymbol{Y}_n \sin(n\omega_1 t)\right] dt.$$

Since

$$\frac{1}{T}\int_0^T \cos(n\omega_1 t)\sin(n\omega_1 t)\,dt = 0$$

thus

$$(x_n, y_n) = 2\frac{1}{T}\int_0^T \left[\mathrm{Re}\boldsymbol{X}_n\mathrm{Re}\boldsymbol{Y}_n\cos^2(n\omega_1 t) + \mathrm{Im}\boldsymbol{X}_n\mathrm{Im}\boldsymbol{Y}_n\sin^2(n\omega_1 t)\right] dt =$$

$$= \mathrm{Re}\boldsymbol{X}_n\mathrm{Re}\boldsymbol{Y}_n + \mathrm{Im}\boldsymbol{X}_n\mathrm{Im}\boldsymbol{Y}_n = \mathrm{Re}\{\boldsymbol{X}_n\boldsymbol{Y}_n^*\}$$

and eventually,

$$(x, y) = \mathrm{Re}\sum_{n\in N_0} \boldsymbol{X}_n\boldsymbol{Y}_n^* \tag{3.23}$$

where the asterisk denotes the conjugate complex number.

This result means that the integration needed for the scalar product calculation according to its definition can be superseded by a sum of products of crms values of harmonics of these two quantities. It usually reduces the number of calculations substantially. There are also other benefits from the frequency domain approach to scalar product calculation. It is clear that components of different frequencies do not contribute to this product, nor to components of the same frequency but shifted by 90°.

Illustration 3.19 *If the voltage and current at the load terminals have the waveform,*

$$u(t) = 20 + \sqrt{2}\,\mathrm{Re}\{120\,e^{j\omega_1 t} + 10\,e^{-j30°}e^{j3\omega_1 t}\}\,\mathrm{V}$$

$$i(t) = 30 + \sqrt{2}\,\mathrm{Re}\{50\,e^{-j45°}e^{j\omega_1 t} + 25\,e^{-j60°}e^{j3\omega_1 t}\}\,\mathrm{A}$$

then the load active power, as the scalar product of the load voltage and current, is equal to

$$P = (u, i) = \mathrm{Re}\sum_{n=0,1,3} \boldsymbol{U}_n\boldsymbol{I}_n^* = 20\times 30 + \mathrm{Re}\{120\times 50\,e^{j45°} + 10\,e^{-j30°}\times 25\,e^{j60°}\}$$

$$= 600 + \mathrm{Re}\{6000\,e^{j45°} + 250\,e^{j30°}\} = 5059\ \mathrm{W}.$$

*****

Rms Value in the Frequency Domain: Just like the scalar product, the crms value of periodic quantities can be calculated in the frequency domain when the frequency spectrum of the quantity is known. This is an obvious conclusion since the rms $||x||$ value is defined as the root of the scalar product (x, x), namely

$$||x|| = \sqrt{(x,x)} = \sqrt{\mathrm{Re} \sum_{n \in N_0} \boldsymbol{X}_n \boldsymbol{X}_n^*} = \sqrt{\sum_{n \in N_0} X_n^2} \tag{3.24}$$

and the product of a complex number and its conjugate is real and equal to the square of the number magnitude. It is a convenient and commonly used method of rms value calculation.

Illustration 3.20 *The rms value of the current in Illustration 3.14, due to orthogonality of harmonics, is equal to*

$$||i|| = \sqrt{\sum_{n \in \{0,1,3,7\}} I_n^2} = \sqrt{10^2 + 70.7^2 + 20^2 + 7.07^2} = 74.49 \text{ A.}$$

****

3.5 Properties of Complex Rms Values

The complex rms values of harmonics have several mathematical properties that simplify their calculation. These properties are compiled below.

1. Crms Values of Harmonics of Linear Forms. If quantity $z(t)$ is a linear form in space $\boldsymbol{L}_{\mathrm{T}}^2$, meaning

$$z(t) = \alpha x(t) + \beta y(t)$$

then crms values $\boldsymbol{Z}_n$ of harmonics of this linear form are equal to

$$\boldsymbol{Z}_n = \frac{\sqrt{2}}{T} \int_0^T [\alpha x(t) + \beta y(t)]\, e^{-jn\omega_1 t} dt =$$

$$= \alpha \frac{\sqrt{2}}{T} \int_0^T x(t)\, e^{-jn\omega_1 t} dt + \beta \frac{\sqrt{2}}{T} \int_0^T y(t)\, e^{-jn\omega_1 t} dt = \alpha \boldsymbol{X}_n + \beta \boldsymbol{Y}_n \tag{3.25}$$

thus, they are linear forms of crms values $\boldsymbol{X}_n$, $\boldsymbol{Y}_n$ of the elements which create this form with the same coefficients, α, β.

This is a very useful property that enables the calculation of crms values of harmonics of complex quantities by decomposing them into simple components, for which this calculation can be much more simple, or the crms values of their harmonics are known.

2. Crms Values of Harmonics of Shifted Quantities. If harmonics of periodic quantity $x(t)$ quantity have crms values $\boldsymbol{X}_n$, then harmonics of the quantity $y(t)$ shifted by τ, as shown in Fig. 3.22 meaning quantity $y(t) \equiv x(t - \tau)$, have the crms values equal to

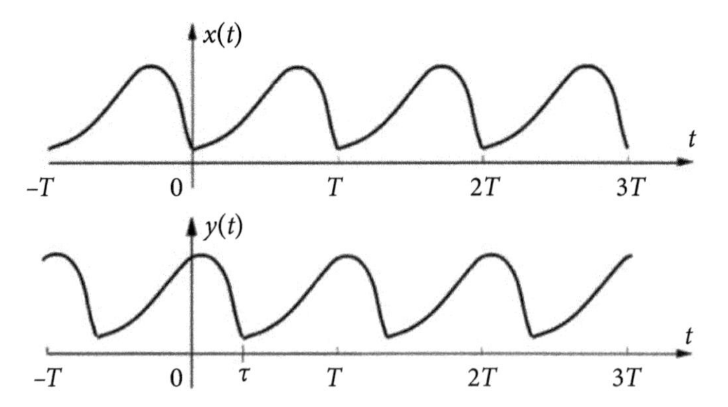

Figure 3.22 An example of mutually shifted quantities.

$$\boldsymbol{Y}_n = \frac{\sqrt{2}}{T}\int_0^T x(t-\tau)\, e^{-jn\omega_1 t} dt = e^{-jn\omega_1 \tau}\frac{\sqrt{2}}{T}\int_0^T x(t-\tau)\, e^{-jn\omega_1 (t-\tau)} d(t-\tau) =$$

$$= e^{-jn\omega_1 \tau}\frac{\sqrt{2}}{T}\int_0^T x(\lambda)\, e^{-jn\omega_1 \lambda}\, d\lambda$$

and consequently,

$$\boldsymbol{Y}_n = e^{-jn\omega_1 \tau}\boldsymbol{X}_n. \tag{3.26}$$

Thus, when the crms values $\boldsymbol{X}_n$ are known, then the crms values of harmonics, $\boldsymbol{Y}_n$, of a quantity shifted by τ are obtained by multiplying these crms values by the coefficient

$$\boldsymbol{z} \stackrel{\text{df}}{=} 1\, e^{-jn\omega_1 \tau}.$$

Observe that the magnitude of this coefficient is 1, thus only phases of crms values are affected by the shift in time, but not their magnitudes.

When the shift of a quantity is not specified in time, τ, but is specified by a phase angle δ, meaning $y(\omega t) = y(\beta) = x(\beta - \delta)$, then

$$\boldsymbol{z} = 1\, e^{-jn\omega_1 \tau} = 1\, e^{-jn\delta}$$

Because

$$\omega_1 \tau = \frac{2\pi}{T}\tau = 2\pi\frac{\tau}{T} = \delta$$

and then

$$\boldsymbol{Y}_n = e^{-jn\delta}\boldsymbol{X}_n. \tag{3.27}$$

Illustration 3.21 *The sequence of periodic rectangular pulses shown in Fig. 3.23 can be regarded as a sequence of pulses in Fig. 3.15, but shifted in time by $\tau = T/2$, meaning*

$$i\,(t) = x(t - T/2).$$

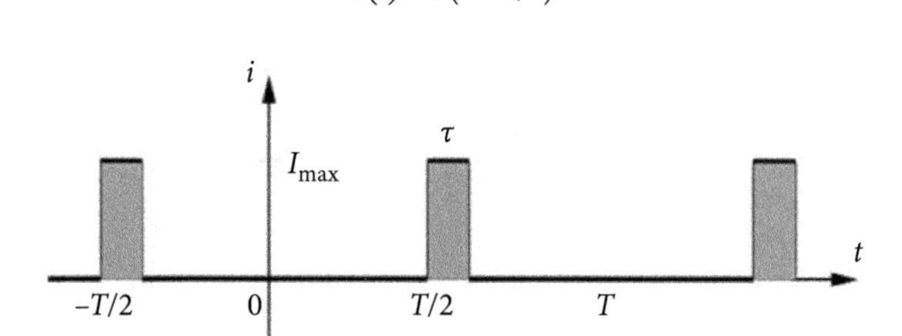

Figure 3.23 A sequence of periodic rectangular pulses.

Thus, the crms values of harmonics of the current pulses in Fig. 3.23 are equal to

$$\boldsymbol{I}_n = e^{-jn\omega_1\frac{T}{2}}\boldsymbol{X}_n = e^{-jn\pi}\boldsymbol{X}_n.$$

crms values $\boldsymbol{X}_n$ were calculated in Illustration 3.15. Thus the crms values of the current harmonics up to harmonic order $n = 5$ are equal to

$$\boldsymbol{I}_0 = 20\ \text{A},\ \boldsymbol{I}_1 = 26.46e^{-j216^\circ}\text{A},\ \boldsymbol{I}_2 = 21.41e^{-j72^\circ}$$
$$\boldsymbol{I}_3 = 14.27e^{-j288^\circ}\text{A},\ \boldsymbol{I}_4 = 6,61e^{-j144^\circ}\text{A},\ \boldsymbol{I}_5 = 0$$

This illustration demonstrates how the known crms values of harmonics of a quantity can be used for calculating these values for another quantity shifted in time.

*****

Illustration 3.22 *TRIAC current in a circuit with purely resistive load has the form shown in Fig. 3.24.*

It can be considered as a sum of two components, namely $x(\beta)$ and $-x\,(\beta - 180^\circ)$, where $x(\beta)$ is the thyristor current of the waveform shown in Fig. 3.17, meaning

$$i(\beta) = x(\beta) - x(\beta - \pi)$$

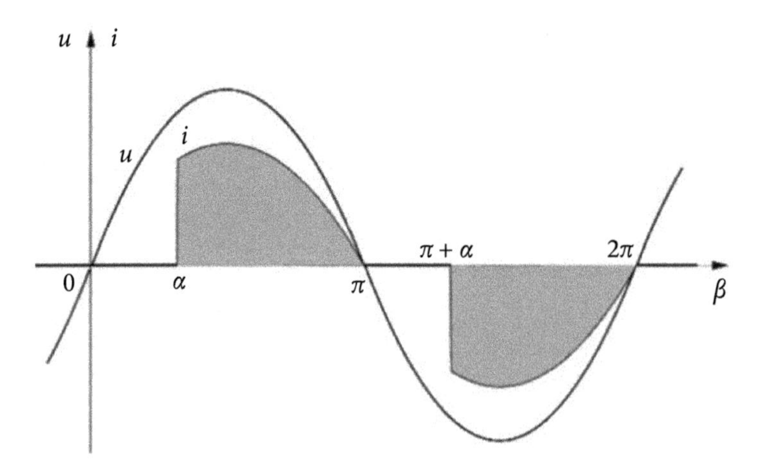

Figure 3.24 TRIAC current in a resistive circuit

Thus, the crms values of the TRIAC current harmonics are equal to

$$\boldsymbol{I}_n = \boldsymbol{X}_n - \boldsymbol{X}_n e^{-jn\pi} = (1 - e^{-jn\pi})\boldsymbol{X}_n = \begin{cases} 2\boldsymbol{X}_n, \text{ for } n = 2k+1 \\ 0, \text{ for } n = 2k \end{cases} \tag{3.29}$$

The crms vales $\boldsymbol{X}_n$ were calculated in Illustration 3.16. It means that the TRIAC current cannot contain harmonics of the even order n. It contains odd order harmonics with the crms values

$$\boldsymbol{I}_n = 2\boldsymbol{X}_n = \frac{I_{\max}}{\sqrt{2}\,\pi(n^2-1)}[(n-1)\,e^{-j(n+1)\alpha} - (n+1)\,e^{-j(n-1)\alpha} + 2].$$

If we assume, as in Illustration 3.16, that $\alpha = 45^\circ$ and $I_{\max} = 100$ A, then

$$\boldsymbol{I}_1 = 63,2\,e^{-j99,9^\circ}\text{A},\ \boldsymbol{I}_3 = 11,26\,e^{j90^\circ}\text{A},$$
$$\boldsymbol{I}_5 = 8,38\,e^{j26,6^\circ}\text{A},\ \boldsymbol{I}_7 = 5,30\,e^{-j45^\circ}\text{A},$$
$$\boldsymbol{I}_9 = 3,18\,e^{-j135^\circ}\text{A},\ \boldsymbol{I}_{11} = 2,70\,e^{j124^\circ}\text{A},$$

*****

3. Crms Values of Harmonics of a Quantity Derivative. If the crms values of harmonics of quantity $x(t)$ are $\boldsymbol{X}_n$, and the derivative of this quantity

$$y(t) = \frac{dx(t)}{dt} \in \boldsymbol{L}_{\mathrm{T}}^2$$

then the crms values of harmonics of the derivative $y(t)$ are

$$\boldsymbol{Y}_n = jn\omega_1\boldsymbol{X}_n.$$

For example, since an inductor voltage is

$$u = L\frac{di}{dt}$$

thus the rms values of the inductor voltage harmonics are equal to

$$\boldsymbol{U}_n = jn\omega_1 L\,\boldsymbol{I}_n. \tag{3.30}$$

4. Crms Values of Harmonics of Reflected Quantity. A quantity reflected in time, $y(t)$, is a quantity that changes in time like the original quantity in the negative time, $x(-t)$, meaning

$$y(t) \equiv x(-t).$$

A pair of such quantities is shown in Fig. 3.25. The crms values of harmonics of the reflected quantity are equal to

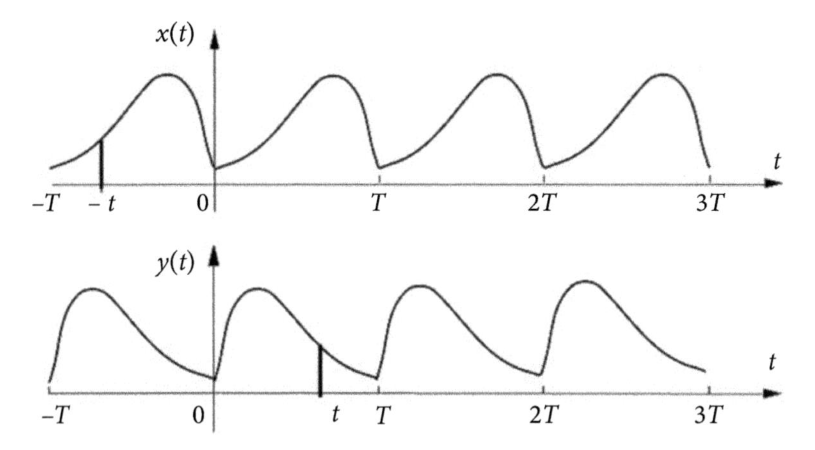

Figure 3.25 Quantity and reflected quantity.

$$\boldsymbol{Y}_n = \frac{\sqrt{2}}{T}\int_0^T x(-t)\, e^{-jn\omega_1 t} dt = -\frac{\sqrt{2}}{T}\int_0^{-T} x(-t)\, e^{jn\omega_1(-t)}\, d(-t) =$$

$$= \frac{\sqrt{2}}{T}\int_0^T x(\lambda)\, e^{jn\omega_1\lambda}\, d\lambda = \boldsymbol{X}_n^*. \tag{3.31}$$

The crms values of harmonics of the reflected quantity are equal to

$$Y_n = \frac{\sqrt{2}}{T}\int_0^T x(-t)\, e^{-jn\omega_1 t} dt = -\frac{\sqrt{2}}{T}\int_0^{-T} x(-t)\, e^{jn\omega_1(-t)}\, d(-t) =$$

$$= \frac{\sqrt{2}}{T}\int_0^T x(\lambda)\, e^{jn\omega_1\lambda}\, d\lambda = X_n^*. \tag{3.32}$$

Thus, the crms values of harmonics of the reflected quantity are conjugate to the crms values of the original quantity. It is enough to change the sign of the imaginary part of the crms value $\boldsymbol{X}_n$ or its argument to the opposite to obtain $\boldsymbol{Y}_n$ values.

Illustration 3.23 *The current of the waveform shown in Fig. 3.26 can be regarded as reflected against the waveform shown in Fig. 3.19, meaning*

$$i(t) \equiv x(-t).$$

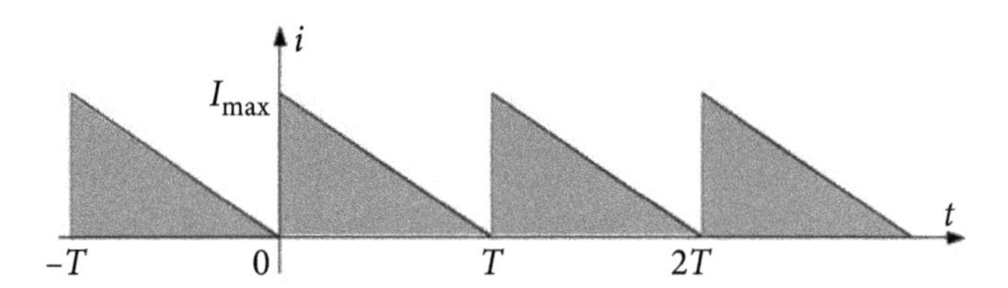

Figure 3.26 Saw-like current waveform.

The crms values $\boldsymbol{X}_n$ *of harmonics of that quantity, calculated in Illustration 3.17 are known, thus*

$$\boldsymbol{I}_n = \boldsymbol{X}_n^* = \frac{I_{\max}}{\sqrt{2}\pi}\frac{1}{n}\, e^{-j90^\circ}.$$

*****

5. Crms Values of Harmonics of Even Quantities. When a quantity has symmetry against $t = 0$, i.e., $x(t) = x(-t)$, meaning it is an *even quantity*, then the crms values of its harmonics have to fulfill the identity

$$\boldsymbol{X}_n \equiv \boldsymbol{X}_n^*.$$

It is possible only on the condition that

$$\mathrm{Im}\{\boldsymbol{X}_n\} \equiv 0$$

thus, the crms values of harmonics of even quantities have to be real numbers.

6. Crms Values of Harmonics of Odd Quantities. When a quantity has a negative symmetry against $t = 0$, that is, $x(t) = -x(-t)$, meaning it is an *odd quantity*, then the crms values of its harmonics fulfill the identity

$$\boldsymbol{X}_n \equiv -\boldsymbol{X}_n^*.$$

It is possible only on the condition that

$$\mathrm{Re}\{\boldsymbol{X}_n\} \equiv 0$$

thus, the crms values of harmonics of odd quantities have to be imaginary numbers. They cannot have, of course, any DC component, meaning the zero-order harmonic.

7. Crms Values of Harmonics of Quantities With Half-Wave Negative Symmetry. When a quantity has a negative symmetry versus values shifted by half of the period, that is,

$$x(t - T/2) \equiv -x(t)$$

as illustrated in Fig. 3.27,

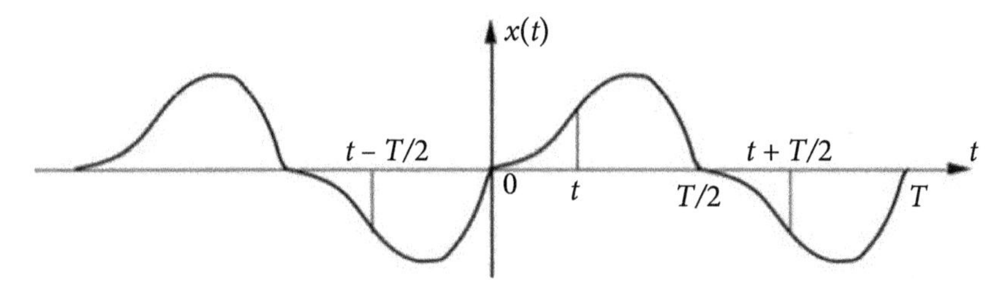

Figure 3.27 A quantity with half-wave negative symmetry.

then the crms values of its harmonics have to fulfill the identity

$$\boldsymbol{X}_n\, e^{-jn\omega_1 \frac{T}{2}} \equiv -\boldsymbol{X}_n.$$

Since

$$n\omega_1 \frac{T}{2} = n\pi$$

thus the identity

$$\boldsymbol{X}_n\, e^{-jn\pi} \equiv -\boldsymbol{X}_n$$

is fulfilled for each harmonic order n, only if the crms value $\boldsymbol{X}_n$ is zero for each even order harmonic. It means that quantities with half-wave symmetry cannot have any even order harmonic. A great majority of voltages and currents in electrical power systems have this half-wave negative symmetry, therefore, the contents of even order harmonics in such s are much lower than the contents of the odd order harmonics. The reasons for which voltages and currents in electrical power s usually have this half-wave negative symmetry were discussed in Chapter 2. According to IEEE Recommended Practices, IEEE Std. 519, the contents of even order harmonics should not be higher than one-quarter of the contents of the odd order harmonics.

This property of quantities with half-wave negative symmetry also substantially simplifies the calculation of the crms values of harmonics; namely a quantity with a half-wave negative symmetry can be decomposed into two identical components, as in the case of the quantity shown in Fig. 3.27. Such components, shown in Fig. 3.28, are only shifted mutually by $T/2$ so that

$$x(t) = y(t) - y(t - T/2).$$

Consequently

$$\boldsymbol{X}_n = \boldsymbol{Y}_n - e^{-jn\omega_1 \frac{T}{2}} \boldsymbol{Y}_n = (1 - e^{-jn\pi})\boldsymbol{Y}_n = \begin{cases} 2\boldsymbol{Y}_n, & n = 2k+1 \\ 0, & n = 2k \end{cases}$$

Thus, it is enough to calculate the crms values only of the odd order harmonics for the quantity component $y(t)$. This reduces the amount of calculation.

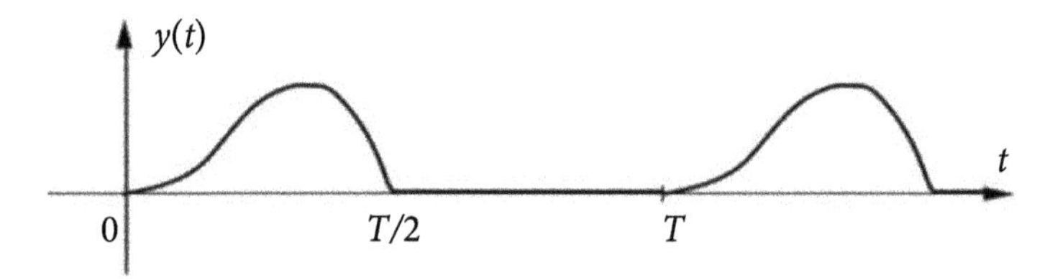

Figure 3.28 Component of quantity $x(t)$, shown in Figure 3.27.

8. Crms Values of Harmonics of a Quantity Derivative. If the crms values of harmonics of quantity $x(t)$ are $\boldsymbol{X}_n$, and the derivative of this quantity

$$y(t) = \frac{dx(t)}{dt} \in \boldsymbol{L}_{\mathrm{T}}^2,$$

then the crms values of harmonics of the derivative $y(t)$ are

$$\boldsymbol{Y}_n = jn\omega_1 \boldsymbol{X}_n.$$

For example, since an inductor voltage is

$$u = L\frac{di}{dt},$$

thus the rms values of the inductor voltage harmonics are equal to

$$\boldsymbol{U}_n = jn\omega_1 L\, \boldsymbol{I}_n. \tag{3.33}$$

9. Crms Values of Harmonics of a Quantity Integral. If the crms values of harmonics of quantity $x(t)$ are $\boldsymbol{X}_n$, and the integral of this quantity

$$y(t) = \int x(t)\, dt \in \boldsymbol{L}^2_{\mathrm{T}},$$

then the crms values of harmonics of the integral $y(t)$ are

$$\boldsymbol{Y}_n = \frac{1}{jn\omega_1} \boldsymbol{X}_n. \tag{3.34}$$

Observe that an integral of a quantity can be a periodic quantity on the condition that the mean value of such a quantity is zero.

Illustration 3.24 *Calculation of the crms values of the current shown in Fig. 3.29a directly from the definition requires integration over three intervals, where the current waveform is described differently.*
It could be a time-consuming process. However, it can be regarded as an integral of the $x(t)$ quantity, shown in Fig. 3.29b, meaning

$$i(t) = k \int x(t)\, dt$$

where the coefficient k has to fulfill the condition

$$k \int_0^{T/4} dt = I_{max}, \text{thus, } k = \frac{4I_{\max}}{T}.$$

Consequently, the crms values of the current harmonics can be expressed in terms of the crms values of harmonics of the $x(t)$ quantity as follows:

$$\boldsymbol{I}_n = k\frac{1}{jn\omega_1}\boldsymbol{X}_n = \frac{4I_{\max}}{T}\frac{1}{jn\omega_1}\boldsymbol{X}_n = \frac{2I_{\max}}{jn\pi}\boldsymbol{X}_n.$$

Since quantity $x(t)$ has a half-wave negative symmetry, it has only odd order harmonics of the crms value $\boldsymbol{X}_n = 2\boldsymbol{Y}_n$, where

$$\boldsymbol{Y}_n = \frac{\sqrt{2}}{T}\int_{-T/4}^{T/4} 1\, e^{-jn\omega_1 t} dt = \frac{\sqrt{2}}{n\pi}\sin\left(n\frac{\pi}{2}\right).$$

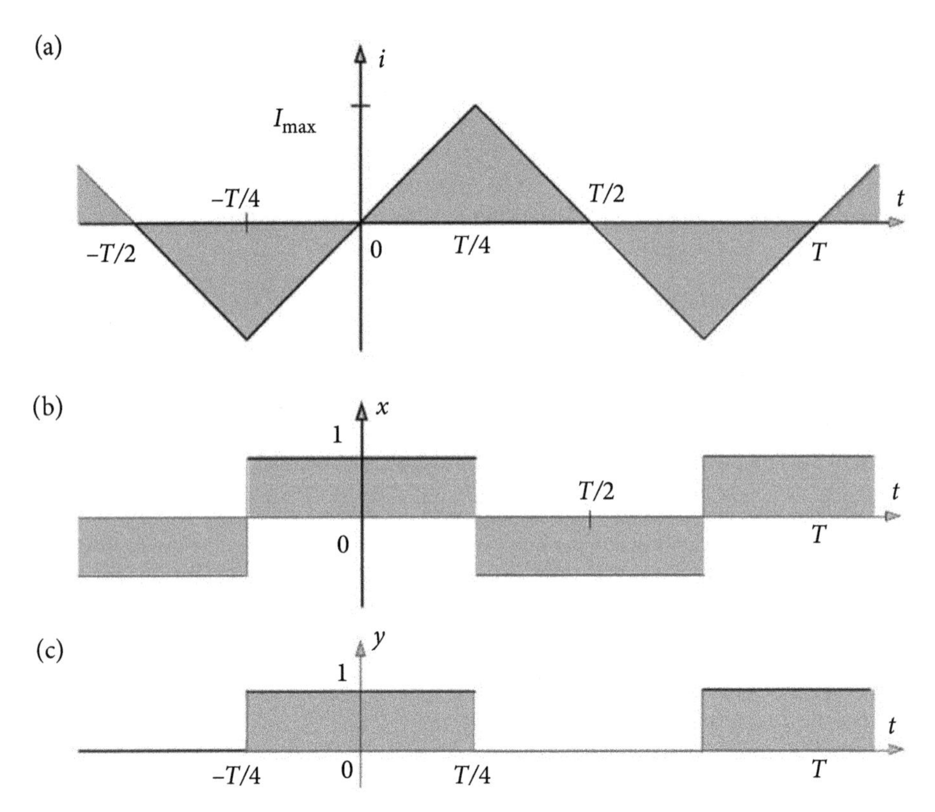

Figure 3.29 A current with saw-like waveform (a), and related quantities, (b) and (c).

Eventually, the crms values of the current harmonics are

$$\boldsymbol{I}_n = \frac{2I_{\max}}{jn\pi}2\boldsymbol{Y}_n = \frac{4\sqrt{2}I_{\max}}{j(n\pi)^2}\,\sin\left(n\frac{\pi}{2}\right).$$

In particular, crms values of harmonics up to $n = 7$ are equal to

$$\boldsymbol{I}_1 = 0.57I_{\max}e^{-j90^\circ}, \boldsymbol{I}_3 = 0.06\,I_{\max}\,e^{j90^\circ}, \boldsymbol{I}_5 = 0.02\,I_{\max}\,e^{-j90^\circ}, \boldsymbol{I}_7 = 0.01\,I_{\max}\,e^{j90^\circ}.$$

This illustration demonstrates how some properties of the crms values of harmonics can be used for effective reduction of the amount of calculation. The benefits of such an approach become more and more visible with an increase in waveform complexity.

*****

Calculation of the crms values of harmonics of the trapezoidal approximation of the supply current of a three-phase AC/DC converter illustrates these benefits. Taking into account that in one period of time there are seven intervals where the current is described differently, calculation of the crms values of harmonics directly from definition can be very toilsome and prone to a calculation error, which is more probable with the amount of calculation increase.

Illustration 3.25 *A three-phase AC/DC converter supply current, with trapezoidal approximation, is shown in Fig. 3.30.*

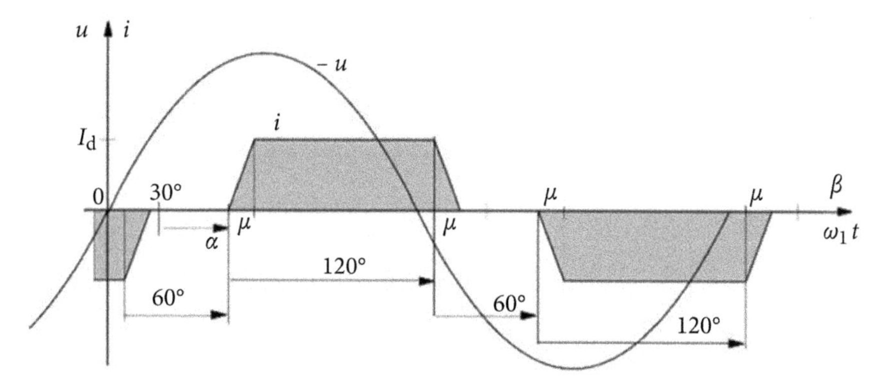

Figure 3.30 A trapezoidal approximation of the supply current of three-phase AC/DC converter.

It can be specified in terms of a few simplified components, shown in Fig. 3.31,

$$i(\beta) = x(\beta) - x(\beta - \pi)$$

$$x(\beta) = y[\beta - (30° + \alpha + 60° + \mu/2)]$$

$$y(\beta) = k \int w(\beta)\, d\beta$$

where coefficient k should fulfill the condition

$$k \int_0^{\mu} 1\, d\beta = I_d, \quad ==> \quad k = \frac{I_d}{\mu}$$

and eventually,

$$w(\beta) = z(\beta + 60°) - z(\beta - 60°).$$

The current has only odd order harmonics of the crms value

$$\boldsymbol{I}_n = 2\boldsymbol{X}_n, \quad n = 2k - 1$$

where $\mathbf{X}_n$ is the crms value of harmonics of x(t) component, shown in Fig. 3.31b. It is

$$\mathbf{X}_n = e^{-jn\,(90° + \alpha + \mu/2)}\,\mathbf{Y}_n.$$

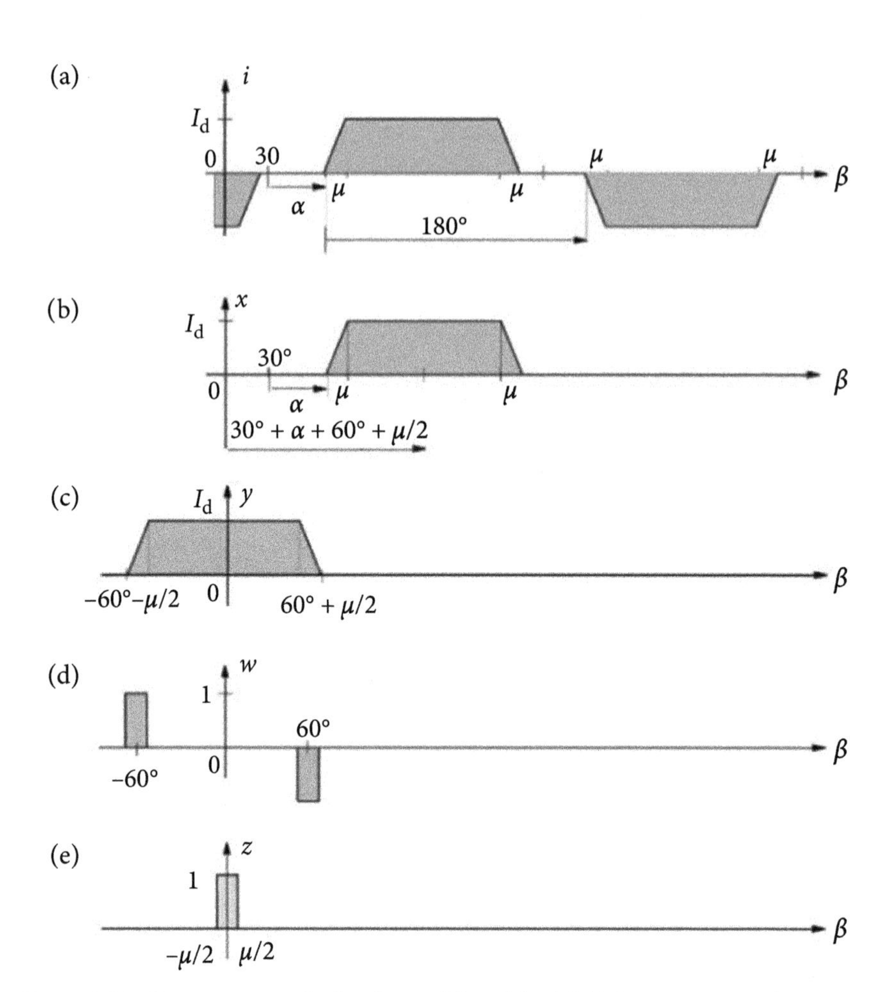

Figure 3.31 Components and related quantities of the AC/DC converter supply current.

These components are related as follows. Quantity y(t) is integral of w(t), thus its crms values are

$$\mathbf{Y}_n = k\frac{1}{jn}\mathbf{W}_n = \frac{I_\mathrm{d}}{\mu}\frac{1}{jn}\mathbf{W}_n$$

while w(t) is a difference between two shifted pulses z(t), hence

$$\mathbf{W}_n = (e^{jn60°} - e^{-jn60°})\,\mathbf{Z}_n = 2j\sin(n\,60°)\,\mathbf{Z}_n.$$

Consequently

$$\boldsymbol{Z}_n = \frac{1}{\sqrt{2\pi}} \int_{-\mu/2}^{\mu/2} e^{-jn\beta} d\beta = \frac{\sqrt{2}}{n\pi} \sin\left(n\frac{\mu}{2}\right).$$

All these formulae, when combined, result in

$$\boldsymbol{I}_n = 2\, e^{-jn\,(90^\circ + \alpha + \mu/2)} \times \frac{I_\mathrm{d}}{\mu} \frac{1}{jn} \times 2j \sin(n\, 60^\circ) \times \frac{\sqrt{2}}{n\pi} \sin\left(n\frac{\mu}{2}\right)$$

and, after some rearrangements, in

$$\boldsymbol{I}_n = \frac{4\sqrt{2}\, I_\mathrm{d}}{\pi \mu n^2} \sin(n\, 60^\circ) \times \sin\left(n\frac{\mu}{2}\right) e^{-jn\,(90^\circ + \alpha + \mu/2)}$$

Observe that for n = 3k, the value of sin (n 60°) is zero. Thus, the converter supply current cannot contain harmonics of the order n which is a multiplicity of three. It contains, apart from the fundamental, only harmonics of the order n = 5, 7, 11, in general, n = 6k+/−1, k = 1, 2, 3,. ... Since for these order harmonics

$$\sin(n \times 60^\circ) = \pm\frac{\sqrt{3}}{2}$$

the rms value of the supply current harmonics is equal to

$$I_n = \frac{2\sqrt{2}\, I_\mathrm{d}}{\pi\, n} \frac{\sqrt{3}}{2} \frac{\sin\left(n\frac{\mu}{2}\right)}{n\frac{\mu}{2}} = \frac{\sqrt{6}\, I_\mathrm{d}}{\pi\, n} \frac{\sin\left(n\frac{\mu}{2}\right)}{n\frac{\mu}{2}}. \tag{3.35}$$

When the commutation angle μ approaches zero, then

$$\frac{\sin(n\,\mu/2)}{n\,\mu/2} \to 1$$

and the rms values of the converter supply current harmonics approach values

$$I_n = \frac{I_1}{n}, \quad \text{with} \quad I_1 = \frac{\sqrt{6}\, I_\mathrm{d}}{\pi\, n},$$

meaning they decline with the harmonic order n.

*****

Illustration 3.26 *A trapezoidal approximation of the supply current of a twelve-pulse AC/DC converter is shown in Fig. 3.32.*

There are twenty-one intervals in a single period where the quantity has a different analytic form. Therefore, calculation of the crms values of harmonics for such a current from the definition can be very toilsome. Decomposition of the current into elementary and related components enables a substantial reduction of the amount of these calculations. Such decomposition is shown in Fig. 3.33.

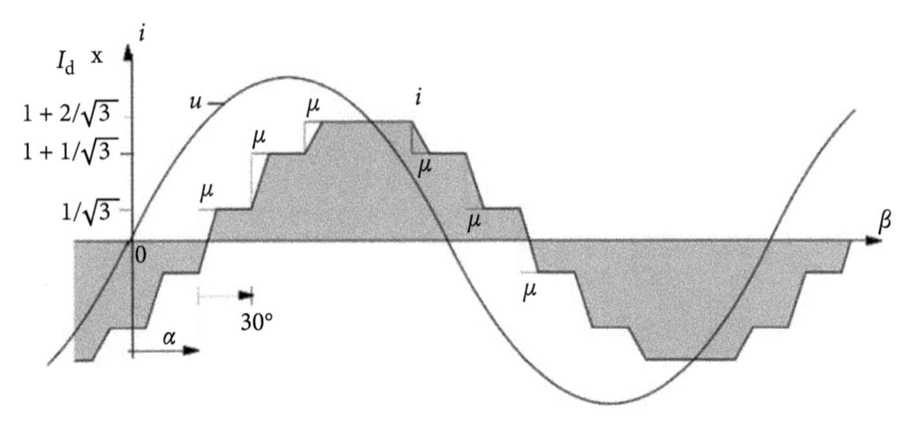

Figure 3.32 Trapezoidal approximation of the supply current of 12-pulse AC/DC converter.

The supply current is equal to the difference of mutually shifted x(β) components, namely

$$i(\beta) = x(\beta) - x(\beta - \pi),$$

thus,

$$\boldsymbol{I}_n = \boldsymbol{X}_n - e^{-jn\pi}\boldsymbol{X}_n = (1 - e^{-jn\pi})\boldsymbol{X}_n = \begin{cases} 2\boldsymbol{X}_n, & \text{for odd } n \\ 0 & \text{for even } n \end{cases}$$

while x(β) is shifted versus y(β) by the angle shown in Fig. 3.33b, thus

$$x(\beta) = y\left[\beta - \left(\frac{\pi}{2} + \alpha + \frac{\mu}{2}\right)\right].$$

Hence

$$\boldsymbol{X}_n = e^{-jn\left(\frac{\pi}{2} + \alpha + \frac{\mu}{2}\right)}\boldsymbol{Y}_n.$$

Quantity y(β) is integral of pulses z(β) shown in Fig. 3.33d, namely

$$y(\beta) = \int z(\beta)\, d\beta, \quad \text{hence } \boldsymbol{Y}_n = \frac{1}{jn}\boldsymbol{Z}_n.$$

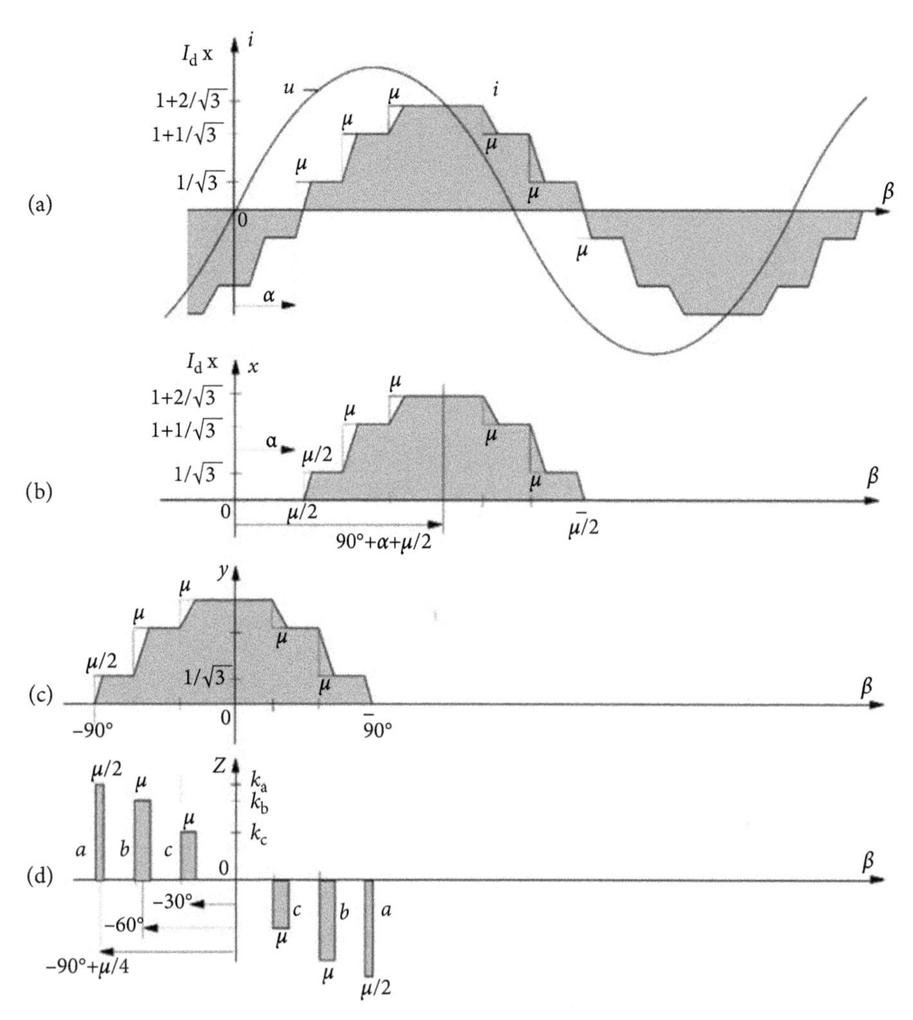

Figure 3.33 Components and related quantities of supply current of 12-pulse AC/DC converter.

Quantity $z(\beta)$ is composed of rectangular pulses of positive and negative polarity as shown in Fig. 3.33, namely pulse $a(\beta)$ of duration $\mu/2$, pulses $b(\beta)$ and $c(\beta)$ of duration μ. They have magnitude k_a, k_b, and k_c, respectively, such that

$$\int_0^{\mu/2} k_a \, d\beta = \frac{I_d}{\sqrt{3}}, \quad \text{hence} \quad k_a = \frac{2\, I_d}{\sqrt{3}\,\mu}$$

$$\int_0^{\mu} k_b \, d\beta = I_d, \quad \text{hence} \quad k_b = \frac{I_d}{\mu}$$

$$\int_0^{\mu} k_c \, d\beta = \frac{I_d}{\sqrt{3}}, \quad \text{hence} \quad k_c = \frac{I_d}{\sqrt{3}\,\mu}.$$

The quantity $z(\beta)$ can be expressed in terms of these pulses as

$$z(\beta) = k_{\rm a}\left[a\left(\beta + \frac{\pi}{2} - \frac{\mu}{4}\right) - a\left(\beta - \frac{\pi}{2} + \frac{\mu}{4}\right)\right] + \\ + k_{\rm b}\left[b\left(\beta + \frac{\pi}{3}\right) - b\left(\beta - \frac{\pi}{3}\right)\right] + k_{\rm c}\left[c\left(\beta + \frac{\pi}{6}\right) - c\left(\beta - \frac{\pi}{6}\right)\right].$$

Hence

$$\boldsymbol{Z}_n = 2jk_{\rm a}\boldsymbol{A}_n \sin\left[n\left(\frac{\pi}{2} - \frac{\mu}{4}\right)\right] + 2jk_{\rm b}\boldsymbol{B}_n \sin\left(n\frac{\pi}{3}\right) + 2jk_{\rm c}\boldsymbol{C}_n \sin\left(n\frac{\pi}{6}\right)$$

where

$$\boldsymbol{A}_n = \frac{1}{\sqrt{2}\pi}\int_{-\mu/4}^{\mu/4} 1\, e^{-jn\beta}\, d\beta = \frac{\sqrt{2}}{n\pi}\sin\left(n\frac{\mu}{4}\right)$$

$$\boldsymbol{B}_n = \frac{1}{\sqrt{2}\pi}\int_{-\mu/2}^{\mu/2} 1\, e^{-jn\beta}\, d\beta = \frac{\sqrt{2}}{n\pi}\sin\left(n\frac{\mu}{2}\right) = \boldsymbol{C}_n.$$

These partial results can be combined as follows. For odd n

$$\boldsymbol{I}_n = 2\boldsymbol{X}_n = 2\,e^{-jn\left(\frac{\pi}{2} + \alpha + \frac{\mu}{2}\right)}\boldsymbol{Y}_n = 2\,e^{-jn\left(\frac{\pi}{2} + \alpha + \frac{\mu}{2}\right)}\frac{1}{jn}\boldsymbol{Z}_n =$$

$$= 2\,e^{-jn\left(\frac{\pi}{2} + \alpha + \frac{\mu}{2}\right)}\frac{2}{n}\left\{k_{\rm a}\boldsymbol{A}_n \sin\left[n\left(\frac{\pi}{2} - \frac{\mu}{4}\right)\right] + k_{\rm b}\boldsymbol{B}_n \sin\left(n\frac{\pi}{3}\right) + k_{\rm c}\boldsymbol{C}_n \sin\left(n\frac{\pi}{6}\right)\right\} =$$

$$= \frac{2\sqrt{2}}{\sqrt{3}\,n\pi}I_{\rm d}\left\{\sin\left(n\frac{\pi}{2}\right) + \sqrt{3}\sin\left(n\frac{\pi}{3}\right) + \sin\left(n\frac{\pi}{6}\right)\right\}\frac{\sin\left(n\frac{\mu}{2}\right)}{n\frac{\mu}{2}}\,e^{-jn\left(\frac{\pi}{2} + \alpha + \frac{\mu}{2}\right)}.$$

At instantaneous commutation, that is, $\mu = 0$, the current harmonics crms values are equal to

$$\boldsymbol{I}_n = \frac{2\sqrt{2}}{\sqrt{3}\,n\pi}I_{\rm d}\left\{\sin\left(n\frac{\pi}{2}\right) + \sqrt{3}\sin\left(n\frac{\pi}{3}\right) + \sin\left(n\frac{\pi}{6}\right)\right\}\,e^{-jn\left(\frac{\pi}{2} + \alpha\right)}.$$

The rms value of the current fundamental harmonic is equal to

$$I_1 = \frac{2\sqrt{2}}{\sqrt{3}\,\pi}I_{\rm d}\left\{1 + \sqrt{3}\frac{\sqrt{3}}{2} + \frac{1}{2}\right\} = \frac{2\sqrt{6}}{\pi}I_{\rm d}$$

thus, it is twice as much as this harmonic for a six-pulse converter with the same DC output current $I_{\rm d}$. Let us calculate the rms value of the fifth and the seventh order harmonics,

$$I_5 = \frac{2\sqrt{2}}{\sqrt{3}\,5\pi} I_\mathrm{d} \left\{\sin\left(\frac{5\pi}{2}\right) + \sqrt{3}\sin\left(\frac{5\pi}{3}\right) + \sin\left(\frac{5\pi}{6}\right)\right\} = \frac{2\sqrt{2}}{\sqrt{3}\,5\pi} I_\mathrm{d} \left\{1 - \sqrt{3}\frac{\sqrt{3}}{2} + \frac{1}{2}\right\} = 0$$

$$I_7 = \frac{2\sqrt{2}}{\sqrt{3}\,7\pi} I_\mathrm{d} \left\{\sin\left(\frac{7\pi}{2}\right) + \sqrt{3}\sin\left(\frac{7\pi}{3}\right) + \sin\left(\frac{7\pi}{6}\right)\right\} = \frac{2\sqrt{2}}{\sqrt{3}\,5\pi} I_\mathrm{d} \left\{-1 + \sqrt{3}\frac{\sqrt{3}}{2} - \frac{1}{2}\right\} = 0$$

thus these harmonics are eliminated from the supply current. The rms values of the eleventh and thirteenth order harmonics are

$$I_{11} = \frac{2\sqrt{2}}{\sqrt{3}\,11\pi} I_\mathrm{d} \left\{\sin\left(\frac{11\pi}{2}\right) + \sqrt{3}\sin\left(\frac{11\pi}{3}\right) + \sin\left(\frac{11\pi}{6}\right)\right\} =$$

$$= \frac{2\sqrt{2}}{\sqrt{3}\,11\pi} I_\mathrm{d} \left\{-1 + \sqrt{3}\frac{-\sqrt{3}}{2} - \frac{1}{2}\right\} = \frac{1}{11} I_1$$

$$I_{13} = \frac{2\sqrt{2}}{\sqrt{3}\,13\pi} I_\mathrm{d} \left\{\sin\left(\frac{13\pi}{2}\right) + \sqrt{3}\sin\left(\frac{13\pi}{3}\right) + \sin\left(\frac{13\pi}{6}\right)\right\} =$$

$$= \frac{2\sqrt{2}}{\sqrt{3}\,13\pi} I_\mathrm{d} \left\{1 + \sqrt{3}\frac{\sqrt{3}}{2} + \frac{1}{2}\right\} = \frac{1}{13} I_1.$$

When commutation is taken into account, these values are reduced by the factor

$$k_{\mu n} \stackrel{\mathrm{df}}{=} \frac{\sin(n\,\mu/2)}{n\,\mu/2}.$$

*****

3.6 Single-Phase LTI Circuit Analysis

Nonsinusoidal but periodic voltages and currents occur in electrical circuits or systems only in a steady state of such systems. Voltages and currents are not periodic, apart from some idealized cases, in transient states. Consequently, analysis of electrical circuits in transient states is not the subject of this chapter, and nor is waveform distortion caused by nonlinearity of the elements or their time variance. It is confined to the analysis of linear, time-invariant (LTI) circuits in a steady state.

Kirchhoff's laws and voltage–current relations for circuit elements provide the ground for the LTI circuit analysis. The Superposition Principle, Thevenen, and Norton concepts of equivalent circuits, as well as the nodal and mash equations, provide major theoretical tools for such analysis. All these methods remain valid for the analysis of LTI circuits with nonsinusoidal but periodic voltages and currents, with only some modifications and adaptations to the harmonic approach to this analysis.

Voltage–current relations for circuit elements, apart from resistors, are differential relations, thus electrical circuits are described by sets of differential equations. In the case of LTI circuits with lumped RLC elements, these are sets of ordinary differential equations with constant coefficients. Unfortunately, analytical solutions can be found only at the low complexity of a circuit. Numerical integration of sets of differential equations is needed for more complex circuits. Just such a numerical integration is performed by common circuit analysis programs such as PSpice or MatLab and Simulink.

In the case of LTI circuits with sinusoidal voltages currents, the set of differential equations can be converted into a set of algebraic equations by the association of complex rms values, known commonly as "phasors," with sinusoidal voltages and currents. This association is based on the relation between time-varying $x(t)$ sinusoidal quantity and its crms value, $\boldsymbol{X}$.

$$x(t) = \sqrt{2}\,\mathrm{Re}\{\boldsymbol{X}e^{j\omega_1 t}\}.$$

The set of circuit differential equations, transformed in such a way into the set of algebraic equations, can be solved with algebraic methods. Integration of differential equations is not needed for such circuit analysis. It is one of the main reasons for the importance of complex algebra in electrical engineering.

As long as nonsinusoidal voltages and currents in a circuit are expressed in terms of Fourier series in a traditional form, such a circuit can be described by a set comprising only differential equations and numerical integration is needed for their solution. Such a set of differential equations can be converted to a set of algebraic equations on the condition that the voltages and currents are expressed in terms of Fourier series in a complex form, meaning in terms of the crms values associated with harmonics of voltages and currents in the circuit. It means that Fourier series in a complex form enables the conversion of differential equations of LTI circuits into algebraic ones, which can be analyzed and solved with algebraic methods. This section presents the fundamentals of such a conversion. Conversion of basic laws of LTI circuits from a differential to an algebraic form is needed for that.

1. Kirchhoff's Current Law (KCL) says that the sum of currents into any node with K branches, shown in Fig. 3.34, equals zero at any instant of time. Since currents in the node satisfy the identity

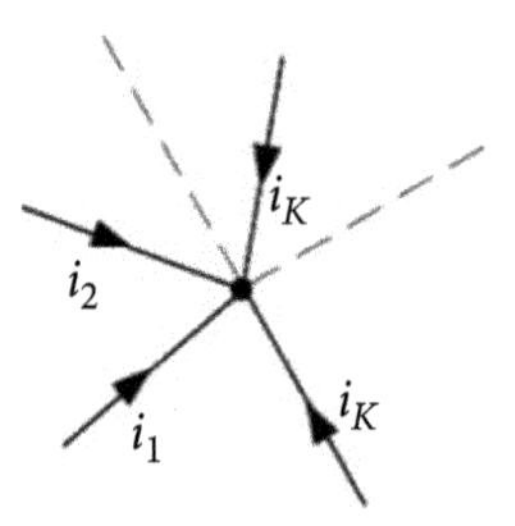

Figure 3.34 A circuit node with currents.

$$\sum_{k=1}^{K} i_k(t) \equiv 0$$

belong to space L_T^2, meaning they can be expressed as the sum of harmonics, thus they have the Fourier series and are sums of harmonics, that is,

$$i_k(t) = \sum_{n \in N_0} i_{kn}(t)$$

then KCL can be written in the form

$$\sum_{k=1}^{K} i_k(t) \equiv \sum_{k=1}^{K} \left[\sum_{n \in N_0} i_{kn}(t) \right] \equiv \sum_{n \in N_0} \left[\sum_{k=1}^{K} i_{kn}(t) \right] \equiv 0.$$

Since the terms

$$\sum_{k=1}^{K} i_{kn}(t)$$

for each non-zero harmonic are quantities with different frequencies, their sum can be equal to zero at any instant of time only if for each harmonic separately

$$\sum_{k=1}^{K} i_{kn}(t) \equiv 0.$$

It means that Kirchhoff's current law has to be fulfilled for each current harmonic separately. Thus, the current harmonics in any node are independent of each other. Consequently, a node diagram, as shown in Fig. 3.35, can be drawn for each current harmonic.

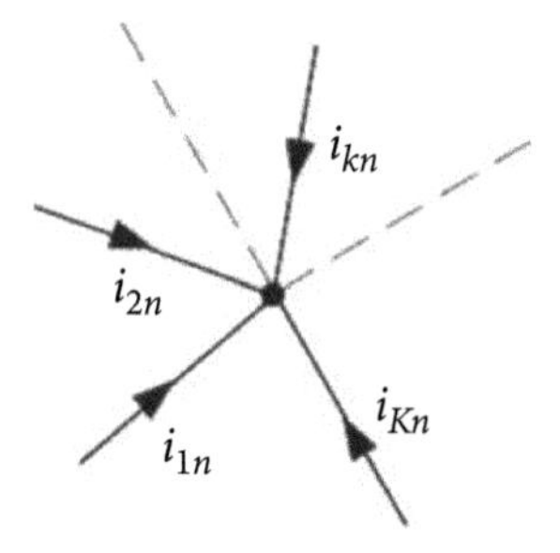

Figure 3.35 A node with current harmonics.

When the node currents belong to space L^2_T, thus they have the Fourier series

$$i_k(t) = I_{k0} + \sqrt{2}\,\mathrm{Re} \sum_{n \in N} \boldsymbol{I}_{kn}\, e^{jn\omega_1 t}$$

then KCL can be expressed, after switching the sequence of harmonic order n summation with the branch number, in the form

$$\sum_{k=1}^{K} \left(I_{k0} + \sqrt{2}\,\mathrm{Re} \sum_{n \in N} \boldsymbol{I}_{kn}\, e^{jn\omega_1 t} \right) \equiv \sum_{k=1}^{K} I_{k0} + \sqrt{2}\,\mathrm{Re} \sum_{n \in N} \left(\sum_{k=1}^{K} \boldsymbol{I}_{kn} \right) e^{jn\omega_1 t} \equiv 0.$$

The function $e^{jn\omega_1 t}$ can have in a period T all values on the unit circle on a plane of complex numbers for each harmonic order n. Thus, KCL can be fulfilled only on the condition that for each harmonic order n from the set N, along with $n = 0$, the coefficient at these functions is zero, that is,

$$\sum_{k=1}^{K} \boldsymbol{I}_{kn} \equiv 0. \tag{3.36}$$

Thus, KCL has to be fulfilled for the crms value for each current harmonic separately. Therefore, a node on circuit diagrams can be drawn as in Fig. 3.36.

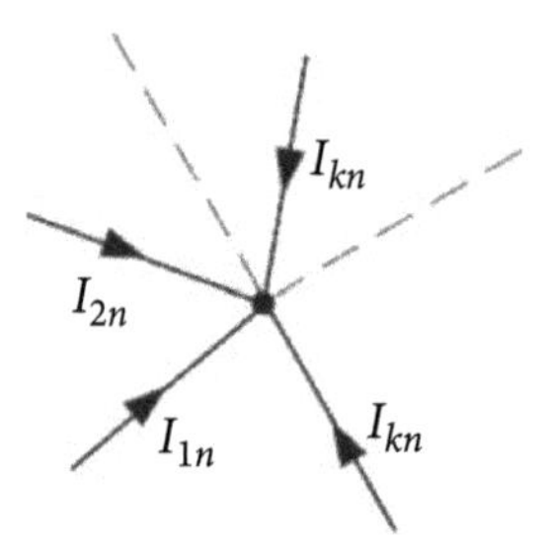

Figure 3.36 A node with the crms values of current harmonics.

Observe an important difference between KCL in the original form, which applies to K time-varying values at each instant of time, t, and KCL for current harmonic crms values, which applies to K complex numbers.

KCL for crms values of current harmonics applies not only to a physical node, meaning a physical connection of a few branches, but also to any closed space, sometimes referred to as a *super-node*. It means that the sum of crms values of individual harmonics of branch currents that flow to a confined area equals zero, even if the branches are not physically connected.

Illustration 3.27 *Currents in node shown in Fig. 3.37 are equal to, respectively,*

$$i_1(t) = 10 + \sqrt{2}\,\mathrm{Re}\{50\, e^{j\omega_1 t} + 12\, e^{-j45^\circ} e^{j3\omega_1 t}\}\mathrm{A}$$

$$i_2(t) = 20 + \sqrt{2}\,\mathrm{Re}\{40\, e^{j60^\circ} e^{j\omega_1 t} + 15\, e^{j5\omega_1 t}\}\mathrm{A}.$$

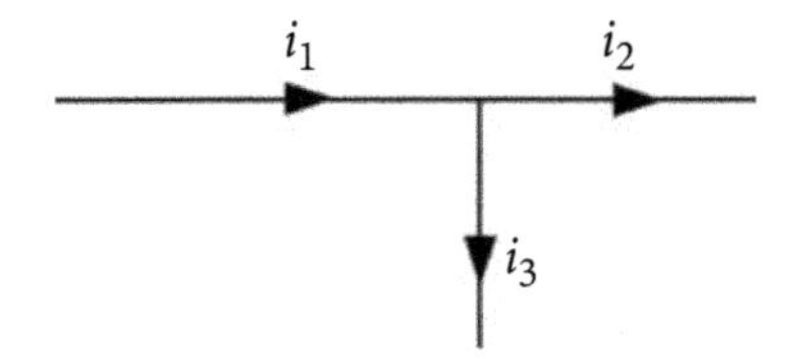

Figure 3.37 A circuit node.

The current of branch #3 can be calculated by applying KCL for crms values of current harmonics. To avoid conclusion as to indices, the first index denotes the number of a branch and is separated with a comma from the second index which denotes the harmonic order. Written without a comma index, for example, 113 can denote thirteenth order harmonic in branch 1, or third order harmonic in branch 11.

$$I_{3,0} = I_{1,0} - I_{2,0} = 10 - 20 = -10\,\text{A}$$

$$\boldsymbol{I}_{3,1} = \boldsymbol{I}_{1,1} - \boldsymbol{I}_{2,1} = 50 - 40\,e^{j60^\circ} = 45.8\,e^{-j49.1^\circ}\,\text{A}$$

$$\boldsymbol{I}_{3,3} = \boldsymbol{I}_{1,3} - \boldsymbol{I}_{2,3} = 12\,e^{-j45^\circ} = 12\,e^{-j45^\circ}\,\text{A}$$

$$\boldsymbol{I}_{3,5} = \boldsymbol{I}_{1,5} - \boldsymbol{I}_{2,5} = -15 = 15\,e^{j180^\circ}\,\text{A}$$

thus the current $i_3(t)$ is equal to

$$i_3(t) = -10 + \sqrt{2}\,\text{Re}\left\{45.9\,e^{-j49.1^\circ}e^{j\omega_1 t} + 12\,e^{-j45^\circ}e^{j3\omega_1 t} + 15\,e^{j180^\circ}e^{j5\omega_1 t}\right\}\text{A}$$

and its rms value is

$$||i_3|| = \sqrt{10^2 + 49.5^2 + 12^2 + 15^2} = 54.0\,\text{A}.$$

*****

2. Kirchhoff's Voltage Law (KVL) says that the sum of voltages along any closed path in a circuit, as shown in Fig. 3.38, equals zero at any instant of time, that is,

$$\sum_{m=1}^{M} u_m(t) \equiv 0.$$

If voltages $u_{\text{m}}(t)$ belong to space L_{T}^2, meaning they can be expressed as the sum of harmonics

$$u_m(t) = \sum_{n \in N_0} u_{m\,n}(t)$$

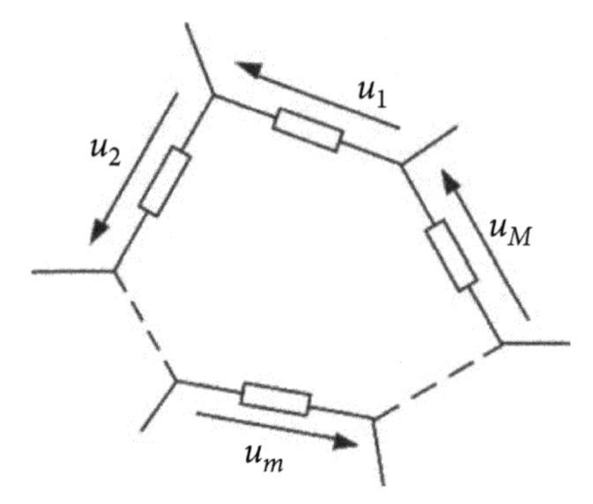

Figure 3.38 A circuit path.

then KVL for the circuit path can be written in the form

$$\sum_{m=1}^{M} u_m(t) \equiv \sum_{m=1}^{M}\left[\sum_{n\in N_0} u_{mn}(t)\right] \equiv \sum_{n\in N_0}\left[\sum_{m=1}^{M} u_{mn}(t)\right] \equiv 0.$$

Terms $u_{mn}(t)$, if non-zero, are quantities of a different frequency. Their sum at each instant of time can equal zero only on the condition that for each harmonic order, separately

$$\sum_{m=1}^{M} u_{mn}(t) \equiv 0.$$

It means that KVL has to be fulfilled along any closed path for each voltage harmonic separately. The distribution of the voltage harmonics along a closed path is independent of each other harmonic. A diagram of a circuit closed path can be drawn for each separate voltage harmonic as shown in Fig. 3.39

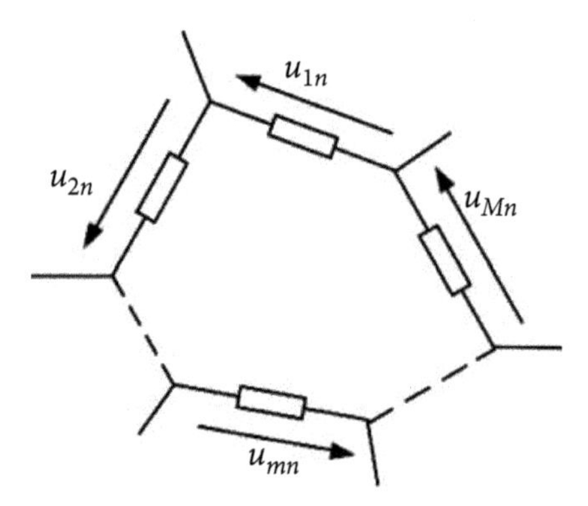

Figure 3.39 A circuit path for voltage harmonics.

If branch voltages are expressed in terms of Fourier series,

$$u_m(t) = U_{m0} + \sqrt{2}\,\mathrm{Re} \sum_{n \in N} \boldsymbol{U}_{mn}\, e^{jn\omega_1 t},$$

then after the sequence of summation over branch number m, and harmonic order n, are switched, KVL can be presented in the form

$$\sum_{m=1}^{M} u_m(t) = \sum_{m=1}^{M} U_{0m} + \sqrt{2}\,\mathrm{Re} \sum_{n \in N} \left(\sum_{m=1}^{M} \boldsymbol{U}_{mn} \right) e^{jn\omega_1 t} \equiv 0.$$

This can be fulfilled only on the condition that the coefficient at each function $e^{jn\omega_1 t}$ is equal to zero, meaning if

$$\sum_{m=1}^{M} \boldsymbol{U}_{mn} \equiv 0. \tag{3.37}$$

Thus KVL has to be fulfilled for the crms values of voltage harmonics along any closed path of a circuit for each harmonic separately. Circuit diagrams of such paths for the crms values of voltage harmonics can be drawn as shown in Fig. 3.40.

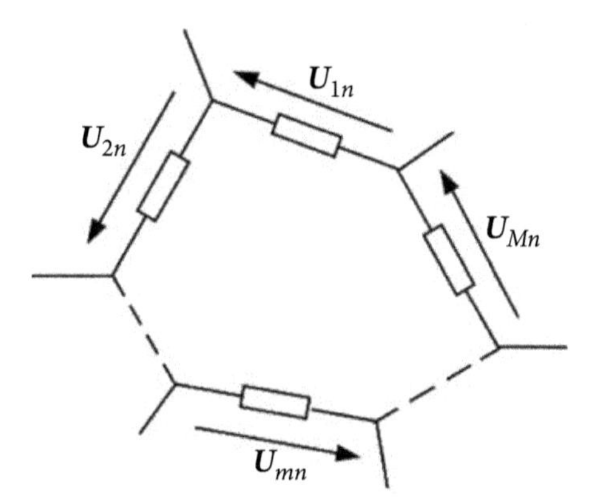

Figure 3.40 A circuit path for the crms values of voltage harmonics.

Illustration 3.28 *Voltages u_1 and u_3 in the circuit shown in Fig. 3.41 are equal to*

$$u_1(t) = \sqrt{2}\,\mathrm{Re}\{120\, e^{j\omega_1 t}\}\mathrm{V}$$

$$u_3(t) = 8 + \sqrt{2}\,\mathrm{Re}\{110\, e^{-j5^\circ} e^{j\omega_1 t} + 5\, e^{j90^\circ} e^{j3\omega_1 t}\}\mathrm{V}.$$

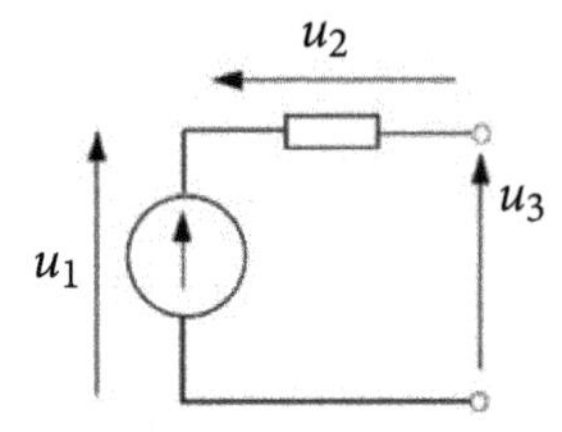

Figure 3.41 A circuit example.

Voltage $u_2(t)$ can be found applying KVL to crms values of voltage harmonics, namely

$$U_{2,0} = U_{1,0} - U_{3,0} = -8\ V$$

$$\boldsymbol{U}_{2,1} = \boldsymbol{U}_{1,1} - \boldsymbol{U}_{3,1} = 120 - 110\, e^{-j5°} = 14.1\, e^{j42.6°}\,\mathrm{V}$$

$$\boldsymbol{U}_{2,3} = \boldsymbol{U}_{1,3} - \boldsymbol{U}_{3,3} = -5\, e^{j90°} = 5\, e^{-j90°}\,\mathrm{V}.$$

Thus, the voltage $u_2(t)$ is

$$u_2(t) = -8 + \sqrt{2}\,\mathrm{Re}\{14.1\, e^{j42.6°} e^{j\omega_1 t} + 5\, e^{-j90°} e^{j3\omega_1 t}\}\mathrm{V}.$$

*****

3.7 Voltage–Current Relations of LTI One-Ports

Let us consider a linear time-invariant one-port as shown in Fig. 3.42.

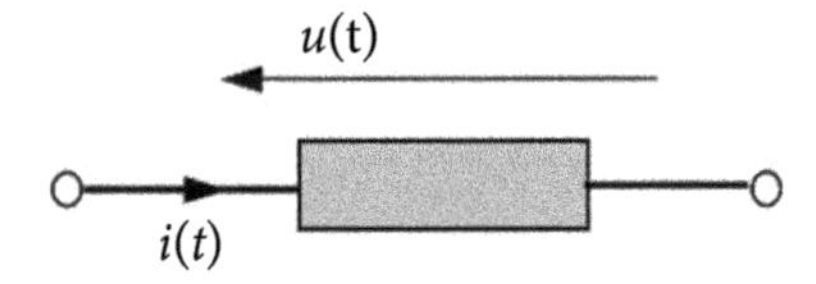

Figure 3.42 A one-port.

Relations between the voltage and current harmonics for such one-ports can be specified in terms of the one-port impedance $\boldsymbol{Z}_n$ or admittance $\boldsymbol{Y}_n$ for harmonic frequencies, defined as

$$\frac{\boldsymbol{U}_n}{\boldsymbol{I}_n} \overset{\mathrm{df}}{=} \boldsymbol{Z}_n = Z_n\, e^{j\varphi_n} = R_n + jX_n \tag{3.38}$$

$$\frac{\boldsymbol{I}_n}{\boldsymbol{U}_n} \overset{\mathrm{df}}{=} \boldsymbol{Y}_n = \frac{1}{\boldsymbol{Z}_n} = Y_n\, e^{-j\varphi_n} = G_n + jB_n \tag{3.39}$$

If the one-port voltage has Fourier Series in a complex form

$$u = U_0 + \sqrt{2}\,\mathrm{Re} \sum_{n \in N} \boldsymbol{U}_n\, e^{jn\omega_1 t}$$

then the one-port current is

$$i = I_0 + \sqrt{2}\,\mathrm{Re} \sum_{n \in N} \boldsymbol{I}_n\, e^{jn\omega_1 t} = Y_0 U_0 + \sqrt{2}\mathrm{Re} \sum_{n \in N} \boldsymbol{Y}_n \boldsymbol{U}_n\, e^{jn\omega_1 t}. \tag{3.40}$$

One should observe, however, that admittance $\boldsymbol{Y}_n$ of an idealized one-port can approach infinity for some harmonic frequencies. There could be a short-circuit for a DC current or there could be a voltage resonance for some harmonic frequencies. If there is a supply voltage harmonic of the frequency where $\boldsymbol{Y}_n$, is infinity high, the one-port current harmonic of that frequency will approach infinity. The current remains limited and periodic in such a case if there is no voltage of the frequency where $\boldsymbol{Y}_n$ is infinitely high.

When the current is a primary quantity, meaning it is exerted by a current source, that is,

$$i = I_0 + \sqrt{2}\,\mathrm{Re} \sum_{n \in N} \boldsymbol{I}_n\, e^{jn\omega_1 t}$$

then the one-port response, meaning the voltage equals to

$$u = U_0 + \sqrt{2}\,\mathrm{Re} \sum_{n \in N} \boldsymbol{U}_n\, e^{jn\omega_1 t} = Z_0 I_0 + \sqrt{2}\,\mathrm{Re} \sum_{n \in N} \boldsymbol{Z}_n \boldsymbol{I}_n\, e^{jn\omega_1 t}. \tag{3.41}$$

As in the previous case, there is a possibility that for some harmonic order n the product $\boldsymbol{I}_n \boldsymbol{Z}_n$ can be infinite. The one-port could be open for a DC current or can be a current resonance at harmonic frequencies. Such situations have to be excluded.

Resistor: The symbol of a resistor with the voltage and current orientation is shown in Fig. 3.43.

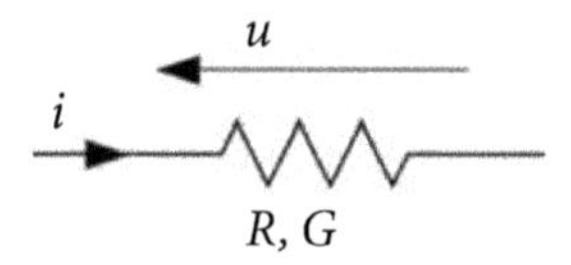

Figure 3.43 Symbol of a resistor and voltage and current orientation.

The current of the resistor is proportional to the resistor voltage, meaning

$$u(t) = Ri(t), \quad or \quad i(t) = Gu(t), \quad G = 1/R$$

thus

$$i(t) = G\,u(t) = GU_0 + \sqrt{2}\,\mathrm{Re}\sum_{n\in N} G\,\boldsymbol{U}_n\, e^{jn\omega_1 t}. \tag{3.42}$$

The crms values of harmonics of resistor voltage and current satisfy the relationship

$$\boldsymbol{I}_n = G\boldsymbol{U}_n \quad or \quad \boldsymbol{U}_n = R\boldsymbol{I}_n.$$

Inductor. The symbol of an inductor with the voltage and current orientation is shown in Fig. 3.44.

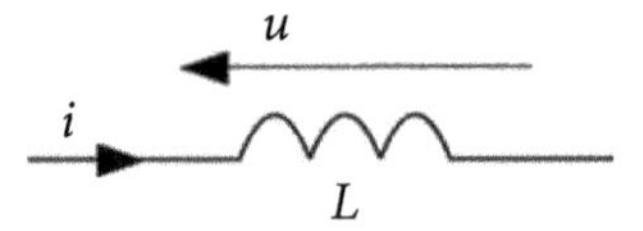

Figure 3.44 The symbol of an inductor and the voltage and current orientations.

The inductor voltage and current satisfy the relations

$$u(t) = L\frac{di(t)}{dt}, \quad i(t) = \frac{1}{L}\int u(t)\,dt + B$$

where B is any constant. When the inductor's current

$$i(t) = I_0 + \sqrt{2}\,\mathrm{Re}\sum_{n\in N} \boldsymbol{I}_n\, e^{jn\omega_1 t}$$

is the primary inductor quantity, then its voltage is

$$u(t) = L\frac{di(t)}{dt} = \sqrt{2}\,\mathrm{Re}\sum_{n\in N} jn\omega_1 L\,\boldsymbol{I}_n\, e^{jn\omega_1 t} = \sqrt{2}\,\mathrm{Re}\sum_{n\in N} \boldsymbol{U}_n\, e^{jn\omega_1 t}.$$

Complex rms values of harmonics of the inductor voltage and current satisfy the relationship

$$\boldsymbol{U}_n = jn\omega_1 L\,\boldsymbol{I}_n. \tag{3.43}$$

The voltage DC component, U_0, has to be equal, of course, to zero, while the current DC component, I_0, can have any value. Its value is not determined by the inductor but by the structure and other elements of the circuit. Since ideal inductors for DC currents are equivalent to a short-circuit, while capacitors are equivalent to an open circuit, the inductor current DC component can be found when all inductors in the analyzed circuit are short-circuited and capacitors are removed.

Illustration 3.29 *Let us assume that the inductor's current is*

$$i(t) = 8 + \sqrt{2}\,\mathrm{Re}\{10\, e^{j\omega_1 t} + 2e^{j5\omega_1 t}\}\ \mathrm{A}$$

and the inductor reactance for the fundamental frequency is $\omega_1 L = 1.5\ \Omega$. *The inductor voltage is*

$$u(t) = \sqrt{2}\,\mathrm{Re}\{15\, e^{j90^\circ} e^{j\omega_1 t} + 15\, e^{j90^\circ} e^{j5\omega_1 t}\}\ \mathrm{V}.$$

****

Capacitor: The symbol of a capacitor and voltage and current orientation are shown in Fig. 3.45.

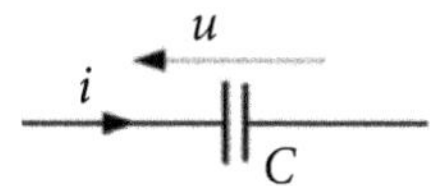

Figure 3.45 The symbol of a capacitor and the voltage And current orientation.

The capacitor voltage and current satisfy the relationship

$$i(t) = C\frac{du(t)}{dt}, \qquad u(t) = \frac{1}{L}\int i(t)\,dt + B$$

where B is any constant. If the voltage on the capacitor is the primary quantity, then the capacitor current is

$$i(t) = C\frac{du(t)}{dt} = \sqrt{2}\,\mathrm{Re}\sum_{n\in N} jn\omega_1 C U_n\, e^{jn\omega_1 t}.$$

The crms values of harmonics of the capacitor voltage and current satisfy the relationship

$$\boldsymbol{I}_n = jn\omega_1 C \boldsymbol{U}_n. \tag{3.44}$$

The DC component of the capacitor current has to be equal to zero, while the DC component of the capacitor voltage can have any value. This value is not determined by capacitance but by other elements of the circuit. Since a capacitor is equivalent at a DC voltage to an open circuit, the DC component can be found by removing all capacitors from the analyzed circuit.

Series RLC Branch: The voltage on a series RLC branch, shown in Fig. 3.46, is the sum of voltages on the resistor, inductor, and capacitor at the common branch current $i(t)$. Since the capacitor current cannot have any DC component, it cannot be in the branch current.

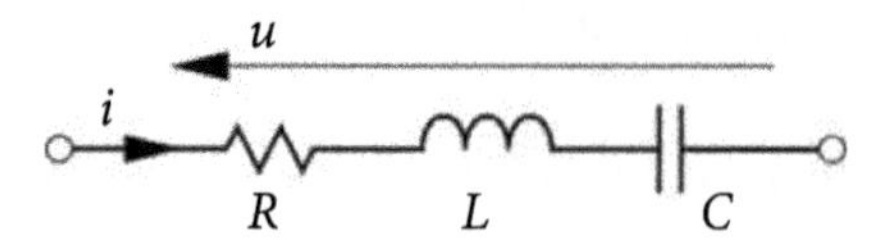

Figure 3.46 A series RLC branch.

Thus, the branch current can have the form

$$i(t) = \sqrt{2}\,\mathrm{Re} \sum_{n\in N} \boldsymbol{I}_n\, e^{jn\omega_1 t}$$

so that, the branch voltage is

$$u(t) = \sqrt{2}\,\mathrm{Re} \sum_{n\in N} \left(R + jn\omega_1 L + \frac{1}{jn\omega_1 C}\right) \boldsymbol{I}_n\, e^{jn\omega_1 t} + B$$

where B is any constant. This constant stands for the DC component of the capacitor voltage and cannot be calculated for the branch isolated from the entire circuit.

The impedance of a series RLC branch is

$$\boldsymbol{Z}_n = Z_n\, e^{j\varphi_n} = R_n + jX_n = R + jn\omega_1 L + \frac{1}{jn\omega_1 C} \tag{3.45}$$

where R_n is the branch resistance and X_n is the branch reactance. The admittance of such a branch is

$$\frac{1}{\boldsymbol{Z}_n} = \boldsymbol{Y}_n = Y_n e^{-j\varphi_n} = G_n + jB_n = \frac{1}{R + jn\omega_1 L + \dfrac{1}{jn\omega_1 C}} \tag{3.46}$$

where G_n is the series branch conductance and B_n is the series branch susceptance.

Illustration 3.30 *Let us calculate the current of a series RLC branch with the branch voltage*

$$u(t) = 25 + \sqrt{2}\,\mathrm{Re}\{220\, e^{j\omega_1 t} + 15\, e^{j90^\circ} e^{j5\omega_1 t}\}\mathrm{V}, \quad \omega_1 = 314\ \mathrm{rad/s}$$

Assuming that $R = 2.5\ \Omega$, $L = 1.5$ mH, $C = 1.2$ mF.

The branch admittance for DC voltage equals zero because of the capacitance. For the two remaining harmonics, the fundamental and the fifth order

$$\boldsymbol{Y}_1 = \frac{1}{R + j\omega_1 L + \dfrac{1}{j\omega_1 C}} = \frac{1}{1.5 + j0.47 - j2.65} = 0.38\, e^{j55.5^\circ}\ \mathrm{S}$$

$$\boldsymbol{Y}_5 = \frac{1}{R + j5\omega_1 L + \dfrac{1}{j5\omega_1 C}} = \frac{1}{1.5 + j5 \times 0.47 - j\frac{2.65}{5}} = 0.42 e^{-j50.6^\circ}\ \mathrm{S}.$$

Thus, the branch current is

$$i(t) = \sqrt{2}\,\mathrm{Re}\{0.38\, e^{j55.4^\circ} \times 220\, e^{j\omega_1 t} + 0.42\, e^{-j50.6^\circ} \times 15\, e^{j90^\circ} e^{j5\omega_1 t}\} =$$
$$= \sqrt{2}\,\mathrm{Re}\{83.6\, e^{j55.4^\circ} e^{j\omega_1 t} + 6.3\, e^{j.39.4^\circ} e^{j5\omega_1 t}\}\mathrm{A}.$$

*****

Parallel RLC Branch: The current of the RLC branch, shown in Fig. 3.47, is the sum of the resistor, inductor, and capacitor currents at the same voltage u(t).

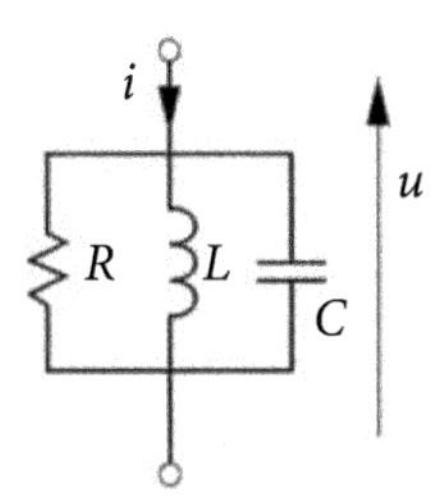

Figure 3.47 A parallel RLC branch.

Since the DC component of the inductor voltage has to be equal to zero, the branch voltage cannot have any non-zero DC component. It can have the form

$$u(t) = \sqrt{2}\,\mathrm{Re} \sum_{n \in N} \boldsymbol{U}_n\, e^{jn\omega_1 t}.$$

The branch current, as the sum of individual elements' current, is

$$i(t) = \sqrt{2}\,\mathrm{Re} \sum_{n \in N} \left(G + jn\omega_1 C + \frac{1}{jn\omega_1 L}\right) \boldsymbol{U}_n\, e^{jn\omega_1 t} + B$$

where B is any constant. It stands for a DC component of the inductor current, which cannot be determined without the knowledge of the circuit structure and resistance values.

The admittance of such a branch for the nth order harmonic is

$$\frac{\boldsymbol{I}_n}{\boldsymbol{U}_n} \stackrel{\mathrm{df}}{=} \boldsymbol{Y}_n \stackrel{\mathrm{df}}{=} Y_n e^{-j\varphi_n} \stackrel{\mathrm{df}}{=} G_n + jB_n = G + jn\omega_1 C + \frac{1}{jn\omega_1 L} \tag{3.47}$$

where G_n denotes the branch conductance and B_n denotes the branch susceptance.

The negative sign of the admittance argument, $-\varphi_n$, results from a convention. The argument of impedance was selected to be positive, meaning

$$\frac{\boldsymbol{U}_n}{\boldsymbol{I}_n} \stackrel{\text{df}}{=} \boldsymbol{Z}_n \stackrel{\text{df}}{=} Z_n\, e^{j\varphi_n} = \frac{1}{\boldsymbol{Y}_n}.$$

The impedance of the parallel branch for the n-th order harmonic is

$$\frac{1}{\boldsymbol{Y}_n} = \boldsymbol{Z}_n = \boldsymbol{Z}_n\, e^{j\varphi_n} = R_n + jX_n = \frac{1}{G + jn\omega_1 C + \dfrac{1}{jn\omega_1 L}} \tag{3.48}$$

where R_n is the branch resistance and X_n is the branch reactance.

Kirchhoff's laws, along with voltage–current relations of circuit elements, provide fundamental for circuit analysis. When the circuit is LTI, such a set of algebraic equations, as was demonstrated above, can be written for each harmonic order n separately, expressed in terms of the crms values of harmonics of the circuit voltages and currents.

At high circuit complexity, computer methods are needed for generating such sets of equations and their solution. At low complexity, such analysis is possible without computers, and such analysis is needed. Analysis of very simple circuits can be needed and sufficient for understanding general properties, especially power properties of electrical systems.

Illustration 3.31 *Let us calculate the supply current in the circuit shown in Fig. 3.48, with parameters*

$$R_\mathrm{a} = 0.3\Omega, \quad R_\mathrm{b} = 0.2\Omega,\ \omega_1 L_\mathrm{b} = 0.5\Omega,\ \omega_1 C_\mathrm{a} = 2.5\ \mathrm{S}$$

if the supply voltage is

$$u(t) = 50 + \sqrt{2}\,\mathrm{Re}\{120\, e^{j\omega_1 t} + 6\, e^{j30^\circ} e^{j5\omega_1 t} + 2\, e^{-j30^\circ} e^{j7\omega_1 t}\}\mathrm{V}.$$

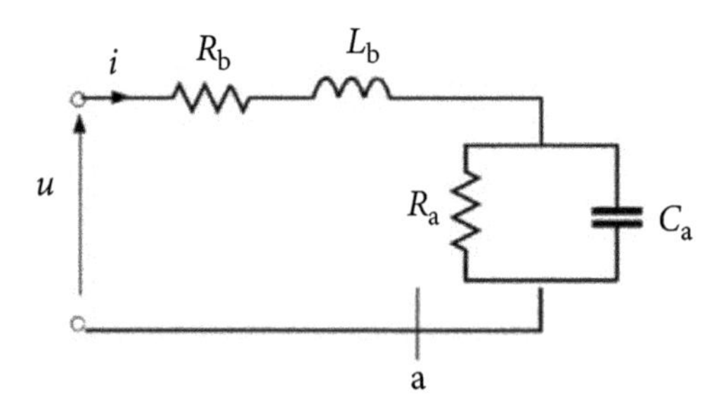

Figure 3.48 An example of a circuit.

The DC component of the supply current, calculated with a short-circuited inductor and removed capacitor, is equal to

$$I_0 = \frac{U_0}{R_a + R_b} = \frac{50}{0.3 + 0.2} = 100\,\mathrm{A}.$$

The admittance in the cross-section "a" is

$$\boldsymbol{Y}_{a\,n} = G_1 + jn\omega_1 C_1$$

thus, for harmonic orders, n = 1, 5, and 7, its numerical value is

$$\boldsymbol{Y}_{a1} = 1/0.3 + j2.5 = 3.33 + j2.5\,\mathrm{S}, \qquad \boldsymbol{Z}_{a1} = 1/\boldsymbol{Y}_{a1} = 0.192 - j0.144\,\Omega,$$

$$\boldsymbol{Y}_{a5} = 1/0.3 + j5 \times 2.5 = 3.33 + j12.5\,\mathrm{S}, \quad \boldsymbol{Z}_{a5} = 1/\boldsymbol{Y}_{a5} = 0.020 - j0.075\,\Omega,$$

$$\boldsymbol{Y}_{a7} = 1/0.3 + j7 \times 2.5 = 3.33 + j17.5\,\mathrm{S}, \quad \boldsymbol{Z}_{a7} = 1/\boldsymbol{Y}_{a7} = 0.011 - j0.055\,\Omega.$$

The circuit impedance is equal to

$$\boldsymbol{Z}_n = R_b + jn\omega_1 L_b + \boldsymbol{Z}_{an}$$

hence

$$\boldsymbol{Z}_1 = 0.2 + j0.5 + (0.192 - j0.144) = 0.392 + j0.356\,\Omega, \qquad \boldsymbol{Y}_1 = 1/\boldsymbol{Z}_1 = 1.89\,e^{-j42.2^\circ}\,\mathrm{S}$$

$$\boldsymbol{Z}_5 = 0.2 + j5 \times 0.5 + (0.020 - j0.075) = 0.220 + j2.425\,\Omega, \quad \boldsymbol{Y}_5 = 1/\boldsymbol{Z}_5 = 0.41\,e^{-j84.8^\circ}\,\mathrm{S}$$

$$\boldsymbol{Z}_7 = 0.2 + j7 \times 0.5 + (0.011 - j0.055) = 0.211 + j3.445\,\Omega, \quad \boldsymbol{Y}_7 = 1/\boldsymbol{Z}_7 = 0.29\,e^{-j86.5^\circ}\,\mathrm{S}.$$

The crms values of the supply current harmonics are equal to, respectively,

$$\boldsymbol{I}_1 = \boldsymbol{Y}_1 \boldsymbol{U}_1 = 1.89\,e^{-j42.2^\circ} \times 120 = 226.8\,e^{-j42.2^\circ}\,\mathrm{A}$$

$$\boldsymbol{I}_5 = \boldsymbol{Y}_5 \boldsymbol{U}_5 = 0.41\,e^{-j84.8^\circ} \times 6\,e^{j30^\circ} = 1.4\,e^{-j54.8^\circ}\,\mathrm{A}$$

$$\boldsymbol{I}_7 = \boldsymbol{Y}_7 \boldsymbol{U}_7 = 0.29\,e^{-j86.5^\circ} \times 2\,e^{-j30^\circ} = 0.54\,e^{-j116.5^\circ}\,\mathrm{A}.$$

Thus the supply current has the waveform

$$i = 100 + \sqrt{2}\,\mathrm{Re}\{226.8\,e^{-j42.2^\circ} e^{j\omega_1 t} + 1.4\,e^{-j54.8^\circ} e^{j5\omega_1 t} + 0.54\,e^{-j116.5^\circ} e^{j7\omega_1 t}\}\,\mathrm{A}.$$

****

Such an approach to circuit analysis as presented above is not, however, universal. It can be used only when a circuit does not have high complexity. An individual approach is needed for such circuit analysis, dependent on the circuit's particular structure. At the same time, such an approach provides a deeper inside into circuit properties than computerized methods.

A global approach is needed for circuits with high complexity, meaning the circuit is not analyzed step by step, but the circuit as a whole is described by sets of equations.

Analysis of LTI circuits with nonsinusoidal but periodic voltages and currents do not differ from such circuit analysis at sinusoidal quantities, only a set of equations has to be developed and solved for each order harmonic that can exist in the circuit voltages.

Resistance, Conductance, Reactance, and Susceptance: Resistance, conductance, reactance, and susceptance of one-ports are sometimes mistaken with these parameters for circuit RLC elements. In particular, it is important to observe, since this is a common error, that the branch conductance G_n is not an inversion of the branch resistance, R_n, meaning

$$G_n \neq \frac{1}{R_n}.$$

Similarly, the branch susceptance B_n is not an inversion of the branch reactance, X_n, meaning

$$B_n \neq \frac{1}{X_n}.$$

One-ports are described in terms of the following parameters for harmonic frequencies,

- One-port resistance,

$$R_n = \text{Re}\{\boldsymbol{Z_n}\}$$

- One-port reactance

$$X_n = \text{Im}\{\boldsymbol{Z_n}\}$$

- One-port conductance

$$G_n = \text{Re}\{\boldsymbol{Y_n}\}$$

- One-port susceptance

$$B_n = \text{Im}\{\boldsymbol{Y_n}\}.$$

These four one-port parameters are mutually related as follows

$$\boldsymbol{Y}_n = G_n + jB_n = \frac{1}{\boldsymbol{Z}_n} = \frac{1}{R_n + jX_n} = \frac{R_n - jX_n}{R_n^2 + X_n^2}$$

hence

$$G_n = \frac{R_n}{R_n^2 + X_n^2}, \quad B_n = -\frac{X_n}{R_n^2 + X_n^2}. \tag{3.49}$$

It means that the one-port conductance depends not only on the one-port resistance but also on the one-port reactance. Also, the one-port susceptance depends not only on the one-port reactance but also on the one-port resistance. Similarly,

$$\boldsymbol{Z}_n = R_n + jX_n = \frac{1}{\boldsymbol{Y}_n} = \frac{1}{G_n + jB_n} = \frac{G_n - jB_n}{G_n^2 + B_n^2}$$

hence

$$R_n = \frac{G_n}{G_n^2 + B_n^2}, \quad X_n = -\frac{B_n}{G_n^2 + B_n^2}. \tag{3.50}$$

Thus, the one-port resistance depends not only on the one-port conductance but also, on the one-port susceptance. The same is with the one-port susceptance.

Illustration 3.32 *Let us calculate the conductance and susceptance of the one-port shown in Fig. 3.49, for the fundamental harmonic and the third order harmonic, assuming that $R = 1\ \Omega$ and $\omega_1 L = 1\Omega$.*

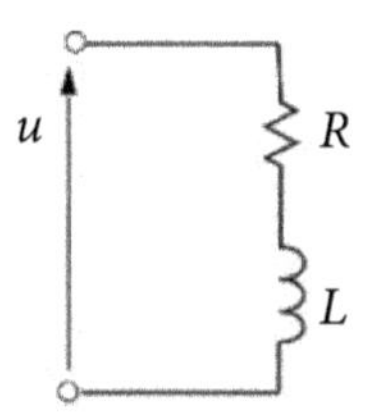

Figure 3.49 An RL one-port.

The one-port admittance for the fundamental harmonic is

$$\boldsymbol{Y}_1 = \frac{1}{R + j\omega_1 L} = \frac{1}{1 + j1} = 0.5 - j0.5\ \text{S}$$

thus the conductance and susceptance for fundamental frequency are equal to

$$G_1 = \text{Re}\{\boldsymbol{Y}_1\} = 0.5\ \text{S}, \qquad B_1 = \text{Im}\{\boldsymbol{Y}_1\} = -0.5\ \text{S}.$$

The admittance for the third order harmonic is

$$Y_3 = \frac{1}{R + j3\omega_1 L} = \frac{1}{1 + j3} = 0.1 - j0.3\ \mathrm{S}$$

thus, the conductance and susceptance for the third order harmonic are equal to

$$G_3 = \mathrm{Re}\{Y_3\} = 0.1\ \mathrm{S}$$
$$B_3 = \mathrm{Im}\{Y_3\} = -0.3\ \mathrm{S}.$$

Observe that despite the constant branch resistance R, the conductance G_n has different values for different frequencies.

*****

Illustration 3.33 *Let us calculate the resistance and reactance of the one-port in Fig. 3.50, for the fundamental and the third order harmonics, for $R = 0.5\ \Omega$ and $\omega_1 L = 0.5\ \Omega$.*

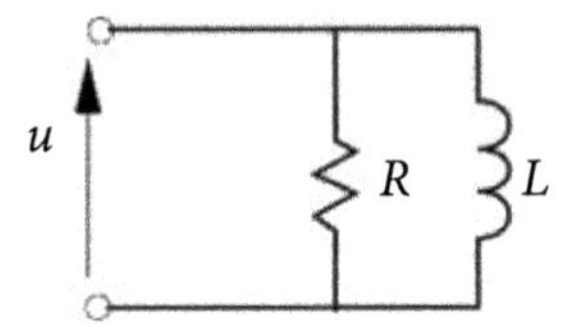

Figure 3.50 An RL one-port.

The one-port impedance for the fundamental harmonic is

$$Z_1 = \frac{1}{G + \dfrac{1}{j\omega_1 L}} = \frac{1}{2 - j2} = 0.25 + j0.25\Omega$$

thus

$$R_1 = \mathrm{Re}\{Z_1\} = 0.25\Omega$$
$$X_1 = \mathrm{Im}\{Z_1\} = 0.25\Omega.$$

The impedance for the third order harmonic is

$$Z_3 = \frac{1}{G + \dfrac{1}{j3\omega_1 L}} = \frac{1}{2 - j\dfrac{2}{3}} = 0.45 + j0.15$$

thus

$$R_3 = \mathrm{Re}\{Z_3\} = 0.45\Omega,\ X_3 = \mathrm{Im}\{Z_3\} = 0.15\Omega.$$

****

3.8 Node and Mash Equations

Node Equations: A circuit branch that connects node j and node k of a circuit with K+1 nodes is shown in Fig. 3.51.

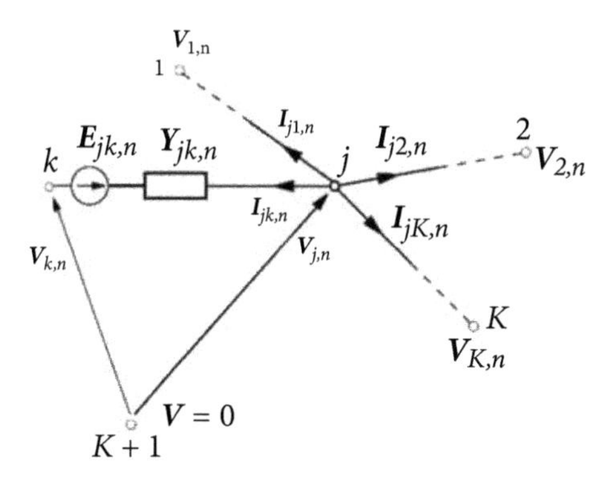

Figure 3.51 A node j of a circuit.

The crms value of the nth order harmonic, $\boldsymbol{I}_{jk,\,n}$, of the branch current can be expressed by the crms value of node j and node k voltage harmonics, the crms value of the source voltage source harmonic in this branch, and the branch admittance for the nth order harmonic, namely

$$\boldsymbol{I}_{jk,n} = \boldsymbol{Y}_{jk,n}(\boldsymbol{V}_{j,n} - \boldsymbol{E}_{jk,n} - \boldsymbol{V}_{k,n}).$$

If the circuit has K+1 nodes, then KCL for the crms values of current harmonics in node j has the form

$$\sum_{k=1}^{K} \boldsymbol{I}_{jk,n} = \sum_{k=1}^{K} \boldsymbol{Y}_{jk,n}(\boldsymbol{V}_{j,n} - \boldsymbol{E}_{jk,n} - \boldsymbol{V}_{k,n}) = 0.$$

This equation can be rearranged as follows

$$\boldsymbol{V}_{j,n} \sum_{k=1}^{K} \boldsymbol{Y}_{jk,n} - \sum_{k=1}^{K} \boldsymbol{Y}_{jk,n}\, \boldsymbol{V}_{k,n} = \sum_{k=1}^{K} \boldsymbol{Y}_{jk,n}\, \boldsymbol{E}_{jk,n} \tag{3.51}$$

and it is called *node equation* for node j, for the nth order harmonic. K+1 of such equations can be written for a circuit with K+1 nodes, however only K equations are independent. One of them is a linear form of the remaining ones. The number of K crms values of node voltage harmonics $\boldsymbol{V}_{j,\,n}$ are unknown variables. Therefore, one of K+1 voltages can have any value. Usually the voltage of the node K+1 is assumed to be zero, and this node disappears from the set of node equations.

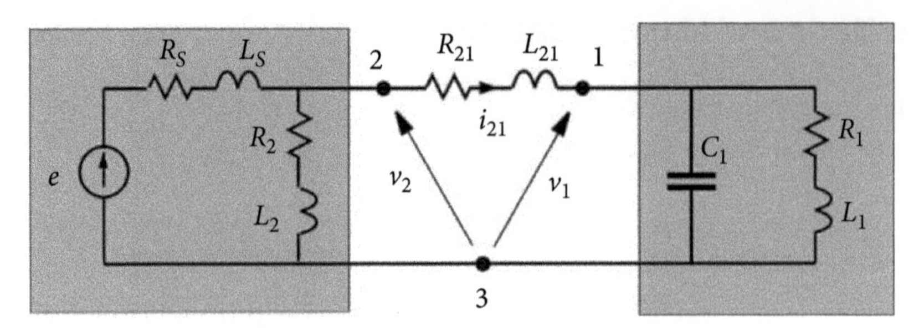

Figure 3.52 An example of a circuit.

Illustration 3.34 *Let us calculate current* i_{21} *in the circuit shown in Fig. 3.52, assuming that the supply voltage is*

$$e(t) = \sqrt{2}\,\mathrm{Re}\{220\, e^{j\omega_1 t} + 11\, e^{j5\omega_1 t}\}\mathrm{V}$$

and the circuit parameters are

$$R_1 = 2\Omega,\ \omega_1 L_1 = 2\Omega,\ \omega_1 C_1 = 0.25\ \mathrm{S},\ R_{21} = 0.1\Omega,\ \omega_1 L_{21} = 0.1\Omega$$

$$R_2 = 1\Omega,\ \omega_1 L_2 = 1\Omega,\ R_s = 0.05\Omega,\ \omega_1 L_s = 0.05\Omega.$$

The set of node equations can be arranged into a matrix form as follows:

$$\begin{bmatrix} \mathbf{Y}_{1,n} + \mathbf{Y}_{21,n}, & -\mathbf{Y}_{21,n} \\ -\mathbf{Y}_{21,n}, & \mathbf{Y}_{2,n} + \mathbf{Y}_{s,n} + \mathbf{Y}_{21,n} \end{bmatrix} \begin{bmatrix} \mathbf{V}_{1,n} \\ \mathbf{V}_{2,n} \end{bmatrix} = \begin{bmatrix} 0 \\ \mathbf{Y}_{s,n}\mathbf{E}_n \end{bmatrix}$$

with two unknowns, namely, crms values $\mathbf{V}_{1,n}$ *and* $\mathbf{V}_{2,n}$ *of node 1 and 2 voltage harmonics.*

Admittances in this equation for the fundamental frequency are

$$\mathbf{Y}_{1,1} = j\omega_1 \mathbf{C}_1 + \frac{1}{R_1 + j\omega_1 L_1} = j0.25 + \frac{1}{2 + j2} = 0.25\ \mathrm{S}$$

$$\mathbf{Y}_{21,1} = \frac{1}{R_{21} + j\omega_1 L_{21}} = \frac{1}{0.1 + j0.1} = 5 - j5 = 7.07\, e^{-j45^\circ}\ \mathrm{S}$$

$$\mathbf{Y}_{2,1} = \frac{1}{R_2 + j\omega_1 L_2} = \frac{1}{1 + j1} = 0.5 - j0.5 = 0.707\, e^{-j45^\circ}\ \mathrm{S}$$

$$\mathbf{Y}_{s,1} = \frac{1}{R_s + j\omega_1 L_s} = \frac{1}{0.05 + j0.05} = 10 - j10 = 14.14\, e^{-j45^\circ}\ \mathrm{S}.$$

The node admittances in this set of equations are equal to, respectively,

$$\mathbf{Y}_{1,1} + \mathbf{Y}_{21,1} = 0.25 + (5 - j5) = 7.25\, e^{-j43.6^\circ}\ \mathrm{S}$$

$$\mathbf{Y}_{2,1} + \mathbf{Y}_{s,1} + \mathbf{Y}_{21,1} = (0.5 - j0.5) + (10 - j10) + (5 - j5) = 21.82\, e^{-j43.6^\circ}\ \mathrm{S}$$

Thus the set of node equations for the fundamental harmonic is

$$\begin{bmatrix} 7.25\, e^{-j43.6^\circ}, & -7.07\, e^{-j45^\circ} \\ -7.07\, e^{-j45^\circ}, & 21.82\, e^{-j43.6^\circ} \end{bmatrix} \begin{bmatrix} V_{1,1} \\ V_{2,1} \end{bmatrix} = \begin{bmatrix} 0 \\ 14,14\, e^{-j45^\circ} \times 220 \end{bmatrix}.$$

It has the solution

$$\begin{bmatrix} V_{1,1} \\ V_{2,1} \end{bmatrix} = \frac{1}{\det\{Y_1\}} \begin{bmatrix} 21.82\, e^{-j45^\circ}, & 7.07\, e^{-j45^\circ} \\ 7.07\, e^{-j45^\circ}, & 7.25\, e^{-j43,6^\circ} \end{bmatrix} \begin{bmatrix} 0 \\ 3110.8\, e^{-j45^\circ} \end{bmatrix}$$

where

$$\det\{Y_1\} = 7.25\, e^{-j43.6^\circ} \times 21.82\, e^{-j45^\circ} - (7.07\, e^{-j45^\circ})^2 = 108.94\, e^{-j88.0^\circ}\ \mathrm{S}^2$$

hence

$$V_{1,1} = \frac{1}{\det\{Y_1\}} \times 7.07\, e^{-j45^\circ} \times 3110.8\, e^{-j45^\circ} = 201.94\, e^{-j2.0^\circ}\ \mathrm{V}$$

$$V_{2,1} = \frac{1}{\det\{Y_1\}} \times 7.25\, e^{-j43.6^\circ} \times 3110.8\, e^{-j45^\circ} = 207.05\, e^{-j0.6^\circ}\ \mathrm{V}$$

The crms value of the current i_{21} fundamental harmonic is equal to

$$I_{21,1} = Y_{21,1}(V_{2,1} - V_{1,1}) = 7.07\, e^{-j45^\circ} \times (207.05\, e^{-j0.6^\circ} - 201.94\, e^{-j2.0^\circ}) = 50.48\, e^{-j2.0^\circ}\ \mathrm{A}.$$

Admittances of the node equations for the fifth order harmonics are equal to

$$Y_{1,5} = j5\omega_1 C_1 + \frac{1}{R_1 + j5\omega_1 L_1} = j1.25 + \frac{1}{2 + j10} = 0.019 + j1.154\ \mathrm{S}$$

$$Y_{21,5} = \frac{1}{R_{21} + j5\omega_1 L_{21}} = \frac{1}{0.1 + j0.5} = 0.38 - j1.92 = 1.96\, e^{-j78.7}\ \mathrm{S}$$

$$Y_{2,5} = \frac{1}{R_2 + j5\omega_1 L_2} = \frac{1}{1 + j5} = 0.038 - j0.192\ \mathrm{S}$$

$$Y_{s,5} = \frac{1}{R_s + j5\omega_1 L_s} = \frac{1}{0.05 + j0.25} = 0.77 - j3.85 = 3.92\, e^{-j78.7^\circ}\ \mathrm{S}.$$

The node admittances equal to

$$Y_{1,5} + Y_{21,5} = (0.019 + j1.154) + (0.38 - j1.92) = 0.864\, e^{-j62.5^\circ}\ \mathrm{S}$$

$$Y_{2,5} + Y_{s,5} + Y_{21,5} = (0.038 - j0.192) + (0.77 - j3.85) + (0.38 - j1.92) = 6.08\, e^{-j78.7^\circ}\ \mathrm{S}$$

The set of node equations for the fifth order harmonic has the form

$$\begin{bmatrix} 0.864\, e^{-j62.5^\circ}, & -1.96\, e^{-j78.7^\circ} \\ -1.96\, e^{-j78,7^\circ}, & 6.08\, e^{-j78.7^\circ} \end{bmatrix} \begin{bmatrix} V_{1,5} \\ V_{2,5} \end{bmatrix} = \begin{bmatrix} 0 \\ 3.92\, e^{-j78.7^\circ} \times 11 \end{bmatrix}.$$

It has the solution

$$\begin{bmatrix} V_{1,5} \\ V_{2,5} \end{bmatrix} = \frac{1}{\det\{\boldsymbol{Y}_5\}} \begin{bmatrix} 6.08\, e^{-j78.7^\circ}, & 1.96\, e^{-j78.7^\circ} \\ 1.96\, e^{-j78.7^\circ}, & 0.864\, e^{-j62.5^\circ} \end{bmatrix} \begin{bmatrix} 0 \\ 43.12\, e^{-j78.7^\circ} \end{bmatrix}.$$

where

$$\det\{\boldsymbol{Y}_5\} = 0.864\, e^{-j62.5^\circ} \times 6.08\, e^{-j78.7^\circ} - (1.96\, e^{-j78.7^\circ})^2 = 1.93\, e^{-j106.7^\circ}\ \mathrm{S}^2$$

hence,

$$V_{1,5} = \frac{1}{\det\{\boldsymbol{Y}_5\}} \times 1.96\, e^{-j78,7^\circ} \times 43.12\, e^{-j78.7^\circ} = 43.91\, e^{-j50.7^\circ}\ \mathrm{V}$$

$$V_{2,5} = \frac{1}{\det\{\boldsymbol{Y}_5\}} \times 0.864\, e^{-j62.5^\circ} \times 43.12\, e^{-j78.7^\circ} = 19.45\, e^{-j34.3^\circ}\ \mathrm{V}.$$

Thus, the crms value of the current i_{21} the fifth order harmonic is equal to

$$\boldsymbol{I}_{21,5} = \boldsymbol{Y}_{21,5}(\boldsymbol{V}_{2,5} - \boldsymbol{V}_{1,5}) = 1.96\, e^{-j78.7^\circ} \times (19.45\, e^{-j34.3^\circ} - 43.91\, e^{-j50.7^\circ}) = 50.68\, e^{j38.4^\circ}\mathrm{A}.$$

Current i_{21} has the waveform

$$i_{21} = \sqrt{2}\,\mathrm{Re}\{50.48\, e^{-j2.0^\circ} e^{j\omega_1 t} + 50.68\, e^{j38.4^\circ} e^{j5\omega_1 t}\}\ \mathrm{A}.$$

*****

This illustration demonstrates, in an example of analysis of a circuit with three nodes, how node equations are used for circuit analysis at nonsinusoidal voltage. It was the analysis of simple a three-node circuit, with the voltage composed of only two harmonics, nonetheless, that analysis was toilsome. Therefore, such a method provides fundamentals for computer-based analysis rather than for "by hand" calculations.

Even if the circuit analyzed was relatively simple, this was an equivalent circuit of an RL load with a capacitive compensator. The load was supplied from a distorted voltage source with RL internal impedance. A shunt branch, which can be regarded as an equivalent magnetizing branch of a transformer, and RL parameters of the supply line were taken into account in that circuit. Thus, the main components of a distribution system and compensated load were taken into account.

Looking into the results of that analysis, it is worth observing, that despite the supply voltage fifth order harmonic being at the level of only 5%, this harmonics in the supply current is comparable with the fundamental. It is because of a voltage resonance located in the vicinity of the 5th order harmonic.

Mesh Equations: Let us assume that the circuit is flat, meaning it can be drawn on a plane, as shown in Fig. 3.53. Its branches divide the whole plane into $M + 1$ areas referred to as *meshes*. They can be numbered as $1, 2, \ldots k, \ldots M + 1$. Let the last, $M + 1$ number, be allocated to the mesh which encloses the entire circuit, meaning an infinity point of the plane is in the last mesh.

Voltages along meshes have to satisfy KVL, thus there are $M+1$ voltage equations. One of these equations is a linear form of the remaining one; thus, it is a dependent equation. Usually, we assume, that the voltage equation of the last, M + 1 mesh, is the dependent equation.

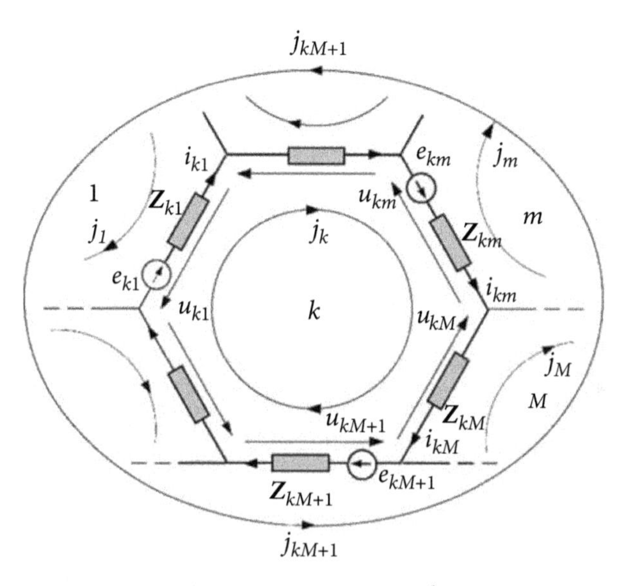

Figure 3.53 A circuit mesh.

A hypothetical current, j_k, can be allocated to each mesh. Such current is referred to as a *mesh current*. There are M + 1 mesh currents and one of them is dependent, meaning it can be assumed as known. Usually we assume that the mesh current of the last mesh, j_{K+1}, = 0. Currents of branches common for two meshes are differences in mesh currents of these meshes.

The crms value of the nth order voltage harmonic on a branch that is common for mesh k and mesh m can be expressed as

$$\boldsymbol{U}_{km,n} = \boldsymbol{Z}_{km,n}\, \boldsymbol{I}_{km,n} - \boldsymbol{E}_{km,n} = \boldsymbol{Z}_{km,n}(\boldsymbol{J}_{k,n} - \boldsymbol{J}_{m,n}) - \boldsymbol{E}_{km,n}$$

thus from the KVL for crms values for mesh k results

$$\sum_{m=1}^{M+1} \boldsymbol{U}_{km,n} = \sum_{m=1}^{M+1} [\boldsymbol{Z}_{km,n}(\boldsymbol{J}_{k,n} - \boldsymbol{J}_{m,n}) - \boldsymbol{E}_{km,n}] \equiv 0.$$

It can be rearranged as follows

$$\boldsymbol{J}_{k,n} \sum_{m=1}^{M+1} \boldsymbol{Z}_{km,n} - \sum_{m=1}^{M+1} \boldsymbol{Z}_{km,n}\, \boldsymbol{J}_{m,n} \equiv \sum_{m=1}^{M+1} \boldsymbol{E}_{km,n}. \tag{3.52}$$

In this equation

$$\sum_{m=1}^{M+1} \boldsymbol{Z}_{km,n} \stackrel{\text{df}}{=} \boldsymbol{Z}_{k,n}$$

is the impedance of mesh k for the nth order harmonic, and

$$\sum_{m=1}^{M+1} \boldsymbol{E}_{km,n} \stackrel{\text{df}}{=} \boldsymbol{E}_{k,n}$$

is the crms value of the voltage source of the nth order harmonic in mesh k. Assuming that $\boldsymbol{J}_{M+1,n} = 0$, the equation for the mesh k can be written in the form

$$\boldsymbol{Z}_{k,n}\, \boldsymbol{J}_{k,n} - \sum_{m=1}^{M} \boldsymbol{Z}_{km,n}\, \boldsymbol{J}_{m,n} \equiv \boldsymbol{E}_{k,n}. \tag{3.53}$$

Such an equation is referred to as a *mesh equation* for the nth order harmonic. Thus, the circuit is described for each harmonic order with M mesh equations with M unknown crms values $\boldsymbol{J}_{k,\,n}$ of mesh current harmonics.

Illustration 3.35 *Let's calculate the same current calculated in the previous illustration using node equations, but now using mesh equations. The supply voltage and the circuit parameters are the same as in Illustration 3.32.*

Let the mesh currents be selected and numbered as shown in Fig. 3.54.

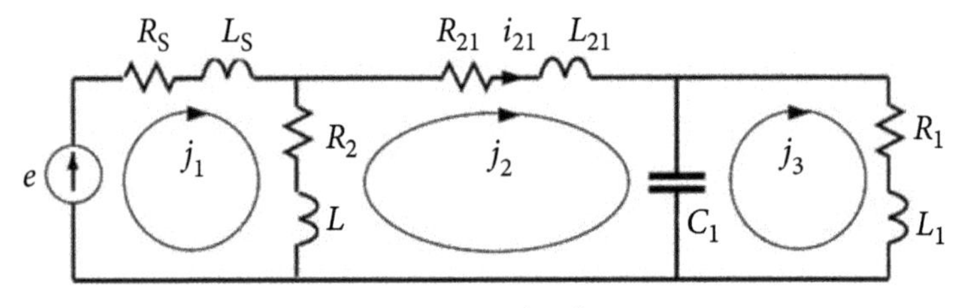

Figure 3.54 Example of a circuit.

The set of mesh equations can be arranged in a matrix form for the fundamental harmonic is

$$\begin{bmatrix} \boldsymbol{Z}_{1,1} & -\boldsymbol{Z}_{12,1} & 0 \\ -\boldsymbol{Z}_{12,1} & \boldsymbol{Z}_{2,1} & -\boldsymbol{Z}_{23,1} \\ 0 & -\boldsymbol{Z}_{23,1} & \boldsymbol{Z}_{3,1} \end{bmatrix} \begin{bmatrix} \boldsymbol{J}_{1,1} \\ \boldsymbol{J}_{2,1} \\ \boldsymbol{J}_{3,1} \end{bmatrix} = \begin{bmatrix} \mathrm{E}_1 \\ 0 \\ 0 \end{bmatrix}.$$

Impedances in this equation are

$$\mathbf{Z}_{1,1} = (R_s + j\omega_1 L_s) + (R_2 + j\omega_1 L_2) = (0.05 + j0.05) + (1 + j1) = 1.05 + j1.05\ \Omega$$

$$\mathbf{Z}_{12,1} = (R_2 + j\omega_1 L_2) = 1 + j1\ \Omega$$

$$\mathbf{Z}_{2,1} = (R_2 + j\omega_1 L_2) + (R_{21} + j\omega_1 L_{21}) + \frac{1}{j\omega_1 C_1} = (1 + j1) + (0.1 + j0.1) +$$

$$+ \frac{1}{j0.25} = 1.1 - j2.9\ \Omega$$

$$\mathbf{Z}_{23,1} = \frac{1}{j\omega_1 C_1} = \frac{1}{j0.25} = -j4.0\ \Omega$$

$$\mathbf{Z}_{3,1} = \frac{1}{j\omega_1 C_1}(R_1 + j\omega_1 L_1) = \frac{1}{j0.25} + (2 + j2) = 2.0 - j2.0\ \Omega.$$

Since in the considered circuit $\boldsymbol{I}_{21,1} = \boldsymbol{J}_{2,1}$ *it is enough to solve mesh equations only for the mesh current* $\boldsymbol{J}_{2,1}$. *The mesh equations for the crms values have complex coefficients, which makes calculations more toilsome. Computer methods should be used for it. The MATLAB program results in*

$$\boldsymbol{J}_{1,2} = \boldsymbol{I}_{1,21} = 50.48\, e^{-j2.0^\circ}\ \text{A}.$$

Mash equations for the fifth order harmonic in the matrix form are

$$\begin{bmatrix} \mathbf{Z}_{1,5} & -\mathbf{Z}_{12,5} & 0 \\ -\mathbf{Z}_{12,5} & \mathbf{Z}_{2,5} & -\mathbf{Z}_{23,5} \\ 0 & -\mathbf{Z}_{23,5} & \mathbf{Z}_{3,5} \end{bmatrix} \begin{bmatrix} \boldsymbol{J}_{1,5} \\ \boldsymbol{J}_{2,5} \\ \boldsymbol{J}_{3,5} \end{bmatrix} = \begin{bmatrix} \boldsymbol{E}_5 \\ 0 \\ 0 \end{bmatrix}.$$

Impedances in this equation are equal to

$$\mathbf{Z}_{1,5} = (R_s + j5\omega_1 L_s) + (R_2 + j5\omega_1 L_2) = (0.05 + j5 \times 0.05) +$$

$$+ (1 + j5 \times 1) = 1.05 + j5.25\ \Omega$$

$$\mathbf{Z}_{12,5} = (R_2 + j5\omega_1 L_2) = 1 + j5\ \Omega$$

$$\mathbf{Z}_{2,5} = (R_2 + j5\omega_1 L_2) + (R_{21} + j5\omega_1 L_{21}) + \frac{1}{j5\omega_1 C_1} = 1.1 + j4.7\ \Omega$$

$$\mathbf{Z}_{23,5} = \frac{1}{j5\omega_1 C_1} = -j0.8\ \Omega$$

$$\mathbf{Z}_{3,5} = \frac{1}{j5\omega_1 C_1} + (R_1 + j5\omega_1 L_1) = 2.0 + j9.2\ \Omega.$$

The solution of these equations for $\boldsymbol{J}_{2,5}$ *results in*

$$\boldsymbol{I}_{21,5} = \boldsymbol{J}_{2,5} = 50.68\, e^{j38.4^\circ}\ \text{A}.$$

It is, of course, the same crms value obtained using node equations.

*****

A set of three equations had to be solved, however. The same circuit was described in terms of three mesh equations, instead of only two nodes equations. It does not mean that the node analysis results in a lower number of equations. It depends entirely on the circuit structure.

3.9 Three-Phase, Three-Wire Circuits

Electric energy in large amounts is delivered mainly by three-phase, three-wire (3p3w) circuits, used for energy transmission and distribution in situations where only three-phase loads, such as three-phase motors, high power rectifiers, AC/DC converters, or arc furnaces are supplied. Four-wire circuits are used when also single-phase loads have to be energized from three-phase sources which is a common situation, for example, in commercial buildings.

Three-phase, four-wire circuits, meaning with a neutral conductor, can be regarded, when loads of individual lines are not mutually coupled, as three single-phase circuits, with only a secondary importance interaction through the impedance of the neutral conductor. Their power properties can be interpreted essentially in terms of the properties of single-phase circuits.

The 3p3w circuit cannot be regarded as three single-phase, however. Even if individual line-to-line loads are not coupled mutually, meaning these are single-phase loads, such loads are coupled by the impedance of the supply source and supply lines. Any change in power of one line-to-line load affects the voltage and power of other loads. Only at an ideal supply, meaning with zero impedance, the line-to-line loads do not affect each other.

Transformers configured in Δ/Y are used, as shown in Fig. 3.55, to connect three wires and four wires.

Three-phase terminals and quantities are denoted according to various conventions, in particular, as (a, b, c), (L1, L2, L3), (R, S, T), or (R, B, Y) after "red," blue," and "yellow." Since index "a" is used in this book for "active" current, and "b" is used for a "fictitious" current, the first convention could be misleading. The second convention, (L1, L2, L3), results in very long indices, inconvenient in mathematical formulae, in particular, when also harmonic order, n, has to be specified. Instead, the convention (R, S, T) is used in this book, also because this convention is used in a great majority of the author's publications.

Description of power properties of 3p3w circuits with nonsinusoidal but periodic voltages and currents in the frequency domain is mathematically complex since, on the structural complexity of three-phase circuits, harmonic description has to be superimposed. Therefore, a set of compact and clear symbols and effective mathematical operations are needed for the description of the power properties of 3p3w circuits. Such a set of symbols and operations for systems with nonsinusoidal voltages and currents were developed in Ref. [93], where the concept of *three-phase vectors*, meaning vectors that have three-phase quantities as the vector elements; *three-phase*

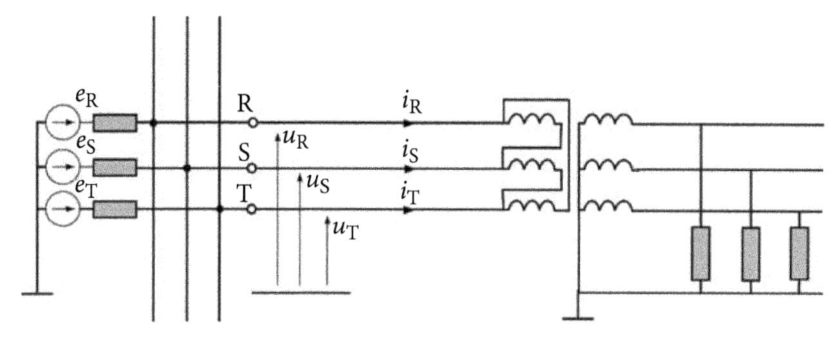

Figure 3.55 A junction of a three-wire, and a four-wire circuit.

rms value, and *three-phase scalar product* for nonsinusoidal asymmetrical quantities were introduced.

The main property that differentiates three-wire circuits from four-wire ones is the sum of line currents. This sum for three-wires

$$i_R(t) + i_S(t) + i_T(t) \equiv 0$$

while such a sum for four-wire circuits can have any value. It means that line currents in three-wire circuits cannot have any common component. It cannot flow in such circuits. Such a common component $u^0(t)$ can be present in line-to-ground voltages, denoted in Fig. 3.56 with upper index "comma," however. Thus

$$u'_R(t) + u'_S(t) + u'_T(t) = 3\,u^{\circ}(t).$$

This supply voltage common component, $u^0(t)$, cannot cause, however, any current flow in three-wire circuits. Its presence is entirely irrelevant to power phenomena in the load.

This common component occurs between the ground of the circuit and the common point of three identical resistors, connected as shown in Fig. 3.56, referred to as an *artificial zero*. The line voltages measured versus the artificial zero do not contain the common component $u^0(t)$, and such voltages hold the relation

$$u_R(t) + u_S(t) + u_T(t) \equiv 0.$$

The presence of the component, $u^0(t)$, in the supply voltage does not affect power phenomena in the circuit but it affects the voltage and its rms value. Therefore, as long as three-wire s are discussed, it will be assumed that the supply voltage does not contain any common component or it is referenced to the artificial zero.

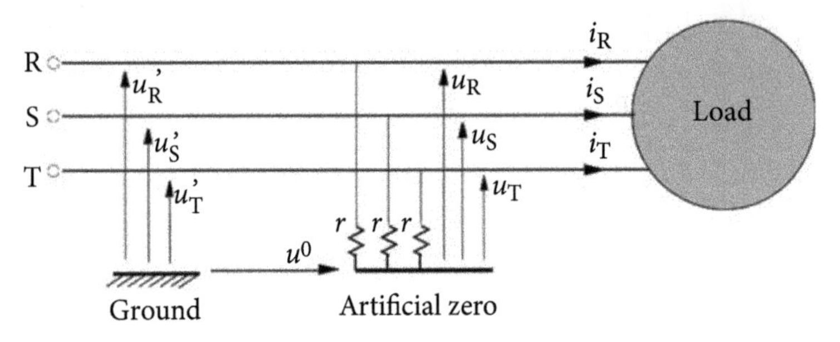

Figure 3.56 A ground and artificial zero.

3.10 Three-Phase Vectors and Their Rms Value

Let us introduce this concept first for 3p3w s with sinusoidal voltages and currents. It will be next extended for s with nonsinusoidal quantities.

Sinusoidal line-to-artificial zero voltages u_R, u_S, u_T, and line currents i_R, i_S, i_T, denoted generally as x_R, x_S, x_T, in three-wire circuits can be arranged into the following vector

$$\boldsymbol{x} \stackrel{\text{df}}{=} \boldsymbol{x}(t) \stackrel{\text{df}}{=} \begin{bmatrix} x_R(t) \\ x_S(t) \\ x_T(t) \end{bmatrix} \stackrel{\text{df}}{=} \begin{bmatrix} x_R \\ x_S \\ x_T \end{bmatrix} = \sqrt{2}\,\text{Re} \begin{bmatrix} X_R \\ X_S \\ X_T \end{bmatrix} e^{j\omega_1 t} = \sqrt{2}\,\text{Re}\,\boldsymbol{X}\,e^{j\omega_1 t} \tag{3.54}$$

where X_R, X_S, and X_T are crms values of quantities x_R, x_S, and x_T, while

$$\boldsymbol{X} \stackrel{\text{df}}{=} \begin{bmatrix} X_R \\ X_S \\ X_T \end{bmatrix}$$

is a three-phase vector of crms values of these quantities.

Observe that the association of vectors $\boldsymbol{x}(t)$ and $\boldsymbol{X}$ in 3p3w circuits is like the association of sinusoidal quantity $x(t)$ and its crms value X in single-phase, namely

$$x = x(t) = \sqrt{2}\,\text{Re}\,X\,e^{j\omega_1 t}.$$

Three-phase scalar product of three-phase vectors $\boldsymbol{x}$ and $\boldsymbol{y}$ is defined as

$$(\boldsymbol{x}, \boldsymbol{y}) \stackrel{\text{df}}{=} \frac{1}{T}\int_0^T \boldsymbol{x}^{\text{T}}(t)\,\boldsymbol{y}(t)\,dt = \frac{1}{T}\int_0^T [x_R(t)y_R(t) + x_S(t)y_S(t) + x_T(t)y_T(t)]dt \tag{3.55}$$

Observe, that the three-phase scalar product can be expressed as

$$(\boldsymbol{x}, \boldsymbol{y}) = (x_R, y_R) + (x_S, y_S) + (x_T, y_T) \tag{3.56}$$

meaning it can be calculated as the sum of the scalar products of individual phase quantities.

When $\boldsymbol{u}$ and $\boldsymbol{i}$ are vectors of voltages and currents of a three-phase load, then their scalar product

$$(\boldsymbol{u}, \boldsymbol{i}) = \frac{1}{T}\int_0^T \boldsymbol{u}^{\mathrm{T}}\boldsymbol{i}\, dt = \frac{1}{T}\int_0^T (u_{\mathrm{R}}\, i_{\mathrm{R}} + u_{\mathrm{S}}\, i_{\mathrm{S}} + u_{\mathrm{T}}\, i_{\mathrm{T}})\, dt = P \tag{3.57}$$

is the load active power. This relation has a close analogy to the scalar product of the voltage and current in single-phases,

$$(u, i) = \frac{1}{T}\int_0^T u\, i\, dt = P.$$

The *three-phase rms value* of vector $\boldsymbol{x}(t)$ is defined as

$$||\boldsymbol{x}|| = \sqrt{(\boldsymbol{x}, \boldsymbol{x})} = \sqrt{\frac{1}{T}\int_0^T \boldsymbol{x}^{\mathrm{T}}\boldsymbol{x}\, dt} = \sqrt{\frac{1}{T}\int_0^T (x_{\mathrm{R}}^2 + x_{\mathrm{S}}^2 + x_{\mathrm{T}}^2)\, dt} = \sqrt{||x_{\mathrm{R}}||^2 + ||x_{\mathrm{S}}||^2 + ||x_{\mathrm{T}}||^2}. \tag{3.55}$$

Thus the three-phase rms value can be calculated as the root of the sum of squares of the rms values of phase quantities.

Both functionals, the three-phase scalar product and the three-phase rms value, originally defined in the time domain, meaning by integrals of time quantities, can be also expressed in terms of three-phase vectors of crms values. Namely, the scalar product of three-phase sinusoidal quantities $\boldsymbol{x}(t)$ and $\boldsymbol{y}(t)$ is

$$(\boldsymbol{x}, \boldsymbol{y}) = (x_{\mathrm{R}}, y_{\mathrm{R}}) + (x_{\mathrm{S}}, y_{\mathrm{S}}) + (x_{\mathrm{T}}, y_{\mathrm{T}}) = \mathrm{Re}\{X_{\mathrm{R}}Y_{\mathrm{R}}^* + X_{\mathrm{S}}Y_{\mathrm{S}}^* + X_{\mathrm{T}}Y_{\mathrm{T}}^*\} = \mathrm{Re}\{\boldsymbol{X}^{\mathrm{T}}\boldsymbol{Y}^*\}. \tag{3.56}$$

Similarly, the three-phase rms value

$$||\boldsymbol{x}|| = \sqrt{(\boldsymbol{x}, \boldsymbol{x})} = \sqrt{\mathrm{Re}\{\boldsymbol{X}^{\mathrm{T}}\boldsymbol{Y}^*\}} = \sqrt{X_{\mathrm{R}}^2 + X_{\mathrm{S}}^2 + X_{\mathrm{T}}^2}. \tag{3.57}$$

The three-phase rms value, as defined above, is only a mathematical concept. One can ask the question: has it any physical or technical merit or not?

This question was answered in Ref. [93] as follows. Let us consider a three-phase purely resistive device, shown in Fig. 3.57, with three currents i_{R}, i_{S}, i_{T} of the same period T.

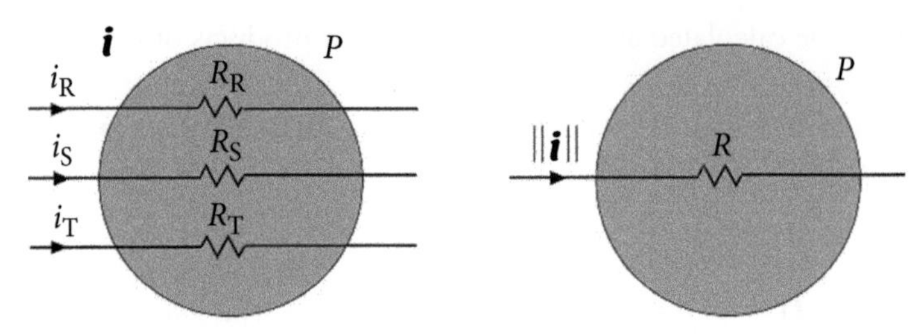

Figure 3.57 A physical interpretation of three-phase current rms value.

The heat released in this device, meaning energy dissipated by these currents in this device in one period T, is equal to

$$W = \int_0^T (R_R\, i_R^2 + R_S\, i_S^2 + R_T\, i_T^2)\, dt.$$

Three-phase devices are built in such a way that their symmetry is as high as possible, meaning that

$$R_R = R_S = R_T \overset{\text{df}}{=} R$$

thus the active power of the device, meaning the average value of energy dissipation, is

$$P = \frac{W}{T} = R\frac{1}{T}\int_0^T (i_R^2 + i_S^2 + i_T^2)\, dt = R\,||\boldsymbol{i}||^2.$$

Therefore, the three-phase rms value $||\boldsymbol{i}||$ of the current vector, $\boldsymbol{i}$, is equivalent to a DC current that causes the same dissipation of energy on the resistance R of a single-phase device as the current vector $\boldsymbol{i}$, meaning three currents i_R, i_S, i_T on resistances R of symmetrical three-phase device.

Orthogonality of Three-Phase Vectors: The rms value of a sum of three-phase vectors $\boldsymbol{x}(t)$ and $\boldsymbol{y}(t)$ is equal to

$$||\boldsymbol{x} + \boldsymbol{y}|| = \sqrt{\frac{1}{T}\int_0^T [\boldsymbol{x}(t) + \boldsymbol{y}(t)]^{\mathrm{T}}[\boldsymbol{x}(t) + \boldsymbol{y}(t)]\, dt} = \sqrt{||\boldsymbol{x}||^2 + 2\,(\boldsymbol{x}, \boldsymbol{y}) + ||\boldsymbol{y}||^2} \tag{3.58}$$

It can be calculated as a root of the sum of squares of the rms values of individual vectors, meaning

$$||\boldsymbol{x} + \boldsymbol{y}|| = \sqrt{||\boldsymbol{x}||^2 + ||\boldsymbol{y}||^2} \tag{3.59}$$

only if the scalar product of these vectors is equal to zero, that is,

$$(\boldsymbol{x}, \boldsymbol{y}) = 0.$$

Such vectors are referred to as orthogonal three-phase vectors.

According to relation (3.56), the scalar product of three-phase vectors is equal to the sum of scalar products of phase quantities. Situations where single-phase quantities are orthogonal were discussed at the beginning of this chapter. When one of these situations occurs then the three-phase quantities are also orthogonal. However, the orthogonality of phase quantities is not the necessary condition for three-phase vectors orthogonality. Two three-phase quantities can be orthogonal even if the phase quantities which form these vectors are not orthogonal.

3.11 Three-Phase Equivalent Load in Δ Configuration

Having the crms values of the voltage and current harmonics at the load terminals, meaning at terminals of the load shown in Fig. 3.58(a), the following question can be asked: is it possible to calculate line-to-line admittances load in Δ configuration, that is, of the load shown in Fig. 3.58(b), which would be equivalent to the original load with regard to the supply currents?

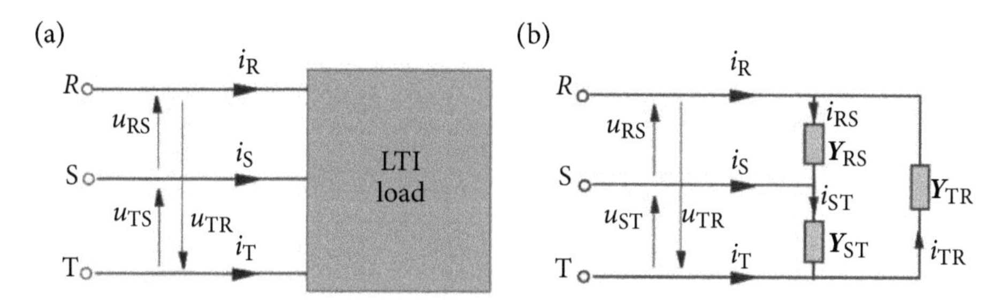

Figure 3.58 A three-phase LTI load and its equivalent Δ structure.

If P and Q are the active and reactive powers measured at the load terminals, then the complex power of the load in Fig. 3.58(b) is

$$C = P + jQ = \boldsymbol{U}_{\mathrm{RS}}\,\boldsymbol{I}_{\mathrm{RS}}^{*} + \boldsymbol{U}_{\mathrm{ST}}\,\boldsymbol{I}_{\mathrm{ST}}^{*} + \boldsymbol{U}_{\mathrm{TR}}\,\boldsymbol{I}_{\mathrm{TR}}^{*} = \boldsymbol{Y}_{\mathrm{RS}}^{*}\,U_{\mathrm{RS}}^{2} + \boldsymbol{Y}_{\mathrm{ST}}^{*}\,U_{\mathrm{ST}}^{2} + \boldsymbol{Y}_{\mathrm{TR}}^{*}\,U_{\mathrm{TR}}^{2}$$

while the crms values of the line-to-line currents in the equivalent load have to satisfy node equations:

$$\boldsymbol{I}_{\mathrm{R}} = \boldsymbol{Y}_{\mathrm{RS}}\,\boldsymbol{U}_{\mathrm{RS}} - \boldsymbol{Y}_{\mathrm{TR}}\,\boldsymbol{U}_{\mathrm{TR}}$$

$$\boldsymbol{I}_{\mathrm{S}} = \boldsymbol{Y}_{\mathrm{ST}}\,\boldsymbol{U}_{\mathrm{ST}} - \boldsymbol{Y}_{\mathrm{RS}}\,\boldsymbol{U}_{\mathrm{RS}}.$$

These three equations can be written in the matrix form

$$\begin{bmatrix} U_{RS}^2 & U_{ST}^2 & U_{TR}^2 \\ U_{RS} & 0 & -U_{TR} \\ -U_{RS} & U_{ST} & 0 \end{bmatrix} \begin{bmatrix} Y_{RS} \\ Y_{ST} \\ Y_{TR} \end{bmatrix} = \begin{bmatrix} C^* \\ I_R \\ I_S \end{bmatrix} \tag{3.60}$$

with three unknown line-to-line admittances Y_{RS}, Y_{ST}, and Y_{TR}. Observe that the conjugate complex power of the load $C^* = P - jQ$ is a known quantity. It can be measured at the load terminals. Let us denote the square matrix on the left side by $\boldsymbol{U}$. Equation (3.60) has a unique solution on the condition that the determinant of matrix $\boldsymbol{U}$ is not equal to zero. Let us calculate this determinant

$$\begin{aligned} \text{Det}\{\boldsymbol{U}\} &= U_{ST}^2\, U_{TR}\, U_{RS} + U_{TR}^2\, U_{RS}\, U_{ST} + U_{RS}^2\, U_{TR} = \\ &= U_{RS}\, U_{ST}\, U_{TR}\, (U_{ST}^* + U_{TR}^* + U_{RS}^*) = 0 \end{aligned}$$

which means that (3.60) does not a have unique solution. It has an infinite number of solutions. One of three line-to-line admittances, Y_{RS}, Y_{ST}, or Y_{TR} can have any value. Thus, there is an infinite number of loads of Δ structure equivalent to the original load with regard to currents at the load terminals and consequently, with regard to the load active and reactive powers P and Q.

Illustration 3.36 *The load shown in Fig. 3.59, at the supply voltage of the crms values $U_{RS} = 100$V, $U_{ST} = j100$V, and $U_{TR} = (-100 - j100)$ V has the active power $P = 40$kW. The crms values of the line currents are:*

$$I_R = (200 + j100)\ \text{A}, \quad I_S = (-100 + j100)\ \text{A}, \quad I_T = (-100 - j200)\ \text{A}.$$

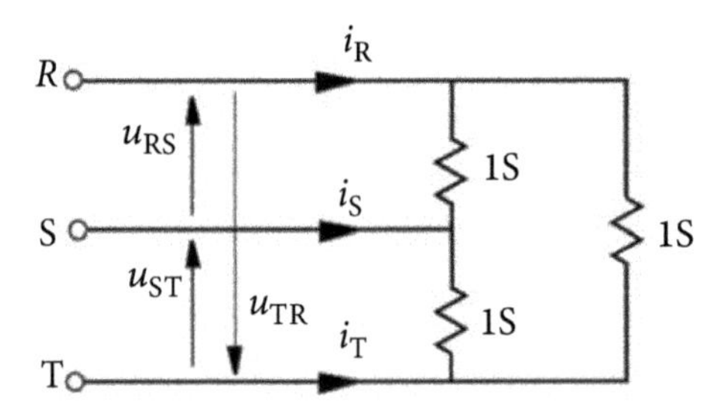

Figure 3.59 An example of a load

We can find a load equivalent to that in Fig. 3.59, assuming any value for one of the three line-to-line admittances. Let's assume that $Y_{RS} = 0$, then

$$Y_{ST} = (1.0 + j1.0)\ \text{S}, \quad Y_{TR} = (1.5 - j0.5)\ \text{S}$$

that is, equivalent load, as shown in Fig. 3.60.

This load at the same supply voltages has identical line currents as that in Fig. 3.59 and the same active and reactive powers, $P = 40$kW, $Q = 0$.

*****

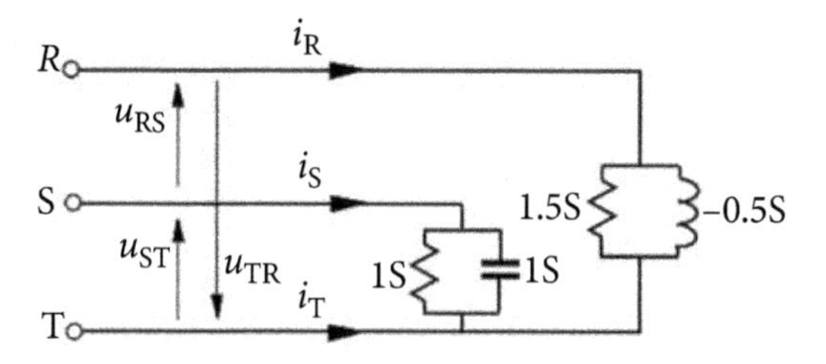

Figure 3.60 A load equivalent to that in Fig. 3.59.

3.12 Three-Phase Reduced Vectors

Since sums of line currents in three-wire s are equal to zero, one of them can be expressed by the remaining two. Thus, only two currents are needed to specify a three-phase current vector. Such two currents of line R and line S, arranged into a two-element vector denoted by a bold, not filled font, as follows

$$\mathring{i} \overset{\text{df}}{=} \begin{bmatrix} i_R \\ i_S \end{bmatrix}$$

will be referred to in this book as a current *three-phase reduced vector*. The whole three-phase vector $\boldsymbol{i}$, if needed, can be reconstructed from the reduced vector as follows

$$\boldsymbol{i} \overset{\text{df}}{=} \begin{bmatrix} i_R \\ i_S \\ i_T \end{bmatrix} = \begin{bmatrix} 1 \\ -[1,\ 1] \end{bmatrix} \begin{bmatrix} i_R \\ i_S \end{bmatrix} \overset{\text{df}}{=} \begin{bmatrix} 1 \\ -1^T \end{bmatrix} \mathring{i}, \quad \text{where} \quad 1 \overset{\text{df}}{=} \begin{bmatrix} 1 \\ 1 \end{bmatrix}. \tag{3.61}$$

Similarly, a voltage three-phase reduced vector can be defined, namely

$$\mathring{u} \overset{\text{df}}{=} \begin{bmatrix} u_R \\ u_S \end{bmatrix}.$$

When a circuit is described in terms of reduced vectors, $\mathring{u}$ and $\mathring{i}$, only two crms values of the voltage and currents are needed to specify these vectors, namely

$$\mathring{u} \overset{\text{df}}{=} \begin{bmatrix} u_R \\ u_S \end{bmatrix} = \sqrt{2}\,\text{Re}\begin{bmatrix} U_R \\ U_S \end{bmatrix} e^{j\omega_1 t} = \sqrt{2}\,\text{Re}\{\mathring{U} e^{j\omega_1 t}\}, \quad \mathring{U} \overset{\text{df}}{=} \begin{bmatrix} U_R \\ U_S \end{bmatrix}.$$

Similarly,

$$\mathring{i} \overset{\text{df}}{=} \begin{bmatrix} i_R \\ i_S \end{bmatrix} = \sqrt{2}\,\text{Re}\begin{bmatrix} I_R \\ I_S \end{bmatrix} e^{j\omega_1 t} = \sqrt{2}\,\text{Re}\{\mathring{I} e^{j\omega_1 t}\}, \quad \mathring{I} \overset{\text{df}}{=} \begin{bmatrix} I_R \\ I_S \end{bmatrix}.$$

Use of such reduced vectors can benefit in a reduction of calculation amount at three-wire s description and the analysis since such s at a distinctive cross-section are specified by only four, not by six, quantities, namely, only by two line-to-ground voltages and two lines currents.

3.13 Symmetrical Components

When a three-phase voltage is applied to a motor, it can rotate in a clockwise direction or a counterclockwise direction. It depends on the sequence with which the voltage of the same value is observed at the motor terminals, R, S, and T. It is referred to as the voltage *sequence*. When labels R, S, and T are allocated to a motor's terminals, then when some phase of the voltage, say maximum, occurs at these terminals in such a sequence that after maximum on the R terminal, the maximum is observed at terminal S and finally at terminal T, as shown in Fig. 3.61a, the motor rotates in a clockwise direction, and the voltage is referred to as a voltage of *positive sequence.*

If, after the maximum value is observed on the terminal R, the maximum is observed at terminal T and finally at terminal S, as shown in Fig. 3.61b, the motor rotates in a counter-clockwise direction, and the voltage is referred to as a voltage of the *negative sequence.*

The same concept applies, of course, also to three-phase currents, in general, to x_R, x_S, and x_T quantities. They are of the positive sequence if

$$x_S\,(t) \equiv x_R(t - T/3), \quad x_T\,(t) \equiv x_R(t + T/3)$$

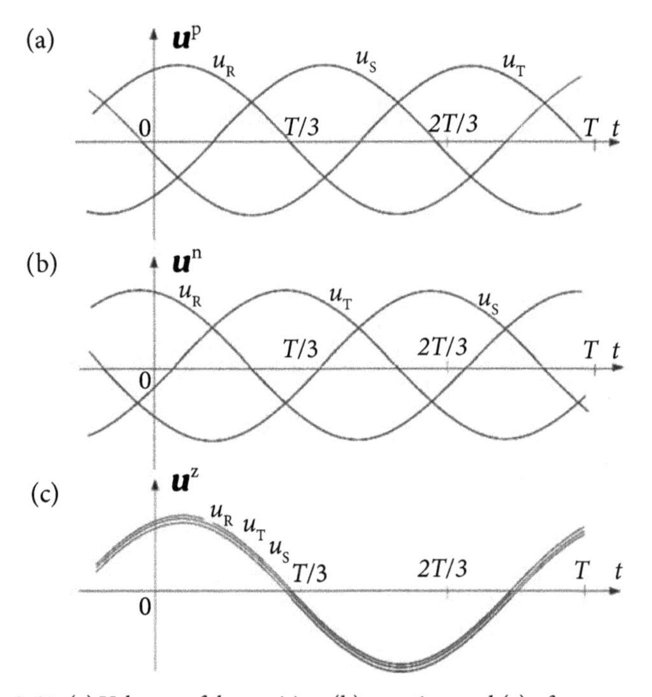

Figure 3.61 (a) Voltages of the positive, (b) negative, and (c) of zero sequences.

and of negative sequence if

$$x_S(t) \equiv x_R(t + T/3), \quad x_T(t) \equiv x_R(t - T/3).$$

Voltages and currents in three-wire circuits with a neutral conductor can also be in-phase. If an induction motor is supplied with such in-phase voltages it will not, of course, rotate. In-phase currents in the motor windings do not create a rotating magnetic field. Such in-phase, but symmetrical quantities

$$x_R(t) \equiv x_S(t) \equiv x_T(t)$$

are referred to as *zero-sequence* quantities.

When these quantities are sinusoidal and have the crms values $\boldsymbol{X}_R$, $\boldsymbol{X}_S$, and $\boldsymbol{X}_T$, then these values for both sequences can be expressed in terms of the crms value at terminal R, $\boldsymbol{X}_R$, and complex rotation coefficients

$$\alpha \stackrel{\text{df}}{=} 1\, e^{j120°}, \quad \alpha^2 = 1\, e^{-j120°} = \alpha^*.$$

The line R is regarded usually as a reference line and the line index R is omitted, namely

$$\boldsymbol{X}_R^p \stackrel{\text{df}}{=} \boldsymbol{X}^p$$

while the crms values of the positive sequence quantity for the remaining lines S and T are

$$\boldsymbol{X}_S^p = \alpha^* \boldsymbol{X}^p$$
$$\boldsymbol{X}_T^p = \alpha \boldsymbol{X}^p.$$

The diagram of these crms values is shown in Fig. 3.62.

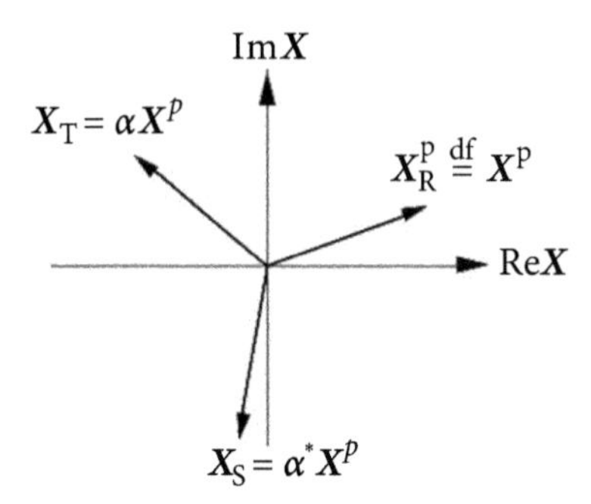

Figure 3.62 A diagram of the crms values of a quantityof positive sequence.

Similarly, for the negative sequence, the line index R is omitted, namely

$$X_R^n \stackrel{df}{=} X^n$$

while the crms values of the negative sequence quantity for the remaining lines S and T are

$$X_S^n = \alpha X^n$$

$$X_T^n = \alpha^* X^n.$$

These relations are illustrated in the diagram shown in Fig. 3.63.

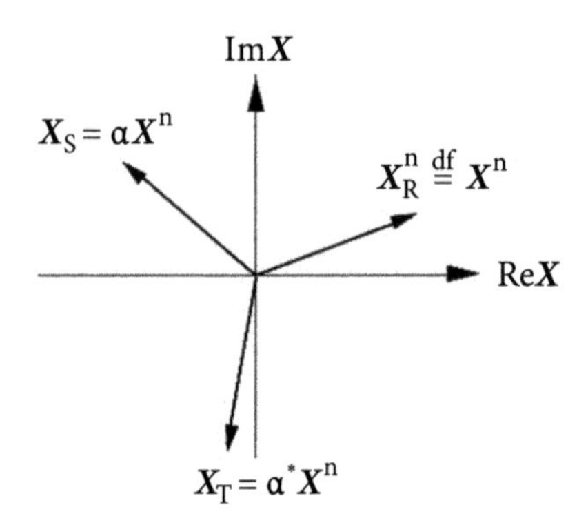

Figure 3.63 A diagram of the crms values of a quantity of negative sequence.

Asymmetrical but sinusoidal voltages and currents in three-wire circuits with a neutral conductor can be expressed, according to Ref. [8], as a sum of vectors of symmetrical quantities of the positive, negative, and zero-sequence components, namely

$$\boldsymbol{x} = \boldsymbol{x}^p + \boldsymbol{x}^n + \boldsymbol{x}^z$$

where

$$\boldsymbol{x}^p \stackrel{df}{=} \begin{bmatrix} x_R^p \\ x_S^p \\ x_T^p \end{bmatrix} = \sqrt{2}\,\mathrm{Re} \begin{bmatrix} X_R^p \\ X_S^p \\ X_S^p \end{bmatrix} e^{j\omega_1 t} = \sqrt{2}\,\mathrm{Re} \begin{bmatrix} 1 \\ \alpha^* \\ \alpha \end{bmatrix} X^p e^{j\omega_1 t} = \sqrt{2}\,\mathrm{Re}\,\boldsymbol{X}^p e^{j\omega_1 t} \quad (3.62)$$

$$\boldsymbol{x}^n \stackrel{df}{=} \begin{bmatrix} x_R^n \\ x_S^n \\ x_T^n \end{bmatrix} = \sqrt{2}\,\mathrm{Re} \begin{bmatrix} X_R^n \\ X_S^n \\ X_S^n \end{bmatrix} e^{j\omega_1 t} = \sqrt{2}\,\mathrm{Re} \begin{bmatrix} 1 \\ \alpha \\ \alpha^* \end{bmatrix} X^n e^{j\omega_1 t} = \sqrt{2}\,\mathrm{Re}\,\boldsymbol{X}^n e^{j\omega_1 t}. \quad (3.63)$$

$$\boldsymbol{x}^z \stackrel{df}{=} \begin{bmatrix} x_R^z \\ x_S^z \\ x_T^z \end{bmatrix} = \sqrt{2}\,\mathrm{Re} \begin{bmatrix} X_R^z \\ X_S^z \\ X_S^z \end{bmatrix} e^{j\omega_1 t} = \sqrt{2}\,\mathrm{Re} \begin{bmatrix} 1 \\ 1 \\ 1 \end{bmatrix} X^z e^{j\omega_1 t} = \sqrt{2}\,\mathrm{Re}\,\boldsymbol{X}^z e^{j\omega_1 t}. \quad (3.64)$$

For three-wire circuits, this sum is reduced to only the positive and negative sequence components, that is,

$$\boldsymbol{x} = \boldsymbol{x}^{\mathrm{p}} + \boldsymbol{x}^{\mathrm{n}}$$

Let us define the unit three-phase symmetrical vectors of the positive, negative, and zero sequence

$$\mathbf{1}^{\mathrm{p}} \stackrel{\mathrm{df}}{=} \begin{bmatrix} 1 \\ \alpha^* \\ \alpha \end{bmatrix} = \begin{bmatrix} 1 \\ 1e^{-j2\pi/3} \\ 1e^{j2\pi/3} \end{bmatrix}, \quad \mathbf{1}^{\mathrm{n}} \stackrel{\mathrm{df}}{=} \begin{bmatrix} 1 \\ \alpha \\ \alpha^* \end{bmatrix} = \begin{bmatrix} 1 \\ 1e^{j2\pi/3} \\ 1e^{-j2\pi/3} \end{bmatrix}, \quad \mathbf{1}^{\mathrm{z}} \stackrel{\mathrm{df}}{=} \begin{bmatrix} 1 \\ 1 \\ 1 \end{bmatrix}. \tag{3.65}$$

Unit three-phase vectors of the positive and negative sequence are shown in Fig. 3.64.

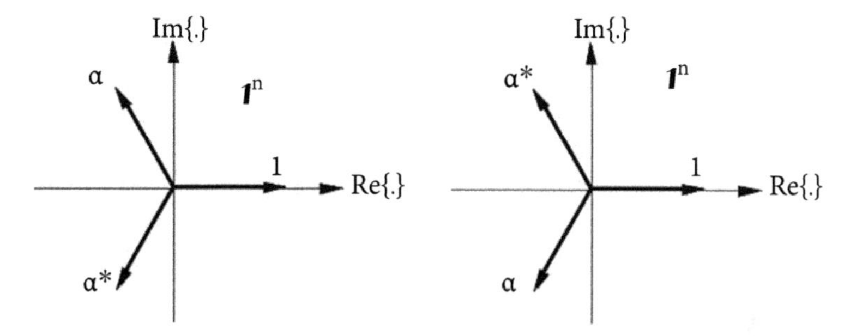

Figure 3.64 Unit three-phase sequence vectors $\mathbf{1}^{\mathrm{p}}$ and $\mathbf{1}^{\mathrm{n}}$.

With these unit vectors

$$\boldsymbol{x}^{\mathrm{p}} = \sqrt{2}\,\mathrm{Re}\,\boldsymbol{X}^{\mathrm{p}}\, e^{j\omega_1 t} = \sqrt{2}\,\mathrm{Re}\,\mathbf{1}^{\mathrm{p}} X^{\mathrm{p}}\, e^{j\omega_1 t} \tag{3.66}$$

$$\boldsymbol{x}^{\mathrm{n}} = \sqrt{2}\,\mathrm{Re}\,\boldsymbol{X}^{\mathrm{n}}\, e^{j\omega_1 t} = \sqrt{2}\,\mathrm{Re}\,\mathbf{1}^{\mathrm{n}} X^{\mathrm{n}}\, e^{j\omega_1 t} \tag{3.67}$$

$$\boldsymbol{x}^{\mathrm{z}} = \sqrt{2}\,\mathrm{Re}\,\boldsymbol{X}^{\mathrm{z}}\, e^{j\omega_1 t} = \sqrt{2}\,\mathrm{Re}\,\mathbf{1}^{\mathrm{z}} X^{\mathrm{z}}\, e^{j\omega_1 t} \tag{3.68}$$

Thus, the vector of crms values $\boldsymbol{X}$ of sinusoidal asymmetrical quantities can be expressed as

$$\begin{bmatrix} X_{\mathrm{R}} \\ X_{\mathrm{S}} \\ X_{\mathrm{T}} \end{bmatrix} = \mathbf{1}^{\mathrm{p}} X^{\mathrm{p}} + \mathbf{1}^{\mathrm{n}} X^{\mathrm{n}} + \mathbf{1}^{\mathrm{z}} X^{\mathrm{z}}.$$

This equation can be solved for X^{p}, X^{n}, and X^{z}, resulting in

$$\begin{bmatrix} X^{\mathrm{p}} \\ X^{\mathrm{n}} \\ X^{\mathrm{z}} \end{bmatrix} = \frac{1}{3} \begin{bmatrix} 1, & \alpha, & \alpha^* \\ 1, & \alpha^*, & \alpha \\ 1, & 1, & 1 \end{bmatrix} \begin{bmatrix} X_{\mathrm{R}} \\ X_{\mathrm{S}} \\ X_{\mathrm{T}} \end{bmatrix}. \tag{3.69}$$

A three-phase sinusoidal asymmetrical quantity in three-wire s can be decomposed with this formula into symmetrical components of the positive and negative sequence. Since

$$\boldsymbol{X}_{\mathrm{T}} = -\boldsymbol{X}_{\mathrm{R}} - \boldsymbol{X}_{\mathrm{S}}$$

this formula can be simplified using the three-phase reduced vector

$$\mathcal{X} \stackrel{\mathrm{df}}{=} \begin{bmatrix} \boldsymbol{X}_{\mathrm{R}} \\ \boldsymbol{X}_{\mathrm{S}} \end{bmatrix}$$

to the following form, namely

$$\mathcal{X}^{\mathrm{s}} \stackrel{\mathrm{df}}{=} \begin{bmatrix} \boldsymbol{X}^{\mathrm{p}} \\ \boldsymbol{X}^{\mathrm{n}} \end{bmatrix} = \frac{1}{3}\begin{bmatrix} 1, & \alpha, & \alpha^* \\ 1, & \alpha^*, & \alpha \end{bmatrix}\begin{bmatrix} \boldsymbol{X}_{\mathrm{R}} \\ \boldsymbol{X}_{\mathrm{S}} \\ -\boldsymbol{X}_{\mathrm{R}} - \boldsymbol{X}_{\mathrm{S}} \end{bmatrix} = \frac{1}{j\sqrt{3}}\begin{bmatrix} \alpha, & -1 \\ -\alpha^*, & 1 \end{bmatrix}\begin{bmatrix} \boldsymbol{X}_{\mathrm{R}} \\ \boldsymbol{X}_{\mathrm{S}} \end{bmatrix} \stackrel{\mathrm{df}}{=} \mathbb{S}\mathcal{X}$$

or

$$\mathcal{X}^{\mathrm{s}} = \mathbb{S}\mathcal{X}. \tag{3.70}$$

Illustration 3.37 *Let line T of a three-wire circuit be open, while the crms value of the current of line* R *is* $\boldsymbol{I}_{\mathrm{R}} = I_{\mathrm{R}} = 100$ A. *Thus,* $\boldsymbol{I}_{\mathrm{S}} = -\boldsymbol{I}_{\mathrm{R}} = 100\, e^{j\pi}$ A. *The symmetrical components of the line currents are*

$$\mathcal{I}^{\mathrm{s}} \stackrel{\mathrm{df}}{=} \begin{bmatrix} \boldsymbol{I}^{\mathrm{p}} \\ \boldsymbol{I}^{\mathrm{n}} \end{bmatrix} = \mathbb{S}\mathcal{I} = \frac{1}{j\sqrt{3}}\begin{bmatrix} \alpha, & -1 \\ -\alpha^*, & 1 \end{bmatrix}\begin{bmatrix} 100 \\ -100 \end{bmatrix} = \begin{bmatrix} 57.7\, e^{-j30^\circ} \\ 57.7\, e^{j30^\circ} \end{bmatrix} \mathrm{A}.$$

Verification:

$$\boldsymbol{I}^{\mathrm{p}} + \boldsymbol{I}^{\mathrm{n}} = \boldsymbol{I}_{\mathrm{R}}$$

and indeed

$$\boldsymbol{I}^{\mathrm{p}} + \boldsymbol{I}^{\mathrm{n}} = 57.7\, e^{-j30^\circ} + 57.7\, e^{j30^\circ} = 100\,\mathrm{A} = \boldsymbol{I}_{\mathrm{R}}.$$

*****

The crms values of line quantities in a three-wire circuit can be restored from the crms values of their symmetrical components with the equation

$$\mathcal{X} = \mathbb{S}^{-1}\mathcal{X}^{\mathrm{s}}$$

where

$$\mathbb{S}^{-1} = \begin{bmatrix} 1, & 1 \\ \alpha^*, & \alpha \end{bmatrix}. \tag{3.71}$$

3.14 Orthogonality of Symmetrical Components

One of the main properties of symmetrical components, crucial for power properties of three-phase circuits, operated at asymmetrical supply voltage, is their orthogonality.

Let us consider positive and negative sequence quantities $\boldsymbol{x}$ and $\boldsymbol{y}$, with the crms values drawn in diagrams in Fig. 3.65.

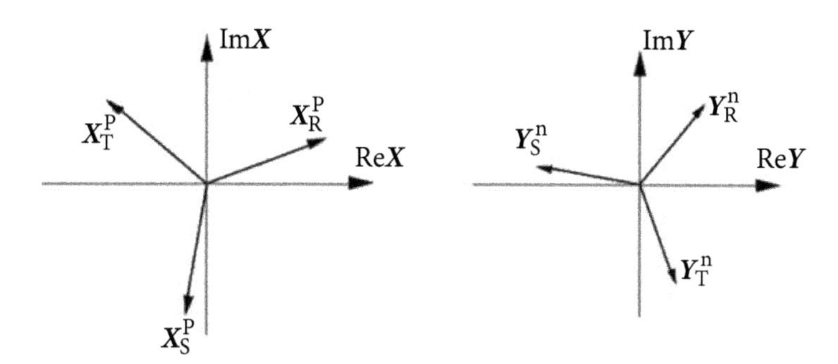

Figure 3.65 A diagram of the crms values of positive $\boldsymbol{x}$ and negative sequence $\boldsymbol{y}$ quantities

The scalar product of symmetrical quantities of the opposite sequence, $\boldsymbol{x}^p$ and $\boldsymbol{y}^n$, is equal to zero, since

$$(\boldsymbol{x}^p, \boldsymbol{y}^n) = \mathrm{Re}\{\mathbf{X}^{pT}\mathbf{Y}^{n*}\} = \mathrm{Re}\{\mathbf{1}^{pT}X^p\,(\mathbf{1}^n Y^n)^*\}$$

$$= \mathrm{Re}\{[1,\ \alpha^*, \alpha]\begin{bmatrix} 1 \\ \alpha^* \\ \alpha \end{bmatrix} X^p Y^{n*}\} = \mathrm{Re}\{(1+\alpha+\alpha^*)\, X^p Y^{n*}\} = 0.$$

The scalar product of the zero and positive sequence symmetrical quantities $\boldsymbol{x}^z$ and $\boldsymbol{y}^p$ is equal to zero, since

$$(\boldsymbol{x}^z, \boldsymbol{y}^p) = \mathrm{Re}\{\mathbf{X}^{zT}\mathbf{Y}^{p*}\} = \mathrm{Re}\{\mathbf{1}^{zT}X^z\,(\mathbf{1}^p Y^p)^*\}$$

$$= \mathrm{Re}\{[1,\ 1, 1]\begin{bmatrix} 1 \\ \alpha \\ \alpha^* \end{bmatrix} X^z Y^{p*}\} = \mathrm{Re}\{(1+\alpha+\alpha^*)\, X^z Y^{p*}\} = 0.$$

The scalar product of the zero and negative sequence symmetrical quantities $\boldsymbol{x}^z$ and $\boldsymbol{y}^n$ is equal to zero, since

$$(\boldsymbol{x}^z, \boldsymbol{y}^n) = \mathrm{Re}\{\mathbf{X}^{zT}\mathbf{X}^{n*}\} = \mathrm{Re}\{\mathbf{1}^{zT}X^z\,(\mathbf{1}^n Y^n)^*\}$$

$$= \mathrm{Re}\{[1,\ 1, 1]\begin{bmatrix} 1 \\ \alpha^* \\ \alpha \end{bmatrix} X^z Y^{n*}\} = \mathrm{Re}\{(1+\alpha^*+\alpha)\, X^z Y^{n*}\} = 0.$$

Thus, symmetrical components of an asymmetrical quantity are mutually orthogonal and consequently, the three-phase rms values of these quantities satisfy the relationship

$$||\boldsymbol{x}^{\mathrm{p}} + \boldsymbol{x}^{\mathrm{n}} + \boldsymbol{x}^{\mathrm{z}}||^2 = ||\boldsymbol{x}^{\mathrm{p}}||^2 + ||\boldsymbol{x}^{\mathrm{n}}||^2 + ||\boldsymbol{x}^{\mathrm{z}}||^2. \tag{3.72}$$

This property has important implications. For example, when the supply voltage is sinusoidal and symmetrical, thus, $\boldsymbol{u} = \boldsymbol{u}^{\mathrm{p}}$, while the load current of LTI load is asymmetrical, thus $\boldsymbol{i} = \boldsymbol{i}^{\mathrm{p}} + \boldsymbol{i}^{\mathrm{n}}$, then, the load active power is

$$P = (\boldsymbol{u}, \boldsymbol{i}) = (\boldsymbol{u}^{\mathrm{p}}, \boldsymbol{i}^{\mathrm{p}} + \boldsymbol{i}^{\mathrm{n}}) = (\boldsymbol{u}^{\mathrm{p}}, \boldsymbol{i}^{\mathrm{p}}) \tag{3.73}$$

thus, only the positive sequence component of the current contributes to the active power.

Since any three-phase vector of a sinusoidal asymmetrical quantity can be expressed as

$$\boldsymbol{x} = \boldsymbol{x}^{\mathrm{p}} + \boldsymbol{x}^{\mathrm{n}},$$

therefore, three-phase rms values of these components satisfy the relation

$$||\boldsymbol{x}||^2 = ||\boldsymbol{x}^{\mathrm{p}}||^2 + ||\boldsymbol{x}^{\mathrm{n}}||^2$$

where

$$||\boldsymbol{x}^{\mathrm{p}}|| = \sqrt{3}\,X^{\mathrm{p}}, \qquad ||\boldsymbol{x}^{\mathrm{n}}|| = \sqrt{3}\,X^{\mathrm{n}}. \tag{3.74}$$

When this is applied to a load with asymmetrical current, its three-phase rms value is

$$||\boldsymbol{i}|| = \sqrt{||\boldsymbol{i}^{\mathrm{p}}||^2 + ||\boldsymbol{i}^{\mathrm{n}}||^2} = \sqrt{3}\sqrt{I^{\mathrm{p}2} + I^{\mathrm{n}2}}. \tag{3.75}$$

Thus, the negative sequence component of the load current contributes to the rms value increase, but not to the load active power.

Illustration 3.38 *Let us consider the circuit shown in Fig. 3.66, with symmetrical sinusoidal supply voltage with*

$$u_{\mathrm{R}} = 120\sqrt{2}\cos\omega_1 t\,\mathrm{V}$$

and ideal transformer with 1:1 turn ratio. The crms values of the supply voltages are

$$U_{\mathrm{R}} = 120\,\mathrm{V}, \quad U_{\mathrm{S}} = 120e^{-j120^\circ}\mathrm{V}, \quad U_{\mathrm{T}} = 120e^{j120^\circ}\,\mathrm{V}.$$

Let us assume that and the load line-to-ground resistance on the transformer secondary side are $R = 2\,\Omega$.

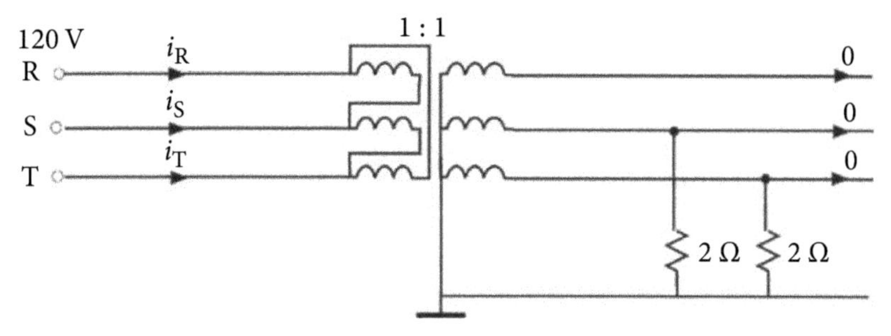

Figure 3.66 An example of a three-phase circuit.

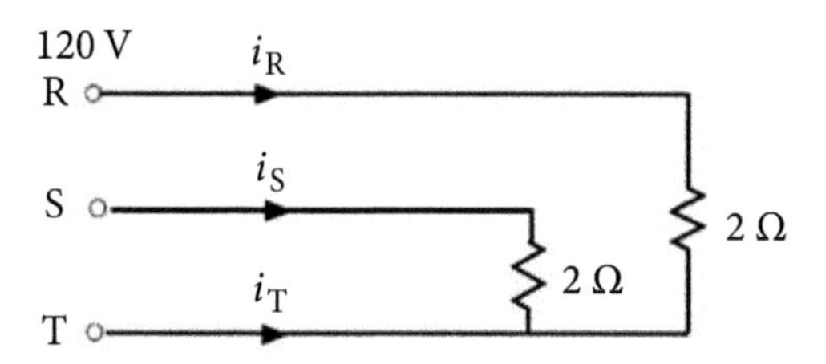

Figure 3.67 An equivalent circuit to the circuit shown in Fig. 3.66.

Such a circuit is equivalent to the circuit without a transformer shown in Fig. 3.67. Hence,

$$\boldsymbol{I}_R = \frac{\boldsymbol{U}_R - \boldsymbol{U}_T}{R} = 103.9 e^{-j30^\circ} \text{A}$$

$$\boldsymbol{I}_S = \frac{\boldsymbol{U}_S - \boldsymbol{U}_T}{R} = 103.9 e^{-j90^\circ} \text{A}$$

$$\boldsymbol{I}_T = -(\boldsymbol{I}_R + \boldsymbol{I}_S) = 180.0\, e^{j120^\circ} \text{A}.$$

The three-phase rms value of this current is

$$||\boldsymbol{i}|| = \sqrt{I_R^2 + I_S^2 + I_T^2} = \sqrt{103.9^2 + 103.9^2 + 180.0^2} = 232.4\text{A}.$$

The symmetrical components of line currents are equal to

$$\begin{bmatrix} I^p \\ I^n \end{bmatrix} = \frac{1}{j\sqrt{3}} \begin{bmatrix} \alpha, & -1 \\ -\alpha^*, & 1 \end{bmatrix} = \begin{bmatrix} 103.9\, e^{-j30^\circ} \\ 103.9\, e^{-j90^\circ} \end{bmatrix} = \begin{bmatrix} 120 \\ 60\, e^{-j120^\circ} \end{bmatrix} \text{A}.$$

The three-phase rms value of the load current, calculated in terms of rms values of symmetrical components, is

$$||\boldsymbol{i}|| = \sqrt{||\boldsymbol{i}^p||^2 + ||\boldsymbol{i}^n||^2} = \sqrt{3}\sqrt{I^{p2} + I^{n2}} = \sqrt{3}\sqrt{120^2 + 60^2} = 232.4\text{A}.$$

*****

3.15 Asymmetry Propagation

Asymmetry occurs in some points of distribution systems due to structural asymmetry of power system equipment or due to load imbalance. This asymmetry can next propagate across the system. The following important question could be asked: *Is the circuit capable of reducing this asymmetry, or quite opposite, can this asymmetry be amplified, assuming that the circuit is built as symmetrical as possible?* To answer this question, let us consider a segment of a three-phase line, shown in Fig. 3.68, with asymmetrical but sinusoidal currents.

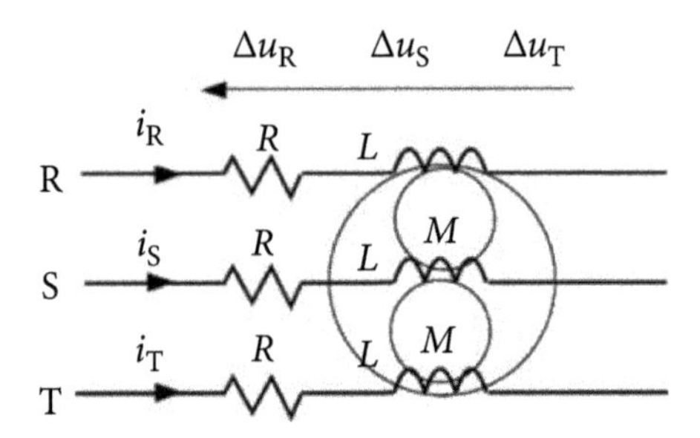

Figure 3.68 A segment of a three-phase line.

The crms values of the voltage drop at such a segment are

$$\begin{bmatrix} \Delta \boldsymbol{U}_R \\ \Delta \boldsymbol{U}_S \end{bmatrix} = \Delta U = \begin{bmatrix} R+j\omega_1 L & j\omega_1 M & j\omega_1 M \\ j\omega_1 M & R+j\omega_1 L & j\omega_1 M \end{bmatrix} \begin{bmatrix} \boldsymbol{I}_R \\ \boldsymbol{I}_S \\ \boldsymbol{I}_T \end{bmatrix} = \begin{bmatrix} \boldsymbol{Z} & 0 \\ 0 & \boldsymbol{Z} \end{bmatrix} \begin{bmatrix} \boldsymbol{I}_R \\ \boldsymbol{I}_S \end{bmatrix} = \mathbf{Z} I \tag{3.76}$$

where

$$\boldsymbol{Z} = R + j\omega_1(L - M).$$

The crms values of the line currents and the voltage drops can be expressed in terms of their symmetrical components, namely

$$S^{-1}\Delta U^s = \mathbf{Z}(S^{-1}I^s)$$

and consequently

$$\Delta U^s = S\mathbf{Z}S^{-1}I^s.$$

The product of matrices in this formula is

$$S\mathbf{Z}S^{-1} = \frac{1}{j\sqrt{3}} \begin{bmatrix} \alpha, & -1 \\ -\alpha^*, & 1 \end{bmatrix} \begin{bmatrix} \boldsymbol{Z} & 0 \\ 0 & \boldsymbol{Z} \end{bmatrix} \begin{bmatrix} 1, & 1 \\ \alpha^*, & \alpha \end{bmatrix} = \begin{bmatrix} \boldsymbol{Z} & 0 \\ 0 & \boldsymbol{Z} \end{bmatrix} = \mathbf{Z}$$

hence the equation of a segment of a symmetrical three-phase line, expressed in terms of the symmetrical components, has the form

$$\Delta U^{s} = \mathbf{Z} I^{s}. \tag{3.77}$$

Since the impedance matrix is diagonal, the symmetrical components of the line voltages and currents are mutually independent. The most important conclusion which can be drawn from this relation is the conclusion that

$$\frac{\Delta \boldsymbol{U}^{\mathrm{p}}}{\boldsymbol{I}^{\mathrm{p}}} = \frac{\Delta \boldsymbol{U}^{\mathrm{n}}}{\boldsymbol{I}^{\mathrm{n}}} = \boldsymbol{Z}$$

thus, asymmetry propagates through three-phase distribution lines without any change.

3.16 Nonsinusoidal Voltages and Currents in Three-Phase Circuits

Nonsinusoidal but periodic line-to-ground voltages and line currents, denoted in general by $x(t)$ or by x, that belong to space $\boldsymbol{L}_{\mathrm{T}}^{2}$, can be arranged into three-phase vectors $\mathbf{x}$, composed of three-phase vectors of harmonics, $\mathbf{x}_n$, namely

$$\mathbf{x} \stackrel{\mathrm{df}}{=} \begin{bmatrix} x_{\mathrm{R}} \\ x_{\mathrm{S}} \\ x_{\mathrm{T}} \end{bmatrix} = \sum_{n\in N} \mathbf{x}_n = \sqrt{2}\,\mathrm{Re} \sum_{n\in N} \begin{bmatrix} X_{\mathrm{R}\,n} \\ X_{\mathrm{S}\,n} \\ X_{\mathrm{T}\,n} \end{bmatrix} e^{jn\omega_1 t} \stackrel{\mathrm{df}}{=} \sqrt{2}\,\mathrm{Re} \sum_{n\in N} \mathbf{X}_n\, e^{jn\omega_1 t} \tag{3.78}$$

where

$$\mathbf{X}_n \stackrel{\mathrm{df}}{=} \begin{bmatrix} X_{\mathrm{R}\,n} \\ X_{\mathrm{S}\,n} \\ X_{\mathrm{T}\,n} \end{bmatrix}$$

is a three-phase vector of crms values of the nth order harmonic. Observe that a three-phase voltage $\mathbf{u}(t)$, measured versus artificial zero, cannot contain any DC component. A line current of a three-wire system can contain such a DC component, but this is rather an unusual situation. Its presence increases the complexity of various formulae which describe three-phase systems. Therefore, for a sake of reduction of this complexity, it is assumed here that three-phase quantities do not contain any DC component.

Scalar Product of Vectors of Nonsinusoidal Quantities: At the assumption that three-phase quantities do not have a DC component, and

$$\mathbf{x} = \sum_{n\in N} \mathbf{x}_n,\ \mathbf{y} = \sum_{n\in N} \mathbf{y}_n$$

their scalar product, defined in Ref. [93] can be expressed as

$$(\boldsymbol{x}, \boldsymbol{y}) \stackrel{\text{df}}{=} \frac{1}{T}\int_0^T \boldsymbol{x}^{\mathrm{T}}\boldsymbol{y}\, dt = \frac{1}{T}\int_0^T \sum_{r\in N} \boldsymbol{x}_r^{\mathrm{T}} \sum_{s\in N} \boldsymbol{y}_s\, dt = \sum_{r\in N}\sum_{s\in N} (\boldsymbol{x}_r, \boldsymbol{y}_s).$$

Harmonics of different order r, s, are orthogonal, meaning $(\boldsymbol{x}_r, \boldsymbol{y}_s) = 0$, thus

$$(\boldsymbol{x}, \boldsymbol{y}) = \sum_{n\in N} (\boldsymbol{x}_n, \boldsymbol{y}_n) \tag{3.79}$$

meaning, that the scalar product $(\boldsymbol{x}, \boldsymbol{y})$ can be calculated harmonic by harmonic.

When vectors of crms values of harmonics $\boldsymbol{X}_n$ and $\boldsymbol{Y}_n$ are known, then the scalar product of three-phase vectors of nonsinusoidal quantities can be calculated in the frequency domain:

$$(\boldsymbol{x}, \boldsymbol{y}) = \frac{1}{T}\int_0^T \boldsymbol{x}^{\mathrm{T}}\boldsymbol{y}\, dt = \sum_{n\in N} (\boldsymbol{x}_n, \boldsymbol{y}_n) = \mathrm{Re}\sum_{n\in N} \boldsymbol{X}_n^{\mathrm{T}}\boldsymbol{Y}_n^{*}. \tag{3.80}$$

In particular, if the vectors of nonsinusoidal voltages and currents at load terminals are

$$\boldsymbol{u} = \sum_{n\in N} \boldsymbol{u}_n, \quad \boldsymbol{i} = \sum_{n\in N} \boldsymbol{i}_n$$

then the scalar product of the load voltage and current vectors is equal to the load active power.

$$(\boldsymbol{u}, \boldsymbol{i}) = \frac{1}{T}\int_0^T \boldsymbol{u}^{\mathrm{T}}\boldsymbol{i}\, dt = \frac{1}{T}\int_0^T (u_{\mathrm{R}}\, i_{\mathrm{R}} + u_{\mathrm{S}}\, i_{\mathrm{S}} + u_{\mathrm{T}}\, i_{\mathrm{T}})\, dt = P. \tag{3.81}$$

which can be calculated algebraically, harmonic by harmonic, namely

$$P = (\boldsymbol{u}, \boldsymbol{i}) = \sum_{n\in N} (\boldsymbol{u}_n, \boldsymbol{i}_n) = \mathrm{Re}\sum_{n\in N} \boldsymbol{U}_n^{\mathrm{T}}, \boldsymbol{I}_n^{*}. \tag{3.82}$$

3.17 Orthogonality of Three-Phase Nonsinusoidal Quantities

Two three-phase vectors of nonsinusoidal quantities are orthogonal if

$$(\boldsymbol{x}, \boldsymbol{y}) = \frac{1}{T}\int_0^T \boldsymbol{x}^{\mathrm{T}}\boldsymbol{y}\, dt = \sum_{n\in N} (\boldsymbol{x}_n, \boldsymbol{y}_n) = \mathrm{Re}\sum_{n\in N} X_n^{\mathrm{T}} Y_n^{*} = 0.$$

Observe that it is sufficient for vectors $\boldsymbol{x}$ and $\boldsymbol{y}$ to be orthogonal to have mutually orthogonal all harmonic components $\boldsymbol{x}_n$ and $\boldsymbol{y}_n$. This is not the necessary condition, however, because scalar products $(\boldsymbol{x}_n, \boldsymbol{y}_n)$ do not have to be positive numbers. Positive and negative terms can cancel mutually to zero. Since

$$(\boldsymbol{x}_n, \boldsymbol{y}_n) = (x_{\mathrm{R}\,n}, y_{\mathrm{R}\,n}) + (x_{\mathrm{S}\,n}, y_{\mathrm{S}\,n}) + (x_{\mathrm{T}\,n}, y_{\mathrm{T}\,n}) \tag{3.83}$$

a similar conclusion applies to the orthogonality of individual harmonics of line quantities. For the orthogonality of vectors $\boldsymbol{x}_n$ and $\boldsymbol{y}_n$ it is sufficient that harmonics of line quantities in all three phases are orthogonal, but it is not necessary.

Observe, that vectors of each harmonic, $\boldsymbol{x}_n$ and $\boldsymbol{y}_n$, can be decomposed, similarly as vectors of sinusoidal voltages and currents, into symmetrical components of the positive, negative, and, in four-wire circuits, into zero sequence, namely

$$\boldsymbol{x}_n = \boldsymbol{x}_n^{\mathrm{p}} + \boldsymbol{x}_n^{\mathrm{n}} + \boldsymbol{x}_n^{\mathrm{z}}, \qquad \boldsymbol{y}_n = \boldsymbol{y}_n^{\mathrm{p}} + \boldsymbol{y}_n^{\mathrm{n}} + \boldsymbol{y}_n^{\mathrm{z}}.$$

Symmetrical components of different sequences are mutually orthogonal, thus

$$(\boldsymbol{x}_n, \boldsymbol{y}_n) = (\boldsymbol{x}_n^{\mathrm{p}}, \boldsymbol{y}_n^{\mathrm{p}}) + (\boldsymbol{x}_n^{\mathrm{n}}, \boldsymbol{y}_n^{\mathrm{n}}) + (\boldsymbol{x}_n^{\mathrm{z}}, \boldsymbol{y}_n^{\mathrm{z}}). \tag{3.84}$$

Rms Value of Three-Phase Vectors of Nonsinusoidal Quantities: The concept of the rms value of a three-phase vector $||\boldsymbol{x}||$, explained in Fig. 3.57, and formally defined with formula (3.74), is not confined to only sinusoidal quantities. It can be applied as well to three-phase vectors of nonsinusoidal quantities. In such a case

$$||\boldsymbol{x}|| = \sqrt{(\boldsymbol{x},\boldsymbol{x})} = \sqrt{\mathrm{Re}\sum_{n\in N} \boldsymbol{X}_n^{\mathrm{T}} \boldsymbol{X}_n^{*}} = \sqrt{\sum_{n\in N} (||x_{\mathrm{R}\,n}||^2 + ||x_{\mathrm{S}\,n}||^2 + ||x_{\mathrm{T}\,n}||^2)}. \tag{3.85}$$

The rms value of a three-phase vector is expressed with this formula in terms of rms values of line harmonics. This is a sort of "line-by-line" approach. In some situations, a "symmetrical component" approach might be more useful. The three-phase rms value $||\boldsymbol{x}||$, in a three-wire circuit, with decomposition (3.106), due to orthogonality of the positive and negative sequence components, can be expressed as

$$||\boldsymbol{x}|| = \sqrt{\sum_{n\in N} (\boldsymbol{x}_n, \boldsymbol{x}_n)} = \sqrt{\sum_{n\in N} ||x_n^{\mathrm{p}}||^2 + ||x_n^{\mathrm{n}}||^2}. \tag{3.86}$$

3.18 The Sequence of Harmonic Symmetrical Components

Let us assume that concerning the supply voltage

$$u_{\mathrm{S}}(t) = u_{\mathrm{R}}(t - T/3), \qquad u_{\mathrm{T}}(t) = u_{\mathrm{R}}(t + T/3).$$

When the shifting property is applied for the calculation of the crms value of the voltage harmonics of line S voltage, we obtain

$$\boldsymbol{U}_{\mathrm{S}\,n} = e^{-jn\omega_1 \frac{T}{3}}\, \boldsymbol{U}_{\mathrm{R}\,n} = e^{-jn120^{\circ}}\, \boldsymbol{U}_{\mathrm{R}\,n} = \begin{cases} \alpha^{*} \boldsymbol{U}_{\mathrm{R}\,n}, & n = 1, 4, 7 \ldots 3\,k + 1 \\ \boldsymbol{U}_{\mathrm{R}\,n}, & n = 3, 6, 9 \ldots 3\,k \\ \alpha\, \boldsymbol{U}_{\mathrm{R}\,n}, & n = 2, 5, 8 \ldots 3\,k - 1 \end{cases}$$

and for line T voltage

$$\boldsymbol{U}_{\mathrm{T}n} = e^{jn\omega_1 \frac{T}{3}}\, \boldsymbol{U}_{\mathrm{R}n} = e^{jn120^\circ}\, \boldsymbol{U}_{\mathrm{R}n} = \begin{cases} \alpha\, \boldsymbol{U}_{\mathrm{R}n}, & n = 1,\ 4, 7 ... 3\,k + 1 \\ \boldsymbol{U}_{\mathrm{R}n}, & n = 3,\ 6, 9 ... 3\,k \\ \alpha^{*} \boldsymbol{U}_{\mathrm{R}n}, & n = 2,\ 5, 8 ... 3\,k - 1 \end{cases} .$$

Consequently, the voltage harmonics of the order $n = 3k{+}1$ can be presented in the form

$$\boldsymbol{u}_n \stackrel{\mathrm{df}}{=} \begin{bmatrix} u_{\mathrm{R}n} \\ u_{\mathrm{S}n} \\ u_{\mathrm{T}n} \end{bmatrix} = \sqrt{2}\,\mathrm{Re} \begin{bmatrix} \boldsymbol{U}_{\mathrm{R}n} \\ \boldsymbol{U}_{\mathrm{S}n} \\ \boldsymbol{U}_{\mathrm{T}n} \end{bmatrix} e^{jn\omega_1 t} = \sqrt{2}\,\mathrm{Re}\{\mathbf{1}^{\mathrm{p}} \boldsymbol{U}_n^{\mathrm{p}}\, e^{jn\omega_1 t}\} \stackrel{\mathrm{df}}{=} \boldsymbol{u}_n^{\mathrm{p}} \tag{3.87}$$

where

$$\boldsymbol{U}_n^{\mathrm{p}} \stackrel{\mathrm{df}}{=} \boldsymbol{U}_{\mathrm{R}n}^{\mathrm{p}}$$

thus they have a positive sequence. Harmonics of the order $n = 3k{-}1$ can be presented in the form

$$\boldsymbol{u}_n \stackrel{\mathrm{df}}{=} \begin{bmatrix} u_{\mathrm{R}n} \\ u_{\mathrm{S}n} \\ u_{\mathrm{T}n} \end{bmatrix} = \sqrt{2}\,\mathrm{Re} \begin{bmatrix} \boldsymbol{U}_{\mathrm{R}n} \\ \boldsymbol{U}_{\mathrm{S}n} \\ \boldsymbol{U}_{\mathrm{T}n} \end{bmatrix} e^{jn\omega_1 t} = \sqrt{2}\,\mathrm{Re}\{\mathbf{1}^{\mathrm{n}} \boldsymbol{U}_n^{\mathrm{n}}\, e^{jn\omega_1 t}\} \stackrel{\mathrm{df}}{=} \boldsymbol{u}_n^{\mathrm{n}} \tag{3.88}$$

where

$$\boldsymbol{U}_n^{\mathrm{n}} \stackrel{\mathrm{df}}{=} \boldsymbol{U}_{\mathrm{R}n}^{\mathrm{p}}$$

thus they have the negative sequence, while harmonics of the order $n = 3k$

$$\boldsymbol{u}_n \stackrel{\mathrm{df}}{=} \begin{bmatrix} u_{\mathrm{R}n} \\ u_{\mathrm{S}n} \\ u_{\mathrm{T}n} \end{bmatrix} = \sqrt{2}\,\mathrm{Re} \begin{bmatrix} \boldsymbol{U}_{\mathrm{R}n} \\ \boldsymbol{U}_{\mathrm{S}n} \\ \boldsymbol{U}_{\mathrm{T}n} \end{bmatrix} e^{jn\omega_1 t} = \sqrt{2}\,\mathrm{Re}\{\mathbf{1}^{\mathrm{z}} \boldsymbol{U}_n^{\mathrm{z}}\, e^{jn\omega_1 t}\} \stackrel{\mathrm{df}}{=} \boldsymbol{u}_n^{\mathrm{z}} \tag{3.89}$$

are of the zero sequence, meaning they are in phase, where $\boldsymbol{U}_{\mathrm{n}}^{\mathrm{z}} \stackrel{\mathrm{df}}{=} \boldsymbol{U}_{\mathrm{Rn}}^{\mathrm{z}}$ and

$$\mathbf{1}^{\mathrm{z}} \stackrel{\mathrm{df}}{=} [\ 1,\ \ 1,\ \ 1\ ]^{\mathrm{T}}.$$

Observe, however, that the sum of line-to-artificial zero voltages is zero, thus the supply voltage cannot contain the zero order sequence component, meaning harmonics of the order $n = 3k$. Such harmonics can occur, however, in the line-to-ground voltages. Nonetheless, when such harmonics are symmetrical they do not cause any currents in three-wires.

It means that set N of the voltage harmonic orders can be decomposed into a set of orders of harmonics that are of the positive sequence N^{p} and harmonics that are of negative sequence N^{n}, namely,

$$N = N^{\mathrm{p}} \oplus N^{\mathrm{n}}$$

where $\oplus$ is the symbol of the logical sum of sets, and

$$N^{\mathrm{p}} \stackrel{\mathrm{df}}{=} \{1, 4, 7, \ldots .3k + 1\}, \quad N^{\mathrm{n}} \stackrel{\mathrm{df}}{=} \{2, 5, 8, \ldots .3k - 1\},$$

and consequently, a symmetrical line-to-artificial zero voltage $\boldsymbol{u}$ can be expressed as the sum of the positive and negative sequence components

$$\boldsymbol{u} = \sum_{n \in N} \boldsymbol{u}_n = \sum_{n \in N^{\mathrm{p}}} \boldsymbol{u}_n + \sum_{n \in N^{\mathrm{n}}} \boldsymbol{u}_n \stackrel{\mathrm{df}}{=} \boldsymbol{u}^{\mathrm{p}} + \boldsymbol{u}^{n}.$$

A similar conclusion can be drawn concerning the line currents in three-wire systems, on the condition, however, that these currents are symmetrical, meaning at approximation with the exemption of the zero-sequence components. Since, irrelevant of the harmonic order,

$$i_{\mathrm{R}n}(t) + i_{\mathrm{S}n}(t) + i_{\mathrm{T}n}(t) \equiv 0$$

the zero-sequence component $\boldsymbol{i}^{\mathrm{z}}$ cannot exist in supply lines of three-wire systems, even if line currents are not symmetrical. When these currents are not symmetrical, harmonics of the order $n = 3k$ can occur in line currents, but these harmonics do not create the zero-sequence component. Their sum has to be equal to zero. It means that the third order harmonic is not the zero sequence. The set N of the current harmonics cannot be decomposed into sets N^{p} and N^{n}, however. Each current harmonic can have both the positive and negative sequence components, $\boldsymbol{i}^{\mathrm{p}}$ and $\boldsymbol{i}^{\mathrm{n}}$. This usually happens when there are different single-phase harmonic generating loads in the distribution system, as shown in the following illustration.

Illustration 3.38 *A lighting system, composed mainly of fluorescent bulbs, meaning the current harmonic generating load, is supplied from a single phase of Δ/Y transformer as shown in Figure 3.69, assuming that transformer is ideal of 1:1 turn ratio.*

Let us assume that the lightning system draws the current:

$$i_{\mathrm{L}} = 100\sqrt{2}\sin \omega_1 t - 60\sqrt{2}\sin 3\omega_1 t + 40\sqrt{2}\sin 5\omega_1 t\,\mathrm{A}.$$

The reduced vectors of the crms values of the line current harmonics are

$$I_1 \stackrel{\mathrm{df}}{=} \begin{bmatrix} \boldsymbol{I}_{\mathrm{R}1} \\ \boldsymbol{I}_{\mathrm{S}1} \end{bmatrix} = \begin{bmatrix} 100\, e^{-j90^\circ} \\ 100\, e^{j90^\circ} \end{bmatrix} \mathrm{A}, \quad I_3 \stackrel{\mathrm{df}}{=} \begin{bmatrix} \boldsymbol{I}_{\mathrm{R}3} \\ \boldsymbol{I}_{\mathrm{S}3} \end{bmatrix} = \begin{bmatrix} 60\, e^{j90^\circ} \\ 60\, e^{-j90^\circ} \end{bmatrix} \mathrm{A},$$

$$I_5 \stackrel{\mathrm{df}}{=} \begin{bmatrix} \boldsymbol{I}_{\mathrm{R}5} \\ \boldsymbol{I}_{\mathrm{S}5} \end{bmatrix} = \begin{bmatrix} 40\, e^{-j90^\circ} \\ 40\, e^{j90^\circ} \end{bmatrix} \mathrm{A}.$$

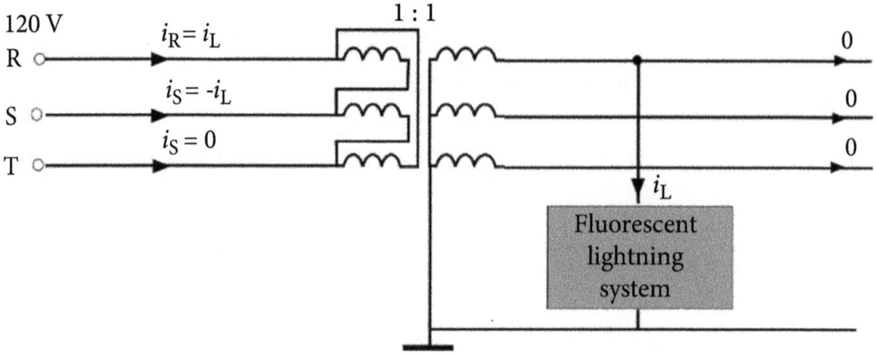

Figure 3.69 An example of a 3p3w lighting system supply circuit.

The crms values of symmetrical components of the current harmonics are equal to

$$\boldsymbol{I}_1^{\rm s} = \begin{bmatrix} I_1^{\rm p} \\ I_1^{\rm n} \end{bmatrix} = \boldsymbol{S}\boldsymbol{I}_1 = \frac{1}{j\sqrt{3}} \begin{bmatrix} \alpha, & -1 \\ -\alpha^*, & 1 \end{bmatrix} \begin{bmatrix} 100\, e^{-j90^\circ} \\ 100\, e^{j90^\circ} \end{bmatrix} = \begin{bmatrix} 57.7\, e^{-j120^\circ} \\ 57.7\, e^{-j60^\circ} \end{bmatrix} \mathrm{A}$$

$$\boldsymbol{I}_3^{\rm s} = \begin{bmatrix} I_3^{\rm p} \\ I_3^{\rm n} \end{bmatrix} = \boldsymbol{S}\boldsymbol{I}_3 = \frac{1}{j\sqrt{3}} \begin{bmatrix} \alpha, & -1 \\ -\alpha^*, & 1 \end{bmatrix} \begin{bmatrix} 60\, e^{j90^\circ} \\ 60\, e^{-j90^\circ} \end{bmatrix} = \begin{bmatrix} 34.6\, e^{j60^\circ} \\ 34.6\, e^{j120^\circ} \end{bmatrix} \mathrm{A}$$

$$\boldsymbol{I}_5^{\rm s} = \begin{bmatrix} I_5^{\rm p} \\ I_5^{\rm n} \end{bmatrix} = \boldsymbol{S}\boldsymbol{I}_5 = \frac{1}{j\sqrt{3}} \begin{bmatrix} \alpha, & -1 \\ -\alpha^*, & 1 \end{bmatrix} \begin{bmatrix} 40\, e^{-j90^\circ} \\ 40\, e^{j90^\circ} \end{bmatrix} = \begin{bmatrix} 23.1\, e^{-j120^\circ} \\ 23.1\, e^{-j60^\circ} \end{bmatrix} \mathrm{A}.$$

Thus both the positive and negative sequence currents on the primary side of the distribytion transformer are composed of harmonics of the same order, from the set $N = \{1, 3, 5\}$, *namely*

$$\boldsymbol{i}^{\rm p} = \sum_{n=1,3,5} \boldsymbol{i}_n^{\rm p} = \sqrt{2}\,\mathrm{Re} \sum_{n=1,3,5} \begin{bmatrix} 1 \\ \alpha^* \\ \alpha \end{bmatrix} I_n^{\rm p}\, e^{jn\omega_1 t} = \sqrt{2}\,\mathrm{Re} \sum_{n=1,3,5} \mathbf{1}^{\rm p} \begin{bmatrix} 57.7\, e^{-j120^\circ} \\ 34.6\, e^{j60^\circ} \\ 23.1\, e^{-j120^\circ} \end{bmatrix} e^{jn\omega_1 t} \mathrm{A}$$

$$\boldsymbol{i}^{\rm n} = \sum_{n=1,3,5} \boldsymbol{i}_n^{\rm n} = \sqrt{2}\,\mathrm{Re} \sum_{n=1,3,5} \begin{bmatrix} 1 \\ \alpha \\ \alpha^* \end{bmatrix} I_n^{\rm n}\, e^{jn\omega_1 t} = \sqrt{2}\,\mathrm{Re} \sum_{n=1,3,5} \mathbf{1}^{\rm n} \begin{bmatrix} 57.7\, e^{-j60^\circ} \\ 34.6\, e^{j120^\circ} \\ 23.1\, e^{-j60^\circ} \end{bmatrix} e^{jn\omega_1 t} \mathrm{A}.$$

*****

Summary

The voltage and current harmonics, as well as three-phase structure, add up two levels of complexity to the three-phase circuits with distorted voltages and current analysis as compared to the analysis of single-phase circuits with sinusoidal quantities. Despite that, by a selection of compact algebraic symbols, as was done in this chapter, it is still possible to preserve the complexity of the mathematical expressions needed for three-phase circuits with nonsinusoidal voltages and currents analysis at a reasonable level.

Chapter 4
Semi-Periodic Voltages and Currents

4.1 Roots of Non-Periodicity and Its Consequences

There are situations in electrical distribution systems where voltages and currents cannot be regarded as periodic quantities. They could be non-periodic. Any switching operation in distribution systems disturbs periodicity. Moreover, loads with a random behavior draw nonperiodic currents. At the high power of a load, they can affect the voltage and current in the distribution system significantly. Rock crushers, arc furnaces, and mills in metallurgic plants are examples of such loads. Suppliers of pulsing loads, such as X-ray generators in medical facilities, high-power laser beam generators, electromagnetic guns, or aircraft launchers in military facilities do not draw periodic currents. The same can apply to power electronics-based variable speed drives. An energy pump, of the structure shown in Fig. 4.1, which supplies a pulsing load is an example of a load that draws a non-periodic current.

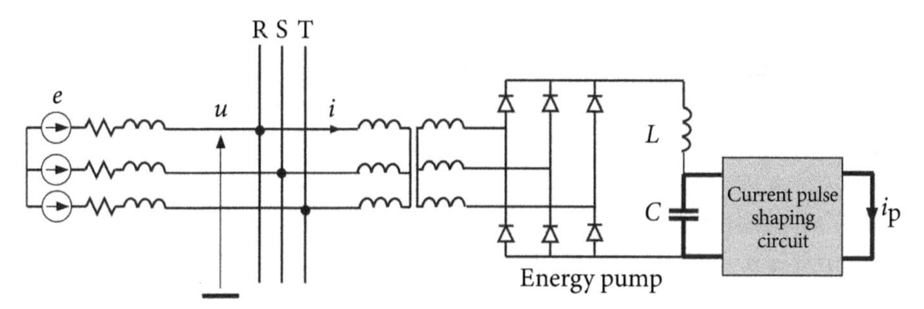

Figure 4.1 An example of an energy pump for a pulsing load.

The supply current of the pump increases over several cycles of the supply voltage and declines to zero when the capacitor voltage approaches its maximum value, to be released next by a current of the order of hundreds kA and only a few milliseconds duration, and the process is repeated. In a response to supply current variation, there is also a variation of the bus voltage and its distortion. Waveforms of the supply current and voltage of such energy pump are shown in Fig. 4.2.

The meaning of the adjective "periodic" is clear and well-defined. At the same time, the adjective "non-periodic" classifies quantities in a rather trivial manner: a quantity is non-periodic if it does not have the periodic property. It does not provide much information on the nature of this non-periodicity, however.

Powers and Compensation in Circuits with Nonsinusoidal Current. Leszek S. Czarnecki, Oxford University Press.
 DOI: 10.1093/oso/9780198879206.003.0004

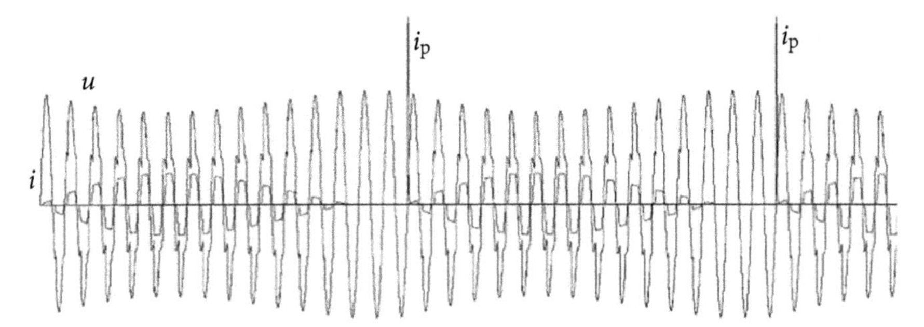

Figure 4.2 Waveforms of the supply current and voltage of an energy pump.

The nature of this not-periodicity can be very diverse and depending on it, non-periodic quantities are categorized into several groups. Only some of them can occur in distribution systems and have importance for electrical engineering.

There is always some level of randomness in electrical quantities and consequently non-periodicity. They are approximated mathematically by describing the most dominating features and just such an approximation provides a ground for their classification. Unfortunately, there is some level of ambiguity as to the meaning of some terms related to non-periodicity. The term "quasi-periodic" was used in Ref. [19] to describe quantities that deviate from periodic ones to a measurable degree. The same term is used in Ref. [188] for quantities composed of components of different frequencies, meaning that their ratio is not a rational number. Quantities of finite energy [94] form another set of non-periodic quantities.

4.2 Frequency Spectra of Periodic and Non-Periodic Quantities

When the quantity $x(t)$ is periodic, then its frequency spectrum, meaning the result of the Fourier transform

$$\mathscr{F}\{x(t)\} = \mathbf{X}(j\omega) = \int_{-\infty}^{\infty} x(t)\, e^{-j\omega t}\, dt$$

has the form of a sequence of Dirac's pulses at harmonic frequencies with pulses area equal to $\sqrt{2}\pi X_n$. Assuming that the mean value of $x(t)$ is zero so that it has the Fourier series

$$x(t) = \sqrt{2}\,\mathrm{Re}\sum_{n=1}^{\infty} \mathbf{X}_n e^{jn\omega_1 t}$$

then [205] the frequency spectrum of $x(t)$ is

$$\mathbf{X}(j\omega) = \mathscr{F}\left\{\sqrt{2}\,\mathrm{Re}\sum_{n=1}^{\infty}\mathbf{X}_n e^{jn\omega_1 t}\right\} = \mathscr{F}\left\{\frac{1}{\sqrt{2}}\sum_{-\infty}^{\infty}\mathbf{X}_n e^{jn\omega_1 t}\right\} =$$

$$= \sqrt{2}\,\pi\sum_{-\infty}^{\infty}\mathbf{X}_n\delta(\omega - n\omega_1). \tag{4.1}$$

It is shown in Fig. 4.3.

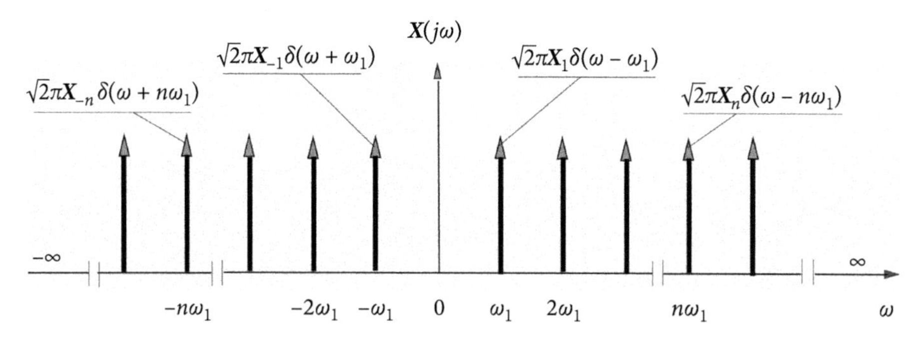

Figure 4.3 A frequency spectrum $X(j\omega)$ of a periodic quantity $x(t)$.

Non-periodic quantities are a result of random processes, effects of the addition of independent processes, or a sort of modulation by changing the circuit parameters. These mechanisms affect the frequency spectrum properties.

When non-periodicity of the quantity $x(t)$ is the effect of random physical processes, its spectrum is a continuous function of frequency as shown in Fig. 4.4.

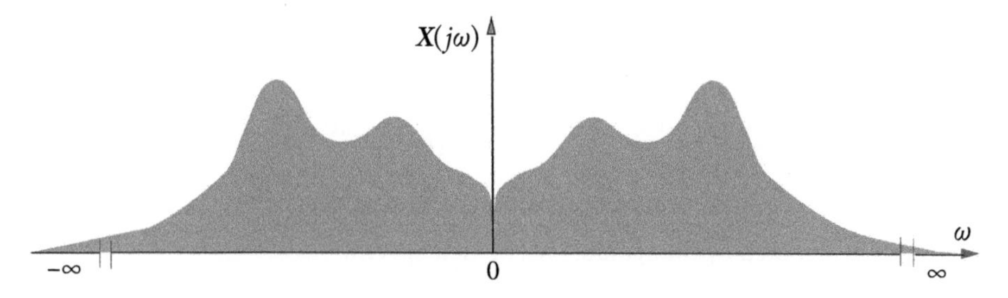

Figure 4.4 A frequency spectrum $X(j\omega)$ of a non-periodic quantity $x(t)$ that is the effect of a random process.

Because the Fourier transform is the form of a linear mapping, thus when the non-periodicity is the effect of an addition of a few processes, the frequency spectrum is the sum of the spectra of these individual processes. They could be continuous or be a sequence of Dirac's pulses. Let us assume, for example, that quantity $x(t)$ has the form

$$x = \sqrt{2}X_1\sin\omega_1 t + \sqrt{2}X_2\sin(\sqrt{2}\,\omega_1)t. \tag{4.2}$$

The frequency ratio of sinusoidal components of this quantity is not a rational number so $x(t)$ is non-periodic despite its components being periodic. Its frequency spectrum is shown in Fig. 4.5.

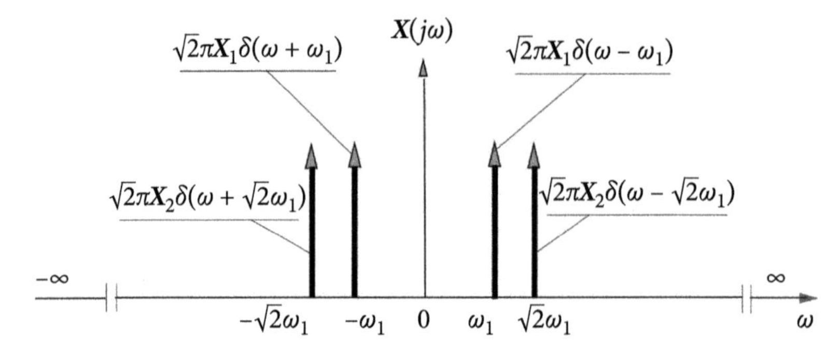

Figure 4.5 A frequency spectrum $X(j\omega)$ of a periodic quantity $x(t)$, specified by (4.2).

It means that non-periodic quantities do not have to have a continuous spectrum. Now let us assume that quantity $y(t)$ is a product of periodic quantity $x(t)$ and a non-periodic quantity $\Delta(t)$ shown in Fig. 4.6.

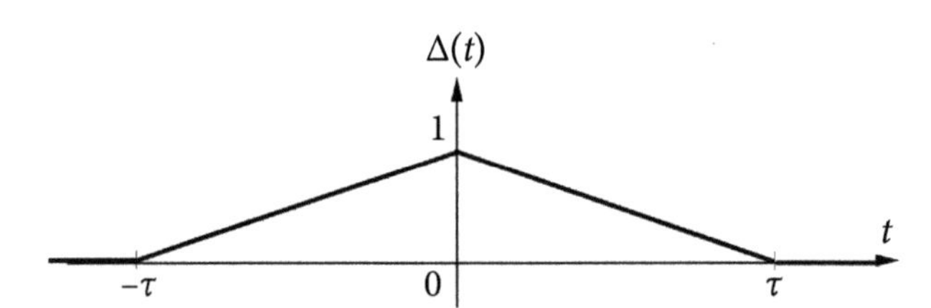

Figure 4.6 A time variance of $\Delta(t)$ function.

Thus

$$y(t) = \Delta(t) \times x(t) = \Delta(t) \times \frac{1}{\sqrt{2}} \sum_{-\infty}^{\infty} \mathbf{X}_{nk} e^{jn\omega_1 t}.$$

The frequency spectrum of the product of two functions in time domain is equal to convolution, denoted by the symbol "⊗", of their spectra

$$\mathscr{F}\{\Delta(t) \times x(t)\} = \mathscr{F}\{\Delta(t)\} \otimes \mathscr{F}\{x(t)\} \tag{4.3}$$

where the spectrum of $\Delta(t)$ function has the form

$$\mathscr{F}\{\Delta(t)\} \overset{\text{df}}{=} \mathbf{D}(j\omega) = \int_{-\tau}^{0} \left(1+\frac{t}{\tau}\right) e^{-j\omega t} dt + \int_{0}^{\tau} \left(1-\frac{t}{\tau}\right) e^{-j\omega t} dt$$

$$= \tau \frac{\sin^2(\frac{\omega\tau}{2})}{(\frac{\omega\tau}{2})^2} = \tau \operatorname{sinc}^2\left(\frac{\omega\tau}{2}\right) \tag{4.4}$$

shown in Fig. 4.7. This plot shows that the spectrum of such a function is concentrated in a narrow band between $(-\pi/\tau < \omega < \pi/\tau)$.

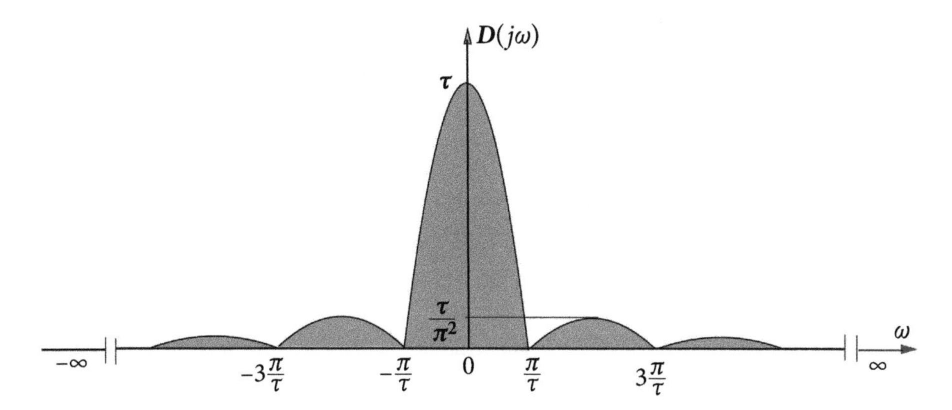

Figure 4.7 Spectrum $D\,(j\omega)$ of $\Delta(t)$ function.

Since the spectrum of periodic function $x(t)$ is given by formula (4.1), we obtain

$$\boldsymbol{Y}(j\omega) = \mathcal{F}\{y(t)\} = \mathcal{F}\{\Delta(t)\} \otimes \mathcal{F}\{x(t)\} = \frac{1}{2\pi}\int_{-\infty}^{\infty} \boldsymbol{D}(j\lambda)\,\boldsymbol{X}(j\lambda - j\omega)\,d\lambda =$$

$$= \frac{1}{2\pi}\int_{-\infty}^{\infty} \tau\,\mathrm{sinc}^2\left(\frac{\lambda\tau}{2}\right)\sqrt{2}\pi\sum_{-\infty}^{\infty}\boldsymbol{X}_{nk}\delta[\lambda - (\omega - n\omega_1)]\,d\lambda = \tag{4.5}$$

$$= \frac{\tau}{\sqrt{2}}\sum_{-\infty}^{\infty}\boldsymbol{X}_{nk}\,\mathrm{sinc}^2\left[(\omega - n\omega_1)\,\frac{\tau}{2}\right].$$

This spectrum is plotted in Fig. 4.8. It is a continuous function of frequency, with the radial frequency components concentrated around harmonic radial frequencies $\omega = n\omega_1$ in bands of width $\Delta\omega = 2\pi/\tau$, or the plain frequency $\Delta f = 1/\tau$.

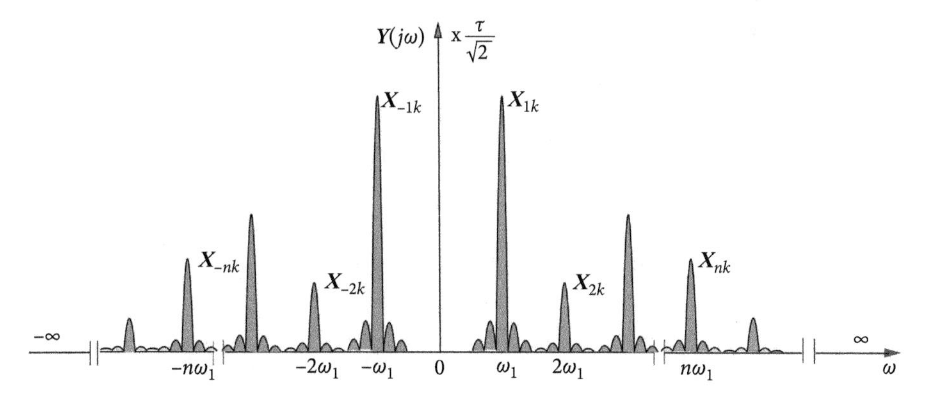

Figure 4.8 The spectrum of an amplitude-modulated periodic quantity $y(t)$.

In general, the frequency spectrum of a non-periodic quantity, which is an effect of modulation, can be confined by a profile shown in Fig. 4.9.

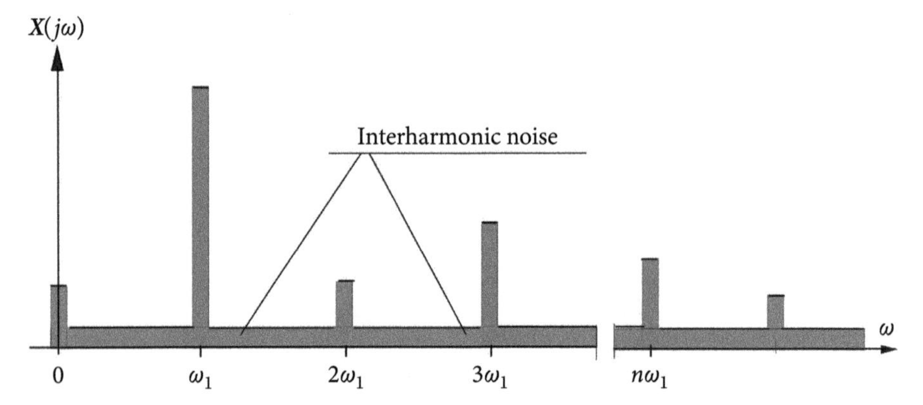

Figure 4.9 A profile of the frequency spectrum of a modulated periodic quantity.

The components of the quantity $y(t)$ of frequency between harmonic frequencies of the modulated quantity $x(t)$ are referred to as an interharmonic noise.

4.3 Concept of Semi-Periodic Currents and Voltages

Non-periodic voltages and currents do not have period T and consequently, the major features of electrical loads used at their supply with periodic voltage, such as the active power

$$P \stackrel{\text{df}}{=} \frac{1}{T}\int_0^T u(t)\, i(t)\, dt = UI\, \cos\varphi$$

and the rms values of voltages and currents

$$||u|| \stackrel{\text{df}}{=} \sqrt{\frac{1}{T}\int_0^T u^2(t)\, dt}, \qquad ||i|| \stackrel{\text{df}}{=} \sqrt{\frac{1}{T}\int_0^T i^2(t)\, dt}$$

cannot be used. The same applies to the concept of harmonics and their crms value

$$\mathbf{X}_n = X_n\, e^{j\alpha_n} = \frac{\sqrt{2}}{T}\int_0^T x(t) e^{-jn\omega_1 t} dt.$$

Thus, the non-periodicity, that is, the lack of period T, causes the main tools for energy flow description have gone. Also, the voltage, current, and active power meters cannot operate according to the above-specified definitions.

Linear circuits with non-periodic supply voltage can be analyzed using the Fourier transform, defined for quantity $x(t)$

$$\mathscr{F}\{x(t)\} = \mathbf{X}(j\omega) = \int_{-\infty}^{\infty} x(t)\, e^{-j\omega t} dt$$

and the circuit frequency transmittances. When a quantity $y(t)$ is related to $x(t)$ by the frequency transmittance $\boldsymbol{H}(j\omega)$, so that $\boldsymbol{Y}(j\omega) = \boldsymbol{H}(j\omega)\,\boldsymbol{X}(j\omega)$, then the response of the circuit to this $x(t)$ quantity is

$$y(t) = \mathscr{F}^{-1}\{\mathbf{Y}(j\omega)\} = \frac{1}{2\pi}\int_{-\infty}^{\infty} \mathbf{Y}(j\omega)\, e^{j\omega t}\, d\omega.$$

Implementation of such an approach to a circuit analysis requires that the excitation $x(t)$ be known over an infinite interval of time. Thus, the response $y(t)$ cannot be obtained in real time, meaning, at a sequence of instants, t_1, t_2, t_3,... t_k. It provides information on a static state of the circuit rather than on a dynamic one. Thus, the Fourier transform approach does not provide mathematical tools for compensation, in particular tools for developing adaptive compensation of time-varying loads.

Non-periodic voltages and currents in electrical distribution systems have some properties that differentiate them from other non-periodic quantities. In other words, energy to such systems is delivered mainly from three-phase synchronous generators which produce almost ideally sinusoidal voltage.

Due to switching operations that cause transients, nonlinearity, and time variation of electrical loads, the current in a circuit could be non-periodic. In a response to the non-periodicity of the distribution currents, also the distribution voltages could lose periodicity, as is illustrated for the single-phase circuit in Fig. 4.10. Eventually, both the distribution voltage and the current are non-periodic.

Nonetheless, the period T of the voltage produced by generators can be detected by filtering and be used as a sort of "time frame" for the energy flow analysis.

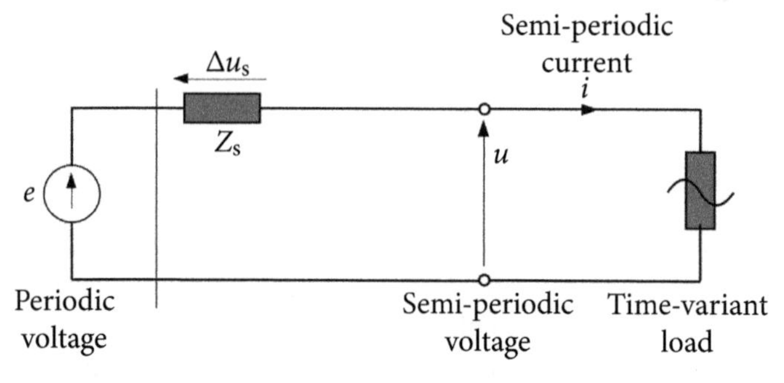

Figure 4.10 Illustration of the mechanism of semi-periodic distortion of the load voltage.

Non-periodic voltages and currents in circuits with time-varying loads but with energy delivered by generators of a sinusoidal voltage, are referred to in this book as *semi-periodic voltages and currents.*

Fundamental harmonic u_1 of the voltage periodic component, separated by a filter, can be used for detecting period T. A time interval of the period T duration will be referred to as an *observation window.* It starts at some instant of time $t = t_k - T$ and ends at instant t_k, as it is shown in Fig. 4.11.

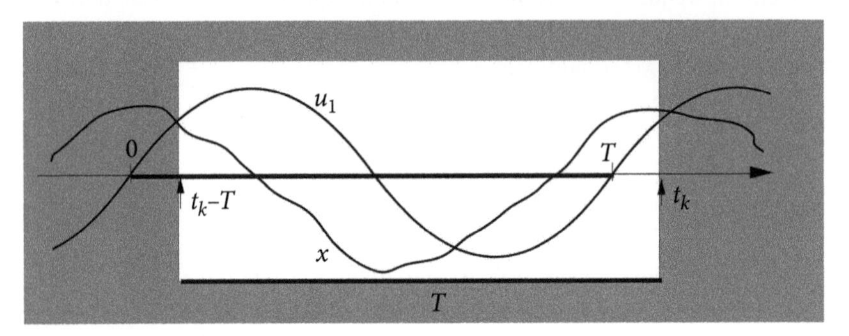

Figure 4.11 An observation window.

If the quantity $x(t)$ as observed in the window shown in Fig. 4.11, is reproduced every period T from minus to plus infinity, then *a periodic extension*

$$x^{\approx}(t) \stackrel{\text{df}}{=} \sum_{n=-\infty}^{n=\infty} x(t - nT) \tag{4.6}$$

is created, which is a fictitious periodic quantity, identical with quantity $x(t)$ only in the observation window but not outside of it.

4.4 Running Active Power and Rms Value

The energy delivered to a load during the observation window preceding the instant t_k is equal to

$$W(t_k) = \int_{t_k-T}^{t_k} u(t)\, i(t)\, dt.$$

When the voltage and current at the load terminals are periodic with period T, this energy is independent of the beginning of integration. It is not constant when the voltage and current are semi-periodic.

Although interval T is not a period of semi-periodic quantities, it can be used for calculating the value of the average rate of energy flow over the observation window. At the end of this flow observation, at instant t_k, this average value is equal to

$$\frac{W(t_k)}{T} = \frac{1}{T}\int_{t_k-T}^{t_k} u(t)\, i(t)\, dt = P^{\approx}(t_k). \tag{4.7}$$

The power calculated in such a way can be regarded as the active power of a load with a semi-periodic voltage and current. When the voltage and current are not periodic then the active power defined in such a way is not constant but a function of time, which is emphasized with the wave symbol "≈". It will be referred to as *running active power.*

The periodic extension $x^{\approx}(t)$ of what is observed in the observation window $x(t)$ has the rms value, which can be calculated at the instant $t = t_k$, namely

$$||x^{\approx}||_{t_k} \stackrel{\text{df}}{=} \sqrt{\frac{1}{T}\int_{t_k-T}^{t_k} x^2(t)\, dt}. \tag{4.8}$$

If the quantity is not periodic, then this value is not constant but changes with the observation window. It will be referred to as the running rms value of semi-periodic quantity. When semi-periodic current $i(t)$ flows through a resistor of resistance R, then the running active power of this resistor is at the instant $t = t_k$, is equal to

$$P_k^{\approx} = \frac{1}{T}\int_{t_k-T}^{t_k} u(t)\, i(t)\, dt = \frac{1}{T}\int_{t_k-T}^{t_k} R\, i^2(t)\, dt = R\, ||i^{\approx}||_{t_k}^2. \tag{4.9}$$

Thus, the running rms value has the same meaning as the conventional rms value. It enables the calculation of energy loss on the resistor in the preceding observation window. It does not allow to calculate it outside of this window, however.

4.5 Running Scalar Product of Semi-Periodic Quantities

For two semi-periodic quantities $x(t)$ and $y(t)$ which have periodic extensions with the same period T, a scalar product can be defined and calculated at the end of the observation window, namely

$$(x^{\approx}, y^{\approx})_{t_k} \stackrel{\text{df}}{=} \frac{1}{T}\int_{t_k-T}^{t_k} x(t)\, y(t)\, dt. \tag{4.10}$$

The running rms value of the sum of two semi-periodic quantities $x(t)$ and $y(t)$ is equal to

$$||x^{\approx} + y^{\approx}||_{t_k} = \sqrt{\frac{1}{T}\int_{t_k-T}^{t_k} [x(t) + y(t)]^2 dt} = \sqrt{||x^{\approx}||_{t_k}^2 + 2(x^{\approx}, y^{\approx})_{t_k} + ||y^{\approx}||_{t_k}^2}.$$

This rms value can be expressed in terms of only the running rms value of this sum components, namely

$$||x^{\approx} + y^{\approx}||_{t_k} = \sqrt{||x^{\approx}||_{t_k}^2 + ||y^{\approx}||_{t_k}^2} \tag{4.11}$$

on the condition that

$$(x^{\approx}, y^{\approx})_{t_k} = 0 \tag{4.12}$$

that is, when these quantities are mutually orthogonal. Property (4.11), at condition (4.12), applies only to the observation window of T duration just preceding instant $t = t_k$.

4.6 Quasi-Harmonics

The periodic extension $x^{\approx}(t)$ can be expressed by the Fourier Series

$$x^{\approx}(t) = X_0^{\approx} + \sqrt{2}\,\mathrm{Re}\sum_{n=0}^{\infty} \mathbf{X}_n^{\approx} e^{jn\omega_1 t} \tag{4.13}$$

with the crms values of its harmonics

$$\mathbf{X}_n^{\approx} = \mathbf{X}_n^{\approx}(t_k) \stackrel{\mathrm{df}}{=} \frac{\sqrt{2}}{T} \int_{t_k-T}^{t_k} x(t)\, e^{-jn\omega_1 t} dt \tag{4.14}$$

calculated at the instant of time t_k. It means that quantity $x(t)$ can be reconstructed from its harmonics, assuming that it does not have points of discontinuity, with full accuracy. This reconstruction applies only to the observation window of T duration, shown in Fig. 4.12, however, but not beyond it.

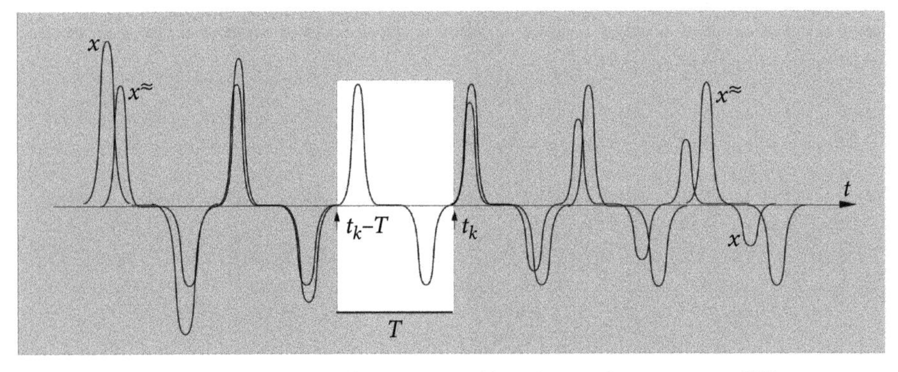

Figure 4.12 Semi periodic quantity $x(t)$ and periodic extension $x^{\approx}(t)$.

When the observed quantity $x(t)$ is not periodic but semi-periodic, then at the next instant of observation, at $t_k + \Delta t$, the observation window moves by Δt, so that its periodic extension $x^{\approx}(t)$ changes. Consequently, the crms values $\boldsymbol{X}_n^{\approx}$ are not constant but functions of time. Inside of the observation window, the term

$$\sqrt{2}\,\mathrm{Re}\{\boldsymbol{X}_{nk}^{\approx} e^{jn\omega_1 t}\} = x_n(t) \tag{4.15}$$

is the nth order harmonic of quantity $x(t)$. Outside of the observation window, this term is the nth order harmonic of the periodic extension, $x^{\approx}(t)$, but it is not a harmonic of quantity $x(t)$, since it is not periodic, so that it cannot be expressed as a sum of harmonics. It means that having all values $\boldsymbol{X}_n^{\approx}$ calculated at instant $t = t_k$, the continuous quantity $x(t)$ can be entirely reconstructed in the preceding observation window, that is, in the interval $t_k - T < t < t_k$, but not outside it. Since the crms value $\boldsymbol{X}_{nk}^{\approx}$ is a function of time, that is,

$$\boldsymbol{X}_{nk}^{\approx} = \boldsymbol{X}_n^{\approx}(t) = X_n^{\approx}(t) e^{j\alpha(t)}$$

the term can be presented in the form

$$\sqrt{2}\,\mathrm{Re}\left\{\boldsymbol{X}_{nk}^{\approx} e^{jn\omega_1 t}\right\} = \sqrt{2} X_n^{\approx}(t)\ \cos[n\omega_1 t + \alpha(t)] \stackrel{\text{df}}{=} x_n^{\approx}(t). \tag{4.16}$$

It is of a sine-like waveform quantity, but with the amplitude modulated by the variability of the rms value $X_n^{\approx}(t)$ and the frequency $\omega = n\omega_1$ modulated by the variability of the phase-angle $\alpha(t)$. An increase α over interval T is equivalent to an increase in the frequency of the component $x_n^{\approx}(t)$. Such a quantity will be referred to as *a quasi-harmonic* of the nth order. A quasi-harmonic is a sinusoidal-like quantity, with variable amplitude and variable frequency in the vicinity of harmonic frequency $n\omega_1$. The crms value $\boldsymbol{X}_{nk}^{\approx}$ will be referred to *as a running crms value* of such a quasi-harmonic.

4.7 Digital Processing of Semi-Periodic Quantities

The running active power and running rms value cannot be measured by conventional analog meters, however. Averaging by the active power meters is the effect of mechanical inertia of the current coil, which rotates in the magnetic field created by the voltage coil. This is not the result of division by period T, as this is suggested by the conventional definition of active power. Digital meters are needed for that. Such a meter performs arithmetic operations on samples of the load voltage and current, provided by analog to digital (A/D) converters. To avoid confusion with symbols of harmonics, denoted in this book by symbols u_n and i_n, these samples will be denoted by u_m and i_m. Assuming that there are M samples of the load current in the observation window, as shown in Fig. 4.13, and M at the same instances taken samples of the load voltage, then such a digital meter can calculate at the instant $t = t_k$ the running active power according to the formula

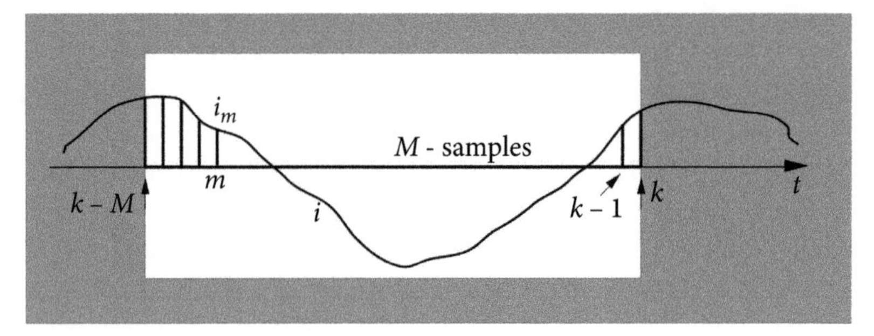

Figure 4.13 Observation window and current samples.

$$P(t_k) \stackrel{\text{df}}{=} P_k^{\approx} = \frac{1}{M} \sum_{m=k-M+1}^{m=k} u_m\, i_m. \tag{4.17}$$

To avoid the spectrum aliasing and consequently an error, the number of samples M has to obey the Nyquist criterion. If $n_{\max}$ is the highest harmonic order of harmonics of the periodic extension $x(t)$, then the minimum number of samples in the period T has to be higher than the double value of $n_{\max}$, that is, $M > 2n_{\max}$. This is the condition for preserving full information on the waveform of periodic quantity when it is specified by the sequence of its discrete values, at instants $t = t_m$.

The voltage and load current sensors have to provide voltage signals fitted to the conversion range of A/D simultaneous sampling converters. The voltage signal has to be used for detecting the period T, which next has to be converted into the number of samples, M, in the observation window. The structure of a digital meter of the running active power is shown in Fig. 4.14.

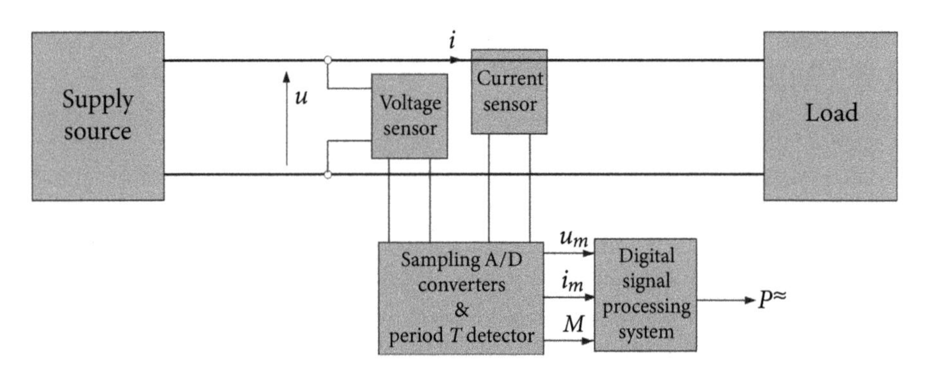

Figure 4.14 A structure of the active power digital meter.

Like the running active power, the running rms value cannot be measured by conventional analog meters but by digital meters. Having M samples x_m taken in the observation window, such a meter calculates the running rms value at instant t_k according to the following algorithm:

$$||x||_k^{\approx} = \sqrt{\frac{1}{M} \sum_{m=k-M+1}^{m=k} x_m^2}. \tag{4.18}$$

The scalar product can be calculated digitally as

$$(x, y)_k^{\approx} = \frac{1}{M} \sum_{m=k-M+1}^{m=k} x_m \, y_m \tag{4.19}$$

while the crms values of quasi-harmonics

$$\boldsymbol{X}_{nk}^{\approx} = \frac{\sqrt{2}}{M} \sum_{m=k-M+1}^{m=k} x_m e^{-jn\frac{2\pi}{M}m}. \tag{4.20}$$

*****

Summary

Voltages and currents in circuits with variable parameters very often are not periodic but can be classified as semi-periodic. This chapter shows how such quantities can be expressed in an analytical form, and how the Fourier series, the concept of harmonics, the scalar product, and the crms value can be generalized to this kind of semi-periodic quantities.

Chapter 5
Historical Overview

5.1 Emergence of Power Terms and Power Theory

Our knowledge of the power properties of electrical circuits and consequently the power theory (PT) in the present form is a product of investigations and discussions of a few generations of scientists and engineers. The main concepts, which have contributed to the development of the PT are compiled in this chapter in chronological order.

Power theory development is not yet completed. This overview, therefore, is important not only for historical reasons. We can recognize the contributions of our predecessors. A return to some ideas formulated long ago could still provide a starting point for the present investigations on the power properties and physical phenomena in circuits with nonsinusoidal voltages and currents.

Physical fundamentals for electrical power systems development were completed in the first half of the nineteen century, due to Oersted's and Faraday's discoveries of relations between electric and magnetic fields, relations that have enabled electromechanical energy conversion. By 1880 all major components of AC power systems, such as the transformer and the generator, which enabled the construction of the first AC power system, were invented. The invention of three-phase generators by Dolivo-Dobrowolski in 1888 made the development of three-phase power circuits possible.

The concepts of the active and apparent powers, P and S, as well as the power factor, $\lambda = P/S$, and the power equation

$$P^2 + Q^2 = S^2$$

occurred at that time. There was also awareness [3, 4] of the possibility of a phase shift between the load voltage and current, regarded as a cause of energy oscillations and the reactive power Q.

This development has applied essentially to circuits with sinusoidal voltages and currents, but as early as 1892, in Ref. [2] Steinmetz asked the question: "does the phase displacement occur in the current of electric arcs?", meaning loads with nonsinusoidal currents had become a matter of concern. He concluded [7] in 1907 that even without the phase shift the power factor of such loads is lower than 1. We should be aware that with the invention of an electromagnetic oscillograph in 1893 and a cathode ray tube oscilloscope in 1897, observations of the voltage and current waveform started to be possible.

Powers and Compensation in Circuits with Nonsinusoidal Current. Leszek S. Czarnecki, Oxford University Press.
 DOI: 10.1093/oso/9780198879206.003.0005

The main focus at the end of the nineteenth century was on the ratio of the active and apparent powers, meaning the power factor, its practical importance [4], physical meaning [6, 7], and measurement [5]. A more detailed history of the development of the power factor measurement methods can be found in Ref. [167].

Although the term "power theory" (PT) occurred for the first time in 1931 in Fryze's paper [17], electrical engineering embarked on its development at the turn of the nineteenth century, when the first concepts were established and the first questions were asked. Steinmetz's observation, reported in Ref. [2], that the power factor in a circuit with an electrical arc is lower than 1, started one of the longest and the most controversial debates in electrical engineering. Even now, more than a hundred years later, this debate is not over. There is a lack of clarity in the electrical engineering community about how electrical powers should be defined and how power phenomena should be interpreted. Explanation of power phenomena in electrical circuits has become one of the most confusing and controversial issues in electrical engineering.

The literature on the subject is above of a few thousands. Even now, after a century of investigations, each year brings several new publications: a vivid proof that the power properties of electrical circuits are only apparently simple.

The sheer number of publications over such a lengthy period, and published in many languages has made the investigation of the history of the PT development enormously difficult and the results of such investigations unreliable. Therefore, the historical overview of this development as presented in this chapter can be rightfully challenged as inaccurate and subjective.

The term "power theory" is not well defined and can be comprehended in various ways. Fryze used this term, in [17], for his explanation of the power properties of electrical circuits, as *Fryze's power theory* (PT). Therefore, other explanations of these properties, as suggested, for example, by Budeanu [14], Shepherd and Zakikhani [39], Kusters and Moore [59], Depenbrock [90], Akagi and Nabae [72] can be referred to, respectively, as Budeanu's PT, Shepherd and Zakikhani's, PT or Czarnecki's PT. On the one hand, this is a convenient approach because it directly identifies a particular concept. Some concepts overlap, however. For example, the concept of the active current is common in Fryze's, Kusters and Moore's, and Czarneck's PTs. Consequently, the association of the PT with names can be confusing. The term "*power theory*" can also be used in a different way, namely as our collective knowledge of the power properties of electrical circuits, a kind of database of true sentences or conclusions about these properties. When used in this way, the power theory is a common product of all those who have contributed to its development. Depending on the situation, the term "*power theory*" is used in this book in both ways, both in regard to a specific theory or as collective knowledge on electrical powers. Sometimes is more appropriate to use it in the traditional meaning, sometimes in the meaning suggested, as a database of our knowledge on electrical powers.

Power theory has developed for two different reasons. The first one cognitive. As humans, we have a curiosity and desire to understand the world around us. With regard to electrical circuits, we try to answer the question: "why does the load with the active power P need a supply source with the apparent power S usually higher than the load power P?" This question involves a need for the interpretation of power phenomena associated with energy transmission. The second reason is very practical. The power theory should provide us with answers to questions: "how we can reduce the required apparent power S of the supply source which supplies the load with the power P?" "How energy loss on delivery can be reduced?" It would be trivial to say that these two cognitive and practical questions are tightly bound. It is hard to imagine that practical issues could be solved without a sufficient level of comprehension of power phenomena.

The apparent power of a supply source can be reduced with a compensator and its parameters can be found by trial and error or using some optimization methods. It seems, however, that such approaches, even if they have sometimes provided useful results, have very little in common with power theory because they do not provide any general knowledge on power properties and the compensation of electrical circuits.

It seems that the development of PT can be divided into a preliminary phase and the current one. The border between them is specified by the capability of the interpretation of energy flow phenomena and the conclusions as to compensation of such a simple load as shown in Fig. 5.1, when the supply voltage is nonsinusoidal.

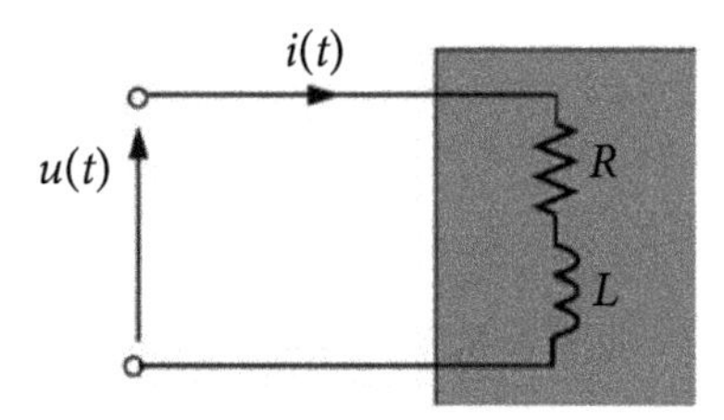

Figure 5.1 A circuit with RL load.

None of the individual PTs as suggested by Budeanu, Fryze, Shepherd and Zakikhani, Kusters and Moore, Depenbrock, Nabae and Akagi, or Tenti has provided the right explanation of the power factor decline in the circuit shown in Fig.5.1 and a method of its improvement. Any generalization of these PTs to three-phase circuits was not capable of explaining power-related phenomena in such circuits. Therefore, all these PTs belong to the preliminary phase of the PT development.

Explanation of power properties of linear loads such as that in Fig. 5.1, along with a method of their compensation was provided in 1983, in Polish [71], and English [75], in the frame of the Currents' Physical Components (CPC)-based PT. This step made the studies on power properties and compensation of nonlinear loads and three-phase circuits possible. This constitutes the present phase of studies on power theory development.

The development of power theory at the beginning of twentieth century started with a quest for the reactive power Q definition, that in the presence of supply voltage distortion would satisfy the power equation (5.1). The literature on the subject by the end of the 1920s amounts to a couple of dozen research papers but no answer was provided. Quite the opposite: Weber concluded in [15] that such a definition is not possible. In such a situation, questions on the reactive power compensation were not even asked.

5.2 Powers in Single-Phase Circuits with Sinusoidal Current

Power properties of single-phase loads, as shown in Fig. 5.2, with a sinusoidal supply voltage and a sinusoidal current were described in terms of powers at the end of nineteenth century.

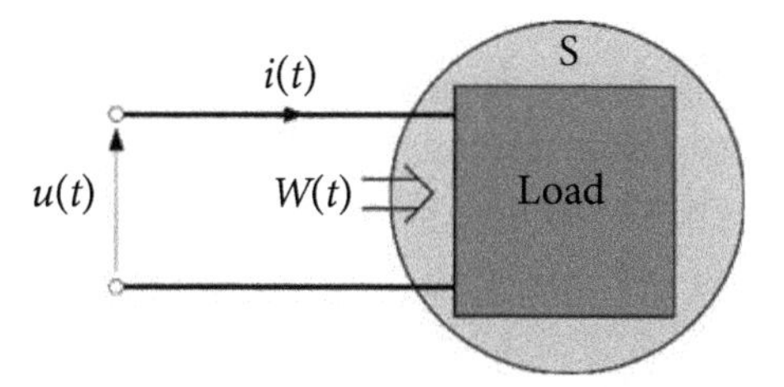

Figure 5.2 A single-phase load.

The rate of energy $W(t)$ flow from the supply source to the load is specified as the *instantaneous power* $p(t)$

$$\frac{d}{dt}\,W(t) \overset{\text{df}}{=} p(t) = u(t)\,i(t). \tag{5.1}$$

Assuming that

$$u(t) = \sqrt{2}\,U\cos\omega_1 t, \quad i(t) = \sqrt{2}\,I\cos(\omega_1 t - \varphi),$$

power properties of the loads supplied with a sinusoidal voltage and current are specified in terms of the active, apparent, and reactive powers, namely.

$$P \overset{\text{df}}{=} \frac{1}{T}\int_0^T u(t)\,i(t)\,dt = UI\cos\varphi \tag{5.2}$$

is the active power,

$$S \overset{\text{df}}{=} \sqrt{\frac{1}{T}\int_0^T u^2(t)\,dt}\,\sqrt{\frac{1}{T}\int_0^T i^2(t)\,dt} = UI \tag{5.3}$$

is the apparent power, and

$$Q \overset{\text{df}}{=} \pm\sqrt{S^2 - P^2} = UI\sin\varphi \tag{5.4}$$

is *the reactive power*. These three powers fulfill the following relationship

$$P^2 + Q^2 = S^2 \tag{5.5}$$

known as the *power equation*.

The active power P is the mean value of the instantaneous power $p(t)$, calculated over the period T of the supply voltage. It specifies the average rate, over the period T, of the electric energy flow to the load.

The apparent power S is the product of the supply voltage and load current rms values, U and I. There is no physical phenomenon in the circuit that could be characterized by the apparent power S. It has a clear technical meaning, however. To operate a load with the active power P, the supply source has to provide the voltage of the rms value U and the current of the rms value I.

The reactive power Q was defined to fill the gap between the apparent power S and the active power P. It accompanies the active power when there is a phase shift between the supply voltage and the load current. Its sign was selected, by convention, in such a way that for the resistive-inductive (RL) loads, dominating in distribution systems, this power is positive.

Definitions of all these powers were fitted to the instrumentation technology available at the end of the nineteenth century, that is, analog instrumentation. Analog meters connected as shown in Fig. 5.3, at terminals of an RL load provided values of all defined above powers.

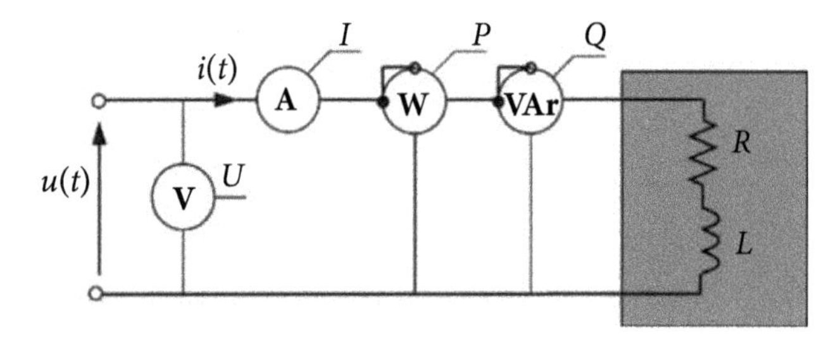

Figure 5.3 An RL load with analog meters.

5.3 Illovici's Reactive Power Definitions

The first definition of the reactive power Q at nonsinusoidal supply voltage can be credited to M.A. Illovici. In 1925 he suggested [13] that two quantities

$$Q_{\mathrm{IC}} = \sum_{n=1}^{\infty} n\, U_n I_n \sin\varphi_n, \quad Q_{\mathrm{IL}} = \sum_{n=1}^{\infty} \frac{1}{n}\, U_n I_n \sin\varphi_n. \tag{5.6}$$

can be regarded as the reactive power. These are quantities measured by a wattmeter as shown in Fig. 5.4a, with the resistor in the branch with the voltage coil replaced by a capacitor, as shown in Fig. 5.4b, or by an inductor.

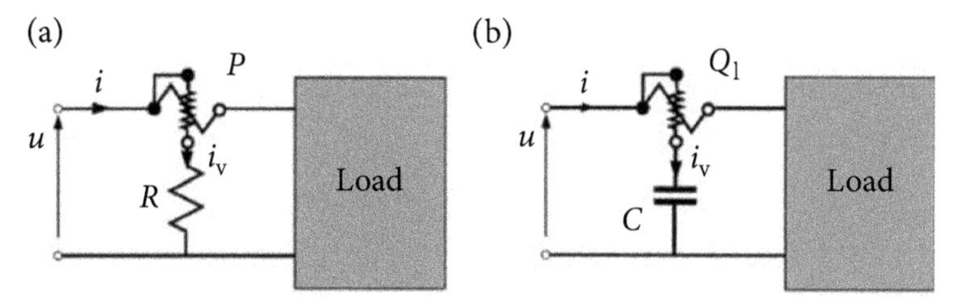

Figure 5.4 (a) A wattmeter and (b) a meter of the reactive power Q_{IC}.

Indeed, the voltage coil current, when the resistor is replaced by a capacitor, is

$$i_v = C\frac{du}{dt}.$$

The mean value of the mechanical torque that deflects the meter needle in the case of sinusoidal supply voltage

$$T_d = k_m \frac{1}{T}\int_0^T i_v(t)\, i(t)\, dt = k_m \omega_1 C\, UI \sin\varphi = kQ, \qquad k = k_m \omega_1 C$$

is proportional to the load reactive power Q. When the supply voltage is nonsinusoidal,

$$u(t) = \sqrt{2}\sum_{n\in N} U_n \sin n\omega_1 t$$

then the load current is

$$i(t) = \sqrt{2}\sum_{n\in N} I_n \sin(n\omega_1 t - \varphi_n)$$

and the mean value of the mechanical torque changes to

$$T_d = k_m \frac{1}{T}\int_0^T i_v(t)\, i(t)\, dt = k_m \sum_{n\in N} n\,\omega_1 CUI \sin\varphi = k\sum_{n\in N} n\, U_n I_n \sin\varphi = k\, Q_{IC}$$

thus, it is proportional to the reactive power Q_{IC} as defined by Illovici.

When the resistor is replaced by an inductor of inductance L, then the mean value of the mechanical torque produced by the meter coils.

$$T_d = \sum_{n\in N} \frac{1}{n} U_n I_n \sin\varphi_n = k\, Q_{IL}, \quad \text{with} \quad k = \frac{k_m}{\omega_1 L}$$

is proportional to the reactive power defined according to Illovici's second suggestion.

Let us check, however, whether the power equation (5.1) is satisfied by Illovici's reactive powers. Let us assume that the load is purely inductive, thus, $P = 0$, and consequently

$$Q_{\mathrm{IC}} = \sum_{n\in N} n\, U_n I_n, \quad Q_{\mathrm{IL}} = \sum_{n\in N} \tfrac{1}{n}\, U_n I_n$$

while the apparent power is

$$S \stackrel{\mathrm{df}}{=} ||u||\,||i|| = \sqrt{\sum_{n\in N} U_n^2}\ \sqrt{\sum_{n\in N} I_n^2}.$$

Let us suppose that the supply voltage is composed of only one harmonic of the order different from the fundamental, meaning, the set $N = \{n\}$ and $n \neq 1$.

The apparent power is $S = U_n I_n$, while

$$Q_{\mathrm{IC}} = \sum_{n\in N} n\, U_n I_n = n\, U_n I_n > S$$

$$Q_{\mathrm{IL}} = \sum_{n\in N} \frac{1}{n}\, U_n I_n = \frac{1}{n}\, U_n I_n < S.$$

Thus, reactive powers defined by Illovici do not satisfy the power equation (5.1). Consequently, Illovici's definitions of reactive power do not contribute to the answer to the question: "why can the apparent power S be higher than the active power P?" Moreover, Illovici's reactive powers do not characterize any physical phenomenon in the circuit. The only justification for these two definitions was the possibility of their measurement. However, measurement issues, of course important, are only secondary to physical or technical reasons for which a quantity should be introduced. Illovici's switched this hierarchy of importance.

Summary: Illovici suggested new definitions of reactive power, but not a new form of the power equation. The traditional power equation is not fulfilled with Illovici's reactive powers.

5.4 Budeanu's Power Theory

In 1927 C.I. Budeanu, a Romanian professor at Bucharest University, suggested a definition of reactive power [14] in the form that has an analogy to the formula for calculation of the active power P in the presence of the supply voltage harmonics,

$$P = \sum_{n=0}^{\infty} U_n I_n \cos\varphi_n$$

namely, as

$$Q = Q_{\mathrm{B}} \stackrel{\mathrm{df}}{=} \sum_{n=1}^{\infty} U_n I_n \sin\varphi_n. \tag{5.7}$$

At such a definition of reactive power, the power equation (5.1) is not satisfied, since

$$P^2 + Q_B^2 \leq S^2,$$

therefore, Budeanu suggested that the power equation should contain also another power quantity, defined as

$$D \stackrel{\text{df}}{=} \sqrt{S^2 - (P^2 + Q_B^2)}. \tag{5.8}$$

Budeanu concluded that this power occurs only when the load voltage and current are distorted. Therefore, he suggested referring to this power as a *distortion power.*

Although the International Electrotechnical Commission (IEC), which debated Budeanu's proposal in 1930, did not endorse it, this equation was supported by the German Standard DIN [37] and the IEEE Standard Dictionary of Electrical and Electronics Terms [154]. It is spread by academic teaching, textbooks, and journal papers. The power equation introduced in 1927 by Budeanu

$$P^2 + Q^2 + D^2 = S^2 \tag{5.9}$$

has become the best-known power equation of single phase-phase loads supplied with non-sinusoidal voltage. Budeanu's reactive power Q was interpreted as the effect of energy oscillations upon the load apparent power S, while Budeanu's distortion power D was interpreted as the effect of the voltage and current distortion upon this power.

Budeanu's reactive power Q measurement. Unlike the reactive powers defined by Illovici, the quantity introduced by Budeanu as the reactive power has turned out to be very difficult for measuring, especially with analog metering technology. Digital signal processing was not available when Budeanu developed his definition of reactive power. Until the 1970s, meaning forty years after Budeanu's reactive power was defined there were no published reports on any method of this power measurement. A report [34] on using an analog computer for its measurement was published in 1967. A few teams have worked on the issue and published results in Refs. [48] and [49]. First patents for a meter of the Budeanu's reactive power were granted [38], [41], and [43] to the author of this book, in 1972 and 1975. Fundamentals of their operation are presented in Refs. [64] and [79].

The measurement of the reactive power Q at sinusoidal voltages and current requires that in the wattmeter used for the active power P measurement, as shown in Fig. 5.4(a), the current in the voltage coil must be shifted by 90° versus the supply voltage. A capacitor, as shown in Fig. 5.4(b), or an inductor can be used for this shift. Such an approach cannot be applied, however, when the voltage is nonsinusoidal, because this 90° phase shift is accompanied by a change in the current harmonic rms value, I_{Vn}, of the voltage coil current. Namely, when a capacitor is connected to the voltage coil branch, then

$$I_{Vn} = n\omega_1 C U_n$$

or in the case of an inductor in that branch

$$I_{Vn} = \frac{U_n}{n\omega_1 L}.$$

In effect, the meter measures the reactive power according to Illovici but not Budeanu's definitions. To measure the Budeanu's reactive power, the 90° phase shift cannot be accompanied by the change in the current harmonic rms value.

Such a relation between input $x(t)$ and output $y(t)$ quantities, meaning the 90° phase shift without change in the rms value with frequency has the Hilbert transform,

$$y(t) \stackrel{\text{df}}{=} \frac{1}{\pi}\int_{-\infty}^{\infty} \frac{x(\tau)}{\tau - t}\, d\tau \stackrel{\text{df}}{=} \mathscr{H}\{x(t)\} \tag{5.10}$$

since the frequency spectra of these two quantities satisfy the relationship

$$Y(j\omega) = j\,\text{sgn}(\omega) X(j\omega)$$

meaning, they are shifted by 90° and have the same harmonic rms value. Therefore, Budeanu's reactive power can be expressed as

$$Q_\text{B} \stackrel{\text{df}}{=} \sum_{n=1}^{\infty} U_n I_n \sin\varphi_n = -\frac{1}{T}\int_0^T i(t)\, \mathscr{H}\{u(t)\}\, dt. \tag{5.11}$$

Thus, when the resistor in the voltage branch of a wattmeter is replaced by a one-port, such that its current, $i_\text{v}(t)$, is equal to the Hilbert transform of the branch voltage, $u(t)$, then the meter measures Budeanu's reactive power. Therefore, conversion of the load voltage to a current which is equal to the Hilbert transform of this voltage opens the path to Budeanu's reactive power measurement.

The Hilbert transform can be regarded as a convolution of the input signal $x(t)$ with an impulse response $h(t)$ of a linear system shown in Fig. 5.5(a), since

$$y(t) \stackrel{\text{df}}{=} \frac{1}{\pi}\int_{-\infty}^{\infty} \frac{x(\tau)}{\tau - t}\, d\tau = \int_{-\infty}^{\infty} x(\tau)\frac{1}{\pi(\tau - t)}\, d\tau = \int_{-\infty}^{\infty} x(\tau)\, h(t-\tau)\, d\tau \stackrel{\text{df}}{=} x(t) * h(t)$$

where the impulse response is equal to

$$h(t) \stackrel{\text{df}}{=} -\frac{1}{\pi t}.$$

Observe, however, that the impulse response is non-zero, even before the system is excited by impulse $\delta(t)$, as it is shown in Fig. 5.5(b).

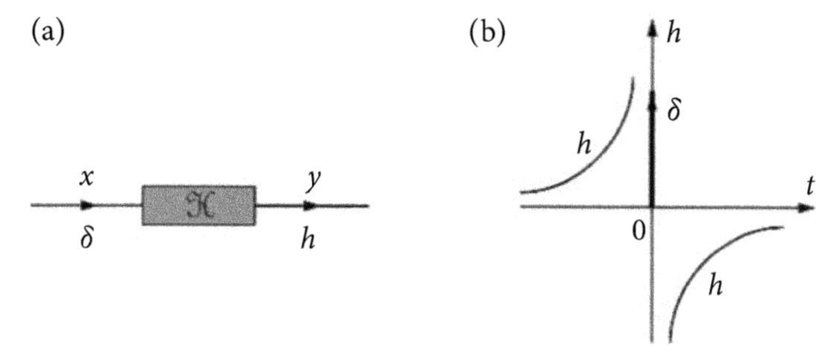

Figure 5.5 (a) A system that converts signals to Hilbert transforms and (b) its impulse response.

It means that a hypothetical system that would convert a signal into a Hilbert transform will violate the *Causality Principle.* Any physical system cannot have such impulse response, meaning it cannot be built.

It was concluded in Ref. [35] however that this constraint does not apply to periodic quantities because such quantities do not convey information. Their values in the future are determined entirely by the past. Therefore, systems that can convert periodic quantities into Hilbert transforms can be built. As proved in Ref. [35], they can be built as reactance one-ports, referred to as *orthonormal one-ports* [53]. A wattmeter with the voltage coil branch resistor replaced by such an orthonormal one-port enables [79] measurement of Budeanu's reactive power, Q_B.

5.5 Fryze's Power Theory

S. Fryze, a Polish professor at Lwów Polytechnic, suggested [17] the reactive power definition that satisfies the power equation (5.1). At the same time, Fryze introduced several important and permanent ideas to the power theory, as well as the term "*power theory*" itself. The most important of these ideas were: the concept of decomposition of the load current into *orthogonal components*; the concept of *active current* and the need of defining power quantities in the *time domain*, without using the Fourier series.

The observation that a purely resistive load is the best load from the point of view of energy conversion was the starting point for Fryze's PT. Such a purely resistive load, shown in Fig. 5.6b, is equivalent considering the active power P at the same supply voltage $u(t)$, to the original load, shown in Fig 5.6a, when

$$G_e||u||^2 = P$$

meaning when its conductance is equal to

$$G_e = \frac{P}{||u||^2}. \tag{5.12}$$

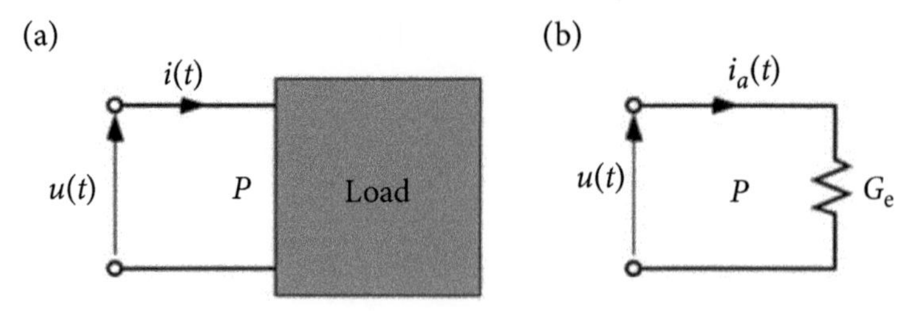

Figure 5.6 (a) A load and (b) a resistive equivalent load of the same active power P.

It is referred to as *equivalent conductance.* The current of such a resistive load is

$$i_a(t) = G_e\, u(t). \tag{5.13}$$

This current is referred to as the *active current.* Its rms value is equal to

$$||i_a|| = \frac{P}{||u||^2}||u|| = \frac{P}{||u||}$$

and this is the minimum value of the load current for a load of active power P. The remaining current is *Fryze's reactive current*

$$i_{rF} \overset{\text{df}}{=} i - i_a. \tag{5.14}$$

The scalar product of these two current components is

$$(i_a, i_{rF}) = (i_a, i - i_a) = (i_a, i) - (i_a, i_a) =$$

$$= \left(\frac{P}{||u||^2}u, i\right) - ||i_a||^2 = \frac{P}{||u||^2}(u, i) - ||i_a||^2 = \frac{P^2}{||u||^2} - \left(\frac{P}{||u||}\right)^2 = 0;$$

thus, these two currents are mutually orthogonal and consequently, their rms values satisfy the relationship

$$||i||^2 = ||i_a||^2 + ||i_{rF}||^2. \tag{5.15}$$

When both sides of this equation are multiplied by the square of the voltage rms value $||u||$, Fryze's power equation is obtained:

$$S^2 = P^2 + Q_F^2 \tag{5.16}$$

with Fryze's definition of the reactive power

$$Q_F \overset{\text{df}}{=} ||u||\, ||i_{rF}||.$$

Fryze's PT, originally published in Polish in a series of papers [17], has become known internationally mainly through its German translation [18] and the German Standard DIN 40110 [37].

Fryze was aware that the endowment of harmonics with physical existence, while they are only mathematical entities, could lead to major physical misinterpretations of power phenomena in electrical circuits. One such misinterpretation was demonstrated in Chapter 1.

Therefore, while developing his concept of the power theory, Fryze strongly emphasized the need for defining power quantities in the time domain, without any use of the Fourier Series and harmonics. Indeed, before computer-based digital signal processing methods were developed, this was a very attractive feature of Fryze's approach. In Budeanu's and Fryze's time, it was not possible practically to measure phases of harmonics. Harmonic analyzers built at that time, composed of analog filters, were capable of measuring only the rms value of harmonics, but not their phase. Nonetheless, Fryze's concept was dominated by the theory developed by Budeanu.

Summary: Fryze's PT introduced a few important concepts. First, it was the concept of the active current and decomposition on the current level, rather than on the power level, as well as the concept of current components orthogonality. Equally important was his conclusion that only apparently the power factor declines because of energy oscillations. His warning that mathematical entities such as harmonics should not be regarded as physical ones, was also very important for further studies.

Let us summarize the first phase of the power theory-development. Two substantially different concepts of PT were developed by Budeanu and by Fryze in this phase. By the end of this phase, Milic, [36], Nowomiejski [65], and Fisher [67] provided very advanced and sophisticated mathematical forms for Budeanu's PT, but unfortunately without any progress on the issue of power factor improvement, even in single-phase circuits with LTI loads.

5.6 Shepherd and Zakikhani's Power Theory

In 1972, W. Shepherd and P. Zakikhani at Bradford University, England, suggested [39] a power equation based on the load current decomposition into a component that is *in phase* with the supply voltage and a component with harmonics shifted by $\pi/2$ versus the supply voltage harmonics. This approach will be denoted, for simplicity, as *S&Z power theory*. At the supply voltage

$$u = U_0 + \sqrt{2} \sum_{n=1}^{\infty} U_n \cos(n\omega_1 t + \alpha_n)$$

the current of an LTI load

$$i = I_0 + \sqrt{2} \sum_{n=1}^{\infty} I_n \cos(n\,\omega_1 t + \alpha_n - \varphi_n)$$

can be decomposed as follows

$$i = I_0 + \sqrt{2}\sum_{n=1}^{\infty} I_n \cos\varphi_n \,\cos(n\,\omega_1 t + \alpha_n) + \sqrt{2}\sum_{n=1}^{\infty} I_n \sin\varphi_n \,\sin(n\,\omega_1 t + \alpha_n) = i_{\mathrm{R}} + i_{\mathrm{rS}}$$

where

$$i_{\mathrm{R}} \stackrel{\mathrm{df}}{=} I_0 + \sqrt{2}\sum_{n=1}^{\infty} I_n \cos\varphi_n \,\cos(n\omega_1 t + \alpha_n) \tag{5.18}$$

is the load current component in phase with the supply voltage, referred to as a *resistive current*, and a component shifted versus the voltage harmonics by $\pi/2$, namely

$$i_{\mathrm{rS}} \stackrel{\mathrm{df}}{=} \sqrt{2}\sum_{n=1}^{\infty} I_n \sin\varphi_n \,\sin(n\omega_1 t + \alpha_n) \tag{5.19}$$

referred to as a *quadrature or reactive current*, Thus, according to S&Z PT,

$$i(t) = i_{\mathrm{R}}(t) + i_{\mathrm{rS}}(t)\,. \tag{5.20}$$

Since harmonics of the resistive and reactive currents are shifted mutually by $\pi/2$, their scalar product

$$(i_{\mathrm{R}}, i_{r\mathrm{S}}) = 0.$$

Thus, they are mutually orthogonal, and hence, their rms values satisfy the relationship

$$||i||^2 = ||i_{\mathrm{R}}||^2 + ||i_{\mathrm{rS}}||^2 \tag{5.21}$$

where

$$||i_{\mathrm{R}}|| = \sqrt{\sum_{n=0}^{\infty} (I_n \cos\varphi_n)^2} \tag{5.22}$$

$$||i_{\mathrm{rS}}|| = \sqrt{\sum_{n=1}^{\infty} I_n^2 \sin^2\varphi_n}. \tag{5.23}$$

Formula (5.21) multiplied by the square of the voltage rms value, results in the S&Z power equation of single-phase LTI loads in the form

$$S^2 = S_R^2 + Q^2 \tag{5.24}$$

with

$$S_R \stackrel{\text{df}}{=} ||u||\,||i_R|| \tag{5.25}$$

referred to as a *resistive power* and

$$Q \stackrel{\text{df}}{=} ||u||\,||i_{rS}|| \tag{5.26}$$

which is Shepherd and Zakikhani's definition of reactive power.

Observe that there is a lack of the most important power, namely the active power P in this equation. Although the resistive power S_R occurs in the effect of the current component which is in line with the supply voltage, it is not equal to the active power but fulfills the inequality

$$S_R \geq P.$$

This lack of active power in this equation caused the S&Z power equation to be met with a cold reception.

5.7 Optimal Capacitance

Despite the cognitive deficiencies of Shepherd and Zakikhani's approach, they obtained the first practical result on power factor improvement in circuits with nonsinusoidal voltages and currents.

They concluded that harmonics of the current i_C of a shunt capacitor connected as shown in Fig. 5.7 cannot have components in phase with the supply voltage, thus this current does not affect the resistive current. Only the reactive component of the supply current is affected by the capacitor.

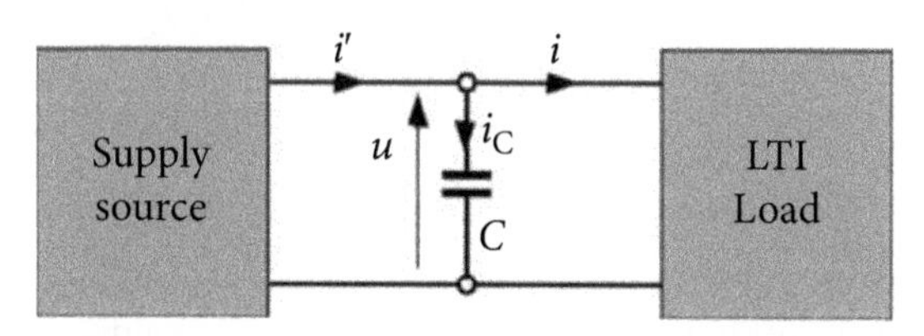

Figure 5.7 An LTI load with compensating capacitor.

The capacitor changes the reactive current to

$$i'_{rS} = C\frac{du}{dt} + i_{rS}.$$

It can be expressed in the form

$$\begin{aligned} i'_{rS} &= -\sqrt{2}\sum_{n=1}^{\infty} n\omega_1 C\, U_n \sin(n\omega_1 t + \alpha_n) + \sqrt{2}\sum_{n=1}^{\infty} I_n \sin\varphi_n \sin(n\omega_1 t + \alpha_n) = \\ &= \sqrt{2}\sum_{n=1}^{\infty} (I_n \sin\varphi_n - n\omega_1 C\, U_n)\sin(n\,\omega_1 t + \alpha_n) \end{aligned}$$

and it has the rms value

$$||\, i'_{rS}\, || = \sqrt{\sum_{n=1}^{\infty} (I_n \sin\varphi_n - n\omega_1 C\, U_n)^2}$$

dependent, of course, on the capacitance C. This formula can be used for calculating the capacitance C that minimizes the reactive current rms value. It is necessary that at such capacitance, the derivative

$$\frac{d}{dC}||i'_{rS}|| = 0.$$

Shepherd and Zakikhani calculated this value assuming that rms values of harmonics of the load voltage and current I_n and U_n are not affected by the capacitance. In such a case

$$\begin{aligned} \frac{d}{dC}||i'_{rS}||^2 &= \frac{d}{dC}\sum_{n=1}^{\infty} (I_n \sin\varphi_n - n\omega_1 C\, U_n)^2 = \\ &= -\sum_{n=1}^{\infty} 2(I_n \sin\varphi_n - n\,\omega_1 C\, U_n)\, n\,\omega_1\, U_n = 0. \end{aligned}$$

The last formula can be solved versus the capacitance C. Thus, condition (5.40) is fulfilled at capacitance equal to

$$C = C_{opt} = \frac{\sum_{n=1}^{\infty} n\, U_n I_n \sin\varphi_n}{\omega_1 \sum_{n=1}^{\infty} n^2 U_n^2} = \frac{\sum_{n=1}^{\infty} n\, Q_n}{\omega_1 \sum_{n=1}^{\infty} n^2 U_n^2}. \qquad (5.27)$$

This capacitance was called by Shepherd and Zakikhani the *optimal capacitance*. According to S&Z, it minimizes the reactive current rms value, and consequently, it minimizes the rms value of the supply current.

Illustration 5.1 *The RL load of the structure shown in Fig. 5.8, at a sinusoidal voltage of the rms value* $U_1 = 230$ V, *has the active power* $P_1 = 35$ kW *with the power factor* $\lambda_1 = 0.5$.

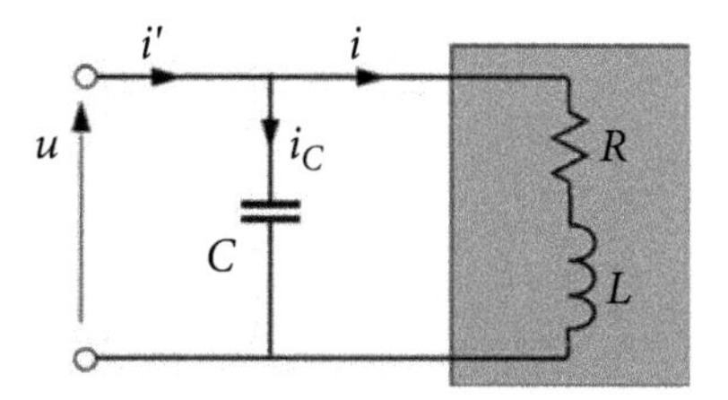

Figure 5.8 An RL load with compensating capacitor.

Let us calculate the optimum capacitance C_{opt} *for the load compensation if the supply voltage is distorted by harmonics of the* 5th, 7th, *and* 11th *order of the relative rms value*

$$U_5 = 6.0\% U_1, \quad U_7 = 4.0\% U_1, \quad U_{11} = 2.0\% U_1.$$

Calculation of the optimum capacitance according to formula (5.27) requires that the reactive power of harmonics of the 5th, 7th, *and* 11th *order is known. It requires that the load susceptance* B_n *for these harmonics is calculated.*

Since the phase shift for the fundamental frequency is

$$\varphi_1 = \text{arc} \ \cos(\lambda_1) = 60^{\circ}$$

the load conductance and susceptance for the fundamental frequency are equal to

$$G_1 = \frac{P_1}{U_1^2} = \frac{35 \times 10^3}{230^2} = 0.661 \text{ S}, \qquad B_1 = -\frac{Q_1}{U_1^2} = -G_1 \ \text{tg}\varphi_1 = -1.146 \text{ S}.$$

Thus, the load resistance and reactance for this frequency are

$$R_1 = \frac{G_1}{G_1^2 + B_1^2} = \frac{0.661}{0.661^2 + 1.146^2} = 0.378 \ \Omega, \quad X_1 = -\frac{B_1}{G_1^2 + B_1^2} = 0.654 \ \Omega.$$

The load susceptance for the supply voltage harmonics is equal to

$$B_1 = -1.146 \text{ S}, \quad B_5 = -0.301 \text{ S}, \quad B_7 = -0.217 \text{ S}, \quad B_{11} = -0.138 \text{ S}$$

and hence, the reactive power of the supply voltage harmonics

$$Q_n = -B_n \ U_n^2$$

are equal to

$$Q_1 = 60619 \text{ VAr}, \ Q_5 = 57.4 \text{ VAr}, \quad Q_7 = 19.3 \text{ VAr}, \quad Q_{11} = 11.7 \text{ VAr}.$$

The optimum capacitance calculated with the S&Z formula (5.27), assuming that the frequency is normalized to $\omega_1 = 1$ *rad/s, is equal to*

$$C_{\text{opt}} = \frac{\sum\limits_{n=1,5,7,11} n\,Q_n}{\sum\limits_{n=1,5,7,11} n^2 U_n^2} = 0.85 \text{ F}$$

Observe, that at sinusoidal supply voltage this capacitance, denoted as C_0, *has to satisfy the relation* $\omega_1 C_0 = -B_1$ *meaning, at* $\omega_1 = 1$ rad/s, *this capacitance is equal to* $C_0 = 1.15$ F. *Thus, the optimum capacitance at distorted voltage is substantially lower than this capacitance in the lack of the supply voltage distortion. These results are confirmed by the plot of the power factor* λ *versus compensating capacitance C, shown in Fig. 5.9.*

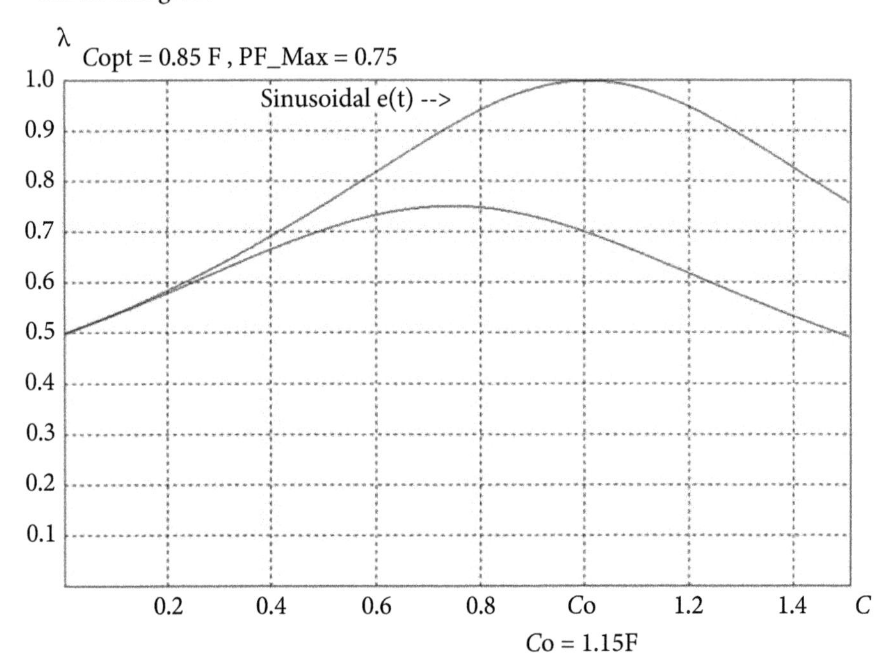

Figure 5.9 The plot of power factor λ versus compensating capacitance C.

This plot shows how the power factor λ *changes with the capacitance at a sinusoidal supply voltage and at the voltage distorted by harmonics of the order and the rms as assumed in this example. It shows that the power factor in the presence of the supply voltage distortion cannot reach the unity value however, even if the capacitance has the optimum value. In the situation considered, the power factor cannot be higher than* $\lambda_{\text{max}} = 0.75$.

*****

Summary: Shepherd and Zakikhani introduced a new definition of the reactive current for circuits with nonsinusoidal supply voltage. This reactive current has the right physical interpretation as the result of the phase shift between the voltage and current harmonics. The method of compensating capacitance calculation, as developed by S&Z, was the first practical result of the power theory development. It requires that the load voltage and current harmonics rms values and their mutual phase shifts are known. Thus, the optimum capacitance is calculated in the frequency domain.

5.8 Depenbrock's Power Theory

In 1979, M. Depenbrock, Professor of Aachen University, suggested [55] the load current and voltage decomposition, which emphasizes the importance of the fundamental harmonic, i_1 and u_1, in the process of energy transmission, namely

$$u = u_1 + u_h, \qquad i = i_1 + i_h. \tag{5.28}$$

The fundamental harmonic of the load current, i_1, is next decomposed into the active and reactive components, i_{1a}, i_{1r}, defined as

$$i_{1a} \stackrel{\text{df}}{=} \frac{P_1}{||u_1||^2} u_1, \quad i_{1r} \stackrel{\text{df}}{=} \frac{Q_1}{||u_1||^2} u_1\left(t - \frac{T}{4}\right) \tag{5.29}$$

Symbols P_1 and Q_1 in this equation denote the active and reactive power of the fundamental harmonic. Apart from the active and reactive components, the current fundamental harmonic contains also the component

$$i_{1V} \stackrel{\text{df}}{=} \left(\frac{P_1}{||u_1||^2} - \frac{P}{||u||^2}\right) u_1. \tag{5.30}$$

The distortion current, i_h, can be decomposed into components

$$i_{ha} \stackrel{\text{df}}{=} \frac{P_h}{||u_h||^2} u_h, \qquad i_{hV} \stackrel{\text{df}}{=} \left(\frac{P_h}{||u_h||^2} - \frac{P}{||u_h||^2}\right) u_h \tag{5.31}$$

where P_h is the active power of harmonics. The sum of such defined currents can, however, be lower than the load current. Their difference is

$$i_N \stackrel{\text{df}}{=} i - (i_{1a} + i_{1r} + i_{1V} + i_{ha} + i_{hV}). \tag{5.32}$$

Thus, according to Depenbrock, the load current can be decomposed as follows:

$$i = i_{1a} + i_{1r} + i_{1V} + i_{ha} + i_{hV} + i_N. \tag{5.33}$$

Unfortunately, apart from the physical meanings of the active and reactive components of the current fundamental harmonic, the physical meanings of the remaining components remain unclear. The power equation can be written in the form

$$S^2 = P^2 + Q_1^2 + V^2 + N^2 \tag{5.34}$$

with

$$V \stackrel{\text{df}}{=} ||u|| \sqrt{||i_{1V}||^2 + ||i_{hV}||^2}, \quad N \stackrel{\text{df}}{=} ||u||\ ||i_N||. \tag{5.35}$$

Summary: Depenbrock's observation that the fundamental harmonics of the load voltage and current have particular importance for energy transmission, and consequently, these harmonics should be discriminated in the voltage and current decomposition, was an important contribution to PT.

5.9 Kusters and Moore's Power Theory

N.L. Kusters and W.J.M. Moore of the National Research Council (NRC) of Canada suggested [59] in 1980 a decomposition of the load current in the time domain, oriented at capacitive compensation of RL loads or inductive compensation of RC loads. Thus, their PT, denoted shortly as *K&M's power theory,* has two different formulations.

For RL loads, much more common in distribution systems than resistive-capacitive loads, Kusters and Moore decomposed the load current into the active current, a *reactive capacitive current,* i_{qC}, and a *residual reactive capacitive current,* i_{qCr}, defined as follows:

$$i_{qC}(t) \stackrel{\text{df}}{=} \frac{(\dot{u}, i)}{||\dot{u}||^2} \dot{u}(t), \quad i_{qCr} \stackrel{\text{df}}{=} i - i_a - i_{qC} \tag{5.36}$$

where $\dot{u} = du/dt$ denotes the derivative of the supply voltage. Thus,

$$i = i_a + i_{qC} + i_{qCr} \tag{5.37}$$

According to [59] the reactive capacitive current, i_{qC}, is the current that can be entirely compensated by the optimum capacitance, C_{opt}, that is,

$$i_{qC}(t) = -C_{opt}\, \dot{u}(t) = \frac{(\dot{u}, i)}{||\dot{u}||^2}\, \dot{u}(t). \tag{5.38}$$

The active current is proportional to the supply voltage, $i_a(t) = G_e\, u(t)$, while the reactive capacitive current, i_{qC}, is proportional to this voltage derivative, thus these two currents are mutually orthogonal, since

$$(i_a, i_{qC}) = G_e \frac{(\dot{u}, i)}{||\dot{u}||^2} (u, \dot{u}) = 0.$$

Let us calculate the scalar products of the remaining components of the current.

$$\begin{aligned}(i_a, i_{qCr}) &= (i_a, [i - i_a - i_{qC}]) = (i_a, i) - ||i_a||^2 \\ &= G_e (u, i) - G_e{}^2||u||^2 = G_e(P - P) = 0\end{aligned}$$

thus, they are orthogonal. Similarly,

$$\begin{aligned}(i_{qC}, i_{qCr}) &= (i_{qC}, [i - i_a - i_{qC}]) = (i_{qC}, i) - ||i_{qC}||^2 \\ &= -C_{opt}(\dot{u}, i) - C^2_{opt} ||\dot{u}||^2 = C^2_{opt} ||\dot{u}||^2 - C^2_{opt} ||\dot{u}||^2 = 0.\end{aligned}$$

Thus, all current components defined in K&M's power theory are mutually orthogonal. Therefore, their rms values satisfy the relationship

$$||i||^2 = ||i_a||^2 + ||i_{qC}||^2 + ||i_{qCr}||^2.$$

When this equation is multiplied by the square of the supply voltage rms value, Kusters and Moore's power equation is obtained

$$S^2 = P^2 + Q^2_C + Q^2_{Cr} \tag{5.39}$$

with

$$Q_C \overset{df}{=} ||u|| \, ||i_{qC}|| \, \text{sgn}(\dot{u}, i), \quad Q_{Cr} \overset{df}{=} ||u|| \, ||i_{qCr}||. \tag{5.40}$$

where the symbol $\text{sgn}(\dot{u},i)$ denotes the sign of the scalar product $(\dot{u},i)$.

This theory introduces the concept of a *capacitive reactive power* Q_C, which is compensated by the optimum capacitance, C_{opt},

$$C_{opt} = -\frac{Q_C}{||u|| \, ||\dot{u}||} \tag{5.41}$$

Observe that the crms values of the voltage and current harmonics are not needed for the optimum capacitance calculation. This was substantial progress as compared to the S&Z's PT-based approach to this capacitance calculation. Therefore, Kusters and Moore's PT was recommended [57] in 1979 by the International Electrotechnical Commission (IEC). More sophisticated mathematical fundamentals for this concept were provided by Fodor in Ref [68]. Its generalization for all linear time-invariant (LTI) loads was developed by Page in Ref. [60].

5.9.1 Optimum capacitance converter

An electronic converter that converts, in real time, the load voltage $u(t)$ and current $i(t)$ into a DC voltage U proportional to the optimum capacitance C_{opt}, was proposed in Ref. [63] by the author of this book and is patented. It operates according to K&M's concept, in the time domain. Its structure is shown in Fig. 5.10.

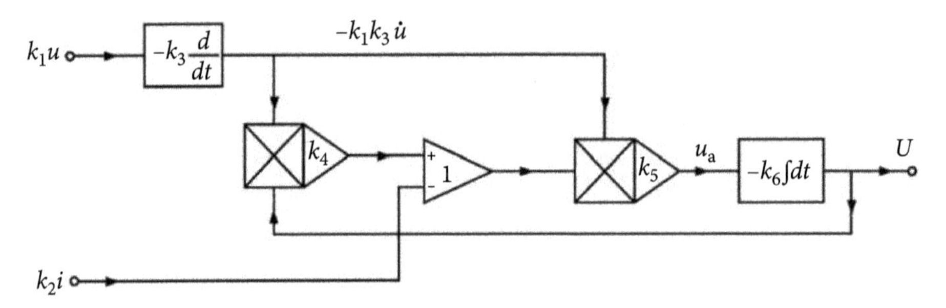

Figure 5.10 A structure of a converter of the load voltage and current into a DC voltage proportional to C_{opt}.

The converter is composed of two analog multipliers, differentiating circuit and an integrator. The input signals, $k_1 u$ and $k_2 i$, are proportional to the load voltage and current, respectively.

The converter output voltage U is constant on the condition that the mean value of the input voltage of the integrator is zero, meaning

$$\frac{1}{T}\int_0^T u_a\, dt = \frac{1}{T}\int_0^T k_5\,(-k_1 k_3\, \dot{u}) \times k_4(-k_1 k_3\, \dot{u} U - k_2\, i)\, dt = 0.$$

This formula can be rearranged to

$$k_1 k_3\, k_4 k_5 \frac{1}{T}\int_0^T (k_1 k_3\, \dot{u}^2\, U + k_2\, \dot{u}\, i)\, dt = 0.$$

It is fulfilled when

$$k_1 k_3\, ||\dot{u}||^2\, U + k_2(\dot{u}, i) = 0,$$

meaning when the converter output voltage U is equal to

$$U = -\frac{k_2(\dot{u}, i)}{k_1 k_3\, ||\dot{u}||^2} = \frac{k_2}{k_1 k_3} C_{opt} \tag{5.42}$$

that is, when it is proportional to the optimum compensating capacitance C_{opt}.

5.10 Czarnecki's Power Theory of Single-Phase LTI Circuits

For linear loads supplied with a nonsinusoidal voltage

$$u \stackrel{\text{df}}{=} U_0 + \sqrt{2}\,\text{Re} \sum_{n \in N} \boldsymbol{U}_n\, e^{jn\,\omega_1 t}$$

L.S. Czarnecki introduced, in 1983 [71], the following decomposition of the supply current

$$i(t) = i_{\text{a}}(t) + i_{\text{s}}(t) + i_{\text{r}}(t) \tag{5.43}$$

where $i_a(t)$ denotes the active current defined in Fryze's PT

$$i_{\text{a}}(t) = G_{\text{e}}\, u(t), \qquad G_{\text{e}} = \frac{P}{||u||^2}. \tag{5.44}$$

When the load for harmonic frequencies has the admittance

$$\boldsymbol{Y}_n = G_n + jB_n$$

then

$$i_{\text{s}}(t) \stackrel{\text{df}}{=} (G_0 - G_{\text{e}})\boldsymbol{U}_n + \sqrt{2}\,\text{Re} \sum_{n \in N} (G_n - G_{\text{e}})\, \boldsymbol{U}_n\, e^{jn\,\omega_1 t} \tag{5.45}$$

is the scattered current, and

$$i_{\text{r}}(t) \stackrel{\text{df}}{=} \sqrt{2}\,\text{Re} \sum_{n \in N} jB_n\, \boldsymbol{U}_n\, e^{jn\,\omega_1 t} \tag{5.46}$$

is the reactive current.

The current components in decomposition (5.43) are mutually orthogonal so that their rms values satisfy the relationship

$$||i||^2 = ||i_{\text{a}}||^2 + ||i_{\text{s}}||^2 + ||i_{\text{r}}||^2. \tag{5.47}$$

Thus, the load current rms value is specified by the phenomenon of the permanent energy transmission which requires that the supply has the active component i_{a}, and by the phenomenon of the phase shift of the load current harmonics versus the voltage harmonics, which creates the reactive current i_{r}. Decomposition (5.43) reveals a new phenomenon that causes an increase of the load current rms value, namely the change of the load conductance for harmonic frequencies, G_n, around the load equivalent conductance G_{e}. The scattered current $i_{\text{s}}(t)$ occurs in the load current due to this phenomenon. Consequently, decomposition (5.43) along with relationship (5.47) explain all causes for which the power factor of linear loads with a nonsinusoidal supply voltage can be lower than 1. This decomposition was a turning point in studies on the power theory development.

The current components in decomposition (5.43) are associated with distinctive energy flow-related phenomena, therefore, these components were later, from paper [195], referred to as *Currents' Physical Components* (CPC).

Decomposition (5.43) also solves the issue of reactive compensation in circuits with linear loads and nonsinusoidal supply voltage. Namely, the reactive current can be compensated entirely by a shunt reactance compensator of susceptance B_{Cn} such that for each harmonic of the order n from set N,

$$B_{Cn} = -B_n. \tag{5.48}$$

At the same time, such a compensator does not affect the scattered current, so that cannot be compensated. It means, that the power factor cannot be improved to a value higher than

$$\lambda_{\max} = \frac{||i_a||}{\sqrt{||i_a||^2 + ||i_s||^2}}. \tag{5.49}$$

This result has provided the answer to the question asked in the PT of circuits with nonsinusoidal voltages and currents: *can any LTI load be compensated to the unity power factor?* The answer is no. When the load conductance G_n changes with harmonic order so that the load draws the scattered current, then the load cannot be compensated by a shunt reactance compensator to unity.

5.11 Instantaneous Reactive Power p-q Theory

H. Akagi, Y. Kanazawa, and A. Nabae, Professors at the Tokyo Institute of Technology, suggested [72] in 1983, a power theory of three-phase circuits with nonsinusoidal voltages and currents, referred to as the *Instantaneous Reactive Power* (IRP) *p-q Theory.*

Three-phase voltages and currents in the IRP p-q Theory are transformed into orthogonal α, β, and 0 coordinates with the Clarke Transform, and two instantaneous powers, active and reactive, are defined. Line currents and line-to-artificial zero voltages in three-wire circuits, denoted by common symbols $x_R(t), x_S(t)$, and $x_T(t)$, satisfy the relation

$$x_R(t) + x_S(t) + x_T(t) \equiv 0$$

thus, only two quantities are independent. Therefore, the Clarke Transform, which in general is more complex, for such three-phase mutually dependent quantities has the form

$$\begin{bmatrix} x_\alpha \\ x_\beta \end{bmatrix} \stackrel{\text{df}}{=} \begin{bmatrix} \sqrt{3/2}, & 0 \\ 1/\sqrt{2}, & \sqrt{2} \end{bmatrix} \begin{bmatrix} x_\mathrm{R} \\ x_\mathrm{S} \end{bmatrix} \stackrel{\text{df}}{=} \mathbf{C} \begin{bmatrix} x_\mathrm{R} \\ x_\mathrm{S} \end{bmatrix}. \tag{5.50}$$

Having voltages and currents in α and β coordinates, the *instantaneous active power*

$$p \stackrel{\text{df}}{=} u_\alpha i_\alpha + u_\beta i_\beta \tag{5.51}$$

and the instantaneous reactive power

$$q \stackrel{\text{df}}{=} u_\alpha i_\beta - u_\beta i_\alpha \tag{5.52}$$

are defined. According to Akagi and Nabae, [77, 111] these two powers can instantaneously specify the power properties of three-phase loads. They stated that the development of the IRP p-q Theory, was a response to “the demand to instantaneously compensate the reactive power”. The adverb “instantaneous” in the name of this theory and definitions of p and q powers in terms of the instantaneous value of voltages and currents suggest the possibility of instantaneous identification and compensation of the reactive power of a three-phase load. This is one of the main reasons for this theory’s attractiveness, both as a theoretical fundamental of control algorithms and as a power theory.

The instantaneous active power has a constant and an oscillating component

$$p = \bar{p} + \tilde{p}. \tag{5.53}$$

According to IRP p-q Theory, a compensator, connected as shown in Fig. 5.11

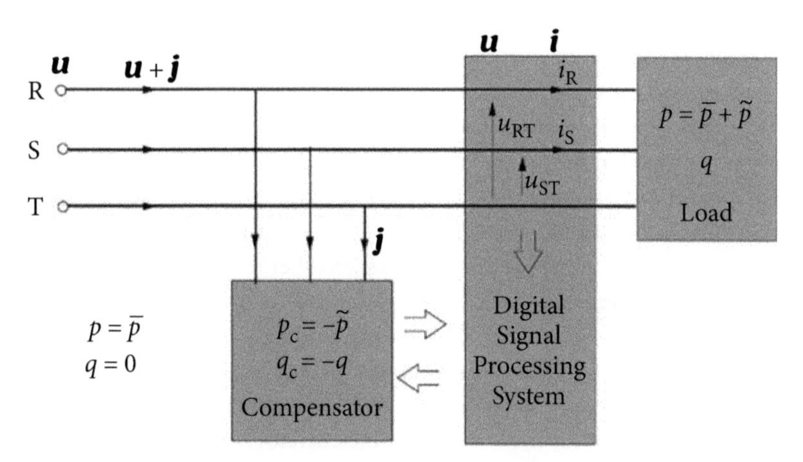

Figure 5.11 The compensation principle according to IRP p-q theory.

should be controlled in such a way, that it burdens the supply source with the negative values of the load instantaneous reactive power and the oscillating component of the load instantaneous active power. Thus, in the effect of such control, the instantaneous active power of the supply source after compensation should be constant.

Instantaneous powers, p and q, can be obtained almost instantaneously. Only a few multiplications and additions are needed for that, which means these quantities can provide data for instantaneous compensation. Indeed, Nabae and Akagi asserted in Ref. [111] that this concept was developed to enable instantaneous compensation of fast varying loads.

5.12 CPC in Single-Phase Circuits with Harmonics Generating Loads

Decomposition (5.43) of linear loads current into CPC was generalized in 1983 [73] to circuits that, due to nonlinearity or periodically time-varying parameters, are the sources of current harmonics, later [99] referred to as *Harmonics Generating Loads* (HGLs).

Unfortunately, the power theories as suggested by Budeanu, Fryze, S&Z, K&M, and Depenbrock did not provide answers to Steinmetz's question. In his experiment, the supply voltage was sinusoidal, while the current was distorted only because of the nonlinearity of the arc bulb. Just such a distortion disturbed the traditional power equation and motivated his question. In theories referenced above and developed by 1983, the distortion of the supply voltage disturbs the traditional power equation, while the load nonlinearity is not a matter of concern in these theories.

Load nonlinearity can disturb the traditional power equation because such a nonlinearity can originate the current harmonics. The same effect can have a periodic time variance of the load parameters, usually caused by the power switches, mainly power transistors or thyristors. These are just the HGLs. Steinmetz's question occurred as a conclusion drawn from measurements of powers in a very simple circuit with an HGL.

When the supply source is ideal, meaning with zero internal impedance, and has a sinusoidal voltage, the current harmonics originated in an HGL do not permanently transfer energy between the supply source and the load because they are orthogonal to the sinusoidal supply voltage. Nonetheless, they contribute to an increase in the supply current rms value, so to an increase in the apparent power S at the load terminals. This is a new component of this power, other than the active and reactive powers, p and q. The presence of the load-generated harmonics can be recognized in such a situation [73] only by the comparison of the set of harmonic orders n of the voltage harmonics with the set of orders of the current harmonics. New orders of the current harmonics can occur only because of their generation by the load. Some orders could be common for the voltage and the current, however.

This problem can be partially solved in real circuits where the supply source has some resistance as shown in Fig. 5.12.

In such a situation, when the supply voltage harmonic causes a permanent flow of energy, then the active power of that harmonic P_n is positive. When this flow is caused by the load-originated harmonic, then the harmonic active power P_n is negative. Thus, the detection, by measurement at the load terminals, of the sign of

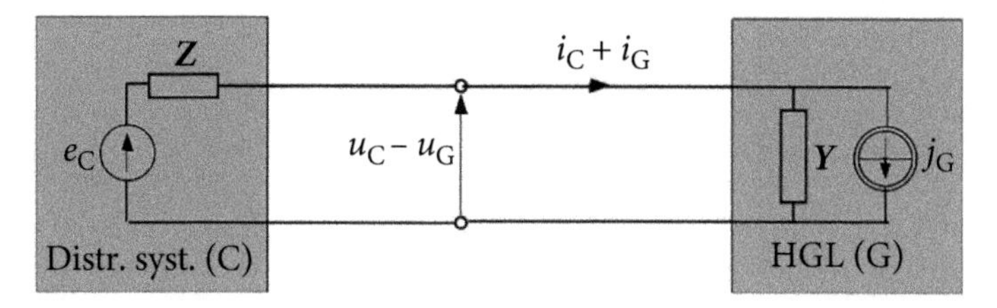

Figure 5.12 A cross-section between a distribution system and an HGL.

P_n enables the decomposition of the current into a component i_C responsible for the transfer of energy to the load and a component i_G responsible for its transfer from the load to the supply source. Thus, the current of an HGL can be decomposed (see Section 6.11),

$$i(t) = i_C(t) + i_G(t) \tag{5.54}$$

and the voltage

$$u(t) = u_C(t) - u_G(t). \tag{5.55}$$

The voltage u_G enters the last formula with the negative sign because this voltage is the supply source response for the load-generated current i_G.

The current component i_G in the supply current is the effect of the harmonics generated by the HGL. When this effect is ignored then the remaining current component i_C can be regarded as a linear load response to the supply voltage, thus it can be decomposed into the active, scattered, and reactive currents, so the current of an HGL can be decomposed into four Current's Physical Components:

$$i = i_{Ca} + i_{Cs} + i_{Cr} + i_G. \tag{5.56}$$

These components are mutually orthogonal, hence

$$||i||^2 = ||i_{Ca}||^2 + ||i_{Cs}||^2 + ||i_{Cr}||^2 + ||i_G||^2. \tag{5.57}$$

This formula reveals four physical phenomena responsible for the rms value of the supply current in single-phase circuits with an HGL.

Very often, the HGL-originated current harmonics are much higher than those caused by the supply voltage harmonics. In such a situation, the supply voltage harmonics can be ignored and assumed that the supply voltage is sinusoidal. Detection of the sign of the harmonic active power is no longer needed for the identification of the HGL-originated current i_G. All current harmonics form this current, while the current i_C is identical to the current fundamental harmonic i_1. It means that the circuit with an HGL can be approximated by an equivalent circuit shown in Fig. 5.13.

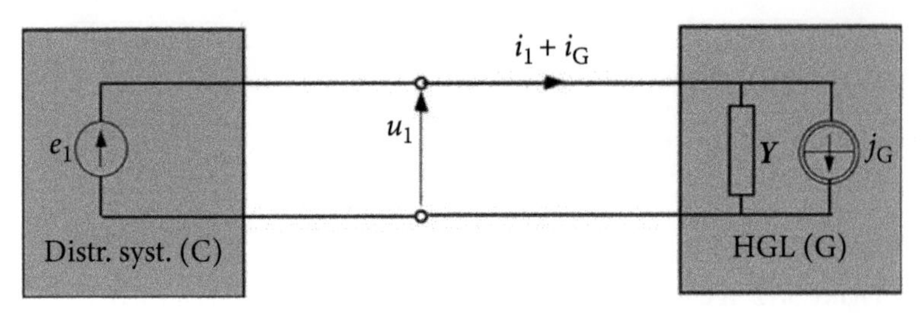

Figure 5.13 An equivalent circuit of an HGL supplied from an ideal source of a sinusoidal voltage.

The current decomposition (5.56) can be reduced in such a case to

$$i = i_{1a} + i_{1r} + i_G. \tag{5.58}$$

In the case of Steinmetz's experiment, because the inductance of the arc bulb is zero, this equation can be reduced even further, namely to

$$i = i_{1a} + i_G.$$

so the square of the rms value of the supply current is

$$||i||^2 = ||i_{1a}||^2 + ||i_G||^2. \tag{5.59}$$

and the power equation

$$S^2 = P^2 + D_G^2. \tag{5.60}$$

with

$$S = U_1||i||, \quad P = U_1||i_{1a}||, \quad D_G = U_1||i_G||.$$

Thus, the answer to Steinmetz's question on how to describe his experiment with the electric bulb in power terms is quite simple. It was found, however, in terms of the CPC-based approach to the power theory but not with any other PT as suggested by Budeanu, Fryze, S&Z, K&M, or Depenbrock.

5.13 CPC in Three-Phase Circuits

The lack of a credible power theory of single-phase circuits and confusion as to the definition of the apparent power S were the main obstacles to the development of the power theory of three-phase circuits.

The CPC-based PT of single-phase circuits was generalized [93] in 1988 to three-phase circuits. Assuming that the supply voltage is nonsinusoidal but symmetrical and it is expressed in the form of the three-phase vector

$$\boldsymbol{u} = \sqrt{2}\mathrm{Re}\sum_{n\in N}\begin{bmatrix} U_{\mathrm{R}n} \\ U_{\mathrm{S}n} \\ U_{\mathrm{T}n} \end{bmatrix} e^{jn\omega_1 t} = \sqrt{2}\mathrm{Re}\sum_{n\in N} \boldsymbol{U}_n\, e^{jn\omega_1 t}$$

then the vector of load line currents

$$\boldsymbol{i} = \sqrt{2}\mathrm{Re}\sum_{n\in N}\begin{bmatrix} I_{\mathrm{R}n} \\ I_{\mathrm{S}n} \\ I_{\mathrm{T}n} \end{bmatrix} e^{jn\omega_1 t} = \sqrt{2}\mathrm{Re}\sum_{n\in N} \boldsymbol{I}_n\, e^{jn\omega_1 t}$$

can be decomposed into CPC of the form

$$\boldsymbol{i} = \boldsymbol{i}_{\mathrm{Ca}} + \boldsymbol{i}_{\mathrm{Cs}} + \boldsymbol{i}_{\mathrm{Cr}} + \boldsymbol{i}_{\mathrm{Cu}} + \boldsymbol{i}_{\mathrm{G}}. \tag{5.61}$$

In this decomposition, the active, scattered, reactive, and the load-generated currents have the same meaning as in single-phase circuits, only they have the form of three-phase vectors. A new physical component

$$\boldsymbol{i}_{\mathrm{Cu}} = \sqrt{2}\mathrm{Re}\sum_{n\in N} Y_{\mathrm{u}n}\boldsymbol{U}_n^{\#}\, e^{jn\omega_1 t} \tag{5.62}$$

referred to as the *unbalanced current*, occurs in this decomposition, however. The symbol $\boldsymbol{Y}_{\mathrm{u}n}$ denotes the unbalanced admittance of the load equal to

$$Y_{\mathrm{u}n} = -(Y_{\mathrm{ST}n} + \alpha_n Y_{\mathrm{TR}n} + \alpha_n^* Y_{\mathrm{RS}n}) \tag{5.63}$$

for the positive sequence harmonics, that is, of the order $n = 3k+1$, $k = 0, 1, 2, \ldots$, and

$$Y_{\mathrm{u}n} = -(Y_{\mathrm{ST}n} + \alpha_n^* Y_{\mathrm{TR}n} + \alpha_n Y_{\mathrm{RS}n}) \tag{5.64}$$

for the negative sequence harmonics, that is, of the order $n = 3k-1$, $k = 1, 2, \ldots$. The upper index "#" in formula (5.62) denotes the vector of crms values of line voltages with switched items $\boldsymbol{U}_{\mathrm{S}n}$ and $\boldsymbol{U}_{\mathrm{T}n}$, while $\boldsymbol{Y}_{\mathrm{ST}n}$, $\boldsymbol{Y}_{\mathrm{TR}n}$, and $\boldsymbol{Y}_{\mathrm{RS}n}$ are line-to-line admittances of the load for harmonic frequencies.

The paper [93] also introduced the concept of *three-phase rms* value $||\boldsymbol{i}||$ of three-phase currents. This is a DC current value that is equivalent concerning the heat released on a resistor of resistance R to the heat released by three-phase currents i_{R}, i_{S}, and i_{T} on a three-phase symmetrical device with the line resistance R. It is equal to

$$||\boldsymbol{i}|| = \sqrt{||i_{\mathrm{R}}||^2 + ||i_{\mathrm{S}}||^2 + ||i_{\mathrm{T}}||^2}. \tag{5.65}$$

Similarly, the three-phase rms value of three-phase voltages was defined, namely

$$||\boldsymbol{u}|| = \sqrt{||u_{\mathrm{R}}||^2 + ||u_{\mathrm{S}}||^2 + ||u_{\mathrm{T}}||^2}. \tag{5.66}$$

Having the three-phase rms value of the voltage and current at the load supply terminals, the apparent power was defined in [93] as

$$S \overset{\mathrm{df}}{=} ||\boldsymbol{u}||\ ||\boldsymbol{i}||. \tag{5.67}$$

5.14 FBD Method

In 1993 M. Depenbrock presented [113] the *FBD Method*, which is an acronym derived from the names of scientists who created fundamentals for that method: Fryze, Buchholz, and Depenbrock. The FBD Method generalizes the PT as developed by Fryze for single-phase circuits in [17] to m-phase circuits, with the definition of the apparent power S as suggested by Buchholz.

The FBD Method is based on the assumption [113] that "none of the conductors is looked at as a special one". It is formulated both on the instantaneous level and the averaging over the period T level, in the time domain.

Some power-related quantities and their symbols in the FBD Method are confined only to that method. Some of them are common to other power theories. Therefore, to reduce ambiguity as to their meaning, these quantities defined in terms of FBD symbols, will be also expressed in terms of common symbols.

If $u_k(t)$ is a voltage of line k, measured versus an artificial zero, and $i_k(t)$ is the current of that line, then FBD defines a *collective instantaneous value* of the voltage and current, which for a three-phase circuit has the form

$$i_{\Sigma}(t) \overset{\mathrm{df}}{=} \sqrt{\sum_{k=1}^{3} i_k^2(t)}, \qquad u_{\Sigma}(t) \overset{\mathrm{df}}{=} \sqrt{\sum_{k=1}^{3} u_k^2(t)}. \tag{5.68}$$

It also defines a *collective rms value*, which is identical to the three-phase rms value, defined in [113], namely

$$||i_{\Sigma}|| \overset{\mathrm{df}}{=} \sqrt{\sum_{k=1}^{3} ||\,i_k\,||^2} \qquad ||\,u_{\Sigma}\,|| \overset{\mathrm{df}}{=} \sqrt{\sum_{k=1}^{3} ||\,u_k\,||^2} \tag{5.69}$$

and the *collective apparent power* S_{Σ}, which is identical to the apparent power S defined in 1988 in paper [93], but denoted by symbols $||\boldsymbol{i}||$ and $||\boldsymbol{u}||$

$$S_{\Sigma} \overset{\mathrm{df}}{=} ||u_{\Sigma}||\ ||i_{\Sigma}|| = ||\boldsymbol{u}||\ ||\boldsymbol{i}|| = S. \tag{5.70}$$

The collective instantaneous power

$$p_{\Sigma}(t) = \sum_{k=1}^{3} u_k(t)\, i_k(t) = p(t) = \frac{dW}{dt} \tag{5.71}$$

is nothing else than the instantaneous power $p(t)$ of the load, that is, the rate of energy W flow from the source to the load, known since the nineteenth century.

Having the collective instantaneous value of the voltage and power, the *power current* in the k-line is defined, namely

$$i_{kp}(t) \stackrel{\text{df}}{=} G_{\text{p}}(t) u_k(t), \quad G_{\text{p}}(t) \stackrel{\text{df}}{=} \frac{p_{\Sigma}(t)}{u_{\Sigma}^2(t)} = \frac{p(t)}{u_{\Sigma}^2(t)}. \tag{5.72}$$

The power current $i_{kp}(t)$ is a time-variant quantity for two reasons. First, the conductance $G_{\text{p}}(t)$ can change in time and it changes with line voltage $u_{\text{k}}(t)$ change. When the load voltage is symmetrical, so $u_{\Sigma}(t) = Const.$ and the load is balanced and purely resistive so that $p(t) = Const. = P$, then conductance $G_{\text{p}}(t) = Const.$ and the power current is proportional to the line voltage $u_{\text{k}}(t)$. The power current can occur in circuits with purely reactive load, therefore it should not be interpreted as a kind of an active current, that is, the current which exists in the presence of the active power P. The active current $i_{\text{ka}}(t)$, that is, the current associated with permanent energy transfer from the source to the load, with the average rate equal to the active power P, is only a component of the power current $i_{kp}(t)$.

A *zero-power current* is next defined in the FBD Method. namely

$$i_{kz}(t) \stackrel{\text{df}}{=} i_k(t) - i_{kp}(t). \tag{5.73}$$

It represents the k-line current component not associated with the energy flow. The reactive current vector $\mathbf{i}_{\text{r}}(t)$, defined in [113], is such a component in sinusoidal situations at symmetrical supply because the instantaneous power associated with this component is

$$p_{\text{r}}(t) = \mathbf{u}^{\text{T}}(t)\, \mathbf{i}_{\text{r}}(t) \equiv 0. \tag{5.74}$$

Thus, the zero power current $i_{kz}(\text{t})$ in the k-line can be regarded as a component of the reactive current vector $\mathbf{i}_{\text{r}}(t)$.

The zero-power current $i_{kz}(\text{t})$ can be used as a reference signal for control of a switching compensator that compensates this current, practically without any time delay. The power current $i_{kp}(t)$ does not provide such a reference. The active component $i_{\text{ka}}(t)$ has to be first extracted from the power current $i_{kp}(t)$ to use next to the remainder as the reference signal. At least one period T of the voltage variability is needed for that.

In the FBD Method, some quantities are defined not only as instantaneous but also based on values averaged over the period T. This is, first of all, an *active current* in the k-line

$$i_{ka}(t) \stackrel{\text{df}}{=} G u_k(t), \quad G \stackrel{\text{df}}{=} \frac{P_\Sigma}{||u_\Sigma||^2} = \frac{P}{||\boldsymbol{u}||^2} = G_e. \tag{5.75}$$

which in the CPC power theory is referred to [71] as the equivalent conductance of the LTI loads. It is identical to the equivalent conductance as defined by Fryze [17] for single-phase loads. The remaining current in the k-line, after the active current is subtracted

$$i_{kn}(t) \stackrel{\text{df}}{=} i_k(t) - i_{ka}(t) \tag{5.76}$$

is a *nonactive current* of the k-line. It is identical to the reactive current defined by Fryze.

Summary: When applied to single-phase circuits, the FBD Method does not go beyond the results obtained by Fryze. When applied to three-phase circuits, it introduces the right definition of the apparent power, as suggested by Buchholz. The assumption that all lines in an m-line circuit are identical is questionable, however. The neutral conductor in four-wire systems is connected to the ground, so the transformer line impedance does not contribute to the neutral impedance. It is also not loaded as the remaining lines. Thus the FBD Method does not apply to three-phase circuits with the neutral conductor.

5.15 Apparent Power in Three-Phase Circuits

The confusion as to the power theory of three-phase circuits was caused to a great degree by confusion as to the definition of the apparent power S.

According to the IEEE Standard Dictionary of Electrical and Electronics Terms [154], and the conclusion of a debate in the 1930s, summarized by Curtis and Silsbee in Ref. [21], apparent power can be defined in two different ways. It can be defined as *arithmetical apparent power*, meaning the sum of apparent powers of individual lines,

$$S_A \stackrel{\text{df}}{=} U_R I_R + U_S I_S + U_T I_T$$

or as the geometrical apparent power.

$$S_G \stackrel{\text{df}}{=} \sqrt{P^2 + Q^2}.$$

Another definition but not disseminated in the electrical engineering community was suggested by Buchholz [11]:

$$S_{\mathrm{B}} \stackrel{\mathrm{df}}{=} \sqrt{U_{\mathrm{R}}^2 + U_{\mathrm{S}}^2 + U_{\mathrm{T}}^2}\,\sqrt{I_{\mathrm{R}}^2 + I_{\mathrm{S}}^2 + I_{\mathrm{T}}^2}.$$

The conclusion on which of these three definitions results in the right value of the power factor was presented [165] in 1999.

The amount of energy delivered and the value of the power factor at its delivery are two main data for energy accounts between utilities and customers. The importance of the power factor stems from the fact that it characterizes the level of utilization of the transmission equipment and the level of energy loss at its delivery. At a fixed active power P, this loss increases with the power factor decline.

While the level of utilization of the transmission equipment is rather a vague measure, the level of energy loss is more distinctively related to the power factor. Therefore, the level of energy loss on its delivery can be used for the evaluation of whether the power factor is calculated correctly or not, meaning whether the apparent power definition is correctly selected.

The issue of selection of the apparent power definition boils down to the question: which value, λ_{A}, λ_{G}, or λ_{B}, provides the true value of the power factor of an unbalanced load if this power factor is to characterize energy loss on its delivery?

The answer to this question is based on the following reasoning. At first, a circuit with a balanced resistive load was designed, a circuit that at the load active power P = 100 kW has 5% of the energy loss at delivery, that is, ΔP = 5 kW. Parameters of such a circuit, along with results of the circuit analysis, are shown in Fig. 5.14.

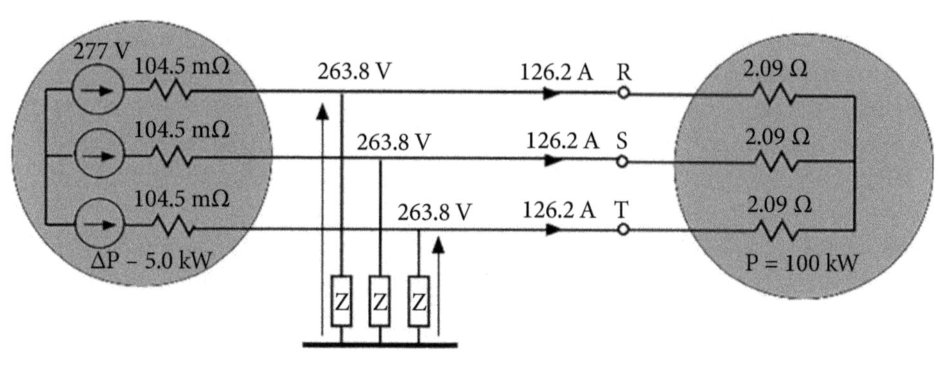

Figure 5.14 A circuit with a balanced resistive load.

The load is purely resistive and balanced thus, the apparent power is independent on its definition and is $S = 100$ kVA.

In the next step, it was assumed that the same source supplies an unbalanced resistive load, with the same active power $P = 100$ kW. The load parameters and results of the circuit analysis are shown in Fig. 5.15.

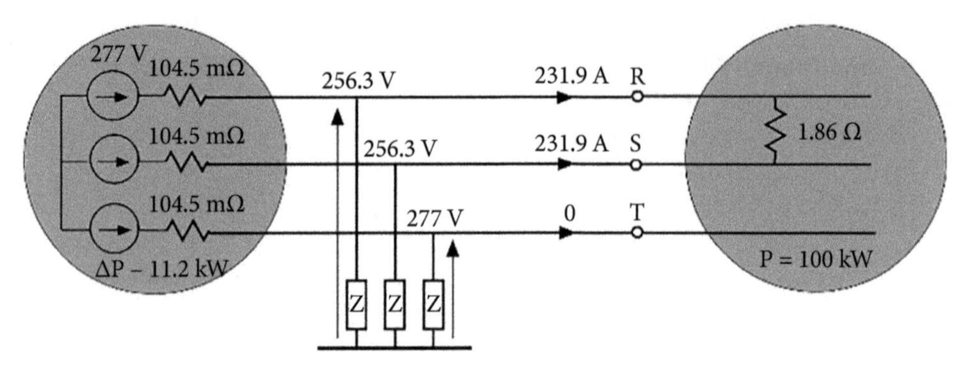

Figure 5.15 A circuit with an unbalanced load.

Depending on definition of the apparent power, it is equal to

$$S_A = 119 \text{ kVA}, \quad S_G = 100 \text{ kVA}, \quad S_B = 149 \text{ kVA}.$$

Because the apparent power S value depends and the selected definition, the power factor is equal to, respectively,

$$\lambda_A = 0.84, \quad \lambda_G = 1, \quad \lambda_B = 0.67.$$

Observe, that despite the same load active power, the power loss on energy delivery has increased in the circuit with the unbalanced load from $\Delta P_s = 5.0$ kW to $\Delta P_s = 11.2$ kW. It means that the load shown in Fig. 5.15 is not a load with a unity power factor, meaning the power factor, λ_G, is calculated using the geometrical definition of the apparent power, S_G. This conclusion disqualifies the geometrical definition.

There is still an open question, however: which value, λ_A or λ_B, provides the right value of the power factor? To find the answer, let us calculate the power factor of a balanced load that at power $P = 100$ kW would cause the same power loss in the supply source as the unbalanced load in Fig. 7.5, that is, $\Delta P_s = 11.2$ kW. To be balanced and have the power factor lower than 1, such a load has to be an RL or RC load.

Such an RL balanced load has parameters shown, along with results of the circuit analysis, in Fig. 5.16.

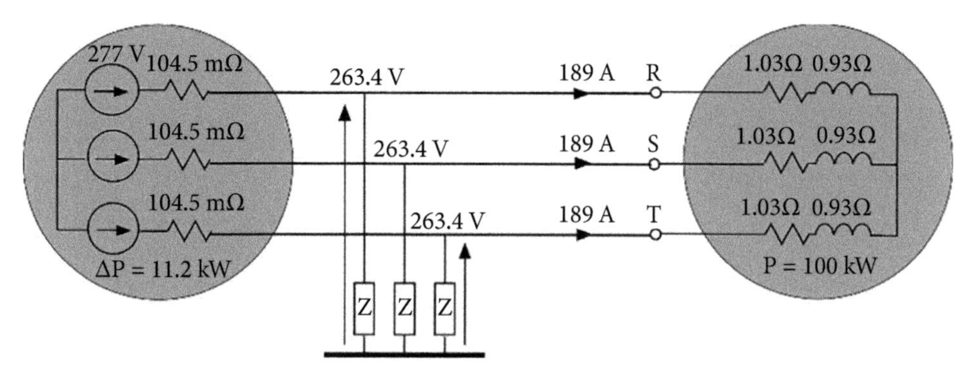

Figure 5.16 A balanced load that is equivalent to an unbalanced load in Fig. 5.15 with regard to the power loss in the source.

Since the load is balanced, the apparent power does not depend on the selected definition. It is equal to

$$S_A = S_B = 149 \text{ kVA}.$$

Thus, the power factor is $\lambda_B = \lambda = 0.67$. It means that the power factor λ in the circuit with the unbalanced load shown in Fig. 5.15 has the right value, only if the apparent power S is calculated according to Buchholz' definition, or Czarnecki's definition (5.67). Arithmetical and geometrical definitions of the apparent power, resulting in an erroneous value of the power factor. When the apparent power S is calculated according to Buchholtz's definition, the power equation

$$S^2 = P^2 + Q^2 \tag{5.77}$$

is not fulfilled, however. For example, in the circuit shown in Fig. 5.15, $P = 100$ kW, $Q = 0$, $S = 149$ kVA, and evidently $149^2 \neq 100^2 + 0^2$. Thus, the power equation (5.77) is erroneous even for sinusoidal voltages and currents. It is true only for balanced LTI loads supplied with a symmetrical voltage. Power properties of such circuits are trivial, however, and can be described phase by phase as properties of single-phase circuits.

5.16 Tenti's Power Theory

In 2003, Tenti and co-workers developed *Conservative Power Theory* (CPT), originally formulated for single-phase circuits and generalized next [191] to three-phase circuits. It follows the Fryze approach as to the definition of the active current $i_a(t)$, but the reactive current definition, denoted here by $i_{rT}(t)$, is based on the concept of "*reactive energy W*". This "energy" is defined as

$$W \stackrel{\text{df}}{=} (\widehat{u}, i) \stackrel{\text{df}}{=} \frac{1}{T}\int_0^T \widehat{u}(t)\, i(t)dt \tag{5.78}$$

where the symbol $\widehat{u}$ denotes unbiased voltage integral:

$$\widehat{u}(t) = \int_0^t u(\tau)d\tau - \frac{1}{T}\int_0^T \left[\int_0^t u(\tau)d\tau\right] dt. \tag{5.79}$$

The name "reactive energy W" is written here in quotation marks because its value for capacitors is negative, thus the W quantity cannot be regarded as the energy.

The reactive current is defined in the CPT as

$$i_{\mathrm{rT}}(t) \stackrel{\text{df}}{=} \frac{W}{||\widehat{u}\,||^2}\widehat{u}(t). \tag{5.80}$$

The remaining component of the load current

$$i_{\mathrm{v}}(t) \stackrel{\text{df}}{=} i(t) - i_{\mathrm{a}}(t) - i_{\mathrm{rT}}(t) \tag{5.81}$$

is referred to as *a void current*. The theory was called *Conservative* because the "reactive energy" W satisfies the conservative property.

According to CPT the current of a single-phase load can be decomposed as follows

$$i(t) = i_{\mathrm{a}}(t) + i_{\mathrm{rT}}(t) + i_{\mathrm{v}}(t). \tag{5.82}$$

The current CPT components are mutually orthogonal so that their rms values satisfy the relationship

$$||i||^2 = ||i_{\mathrm{a}}||^2 + ||i_{\mathrm{rT}}||^2 + ||i_{\mathrm{v}}||^2. \tag{5.83}$$

Multiplying this equation by the square of the voltage rms value $||u||$, the power equation is obtained

$$S^2 = P^2 + Q_{\mathrm{T}}^2 + D_{\mathrm{T}}^2 \tag{5.84}$$

where

$$Q_{\mathrm{rT}} \stackrel{\text{df}}{=} ||u|| \cdot ||i_{\mathrm{rT}}|| \tag{5.85}$$

is the reactive power, while

$$D_{\mathrm{T}} \stackrel{\mathrm{df}}{=} ||u|| \cdot ||i_{\mathrm{v}}|| \tag{5.86}$$

is the distortion power.

The CPT was also generalized for three-phase circuits. The most crucial question for that theory is how it describes and interprets the power properties of single-phase circuits, however.

5.17 CPC-Based PT of Three-Phase LTI Circuits with Neutral

The decomposition of the supply current in three-phase, three-wire circuits with a nonsinusoidal supply voltage into Physical Components was generalized [273] in 2015 to three-phase circuits with LTI loads and a neutral conductor.

It was shown there that the neutral conductor does not change the structure of the CPC decomposition. It is still composed of active, scattered, reactive, and unbalanced currents,

$$\boldsymbol{i} = \boldsymbol{i}_{\mathrm{a}} + \boldsymbol{i}_{\mathrm{s}} + \boldsymbol{i}_{\mathrm{r}} + \boldsymbol{i}_{\mathrm{u}} \tag{5.87}$$

only the unbalanced current $\boldsymbol{i}_{\mathrm{u}}$ can be decomposed into symmetrical components of the positive, negative, and zero sequence components so that the decomposition of the supply current into Current's Physical Components has the form

$$\boldsymbol{i} = \boldsymbol{i}_{\mathrm{a}} + \boldsymbol{i}_{\mathrm{s}} + \boldsymbol{i}_{\mathrm{r}} + \boldsymbol{i}_{\mathrm{u}}^{\mathrm{p}} + \boldsymbol{i}_{\mathrm{u}}^{\mathrm{n}} + \boldsymbol{i}_{\mathrm{u}}^{\mathrm{z}}. \tag{5.88}$$

As shown in Section 8.7, all current components can be expressed in terms of the equivalent parameters of the load which can be calculated having the crms values of voltages and currents measured at the load supply terminals. They can be endowed with clear physical interpretation.

These six physical components of the load current are mutually orthogonal so that their three-phase rms value satisfies the relationship

$$||\boldsymbol{i}||^2 = ||\boldsymbol{i}_{\mathrm{a}}||^2 + ||\boldsymbol{i}_{\mathrm{s}}||^2 + ||\boldsymbol{i}_{\mathrm{r}}||^2 + ||\boldsymbol{i}_{\mathrm{u}}^{\mathrm{p}}||^2 + ||\boldsymbol{i}_{\mathrm{u}}^{\mathrm{n}}||^2 + ||\boldsymbol{i}_{\mathrm{u}}^{\mathrm{z}}||^2. \tag{5.89}$$

As shown in Chapter 15, two of three unbalanced currents, along with the reactive current can be entirely compensated by a reactance compensator. One of three unbalanced currents and the scattered current by such a compensator cannot be compensated.

5.18 The State of the CPC-Based PT Development

The Current's Physical Components -based power theory currently (2022) provides the description and interpretation of the energy flow-related phenomena in single-phase and three-phase three-wire, and four-wire circuits with nonsinusoidal supply voltage and LTI loads. The CPC-based PT can also describe the energy flow-related phenomena in single-phase and three-phase three-wire circuits with HGLs. Moreover, in the case of three-phase three-wire circuits, the supply voltage can be asymmetrical. The effect of the supply voltage asymmetry on three-phase circuits with the neutral conductor still has to be investigated. The same applies to three-phase circuits with the neutral conductor and HGLs.

These results were obtained using the frequency domain approach to the power theory development. Albeit the time domain and frequency domain are, due to Fourier transforms, mutually equivalent, there is not a CPC-based power theory in the time domain. Each CPC, defined in the frequency domain, can be recalculated, using the Fourier transform, to the time domain but its physical meaning will be hidden. For example, the presence of the scattered current shows that the change of the load conductance G_n with harmonic order increases the load current rms value. This is an existing phenomenon that is specified in terms of the frequency properties of the load. Unfortunately, the terms such as "conductance G_n" or harmonic" do not even exist in the time domain's vocabulary.

The relationship of the CPC-based PT to other approaches of its development is illustrated in the diagram shown in Fig. 5.17.

Summary

Studies on power properties of electrical circuits with nonsinusoidal voltages and currents and PT development have a long history of well over a century, with hundreds of scientists involved. This chapter has only been able to sketch this history briefly and only a few contributors to this development were referenced.

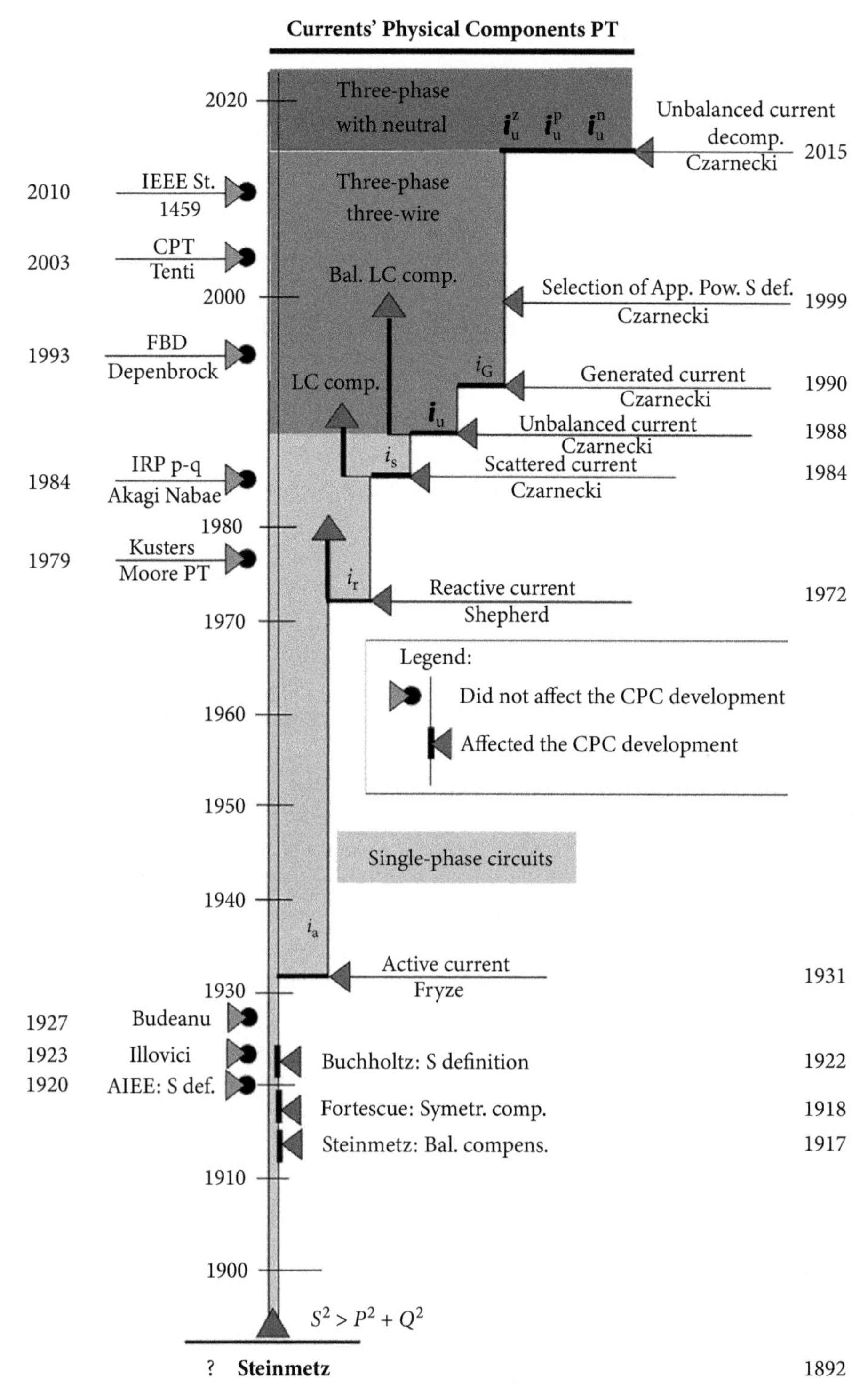

Figure 5.17 A diagram of the CPC-based power theory development.

Chapter 6
Powers in Single-Phase Circuits

6.1 Powers and Currents' Physical Components

Energy transfer in electrical circuits and circuits is commonly described in terms of various powers. Several investigators of this transfer, starting with Fryze, have concluded that powers are secondary to various components of the supply current, and just these current components rather than powers affect the energy transfer effectiveness. This conclusion is followed in this book. It presents the supply current decompositions into various components that affect energy transfer in circuits of different structures and properties.

The decomposition into Currents' Physical Components (CPC) is the very core of this approach. It is referred to as the CPC-based Power Theory (PT). These components are associated with physical phenomena or structural properties of the circuit that are responsible for the energy transfer effectiveness. This approach provides the answer to the question formulated at the very beginning of studies on the power properties of electrical circuits: "what physical phenomena in the circuit are responsible for the supply current increase?"

Although most of the electric energy is transferred and processed in three-phase circuits so that three-phase circuits are the most important objects of the power theory, the energy transfer phenomena in such circuits cannot be comprehended before they are well comprehended in single-phase circuits. Therefore, investigations on power theory development were originally focused on single-phase circuits. This approach is followed in this book as well. The CPC-based power theory, originally developed for single-phase circuits with nonsinusoidal supply voltage and linear time-invariant (LTI) loads, will be gradually extended to circuits with harmonics-generating loads (HGLs), three-phase, three-wire circuits with LTI loads, and HGLs, and finally to circuits with the supply voltage asymmetry.

6.2 CPC of LTI Loads with Nonsinusoidal Voltage

Let us consider an LTI load, shown in Fig. 6.1a, of admittance for harmonic frequencies

$$\boldsymbol{Y}_n = Y_n e^{-j\varphi_n} \stackrel{\text{df}}{=} G_n + jB_n.$$

Powers and Compensation in Circuits with Nonsinusoidal Current. Leszek S. Czarnecki, Oxford University Press.
 DOI: 10.1093/oso/9780198879206.003.0006

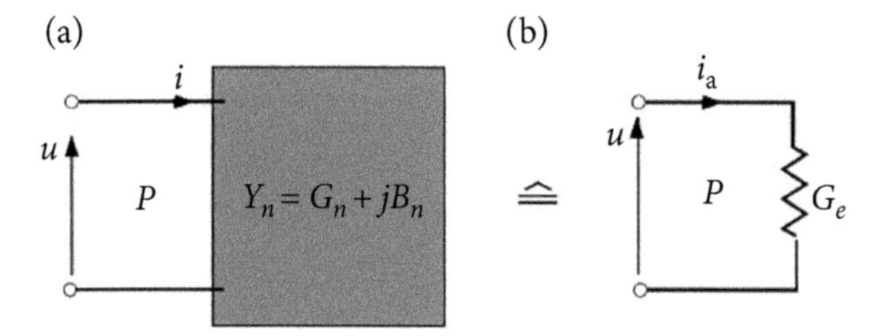

Figure 6.1 (a) A single-phase circuit with LTI load and (b) a resistive equivalent circuit.

If the load is supplied from a source of nonsinusoidal voltage

$$u(t) = U_0 + \sqrt{2}\mathrm{Re}\sum_{n\in N} \boldsymbol{U}_n e^{jn\omega_1 t}, \quad \boldsymbol{U}_n = U_n e^{j\alpha_n}$$

then the load current can be expressed as

$$i(t) = Y_0\, U_0 + \sqrt{2}\mathrm{Re}\sum_{n\in N} \boldsymbol{Y}_n\, \boldsymbol{U}_n\, e^{jn\omega_1 t}.$$

The active current, $i_a(t)$, meaning the current component proportional to the supply voltage $u(t)$ and of the minimum value needed for energy permanent delivery to the load with the average rate equal to the active power P, can be expressed as

$$i_a(t) \stackrel{\text{df}}{=} G_e u(t) = G_e U_0 + \sqrt{2}\mathrm{Re}\sum_{n\in N} G_e\, \boldsymbol{U}_n e^{jn\omega_1 t}. \tag{6.1}$$

This is a current of a purely resistive load, shown in Fig. 6.1(b), that at the supply voltage $u(t)$ has the active power P. Since the active power, P, of the resistive load is equal to

$$P = G\, ||u||^2,$$

this load is equivalent concerning the active power at voltage $u(t)$ to the original load, shown in Fig. 6.1a, if its conductance has the value

$$G \stackrel{\text{df}}{=} G_e = \frac{P}{||u||^2} \tag{6.2}$$

referred to as *equivalent conductance*. The remaining component of the load current, after subtracting the active current, is equal to

$$i(t) - i_a(t) = (Y_0 - G_e)U_0 + \sqrt{2}\,\mathrm{Re}\sum_{n\in N} (\boldsymbol{Y}_n - G_e)\boldsymbol{U}_n\, e^{jn\omega_1 t} =$$

$$= (Y_0 - G_e)U_0 + \sqrt{2}\,\mathrm{Re}\sum_{n\in N} (G_n + jB_n - G_e)\boldsymbol{U}_n\, e^{jn\omega_1 t}.$$

It can be decomposed into the following components

$$\sqrt{2}\,\mathrm{Re}\sum_{n\in N} jB_n\,\boldsymbol{U}_n\,e^{jn\omega_1 t} \stackrel{\mathrm{df}}{=} i_{\mathrm{r}}(t) \tag{6.3}$$

$$(G_0 - G_{\mathrm{e}})U_0 + \sqrt{2}\,\mathrm{Re}\sum_{n\in N} (G_n - G_{\mathrm{e}})\boldsymbol{U}_n\,e^{jn\omega_1 t} \stackrel{\mathrm{df}}{=} i_{\mathrm{s}}(t). \tag{6.4}$$

This decomposition reveals that two entirely different phenomena contribute to a useless component of the supply current of LTI loads at nonsinusoidal supply voltage, namely:

1. The presence of the current harmonic components shifted by 90° versus the voltage harmonics. Their sum is the *reactive current*, $i_{\mathrm{r}}(t)$. It is identical to the Shepherd and Zakikhani's quadrature current. Indeed,

$$\mathrm{Re}\left\{jB_n\,\boldsymbol{U}_n\,e^{jn\omega_1 t}\right\} = -B_n\,U_n\sin(n\omega_1 t + \alpha_n)$$

where

$$B_n = \mathrm{Im}\,\boldsymbol{Y}_n = \mathrm{Im}\{Y_n\cos\varphi_n - jY_n\sin\varphi_n\} = -Y_n\sin\varphi_n, \quad \text{while} \quad Y_n U_n = I_n$$

hence

$$\sqrt{2}\,\mathrm{Re}\sum_{n\in N} jB_n\,\boldsymbol{U}_n\,e^{jn\omega_1 t} = \sqrt{2}\sum_{n\in N} I_n\sin\varphi_n\sin\,(n\omega_1 t + \alpha_n) = i_r(t).$$

2. The difference of the load conductance G_n for harmonic frequency from the load equivalent conductance, G_{e}. In effect of this difference, each current harmonic has a component that is *in phase* with the supply voltage harmonics. The sum of these components creates the current $i_{\mathrm{s}}(t)$. Since this current occurs when conductances G_n are scattered around the equivalent conductance G_{e}, in Ref. [71] this current was called a *scattered current*. The presence of the scattered current in the supply current of LTI loads is a new phenomenon revealed by this decomposition.

Thus, the supply current can be expressed as a sum of three components

$$i(t) = i_{\mathrm{a}}(t) + i_{\mathrm{s}}(t) + i_{\mathrm{r}}(t) \tag{6.5}$$

associated with three distinctively different physical phenomena in the load:

- permanent unidirectional energy conversion, associated with the *active current*, $i_{\mathrm{a}}(t)$,
- change of the load conductance G_n with the harmonic order n, associated with the *scattered current*, $i_{\mathrm{s}}(t)$,

- phase shift between the voltage and current harmonics, thus non-zero load susceptances B_n are associated with the *reactive current*, $i_r(t)$.

These currents are referred to as *Currents' Physical Components* (CPC) of LTI loads. The adjective "physical" does not mean, however, that these currents exist physically. Quite the opposite: they do not exist as physical entities, but like harmonics, only as mathematical ones. These currents are only a product of mathematical decomposition. The adjective "physical" only means the association of these currents with distinctive physical phenomena.

They are, moreover, relative quantities that depend on the reference voltage. As illustrated in Fig. 6.2, the same current in a supply line, with the change of the reference voltage, can have entirely different physical components.

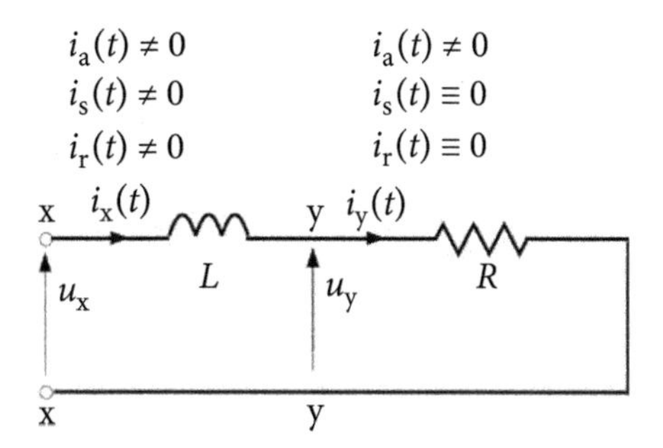

Figure 6.2 A circuit line with the supply current.

There are both non-zero reactive and scattered currents $i_r(t)$ and $i_s(t)$ in the cross-section xx but not in the cross-section yy. Only the active current $i_a(t)$ is non-zero in the last cross-section. It means that Kirchoff's Current Law for the CPC, even for a simple conductor shown in Fig. 6.3, is not satisfied. The same applies to the Ohm Law. With these observations, one might have major doubts about the correctness of the CPC concept. The decomposition into the CPC is not a tool for circuit analysis, however. It describes the circuit from the perspective of energy transfer only at the supply terminals and provides the method for the development and control of a compensator installed just at these terminals.

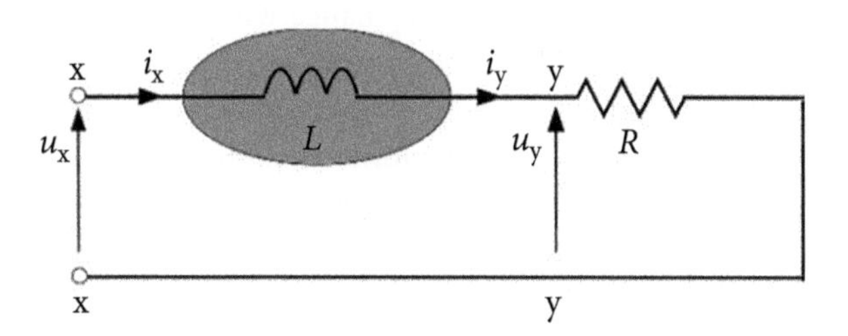

Figure 6.3 A node (shadowed) area in the supply line.

Decomposition (6.5) reveals a new phenomenon that affects the energy transfer, namely the change of the load conductance with harmonic order, *n*. This is not an "exotic" property of electrical loads but a common property of linear, passive, time-invariant loads, as can be demonstrated in the following illustration.

Illustration 6.1 *Let us calculate the conductance G_n for a few harmonic orders, n, of a common RL load shown in Fig. 6.4, assuming that the load is composed of a resistor with resistance $R = 1\Omega$ and an inductor with reactance for the fundamental harmonic $\omega_1 L = 1\Omega$.*

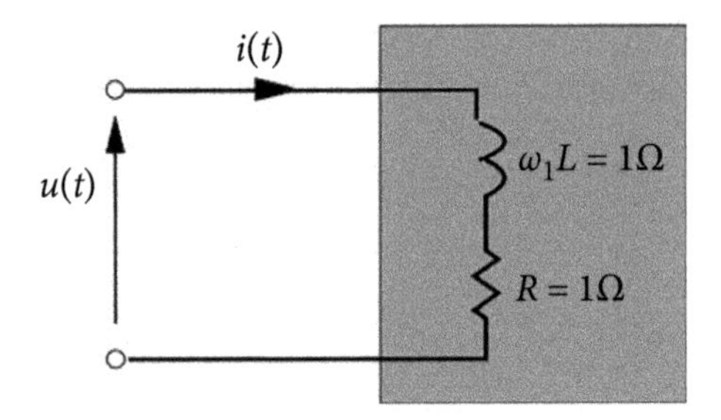

Figure 6.4 An RL load.

The conductance of such a load for harmonic frequencies is equal to

$$G_n = \text{Re}\{\mathbf{Y}_n\} = \text{Re}\frac{1}{R + jn\omega_1 L} = \frac{R}{R^2 + (n\omega_1 L)^2}.$$

For parameters given in the illustration, this conductance has values

$$G_0 = 1\,\text{S}, \quad G_1 = 0.5\,\text{S}, \quad G_2 = 0.2\,\text{S}, \quad G_3 = 0.1\,\text{S}, \quad G_4 = 0.06\,\text{S},$$

thus it changes with harmonic order. Consequently, the current of a common RL load supplied with a nonsinusoidal voltage must contain a scattered current.

****

The phenomenon of the load conductance change with harmonic order and, as a consequence, the supply current increase by a scattered current, is a phenomenon that can be easily identified in the frequency domain, but not in the time domain. This could be a reason why Fryze power theory, formulated in the time domain has failed to explain the power phenomena in single-phase circuits with LTI loads. This shows that as long as power phenomena are a matter of concern, the frequency domain approach is advantageous over the time domain approach.

Illustration 6.2 *Let us assume that the load shown in Fig. 6.2, is supplied with a voltage*

$$u = U_0 + \sqrt{2}\,\mathrm{Re}\,(\boldsymbol{U}_1\, e^{j\omega_1 t} + \boldsymbol{U}_3\, e^{j3\omega_1 t}) = 50 + \sqrt{2}\,\mathrm{Re}\,(100\, e^{j\omega_1 t} + 20\, e^{j3\omega_1 t})\,\mathrm{V}.$$

The load admittance $\boldsymbol{Y}_n$ for the supply voltage harmonics of the order $n = 0$, 1, and 3 is

$$Y_0 = G_0 = 1\ \mathrm{S}$$

$$\boldsymbol{Y}_1 = G_1 + jB_1 = \frac{1}{1+j1} = (0.5 - j0.5)\ \mathrm{S}$$

$$\boldsymbol{Y}_3 = G_3 + jB_3 = \frac{1}{1+j3} = (0.1 - j0.3)\ \mathrm{S}.$$

Since the load conductance changes with the harmonic order, the supply current contains the scattered current. Calculation of this current requires that the equivalent conductance G_e is calculated, meaning the load active power and the supply voltage rms value are known.

The load active power is equal to

$$P = \sum_{n=0,1,3} G_n U_n^2 = 1 \times 50^2 + 0.5 \times 100^2 + 0.1 \times 20^2 = 7540\ \mathrm{W}$$

while the supply voltage rms value

$$||\,u\,|| = \sqrt{\sum_{n=0,1,3} U_n^2} = \sqrt{50^2 + 100^2 + 20^2} = 113.58\ \mathrm{V}.$$

Thus,

$$G_\mathrm{e} = \frac{P}{||\,u||^2} = \frac{7540}{113.58^2} = 0.584\ \mathrm{S}$$

and hence, the Current Physical Components have the following waveforms:

$$i_\mathrm{a} = G_\mathrm{e} u = G_\mathrm{e} U_0 + \sqrt{2}\,\mathrm{Re}\,\{G_\mathrm{e}\boldsymbol{U}_1\, e^{j\omega_1 t} + G_\mathrm{e}\boldsymbol{U}_3\, e^{j3\omega_1 t}\} =$$

$$= 29.2 + \sqrt{2}\,\mathrm{Re}\,\{58.4\, e^{j\omega_1 t} + 11.7\, e^{j3\omega_1 t}\}\ \mathrm{A}$$

$$i_\mathrm{s} = (G_0 - G_\mathrm{e})\, U_0 + \sqrt{2}\,\mathrm{Re}\,\{(G_1 - G_\mathrm{e})\,\boldsymbol{U}_1\, e^{j\omega_1 t} + (G_3 - G_\mathrm{e})\,\boldsymbol{U}_3 e^{j3\omega_1 t}\} =$$

$$= 20.8 + \sqrt{2}\,\mathrm{Re}\,\{8.4\, e^{j\pi} e^{j\omega_1 t} + 9.7 e^{j\pi} e^{j3\omega_1 t}\}\ \mathrm{A}$$

$$i_\mathrm{r} = \sqrt{2}\,\mathrm{Re}\,\{jB_1\boldsymbol{U}_1\, e^{j\omega_1 t} + jB_3\boldsymbol{U}_3\, e^{j3\omega_1 t}\} =$$

$$= \sqrt{2}\,\mathrm{Re}\,\{50\, e^{-j\pi/2} e^{j\omega_1 t} + 6\, e^{-j\pi/2} e^{j3\omega_1 t}\}\ \mathrm{A}.$$

*****

Observe that each current harmonic has the active, scattered, and reactive component, namely

$$i_{\rm a} = i_{\rm a0} + \sum_{n\in N} i_{{\rm a}n} = I_{\rm a0} + \sqrt{2}\,{\rm Re} \sum_{n\in N} \boldsymbol{I}_{{\rm a}n}\, e^{jn\omega_1 t},\ \text{with}\ \boldsymbol{I}_{{\rm a}n} \stackrel{\rm df}{=} G_{\rm e}\boldsymbol{U}_n$$

$$i_{\rm s} = i_{\rm s0} + \sum_{n\in N} i_{{\rm a}n} = I_{\rm s0} + \sqrt{2}\,{\rm Re} \sum_{n\in N} \boldsymbol{I}_{{\rm s}n}\, e^{jn\omega_1 t},\ \text{with}\ \boldsymbol{I}_{{\rm s}n} \stackrel{\rm df}{=} (G_n - G_{\rm e})\boldsymbol{U}_n$$

$$i_{\rm r} = \sum_{n\in N} i_{{\rm r}n} = \sqrt{2}\,{\rm Re} \sum_{n\in N} \boldsymbol{I}_{{\rm r}n}\, e^{jn\omega_1 t},\ \text{with}\ \boldsymbol{I}_{{\rm r}n} \stackrel{\rm df}{=} jB_n\boldsymbol{U}_n.$$

These components satisfy for each harmonic the relationship

$$i_n = i_{{\rm a}n} + i_{{\rm s}n} + i_{{\rm r}n}$$

and

$$\boldsymbol{I}_n = \boldsymbol{I}_{{\rm a}n} + \boldsymbol{I}_{{\rm s}n} + \boldsymbol{I}_{{\rm r}n}.$$

Current Physical Components have the rms value, respectively, equal to

$$||i_{\rm a}|| = G_{\rm e}\,||u|| = \frac{P}{||u||} \tag{6.6}$$

$$||i_{\rm s}|| = \sqrt{\sum_{n\in N_0} (G_n - G_{\rm e})^2 U_n^2} \tag{6.7}$$

$$||i_{\rm r}|| = \sqrt{\sum_{n\in N} B_n^2\, U_n^2} = \sqrt{\sum_{n\in N} \left(\frac{Q_n}{U_n}\right)^2}. \tag{6.8}$$

6.3 Orthogonality of CPC

The possibility of calculating the supply current rms value, having rms values of currents $i_{\rm a}$, $i_{\rm s}$, and $i_{\rm r}$, depends on the orthogonality of these currents, meaning the answer to the question of whether their scalar products are equal to zero or not.

The scalar product of periodic quantities $x(t)$ and $y(t)$ from a space $\boldsymbol{L}_{\rm T}{}^2$, specified in the frequency domain, according to Section 3.4, is equal to

$$(x, y) = {\rm Re} \sum_{n\in N_0} \boldsymbol{X}_n \boldsymbol{Y}_n^*$$

hence

$$(i_{\rm r}, i_{\rm a}) = {\rm Re} \sum_{n \in N} jB_{\rm e}\, \boldsymbol{U}_n\, G_{\rm e}\, \boldsymbol{U}_n^* = {\rm Re} \sum_{n \in N} jB_{\rm e}\, G_{\rm e}\, U_n^2 = 0$$

$$(i_{\rm r}, i_{\rm s}) = {\rm Re} \sum_{n \in N} jB_{\rm e} \boldsymbol{U}_n (G_n - G_{\rm e})\, \boldsymbol{U}_n^* = {\rm Re} \sum_{n \in N} jB_{\rm e} (G_n - G_{\rm e})\, U_n^2 = 0$$

$$(i_{\rm s}, i_{\rm a}) = {\rm Re} \sum_{n \in N_0} (G_n - G_{\rm e}) \boldsymbol{U}_n\, G_{\rm e} \boldsymbol{U}_n^* = G_{\rm e} \sum_{n \in N_0} (G_n - G_{\rm e})\, U_n^2 =$$

$$= G_{\rm e} \left(\sum_{n \in N_0} G_n\, U_n^2 - G_{\rm e} \sum_{n \in N_0} U_n^2 \right) = G_{\rm e}(P - G_{\rm e}\, ||u||^2) = 0.$$

Thus, all Current's Physical Components are mutually orthogonal and consequently

$$||i||^2 = ||i_{\rm a}||^2 + ||i_{\rm s}||^2 + ||i_{\rm r}||^2. \tag{6.9}$$

This relationship explains the physical causes of the load current rms value increase in single-phase circuits with LTI loads, above the rms value $||i_{\rm a}||$ of the active current. There are two causes for that. The first is the phase shift of the current harmonics versus the supply voltage harmonics, thus the presence of the reactive current $i_{\rm a}$. The change of the load conductance for harmonics G_n with the harmonic order, thus the presence of the scattered current $i_{\rm s}$ is the second cause.

The relationship (6.9) can be visualized with a rectangular box shown in Fig. 6.5.

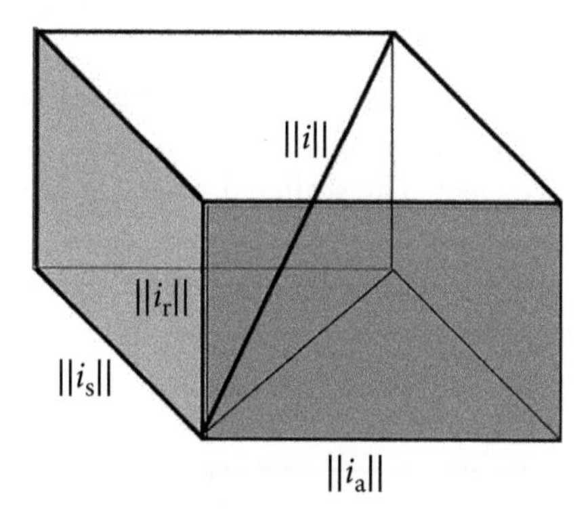

Figure 6.5 A rectangular box of rms values of CPC.

If the length of the box edges is proportional to rms values of Current's Physical Components, then the box diagonal is proportional to the load current rms value, $||i||$.

Observe that if the scattered current is orthogonal to the active current, it is also orthogonal to the supply voltage, since

$$(i_{\rm s},\ u) = \frac{1}{G_{\rm e}}(i_{\rm s},\ i_{\rm a}) = 0.$$

Thus, although the scattered current contains components in phase or counter phase with the supply voltage harmonics, it does not contribute to the load active power.

6.4 Power Equation of LTI Loads with Nonsinusoidal Voltage

Multiplying formula (6.9) by the square of the voltage rms value,

$$||i||^2 = ||i_a||^2 + ||i_s||^2 + ||i_r||^2 \quad | \times \; ||u||^2$$

the power equation of LTI loads at nonsinusoidal supply voltage is obtained:

$$S^2 = P^2 + D_s^2 + Q^2 \tag{6.10}$$

where

$$\begin{aligned} S &= ||u|| \; ||i|| \\ P &= ||u|| \; ||i_a|| \\ D_s &= ||u|| \; ||i_s|| \\ Q &= ||u|| \; ||i_r||. \end{aligned} \tag{6.11}$$

Thus, the CPC-based power theory reveals a new power quantity in circuits with LTI loads, namely the *scattered power*, D_s.

Power Eqn. (6.10) only apparently resembles Budeanu's power equation. The reactive power Q in the CPC-based PT is an entirely different quantity than the reactive power Q_B in Budeanu's concept. The scattered power D_s has nothing in common with Budeanu's distortion power D_B. It can be interpreted as the effect of the load conductance change with harmonic order upon the apparent power of the supply source. The relation between these powers can be visualized geometrically with a rectangular box in Fig. 6.6.

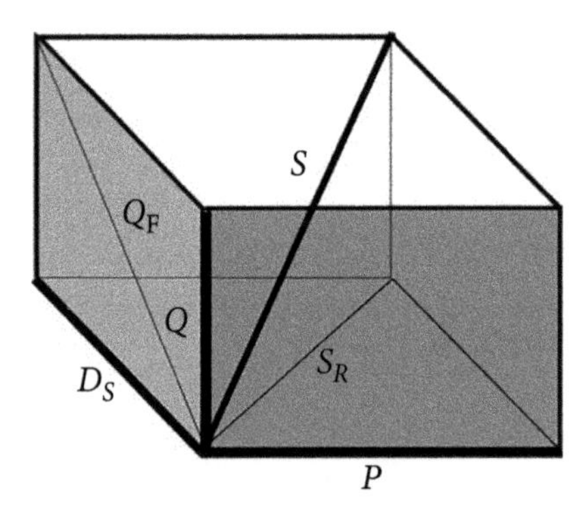

Figure 6.6 Power rectangular box.

Illustration 6.3 *Let us calculate the rms value of the Current's Physical Components of the load shown in Fig. 6.7, if the supply voltage is*

$$u = 50 + \sqrt{2}\,\text{Re}\{100\, e^{j\omega_1 t} + 20\, e^{j5\omega_1 t}\}\,\text{V}, \quad \omega_1 = 1\ \text{rad/s}, \quad ||u|| = 113.58\ \text{V}.$$

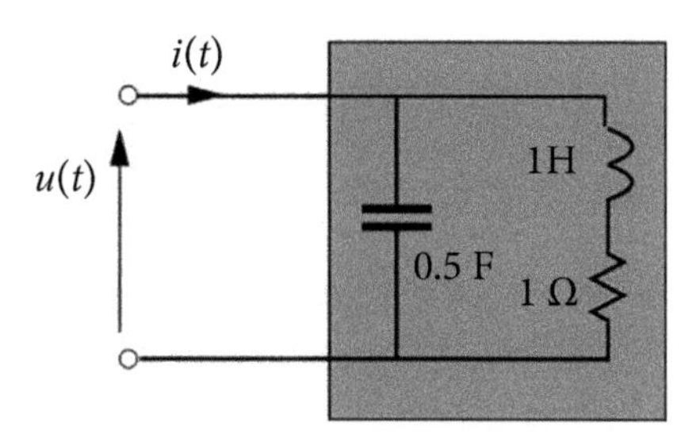

Figure 6.7 An example of a load.

The set of voltage harmonics is $N_0 = \{0,1,5\}$ *and the load admittance for harmonic orders from this set is equal to*

$$Y_0 = 1\ \text{S}, \ \boldsymbol{Y}_1 = 0.5\ \text{S}, \ \boldsymbol{Y}_5 = 0.04 + j2.31\ \text{S}$$

and consequently, the load current is

$$i = 50 + \sqrt{2}\ \text{Re}\{50\, e^{j\omega_1 t} + 46.2\, e^{j89^\circ} e^{j5\omega_1 t}\}\ \text{A}$$

and its rms value is equal to

$$||i|| = \sqrt{I_0^2 + I_1^2 + I_5^2} = \sqrt{50^2 + 50^2 + 46.2^2} = 84.47\ \text{A}.$$

To decompose the current into physical components and calculate their rms value, the active power and equivalent conductance of the load must be calculated. The active power is

$$P = \sum_{n\in 0,1,5} G_n\, U_n^2 = 7.516\ \text{kW}$$

so that, the equivalent conductance of the load has the value

$$G_e = \frac{P}{||u||^2} = \frac{7516}{113.58^2} = 0.5826\ \text{S}.$$

The rms value of the load current physical components is equal to

$$||i_{\rm a}|| = G_{\rm a}\,||u|| = 66.17\ \text{A}$$

$$||i_{\rm s}|| = \sqrt{\sum_{n\in 0,1,5} (G_n - G_{\rm e})^2\, U_n^2} = 24.93\ \text{A}$$

$$||i_{\rm r}|| = \sqrt{\sum_{n\in 1,5} B_n^2\, U_n^2} = 46.2\ \text{A.}$$

It is easy to verify that calculated rms values satisfy the relationship (6.8), and indeed

$$||i|| = \sqrt{||i_{\rm a}||^2 + ||i_{\rm s}||^2 + ||\,i_{\rm r}||^2} = \sqrt{66.17^2 + 24.93^2 + 46.2^2} = 84.47\text{A.}$$

Thus, decomposition (6.5) is satisfied even for very high distortion. This conclusion is trivial to some degree because decomposition (6.4) was not founded on any approximations. The scattered and reactive powers of the load shown in Fig. 6.4 are equal to

$$D_{\rm s} = 2.83\ \text{kVA} \quad \textit{and} \quad Q = 5.25\ \text{kVAr.}$$

*****

The scattered power $D_{\rm s}$ at a distortion level more realistic than that assumed in Illustration 6.2, is usually much smaller than the active power. However, the power equation of loads with nonsinusoidal supply voltage cannot be fulfilled without the scattered power.

6.5 CPC-Based Power Theory Reactance Compensability

Compensation in circuits with sinusoidal voltages and currents has a clear meaning. Reduction of energy loss at its delivery (*i*) is the major objective of compensation. At the same time, compensation reduces the voltage drop and investment cost of equipment needed for energy delivery.

The objective of compensation in circuits with nonsinusoidal voltages and currents could be different. It could be the same as previously (i), or it could be focused on preventing customer (ii) or power utility (iii) equipment from disturbances caused by harmonics, which requires reduction of waveform distortion. These different objectives may require different compensating or filtering equipment or some sort of trade-off between different compensation goals. Objectives (ii) and (iii) are subjects of studies on supply and loading quality. In this section, we are focused only on the power factor improvement.

Power factor, that is, the ratio of the active and apparent powers, can be expressed as

$$\lambda \stackrel{\text{df}}{=} \frac{P}{S} = \frac{||i_{\text{a}}||}{\sqrt{||i_{\text{a}}||^2 + ||i_{\text{s}}||^2 + ||i_{\text{r}}||^2}} \tag{6.12}$$

thus both scattered and reactive currents contribute to power factor degradation and could be improved by a reduction of these currents.

There are two kinds of compensators. The most common are reactance devices, such as capacitor banks or filters. Less common are *switching compensators*, known as active power filters, with compensating currents shaped by fast switching transistors. Some hybrid structures composed of reactance and switching compensators are possible as well. Reactance compensators usually are shunt devices, that is, connected as shown in Fig. 6.8.

An ideal, lossless *reactance compensator* has susceptance B_{Cn} for n-order harmonics and zero conductance for all harmonic frequencies. When a load is supplied with a fixed voltage, then, the active power P and equivalent conductance G_{e} are not affected by the compensator. Consequently, such a compensator does not affect the active and scattered currents. It modifies only the reactive current rms to the value

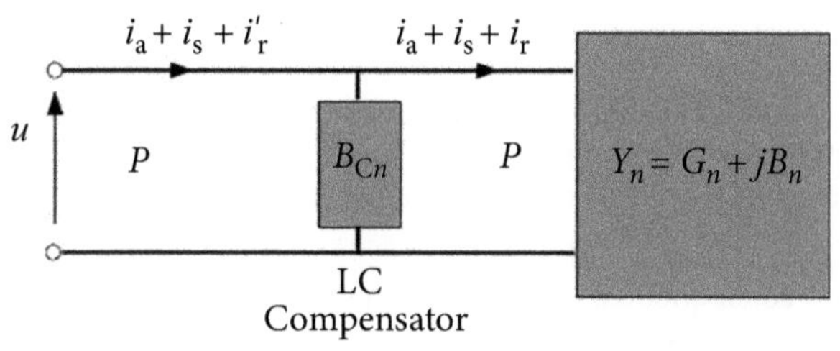

Figure 6.8 A load with a shunt reactance compensator.

$$||i'_{\text{r}}|| = \sqrt{\sum_{n \in N} (B_n + B_{Cn})^2\, U_n^2}. \tag{6.13}$$

In particular, if for each harmonic of the order n from set N,

$$B_{Cn} = -B_n \tag{6.14}$$

then the reactive current is entirely compensated and the power factor reaches its maximum, possible to obtain with a shunt reactive compensator:

$$\lambda_{\text{max}} = \frac{||i_{\text{a}}||}{\sqrt{||i_{\text{a}}||^2 + ||i_{\text{s}}||^2}} \tag{6.15}$$

since the scattered current is not affected by such a compensator.

A reader can observe an essential difference between the frequency properties of a circuit with a reactance compensator that satisfies condition (6.14) and a capacitive compensator, in a situation when the supply source has inductive impedance, as shown in Fig. 6.9. This is a common situation in power circuits, where the reactance of the supply circuit X_s is usually a few times higher than the resistance R_s.

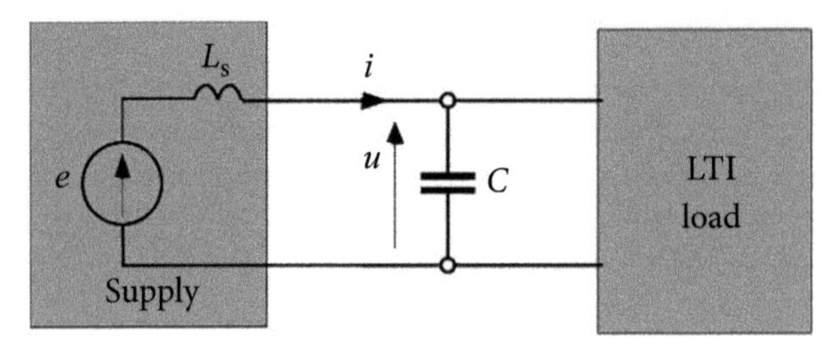

Figure 6.9 An equivalent circuit of a circuit with a capacitive compensator for $\omega >> \omega_1$.

In the first case, the load with compensator is purely resistive for all supply voltage harmonics, thus resonance of such a load with the supply source cannot occur. In the second case, the load is capacitive for all harmonics and consequently such resonance cannot be avoided. Such a resonance has to cause an increase in harmonic distortion of both the supply current and load voltage. This is not possible when instead of the capacitor a reactance compensator that satisfies condition (6.14) is connected.

One should observe, however, that the active and reactive currents are not affected by the reactance compensator only if the load voltage is not affected by the compensator. It is true only at supply source impedance zero value. This voltage is affected in a real circuit and consequently, these two currents are affected as well. The change of these currents is of the order, however, comparable with the ratio of the load apparent power S to the short-circuit power S_{sc} at the bus where the compensator is installed, meaning only a few percent.

6.6 Fryze's Decomposition in Terms of CPC

Comparison of Fryze's decomposition of the supply current in single-phase LTI loads, which has the form

$$i = i_a + i_{rF}$$

with decomposition into Current's Physical Components, reveals the meaning of the reactive current i_{rF} in Fryze power theory PT, namely

$$i_{rF} = i_s + i_r.$$

Thus, this current is associated with two distinctively different phenomena. Therefore, Fryze's power theory has failed to provide any physical interpretation of the reactive current i_{rF}. The decomposition into CPC also explains the meaning of Fryze's reactive power Q_F, namely it is equal to

$$Q_F = \sqrt{D_s^2 + Q^2},$$

thus it occurs not only in effect of the voltage and current phase shift but also because of a change of the load conductance with harmonic order. It explains why compensation of the reactive power Q_F to zero value in some situations is not possible. This power is shown on the power box in Fig. 6.6 as a diagonal on one side of that box.

6.7 Shepherd and Zakikhani's Decomposition in Terms of CPC

Comparison of Shepherd and Zakikhani decomposition of the supply current in single-phase LTI loads, which has the form

$$i = i_R + i_r$$

with decomposition into CPC, reveals the meaning of the resistive current i_R in the S&Z power theory, namely it is equal to

$$i_R = i_a + i_s.$$

Thus, the resistive current is composed of active and scattered currents. Although it is in phase with the supply voltage, it contains a useless component. It clarifies the meaning of the resistive power S_R, namely it is equal to

$$S_R = \sqrt{D_s^2 + P^2}.$$

This power is shown in the power box in Fig. 6.6 as a diagonal of one side of that box.

The association of the resistive power S_R with two different phenomena has created a major obstacle for Shepherd and Zakikhani for providing a physical interpretation of this power.

These two comparisons of decomposition of the supply current into Current's Physical Components with Fryze as well with Shepherd and Zakikhani decompositions show that as long as the presence of the scattered current was not revealed, these two decompositions were not capable of providing the physical interpretation of power phenomena in circuits with LTI loads and nonsinusoidal supply voltage.

6.8 Active, Scattered, and Reactive Voltage

At the load current decomposition into the active, reactive, and scattered components, the voltage at load terminals is a reference. It serves for calculating the equivalent conductance G_e and all CPCs are expressed in terms of load admittances for harmonics and the load terminals. It can be assumed, however, that instead of the load voltage, the load current is such a reference quantity.

Concerning the active power P at current i, any branch of a circuit as shown in Fig. 6.10a, is equivalent to a purely resistive branch of equivalent resistance R_e.

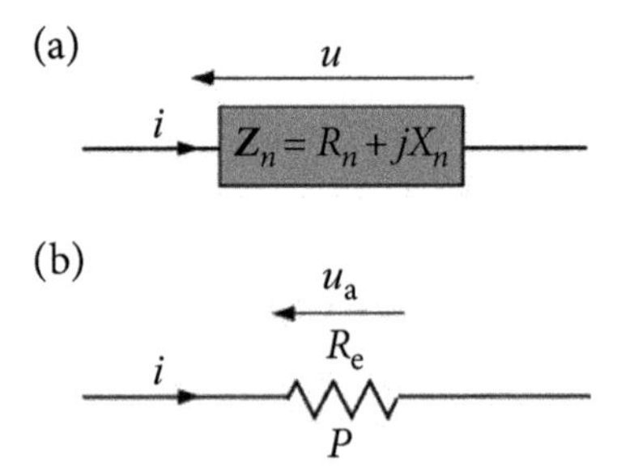

Figure 6.10 (a) A circuit branch and (b) a resistive branch equivalent as to active power P.

The voltage at such equivalent resistance at current i will be referred to as an *active voltage* u_a. Indeed, the scalar product of the active voltage and the resistor current is

$$(u_a, i) = R(i, i) = R\,||i||^2 = P$$

and hence the resistor is equivalent to the original branch when

$$R = R_e \overset{\text{df}}{=} \frac{P}{||i||^2}. \tag{6.16}$$

Let us assume that the brunch current has the Fourier Series

$$i = I_0 + \sqrt{2}\,\text{Re} \sum_{n \in N} \boldsymbol{I}_n\, e^{jn\omega_1 t}$$

then the active voltage has the waveform of the branch current and can be expressed in the form

$$u_a \overset{\text{df}}{=} R_e\, i = R_e I_0 + \sqrt{2}\,\text{Re} \sum_{n \in N} R_e\, \boldsymbol{I}_n\, e^{jn\omega_1 t}. \tag{6.17}$$

The voltage on the branch is equal to

$$u = R_0 I_0 + \sqrt{2}\,\text{Re} \sum_{n \in N} (R_n + jX_n)\, \boldsymbol{I}_n\, e^{jn\omega_1 t}$$

thus, the voltage which does not contribute to the branch active power P is equal to the difference between the actual voltage u on the branch and its active component

u_{a}, namely

$$u - u_{\mathrm{a}} = (R_0 - R_{\mathrm{e}})I_0 + \sqrt{2}\mathrm{Re}\sum_{n\in N}(R_n + jX_n - R_{\mathrm{e}})\,\boldsymbol{I}_n\,e^{jn\omega_1 t}.$$

This voltage can be decomposed into two components

$$(R_0 - R_{\mathrm{e}})\,I_0 + \sqrt{2}\;\mathrm{Re}\sum_{n\in N}(R_n - R_{\mathrm{e}})\,\boldsymbol{I}_n\,e^{jn\omega_1 t} \stackrel{\mathrm{df}}{=} u_{\mathrm{s}} \tag{6.18}$$

$$\sqrt{2}\;\mathrm{Re}\sum_{n\in N} jX_n\,\boldsymbol{I}_n\,e^{jn\omega_1 t} \stackrel{\mathrm{df}}{=} u_{\mathrm{r}}. \tag{6.19}$$

In such a way the branch voltage was decomposed into three components, namely

$$u = u_{\mathrm{a}} + u_{\mathrm{s}} + u_{\mathrm{r}} \tag{6.20}$$

of entirely different physical meaning. The active voltage occurs only when there is permanent energy conversion in the branch at a rate equal to the active power P. The voltage u_{s} occurs only when the resistance of the branch for harmonic frequencies R_n is not constant and equal to equivalent resistance R_{e}, but when it is scattered around that value. Therefore, this component was called a *scattered voltage* in Ref. [105]. The voltage u_{r} occurs when voltage harmonics are shifted versus the current harmonics. Therefore, this voltage is referred to as a *reactive voltage.*

These three voltages are associated with three entirely different phenomena in the branch, therefore, they can be referred to as the *voltage physical components* (VPC). Their rms values are equal to, respectively

$$||u_{\mathrm{a}}|| = \frac{P}{||i||} \tag{6.21}$$

$$||u_{\mathrm{s}}|| = \sqrt{\sum_{n\in N_0}[(R_n - R_{\mathrm{e}})\,I_n]^2} \tag{6.22}$$

$$||u_{\mathrm{r}}|| = \sqrt{\sum_{n\in N}(X_n I_n)^2}. \tag{6.23}$$

6.9 Orthogonality of the Voltage Physical Components

Orthogonality of the active and reactive voltages is evident due to the phase shift by 90° of their harmonic components. The same applies to scattered and reactive voltages. Thus

$$(u_{\rm a}, u_{\rm r}) = (u_{\rm s}, u_{\rm r}) = 0.$$

The orthogonality of the active and scattered voltages is not so evident, however. Let us calculate their scalar product

$$(u_{\rm s}, u_{\rm a}) = {\rm Re} \sum_{n\in N_0} \boldsymbol{U}_{{\rm s}n} \boldsymbol{U}^*_{{\rm a}n} = {\rm Re} \sum_{n\in N_0} (R_n - R_{\rm e}) \boldsymbol{I}_n R_{\rm e} \boldsymbol{I}^*_n =$$

$$= R_{\rm e} \sum_{n\in N_0} (R_n - R_{\rm e})\, I_n^2 = R_{\rm e} \left[\sum_{n\in N_0} R_n I_n^2 - R_{\rm e} \sum_{n\in N_0} I_n^2 \right] = R_{\rm e}[P - P] = 0.$$

Thus, active, scattered, and reactive components of the branch voltage are mutually orthogonal and hence, their rms values have to satisfy the relationship

$$||u||^2 = ||u_{\rm a}||^2 + ||u_{\rm s}||^2 + ||u_{\rm r}||^2. \tag{6.24}$$

When this equation is multiplied by the square of the branch current rms value, the power equation is obtained

$$S^2 = P^2 + D_{\rm si}^2 + Q_{\rm i}^2 \tag{6.25}$$

where $D_{\rm si}$ and $Q_{\rm i}$ denote the *scattered and reactive powers*, defined as

$$D_{\rm si} \stackrel{\rm df}{=} ||u_{\rm s}||\ ||i||, \quad Q_{\rm i} \stackrel{\rm df}{=} ||u_{\rm r}||\ ||i||. \tag{6.26}$$

Index "i" in power symbols means that these powers were defined with the branch current taken as the reference quantity. Using this convention, the powers defined at the load voltage taken as the reference should be denoted as $D_{\rm su}$ and $Q_{\rm u}$, respectively. Such differentiation of the power symbols is not objectless since their numerical value might depend on the reference quantity.

Illustration 6.4 *Let us calculate the rms values of the active, scattered, and reactive voltages and powers for an RC branch shown in Fig. 6.11, assuming that the branch current is equal to*

$$i = 50 + \sqrt{2}\,{\rm Re}\,\{200\, e^{j\omega_1 t} + 100\, e^{j5\omega_1 t}\}{\rm A}.$$

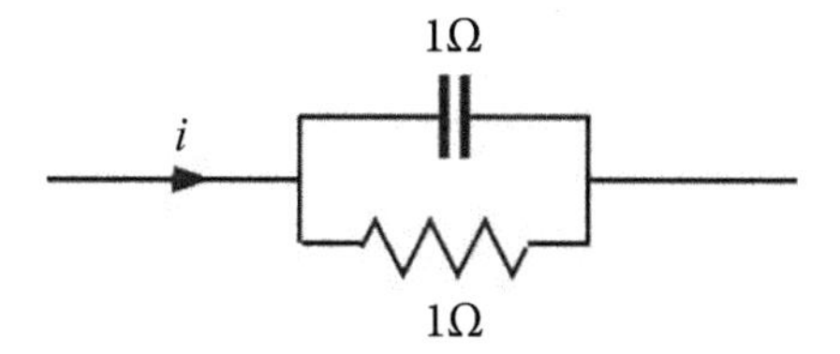

Figure 6.11 An RC branch.

The current rms value is $||\mathrm{i}|| = 229.13A$. *The branch impedance for the* n^{th} *order harmonic is*

$$\boldsymbol{Z}_n = R_n + jX_n = \frac{1}{G + jn\omega_1 C} = \frac{G}{G^2 + (n\omega_1 C)^2} - j\frac{n\omega_1 C}{G^2 + (n\omega_1 C)^2}$$

thus, the branch impedance for the zero, the first, and the fifth-order harmonics of the supply voltage is equal to $\mathrm{R}_0 = 1\ \Omega$,

$$\boldsymbol{Z}_1 = R_1 + jX_1 = \frac{1}{1 + j1} = 0.5 - j0.5 = 0.707e^{-j45^\circ}\Omega$$

$$\boldsymbol{Z}_5 = R_5 + jX_5 = \frac{1}{1 + j5} = 0.038 - j0.192 = 0.196\, e^{-j78.7^\circ}\Omega.$$

The branch voltage at the current assumed in this illustration has the waveform

$$u = 50 + \sqrt{2}\,\mathrm{Re}\,\{141.42\, e^{-j45^\circ} e^{j\omega_1 t} + 19.61 e^{-j78.7^\circ} e^{j5\omega_1 t}\}\mathrm{V}, \text{ with } ||u|| = 151.27\ \mathrm{V}.$$

The branch's active power is

$$P = \mathrm{G}\,||u||^2 = 1 \times 151.27^2 = 22\,884\ \mathrm{W}$$

thus, the equivalent resistance is equal to,

$$R_\mathrm{e} = \frac{P}{||i||^2} = \frac{22\,884}{229.13^2} = 0.436\ \Omega.$$

The voltage physical components u_a, u_s *and* u_r *have the rms values*

$$||u_\mathrm{a}|| = \frac{P}{||i||} = 99.87\ \mathrm{V}$$

$$||u_\mathrm{s}|| = \sqrt{\sum_{n=0,1,5} [(R_n - R_\mathrm{e})\, I_n]^2} =$$

$$= \sqrt{[(1 - 0.436)\,50]^2 + [(0.5 - 0.436)\,200]^2 + [(0.038 - 0.436)\,100]^2} = 50.41\ \mathrm{V}$$

$$||u_\mathrm{r}|| = \sqrt{\sum_{n=1,5} (X_n\, I_n)^2} = \sqrt{(0.5 \times 200)^2 + (0.192 \times 100)^2} = 101.83\ \mathrm{V}.$$

This decomposition of the branch voltage into physical components can be verified numerically by comparison of the rms value calculated with the formula (6.23) with that calculated directly from the harmonic contents. Indeed,

$$\sqrt{||u_a||^2 + ||u_s||^2 + ||u_r||^2} = \sqrt{99.87^2 + 50.41^2 + 101.83^2} = 151.27 \text{ V} = ||u||.$$

The scattered and reactive powers are equal to

$$D_{si} = ||u_s|| \, ||i|| = 50.41 \times 229.13 = 11.55 \text{ kVA}$$
$$Q_i = ||u_r|| \, ||i|| = 101.83 \times 229.13 = 23.33 \text{ kvar}.$$

The apparent power of the branch has the value

$$S = ||u|| \, ||i|| = 151.27 \times 229.13 = 34.66 \text{ kVA}.$$

It might be interesting to observe that unlike resistance R_n of the branch in this example, the conductance G_n does not change with the harmonic order, that is, $G_0 = G_1 = G_5 = G_e, = 1$ S. *It means that the load current cannot contain any scattered component i_s. Consequently, when the powers of the load are calculated based on the current decomposition, then the scattered power $D_s = D_{su}$, must be equal to zero, while it is not equal to zero when powers are calculated using the voltage decomposition. Thus*

$$D_{su} \neq D_{si}. \tag{6.27}$$

Since the active and apparent powers, P and S, are independent of decomposition, inequality (6.27) results in an inequality of reactive powers, that is,

$$Q_u \neq Q_i. \tag{6.28}$$

In particular, the reactive power calculated based on the current decomposition into CPCs, due to zero scattered power, D_{su}, is equal to

$$Q_u = \sqrt{S^2 - P^2} = \sqrt{34.66^2 - 22.88^2} = 26.03 \text{ kvar}$$

while $Q_i = 23.33$ kVAr.

*****

The difference in numerical values between reactive power Q_u and Q_1 challenges the straightforward interpretation of the reactive power as a component of the apparent power caused by the phase shift between the voltage and current harmonics. This interpretation cannot explain this difference. Observe that the reactive power Q_u

is the power that can be entirely compensated by a shunt reactance compensator. Nothing else but Q_u can be compensated by such a compensator. Thus, this power provides us with information on the possibility of shunt compensation.

6.10 Series Reactance Compensability

The power factor specified by formula (6.12) in terms of Current's Physical Components can be also specified in terms of Voltage Physical Components, namely

$$\lambda \stackrel{\text{df}}{=} \frac{P}{S} = \frac{||u_a||}{\sqrt{||u_a||^2 + ||u_s||^2 + ||u_r||^2}}, \tag{6.29}$$

thus it is reduced by the presence of reactive and scattered voltages. It can be improved by reduction of the rms value of these voltages, which requires that the compensator be connected in series with the load. The load preserves its active power P on the condition, however, that the load current is not affected by the compensator, which is possible only when the load is supplied from a current source. When a load is supplied from a voltage source, a series compensator affects the load voltage and its power. Therefore, series compensation is not used in common distribution systems. It can be used only for special purposes, for example when an increase in the load voltage would be appropriate. Such compensation is regarded in this book not for practical but only for some cognitive reasons.

Let us consider a series lossless reactance compensator with impedance for harmonic frequencies

$$Z_{Cn} = jX_{Cn}$$

connected in series with the load, as shown in Fig. 6.12.

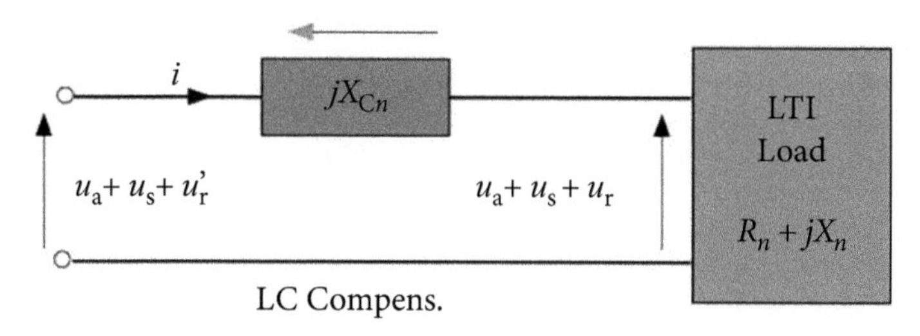

Figure 6.12 An LTI load with a series reactance compensator.

The load impedance for harmonic frequencies is

$$Z_n = R_n + jX_n,$$

thus the impedance as seen from the supply is equal to

$$\boldsymbol{Z}'_n = R_n + j(X_n + jX_{\mathrm{C}n})$$

Such a compensator, at the assumption that the load current is not affected, meaning the load is supplied from a current source, does not change the active power, meaning the equivalent resistance R_{e} and harmonic active powers, meaning resistance for harmonics R_n. Consequently, neither the active voltage nor the scattered voltage is affected by such a compensator. Only the reactive voltage is modified to

$$u'_{\mathrm{r}} = \sqrt{2}\,\mathrm{Re}\sum_{n\in N} j(X_{\mathrm{C}n} + X_n)\,\boldsymbol{I}_n\, e^{jn\omega_1 t}.$$

In particular, the reactive voltage is equal to zero when for each harmonic in the load current

$$X_{\mathrm{C}n} + X_n = 0. \tag{6.30}$$

When this condition is fulfilled then the load reactive power Q_{i} is entirely compensated. Thus, the difference in numerical values of reactive powers Q_{u} and Q_1 for the same load at the same voltage and current shows that different values of reactive power can be compensated by a shunt and by a series reactance compensator. In particular, observe that a common series RL load as shown in Fig. 6.13 has a constant resistance $R_n = R$, thus its scattered power $D_{\mathrm{si}} = 0$. Hence, such a load can be compensated with a reactance series compensator to the unity power factor. At the same time, the conductance of such a load, G_n, changes with harmonic order, thus at nonsinusoidal supply voltage such a load has a scattered power D_{su}, which cannot be compensated by a shunt reactance compensator. Thus, such a load cannot be compensated by a shunt compensator to the unity power factor. It means that the difference in values of reactive powers Q_{u} and Q_1 demonstrates a difference in this power value that can be compensated by a shunt and by a series compensator.

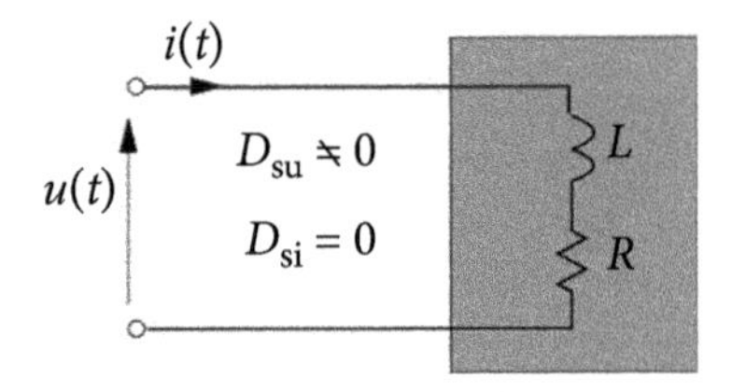

Figure 6.13 A series RL load.

Observe that when a load is compensated by a series compensator that compensates entirely the reactive power, then the voltage rms value is reduced from

$$||u|| = \sqrt{||u_{\mathrm{a}}||^2 + ||u_{\mathrm{s}}||^2 + ||u_{\mathrm{r}}||^2}$$

to

$$||u'|| = \sqrt{||u_a||^2 + ||u_s||^2}.$$

When such a load is supplied from a voltage source, meaning from a source with a fixed voltage rms value, the series compensator increases the load current rms value and consequently its active power. To keep the active power of the load after compensation unchanged, a transformer is connected as shown in Fig. 6.14

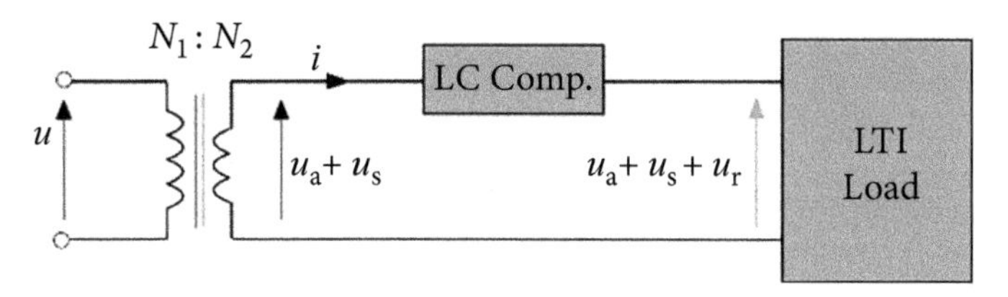

Figure 6.14 A series compensator with a transformer.

with the turn ratio

$$\frac{N_2}{N_1} = \frac{\sqrt{||u_a||^2 + ||u_s||^2}}{\sqrt{||u_a||^2 + ||u_s||^2 + ||u_r||^2}} \tag{6.31}$$

would be needed. It increases, of course, the cost of the compensation, especially since there is also an additional loss of energy in the transformer. Therefore, series compensation, even if possible, is usually technically and economically inferior to shunt compensation.

6.11 CPC in Circuits with Harmonics-Generating Loads

Linear time-invariant loads cannot be sources of current harmonics. Current harmonics can be generated by the load nonlinearity or by the periodic variance of the load parameters, on the condition that the period of this variance is equal to the period of the supply voltage. Such loads can be referred to as *harmonic generating loads* (HGLs). The following illustration demonstrates how the load-generated current harmonics affect the power properties of a circuit.

Illustration 6.5 *Let us consider a purely resistive circuit shown in Fig. 6.15. It is assumed that at the supply voltage*

$$e = e_1 = 100\sqrt{2}\,\sin\omega_1 t\ \text{V}$$

the third order current harmonic

$$j = j_3 = 50\sqrt{2}\sin 3\omega_1 t\ \text{A}$$

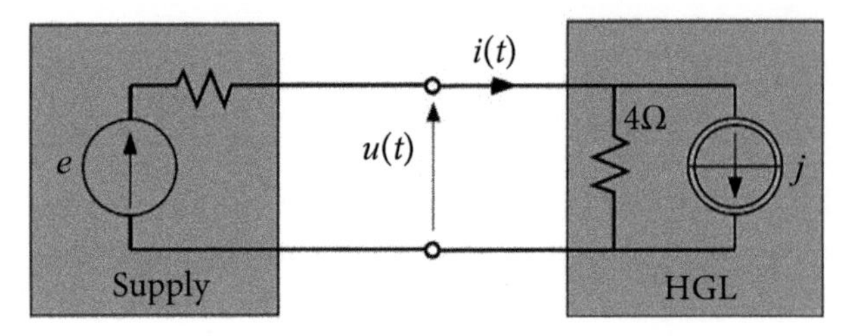

Figure 6.15 A circuit with a harmonic generating load.

is generated in the load. The voltage and current at the load terminals are equal to

$$u = u_1 + u_3 = 80\sqrt{2}\sin\omega_1 t - 40\sqrt{2}\,\sin\,3\omega_1 t\ \text{V}$$

$$i = i_1 + i_3 = 20\sqrt{2}\sin\omega_1 t + 40\sqrt{2}\sin 3\omega_1 t\ \ \text{A},$$

thus the active power is zero, since

$$P = \frac{1}{T}\int_0^T u\,i\,dt = \sum_{n=1,3} U_n\,I_n\cos\varphi_n = 1600 - 1600 = 0.$$

Observe that there is neither a phase shift between voltage and current harmonics nor any change of the load conductance with harmonic order, thus no reactive and scattered currents and powers. The apparent power S at the supply terminals is equal to

$$S = ||u||\,||i|| = 89.44 \times 44.72 = 4\,000\ \text{VA}$$

but we are not able to write the power equation in form (6.10).

*****

It is easy to find the reason for this. Namely, it is caused by the presence of the active power associated with the third order voltage and current harmonic j_3 generated in the load, of the negative value, $P_3 = -1600$ W. When a current harmonic is generated in the load, it is a source of energy flow from the load to the supply source, where this energy is dissipated on the supply source resistance. Thus, there is a current component in the supply current, which cannot be interpreted as a reactive or a scattered current, but it does not contribute to the load active power P. Quite the opposite: it reduces that power. This component can be associated with energy flow in the opposite direction to the normal flow, meaning back from the load to the supply source. Thus, the generation of current harmonics in the load, due to its nonlinearity or periodic time variance, has to be regarded [99] as a phenomenon that affects the power properties of electric circuits.

The presence of current harmonics generated in the load can be identified by measuring the phase angle, φ_n, between the supply voltage and current harmonics, u_n and i_n, at the cross-section between the distribution system (C) and a harmonic generating load (G), as shown in Fig. 6.16. Since the active power of the nth order harmonic is equal to

$$P_n = U_n I_n \cos \varphi_n$$

then if

$$|\varphi_n| < \pi/2,$$

there is an average component of energy flow at the nth order harmonic from the supply towards the load, and if

$$|\varphi_n| > \pi/2,$$

there is an average component of energy flow at n^{th} order harmonic from the load back to the supply source.

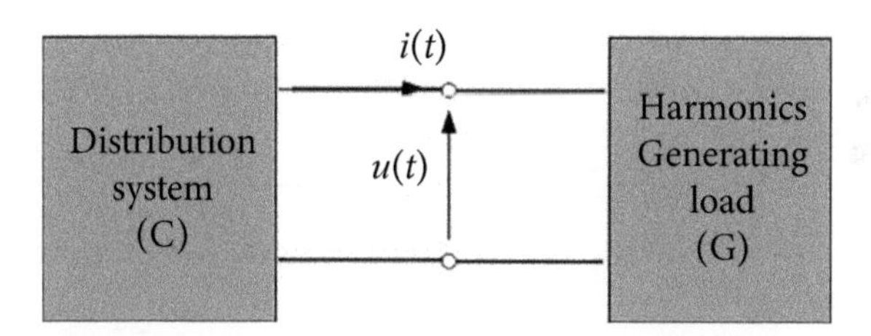

Figure 6.16 A cross-section between a distribution system and a HGL.

With this observation, the set N of all harmonic orders n can be decomposed into subsets N_{C} and N_{G}, as follows:

$$\begin{aligned} &\text{if} \quad |\varphi_n| \leq \pi/2, \text{ then } n \in N_{\mathrm{C}} \\ &\text{if} \quad |\varphi_n| > \pi/2, \text{ then } n \in N_{\mathrm{G}}. \end{aligned} \tag{6.32}$$

It enables the voltage and current decomposition into components with harmonics from subsets N_{C} and N_{G},

$$i = \sum_{n \in N} i_n = \sum_{n \in N_{\mathrm{C}}} i_n + \sum_{n \in N_{\mathrm{G}}} i_n \stackrel{\text{df}}{=} i_{\mathrm{C}} + i_{\mathrm{G}} \tag{6.33}$$

$$u = \sum_{n \in N} u_n = \sum_{n \in N_{\mathrm{C}}} u_n + \sum_{n \in N_{\mathrm{G}}} u_n \stackrel{\text{df}}{=} u_{\mathrm{C}} - u_{\mathrm{G}}. \tag{6.34}$$

The voltage u_C is defined as the negative sum of voltage harmonics because as a supply source response to load-generated current, i_C, it has the opposite sign as compared to the sign of the distribution system-originated voltage harmonics. The same applies to harmonic active power, thus

$$P = \sum_{n \in N} P_n = \sum_{n \in N_C} P_n + \sum_{n \in N_G} P_n \stackrel{\text{df}}{=} P_C - P_G. \tag{6.35}$$

Subsets N_C and N_G do not contain common harmonic orders n, thus currents i_D and i_C are mutually orthogonal. Hence, their rms values satisfy the relationship

$$||i||^2 = ||i_C||^2 + ||i_G||^2.$$

The same applies to the voltage rms values, namely

$$||u||^2 = ||u_C||^2 + ||u_G||^2.$$

Decomposition (6.32) of harmonic orders and the voltage and current according to (6.33) and (6.34) mean that the circuit, as presented in Fig. 6.16, can be regarded as a superposition of two circuits. The first, shown in Fig. 6.17a, has an LTI load and the second, shown in Fig. 6.17b, has only a current source on the customer side while the distribution system is a passive energy receiver.

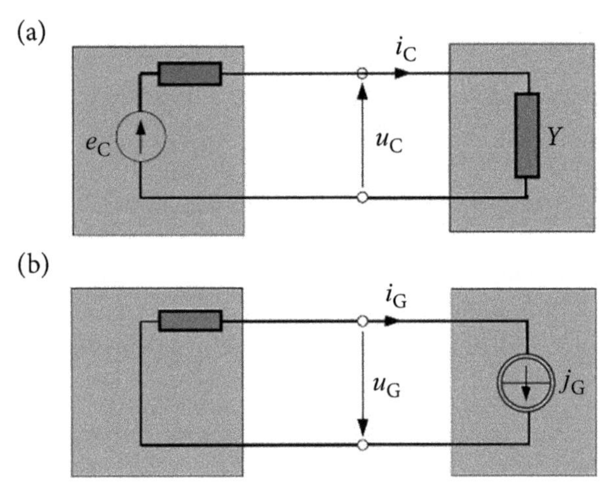

Figure 6.17 (a) Equivalent circuit for harmonics $n \in N_C$ (b) equivalent circuit for harmonics $n \in N_G$.

The circuit in Fig. 6.17a, as a circuit with an LTI load, can be described according to the CPC approach. Namely, if the equivalent admittance is

$$\boldsymbol{Y}_n = G_n + jB_n = \frac{\boldsymbol{I}_n}{\boldsymbol{U}_n}$$

and the equivalent conductance of the HGL according to the CPC power theory

$$G_{\mathrm{Ce}} \stackrel{\mathrm{df}}{=} \frac{P_{\mathrm{C}}}{||u_{\mathrm{C}}||^2}$$

then the current i_{C} can be decomposed into the active, scattered, and reactive components, with the active current, defined as

$$i_{\mathrm{Ca}} \stackrel{\mathrm{df}}{=} G_{\mathrm{Ce}}\, u_{\mathrm{C}}. \tag{6.36}$$

Definitions of the equivalent conductance G_{Ce} of harmonic generating loads and the active current i_{Ca} differ from Fryze's definitions. To differentiate these definitions, index "F" is added to Fryze's definitions, namely they are defined as

$$G_{\mathrm{Fe}} \stackrel{\mathrm{df}}{=} \frac{P}{||u||^2}$$

and

$$i_{\mathrm{Fa}} \stackrel{\mathrm{df}}{=} G_{\mathrm{Fe}}\, u.$$

The whole load voltage $u(t)$ is taken as a reference for the active current definition in Fryze's decomposition, while only component $u_{\mathrm{C}}(t)$ is regarded as the reference voltage in the CPC-based current decomposition. It has the form

$$i_{\mathrm{C}} = i_{\mathrm{Ca}} + i_{\mathrm{Cs}} + i_{\mathrm{Cr}},$$

and eventually, the load current can be decomposed into four physical components,

$$i = i_{\mathrm{Ca}} + i_{\mathrm{Cs}} + i_{\mathrm{Cr}} + i_{\mathrm{G}}. \tag{6.37}$$

These currents are mutually orthogonal, hence

$$||i||^2 = ||i_{\mathrm{Ca}}||^2 + ||i_{\mathrm{Cs}}||^2 + ||i_{\mathrm{Cr}}||^2 + ||i_{\mathrm{G}}||^2. \tag{6.38}$$

This relation can be visualized in the diagram shown in Fig. 6.18.

Decomposition (6.36) reveals a new current component, referred to as the *load-generated current* i_{G}, orthogonal to the remaining current components. It is associated with the phenomenon of the generation of harmonic current in the load.

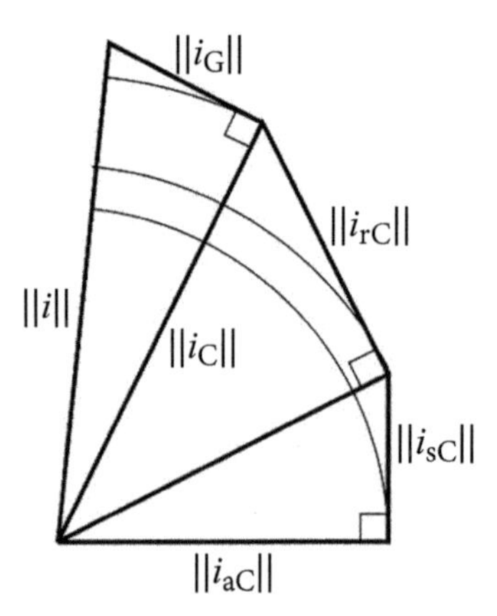

Figure 6.18 A diagram of CPC rms values in a circuit with HGL.

Illustration 6.6 *Let us consider a circuit shown in Fig. 6.19, with the internal voltage of the supply source equal to*

$$e = 220\sqrt{2}\cos\omega_1 t + 15\sqrt{2}\cos 3\omega_1 t\,\mathrm{V}$$

assuming that the load generates the 5th and 7th order harmonics such that

$$j = 20\sqrt{2}\cos 5\omega_1 t + 15\sqrt{2}\cos 7\omega_1 t\,\mathrm{A}.$$

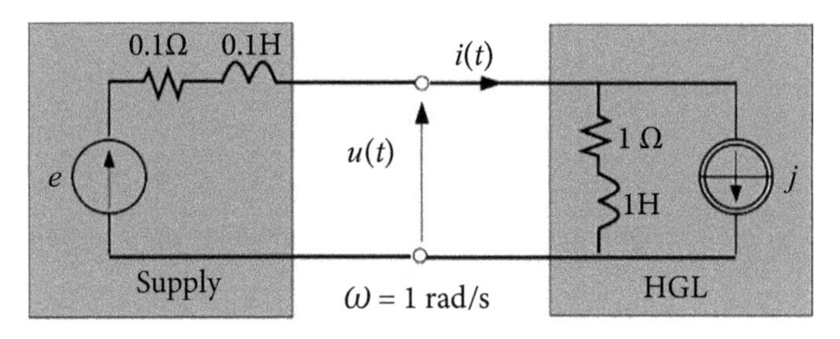

Figure 6.19 Example of a circuit with HGL.

At such assumption, the load voltage and current have harmonics of the crms value equal to, respectively

$$\begin{aligned}
\boldsymbol{I}_1 &= 141.42e^{-j45^\circ}\,\mathrm{A}, & \boldsymbol{U}_1 &= 200.0\,\mathrm{V}\\
\boldsymbol{I}_3 &= 4.31e^{-j71.6^\circ}\,\mathrm{A}, & \boldsymbol{U}_3 &= 13.64e^{j8.4^\circ}\,\mathrm{V}\\
\boldsymbol{I}_5 &= 18.18\mathrm{A}, & \boldsymbol{U}_5 &= 9.27e^{-j101.3^\circ}\,\mathrm{V}\\
\boldsymbol{I}_7 &= 13.64\mathrm{A}, & \boldsymbol{U}_7 &= 9.64e^{-j98.1^\circ}\,\mathrm{V}.
\end{aligned}$$

The active power of the fundamental and the third order harmonics in this circuit are positive, thus $N_C = \{1, 3\}$*, hence*

$$P_C = \mathrm{Re}\{\boldsymbol{U}_1\boldsymbol{I}_1^*\} + \mathrm{Re}\{\boldsymbol{U}_3\boldsymbol{I}_3^*\} = 20010\ \mathrm{W}.$$

$$||u_C|| = \sqrt{U_1^2 + U_3^2} = 200.46\ \mathrm{V}$$

$$G_{Ce} = \frac{P_C}{||u_C||^2} = 0.4979\ \mathrm{S}.$$

The load admittance for harmonics of the order from the subset N_C *is equal to*

$$\mathbf{Y}_1 = 0.50 - j0.50\ \mathrm{S}$$

$$\mathbf{Y}_3 = 0.10 - j0.30\ \mathrm{S}$$

thus, the CPC of the load have the rms values equal to

$$||i_{Ca}|| = \frac{P_C}{||u_C||} = 99.84\ \mathrm{A}$$

$$||i_{Cs}|| = \sqrt{\sum_{n=1,3} [(G_n - G_e)\ U_n]^2} = 5.44\ \mathrm{A}$$

$$||i_{Cr}|| = \sqrt{\sum_{n=1,3} (B_n\ U_n)^2} = 100.08\ \mathrm{A}$$

$$||i_G|| = \sqrt{\sum_{n=5,7} I_n^2} = 22.73\ \mathrm{A}.$$

The sum, with square, of the rms values of the Current's Physical Components results in

$$\sqrt{||i_{Ca}||^2 + ||i_{Cs}||^2 + ||i_{Cr}||^2 + ||i_G||^2} = \sqrt{99.84^2 + 5.44^2 + 100.08^2 + 22.73^2} =$$

$$= 143.29\ \mathrm{A}$$

and it should be equal to the supply current rms value. Indeed,

$$||i|| = \sqrt{\sum_{n\in N} I_n^2} = \sqrt{141.42^2 + 4.31^2 + 18.18^2 + 13.64^2} = 143.29\ \mathrm{A}.$$

This value verifies the numerical correctness of the current decomposition into physical components.

*****

6.12 Power Equation of Circuits with HGL

Multiplication of the square of the supply current rms value of a HGL by the square of its voltage rms value, namely

$$||u||^2||i||^2 = (||u_{\mathrm{C}}||^2 + ||u_{\mathrm{G}}||^2)(||i_{\mathrm{C}}||^2 + ||i_{\mathrm{G}}||^2)$$

results in the power equation of HGLs in the form

$$S^2 = S_{\mathrm{C}}^2 + S_{\mathrm{CG}}^2 + S_{\mathrm{G}}^2. \tag{6.39}$$

In this equation

$$S_{\mathrm{C}} \stackrel{\mathrm{df}}{=} ||u_{\mathrm{C}}||\,||i_{\mathrm{C}}|| = ||u_{\mathrm{C}}||\sqrt{||i_{\mathrm{Ca}}||^2 + ||i_{\mathrm{Cs}}||^2 + ||i_{\mathrm{Cr}}||^2} = \sqrt{P_{\mathrm{C}}^2 + D_{\mathrm{Cs}}^2 + Q_{\mathrm{C}}^2} \tag{6.40}$$

is the apparent power of an LTI load, obtained from the actual HGL load by neglecting current harmonics generated in this load. The term

$$S_{\mathrm{G}} \stackrel{\mathrm{df}}{=} ||u_{\mathrm{G}}||\,||i_{\mathrm{G}}|| \tag{6.41}$$

represents apparent power that occurs due to the generation of current harmonics in the load and the voltage response of the supply source to these generated harmonics. This response can be found neglecting the internal voltage e of the supply, meaning regarding the supply as a passive one-port. This power will be referred to as a *load-generated apparent power.* Observe that some active, reactive, and scattered powers are associated with current i_{G} and voltage u_{G}, meaning some physical phenomena can contribute to this power. These powers in common circuits are small, however, as compared to components of the load apparent power, P_{C}, D_{s}, and Q_{C}. Their separation in the power equation would only make this equation more complex without providing any practical benefits.

The apparent power in the cross-section between the source and the load is elevated additionally by the power

$$S_{\mathrm{CG}} \stackrel{\mathrm{df}}{=} \sqrt{||u_{\mathrm{C}}||^2||i_{\mathrm{G}}||^2 + ||u_{\mathrm{G}}||^2||i_{\mathrm{C}}||^2}. \tag{6.42}$$

Quantities in both products in this formula do not have common harmonics and consequently, there cannot be any physical phenomenon associated with power S_{CG}. It will be referred to as *apparent traverse power.* This power occurs in the effect of the presence of mutually orthogonal components in the supply current and load voltage. These components contribute to the apparent power increase, but due to their orthogonality are not associated with active, reactive, or scattered powers.

Consequently, with definitions (6.40–6.42), the power Eqn. (6.39) of HGLs can be presented in a more explicit form:

$$S^2 = P_{\mathrm{C}}^2 + D_{\mathrm{s}}^2 + Q_{\mathrm{C}}^2 + S_{\mathrm{CG}}^2 + S_{\mathrm{G}}^2. \tag{6.43}$$

Thus, four different powers associated with different power phenomena contribute to an increase of the apparent power S above power P_{C}. Observe, that the common active power P is not visible in this equation. Instead, the active power originated in the distribution system P_{C}, but not the active power P_{G} originated in the HGL is present in this equation.

6.13 Power Factor of HGLs

The power factor of HGLs can be expressed in the form

$$\lambda \stackrel{\mathrm{df}}{=} \frac{P}{S} = \frac{P_{\mathrm{C}} - P_{\mathrm{G}}}{\sqrt{P_{\mathrm{C}}^2 + D_{\mathrm{s}}^2 + Q_{\mathrm{C}}^2 + S_{\mathrm{CG}}^2 + S_{\mathrm{G}}^2}} \tag{6.44}$$

and this formula reveals all power components that contribute to the degradation of the power factor. Observe that in the case of LTI loads the power factor is reduced below unity only by an increase of the denominator versus to the numerator. In the case of HGLs, a negative active power P_{C} occurs in the circuit, and it reduces the numerator.

A particularly common and important situation occurs when the distortion of the supply voltage can be neglected, meaning it can be assumed that the distribution voltage e is sinusoidal. In such a case

$$e = e_1, \quad u_{\mathrm{C}} = u_1, \quad i_{\mathrm{C}} = i_1$$

and consequently

$$P_{\mathrm{C}} = P_1, \; D_{\mathrm{s}} = 0, \; Q_{\mathrm{C}} = Q_1, \; S_{\mathrm{C}} = S_1.$$

The power equation of an HGL supplied from a sinusoidal voltage source is reduced to the form

$$S^2 = P_1^2 + Q_1^2 + S_{\mathrm{CG}}^2 + S_{\mathrm{G}}^2 \tag{6.45}$$

with

$$S_{\mathrm{CG}} = \sqrt{||u_1||^2||i_{\mathrm{G}}||^2 + ||u_{\mathrm{G}}||^2 + ||i_1||^2} = S_1\sqrt{\left(\frac{||i_{\mathrm{G}}||}{||i_1||}\right)^2 + \left(\frac{||u_{\mathrm{G}}||}{||u_1||}\right)^2}.$$

Since

$$\frac{||i_G||}{||i_1||} = \frac{||i_G||}{I_1} \overset{\text{df}}{=} \delta_i$$

is the *total harmonic distortion* (THD) of the supply current, while

$$\frac{||u_G||}{||u_1||} = \frac{||u_G||}{U_1} \overset{\text{df}}{=} \delta_u$$

is the total harmonic distortion of the load voltage, thus

$$S_{CG} = S_1\sqrt{\left(\frac{||i_G||}{||i_1||}\right)^2 + \left(\frac{||u_G||}{||u_1||}\right)^2} = S_1\sqrt{\delta_i^2 + \delta_u^2}.$$

The load-generated apparent power, S_C, can be expressed in the form

$$S_G = ||u_G||\;||i_G|| = \frac{||u_G||}{U_1}\frac{||i_G||}{I_1}S_1 = \delta_u\delta_i S_1.$$

Since in real circuits $\delta_u \ll 1$, thus $S_C \ll S_1$ so that it could be neglected. Thus,

$$S^2 = (P_1^2 + Q_1^2)(1 + \delta_i^2 + \delta_u^2) \tag{6.46}$$

and the power factor is

$$\lambda = \frac{P}{S} \approx \frac{P_1 - P_G}{S_1\sqrt{1 + \delta_i^2 + \delta_u^2}} = \frac{1 - P_G/P_1}{\sqrt{1 + \delta_i^2 + \delta_u^2}}\lambda_1 \tag{6.47}$$

where λ_1 is the power factor of the load for the fundamental frequency. Sometimes it is referred to as a *displacement power factor*, although this seems to be rather misleading jargon. This formula reveals three factors that contribute to the power factor decline in circuits with HGLs and sinusoidal internal voltage in the distribution system.

The total harmonic distortion of the load voltage δ_u is usually small, confined by standards to a few percent. Moreover, the load-generated active power P_C is usually much smaller than the active power of the fundamental harmonic P_1. The power factor of HGLs supplied from a sinusoidal voltage source can be in such a situation approximated by

$$\lambda \approx \frac{1}{\sqrt{1 + \delta_i^2}}\lambda_1 \tag{6.48}$$

thus, the current distortion is the main factor that contributes to the power factor decline below its value for the fundamental harmonic.

When an HGL is supplied from an ideal voltage source, then

$$u_{\mathrm{C}}(t) \equiv 0$$

and consequently, there is no negative active power, P_{C}. Moreover, the load-generated apparent power S_{C} is equal to zero. Power phenomena in such situations are discussed in the following example.

Illustration 6.7 *The current of a resistor in the circuit shown in Fig. 6.20, with sinusoidal supply voltage*

$$u = \sqrt{2}\, U \sin \omega_1 t$$

is controlled by a TRIAC, with $U = 220\mathrm{V}$, $R = 1\ \Omega$, *and TRIAC firing angle* $\alpha = 135^{\circ} = ¾\,\pi$.

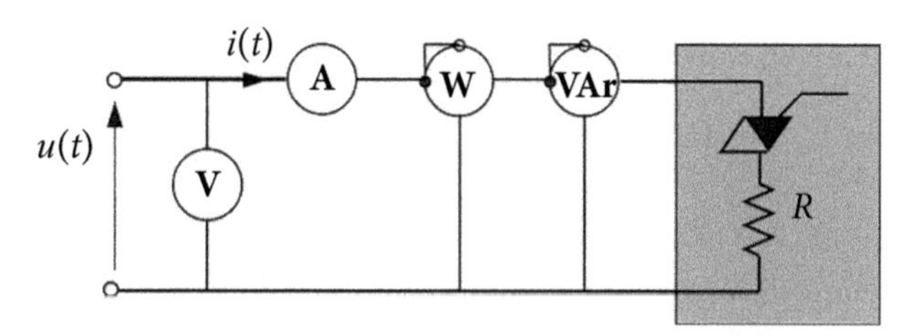

Figure 6.20 A resistive load controlled with a TRIAC.

The resistor with the TRIAC represents an HGL. Let us describe this load in power terms and answer an intriguing question: what is the reading of the varmeter?

Because the load is purely resistive, meaning without any capability of energy storage, one might answer that the varmeter reading should be zero. This is because the reactive power Q in the teaching of electrical engineering is commonly associated with energy oscillation between the load and the supply source, due to energy storage in reactive elements of the load. This would be a wrong answer, however.

The supply current has the waveform shown in Fig. 6.21.

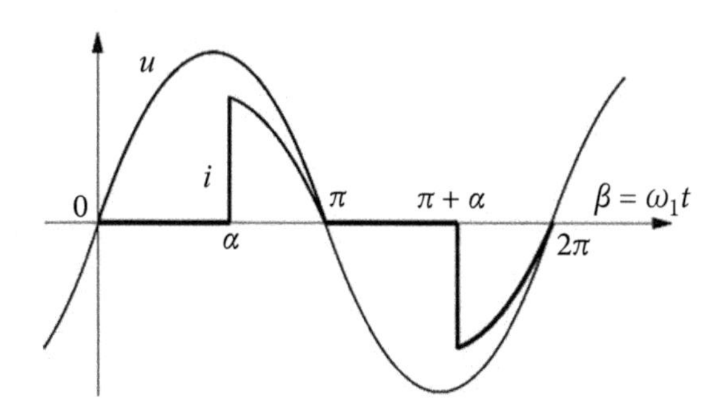

Figure 6.21 The load current waveform.

The general formula for the rms value of such waveform calculation is

$$||i|| \stackrel{\text{df}}{=} \sqrt{\frac{1}{2\pi}\int_0^{2\pi} i^2 d\beta} = \sqrt{\frac{1}{\pi}\int_0^{\pi}\left(\sqrt{2}\frac{U}{R}\sin\beta\right)^2 d\beta} =$$

$$= \sqrt{\frac{2}{\pi}}\frac{U}{R}\sqrt{\int_{\alpha}^{\pi}\frac{1}{2}(1-\cos 2\beta)\, d\beta} = \sqrt{\frac{1}{\pi}}\frac{U}{R}\sqrt{\pi-\alpha-\frac{1}{2}\sin 2\beta\,|_{\alpha}^{\pi}} =$$

$$= \frac{U}{R}\sqrt{1-\frac{\alpha}{\pi}+\frac{\sin 2\alpha}{2\pi}}.$$

For $U = 220$ V, $R = 1\Omega$, *and* $\alpha = 135^{\circ} = \frac{3}{4}\pi$, *the supply current rms value is*

$$||i|| = \frac{U}{R}\sqrt{1-\frac{\alpha}{\pi}+\frac{\sin 2\alpha}{2\pi}} = 220\sqrt{1-\frac{3}{4}+\frac{\sin 270^{\circ}}{2\pi}} = 66.307\text{A}.$$

The apparent power S of the supply source is equal to

$$S = ||u||\,||i|| = 220 \times 66.307 = 14.588 \text{ kVA}.$$

The active power P of the load is

$$P = ||i||^2 R = 66.307^2 \times 1 = 4.397 \text{ kW}.$$

Consequently, the power factor is

$$\lambda = \frac{P}{S} = \frac{4.397}{14.588} = 0.301.$$

Readings of all meters, apart from the varmeter, are shown in Fig. 6.22.

The supply voltage is sinusoidal, thus a common reactive power meter can be used for measuring the reactive power Q. Reading of such a meter is proportional to the mean value of currents in the voltage and current coils, meaning the value

$$Q = k\frac{1}{T}\int_0^T i_V(t)\, i(t)\, dt \stackrel{\text{df}}{=} k\,(i_V, i).$$

If the varmeter is a meter with an inductor in the voltage coil, as shown in Fig. 6.23, then assuming that the voltage coil is purely inductive

$$i_V = \frac{1}{L}\int u(t)\, dt = -\sqrt{2}\frac{U}{\omega_1 L}\cos\omega_1 t \stackrel{\text{df}}{=} i_{V1}$$

thus the current of the voltage coil is sinusoidal.

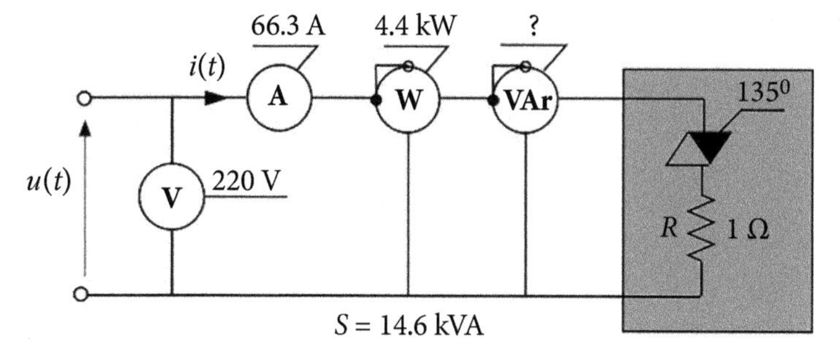

Figure 6.22 Meter readings.

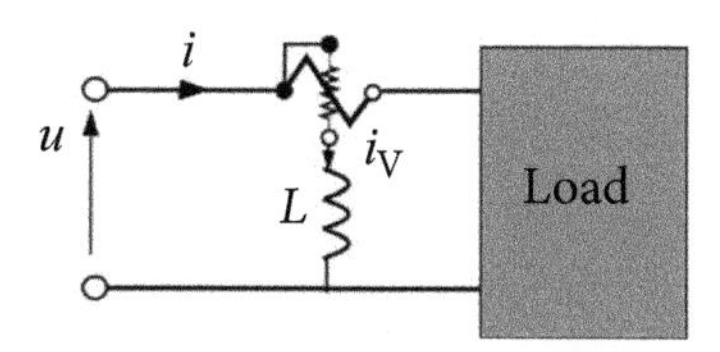

Figure 6.23 A reactive power meter.

If the load current is nonsinusoidal, meaning composed of harmonics

$$i = \sum_{n \in N} i_n$$

then due to the orthogonality of harmonics of a different order, the meter reading is

$$Q = k\,(i_{\rm V}, i) = k\left(i_{\rm V1}, \sum_{n \in N} i_n\right) = k\,(i_{\rm V1}, i_1) \stackrel{\rm df}{=} Q_1.$$

The reactive power Q_1 can be calculated if the crms value $\boldsymbol{I}_1$ of the supply current fundamental harmonic is known. It is equal to

$$\boldsymbol{I}_1 = \frac{1}{\sqrt{2}\,\pi} \int_0^{2\pi} i(\beta)\, e^{-j\beta}\, d\beta.$$

The load current can be presented as

$$i\,(\beta) = x\,(\beta) - x\,(\beta - \pi)$$

where the current component $x(\beta)$ is shown in Fig. 6.24, and using shifting property, the crms value of the current fundamental harmonic $\boldsymbol{I}_1$ can be expressed as

$$\boldsymbol{I}_1 = \boldsymbol{X}_1 - \boldsymbol{X}_1\, e^{-j\pi} = 2\boldsymbol{X}_1$$

where

$$\boldsymbol{X}_1 = \frac{1}{\sqrt{2}\,\pi}\int_0^{2\pi} x(\beta)\, e^{-j\beta}\, d\beta = \frac{U}{R}\frac{1}{\pi}\int_\alpha^{\pi} \sin\beta\, e^{-j\beta}\, d\beta.$$

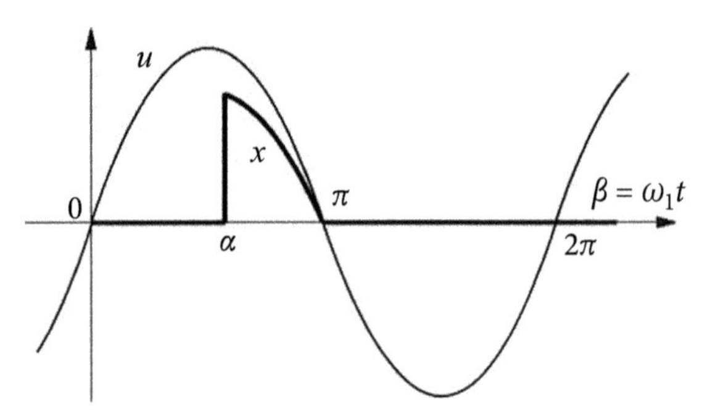

Figure 6.24 Quantity $x(\beta)$ waveform.

Thus,

$$\boldsymbol{I}_1 = 2\frac{U}{R}\frac{1}{\pi}\int_\alpha^{\pi} \sin\beta\, e^{-j\beta}\, d\beta = 2\frac{U}{R}\frac{1}{\pi}\int_\alpha^{\pi} \frac{e^{j\beta} - e^{-j\beta}}{2j} e^{-j\beta}\, d\beta = -j\frac{U}{R}\frac{1}{\pi}\int_\alpha^{\pi} (1 - e^{-j2\beta})\, d\beta =$$

$$= -j\frac{U}{R}\frac{1}{\pi}\left[(\pi - \alpha) - \frac{1}{-2j}(e^{-j2\pi} - e^{-j2\alpha})\right] = -j\frac{U}{R}\left[1 - \frac{\alpha}{\pi} - j\frac{1}{2\pi}(1 - e^{-j2\alpha})\right] =$$

$$= -j\frac{U}{R}\left[1 - \frac{\alpha}{\pi} + \frac{\sin 2\alpha}{2\pi} - j\frac{1}{2\pi}(1 - \cos 2\alpha)\right].$$

The crms value of the supply voltage is $\boldsymbol{U} = -jU$. *Indeed,*

$$u(t) = \sqrt{2}\,\mathrm{Re}\{\boldsymbol{U} e^{j\omega_1 t}\} = \sqrt{2}\,\mathrm{Re}\{-j\,U e^{j\omega_1 t}\} = \sqrt{2}\,U \sin\omega_1 t,$$

therefore the crms value of the load current fundamental harmonic $\boldsymbol{I}_1$ *has a component that is in phase with the voltage crms value, equal to*

$$-j\frac{U}{R}\left(1 - \frac{\alpha}{\pi} + \frac{\sin 2\alpha}{2\pi}\right) \stackrel{\mathrm{df}}{=} \boldsymbol{I}_{1\mathrm{a}}$$

and a component that is shifted against the voltage by $\pi/2$, *equal to*

$$-j\frac{U}{R}\left[-j\frac{1}{2\pi}(1 - \cos 2\alpha)\right] = -\frac{U}{R}\left[\frac{1}{2\pi}(1 - \cos 2\alpha)\right] \stackrel{\mathrm{df}}{=} \boldsymbol{I}_{1\mathrm{r}}.$$

For $U = 220$ V, $R = 1\Omega$, *and* $\alpha = 135^\circ = \frac{3}{4}\pi$

$$\boldsymbol{I}_{1\mathrm{a}} = -j\frac{U}{R}\left(1 - \frac{\alpha}{\pi} + \frac{\sin 2\alpha}{2\pi}\right) = -j\frac{220}{1}\left(1 - \frac{3}{4} + \frac{-1}{2\pi}\right) = -j19.985\ \mathrm{A}.$$

Verification. The active power P at the load terminal should satisfy the relation

$$P = \mathrm{Re}\{\boldsymbol{U}\boldsymbol{I}_{1a}^{*}\} = \mathrm{Re}\{-j220 \times (-j19.985)^{*}\} = 4.397 \text{ kW} = ||i||^2 R.$$

The crms value of the reactive current

$$\boldsymbol{I}_{1r} = -\frac{U}{R}\left[\frac{1}{2\pi}(1 - \cos 2\alpha)\right] = -\frac{220}{1}\left[\frac{1}{2\pi}(1 - \cos 270°)\right] = -35.015 \text{ A}.$$

The total value of the load current fundamental harmonic crms is

$$\boldsymbol{I}_1 = \boldsymbol{I}_{1a} + \boldsymbol{I}_{1r} = -j19.985 - 35.015 = 40.317\, e^{-j150.28°} \text{ A}$$

and its time variance

$$i_1(t) = \sqrt{2}\,\mathrm{Re}\{\boldsymbol{I}_1\, e^{j\omega_1 t}\} = \sqrt{2}\,\mathrm{Re}\{40.317\, e^{-j150.28°} e^{j\omega_1 t}\} = 40.317\sqrt{2}\cos(\omega_1 t - 150.28°) =$$
$$= 40.317\sqrt{2}\cos(\omega_1 t - 150.28°) = 40.317\sqrt{2}\sin(\omega_1 t - 60.28°) \text{ A}.$$

Thus, the current fundamental harmonic i_1 is delayed versus the supply voltage. Because this fundamental harmonic of the supply current is shifted versus the load voltage, it has an in-phase component, meaning the active power, and a component shifted by $\pi/2$, meaning a reactive component. These components are equal to

$$i_1 = i_{1a} + i_{1r} = \sqrt{2}\, I_{1a} \sin \omega_1 t + \sqrt{2}\, I_{1r} \sin(\omega_1 t - 90°) = 40.317\sqrt{2}\sin(\omega_1 t - 60.28°) \text{ A}$$

with

$$I_{1a} = I_1 \cos \varphi_1 = 40.317 \times 0.496 = 19.987 \text{ A}$$

$$I_{1r} = I_1 \sin \varphi_1 = 40.317 \times 0.868 = 35.014 \text{ A}.$$

Thus, despite the lack of energy oscillation, there is a current component shifted against the supply voltage by $\pi/2$, meaning the reactive current.

The meter of the reactive power reading is equal to

$$Q = k\,(i_{V1}, i_1) = k\left(-\sqrt{2}\frac{U}{\omega_1 L}\cos \omega_1 t, \sqrt{2}\, I_{1a} \sin \omega_1 t + \sqrt{2}\, I_{1r} \sin\,(\omega_1 t - 90°)\right) =$$
$$= k\, 2\frac{U}{\omega_1 L} I_{1r}(\cos \omega_1 t,\ \cos \omega_1 t) = \frac{k}{\omega_1 L} U I_{1r}.$$

Assuming that the coefficient is selected such that $k/\omega_1 L = 1$, the meter reading is

$$Q = U I_{1r} = 220 \times 35.014 = 7.703 \text{ kVAr}.$$

The reactive power Q occurs in this circuit despite the lack of energy oscillation between the supply source and the load. TRIAC switching causes a phase shift of the load current fundamental harmonic i_1 versus the supply voltage.

The supply current i in the circuit considered can be decomposed as follows

$$i = \sum_{n=1}^{\infty} i_n = i_1 + \sum_{n=2}^{\infty} i_n = i_{1a} + i_{1r} + i_G, \quad \text{with} \quad i_G = \sum_{n=2}^{\infty} i_n.$$

These three components are mutually orthogonal, so that their rms values satisfy the relationship

$$||i||^2 = ||i_{1a}||^2 + ||i_{1r}||^2 + ||i_G||^2.$$

In the circuit considered $u_C = u$, $u_G = 0$. Multiplying this equation by $||u||^2$, we obtain the power equation

$$S^2 = P^2 + Q^2 + S_{CG}^2.$$

Since

$$||i_G|| = \sqrt{||i||^2 - (||i_{1a}||^2 + ||i_{1r}||^2)} =$$

$$= \sqrt{66.307^2 - (19.985^2 + 35.015^2)} = 52.642 \text{ A}$$

$$S_{CG} = ||i_G||\,||u|| = 52.642 \times 220 = 11.581 \text{ kVA}.$$

*****

6.14 Working, Reflected, and Detrimental Active Powers

The active power P is one of the best-established power quantities, commonly regarded as useful power. There is awareness in the electrical engineering community, however, that some components of this power in circuits with nonsinusoidal and asymmetrical voltages and currents are not useful for customers but are even harmful. For example, the active power of the negative sequence symmetrical component in the inductor motor supply voltage causes only the motor overheating. Therefore, the concept of active power should be revised.

The active power P can be decomposed [262, 279] into three components of different usefulness, referred to as working, reflected, and detrimental active powers. Let us present a rationale for such a decomposition.

To reduce the risk that circuit complexity may obscure the rationale, let us analyze the simplest possible single-phase resistive circuit with an HGL, shown in Fig. 6.25, assuming that the distribution voltage e is sinusoidal, while the HGL is a source of current harmonics j.

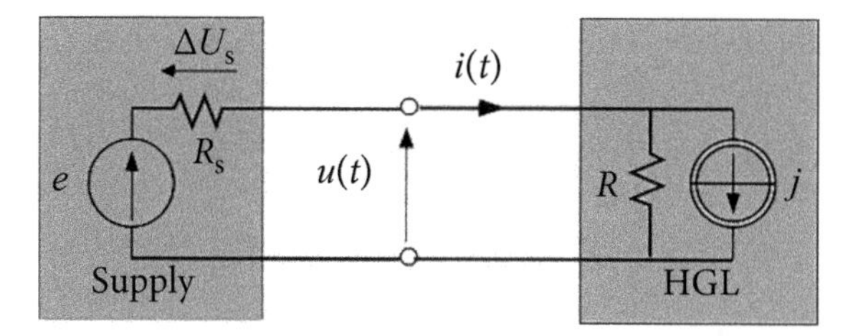

Figure 6.25 A resistive circuit with HGL.

Let us assume for simplicity's sake, that the load-generated current j does not contain a DC component, thus the load current can be expressed in the form

$$i(t) = \sum_{n \in N} i_n = i_1 + i_G.$$

In a response to the load distorted current, the load voltage is distorted and has the form

$$u(t) = \sum_{n \in N} u_n = u_1 + u_G.$$

The active power at the load terminals is equal to

$$P = \frac{1}{T}\int_0^T u(t)\, i(t)\, dt = \sum_{n \in N} P_n = P_1 + P_2 + P_3 + P_4 + \ldots \tag{6.49}$$

The load-generated current j is composed of harmonic of the order $n > 1$, thus for the fundamental harmonic the load is passive, hence

$$P_1 = U_1 I_1 > 0.$$

The active power of other harmonics, for $n > 1$, is equal to

$$P_n = U_n I_n = (-R_s I_n) I_n = -R_s I_n^2 < 0,$$

thus all of them are negative. Thus

$$\sum_{n \notin 1} P_n = P_2 + P_3 + P_4 + \ldots \stackrel{\text{df}}{=} P_G < 0.$$

Therefore, formula (6.49) can be rewritten in the form

$$P = P_1 + P_G.$$

The last formula shows that the active power P of HGLs contains two substantially different components. The first of them, P_1, stands for the average rate of energy flow from the supply source to the load. This component can be referred to as a *working active power* and denoted as P_{w}. The second term, P_{G}, stands for the average rate of energy flow back from the load to the supply source, where this energy is dissipated on the source's internal resistance. Therefore, the power P_{G} can be referred to as the load-generated active power. It means that a part of the energy delivered to the load is sent back to the supply source as a *reflected active power* and denoted as P_{r}. Since power P_{G} is negative, thus $P_1 > P$, which means that the HGLs have to be supplied with higher power than measured at the load terminals, active power P. It has to be supplied with power P_1. Thus, to operate such a load, it has to be supplied to the load by the voltage, and current fundamental harmonics are transformed, due to the load nonlinearity, into the energy of harmonics and sent back to the supply circuit, where it is dissipated on the supply source resistance.

The relation between the working and reflected active powers, P_{w}, P_{r}, and the active power P in a single-phase, resistive circuit with HGLs is visualized in Fig. 6.26.

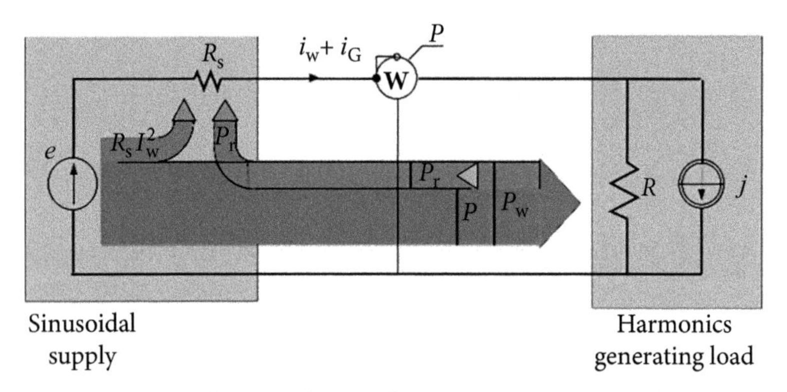

Figure 6.26 A diagram of energy flow in a resistive circuit with HGL.

Symbol i_{w} in this figure denotes a *working active current*, meaning the minimum current needed for providing the working active power P_{w} to the load. In single-phase circuits, this current is identical to the component of the fundamental harmonic which is in phase with the voltage fundamental harmonic, $u_1(t)$. Part of the energy delivered to such a load by the fundamental harmonic is converted to energy of higher harmonics and reflected into the supply source. This energy is not recovered in the supply source, since its voltage e is sinusoidal, but it is dissipated on the supply source resistance.

Illustration 6.8 *Figure 6.27 shows a structure of a single-phase DC battery charger, simplified to a purely resistive circuit with ideal lossless diodes. The rectifier is supplied with a sinusoidal voltage e of the rms value* $E = 230$ V *and provides the active power* $P = 5000$ W *to the DC side.*

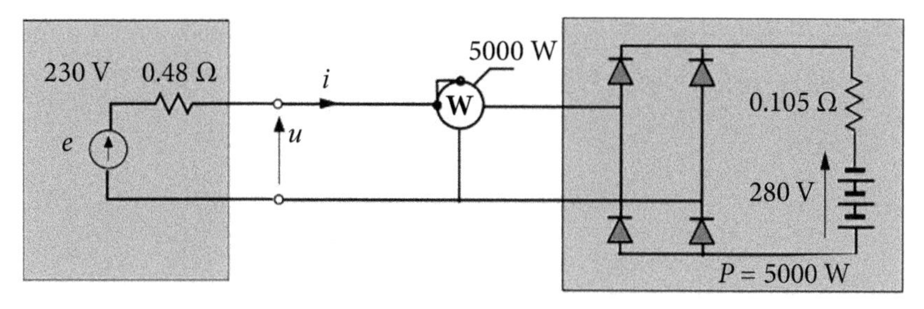

Figure 6.27 The structure of a battery charger simplified to a resistive circuit.

Analysis of this circuit results in the crms value of the supply voltage and current fundamental harmonic equal to

$$U_1 = 218.5 \text{ V}, \qquad I_1 = 24.99 \; A$$

which are mutually in phase, so that the active power of the fundamental harmonic, meaning the working power, is equal to $P_w = 5242$ W. *The reflected active power is equal to*

$$P_r = P_1 - P = 242 \text{ W}.$$

Thus the rectifier has to be supplied with a power, which is a few percent higher than the active power P measured by the wattmeter at the load terminals.

****

It was assumed above that the internal voltage e of the supply source is sinusoidal. It can be distorted, however, and the voltage and current harmonics contribute to energy transfer in the circuit. When the load is, on purpose, resistive, as in a case of a heater, then the energy transferred to such a load by these harmonics is useful. It is not so evident in the case of electronic kinds of loads, mainly rectifiers in video and/or computer-like appliances, motors, or LED-based lighting systems. Harmonics disturb the performance of such devices rather than contribute to energy transfer. Therefore, the active power of such harmonics should be regarded as a *detrimental active power* and denoted as P_d.

Customers are charged currently for the active energy equal to in a payoff interval (suppose one month), meaning the integral of the active power P,

$$\int_0^{\text{month}} P \, dt = W$$

and measured by common energy meters. Unfortunately, this energy is lower than the working energy needed to run loads that generate current harmonics

$$\int_{0}^{\text{month}} P_{\mathrm{W}}\, dt = W_{\mathrm{W}}.$$

Financial accounts between energy providers and its user would be fairer if this account would be based on the working energy delivered to the customer. Moreover, when charged for the active energy, the customer with current harmonics generating loads is not financially responsible for negative effects caused by harmonics generated by his loads on the energy provider side. Such an energy user does not now have financial motivations for reducing harmonics. Unfortunately, these conclusions challenge deeply rooted fundamentals, regulations, and standards, all based on the concept of active power.

Summary

The CPC-based power theory of single-phase circuits with nonsinusoidal voltages and currents reveals all power-related phenomena in such circuits and describes them in mathematical terms. It provides, moreover, a starting point for the development of the power theory of three-phase circuits with nonsinusoidal voltages and currents, as presented in Chapters 7 and 8.

The CPC-based power theory creates fundamentals for the power factor improvement in electrical circuits by reactance compensators, discussed in detail in Chapters 13–15. This theory also provides a reference for studies on other power theories presented in Chapter 5 and Chapters 18–22.

Chapter 7
CPC in Three-Phase Three-Wire Circuits

7.1 Troubles with the Power Equation

Energy in electrical power systems is delivered to customers mainly by three-phase three-wire (3p3w) lines. Therefore, the knowledge of the power properties of 3p3w circuits is of the utmost importance for power systems engineering. Power properties and compensation of such circuits are central issues of the power theory.

On the customer's side, a neutral conductor is usually added to enable the supply of single-phase loads, as shown in Fig. 7.1, so that the customers' distribution system has the form of a three-phase with a neutral circuit or three-phase four-wire (3p4w) circuit.

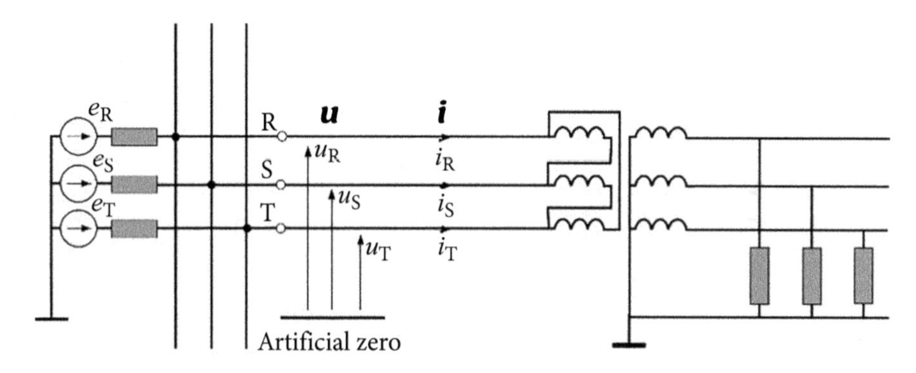

Figure 7.1 A three-phase, three-wire (3p3w) circuit supplying single-phase loads.

Three-phase three-wire circuits can be regarded as the next step in the complexity to single-phase circuits. Consequently, clarification of their properties was not possible before clarification of these properties in single-phase circuits.

There also was another crucial factor that has contributed to a failure of clarification power properties of 3p3w circuits with nonsinusoidal voltages and currents (nv&c) for a long time. While the power equation of a single-phase circuit with sinusoidal voltages and current (sv&c)

$$S^2 = P^2 + Q^2 \tag{7.1}$$

provides a good fundamental for developing power equation at sv&c, this equation, as explained in Chapter 1, does not describe the relationship between powers in 3p3w circuits correctly, even at sinusoidal voltages and currents and a symmetrical supply

Powers and Compensation in Circuits with Nonsinusoidal Current. Leszek S. Czarnecki, Oxford University Press.
 DOI: 10.1093/oso/9780198879206.003.0007

voltage. It is valid only in circuits with balanced loads, that is in circuits with symmetrical currents. The power properties of such circuits are trivial, however. There are not any other powers than only the active, reactive, and apparent power, which satisfy Eqn. (7.1) at any definition of apparent power. Power properties in such circuits can be explained in terms of the power theory of single-phase circuits. This is not possible, however, when, due to the load imbalance, the load supply currents are not symmetrical.

It involves two issues. The first of them is the selection of the right definition of the apparent power S. However, even with the right definition of the apparent power, Eqn. (7.1) is not valid in the presence of voltage and current distortion. Thus, a right relation between powers, meaning the power equation of 3p3w circuits, has to be formulated. Therefore, this chapter is confined to the identification of power phenomena in 3p3w circuits that contribute to an increase of the supply current rms value. Eqn. (7.1) is not valid, however, even if the supply voltage is sinusoidal but the load is unbalanced. Therefore, the explanation of the power properties of circuits with nv&c has to be preceded with the explanation of these properties of circuits with unbalanced loads with sv&c.

The basic structure of a three-phase circuit with LTI load and symbols of voltages and currents, as well their three-phase vectors $\boldsymbol{u}$ and $\boldsymbol{i}$, used in this book are shown in Fig. 7.2.

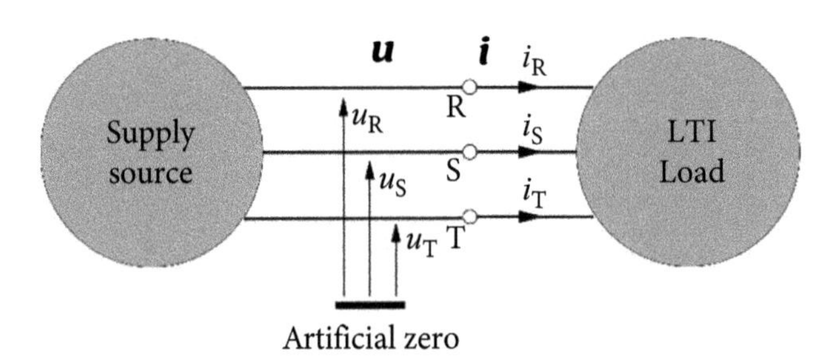

Figure 7.2 A structure of a three-phase, three-wire circuit.

The instantaneous power $p(t)$ of the load, meaning the rate of energy flow from the supply source to the load, is equal to

$$p(t) \stackrel{\text{df}}{=} \frac{dW(t)}{dt} = u_R(t)\, i_R(t) + u_S(t)\, i_S(t) + u_T(t)\, i_T(t)$$

and can be expressed in a more compact form as a dot product of voltage and current vectors:

$$p(t) = \boldsymbol{u}^T(t)\boldsymbol{i}(t) = \boldsymbol{u}^T\boldsymbol{i}.$$

According to these formulae, the instantaneous power is specified by three line-to-artificial zero voltages and three supply currents, meaning six quantities. In three-wire circuits that are supplied with a symmetrical voltage, meaning such that

$$u_{\mathrm{R}}(t) + u_{\mathrm{S}}(t) + u_{\mathrm{T}}(t) \equiv 0$$

$$i_{\mathrm{R}}(t) + i_{\mathrm{S}}(t) + i_{\mathrm{T}}(t) \equiv 0$$

the instantaneous power can be expressed in terms of only two voltages and two currents,

$$p(t) = \mathbf{u}^{\mathrm{T}}\mathbf{i} = (u_{\mathrm{R}} - u_{\mathrm{T}})\, i_{\mathrm{R}} + (u_{\mathrm{S}} - u_{\mathrm{T}})\, i_{\mathrm{S}} = u_{\mathrm{RT}}\, i_{\mathrm{R}} + u_{\mathrm{ST}}\, i_{\mathrm{S}} = \mathscr{u}^{\mathrm{T}}\mathscr{i}$$

where

$$\mathscr{u} \stackrel{\mathrm{df}}{=} \begin{bmatrix} u_{\mathrm{RT}} \\ u_{\mathrm{ST}} \end{bmatrix}, \quad \mathscr{i} \stackrel{\mathrm{df}}{=} \begin{bmatrix} i_{\mathrm{R}} \\ i_{\mathrm{S}} \end{bmatrix}$$

are reduced vectors of line-to-line voltages and supply currents of only two lines.

The active power P in 3p3w circuits is defined as a mean value of the instantaneous power over a period T.

$$P = \frac{1}{T}\int_0^T (u_{\mathrm{R}}i_{\mathrm{R}} + u_{\mathrm{S}}i_{\mathrm{S}} + u_{\mathrm{T}}i_{\mathrm{T}})\, dt = \frac{1}{T}\int_0^T \mathbf{u}^{\mathrm{T}}\mathbf{i}\, dt \stackrel{\mathrm{df}}{=} (\mathbf{u}, \mathbf{i}) = (\mathscr{u}, \mathscr{i}),$$

thus it is equal to a scalar product of the supply voltage and the supply currents' three-phase vectors $\mathbf{u}$ and $\mathbf{i}$, or reduced vectors $\mathscr{u}$ and $\mathscr{i}$. This formula, valid also for nonsinusoidal voltages and currents, in sinusoidal conditions has the form

$$P = (\mathbf{u}, \mathbf{i}) = \sum_{p=\mathrm{R,S,T}} U_p\, I_p \cos\varphi_p.$$

The reactive power Q in three-phase three-wire circuits with sinusoidal voltages and currents is defined, by analogy to the active power, as

$$Q \stackrel{\mathrm{df}}{=} \sum_{p=\mathrm{R,S,T}} U_p\, I_p \sin\varphi_p$$

meaning as a sum of reactive powers of individual phases. This is a misleading analogy, however, because energy or powers cannot be associated with individual supply lines of three-phase circuits, as was explained in Chapter 1.

The reactive power Q occurs as a result of a phase shift φ between the supply voltage and line currents. Unlike the reactive power in single-phase circuits, the reactive power Q in 3p3w circuits cannot be associated, however, and as discussed in [324], with energy oscillation between the supply source and the load. At symmetrical line currents, the instantaneous power $p(t)$ is constant, meaning it does not contain any oscillating component, even if the reactive power Q is not equal to zero.

Traditional definitions of apparent power. The apparent power in 3p3w circuits with sv&c is defined according to IEEE Standard Dictionary of Electrical and Electronics Terms [154] in two different ways. It is defined as arithmetical apparent power, meaning the sum of apparent powers of individual phases,

$$S_{\mathrm{A}} \stackrel{\mathrm{df}}{=} U_{\mathrm{R}}I_{\mathrm{R}} + U_{\mathrm{S}}I_{\mathrm{S}} + U_{\mathrm{T}}I_{\mathrm{T}},$$

or as the geometrical apparent power.

$$S_{\mathrm{G}} \stackrel{\mathrm{df}}{=} \sqrt{P^2 + Q^2}.$$

The definition of the apparent power suggested by Buchholz [11]

$$S_{\mathrm{B}} \stackrel{\mathrm{df}}{=} \sqrt{U_{\mathrm{R}}^2 + U_{\mathrm{S}}^2 + U_{\mathrm{T}}^2}\sqrt{I_{\mathrm{R}}^2 + I_{\mathrm{S}}^2 + I_{\mathrm{T}}^2}$$

is not included in the IEEE Standard Dictionary thus it is not used in IEEE Standard-supported documents. The same applies to the definition of the apparent power S_{C} introduced in 1988 [93] for 3p3w circuits with nv&c. This last definition was based on the following reasoning. The single-phase supply or transmission equipment is rated by the product of the rms value of the voltage and current provided, as shown in Fig. 7.3a, by such a device to loads. Such a formal product is referred to as the apparent power S of single-phase circuits sources

$$S \stackrel{\mathrm{df}}{=} ||u||\,||i||.$$

The three-phase rms voltage and current values $||\boldsymbol{u}||$ and $||\boldsymbol{i}||$ of a three-phase supply source, shown in Fig. 7.2b, are equivalent to rms values in single-phase circuits $||u||$ and $||i||$ with regard to energy dissipation. Therefore, it is reasonable to assume, as in paper [93], that the apparent power of three-phase devices be defined as the product of the three-phase rms value of the supply voltage and current, namely:

$$S = S_{\mathrm{C}} \stackrel{\mathrm{df}}{=} ||\boldsymbol{u}||\,||\boldsymbol{i}||. \tag{7.2}$$

This definition at sv&c is identical to that suggested by Buchholz in Ref. [11], meaning $S_{\mathrm{B}} = S_{\mathrm{C}}$. However, definition (7.2) was introduced for 3p3w circuits with nv&c, while Buchholtz's definition was introduced only for circuits with sv&c. Thus, it is more general.

When the apparent power is calculated in symmetrical 3p3w circuits with sv&c then the numerical value of the apparent power S is independent of the definition

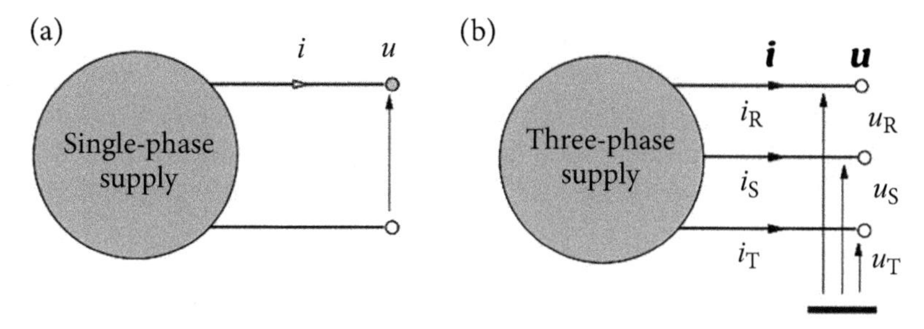

Figure 7.3 (a) A single-phase, and (b) a three-phase source.

used for this power calculation. In circuits with unbalanced loads, these definitions provide different values of apparent power and the power factor, as was shown in Chapter 1. Unfortunately, the IEEE Standard Dictionary of Electrical and Electronics Terms [154] provides no hints for selecting the right definition of the apparent power S even in sinusoidal situations. In this situation, the power factor can have three different values. This was illustrated in Section 1.9.

Reasoning that enables such a selection was presented in Ref. [165] in 1999. It can be found in Section 5.15.

7.2 Currents' Physical Components in Circuits with sv&c

Apparent power S in single-phase circuits and in balanced three-phase circuits with sv&c is equal to the magnitude of the *complex apparent power* $\boldsymbol{S}$, which for single-phase circuits is defined as

$$\boldsymbol{S} = \boldsymbol{U}\boldsymbol{I}^* = Se^{j\varphi} = P + jQ.$$

When the load is unbalanced and/or voltages are asymmetrical or nonsinusoidal then the apparent power S is no longer the magnitude of the complex apparent power $\boldsymbol{S}$. Unfortunately, the similarity of symbols for both powers could confuse and even could lead to errors. Since it is a very common custom of denoting the apparent power by S, a different symbol will be used in this book for the power defined as

$$\boldsymbol{U}^{\mathrm{T}}\boldsymbol{I}^* = P + jQ = \boldsymbol{C} = C\,e^{j\varphi} \tag{7.3}$$

where

$$\boldsymbol{U} \stackrel{\mathrm{df}}{=} [\boldsymbol{U}_{\mathrm{R}}, \boldsymbol{U}_{\mathrm{S}}, \boldsymbol{U}_{\mathrm{T}}]^{\mathrm{T}},\ \boldsymbol{I} \stackrel{\mathrm{df}}{=} [\boldsymbol{I}_{\mathrm{R}}, \boldsymbol{I}_{\mathrm{S}}, \boldsymbol{I}_{\mathrm{T}}]^{\mathrm{T}}.$$

Also, the adjective "apparent" will not be used. The quantity defined by (7.3) will be referred to as a *complex power*, but not a complex apparent power.

Having selected the right definition of the apparent power S, we can look for the true power equation of 3p3w circuits with sv&c. The concept of the Currents' Physical Components (CPC), meaning identification of currents components, associated with distinctive power-related physical phenomena in the circuit shall be used for it.

Let us consider a three-phase load supplied by a three-wire line as shown in Fig. 7.4.

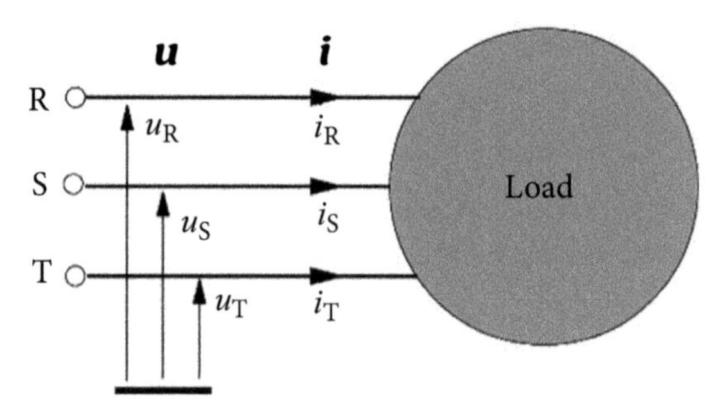

Figure 7.4 A three-phase load.

If the supply voltage is symmetrical of the positive sequence, i.e. $u_S(t) = u_R(t - T/3)$ and $u_T(t) = u_R(t + T/3)$, and the load is linear and time invariant (LTI) then, according to Ref. [93], such a load has an equivalent load of Δ structure shown in Fig. 7.5, with line-to-line admittances $\boldsymbol{Y}_{RS}$, $\boldsymbol{Y}_{ST}$ and $\boldsymbol{Y}_{TR}$.

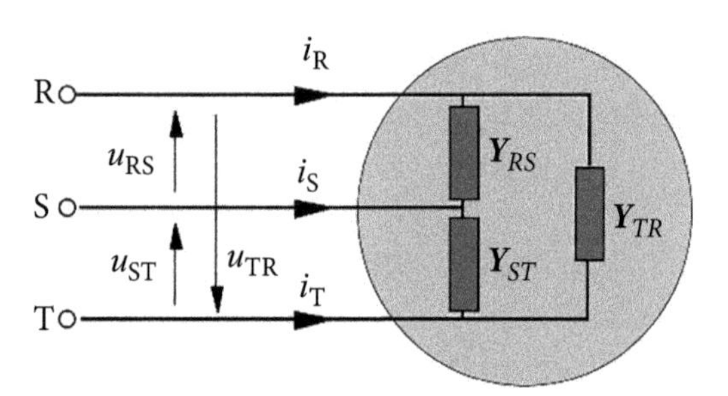

Figure 7.5 An equivalent load in Δ structure.

Let us express the vector of the load line currents

$$\boldsymbol{i} = \begin{bmatrix} i_R \\ i_S \\ i_T \end{bmatrix} = \sqrt{2}\mathrm{Re}\begin{bmatrix} \boldsymbol{I}_R \\ \boldsymbol{I}_S \\ \boldsymbol{I}_T \end{bmatrix} e^{j\omega t} = \sqrt{2}\mathrm{Re}\,\{\boldsymbol{I}e^{j\omega t}\}$$

in terms of the load admittances as

$$\begin{aligned} \boldsymbol{I}_R &= \boldsymbol{Y}_{RS}(\boldsymbol{U}_R - \boldsymbol{U}_S) - \boldsymbol{Y}_{TR}(\boldsymbol{U}_T - \boldsymbol{U}_R) \\ \boldsymbol{I}_S &= \boldsymbol{Y}_{ST}(\boldsymbol{U}_S - \boldsymbol{U}_T) - \boldsymbol{Y}_{RS}(\boldsymbol{U}_R - \boldsymbol{U}_S) \\ \boldsymbol{I}_T &= \boldsymbol{Y}_{TR}(\boldsymbol{U}_T - \boldsymbol{U}_R) - \boldsymbol{Y}_{ST}(\boldsymbol{U}_S - \boldsymbol{U}_T). \end{aligned}$$

For the supply voltage of the positive sequence

$$\boldsymbol{U}_{\mathrm{S}} = \alpha^* \boldsymbol{U}_{\mathrm{R}}, \quad \boldsymbol{U}_{\mathrm{T}} = \alpha\, \boldsymbol{U}_{\mathrm{R}}, \quad \alpha \stackrel{\mathrm{df}}{=} 1 e^{j\frac{2\pi}{3}}$$

where α is the complex rotation coefficient, and α^* is its conjugate. Thus, these crms values can be rearranged to the form

$$\boldsymbol{I}_{\mathrm{R}} = \boldsymbol{Y}_{\mathrm{RS}}(\boldsymbol{U}_{\mathrm{R}} - \boldsymbol{U}_{\mathrm{S}}) - \boldsymbol{Y}_{\mathrm{TR}}(\boldsymbol{U}_{\mathrm{T}} - \boldsymbol{U}_{\mathrm{R}}) = (\boldsymbol{Y}_{\mathrm{RS}} + \boldsymbol{Y}_{\mathrm{ST}} + \boldsymbol{Y}_{\mathrm{TR}})\boldsymbol{U}_{\mathrm{R}} - (\boldsymbol{Y}_{\mathrm{ST}} + \alpha\boldsymbol{Y}_{\mathrm{TR}} + \alpha^*\boldsymbol{Y}_{\mathrm{RS}})\boldsymbol{U}_{\mathrm{R}}$$

$$\boldsymbol{I}_{\mathrm{S}} = \boldsymbol{Y}_{\mathrm{ST}}(\boldsymbol{U}_{\mathrm{S}} - \boldsymbol{U}_{\mathrm{T}}) - \boldsymbol{Y}_{\mathrm{RS}}(\boldsymbol{U}_{\mathrm{R}} - \boldsymbol{U}_{\mathrm{S}}) = (\boldsymbol{Y}_{\mathrm{RS}} + \boldsymbol{Y}_{\mathrm{ST}} + \boldsymbol{Y}_{\mathrm{TR}})\boldsymbol{U}_{\mathrm{S}} - (\boldsymbol{Y}_{\mathrm{ST}} + \alpha\boldsymbol{Y}_{\mathrm{TR}} + \alpha^*\boldsymbol{Y}_{\mathrm{RS}})\boldsymbol{U}_{\mathrm{T}}$$

$$\boldsymbol{I}_{\mathrm{T}} = \boldsymbol{Y}_{\mathrm{TR}}(\boldsymbol{U}_{\mathrm{T}} - \boldsymbol{U}_{\mathrm{R}}) - \boldsymbol{Y}_{\mathrm{ST}}(\boldsymbol{U}_{\mathrm{S}} - \boldsymbol{U}_{\mathrm{T}}) = (\boldsymbol{Y}_{\mathrm{RS}} + \boldsymbol{Y}_{\mathrm{ST}} + \boldsymbol{Y}_{\mathrm{TR}})\boldsymbol{U}_{\mathrm{T}} - (\boldsymbol{Y}_{\mathrm{ST}} + \alpha\boldsymbol{Y}_{\mathrm{TR}} + \alpha^*\boldsymbol{Y}_{\mathrm{RS}})\boldsymbol{U}_{\mathrm{S}}.$$

If we denote

$$\boldsymbol{Y}_{\mathrm{RS}} + \boldsymbol{Y}_{\mathrm{ST}} + \boldsymbol{Y}_{\mathrm{TR}} \stackrel{\mathrm{df}}{=} \boldsymbol{Y}_{\mathrm{e}} \stackrel{\mathrm{df}}{=} G_{\mathrm{e}} + jB_{\mathrm{e}} \tag{7.4}$$

$$-(\boldsymbol{Y}_{\mathrm{ST}} + \alpha\, \boldsymbol{Y}_{\mathrm{TR}} + \alpha^*\boldsymbol{Y}_{\mathrm{RS}}) \stackrel{\mathrm{df}}{=} \boldsymbol{Y}_{\mathrm{u}} \stackrel{\mathrm{df}}{=} Y_{\mathrm{u}} e^{j\psi} \tag{7.5}$$

and assuming that $\boldsymbol{U}_{\mathrm{R}} = \boldsymbol{U}$, then the vector of the supply currents can be expressed as follows

$$\boldsymbol{i} = \sqrt{2}\mathrm{Re}\{\boldsymbol{Y}_{\mathrm{e}} \begin{bmatrix} \boldsymbol{U}_{\mathrm{R}} \\ \boldsymbol{U}_{\mathrm{S}} \\ \boldsymbol{U}_{\mathrm{T}} \end{bmatrix} + \boldsymbol{Y}_{\mathrm{u}} \begin{bmatrix} \boldsymbol{U}_{\mathrm{R}} \\ \boldsymbol{U}_{\mathrm{T}} \\ \boldsymbol{U}_{\mathrm{S}} \end{bmatrix} e^{j\omega t}\} = \sqrt{2}\mathrm{Re}\{(G_{\mathrm{e}}\, \mathbf{1}^{\mathrm{p}}\boldsymbol{U} + jB_{\mathrm{e}}\, \mathbf{1}^{\mathrm{p}}\boldsymbol{U} + \boldsymbol{Y}_{\mathrm{u}}\, \mathbf{1}^{\mathrm{n}}\boldsymbol{U})\, e^{j\omega t}\}.$$

This formula shows that the supply current of a three-phase LTI load supplied with symmetrical sinusoidal voltage has three components:

$$\boldsymbol{i}_{\mathrm{a}} \stackrel{\mathrm{df}}{=} \sqrt{2}\, \mathrm{Re}\{G_{\mathrm{e}}\mathbf{1}^{\mathrm{p}}\boldsymbol{U}\, e^{j\omega t}\} \tag{7.6}$$

$$\boldsymbol{i}_{\mathrm{r}} \stackrel{\mathrm{df}}{=} \sqrt{2}\, \mathrm{Re}\{jB_{\mathrm{e}}\mathbf{1}^{\mathrm{p}}\boldsymbol{U}\, e^{j\omega t}\} \tag{7.7}$$

$$\boldsymbol{i}_{\mathrm{u}} \stackrel{\mathrm{df}}{=} \sqrt{2}\, \mathrm{Re}\{\boldsymbol{Y}_{\mathrm{u}}\mathbf{1}^{\mathrm{n}}\boldsymbol{U}\, e^{j\omega t}\} \tag{7.8}$$

such that

$$\boldsymbol{i} = \boldsymbol{i}_{\mathrm{a}} + \boldsymbol{i}_{\mathrm{r}} + \boldsymbol{i}_{\mathrm{u}}. \tag{7.9}$$

Since the three-phase rms value of a three-phase vector

$$\boldsymbol{x} = \begin{bmatrix} x_{\mathrm{R}} \\ x_{\mathrm{S}} \\ x_{\mathrm{T}} \end{bmatrix} = \sqrt{2}\mathrm{Re}\begin{bmatrix} X_{\mathrm{R}} \\ X_{\mathrm{S}} \\ X_{\mathrm{T}} \end{bmatrix} e^{j\omega t} = \sqrt{2}\mathrm{Re}\{\boldsymbol{X}\, e^{j\omega t}\}$$

is equal to

$$||\boldsymbol{x}|| = \sqrt{\boldsymbol{X}^{\mathrm{T}}\boldsymbol{X}^{*}} = \sqrt{X_{\mathrm{R}}^2 + X_{\mathrm{S}}^2 + X_{\mathrm{T}}^2} \tag{7.10}$$

thus the supply current components have the three-phase rms values

$$||\boldsymbol{i}_a|| = G_e||\boldsymbol{u}|| \tag{7.11}$$

$$||\boldsymbol{i}_r|| = |B_e|\ ||\boldsymbol{u}|| \tag{7.12}$$

$$||\boldsymbol{i}_u|| = Y_u||\boldsymbol{u}||. \tag{7.13}$$

Let us clarify the physical meaning of the supply current components. The LTI load with the equivalent circuit shown in Fig. 7.5 has the active power

$$\begin{aligned} P &= \mathrm{Re}\,\{Y_{\mathrm{RS}}\}U_{\mathrm{RS}}^2 + \mathrm{Re}\,\{Y_{\mathrm{ST}}\}U_{\mathrm{ST}}^2 + \mathrm{Re}\,\{Y_{\mathrm{TR}}\}U_{\mathrm{TR}}^2 = \\ &= \mathrm{Re}\,\{Y_{\mathrm{RS}} + Y_{\mathrm{ST}} + Y_{\mathrm{TR}}\}3U_{\mathrm{R}}^2 = G_{\mathrm{e}}||\boldsymbol{u}||^2 = ||\boldsymbol{u}||\ ||\boldsymbol{i}_{\mathrm{a}}|| \end{aligned} \tag{7.14}$$

thus the current $\boldsymbol{i}_{\mathrm{a}}$ is an active component of the supply current. It will be referred to as the *active current.* The reactive power of the load is equal to

$$\begin{aligned} Q &= \mathrm{Im}\{Y_{\mathrm{RS}}^{*}\}U_{\mathrm{RS}}^2 + \mathrm{Im}\{Y_{\mathrm{ST}}^{*}\}U_{\mathrm{ST}}^2 + \mathrm{Im}\{Y_{\mathrm{TR}}^{*}\}U_{\mathrm{TR}}^2 = \\ &= \mathrm{Im}\{Y_{\mathrm{RS}}^{*} + Y_{\mathrm{ST}}^{*} + Y_{\mathrm{TR}}^{*}\}3U_{\mathrm{R}}^2 = -B_{\mathrm{e}}||\boldsymbol{u}||^2 = \pm\ ||\boldsymbol{u}||\ ||\boldsymbol{i}_{\mathrm{r}}|| \end{aligned} \tag{7.15}$$

thus current $\boldsymbol{i}_{\mathrm{r}}$ is a reactive component of the supply current. It will be referred to as the *reactive current.* Both the active and reactive currents have the same sequence as the supply voltage. These are currents of balanced purely resistive and purely reactive loads shown in Fig. 7.6b and c, equivalent to the original load, shown in Fig. 7.6a, concerning the active and reactive power P and Q.

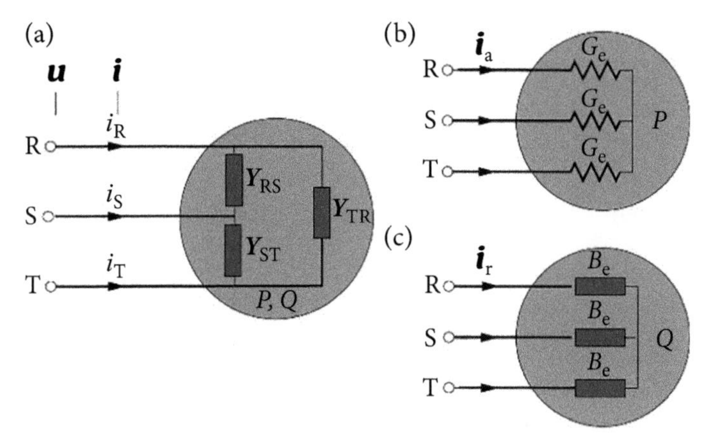

Figure 7.6 (a) An unbalanced load and balanced loads (b) equivalent to the active power, and (c) equivalent to the reactive power.

The sum of the active and reactive currents, $\boldsymbol{i}_a + \boldsymbol{i}_r = \boldsymbol{i}_b$, is the supply current of a balanced load with admittance per phase

$$Y_e = G_e + jB_e$$

equivalent to the active and reactive powers of the original unbalanced load. Therefore,

$$\boldsymbol{Y}_{RS} + \boldsymbol{Y}_{ST} + \boldsymbol{Y}_{TR} \stackrel{df}{=} \boldsymbol{Y}_e \stackrel{df}{=} G_e + jB_e$$

was called in Ref. [93] as an *equivalent admittance* of three-phase loads. The real part of this admittance, G_e, is referred to as *equivalent conductance*, while its imaginary part, B_e, is referred to as an *equivalent susceptance* of three-phase loads. When the load is balanced, meaning

$$\boldsymbol{Y}_{ST} = \boldsymbol{Y}_{TR} = \boldsymbol{Y}_{RS} \stackrel{df}{=} \boldsymbol{Y}$$

then

$$\boldsymbol{Y}_u \stackrel{df}{=} -(\boldsymbol{Y}_{ST} + \alpha \boldsymbol{Y}_{TR} + \alpha^* \boldsymbol{Y}_{RS}) = -(1 + \alpha + \alpha^*)\, \boldsymbol{Y} = 0,$$

thus the component $\boldsymbol{i}_u$ does not occur in the supply current. It occurs only when, due to the load imbalance, the term $\boldsymbol{A}$ is not equal to zero. Therefore, the term

$$-(\boldsymbol{Y}_{ST} + \alpha \boldsymbol{Y}_{TR} + \alpha^* \boldsymbol{Y}_{RS}) \stackrel{df}{=} \boldsymbol{Y}_u \stackrel{df}{=} Y_u e^{j\psi}$$

was called an *unbalanced admittance* of three-phase loads in Ref. [93].

Elements $\boldsymbol{U}_{\mathrm{S}}$ and $\boldsymbol{U}_{\mathrm{T}}$ of the vector a $\mathbf{1}^{\mathrm{n}}\boldsymbol{U}$ in formula (7.8) are switched as compared to the vector $\mathbf{1}^{\mathrm{p}}\boldsymbol{U}$, thus the current $\boldsymbol{i}_{\mathrm{u}}$ has the opposite sequence to the sequence of the supply voltage. It occurs due to the load imbalance and causes the supply current asymmetry. Therefore, it was called an *unbalanced current.* Observe, that the crms value of the unbalanced current in line S is proportional to the crms value of the line voltage u_{T}, namely

$$\boldsymbol{I}_{\mathrm{uS}} = \boldsymbol{Y}_{\mathrm{u}}\boldsymbol{Y}_{\mathrm{T}} + \alpha^{*}\boldsymbol{Y}_{\mathrm{u}}\boldsymbol{U}_{\mathrm{S}} \quad \text{and} \quad \boldsymbol{I}_{\mathrm{uT}} = \boldsymbol{Y}_{\mathrm{u}}\boldsymbol{U}_{\mathrm{S}} = \alpha\boldsymbol{Y}_{\mathrm{u}}\boldsymbol{U}_{\mathrm{T}},$$

thus the unbalanced current $\boldsymbol{i}_{\mathrm{u}}$ can be regarded as the current of a load, shown in Fig. 7.7, with line admittances $\boldsymbol{Y}_{\mathrm{u}}$, $\alpha^{*}\boldsymbol{Y}_{\mathrm{u}}$, and $\alpha\boldsymbol{Y}_{\mathrm{u}}$, respectively.

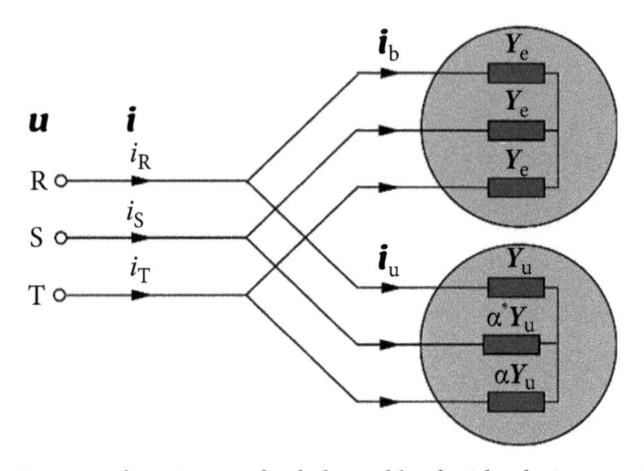

Figure 7.7 An equivalent circuit of unbalanced load with admittances $\boldsymbol{Y}_{\mathrm{e}}$ and $\boldsymbol{Y}_{\mathrm{u}}$.

This decomposition demonstrates that the supply current of LTI unbalanced load supplied with symmetrical sinusoidal voltage is composed of three components associated with three distinctively different power-related physical phenomena.

- active current, $\boldsymbol{i}_{\mathrm{a}}$, associated with permanent energy flow from the supply to the load;
- reactive current, $\boldsymbol{i}_{\mathrm{r}}$, associated with phase shift between the supply voltage and current; and
- unbalanced current, $\boldsymbol{i}_{\mathrm{u}}$, associated with the supply current asymmetry, caused by the load imbalance.

These are Current's Physical Components of LTI three-phase loads supplied with symmetrical sinusoidal voltage. Let us check their orthogonality, meaning let us check whether their scalar product is equal to zero or not.

7.3 Orthogonality of CPC in Circuits with sv&c

The scalar product of the active and reactive currents is equal to

$$(\boldsymbol{i}_{\rm a}, \boldsymbol{i}_{\rm r}) = {\rm Re}\,\{\boldsymbol{I}_{\rm a}^{\rm T}\boldsymbol{I}_{\rm r}{}^{*}\} = {\rm Re}\,\{G_{\rm e}\,\mathbf{1}^{\rm pT}\boldsymbol{U}\,(jB_{\rm e}\mathbf{1}^{\rm p}U)^{*}\} = {\rm Re}\,\{-jB_{\rm e}G_{\rm e}||\boldsymbol{u}||^{2}\} = 0.$$

The scalar product of the active and unbalanced currents is equal to

$$(\boldsymbol{i}_{\rm a},\boldsymbol{i}_{\rm u}) = {\rm Re}\{\boldsymbol{I}_{\rm a}^{\rm T}\boldsymbol{I}_{\rm u}^{*}\} = {\rm Re}\{G_{\rm e}\,\mathbf{1}^{\rm pT}\boldsymbol{U}(Y_{\rm u}\mathbf{1}^{\rm n}U)^{*}\} =$$

$$= {\rm Re}\{G_{\rm e}\boldsymbol{Y}_{\rm u}^{*}\}[1, \alpha^{*}, \alpha]\begin{bmatrix} 1 \\ \alpha^{*} \\ \alpha \end{bmatrix} U_{\rm R}^{2}\} = {\rm Re}\{G_{\rm e}\boldsymbol{Y}_{\rm u}^{*}(1+\alpha+\alpha^{*})U_{\rm R}^{2}\} = 0.$$

The scalar product of the reactive and unbalanced currents is equal to

$$(\boldsymbol{i}_{\rm r}, \boldsymbol{i}_{\rm u}) = {\rm Re}\,\{\boldsymbol{I}_{\rm r}^{\rm T}\boldsymbol{I}_{\rm u}^{*}\} = {\rm Re}\,\{jB_{\rm e}\mathbf{1}^{\rm pT}\boldsymbol{U}\,(Y_{\rm u}\mathbf{1}^{\rm n}U)^{*}\} = {\rm Re}\,\{jB_{\rm e}\,\boldsymbol{Y}_{\rm u}^{*}(1+\alpha^{*}+\alpha)U_{\rm R}^{2}\} = 0.$$

Thus physical components of the supply current of three-phase LTI loads supplied with sinusoidal symmetrical voltage are mutually orthogonal. Hence, their rms values satisfy the relationship

$$||\boldsymbol{i}||^{2} = ||\,\boldsymbol{i}_{\rm a}||^{2} + ||\,\boldsymbol{i}_{\rm r}||^{2} + ||\,\boldsymbol{i}_{\rm u}||^{2}. \tag{7.16}$$

Relation (7.16) reveals that three but not two phenomena in the load affect the supply current rms value.

It is assumed traditionally that only the active and reactive currents contribute to the supply current three-phase rms $||\boldsymbol{i}||$ value. The presented decomposition reveals the load imbalance as the additional cause of this rms value increase. The unbalanced current, $\boldsymbol{i}_{\rm u}$, as the current component of the negative sequence, causes the supply current asymmetry, which in turn, due to the voltage drop on the circuit impedance, causes the distribution voltage asymmetry. The unbalanced current, as the current of the negative sequence, can be found by the supply current decomposition into symmetrical components. This method was not used in the presented current decomposition into CPC. It was revealed by analyzing the power properties of unbalanced loads.

Relation (7.16) means that each phenomenon that affects the supply current rms value (i.e., permanent flow of energy from the supply source to loads, phase shift between voltage and current, and the supply current asymmetry) affects this value independently of each other. This relationship can be illustrated geometrically with a rectangular box, shown in Fig. 7.8.

The length of the edges of this box is proportional to the rms values of the active, reactive, and unbalanced currents, while its diagonal is proportional to the supply current rms value.

An important property of the presented decomposition of the supply current $\boldsymbol{i}$ into physical components, $\boldsymbol{i}_{\rm a}$, $\boldsymbol{i}_{\rm r}$, and $\boldsymbol{i}_{\rm u}$, is the possibility of relating them to the load

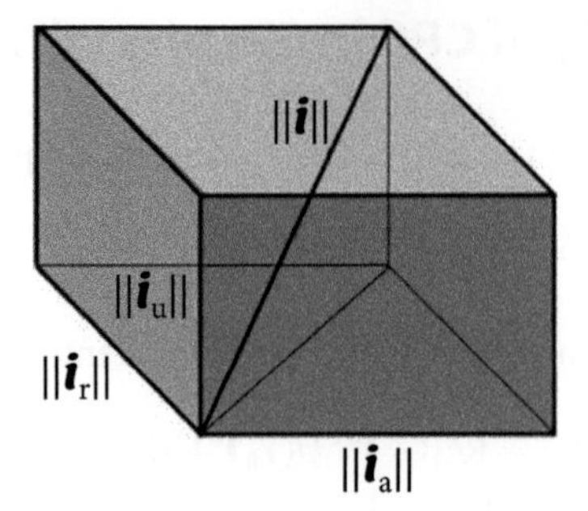

Figure 7.8 A rectangular box of active, reactive, and unbalanced currents rms value.

parameters, namely to the load equivalent conductance and susceptance, G_e and B_e, respectively, and to the load unbalanced admittance $\boldsymbol{Y}_u$ as well as, to admittances $\boldsymbol{Y}_{RS}$, $\boldsymbol{Y}_{ST}$ and $\boldsymbol{Y}_{TR}$.

The reader should be aware, however, that despite the name *Physical Components*, currents $\boldsymbol{i}_a$, $\boldsymbol{i}_r$, and $\boldsymbol{i}_u$ do not exist physically. They are only a product of mathematical decomposition and consequently, they exist as mathematical but not physical entities. The term *Physical Components* emphasizes only their association with distinctive physical phenomena but not their physical existence.

Illustration 7.1 *Let us calculate the Physical Components of the supply current in the circuit shown in Fig. 7.9, assuming that the supply voltage is symmetrical, with the line R-to-ground voltage*

$$u_R(t) = 220\sqrt{2}\cos\omega t \text{ V}$$

and the transformer is ideal with a turn ratio 1:1.

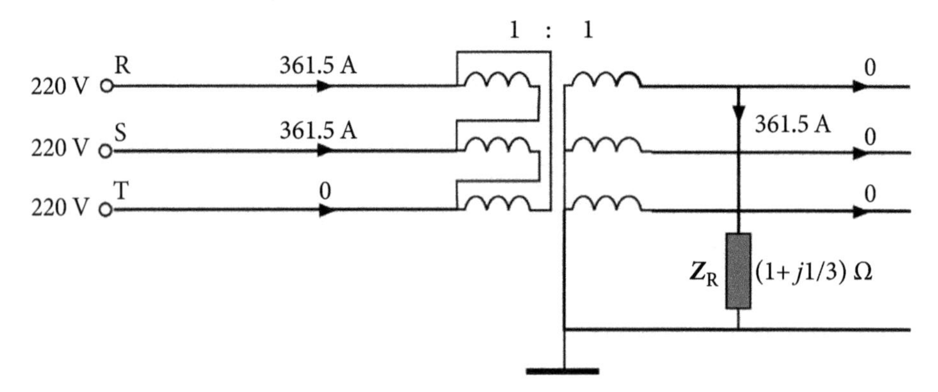

Figure 7.9 An example of a circuit.

The three-phase rms value of the supply voltage is

$$||\boldsymbol{u}|| = \sqrt{U_R^2 + U_S^2 + U_T^2} = \sqrt{3 \times 220^2} = 220\sqrt{3} = 381 \text{ V}.$$

The supply current rms value in lines R and S are equal to $I_{\mathrm{R}} = I_{\mathrm{S}} = 361.5$ A *and* $I_{\mathrm{T}} = 0$, *hence*

$$||\boldsymbol{i}|| = \sqrt{I_{\mathrm{R}}^2 + I_{\mathrm{S}}^2 + I_{\mathrm{T}}^2} = \sqrt{2 \times 361,5^2} = 361,5\sqrt{2} = 511\,\mathrm{A}.$$

The load as seen from the supply source has only one non-zero admittance, $\boldsymbol{Y}_{\mathrm{RS}}$, *equal to*

$$\boldsymbol{Y}_{\mathrm{RS}} = \frac{1}{\boldsymbol{Z}_{\mathrm{R}}} = \frac{1}{1 + j0.333} = 0.90 - j0.30\mathrm{S}, \quad \boldsymbol{Y}_{\mathrm{ST}} = 0, \qquad \boldsymbol{Y}_{\mathrm{TR}} = 0,$$

thus the equivalent admittance of the load is equal to

$$Y_{\mathrm{e}} = G_{\mathrm{e}} + jB_{\mathrm{e}} = \boldsymbol{Y}_{\mathrm{RS}} + \boldsymbol{Y}_{\mathrm{ST}} + \boldsymbol{Y}_{\mathrm{TR}} = \boldsymbol{Y}_{\mathrm{RS}} = 0.90 - j0.30 = 0.95\, e^{-j18^\circ}\ \mathrm{S}$$

while the unbalanced admittance

$$\boldsymbol{Y}_{\mathrm{u}} = -(\boldsymbol{Y}_{\mathrm{ST}} + \alpha\, \boldsymbol{Y}_{\mathrm{TR}} + \alpha^* \boldsymbol{Y}_{\mathrm{RS}}) = -\alpha^* \boldsymbol{Y}_{\mathrm{RS}} = -1\, e^{-j120^\circ} \times 0.95\, e^{-j18^\circ} = 0.95\, e^{j42^\circ}\ \mathrm{S}.$$

The active current, according to definition (11) is equal to

$$\boldsymbol{i}_{\mathrm{a}} = \sqrt{2}\mathrm{Re}\left\{0.90\begin{bmatrix} 220 \\ 220\, e^{-j120^\circ} \\ 220\, e^{j120^\circ} \end{bmatrix} e^{j\omega t}\right\} = \sqrt{2}\mathrm{Re}\left\{\begin{bmatrix} 198 \\ 198\, e^{-j120^\circ} \\ 198\, e^{j120^\circ} \end{bmatrix} e^{j\omega t}\right\}\ \mathrm{A}$$

the reactive current

$$\boldsymbol{i}_{\mathrm{r}} = \sqrt{2}\mathrm{Re}\left\{j\,0.30\begin{bmatrix} 220 \\ 220\, e^{-j120^\circ} \\ 220\, e^{j120^\circ} \end{bmatrix} e^{j\omega t}\right\} = \sqrt{2}\mathrm{Re}\left\{\begin{bmatrix} 66\, e^{j90^\circ} \\ 66\, e^{-j30^\circ} \\ 66\, e^{j210^\circ} \end{bmatrix} e^{j\omega t}\right\}\ \mathrm{A}$$

while the unbalanced current is equal to

$$\boldsymbol{i}_{\mathrm{u}} = \sqrt{2}\mathrm{Re}\left\{0.95\, e^{j42^\circ}\begin{bmatrix} 220 \\ 220\, e^{+j120^\circ} \\ 220\, e^{-j120^\circ} \end{bmatrix} e^{j\omega t}\right\} = \sqrt{2}\mathrm{Re}\left\{\begin{bmatrix} 209\, e^{j42^\circ} \\ 209\, e^{j162^\circ} \\ 209\, e^{-j78^\circ} \end{bmatrix} e^{j\omega t}\right\}\ \mathrm{A}.$$

Three-phase rms values of these currents are

$$||\boldsymbol{i}_{\mathrm{a}}|| = G_{\mathrm{e}}||\boldsymbol{u}|| = 0.90 \times 381 = 343\ A$$

$$||\boldsymbol{i}_{\mathrm{r}}|| = |B_{\mathrm{e}}|\ ||\boldsymbol{u}|| = 0.3 \times 381 = 114\,\mathrm{A}$$

$$||\boldsymbol{i}_{u}|| = Y_{u}||\boldsymbol{u}|| = 0.95 \times 381 = 361\,\mathrm{A},$$

thus the unbalanced current is the most dominating component of the supply current. The sum of these currents square rms values results in

$$||\boldsymbol{i}|| = \sqrt{||\boldsymbol{i}_{\mathrm{a}}||^2 + ||\boldsymbol{i}_{\mathrm{r}}||^2 + ||\boldsymbol{i}_{\mathrm{u}}||^2} = \sqrt{343^2 + 114^2 + 361^2} = 511\,\mathrm{A}$$

which verifies the numerical correctness of this decomposition.

*****

7.4 Power Equation in Circuits with sv&c

Multiplying relationship (7.16) by the square of the supply voltage rms value$||\boldsymbol{u}||$, the power equation of three-phase three-wire circuits with LTI loads supplied with symmetrical sinusoidal voltage

$$S^2 = P^2 + Q^2 + D_{\mathrm{u}}^2 \tag{7.17}$$

is obtained, where

$$S \overset{\mathrm{df}}{=} ||\boldsymbol{u}||\ ||\boldsymbol{i}||$$

$$P \overset{\mathrm{df}}{=} ||\boldsymbol{u}|| \cdot ||\boldsymbol{i}_{\mathrm{a}}|| = G_{\mathrm{e}}||\boldsymbol{u}||^2 \tag{7.18}$$

$$Q \overset{\mathrm{df}}{=} \pm||\boldsymbol{u}|| \cdot ||\boldsymbol{i}_{\mathrm{r}}|| = -B_{\mathrm{e}}||\boldsymbol{u}||^2 \tag{7.19}$$

$$D_{\mathrm{u}} \overset{\mathrm{df}}{=} ||\boldsymbol{u}|| \cdot ||\boldsymbol{i}_{\mathrm{u}}|| = Y_{\mathrm{u}}||\boldsymbol{u}||^2. \tag{7.20}$$

The sign of reactive power is based on the convention that resistive-inductive (RL) loads have positive reactive power, while the reactive power of resistive-capacitive (RC) loads is negative.

This equation provides the right relationship between powers in 3p3w circuits and should supersede the common power Eqn. (7.1). It reveals a new power quantity, referred to as an *unbalanced power*, D_{u}. This power enables us to express quantitatively the effect of the load imbalance on the apparent power S increase. The power Eqn. (7.17) emphasizes the fact that the apparent power value S is determined not by two but by three distinctively different phenomena. It is important to observe that the active, reactive, and unbalanced powers can be expressed in terms of the load parameters, G_{e}, B_{e}, and Y_{u}. The relationship between these four powers can be illustrated geometrically with the rectangular box, shown in Fig. 7.10.

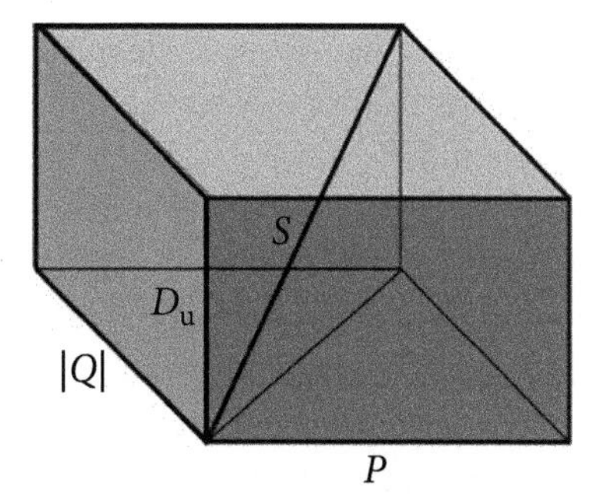

Figure 7.10 A rectangular box of active, reactive, and unbalanced powers.

The power factor of LTI loads supplied with three-phase symmetrical and sinusoidal voltage is equal to

$$\lambda \overset{\text{df}}{=} \frac{P}{S} = \frac{P}{\sqrt{P^2 + Q^2 + D_u^2}}$$

thus, the unbalanced power D_u contributes to the power factor reduction to the same degree as the reactive power Q. Observe, that the traditional power equation does not provide the power factor value for unbalanced loads.

The power factor declines due to a useless increase in the supply current rms value, which is more visible when it is expressed, due to relations (7.18–7.20) between powers and the CPC rms value, in the form

$$\lambda \overset{\text{df}}{=} \frac{P}{S} = \frac{P}{\sqrt{P^2 + Q^2 + D_u^2}} = \frac{||\boldsymbol{i}_a||}{\sqrt{||\boldsymbol{i}_a||^2 + ||\boldsymbol{i}_r||^2 + ||\boldsymbol{i}_u||^2}},$$

thus the reactive and unbalanced currents are responsible for the power factor degradation. However, even this last formula hides the very true mechanism of its degradation, meaning the phase shift between the supply voltage and currents and the line currents asymmetry, specified entirely by the load parameters G_e, B_e, and Y_u, which is visible when the power factor is expressed in the form

$$\lambda = \frac{||\boldsymbol{i}_a||}{\sqrt{||\boldsymbol{i}_a||^2 + ||\boldsymbol{i}_r||^2 + ||\boldsymbol{i}_u||^2}} = \frac{G_e}{\sqrt{G_e^2 + B_e^2 + Y_u^2}}. \tag{7.21}$$

This formula emphasizes the fact that the power factor is not dependent on the voltages and currents but only on the load properties, and these are entirely specified by equivalent conductance, susceptance, and unbalanced admittance. The traditional power Eqn. (7.1) does not relate the power factor value to the load parameters in a way that might have an analogy to the formula (7.21).

One should observe, however, that even if definitions of the reactive and unbalanced powers, Q and D_{u}, are distinctively related to Current's Physical Components, unlike the instantaneous and the active powers, they do not have clear physical interpretations. Like the apparent power S, they are only formal products of the supply voltage and reactive, or unbalanced currents, $\boldsymbol{i}_{\mathrm{r}}$ and $\boldsymbol{i}_{\mathrm{u}}$, three-phase rms values. These products are some measures of the effect of the phase shift between the supply voltage and current and the line currents asymmetry on the apparent power of the load. These powers are certainly associated with distinctive power-related phenomena, but this does not mean that they can be interpreted physically.

7.5 CPC and the Instantaneous Power

The instantaneous power $p(t)$ is a power quantity with the most distinctive physical interpretation, meaning the rate of energy flow between the supply source and the load. The relationship between this power and the Current's Physical Components can get deeper, therefore, our understanding of the power-related phenomena in 3p3w circuits.

The instantaneous power $p(t)$ can be expressed as follows:

$$p(t) \stackrel{\mathrm{df}}{=} \frac{dW(t)}{dt} = \boldsymbol{u}^{\mathrm{T}}\boldsymbol{i} = \boldsymbol{u}^{\mathrm{T}}(\boldsymbol{i}_{\mathrm{a}} + \boldsymbol{i}_{\mathrm{r}} + \boldsymbol{i}_{\mathrm{u}}) \stackrel{\mathrm{df}}{=} p_{\mathrm{a}}(t) + p_{\mathrm{r}}(t) + p_{\mathrm{u}}(t), \tag{7.22}$$

thus it has three additive components associated separately with the CPC, that is with the active, reactive, and unbalanced currents.

The component of the instantaneous power associated with the active current is equal to

$$p_{\mathrm{a}}(t) \stackrel{\mathrm{df}}{=} \boldsymbol{u}^{\mathrm{T}}\boldsymbol{i}_{\mathrm{a}} = \boldsymbol{u}^{\mathrm{T}}G_{\mathrm{e}}\boldsymbol{u} = G_{\mathrm{e}}||\boldsymbol{u}||^2 = P \tag{7.23}$$

meaning it does not change in time. Assuming that

$$u_{\mathrm{R}}(t) = \sqrt{2}\,U\cos\omega t$$

then

$$\boldsymbol{u} \stackrel{\mathrm{df}}{=} \begin{bmatrix} u_{\mathrm{R}} \\ u_{\mathrm{S}} \\ u_{\mathrm{T}} \end{bmatrix} = \sqrt{2}\,U_{\mathrm{R}} \begin{bmatrix} \cos\omega t \\ \cos(\omega t - 120^{\circ}) \\ \cos(\omega t + 120^{\circ}) \end{bmatrix}$$

hence the reactive current changes as

$$\boldsymbol{i}_{\mathrm{r}} = \sqrt{2}\mathrm{Re}\,\{jB_{\mathrm{e}}\,U_{\mathrm{R}} \begin{bmatrix} 1 \\ e^{-j120^{\circ}} \\ e^{j120^{\circ}} \end{bmatrix} e^{j\omega t}\} = -\sqrt{2}\,B_{\mathrm{e}}U_{\mathrm{R}} \begin{bmatrix} \sin\omega t \\ \sin(\omega t - 120^{\circ}) \\ \sin(\omega t + 120^{\circ}) \end{bmatrix}.$$

The component of the instantaneous power associated with the reactive current is equal to

$$p_{\mathrm{r}}(t) \stackrel{\mathrm{df}}{=} \boldsymbol{u}^{\mathrm{T}}\boldsymbol{i}_{\mathrm{r}} = -2B_{\mathrm{e}}U_{\mathrm{R}}^{2}[\cos\omega t \sin\omega t +$$

$$+ \cos(\omega t - 120°)\sin(\omega t - 120°) + \cos(\omega t + 120°)\sin(\omega t + 120°)] =$$

$$= - B_{\mathrm{e}}U_{\mathrm{R}}^{2}[\sin 2\omega t + \sin(2\omega t + 120°) + \sin(2\omega t - 120°) \equiv 0. \tag{7.24}$$

Thus the reactive current does not contribute to the energy flow between the supply source and the load.

The unbalanced current changes as

$$\boldsymbol{i}_{\mathrm{u}} = \sqrt{2}\mathrm{Re}\,\{Y_{\mathrm{u}}e^{j\Psi}\,U_{\mathrm{R}}\begin{bmatrix}1\\ e^{+j120°}\\ e^{-j120°}\end{bmatrix}e^{j\omega t}\} = \sqrt{2}Y_{\mathrm{u}}U_{\mathrm{R}}\begin{bmatrix}\cos(\omega t + \Psi)\\ \cos(\omega t + \Psi + 120°)\\ \cos(\omega t + \Psi - 120°)\end{bmatrix},$$

thus the component of the instantaneous power associated with the unbalanced current is equal to

$$p_{\mathrm{u}}(t) \stackrel{\mathrm{df}}{=} \boldsymbol{u}^{\mathrm{T}}\boldsymbol{i}_{\mathrm{u}} = 2Y_{\mathrm{u}}U_{\mathrm{R}}^{2}[\cos\omega t\cos(\omega t + \Psi) +$$

$$+ \cos(\omega t - 120°)\cos(\omega t + \Psi + 120°)+$$

$$+ \cos(\omega t + 120°)\cos(\omega t + \Psi - 120°)] =$$

$$= 3Y_{\mathrm{u}}\,U_{\mathrm{R}}^{2}\cos(2\omega t + \Psi) = Y_{\mathrm{u}}||\boldsymbol{u}||^{2}\cos(2\omega t + \Psi). \tag{7.25}$$

Consequently, the instantaneous power of LTI three-phase loads supplied with symmetrical sinusoidal voltage can be expressed as

$$p(t) = p_{\mathrm{a}}(t) + p_{\mathrm{r}}(t) + p_{\mathrm{u}}(t) = P + D_{\mathrm{u}}\cos(2\omega t + \Psi). \tag{7.26}$$

This result can be surprising for those who believe that reactive power is caused by energy oscillation between the supply source and the load. Such oscillations are associated only with the presence of the unbalanced power D_{u}, which occurs in the effect of the supply current asymmetry.

This illustration shows that the CPC-based approach to power theory provides a very simple and transparent method of calculation of the balancing compensator parameters. It confirms the applicability of this approach to reactance balancing compensator design.

7.6 Three-Phase Load Equivalent Δ Circuits

Equivalent and unbalanced admittances, $\boldsymbol{Y}_{\mathrm{e}}$ and $\boldsymbol{Y}_{\mathbf{u}}$, and consequently parameters of a balancing compensator, are specified in terms of admittances $\boldsymbol{Y}_{\mathrm{RS}}$, $\boldsymbol{Y}_{\mathrm{ST}}$, and $\boldsymbol{Y}_{\mathrm{TR}}$ of the load-equivalent circuit of a Δ structure. The crms values of the supply currents for any LTI load have to satisfy the following equations:

$$\boldsymbol{I}_{\mathrm{R}} = \boldsymbol{Y}_{\mathrm{RS}}\boldsymbol{U}_{\mathrm{RS}} - \boldsymbol{Y}_{\mathrm{TR}}\boldsymbol{U}_{\mathrm{TR}}$$

$$\boldsymbol{I}_{\mathrm{S}} = \boldsymbol{Y}_{\mathrm{ST}}\boldsymbol{U}_{\mathrm{ST}} - \boldsymbol{Y}_{\mathrm{RS}}\boldsymbol{U}_{\mathrm{RS}}$$

$$\boldsymbol{I}_{\mathrm{T}} = \boldsymbol{Y}_{\mathrm{TR}}\boldsymbol{U}_{\mathrm{TR}} - \boldsymbol{Y}_{\mathrm{ST}}\boldsymbol{U}_{\mathrm{ST}}$$

with unknown admittances $\boldsymbol{Y}_{\mathrm{RS}}$, $\boldsymbol{Y}_{\mathrm{ST}}$, and $\boldsymbol{Y}_{\mathrm{TR}}$. In three-wire circuits

$$\boldsymbol{I}_{\mathrm{R}} + \boldsymbol{I}_{\mathrm{S}} + \boldsymbol{I}_{\mathrm{T}} = 0,$$

thus these equations are not linearly independent. As proven in Ref. [140], at symmetrical supply voltage, these equations have an infinite number of solutions, meaning any complex number can be selected for one of admittances $\boldsymbol{Y}_{\mathrm{RS}}$, $\boldsymbol{Y}_{\mathrm{ST}}$, and $\boldsymbol{Y}_{\mathrm{TR}}$. In particular, it can be zero. Assuming that $\boldsymbol{Y}_{\mathrm{RS}} = 0$, any LTI load supplied with a three-wire line can have an equivalent Δ circuit in the form shown in Fig. 7.11. It enables very simple calculation of admittances of the equivalent circuit from

$$\boldsymbol{Y}_{\mathrm{ST}} = \frac{\boldsymbol{I}_{\mathrm{S}}}{\boldsymbol{U}_{\mathrm{ST}}}, \qquad \boldsymbol{Y}_{\mathrm{TR}} = \frac{\boldsymbol{I}_{\mathrm{R}}}{\boldsymbol{U}_{\mathrm{RT}}}.$$

Thus the crms values of two line-to-line voltages and two currents of the supply lines enable the calculation of admittances of the Δ equivalent circuit. The reader should be aware, however, that these admittances may not represent the admittance of any real one-port; for example: a real part of these admittances can be negative. This is because one of them, in this

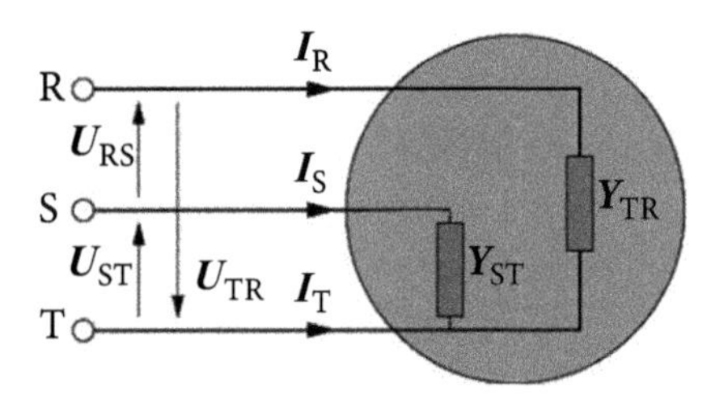

Figure 7.11 An equivalent Δ circuit for any LTI three-phase load.

case $\boldsymbol{Y}_{\mathrm{RS}}$, was assumed to be zero. A Discrete Fourier Transform (DFT) performed on a set of voltage and current samples, taken in one period T, is a common method of calculating the crms values $\boldsymbol{U}_{\mathrm{ST}}$, $\boldsymbol{U}_{\mathrm{RT}}$, $\boldsymbol{I}_{\mathrm{R}}$, and $\boldsymbol{I}_{\mathrm{S}}$ of these voltages and currents. There are also computationally more effective methods, for example those that were presented in Ref. [108]. Only two samples taken at the interval of $T/4$ are needed for calculating the crms value of a sinusoidal quantity.

The equivalent admittance of the load can be expressed as

$$Y_e \overset{\mathrm{df}}{=} G_e + jB_e = Y_{RS} + Y_{ST} + Y_{TR} = \frac{I_S}{U_{ST}} - \frac{I_R}{U_{RT}}$$

and its unbalanced admittance

$$Y_u \overset{\mathrm{df}}{=} -(Y_{ST} + \alpha Y_{TR} + \alpha^* Y_{RS}) = -\frac{I_S}{U_{ST}} + \alpha\frac{I_R}{U_{RT}}.$$

7.7 CPC in Circuits with nv&c and LTI Loads

Voltage and current harmonics occur in three-phase circuits due to high-power three-phase nonlinear or time-variant loads harmonics generating loads (HGLs) or due to such single-phase loads supplied from three-phase circuits. In the last case, harmonics can be accompanied by voltage and current asymmetry, due to load imbalance. This asymmetry occurs mainly at the junctions of single-phase and three-phase circuits. Since energy is delivered in large amounts by three-phase three-wire circuits, just such circuits under non-sinusoidal conditions are the major object of the power theory. Apart from the Instantaneous Reactive Power p-q Theory [72] and the FBD Method [90], power theories of single-phase circuits were starting points for the investigation of power properties of three-phase circuits at nv&c as carried out by Quade [22], Nedelcu [28], and Fryze [81].

Unfortunately, as long as the power properties of single-phase circuits with nv&c were not properly identified, all misinterpretations of these properties in single-phase circuits were naturally conveyed to interpretations of the power properties of three-phase circuits. Moreover, the generalizations of the power theory from single-phase to three-phase circuits were done in a situation where the power properties of unbalanced three-phase circuits with LTI loads were incorrectly described even at sinusoidal voltages and currents, meaning with the power equation $S^2 = P^2 + Q^2$.

Power properties of 3p3w circuits with LTI loads using the CPC concept were eventually explained in Ref. [93] in 1988. The right power equation of unbalanced loads with sv&c was developed in that paper as well, while the right power equation of single-phase LTI loads with nv&c was developed in Refs [71] and [75]. Fundamentals of designing reactance compensators for 3p3w circuits were developed one year later and published in Ref. [96].

Power properties of three-phase unbalanced loads supplied with a symmetrical nonsinusoidal voltage, as shown in Fig. 7.12, will be explained in this chapter using power theory founded on CPC-based power theory of 3p3w circuits with sinusoidal voltages and currents. We will assume, however, that the three-phase load is LTI, meaning it cannot generate current harmonics.

Harmonic sequence. Harmonics of symmetrical three-phase quantities, depending on their order n, form three-phase sets of symmetrical quantities of a different

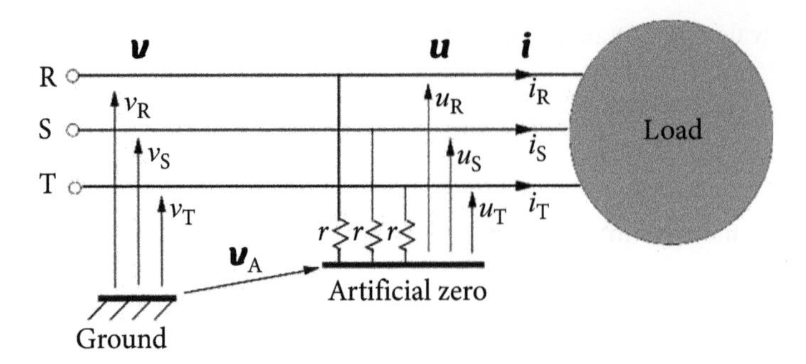

Figure 7.12 A three-phase load supplied from a three-wire line.

order in the sense of the sequence we observe the same phase, for example maximum, at R, S, and T terminals.

Power systems operate in such a way, that the same phase is observed sequentially on terminal R, next on terminal S, and finally on terminal T. This sequence is referred to as the positive sequence. Induction motors supplied with a voltage of such a sequence rotate clockwise, while motors supplied with a voltage of the negative sequence rotate counterclockwise. At a voltage of positive sequence, the fundamental harmonic, and all other harmonics of the order $n = 3k+1$, $k = 0, 1, 2...$, have the same, positive sequence, harmonics of the order $n = 3k-1$, $k = 1, 2, 3...$, have the negative sequence, while harmonics of order n which is a multiplicity three, $n = 3k$, $k = 0, 1, 2...$, are zero-sequence harmonics.

Thus the vector of symmetrical nonsinusoidal voltages, measured versus the ground, as shown in Fig. 7.12, can be decomposed into a sum of three different, considering their sequence, groups of harmonics,

$$\boldsymbol{v}(t) \stackrel{\text{df}}{=} \begin{bmatrix} v_{\mathrm{R}}(t) \\ v_{\mathrm{S}}(t) \\ v_{\mathrm{T}}(t) \end{bmatrix} = \sum_{n \in N} \boldsymbol{v}_n(t) = \boldsymbol{v}^{\mathrm{p}} + \boldsymbol{v}^{\mathrm{n}} + \boldsymbol{v}^{\mathrm{z}}$$

with

$$\boldsymbol{v}^{\mathrm{p}} = \sum_{n=3k+1} \boldsymbol{v}_n = \sqrt{2}\mathrm{Re} \sum_{n=3k+1} \boldsymbol{V}_n\, e^{jn\omega_1 t} = \sqrt{2}\mathrm{Re} \sum_{n=3k+1} \mathbf{1}^{\mathrm{p}} V_n\, e^{jn\omega_1 t}$$

$$\boldsymbol{v}^{\mathrm{n}} = \sum_{n=3k-1} \boldsymbol{v}_n = \sqrt{2}\mathrm{Re} \sum_{n=3k-1} \boldsymbol{V}_n\, e^{jn\omega_1 t} = \sqrt{2}\mathrm{Re} \sum_{n=3k-1} \mathbf{1}^{\mathrm{n}} V_n\, e^{jn\omega_1 t}$$

$$\boldsymbol{v}^{\mathrm{z}} = \sum_{n=3k} \boldsymbol{v}_n = \sqrt{2}\mathrm{Re} \sum_{n=3k} \boldsymbol{V}_n\, e^{jn\omega_1 t} = \sqrt{2}\mathrm{Re} \sum_{n=3k} \mathbf{1}^{\mathrm{z}} V_n\, e^{jn\omega_1 t}.$$

These three different sequence groups of harmonics are specified with three different formulae, which is a disadvantage for the compactness of the mathematical description of such quantities. To remove this disadvantage, let us define a *generalized unit three-phase vector*

$$\mathbf{1}_n \stackrel{\text{df}}{=} \begin{bmatrix} 1 \\ 1\,e^{-jn\frac{2\pi}{3}} \\ 1\,e^{jn\frac{2\pi}{3}} \end{bmatrix} \stackrel{\text{df}}{=} \begin{bmatrix} 1 \\ \beta_n^* \\ \beta_n \end{bmatrix} = \begin{cases} \mathbf{1}^{\text{p}}, & \text{for } n = 3k+1 \\ \mathbf{1}^{\text{n}}, & \text{for } n = 3k-1 \\ \mathbf{1}^{\text{z}}, & \text{for } n = 3k \end{cases}$$

where

$$\beta_n \stackrel{\text{df}}{=} 1e^{jn\frac{2\pi}{3}} = \alpha^n.$$

Thus, with such a unit three-phase vector

$$\boldsymbol{v}(t) = \sum_{n\in N} \boldsymbol{v}_n(t) = \boldsymbol{v}^{\text{p}} + \boldsymbol{v}^{\text{n}} + \boldsymbol{v}^{\text{z}} = \sqrt{2}\text{Re}\sum_{n\in N} \mathbf{1}_n V_n\, e^{jn\omega_1 t}.$$

Ground and Artificial Zero. When a three-phase load is supplied from a source of any nonsinusoidal, but symmetrical voltage $\boldsymbol{v}$ by a three-wire line, as shown in Fig. 7.1, then the zero-sequence voltage harmonics, that is, of the zero, third, or sixth (i.e., $n = 3k$) order, cannot cause any current harmonics in the supply line.

It means that there are no energy flow phenomena associated with such harmonics. At the same time, such harmonics affect the supply voltage three-phase rms value$||\boldsymbol{v}||$.

Voltages in the circuit shown in Fig. 7.12 have to satisfy the relation

$$\boldsymbol{v} = \boldsymbol{v}_{\text{A}} + \boldsymbol{u}$$

where the ground-to-artificial zero voltage $\boldsymbol{u}_{\text{A}}$ contains all zero-sequence harmonics, meaning

$$\boldsymbol{v}_{\text{A}} = \boldsymbol{v}^{\text{z}} = \sum_{n=3k} \boldsymbol{v}_n,$$

while the voltage measured versus the artificial ground does not contain these harmonics. Since the voltage $\boldsymbol{u}$ and $\boldsymbol{u}_{\text{A}}$ do not contain harmonics of the same order, they are mutually orthogonal, thus the supply voltage three-phase rms value is equal to

$$||\boldsymbol{v}|| = \sqrt{||\boldsymbol{v}_{\text{A}}||^2 + ||\boldsymbol{u}||^2}.$$

Let us assume that the load is balanced, resistive of the phase conductance G, thus its active power is $P = G||\boldsymbol{u}||^2$. The power factor λ of such a load is, of course, unity. We obtain such a unity value only if the apparent power is defined as

$$S \stackrel{\text{df}}{=} ||\boldsymbol{u}||\;||\boldsymbol{i}||$$

since, at such definition, the power factor, indeed, is

$$\lambda \stackrel{\text{df}}{=} \frac{P}{S} = \frac{G||\boldsymbol{u}||^2}{||\boldsymbol{u}||\;||\boldsymbol{i}||} = \frac{G||\boldsymbol{u}||^2}{||\boldsymbol{u}||G||\boldsymbol{u}||} = 1.$$

When the supply voltage $\boldsymbol{v}$ contains zero-sequence harmonics, meaning $\boldsymbol{u}_{\text{A}} \neq 0$, then the power factor λ is not calculated correctly under the assumption that the apparent power is defined as $S = ||\boldsymbol{v}||\,||\boldsymbol{i}||$. At such a definition, the power factor even of an ideal load would be equal to

$$\lambda \stackrel{\text{df}}{=} \frac{P}{S} = \frac{G||\boldsymbol{u}||^2}{||\boldsymbol{v}||\,||\boldsymbol{i}||} = \frac{||\boldsymbol{u}||}{\sqrt{||\boldsymbol{v}_\text{A}||^2 + ||\boldsymbol{u}||^2}} \leq 1$$

thus lower than unity. Therefore, only the voltage measured versus the artificial zero, meaning $\boldsymbol{u}$, can be regarded as the voltage applied to three-phase loads in three-wire circuits.

Illustration 7.2 *Let us assume that the voltage of line R-to-ground is*

$$v_\text{R} = 15 + 110\sqrt{2}\sin\omega_1 t + 12\sqrt{2}\sin 3\omega_1 t + 5\sqrt{2}\sin 5\omega_1 t \text{ V}.$$

Assuming that the voltage is symmetrical

$$u_\text{R} = 110\sqrt{2}\sin\omega_1 t + 5\sqrt{2}\sin 5\omega_1 t \text{ V}$$

and the three-phase rms value of this voltage is

$$||\boldsymbol{u}|| = \sqrt{3}\sqrt{110^2 + 5^2} = 190.7\,\text{V}.$$

****

Current decomposition into Currents' Physical Components. Let us start by explaining the power properties of three-phase loads that do not generate current harmonics, meaning LTI loads, supplied with symmetrical nonsinusoidal voltage. Such a voltage can be presented as a three-phase vector and expressed as a sum of harmonics of the order N that does not contain the zero-sequence harmonics $n = 3k$,

$$\boldsymbol{u}(t) = \boldsymbol{v}^\text{p}(t) + \boldsymbol{v}^\text{n}(t) = \sqrt{2}\text{Re}\sum_{n\in N} \mathbf{1}_n U_n\, e^{jn\omega_1 t}.$$

The supply current of the load can be presented in the three-phase vector form:

$$\boldsymbol{i}(t) \stackrel{\text{df}}{=} \begin{bmatrix} i_\text{R}(t) \\ i_\text{S}(t) \\ i_\text{T}(t) \end{bmatrix} = \sum_{n\in N} \boldsymbol{i}_n(t) = \sqrt{2}\text{Re}\sum_{n\in N} \begin{bmatrix} I_{\text{R}n} \\ I_{\text{S}n} \\ I_{\text{T}n} \end{bmatrix} e^{jn\omega_1 t} \stackrel{\text{df}}{=} \sqrt{2}\text{Re}\sum_{n\in N} \boldsymbol{I}_n\, e^{jn\omega_1 t}.$$

The load current of unbalanced loads is asymmetrical, thus each current harmonic can be composed of symmetrical components of the positive and the negative sequence.

Although they are three-phase vectors, with line voltages and currents as these vectors' entries, quantities $\boldsymbol{u}(t)$ and $\boldsymbol{i}(t)$, as well as their harmonics, $\boldsymbol{u}_n(t)$ and $\boldsymbol{i}_n(t)$, although they are three-phase vectors, with line voltages and currents as these vectors' entries, will be referred to, for simplicity's sake, as supply voltage, current, and their harmonics, respectively. Moreover, when there is no need to emphasize that these quantities are functions of time, their symbols will be simplified to the form of $\boldsymbol{u}$, $\boldsymbol{i}$, $\boldsymbol{u}_n$, or $\boldsymbol{i}_n$.

The active power of the load is equal to the scalar product of the supply voltage and current, since

$$P = \frac{1}{T}\int_0^T (u_R i_R + u_S i_S + u_T i_T)\, dt = \frac{1}{T}\int_0^T \boldsymbol{u}^{\mathrm{T}}\boldsymbol{i}\, dt = (\boldsymbol{u}, \boldsymbol{i}) = \mathrm{Re}\sum_{n\in N} \boldsymbol{U}_{\mathrm{n}}^{\mathrm{T}} \boldsymbol{I}_n^*.$$

Concerning the active power P at the same voltage $\boldsymbol{u}$, the load is equivalent to a balanced resistive load, configured in star, Y, as shown in Fig. 7.13, of the phase conductance G_e, referred to as the *equivalent conductance* of three-phase loads, on the condition that this conductance is equal to

$$G_e = \frac{P}{||\boldsymbol{u}||^2}.$$

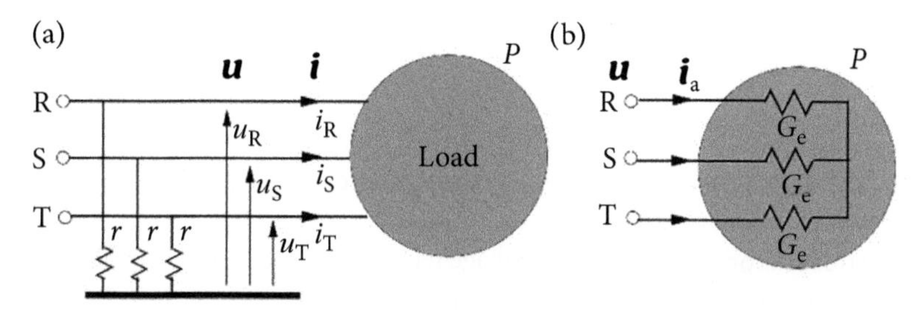

Figure 7.13 (a) A three-phase load and (b) balanced resistive load equivalent to the active power P.

This is because the active power of such an equivalent load is equal to

$$P = G_e(||u_R||^2 + ||u_S||^2 + ||u_T||^2) = G_e||\boldsymbol{u}||^2.$$

The supply current of the equivalent load is proportional to the supply voltage, since

$$\boldsymbol{i}_a(t) = G_e\, \boldsymbol{u}(t) \tag{7.27}$$

and it is the active component of the supply current of the original load, or simply, the *active current*. It is a generalization to three-phase circuits of the active current defined by Fryze for single-phase circuits. The active current $\boldsymbol{i}_a(t)$ is symmetrical and of the same waveform as the supply voltage $\boldsymbol{u}(t)$. Formula (7.27) specifies the active current in the time domain, but it can be calculated also in the frequency domain as

$$\boldsymbol{i}_a \stackrel{\mathrm{df}}{=} \begin{bmatrix} i_{Ra} \\ i_{Sa} \\ i_{Ta} \end{bmatrix} = G_e\, \boldsymbol{u} = \sqrt{2}\mathrm{Re}\sum_{n\in N} \mathbf{1}_n G_e\, U_n\, e^{jn\omega_1 t} = \sqrt{2}\mathrm{Re}\sum_{n\in N} \boldsymbol{I}_{an}\, e^{jn\omega_1 t}.$$

The remaining current component

$$\boldsymbol{i}_{\mathrm{b}} = \boldsymbol{i} - \boldsymbol{i}_{\mathrm{a}}$$

is not associated with the load active power P. It only increases the supply current three-phase rms value. It has an analogy to Fryze's reactive current, generalized to three-phase circuits. Unfortunately, it is not associated with any specific physical phenomenon in the circuit or the load features.

To reveal physical phenomena and load features responsible for this current presence, the time domain approach recommended by Fryze has to be abandoned for the frequency domain approach. It is because the supply current can, as was shown in Chapter 6, depend on the frequency properties of the load. Namely, it depends on the change of the load conductance G_n with the harmonic order n and consequently, the scattered current i_s, and scattered power D_s occur in the circuit. Consequently, the power factor declines due to a physical phenomenon visible in the frequency domain, but not in the time domain. The time domain approach has failed to reveal it.

Current $\boldsymbol{i}_{\mathrm{b}}$ in the frequency domain has the form

$$\boldsymbol{i}_{\mathrm{b}} = \boldsymbol{i} - \boldsymbol{i}_{\mathrm{a}} = \sqrt{2}\mathrm{Re}\sum_{n\in N}(\boldsymbol{I}_n - G_{\mathrm{e}}\,\boldsymbol{U}_n)e^{jn\omega_1 t}.$$

It was assumed in this section that the load is linear and consequently the circuit satisfies the superposition principle, meaning the current response to the supply voltage can be calculated harmonic by harmonic, independently of each other. To calculate this response to the supply voltage harmonic, let us use the load equivalent circuit in Δ configuration, shown in Fig. 7.14, for the harmonic of the nth order.

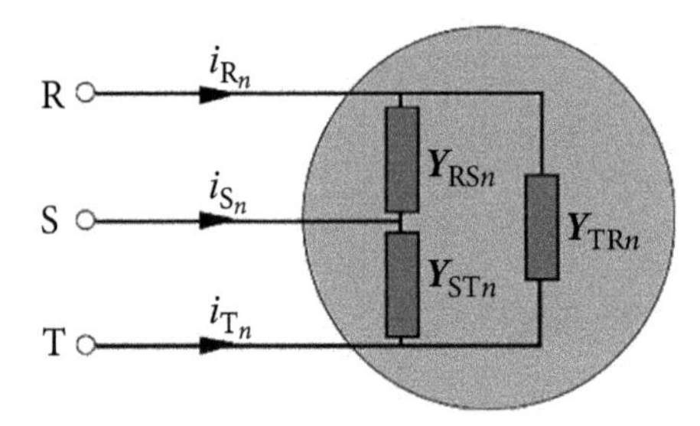

Figure 7.14 An equivalent circuit of LTI load configured in Δ for the n-th order harmonic.

The supply current crms values can be expressed in terms of the line-to-line admittances as follows:

$$\boldsymbol{I}_{\mathrm{R}n} = \boldsymbol{Y}_{\mathrm{RS}n}(\boldsymbol{U}_{\mathrm{R}n} - \boldsymbol{U}_{\mathrm{S}n}) - \boldsymbol{Y}_{\mathrm{TR}n}(\boldsymbol{U}_{\mathrm{T}n} - \boldsymbol{U}_{\mathrm{R}n})$$

$$\boldsymbol{I}_{\mathrm{S}n} = \boldsymbol{Y}_{\mathrm{ST}n}(\boldsymbol{U}_{\mathrm{S}n} - \boldsymbol{U}_{\mathrm{T}n}) - \boldsymbol{Y}_{\mathrm{RS}n}(\boldsymbol{U}_{\mathrm{R}n} - \boldsymbol{U}_{\mathrm{S}n})$$

$$\boldsymbol{I}_{\mathrm{T}n} = \boldsymbol{Y}_{\mathrm{TR}n}(\boldsymbol{U}_{\mathrm{T}n} - \boldsymbol{U}_{\mathrm{R}n}) - \boldsymbol{Y}_{\mathrm{ST}n}(\boldsymbol{U}_{\mathrm{S}n} - \boldsymbol{U}_{\mathrm{T}n}).$$

For the supply voltage harmonic $\boldsymbol{u}_n$ of the positive sequence, meaning of the order $n = 1, 4, 7, 13, \ldots$, i.e., $n = 3k + 1$, with $k = 0, 1, 2\ldots$,

$$\boldsymbol{U}_{\mathrm{S}n} = \alpha^* \boldsymbol{U}_{\mathrm{R}n}, \quad \boldsymbol{U}_{\mathrm{T}n} = \alpha \boldsymbol{U}_{\mathrm{R}n}$$

and the vector of the supply current harmonic can be presented in the form

$$\boldsymbol{I}_n \stackrel{\mathrm{df}}{=} \begin{bmatrix} \boldsymbol{I}_{\mathrm{R}n} \\ \boldsymbol{I}_{\mathrm{S}n} \\ \boldsymbol{I}_{\mathrm{T}n} \end{bmatrix} = \begin{bmatrix} Y_{en} + Y^{\mathrm{p}}_{\mathrm{u}n} \\ Y_{en} + \alpha^* Y^{\mathrm{p}}_{\mathrm{u}n} \\ Y_{en} + \alpha Y^{\mathrm{p}}_{\mathrm{u}n} \end{bmatrix} \begin{bmatrix} \boldsymbol{U}_{\mathrm{R}n} \\ \boldsymbol{U}_{\mathrm{S}n} \\ \boldsymbol{U}_{\mathrm{T}n} \end{bmatrix} = \mathbf{1}^{\mathrm{p}} Y_{en} \boldsymbol{U}_n + \mathbf{1}^{\mathrm{n}} Y^{\mathrm{p}}_{\mathrm{u}n} \boldsymbol{U}_n \tag{7.28}$$

where

$$Y_{en} \stackrel{\mathrm{df}}{=} G_{en} + jB_{en} \stackrel{\mathrm{df}}{=} Y_{\mathrm{RS}n} + Y_{\mathrm{ST}n} + Y_{\mathrm{TR}\,n} \tag{7.29}$$

$$Y^{\mathrm{p}}_{\mathrm{u}n} \stackrel{\mathrm{df}}{=} Y^{\mathrm{p}}_{\mathrm{u}n} e^{j\psi^{\mathrm{p}}_n} \stackrel{\mathrm{df}}{=} -(Y_{\mathrm{ST}n} + \alpha\, Y_{\mathrm{TR}n} + \alpha^* Y_{\mathrm{RS}\,n}). \tag{7.30}$$

is unbalanced admittance for the positive sequence harmonics, while

$$\begin{bmatrix} \boldsymbol{U}_{\mathrm{R}n} \\ \alpha^* \boldsymbol{U}_{\mathrm{S}n} \\ \alpha \boldsymbol{U}_{\mathrm{T}n} \end{bmatrix} = \begin{bmatrix} \boldsymbol{U}_{\mathrm{R}n} \\ \boldsymbol{U}_{\mathrm{T}n} \\ \boldsymbol{U}_{\mathrm{S}n} \end{bmatrix} = \mathbf{1}^{\mathrm{n}} \boldsymbol{U}_n.$$

For the supply voltage harmonics $\boldsymbol{u}_n$ of the negative sequence, meaning $n = 2, 5, 8, 11, \ldots$, that is, of the order $n = 3k-1$, with $k = 1,2\ldots$,

$$\boldsymbol{U}_{\mathrm{S}n} = \alpha \boldsymbol{U}_{\mathrm{R}n}, \quad \boldsymbol{U}_{\mathrm{T}n} = \alpha^* \boldsymbol{U}_{\mathrm{R}n}$$

the vector of the supply current harmonic can be presented in the form

$$\boldsymbol{I}_n = \begin{bmatrix} Y_{en} + Y^{\mathrm{n}}_{\mathrm{u}n} \\ Y_{en} + \alpha Y^{\mathrm{n}}_{\mathrm{u}n} \\ Y_{en} + \alpha^* Y^{\mathrm{n}}_{\mathrm{u}n} \end{bmatrix} \begin{bmatrix} \boldsymbol{U}_{\mathrm{R}n} \\ \boldsymbol{U}_{\mathrm{S}n} \\ \boldsymbol{U}_{\mathrm{T}n} \end{bmatrix} = \mathbf{1}^{\mathrm{n}} Y_{en} \boldsymbol{U}_n + \mathbf{1}^{\mathrm{p}} Y^{\mathrm{n}}_{\mathrm{u}n} \boldsymbol{U}_n \tag{7.31}$$

where

$$Y^{\mathrm{n}}_{\mathrm{u}n} \stackrel{\mathrm{df}}{=} Y^{\mathrm{n}}_{\mathrm{u}n}\, e^{j\psi^{\mathrm{n}}_n} \stackrel{\mathrm{df}}{=} -(Y_{\mathrm{ST}n} + \alpha^* Y_{\mathrm{TR}n} + \alpha\, Y_{\mathrm{RS}n}) \tag{7.32}$$

is unbalanced admittance for the negative sequence harmonics, while

$$\begin{bmatrix} \boldsymbol{U}_{\mathrm{R}n} \\ \alpha\, \boldsymbol{U}_{\mathrm{S}n} \\ \alpha^*\, \boldsymbol{U}_{\mathrm{T}n} \end{bmatrix} = \begin{bmatrix} \boldsymbol{U}_{\mathrm{R}n} \\ \boldsymbol{U}_{\mathrm{T}n} \\ \boldsymbol{U}_{\mathrm{S}n} \end{bmatrix} = \mathbf{1}^{\mathrm{p}} \boldsymbol{U}_n.$$

Formulae (7.28) and (7.31) show that each current harmonic contains a component proportional to the load equivalent admittance of the same sequence as the harmonic sequence and a component proportional to the unbalanced admittance of the opposite sequence.

Instead of using two different unbalanced admittances for harmonics of the positive sequence $Y^{\mathrm{p}}_{\mathrm{u}n}$ and the negative sequence $Y^{\mathrm{n}}_{\mathrm{u}n}$, it would be appropriate to define this admittance in such a way that the harmonic order n would switch its definition, namely

$$Y_{\mathrm{u}n} \stackrel{\mathrm{df}}{=} -(Y_{\mathrm{ST}n} + e^{jn\frac{2\pi}{3}} Y_{\mathrm{TR}n} + e^{-jn\frac{2\pi}{3}} Y_{\mathrm{RS}n}) = \begin{cases} Y^{\mathrm{p}}_{\mathrm{u}n}, & \text{for } n = 3k+1 \\ Y^{\mathrm{n}}_{\mathrm{u}n}, & \text{for } n = 3k-1 \end{cases}. \tag{7.33}$$

Having such a definition of the unbalanced admittance valid for both the positive and the negative sequence harmonic, the load current does not have to be decomposed separately for harmonics of these sequences, but together. Thus, the non-active current $\boldsymbol{i}_{\mathrm{b}}$ with a generalized unit three-phase vector $\mathbf{1}_{\mathrm{n}}$ can be rearranged to the form

$$\boldsymbol{i}_{\mathrm{b}} = \boldsymbol{i} - \boldsymbol{i}_{\mathrm{a}} = \sqrt{2}\mathrm{Re}\sum_{n\in N} [(G_{\mathrm{e}n} + jB_{\mathrm{e}n})\mathbf{1}_n \boldsymbol{U}_n + Y_{\mathrm{u}n}\mathbf{1}^*_n \boldsymbol{U}_n - G_{\mathrm{e}}\mathbf{1}_n \boldsymbol{U}_n]\, e^{jn\omega_1 t}.$$

The non-active current $\boldsymbol{i}_{\mathrm{b}}$ contains a *scattered current*

$$\boldsymbol{i}_{\mathrm{s}}(t) \stackrel{\mathrm{df}}{=} \sqrt{2}\mathrm{Re}\sum_{n\in N} (G_{\mathrm{e}n} - G_{\mathrm{e}})\mathbf{1}_n \boldsymbol{U}_n\, e^{jn\omega_1 t} = \sqrt{2}\mathrm{Re}\sum_{n\in N} \boldsymbol{I}_{\mathrm{s}n}\, e^{jn\omega_1 t} \tag{7.34}$$

a reactive current

$$\boldsymbol{i}_{\mathrm{r}}(t) \stackrel{\mathrm{df}}{=} \sqrt{2}\mathrm{Re}\sum_{n\in N} jB_{\mathrm{e}n}\mathbf{1}_n \boldsymbol{U}_n\, e^{jn\omega_1 t} = \sqrt{2}\mathrm{Re}\sum_{n\in N} \boldsymbol{I}_{\mathrm{r}n}\, e^{jn\omega_1 t} \tag{7.35}$$

and an unbalanced current

$$\boldsymbol{i}_{\mathrm{u}}(t) \stackrel{\mathrm{df}}{=} \sqrt{2}\mathrm{Re}\sum_{n\in N} Y_{\mathrm{u}n}\mathbf{1}^*_n \boldsymbol{U}_n\, e^{jn\omega_1 t} = \sqrt{2}\mathrm{Re}\sum_{n\in N} \boldsymbol{I}_{\mathrm{u}n}\, e^{jn\omega_1 t}. \tag{7.36}$$

Three-phase vectors of crms values of these currents harmonics are

$$\boldsymbol{I}_{\mathrm{s}n} \stackrel{\mathrm{df}}{=} (G_{\mathrm{e}n} - G_{\mathrm{e}})\,\mathbf{1}_n \boldsymbol{U}_n$$

$$\boldsymbol{I}_{\mathrm{r}n} \stackrel{\mathrm{df}}{=} jB_{\mathrm{e}n}\,\mathbf{1}_n \boldsymbol{U}_n$$

$$\boldsymbol{I}_{\mathrm{u}n} \stackrel{\mathrm{df}}{=} Y_{\mathrm{u}n}\,\mathbf{1}^*_n \boldsymbol{U}_n$$

thus these currents are symmetrical, with the sequence specified by unit vectors $\mathbf{1}_{\mathrm{n}}$ and $\mathbf{1}^*_{\mathrm{n}}$. Thus, the three-phase vector of the supply current $\boldsymbol{i}(t)$ of a three-phase load supplied with a nonsinusoidal, but symmetrical voltage $\boldsymbol{u}(t)$, can be decomposed into four components

$$\boldsymbol{i}(t) = \boldsymbol{i}_{\mathrm{a}}(t) + \boldsymbol{i}_{\mathrm{s}}(t) + \boldsymbol{i}_{\mathrm{r}}(t) + \boldsymbol{i}_{\mathrm{u}}(t). \tag{7.37}$$

Each current component in this decomposition is associated with a distinctively different phenomenon in the circuit, thus they can be referred to as Current's

Physical Components of three-phase three-wire circuits with LTI loads supplied with nonsinusoidal symmetrical voltage.

The active current $\boldsymbol{i}_{\mathrm{a}}(t)$ is associated with permanent energy transfer from the supply source to the load at active power P. It is the only current component needed for energy delivery to the load, thus it is the smallest current indispensable for energy delivery with the active power P. Its three-phase rms value is

$$||\boldsymbol{i}_{\mathrm{a}}|| = G_{\mathrm{e}}||\boldsymbol{u}|| = \frac{P}{||\boldsymbol{u}||}.$$

The scattered current $\boldsymbol{i}_{\mathrm{s}}(t)$ of the rms value

$$||\boldsymbol{i}_{\mathrm{s}}|| = \sqrt{3}\sqrt{\sum_{n\in N} I_{\mathrm{s}n}^2} = \sqrt{3}\sqrt{\sum_{n\in N} (G_{\mathrm{e}n} - G_{\mathrm{e}})^2\, U_n^2} \tag{7.38}$$

occurs in the supply current when the load conductance $G_{\mathrm{e}n}$ changes with the harmonic order. The reactive current $\boldsymbol{i}_{\mathrm{r}}(t)$ of the rms value

$$||\boldsymbol{i}_{\mathrm{r}}|| = \sqrt{3}\sqrt{\sum_{n\in N} I_{\mathrm{r}n}^2} = \sqrt{3}\sqrt{\sum_{n\in N} B_{\mathrm{e}n}^2\, U_n^2} \tag{7.39}$$

occurs in the supply current when there is a phase shift between the supply voltage and the supply current harmonics, and consequently the load equivalent susceptance $B_{\mathrm{e}n}$ is not equal to zero. We should observe, however, that the reactive current $\boldsymbol{i}_{\mathrm{r}}$, associated with the phase shift between the supply voltage and current, is not accompanied in three-phase circuits by energy oscillation between the supply and the load. It is because, for an individual harmonic in circuits with balanced loads, the rate of energy flow is

$$p_n(t) = \boldsymbol{u}_n^T(t)\boldsymbol{i}_n(t) = \text{const.}$$

independently of the phase shift and consequently, the presence of the reactive current $\boldsymbol{i}_{\mathrm{r}}(t)$.

The unbalanced current, $\boldsymbol{i}_{\mathrm{u}}(t)$ of the rms value

$$||\boldsymbol{i}_{\mathrm{u}}|| = \sqrt{3}\sqrt{\sum_{n\in N} I_{\mathrm{u}n}^2} = \sqrt{3}\sqrt{\sum_{n\in N} Y_{\mathrm{u}n}^2\, U_n^2} \tag{7.40}$$

occurs in the supply current when there is a supply current asymmetry and consequently, the load unbalanced admittance for harmonic frequencies Y_{un} is not equal to zero.

Equivalent circuit. Decomposition (7.37) along with relations (7.34), (7.35), and (7.36) means that an LTI load can be replaced for each harmonic by an equivalent circuit, composed of four three-phase loads connected in parallel as shown in Fig. 7.15, and responsible for each current component independently of each other. It is important to observe that these currents are specified by equivalent parameters of the load.

The active $\boldsymbol{i}_a(t)$, scattered $\boldsymbol{i}_s(t)$, and reactive $\boldsymbol{i}_r(t)$ currents have identical physical meanings to these currents in single-phase circuits. They are associated with distinctive physical phenomena in the load. Only their mathematical form is more complex, due to the three-phase circuits' complexity. They have a form of three-phase vectors. The unbalanced current $\boldsymbol{i}_u(t)$ can occur only in three-phase circuits. It is similar to this current in circuits with sinusoidal voltage, however. At nonsinusoidal supply

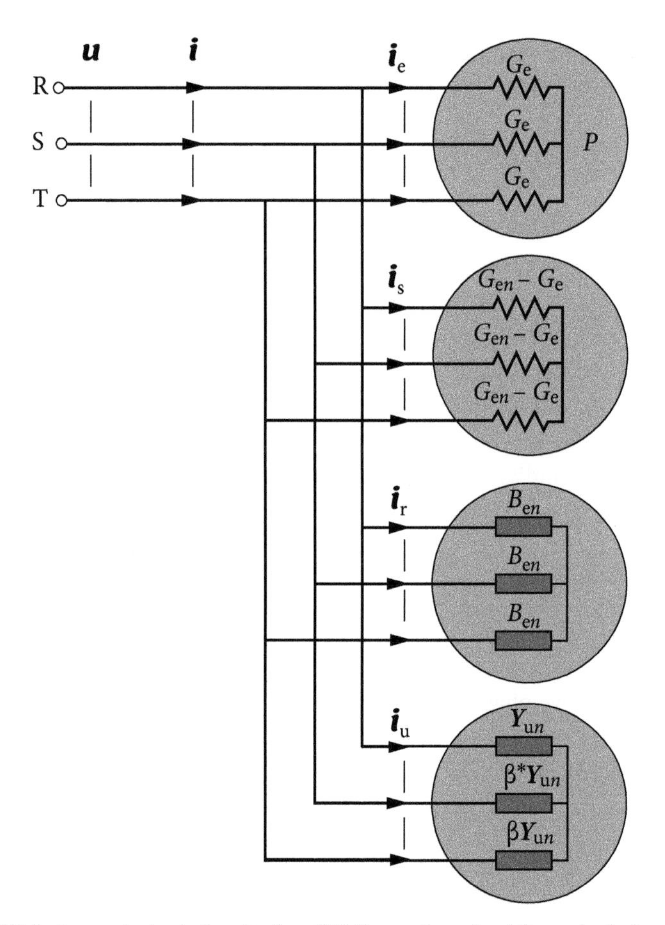

Figure 7.15 An equivalent circuit of an LTI three-phase load for a single harmonic.

voltage, it is a sum of unbalanced currents for each harmonic. All these currents are symmetrical. The active $\mathbf{i}_{\mathrm{a}}(t)$, scattered $\mathbf{i}_{\mathrm{s}}(t)$, and reactive $\mathbf{i}_{\mathrm{r}}(t)$ currents have the same sequence as the supply voltage, in other words they are of positive sequence, while the unbalanced current $\mathbf{i}_{\mathrm{u}}(t)$ has the negative sequence. Therefore, only this current causes the supply current asymmetry.

7.8 Orthogonality of CPCs in Circuits with nv&c and LTI Loads

Physical components $\mathbf{i}_{\mathrm{a}}(t)$, $\mathbf{i}_{\mathrm{s}}(t)$, $\mathbf{i}_{\mathrm{r}}(t)$, and $\mathbf{i}_{\mathrm{u}}(t)$ of the supply current affect the supply current rms value $||\mathbf{i}||$, and consequently, the apparent power S of the supply source independently on each other, on the condition that these currents are mutually orthogonal, meaning their scalar products

$$(\mathbf{i}_{\mathrm{a}}, \mathbf{i}_{\mathrm{s}}),\quad (\mathbf{i}_{\mathrm{a}}, \mathbf{i}_{\mathrm{r}}),\quad (\mathbf{i}_{\mathrm{a}}, \mathbf{i}_{\mathrm{u}}),\quad (\mathbf{i}_{\mathrm{s}}, \mathbf{i}_{\mathrm{r}}),\quad (\mathbf{i}_{\mathrm{s}}, \mathbf{i}_{\mathrm{u}}),\quad (\mathbf{i}_{\mathrm{r}}, \mathbf{i}_{\mathrm{u}})$$

are equal to zero.

The following property of the scalar product $(\mathbf{x}, \mathbf{y})$ of quantities that are a sum of harmonics is very useful for the proof of orthogonality of the current physical components in three-phase circuits. If three-phase vectors $\mathbf{x}$ and $\mathbf{y}$ are sums of harmonics

$$\mathbf{x} = \sum_{n\in N} \mathbf{x}_n,\quad \mathbf{y} = \sum_{n\in N} \mathbf{y}_n,$$

then, due to orthogonality of harmonics of different order n, their scalar product

$$(\mathbf{x}, \mathbf{y}) = \left(\sum_{n\in N} \mathbf{x}_n, \sum_{n\in N} \mathbf{y}_n\right) = \sum_{n\in N} (\mathbf{x}_n, \mathbf{y}_n)$$

is equal to the sum of scalar products of individual harmonics of these quantities.

The orthogonality of the active, reactive, and unbalanced currents, $\mathbf{i}_{\mathrm{a}}$, $\mathbf{i}_{\mathrm{r}}$, and $\mathbf{i}_{\mathrm{u}}$, when these currents are sinusoidal, was proven earlier in this chapter. These proofs remain valid for each harmonic of these quantities. Hence

$$(\mathbf{i}_{\mathrm{a}}, \mathbf{i}_{\mathrm{r}}) = 0,\quad (\mathbf{i}_{\mathrm{a}}, \mathbf{i}_{\mathrm{u}}) = 0,\quad (\mathbf{i}_{\mathrm{r}}, \mathbf{i}_{\mathrm{u}}) = 0.$$

To prove the orthogonality of these three currents with the scattered current observe that

$$(\mathbf{x}, \mathbf{y}) = \mathrm{Re} \sum_{n\in N} \mathbf{X}_n^{\mathrm{T}} \mathbf{Y}_n^{*},$$

hence the scalar product of the scattered and the active currents is

$$(\boldsymbol{i}_{\mathrm{s}}, \boldsymbol{i}_{\mathrm{a}}) = \mathrm{Re} \sum_{n\in N} (G_{en} - G_{\mathrm{e}})\ \mathbf{1}_n^{\mathrm{T}} U_n\ (G_{\mathrm{e}}\, \mathbf{1}_n U_n)^* =$$
$$= G_{\mathrm{e}}(\sum_{n\in N} G_{en}||\boldsymbol{u}_n||^2 - G_{\mathrm{e}} \sum_{n\in N} ||\boldsymbol{u}_n||^2) = G_{\mathrm{e}}(P - P) = 0,$$

thus they are orthogonal. Similarly, the scalar product of the scattered and the reactive currents is equal to

$$(\boldsymbol{i}_{\mathrm{s}}, \boldsymbol{i}_{\mathrm{r}}) = \mathrm{Re} \sum_{n\in N} (G_{en} - G_{\mathrm{e}})\ \mathbf{1}_n^{\mathrm{T}} U_n\ (jB_{en}\, \mathbf{1}_n U_n)^* =$$
$$= \mathrm{Re}\,\{-j \sum_{n\in N} (G_{en} - G_{\mathrm{e}})\, B_{en}||\boldsymbol{u}_n||^2\} = 0$$

so that these two currents are orthogonal. The scattered and unbalanced currents are of the opposite sequence hence their scalar product should be equal to zero. Indeed,

$$(\boldsymbol{i}_{\mathrm{s}}, \boldsymbol{i}_{\mathrm{u}}) = \mathrm{Re} \sum_{n\in N} (G_{en} - G_{\mathrm{e}})\mathbf{1}_n^{\mathrm{T}} U_n (Y_{\mathrm{u}\,n}\mathbf{1}_n^{*} U_n)^* = 0,$$

since

$$\mathbf{1}_n^{\mathrm{T}}\mathbf{1}_n = \left[1,\ 1e^{-jn\frac{2\pi}{3}}, 1e^{jn\frac{2\pi}{3}}\right] \begin{bmatrix} 1 \\ 1\,e^{-jn\frac{2\pi}{3}} \\ 1\,e^{jn\frac{2\pi}{3}} \end{bmatrix} = 1 + 1e^{jn\frac{2\pi}{3}} + 1e^{-jn\frac{2\pi}{3}} = 0.$$

This completes the proof that the supply current physical components $\boldsymbol{i}_{\mathrm{a}}$, $\boldsymbol{i}_{\mathrm{s}}$, $\boldsymbol{i}_{\mathrm{r}}$, and $\boldsymbol{i}_{\mathrm{u}}$ in three-phase circuits with LTI loads supplied with a nonsinusoidal voltage are mutually orthogonnal, hence their three-phase rms values satisfy the relationship

$$||\boldsymbol{i}||^2 = ||\boldsymbol{i}_{\mathrm{a}}||^2 + ||\boldsymbol{i}_{\mathrm{s}}||^2 + ||\boldsymbol{i}_{\mathrm{r}}||^2 + ||\boldsymbol{i}_{\mathrm{u}}||^2. \tag{7.41}$$

It means that these currents affect the supply current rms value independently of each other.

This relationship is illustrated with a polygon shown in Fig. 7.16.

It is drawn in such a way that the edges of the length proportional to the current rms value are perpendicular to the geometrical sum of the previous edges.

This decomposition of the supply current into physical components is the exact decomposition in the sense that it is not founded on any approximations. As long as the circuit satisfies the conditions for which this decomposition is defined, the relations presented above are satisfied independently of the circuit parameters and the level of the supply voltage harmonics, and the level of the load imbalance. It is illustrated with the following example.

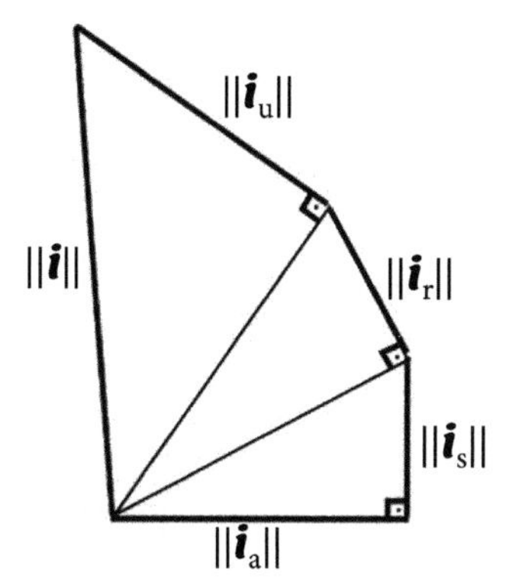

Figure 7.16 A polygon of the CPCs rms values in three-phase circuits with an LTI load with a nonsinusoidal supply voltage.

Illustration 7.3 *The load shown in Fig. 7.17 is supplied with a symmetrical voltage such that the line R-to-artificial zero voltage* u_R *is*

$$u_R = \sqrt{2}\text{Re}\{220\, e^{j\omega_1 t} + 44\, e^{j5\omega_1 t}\}\ \text{V}.$$

The supply transformer is ideal with a turn ratio 1:1.

The three-phase rms values of the supply voltage harmonics are equal to, respectively,

$$||\boldsymbol{u}_1|| = \sqrt{U_{R1}^2 + U_{S1}^2 + U_{T1}^2} = \sqrt{3 \times 220^2} = 220\sqrt{3} = 381.05\ \text{V}$$

and consequently, the supply voltage three-phase rms value is equal to

$$||\boldsymbol{u}|| = \sqrt{||\boldsymbol{u}_1||^2 + ||\boldsymbol{u}_5||^2} = 388.6\ \text{V}.$$

Line-to-line admittances for the fundamental harmonics are equal to

$$Y_{RS1} = \frac{1}{1+j1} = 0.50 - j0.50\ \text{S}, \quad Y_{ST1} = \frac{1}{0.5-j2} = 0.118 + j0.471\ \text{S},$$

hence the equivalent admittance for the fundamental harmonic is equal to

$$Y_{e1} = G_{e1} + jB_{e1} = Y_{ST1} + Y_{TR1} + Y_{RS1} = 0.618 - j0.029\ \text{S}$$

while the unbalanced admittance

$$Y_{u1} = -(Y_{ST1} + 1e^{-j120°}Y_{RS1}) =$$

$$= -[0.118 + j0.471 + 1e^{-j120°}(0.50 - j0.50)] = 0.634e^{-j27°}\text{S}.$$

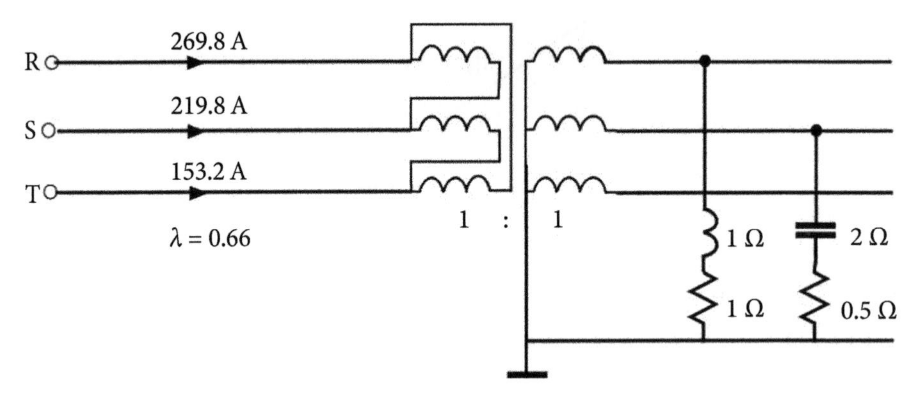

Figure 7.17 An example of a three-phase load.

The line-to-line admittances of the load for the fifth order harmonic have the values

$$Y_{RS5} = \frac{1}{1 + j5 \times 1} = 0.038 - j0.192 \text{ S}$$

$$Y_{ST5} = \frac{1}{0.5 - j2/5} = 1.220 + j0.978 \text{ S},$$

hence the equivalent admittance of the load for the fifth order harmonic is equal to

$$Y_{e5} = G_{e5} + jB_{e5} = Y_{ST5} + Y_{TR5} + Y_{RS5} = 1.258 + j0.783 \text{ S}$$

while the unbalance admittance

$$Y_{u5} = -(Y_{ST5} + 1e^{j120°} Y_{RS5}) =$$

$$= -[1.220 + j0.978 + 1e^{j120°}(0.038 - j0.192)] = 1.759\, e^{-j141°} \text{S}.$$

The load active power is

$$P = G_{e1}||\boldsymbol{u}_1||^2 + G_{e5}||\boldsymbol{u}_5||^2 = 96988.7 \text{ W},$$

thus its equivalent conductance is equal to

$$G_e = \frac{P}{||\boldsymbol{u}||^2} = \frac{96988.7}{388.6^2} = 0.6423 \text{ S}.$$

The rms values of the supply current physical components are

$$||\boldsymbol{i}_{\rm a}|| = G_{\rm e}||\boldsymbol{u}|| = 0.6423 \times 388.6 = 249.6\,\text{A}$$

$$||\boldsymbol{i}_{\rm s}|| = \sqrt{\sum_{n=1,5} (G_{{\rm e}n} - G_{\rm e})^2 ||\boldsymbol{u}_n||^2} =$$

$$= \sqrt{(0.618 - 0.642)^2 \times 381.05^2 + (1.258 - 0.642)^2 \times 76.2^2} = 47.8\,\text{A}$$

$$||\boldsymbol{i}_{\rm r}|| = \sqrt{\sum_{n=1,5} B_{{\rm e}n}^2 ||\boldsymbol{u}_n||^2} = \sqrt{(0.029 \times 381.05)^2 + (0.783 \times 76.2)^2} = 60.7\,\text{A}$$

$$||\boldsymbol{i}_{\rm u}|| = \sqrt{\sum_{n=1,5} Y_{{\rm u}n}^2 ||\boldsymbol{u}_n||^2} = \sqrt{(0.634 \times 381.05)^2 + (1.759 \times 76.2)^2} = 276.2\,\text{A}.$$

The supply current rms value, calculated as the geometrical sum of rms values of the current physical components, is equal to

$$||\boldsymbol{i}|| = \sqrt{||\boldsymbol{i}_{\rm a}||^2 + ||\boldsymbol{i}_{\rm s}||^2 + ||\boldsymbol{i}_{\rm u}||^2 + ||\boldsymbol{i}_{\rm r}||^2} = \sqrt{249.6^2 + 47.8^2 + 60.7^2 + 276.3^2} = 380.2\,\text{A}.$$

It is, of course, equal to the supply current rms value, calculated as a geometrical sum of the line currents' rms values, namely

$$||\boldsymbol{i}|| = \sqrt{||i_{\rm R}||^2 + ||i_{\rm S}||^2 + ||i_{\rm T}||^2} = \sqrt{269.8^2 + 219.8^2 + 153.2^2} = 380.2\,\text{A}$$

and this verifies decomposition and numerical calculations.

*****

7.9 Powers in Circuits with nv&c and LTI Loads

Although power properties are the subject of the power theory, these properties can be known entirely even without introducing the concept of power. These properties are specified entirely by the Current's Physical Components, meaning in terms of currents. Traditionally, electrical engineers are accustomed to specifying these properties in terms of powers rather in terms of current components' rms values.

To satisfy this traditional expectation, let us multiply Eqn. (7.41) by the square of the supply voltage rms value, $||\boldsymbol{u}||^2$. In effect, the power equation of the load introduced originally in Ref. [93]:

$$S^2 = P^2 + D_{\rm s}^2 + Q^2 + D_{\rm u}^2 \tag{7.42}$$

is obtained, with the scattered power

$$D_{\rm s} \stackrel{\rm df}{=} ||\boldsymbol{u}|| \cdot ||\boldsymbol{i}_{\rm s}||, \tag{7.43}$$

the reactive power,

$$Q \stackrel{\mathrm{df}}{=} ||\boldsymbol{u}|| \cdot ||\boldsymbol{i}_{\mathrm{r}}||, \tag{7.44}$$

and the unbalanced power

$$D_{\mathrm{u}} \stackrel{\mathrm{df}}{=} ||\boldsymbol{u}|| \cdot ||\boldsymbol{i}_{\mathrm{u}}||. \tag{7.45}$$

Apart from the active power P, the powers defined by these formulae do not have physical interpretations as explicit as the interpretation of the active power. They are only *associated* with some distinctive physical phenomena. The scattered power D_{s} occurs due to the change of the load conductance with harmonic frequency. The reactive power Q occurs due to the phase shift between the supply voltage and the load current harmonics. The unbalanced power D_{u} occurs due to the load current harmonic asymmetry, which is caused by the load imbalance.

These powers describe quantitatively the effects of these three phenomena on the apparent power of the load, but they are not any measures of these phenomena.

Similarly as with the CPCs three-phase rms values, the relationship between powers in 3p3w circuits with LTI loads and nonsinusoidal voltage can be visualized with a diagram as shown in Fig. 7.18. Projections of the diagram sides of the length proportional to particular powers, on the closing side of the diagram, show the contribution of these powers to the apparent power of the load.

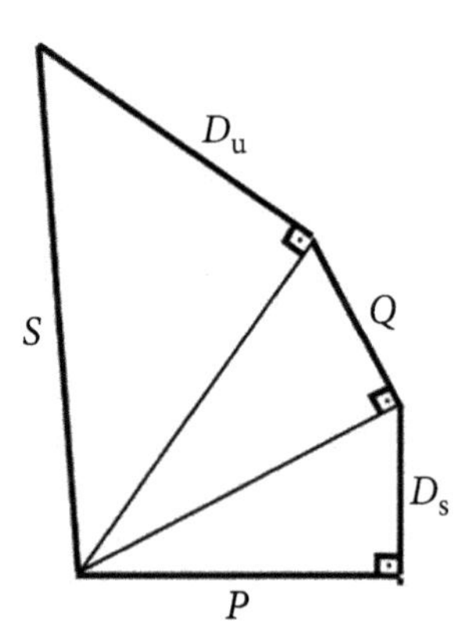

Figure 7.18 A diagram of powers of three-phase, LTI loads supplied with three-phase nonsinusoidal voltage.

The power factor of LTI loads in 3p3w circuits with nonsinusoidal supply voltage is equal to

$$\lambda \stackrel{\mathrm{df}}{=} \frac{P}{S} = \frac{P}{\sqrt{P^2 + D_{\mathrm{s}}^2 + Q^2 + D_{\mathrm{u}}^2}}.$$

It can be expressed in terms of the current components' three-phase rms values

$$\lambda = \frac{P}{S} = \frac{||\boldsymbol{u}||\,||\boldsymbol{i}_\mathrm{a}||}{||\boldsymbol{u}||\,||\boldsymbol{i}||} = \frac{||\boldsymbol{i}_\mathrm{a}||}{\sqrt{||\boldsymbol{i}_\mathrm{a}||^2 + ||\boldsymbol{i}_\mathrm{s}||^2 + ||\boldsymbol{i}_\mathrm{r}||^2 + ||\boldsymbol{i}_\mathrm{u}||^2}}. \tag{7.46}$$

According to traditional interpretations, only the reactive current is responsible for the power factor decline. The unbalanced and scattered currents contribute to this decline similarly to the reactive current.

7.10 CPC in Circuits with nv&c and HGLs

Three-phase nonlinear loads and/or loads with periodic time-variance of their parameters can generate current harmonics of the order not present in the supply voltage.

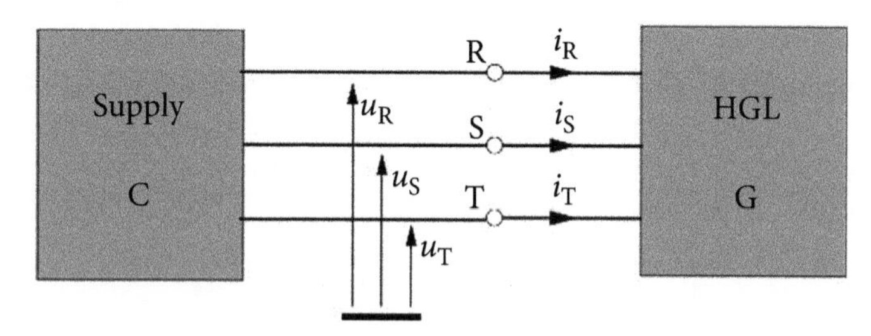

Figure 7.19 A three-phase three-wire circuit with harmonic generating load (HGL).

Three-phase circuits with harmonic generating loads, shown in Fig. 7.19, are increasingly common due to the power electronics equipment, used for energy flow control in industrial and commercial environments. Such power electronics equipment is usually nonlinear and has parameters that vary periodically with the period of the distribution voltage. This causes periodic distortion of the supply current and consequently, current harmonics, $\boldsymbol{j}$, are generated in the load.

The equivalent circuit of such a harmonic-generating load, which emphasizes the phenomenon of generation of these current harmonics is shown in Fig. 7.20.

Similarly, as in single-phase circuits, the load-generated harmonics in three-phase loads can cause the load, normally an energy consumer, to become the source of the energy flow at some harmonic frequencies. It means that the active power P_n for some frequencies becomes negative. For example, if the distribution voltage does not contain the fifth order harmonic, but this harmonic exists in the load current, then the active power of the fifth order harmonic is negative. The flow of energy associated with this harmonic originates in the load. This energy is dissipated in the resistance of the distribution system.

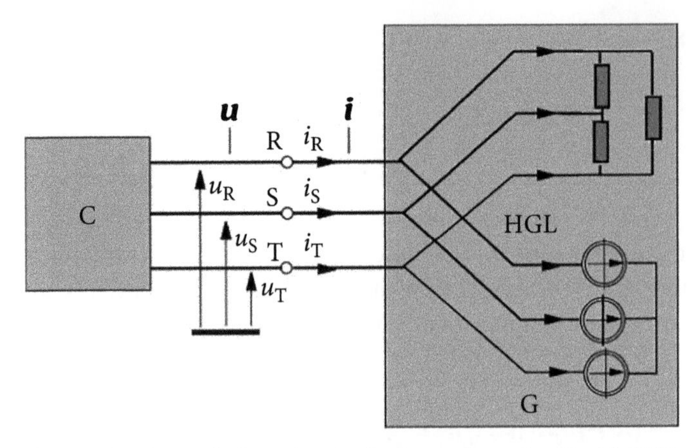

Figure 7.20 An equivalent circuit of a three-phase circuit with HGL.

Taking into account that the load is passive, this energy has to be delivered to the load, of course, by other voltage and current harmonics, mainly by the fundamental. Unlike LTI loads that have to satisfy the Energy Balance Principle, harmonic by harmonic, the HGLs do not obey this principle harmonic by harmonic but only as a whole. Due to nonlinearity, or time variance of the HGLs, the energy can change harmonic carriers. It is a well-known feature of power electronics devices.

The active power of the nth order harmonic is equal to

$$P_n = U_{\text{R}n} I_{\text{R}n} \cos\varphi_{\text{R}n} + U_{\text{S}n} I_{\text{S}n} \cos\varphi_{\text{S}n} + U_{\text{T}n} I_{\text{T}n} \cos\varphi_{\text{T}n}.$$

If the supply voltage is symmetrical, then

$$U_{\text{R}n} = U_{\text{S}n} = U_{\text{T}n} \stackrel{\text{df}}{=} U_n$$

so that the harmonic active power can be expressed as

$$P_n = U_n (I_{\text{R}n} \cos\varphi_{\text{R}n} + I_{\text{S}n} \cos\varphi_{\text{S}n} + I_{\text{T}n} \cos\varphi_{\text{T}n}) \stackrel{\text{df}}{=} U_n I_{\text{a}n}$$

where

$$I_{\text{a}n} \stackrel{\text{df}}{=} I_{\text{R}n} \cos\varphi_{\text{R}n} + I_{\text{S}n} \cos\varphi_{\text{S}n} + I_{\text{T}n} \cos\varphi_{\text{T}n}.$$

This is a generalized (meaning it could be positive or negative) rms value of the active component of the nth order current harmonic. Due to $\cos\varphi$ terms, $I_{\text{a}n}$ value can be positive or negative, which is in conflict with the common meaning of rms value.

As was done in Chapter 6 for single-phase circuits with the HGLs, the set N of all harmonic orders can be decomposed, depending on the sign of the harmonic active power P_n, into two sets, N_{C} and N_{G}. There is one important difference with respect to single-phase circuits, however. It was enough to observe for that purpose in single-phase circuits the value of the phase angle, φ_n, between the voltage and current harmonics. To find the direction of energy flow at some harmonic frequency in a three-phase circuit, the sign of the generalized rms value of the active current I_{an}, has to be calculated.

When the harmonic active power, P_n, that is, the generalized rms value I_{an}, is non-negative, we can assume that the distribution voltage harmonic $\boldsymbol{u}_n$ is the cause of energy flow from the source to the load. When this power, that is, the rms value I_{an}, is negative, we can assume that the current harmonic generated in the load, $\boldsymbol{j}_n$, is the cause of energy flow back from the load to the distribution system. This enables the decomposition of the set N into two subsets, such that

$$\text{when } I_{an} \geq 0, \quad n \in N_C$$

$$\text{when } I_{an} < 0, \quad n \in N_G.$$

Having these sets defined, the current, voltage, and active power component can then be associated with the direction of energy flow. All harmonics with indices from the set N_C specify the current and voltage components, $\boldsymbol{i}_C$ and $\boldsymbol{u}_C$, associated with energy flow from the distribution system (C) to the load, namely

$$\sum_{n\in N_C} \boldsymbol{i}_n \stackrel{\text{df}}{=} \boldsymbol{i}_C, \qquad \sum_{n\in N_C} \boldsymbol{u}_n \stackrel{\text{df}}{=} \boldsymbol{u}_C, \qquad \sum_{n\in N_C} P_n \stackrel{\text{df}}{=} P_C.$$

All harmonics with indices from the set N_G specify the current and voltage components, $\boldsymbol{i}_G$ and $\boldsymbol{u}_G$, associated with energy flow from the customer to the distribution system, namely

$$\sum_{n\in N_G} \boldsymbol{i}_n \stackrel{\text{df}}{=} \boldsymbol{i}_G, \qquad \sum_{n\in N_G} \boldsymbol{u}_n \stackrel{\text{df}}{=} -\boldsymbol{u}_G, \qquad \sum_{n\in N_G} P_n \stackrel{\text{df}}{=} -P_G.$$

Thus, similarly to single-phase circuits with the HGLs, the current, voltage, and active power can be decomposed into components associated with the direction of energy flow

$$\boldsymbol{i} = \sum_{n\in N} \boldsymbol{i}_n = \boldsymbol{i}_C + \boldsymbol{i}_G, \quad \boldsymbol{u} = \sum_{n\in N} \boldsymbol{u}_n = \boldsymbol{u}_C - \boldsymbol{u}_G, \quad \sum_{n\in N} P_n = P_C - P_G. \tag{7.47}$$

These decompositions and their interpretations do not differ from that in single-phase circuits with HGLs and, as previously mentioned, the current and voltage components are orthogonal. It is because these components do not contain harmonics of the same order, thus $(\boldsymbol{i}_C, \boldsymbol{i}_G,) = 0$ and $(\boldsymbol{u}_C, \boldsymbol{u}_G,) = 0$, hence

$$||\boldsymbol{i}||^2 = ||\boldsymbol{i}_C||^2 + ||\boldsymbol{i}_G||^2 \tag{7.48}$$

and

$$||\boldsymbol{u}||^2 = ||\boldsymbol{u}_C||^2 + ||\boldsymbol{u}_G||^2. \tag{7.49}$$

The original circuit can be regarded for the voltage and current harmonics of the order from the set N_{SC} as a circuit with a passive load, shown in Fig. 7.21, of equivalent admittance for harmonic frequencies

$$Y_{en} \stackrel{\text{df}}{=} G_{en} + jB_{en} = \frac{\boldsymbol{C}_n^*}{||\boldsymbol{u}_n||^2} \tag{7.50}$$

where $\boldsymbol{C}_n$ denotes the complex power at the nth order harmonic equal to

$$\boldsymbol{C}_n \stackrel{\text{df}}{=} P_n + jQ_n = \boldsymbol{U}_n^{\text{T}} \boldsymbol{I}_n^*. \tag{7.51}$$

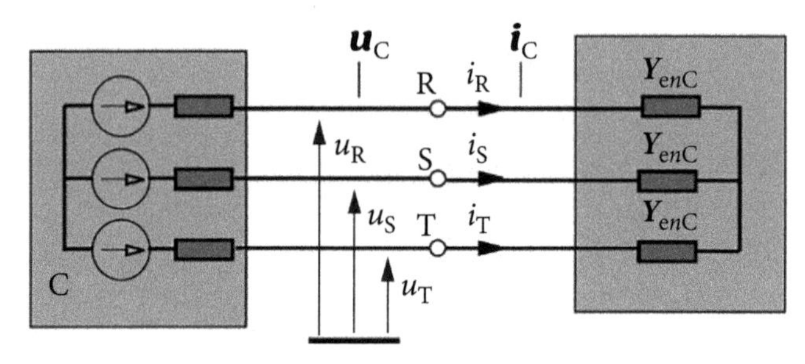

Figure 7.21 An equivalent circuit of the circuit for harmonics of order from sub-set N_{C}.

The same circuit can be regarded for harmonics of the order from the subset N_{G} as a circuit with current sources, $\boldsymbol{j}_{\text{G}}$, on the load side and a passive distribution system, as shown in Fig. 7.22. Observe that the voltage response $\boldsymbol{u}_{\text{G}}$ of the distribution system to the load-generated current, $\boldsymbol{j}_{\text{G}}$, has an opposite direction to the distribution voltage $\boldsymbol{u}_{\text{C}}$.

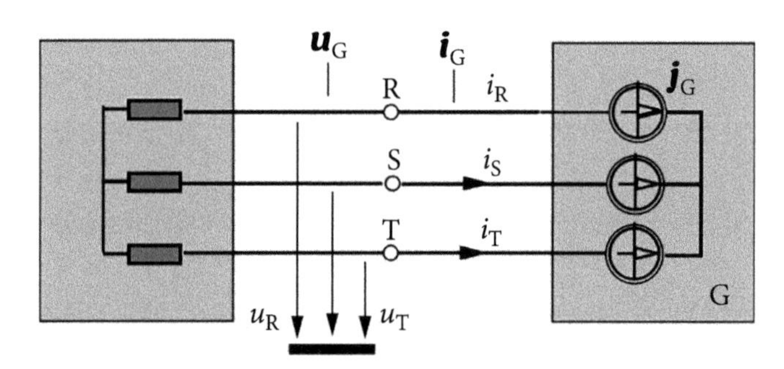

Figure 7.22 An equivalent circuit of the circuit for harmonics of order from sub-set N_{G}.

Concerning active power P_{C} at voltage $\boldsymbol{u}_{\text{C}}$, the load shown in Fig. 7.21, is equivalent to a resistive symmetrical load in Y configuration, and the conductance per phase equal to

$$G_{eC} \stackrel{df}{=} \frac{P_C}{||\boldsymbol{u}_C||^2}. \tag{7.52}$$

Such a load draws the active current

$$i_{aC} = G_{eC}\, u_C \tag{7.53}$$

from the source. The remaining part of the current $\boldsymbol{i}_C$ can be decomposed into the scattered, reactive, and unbalanced currents

$$\boldsymbol{i}_{sC} \stackrel{df}{=} \sqrt{2}\mathrm{Re} \sum_{n\in N_C} (G_{en} - G_{eC})\, \mathbf{1}_n \boldsymbol{U}_n\, e^{jn\,\omega_1 t} \tag{7.54}$$

$$\boldsymbol{i}_{rC} \stackrel{df}{=} \sqrt{2}\mathrm{Re} \sum_{n\in N_C} jB_{en}\, \mathbf{1}_n \boldsymbol{U}_n\, e^{jn\,\omega_1 t} \tag{7.55}$$

$$\boldsymbol{i}_{uC} \stackrel{df}{=} \sqrt{2}\mathrm{Re} \sum_{n\in N_C} Y_{un} \mathbf{1}_n^* \boldsymbol{U}_n\, e^{jn\,\omega_1 t}. \tag{7.56}$$

Thus, taking into account relation (7.47), the supply current of HGLs can be decomposed into five components, namely

$$\boldsymbol{i} = \boldsymbol{i}_{aC} + \boldsymbol{i}_{sC} + \boldsymbol{i}_{rC} + \boldsymbol{i}_{uC} + \boldsymbol{i}_G. \tag{7.57}$$

These five components are mutually orthogonal, hence their three-phase rms values fulfill the relationship

$$||\boldsymbol{i}||^2 = ||\boldsymbol{i}_{aC}||^2 + ||\boldsymbol{i}_{sC}||^2 + ||\boldsymbol{i}_{rC}||^2 + ||\boldsymbol{i}_{uC}||^2 + ||\boldsymbol{i}_G||^2. \tag{7.58}$$

where

$$||\boldsymbol{i}_{aC}|| = G_{eC}||\boldsymbol{u}_C|| = \frac{P_C}{||\boldsymbol{u}_C||} \tag{7.59}$$

$$||\boldsymbol{i}_{sC}|| = \sqrt{3}\sqrt{\sum_{n\in N_C} I_{sn}^2} = \sqrt{3}\sqrt{\sum_{n\in N_C} (G_{en} - G_{eC})^2 U_n^2} \tag{7.60}$$

$$||\boldsymbol{i}_{rC}|| = \sqrt{3}\sqrt{\sum_{n\in N_C} I_{rn}^2} = \sqrt{3}\sqrt{\sum_{n\in N_C} B_{en}^2\, U_n^2} \tag{7.61}$$

$$||\boldsymbol{i}_{uC}|| = \sqrt{3}\sqrt{\sum_{n\in N_C} I_{un}^2} = \sqrt{3}\sqrt{\sum_{n\in N_C} Y_{un}^2\, U_n^2} \tag{7.62}$$

Observe that the scattered, reactive, and unbalanced currents are symmetrical so that their rms value can be calculated for a single line and the three-phase rms value is higher than the line rms value by the root of three. The load-generated current could be asymmetrical, however, so that

$$||\boldsymbol{i}_{\mathrm{G}}|| = \sqrt{\sum_{n\in N_{\mathrm{G}}} ||\boldsymbol{i}_n||^2}. \tag{7.63}$$

This relationship can be visualized with the help of the polygon shown in Fig. 7.23.

According to this decomposition, five distinctively different power phenomena are responsible for the load current three-phase rms value. Therefore, these are the supply Current's Physical Components in circuits with the HGLs. The active, scattered, reactive, unbalanced, and load-generated currents are associated with these phenomena.

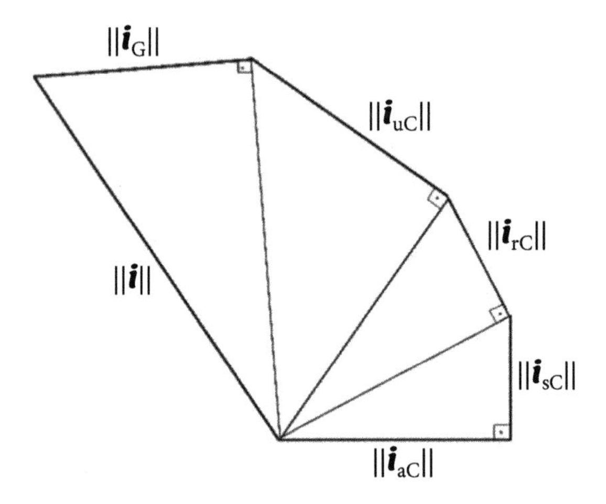

Figure 7.23 A diagram of three-phase rms values of CPC in a three-phase circuit with HGLs.

The phenomena that result in the active, scattered, reactive, and unbalanced currents essentially do not differ from those in systems with LTI loads. Observe, however, that only a part of the supply voltage, $\boldsymbol{u}$, namely, only the voltage $\boldsymbol{u}_{\mathrm{C}}$ affects these currents. These phenomena do not depend on the voltage $\boldsymbol{u}_{\mathrm{G}}$ which occurs as a distribution system response to the load-generated current $\boldsymbol{j}_{\mathrm{G}}$.

Generation of current harmonics in the load and consequently, the presence of the negative harmonic active power P_n, is a new physical phenomenon, as compared to those in circuits with linear time-invariant (LTI) loads, that affects the supply current decomposition. Due to this phenomenon, the energy flows at some harmonic frequencies from the load to the source, meaning in the opposite direction to its transfer at the fundamental frequency.

Since the voltage vectors, $\boldsymbol{u}_{\mathrm{C}}$ and $\boldsymbol{u}_{\mathrm{G}}$, are orthogonal and consequently, they satisfy the relation (7.49), the source apparent power can be expressed as

$$S \stackrel{\mathrm{df}}{=} ||\boldsymbol{u}||\cdot||\boldsymbol{i}|| = \sqrt{||\boldsymbol{u}_{\mathrm{C}}||^2 + ||\boldsymbol{u}_{\mathrm{G}}||^2}\,\sqrt{||\boldsymbol{i}_{\mathrm{C}}||^2 + ||\boldsymbol{i}_{\mathrm{G}}||^2} = \sqrt{S_{\mathrm{C}}^2 + S_{\mathrm{G}}^2 + S_{\mathrm{E}}^2} \tag{7.64}$$

with

$$S_{\mathrm{C}} \stackrel{\mathrm{df}}{=} ||\boldsymbol{u}_{\mathrm{C}}|| \cdot ||\boldsymbol{i}_{\mathrm{C}}|| = \sqrt{P_{\mathrm{C}}^2 + D_{\mathrm{sC}}^2 + Q_{\mathrm{rC}}^2 + D_{\mathrm{uC}}^2} \tag{7.65}$$

and

$$S_{\mathrm{G}} \stackrel{\mathrm{df}}{=} ||\boldsymbol{u}_{\mathrm{G}}|| \cdot ||\boldsymbol{i}_{\mathrm{G}}||, \quad S_{\mathrm{E}} \stackrel{\mathrm{df}}{=} \sqrt{||\boldsymbol{u}_{\mathrm{C}}||^2||\boldsymbol{i}_{\mathrm{G}}||^2 + ||\boldsymbol{u}_{\mathrm{G}}||^2||\boldsymbol{i}_{\mathrm{C}}||^2}. \tag{7.66}$$

The apparent power in circuits with LTI loads $S = S_{\mathrm{C}}$, while in circuits with HGLs, the apparent power contains other new components which contribute to its increase. These are S_{G} and S_{E} powers. The meaning of the apparent power S_{G} does not differ substantially from the meaning of the power S_{C}. It depends not on the load power properties, however, but on features of the distribution system impedance, $\boldsymbol{Z}_{\mathrm{s}n}$, and the load-generated harmonics, $\boldsymbol{j}_{\mathrm{G}}$. There are some physical phenomena in the distribution system, which contribute to this power. These are mainly the energy dissipation on the distribution system resistance, $R_{\mathrm{s}n}$, and the phase shift between the voltage and current harmonics, due to the distribution system reactance, $X_{\mathrm{s}n}$. There could be also some scattered power as well. Three-phase distribution system equipment is built to be as symmetrical as possible so that the unbalanced power on the distribution system is unlikely to contribute to the apparent power S_{G}. Thus, the apparent power S_{G}, referred to as a *load-generated apparent power*, can be decomposed into other powers. This power is much lower than the apparent power S_{C}, however. Consequently, such decomposition, even if possible, does not seem to have technical importance. The apparent power S_{E} differs entirely from powers S_{C} and S_{G}. Its square is composed of products of the rms values of voltages and currents, composed of harmonics of exclusively different orders from subsets N_{C} and N_{G}. There is no physical phenomenon associated with such harmonics of the different order. They only contribute to an increase in the voltage and current rms value. It is referred to as a *cross apparent power*.

7.10.1 Power Factor

The power factor λ of a three-phase unbalanced HGL can be expressed in the form

$$\lambda \stackrel{\mathrm{df}}{=} \frac{P}{S} = \frac{P_{\mathrm{C}} - P_{\mathrm{G}}}{\sqrt{P_{\mathrm{C}}^2 + D_{\mathrm{sC}}^2 + Q_{\mathrm{rC}}^2 + D_{\mathrm{uC}}^2 + S_{\mathrm{G}}^2 + S_{\mathrm{E}}^2}}. \tag{7.67}$$

This formula reveals all power components that contribute to the deterioration of the power factor in three-phase circuits with harmonic-generating loads. Observe, that generation of current harmonics in the load reduces the numerator of the formula (7.62) and increases the denominator, thus it reduces the power factor even more.

7.11 Circuits with Asymmetrical Supply, sv&c, and LTI Loads

The load current decomposition (7.9) into the CPCs was obtained under the assumption that the supply voltage was sinusoidal and symmetrical. This voltage could be asymmetrical, however, and consequently, the current decomposition obtained at this assumption is not valid. Indeed, at the first approximation, the voltage in a three-phase circuit is symmetrical because the advantages of three-phase circuits over one-phase circuits stem mainly from the voltage symmetry. Any deviation from the voltage symmetry is harmful to the power system equipment's performance. Therefore, three-phase power systems are built in such a way that the voltage is symmetrical as much as possible. Despite all efforts aimed at keeping this voltage symmetrical, there is always some level of asymmetry, however.

There are two causes of the distribution voltage asymmetry. These are (i) a difference in the phase impedance of the distribution system and (ii) an asymmetry of line currents. The phase leg of a transformer core has a different geometry than the external legs; the position of line conductors versus the ground, and versus themselves, is different for each conductor. Consequently, due to some difference in phase impedances, even symmetrical currents cause asymmetrical voltage drop and consequently, distribution voltage asymmetry. This asymmetry, caused by the geometry of lines and transformers, is referred to as a *structural or primary asymmetry*. When, due to the load imbalance, the line currents are asymmetrical, the voltage drop is asymmetrical even if the line impedances are mutually equal. This asymmetry of the voltage caused by asymmetrical currents will be referred to in this book as a *secondary asymmetry*. Asymmetrical currents are caused by unbalanced loads and therefore, the load imbalance is the cause of the distribution voltage asymmetry. This asymmetry is a by-product of the load imbalance. It can be observed in the highest degree at junctions of one-phase and three-phase circuits. It declines at buses distant from such junctions, meaning it declines with the bus voltage increase. On the transmission level usually, the primary asymmetry can be observed.

Power phenomena in three-wire circuits were discussed in this chapter under the assumption that the supply voltage was symmetrical. The supply current decomposition into CPCs, as well as power definitions, and the developed power equation are not valid when this voltage is asymmetrical. This equation provides only approximate relations between powers, but the accuracy of this approximation cannot even be assessed. Therefore, one can legitimately ask the question: *how does the voltage asymmetry affect powers and power relations in a three-phase circuit?* Unfortunately, an answer is unavailable, even for the simplest circuits, composed of only LTI loads with a sinusoidal but asymmetrical supply voltage.

Power theory should provide accurate results independently on the level of voltage asymmetry. Therefore, despite the fact that this asymmetry in distribution systems is usually no higher than a few percent, the analysis of power phenomena shall be carried out in this chapter without any limitations on the level of this asymmetry. Just the opposite: it will be assumed that it is much higher than the voltage asymmetry in real circuits. This will enable emphasizing various effects of the voltage asymmetry on the power-related phenomena. At the level of the voltage asymmetry that can occur in real circuits, these effects could not be visible.

The analysis of power phenomena in circuits with asymmetrical voltages is based here on the concept of symmetrical components, developed by Fortescue [8] in 1918, and on the CPC-based power theory of three-phase, three-wire unbalanced circuits as presented above.

Apparent power. There could be some level of discrepancy respective the apparent power S of the supply in the presence of the supply voltage asymmetry. The internal voltage $\boldsymbol{e}$ of the distribution system, as shown in Fig. 7.24, can be composed of all three symmetrical components, meaning

$$\boldsymbol{e} = \boldsymbol{e}^{\mathrm{p}} + \boldsymbol{e}^{\mathrm{n}} + \boldsymbol{e}^{\mathrm{z}}.$$

The apparent power, from the perspective of the supply source, can be defined as

$$S = ||\boldsymbol{e}||\ ||\boldsymbol{i}|| \stackrel{\mathrm{df}}{=} S_{\mathrm{e}}.$$

The zero-sequence symmetrical component $\boldsymbol{e}^{\mathrm{z}}$ of the distribution voltage does not contribute to energy delivery to the load supplied by a three-wire line, however, since the load current $\boldsymbol{i}$ cannot contain the zero-sequence component, $\boldsymbol{i}^{\mathrm{z}}$. The voltage, which affects the energy flow, is composed of only the positive and the negative sequence components of the distribution voltage, and should be measured versus artificial zero, as shown in Fig. 7.24, but not versus the circuit ground. At such measurement,

$$\boldsymbol{u}^{\mathrm{p}} \equiv \boldsymbol{e}^{\mathrm{p}}, \quad \boldsymbol{u}^{\mathrm{n}} \equiv \boldsymbol{e}^{\mathrm{n}}, \quad \boldsymbol{u}^{\mathrm{z}} \equiv 0.$$

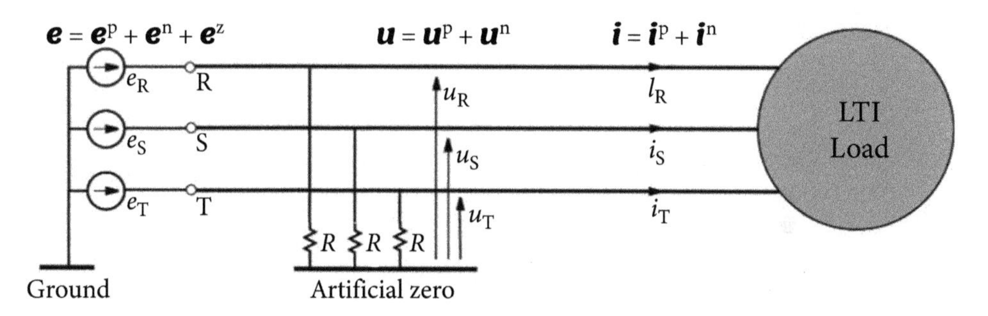

Figure 7.24 A three-phase circuit with asymmetrical distribution voltage.

Therefore, the apparent power of the load should be defined as

$$S \stackrel{\mathrm{df}}{=} ||\boldsymbol{u}||\ ||\boldsymbol{i}|| \leq S_{\mathrm{e}}.$$

If the apparent power is defined as S_{e}, even an ideal, meaning a balanced and resistive load, then in the presence of the zero-sequence component $\boldsymbol{e}^{\mathrm{z}}$, it might have power factor lower than 1.

7.12 Induction Motor Supplied with Asymmetrical Voltage

Induction motors are the most common LTI loads that are inherently balanced. For example, approximately two-thirds of the electric energy produced by power plants in the US power system is used by such motors. When the supply voltage of the motor is asymmetrical, the motor current is also asymmetrical. A schematic circuit of an induction motor supplied with asymmetrical voltage is shown in Fig. 7.25. In a response to the negative sequence symmetrical component in the internal voltage of the supply system $\boldsymbol{e}^{\mathrm{n}}$, the motor current contains a negative sequence symmetrical component, $\boldsymbol{i}^{\mathrm{n}}$. Since rotors of induction motors rotate in the direction of the rotating magnetic field created by the supply voltage component of the positive sequence, the motor equivalent admittance for the positive and the negative sequence component have different values,

$$Y_{\mathrm{e}}^{\mathrm{p}} = G_{\mathrm{e}}^{\mathrm{p}} + jB_{\mathrm{e}}^{\mathrm{p}}, \quad Y_{\mathrm{e}}^{\mathrm{n}} = G_{\mathrm{e}}^{\mathrm{n}} + jB_{\mathrm{e}}^{\mathrm{n}}.$$

The load current is

$$\boldsymbol{i} = \boldsymbol{i}^{\mathrm{p}} + \boldsymbol{i}^{\mathrm{n}} = \sqrt{2}\mathrm{Re}\,\{\mathbf{1}^{\mathrm{p}}\boldsymbol{I}^{\mathrm{p}}\,e^{j\omega_1 t}\} + \sqrt{2}\mathrm{Re}\,\{\mathbf{1}^{\mathrm{n}}\boldsymbol{I}^{\mathrm{n}}\,e^{j\omega_1 t}\}$$

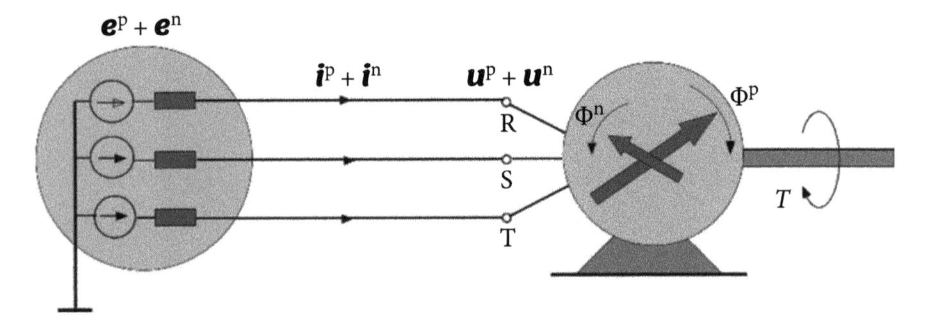

Figure 7.25 An induction motor with asymmetrical supply voltage.

with

$$\boldsymbol{I}^{\mathrm{p}} = Y_{\mathrm{e}}^{\mathrm{p}}\boldsymbol{U}^{\mathrm{p}}, \quad \boldsymbol{I}^{\mathrm{n}} = Y_{\mathrm{e}}^{\mathrm{n}}\boldsymbol{U}^{\mathrm{n}}.$$

The load active power can be expressed, due to orthogonality of the opposite sequence components, in the form

$$P = (\boldsymbol{u}^{\mathrm{p}} + \boldsymbol{u}^{\mathrm{n}}, \boldsymbol{i}^{\mathrm{p}} + \boldsymbol{i}^{\mathrm{n}}) = (\boldsymbol{u}^{\mathrm{p}}, \boldsymbol{i}^{\mathrm{p}}) + (\boldsymbol{u}^{\mathrm{n}}, \boldsymbol{u}^{\mathrm{n}}) = P^{\mathrm{p}} + P^{\mathrm{n}}$$

where

$$P^{\mathrm{p}} = G_{\mathrm{e}}^{\mathrm{p}}||\boldsymbol{u}^{\mathrm{p}}||^2, \quad P^{\mathrm{n}} = G_{\mathrm{e}}^{\mathrm{n}}||\boldsymbol{u}^{\mathrm{n}}||^2. \tag{7.68}$$

This result is rather trivial. It only shows that the load active power is a sum of active powers associated with the positive and negative sequence components of the load voltages and currents. One should notice only that the active power P^{n} is not the useful power. The energy transferred by the negative sequence voltage and current components is entirely dissipated in the motor, thus it only increases its temperature. Therefore, it would not be fair to charge a customer for this energy. Just the opposite: the energy provider should reimburse the customer for the negative effects of the supply voltage asymmetry.

Illustration 7.4 *To give an idea of the level of these two powers, let us observe that a common,* 220 V, 100 kW *induction motor at the rated speed has the phase admittance for the positive and negative sequence voltage equal approximately to*

$$Y_{e}^{p} = 0.6 - j0.4 = 0.7e^{j33^{\circ}}\ \text{S}$$

$$Y_{e}^{n} = 0.4 - j0.13 = 1.4e^{j73^{\circ}}\ \text{S}.$$

*****

Taking into account that $G_{e}^{p} > G_{e}^{n}$, and that the active power is proportional to the square of the supply voltage rms value, thus at the level of the voltage asymmetry that is allowed by standards, the power P^{n} is only a minute part of the motor active power. Observe, however, that most of the power P^{p} is converted to mechanical power on the motor shaft, while only a few percent of this power contributes to the motor heating. Consequently, the share of the positive and negative sequence voltages in the motor heating could be comparable.

Observe that the magnitude of the motor admittance for the negative sequence voltage is higher than this magnitude for the positive sequence. Consequently, the motor current asymmetry has to be higher than the supply voltage asymmetry. Thus, due to the voltage drop on the distribution system's internal impedance, induction motors contribute to an increase in voltage asymmetry in such a circuit.

7.13 Superposition-Based Current Decomposition

Let us investigate power phenomena in the circuit with an unbalanced LTI load shown in Fig. 7.24. Since the circuit considered is linear one might be tempted to use the Superposition Principle for the current decomposition. The current component of the positive sequence, $\boldsymbol{i}^{p}$, is composed of the active and reactive currents, which are load responses to the voltage, $\boldsymbol{u}^{p}$, and unbalanced current, which occurs due to negative sequence voltage, $\boldsymbol{u}^{n}$, namely

$$\boldsymbol{i}^{p} = \boldsymbol{i}_{a}^{p} + \boldsymbol{i}_{r}^{p} + \boldsymbol{i}_{u}^{p}$$

where

$$\begin{aligned}
\boldsymbol{i}_{\mathrm{a}}^{\mathrm{p}} &\stackrel{\mathrm{df}}{=} \sqrt{2}\mathrm{Re}\{\boldsymbol{I}_{\mathrm{a}}^{\mathrm{p}}\, e^{j\omega_1 t}\} = \sqrt{2}\mathrm{Re}\{G_{\mathrm{e}}^{\mathrm{p}}\mathbf{1}^{\mathrm{p}}U^{\mathrm{p}}\, e^{j\omega_1 t}\}\\
\boldsymbol{i}_{\mathrm{r}}^{\mathrm{p}} &\stackrel{\mathrm{df}}{=} \sqrt{2}\mathrm{Re}\{\boldsymbol{I}_{\mathrm{r}}^{\mathrm{p}}\, e^{j\omega_1 t}\} = \sqrt{2}\mathrm{Re}\{jB_{\mathrm{e}}^{\mathrm{p}}\mathbf{1}^{\mathrm{p}}U^{\mathrm{p}}\, e^{j\omega_1 t}\}\\
\boldsymbol{i}_{\mathrm{u}}^{\mathrm{p}} &\stackrel{\mathrm{df}}{=} \sqrt{2}\mathrm{Re}\{\boldsymbol{I}_{\mathrm{u}}^{\mathrm{p}}\, e^{j\omega_1 t}\} = \sqrt{2}\mathrm{Re}\{Y_{\mathrm{u}}^{\mathrm{n}}\mathbf{1}^{\mathrm{p}}U^{\mathrm{n}}\, e^{j\omega_1 t}\}
\end{aligned} \tag{7.69}$$

In the last formula, $Y_{\mathrm{u}}^{\mathrm{n}}$ is the load unbalanced admittance for the negative sequence voltage. When the load is unbalanced, then at the negative sequence voltage, $\boldsymbol{u}^{\mathrm{n}}$, the positive sequence components occur in the supply current.

The current component of the negative sequence, $\boldsymbol{i}^{\mathrm{n}}$, is composed of the active and reactive currents, which are load responses to the voltage, $\boldsymbol{u}^{\mathrm{n}}$, and unbalanced current, which occurs due to positive sequence voltage, $\boldsymbol{u}^{\mathrm{p}}$, namely

$$\boldsymbol{i}^{\mathrm{n}} = \boldsymbol{i}_{\mathrm{a}}^{\mathrm{n}} + \boldsymbol{i}_{\mathrm{r}}^{\mathrm{n}} + \boldsymbol{i}_{\mathrm{u}}^{\mathrm{n}}$$

where

$$\begin{aligned}
\boldsymbol{i}_{\mathrm{a}}^{\mathrm{n}} &\stackrel{\mathrm{df}}{=} \sqrt{2}\mathrm{Re}\{\boldsymbol{I}_{\mathrm{a}}^{\mathrm{n}}\, e^{j\omega_1 t}\} = \sqrt{2}\mathrm{Re}\{G_{\mathrm{e}}^{\mathrm{n}}\mathbf{1}^{\mathrm{n}}U^{\mathrm{n}}\, e^{j\omega_1 t}\}\\
\boldsymbol{i}_{\mathrm{r}}^{\mathrm{n}} &\stackrel{\mathrm{df}}{=} \sqrt{2}\mathrm{Re}\{\boldsymbol{I}_{\mathrm{r}}^{\mathrm{n}}\, e^{j\omega_1 t}\} = \sqrt{2}\mathrm{Re}\{jB_{\mathrm{e}}^{\mathrm{n}}\mathbf{1}^{\mathrm{n}}U^{\mathrm{n}}\, e^{j\omega_1 t}\}\\
\boldsymbol{i}_{\mathrm{u}}^{\mathrm{n}} &\stackrel{\mathrm{df}}{=} \sqrt{2}\mathrm{Re}\{\boldsymbol{I}_{\mathrm{u}}^{\mathrm{n}}\, e^{j\omega_1 t}\} = \sqrt{2}\mathrm{Re}\{Y_{\mathrm{u}}^{\mathrm{p}}\mathbf{1}^{\mathrm{n}}U^{\mathrm{p}}\, e^{j\omega_1 t}\}.
\end{aligned} \tag{7.70}$$

the load unbalanced admittance $Y_{\mathrm{u}}^{\mathrm{p}}$ in the last formula is for the positive sequence voltage, introduced at the beginning of this chapter and denoted without upper index "p". As discussed previously, when the load is unbalanced, then at the positive sequence voltage, $\boldsymbol{u}^{\mathrm{p}}$, the negative sequence components occur in the supply current.

According to the Superposition Principle, the current responses to the positive and the negative sequence voltage components can be added, which results in the load current decomposed as follows

$$\boldsymbol{i} = \boldsymbol{i}_{\mathrm{a}}^{\mathrm{p}} + \boldsymbol{i}_{\mathrm{r}}^{\mathrm{p}} + \boldsymbol{i}_{\mathrm{u}}^{\mathrm{p}} + \boldsymbol{i}_{\mathrm{a}}^{\mathrm{n}} + \boldsymbol{i}_{\mathrm{r}}^{\mathrm{n}} + \boldsymbol{i}_{\mathrm{u}}^{\mathrm{n}}. \tag{7.71}$$

Some of these current components are not mutually orthogonal, however. For example, let us calculate the scalar product

$$\begin{aligned}
(\boldsymbol{i}_{\mathrm{a}}^{\mathrm{p}}, \boldsymbol{i}_{\mathrm{u}}^{\mathrm{p}}) &= \mathrm{Re}\{\boldsymbol{I}_{\mathrm{a}}^{\mathrm{pT}}(\boldsymbol{I}_{\mathrm{u}}^{\mathrm{p}})^*\} = \mathrm{Re}\{(\mathbf{1}^{\mathrm{pT}}G_{\mathrm{e}}^{\mathrm{p}}\, U^{\mathrm{p}})(\mathbf{1}^{\mathrm{p}}Y_{\mathrm{u}}^{\mathrm{n}}U^{\mathrm{n}})^*\} =\\
&= \mathrm{Re}\left\{[1, \alpha^*, \alpha]\begin{bmatrix}1\\ \alpha\\ \alpha^*\end{bmatrix}G_{\mathrm{e}}^{\mathrm{p}}\, U^{\mathrm{p}}Y_{\mathrm{u}}^{\mathrm{n}*}\, U^{\mathrm{n}*}\right\} = 3G_{\mathrm{e}}^{\mathrm{p}}\,\mathrm{Re}\{Y_{\mathrm{u}}^{\mathrm{n}*}U^{\mathrm{p}}U^{\mathrm{n}*}\}.
\end{aligned}$$

The value of this scalar product depends on two mutually independent factors: the load imbalance and the supply voltage asymmetry. Depending on these factors, the scalar product can have any value. Thus, these currents are not orthogonal.

Due to the lack of mutual orthogonality of the current components in decomposition (7.71), even if the rms value of all these components is known, then the rms value of the load current cannot be calculated, since

$$||\boldsymbol{i}_{\mathrm{a}}^{\mathrm{p}}||^2 + ||\boldsymbol{i}_{\mathrm{r}}^{\mathrm{p}}||^2 + ||\boldsymbol{i}_{\mathrm{u}}^{\mathrm{p}}||^2 + ||\boldsymbol{i}_{\mathrm{a}}^{\mathrm{n}}||^2 + ||\boldsymbol{i}_{\mathrm{r}}^{\mathrm{n}}||^2 + ||\boldsymbol{i}_{\mathrm{u}}^{\mathrm{n}}||^2 \neq ||\boldsymbol{i}||^2.$$

Thus, the power equation of unbalanced loads at asymmetrical supply voltage cannot be developed based on decomposition (7.71).

Illustration 7.5 *The LTI unbalanced load shown in Fig. 7.26 is supplied by a strongly asymmetrical voltage. Decompose the load current into active, reactive, and unbalanced currents as defined by formulae (7.69) and (7.70).*

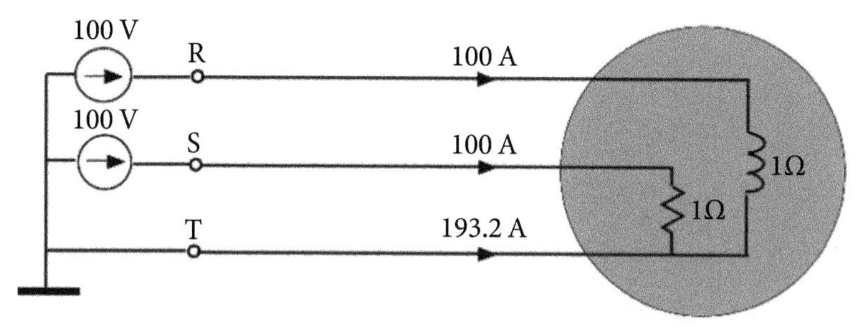

Figure 7.26 An example of an unbalanced load supplied with asymmetrical voltage.

Assuming that the crms value of the voltage at terminal R is $\boldsymbol{U}_{\mathrm{R}} = U_{\mathrm{R}} = 100$ V, *the crms values of the positive and negative sequence components* $\boldsymbol{U}^{\mathrm{p}}$ *and* $\boldsymbol{U}^{\mathrm{n}}$ *of the supply voltage are equal to*

$$\begin{bmatrix} \boldsymbol{U}^{\mathrm{p}} \\ \boldsymbol{U}^{\mathrm{n}} \end{bmatrix} = \frac{1}{3}\begin{bmatrix} 1, \ \alpha, \ \alpha^* \\ 1, \ \alpha^*, \ \alpha \end{bmatrix}\begin{bmatrix} \boldsymbol{U}_{\mathrm{R}} \\ \boldsymbol{U}_{\mathrm{S}} \\ \boldsymbol{U}_{\mathrm{T}} \end{bmatrix} = \frac{1}{3}\begin{bmatrix} 1, \ \alpha, \ \alpha^* \\ 1, \ \alpha^*, \ \alpha \end{bmatrix}\begin{bmatrix} 100 \\ 100e^{-j120^\circ} \\ 0 \end{bmatrix}$$

$$= \begin{bmatrix} 66.66 \\ 33.33e^{j60^\circ} \end{bmatrix} \ \mathrm{V}.$$

Hence, three-phase rms values of these sequence components are equal to

$$||\boldsymbol{u}^{\mathrm{p}}|| = \sqrt{3}\,U^{\mathrm{p}} = \sqrt{3} \times 66.66 = 115.47\,\mathrm{V}, \quad ||\boldsymbol{u}^{\mathrm{n}}|| = \sqrt{3}\,U^{\mathrm{n}} = \sqrt{3} \times 33.33 = 57.73\,\mathrm{V}$$

and consequently, the three-phase rms value of the load voltage is

$$||\boldsymbol{u}|| = \sqrt{||\boldsymbol{u}^{\mathrm{p}}||^2 + ||\boldsymbol{u}^{\mathrm{n}}||^2} = \sqrt{115.47^2 + 57.73^2} = 129.1\,\mathrm{V}.$$

Since the load is static, thus

$$\boldsymbol{Y}_{\mathrm{e}}^{\mathrm{p}} = \boldsymbol{Y}_{\mathrm{e}}^{\mathrm{n}} = \boldsymbol{Y}_{\mathrm{ST}} + \boldsymbol{Y}_{\mathrm{TR}} = 1 - j1\ \mathrm{S}$$

hence

$$G_e^p = G_e^n = 1\,\text{S}, \quad B_e^p = B_e^n = -1\,\text{S}.$$

The unbalanced admittance for the positive sequence voltage is

$$Y_u^p = -(Y_{ST} + \alpha\, Y_{TR} + \alpha^* Y_{RS}) = -[1 + \alpha(-j1)] = 1.932 e^{-j165°}\ \text{S}$$

while this admittance for the negative sequence voltage is

$$Y_u^n = -(Y_{ST} + \alpha^* Y_{TR} + \alpha Y_{RS}) = -[1 + \alpha^*(-j1)] = 0.518 e^{-j105°}\ \text{S}.$$

The three-phase rms values of the current components are equal to

$$\begin{aligned}
||\boldsymbol{i}_a^p|| &= G_e^p||\boldsymbol{u}^p|| = 1 \times 115.47 = 115.47\,\text{A}\\
||\boldsymbol{i}_a^n|| &= G_e^n||\boldsymbol{u}^n|| = 1 \times 57.73 = 57.73\,\text{A}\\
||\boldsymbol{i}_r^p|| &= |B_e^p|\,||\boldsymbol{u}^p|| = 1 \times 115.47 = 115.47\,\text{A}\\
||\boldsymbol{i}_r^n|| &= |B_e^n|\,||\boldsymbol{u}^n|| = 1 \times 57.73 = 57.73\,\text{A}\\
||\boldsymbol{i}_u^p|| &= Y_u^n||\boldsymbol{u}^n|| = 0.518 \times 57.73 = 29.90\,\text{A}\\
||\boldsymbol{i}_u^n|| &= Y_u^n||\boldsymbol{u}^p|| = 1.932 \times 115.47 = 223.09\,\text{A}.
\end{aligned}$$

The three-phase rms value of the load current is

$$||\boldsymbol{i}|| = \sqrt{I_R^2 + I_S^2 + I_T^2} = \sqrt{100^2 + 100^2 + 193.2^2} = 239.43\,\text{A}$$

while

$$\sqrt{||\boldsymbol{i}_a^p||^2 + ||\boldsymbol{i}_r^p||^2 + ||\boldsymbol{i}_u^p||^2 + ||\boldsymbol{i}_a^n||^2 + ||\boldsymbol{i}_r^n||^2 + ||\boldsymbol{i}_u^n||^2} = 289.82\,\text{A} \neq ||\boldsymbol{i}||^2.$$

*****

The lack of orthogonality of the current components obtained using the Superposition Principle raises the question of whether these components can be regarded as the Current's Physical Components.

The association of physical phenomena with distinctive current components is the main core of the CPC concept. Thus, let us check if there is such a distinctive association in the case of current decomposition (7.71) developed using the Superposition Principle.

The scalar product of quantities of the opposite sequence is zero. Moreover,it is equal to zero also the scalar product of voltages and reactive currents. Therefore, the active power at the load terminals can be expressed as

$$P = (\boldsymbol{u},\ \boldsymbol{i}) = (\boldsymbol{u}^{\mathrm{p}} + \boldsymbol{u}^{\mathrm{n}},\ \boldsymbol{i}_{\mathrm{a}}^{\mathrm{p}} + \boldsymbol{i}_{\mathrm{r}}^{\mathrm{p}} + \boldsymbol{i}_{\mathrm{u}}^{\mathrm{p}} + \boldsymbol{i}_{\mathrm{a}}^{\mathrm{n}} + \boldsymbol{i}_{\mathrm{r}}^{\mathrm{n}} + \boldsymbol{i}_{\mathrm{u}}^{\mathrm{n}}) =$$
$$= (\boldsymbol{u}^{\mathrm{p}},\ \boldsymbol{i}_{\mathrm{a}}^{\mathrm{p}}) + (\boldsymbol{u}^{\mathrm{n}},\ \boldsymbol{i}_{\mathrm{a}}^{\mathrm{n}}) + (\boldsymbol{u}^{\mathrm{p}},\ \boldsymbol{i}_{\mathrm{u}}^{\mathrm{p}}) + (\boldsymbol{u}^{\mathrm{n}},\ \boldsymbol{i}_{\mathrm{u}}^{\mathrm{n}}).$$

The scalar product of three-phase symmetrical quantities of the opposite sequence is equal to zero, thus they are orthogonal. The scalar product of quantities of the same sequence in the last formula is equal to

$$P^{\mathrm{p}} \stackrel{\mathrm{df}}{=} (\boldsymbol{u}^{\mathrm{p}},\ \boldsymbol{i}_{\mathrm{a}}^{\mathrm{p}}) = G_{\mathrm{e}}^{\mathrm{p}}||\boldsymbol{u}^{\mathrm{p}}||^2$$

$$P^{\mathrm{n}} \stackrel{\mathrm{df}}{=} (\boldsymbol{u}^{\mathrm{n}},\ \boldsymbol{i}_{\mathrm{a}}^{\mathrm{n}}) = G_{\mathrm{e}}^{\mathrm{n}}||\boldsymbol{u}^{\mathrm{n}}||^2$$

and

$$P_{\mathrm{u}}^{\mathrm{p}} = (\boldsymbol{u}^{\mathrm{p}}, \boldsymbol{i}_{\mathrm{u}}^{\mathrm{p}}) = \mathrm{Re}\,\{\boldsymbol{U}^{\mathrm{pT}}(\boldsymbol{I}_{\mathrm{u}}^{\mathrm{p}})^*\} = \mathrm{Re}\,\{(\mathbf{1}^{\mathrm{pT}}U^{\mathrm{p}})\,(\mathbf{1}^{\mathrm{p}}Y_{\mathrm{u}}^{\mathrm{n}}U^{\mathrm{n}})^*\} =$$
$$= \mathrm{Re}\,\{[1,\ \alpha^*,\ \alpha]\begin{bmatrix} 1 \\ \alpha \\ \alpha^* \end{bmatrix} U^{\mathrm{p}}Y_{\mathrm{u}}^{\mathrm{n}*}U^{\mathrm{n}*}\} = 3\mathrm{Re}\,\{Y_{\mathrm{u}}^{\mathrm{n}*}U^{\mathrm{p}}U^{\mathrm{n}*}\}.$$

This scalar product is in general not equal to zero, meaning the unbalanced current of the positive sequence contributes to a permanent flow of energy between the supply source and the load. Therefore this current cannot be regarded as the load current physical component in the sense of the CPC-based power theory. The same stands for the last scalar product:

$$P_{\mathrm{u}}^{\mathrm{n}} = (\boldsymbol{u}^{\mathrm{n}},\ \boldsymbol{i}_{\mathrm{u}}^{\mathrm{n}}) = \mathrm{Re}\,\{\boldsymbol{U}^{\mathrm{nT}}(\boldsymbol{I}_{\mathrm{u}}^{\mathrm{n}})^*\} = \mathrm{Re}\,\{(\mathbf{1}^{\mathrm{nT}}U^{\mathrm{n}})\,(\mathbf{1}^{\mathrm{n}}Y_{\mathrm{u}}^{\mathrm{p}}U^{\mathrm{p}})^*\} =$$
$$= \mathrm{Re}\,\{[1,\ \alpha,\ \alpha^*]\begin{bmatrix} 1 \\ \alpha^* \\ \alpha \end{bmatrix} U^{\mathrm{n}}Y_{\mathrm{u}}^{\mathrm{p}*}U^{\mathrm{p}*}\} = 3\mathrm{Re}\,\{Y_{\mathrm{u}}^{\mathrm{p}*}U^{\mathrm{n}}U^{\mathrm{p}*}\}.$$

Consequently, the active power of the load is

$$P = G_{\mathrm{e}}^{\mathrm{p}}||\boldsymbol{u}^{\mathrm{p}}||^2 + G_{\mathrm{e}}^{\mathrm{n}}||\boldsymbol{u}^{\mathrm{n}}||^2 + 3\mathrm{Re}\{Y_{\mathrm{u}}^{\mathrm{n}*}U^{\mathrm{p}}U^{\mathrm{n}*}\} + 3\mathrm{Re}\{Y_{\mathrm{u}}^{\mathrm{p}*}U^{\mathrm{n}}U^{\mathrm{p}*}\}.$$

Illustration 7.6 *The load shown in Fig. 7.26 has the active power* P = 10.0 kW. *At the current decomposition specified by formula (7.71), the active power has the components:*

$$P^{\rm p} = G_{\rm e}^{\rm p}||\boldsymbol{u}^{\rm p}||^2 = 1 \times 115.47^2 = 13.33\ {\rm kW}$$

$$P^{\rm n} = G_{\rm e}^{\rm n}||\boldsymbol{u}^{\rm n}||^2 = 1 \times 57.73^2 = 3.33\ {\rm kW}$$

$$P_{\rm u}^{\rm p} = 3{\rm Re}\{Y_{\rm u}^{\rm n*}U^{\rm p}U^{\rm n*}\} = 3{\rm Re}\,\{0.518e^{j105^\circ} \times 66.66 \times 33.33e^{-j60^\circ}\} = 2.44\ {\rm kW}$$

$$P_{\rm u}^{\rm n} = 3{\rm Re}\{Y_{\rm u}^{\rm n*}U^{\rm n}U^{\rm p*}\} = 3{\rm Re}\{1.932e^{j165^\circ} \times 33.33e^{j60^\circ} \times 66.66\} = -9.10\ {\rm kW}$$

and indeed,

$$P = P^{\rm p} + P^{\rm n} + P_{\rm u}^{\rm p} + P_{\rm u}^{\rm n} = 13.33 + 3.33 + 2.44 - 9.10 = 10.0\ {\rm kW}.$$

*****

To summarize the considerations presented above, the Superposition Principle provides a straightforward and intuitively obvious decomposition of the supply current of LTI loads supplied with asymmetrical voltage. Unfortunately, the current components obtained are not mutually orthogonal and cannot be regarded as the CPC. The power equation of LTI cannot be developed based on such decomposition of the load current.

7.14 CPC at Asymmetrical Supply with sv&c and LTI Load

The complex power of the considered unbalanced load supplied with asymmetrical voltage in the 3p3w circuit, as shown in Fig. 7.27 is equal to

$$C = \boldsymbol{U}^{\rm T}\boldsymbol{I}^* = U_{\rm R}I_{\rm R}^* + U_{\rm S}I_{\rm S}^* + U_{\rm T}I_{\rm T}^* = U_{\rm R}(I_{\rm RS}^* - I_{\rm TR}^*) + U_{\rm S}(I_{\rm ST}^* - I_{\rm RS}^*) + U_{\rm T}(I_{\rm TR}^* - I_{\rm ST}^*) =$$
$$= U_{\rm RS}I_{\rm RS}^* + U_{\rm ST}I_{\rm ST}^* + U_{\rm TR}I_{\rm TR}^* = P + jQ.$$

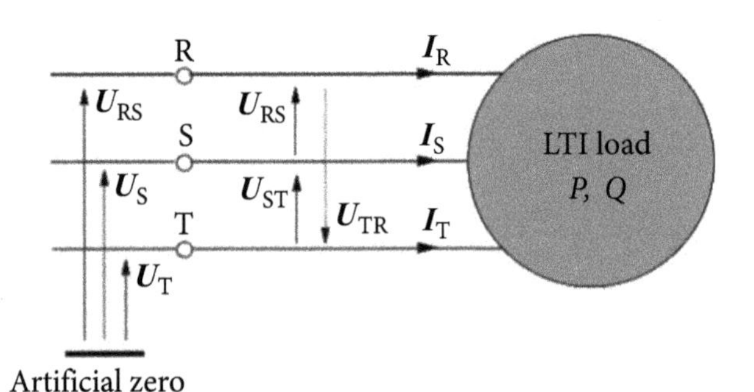

Figure 7.27 A three-phase load supplied from a three-wire line.

Concerning active and reactive powers P and Q at the supply voltage $\boldsymbol{u}$, the unbalanced load, shown in Fig. 7.27, is equivalent to a balanced load shown in Fig. 7.28, on the condition that its phase admittances are equal to

$$Y_b = G_b + jB_b = \frac{P - jQ}{||\boldsymbol{u}||^2} = \frac{C^*}{||\boldsymbol{u}||^2}. \tag{7.72}$$

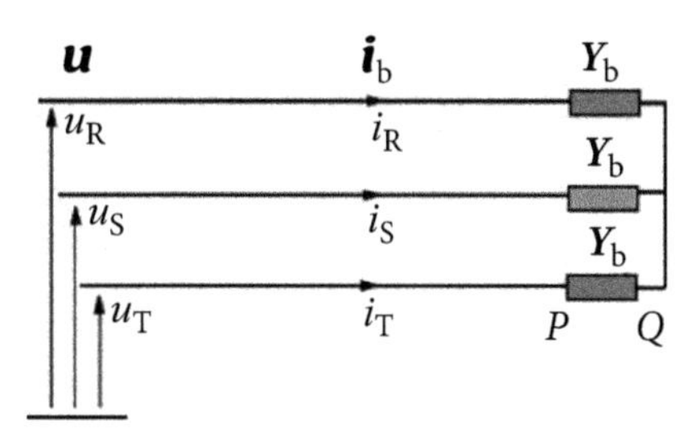

Figure. 7.28 A balanced load that is equivalent to the original load with regard to P and Q powers.

Indeed, the complex power C_b of such a balanced load is

$$C_b = \boldsymbol{U}^T \boldsymbol{I}_b^* = \boldsymbol{U}^T (Y_b \boldsymbol{U})^* = Y_b^* ||\boldsymbol{u}||^2 = P + jQ = C.$$

The supply voltage $\boldsymbol{u}$ is sinusoidal but can be asymmetrical. Thus, it can be decomposed into a sum of symmetrical voltages of the positive $\boldsymbol{u}^p$ and negative $\boldsymbol{u}^n$ sequence, so that

$$\boldsymbol{u} = \boldsymbol{u}^p + \boldsymbol{u}^n = \sqrt{2}\mathrm{Re}\,\{(\mathbf{1}^p U^p + \mathbf{1}^n U^n)\, e^{j\omega_1 t}\}.$$

The zero-sequence symmetrical component $\boldsymbol{u}^z$ cannot cause any current flow in three-wire circuits. Thus it can be neglected, or line voltages should be measured versus an artificial zero of the circuit, so the voltage vector $\boldsymbol{u}$ does not contain any zero-sequence component.

Since Y_b is the admittance of a balanced load, which is equivalent to the original load with regard to the active and reactive powers, it will be referred to as the *equivalent balanced admittance*. Such an equivalent balanced load draws the current

$$\boldsymbol{i}_b = \boldsymbol{i}_a + \boldsymbol{i}_r = \sqrt{2}\mathrm{Re}\,\{\boldsymbol{I}_b\, e^{j\omega_1 t}\} = \sqrt{2}\mathrm{Re}\,\{Y_b\, \boldsymbol{U}\, e^{j\omega_1 t}\} \tag{7.73}$$

composed of the active current

$$\boldsymbol{i}_a = G_b\, \boldsymbol{u} = \sqrt{2}\mathrm{Re}\,\{G_b\, \boldsymbol{U} e^{j\omega_1 t}\} = \sqrt{2}\mathrm{Re}\,\{G_b(\mathbf{1}^p\, U^p + \mathbf{1}^n\, U^n)^{j\omega_1 t}\} \tag{7.74}$$

and the reactive current

$$\boldsymbol{i}_r = \sqrt{2}\mathrm{Re}\,\{jB_b\, \boldsymbol{U} e^{j\omega_1 t}\} = \sqrt{2}\mathrm{Re}\,\{jB_b\, (\mathbf{1}^p\, U^p + \mathbf{1}^n\, U^n) e^{j\omega_1 t}\}. \tag{7.75}$$

The remaining current of the load, after the current of the balanced load is subtracted, is caused by the load imbalance

$$\boldsymbol{i} - \boldsymbol{i}_{\rm b} = \sqrt{2}\,{\rm Re}\,\{(\boldsymbol{I} - \boldsymbol{I}_{\rm b})\, e^{j\omega_1 t}\} \stackrel{\rm df}{=} \boldsymbol{i}_{\rm u} = \sqrt{2}\,{\rm Re}\,\{\boldsymbol{I}_{\rm u}\, e^{j\omega_1 t}\}. \tag{7.76}$$

Consequently, the load current is decomposed into the active, reactive, and unbalanced current components, such that

$$\boldsymbol{i}\,(t) = \boldsymbol{i}_{\rm a}\,(t) + \boldsymbol{i}_{\rm r}\,(t) + \boldsymbol{i}_{\rm u}\,(t)\,. \tag{7.77}$$

Mutual orthogonality of the active and reactive currents results from their mutual phase shift by $\pi/2$. The orthogonality of the balanced and unbalanced current has to be proven. Indeed,

$$\begin{aligned}(\boldsymbol{i}_{\rm b},\, \boldsymbol{i}_{\rm u}) &= {\rm Re}\,\{\boldsymbol{I}_{\rm b}^{\rm T}(\boldsymbol{I} - \boldsymbol{I}_{\rm b})^*\} = {\rm Re}\,\{Y_{\rm b}\, \boldsymbol{U}^{\rm T}\boldsymbol{I}^* - Y_{\rm b}\, \boldsymbol{U}^{\rm T} Y_{\rm b}^*\, \boldsymbol{U}^*\} = \\ &= {\rm Re}\,\{Y_{\rm b}\, (\boldsymbol{U}^{\rm T}\boldsymbol{I}^* - \boldsymbol{U}^{\rm T} Y_{\rm b}^*\, \boldsymbol{U}^*)\} = {\rm Re}\,\{Y_{\rm b}(C - C_{\rm b})\} = 0\end{aligned}$$

thus these currents are mutually orthogonal and consequently

$$||\boldsymbol{i}||^2 = ||\boldsymbol{i}_{\rm a}||^2 + ||\boldsymbol{i}_{\rm r}||^2 + ||\boldsymbol{i}_{\rm u}||^2. \tag{7.78}$$

Each current component in decomposition (7.72) is distinctively associated with a unique physical phenomenon in the circuit, thus they can be regarded as Current's Physical Components.

This decomposition can be performed, and the three-phase rms values of each particular current can be measured or calculated by measurements of active and reactive powers P and Q at the load terminals as well as crms values $\boldsymbol{I}_{\rm R}$, $\boldsymbol{I}_{\rm S}$, and $\boldsymbol{I}_{\rm T}$ of the load currents.

Multiplying Eqn. (7.78) by the square of the three-phase rms value $||\boldsymbol{u}||$ of the supply voltage

$$\{||\boldsymbol{i}||^2 = ||\boldsymbol{i}_{\rm a}||^2 + ||\boldsymbol{i}_{\rm r}||^2 + ||\boldsymbol{i}_{\rm u}||^2\} \times ||\boldsymbol{u}||^2$$

the power equation

$$S^2 = P^2 + Q^2 + D_{\rm u}^2 \tag{7.79}$$

is obtained, with the unbalanced power defined as

$$D_{\rm u} \stackrel{\rm df}{=} ||\boldsymbol{u}||\ ||\boldsymbol{i}_{\rm u}||. \tag{7.80}$$

This power equation is identical to Eqn. (7.17) but developed without the assumption that the supply voltage is symmetrical. Thus, the supply voltage asymmetry does not affect the general form of the power equation of stationary LTI loads with sinusoidal supply voltages.

Power Eqn (7.79) and the values of the active, reactive, and unbalanced powers provide distinctive information on how a permanent flow of energy to the load, the phase shift between the supply voltage and the load current, as well as the load current asymmetry affect the apparent power S.

The unbalanced current $\boldsymbol{i}_{\mathrm{u}}$ in formula (7.76) is not expressed in terms of the load parameters, however. It only fills a gap between the load current and its active and reactive components. The same applies to the unbalanced power D_{u}. Definition (7.80) has no analogy to definition (7.20). It is possible to calculate its value but it cannot be used in a design process of a reactance compensator that would compensate that power. The dependence of the unbalanced power on the circuit parameters is needed for that.

Therefore, let us find how the unbalanced current and power depend on the circuit parameters.

The active and reactive currents in circuits with symmetrical supply voltage are symmetrical currents, such that

$$\boldsymbol{i}_{\mathrm{a}} + \boldsymbol{i}_{\mathrm{r}} = \sqrt{2}\mathrm{Re}\{(G_{\mathrm{e}} + jB_{\mathrm{e}})\boldsymbol{U}e^{j\omega_1 t}\} = \sqrt{2}\mathrm{Re}\{\boldsymbol{Y}_{\mathrm{e}}\,\boldsymbol{U}\,e^{j\omega_1 t}\}$$

where according to formula (7.7)

$$\boldsymbol{Y}_{\mathrm{e}} + G_{\mathrm{e}} + jB_{\mathrm{e}} = \boldsymbol{Y}_{\mathrm{ST}} + \boldsymbol{Y}_{\mathrm{TR}} + \boldsymbol{Y}_{\mathrm{RS}}$$

is the equivalent admittance of the load. It is a phase admittance of a balanced load, which is equivalent to the original one concerning the active and reactive powers, P and Q.

When the supply voltage is asymmetrical then, according to the formula (7.74) and (7.75), the active and reactive currents follow the voltage asymmetry. The phase admittance $\boldsymbol{Y}_{\mathrm{b}}$ of the equivalent balanced load is different than the equivalent admittance $\boldsymbol{Y}_{\mathrm{e}}$, because

$$\boldsymbol{Y}_{\mathrm{b}} = G_{\mathrm{b}} + jB_{\mathrm{b}} = \frac{P - jQ}{||\boldsymbol{u}||^2} = \frac{\boldsymbol{Y}_{\mathrm{RS}}\,U_{\mathrm{RS}}^2 + \boldsymbol{Y}_{\mathrm{ST}}\,U_{\mathrm{ST}}^2 + \boldsymbol{Y}_{\mathrm{TR}}\,U_{\mathrm{TR}}^2}{||\boldsymbol{u}||^2}. \tag{7.81}$$

At symmetrical supply voltage $U_{\mathrm{RS}} = U_{\mathrm{ST}} = U_{\mathrm{TR}} = ||\boldsymbol{u}||$, so that $\boldsymbol{Y}_{\mathrm{b}} = \boldsymbol{Y}_{\mathrm{e}}$. If the voltage is asymmetrical then admittances $\boldsymbol{Y}_{\mathrm{e}}$ and $\boldsymbol{Y}_{\mathrm{b}}$ differ mutually by admittance $\boldsymbol{Y}_{\mathrm{d}}$,

$$\boldsymbol{Y}_{\mathrm{d}} = G_{\mathrm{d}} + jB_{\mathrm{d}} \stackrel{\mathrm{df}}{=} \boldsymbol{Y}_{\mathrm{e}} - \boldsymbol{Y}_{\mathrm{b}} \tag{7.82}$$

dependent on the voltage asymmetry. This asymmetry can be specified quantitatively by a *complex coefficient of the supply voltage asymmetry*, defined as

$$\frac{\boldsymbol{U}^{\mathrm{n}}}{\boldsymbol{U}^{\mathrm{p}}} = \frac{U^{\mathrm{n}}\,e^{j\varphi}}{U^{\mathrm{p}}\,e^{j\phi}} = \frac{U^{\mathrm{n}}}{U^{\mathrm{p}}}e^{j(\varphi-\phi)} \stackrel{\mathrm{df}}{=} \boldsymbol{a} = a\,e^{j\psi}. \tag{7.83}$$

The admittance of the balanced load Y_b can be expressed as

$$Y_b = \frac{C^*}{||\boldsymbol{u}||^2} = \frac{C^*_{RS} + C^*_{ST} + C^*_{TR}}{||\boldsymbol{u}||^2}$$

where the complex power for individual branches can be expressed as follows:

$$C_{RS} = U_{RS} I^*_{RS} = U_{RS} Y^*_{RS} U^*_{RS} = Y^*_{RS}(U^2_R + U^2_S - 2\mathrm{Re}\{U_R U^*_S\}).$$

Since

$$U^2_T = (-U_R - U_S)(-U_R - U_S)^* = U^2_R + U^2_S + 2\mathrm{Re}\{U_R U^*_S\},$$

thus

$$C_{RS} = Y^*_{RS}(2U^2_R + 2U^2_S - U^2_T) = Y^*_{RS}(2||\boldsymbol{u}||^2 - 3U^2_T).$$

Similarly,

$$C_{ST} = U_{ST} Y^*_{ST} U^*_{ST} = Y^*_{ST}(2||\boldsymbol{u}||^2 - 3U^2_R)$$

$$C_{TR} = U_{TR} Y^*_{TR} U^*_{TR} = Y^*_{TR}(2||\boldsymbol{u}||^2 - 3U^2_S).$$

The equivalent balanced admittance of the load Y_b, defined by formula (7.81), can be therefore expressed in the form

$$Y_b = \frac{C^*_{RS} + C^*_{ST} + C^*_{TR}}{||\boldsymbol{u}||^2} = 2Y_e - \frac{3}{||\boldsymbol{u}||^2}(Y_{ST} U^2_R + Y_{TR} U^2_S + Y_{RS} U^2_T) \stackrel{\text{df}}{=} Y_e - Y_d \quad (7.84)$$

which is the equivalent admittance of the load when it is supplied with a symmetrical voltage, and

$$Y_d \stackrel{\text{df}}{=} \frac{3}{||\boldsymbol{u}||^2}(Y_{ST}\, U^2_R + Y_{TR}\, U^2_S + Y_{RS}\, U^2_T) - Y_e. \quad (7.85)$$

Let us express this admittance in terms of crms values of symmetrical components of the positive sequence U^p and the negative sequence U^n. Since

$$U_R = U^p + U^n,\ U_S = \alpha^* U^p + \alpha U^n,\ U_T = \alpha U^p + \alpha^* U^n$$

then

$$U^2_R = U^{p2} + U^{n2} + 2\mathrm{Re}\{U^{p*} U^n\}$$

$$U^2_S = U^{p2} + U^{n2} + 2\mathrm{Re}\{\alpha^* U^{p*} U^n\}$$

$$U^2_T = U^{p2} + U^{n2} + 2\mathrm{Re}\{\alpha U^{p*} U^n\}.$$

The crms values $\boldsymbol{U}^{\mathrm{p}}$ and $\boldsymbol{U}^{\mathrm{n}}$ have the form

$$\boldsymbol{U}^{\mathrm{p}} = U^{\mathrm{p}}\, e^{j\phi}, \qquad \boldsymbol{U}^{\mathrm{n}} = U^{\mathrm{n}}\, e^{j\varphi}$$

therefore, if we denote

$$\boldsymbol{U}^{\mathrm{p}*}\boldsymbol{U}^{\mathrm{n}} = U^{\mathrm{p}}U^{\mathrm{n}}e^{j(\varphi-\phi)} \stackrel{\mathrm{df}}{=} \boldsymbol{W} = We^{j\psi}$$

the admittance $\boldsymbol{Y}_{\mathrm{d}}$, given by formula (7.85), can be expressed as

$$Y_{\mathrm{d}} = 2\frac{Y_{\mathrm{ST}}\mathrm{Re}\,\{\boldsymbol{W}\} + Y_{\mathrm{TR}}\mathrm{Re}\,\{\alpha^{*}\boldsymbol{W}\} + Y_{\mathrm{RS}}\mathrm{Re}\,\{\alpha\boldsymbol{W}\}}{U^{\mathrm{p}2} + U^{\mathrm{n}2}}. \tag{7.86}$$

When the supply voltage asymmetry is specified by complex asymmetry coefficient $\boldsymbol{a}$, then

$$\frac{\mathrm{Re}\,\{\boldsymbol{U}^{\mathrm{p}*}\boldsymbol{U}^{\mathrm{n}}\}}{U^{\mathrm{p}2} + U^{\mathrm{n}2}} = \frac{U^{\mathrm{p}}U^{\mathrm{n}}}{U^{\mathrm{p}2} + U^{\mathrm{n}2}}\mathrm{Re}\{e^{j(\varphi-\phi)}\} = \frac{a}{1 + a^{2}}\cos\psi$$

and consequently, the asymmetry-dependent unbalanced admittance $\boldsymbol{Y}_{\mathrm{d}}$ can be rearranged to the form

$$Y_{\mathrm{d}} = \frac{2a}{1 + a^{2}}[Y_{\mathrm{ST}}\cos\psi + Y_{\mathrm{TR}}\cos(\psi - \frac{2\pi}{3}) + Y_{\mathrm{RS}}\cos(\psi + \frac{2\pi}{3})]. \tag{7.87}$$

Admittance $\boldsymbol{Y}_{\mathrm{d}}$ depends not only on the line-to-line admittances of the load but also on the supply voltage asymmetry. When the load is balanced, that is, $\boldsymbol{Y}_{\mathrm{RS}} = \boldsymbol{Y}_{\mathrm{TR}} = \boldsymbol{Y}_{\mathrm{ST}} = \boldsymbol{Y}_{\mathrm{e}}/3$, then $\boldsymbol{Y}_{\mathrm{d}} = 0$, independently of the supply voltage asymmetry. When the supply voltage is symmetrical and consequently, asymmetry coefficient $a = 0$, then $\boldsymbol{Y}_{\mathrm{d}} = 0$, independently of the load imbalance. Therefore, admittance $\boldsymbol{Y}_{\mathrm{d}}$ is referred to as a *voltage asymmetry-dependent unbalanced admittance* in this chapter. Admittance $\boldsymbol{Y}_{\mathrm{d}}$ can have a non-zero value only if the load is unbalanced and the supply voltage is asymmetrical.

The crms value of line R current is equal to

$$\boldsymbol{I}_{\mathrm{R}} = \boldsymbol{Y}_{\mathrm{RS}}(\boldsymbol{U}_{\mathrm{R}} - \boldsymbol{U}_{\mathrm{S}}) - \boldsymbol{Y}_{\mathrm{TR}}(\boldsymbol{U}_{\mathrm{T}} - \boldsymbol{U}_{\mathrm{R}})$$

and can be rearranged to the form

$$\boldsymbol{I}_{\mathrm{R}} = \boldsymbol{Y}_{\mathrm{e}}\boldsymbol{U}_{\mathrm{R}} - (\boldsymbol{Y}_{\mathrm{ST}}\boldsymbol{U}_{\mathrm{R}} + \boldsymbol{Y}_{\mathrm{TR}}\boldsymbol{U}_{\mathrm{T}} + \boldsymbol{Y}_{\mathrm{RS}}\boldsymbol{U}_{\mathrm{S}})$$

and next expressed in terms of the crms values of the voltage symmetrical components

$$\boldsymbol{I}_{\mathrm{R}} = \boldsymbol{Y}_{\mathrm{e}}(\boldsymbol{U}_{\mathrm{R}}^{\mathrm{p}} + \boldsymbol{U}_{\mathrm{R}}^{\mathrm{n}}) - [\boldsymbol{Y}_{\mathrm{ST}}(\boldsymbol{U}_{\mathrm{R}}^{\mathrm{p}} + \boldsymbol{U}_{\mathrm{R}}^{\mathrm{n}}) + \boldsymbol{Y}_{\mathrm{TR}}(\boldsymbol{U}_{\mathrm{T}}^{\mathrm{p}} + \boldsymbol{U}_{\mathrm{T}}^{\mathrm{n}}) + \boldsymbol{Y}_{\mathrm{RS}}(\boldsymbol{U}_{\mathrm{S}}^{\mathrm{p}} + \boldsymbol{U}_{\mathrm{S}}^{\mathrm{p}})].$$

This formula can be rearranged to the form

$$\boldsymbol{I}_{\mathrm{R}} = Y_{\mathrm{e}}\,\boldsymbol{U}_{\mathrm{R}} + \boldsymbol{Y}_{\mathrm{u}}^{\mathrm{p}}\,\boldsymbol{U}_{\mathrm{R}}^{\mathrm{p}} + \boldsymbol{Y}_{\mathrm{u}}^{\mathrm{n}}\,\boldsymbol{U}_{\mathrm{R}}^{\mathrm{n}}$$

where

$$\boldsymbol{Y}_{\mathrm{u}}^{\mathrm{p}} \stackrel{\mathrm{df}}{=} -(\boldsymbol{Y}_{\mathrm{ST}} + \alpha\boldsymbol{Y}_{\mathrm{TR}} + \alpha^{*}\boldsymbol{Y}_{\mathrm{RS}}) \tag{7.88}$$

$$\boldsymbol{Y}_{\mathrm{u}}^{\mathrm{n}} \stackrel{\mathrm{df}}{=} -(\boldsymbol{Y}_{\mathrm{ST}} + \alpha^{*}\boldsymbol{Y}_{\mathrm{TR}} + \alpha\boldsymbol{Y}_{\mathrm{RS}}) \tag{7.89}$$

are unbalanced admittances for the positive and the negative sequence voltages.

The same rearrangement can be repeated for the crms value of the current in lines S and T, namely

$$\boldsymbol{I}_{\mathrm{S}} = Y_{\mathrm{e}}\,\boldsymbol{U}_{\mathrm{S}} + \boldsymbol{Y}_{\mathrm{u}}^{\mathrm{p}}\,\boldsymbol{U}_{\mathrm{T}}^{\mathrm{p}} + \boldsymbol{Y}_{\mathrm{u}}^{\mathrm{n}}\,\boldsymbol{U}_{\mathrm{T}}^{\mathrm{n}}$$

$$\boldsymbol{I}_{\mathrm{T}} = Y_{\mathrm{e}}\,\boldsymbol{U}_{\mathrm{T}} + \boldsymbol{Y}_{\mathrm{u}}^{\mathrm{p}}\,\boldsymbol{U}_{\mathrm{S}}^{\mathrm{p}} + \boldsymbol{Y}_{\mathrm{u}}^{\mathrm{n}}\,\boldsymbol{U}_{\mathrm{S}}^{\mathrm{n}}$$

and expressed in the vector form

$$\boldsymbol{I} = \begin{bmatrix} \boldsymbol{I}_{\mathrm{R}} \\ \boldsymbol{I}_{\mathrm{S}} \\ \boldsymbol{I}_{\mathrm{T}} \end{bmatrix} = Y_{\mathrm{e}}\,\boldsymbol{U} + \mathbf{1}^{\mathrm{n}}\boldsymbol{Y}_{\mathrm{u}}^{\mathrm{p}}\boldsymbol{U}^{\mathrm{p}} + \mathbf{1}^{\mathrm{p}}\boldsymbol{Y}_{\mathrm{u}}^{\mathrm{n}}\boldsymbol{U}^{\mathrm{n}}.$$

So the vector of unbalanced currents crms values is equal to

$$\boldsymbol{I}_{\mathrm{u}} = \boldsymbol{I} - \boldsymbol{I}_{\mathrm{b}} = (Y_{\mathrm{e}} - Y_{\mathrm{b}})\boldsymbol{U} + \mathbf{1}^{\mathrm{n}}\boldsymbol{Y}_{\mathrm{u}}^{\mathrm{p}}\boldsymbol{U}^{\mathrm{p}} + \mathbf{1}^{\mathrm{p}}\boldsymbol{Y}_{\mathrm{u}}^{\mathrm{n}}\boldsymbol{U}^{\mathrm{n}}$$

or

$$\boldsymbol{I}_{\mathrm{u}} = Y_{\mathrm{d}}\,\boldsymbol{U} + \mathbf{1}^{\mathrm{n}}\boldsymbol{Y}_{\mathrm{u}}^{\mathrm{p}}\boldsymbol{U}^{\mathrm{p}} + \mathbf{1}^{\mathrm{p}}\boldsymbol{Y}_{\mathrm{u}}^{\mathrm{n}}\boldsymbol{U}^{\mathrm{n}}. \tag{7.90}$$

The term

$$\mathbf{1}^{\mathrm{n}}\boldsymbol{Y}_{\mathrm{u}}^{\mathrm{p}}\boldsymbol{U}^{\mathrm{p}} \stackrel{\mathrm{df}}{=} \boldsymbol{J}^{\mathrm{n}} \tag{7.91}$$

in formula (7.90) stands for a vector of crms values of symmetrical currents of negative sequence proportional to the positive sequence voltage, while

$$\mathbf{1}^{\mathrm{p}}\boldsymbol{Y}_{\mathrm{u}}^{\mathrm{n}}\boldsymbol{U}^{\mathrm{n}} \stackrel{\mathrm{df}}{=} \boldsymbol{J}^{\mathrm{p}} \tag{7.92}$$

stands for a vector of crms values of symmetrical currents of positive sequence proportional to the negative sequence voltage. Thus, the vectors of the active, reactive, and unbalanced currents, $\boldsymbol{i}_{a}$, $\boldsymbol{i}_{r}$, and $\boldsymbol{i}_{u}$ can be specified in terms of four admittances, Y_{b}, Y_{d}, Y_{u}^{p}, and Y_{u}^{n}, which can be expressed in terms of line-to-line admittances Y_{RS}, Y_{ST}, and Y_{TR}, line-to-line supply voltage rms values, and the coefficient of its asymmetry, $\boldsymbol{a}$.

Equivalent circuit. The original unbalanced LTI load supplied with asymmetrical voltage can be regarded as a parallel connection of two balanced loads with phase admittance Y_{b} and Y_{d}, and two symmetrical current sources, connected as shown in Fig. 7.29, which inject two three-phase currents $\boldsymbol{j}^{n}$ and $\boldsymbol{j}^{p}$.

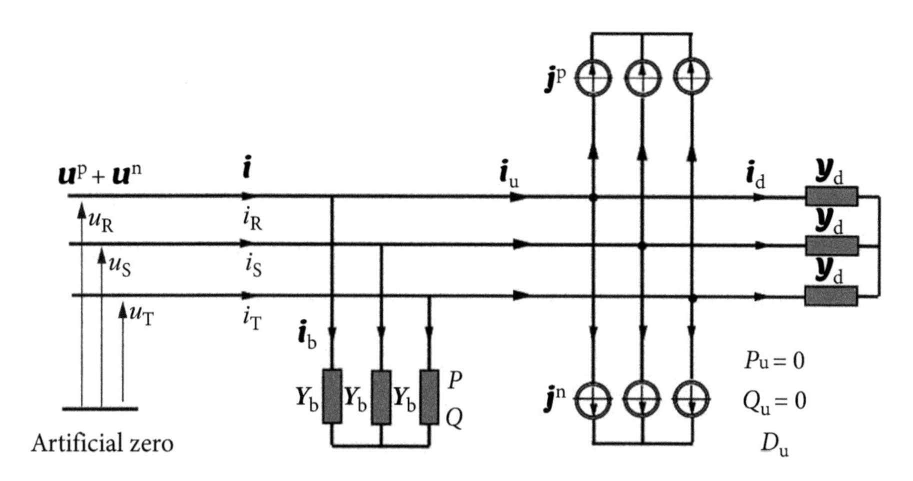

Figure 7.29 An equivalent circuit of an unbalanced load.

The balanced loads draw asymmetrical currents of the crms value proportional to admittance Y_{b} and Y_{d}. Three-phase currents $\boldsymbol{j}^{n}$ and $\boldsymbol{j}^{p}$ are symmetrical currents proportional to the positive and negative sequence components of the supply voltage, $\boldsymbol{u}^{p}$ and $\boldsymbol{u}^{n}$, but of opposite sequence to those voltages, according to formulae (7.91) and (7.92). All parameters of such an equivalent circuit are expressed in terms of line-to-line admittances Y_{RS}, Y_{ST}, and Y_{TR}, of the Δ configured equivalent circuit shown in Fig. 7.29. The balanced branch with admittance Y_{b} has the active and reactive powers equal to P and Q, respectively, because this admittance was calculated, according to the formula (7.81), just to satisfy such a condition.

Such a circuit can be regarded as an *equivalent circuit of unbalanced loads* supplied with sinusoidal but asymmetrical voltage. It visualizes the complex nature of the unbalanced current $\boldsymbol{i}_{u}$. This nature could be irrelevant when only its rms value or/and the unbalanced power have to be known. Knowledge of this nature could yet be crucial for the process of design of a reactance compensator that would be capable to compensate the unbalanced current.

The branch with the unbalanced current $\boldsymbol{i}_{u}$ has to have zero active and reactive powers, P and Q, because these two powers of the original load are equal, according to formula (7.76), to the powers of the branch with current $\boldsymbol{i}_{b}$, while the whole equivalent circuit has to satisfy the balance principle for the active and reactive powers. The only non-zero power of this branch could be the unbalanced power D_{u}.

Illustration 7.7 *Let us calculate the physical components of the load current in the circuit analyzed in Illustration 7.5, shown in Fig. 7.26.*

The crms values of the positive and negative symmetrical components of the supply voltage were calculated in Illustration 7.5 and they are equal to

$$U^{\mathrm{p}} = 66.66\,\mathrm{V}, \quad U^{\mathrm{n}} = 33.33\,e^{j60^\circ}$$

so that while the vector of the supply voltage versus the artificial zero is

$$\boldsymbol{U} = \mathbf{1}^{\mathrm{p}}U^{\mathrm{p}} + \mathbf{1}^{\mathrm{n}}U^{\mathrm{n}} \begin{bmatrix} 1 \\ \alpha^* \\ \alpha \end{bmatrix} 66.7 + \begin{bmatrix} 1 \\ \alpha \\ \alpha^* \end{bmatrix} 33.3\,e^{j60^\circ} = \begin{bmatrix} 88.2\,e^{j19.1^\circ} \\ 88.2\,e^{-j139.1^\circ} \\ 33.3\,e^{j120^\circ} \end{bmatrix} \mathrm{V}$$

while the three-phase rms values of the supply voltage symmetrical components are

$$||\boldsymbol{u}^{\mathrm{p}}|| = \sqrt{3}\,U^{\mathrm{p}} = 115.47\,\mathrm{V}, \quad ||\boldsymbol{u}^{\mathrm{n}}|| = \sqrt{3}\,U^{\mathrm{n}} = 57.73\,\mathrm{V}$$

and the three-phase rms value of the supply voltage is

$$||\boldsymbol{u}|| = \sqrt{||\boldsymbol{u}^{\mathrm{p}}||^2 + ||\boldsymbol{u}^{\mathrm{n}}||^2} = 129.1\,\mathrm{V}.$$

The equivalent balanced admittance Y_{b} *of the load in the circuit shown in Fig. 7.3 is equal to*

$$Y_{\mathrm{b}} = G_{\mathrm{b}} + jB_{\mathrm{b}} = \frac{P - jQ}{||\boldsymbol{u}||^2} = 0.600 - j0.600\,\mathrm{S}$$

hence, the active current vector has the waveform

$$\boldsymbol{i}_{\mathrm{a}} = \sqrt{2}\mathrm{Re}\,\{\boldsymbol{I}_{\mathrm{a}}e^{j\omega_1 t}\} = \sqrt{2}\mathrm{Re}\,\{G_{\mathrm{b}}(\mathbf{1}^{\mathrm{p}}U^{\mathrm{p}} + \mathbf{1}^{\mathrm{n}}U^{\mathrm{n}})e^{j\omega_1 t}\} =$$

$$= \sqrt{2}\mathrm{Re}\,\{0.60(\mathbf{1}^{\mathrm{p}} \times 66.7 + \mathbf{1}^{\mathrm{n}} \times 3.33\,e^{j60^\circ})e^{j\omega_1 t}\} = \sqrt{2}\mathrm{Re}\left\{\begin{bmatrix} 52.9\,e^{j19.1^\circ} \\ 52.9\,e^{-j139.1^\circ} \\ 20.0\,e^{j120^\circ} \end{bmatrix} e^{j\omega_1 t}\right\} \mathrm{A}.$$

The vector of the reactive current is

$$\boldsymbol{i}_{\mathrm{r}} = \sqrt{2}\mathrm{Re}\,\{\boldsymbol{I}_{\mathrm{r}}\,e^{j\omega t}\} = \sqrt{2}\mathrm{Re}\,\{jB_{\mathrm{b}}(\mathbf{1}^{\mathrm{p}}U^{\mathrm{p}} + \mathbf{1}^{\mathrm{n}}U^{\mathrm{n}})e^{j\omega t}\} = \sqrt{2}\mathrm{Re}\left\{\begin{bmatrix} 52.9\,e^{-j70.9^\circ} \\ 52.9\,e^{j130.9^\circ} \\ 20.0\,e^{j30^\circ} \end{bmatrix} e^{j\omega t}\right\} \mathrm{A}.$$

The unbalanced current vector can be presented in the form

$$\boldsymbol{i}_{\mathrm{u}} = \sqrt{2}\mathrm{Re}\{(\boldsymbol{I} - \boldsymbol{I}_{\mathrm{a}} - \boldsymbol{I}_{\mathrm{r}})\, e^{j\omega_1 t}\} = \sqrt{2}\mathrm{Re}\left\{\begin{bmatrix} 95.2\, e^{-j135°} \\ 95.2\, e^{-j75°} \\ 164.9\, e^{j75°} \end{bmatrix} e^{j\omega_1 t}\right\} \mathrm{A}.$$

Three-phase rms values of the current components are equal to

$$||\boldsymbol{i}_{\mathrm{a}}|| = G_{\mathrm{b}}||\boldsymbol{u}|| = 0.60 \times 129.1 = 77.46\,\mathrm{A}$$

$$||\boldsymbol{i}_{\mathrm{r}}|| = |B_{\mathrm{b}}|\;||\boldsymbol{u}|| = 0.60 \times 129.1 = 77.46\,\mathrm{A}$$

$$||\boldsymbol{i}_{\mathrm{u}}|| = \sqrt{I_{\mathrm{uR}}^2 + I_{\mathrm{uS}}^2 + I_{\mathrm{uT}}^2} = \sqrt{95.2^2 + 95.2^2 + 164.9^2} = 212.9\,\mathrm{A}.$$

The supply current has the three-phase rms value

$$||\boldsymbol{i}|| = \sqrt{I_{\mathrm{R}}^2 + I_{\mathrm{S}}^2 + I_{\mathrm{T}}^2} = \sqrt{100^2 + 100^2 + 193.2^2} = 239.4\,\mathrm{A}$$

and indeed

$$\sqrt{||\boldsymbol{i}_{\mathrm{a}}||^2 + ||\boldsymbol{i}_{\mathrm{r}}||^2 + ||\boldsymbol{i}_{\mathrm{u}}||^2} = \sqrt{77.46^2 + 77.46^2 + 212.9^2} = 239.4\,\mathrm{A}$$

which confirms numerical correctness of the current decomposition into physical components. The load power factor $\lambda = P/S = ||\boldsymbol{i}_{\mathrm{a}}||/||\boldsymbol{i}|| = 0.32$. *The unbalanced power is equal to*

$$D_{\mathrm{u}} = ||\boldsymbol{u}|| \times ||\boldsymbol{i}_{\mathrm{u}}|| = 129.1 \times 212.9 = 27.5\,\mathrm{kVA}.$$

To find parameters of the equivalent circuit of the load, let us calculate unbalanced admittances $\boldsymbol{Y}_{\mathrm{u}}^{\mathrm{p}}$ *and* $\boldsymbol{Y}_{\mathrm{u}}^{\mathrm{n}}$, *namely*

$$\boldsymbol{Y}_{\mathrm{u}}^{\mathrm{p}} = -(\boldsymbol{Y}_{\mathrm{ST}} + \alpha\boldsymbol{Y}_{\mathrm{TR}} + \alpha^*\boldsymbol{Y}_{\mathrm{RS}}) = -[1 + \alpha(-j1)] = 1.932e^{-j165°}\,\mathrm{S}$$

$$\boldsymbol{Y}_{\mathrm{u}}^{\mathrm{n}} = -(\boldsymbol{Y}_{\mathrm{ST}} + \alpha^*\boldsymbol{Y}_{\mathrm{TR}} + \alpha\boldsymbol{Y}_{\mathrm{RS}}) = -[1 + \alpha^*(-j1)] = 0.518e^{-j105°}\,\mathrm{S}.$$

The complex coefficient of the supply voltage asymmetry is equal to

$$\boldsymbol{a} = a\, e^{j\psi} = \frac{\boldsymbol{U}^{\mathrm{n}}}{\boldsymbol{U}^{\mathrm{p}}} = \frac{33.33e^{j60°}}{66.66} = 0.50\, e^{j60°}$$

therefore, the asymmetry-dependent unbalanced admittance is equal to

$$\boldsymbol{Y}_{\mathrm{d}} = \frac{2a}{1 + a^2}[\boldsymbol{Y}_{\mathrm{ST}} \cos\psi + \boldsymbol{Y}_{\mathrm{TR}} \cos(\psi - \frac{2\pi}{3}) + \boldsymbol{Y}_{\mathrm{RS}} \cos(\psi + \frac{2\pi}{3})] =$$

$$= \frac{2 \times 0.5}{1 + 0.5^2}[\cos(60°) - j\cos(60° - 120°)] = 0.566\, e^{-j45°}\,\mathrm{S}.$$

With these parameters of the equivalent circuit, the vector of the unbalanced current crms values is equal to

$$\boldsymbol{I}_{\mathrm{u}} = Y_{\mathrm{d}}\, \boldsymbol{U} + \mathbf{1}^{\mathrm{n}} Y_{\mathrm{u}}^{\mathrm{p}}\, U^{\mathrm{p}} + \mathbf{1}^{\mathrm{p}} Y_{\mathrm{u}}^{\mathrm{n}} U^{\mathrm{n}} =$$

$$= 0.566\, e^{-j45^\circ} \begin{bmatrix} U_{\mathrm{R}} \\ U_{\mathrm{S}} \\ U_{\mathrm{T}} \end{bmatrix} + 1.93\, e^{-j165^\circ} \begin{bmatrix} 1 \\ \alpha \\ \alpha^* \end{bmatrix} U^{\mathrm{p}} + 0.518\, e^{-j105^\circ} \begin{bmatrix} 1 \\ \alpha^* \\ \alpha \end{bmatrix}$$

$$\boldsymbol{U}^{\mathrm{n}} = \begin{bmatrix} 95.2 e^{-j135^\circ} \\ 95.2 e^{-j75^\circ} \\ 164.9\, e^{j75^\circ} \end{bmatrix} \mathrm{A}.$$

One could checked that the complex power of the branch of the equivalent circuit with the unbalanced current $\boldsymbol{i}_{\mathrm{u}}$ *is*

$$C_{\mathrm{u}} = \boldsymbol{U}^{\mathrm{T}} \boldsymbol{I}_{\mathrm{u}}^{*} = P_{\mathrm{u}} + jQ_{\mathrm{u}} = 0.$$

This confirms the numerical correctness of the calculation of the unbalanced current, thus the correctness of the calculation of the parameters of the equivalent circuit.

****

Formula (7.90) shows that the unbalanced current is an intricate quantity. Unbalanced admittances $Y_{\mathrm{u}}^{\mathrm{p}}$ and $Y_{\mathrm{u}}^{\mathrm{n}}$ depend only on the load parameters, while the asymmetry-dependent unbalanced admittance Y_{d} depends moreover on the supply voltage asymmetry. Admittances $Y_{\mathrm{u}}^{\mathrm{p}}$ and $Y_{\mathrm{u}}^{\mathrm{n}}$ can have different values, and consequently, the dependence of the unbalanced current on the supply voltage positive and negative sequence components can be different.

Since the equivalent balanced admittance $Y_{\mathrm{b}} = Y_{\mathrm{e}} - Y_{\mathrm{d}}$ and consequently, $G_{\mathrm{b}} = G_{\mathrm{e}} - G_{\mathrm{d}}$ and $B_{\mathrm{b}} = B_{\mathrm{e}} - B_{\mathrm{d}}$, the active and reactive powers can be decomposed into components independent of the voltage asymmetry and dependent on it, namely

$$P = G_{\mathrm{b}} ||\boldsymbol{u}||^2 = (G_{\mathrm{e}} - G_{\mathrm{d}}) ||\boldsymbol{u}||^2 = P_{\mathrm{s}} - P_{\mathrm{d}} \tag{7.93}$$

where P_{s} denotes the load active power at a symmetrical supply voltage, but with the same rms value as the asymmetrical one. The power P_{d} occurs because of the supply voltage asymmetry, but it disappears, independently of this asymmetry, when the load is balanced.

Similarly, the reactive power

$$Q = B_{\mathrm{b}} ||\boldsymbol{u}||^2 = -(B_{\mathrm{e}} - B_{\mathrm{d}}) ||\boldsymbol{u}||^2 = Q_{\mathrm{s}} - Q_{\mathrm{d}} \tag{7.94}$$

where Q_{s} denotes the reactive power at symmetrical supply voltage, while the power Q_{d} occurs because of the supply voltage asymmetry in presence of the load imbalance.

Observe that the unbalanced current contains both positive and negative sequence components, since the vector

$$\mathbf{1}^{\mathrm{n}} Y_{\mathrm{u}}^{\mathrm{p}} U^{\mathrm{p}} + Y_{\mathrm{d}} \boldsymbol{U}^{\mathrm{n}} \stackrel{\mathrm{df}}{=} \boldsymbol{I}_{\mathrm{u}}^{\mathrm{n}} \tag{7.95}$$

is a vector of the crms values of the supply currents of the negative sequence, while the vector

$$\mathbf{1}^{\mathrm{p}} Y_{\mathrm{u}}^{\mathrm{n}} U^{\mathrm{n}} + Y_{\mathrm{d}} \boldsymbol{U}^{\mathrm{p}} \stackrel{\mathrm{df}}{=} \boldsymbol{I}_{\mathrm{u}}^{\mathrm{p}} \tag{7.96}$$

is a vector of crms values of the positive sequence currents. Thus, the unbalanced current can be expressed in the form

$$\boldsymbol{i}_{\mathrm{u}} = \sqrt{2}\mathrm{Re}\,\{(\boldsymbol{I}_{\mathrm{u}}^{\mathrm{p}} + \boldsymbol{I}_{\mathrm{u}}^{\mathrm{n}})e^{j\omega_1 t}\} = \boldsymbol{i}_{\mathrm{u}}^{\mathrm{p}} + \boldsymbol{i}_{\mathrm{u}}^{\mathrm{n}} \tag{7.97}$$

so that, the load current can be decomposed into four components

$$\boldsymbol{i} = \boldsymbol{i}_{\mathrm{a}} + \boldsymbol{i}_{\mathrm{r}} + \boldsymbol{i}_{\mathrm{u}}^{\mathrm{p}} + \boldsymbol{i}_{\mathrm{u}}^{\mathrm{n}}. \tag{7.98}$$

These components are mutually orthogonal so that their three-phase rms values satisfy the relationship

$$||\boldsymbol{i}||^2 = ||\boldsymbol{i}_{\mathrm{a}}||^2 + ||\boldsymbol{i}_{\mathrm{r}}||^2 + ||\boldsymbol{i}_{\mathrm{u}}^{\mathrm{p}}||^2 + ||\boldsymbol{i}_{\mathrm{u}}^{\mathrm{n}}||^2. \tag{7.99}$$

The active current $\boldsymbol{i}_{\mathrm{a}}$ is associated exclusively with permanent energy transfer from the supply source to the load, meaning with the load active power P. The reactive current $\boldsymbol{i}_{\mathrm{r}}$ is associated exclusively with the phase shift between the supply voltage and the load current, meaning with the load reactive power Q. These two currents are asymmetrical currents and their asymmetry reproduces the asymmetry of the supply voltage. Currents $\boldsymbol{i}_{\mathrm{u}}^{\mathrm{p}}$ and $\boldsymbol{i}_{\mathrm{u}}^{\mathrm{n}}$ are symmetrical currents, which occur exclusively due to the load imbalance. They do not contribute to the active and reactive powers P and Q of the load, but only to an increase of its three-phase rms value. Therefore, these four components of the load current can be regarded as the Current's Physical Components.

Multiplying Eqn. (7.99) by the square of the three-phase rms value of the supply voltage, the power equation is obtained in the form

$$S^2 = P^2 + Q^2 + D_{\mathrm{u}}^{\mathrm{p}2} + D_{\mathrm{u}}^{\mathrm{n}2} \tag{7.100}$$

with

$$D_{\mathrm{u}}^{\mathrm{p}} \stackrel{\mathrm{df}}{=} ||\boldsymbol{u}||\,||\boldsymbol{i}_{\mathrm{u}}^{\mathrm{p}}||, \quad D_{\mathrm{u}}^{\mathrm{n}} \stackrel{\mathrm{df}}{=} ||\boldsymbol{u}||\,||\boldsymbol{i}_{\mathrm{u}}^{\mathrm{n}}||. \tag{7.101}$$

7.15 CPC at Asymmetrical Supply with nv&c and LTI Load

The internal voltage $\boldsymbol{e}(t)$ of a distribution system can be not only asymmetrical but also nonsinusoidal. Let it be composed of harmonics of the order n from a set N, namely

$$\boldsymbol{e}(t) \stackrel{\text{df}}{=} \begin{bmatrix} e_{\text{R}}(t) \\ e_{\text{S}}(t) \\ e_{\text{T}}(t) \end{bmatrix} = \sum_{n\in N} \begin{bmatrix} e_{\text{R}n}(t) \\ e_{\text{S}n}(t) \\ e_{\text{T}n}(t) \end{bmatrix} = \sum_{n\in N} \boldsymbol{e}_n(t).$$

Each harmonic of this voltage can contain symmetrical components of the positive, negative, and zero order, so that

$$\boldsymbol{e}(t) = \sum_{n\in N} \boldsymbol{e}_n = \sum_{n\in N} (\boldsymbol{e}_n^{\text{p}} + \boldsymbol{e}_n^{\text{n}} + \boldsymbol{e}_n^{\text{z}}) = \boldsymbol{e}^{\text{p}} + \boldsymbol{e}^{\text{n}} + \boldsymbol{e}^{\text{z}}.$$

The zero-sequence component $\boldsymbol{e}^{\text{z}}(t)$ of the internal voltage cannot cause any current in three-wire circuits but it contributes to an increase of the three-phase rms value of the supply voltage, thus it increases the apparent power S and reduces the load power factor. Even an ideal resistive balanced load supplied with a voltage that contains the symmetrical component of the zero sequence would have power factor lower than 1. To avoid this, the zero-sequence component of the supply voltage has to be eliminated by referencing the voltage to an artificial zero, as shown in Fig. 7.24. Such a voltage can be composed of symmetrical components of only the positive and the negative sequence:

$$\boldsymbol{u}(t) = \sum_{n\in N} \boldsymbol{u}_n = \sum_{n\in N} (\boldsymbol{u}_n^{\text{p}} + \boldsymbol{u}_n^{\text{n}}) = \boldsymbol{u}^{\text{p}} + \boldsymbol{u}^{\text{n}}.$$

The crms values of the symmetrical components of the supply voltage harmonics $\boldsymbol{U}_{\text{n}}^{\text{p}}$, $\boldsymbol{U}_{\text{n}}^{\text{n}}$, and $\boldsymbol{U}_{\text{n}}^{\text{z}}$ can be calculated in the same way as for sinusoidal quantities, namely

$$\begin{bmatrix} \boldsymbol{U}_n^{\text{z}} \\ \boldsymbol{U}_n^{\text{p}} \\ \boldsymbol{U}_n^{\text{n}} \end{bmatrix} = \frac{1}{3} \begin{bmatrix} 1 & 1 & 1 \\ 1 & \alpha & \alpha^* \\ 1 & \alpha^* & \alpha \end{bmatrix} \begin{bmatrix} \boldsymbol{U}_{\text{R}n} \\ \boldsymbol{U}_{\text{S}n} \\ \boldsymbol{U}_{\text{T}n} \end{bmatrix}.$$

When the supply voltage is symmetrical then the voltage harmonics are also symmetrical and of a specified sequence. Harmonics of the order $n = 3k + 1$ are of positive sequence; harmonics of the order $n = 3k - 1$ are of negative sequence, while the voltage harmonics of the order $n = 3k$ are of zero sequence. When the supply voltage is asymmetrical then this property is no longer valid, however. In particular, the third order harmonic can exist both in the supply voltage and in the load current because at the supply voltage asymmetry, the third order harmonic is not a zero sequence harmonic. The third order harmonic can contain symmetrical components both of the positive and the negative sequence.

The supply voltage, referenced to the artificial zero, arranged in a three-phase vector, can be presented in the form

$$\boldsymbol{u}(t) \stackrel{\text{df}}{=} \begin{bmatrix} u_{\text{R}}(t) \\ u_{\text{S}}(t) \\ u_{\text{T}}(t) \end{bmatrix} = \sum_{n\in N} \boldsymbol{u}_n(t) = \sqrt{2}\text{Re}\sum_{n\in N} \begin{bmatrix} U_{\text{R}n} \\ U_{\text{S}n} \\ U_{\text{T}n} \end{bmatrix} e^{jn\omega_1 t} = \sqrt{2}\text{Re}\sum_{n\in N} \boldsymbol{U}_n e^{jn\omega_1 t} =$$

$$= \sqrt{2}\text{Re}\sum_{n\in N} (\boldsymbol{U}_n^{\text{p}} + \boldsymbol{U}_n^{\text{n}}) e^{jn\omega_1 t} = \sqrt{2}\text{Re}\sum_{n\in N} (\mathbf{1}^{\text{p}} U_n^{\text{p}} + \mathbf{1}^{\text{n}} U_n^{\text{n}}) e^{jn\omega_1 t}.$$

The load in Fig. 7.30(a) is equivalent to active power P of a balanced resistive load, shown in Fig. 7.30(b), of conductance

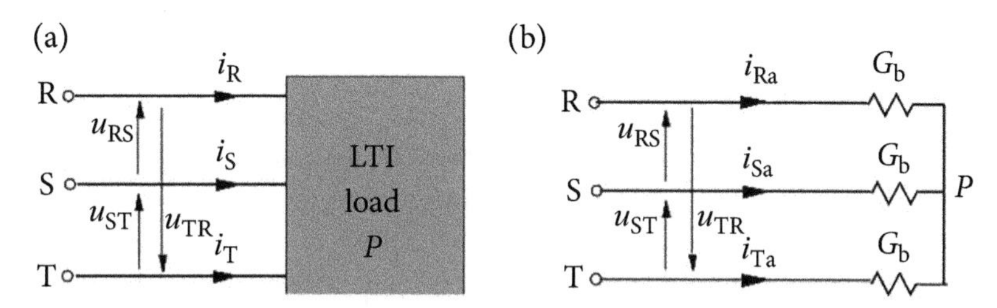

Figure 7.30 (a) A three-phase load and (b) a balanced resistive load equivalent to active power P.

$$G_{\text{b}} = \frac{P}{||\boldsymbol{u}||^2}.$$

The current of such an equivalent load

$$\boldsymbol{i}_{\text{a}} = G_{\text{b}}\, \boldsymbol{u} = \sqrt{2}\text{Re}\sum_{n\in N} G_{\text{b}}(\mathbf{1}^{\text{p}} U_n^{\text{p}} + \mathbf{1}^{\text{n}} U_n^{\text{n}}) e^{jn\omega_1 t} \tag{7.102}$$

is the active current.

According to the assumption in this chapter, the load is linear so that the current response of the load to this voltage can be calculated by using harmonic-by-harmonic approach.

The load at each harmonic frequency has active and reactive powers. For a harmonic of the nth order

$$P_n = \text{Re}\{\boldsymbol{U}_n^{\text{T}} \boldsymbol{I}_n^*\}, \quad Q_n = \text{Im}\{\boldsymbol{U}_n^{\text{T}} \boldsymbol{I}_n^*\}.$$

The load can be unbalanced for the nth order harmonic, but with regard to the active and reactive powers P_n and Q_n at voltage $\boldsymbol{u}_n$, such a load is equivalent to a balanced load of the phase admittance

$$Y_{bn} = G_{bn} + jB_{bn} = \frac{P_n - jQ_n}{||\boldsymbol{u}_n||^2} = \frac{C_n^*}{||\boldsymbol{u}_n||^2} \tag{7.103}$$

where $||\boldsymbol{u}_n||$ denotes the three-phase rms value of the nth order voltage harmonic. Just such an equivalent load is shown in Fig. 7.31.

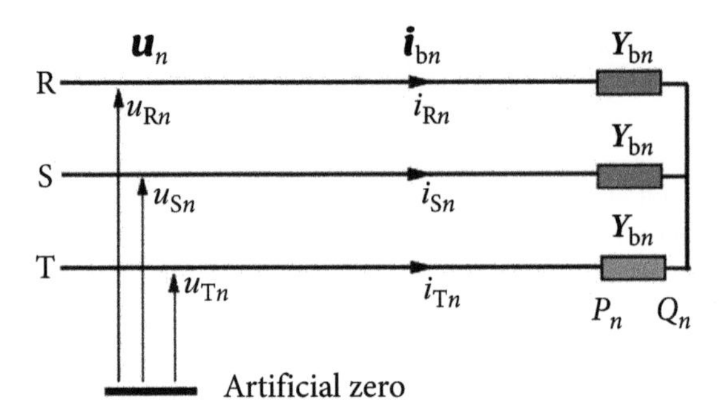

Figure 7.31 A balanced load equivalent to the original with regard to P_n and Q_n powers of the nth order harmonic.

The line current of such an equivalent load is composed of the active current

$$\boldsymbol{i}_{an}(t) = G_{bn}\boldsymbol{u}_n(t) = \sqrt{2}\text{Re}\,\{G_{bn}(\boldsymbol{U}_n^{\text{p}} + \boldsymbol{U}_n^{\text{n}})e^{jn\omega_1 t}\} =$$
$$= \sqrt{2}\text{Re}\,\{G_{bn}(\mathbf{1}^{\text{p}}U_n^{\text{p}} + \mathbf{1}^{\text{n}}U_n^{\text{n}})e^{jn\omega_1 t}\} \tag{7.104}$$

and the reactive current

$$\boldsymbol{i}_{rn}(t) = B_{bn}\boldsymbol{u}_n(t + \frac{T}{4n}) = \sqrt{2}\text{Re}\,\{jB_{bn}(\boldsymbol{U}_n^{\text{p}} + \boldsymbol{U}_n^{\text{n}})e^{jn\omega_1 t}\} =$$
$$= \sqrt{2}\text{Re}\,\{jB_{bn}(\mathbf{1}^{\text{p}}U_n^{\text{p}} + \mathbf{1}^{\text{n}}U_n^{\text{n}})e^{jn\omega_1 t}\}. \tag{7.105}$$

The admittance Y_{bn} is the admittance of an equivalent balanced load, while the load can be unbalanced for the nth order harmonic. Consequently, the nth order harmonic of the line current $\boldsymbol{i}_n$ can contain the unbalanced current

$$\boldsymbol{i}_{un} = \boldsymbol{i}_n - \boldsymbol{i}_{bn} = \boldsymbol{i}_n - (\boldsymbol{i}_{an} + \boldsymbol{i}_{rn}) =$$
$$= \sqrt{2}\text{Re}\,\{\boldsymbol{I}_n - Y_{bn}(\mathbf{1}^{\text{p}}U_n^{\text{p}} + \mathbf{1}^{\text{n}}U_n^{\text{n}})e^{jn\omega_1 t}\}.$$

This means that each current harmonic $\boldsymbol{i}_n$ can be regarded as a sum of three components

$$\boldsymbol{i}_n = \boldsymbol{i}_{an} + \boldsymbol{i}_{rn} = \boldsymbol{i}_{un} \tag{7.106}$$

and consequently, the load current is equal to

$$\boldsymbol{i}(t) = \begin{bmatrix} i_{\mathrm{R}}(t) \\ i_{\mathrm{S}}(t) \\ i_{\mathrm{T}}(t) \end{bmatrix} = \sum_{n\in N} \boldsymbol{i}_n(t) = \sum_{n\in N} (\boldsymbol{i}_{\mathrm{a}n} + \boldsymbol{i}_{\mathrm{r}n} + \boldsymbol{i}_{\mathrm{u}n}). \tag{7.107}$$

The current

$$\sum_{n\in N} \boldsymbol{i}_{\mathrm{r}n} \stackrel{\mathrm{df}}{=} \boldsymbol{i}_{\mathrm{r}} = \sqrt{2}\mathrm{Re} \sum_{n\in N} jB_{\mathrm{b}n}(\mathbf{1}^{\mathrm{p}} U_n^{\mathrm{p}} + \mathbf{1}^{\mathrm{n}} U_n^{\mathrm{n}}) e^{jn\omega_1 t} \tag{7.108}$$

occurs in the load current because of the phase shift of the current harmonics against the supply voltage harmonics. Therefore, it can be regarded as a reactive current of the load.

The current

$$\sum_{n\in N} \boldsymbol{i}_{\mathrm{u}n} \stackrel{\mathrm{df}}{=} \boldsymbol{i}_{\mathrm{u}}$$

occurs in the load current because of the load imbalance for harmonic frequencies.

The current

$$\sum_{n\in N} \boldsymbol{i}_{\mathrm{a}n} = \sqrt{2}\mathrm{Re} \sum_{n\in N} G_{\mathrm{b}n}(\mathbf{1}^{\mathrm{p}} U_n^{\mathrm{p}} + \mathbf{1}^{\mathrm{n}} U_n^{\mathrm{n}}) e^{jn\omega_1 t}$$

is not the active current $\boldsymbol{i}_{\mathrm{a}}$ of the load, however. These two currents differ by

$$\sum_{n\in N} \boldsymbol{i}_{\mathrm{a}n} - \boldsymbol{i}_{\mathrm{a}} = \sqrt{2}\mathrm{Re} \sum_{n\in N} (G_{\mathrm{b}n} - G_{\mathrm{b}})(\mathbf{1}^{\mathrm{p}} U_n^{\mathrm{p}} + \mathbf{1}^{\mathrm{n}} U_n^{\mathrm{n}}) e^{jn\omega_1 t} \stackrel{\mathrm{df}}{=} \boldsymbol{i}_{\mathrm{s}}.$$

This difference occurs when the conductance G_{b} of an equivalent balanced load differs from the conductance $G_{\mathrm{b}n}$ for harmonic frequencies. The current $\boldsymbol{i}_{\mathrm{s}}$ will be referenced as a scattered current. Combining (7.106) to (7.108), the load current can be expressed as follows

$$\boldsymbol{i}(t) = \boldsymbol{i}_{\mathrm{a}}(t) + \boldsymbol{i}_{\mathrm{s}}(t) + \boldsymbol{i}_{\mathrm{r}}(t) + \boldsymbol{i}_{\mathrm{u}}(t). \tag{7.109}$$

Each of these currents is associated with a different phenomenon in the load. The active current $\boldsymbol{i}_{\mathrm{a}}$ is associated with permanent energy delivery with active power P to the load. Other currents do not contribute to this transfer. The scattered current $\boldsymbol{i}_{\mathrm{s}}$ is associated with the phenomenon of a change of the load conductance $G_{\mathrm{b}n}$ with harmonic order n. The reactive current $\boldsymbol{i}_{\mathrm{r}}$ is associated with the phase shift of the load current harmonics versus the supply voltage harmonics. The unbalanced current $\boldsymbol{i}_{\mathrm{u}}$ is associated with the load imbalance for harmonic frequencies.

The three-phase rms values of the load current components are equal to

$$||\boldsymbol{i}_{\mathrm{a}}|| = G_{\mathrm{b}}||\boldsymbol{u}|| = \frac{P}{||\boldsymbol{u}||}$$

$$||\boldsymbol{i}_{\mathrm{s}}|| = \sqrt{\sum_{n\in N} ||\boldsymbol{i}_{\mathrm{s}n}||^2} = \sqrt{\sum_{n\in N} (G_{\mathrm{b}n} - G_{\mathrm{b}})^2||\boldsymbol{u}_n||^2}$$

$$||\boldsymbol{i}_{\mathrm{r}}|| = \sqrt{\sum_{n\in N} ||\boldsymbol{i}_{\mathrm{r}n}||^2} = \sqrt{\sum_{n\in N} B_{\mathrm{b}n}^2||\boldsymbol{u}_n||^2}$$

$$||\boldsymbol{i}_{\mathrm{u}}|| = \sqrt{\sum_{n\in N} ||\boldsymbol{i}_{\mathrm{u}n}||^2}.$$

Orthogonality: The orthogonality of the active, reactive, and unbalanced currents in three-phase circuits with sinusoidal quantities was proven at the beginning of this chapter. This applies, of course, to an individual harmonic of any order n. Therefore, the active, reactive, and unbalanced components of the nth order harmonic of the load current $\boldsymbol{i}_n$ in decomposition (7.106) are mutually orthogonal, that is:

$$(\boldsymbol{i}_{\mathrm{a}n}, \boldsymbol{i}_{\mathrm{r}n}) = (\boldsymbol{i}_{\mathrm{a}n}, \boldsymbol{i}_{\mathrm{u}n}) = (\boldsymbol{i}_{\mathrm{r}n}, \boldsymbol{i}_{\mathrm{u}n}) = 0.$$

Current components in decomposition (7.107) are sums of harmonics. Because harmonics of different order r and s are mutually orthogonal, then the scalar products of two currents, which are sums of harmonics, can be expressed generally as

$$(\boldsymbol{i}_{\mathrm{x}}, \boldsymbol{i}_{\mathrm{v}}) = (\sum_{r\in N} \boldsymbol{i}_{\mathrm{x}r}, \sum_{s\in N} \boldsymbol{i}_{\mathrm{v}s}) = \sum_{n\in N} (\boldsymbol{i}_{\mathrm{x}n}, \boldsymbol{i}_{\mathrm{v}n}).$$

Thus, if harmonics of the same order n of two currents are mutually orthogonal, that is, $(\boldsymbol{i}_{\mathrm{xn}}, \boldsymbol{i}_{\mathrm{vn}}) = 0$, then such currents are orthogonal as well. Consequently, all terms on the right side of (7.107) are mutually orthogonal, hence

$$||\boldsymbol{i}||^2 = ||\sum_{n\in N} \boldsymbol{i}_{\mathrm{a}n}||^2 + ||\boldsymbol{i}_{\mathrm{r}}||^2 + ||\boldsymbol{i}_{\mathrm{u}}||^2.$$

The sum of harmonic active currents is not the active current, because it is equal to

$$\sum_{n\in N} \boldsymbol{i}_{\mathrm{a}n} = \boldsymbol{i}_{\mathrm{a}} + \boldsymbol{i}_{\mathrm{s}}.$$

There is a question to be posed here: are the active and scattered currents mutually orthogonal? Their scalar product is

$$(\boldsymbol{i}_{\mathrm{s}}, \boldsymbol{i}_{\mathrm{a}}) = \mathrm{Re} \sum_{n\in N} (G_{\mathrm{b}n} - G_{\mathrm{b}})\, \mathbf{1}_n U_n \cdot (G_{\mathrm{b}} \mathbf{1}_n U_n)^* =$$

$$= G_{\mathrm{b}} \left(\sum_{n\in N} G_{\mathrm{b}n}||\boldsymbol{u}_n||^2 - G_{\mathrm{b}} \sum_{n\in N} ||\boldsymbol{u}_n||^2 \right) = G_{\mathrm{b}}(P - P) = 0.$$

Thus, the scattered and the active currents are orthogonal even if the supply voltage is asymmetrical. This completes the proof that the supply current physical components $\boldsymbol{i}_a$, $\boldsymbol{i}_s$, $\boldsymbol{i}_r$, and $\boldsymbol{i}_u$ in three-phase circuits with LTI loads supplied with an asymmetrical nonsinusoidal voltage are mutually orthogonal, hence their three-phase rms values satisfy the relationship

$$||\boldsymbol{i}||^2 = ||\boldsymbol{i}_a||^2 + ||\boldsymbol{i}_s||^2 + ||\boldsymbol{i}_r||^2 + ||\boldsymbol{i}_u||^2. \tag{7.110}$$

This relationship enables to develop power equation

$$S^2 = P^2 + D_s^2 + Q^2 + D_u^2 \tag{7.111}$$

with the scattered power defined as

$$D_s \overset{\text{df}}{=} ||\boldsymbol{u}||\ ||\boldsymbol{i}_s||. \tag{7.112}$$

The developed above decomposition of the load current into Physical Components has a major deficiency when reactive compensation is a matter of interest, however. It does not provide information on how the unbalanced current $\boldsymbol{i}_u(t)$ depends on the parameters of the load. Consequently, it does not provide information on how the compensator parameters should be selected to compensate that current.

This issue for a sinusoidal supply voltage was solved earlier in this Section 7.14 with Eqn (7.80).

That equation for symmetrical components of the supply voltage harmonics remains the same, independently of the harmonic order n. For such a harmonic of the nth order the three-phase vector of the crms values of the unbalanced component of that current harmonic is

$$\boldsymbol{I}_{un} = Y_{dn}\,\boldsymbol{U}_n + \boldsymbol{1}^{n}\boldsymbol{Y}_{un}^{p}\,\boldsymbol{U}_n^{p} + \boldsymbol{1}^{p}\boldsymbol{Y}_{un}^{n}\,\boldsymbol{U}_n^{n} \tag{7.113}$$

where

$$Y_{dn} = \frac{2a_n}{1+a_n^2}\left[Y_{STn}\cos\psi_n + Y_{TRn}\cos\left(\psi_n - \frac{2\pi}{3}\right) + Y_{RSn}\cos\left(\psi_n + \frac{2\pi}{3}\right)\right] \tag{7.114}$$

$$\boldsymbol{a}_n \overset{\text{df}}{=} a_n\,e^{j\psi_n} = \frac{\boldsymbol{U}_n^{n}}{\boldsymbol{U}_n^{p}} \tag{7.115}$$

$$\boldsymbol{Y}_{un}^{p} \overset{\text{df}}{=} -(\boldsymbol{Y}_{STn} + \alpha\boldsymbol{Y}_{TRn} + \alpha^{*}\boldsymbol{Y}_{RSn}) \tag{7.116}$$

$$\boldsymbol{Y}_{un}^{n} \overset{\text{df}}{=} -(\boldsymbol{Y}_{STn} + \alpha^{*}\boldsymbol{Y}_{TRn} + \alpha\boldsymbol{Y}_{RSn}). \tag{7.117}$$

Thus, the unbalanced current at asymmetrical and a nonsinusoidal supply voltage can be presented in the form

$$\boldsymbol{i}_{\mathrm{u}} \stackrel{\mathrm{df}}{=} \sum_{n\in N} \boldsymbol{i}_{\mathrm{u}n} = \sqrt{2}\mathrm{Re}\sum_{n\in N} \boldsymbol{I}_{\mathrm{u}n} e^{jn\omega_1 t} =$$

$$= \sqrt{2}\mathrm{Re}\sum_{n\in N} (\boldsymbol{Y}_{\mathrm{d}n}\,\boldsymbol{U}_n + \mathbf{1}^{\mathrm{n}} \mathrm{Y}^{\mathrm{p}}_{\mathrm{u}n} U^{\mathrm{p}}_n + \mathbf{1}^{\mathrm{p}} \mathrm{Y}^{\mathrm{p}}_{\mathrm{u}n} U^{\mathrm{p}}_n) e^{jn\omega_1 t}. \tag{7.118}$$

This means it can be expressed in terms of the load equivalent parameters.

Illustration 7.8 *The presented current decomposition into CPC is verified numerically with the circuit in Fig. 7.32.*

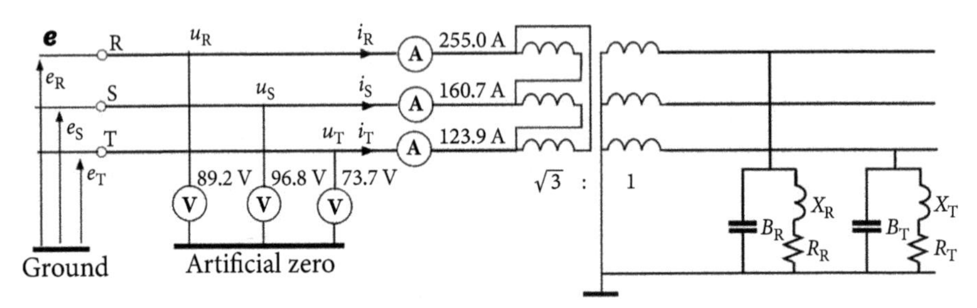

Figure 7.32 Rms values of the line-to-artificial zero voltages and load line currents.

This decomposition was obtained without any restrictions as to the level of the supply voltage asymmetry, its distortion, and the load imbalance. To demonstrate that it is valid independently of the supply voltage asymmetry and distortion, and independently of the load imbalance, a very high level of voltage distortion and asymmetry was assumed for this verification. It was assumed that the internal voltage **e** *fundamental harmonic of the supply circuit is* E_{R1} = 100V, *while the higher-order harmonics, of the order* $n = 3, 5,$ *and* 7 *have the same, very high rms value, equal to* $E_{R3} = E_{R5} = E_{R7} = 20$V. *It was also assumed that the voltage at the supply terminal T is* $E_T = E_R/2$. *Moreover, the load voltage asymmetry was increased by an extra phase shift of voltages at lines S and T, with the angle,* $\Delta\varphi_S = 10$ deg, *and* $\Delta\varphi_T = -10$ deg, *the same for each voltage harmonic. Such assumptions, by making the circuit asymmetry significant, enable reliable numerical verification and can enhance the credibility of the developed current decomposition. The load parameters for the fundamental harmonic are assumed to be equal to*

$$R_R = R_T = 1.0\Omega, \quad X_R = X_T = 1.0\Omega, \quad B_R = B_T = 0.50\ \mathrm{S}.$$

The strongly unbalanced load is supplied from an ideal transformer in Δ/Y *configuration with the turn ratio* $\sqrt{3}$: 1.

The crms values of the zero-sequence symmetrical component of the supply voltage harmonics are equal to

$$U_1^z = 22.56\, e^{-j44.0°}\text{V},\quad U_3^z = 16.52\, e^{j2.0°}\ \text{V}$$
$$U_5^z = 2.56\, e^{j84.6°}\text{V},\quad U_7^z = 4.51\, e^{-j44.0°}\ \text{V}.$$

The crms values of the positive and the negative sequence symmetrical components of load voltage harmonics as referenced to the artificial zero are

$$U_1^p = 82.62\, e^{j2.0°}\text{V},\quad U_3^p = 2.56\, e^{j84.7°}\ \text{V}$$
$$U_5^p = 4.51\, e^{-j44.0°}\text{V},\quad U_7^p = 16.52\, e^{j2.0°}\ \text{V}$$

$$U_1^n = 12.82\, e^{j84.7°}\text{V},\quad U_3^n = 4.51\, e^{-j44.0°}\ \text{V}$$
$$U_5^n = 16.52\, e^{-j2.0°}\text{V},\quad U_7^n = 2.56\, e^{j84.6°}\ \text{V}.$$

The crms values of harmonics of line current are

$$\boldsymbol{I}_1 = \begin{bmatrix} I_{R1} \\ I_{S1} \\ I_{T1} \end{bmatrix} = \begin{bmatrix} 127.8\, e^{j10.6°} \\ 81.9\, e^{-j145.0°} \\ 63.1\, e^{j158.1°} \end{bmatrix} \text{A},\quad \boldsymbol{I}_3 = \begin{bmatrix} I_{R3} \\ I_{S3} \\ I_{T3} \end{bmatrix} = \begin{bmatrix} 12.8\, e^{j75.8°} \\ 4.2\, e^{-j179.8°} \\ 12.4\, e^{-j85.1°} \end{bmatrix} \text{A}$$

$$\boldsymbol{I}_5 = \begin{bmatrix} I_{R5} \\ I_{S5} \\ I_{T5} \end{bmatrix} = \begin{bmatrix} 138.0\, e^{j81.7°} \\ 83.7\, e^{-j115.9°} \\ 63.5\, e^{-j74.8°} \end{bmatrix} \text{A},\quad \boldsymbol{I}_7 = \begin{bmatrix} I_{R7} \\ I_{S7} \\ I_{T7} \end{bmatrix} = \begin{bmatrix} 171.8\, e^{j100.2°} \\ 110.1\, e^{-j55.3°} \\ 85.0\, e^{-j112.2°} \end{bmatrix} \text{A}.$$

The rms values of the line-to-artificial zero voltages and line currents in the circuit are shown in Fig. 7.32. The load unbalanced admittances for the positive sequence are equal to

$$Y_{u1}^p = 0.500\, e^{j0°}\ \text{S},\quad Y_{u3}^p = 1.204\, e^{j85.2°}\text{S},$$
$$Y_{u5}^p = 2.308\, e^{j89.0°}\text{S},\quad Y_{u7}^p = 3.360\, e^{j89.7°}\text{S}.$$

Observe, that for the selected load $Y_{un}^n = Y_{un}^p$.

The results of the circuit analysis with regard to the load equivalent parameters for harmonics and harmonic active and reactive powers are compiled in Table 7.1.

The active power of the load is P = *21,506 W and the three-phase rms value of the supply voltage is* $||\boldsymbol{u}|| = 150.91$*V, so that, the equivalent balanced conductance is*

$$G_b = P/||\boldsymbol{u}||^2 = 0.9443\ \text{S}.$$

The voltage asymmetry-dependent admittances for harmonics are equal to

$$Y_{d1}^p = 0.0194\, e^{j180°}\text{S},\quad Y_{u3}^p = 0.6460\, e^{j85.2°}\text{S},$$
$$Y_{u5}^p = 0.8149\, e^{-j90.0°}\text{S},\quad Y_{u7}^p = 0.1303\, e^{-j90.3°}\text{S}.$$

Table 7.1 Results of the circuit analysis.

	n	1	3	5	7
P_n	W	21,380	12	80	34
Q_n	VAr	0	−142	−4780	−5747
$\|\|\boldsymbol{u}_n\|\|$	V	144.80	8.9	29.7	29.0
G_{bn}	S	1.0194	0.1464	0.0905	0.0408
B_{bn}	S	0	1.7562	5.4302	6.8503

Having values of equivalent parameters of the load as compiled in Table 7.1, three-phase rms values of CPC of the load current can be calculated, namely

$$||\boldsymbol{i}_a|| = G_b||\boldsymbol{u}|| = 142.5\,\text{A}$$

$$||\boldsymbol{i}_r|| = \sqrt{\sum_{n\in N} B_{bn}^2||\boldsymbol{u}_n||^2} = 256.0\,\text{A}$$

$$||\boldsymbol{i}_s|| = \sqrt{\sum_{n\in N} (G_{bn} - G_b)^2||\boldsymbol{u}_n||^2} = 38.7\,\text{A}$$

$$||\boldsymbol{i}_u^p|| = 65.0\,\text{A}, \quad ||\boldsymbol{i}_u^n|| = 121.0\,\text{A}$$

so that the rms value of the unbalanced current is

$$||\boldsymbol{i}_u|| = \sqrt{||\boldsymbol{i}_u^p||^2 + ||\boldsymbol{i}_u^n||^2} = 137.4\,\text{A}.$$

The root of the sum of squares of three-phase rms values of the CPC has to be equal to the three-phase rms value of the load current which is equal to

$$||\boldsymbol{i}|| = \sqrt{||i_R||^2 + ||i_S||^2 + ||i_T||^2} = \\ = \sqrt{255.0^2 + 160.7^2 + 123.9^2} = 325.9\,\text{A}.$$

Indeed,

$$||\boldsymbol{i}|| = \sqrt{||\boldsymbol{i}_a||^2 + ||\boldsymbol{i}_r||^2 + ||\boldsymbol{i}_s||^2 + ||\boldsymbol{i}_u||^2} = \\ = \sqrt{142.5^2 + 256.0^2 + 38.7^2 + 137.4^2} = 325.9\,\text{A}.$$

with a numerical error on the level of 10^{-7}.

*****

These numerical results confirm the correctness of the load current decomposition into the CPC. Thus, CPC-based power theory enables us to associate distinctive physical phenomena in the load with specific components of the load current and express them in terms of the load parameters, the supply voltage harmonics, and asymmetry.

7.16 CPC at Asymmetrical Supply with nv&c and HGLs

It was assumed in the previous section that the load is linear and time invariant. The load could cause the current distortion however, due to non-linearity or periodic switching it could generate current harmonics. Thus, how to decompose the load current into CPC in the circuit with an HGL?

The answer to such a question for the HGLs supplied with a symmetrical voltage was given in Section 7.12. Asymmetry of the supply voltage does not affect the fundamentals of the current decomposition discussed previously. It is based on an approximation that relies on the concept of dominating harmonics. When it is not known where sources of harmonics are located, meaning at the supply side or the load side, then by observing the sign of the harmonic active power P_n for a specific harmonic order n at the load terminals, we can only conclude which source of harmonics is the dominating one. When this power is positive, then the dominating source is on the supply side; when this power is negative then dominating source is on the load side. This approximation provides the ground for the decomposition of the set N of all harmonic orders n into two subsets $N_{\rm C}$ and $N_{\rm G}$. The reasoning that follows this decomposition and formula (7.52) remain unchanged to the point, where the current $\boldsymbol{i}_{\rm C}(t)$ requires decomposition valid for loads with asymmetrical supply voltage, that is:

$$\boldsymbol{i}_{\rm C}(t) = \boldsymbol{i}_{\rm Ca}(t) + \boldsymbol{i}_{\rm Cs}(t) + \boldsymbol{i}_{\rm Cr}(t) + \boldsymbol{i}_{\rm Cu}(t) \tag{7.119}$$

where the active current is defined as

$$\boldsymbol{i}_{\rm Ca} = G_{\rm Cb}\,\boldsymbol{u}_{\rm C} = \sqrt{2}{\rm Re}\sum_{n\in N_{\rm C}} G_{\rm Cb}(\mathbf{1}^{\rm p}\boldsymbol{U}_n^{\rm p} + \mathbf{1}^{\rm n}\boldsymbol{U}_n^{\rm n})e^{jn\omega_1 t} \tag{7.120}$$

with

$$G_{\rm Cb} = \frac{P_{\rm C}}{||\boldsymbol{u}_{\rm C}||^2}. \tag{7.121}$$

The balanced equivalent admittance of the HGL for harmonics of the order n from the subset $N_{\rm C}$ has the values

$$Y_{{\rm b}n} = G_{{\rm b}n} + jB_{{\rm b}n} = \frac{P_n - jQ_n}{||\boldsymbol{u}_n||^2} = \frac{\boldsymbol{C}_n^*}{||\boldsymbol{u}_n||^2} \tag{7.122}$$

so that the scattered, reactive, and unbalanced currents are equal to

$$\boldsymbol{i}_{\mathrm{Cs}}(t) = \sqrt{2}\mathrm{Re}\sum_{n\in N_{\mathrm{C}}} (G_{\mathrm{b}n} - G_{\mathrm{Cb}})(\mathbf{1}^{\mathrm{p}}U_n^{\mathrm{p}} + \mathbf{1}^{\mathrm{n}}U_n^{\mathrm{n}})e^{jn\omega_1 t} \tag{7.123}$$

$$\boldsymbol{i}_{\mathrm{Cr}} = \sqrt{2}\mathrm{Re}\sum_{n\in N_{\mathrm{C}}} jB_{\mathrm{b}n}(\mathbf{1}^{\mathrm{p}}U_n^{\mathrm{p}} + \mathbf{1}^{\mathrm{n}}U_n^{\mathrm{n}})e^{jn\omega_1 t} \tag{7.124}$$

$$\boldsymbol{i}_{\mathrm{Cu}} = \sqrt{2}\mathrm{Re}\sum_{n\in N_{\mathrm{C}}} (Y_{\mathrm{d}n}\boldsymbol{U}_n + \mathbf{1}^{\mathrm{n}}Y_{\mathrm{u}n}^{\mathrm{p}}U_n^{\mathrm{p}} + \mathbf{1}^{\mathrm{p}}Y_{\mathrm{u}n}^{\mathrm{n}}U_n^{\mathrm{n}})e^{jn\omega_1 t}. \tag{7.125}$$

With such components of the current $\boldsymbol{i}_{\mathrm{C}}(t)$, the load current decomposition has the form

$$\boldsymbol{i}(t) = \boldsymbol{i}_{\mathrm{Ca}}(t) + \boldsymbol{i}_{\mathrm{Cs}}(t) + \boldsymbol{i}_{\mathrm{Cr}}(t) + \boldsymbol{i}_{\mathrm{Cu}}(t) + \boldsymbol{i}_{\mathrm{G}}(t) \tag{7.126}$$

where

$$\boldsymbol{i}_{\mathrm{G}}(t) = \sum_{n\in N_{\mathrm{G}}} \boldsymbol{i}_n. \tag{7.127}$$

All these components of the load current are mutually orthogonal so that their three-phase rms values satisfy the relationship

$$||\boldsymbol{i}||^2 = ||\boldsymbol{i}_{\mathrm{Ca}}||^2 + ||\boldsymbol{i}_{\mathrm{Cs}}||^2 + ||\boldsymbol{i}_{\mathrm{Cr}}||^2 + ||\boldsymbol{i}_{\mathrm{Cu}}||^2 + ||\boldsymbol{i}_{\mathrm{G}}||^2 \tag{7.128}$$

where

$$||\boldsymbol{i}_{\mathrm{Ca}}|| = G_{\mathrm{Cb}}||\boldsymbol{u}_{\mathrm{C}}|| = \frac{P_{\mathrm{C}}}{||\boldsymbol{u}_{\mathrm{C}}||} \tag{7.129}$$

$$||\boldsymbol{i}_{\mathrm{Cs}}|| = \sqrt{3}\sqrt{\sum_{n\in N_{\mathrm{C}}} I_{\mathrm{s}n}^2} = \sqrt{3}\sqrt{\sum_{n\in N_{\mathrm{C}}} (G_{\mathrm{b}n} - G_{\mathrm{Cb}})^2\, U_n^2} \tag{7.130}$$

$$||\boldsymbol{i}_{\mathrm{Cr}}|| = \sqrt{3}\sqrt{\sum_{n\in N_{\mathrm{C}}} I_{\mathrm{r}n}^2} = \sqrt{3}\sqrt{\sum_{n\in N_{\mathrm{C}}} B_{\mathrm{b}n}^2\, U_n^2} \tag{7.131}$$

$$||\boldsymbol{i}_{\mathrm{Cu}}|| = \sqrt{3}\sqrt{\sum_{n\in N_{\mathrm{C}}} I_{\mathrm{u}n}^2} \tag{7.132}$$

$$||\boldsymbol{i}_{\mathrm{G}}|| = \sqrt{\sum_{n\in N_{\mathrm{G}}} ||\boldsymbol{i}_n||^2}. \tag{7.133}$$

The power equation of an HGL with asymmetrical supply voltage has the same structure as that of the HGL with a symmetrical supply, namely

$$S \stackrel{\mathrm{df}}{=} ||\boldsymbol{u}||\cdot||\boldsymbol{i}|| = \sqrt{||\boldsymbol{u}_{\mathrm{C}}||^2 + ||\boldsymbol{u}_{\mathrm{G}}||^2}\,\sqrt{||\boldsymbol{i}_{\mathrm{C}}||^2 + ||\boldsymbol{i}_{\mathrm{G}}||^2} = \sqrt{S_{\mathrm{C}}^2 + S_{\mathrm{G}}^2 + S_{\mathrm{E}}^2} \tag{7.134}$$

only the apparent power S_{C} should be calculated as presented in Section 7.12 but for harmonic orders n only from the subset N_{C}.

7.17 Active Power Components in 3p3w Circuits

The concept of the working, reflected, and detrimental active powers, explained for single-phase circuits in Section 6.14 can be extended to 3p3w circuits.

Let us consider for this purpose a purely resistive unbalanced circuit shown in Fig. 7.33, supplied from a voltage source of symmetrical and sinusoidal internal voltage **e**.

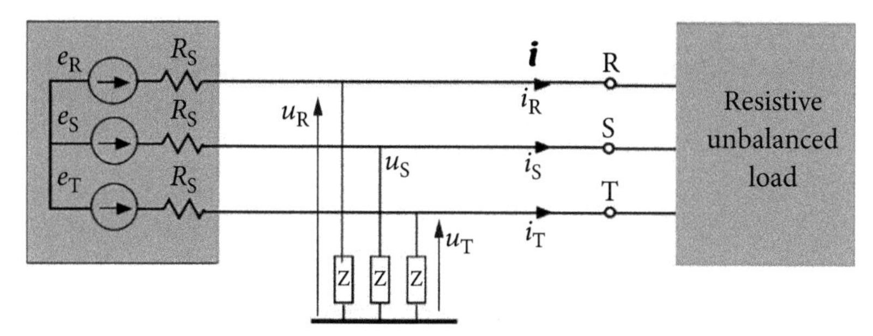

Figure 7.33 A circuit with an unbalanced resistive load.

The vector of the load currents in such a circuit is composed of the vector of the positive and negative sequence symmetrical components, $\boldsymbol{i}^{\mathrm{p}}$ and $\boldsymbol{i}^{\mathrm{n}}$, namely

$$[i_{\mathrm{R}},\, i_{\mathrm{S}},\, i_{\mathrm{T}}]^{\mathrm{T}} \stackrel{\mathrm{df}}{=} \boldsymbol{i} = \boldsymbol{i}^{\mathrm{p}} + \boldsymbol{i}^{\mathrm{n}}.$$

Since the load voltage, due to asymmetrical currents, is asymmetrical, it also contains the positive and negative sequence components

$$[u_{\mathrm{R}},\, u_{\mathrm{S}},\, u_{\mathrm{T}}]^{\mathrm{T}} \stackrel{\mathrm{df}}{=} \boldsymbol{u} = \boldsymbol{u}^{\mathrm{p}} + \boldsymbol{u}^{\mathrm{n}}.$$

The positive and negative sequence quantities are mutually orthogonal, hence the active power at the load terminals is equal to

$$P = (\boldsymbol{u}, \boldsymbol{i}) = (\boldsymbol{u}^{\mathrm{p}}, \boldsymbol{i}^{\mathrm{p}}) + (\boldsymbol{u}^{\mathrm{n}}, \boldsymbol{i}^{\mathrm{n}}) = P^{\mathrm{p}} + P^{\mathrm{n}}.$$

It was assumed that the voltage **e** was symmetrical, therefore the negative sequence symmetrical component of the voltage at the load terminals occurs only due to the voltage drop of the negative sequence current at the supply source resistance, therefore

$$P^{\mathrm{n}} = (\boldsymbol{u}^{\mathrm{n}}, \boldsymbol{i}^{\mathrm{n}}) = (-R_{\mathrm{s}}\, \boldsymbol{i}^{\mathrm{n}}, \boldsymbol{i}^{\mathrm{n}}) = -R_{\mathrm{s}}\, ||\boldsymbol{i}^{\mathrm{n}}||^2 < 0. \tag{7.135}$$

The active power of the negative sequence component of the load voltages and currents is negative. Thus, these components transfer energy back to the supply source. Since the internal voltage **e** is of the positive sequence, while $\boldsymbol{i}^{\mathrm{n}}$ is the current of the negative sequence, the energy cannot be returned to the internal voltage source, but is dissipated, with power P^{n}, specified by (7.135), at the source internal resistance R_{S}. Its negative value will be regarded as a *reflected active power* and denoted by P_{r}, namely

$$P_{\mathrm{r}} \overset{\mathrm{df}}{=} -P^{\mathrm{n}} = -(\boldsymbol{u}^{\mathrm{n}}, \boldsymbol{i}^{\mathrm{n}}). \tag{7.136}$$

Thus, the energy needed for supplying an unbalanced load with the active power P is higher than the integral of that power. Such an unbalanced load has to be supplied with the active power of the positive sequence symmetrical component P^{p}, meaning with the *working active power*, P_{w}.

$$P_{\mathrm{w}} \overset{\mathrm{df}}{=} P^{\mathrm{p}} = P - P^{\mathrm{n}} > P. \tag{7.137}$$

The relation between the working and reflected active powers, in three-phase, resistive circuits with unbalanced loads is visualized in Fig. 7.34.

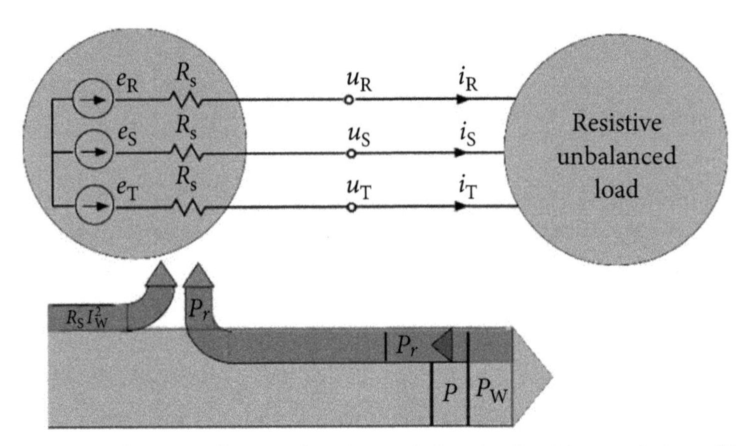

Figure 7.34 A diagram of energy flow in a resistive circuit with an unbalanced load.

Illustration 7.9 *Parameters of the circuit shown in Fig. 7.35 with sinusoidal and symmetrical distribution voltage* **e** *and unbalanced load were selected in such a way that at the line-to-ground voltage of rms value E = 230V, the load active power is P = 100 kW.*

Assuming that the crms value of the distribution voltage is

$$E_{\mathrm{R}} = 230\, e^{j0}\ \mathrm{V}$$

the crms values of the line currents are equal to

$$\boldsymbol{I}_{\mathrm{R}} = -\boldsymbol{I}_{\mathrm{T}} = 279.2\, e^{j30^{\circ}}\ \mathrm{A}$$

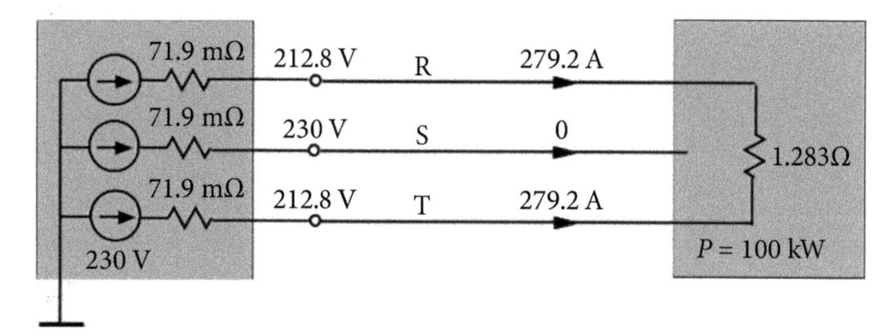

Figure 7.35 A three-phase resistive circuit with an unbalanced load.

and consequently, the crms values of their symmetrical components

$$\begin{bmatrix} \boldsymbol{I}^{\rm p} \\ \boldsymbol{I}^{\rm n} \end{bmatrix} = \frac{1}{3}\begin{bmatrix} 1, \alpha, \alpha^* \\ 1, \alpha^*, \alpha \end{bmatrix}\begin{bmatrix} \boldsymbol{I}_{\rm R} \\ \boldsymbol{I}_{\rm S} \\ \boldsymbol{I}_{\rm T} \end{bmatrix} = \begin{bmatrix} 161.2\, e^{j0^{\rm O}} \\ 161.2\, e^{-j60^{\rm O}} \end{bmatrix} \text{A.}$$

The crms values of the load voltage are

$$\boldsymbol{U}_{\rm R} = \boldsymbol{E}_{\rm R} - R_{\rm s}\,\boldsymbol{I}_{\rm R} = 212.8\, e^{j2.7^{\rm O}}\ \text{V}$$

$$\boldsymbol{U}_{\rm T} = \boldsymbol{E}_{\rm T} - R_{\rm s}\,\boldsymbol{I}_{\rm T} = 212.8\, e^{j117.3^{\rm O}}\ \text{V}$$

and consequently, the crms values of their symmetrical components

$$\begin{bmatrix} \boldsymbol{U}^{\rm p} \\ \boldsymbol{U}^{\rm n} \end{bmatrix} = \frac{1}{3}\begin{bmatrix} 1, \alpha, \alpha^* \\ 1, \alpha^*, \alpha \end{bmatrix}\begin{bmatrix} \boldsymbol{U}_{\rm R} \\ \boldsymbol{U}_{\rm S} \\ \boldsymbol{U}_{\rm T} \end{bmatrix} = \begin{bmatrix} 218.4\, e^{j0^{\rm O}} \\ 11.6\, e^{j120^{\rm O}} \end{bmatrix} \text{V.}$$

The working active power of the load is

$$P_{\rm w} = P^{\rm p} = (\boldsymbol{u}^{\rm p}, \boldsymbol{i}^{\rm p}) = 3U^{\rm p}\, I^{\rm p} = 105.6\ \text{kW}$$

while the reflected active power of the load

$$P_{\rm r} = P^{\rm n} = -(\boldsymbol{u}^{\rm n}, \boldsymbol{i}^{\rm n}) = -3\text{Re}\,\{\boldsymbol{U}^{\rm n}\boldsymbol{I}^{{\rm n}*}\} = 5.6\ \text{kW}.$$

This is the power of energy loss on the supply source resistance.

****

Let us assume that the internal voltage of a distribution system $\boldsymbol{e}$ is sinusoidal but asymmetrical and it supplies an inductor motor as shown in Fig. 7.25.

In response to an asymmetrical supply voltage, the motor current $\boldsymbol{i}$ contains positive and negative sequence components, $\boldsymbol{i}^{\rm p}$ and $\boldsymbol{i}^{\rm n}$. The active power at the motor terminals is

$$P = (\boldsymbol{u}, \boldsymbol{i}) = (\boldsymbol{u}^{\rm p}, \boldsymbol{i}^{\rm p}) + (\boldsymbol{u}^{\rm n}, \boldsymbol{i}^{\rm n}) = P^{\rm p} + P^{\rm n}$$

but only the active power of the positive sequence $P^{\rm p}$ can be converted, with some losses in the motor, into a mechanical power on the motor shaft. Therefore, only this component of the motor active power P can be regarded as the motor working power; that is,

$$P_{\mathrm{w}} \stackrel{\mathrm{df}}{=} P^{\mathrm{p}}.$$

The negative sequence current $\boldsymbol{i}^{\mathrm{n}}$ creates a rotating magnetic field Φ^{n} which rotates in the opposite direction to the direction of the shaft rotation as a sort of braking torque so that it reduces the motor torque T. The energy delivered to the motor by the voltage and current of the negative sequence is dissipated in the motor as the heat, which raises its temperature and reduces its life span. Thus, the active power of the negative sequence symmetrical component P^{n} should be regarded as a *detrimental active power*. It will be denoted by P_{d}.

$$P_{\mathrm{d}} \stackrel{\mathrm{df}}{=} P^{\mathrm{n}}.$$

With new symbols, the active power measured by a wattmeter at the motor terminals is the sum of working and detrimental powers

$$P = P^{\mathrm{p}} + p^{\mathrm{n}} \stackrel{\mathrm{df}}{=} P_{\mathrm{w}} + P_{\mathrm{d}}. \tag{7.138}$$

Let us assume that the internal voltage $\boldsymbol{e}$ of the distribution system is distorted. Suppose it contains a harmonic of the nth order, which is of the positive sequence, for example the seventh order. It creates the rotating magnetic field which rotates at angular velocity versus the rotor, approximately $(n{-}1)\omega_1$, while the rotor rotates at the angular velocity approximately equal to ω_1. At such a velocity of the magnetic field rotation, the motor behaves as if it is in a permanent starting mode. The energy delivered by the harmonic in such a mode is dissipated as heat in the motor winding.

Let us assume that the voltage $\boldsymbol{e}$ contains a harmonic of the negative sequence, for example, the fifth order harmonic. It creates the magnetic flux which rotates in the opposite direction to the rotor at an angular velocity of $(n{+}1)\omega_1$. The energy delivered by such harmonics is entirely converted into heat which increases the motor temperature. Thus, the active power P_n of such voltage harmonic contributes to the detrimental active power P_{d} rather than to the working active power P_{w}. Thus, if P_{h} is the active power of all harmonics in the internal voltage $\boldsymbol{e}$, then the detrimental active power of induction motors increases to

$$P_{\mathrm{d}} = P^{\mathrm{n}} + P_{\mathrm{h}}. \tag{7.139}$$

When a source of sinusoidal voltage supplies a harmonics-generating resistive load, then the active power of the nth order harmonic at the load terminals is

$$P_n = ||\boldsymbol{u}_n||\ ||\boldsymbol{i}_n|| = (-R_{\mathrm{s}}||\boldsymbol{i}_n||)||\boldsymbol{i}_n|| = -R_{\mathrm{s}}||\boldsymbol{i}_n||^2 < 0,$$

thus all of them are negative. Thus, the load-generated active power is

$$\sum_{n\notin 1} P_{\rm n} = P_2 + P_3 + P_4 + \ldots \stackrel{\rm df}{=} P_G < 0.$$

Therefore, the rejected active power $P_{\rm r}$ of unbalanced, harmonics generating loads is composed of the negative sequence active power of the fundamental harmonic and the load-generated active power, namely

$$P_{\rm r} = -P^{\rm n} + P_{\rm G}. \tag{7.140}$$

It increases the working active power $P_{\rm w}$ as compared to the active power P measured at the load terminals.

Decomposition of the active power P into components of different technical meanings

$$P = P_{\rm w} + P_{\rm r} + P_{\rm d} \tag{7.141}$$

has its equivalence in the current decomposition. Each power in the last formula is associated with a component of the active current $\boldsymbol{i}_{\rm Ca}$(t), defined by (7.53), namely

$$\boldsymbol{i}_{\rm Ca}(t) = \boldsymbol{i}_{\rm w}(t) + \boldsymbol{i}_{\rm f}(t) + \boldsymbol{i}_{\rm d}(t) \tag{7.142}$$

where $\boldsymbol{i}_{\rm w}$(t) is the vector of a *working active current*, $\boldsymbol{i}_{\rm f}$(t) is the vector of a *reflected active current*, and $\boldsymbol{i}_{\rm d}$(t) is the vector of a *detrimental active current*. The index "f" was used instead of "r" in the symbol of a reflected current to avoid confusion with the common symbol of a reactive current.

With (7.142), decomposition (7.126) can be rewritten to the form

$$\boldsymbol{i}(t) = \boldsymbol{i}_{\rm w}(t) = \boldsymbol{i}_{\rm f}(t) + \boldsymbol{i}_{\rm d}(t) + \boldsymbol{i}_{\rm Cs}(t) + \boldsymbol{i}_{\rm Cr}(t) + \boldsymbol{i}_{\rm Cu}(t) + \boldsymbol{i}_{\rm G}(t). \tag{7.143}$$

The working active current is defined as follows

$$\boldsymbol{i}_{\rm w}(t) \stackrel{\rm df}{=} \frac{P_{\rm W}}{||\boldsymbol{u}_{\rm w}||^2}\boldsymbol{u}_{\rm w}(t) = \frac{P_1^{\rm p}}{||\boldsymbol{u}_1^{\rm p}||^2}\boldsymbol{u}_1^{\rm p}(t) = G_{\rm w}\boldsymbol{u}_1^{\rm p}(t) \tag{7.144}$$

where

$$G_{\rm W} \stackrel{\rm df}{=} \frac{P_1^{\rm p}}{||\boldsymbol{u}_1^{\rm p}||^2} \tag{7.145}$$

is the *working conductance* of the load.

The reflected active current is defined as follows

$$\boldsymbol{i}_{\mathrm{f}}(t) \stackrel{\mathrm{df}}{=} \frac{P_{\mathrm{r}}}{||\boldsymbol{u}_{\mathrm{r}}||^2}\boldsymbol{u}_{\mathrm{r}}(t) = \frac{P_1^{\mathrm{n}}}{||\boldsymbol{u}_1^{\mathrm{n}}||^2}\boldsymbol{u}_1^{\mathrm{n}}(t) = G_{\mathrm{f}}\,\boldsymbol{u}_1^{\mathrm{n}}(t) \tag{7.146}$$

where

$$G_{\mathrm{f}} \stackrel{\mathrm{df}}{=} \frac{P_1^{\mathrm{n}}}{||\boldsymbol{u}_1^{\mathrm{n}}||^2} \tag{7.147}$$

is the *reflected conductance* of the load.

Summary

This chapter showed that the commonly used power equation of three-phase three-wire circuits does not describe the power properties of such circuits with unbalanced loads correctly, even if the supply voltage is sinusoidal and the load is linear and time invariant. It is because the commonly used definitions of apparent power do not characterize the supply sources concerning energy loss at its delivery adequately. Moreover, the commonly used power equation of such circuits does not take into account the effect of the supply current asymmetry on the apparent power of the supply.

The chapter presents reasoning that enables the selection of the apparent power definition. It demonstrates that the supply current can be decomposed into mutually orthogonal, active, reactive, and unbalanced currents. This decomposition reveals the current asymmetry as the cause of the increase of the supply current three-phase rms value and associates a current component with this phenomenon. This observation enables the development of a true power equation of 3p3w circuits with sv&c but asymmetrical currents with a new power quantity called unbalanced power. The selection of the apparent power definition and the developed power equation provide the right starting point for studies on the power properties of three-phase three-wire circuits with nonsinusoidal voltages and currents. The decomposition of the supply current into the Current's Physical Components was extended next for unbalanced loads supplied with asymmetrical voltage.

Chapter 8
CPC and Powers in Four-Wire Circuits

8.1 Neutral Conductor

Loads in three-phase three-wire circuits (3p3w), such as that discussed in the previous chapter, are supplied with line-to-line voltages. When single-phase loads have to be supplied too, a neutral conductor is needed for that. Therefore, distribution systems that have to supply both three-phase and single-phase loads are built as three-phase circuits with a neutral conductor, meaning, four-wire (3p4w) circuits as shown in Fig. 8.1.

Such circuits are supplied from a transformer with secondary windings configured in Y with a grounded star point, as shown in Fig. 8.1.

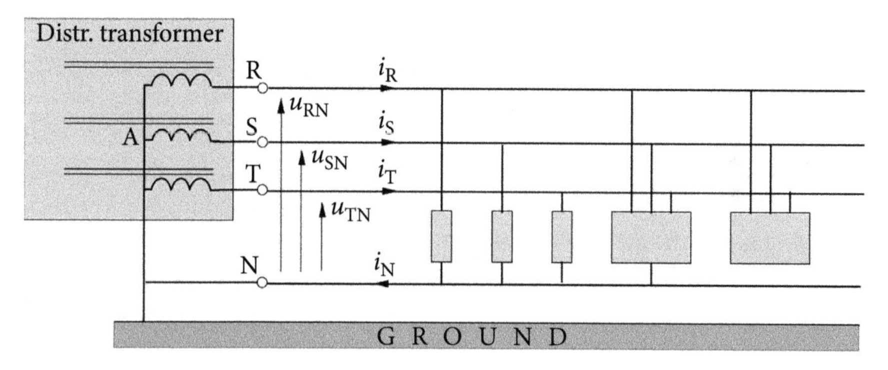

Figure 8.1 A three-phase four-wire (3p4w) circuit.

Lighting and single-phase loads are supplied with line-to-neutral voltages u_{RN}, u_{SN}, and u_{TN}. Three-phase equipment is supplied with line-to-line voltages u_{RS}, u_{ST}, and u_{TR}. The star point of three-phase equipment may or may not be connected to the neutral conductor, referred to in brief as a *neutral.*

The supply transformer is built in such a way that its winding impedances are as similar as possible. The stray inductance and the winding resistance of the transformer contribute mainly to the supply line impedance $\boldsymbol{Z}_{\rm s}$. The impedance of the neutral, $\boldsymbol{Z}_{\rm N}$, is usually much lower than the transformer impedance, so that can be assumed to be zero.

The previous chapter on the Current's Physical Components and powers in 3p3w circuits describes the power properties of three-phase circuits without a neutral conductor. The sum of supply currents in such circuits is equal to zero at each instant of

Powers and Compensation in Circuits with Nonsinusoidal Current. Leszek S. Czarnecki, Oxford University Press.
 DOI: 10.1093/oso/9780198879206.003.0008

time. This is no longer true in 3p3w circuits with a neutral, where

$$i_R(t) + i_S(t) + i_T(t) \equiv i_N(t).$$

Consequently, some definitions of power-related quantities introduced in Chapter 7 and conclusions on power properties are no longer valid.

Loads in 3p4w circuits are supplied with line-to-neutral voltages. However, voltages in three-phase circuits are sometimes referenced or even measured against different points. These could be the ground, neutral, or artificial zero, as shown in Fig. 8.2.

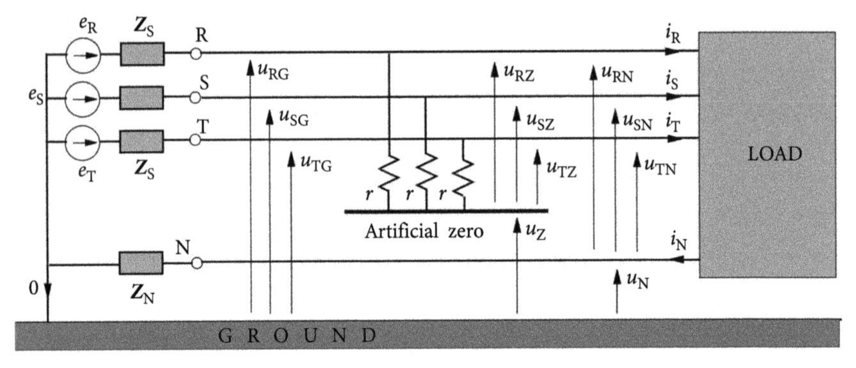

Figure 8.2 Voltages in a 3p4w circuit.

The choice of the reference point does not, of course, affect the energy flow in the circuit. Nonetheless, this choice can confuse. To avoid it, relations between these voltages should be distinctively specified.

It is assumed in this chapter that internal voltages e_R, e_S, and e_T of the supply distribution system are mutually symmetrical and of the positive sequence, meaning

$$e_S(t) = e_R\left(t - \frac{T}{3}\right), \quad e_T(t) = e_R\left(t - 2\frac{T}{3}\right) = e_R\left(t + \frac{T}{3}\right)$$

and without any DC component.

The line voltages measured between R, S, and T lines and the ground, u_{RG}, u_{SG}, and u_{TG}, are reduced as compared to the internal voltages e_R, e_S, and e_T of the distribution system by the voltage drop on the internal impedance, $\boldsymbol{Z}_s$, of the circuit. Consequently, when due to the load imbalance, the supply currents are asymmetrical, and these voltages are asymmetrical as well. These voltages are not applied to the load, however. The load is supplied by line-to-neutral voltages u_{RN}, u_{SN}, and u_{TN} are reduced as compared to voltages u_{RG}, u_{SG}, and u_{TG} by the voltage u_N, meaning by the voltage drop of the current i_N on the neutral impedance $\boldsymbol{Z}_N$. For example, the nth order harmonic of the voltage u_{RN} is equal to

$$\boldsymbol{U}_{RNn} = \boldsymbol{E}_{Rn} - \boldsymbol{Z}_{sn}\boldsymbol{I}_{Rn} - \boldsymbol{Z}_{Nn}\boldsymbol{I}_{Nn}.$$

Since the neutral current is a sum of the supply currents, then at non-zero neutral impedance $\boldsymbol{Z}_{\mathrm{N}}$, the individual line-to-neutral voltages u_{RN}, u_{SN}, and u_{TN} are affected by all supply currents i_{R}, i_{S}, and i_{T}. It means, for example, that a single-phase load in phase S can disturb the load in phase T and the opposite. Therefore, the impedance of the neutral $\boldsymbol{Z}_{\mathrm{N}}$ should be as low as possible.

The voltages specified with the artificial zero as the reference point, u_{RZ}, u_{SZ}, and u_{TZ} are reduced as compared to that specified versus the ground u_{RG}, u_{SG}, and u_{TG}, by the voltage u_{Z}. Its crms value for the nth order harmonic, calculated from the nodal equation, is

$$\boldsymbol{U}_{\mathrm{Z}n} = \frac{G\boldsymbol{U}_{\mathrm{RG}n} + G\boldsymbol{U}_{\mathrm{SG}n} + G\boldsymbol{U}_{\mathrm{TG}n}}{3G} = \frac{1}{3}(\boldsymbol{E}_{\mathrm{R}n} + \boldsymbol{E}_{\mathrm{S}n} + \boldsymbol{E}_{\mathrm{T}n} - \boldsymbol{Z}_{\mathrm{s}n}\boldsymbol{I}_{\mathrm{N}n}), \qquad G = \frac{1}{r}.$$

This formula specifies the difference between the voltage harmonic crms values measured with the artificial zero and the circuit ground as the reference point.

When the internal voltages e_{R}, e_{S}, and e_{T} of the supply distribution system are mutually symmetrical, then the voltage harmonics of the order $n = 3k + 1$ are of the positive sequence and their sum is equal to zero. Similarly, the voltage harmonics of the order $n = 3k - 1$ are of the negative sequence and their sum is also equal to zero. It is only harmonics of the order $n = 3k$ do not add up to zero, because these harmonics are in phase; they are of the zero sequence. Thus,

$$\boldsymbol{E}_{\mathrm{R}n} + \boldsymbol{E}_{\mathrm{S}n} + \boldsymbol{E}_{\mathrm{T}n} = \begin{cases} 0, & \text{for } n \neq 3k \\ 3\boldsymbol{E}_n, & \text{for } n = 3k \end{cases}.$$

Therefore, for the harmonics of the positive and negative sequence

$$\boldsymbol{U}_{\mathrm{Z}n} = -\frac{1}{3}\boldsymbol{Z}_{\mathrm{s}n}\boldsymbol{I}_{\mathrm{N}n}$$

while for the zero sequence

$$\boldsymbol{U}_{\mathrm{Z}n} = \boldsymbol{E}_n - \frac{1}{3}\boldsymbol{Z}_{\mathrm{s}n}\boldsymbol{I}_{\mathrm{N}n}.$$

The load in 3p4w circuits is supplied by line-to-neutral voltages u_{RN}, u_{SN}, and u_{TN}. They depend on the internal voltages e_{R}, e_{S}, and e_{T} of the distribution system and on the voltage drop on the distribution system impedance $\boldsymbol{Z}_{\mathrm{s}}$ and the neutral impedance $\boldsymbol{Z}_{\mathrm{N}}$. The internal voltages and the voltage drops are not available by the measurement at terminals R, S, and T, however. The power properties of fixed-parameter loads should be specified by the load currents i_{R}, i_{S}, and i_{T} measured at the load terminals at voltages measured against the neutral u_{RN}, u_{SN}, and u_{TN}. The index "N" for the sake of simplicity will be ignored.

8.2 Current's Three-Phase Rms Value in 3p4w Circuit

The three-phase current's $\boldsymbol{i}(t)$ three-phase rms value $||\boldsymbol{i}||$ in 3p3w circuits is the rms value of a single-phase current, which results in the single line with the same active power P as the current $\boldsymbol{i}(t)$ in the three-phase symmetrical device of the same line resistance. Its concept was introduced for 3p3w circuits in Ref. [93].

Dissipation of energy in a three-phase transmission device with a neutral conductor is caused not only by line currents i_R, i_S, and i_T but also by the neutral current i_N, however. Thus, the above definition of a three-phase rms value is no longer valid. To distinguish the three-phase rms value $||\boldsymbol{i}||$ in 3p3w circuits from this value in 3p4w circuits, the former will be denoted, if needed, by $||\boldsymbol{i}|| = ||\boldsymbol{i}||_3$, while in 3p4w circuits it will be denoted by $||\boldsymbol{i}||_4$.

To develop the definition of the three-phase rms value in 3p4w circuits, let us consider a resistive device shown in Fig. 8.3a.

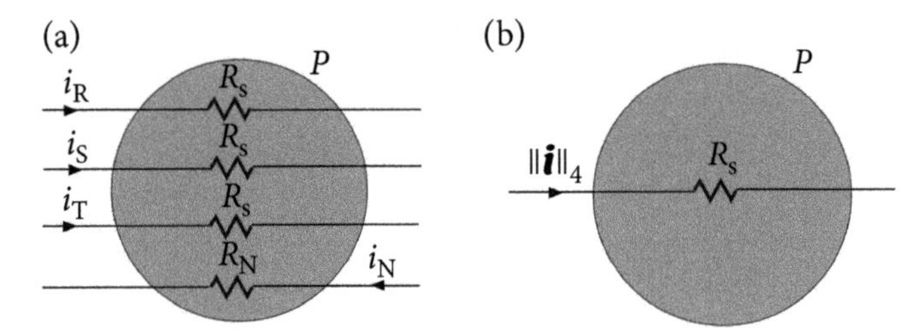

Figure 8.3 (a) A resistive 3p4w device and (b) a single-phase device that is equivalent as to the active power P to 3p4w device as to the active power P.

The active power of this device is equal to

$$P = \left(||i_R||^2 + ||i_S||^2 + ||i_T||^2\right) R_s + ||i_N||^2 R_N$$

or

$$P = ||\boldsymbol{i}||_3^2 R_s + ||i_N||^2 R_N.$$

The first term specifies the active power of the single-phase device with resistance R_s, and line current of the three-phase rms value $I = ||\boldsymbol{i}||_4$

$$P = ||\boldsymbol{i}||_4^2 R_s.$$

Devices in Fig. 8.3a and b are mutually equivalent as to the active power P, if

$$||\boldsymbol{i}||_4^2 R_s = ||\boldsymbol{i}||_3^2 R_s + ||i_N||^2 R_N$$

hence

$$||\boldsymbol{i}||_4 = ||\boldsymbol{i}||_3 \sqrt{1 + \frac{R_N}{R_s}\left(\frac{||i_N||}{||\boldsymbol{i}||_3}\right)^2}. \tag{8.1}$$

This is the three-phase rms value of a three-phase current of a four-wire device. Formula (8.1) takes into account a common fact that 3p4w devices may not have internal symmetry. The resistance of the neutral is usually different from the resistance of the supply R, S, and T lines. Unfortunately, the current three-phase rms value $||\boldsymbol{i}||_4$ cannot be calculated without information on the device resistance asymmetry, meaning the ratio R_N/R_s.

If this would not cause any confusion, this value will be denoted in this chapter without index "4", meaning

$$||\boldsymbol{i}|| \stackrel{\text{df}}{=} ||\boldsymbol{i}||_4.$$

Formula (8.1) was developed without any conditions as to the current waveform. Therefore, it applies not only to circuits with sinusoidal currents but also to circuits with nonsinusoidal currents.

Observe, that when there is no dissipation of energy associated with the neutral current, meaning $||i_N|| = 0$ or $R_N = 0$, then

$$||\boldsymbol{i}||_4 = ||\boldsymbol{i}||_3 = ||\boldsymbol{i}||.$$

Illustration 8.1 *Let us calculate the supply current three-phase rms value $||\boldsymbol{i}||_4$ of 3p4w device shown in Fig. 8.4*

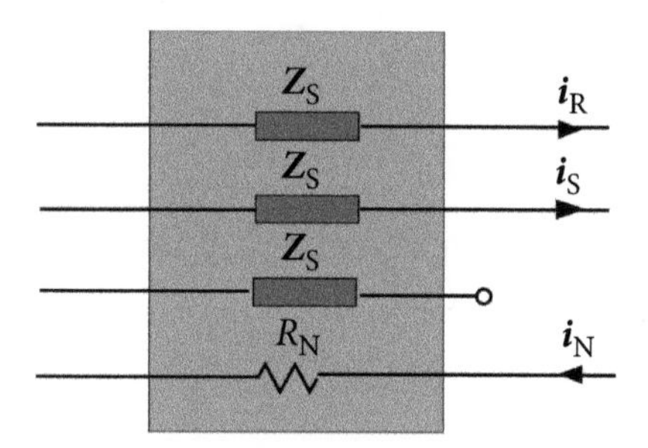

Figure 8.4 An example of a 3p4w device.

if the line currents are: $i_R = 50\sqrt{2}\sin\omega_1 t$ A, $i_S = 50\sqrt{2}\sin(\omega_1 t - 120°)$A, $i_T = 0$, *assuming that* $R_s = \text{Re}\{\boldsymbol{Z}_s\} = 2\ \Omega$, *and* $R_N = 0.2\ \Omega$.

The neutral current is equal to

$$i_N = i_R + i_S + i_T = 50\sqrt{2}\sin\omega_1 t + 50\sqrt{2}\sin(\omega_1 t - 120°) = 50\sqrt{2}\sin(\omega_1 t - 60°)\,\text{A}.$$

The rms value $||\boldsymbol{i}||_3$ is equal to

$$||\boldsymbol{i}||_3 = \sqrt{||i_R||^2 + ||i_S||^2 + ||i_T||^2} = \sqrt{50^2 + 50^2} = 70.7\ \text{A}$$

hence, according to (8.1), the three-phase rms value of the current is

$$||\boldsymbol{i}||_4 = ||\boldsymbol{i}||_3\sqrt{1+\frac{R_N}{R_s}\left(\frac{||i_N||}{||\boldsymbol{i}||_3}\right)^2} = 70.7\sqrt{1+\frac{0.2}{2}\left(\frac{50}{70.7}\right)^2} = 72.4\ \text{A}.$$

Thus, this illustration shows that the effect of the neutral current on the current rms value could not be high.

****

8.3 CPC in 3p4w Circuits with sv&c and LTI Loads

The basic 3p4w circuit for analyzing power phenomena is a circuit with LTI load supplied from an ideal source of symmetrical and sinusoidal voltage and current (sv&c) of the positive sequence.

To simplify notation, the supply voltage specified against the neutral conductor is denoted in this chapter without index N, meaning, for example, it is denoted as u_R but not as u_{RN}.

Any LTI load in 3p4w circuit, as shown in Fig. 8.2, is equivalent to star (Y)-configured load, shown in Fig. 8.5(a), with admittances

$$Y_R = \frac{I_R}{U_R}, \qquad Y_S = \frac{I_S}{U_S}, \qquad Y_T = \frac{I_T}{U_T}$$

and an ideal (meaning with zero impedance) neutral conductor. Concerning the active power P, such a load is equivalent to a balanced resistive load, shown in Fig. 8.5(b), of the line conductance G_e.

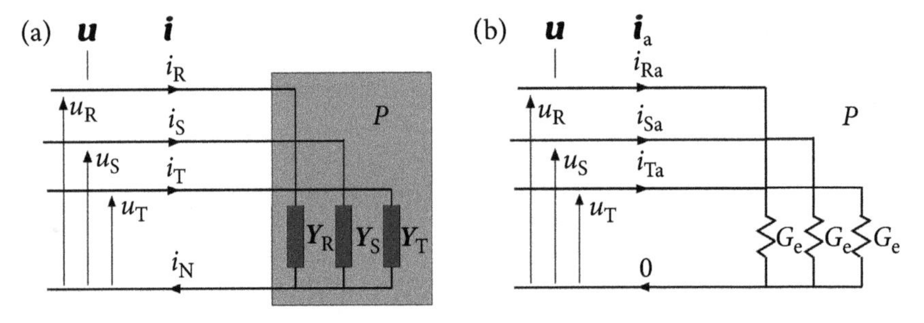

Figure 8.5 (a) Equivalent load configured in Y and (b) resistive balanced load equivalent as to the active power P.

Since the active power of the original load is

$$P = \text{Re}\{Y_R + Y_S + Y_T\}U_R^2 = (G_R + G_S + G_T)\ U_R^2,$$

thus conductance G_e is equal to

$$G_e = \frac{P}{||\mathbf{u}||^2} = \frac{1}{3}\frac{P}{U_R^2} = \frac{1}{3}(G_R + G_S + G_T) \tag{8.2}$$

it will be referred to as an *equivalent conductance* of a load supplied by a 3p4w line. Such an equivalent resistive load draws a current, which is in phase with the supply voltage $\mathbf{u}$ and can be regarded as the active current of the load, namely

$$\mathbf{i}_a(t) = G_e\,\mathbf{u}(t).$$

It is the current of the lowest three-phase rms value $||\mathbf{i}||$ for a load that at voltage $\mathbf{u}$ has active power P. Concerning the reactive power Q, the original load is equivalent to a balanced reactive load, shown in Fig. 8.6b, of susceptance B_e.

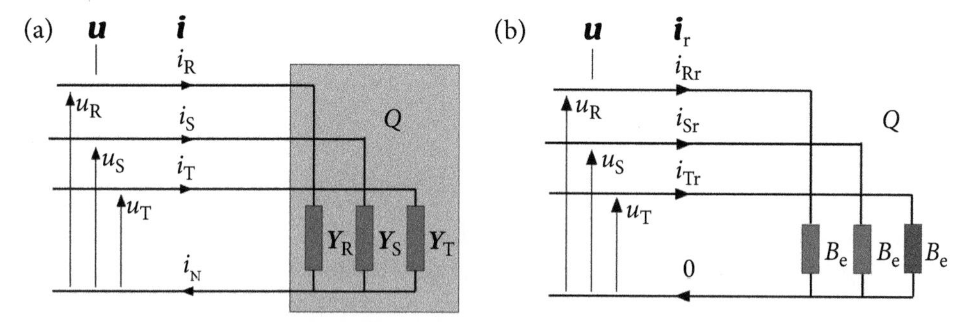

Figure 8.6 (a). Equivalent load configured in Y and (b) reactive balanced load equivalent to the original as to the reactive power Q.

The reactive power Q of the original load is

$$Q = -\mathrm{Im}\{Y_R + Y_S + Y_T\}U_R^2 = -(B_R + B_S + B_T)U_R^2,$$

thus the susceptance of a reactive balanced load, equivalent to the original load in regard to the reactive power Q, is equal to

$$B_e = -\frac{Q}{||\mathbf{u}||^2} = -\frac{1}{3}\frac{Q}{U_R^2} = \frac{1}{3}(B_R + B_S + B_T). \tag{8.3}$$

It will be referred to as the *equivalent susceptance* of loads supplied by a 3p4w line. Such a reactance load draws a reactive current

$$\mathbf{i}_r = B_e\frac{d}{d(\omega_1 t)}\mathbf{u} = \sqrt{2}\mathrm{Re}\left\{jB_e\begin{bmatrix} U_R \\ U_S \\ U_R \end{bmatrix} e^{j\omega_1 t}\right\}. \tag{8.4}$$

This current is symmetrical with the same sequence as the supply voltage, meaning the positive sequence.

The equivalent conductance and the equivalent susceptance can be combined to form the *equivalent admittance*

$$\boldsymbol{Y}_{\mathrm{e}} \stackrel{\mathrm{df}}{=} G_{\mathrm{e}} + jB_{\mathrm{e}} = \frac{1}{3}(\boldsymbol{Y}_{\mathrm{R}} + \boldsymbol{Y}_{\mathrm{S}} + \boldsymbol{Y}_{\mathrm{T}}).$$

The residual component of the current occurs due to the load imbalance and is equal to

$$\boldsymbol{i}_{\mathrm{u}} = \boldsymbol{i} - \boldsymbol{i}_{\mathrm{a}} - \boldsymbol{i}_{\mathrm{r}} = \sqrt{2}\mathrm{Re}\left\{\begin{bmatrix} \boldsymbol{Y}_{\mathrm{R}}\boldsymbol{U}_{\mathrm{R}} \\ \boldsymbol{Y}_{\mathrm{S}}\boldsymbol{U}_{\mathrm{S}} \\ \boldsymbol{Y}_{\mathrm{T}}\boldsymbol{U}_{\mathrm{T}} \end{bmatrix} e^{j\omega_1 t}\right\} - \sqrt{2}\mathrm{Re}\left\{\begin{bmatrix} G_{\mathrm{e}}\boldsymbol{U}_{\mathrm{R}} \\ G_{\mathrm{e}}\boldsymbol{U}_{\mathrm{S}} \\ G_{\mathrm{e}}\boldsymbol{U}_{\mathrm{T}} \end{bmatrix} e^{j\omega_1 t}\right\} =$$

$$- \sqrt{2}\mathrm{Re}\left\{ j \begin{bmatrix} B_{\mathrm{e}}\boldsymbol{U}_{\mathrm{R}} \\ B_{\mathrm{e}}\boldsymbol{U}_{\mathrm{S}} \\ B_{\mathrm{e}}\boldsymbol{U}_{\mathrm{T}} \end{bmatrix} e^{j\omega_1 t}\right\} =$$

$$= \sqrt{2}\mathrm{Re}\left\{\begin{bmatrix} (\boldsymbol{Y}_{\mathrm{R}} - G_{\mathrm{e}} - jB_{\mathrm{e}}) \\ (\boldsymbol{Y}_{\mathrm{S}} - G_{\mathrm{e}} - jB_{\mathrm{e}})\,\alpha^* \\ (\boldsymbol{Y}_{\mathrm{T}} - G_{\mathrm{e}} - jB_{\mathrm{e}})\,\alpha \end{bmatrix} \boldsymbol{U}_{\mathrm{R}}\, e^{j\omega_1 t}\right\} = \sqrt{2}\mathrm{Re}\left\{\begin{bmatrix} \boldsymbol{I}_{\mathrm{Ru}} \\ \boldsymbol{I}_{\mathrm{Su}} \\ \boldsymbol{I}_{\mathrm{Tu}} \end{bmatrix} e^{j\omega_1 t}\right\}. \tag{8.5}$$

The physical nature of the residual current $\boldsymbol{i}_{\mathrm{u}}$ is not clear at this moment. We can only say that this current is associated neither with the active nor with the reactive power. Let us calculate the crms value of the symmetrical component of the positive sequence of this current. It is equal to

$$\begin{aligned} \boldsymbol{I}_{\mathrm{u}}^{\mathrm{p}} &= \frac{1}{3}(\boldsymbol{I}_{\mathrm{Ru}} + \alpha\boldsymbol{I}_{\mathrm{Su}} + \alpha^*\boldsymbol{I}_{\mathrm{Tu}}) = \\ &= \frac{1}{3}[(\boldsymbol{Y}_{\mathrm{R}} - G_{\mathrm{e}} - jB_{\mathrm{e}}) + \alpha(\boldsymbol{Y}_{\mathrm{S}} - G_{\mathrm{e}} - jB_{\mathrm{e}})\,\alpha^* + \alpha^*(\boldsymbol{Y}_{\mathrm{T}} - G_{\mathrm{e}} - jB_{\mathrm{e}})\alpha]\,\boldsymbol{U}_{\mathrm{R}} = \\ &= \frac{1}{3}[(\boldsymbol{Y}_{\mathrm{R}} + \boldsymbol{Y}_{\mathrm{S}} + \boldsymbol{Y}_{\mathrm{T}}) - 3G_{\mathrm{e}} - j3B_{\mathrm{e}}]\boldsymbol{U}_{\mathrm{R}} = 0. \end{aligned}$$

Thus this current does not contain any component of the positive sequence, meaning it occurs due to the supply current asymmetry.

The crms value of the negative sequence component of this current is equal to

$$\begin{aligned} \boldsymbol{I}_{\mathrm{u}}^{\mathrm{n}} &= \frac{1}{3}(\boldsymbol{I}_{\mathrm{Ru}} + \alpha^*\boldsymbol{I}_{\mathrm{Su}} + \alpha\boldsymbol{I}_{\mathrm{Tu}}) = \\ &= \frac{1}{3}[(\boldsymbol{Y}_{\mathrm{R}} - G_{\mathrm{e}} - jB_{\mathrm{e}}) + \alpha^*(\boldsymbol{Y}_{\mathrm{S}} - G_{\mathrm{e}} - jB_{\mathrm{e}})\,\alpha^* + \alpha(\boldsymbol{Y}_{\mathrm{T}} - G_{\mathrm{e}} - jB_{\mathrm{e}})\alpha]\,\boldsymbol{U}_{\mathrm{R}} = \\ &= \frac{1}{3}[(\boldsymbol{Y}_{\mathrm{R}} + \alpha\boldsymbol{Y}_{\mathrm{S}} + \alpha^*\boldsymbol{Y}_{\mathrm{T}})\,\boldsymbol{U}_{\mathrm{R}} \stackrel{\mathrm{df}}{=} \boldsymbol{Y}_{\mathrm{u}}^{\mathrm{n}}\boldsymbol{U}_{\mathrm{R}} \end{aligned}$$

where

$$\boldsymbol{Y}_{\mathrm{u}}^{\mathrm{n}} \stackrel{\mathrm{df}}{=} \frac{1}{3}(\boldsymbol{Y}_{\mathrm{R}} + \alpha\,\boldsymbol{Y}_{\mathrm{S}} + \alpha^*\boldsymbol{Y}_{\mathrm{T}}). \tag{8.6}$$

The crms value of the zero sequence component of the residual current is equal to

$$\begin{aligned} \boldsymbol{I}_{\rm u}^{\rm z} &= \frac{1}{3}(\boldsymbol{I}_{\rm Ru} + \boldsymbol{I}_{\rm Su} + \boldsymbol{I}_{\rm Tu}) = \\ &= \frac{1}{3}[(\boldsymbol{Y}_{\rm R} - G_{\rm e} - jB_{\rm e}) + (\boldsymbol{Y}_{\rm S} - G_{\rm e} - jB_{\rm e})\,\alpha^* + (\boldsymbol{Y}_{\rm T} - G_{\rm e} - jB_{e})\alpha]\,\boldsymbol{U}_{\rm R} = \\ &= \frac{1}{3}[(\boldsymbol{Y}_{\rm R} + \alpha^*\boldsymbol{Y}_{\rm S} + \alpha\boldsymbol{Y}_{\rm T})\,\boldsymbol{U}_{\rm R} \stackrel{\rm df}{=} \boldsymbol{Y}_{\rm u}^{\rm z}\boldsymbol{U}_{\rm R} \end{aligned}$$

where

$$\boldsymbol{Y}_{\rm u}^{\rm z} \stackrel{\rm df}{=} \frac{1}{3}(\boldsymbol{Y}_{\rm R} + \alpha^*\boldsymbol{Y}_{\rm S} + \alpha\boldsymbol{Y}_{\rm T}). \tag{8.7}$$

When phase-to-neutral admittances $\boldsymbol{Y}_{\rm R}$, $\boldsymbol{Y}_{\rm S}$, and $\boldsymbol{Y}_{\rm T}$ are mutually equal, meaning the load is balanced, then admittances $\boldsymbol{Y}_{\rm u}^{\rm n}$ and $\boldsymbol{Y}_{\rm u}^{\rm z}$ are equal to zero and the supply current does not contain $\boldsymbol{i}_{\rm u}$ component. It occurs only when the load is unbalanced. It means that the current $\boldsymbol{i}_{\rm u}$ stands for the unbalanced current. It is composed of the negative and positive sequence components,

$$\boldsymbol{i}_{\rm u} = \boldsymbol{i}_{\rm u}^{\rm n} + \boldsymbol{i}_{\rm u}^{\rm z} \tag{8.8}$$

where

$$\boldsymbol{i}_{\rm u}^{\rm n} \stackrel{\rm df}{=} \sqrt{2}{\rm Re}\left\{\begin{bmatrix} \boldsymbol{I}_{\rm R}^{\rm n} \\ \boldsymbol{I}_{\rm S}^{\rm n} \\ \boldsymbol{I}_{\rm T}^{\rm n} \end{bmatrix} e^{j\omega_1 t}\right\} = \sqrt{2}{\rm Re}\left\{\begin{bmatrix} \boldsymbol{Y}_{\rm u}^{\rm n}\,\boldsymbol{U}_{\rm R} \\ \boldsymbol{Y}_{\rm u}^{\rm n}\,\boldsymbol{U}_{\rm T} \\ \boldsymbol{Y}_{\rm u}^{\rm n}\,\boldsymbol{U}_{\rm S} \end{bmatrix} e^{j\omega_1 t}\right\} = \sqrt{2}{\rm Re}\left\{\boldsymbol{Y}_{\rm u}^{\rm n}\mathbf{1}^{\rm n}\boldsymbol{U}_{\rm R}\,e^{j\omega_1 t}\right\} \tag{8.9}$$

$$\boldsymbol{i}_{\rm u}^{\rm z} \stackrel{\rm df}{=} \sqrt{2}{\rm Re}\left\{\begin{bmatrix} \boldsymbol{I}_{\rm R}^{\rm z} \\ \boldsymbol{I}_{\rm S}^{\rm z} \\ \boldsymbol{I}_{\rm T}^{\rm z} \end{bmatrix} e^{j\omega_1 t}\right\} = \sqrt{2}{\rm Re}\left\{\begin{bmatrix} \boldsymbol{Y}_{\rm u}^{\rm z}\,\boldsymbol{U}_{\rm R} \\ \boldsymbol{Y}_{\rm u}^{\rm z}\,\boldsymbol{U}_{\rm R} \\ \boldsymbol{Y}_{\rm u}^{\rm z}\,\boldsymbol{U}_{\rm R} \end{bmatrix} e^{j\omega_1 t}\right\} = \sqrt{2}{\rm Re}\left\{\boldsymbol{Y}_{\rm u}^{\rm z}\mathbf{1}^{\rm z}\boldsymbol{U}_{\rm R}\,e^{j\omega_1 t}\right\}. \tag{8.10}$$

The symbol $\mathbf{1}^{\rm n}$ in formula (8.9) denotes unit three-phase vectors of the negative sequence, while the symbol $\mathbf{1}^{\rm z}$ in formula (8.10) denotes unit three-phase vectors of the zero sequence. These two currents will be called *negative* and *zero sequence unbalanced currents* of LTI loads, respectively, and consequently, the complex number $\boldsymbol{Y}_{\rm u}^{\rm n}$ will be called *a negative sequence unbalanced admittance*, while $\boldsymbol{Y}_{\rm u}^{\rm z}$ will be called a *zero sequence unbalanced admittance* of such loads.

Unbalanced admittances $\boldsymbol{Y}_{\rm u}^{\rm n}$ and $\boldsymbol{Y}_{\rm u}^{\rm z}$ are equal to zero when admittances $\boldsymbol{Y}_{\rm R}$, $\boldsymbol{Y}_{\rm S}$, and $\boldsymbol{Y}_{\rm T}$ are mutually equal, but this is only a sufficient, but not a necessary condition to have zero unbalanced admittances. They can be zero even if

$$Y_R \neq Y_S \neq Y_T$$

but in such a situation only one of them, Y_u^n or Y_u^z, can be equal to zero. When the load is unbalanced then the supply current has to contain at least one of two unbalanced currents $\boldsymbol{i}_u^n$ or $\boldsymbol{i}_u^z$.

The current decomposition expressed with formula (8.5) can be rewritten with formula (8.8) as

$$\boldsymbol{i} = \boldsymbol{i}_a + \boldsymbol{i}_r + \boldsymbol{i}_u^n + \boldsymbol{i}_u^z. \tag{8.11}$$

Currents $\boldsymbol{i}_a$, $\boldsymbol{i}_r$, $\boldsymbol{i}_u^n$, and $\boldsymbol{i}_u^z$ in this decomposition can be regarded as the Currents' Physical Components (CPC) of three-phase LTI loads with a neutral conductor, supplied from a source of symmetrical sinusoidal voltage. The interpretation of the active and reactive currents, $\boldsymbol{i}_a$, and $\boldsymbol{i}_r$, is the same as in 3p3w circuits. They are associated distinctively with the phenomenon of permanent energy transmission, and consequently, the load active power P, and with the phase shift between the supply voltage and current, thus the load reactive power Q. The negative sequence unbalanced current $\boldsymbol{i}_u^n$ is an effect of the supply current asymmetry due to the load imbalance, but it does not require any neutral conductor for its presence. The zero sequence unbalanced current $\boldsymbol{i}_u^z$ is also an effect of the supply current asymmetry caused by the load imbalance, but it cannot occur in the supply current if the load is not equipped with a neutral conductor.

Equivalent circuit. Decomposition (8.11) means that an LTI load supplied by a 3p4w line with a sinusoidal symmetrical voltage source has an equivalent circuit composed of four parallel three-phase branches that draw $\boldsymbol{i}_a$, $\boldsymbol{i}_r$, $\boldsymbol{i}_u^n$, and $\boldsymbol{i}_u^z$ currents, as shown in Fig. 8.7.

The crms value of the symmetrical components of the unbalanced current of the negative and the zero sequence $\boldsymbol{i}_u^n$ and $\boldsymbol{i}_u^z$ in the line R are proportional, according to (8.9), to the crms value of the voltage u_R, with admittances Y_u^n and Y_u^z as proportionality coefficients. However, the crms value of the negative sequence component $\boldsymbol{i}_u^n$ in the line S is proportional to the crms value $\boldsymbol{U}_T$, and consequently

$$\boldsymbol{I}_{Su}^n = \boldsymbol{Y}_u^n \boldsymbol{U}_T = \boldsymbol{Y}_u^n \alpha^* \boldsymbol{U}_S = (\alpha^* \boldsymbol{Y}_u^n)\, \boldsymbol{U}_S.$$

Similarly, the crms value of the negative sequence component $\boldsymbol{i}_u^n$ in the line T is proportional to the crms value $\boldsymbol{U}_S$, and consequently

$$\boldsymbol{I}_{Tu}^n = \boldsymbol{Y}_u^n \boldsymbol{U}_S = \boldsymbol{Y}_u^n \alpha\, \boldsymbol{U}_T = (\alpha \boldsymbol{Y}_u^n)\, \boldsymbol{U}_T.$$

These two relations are visualized on the equivalent circuit shown in Fig. 8.7.

The crms value of the zero sequence component $\boldsymbol{i}_u^z$ of the supply current in line S is proportional, according to (8.10), to the crms value of the voltage u_R, with

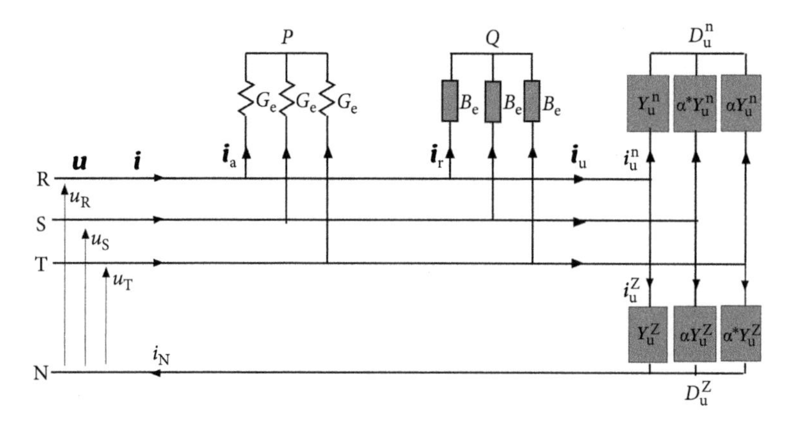

Figure 8.7 Equivalent circuit of three-phase LTI load supplied by a 3p4w line with sv&c.

admittances Y_u^z as the proportionality coefficient and consequently

$$I_{Su}^z = Y_u^z U_R = Y_u^z \alpha\, U_S = (\alpha Y_u^z)\, U_S$$

and similarly, in the line T

$$I_{Tu}^z = Y_u^z U_R = Y_u^z \alpha^* U_T = (\alpha^* Y_u^z)\, U_T.$$

The four physical components of the supply current are mutually orthogonal, hence their rms values fulfill the relationship

$$||\boldsymbol{i}||^2 = ||\boldsymbol{i}_a||^2 + ||\boldsymbol{i}_r||^2 + ||\boldsymbol{i}_u^n||^2 + ||\boldsymbol{i}_u^z||^2. \tag{8.12}$$

Orthogonality of unbalanced currents $\boldsymbol{i}_u^n$ and $\boldsymbol{i}_u^z$ between themselves and other components result from differences in their sequence. Current $\boldsymbol{i}_u^n$ is of negative sequence, current $\boldsymbol{i}_u^z$ is of zero sequence, while currents $\boldsymbol{i}_a$ and $\boldsymbol{i}_r$ are of positive sequence. The relationship (8.12) is illustrated in Fig. 8.8.

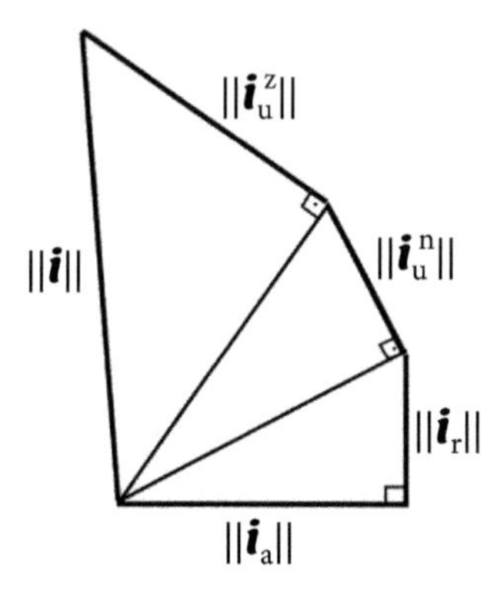

Figure 8.8 A diagram of three-phase rms values of the CPC in 3p4w circuits.

Three-phase rms values of the current particular components are equal to

$$||\boldsymbol{i}_{\rm a}|| = G_{\rm e}||\boldsymbol{u}|| \tag{8.13}$$

$$||\boldsymbol{i}_{\rm r}|| = |B_{\rm e}|\ ||\boldsymbol{u}|| \tag{8.14}$$

$$||\boldsymbol{i}_{\rm u}^{\rm n}|| = Y_{\rm u}^{\rm n}||\boldsymbol{u}|| \tag{8.15}$$

$$||\boldsymbol{i}_{\rm u}^{\rm z}|| = Y_{\rm u}^{\rm z}||\boldsymbol{u}||. \tag{8.16}$$

The active and reactive currents $\boldsymbol{i}_{\rm a}$ and $\boldsymbol{i}_{\rm r}$ are symmetrical and of the positive sequence, while the unbalanced current $\boldsymbol{i}_{\rm u}^{\rm n}$ is symmetrical and of the negative sequence. Thus, these three currents in supply lines R, S, and T add up to zero, not contributing to the neutral current. The neutral current $i_{\rm N}$ is an effect of the presence of the zero sequence component of the unbalanced current, namely

$$i_{\rm N} = 3i_{\rm Ru}^{\rm z} = 3\sqrt{2}{\rm Re}\left\{\boldsymbol{I}_{\rm u}^{\rm z}\, e^{j\omega_1 t}\right\} = 3\sqrt{2}{\rm Re}\left\{\boldsymbol{Y}_{\rm u}^{\rm z}\boldsymbol{U}_{\rm R}\, e^{j\omega_1 t}\right\} \tag{8.17}$$

and its rms value is equal to

$$||i_{\rm N}|| = 3\, Y_{\rm u}^{\rm z}||\boldsymbol{u}||.$$

Illustration 8.2 *Let us calculate the rms values of the supply current physical components for the load shown in Fig. 8.9, assuming that the supply voltage is symmetrical and the rms value of the voltage at terminal* R *is* $U_{\rm R} = 120$ V.

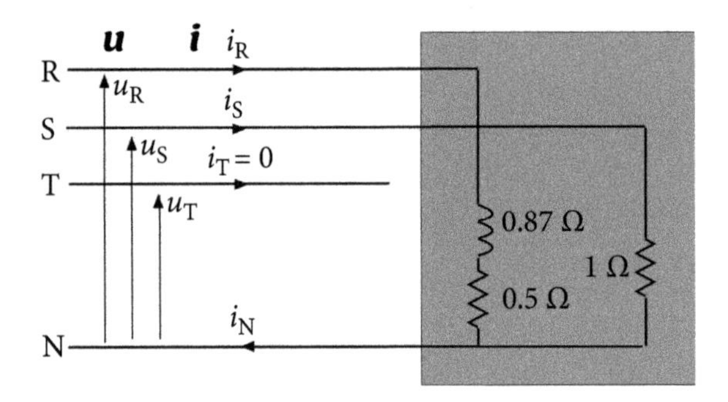

Figure 8.9 An example of an unbalanced load supplied from a 3p4w line.

The line-to-neutral admittances of the load are equal to

$$\boldsymbol{Y}_{\rm R} = \frac{1}{0.5 + j\,0.87} = 0.5 - j\,0.87 \text{ S}, \quad \boldsymbol{Y}_{\rm S} = 1 \text{ S}, \quad \boldsymbol{Y}_{\rm T} = 0.$$

Thus the equivalent admittance of the load equals to

$$\boldsymbol{Y}_{\rm e} = G_{\rm e} + jB_{\rm e} = \frac{1}{3}(\boldsymbol{Y}_{\rm R} + \boldsymbol{Y}_{\rm S} + \boldsymbol{Y}_{\rm T}) = \frac{1}{3}(0.5 - j\,0.87 + 1) = 0.50 - j\,0.29 \text{ S}.$$

The negative sequence unbalanced admittance is equal to zero, since

$$Y_{\rm u}^{\rm n} = \frac{1}{3}(Y_{\rm R} + \alpha Y_{\rm S} + \alpha^* Y_{\rm T}) = \frac{1}{3}(0.5 - j\,0.87 + 1e^{j120^\circ} \times 1 + 0) = 0$$

while the zero sequence unbalanced admittance is equal to

$$Y_{\rm u}^{\rm z} = \frac{1}{3}(Y_{\rm R} + \alpha^* Y_{\rm S} + \alpha\, Y_{\rm T}) = \frac{1}{3}(0.5 - j\,0.87 + 1e^{-j120^\circ} \times 1 + 0) = 0.58\, e^{-j90^\circ}\ {\rm S}.$$

Thus, the supply current contains $\boldsymbol{i}_{\rm u}^{\rm z}$ *but not* $\boldsymbol{i}_{\rm u}^{\rm n}$ *components. Since the three-phase rms value of the supply voltage is*

$$||\boldsymbol{u}|| = \sqrt{3}\, U_{\rm R} = \sqrt{3} \times 120 = 207.8\ {\rm V},$$

three-phase rms values of the supply current physical components are equal to

$$||\boldsymbol{i}_{\rm a}|| = G_{\rm e}||\boldsymbol{u}|| = 0.50 \times 207.8 = 103.9\ {\rm A}$$
$$||\boldsymbol{i}_{\rm r}|| = |B_{\rm e}|\ ||\boldsymbol{u}|| = 0.29 \times 207.8 = 60.3\ {\rm A}$$
$$||\boldsymbol{i}_{\rm u}^{\rm n}|| = Y_{\rm u}^{\rm n}||\boldsymbol{u}|| = 0$$
$$||\boldsymbol{i}_{\rm u}^{\rm z}|| = Y_{\rm u}^{\rm z}||\boldsymbol{u}|| = 0.58 \times 207.8 = 120.5\ {\rm A}.$$

The crms value of the zero sequence unbalanced current in the line R, $i_{\rm Ru}^{\rm z}$, *is equal to*

$$\boldsymbol{I}_{\rm Rn}^{\rm z} = \boldsymbol{Y}_{\rm u}^{\rm z}\boldsymbol{U}_{\rm R} = 0.58\, e^{-j90^\circ} \times 120 = 69.6\, e^{-j90^\circ}\ {\rm A}.$$

The same crms value, of course, has this current in lines S and T. The crms value of the neutral current is

$$\boldsymbol{I}_{\rm N} = 3\boldsymbol{I}_{\rm Rn}^{\rm z} = 3 \times 69.6\, e^{-j90^\circ} = 208.8\, e^{-j90^\circ}\ {\rm A}.$$

****

8.4 Powers and Power Factor

Decomposition of the supply current of LTI load supplied by a symmetrical sinusoidal voltage of the positive sequence into the CPCs leads directly to the power equation of such a load. Namely, by multiplying formula (8.12) by the supply voltage

$\boldsymbol{u}$ three-phase rms value, $||\boldsymbol{u}||$, we obtain

$$S^2 = P^2 + Q^2 + D_u^{n2} + D_u^{z2} \tag{8.18}$$

with

$$P = ||\boldsymbol{i}_a||\ ||\boldsymbol{u}|| = G_e||\boldsymbol{u}||^2 \tag{8.19}$$

$$Q \stackrel{\text{df}}{=} \pm||\boldsymbol{i}_r||\ ||\boldsymbol{u}|| = -B_e||\boldsymbol{u}||^2 \tag{8.20}$$

$$D_u^n \stackrel{\text{df}}{=} ||\boldsymbol{i}_u^n||\ ||\boldsymbol{u}|| = Y_u^n\ ||\boldsymbol{u}||^2 \tag{8.21}$$

$$D_u^z \stackrel{\text{df}}{=} ||\boldsymbol{i}_u^z||\ ||\boldsymbol{u}|| = Y_u^z\ ||\boldsymbol{u}||^2. \tag{8.22}$$

The power Eqn. (8.18) describes the relationship between powers of LTI load supplied from a symmetrical source of sinusoidal voltages in a 3p4w circuit. This power equation contains two new power quantities, D_u^n and D_u^z. These powers are associated with the presence of the negative and zero sequence unbalanced components in the supply current. Therefore, they will be called *negative sequence unbalanced power* and *zero sequence unbalanced power*, respectively. The power equation is illustrated geometrically with the diagram shown in Fig. 8.10.

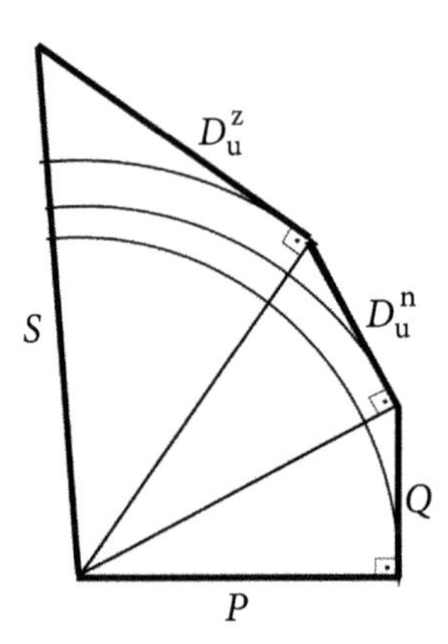

Figure 8.10 Diagram of powers of LTI load supplied with a symmetrical sinusoidal voltage in 3p4w circuit.

Illustration 8.3 *Let us calculate active, reactive, and unbalanced powers for the unbalanced circuit shown in Fig. 8.11, assuming that* $U_R = 120$ V.

For such a load, the equivalent admittance is equal to

$$Y_e = G_e + jB_e = \frac{1}{3}(Y_R + Y_S + Y_T) = \frac{1}{3}(0.50 - j0.87 + 2.0) = 0.833 - j0.290 \text{ S}.$$

The negative sequence unbalanced admittance is

$$Y_u^n = \frac{1}{3}(Y_R + \alpha Y_S + \alpha^* Y_T) = \frac{1}{3}(0.5 - j0.87 + 1e^{-j120^\circ} \times 0.5) = 0.334e^{-120.1^\circ} \text{ S}$$

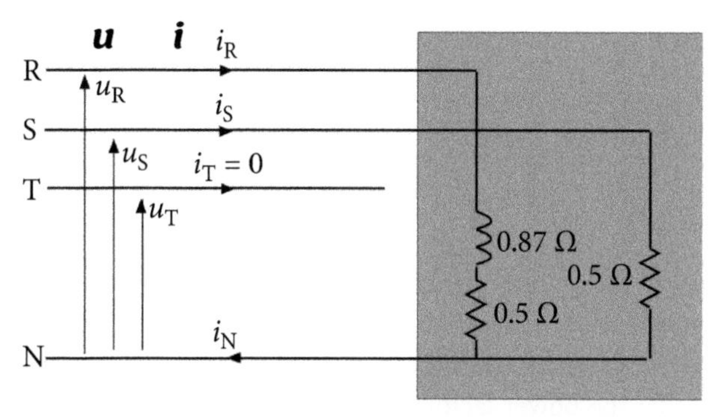

Figure 8.11 Example of an unbalanced load.

while the zero sequence unbalanced admittance has the value

$$Y_u^z = \frac{1}{3}(Y_R + \alpha^* Y_S + \alpha\, Y_T) = \frac{1}{3}(0.5 - j\,0.87 + 1e^{j120^\circ} \times 0.5) = 0.881 e^{-j101^\circ}\ \text{S}.$$

Since

$$||\boldsymbol{u}|| = \sqrt{3}\, U_R = \sqrt{3} \times 120 = 207.8\ \text{V}$$

the particular powers are equal to

$$P = G_e||\boldsymbol{u}||^2 = 0.833 \times (207.8)^2 = 36.0\ \text{kW}$$

$$Q = -B_e||\boldsymbol{u}||^2 = 0.290 \times (207.8)^2 = 12.5\ \text{kVAr}$$

$$D_u^n = Y_u^n||\boldsymbol{u}||^2 = 0.334 \times (207.8)^2 = 7.2\ \text{kVA}$$

$$D_u^z = Y_u^z||\boldsymbol{u}||^2 = 0.881 \times (207.8)^2 = 38.0\ \text{kVA}.$$

The power factor in 3p4w circuits with LTI loads supplied with symmetrical sinusoidal voltage is equal to

$$\lambda \stackrel{\text{df}}{=} \frac{P}{S} = \frac{P}{\sqrt{P^2 + Q^2 + D_u^{n\,2} + D_u^{z\,2}}} \tag{8.23}$$

thus not only the reactive power Q but also both unbalanced powers, D_u^n and D_u^z contribute to the load power factor degradation. It can be expressed not only in terms of powers but also in terms of three-phase rms values of CPCs of the supply current, namely

$$\lambda = \frac{P}{S} = \frac{||\boldsymbol{i}_a||}{||\boldsymbol{i}||} = \frac{||\boldsymbol{i}_a||}{\sqrt{||\boldsymbol{i}_a||^2 + ||\boldsymbol{i}_r||^2 + ||\boldsymbol{i}_u^z||^2 + ||\boldsymbol{i}_u^n||^2}}. \tag{8.24}$$

Especially important is the possibility of expressing the power factor in terms of the load parameters, in particular, in terms of the equivalent conductance,

G_e, susceptance, B_e, and the magnitude of unbalanced admittances Y_u^n and Y_u^z, namely

$$\lambda = \frac{||\boldsymbol{i}_a||}{||\boldsymbol{i}||} = \frac{G_e}{\sqrt{G_e^2 + B_e^2 + A^{n\,2} + A^{z\,2}}}. \tag{8.25}$$

Thus, the power factor of 3p4w circuits declines from unity value because of non-zero equivalent susceptance B_e, negative sequence unbalanced admittance Y_u^n, and the zero sequence unbalanced admittance Y_u^z. This last formula emphasizes the fact that the power factor depends only on the load properties but not on voltages, currents, or powers. It is defined in terms of the active and apparent powers, but eventually only the load properties specify the power factor value. Also, only a change of load parameters using a compensator makes the power factor improvement possible.

8.5 Apparent Power of Δ/Y Transformer in 3p4w Circuits

Energy for 3p4w loads is delivered usually by three-wire lines by transformers in Δ/Y configuration, shown in Fig. 8.12. The connection point of the secondary windings of the transformer is connected to the neutral conductor and grounded. The Δ configuration of primary windings prevents the supply system from being affected by the third order current harmonics, which are created inside of the transformer due to the iron core nonlinearity.

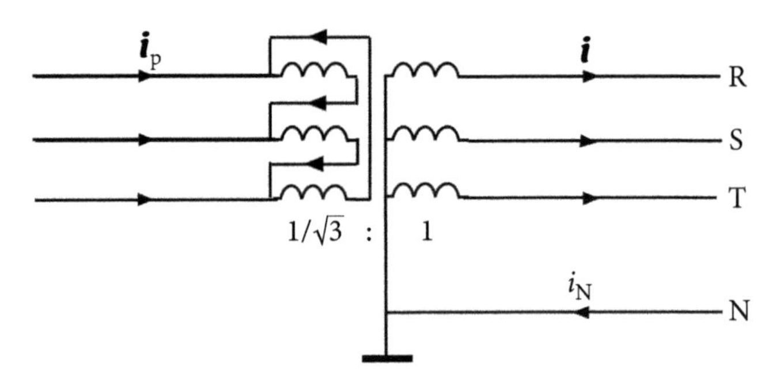

Figure 8.12 A transformer of Δ/Y structure.

The transformer's primary side current $\boldsymbol{i}_p$ can contain the active, the reactive, and the negative sequence symmetrical component of the unbalanced currents. It cannot contain the zero sequence component. Assuming, for simplicity that the transformer is ideal, the zero sequence symmetrical component of the load current circulates in primary windings, so that it does not affect the line current $\boldsymbol{i}_p$ on the transformer supply side. Consequently, the apparent power S_p at the transformer supply terminals

$$S_p = ||\boldsymbol{u}_p||\ ||\boldsymbol{i}_p||$$

is lower than the apparent power on the secondary side

$$S = ||\boldsymbol{u}||\ ||\boldsymbol{i}||$$

since the voltage three-phase rms value for the ideal transformer remains unchanged. Thus, the powers measured on the primary side of the transformer may not specify the true power loading of the transformer. These measurements result in the apparent power

$$S_{\mathrm{p}} = \sqrt{P^2 + Q^2 + D_{\mathrm{u}}^{\mathrm{n}\,2}} \tag{8.26}$$

while the transformer is loaded by the power

$$S = \sqrt{P^2 + Q^2 + D_{\mathrm{u}}^{\mathrm{n}\,2} + D_{\mathrm{u}}^{\mathrm{z}\,2}}. \tag{8.27}$$

Observe moreover that the power factor measured on the primary and the secondary side of the transformer could have different values, as well.

Illustration 8.4 *Let us calculate the apparent power and the power factor at the primary and secondary sides of Δ/Y transformer which supplies the load analyzed in Illustration 8.2, assuming that the transformer is ideal and the load line voltage U_{R} = 120 V. The load with the transformer is shown in Fig. 8.13.*

Equivalent and unbalanced admittances calculated in Illustration 8.2 for the load under consideration are

$$Y_{\mathrm{e}} = G_{\mathrm{e}} + jB_{\mathrm{e}} = 0.50 - j0.29\ \mathrm{S}, \qquad Y_{\mathrm{u}}^{\mathrm{n}} = 0, \qquad Y_{\mathrm{u}}^{\mathrm{z}} = 0.58e^{-j90^\circ}\ \mathrm{S}.$$

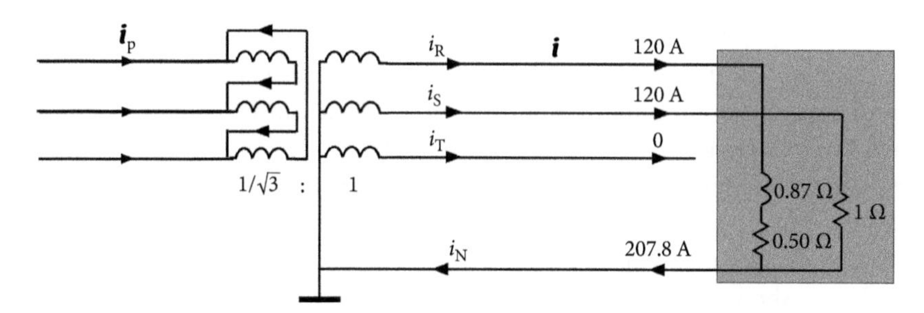

Figure 8.13 Unbalanced load supplied from Δ/Y transformer.

Thus, at the supply voltage of the three-phase rms value

$$||\boldsymbol{u}|| = \sqrt{3}\,U_{\mathrm{R}} = \sqrt{3} \times 120 = 207.8\ \mathrm{V}$$

particular powers have the value

$$P = G_e||\boldsymbol{u}||^2 = 0.50 \times (207.8)^2 = 21.6 \text{ kW}$$

$$Q = -B_e||\boldsymbol{u}||^2 = 0.290 \times (207.8)^2 = 12.5 \text{ kVAr}$$

$$D_u^n = Y_u^n||\boldsymbol{u}||^2 = 0$$

$$D_u^z = Y_u^z||\boldsymbol{u}||^2 = 0.58 \times (207.8)^2 = 25.1 \text{ kVA.}$$

The apparent power on the transformer's secondary side is

$$S = \sqrt{P^2 + Q^2 + D_u^{n\,2} + D_u^{z\,2}} = \sqrt{21.6^2 + 12.5^2 + 25.1^2} = 35.4 \text{ kVA.}$$

The apparent power on the transformer's primary side is

$$S_p = \sqrt{P^2 + Q^2 + D_u^{n\,2}} = \sqrt{21.6^2 + 12.5^2} = 25.0 \text{ kVA.}$$

The power factor on the secondary side has the value

$$\lambda = \frac{P}{S} = \frac{21.6}{35.4} = 0.61$$

while on the primary side

$$\lambda_p = \frac{P}{S_p} = \frac{21.6}{25.0} = 0.86.$$

It means, that Δ/Y transformers, by eliminating the zero sequence current, can improve the power factor on the three-wire side of such transformers.

*****

8.6 Line-to-neutral Admittances

The active and reactive currents, $\boldsymbol{i}_a$ and $\boldsymbol{i}_r$, are specified by equivalent conductance G_e and equivalent susceptance B_e of the load, which can be combined into the equivalent admittance, $\boldsymbol{Y}_e = G_e + jB_e$. Currents $\boldsymbol{i}_u^n$ and $\boldsymbol{i}_u^z$ are specified in terms of unbalanced admittances $\boldsymbol{Y}_u^n$ and $\boldsymbol{Y}_u^z$. To calculate all these admittances, the line-to-neutral admittances $\boldsymbol{Y}_R$, $\boldsymbol{Y}_S$, and $\boldsymbol{Y}_T$ of the load have to be known.

Assume that a load in a 3p4w circuit is composed of K three-phase line-to-neutral branches, as shown in Fig. 8.14, and their line-to-neutral admittances $\boldsymbol{Y}_{Rk}$, $\boldsymbol{Y}_{Sk}$, and $\boldsymbol{Y}_{Tk}$ are known, these admittances should be added to find admittances $\boldsymbol{Y}_R$, $\boldsymbol{Y}_S$, and $\boldsymbol{Y}_T$,

$$\boldsymbol{Y}_R = \sum_{k=1}^{K} \boldsymbol{Y}_{Rk}, \quad \boldsymbol{Y}_S = \sum_{k=1}^{K} \boldsymbol{Y}_{Sk}, \quad \boldsymbol{Y}_T = \sum_{k=1}^{K} \boldsymbol{Y}_{Tk}.$$

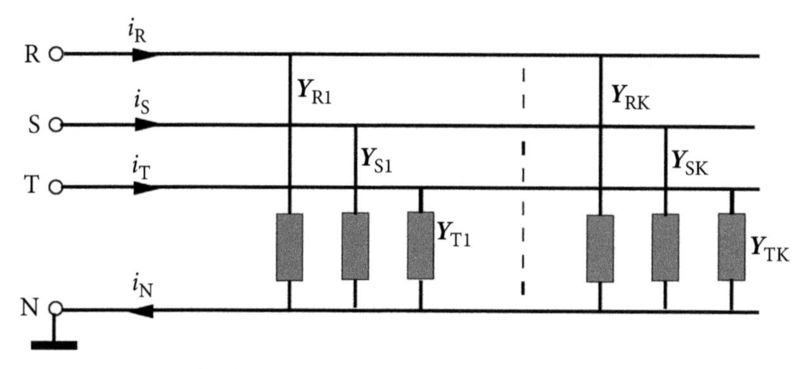

Figure 8.14 A load composed of K three-phase line-to-neutral branches.

Apart from loads connected to the neutral conductor, loads connected in **Δ**, as is shown in Fig. 8.15, can also be supplied by a 3p4w line with line-to-line voltages. These **Δ**-configured loads can be balanced or unbalanced. A three-phase induction motor is an example of a balanced load. Also, single-phase loads can be supplied with line-to-line voltages in such a circuit. Compensators used for load balancing are distinctively unbalanced devices that can be supplied with line-to-line voltages.

Any LTI load in **Δ** configuration. meaning connected as shown in Fig. 8.15(a), has an equivalent circuit in star ("Y") configuration, as shown in Fig. 8.15(b).

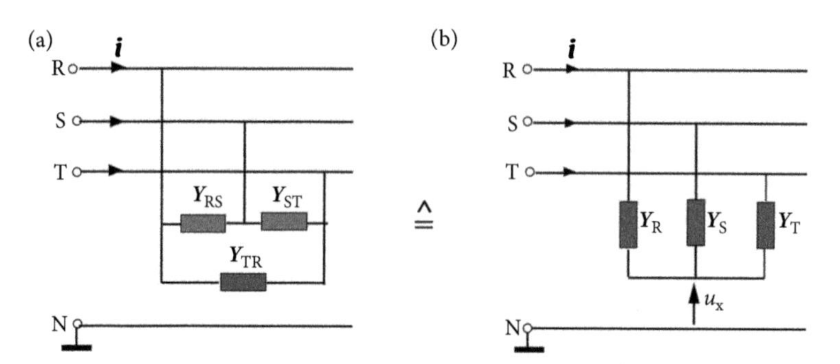

Figure 8.15 A **Δ**-configured load and its Y equivalent.

Admittances of the equivalent circuit can be calculated according to formulae

$$Y_R = Y_{RS} + Y_{TR} + \frac{Y_{RS}Y_{TR}}{Y_{ST}} \tag{8.28}$$

$$Y_S = Y_{ST} + Y_{RS} + \frac{Y_{ST}Y_{RS}}{Y_{TR}} \tag{8.29}$$

$$Y_T = Y_{TR} + Y_{ST} + \frac{Y_{TR}Y_{ST}}{Y_{RS}}. \tag{8.30}$$

This transformation from Δ to Y structure can be a trap, however. The equivalent load, shown in Fig. 8.15(b), is not a 3p4w load, since equivalent one-ports with admittances $\boldsymbol{Y}_{\rm R}$, $\boldsymbol{Y}_{\rm S}$, and $\boldsymbol{Y}_{\rm T}$ are not connected to the neutral conductor. The voltage of the star point, $u_{\rm x}$, could not be equal to zero. Consequently, admittances $\boldsymbol{Y}_{\rm R}$, $\boldsymbol{Y}_{\rm S}$, and $\boldsymbol{Y}_{\rm T}$ are not line-to-neutral admittances. When the voltage $u_{\rm x}$ is not equal to zero, then these one-ports cannot be regarded as connected in parallel to loads that are connected between supply lines and the neutral conductor. Only when the load configured in Δ is balanced, then the star point voltage is zero. Consequently, branches of the equivalent circuit in Y configuration can be regarded as connected in parallel to loads connected to the neutral conductor.

Since loads configured in Δ do not affect the zero sequence current, they do not affect the zero sequence unbalanced admittance $\boldsymbol{Y}_{\rm u}^{\rm z}$. Such loads can affect only the active, reactive, and negative sequence unbalanced currents and consequently equivalent admittance $\boldsymbol{Y}_{\rm e}$ and negative sequence unbalanced admittance $\boldsymbol{Y}_{\rm u}^{\rm n}$ of 3p4w loads. These admittances, as shown in Fig. 8.7, are equivalent parameters of loads that do not need a connection to the neutral conductor, therefore any load in Δ configuration can be regarded as connected in parallel (in three-wire circuit sense) to equivalent circuits specified in terms of only $\boldsymbol{Y}_{\rm e}$ and $\boldsymbol{Y}_{\rm u}^{\rm n}$ admittances. Consequently, the equivalent admittance as introduced in Chapter 7 for three-wire circuits

$$\boldsymbol{Y}_{\rm e} = G_{\rm e} + jB_{\rm e} = \boldsymbol{Y}_{\rm ST} + \boldsymbol{Y}_{\rm TR} + \boldsymbol{Y}_{\rm RS}$$

and the unbalanced admittance

$$-(\boldsymbol{Y}_{\rm ST} + \alpha\,\boldsymbol{Y}_{\rm TR} + \alpha^*\,\boldsymbol{Y}_{\rm RS}) = \boldsymbol{Y}_{\rm u} = Y_{\rm u}e^{j\psi}$$

can be regarded as parameters of parallel circuits and add up for calculating equivalent admittance and negative sequence admittance of loads supplied from 3p4w line and Δ-configured loads.

When the load structure and parameters are not known but the supply voltages and line currents are available for measurement of their crms values, then these admittances can be easily calculated. If the crms values of the line-to-neutral voltages are $\boldsymbol{U}_{\rm R}$, $\boldsymbol{U}_{\rm S}$, and $\boldsymbol{U}_{\rm T}$ and the supply currents $\boldsymbol{I}_{\rm R}$, $\boldsymbol{I}_{\rm S}$, and $\boldsymbol{I}_{\rm T}$, then line-to-neutral admittances can be directly calculated as

$$\boldsymbol{Y}_{\rm R} = \frac{\boldsymbol{I}_{\rm R}}{\boldsymbol{U}_{\rm R}}, \quad \boldsymbol{Y}_{\rm S} = \frac{\boldsymbol{I}_{\rm S}}{\boldsymbol{U}_{\rm S}}, \quad \boldsymbol{Y}_{\rm T} = \frac{\boldsymbol{I}_{\rm T}}{\boldsymbol{U}_{\rm T}}.$$

This approach assumes that one-ports with admittances $\boldsymbol{Y}_{\rm R}$, $\boldsymbol{Y}_{\rm S}$, and $\boldsymbol{Y}_{\rm T}$ are connected between R, S, and T lines and the neutral conductor. This might not be true, however, when some loads have Δ configuration. In such a case, obtained admittances do not stand for admittance of any physical one-port, but for a fictitious one, which is connected between supply lines and the neutral conductor. Nonetheless, obtained admittances specify correctly the circuit performance concerning the current components and powers. This is demonstrated with the following illustration.

Illustration 8.5 *Let us calculate line-to-neutral equivalent admittances of the load shown in Fig. 8.16(a) and compare equivalent and unbalanced admittances of the original load shown in Fig. 8.16(a) and the fictitious one, shown in Fig. 8.16(b), assuming that the voltage of line R rms value is $U_R = 100$ V.*

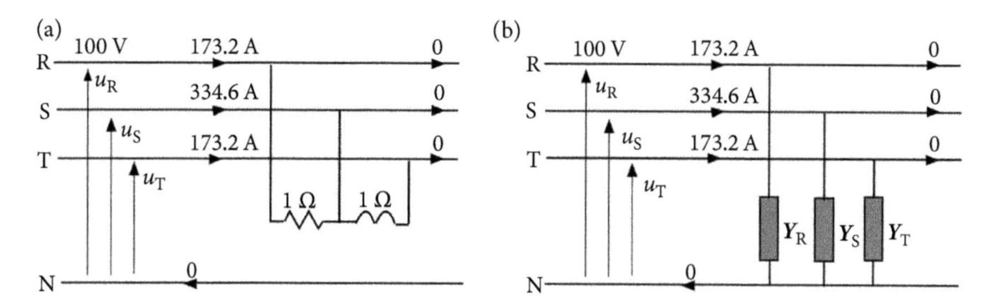

Figure 8.16 A Δ-configured load and its equivalent fictitious load.

The equivalent and unbalanced admittances of the load shown in Fig. 8.16(a) are

$$Y_e = G_e + jB_e = \boldsymbol{Y}_{ST} + \boldsymbol{Y}_{TR} + \boldsymbol{Y}_{RS} = 1 - j1 \text{ S}$$

$$\boldsymbol{Y}_u = -(\boldsymbol{Y}_{ST} + \alpha \boldsymbol{Y}_{TR} + \alpha^* \boldsymbol{Y}_{RS}) = -(-j1 + \alpha^* 1) = 1.932\, e^{j75^\circ} \text{ S.}$$

Since the three-phase rms value of the supply voltage is

$$||\boldsymbol{u}|| = \sqrt{3}\, U_R = \sqrt{3} \times 100 = 173.2 \text{ V,}$$

the rms value of the CPC of the load currents are equal to

$$||\boldsymbol{i}_a|| = G_e||\boldsymbol{u}|| = 1 \times 173.2 = 173.2 \text{ A}$$

$$||\boldsymbol{i}_r|| = |B_e|\ ||\boldsymbol{u}|| = 1 \times 173.2 = 173.2 \text{ A}$$

$$||\boldsymbol{i}_u|| = Y_u\ ||\boldsymbol{u}|| = 1.93 \times 173.2 = 334.5 \text{ A.}$$

Assuming that the crms value of the voltage at terminal R is $\boldsymbol{U}_R = U_R = 100$ V, the line currents of the load shown in Fig. 8.16(a) are equal to

$$\boldsymbol{I}_R = 173.2 e^{j30^\circ} \text{ A}, \quad \boldsymbol{I}_S = 334.6 e^{-j165^\circ} \text{ A}, \quad \boldsymbol{I}_T = 173.2 e^{j0^\circ} \text{ A.}$$

Admittances of a fictitious load connected to neutral at such currents are equal to

$$\boldsymbol{Y}_R = \frac{\boldsymbol{I}_R}{\boldsymbol{U}_R} = \frac{173.2\, e^{j30^\circ}}{100} = 1.732\, e^{j30^\circ} \text{ S}$$

$$\boldsymbol{Y}_S = \frac{\boldsymbol{I}_S}{\boldsymbol{U}_S} = \frac{334.6\, e^{-j165^\circ}}{100\, e^{-j120^\circ}} = 3.346\, e^{-j45^\circ} \text{ S}$$

$$\boldsymbol{Y}_T = \frac{\boldsymbol{I}_T}{\boldsymbol{U}_T} = \frac{173.2\, e^{j0^\circ}}{100\, e^{j120^\circ}} = 1.732\, e^{-j120^\circ} \text{ S.}$$

The equivalent admittance of the fictitious load is

$$Y_e = G_e + jB_e = \frac{1}{3}(Y_R + Y_S + Y_T) = 1.0 - j\,1.0 \text{ S}.$$

and the negative sequence unbalanced admittance

$$Y_u^n = \frac{1}{3}(Y_R + \alpha\, Y_S + \alpha^*\, Y_T) = 1.932\, e^{j75°} \text{ S}$$

thus, loads in Figs. 8.16(a) and 8.16(b) have the same equivalent and unbalanced admittances and consequently the same values of the active, reactive, and unbalanced currents.

*****

Another method of calculation of line-to-neutral admittances of an unknown load, which does not require crms value measurement of voltages and currents, can be based on the measurement of the active and reactive power of individual supply lines. Wattmeters and varmeters connected as shown in Fig. 8.17 are needed for that.

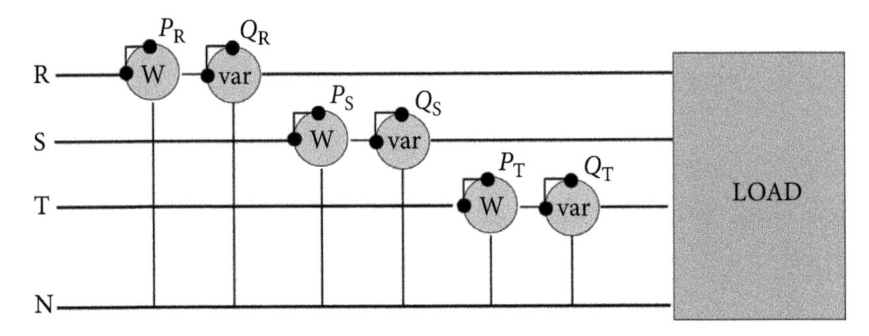

Figure 8.17 Measurement of the active and reactive powers of 3p4w load.

The black terminal of wattmeters and varmeters marked in Fig. 8.17 emphasize the orientation of the voltage and current. To obtain the right sign of the active and reactive powers, these meters should be connected in such a way that the voltage and current are oriented toward marked terminals. The complex apparent power $\boldsymbol{C}_k$ of line k–to-neutral is

$$\boldsymbol{C}_k \stackrel{\text{df}}{=} P_k + jQ_k = \boldsymbol{U}_k \boldsymbol{I}_k^* = \boldsymbol{Y}_k^*\, U_k^2 \tag{8.31}$$

and hence, the line k-to-neutral admittance is equal to

$$\boldsymbol{Y}_k = G_k + jB_k = \frac{\boldsymbol{C}_k^*}{U_k^2} = \frac{P_k - jQ_k}{U_k^2}. \tag{8.32}$$

When some loads have **Δ** configuration then obtained admittances stand for admittances of fictitious loads connected to the neutral conductor.

8.7 CPC in 3p4w Circuits with nv&c and LTI Loads

Currents and powers of LTI loads supplied by a 3p4w line were analyzed to this point under the assumption that the supply voltage was symmetrical and sinusoidal. This voltage is usually distorted, however. When the load under consideration is LTI, the supply voltage distortion is caused by other harmonics generating loads (HGLs), or other sources of distortion inside the power system. The voltage harmonics could be of different sequences, not only the positive and the negative but even of the zero sequence. The third order harmonic is usually the most dominating harmonic of the zero sequence order.

Let us assume that the supply voltage is periodic of period T and contains harmonics of order n from a set of orders N. It is assumed also that this voltage does not contain DC component, that is, the harmonics of the zero order. It is a realistic assumption. The DC component cannot be transferred from the primary side of the supply transformer. It can occur only as a voltage response of the supply system to a DC current on the load side. The LTI load cannot generate such a current, however.

The supply voltage can be regarded as a three-phase vector and expressed as

$$\boldsymbol{u}(t) \stackrel{\mathrm{df}}{=} \begin{bmatrix} u_{\mathrm{R}}(t) \\ u_{\mathrm{S}}(t) \\ u_{\mathrm{T}}(t) \end{bmatrix} = \sum_{n\in N} \boldsymbol{u}_n(t) = \sqrt{2}\mathrm{Re}\sum_{n\in N} \begin{bmatrix} U_{\mathrm{R}n} \\ U_{\mathrm{S}n} \\ U_{\mathrm{T}n} \end{bmatrix} e^{jn\omega_1 t} \stackrel{\mathrm{df}}{=} \sqrt{2}\mathrm{Re}\sum_{n\in N} \boldsymbol{U}_n\, e^{jn\omega_1 t}$$

and the supply current can be expressed in the form of the vector

$$\boldsymbol{i}(t) \stackrel{\mathrm{df}}{=} \begin{bmatrix} i_{\mathrm{R}}(t) \\ i_{\mathrm{S}}(t) \\ i_{\mathrm{T}}(t) \end{bmatrix} = \sum_{n\in N} \boldsymbol{i}_n(t) = \sqrt{2}\mathrm{Re}\sum_{n\in N} \begin{bmatrix} I_{\mathrm{R}n} \\ I_{\mathrm{S}n} \\ I_{\mathrm{T}n} \end{bmatrix} e^{jn\omega_1 t} \stackrel{\mathrm{df}}{=} \sqrt{2}\mathrm{Re}\sum_{n\in N} \boldsymbol{I}_n\, e^{jn\omega_1 t}.$$

Quantities $\boldsymbol{u}(t)$ and $\boldsymbol{i}(t)$, as well as their harmonics, $\boldsymbol{u}_n(t)$ and $\boldsymbol{i}_n(t)$, although they are three-phase vectors, with line-to-neutral voltages and currents as these vectors entries, will be referred to, for simplicity's sake, as supply voltage, current, and their harmonics, respectively. Moreover, when there is no need to emphasize that these quantities are functions of time, their symbols will be simplified to the form of $\boldsymbol{u}$, $\boldsymbol{i}$, $\boldsymbol{u}_n$, or $\boldsymbol{i}_n$.

Let the voltage given by formula (8.33) be applied to an LTI load supplied from a 3p4w line as shown in Fig. 8.18.

If line-to-neutral admittances of the load for harmonic frequencies are $Y_{\mathrm{R}n}$, $Y_{\mathrm{S}n}$, and $Y_{\mathrm{T}n}$ then the supply current is

$$\boldsymbol{i} = \sqrt{2}\mathrm{Re}\sum_{n\in N} \begin{bmatrix} I_{\mathrm{R}n} \\ I_{\mathrm{S}n} \\ I_{\mathrm{T}n} \end{bmatrix} e^{jn\,\omega_1 t} = \sqrt{2}\mathrm{Re}\sum_{n\in N} \begin{bmatrix} Y_{\mathrm{R}n}\, U_{\mathrm{R}n} \\ Y_{\mathrm{S}n}\, U_{\mathrm{S}n} \\ Y_{\mathrm{T}n}\, U_{\mathrm{T}n} \end{bmatrix} e^{jn\,\omega_1 t}.$$

When the load contains devices configured in Δ, admittances in this formula should be regarded as admittances of a fictitious load connected between supply lines and the neutral conductor.

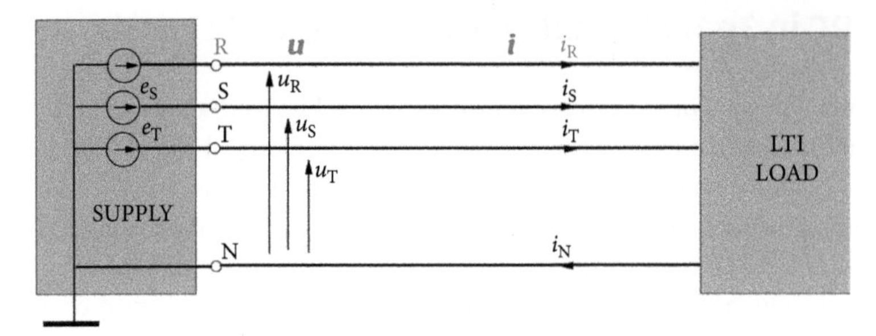

Figure 8.18 A circuit with an LTI load supplied by a 3p4w line.

The active power of the load is equal to the scalar product of the supply voltage and current, since as in three-wire circuits

$$P = \frac{1}{T}\int_0^T (u_R i_R + u_S i_S + u_T i_T)\, dt = \frac{1}{T}\int_0^T \mathbf{u}^{\mathrm{T}}\mathbf{i}\, dt = (\mathbf{u}, \mathbf{i}).$$

When the crms values of the voltage and current harmonics are known, then the active power can be calculated in the frequency domain, namely

$$P = (\mathbf{u}, \mathbf{i}) = \mathrm{Re} \sum_{n\in N} \mathbf{U}_n^{\mathrm{T}} \mathbf{I}_n^{*}.$$

Concerning the active power P, the LTI load in the circuit shown in Fig. 8.18 is equivalent to a purely resistive, symmetrical load of equivalent conductance

$$G_e = \frac{P}{||\mathbf{u}||^2}$$

which draws the active current, meaning the current that is proportional to the supply voltage

$$\mathbf{i}_a \stackrel{\mathrm{df}}{=} \begin{bmatrix} i_{Ra} \\ i_{Sa} \\ i_{Ta} \end{bmatrix} = G_e\, \mathbf{u} = G_e \sum_{n\in N} \mathbf{u}_n = \sqrt{2}\mathrm{Re} \sum_{n\in N} G_e\, \mathbf{U}_n\, e^{jn\omega_1 t}. \tag{8.33}$$

When the voltage is distorted, it is also a distorted but symmetrical current. If the supply voltage contains the zero sequence harmonic, say the third order, the sum of line currents is not equal to zero however, and the active current will have the return pass through the neutral conductor.

The active power of the nth order harmonic, P_n, of the load is

$$P = (\boldsymbol{u}, \boldsymbol{i}) = \text{Re} \sum_{n \in N} \boldsymbol{U}_n^{\text{T}} \boldsymbol{I}_n^*$$

where

$$G_{en} \stackrel{\text{df}}{=} \frac{1}{3}(G_{\text{R}n} + G_{\text{S}n} + G_{\text{T}n}) \tag{8.34}$$

is equivalent conductance of the load for the n-th order harmonic and

$$|\boldsymbol{u}_n| \stackrel{\text{df}}{=} \sqrt{U_{\text{R}n}^2 + U_{\text{S}n}^2 + U_{\text{T}n}^2} = \sqrt{3}\, U_{\text{R}n}^2 \tag{8.35}$$

is the three-phase rms value of the nth order voltage harmonic.

The equivalent conductance of LTI loads, G_{e}, is not equal, apart from special cases, to the equivalent admittance of the load for harmonic frequencies, G_{en}. Therefore, similarly as it was in single-phase and three-wire circuits with nonsinusoidal supply voltage, the supply current contains a scattered component, $\boldsymbol{i}_{\text{s}}$, defined as

$$\boldsymbol{i}_{\text{s}} \stackrel{\text{df}}{=} \sqrt{2}\text{Re} \sum_{n \in N} (G_{en} - G_{\text{e}})\, \boldsymbol{U}_n\, e^{jn\omega_1 t} \tag{8.36}$$

and the reactive current

$$\boldsymbol{i}_{\text{r}} \stackrel{\text{df}}{=} \sqrt{2}\text{Re} \sum_{n \in N} jB_{en}\, \boldsymbol{U}_n\, e^{jn\omega_1 t} \tag{8.37}$$

where

$$B_{en} \stackrel{\text{df}}{=} \frac{1}{3}(B_{\text{R}n} + B_{\text{S}n} + B_{\text{T}n}) \tag{8.38}$$

is equivalent susceptance of the load for the nth order harmonic.

The active, scattered, and reactive currents, $\boldsymbol{i}_{\text{a}}$, $\boldsymbol{i}_{\text{s}}$, and $\boldsymbol{i}_{\text{r}}$, are symmetrical currents of the positive sequence, while the supply current $\boldsymbol{i}$, due to the load potential imbalance can be asymmetrical. Therefore, the residual current, after subtracting the active, scattered, and reactive components,

$$\boldsymbol{i} - (\boldsymbol{i}_{\text{a}} + \boldsymbol{i}_{\text{s}} + \boldsymbol{i}_{\text{r}}) \stackrel{\text{df}}{=} \boldsymbol{i}_{\text{u}}$$

is asymmetrical. It can be expressed as

$$\boldsymbol{i}_{\mathrm{u}} = \sum_{n\in N} \boldsymbol{i}_{\mathrm{u}n} = \sqrt{2}\mathrm{Re}\sum_{n\in N}\begin{bmatrix} \boldsymbol{I}_{\mathrm{Ru}n} \\ \boldsymbol{I}_{\mathrm{Su}n} \\ \boldsymbol{I}_{\mathrm{Tu}n} \end{bmatrix} e^{jn\omega_1 t} =$$

$$= \sqrt{2}\mathrm{Re}\sum_{n\in N}\begin{bmatrix} [Y_{\mathrm{R}n} - G_{\mathrm{e}} - (G_{\mathrm{e}n} - G_{\mathrm{e}}) - jB_{\mathrm{e}n}]\boldsymbol{U}_{\mathrm{R}n} \\ [Y_{\mathrm{S}n} - G_{\mathrm{e}} - (G_{\mathrm{e}n} - G_{\mathrm{e}}) - jB_{\mathrm{e}n}]\boldsymbol{U}_{\mathrm{S}n} \\ [Y_{\mathrm{T}n} - G_{\mathrm{e}} - (G_{\mathrm{e}n} - G_{\mathrm{e}}) - jB_{\mathrm{e}n}]\boldsymbol{U}_{\mathrm{T}n} \end{bmatrix} e^{jn\omega_1 t} =$$

$$= \sqrt{2}\mathrm{Re}\sum_{n\in N}\begin{bmatrix} (Y_{\mathrm{R}n} - Y_{\mathrm{e}n})\boldsymbol{U}_{\mathrm{R}n} \\ (Y_{\mathrm{S}n} - Y_{\mathrm{e}n})\boldsymbol{U}_{\mathrm{S}n} \\ (Y_{\mathrm{T}n} - Y_{\mathrm{e}n})\boldsymbol{U}_{\mathrm{T}n} \end{bmatrix} e^{jn\omega_1 t} \stackrel{\mathrm{df}}{=} \sqrt{2}\mathrm{Re}\sum_{n\in N}\begin{bmatrix} Y_{\mathrm{Ru}n}\boldsymbol{U}_{\mathrm{R}n} \\ Y_{\mathrm{Su}n}\boldsymbol{U}_{\mathrm{S}n} \\ Y_{\mathrm{Tu}n}\boldsymbol{U}_{\mathrm{T}n} \end{bmatrix} e^{jn\omega_1 t}.$$

The waveform of the unbalanced current $\boldsymbol{i}_{\mathrm{u}}$ depends on the load imbalance and it is affected by the fact that voltage harmonics can have, depending on the harmonic order, different sequences: positive, negative, and the zero.

As was discussed in Chapter 3, harmonics of the order $n = 3k + 1$ have a positive sequence; harmonics of the order $n = 3k-1$ have a negative sequence, and harmonics of the order $n = 3k$, have the zero sequence. Mutual orientation on the complex plane of the crms value of lines R, S, and T voltage harmonics is shown in Fig. 8.19.

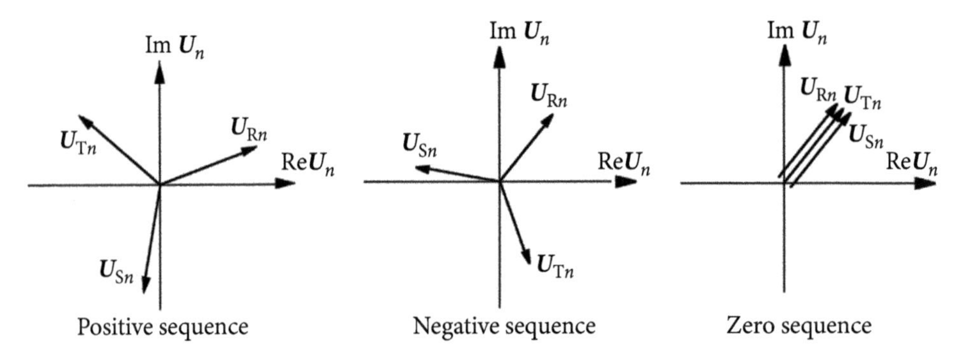

Figure 8.19 Mutual relations of line voltage crms value for harmonics of different sequences.

If the voltage harmonic is of the positive sequence order, that is, $n = 1, 4, 7....$, then

$$\boldsymbol{U}_{\mathrm{S}n} = \alpha^{*}\boldsymbol{U}_{\mathrm{R}n}, \qquad \boldsymbol{U}_{\mathrm{T}n} = \alpha\boldsymbol{U}_{\mathrm{R}n}$$

and the harmonic $\boldsymbol{i}_{\mathrm{u}n}$ of the unbalanced current can be presented in the form

$$\boldsymbol{i}_{\mathrm{u}n} = \sqrt{2}\mathrm{Re}\begin{bmatrix} Y_{\mathrm{Ru}n} \\ Y_{\mathrm{Su}n}\alpha^{*} \\ Y_{\mathrm{Tu}n}\alpha \end{bmatrix} \boldsymbol{U}_{\mathrm{R}n}\, e^{jn\omega_1 t} = \sqrt{2}\mathrm{Re}\begin{bmatrix} \boldsymbol{I}_{\mathrm{Ru}n} \\ \boldsymbol{I}_{\mathrm{Su}n} \\ \boldsymbol{I}_{\mathrm{Tu}n} \end{bmatrix} e^{jn\omega_1 t}.$$

Since this current harmonic can be not symmetrical, it can have positive, negative, and zero sequence components. The crms value of the positive sequence component is equal to

$$\boldsymbol{I}^{\mathrm{p}}_{\mathrm{u}n} = \frac{1}{3}(\boldsymbol{I}_{\mathrm{Ru}n} + \alpha\boldsymbol{I}_{\mathrm{Su}n} + \alpha^*\boldsymbol{I}_{\mathrm{Tu}n}) = \frac{1}{3}(\boldsymbol{Y}_{\mathrm{Ru}n} + \alpha\boldsymbol{Y}_{\mathrm{Su}n}\alpha^* + \alpha^*\boldsymbol{Y}_{\mathrm{Tu}n}\alpha)\boldsymbol{U}_{\mathrm{R}n} =$$

$$= \frac{1}{3}(\boldsymbol{Y}_{\mathrm{Ru}n} + \boldsymbol{Y}_{\mathrm{Su}n} + \boldsymbol{Y}_{\mathrm{Tu}n})\boldsymbol{U}_{\mathrm{R}n} = 0$$

since

$$\boldsymbol{Y}_{\mathrm{Ru}n} + \boldsymbol{Y}_{\mathrm{Su}n} + \boldsymbol{Y}_{\mathrm{Tu}n} = \boldsymbol{Y}_{\mathrm{R}n} + \boldsymbol{Y}_{\mathrm{S}n} + \boldsymbol{Y}_{\mathrm{T}n} - 3\,\boldsymbol{Y}_{en} = 0.$$

The crms value of the negative sequence component is equal to

$$\boldsymbol{I}^{\mathrm{n}}_{\mathrm{u}\,n} = \frac{1}{3}(\boldsymbol{I}_{\mathrm{Ru}n} + \alpha^*\boldsymbol{I}_{\mathrm{Su}n} + \alpha\boldsymbol{I}_{\mathrm{Tu}n}) = \frac{1}{3}(\boldsymbol{Y}_{\mathrm{Ru}n} + \alpha^*\boldsymbol{Y}_{\mathrm{Su}n}\alpha^* + \alpha\,\boldsymbol{Y}_{\mathrm{Tu}n}\alpha)\boldsymbol{U}_{\mathrm{R}n} =$$

$$= \frac{1}{3}(\boldsymbol{Y}_{\mathrm{Ru}n} + \alpha\,\boldsymbol{Y}_{\mathrm{Su}n} + \alpha^*\boldsymbol{Y}_{\mathrm{Tu}n})\boldsymbol{U}_{\mathrm{R}n} = \frac{1}{3}(\boldsymbol{Y}_{\mathrm{R}n} + \alpha\,\boldsymbol{Y}_{\mathrm{S}n} + \alpha^*\boldsymbol{Y}_{\mathrm{T}n})\boldsymbol{U}_{\mathrm{R}n} = \boldsymbol{Y}^{n}_{\mathrm{u}\,n}\boldsymbol{U}_{\mathrm{R}n}$$

where

$$\boldsymbol{Y}^{n}_{\mathrm{u}\,n} \stackrel{\mathrm{df}}{=} \frac{1}{3}(\boldsymbol{Y}_{\mathrm{R}n} + \alpha\,\boldsymbol{Y}_{\mathrm{S}n} + \alpha^*\boldsymbol{Y}_{\mathrm{T}n}). \tag{8.39}$$

And finally, let us calculate the crms value of the zero sequence component

$$\boldsymbol{I}^{\mathrm{z}}_{\mathrm{u}\,n} = \frac{1}{3}(\boldsymbol{I}_{\mathrm{Ru}n} + \boldsymbol{I}_{\mathrm{Su}n} + \boldsymbol{I}_{\mathrm{Tu}n}) = \frac{1}{3}(\boldsymbol{Y}_{\mathrm{Ru}n} + \boldsymbol{Y}_{\mathrm{Su}n}\alpha^* + \boldsymbol{Y}_{\mathrm{Tu}n}\alpha)\boldsymbol{U}_{\mathrm{R}n} =$$

$$= \frac{1}{3}(\boldsymbol{Y}_{\mathrm{R}n} + \alpha^*\boldsymbol{Y}_{\mathrm{S}n} + \alpha\,\boldsymbol{Y}_{\mathrm{T}n})\boldsymbol{U}_{\mathrm{R}n} \stackrel{\mathrm{df}}{=} \boldsymbol{Y}^{z}_{\mathrm{u}\,n}\,\boldsymbol{U}_{\mathrm{R}n}$$

where

$$\boldsymbol{Y}^{\mathrm{z}}_{\mathrm{u}n} \stackrel{\mathrm{df}}{=} \frac{1}{3}(\boldsymbol{Y}_{\mathrm{R}n} + \alpha^*\boldsymbol{Y}_{\mathrm{S}n} + \alpha\,\boldsymbol{Y}_{\mathrm{T}n}). \tag{8.40}$$

It means that the unbalanced current harmonic $\boldsymbol{i}_{\mathrm{u}n}$ of a harmonic of the order of positive sequence contains two components of the negative sequence and the zero sequence, but not the positive sequence

$$\boldsymbol{i}_{\mathrm{u}\,n} = \boldsymbol{i}^{\mathrm{n}}_{\mathrm{u}\,n} + \boldsymbol{i}^{\mathrm{z}}_{\mathrm{u}\,n} = \sqrt{2}\mathrm{Re}\begin{bmatrix} \boldsymbol{Y}^{\mathrm{n}}_{\mathrm{u}\,n}\boldsymbol{U}_{\mathrm{R}n} \\ \boldsymbol{Y}^{\mathrm{n}}_{\mathrm{u}\,n}\boldsymbol{U}_{\mathrm{T}n} \\ \boldsymbol{Y}^{\mathrm{n}}_{\mathrm{u}\,n}\boldsymbol{U}_{\mathrm{S}n} \end{bmatrix} e^{jn\omega_1 t} + \sqrt{2}\mathrm{Re}\begin{bmatrix} \boldsymbol{Y}^{\mathrm{z}}_{\mathrm{u}\,n}\boldsymbol{U}_{\mathrm{R}n} \\ \boldsymbol{Y}^{\mathrm{z}}_{\mathrm{u}\,n}\boldsymbol{U}_{\mathrm{R}n} \\ \boldsymbol{Y}^{\mathrm{z}}_{\mathrm{u}\,n}\boldsymbol{U}_{\mathrm{R}n} \end{bmatrix} e^{jn\omega_1 t} =$$

$$= \sqrt{2}\mathrm{Re}\{(\mathbf{1}^{\mathrm{n}}\boldsymbol{Y}^{\mathrm{n}}_{\mathrm{u}\,n} + \mathbf{1}\,\boldsymbol{Y}^{\mathrm{z}}_{\mathrm{u}\,n})\boldsymbol{U}_{\mathrm{R}n}\, e^{jn\omega_1 t}\}. \tag{8.41}$$

where

$$\mathbf{1} \stackrel{\mathrm{df}}{=} [\ 1,\ \ 1,\ \ 1\ ]^{\mathrm{T}}, \quad \mathbf{1}^{\mathrm{p}} \stackrel{\mathrm{df}}{=} [\ 1,\ \ \alpha^*,\ \ \alpha\ ]^{\mathrm{T}}, \quad \mathbf{1}^{\mathrm{n}} \stackrel{\mathrm{df}}{=} [\ 1,\ \ \alpha,\ \ \alpha^*\ ]^{\mathrm{T}}.$$

If the voltage harmonic is of the negative sequence order, that is, $n = 2, 5, 8....$, then

$$U_{\mathrm{S}n} = \alpha U_{\mathrm{R}n}, \qquad U_{\mathrm{T}n} = \alpha^* U_{\mathrm{R}n}$$

and the harmonic $\boldsymbol{i}_{\mathrm{u}n}$ of the unbalanced current can be expressed in the form

$$\boldsymbol{i}_{\mathrm{u}n} = \sqrt{2}\mathrm{Re}\begin{bmatrix} Y_{\mathrm{Ru}n} \\ Y_{\mathrm{Su}n}\alpha \\ Y_{\mathrm{Tu}n}\alpha^* \end{bmatrix} U_{\mathrm{R}n}\, e^{jn\omega_1 t} = \sqrt{2}\mathrm{Re}\begin{bmatrix} I_{\mathrm{Ru}n} \\ I_{\mathrm{Su}n} \\ I_{\mathrm{Tu}n} \end{bmatrix} e^{jn\omega_1 t}.$$

Since this current harmonic is not symmetrical, it can have positive, negative, and zero sequence components. The crms value of the positive sequence component is equal to

$$\begin{aligned} \boldsymbol{I}^{\mathrm{p}}_{\mathrm{u}n} &= \frac{1}{3}(\boldsymbol{I}_{\mathrm{Ru}n} + \alpha \boldsymbol{I}_{\mathrm{Su}n} + \alpha^* \boldsymbol{I}_{\mathrm{Tu}n}) = \frac{1}{3}(\boldsymbol{Y}_{\mathrm{Ru}n} + \alpha\, \boldsymbol{Y}_{\mathrm{Su}n}\alpha + \alpha^* \boldsymbol{Y}_{\mathrm{Tu}n}\alpha^*)\boldsymbol{U}_{\mathrm{R}n} = \\ &= \frac{1}{3}(\boldsymbol{Y}_{\mathrm{Ru}n} + \alpha^* \boldsymbol{Y}_{\mathrm{Su}n} + \alpha\, \boldsymbol{Y}_{\mathrm{Tu}n})\boldsymbol{U}_{\mathrm{R}n} = \frac{1}{3}(\boldsymbol{Y}_{\mathrm{R}n} + \alpha^* \boldsymbol{Y}_{\mathrm{S}n} + \alpha\, \boldsymbol{Y}_{\mathrm{T}n})\boldsymbol{U}_{\mathrm{R}n} \stackrel{\mathrm{df}}{=} \boldsymbol{Y}^{\mathrm{p}}_{\mathrm{u}n}\boldsymbol{U}_{\mathrm{R}n}. \end{aligned}$$

where

$$\boldsymbol{Y}^{\mathrm{p}}_{\mathrm{u}n} \stackrel{\mathrm{df}}{=} \frac{1}{3}(\boldsymbol{Y}_{\mathrm{R}n} + \alpha^* \boldsymbol{Y}_{\mathrm{S}n} + \alpha \boldsymbol{Y}_{\mathrm{T}n}). \tag{8.42}$$

The crms value of the negative sequence component is equal to

$$\begin{aligned} I^{\mathrm{n}}_{\mathrm{u}n} &= \frac{1}{3}(I_{\mathrm{Ru}n} + \alpha^* I_{\mathrm{Su}n} + \alpha I_{\mathrm{Tu}n}) = \frac{1}{3}(\boldsymbol{Y}_{\mathrm{Ru}n} + \alpha^* \boldsymbol{Y}_{\mathrm{Su}n}\alpha + \alpha \boldsymbol{Y}_{\mathrm{Tu}n}\alpha^*)\boldsymbol{U}_{\mathrm{R}n} = \\ &= \frac{1}{3}(\boldsymbol{Y}_{\mathrm{Ru}n} + \boldsymbol{Y}_{\mathrm{Su}n} + \boldsymbol{Y}_{\mathrm{Tu}n})\boldsymbol{U}_{\mathrm{R}n} = 0. \end{aligned}$$

And finally, let us calculate the crms value of the zero sequence component

$$\begin{aligned} \boldsymbol{I}^{\mathrm{z}}_{\mathrm{u}n} &= \frac{1}{3}(\boldsymbol{I}_{\mathrm{Ru}n} + \boldsymbol{I}_{\mathrm{Su}n} + \boldsymbol{I}_{\mathrm{Tu}n}) = \frac{1}{3}(\boldsymbol{Y}_{\mathrm{Ru}n} + \boldsymbol{Y}_{\mathrm{Su}n}\alpha + \boldsymbol{Y}_{\mathrm{Tu}n}\alpha^*)\boldsymbol{U}_{\mathrm{R}n} = \\ &= \frac{1}{3}(\boldsymbol{Y}_{\mathrm{R}n} + \alpha\, \boldsymbol{Y}_{\mathrm{S}n} + \alpha^* \boldsymbol{Y}_{\mathrm{T}n})\boldsymbol{U}_{\mathrm{R}n} \stackrel{\mathrm{df}}{=} \boldsymbol{Y}^{z}_{\mathrm{u}n}\, \boldsymbol{U}_{\mathrm{R}n}. \end{aligned}$$

where

$$\boldsymbol{Y}^{\mathrm{z}}_{\mathrm{u}n} \stackrel{\mathrm{df}}{=} \frac{1}{3}(\boldsymbol{Y}_{\mathrm{R}n} + \alpha \boldsymbol{Y}_{\mathrm{S}n} + \alpha^* \boldsymbol{Y}_{\mathrm{T}n}). \tag{8.43}$$

It means that the unbalanced current harmonic $\boldsymbol{i}_{un}$ of the negative sequence order n contains two components of the positive sequence and the zero sequence,

$$\boldsymbol{i}_{un} = \boldsymbol{i}^{p}_{un} + \boldsymbol{i}^{z}_{un} = \sqrt{2}\mathrm{Re}\begin{bmatrix} \boldsymbol{Y}^{p}_{un}\boldsymbol{U}_{Rn} \\ \boldsymbol{Y}^{p}_{un}\boldsymbol{U}_{Sn} \\ \boldsymbol{Y}^{p}_{un}\boldsymbol{U}_{Tn} \end{bmatrix} e^{jn\omega_1 t} + \sqrt{2}\mathrm{Re}\begin{bmatrix} \boldsymbol{Y}^{z}_{un}\boldsymbol{U}_{Rn} \\ \boldsymbol{Y}^{z}_{un}\boldsymbol{U}_{Rn} \\ \boldsymbol{Y}^{z}_{un}\boldsymbol{U}_{Rn} \end{bmatrix} e^{jn\omega_1 t} =$$

$$= \sqrt{2}\mathrm{Re}\{(\mathbf{1}^{p}\boldsymbol{Y}^{p}_{un} + \mathbf{1}^{z}\boldsymbol{Y}^{z}_{un})\boldsymbol{U}_{Rn}\, e^{jn\omega_1 t}\}. \tag{8.44}$$

If the voltage harmonic is of the zero sequence order, that is, $n = 3, 6, 9....$, then

$$U_{Sn} = U_{Rn} = U_{Tn}$$

and the harmonic $\boldsymbol{i}_{un}$ of the unbalanced current can be presented in the form

$$\boldsymbol{i}_{un} = \sqrt{2}\mathrm{Re}\begin{bmatrix} \boldsymbol{Y}_{Run} \\ \boldsymbol{Y}_{Sun} \\ \boldsymbol{Y}_{Tun} \end{bmatrix} \boldsymbol{U}_{Rn}\, e^{jn\omega_1 t} = \sqrt{2}\mathrm{Re}\begin{bmatrix} \boldsymbol{I}_{Run} \\ \boldsymbol{I}_{Sun} \\ \boldsymbol{I}_{Tun} \end{bmatrix} e^{jn\omega_1 t}.$$

It should be observed, however, that both line-to-neutral admittances $\boldsymbol{Y}_{Rn}$, $\boldsymbol{Y}_{Sn}$, and $\boldsymbol{Y}_{Tn}$ as well as equivalent admittance $\boldsymbol{Y}_{en}$ for the zero sequence harmonics can differ from these admittances for the positive and the negative sequence harmonics since loads in **Δ** configuration behave for zero sequence harmonics as open circuits. Thus, only loads connected to the neutral conductor should be taken into account when admittances $\boldsymbol{Y}_{Rn}$, $\boldsymbol{Y}_{Sn}$, $\boldsymbol{Y}_{Tn}$, $\boldsymbol{Y}_{en}$, and consequently $\boldsymbol{Y}_{Run}$, $\boldsymbol{Y}_{Sun}$, $\boldsymbol{Y}_{Tun}$ for the zero order harmonics are calculated.

The crms value of the positive and negative sequence components are equal to

$$\boldsymbol{I}^{p}_{un} = \frac{1}{3}(\boldsymbol{I}_{Run} + \alpha\boldsymbol{I}_{Sun} + \alpha^{*}\boldsymbol{I}_{Tun}) = \frac{1}{3}(\boldsymbol{Y}_{Run} + \alpha\boldsymbol{Y}_{Sun} + \alpha^{*}\boldsymbol{Y}_{Tun})\boldsymbol{U}_{Rn} = \boldsymbol{Y}^{p}_{un}\boldsymbol{U}_{Rn}$$

$$\boldsymbol{I}^{n}_{un} = \frac{1}{3}(\boldsymbol{I}_{Run} + \alpha^{*}\boldsymbol{I}_{Sun} + \alpha\boldsymbol{I}_{Tun}) = \frac{1}{3}(\boldsymbol{Y}_{Run} + \alpha^{*}\boldsymbol{Y}_{Sun} + \alpha\boldsymbol{Y}_{Tun})\boldsymbol{U}_{Rn} = \boldsymbol{Y}^{n}_{un}\boldsymbol{U}_{Rn}$$

and the zero sequence

$$\boldsymbol{I}^{z}_{un} = \frac{1}{3}(\boldsymbol{I}_{Run} + \boldsymbol{I}_{Sun} + \boldsymbol{I}_{Tun}) = \frac{1}{3}(\boldsymbol{Y}_{Run} + \boldsymbol{Y}_{Sun} + \boldsymbol{Y}_{Tun})\boldsymbol{U}_{Rn} = 0$$

since

$$\boldsymbol{Y}_{Run} + \boldsymbol{Y}_{Sun} + \boldsymbol{Y}_{Tun} = \boldsymbol{Y}_{Rn} + \boldsymbol{Y}_{Sn} + \boldsymbol{Y}_{Tn} - 3\,\boldsymbol{Y}_{en} = 0.$$

It means that the unbalanced current harmonic $\boldsymbol{i}_{\mathrm{u}n}$ of the zero sequence order n contains components of the positive sequence and the negative sequence,

$$\boldsymbol{i}_{\mathrm{u}n} = \boldsymbol{i}_{\mathrm{u}n}^{\mathrm{p}} + \boldsymbol{i}_{\mathrm{u}n}^{\mathrm{n}} = \sqrt{2}\mathrm{Re}\begin{bmatrix} Y_{\mathrm{u}n}^{\mathrm{p}} U_{\mathrm{R}n} \\ Y_{\mathrm{u}n}^{\mathrm{p}} U_{\mathrm{S}n} \\ Y_{\mathrm{u}n}^{\mathrm{p}} U_{\mathrm{T}n} \end{bmatrix} e^{jn\omega_1 t} + \sqrt{2}\mathrm{Re}\begin{bmatrix} Y_{\mathrm{u}n}^{\mathrm{n}} U_{\mathrm{R}n} \\ Y_{\mathrm{u}n}^{\mathrm{n}} U_{\mathrm{R}n} \\ Y_{\mathrm{u}n}^{\mathrm{n}} U_{\mathrm{R}n} \end{bmatrix} e^{jn\omega_1 t} =$$

$$= \sqrt{2}\mathrm{Re}\{(\mathbf{1}^{\mathrm{p}} Y_{\mathrm{u}n}^{\mathrm{p}} + \mathbf{1}^{\mathrm{n}} Y_{\mathrm{u}n}^{\mathrm{n}}) U_{\mathrm{R}n}\, e^{jn\omega_1 t}\}. \tag{8.45}$$

Thus, zero sequence order, $n = 3, 6, 9...$, voltage harmonics result only in positive and negative sequence components in the unbalanced current. The unbalanced current does not contain the zero sequence component.

In general, independently of the harmonic sequence, the unbalanced current of the nth order harmonic is equal to

$$\boldsymbol{i}_{\mathrm{u}n} = \boldsymbol{i}_{\mathrm{u}n}^{\mathrm{z}} + \boldsymbol{i}_{\mathrm{u}n}^{\mathrm{p}} + \boldsymbol{i}_{\mathrm{u}n}^{\mathrm{n}} = \sqrt{2}\mathrm{Re}\{(\mathbf{1}^{\mathrm{z}} Y_{\mathrm{u}n}^{\mathrm{z}} + \mathbf{1}^{\mathrm{p}} Y_{\mathrm{u}n}^{\mathrm{p}} + \mathbf{1}^{\mathrm{n}} Y_{\mathrm{u}n}^{\mathrm{n}}) U_{\mathrm{R}n}\, e^{jn\omega_1 t}\} \tag{8.46}$$

with

$$Y_{\mathrm{u}n}^{\mathrm{z}} = \begin{cases} 0, & \text{for } n = 3k \\ \frac{1}{3}(Y_{\mathrm{R}n} + \alpha^* Y_{\mathrm{S}n} + \alpha Y_{\mathrm{T}n}), & \text{for } n = 3k+1 \\ \frac{1}{3}(Y_{\mathrm{R}n} + \alpha Y_{\mathrm{S}n} + \alpha^* Y_{\mathrm{T}n}), & \text{for } n = 3k-1 \end{cases} \tag{8.47}$$

$$Y_{\mathrm{u}n}^{\mathrm{p}} = \begin{cases} \frac{1}{3}(Y_{\mathrm{R}n} + \alpha Y_{\mathrm{S}n} + \alpha^* Y_{\mathrm{T}n}), & \text{for } n = 3k \\ 0, & \text{for } n = 3k+1 \\ \frac{1}{3}(Y_{\mathrm{R}n} + \alpha^* Y_{\mathrm{S}n} + \alpha Y_{\mathrm{T}n}), & \text{for } n = 3k-1 \end{cases} \tag{8.48}$$

$$Y_{\mathrm{u}n}^{\mathrm{n}} = \begin{cases} \frac{1}{3}(Y_{\mathrm{R}n} + \alpha^* Y_{\mathrm{S}n} + \alpha Y_{\mathrm{T}n}), & \text{for } n = 3k \\ \frac{1}{3}(Y_{\mathrm{R}n} + \alpha Y_{\mathrm{S}n} + \alpha^* Y_{\mathrm{T}n}), & \text{for } n = 3k+1 \\ 0, & \text{for } n = 3k-1. \end{cases} \tag{8.49}$$

The nth order harmonic of the unbalanced current $\boldsymbol{i}_{\mathrm{un}}$ is composed, as shown by (8.46), of components of the zero, positive, and negative sequence, thus they are mutually orthogonal. Hence, the three-phase rms value of such harmonic, in a square, is equal to

$$||\boldsymbol{i}_{\mathrm{u}n}||^2 = ||\boldsymbol{i}_{\mathrm{u}n}^{\mathrm{z}}||^2 + ||\boldsymbol{i}_{\mathrm{u}n}^{\mathrm{p}}||^2 + ||\boldsymbol{i}_{\mathrm{u}n}^{\mathrm{n}}||^2. \tag{8.50}$$

Eventually, the total unbalanced component of the supply current of unbalanced LTI load is

$$\boldsymbol{i}_{\mathrm{u}} = \sum_{n\in N} \boldsymbol{i}_{\mathrm{u}n} = \sum_{n\in N} (\boldsymbol{i}_{\mathrm{u}n}^{\mathrm{z}} + \boldsymbol{i}_{\mathrm{u}n}^{\mathrm{p}} + \boldsymbol{i}_{\mathrm{u}n}^{\mathrm{n}}) \stackrel{\mathrm{df}}{=} \boldsymbol{i}_{\mathrm{u}}^{\mathrm{z}} + \boldsymbol{i}_{\mathrm{u}}^{\mathrm{p}} + \boldsymbol{i}_{\mathrm{u}}^{\mathrm{n}} \tag{8.51}$$

with

$$\boldsymbol{i}_{\mathrm{u}}^{\mathrm{z}} \stackrel{\mathrm{df}}{=} \sum_{n\in N} \boldsymbol{i}_{\mathrm{u}n}^{\mathrm{z}} = \sqrt{2}\mathrm{Re} \sum_{n\in N} \mathbf{1}^{\mathrm{z}}\, Y_{\mathrm{u}n}^{\mathrm{z}}\, \boldsymbol{U}_{\mathrm{R}n}\, e^{jn\omega_1 t} \tag{8.52}$$

$$\boldsymbol{i}_{\mathrm{u}}^{\mathrm{p}} \stackrel{\mathrm{df}}{=} \sum_{n\in N} \boldsymbol{i}_{\mathrm{u}n}^{\mathrm{p}} = \sqrt{2}\mathrm{Re} \sum_{n\in N} \mathbf{1}^{\mathrm{p}}\, Y_{\mathrm{u}n}^{\mathrm{p}}\, \boldsymbol{U}_{\mathrm{R}n}\, e^{jn\omega_1 t} \tag{8.53}$$

$$\boldsymbol{i}_{\mathrm{u}}^{\mathrm{n}} \stackrel{\mathrm{df}}{=} \sum_{n\in N} \boldsymbol{i}_{\mathrm{u}n}^{\mathrm{n}} = \sqrt{2}\mathrm{Re} \sum_{n\in N} \mathbf{1}^{\mathrm{n}}\, Y_{\mathrm{u}n}^{\mathrm{n}}\, \boldsymbol{U}_{\mathrm{R}n}\, e^{jn\omega_1 t}. \tag{8.54}$$

The unbalanced current $\boldsymbol{i}_{\mathrm{u}}$ definition (8.51) with (8.52–8.54) could be reduced to just one formula by introducing a *generalized complex rotation coefficient*, defined as

$$\beta \stackrel{\mathrm{df}}{=} (\alpha^*)^n = \begin{cases} 1, & \text{for} \quad n = 3k \\ \alpha^*, & \text{for} \quad n = 3k+1 \\ \alpha, & \text{for} \quad n = 3k-1 \end{cases}. \tag{8.55}$$

With this coefficient, the crms values of symmetrical components of the unbalanced current harmonics $\boldsymbol{i}_{\mathrm{u}n}$ can be rewritten as

$$\begin{bmatrix} \boldsymbol{I}_{\mathrm{u}n}^{\mathrm{z}} \\ \boldsymbol{I}_{\mathrm{u}n}^{\mathrm{p}} \\ \boldsymbol{I}_{\mathrm{u}n}^{\mathrm{n}} \end{bmatrix} = \frac{1}{3} \begin{bmatrix} 1 & 1 & 1 \\ 1 & \alpha & \alpha^* \\ 1 & \alpha^* & \alpha \end{bmatrix} \begin{bmatrix} (\boldsymbol{Y}_{\mathrm{R}n} - \boldsymbol{Y}_{\mathrm{e}n}) \\ (\boldsymbol{Y}_{\mathrm{S}n} - \boldsymbol{Y}_{\mathrm{e}n})\beta \\ (\boldsymbol{Y}_{\mathrm{T}n} - \boldsymbol{Y}_{\mathrm{e}n})\beta^* \end{bmatrix} \boldsymbol{U}_{\mathrm{R}n} = \begin{bmatrix} \boldsymbol{Y}_{\mathrm{u}n}^{\mathrm{z}} \\ \boldsymbol{Y}_{\mathrm{u}n}^{\mathrm{p}} \\ \boldsymbol{Y}_{\mathrm{u}n}^{\mathrm{n}} \end{bmatrix} \boldsymbol{U}_{\mathrm{R}n} \tag{8.56}$$

where

$$\boldsymbol{Y}_{\mathrm{u}n}^{\mathrm{z}} = \frac{1}{3}[(\boldsymbol{Y}_{\mathrm{R}n} + \beta \boldsymbol{Y}_{\mathrm{S}n} + \beta^* \boldsymbol{Y}_{\mathrm{T}n}) - \boldsymbol{Y}_{\mathrm{e}n}(1 + \beta + \beta^*)] \tag{8.57}$$

$$\boldsymbol{Y}_{\mathrm{u}n}^{\mathrm{p}} = \frac{1}{3}[(\boldsymbol{Y}_{\mathrm{R}n} + \alpha\beta \boldsymbol{Y}_{\mathrm{S}n} + \alpha^*\beta^* \boldsymbol{Y}_{\mathrm{T}n}) - \boldsymbol{Y}_{\mathrm{e}n}(1 + \alpha\beta + \alpha^*\beta^*)] \tag{8.58}$$

$$\boldsymbol{Y}_{\mathrm{u}n}^{\mathrm{n}} = \frac{1}{3}[(\boldsymbol{Y}_{\mathrm{R}n} + \alpha^*\beta \boldsymbol{Y}_{\mathrm{S}n} + \alpha\beta^* \boldsymbol{Y}_{\mathrm{T}n}) - \boldsymbol{Y}_{\mathrm{e}n}(1 + \alpha^*\beta + \alpha\beta^*)]. \tag{8.59}$$

These last formulae for unbalanced admittance of the zero, positive, and negative sequence are equivalent to those specified by (8.39), (8.40), (8.42), and (8.43).

Components of the unbalanced current $\boldsymbol{i}_{\mathrm{u}}$ as specified by (8.51) are mutually orthogonal so that their three-phase rms values fulfill the relationship

$$||\boldsymbol{i}_{\mathrm{u}}||^2 = ||\boldsymbol{i}_{\mathrm{u}}^{\mathrm{z}}||^2 + ||\boldsymbol{i}_{\mathrm{u}}^{\mathrm{p}}||^2 + ||\boldsymbol{i}_{\mathrm{u}}^{\mathrm{n}}||^2. \tag{8.60}$$

where

$$||\boldsymbol{i}_{\mathrm{u}}^{\mathrm{z}}|| = \sqrt{3 \sum_{n\in N} (Y_{\mathrm{u}n}^{\mathrm{z}}\, U_{\mathrm{R}n})^2} = \sqrt{\sum_{n\in N} Y_{\mathrm{u}n}^{\mathrm{z}\,2} ||\boldsymbol{u}_n||^2} \tag{8.61}$$

$$||\boldsymbol{i}_{\mathrm{u}}^{\mathrm{p}}|| = \sqrt{3\sum_{n\in N}(Y_{\mathrm{u}n}^{\mathrm{p}}\,U_{\mathrm{R}n})^2} = \sqrt{\sum_{n\in N} Y_{\mathrm{u}n}^{\mathrm{p}2}||\boldsymbol{u}_n||^2} \tag{8.62}$$

$$||\boldsymbol{i}_{\mathrm{u}}^{\mathrm{n}}|| = \sqrt{3\sum_{n\in N}(Y_{\mathrm{u}n}^{\mathrm{n}}\,U_{\mathrm{R}n})^2} = \sqrt{\sum_{n\in N} Y_{\mathrm{u}n}^{\mathrm{n}2}||\boldsymbol{u}_n||^2}. \tag{8.63}$$

Taking into account the decomposition of the unbalanced current into the zero, positive, and negative sequence symmetrical components as presented above, the supply current of LTI loads supplied by a three-wire line with a neutral conductor can be decomposed into CPC as follows

$$\boldsymbol{i} = \boldsymbol{i}_{\mathrm{a}} + \boldsymbol{i}_{\mathrm{s}} + \boldsymbol{i}_{\mathrm{r}} + \boldsymbol{i}_{\mathrm{u}}^{\mathrm{z}} + \boldsymbol{i}_{\mathrm{u}}^{\mathrm{p}} + \boldsymbol{i}_{\mathrm{u}}^{\mathrm{n}}. \tag{8.64}$$

All these physical components of the supply current are mutually orthogonal, so that

$$||\boldsymbol{i}||^2 = ||\boldsymbol{i}_{\mathrm{a}}||^2 + ||\boldsymbol{i}_{\mathrm{s}}||^2 + ||\boldsymbol{i}_{\mathrm{r}}||^2 + ||\boldsymbol{i}_{\mathrm{u}}^{\mathrm{z}}||^2 + ||\boldsymbol{i}_{\mathrm{u}}^{\mathrm{p}}||^2 + ||\boldsymbol{i}_{\mathrm{u}}^{\mathrm{n}}||^2. \tag{8.65}$$

Six components of the supply current in decomposition (8.64), associated with distinctive physical phenomena in the load or, as is the case with the unbalanced current, different sequences, specify the supply current three-phase rms value. All of them can be expressed in terms of the load admittances for harmonic frequencies.

8.8 Powers and Power Factor

Multiplication of equation by (8.65) by the square of the supply voltage three-phase rms value, $||\boldsymbol{u}||$, results in the power equation of LTI loads in 3p4w circuits supplied with a nonsinusoidal voltage

$$S^2 = P^2 + D_{\mathrm{s}}^2 + Q^2 + D_{\mathrm{u}}^{\mathrm{z}2} + D_{\mathrm{u}}^{\mathrm{p}2} + D_{\mathrm{u}}^{\mathrm{n}2}. \tag{8.66}$$

In this equation $D_{\mathrm{u}}^{\mathrm{z}}$, $D_{\mathrm{u}}^{\mathrm{p}}$, and $D_{\mathrm{u}}^{\mathrm{n}}$ stand for the unbalanced powers associated with the zero, positive, and negative sequence symmetrical components of the unbalanced current.

This equation can be written in a simplified form as

$$S^2 = P^2 + D_{\mathrm{s}}^2 + Q^2 + D_{\mathrm{u}}^2 \tag{8.67}$$

with

$$D_{\mathrm{u}}^2 = D_{\mathrm{u}}^{\mathrm{z}2} + D_{\mathrm{u}}^{\mathrm{p}2} + D_{\mathrm{u}}^{\mathrm{n}2}. \tag{8.68}$$

Power factor. The power factor of LTI loads supplied with symmetrical nonsinusoidal voltage in 3p4w circuits is equal to

$$\lambda \stackrel{\mathrm{df}}{=} \frac{P}{S} = \frac{P}{\sqrt{P^2 + D_{\mathrm{s}}^2 + Q^2 + D_{\mathrm{u}}^{\mathrm{z}\,2} + D_{\mathrm{u}}^{\mathrm{n}\,2} + D_{\mathrm{u}}^{\mathrm{p}\,2}}} \tag{8.69}$$

thus, not only the reactive power Q, but also the scattered power D_{s} and three unbalanced powers, $D_{\mathrm{u}}^{\mathrm{z}}$, $D_{\mathrm{u}}^{\mathrm{n}}$, and $D_{\mathrm{u}}^{\mathrm{p}}$ contribute to the load power factor degradation. It can be expressed not only in terms of powers but also in terms of three-phase rms values of CPCs of the supply current, namely

$$\lambda = \frac{P}{S} = \frac{||\boldsymbol{i}_{\mathrm{a}}||}{||\boldsymbol{i}||} = \frac{||\boldsymbol{i}_{\mathrm{a}}||}{\sqrt{||\boldsymbol{i}_{\mathrm{a}}||^2 + ||\boldsymbol{i}_{\mathrm{s}}||^2 + ||\boldsymbol{i}_{\mathrm{r}}||^2 + ||\boldsymbol{i}_{\mathrm{u}}^{\mathrm{z}}||^2 + ||\boldsymbol{i}_{\mathrm{u}}^{\mathrm{n}}||^2 + ||\boldsymbol{i}_{\mathrm{u}}^{\mathrm{p}}||^2}} \tag{8.70}$$

thus the current decomposition presented above and the power equation developed provide information on all factors that contribute to the power factor degradation.

8.9 Neutral Conductor Current

The neutral conductor current, $i_{\mathrm{N}}(t)$, is equal to the sum of line currents, i.e.,

$$i_{\mathrm{N}}(t) \equiv i_{\mathrm{R}}(t) + i_{\mathrm{S}}(t) + i_{\mathrm{T}}(t).$$

Since the interpretation of power phenomena in 3p4w circuits requires that the line currents are decomposed into CPC, let us find out how these components affect the neutral current.

The neutral conductor current, $i_{\mathrm{N}}(t)$, is composed of triple zero sequence components of the line currents. According to Eqn. (8.51), this component is a part of the unbalanced current. The zero sequence component is also a part of the active, reactive, and scattered currents. When the supply voltage contains the zero sequence harmonics, $\boldsymbol{u}_n = \boldsymbol{u}_{3k}$, then these three currents contain the same zero sequence harmonics. This is the zero sequence component of the current $\boldsymbol{i}_{\mathrm{b}}$ of a balanced load, composed of equivalent admittances Y_{en}, with a neutral conductor, namely

$$\boldsymbol{i}_{\mathrm{b}}^{\mathrm{z}} \stackrel{\mathrm{df}}{=} \sqrt{2}\mathrm{Re}\sum_{n\in N_{\mathrm{Z}}} [G_e + jB_{en} + (G_{en} - G_{\mathrm{e}})]\,\boldsymbol{U}_n\, e^{jn\omega_1 t} = \sqrt{2}\mathrm{Re}\sum_{n\in N_{\mathrm{Z}}} \mathbf{1}^{\mathrm{z}}\, Y_{en}\, U_{\mathrm{R}n}\, e^{jn\omega_1 t} \tag{8.71}$$

where N_{Z} is the set of orders of zero sequence harmonics, $N_{\mathrm{Z}} = \{3, 6,\ldots 3k\}$.

Since line current components of zero sequence are in phase, the neutral conductor current has triple the value of the zero sequence components of line currents, namely

$$i_N = 3(i_b^z + i_u^z) = 3\sqrt{2}\mathrm{Re}\left\{\sum_{n\in N_Z} Y_{en}\, \boldsymbol{U}_{Rn}\, e^{jn\omega_1 t} + \sum_{n\in N} \boldsymbol{Y}_{un}^{z}\, \boldsymbol{U}_{Rn}\, e^{jn\omega_1 t}\right\} \tag{8.72}$$

and its rms value is

$$||i_N|| = 3\sqrt{\sum_{n\in N_Z} (Y_{en} U_{Rn})^2 + \sum_{n\in N} (Y_{un}^{z}\, U_{Rn})^2}. \tag{8.73}$$

Illustration 8.10 *Let us calculate Current's Physical Components of the supply current for the load shown in Fig. 8.20, assuming that it is supplied with the voltage such that*

$$\boldsymbol{U}_{R1} = 100\, e^{j0}\ \mathrm{V}, \quad \boldsymbol{U}_{R3} = 3.0\, e^{j0}\ \mathrm{V}, \quad \boldsymbol{U}_{R5} = 4.0\, e^{j0}\ \mathrm{V}, \quad \boldsymbol{U}_{R5} = 2.0\, e^{j0}\ \mathrm{V}.$$

Impedances of the load individual components in Fig. 8.20 are specified for the fundamental frequency. Individual loads are compensated by shunt capacitors to the unity power factor.

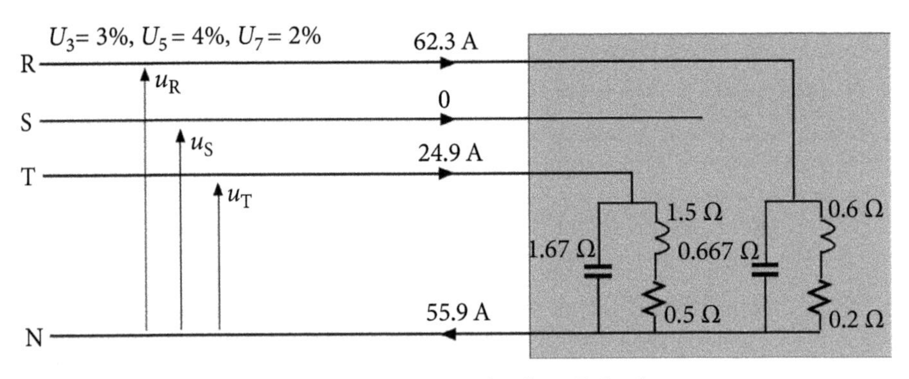

Figure 8.20 An example of an LTI load

Admittances of the load for the supply voltage harmonics are

$$\boldsymbol{Y}_{R1} = 0.50e^{j0}\ \mathrm{S}, \quad \boldsymbol{Y}_{R3} = 3.95e^{j89.1^\circ}\ \mathrm{S}, \quad \boldsymbol{Y}_{R5} = 7.17e^{j89.8^\circ}\ \mathrm{S}, \quad \boldsymbol{Y}_{R7} = 10.26\, e^{j89.9^\circ}\ \mathrm{S}.$$

$$\boldsymbol{Y}_{T1} = 0.20\, e^{j0}\ \mathrm{S}, \quad \boldsymbol{Y}_{T3} = 1.58\, e^{j89.1^\circ}\ \mathrm{S}, \quad \boldsymbol{Y}_{T5} = 2.87e^{j89.8^\circ}\ \mathrm{S}, \quad \boldsymbol{Y}_{T7} = 4.10\, e^{j89.9^\circ}\ \mathrm{S}.$$

At such admittances, the line currents rms values are as shown in Fig. 8.20. The three-phase rms value of the load is

$$||\boldsymbol{i}|| = \sqrt{||i_R||^2 + ||i_S||^2 + ||i_T||^2} = \sqrt{62.3^2 + 24.9^2} = 67.1\ \mathrm{A}.$$

At the voltage distortion as assumed, the three-phase rms value of the supply voltage harmonics are

$$||\boldsymbol{u}_1|| = 173.2\,\mathrm{V}, \quad ||\boldsymbol{u}_3|| = 5.20\,\mathrm{V}, \quad ||\boldsymbol{u}_5|| = 6.93\,\mathrm{V}, \quad ||\boldsymbol{u}_7|| = 3.46\,\mathrm{V}$$

and the three-phase rms value of the total voltage

$$||\boldsymbol{u}|| = \sqrt{\sum_{n=1,3,5,7} ||\boldsymbol{u}_n||^2} = 173.46\,\text{V}.$$

Consequently, the apparent power of the load is

$$S = ||\boldsymbol{u}||\,||\boldsymbol{i}|| = 173.46 \times 67.1 = 11.640\ \text{kVA}.$$

The load active power

$$P = P_{\text{R}} + P_{\text{S}} + P_{\text{T}} = \sum_{n=1,3,5,7} (G_{\text{R}n}\,U_{\text{R}n}^2 + G_{\text{T}n}\,U_{\text{T}n}^2) = 7.001\ \text{kW}$$

hence, the load power factor is

$$\lambda = \frac{P}{S} = 0.601.$$

Observe that despite the reactive power compensation of individual loads for the fundamental harmonic, the power factor is low. Both harmonics and asymmetry contribute to its degradation.

Let us calculate the physical components of the supply current. The equivalent admittance of the load is equal to

$$G_{\text{e}} = \frac{P}{||\boldsymbol{u}||^2} = 0.2327\ \text{S}$$

hence the three-phase rms value of the active current is

$$||\boldsymbol{i}_{\text{a}}|| = G_{\text{e}}||\boldsymbol{u}|| = 40.36\ \text{A}.$$

The equivalent conductance of the load for harmonics is calculated from the formula

$$G_{\text{e}n} = \frac{P_n}{||\boldsymbol{u}_n||^2} = \frac{1}{3}(G_{\text{R}n} + G_{\text{S}n} + G_{\text{T}n})$$

and it is equal to

$$G_{\text{e}1} = 0.2333\ \text{S}, \quad G_{\text{e}3} = 0.0285\ \text{S}, \quad G_{\text{e}5} = 0.0103\ \text{S}, \quad G_{\text{e}7} = 0.0053\ \text{S}$$

hence the three-phase rms value of the scattered current is

$$||\boldsymbol{i}_{\text{s}}|| = \sqrt{\sum_{n=1,3,5,7} (G_{\text{e}n} - G_{\text{e}})^2||\boldsymbol{u}_n||^2} = 2.03\ \text{A}.$$

The equivalent susceptance of the load for harmonics is calculated from the formula

$$B_{en} = \frac{1}{3}(B_{Rn} + B_{Sn} + B_{Tn})$$

and it is equal to

$$B_{e1} = 0, \quad B_{e3} = 1.844 \text{ S}, \quad B_{e5} = 3.345 \text{ S}, \quad B_{e7} = 4.789 \text{ S}$$

hence the three-phase rms value of the reactive current is

$$||\boldsymbol{i}_{\mathrm{r}}|| = \sqrt{\sum_{n=1,3,5,7} B_{en}^2 ||\boldsymbol{u}_n||^2} = 30.07 \text{ A}.$$

The zero sequence unbalanced admittance of the load for particular harmonics has the value

$$\boldsymbol{Y}_{\mathrm{u1}}^{\mathrm{z}} = 0.145\, e^{j23.4^\circ} \text{ S}, \quad \boldsymbol{Y}_{\mathrm{u3}}^{\mathrm{z}} = 0, \quad \boldsymbol{Y}_{\mathrm{u5}}^{\mathrm{z}} = 2.08\, e^{j66.4^\circ} \text{ S}, \quad \boldsymbol{Y}_{\mathrm{u7}}^{\mathrm{z}} = 2.98\, e^{j113.5^\circ} \text{ S}$$

hence, the zero sequence component of the unbalanced current is

$$\boldsymbol{i}_{\mathrm{u}}^{\mathrm{z}} = \sqrt{2}\mathrm{Re}\{\mathbf{1}^{\mathrm{z}}\ (14.5\, e^{j\,23.4^\circ} e^{j\omega_1 t} + 8.32\, e^{j66.4^\circ} e^{j5\omega_1 t} + 5.96\, e^{j113.5^\circ} e^{j7\omega_1 t})\} \text{ A}$$

and its three-phase rms value is

$$||\boldsymbol{i}_{\mathrm{u}}^{\mathrm{z}}|| = \sqrt{\sum_{n=1,5,7} Y_{\mathrm{u}n}^{\mathrm{z}2}\, ||\boldsymbol{u}_n||^2} = 30.79 \text{ A}.$$

The positive sequence unbalanced admittance of the load for particular harmonics has the value

$$\boldsymbol{Y}_{\mathrm{u1}}^{\mathrm{p}} = 0, \quad \boldsymbol{Y}_{\mathrm{u3}}^{\mathrm{p}} = 1.15\, e^{j65.7^\circ} \text{ S}, \quad \boldsymbol{Y}_{\mathrm{u5}}^{\mathrm{p}} = 2.08\, e^{j113.2^\circ} \text{ S}, \quad \boldsymbol{Y}_{\mathrm{u7}}^{\mathrm{p}} = 0,$$

hence the positive sequence component of the unbalanced current is

$$\boldsymbol{i}_{\mathrm{u}}^{\mathrm{p}} = \sqrt{2}\mathrm{Re}\{\mathbf{1}^{\mathrm{p}}(3.45\, e^{j65.7^\circ} e^{j3\omega_1 t} + 8.32\, e^{j113.2^\circ} e^{j5\omega_1 t})\} \text{ A}$$

and its three-phase rms value is

$$||\boldsymbol{i}_{\mathrm{u}}^{\mathrm{p}}|| = \sqrt{\sum_{n=3,5} Y_{\mathrm{u}n}^{\mathrm{p}2}\, ||\boldsymbol{u}_n||^2} = 15.61 \text{ A}.$$

The negative sequence unbalanced admittance of the load for particular harmonics has the value

$$\boldsymbol{Y}_{\mathrm{u1}}^{\mathrm{n}} = 0.145\, e^{j23.4^\circ} \text{ S}, \quad \boldsymbol{Y}_{\mathrm{u3}}^{\mathrm{n}} = 1.45\, e^{j65.7^\circ} \text{ S}, \quad \boldsymbol{Y}_{\mathrm{u5}}^{\mathrm{n}} = 0, \quad \boldsymbol{Y}_{\mathrm{u7}}^{\mathrm{n}} = 2.98\, e^{j113.3^\circ} \text{ S},$$

hence the negative sequence component of the unbalanced current has the waveform

$$\boldsymbol{i}_{\mathrm{u}}^{\mathrm{n}} = \sqrt{2}\mathrm{Re}\{\mathbf{1}^{\mathrm{n}}(14.5\, e^{j23.4^{\circ}} e^{j\omega_1 t} + 4.35\, e^{j65.7^{\circ}} e^{j3\omega_1 t} + 5.96\, e^{j113.3^{\circ}} e^{j7\omega_1 t})\}\ \mathrm{A}.$$

Its three-phase rms value is

$$||\boldsymbol{i}_{\mathrm{u}}^{\mathrm{n}}|| = \sqrt{\sum_{n=1,3,7} Y_{\mathrm{u}n}^{\mathrm{n}2} ||\boldsymbol{u}_n||^2} = 27.85\ \mathrm{A}.$$

Thus the three-phase rms value of the unbalanced current is

$$||\boldsymbol{i}_{\mathrm{u}}|| = \sqrt{||\boldsymbol{i}_{\mathrm{u}}^{\mathrm{z}}||^2 + ||\boldsymbol{i}_{\mathrm{u}}^{\mathrm{p}}||^2 + ||\boldsymbol{i}_{\mathrm{u}}^{\mathrm{n}}||^2} = \sqrt{30.79^2 + 15.61^2 + 27.85^2} = 44.36\,\mathrm{A}.$$

Verification *the rms value of all calculated above physical components of the supply currents should be equal to the three-phase rms value of this current calculated directly. Indeed,*

$$||\boldsymbol{i}|| = \sqrt{||\boldsymbol{i}_{\mathrm{a}}||^2 + ||\boldsymbol{i}_{\mathrm{s}}||^2 + ||\boldsymbol{i}_{\mathrm{r}}||^2 + ||\boldsymbol{i}_{\mathrm{u}}||^2} = \sqrt{40.36^2 + 2.03^2 + 30.07^2 + 44.36^2} = 61.12\,\mathrm{A}.$$

The neutral conductor current has the waveform

$$i_{\mathrm{N}} = 3\sqrt{2}\mathrm{Re}\{\boldsymbol{Y}_{\mathrm{e}3}\, \boldsymbol{U}_{\mathrm{R}3}\, e^{j3\omega_1 t} + Y_{\mathrm{u}1}^{\mathrm{z}}\, \boldsymbol{U}_{\mathrm{R}1}\, e^{j\omega_1 t} + Y_{\mathrm{u}5}^{\mathrm{z}}\, \boldsymbol{U}_{\mathrm{R}5}\, e^{j5\omega_1 t} + Y_{\mathrm{u}7}^{\mathrm{z}}\, \boldsymbol{U}_{\mathrm{R}7}\, e^{j7\omega_1 t}\} =$$

$$= \sqrt{2}\mathrm{Re}\{43.6\, e^{j23.4^{\circ}} e^{j\omega_1 t} + 16.6\, e^{j89.1^{\circ}} e^{j3\omega_1 t} + 25.0\, e^{j66.4^{\circ}} e^{j5\omega_1 t} + 17.9 e^{j113.3^{\circ}}\, e^{j7\omega_1 t}\}\,\mathrm{A}$$

and its rms value is $||\mathrm{i}_{\mathrm{N}}|| = 55.9\,\mathrm{A}$.

****

8.10 CPC in 3p4w Circuits with nv&c and HGLs

The current and power equations of 3p4w circuits were developed to this point under the assumption that the load is linear, time-invariant, meaning it is not a source of current harmonics. Unfortunately, this is usually only a very rough approximation of customer load properties. Waveform distortion and consequently current harmonics originate mainly on the load side. When harmonics originate in the load, then similarly as it was in single-phase circuits, current and power decomposition obtained for LTI loads could be erroneous. It is because current harmonics generated in the load can cause a permanent flow of energy from the load back to the supply source.

To decompose the load current into physical components and develop a power equation, the sign of harmonic active power, meaning the direction of permanent energy flow for harmonic frequency, has to be detected, similarly as was discussed in Chapter 6 for single-phase circuits and in Chapter 7 for three-wire circuits.

The load current decomposition into CPC of three-phase harmonic generating loads with a neutral conductor does not differ from such decomposition of such loads without this conductor. Only harmonics of the zero sequence order can occur in such a decomposition, along with zero sequence components in the unbalanced current.

Decomposition of the set N of all harmonic orders into set N_C of orders of these harmonics that convey the energy from the supply source (C) to the load and set N_G of orders of these harmonics that convey this energy from the HGL (G) back to the supply is identical to that presented in Chapter 6 for single-phase circuits and in Chapter 7 for three-phase three-wire circuits. The whole reasoning presented in Section 7.12 can be applied without any changes to 3p4w circuits, as shown in Fig. 8.21.

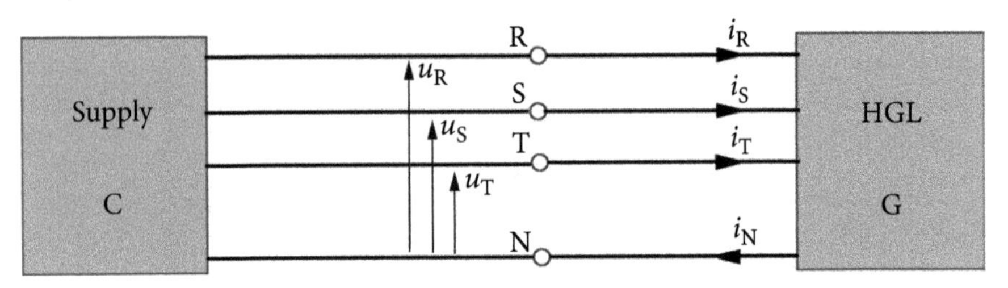

Figure 8.21 A 3p4w circuit with a harmonic generating load (HGL).

Depending on the sign of the active power P_n of the nth order harmonic

$$P_n = U_n(I_{Rn}\cos\varphi_{Rn} + I_{Sn}\cos\varphi_{Sn} + I_{Tn}\cos\varphi_{Tn}) \stackrel{\text{df}}{=} U_n I_{an} \tag{8.74}$$

where:

$$I_{an} \stackrel{\text{df}}{=} I_{Rn}\cos\varphi_{Rn} + I_{Sn}\cos\varphi_{Sn} + I_{Tn}\cos\varphi_{Tn} \tag{8.75}$$

the set N of harmonic orders n is divided into subsets N_C and N_G and the active power P is decomposed into powers P_C and P_G as explained in Section 7.12. Next, the load current vector $\boldsymbol{i}$ is decomposed into currents $\boldsymbol{i}_C$ and $\boldsymbol{i}_G$ and the supply voltage vector $\boldsymbol{u}$ is decomposed into voltages $\boldsymbol{u}_C$ and $\boldsymbol{u}_G$. Eventually, similarly, as was done in Section 7.12, the load current can be decomposed into CPC, namely

$$\boldsymbol{i} = \boldsymbol{i}_{aC} + \boldsymbol{i}_{sC} + \boldsymbol{i}_{rC} + \boldsymbol{i}_{uC}^{z} + \boldsymbol{i}_{uC}^{p} + \boldsymbol{i}_{uC}^{n} + \boldsymbol{i}_G \tag{8.76}$$

only the unbalanced current $\boldsymbol{i}_{uC}$ has three components of the zero, positive and negative sequence, specified by formulae (8.52)–(8.54) of three-phase rms values specified by formulae (8.61)–(8.63). It means that the load current is composed of seven different physical components. The same applies to the relationship between three-phase rms values of the load current physical components, Eqn. (8.65), which preserves its validity for 3p4w circuits with HGLs, namely

$$||\boldsymbol{i}||^2 = ||\boldsymbol{i}_{\mathrm{aC}}||^2 + ||\boldsymbol{i}_{\mathrm{sC}}||^2 + ||\boldsymbol{i}_{\mathrm{rC}}||^2 + ||\boldsymbol{i}^{\mathrm{z}}_{\mathrm{uC}}||^2 + ||\boldsymbol{i}^{\mathrm{p}}_{\mathrm{uC}}||^2 + ||\boldsymbol{i}^{\mathrm{n}}_{\mathrm{uC}}||^2 + ||\boldsymbol{i}_{\mathrm{G}}||^2 \tag{8.78}$$

and the power equation, Eqn. (7.79)

$$S = ||\boldsymbol{u}|| \cdot ||\boldsymbol{i}|| = \sqrt{||\boldsymbol{u}_{\mathrm{C}}||^2 + ||\boldsymbol{u}_{\mathrm{G}}||^2}\sqrt{||\boldsymbol{i}_{\mathrm{C}}||^2 + ||\boldsymbol{i}_{\mathrm{G}}||^2} = \sqrt{S_{\mathrm{C}}^2 + S_{\mathrm{G}}^2 + S_{\mathrm{E}}^2} \tag{8.79}$$

with

$$S_{\mathrm{C}} \overset{\mathrm{df}}{=} ||\boldsymbol{u}_{\mathrm{C}}|| \cdot ||\boldsymbol{i}_{\mathrm{C}}|| = \sqrt{P_{\mathrm{C}}^2 + D_{\mathrm{sC}}^2 + Q_{\mathrm{rC}}^2 + D_{\mathrm{uC}}^{\mathrm{z}\,2} + D_{\mathrm{uC}}^{\mathrm{n}\,2} + D_{\mathrm{uC}}^{\mathrm{p}\,2}} \tag{8.80}$$

and

$$S_{\mathrm{C}} \overset{\mathrm{df}}{=} ||\boldsymbol{u}_{\mathrm{G}}|| \cdot ||\boldsymbol{i}_{\mathrm{G}}||, \quad S_{\mathrm{E}} \overset{\mathrm{df}}{=} \sqrt{||\boldsymbol{u}_{\mathrm{C}}||^2 + ||\boldsymbol{i}_{\mathrm{G}}||^2 + ||\boldsymbol{u}_{\mathrm{G}}||^2 + ||\boldsymbol{i}_{\mathrm{C}}||^2}. \tag{8.81}$$

The power factor λ of three-phase unbalanced circuits with neutral conductor and harmonic generating loads can be expressed in the form

$$\lambda \overset{\mathrm{df}}{=} \frac{P}{S} = \frac{P_{\mathrm{C}} - P_{\mathrm{G}}}{\sqrt{P_{\mathrm{C}}^2 + D_{\mathrm{sC}}^2 + Q_{\mathrm{rC}}^2 + D_{\mathrm{uC}}^{\mathrm{z}\,2} + D_{\mathrm{uC}}^{\mathrm{n}\,2} + D_{\mathrm{uC}}^{\mathrm{p}\,2} + S_{\mathrm{G}}^2 + S_{\mathrm{E}}^2}}. \tag{8.82}$$

This formula reveals all power components that contribute to the deterioration of the power factor in three-phase circuits with harmonic generating loads and the neutral conductor.

Summary

The Current's Physical Components-based approach to power theory enables the explanation of all power-related phenomena in three-phase circuits with the neutral conductor and the development of the load current decomposition into the physical components associated with these phenomena.

PART B

FILTERS AND COMPENSATORS

Introduction

Compensators are traditionally used in power systems for reducing the apparent power S toward the value of the active power P, in other words for reducing the supply current. When the supply voltage and current are sinusoidal, this could be done by a capacitor connected to the load terminals. The ratio of the active-to-apparent powers, meaning the power factor $\lambda_1 = P/S$, can change as this is shown in Figure B.1. This plot was obtained for an LTI load of the active power $P = 10$ kW, power factor for the fundamental harmonic $\lambda_1 = 0.5$, supplied from a distribution transformer with the power ratings of $S = 400$ kVA, and the reactance three times higher than the resistance.

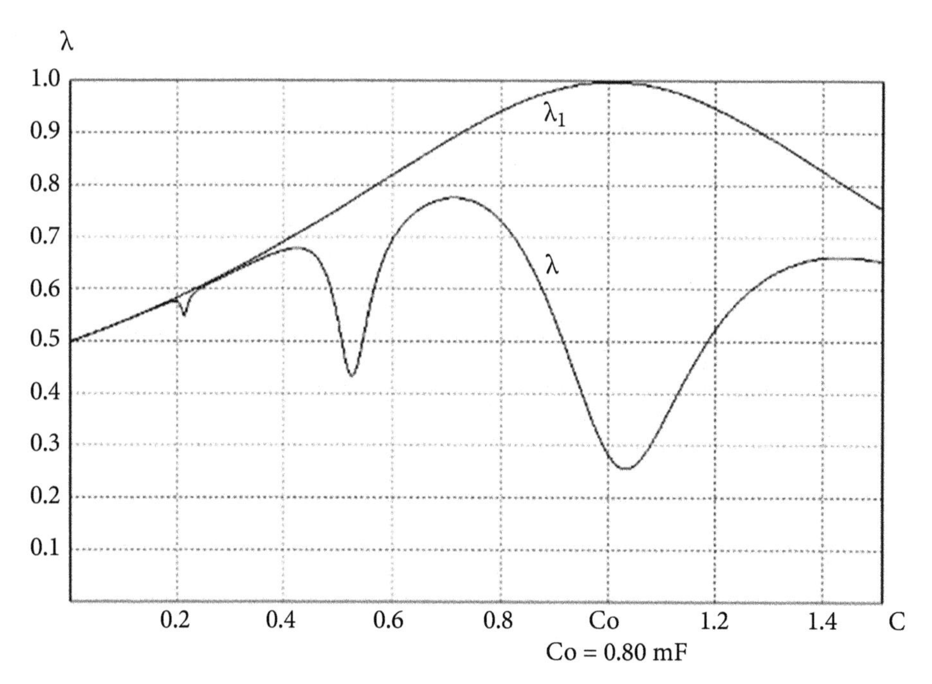

Figure B.1 Dependence of the power factor λ and λ_1 on the compensating capacitance C.

When the supply voltage is sinusoidal then at the capacitance, dependent on the load reactive power Q, and the supply voltage rms value U

$$C = \frac{Q}{\omega_1 U^2}$$

the apparent power S can be reduced entirely to the active power P value. It is enough that the supply voltage e of the transformer is distorted, for example, by harmonics of the relative value

$$E_3/E_1 = 1.0\ \%,\ E_5/E_1 = 3.0\ \%,\ E_7/E_1 = 1.5\ \%,\ E_{11}/E_1 = 0.5\ \%$$

and the power factor declines to $\lambda = 0.28$.

There are long-lasting efforts in electrical engineering aimed at finding methods of filtering and compensation in the presence of waveform distortion and asymmetry. This book presents the methods of compensation founded on the Current's Physical Components-based power theory. These are currently the most advanced methods of reactance compensation, compensation with switching, and with hybrid compensators. They could be used in small micro-grids and distribution grids of ultra-high-power manufacturing plants.

Chapter 9
Overview of Compensation Issues

9.1 Supply Quality and Loading Quality

The effectiveness of energy delivery to customers and its utilization by customer loads could be hampered by various agents. The reactive current which is taken by resistive-inductive (RL) loads, and/or harmonics are the most common example of such an agent. To reduce them, compensators or filters are installed in distribution systems. Providing theoretical fundamentals for the design of filters and compensators is one of the main practical goals of power theory development. Fundamentals of compensation and filtering in distribution systems are discussed in this chapter. When voltages and currents in distribution systems are sinusoidal and symmetrical, then the objective of compensation is very simple. However, when these quantities are nonsinusoidal and asymmetrical, then compensation and filtering can have a variety of different objectives. Explanation and clarification of these objectives are the main subjects of this chapter.

Traditional power systems are composed of electric energy providers and energy consumers. An individual consumer is connected to the provider as shown in Fig. 9.1.

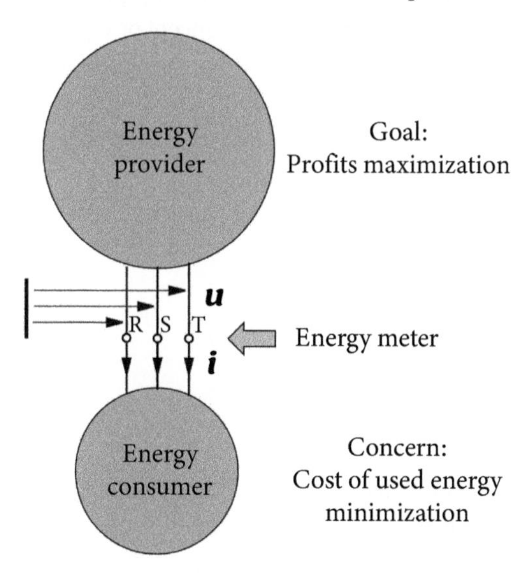

Figure 9.1 A diagram of a power system with economic concerns of energy provider and its consumer.

Powers and Compensation in Circuits with Nonsinusoidal Current. Leszek S. Czarnecki, Oxford University Press.
 DOI: 10.1093/oso/9780198879206.003.0009

Currently, power systems also include distributed generation, meaning power systems where functions of energy production and consumption are not as clearly separated as they are in traditional systems.

When the energy provider and its consumer are distinctive economic entities, they are separated mutually by energy meters. Bills for electric energy paid by its consumers are the source of revenue for an energy provider in electrical power systems. To obtain these revenues the energy provider has to produce or/and buy electric energy, build and run facilities for its distribution, and operate the whole system in an efficient and economical way. It also has to develop the system to meet future energy demands from its customers. The main economic objective of energy providers is to obtain as great a profit as possible from energy delivery. Efficient use of electric energy, thus minimization of energy cost, is one of the major concerns of energy consumers.

Excessive currents and current harmonics generated by customer loads, as well as supply current asymmetry, can increase the energy loss at its delivery, causing overheating of the transmission equipment and coercing the energy provider for equipment power rating increase. Transients, harmonics, and high-frequency (HF) noise injected into the system by customer loads can disturb instrumentation, communication, and control equipment. Energy providers may also need equipment and be burdened with an extra operational cost to keep the supply voltage parameters, such as its waveform, rms value, and symmetry, at a level required by relevant standards. This is because the supply voltage asymmetry, harmonics, HF noise, or transients affect customer loads, causing overheating, performance deterioration, and extra costs. These observations lead to the concept of supply quality and loading quality.

The Supply Quality (SQ). This term provides a qualitative description of the supply source from the point of view of its capability of providing the right conditions for a load operation, meaning from the point of view of the energy consumer. It can be specified as follows.

If the internal voltage $\boldsymbol{e}(t)$ of a three-phase supply source is

(i) symmetrical;
(ii) sinusoidal, meaning without
(iii) harmonic distortion,
(iv) transients,
(v) high-frequency (HF) noise;
(vi) time-invariant rms value;
(vii) constant frequency;
and if, moreover, the supply voltage is
(viii) independent of the load current,

then such a source is regarded as a source with the best SQ possible, or with an *ideal supply quality*. If any of these conditions are not met, then the source is regarded as a source with a *degraded supply quality*. When a load is supplied from such a source, then some undesirable effects can occur. Unfortunately, since quite different features of the voltage can contribute to supply quality degradation, it is not measurable. Two sources can be compared mutually concerning their supply quality only if they differ by one component, for example, by the level of asymmetry, while other features, as compiled above, are identical. Observe that all conditions, apart from (vii) refer to the internal voltage $\boldsymbol{e}(t)$ of the source, but not to the voltage at load terminals. The condition (vii) means that the source with an ideal supply quality has infinite power.

The internal voltage of such a source with a degraded supply quality differs by $\boldsymbol{e}_{\rm d}(t)$ from the voltage $\boldsymbol{e}_{\rm i}(t)$ of the source with an ideal supply quality as is shown in Fig. 9.2.

Voltage $\boldsymbol{e}_{\rm i}(t)$ is a sinusoidal and symmetrical voltage with a constant rms value, while voltage $\boldsymbol{e}_{\rm d}(t)$ can be composed of the negative sequence component $\boldsymbol{e}^{\rm n}(t)$, voltage harmonics $\boldsymbol{e}_{\rm h}(t)$, a high-frequency noise $\boldsymbol{e}_{\rm HF}(t)$, transient components $\boldsymbol{e}_{\rm t}(t)$, or components with slowly varying rms value $\boldsymbol{e}_{\rm v}(t)$.

$$\boldsymbol{e}_d(t) = \boldsymbol{e}^n(t) + \boldsymbol{e}_h(t) + \boldsymbol{e}_{HF}(t) + \boldsymbol{e}_t(t) + \boldsymbol{e}_v(t). \tag{9.1}$$

Each of these components occurs in the internal voltage $\boldsymbol{e}(t)$ of the distribution system due to different reasons and can affect the load performance in a different way.

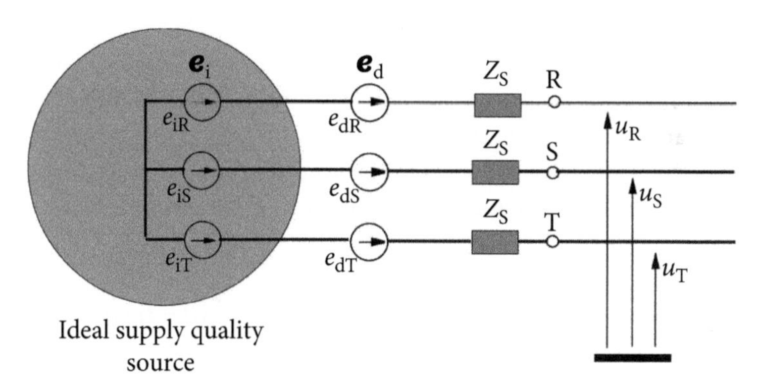

Figure 9.2 A supply source with degraded supply quality (SQ).

The Loading Quality (LQ). This term provides a qualitative description of the load from the point of view of the energy provider. It can be specified as follows.

The best load from the point of view of the energy provider is a resistive, linear, LTI load as shown in Fig. 9.3.

It is a load with the best LQ possible or with an *ideal loading quality*. When supplied from a source with an ideal SQ, such a load has a sinusoidal and symmetrical current, in phase with the supply voltage and of a constant rms value, meaning the load current is proportional to the supply voltage, meaning

$$\boldsymbol{i}_i(t) = G\boldsymbol{e}_i(t)$$

If any of these conditions are not fulfilled, then the load has a *degraded loading quality*. In particular, when the load

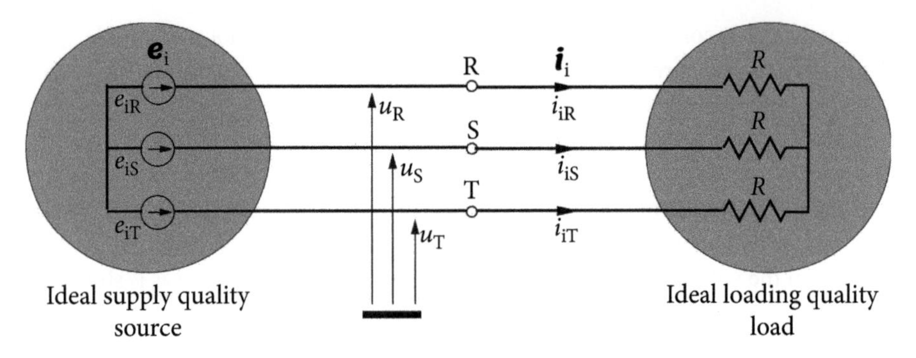

Figure 9.3 A circuit with an ideal supply quality source and ideal loading quality load.

- is not resistive, then the load current contains a reactive component $\boldsymbol{i}_{1r}(t)$;
- is not balanced, then the load current contains an unbalanced component $\boldsymbol{i}_{1u}(t)$;
- is not linear or/and time-invariant, then the load current contains harmonics $\boldsymbol{i}_{h}(t)$;
- is not time-invariant, then the load current contains a component with time-varying rms value $||\boldsymbol{i}_{v}||$.

The current of a load with degraded loading quality can contain, moreover, a scattered current $\boldsymbol{i}_{s}$ and, in the presence of fast switching devices, HF noise $\boldsymbol{i}_{HF}(t)$, and also, due to random switching, transient currents $\boldsymbol{i}_{t}(t)$. Thus, the current of a load with a degraded loading quality differs from the current of a load with an ideal loading quality $\boldsymbol{i}_{i}(t)$ by a distorting current component $\boldsymbol{i}_{d}(t)$, composed in general of seven components of a different nature:

$$\boldsymbol{i}_{d}(t) = \boldsymbol{i}_{s}(t) + \boldsymbol{i}_{1r}(t) + \boldsymbol{i}_{1u}(t) + \boldsymbol{i}_{h}(t) + \boldsymbol{i}_{v}(t) + \boldsymbol{i}_{t}(t) + \boldsymbol{i}_{HF}(t). \tag{9.2}$$

At a fixed supply, voltage $\boldsymbol{u}(t)$ and a fixed load with degraded quality, supplied as shown in Fig. 9.4(a), can be regarded as a load of ideal loading quality connected in parallel as shown in Fig. 9.3(b) to a current source of a distorting current specified by (9.2):

$$\boldsymbol{j}_{d}(t) = \boldsymbol{j}_{1r}(t) + \boldsymbol{j}_{1u}(t) + \boldsymbol{j}_{h}(t) + \boldsymbol{j}_{v}(t) + \boldsymbol{j}_{t}(t) + \boldsymbol{j}_{HF}(t). \tag{9.3}$$

Observe that due to the lack of the internal impedance in the source with an ideal supply quality, the current $\boldsymbol{i}_{d}(t)$ increases the supply current of the source as compared to the current of the load with an ideal loading quality, $\boldsymbol{i}_{i}(t)$. Its rms value increases as well. These two currents are mutually orthogonal. The three-phase rms value of the load current with a degraded loading quality can be expressed as follows

$$||\boldsymbol{i}|| = \sqrt{||\boldsymbol{i}_{1a}||^2 + ||\boldsymbol{i}_{d}||^2}. \tag{9.4}$$

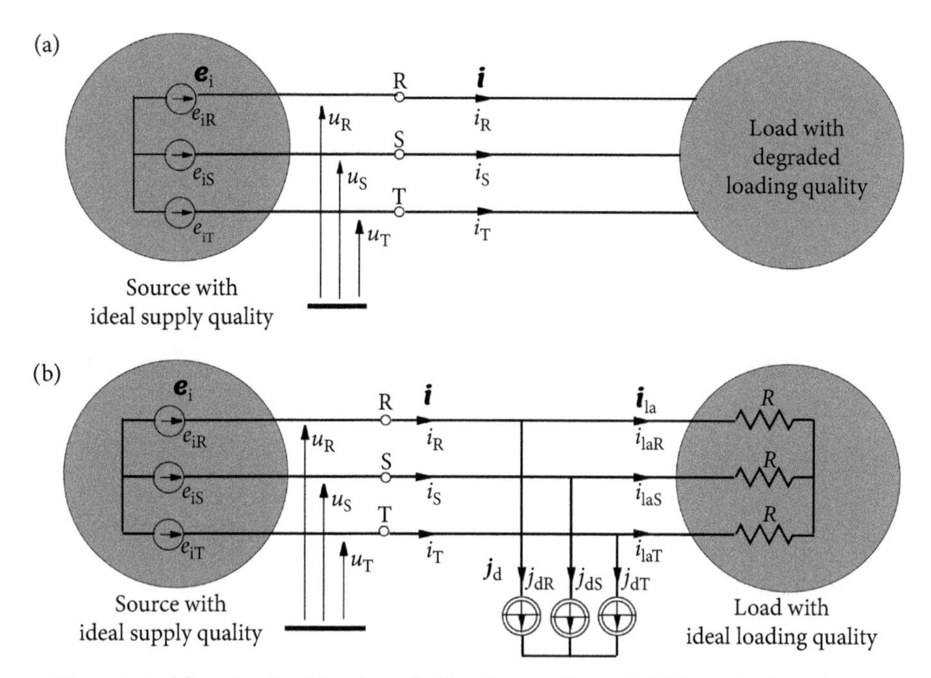

Figure 9.4 (a) A circuit with a degraded loading quality and (b) its equivalent circuit.

9.2 Negative Effects of Degraded LQ and SQ

Effects of degraded LQ upon the supply source. Degradation of the loading quality affects conditions of energy delivery. Some agents, such as the reactive, scattered, or unbalanced currents as well as current harmonics, contribute to the supply current rms value increase. This may elevate the temperature of transmission equipment or increase its cost associated with an increase in the power rating of this equipment. These currents contribute to a reduction of the supply voltage, its asymmetry and distortion HF noise, transients, and the voltage rms value variation. These factors degrading SQ propagate through the point of the common supply (PCS) to other loads, as shown in Fig. 9.5. Some measures might be necessary to confine them, which involves some extra cost. All of these contribute to financial profit reduction on the energy provider side.

The effect of degraded LQ upon the supply source depends on its short circuit power. When the load is supplied from an infinite power source, then due to the zero internal impedance of such a source, there is no extra energy loss on the energy provider side, and the terminal voltage is not affected by the load current. However, the energy provider could be concerned with the presence of current harmonics and HF noise, meaning $\boldsymbol{i}_{\rm h}(t)$ and $\boldsymbol{i}_{\rm HF}$ currents in the supply current. These currents can disturb digital instrumentation, such as protection and control devices, as well as energy meters.

The situation changes when the supply source has finite power. It is enough that the source has a non-zero internal impedance, as shown in Fig. 9.6, and the voltage

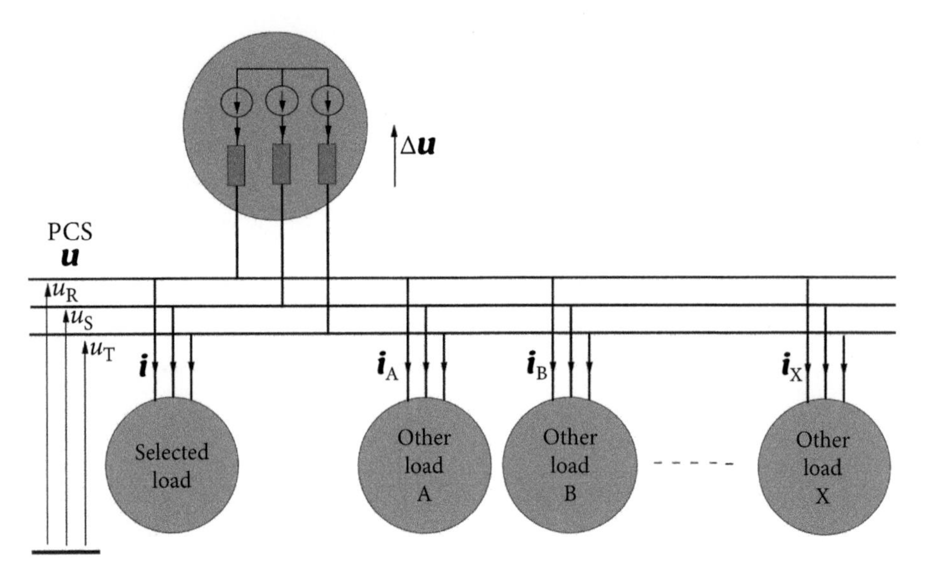

Figure 9.5 Loads connected in the point of common supply (PCS).

as observed at the supply source terminals $\boldsymbol{u}(t)$ is affected by the voltage drop $\Delta\boldsymbol{u}(t)$ of the load current on this impedance.

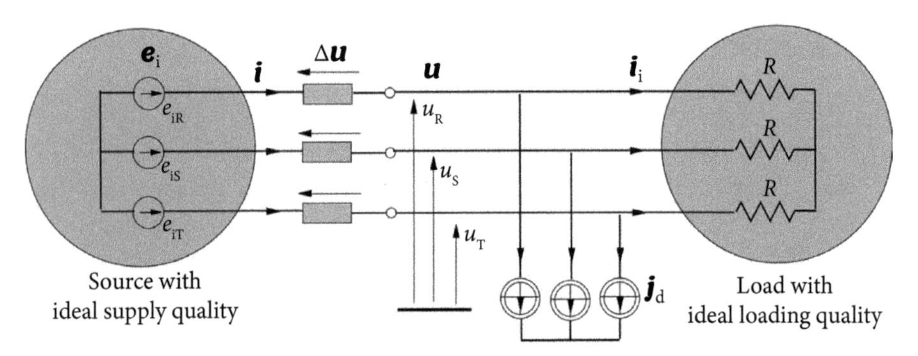

Figure 9.6 A load with degraded loading quality supplied from a source with non-zero internal impedance.

Consequently, the supply voltage can contain the same features as the load current. It can be asymmetrical, distorted by harmonics, and contain slowly varying components, transients, and HF noise,

$$\boldsymbol{u}(t) = \boldsymbol{u}_1(t) + \boldsymbol{u}_n(t) + \boldsymbol{u}_h(t) + \boldsymbol{u}_v(t) + \boldsymbol{u}_t(t) + \boldsymbol{u}_{HF}(t). \tag{9.5}$$

A selected customer may not have any knowledge than that he is supplied from a source as shown in Fig. 9.7. In general, this is an equivalent source, composed of the source of the energy provider, but affected by all other customer loads.

Symbol $\Delta\boldsymbol{u}'(t)$ in this diagram represents the voltage drop inside of the equivalent supply source caused by the current $\boldsymbol{i}(t)$ of the selected load. It is equal to zero when

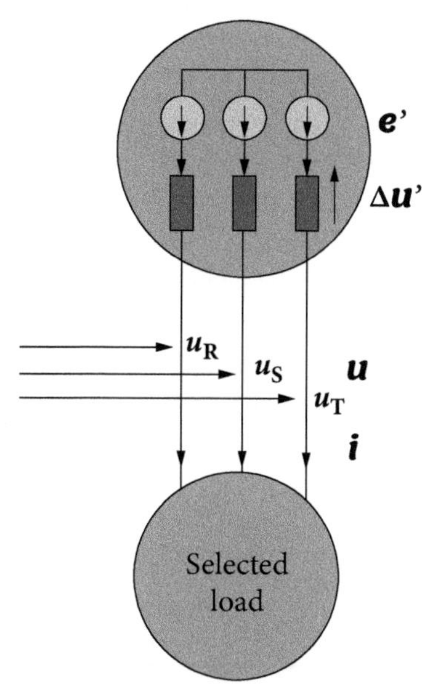

Figure 9.7 A distribution system from the point of view of a selected individual customer.

this load is disconnected. The internal voltage $\mathbf{e}'(t)$ is an equivalent voltage at PCS when the selected load is disconnected. It is composed of the internal voltage of the original source and the voltage response of this source to all currents loads connected at the PCS. It can contain all components which contribute to the supply quality degradation, denoted as $\mathbf{e}_{\mathrm{d}}(t)$, namely

$$\mathbf{e}'(t) = \mathbf{e}_1(t) + \mathbf{e}^{\mathrm{n}}(t) + \mathbf{e}_{\mathrm{h}}(t) + \mathbf{e}_{\mathrm{v}}(t) + \mathbf{e}_{\mathrm{t}}(t) + \mathbf{e}_{\mathrm{HF}}(t) = \mathbf{e}_1(t) + \mathbf{e}_{\mathrm{d}}(t). \qquad (9.6)$$

This internal voltage can be visible on terminals of the selected load only if this load can be disconnected. If not, only the voltage

$$\mathbf{u}(t) = \mathbf{e}'(t) - \Delta\mathbf{u}'(t) = \mathbf{u}_1(t) + \mathbf{u}^n(t) + \mathbf{u}_h(t) + \mathbf{u}_v(t) + \mathbf{u}_t(t) + \mathbf{u}_{HF}(t)$$
$$= \mathbf{u}_1(t) + \mathbf{u}_d(t) \qquad (9.7)$$

can be observed. Unfortunately, without additional information, modeling, or identification procedures there is no ground for a conclusion that can be reached on the origins of the degrading component of the supply voltage, $\mathbf{u}_{\mathrm{d}}(t)$. Is the original source, other loads, or the selected load responsible for the voltage $\mathbf{u}_{\mathrm{d}}(t)$ presence? The same is true for the load current. It can contain all components that contribute to the loading quality degradation:

$$\mathbf{i}(t) = \mathbf{i}_{1a}(t) + \mathbf{i}_{1r}(t) + \mathbf{i}^n(t) + \mathbf{i}_h(t) + \mathbf{i}_v(t) + \mathbf{i}_t(t) + \mathbf{i}_{HF}(t) = \mathbf{i}_{1a}(t) + \mathbf{i}_d(t).$$

As with voltage components that degrade supply quality, similarly without additional information on the distribution system, without modeling or identification processes there is no ground for a conclusion to be drawn about the cause of the current $\boldsymbol{j}_{\mathrm{d}}(t)$ component. We do not know whether it occurs because of degraded loading quality, degraded supply quality, or both. The voltage at the load terminals $\boldsymbol{u}(t)$ and its current $\boldsymbol{i}(t)$ are a result of the distribution internal voltage $\boldsymbol{e}_1(t) + \boldsymbol{e}_{\mathrm{d}}(t)$ and the load current $\boldsymbol{i}_{\mathrm{la}}(t) + \boldsymbol{j}_{\mathrm{d}}(t)$. The equivalent circuit of the system is shown in Fig. 9.8.

Looking at Fig. 9.8 the reader should be aware, however, that accuracy of this equivalent circuit is confined by some approximations. First of all, the current $\boldsymbol{j}_{\mathrm{d}}(t)$ is assumed to be fixed, while in fact it is a response to the supply voltage. It can be regarded as fixed only if the supply voltage is fixed concerning the rms value and the waveform. Any change of this voltage of course changes this current.

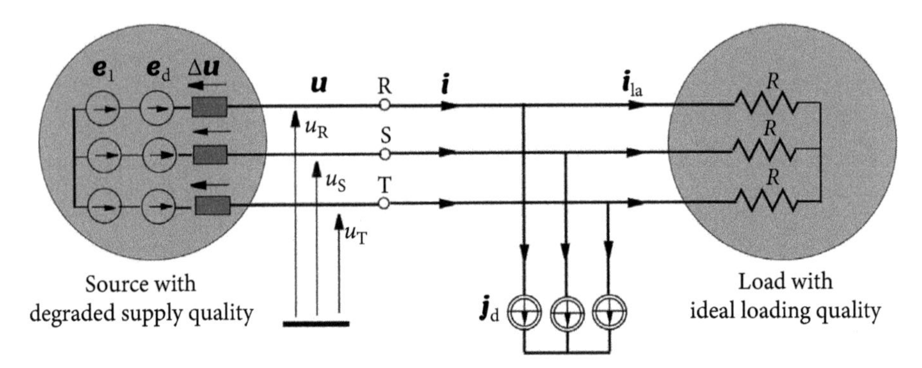

Figure 9.8 An equivalent circuit of a distribution system and a load.

Since this voltage rms value in common power systems does not change more than a few percent, however, we can also expect that the rms value of the current $\boldsymbol{j}_{\mathrm{d}}(t)$ can change only within these limits. The same applies to the current $\boldsymbol{j}_{\mathrm{d}}(t)$ waveform, although it is not easy to express this quantitatively.

Effects of degraded SQ. Degraded supply quality can have several detrimental effects on customer loads. These effects depend on particular features of the supply voltage and the load. Transients, HF noise, and harmonics can affect computer-like devices and, in particular, digital control and instrumentation, as well as power electronics loads such as variable speed drives. Variation of the supply voltage rms value can cause annoying light flickers. Transients in the supply voltage, such as notches or spikes, can disrupt production processes, causing equipment damage and a huge loss of revenue. If rotating machines dominate in the load, then energy conveyed by negative sequence voltage and harmonics do not contribute to energy conversion to mechanical energy by such machines but only to their temperature increase. On the other hand, such machines are rather not sensitive to HF noise or transients. In general, the efficiency of energy conversion by customer loads and customer revenues are reduced by the SQ degradation. All of these effects are illustrated in Fig. 9.9.

The concept of quality immediately involves the issue of its measurement. Unfortunately, the loading and supply qualities are affected by various agents and their effects usually defy any quantitative comparison. How do we compare, for example, the harmful effects of the supply voltage asymmetry with the effects of random spikes? Voltage asymmetry can contribute to an increase in a motor's temperature, while voltage spikes can interrupt control and stop manufacturing processes. Such comparison is possible only on the condition that these harmful effects can be expressed in terms of money, but this is not an easy task, and it is dependent on a specific system situation. Consequently, the measures of the loading or supply qualities can be based only on conventions or standards. Moreover, it is much easier to talk about these two qualities in terms of their *degradation* or *improvement*, regarded only as qualitative, rather than quantitative terms.

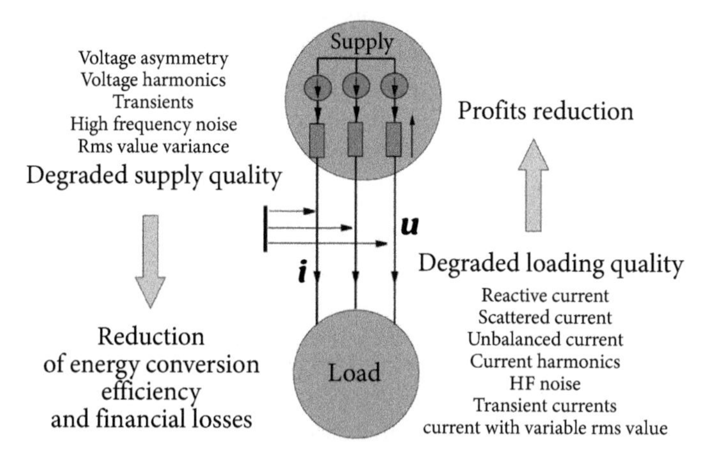

Figure 9.9 A diagram of the effects of degraded supply quality and degraded loading quality.

However, instead of terms the loading quality and the supply quality, the term *power quality* is commonly used. This term is vague, however. For example, if the voltage and current in a cross-section of a three-phase supply line between a load and the distribution system are asymmetrical, one can say that the power quality in that cross-section is degraded. Such a conclusion is not very informative, however. This asymmetry could be the effect of two, distinctively different causes, and, depending on the causes it can have distinctively different effects.

The voltage asymmetry at terminals of a load can occur because of the asymmetry of the distribution voltage. The supply voltage asymmetry is the primary cause in such a case, while the current asymmetry is only a response. A part of energy delivered at such a supply is not converted to mechanical energy on the shaft of induction motors. This energy elevates the temperature of such motors, accelerating their aging, thus the extra cost of such a supply is on the customer side. It increases the cost of energy utilization. The same asymmetry can occur, however, as a result of the load imbalance and consequently, asymmetry of the supply current. It causes an increase in its rms value and extra energy loss at energy delivery. The current asymmetry is the primary cause in such a situation, while the voltage asymmetry is only the response.

Thus it is important to know whether degradation of the quality refers to the supply, the load, or both. This knowledge is crucial for evaluating the technical and economic effects of such degradation. This knowledge is also crucial for developing filters or compensators for these qualities improvement. First of all, compensators for improving loading quality have to be connected in the system in different ways than compensators for improving the supply quality.

9.3 Objectives of Compensation

The objective of compensation in distribution systems supplied with a symmetrical voltage and sinusoidal voltages and currents (sv&c) is simple. Compensators should improve, by reduction of the reactive current, the power factor to such a value that the cost of energy delivery to a load would be minimal. This cost should include, of course, the cost of compensation.

When voltages and currents are nonsinusoidal and asymmetrical, then the objectives of compensation are not unique. Voltages and currents in such situations can contain several harmful components that can be reduced by compensators or filters. It is a matter of hierarchy of their importance and the designer's preferences which of them should be reduced.

The ultimate objective of compensation is a minimization of the cost of energy delivery to the load and utilization of this energy. The extra cost of energy delivery and utilization, caused by the reduced quality of the load and/or the supply, has several components. Some of them, such as the cost of extra loss of energy at delivery, can be evaluated relatively easily. However, it can be very difficult to predict the cost of accelerated aging of equipment, the cost of disturbances, and the reduced reliability of the supply. Compensation itself involves some costs as well. It is the investment cost of the compensator but also the cost of energy needed to run the compensator, as well as its operation maintenance. Therefore, due to the lack of data, and minimization of the cost of energy delivery and utilization, which includes the harmful effects of reduced LQ and SQ, as well as the cost of the compensation, the ultimate goal of compensation is very difficult to achieve. Moreover, from an economic perspective, energy providers and customers are different entities and they do not share the same goals. Therefore, compensation goals are not formulated in terms of minimization of the cost of energy delivery and its utilization but rather by minimization of some current or voltage components that degrade the LQ or SQ. Even more importantly, instead of minimization it could be sufficient that some harmful components of the current or the voltage are kept by compensation within limits accepted by some standards or agreements between energy providers and consumers.

Protection of the supply system against loads with degraded LQ requires that the compensator or a filter is connected in parallel to the load, as shown in Fig. 9.10(a).

Protection of the load against harmful components in the supply voltage, supplied from a source with a degraded SQ, requires that the compensator is connected in series to the load, as shown in Fig. 9.10(b). This means that the compensator's topologies for SQ and LQ improvement are substantially different. Improvement of both of them is possible by a sort of hybrid compensator composed of shunt and series branches shown in Fig. 9.11.

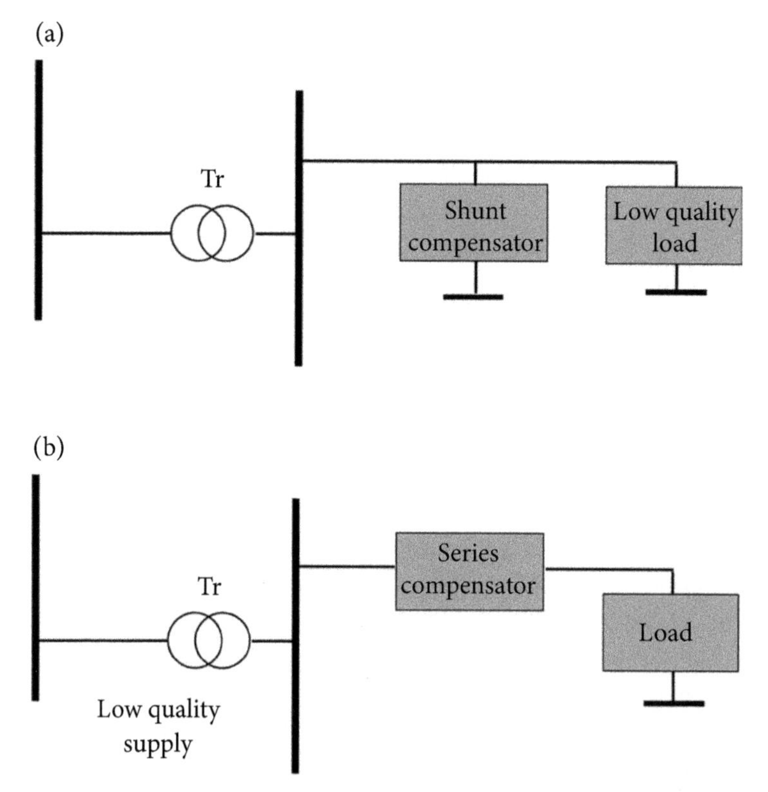

Figure 9.10 (a) A system with a shunt compensator and (b) with a series compensator.

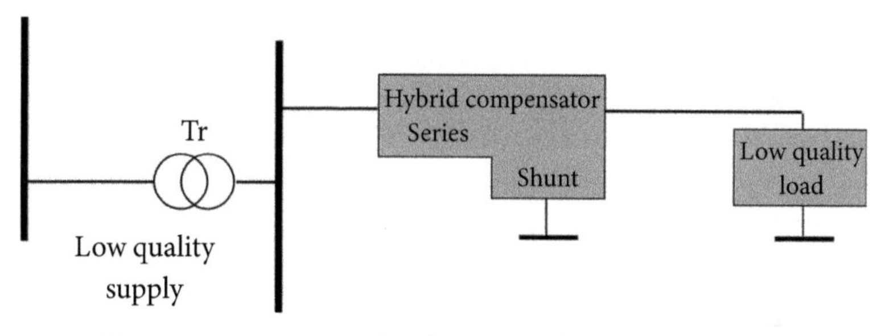

Figure 9.11 A system with a shunt/series hybrid compensator.

Such compensators are not common, however. Most compensators are shunt devices used for the protection of the supply systems against loads with degraded LQ. Therefore, this book confines itself entirely to shunt compensation only.

The smallest current needed for energy delivery to the single-phase LTI load with the power P, according to Section 6.5, is the active current $i_{\rm a}(t)$. In the case of circuits with harmonics generating loads, according to Section 6.11, the smallest is the active current $i_{\rm aC}(t)$. In three-phase circuits, according to Sections 7.7 and 7.10, these are currents $\boldsymbol{i}_{\rm a}(t)$ and $\boldsymbol{i}_{\rm aC}(t)$ respectively. Reduction of the supply current to the active current is often regarded as the ultimate objective of the compensation. Observe, however, that the supply current after compensation, having the lowest rms value, reproduces the supply voltage waveform, thus it is distorted. This current could be sinusoidal if harmonics of the active current are compensated as well.

The most common objective of compensation is the reduction of the reactive power, which means the reduction of the rms value of the reactive component in the supply current. Reduction of the supply current distortion, meaning the current harmonics, is usually the second objective of compensation. The supply current symmetrization, reduction of transients, and/or the supply current rms value variation are much less frequent objectives.

9.4 Compensation Tools

The objectives of compensation dictate what tools can be used. At very high power only reactance devices or under-excited synchronous machines can be used as compensators. Capacitor banks or synchronous machines used as compensators confine compensation only to fundamental harmonics, however. Moreover, this confines compensation to circuits with sinusoidal voltages and currents only. Compensators that are more complex than capacitor bank compensators are needed for compensation of the reactive current in the presence of the supply voltage distortion.

Reactance compensators (RCs) are fixed-parameter devices so they are not well suited to compensation of loads with variable power. *Adaptive compensators* are needed for that. Switches or saturating the magnetic core of inductors can be used to change the compensator parameters. Thyristor switched inductors (TSIs) are commonly used for that.

The current harmonics generated by the harmonic-generating loads (HGLs), along with the reactive current, can be reduced in the supply current of such high-power systems by *resonant harmonic filters* (RHFs). Unfortunately, in the presence of the distribution voltage distortion, the effectiveness of RHFs can decline drastically, as is shown in Ref. [200]. This can occur even if the distortion is within the limits specified by the relevant standards. This confines the use of the RHFs as a tool for reducing current harmonics originated in the HGLs.

At lower load power, up to the order of a few megawatts (MW), compensation is possible by *switching compensators* (SCs), known as active power filters or active harmonic filters. These devices operate as controlled current sources and can generate currents of any waveform specified by the compensator control algorithm. This current, when injected into the distribution system, can compensate the undesirable component of the supply current. This could be the reactive and unbalanced currents as well as the current harmonics or transients. Switching compensators shape the current waveform by fast switching power transistors, with the frequency of hundreds or several kHz, in an inverter structure. Therefore, they have natural adaptability. Relatively low power is a disadvantage of SCs, however. This disadvantage can be overcome by combining switching compensators with reactance compensators which can be built for a much higher power. Such compensators can be referred to as *hybrid compensators*.

9.5 Compensation at Sinusoidal Voltage and Current

Compensators in distribution systems with sinusoidal voltages and currents are installed for only one purpose, namely for improving the power factor which due to the reactive current of RL loads could be unacceptably low. The reactive power compensation increases the distribution voltage rms value and reduces its change with the change of the load power. Compensation improves energy flow control. This subject is beyond the scope of this book, however.

Compensation in a traditional sense has a clear and very distinctive meaning. This term is used for an action aimed at reducing the reactive power Q at terminals of voltage sources that supply inductive loads. Reduction of the supply current rms value and, consequently, enhancement of the power factor λ, are the effects of such compensation.

Capacitor banks or under-excited synchronous machines are used for reactive power reduction. The change of the supply current crms value of RL load, in Fig. 9.12(a), in the effect of compensation is as shown in Fig. 9.12(b). The current of the capacitor reduces the reactive component of the load current and consequently, the supply current i_s rms value. Compensators in the form of a capacitor bank, installed both by large customers and by utilities, are very common devices for reducing reactive currents in distribution systems.

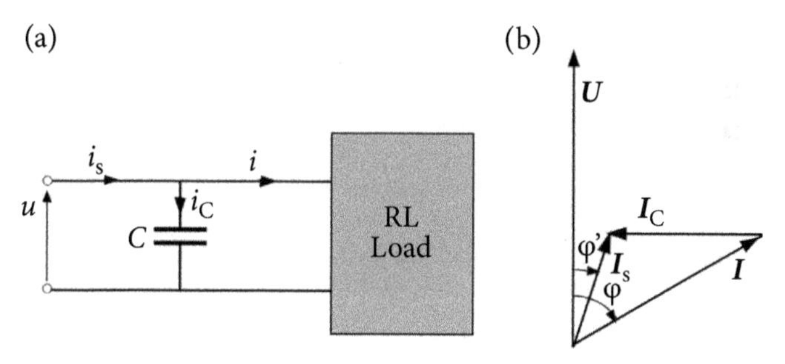

Figure 9.12 (a) An RL load with a capacitive compensator and (b) diagram of the currents' crms values.

Capacitor banks in distribution systems are installed directly at terminals of large loads with low power factor, and/or at power station buses and even along distribution lines, as shown in Fig. 9.13.

Compensators at terminals of large loads are installed usually at the cost of the energy consumers. Compensators along distribution lines and at power stations are installed at the cost of energy providers.

Individual capacitor banks are sources of a relatively small amount of reactive power needed for compensation. Lager amount can be provided by under-excited synchronous machines, and in particular, under-excited synchronous generators. Briefly, the operation of such a generator as a source of capacitive reactive power can be explained as follows.

A simplified equivalent circuit of a generator, turbine, supply bus, and power system is shown in Fig. 9.14. The generator considered is connected to a bus of a power

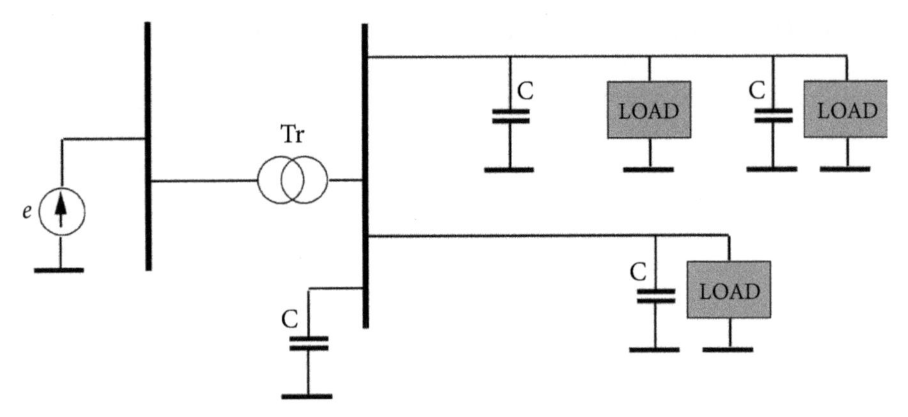

Figure 9.13 An example of capacitor banks distribution.

system, composed of a large number of other generators. Due to this large number, the voltage at the bus and the frequency are independent of the generator considered, but they are a result of the operation of all other generators and their control. Such a bus is referred to as an infinite bus.

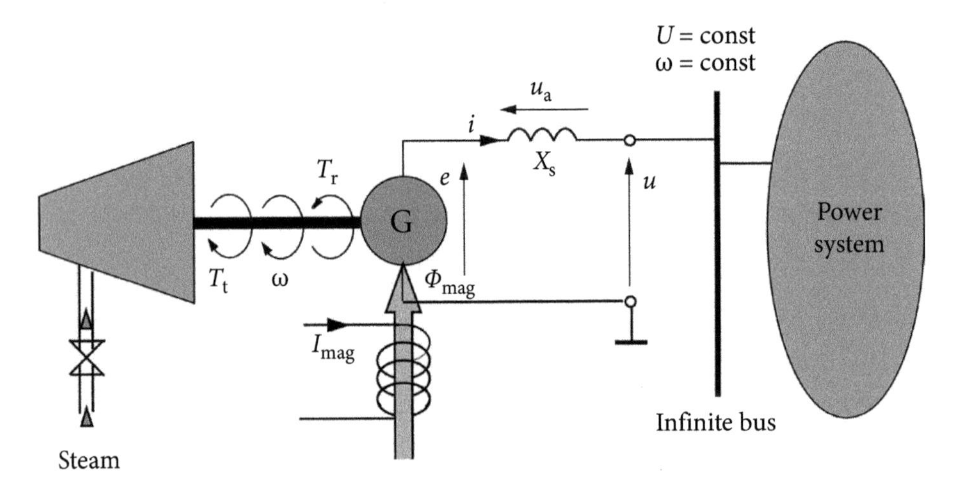

Figure 9.14 A synchronous generator connected to an infinite bus.

The rms value E of the internal voltage $\boldsymbol{e}$ produced by the generator is proportional to the magnetic flux Φ_{mag} produced by the exciting winding of the generator. Assuming that the resistance of the armature winding is much lower than its reactance X_s, the crms values of the bus voltage $\boldsymbol{U}$, internal voltage $\boldsymbol{E}$ of the generator, and the voltage drop $\boldsymbol{U}_a$ on the reactance X_s have to satisfy the diagram shown in Fig. 9.15.

As is shown in the diagram in Fig. 9.15(a), at the constant torque T_r, at the generator shaft, that is at constant active power $P = \omega\, T_r$, the current of the over-excited generator is delayed versus the bus voltage. When by reduction of the magnetic flux, the internal voltage of the generator is substantially reduced, as is shown in

Fig. 9.15(b), the generator current is leading the bus voltage, the phase angle φ, and the reactive power Q of the generator are negative. Concerning the reactive power, the generator behaves as a capacitor.

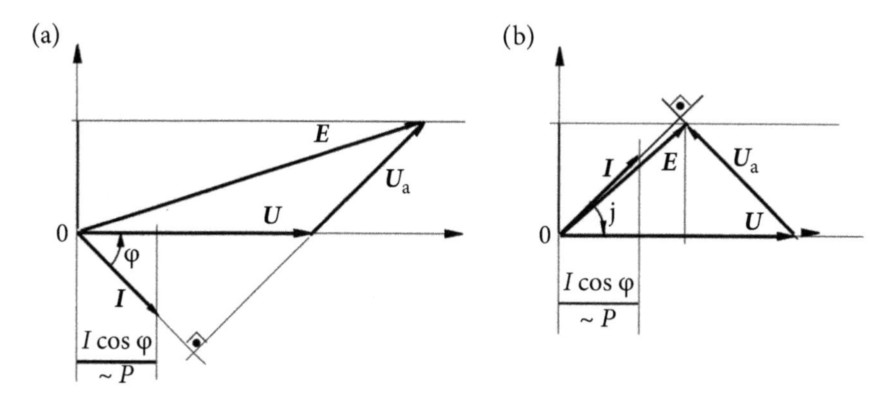

Figure 9.15 Diagrams of the crms values of voltages and current of (a) over-excited and (b) under-excited synchronous generator.

9.6 Reactance Compensation at Nonsinusoidal Voltage

When the distribution voltage is distorted or the load generates current harmonics, then the capacitor banks installed for the power factor improvement can cause amplification of the voltage and current distortion. In the effect of this distortion the effectiveness of capacitive compensation declines. Even worse, instead of improving the power factor, capacitive compensators can cause its decline. A resonance at a harmonic frequency can increase the voltage and current harmonics to such a level that the capacitor bank can be damaged or at least overcurrent protection activated. This could be avoided if instead of capacitors more complex devices, known as reactance compensators (RC), are used for the reactive power compensation.

Parameters of the reactance compensators can be selected in such a way that for selected distribution voltage harmonics the admittance of the load with a shunt RC, connected as shown in Fig. 9.16, is purely resistive.

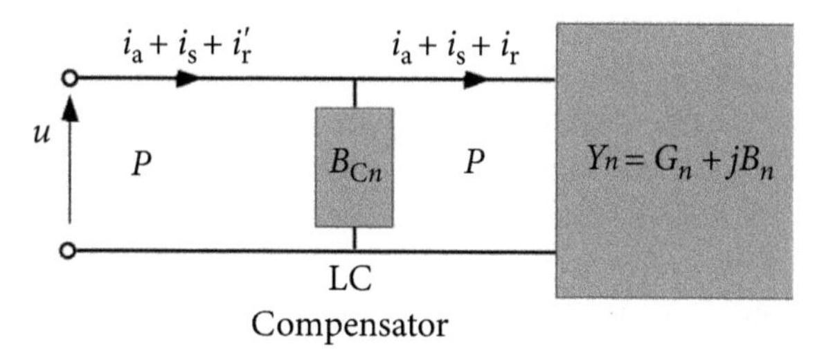

Figure 9.16 A load with reactance compensator.

Indeed, if the compensator admittance for the nth order harmonic $\boldsymbol{Y}_{\mathrm{C}n}$ is equal, with the opposite sign, to the load susceptance B_n, then the admittance as seen from the supply terminals is

$$Y_n = Y_{\mathrm{C}n} + Y_{\mathrm{L}n} = -jB_n + (G_n + jB_n) = G_n \tag{9.9}$$

It means compensation of the reactive current of all harmonics that satisfy (9.9). Such a compensator could be very complex, however. Its complexity can be reduced if instead of whole compensation the reactive current is only minimized.

The reactance compensators in three-phase circuits enable not only the compensation of the reactive current but also the unbalanced current. Parameters of a reactance compensator can be selected in such a way that for selected distribution voltage harmonics, not only the reactive component $\boldsymbol{i}_{\mathrm{r}}$ of the supply current is compensated but also its unbalanced component $\boldsymbol{i}_{\mathrm{u}}$ is compensated. The unbalanced load with such a compensator, connected as shown in Fig. 9.17, is seen by the supply source as a balanced load. The compensator operates as a *balancing compensator.*

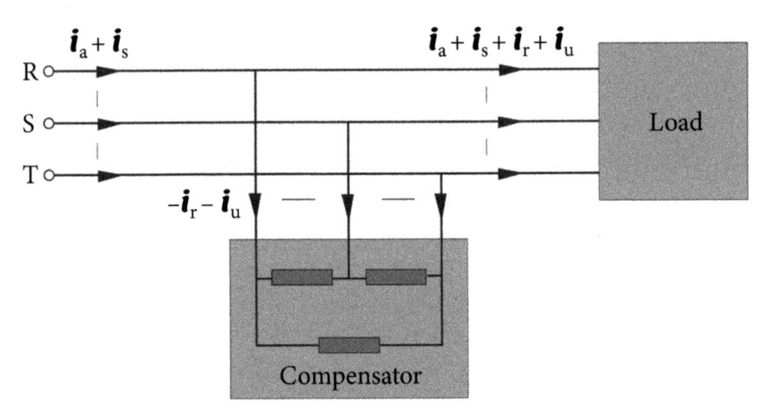

Figure 9.17 An unbalanced load in 3p3w circuit with a balancing reactance compensator.

The reactance balancing compensator in 3p4w circuits has to be composed of two sub-compensators. One of them, connected between supply lines, can have a Δ-configuration, the other Y-configuration, and be connected between supply lines and the neutral, as shown in Fig. 9.18.

The number of reactance elements needed for complete compensation of the reactive and unbalanced currents at a high number of harmonics in the supply voltage can make reactance compensators very complex and expensive. It is possible, however, to reduce the complexity of the compensator to a compensator with branches that have no more than only two reactance devices and almost preserved effectiveness. Details of the procedure that enables a reduction of the compensator complexity are presented in Sections 14.5 and 15.5.

A reactance compensator is a fixed parameters device thus its parameters are selected to fit a fixed load. It can be converted to an adaptive compensator by replacing its fixed inductors and/or capacitors with controllable ones. A thyristor switched

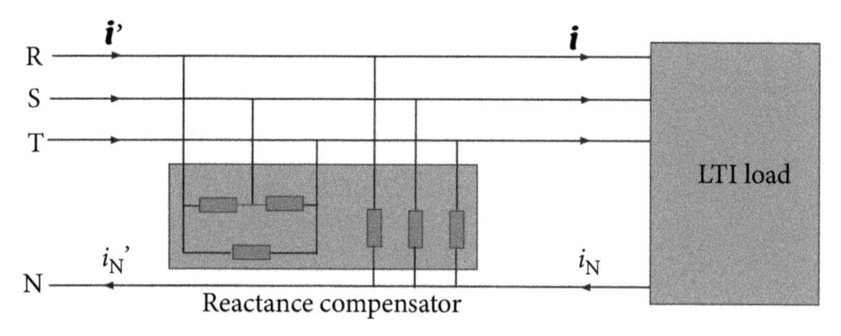

Figure 9.18 An unbalanced load in 3p4w circuit with balancing reactance compensator.

inductors (TSI) in a branch with a shunt third order harmonic filter and series inductor, shown in Fig. 9.19, can be used for that.

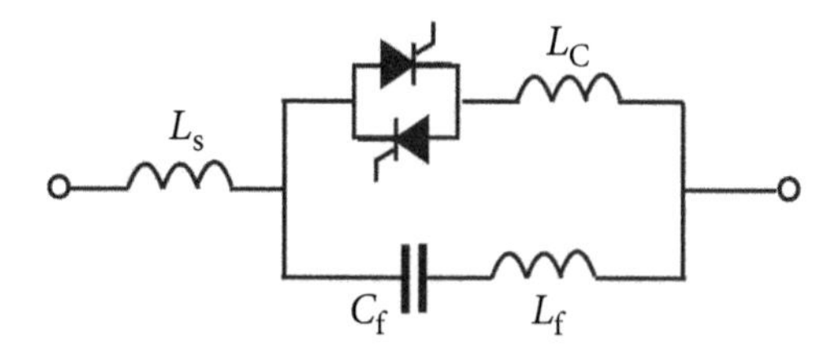

Figure 9.19 A controlled susceptance branch with TSI.

The filter reduces the third order harmonic generated by TSI, while the series inductor keeps the branch inductive to avoid resonances of the compensator with the distribution system inductance.

Such a branch could have the susceptance for the fundamental harmonic B_1 controlled by the thyristor's firing angle in some range from a negative to a positive value. It is referred to as *thyristor-controlled susceptance* (TCS).

9.7 Resonant Harmonic Filters

Reactance compensators can reduce the reactive and unbalanced current components in the supply current but not the current harmonics caused by harmonics generating loads. It can be done by resonant harmonic filters (RHFs).

An RHF is a reactance device built of a few series LC branches, connected in parallel, as shown in Fig. 9.20, tuned to or in a vicinity of the frequency of the load-originated dominating current harmonics. In three-phase circuits these are usually the fifth and the seventh order, sometimes eleventh and thirteenth order harmonics. Such a filter, along with the impedance of the supply source, forms a current divider for the load-originated current harmonics. Since at the frequency to which the LC branch is tuned its impedance declines to the lowest value, the low impedance path is created at such a frequency.

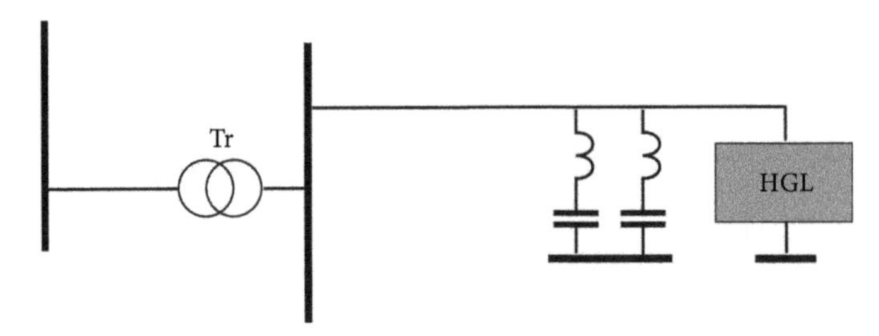

Figure 9.20 A two-branch resonant harmonic filter in a circuit with a harmonic generating load.

The reactance of RHFs for the fundamental harmonic is capacitive, therefore, apart from reducing the harmonic distortion of the supply current, the RHF can also compensate the reactive power of the fundamental. Thus, these devices are both filters of harmonics and compensators of reactive power.

Unfortunately, the low impedance of an LC branch tuned to the frequency of load-originated harmonics causes an increase of that harmonic component in the supply current when this component is caused by the distribution voltage harmonics of the same frequency. It means that an RHF can reduce the load-originated current harmonics at the cost of increasing the distribution system-originated current harmonic. Therefore, with an increase in the voltage distortion in the distribution system, the RHFs effectiveness in reducing current distortion declines [200].

9.8 Harmonics Blocking Compensator

When distribution voltage distortion has such a level that RHFs lose their effectiveness, a *harmonic blocking compensator* (HBC) can be used for both reduction of the current harmonics in the supply line and reactive power compensation. Apart from a shunt capacitor C_0 of the capacitance needed for the reactive power of the fundamental harmonic compensation, it has a series LC branch tuned to the fundamental harmonic and coupled with the circuit by a current transformer, as shown in Fig. 9.21.

Because of the resonance in the series circuit for the fundamental harmonics, this circuit injects only a small impedance to the supply line, while this impedance increases for a higher order harmonics and is inductive. Since the impedance of the capacitor C_0 declines with the harmonic order, this capacitor provides a low impedance path for the load-originated current harmonics. At the same time, the series impedance which increases with the harmonic order blocks the effect of the distribution system voltage harmonics upon the load. It protects the load against the distribution voltage distortion.

This kind of compensator can be used when both the loading quality and the supply quality are a matter of concern.

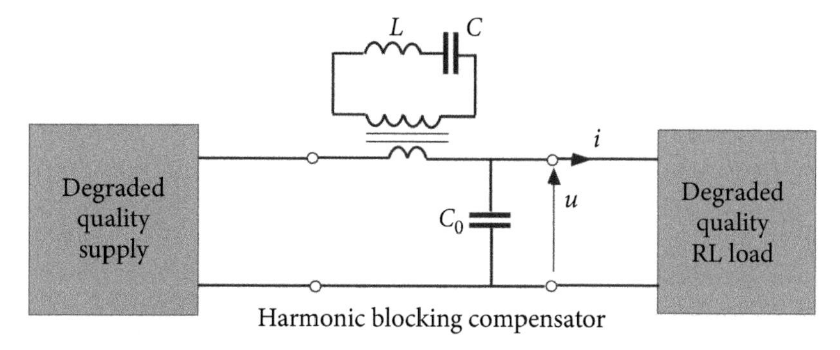

Figure 9.21 A circuit with a harmonic blocking compensator.

9.9 Switching Compensators

Switching compensators (SCs) are power electronics devices, composed of a power transistor-build inverter and a filter. A structure of a three-phase switching compensator with an inductive filter is shown in Fig. 9.22.

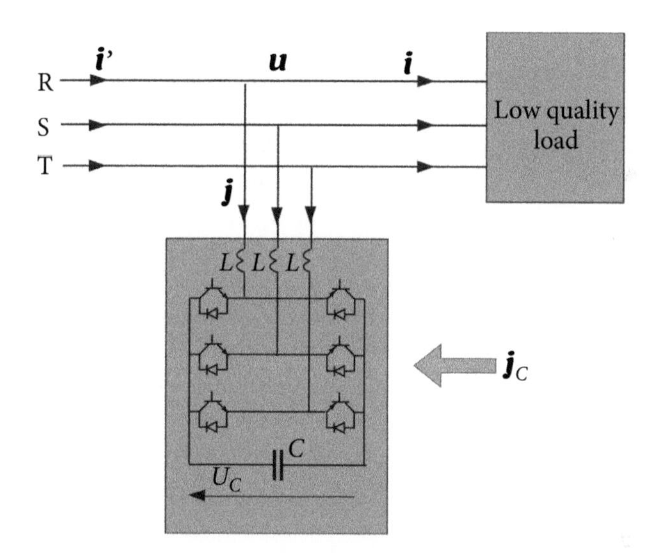

Figure 9.22 A shunt switching compensator.

A switching compensator reproduces a reference signal $\boldsymbol{j}_C$ as a three-phase compensating current $\boldsymbol{j}$ by switching the inverter's transistors with the frequency of hundreds or even up to several kHz. A controller of transistors switching (CTS) is needed for shaping the output current. The reference signal $\boldsymbol{j}_C$ is produced, according to a *compensation algorithm*, by a digital signal processing (DSP) system, based on the measurement of voltages and currents at the load terminals. When all transistors of the inverter are in the OFF state, the inverter operates as a six-pulse three-phase rectifier with a capacitive filter. The capacitor C is charged almost to line-to-line voltage maximum value. This voltage enables the operation of the inverter.

Switching compensators are referred to in the literature on the subject under different names, such as "*active power filters*", "*power conditioners*", or "*active harmonic filters*". This difference in names can be irrelevant for a person who works in the area but probably not for a novice who will need some time to realize that these different names refer to the same device. Such a novice will need some time to comprehend that despite the adjective "*active*" in the name, this is not a source of energy, but dissipates it, meaning it is not an active, but a *passive* device. Such a novice will need time to comprehend that this device does not eliminate unwanted components of the supply current by filtering but by injection of a compensating current, meaning it is not a filter but a compensator. Thus, the "active power filter" is neither an active device nor a filter. It is a passive compensator. These common names do not characterize the device adequately to its properties. Observe also that the phrase "power filter" suggests filtering of power which does not have much sense. Moreover, the phrase "harmonic filter" does not characterize adequately the field of the device applications. It can compensate not only harmonics but also the reactive current. It can balance the load and can compensate non-harmonic deviations of the supply current from a sinusoidal waveform, such as single spikes, notches, or HF noise. Even more, the concept of "harmonics" is not necessary for such devices' control. It is enough to separate the fundamental harmonic from the load voltages and currents. The switching compensator also does not "*condition*" the power. The name "*power conditioner*" is misleading. Thus, although the terms "active power filters", "power conditioners", or "active harmonic filters" are very common in electrical engineering, a name that would better describe the main features of these devices is needed.

These are, of course, passive compensators, but the term "passive compensator" does not differentiate such devices from reactance compensators, even such as capacitor banks. Fast switching of the inverter's switches, which enables shaping the compensating current waveform according to the waveform of the reference signal, is the most dominating feature of such a compensator. Therefore, these kinds of compensators are referred to as switching compensators.

Switching compensators can also be used as series compensators to reduce undesirable components of the supply voltage, thus improving the supply quality for customer loads. A switching compensator has to be connected to the distribution system for such a purpose by three line transformers, as shown in Fig. 9.23.

The output voltage, $\boldsymbol{u}$, of these transformers, which modifies the distribution voltage $\boldsymbol{u}$ to the voltage at the load terminals, $\boldsymbol{u}'$, should reproduce the reference signal, $\boldsymbol{v}_{\mathrm{C}}$.

The reference signal should be composed of undesirable components of the supply voltage, such as voltage harmonics and/or the negative sequence component of this voltage, to provide a symmetrical supply for the load. The reference signal can also contain distribution voltage notches or spikes and anything else that can be detected in the distribution voltage and can hamper the load performance.

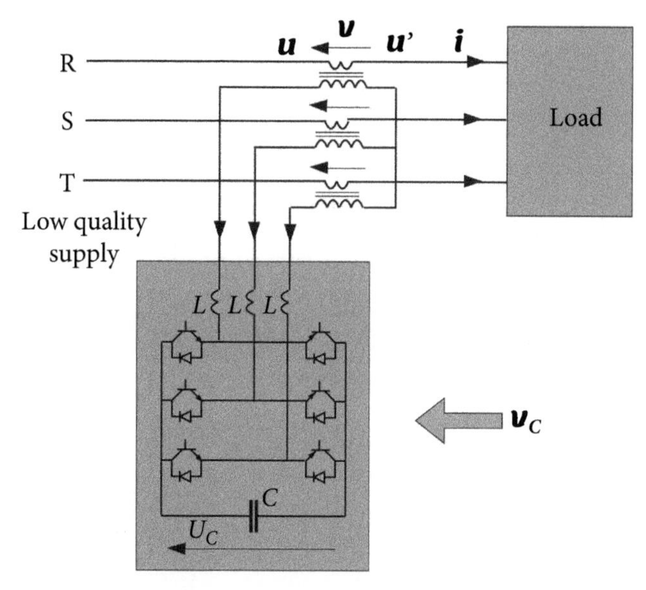

Figure 9.23 A series switching compensator.

9.10 Hybrid Compensators

The limited switching power of the inverter transistors is the main obstacle for high-power applications of shunt switching compensators. Transistors in the inverter operate only as switches between the ON and OFF state, and the energy is dissipated mainly during changing of the transistor state. Thus, the power rating of the compensator can be elevated by a reduction of the number of switching operations per second, meaning by reducing switching frequency. This frequency has to be sufficiently high to shape the current waveform distorted by high order harmonics or short-lasting spikes, however.

One could observe, however, that high order current harmonics that decide on the needed switching frequency are not the main components that contribute to the compensating current rms value $||\boldsymbol{j}||$. Usually, the reactive and/or unbalanced current components of the fundamental frequency contribute mainly to this rms value. These currents are slowly varying components, however. They can be compensated by an SC with low switching frequency and consequently, higher power ratings, in a hybrid structure shown in Fig. 9.24.

Such a hybrid compensator is composed of a sub-compensator of the reactive and unbalanced currents with low switching frequency but high power ratings, and a second sub-compensator of harmonic distortion with high switching frequency but low power ratings. Such decomposition of the compensator into two sub-compensators enables enhancement of the power rating of the compensator using the same power transistors.

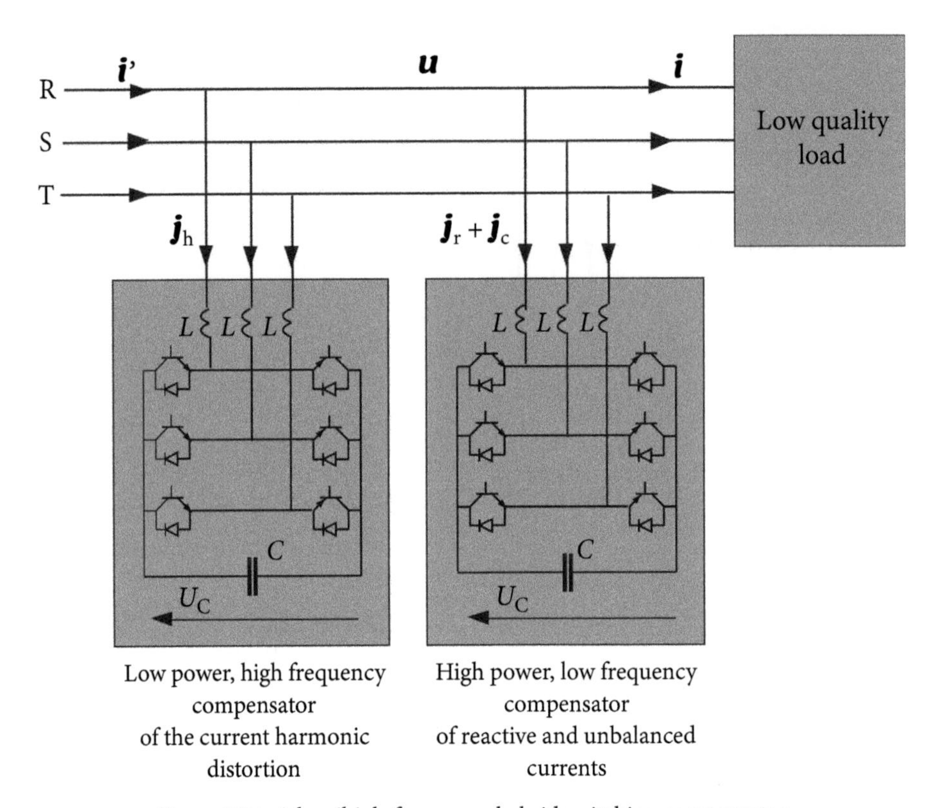

Figure 9.24 A low/high-frequency hybrid switching compensator.

Such a hybrid compensator has one additional important advantage over common switching compensators. ON-OFF switching of the current is the source of electromagnetic interference (EMI). Its intensity increases with the switching frequency. Thus separation of the reactive and unbalanced currents and compensation of these currents with a low-frequency SC reduces the level of EMI.

When the enhancement in the compensator power ratings using the low-/high-frequency hybrid structure is not sufficient, then it can be increased even more by compensating the reactive and unbalanced current by a reactance balancing compensator (RBC), while the switching compensation is confined only to harmonics, which requires much lower power ratings of the compensator. The structure of such a compensator is shown in Fig. 9.25.

Since the switching power of thyristors is approximately one order higher than the switching power of transistors, the hybrid compensator, composed of reactance balancing and switching sub-compensators, can have the power ratings by one order higher than the power rating of common switching compensators. At the same time, only six switching per period is needed for the RBC operation, which substantially reduces electromagnetic interference produced by the compensator.

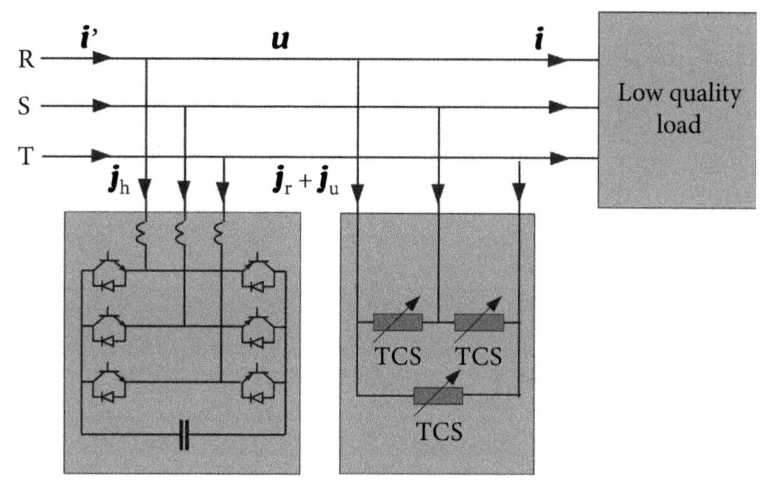

Figure 9.25 A hybrid compensator composed of a reactance balancing compensator and a switching compensator (SC).

These kinds of compensators can be built for very high power loads that substantially degrade the power factor and are the main sources of the current distortion. AC arc furnaces are examples of these kinds of loads.

Hybrid Series/Shunt Compensators. When both loading quality and supply quality require improvement by a compensator, then a hybrid series/shunt compensator can be used. Such compensator injects both the series voltage $\boldsymbol{v}$ and a shunt current $\boldsymbol{j}$ into distribution system lines, as shown in Fig. 9.26.

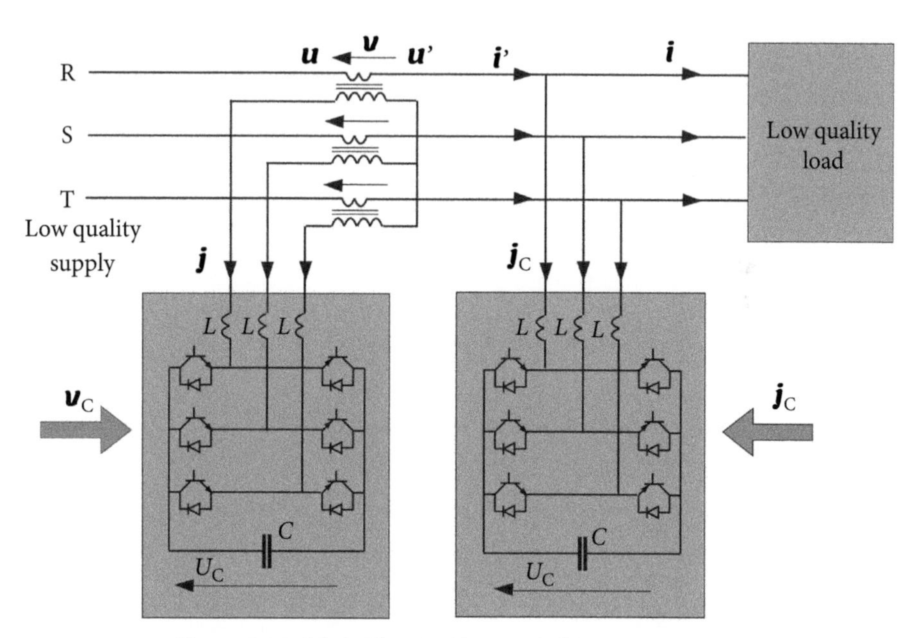

Figure 9.26 A hybrid series/shunt switching compensator.

The voltages and currents are shaped separately by two inverters that reproduce two reference signals, $\boldsymbol{v}_{C}$ and $\boldsymbol{j}_{C}$ generated by the DSP systems according to the compensator control algorithms. For high power loads, the shunt compensator can be built, of course, as the adaptive reactance compensator.

Summary

The possibility of compensation of various detrimental effects of electrical loads and supply sources is an important practical outcome of power theory development. These effects include not only extra energy loss but also degradation of the LQ and SQ qualities. The CPC-based PT provides solid fundamentals for the development of various compensators for that purpose.

Chapter 10
Reactance Compensators Synthesis

10.1 Circuit Synthesis Versus Analysis

The impedance of a reactance, meaning a inductive-capacitive (LC), compensator for circuits with nonsinusoidal voltages and currents should change with frequency in a specific way. It should have specified susceptances B_n for harmonic frequencies $n\omega_1$. Consequently, both its structure and the LC parameters have to be adequately selected. Unfortunately, only in very simple situations can it be done by guesswork.

Let us suppose, for example, that a reactance compensator, shown in Fig. 10.1a, with the susceptance for the fundamental frequency B_1 = 2S and for the third order harmonic B_3 = 1S, as shown in Fig. 10.1b, is needed. What kind of circuit concerning structure and parameters can have such susceptances?

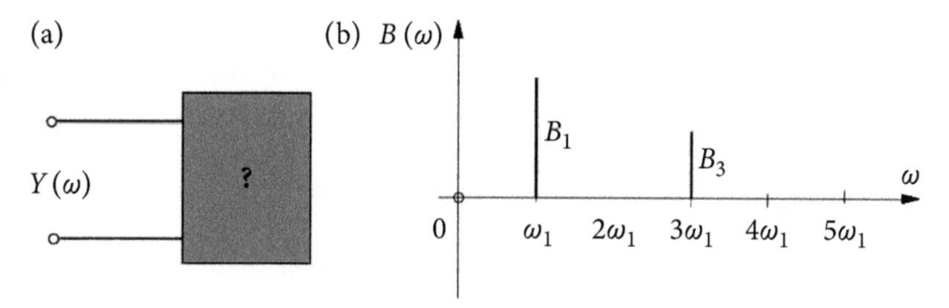

Figure 10.1 (a) A compensator and (b) needed susceptances for the 1st and 3rd order harmonics.

Practically, the structure and parameters of such a compensator cannot be predicted. A systematic approach is necessary for that. Fortunately, mathematical tools have been developed for solving this kind of problems by circuit theory and these tools are known as circuit synthesis theory.

To summarize this introduction, reactance compensators for circuits with nonsinusoidal voltages and currents are much more sophisticated devices compared to compensators used in circuits with sinusoidal voltages and currents. Unfortunately, engineers usually do not have the knowledge which would enable the design and build of such a compensator. Some fundamentals of electrical circuit synthesis are needed for that. This chapter provides the fundamentals of reactance compensator synthesis.

The subject of circuit synthesis is entirely different from that of circuit analysis. In the case of circuit analysis, the circuit, denoted symbolically by **C**, with regard to its

Powers and Compensation in Circuits with Nonsinusoidal Current. Leszek S. Czarnecki, Oxford University Press.
 DOI: 10.1093/oso/9780198879206.003.0010

structure and parameters is known. We are looking for the circuit response, denoted by $\boldsymbol{Y}$, to a given excitation, denoted by $\boldsymbol{X}$. Thus, circuit analysis should provide a solution to a symbolic equation:

$$Y = \{C, X\}.$$

For example, the circuit and the supply voltage are known but we need to find the circuit currents. The answer to this problem, of course, exists and it is unique. In the case of synthesis, we need to obtain a specific response to a specific excitation but the circuit structure and its parameters have to be found first. The synthesis should provide a solution to a symbolic equation

$$C = \{X, Y\}.$$

Such a synthesis problem may not have a solution, however. So the first question in the circuit synthesis is related to the existence of a solution for a specific synthesis problem. When we know that there exists a solution to this problem, that solution can be looked for, meaning the circuit structure has to be found and its RLC parameters have to be calculated. When a solution exists, it is usually not unique, however. A lot of circuits with different structures and parameters can provide to the same excitation the same response. The issue of the selection of the best solution to the synthesis problem arises. Therefore, the analysis of electrical circuits and their synthesis are substantially different fields of circuit theory.

One should observe that synthesis can be regarded as the primary subject of the activity of electrical engineers as compared to circuit analysis. Before a circuit can be analyzed, its structures and parameters have to be designed.

10.2 Positive Real Functions

The concept [16] of *Positive Real* (PR) *functions* is a core concept for circuit synthesis. It is because the *immittance* (the common term for impedance and admittance) of any circuit built of lumped resistive-inductive-capacitive (RLC) elements is a PR function. Also, any function that is PR can be realized physically as an immittance of a circuit built of lumped RLC elements.

Positive Real function $F(s) = \mathrm{Re}\{F(s)\} + j\mathrm{Im}\{F(s)\}$ is a function of the complex variable, $s = \sigma + j\omega$, such that

$$\begin{aligned} &\text{for} \quad \omega = 0, \quad \mathrm{Im}\{F(s)\} = 0 \\ &\text{for} \quad \sigma \geq 0, \quad \mathrm{Re}\{F(s)\} \geq 0. \end{aligned}$$

The first condition says that for a real s, meaning $s = \sigma$, the function $F(s)$ has real values. The second condition says that for s in the right-half plane, the function $F(s)$ has values in the right-half plane of $F(s)$. This mapping of the s-plane into the $F(s)$-plane for PR functions is illustrated in Fig. 10.2.

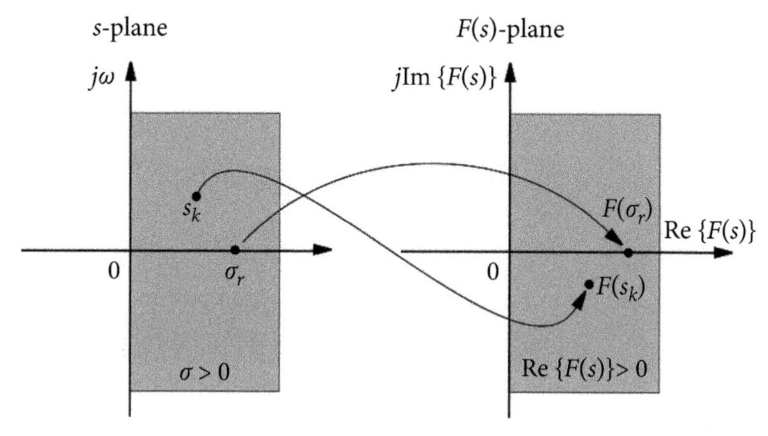

Figure 10.2 The s-plane and the Positive Real Function $F(s)$-plane.

The most general property of the impedance of passive and lumped RLC circuits is expressed by the following Theorem.

Theorem 1: *The impedance of any passive one-port built of lumped RLC elements is a Positive Real function.*

This property is a conclusion from the observation that the energy stored in the electric and magnetic fields and dissipated as heat in such a one-port cannot be negative.

To prove Theorem 1, let us consider one loop of a general structure, as shown in Fig. 10.3, of an RLC one-port. Let us assume that the one-port has N loops.

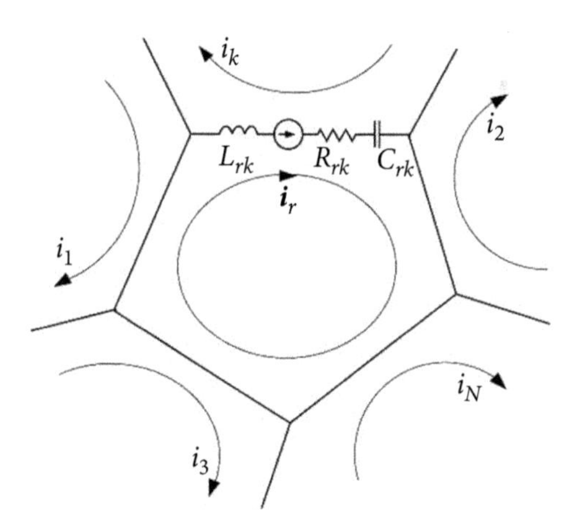

Figure 10.3 A loop r of one-port with N loops.

The capacitor C_{rk} voltage response to the loop current i_k at instant t can be expressed as

$$u_{rk} = \frac{1}{C_{rk}} \int_{-\infty}^{t} i_k \, dt = S_{rk} \, q_k, \qquad \frac{1}{C_{rk}} \overset{\text{df}}{=} S_{rk}$$

where q_k is the capacitor charge.

The loop equation for the loop r can be written in the form

$$\sum_{k=1}^{N} L_{rk} \frac{di_k}{dt} + \sum_{k=1}^{N} R_{rk} \, i_k + \sum_{k=1}^{N} S_{rk} \, q_k = e_r$$

where e_r is the sum of source voltages in loop r. Observe that all currents in this equation are entered with the positive sign, while the common loop currents of the loops other than r enter the loop equation for loop r with the negative sign. This convention is changed here in such a way that instead of the negative current, the circuit parameters, common for loops r and k, enter the loop equation with the opposite sign as compared to their sign in the common form of the loop equation. For example, terms such as this in the traditional loop equation were modified as follows:

$$R_{rk}(-i_k) = (-R_{rk}) i_k,$$

thus only parameters L_{rr}, R_{rr}, and S_{rr} are positive, while parameters R_{rk} and S_{rk} are negative. Mutual inductance, L_{rk}, can be, depending on the coupling, positive or negative.

The instantaneous power of the voltage sources in loop r is

$$p_r(t) \overset{\text{df}}{=} e_r \, i_r = \sum_{k=1}^{N} L_{rk} \frac{di_k}{dt} i_r + \sum_{k=1}^{N} R_{rk} \, i_k \, i_r + \sum_{k=1}^{N} S_{rk} \, q_k \, i_r,$$

thus the instantaneous power of all voltage sources in the circuit, meaning the instantaneous power of all passive elements, is

$$p(t) = \sum_{r=1}^{N} p_r(t) = \sum_{r=1}^{N} \sum_{k=1}^{N} L_{rk} \frac{di_k}{dt} i_r + \sum_{r=1}^{N} \sum_{k=1}^{N} R_{rk} \, i_k \, i_r + \sum_{r=1}^{N} \sum_{k=1}^{N} S_{rk} \, q_k \, i_r$$

Let us denote the instantaneous power of all inductors in the circuit

$$\sum_{r=1}^{N} \sum_{k=1}^{N} L_{rk} \frac{di_k}{dt} i_r \overset{\text{df}}{=} p_{\mathrm{L}}(t) = \frac{dT}{dt}$$

where T is energy stored in the magnetic fields of these inductors.

Let us denote the instantaneous power of resistors, meaning the rate of energy dissipation,

$$\frac{1}{2}\sum_{r=1}^{N}\sum_{k=1}^{N} R_{rk}\, i_k\, i_r \stackrel{\text{df}}{=} p_{\mathrm{R}}(t) = F. \tag{10.1}$$

Let us denote the instantaneous power of all capacitors in the circuit

$$\sum_{r=1}^{N}\sum_{k=1}^{N} S_{rk}\, q_k\, i_r \stackrel{\text{df}}{=} p_{\mathrm{C}}(t) = \frac{dV}{dt} \tag{10.2}$$

where V is energy stored in the electric fields of capacitors. Thus the instantaneous power of all passive elements of the one-port can be expressed as

$$p\,(t) = 2\,F + \frac{d}{dt}(T + V). \tag{10.3}$$

The energy stored in the magnetic fields of inductors is

$$T = \int_{-\infty}^{t} p_{\mathrm{T}}\, dt = \sum_{r=1}^{N}\sum_{k=1}^{N} L_{rk} \int_{-\infty}^{t} i_r\, di_k = \sum_{r=1}^{N}\sum_{k=1}^{N} L_{rk}\, i_r\, i_k - \sum_{r=1}^{N}\sum_{k=1}^{N} L_{rk} \int_{-\infty}^{t} i_k\, di_r. \tag{10.4}$$

The indices r and k of the loop currents can, of course, be switched without any effect on the result, similarly to the sequence of summation. Moreover, the symmetry of the mutual inductance, meaning equality $L_{rk} = L_{kr}$, is a fundamental property of magnetic coupling. It enables rewriting the result of integration in the form

$$\sum_{r=1}^{N}\sum_{k=1}^{N} L_{rk} \int_{-\infty}^{t} i_r\, di_k = \sum_{r=1}^{N}\sum_{k=1}^{N} L_{rk}\, i_r\, i_k - \sum_{r=1}^{N}\sum_{k=1}^{N} L_{rk} \int_{-\infty}^{t} i_r\, di_k$$

and consequently, the energy stored in magnetic fields of inductors is

$$T = \frac{1}{2}\sum_{r=1}^{N}\sum_{k=1}^{N} L_{rk}\, i_r\, i_k. \tag{10.5}$$

Similarly, the energy stored in the electric fields of capacitors is

$$V = \frac{1}{2}\sum_{r=1}^{N}\sum_{k=1}^{N} S_{rk}\, q_r\, q_k. \tag{10.6}$$

Energies T and V, as well the power dissipated in resistors, $2F$, cannot be negative for any loop currents i and their integrals q. This physical feature of energy and resistor power imposes particular requirements on the circuit parameters. The quantities T, V, and F have the same mathematical form,

$$\sum_{r=1}^{N}\sum_{k=1}^{N} a_{rk}\, x_r\, x_k$$

referred to in mathematics as a *quadratic form.* Such a form that is non-negative for any real numbers x_r and x_k is called a *positively semi-defined quadratic form.* Thus, energies in electric circuits and the instantaneous power of all circuit resistors are specified by positively semi-defined quadratic forms.

A quadratic form is positively semi-defined on the condition that coefficients a_{rk} have some specific property. If these coefficients are arranged into the matrix,

$$\mathbf{A} \stackrel{\text{df}}{=} \begin{bmatrix} a_{11}, & a_{12}, & \dots & a_{1N} \\ a_{21}, & a_{22}, & \dots & a_{2N} \\ \dots & \dots & \dots & \dots \\ a_{N1}, & a_{N2}, & \dots & a_{NN} \end{bmatrix}$$

then all main determinants of this matrix cannot be negative, meaning

$$a_{11} \geq 0, \qquad \begin{vmatrix} a_{11}, & a_{12} \\ a_{21}, & a_{22} \end{vmatrix} \geq 0, \; \dots\dots\dots, \; |\mathbf{A}| \geq 0.$$

This condition explains, for example, the known requirement related to mutual inductances, namely they have to satisfy the following condition

$$L_{11}L_{22} - L_{12}L_{21} \geq 0.$$

Let the same circuit is described in terms of the Laplace Transform of voltages and currents. Assuming that

$$\mathscr{L}\{i_r\} \stackrel{\text{df}}{=} I_r(s), \quad \mathscr{L}\{i_k\} \stackrel{\text{df}}{=} I_k(s), \quad \mathscr{L}\{e_r\} \stackrel{\text{df}}{=} E_r(s)$$

the loop equation for loop r, at zero initial conditions, can be written in the form

$$s\sum_{k=1}^{N} L_{rk}\, I_k(s) + \sum_{k=1}^{N} R_{rk}\, I_k(s) + \frac{1}{s}\sum_{k=1}^{N} S_{rk}\, I_k(s) = E_r(s)$$

Let us multiply this equation by the conjugate Laplace Transform of the loop current i_r, namely

$$E_r(s)\, I_r^*(s) \stackrel{\text{df}}{=} S_r(s) = s\sum_{k=1}^{N} L_{rk}\, I_k(s)\, I_r^*(s) + \sum_{k=1}^{N} R_{rk}\, I_k(s)\, I_r^*(s) + \frac{1}{s}\sum_{k=1}^{N} S_{rk}\, I_k(s)\, I_r^*(s)$$

Such a product at sinusoidal quantities, when $s = j\omega$, specifies the complex power $\mathbf{S}_r$ of the loop current. Let us calculate the sum of such "pseudo-powers" of the entire circuit, meaning the quantity

$$S(s) \overset{\text{df}}{=} \sum_{r=1}^{N} S_r(s).$$

It is equal to

$$\sum_{r=1}^{N} E_r(s)\, I_r^*(s) = s \sum_{r=1}^{N}\sum_{k=1}^{N} L_{rk}\, I_k(s)\, I_r^*(s) + \sum_{r=1}^{N}\sum_{k=1}^{N} R_{rk}\, I_k(s)\, I_r^*(s) + \frac{1}{s}\sum_{r=1}^{N}\sum_{k=1}^{N} S_{rk}\, I_k(s)\, I_r^*(s).$$

This formula, with symbols

$$\sum_{r=1}^{N}\sum_{k=1}^{N} L_{rk}\, I_k(s)\, I_r^*(s) \overset{\text{df}}{=} V'(s)$$

$$\sum_{r=1}^{N}\sum_{k=1}^{N} S_{rk}\, I_k(s)\, I_r^*(s) \overset{\text{df}}{=} T'(s)$$

$$\sum_{r=1}^{N}\sum_{k=1}^{N} R_{rk}\, I_k(s)\, I_r^*(s) \overset{\text{df}}{=} F'(s)$$

can be presented in the form

$$\sum_{r=1}^{N} E_r(s)\, I_r^*(s) = sV'(s) + F'(s) + \frac{1}{s}T'(s).$$

Quantities $V'(s)$, $T'(s)$, and $F'(s)$, referred to as *pseudo-energy* functions, are functions of complex variable s, and they have mathematically the same quadratic form as V, T, and F. They are functions of a complex variable z, however, namely they have the form

$$\sum_{r=1}^{N}\sum_{k=1}^{N} a_{rk}\, z_k\, z_r^*.$$

Let us denote

$$z_k = x_k + jy_k, \quad z_r = x_r + jy_r$$

then

$$\sum_{r=1}^{N}\sum_{k=1}^{N} a_{rk}\, z_k\, z_r^* = \sum_{r=1}^{N}\sum_{k=1}^{N} a_{rk}\,(x_k + jy_k)(x_r - jy_r) =$$

$$= \sum_{r=1}^{N}\sum_{k=1}^{N} a_{rk}\,(x_k\, x_r + y_k\, y_r) + j\sum_{r=1}^{N}\sum_{k=1}^{N} a_{rk}\,(x_k\, y_r - y_k\, x_r).$$

Indices k and r can be switched and coefficients in electrical circuits $a_{rk} = a_{kr}$, consequently the imaginary part of forms $V'(s)$, $T'(s)$, and $F'(s)$ is always equal to zero. Thus, these forms can have only a real non-negative value.

Let us assume that there is only one voltage source in the circuit and that source is located in the loop selected as loop number $r = 1$, as shown in Fig. 10.4.

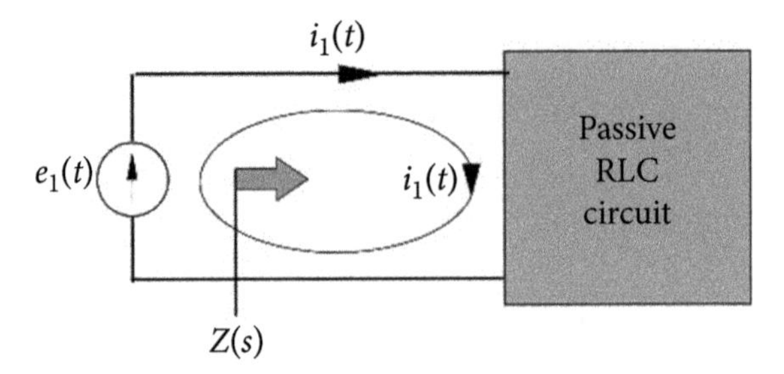

Figure 10.4 An RLC one-port.

The pseudo-power for such a one-port has the form

$$E_1(s)\, I_r^*(s) = sV'(s) + F'(s) + \frac{1}{s}T'(s). \tag{10.7}$$

Let us divide this equation by

$$I_1(s)I_1^*(s) = |I_1(s)|^2$$

which is a real, non-negative number. The result,

$$\frac{E_1(s)}{I_1(s)} \overset{\text{df}}{=} Z(s) = s\frac{V'(s)}{|I(s)|^2} + \frac{F'(s)}{|I(s)|^2} + \frac{1}{s}\frac{T'(s)}{|I(s)|^2}$$

is the one-port impedance, which eventually can be written in the form

$$Z(s) = \text{Re}\{Z(s)\} + j\,\text{Im}\{Z(s)\} = sV(s) + F(s) + \frac{1}{s}T(s) \tag{10.8}$$

with pseudo-energy functions $V(s)$, $T(s)$, and $F(s)$ that have real, non-negative values. This impedance on the real axis of the s-plane, $s = \sigma$

$$\mathbf{Z}(\sigma) = \sigma V(\sigma) + F(\sigma) + \frac{1}{\sigma}T(\sigma)$$

is real. The real part of this impedance in the right-half plane of $s = \sigma + j\omega$, $\sigma > 0$, is

$$\text{Re}\{Z(s)\} = \text{Re}\left\{(\sigma + j\omega)V(s) + F(s) + \frac{1}{\sigma + j\omega}T(s)\right\} =$$
$$= \sigma V(s) + F(s) + \frac{\sigma}{\sigma^2 + \omega^2}T(s) > 0$$

thus impedance $Z(s)$ of a passive RLC one-port is a real positive function. This completes the proof of Theorem 1.
Let us observe that if the "pseudo-power" is divided by

$$E_1(s)E_1^*(s) = |E_1(s)|^2$$

which is a real, non-negative, then the result,

$$\frac{I_1^*(s)}{E_1^*(s)} \stackrel{\text{df}}{=} Y^*(s) = s\frac{V'(s)}{|E(s)|^2} + \frac{F'(s)}{|E(s)|^2} + \frac{1}{s}\frac{T'(s)}{|E(s)|^2}$$

is the conjugate admittance of the one-port. The admittance can be presented in the form

$$Y(s) = s\,W(s) + D(s) + \frac{1}{s}A(s) \tag{10.9}$$

with real, non-negative pseudo-energy functions $W(s)$, $D(s)$, and $A(s)$.

10.3 Properties of Positive Real Functions

Since impedances and admittances are PR functions, according to Refs [26, 27], properties of PR functions specify the most general properties of impedances and admittances, independent of structure and parameters. Thus, let us compile the main features of PR functions.

1. Functions $F(s) = 1$ and $F(s) = s$ are PR functions.
2. Linear form with positive coefficients α, β of PR functions is a PR function, meaning that if $F_1(s)$ and $F_2(s)$ are PR functions then

$$F(s) = \alpha F_1(s) + \beta F_2(s)$$

   is a PR function.
3. If $F_1(s)$ is a PR function, then its inversion

$$F_2(s) \stackrel{\text{df}}{=} \frac{1}{F_1(s)}$$

   is a PR function. To prove it, let us present function $F_1(s)$ as follows

$$F_2(s) = \text{Re}\{F_2(s)\} + j\,\text{Im}\{F_2(s)\} = \frac{1}{F_1(s)} = \frac{1}{\text{Re}\{F_1(s)\} + j\,\text{Im}\{F_1(s)\}} =$$
$$= \frac{\text{Re}\{F_1(s)\}}{[\text{Re}\{F_1(s)\}]^2 + [\text{Im}\{F_1(s)\}]^2} - j\frac{\text{Im}\{F_1(s)\}}{[\text{Re}\{F_1(s)\}]^2 + [\text{Im}\{F_1(s)\}]^2}.$$

Since $F_1(s)$ is real, meaning for $\omega = 0$, $\mathrm{Im}\{F_1(s)\} = 0$, thus $\mathrm{Im}\{F_2(s)\} = 0$, meaning $F_2(s)$ is real. Since $F_1(s)$ is positive, meaning for $\sigma > 0$, $\mathrm{Re}\{F_1(s)\} > 0$, thus $\mathrm{Re}\{F_2(s)\} > 0$, meaning $F_2(s)$ is positive.

4. If $F_1(s)$ and $F_2(s)$ are PR functions then

- the difference,

$$F_3(s) \stackrel{\mathrm{df}}{=} F_1(s) - F_2(s)$$

- the product,

$$F_3(s) \stackrel{\mathrm{df}}{=} F_1(s) \times F_2(s)$$

- and the quotient,

$$F_3(s) \stackrel{\mathrm{df}}{=} F_1(s) \,/\, F_2(s)$$

can, but do not have to be PR functions.

Illustration 10.1 $F_1(s) = 1$ and $F_2(s) = s$ are PR, but

$$F_3(s) \stackrel{\mathrm{df}}{=} 1 - s, \text{ is not a PR.}$$

$$F_3(s) \stackrel{\mathrm{df}}{=} s^2, \text{ is not a PR.}$$

$$F_3(s) \stackrel{\mathrm{df}}{=} 1/s^2, \text{ is not a PR.}$$

*****

5. Positive Real functions are rational functions, meaning there are ratios of two polynomials of complex a variable s, of the general form

$$F(s) = \frac{a_n s^n + \ldots\ldots + a_k s^k + \ldots\ldots + a_0}{b_m s^m + \ldots\ldots + b_r s^r + \ldots\ldots + b_0} \tag{10.10}$$

with real, positive coefficients.

6. Degrees of the numerator and denominator polynomials, n and m, cannot differ more than by 1, meaning

$$|n - m| \leq 1.$$

To prove this property, let us observe that as s approaches infinity, function $F(s)$ approaches

$$F(s) \to \frac{a_n}{b_m} s^{n-m} = \frac{a_n}{b_m} (r\, e^{j\alpha})^{n-m}$$

When the complex variable s is in the right-half plane, as shown in Fig. 10.5, then the angle α is confined to

$$-\frac{\pi}{2} < \alpha < \frac{\pi}{2},$$

thus the real part of $F(s)$ changes as

$$\mathrm{Re}\{F(s)\} = \frac{a_n}{b_m} r^{n-m} \cos(n-m)\alpha$$

and has to be positive in the entire right-half plane. The cosine function in this formula is positive on the condition that $|n-m| \leq 1$. Thus only at this condition $\mathrm{Re}\{F(s)\}$ is positive for positive σ.

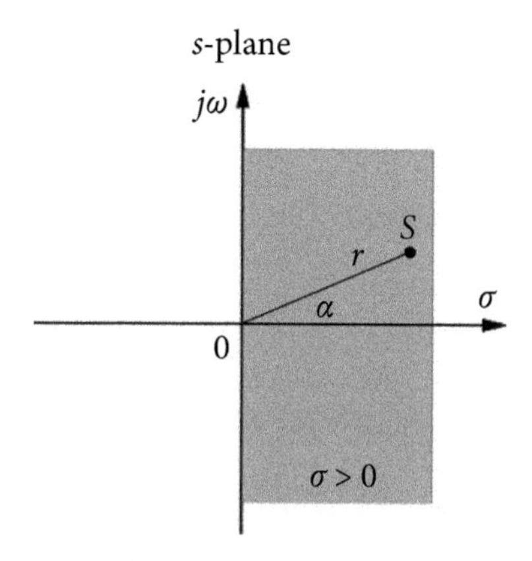

Figure 10.5 The variable s in the right-half s-plane.

7. The immittance $F(s)$ on the imaginary axis, $s = j\omega$, of linear passive RLC one-ports is a Hermitian function of frequency, meaning that

$$F(-j\omega) = F^*(j\omega). \tag{10.11}$$

To prove this property, let us observe that the supply current of any passive linear one-port can be expressed as a convolution of the supply voltage and the circuit pulse response $y(t)$, namely

$$i(t) = \int_{-\infty}^{\infty} y(\tau)\, u(t-\tau)\, d\tau. \tag{10.12}$$

This fundamental relation in the time-domain has its equivalent in the frequency domain,

$$I(j\omega) = Y(j\omega)\, U(j\omega)$$

where $Y(j\omega)$ is the Fourier transform of the pulse response $y(t)$, namely

$$Y(j\omega) = \frac{1}{2\pi} \int_{-\infty}^{\infty} y(t)\, e^{-j\omega t}\, dt. \tag{10.13}$$

This admittance for the negative frequency is

$$Y(-j\omega) = \frac{1}{2\pi} \int_{-\infty}^{\infty} y(t)\, e^{j\omega t}\, dt = Y^{*}(j\omega)$$

meaning it is a Hermitian function. The same can be proven, of course, for the one-port impedance. This property specifies the behavior of the real and imaginary parts of PR functions on the imaginary axis. Since

$$F(j\omega) = \mathrm{Re}\{F(j\omega)\} + j\,\mathrm{Im}\{F(j\omega)\}$$

and

$$F(-j\omega) = \mathrm{Re}\{F(-j\omega)\} + j\,\mathrm{Im}\{F(-j\omega)\} = F^{*}(j\omega) = \mathrm{Re}\{F(j\omega)\} - j\,\mathrm{Im}\{F(j\omega)\},$$

thus

$$\mathrm{Re}\{F(-j\omega)\} = \mathrm{Re}\{F(j\omega)\}$$
$$\mathrm{Im}\{F(-j\omega)\} = -\mathrm{Im}\{F(j\omega)\}.$$

It means that the real part of a PR function on the imaginary axis of the s-plane is an even function of frequency, while the imaginary part of such function is an odd function. Therefore, the reactance of linear passive RLC one ports

$$X(\omega) \overset{\mathrm{df}}{=} \mathrm{Im}\{Z(j\omega)\}$$

and their susceptance

$$B(\omega) \overset{\mathrm{df}}{=} \mathrm{Im}\{Y(j\omega)\}$$

are odd functions of frequency.

The main theorem for the circuit synthesis which makes the concept of PR functions so important is *Brune's Theorem* [16]: *Function F(s) can be realized as immittance of a passive RLC circuit, if and only if it is a PR function. Thus, this property is both the necessary and the sufficient condition for a function to be realized as a passive RLC circuit immittance.* The proof of this theorem can be found in Refs [26, 27].

10.4 Reactance Functions and Their Properties

Reactance filters and compensators are built of inductors and capacitors. These elements are regarded, at the first approximation, as lossless. Therefore, such LC devices are a subset of RLC circuits with all $R_{rk} = 0$. The immittance of such LC one-ports is referred to as a *reactance function*. Such functions form a sub-set of the set of Positive Real functions.

Since there are no resistances in the reactance circuit, thus the pseudo-power function

$$\sum_{r=1}^{N}\sum_{k=1}^{N} R_{rk}\, I_k(s)\, I_r^*(s) = 0$$

and the impedance of passive reactance one-ports has the general form

$$Z(s) = \mathrm{Re}\{Z(s)\} + j\,\mathrm{Im}\{Z(s)\} = sV(s) + \frac{1}{s}T(s) \tag{10.14}$$

and the admittance

$$Y(s) = \mathrm{Re}\{Y(s)\} + j\,\mathrm{Im}\{Y(s)\} = sW(s) + \frac{1}{s}A(s). \tag{10.15}$$

The reactance functions $F(s)$, have POLEs p_r and ZEROs z_k on the s-plane, meaning values of $s = p_r$, where function $F(s)$ approaches infinity and $s = z_k$, where the function $F(s)$ is equal to zero, respectively.

1. Reactance functions have POLEs and ZEROs exclusively on the imaginary axis of the s-plane. To prove it, observe that the pseudo-energy functions $V(s)$ and $T(s)$ are real and positive, thus the impedance $Z(s)$ has ZEROs, meaning

$$z_k\, V(z_k) + \frac{1}{z_k}T(z_k) = 0,$$

thus at points

$$s = z_k = \pm j\sqrt{\frac{T(z_k)}{V(z_k)}} = \pm j\omega_k$$

located on the imaginary axis, as shown in Fig. 10.6.

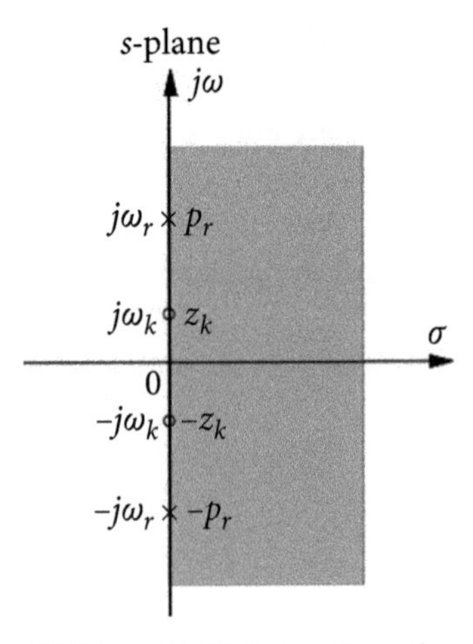

Figure 10.6 The ZEROs and POLEs location of reactance one-port.

The impedance $Z(s)$ has POLEs where its inversion, the admittance has ZEROs, meaning

$$p_r\, W(p_r) + \frac{1}{p_r} A(p_r) = 0,$$

thus at points

$$s = p_r = \pm j\sqrt{\frac{A(p_r)}{W(p_r)}} = \pm j\omega_r.$$

2. The real part of reactance functions on the imaginary axis $s = j\omega$ is zero, thus

$$F(j\omega) = j\,\mathrm{Im}\{F(j\omega)\},$$

   but this imaginary part of $F(j\omega)$ has to be an odd function of the frequency ω. To be an odd function of ω the function $F(s)$ has to be an odd function of s.
3. A reactance function as a rational function is a ratio of two polynomials, namely

$$F(s) = \frac{N(s)}{D(s)} = \frac{W_n(s)}{W_m(s)} = \frac{a_n s^n + a_{n-1} s^{n-1} + \ldots . a_1 s + a_0}{b_m s^m + b_{m-1} s^{m-1} + \ldots . b_1 s + b_0}. \tag{10.16}$$

   Such a function is an odd function of s on the condition that one of the polynomials in this ratio is odd while another is even, meaning one of them has only odd powers of s, while another has only even powers. It means, moreover, that these polynomials cannot have the same orders, n and m.

4. Since these orders for PR functions cannot differ more than by 1, thus reactance functions have to fulfill the condition, that

$$|n - m| = 1.$$

Illustration 10.2 *A reactance function can have the following forms*

$$F(s) = \frac{a_8 s^8 + a_6 s^6 + a_4 s^4 + a_2 s^2 + a_0}{b_7 s^7 + b_5 s^5 + b_3 s^3 + b_1 s}$$

$$F(s) = \frac{a_5 s^5 + a_3 s^3 + a_1 s}{b_6 s^6 + b_4 s^4 + b_2 s^2 + b_0}.$$

*****

5. The derivative of reactance functions versus frequency on the imaginary axis is always positive, that is,

$$\frac{d}{d(j\omega)} F(j\omega) > 0. \tag{10.17}$$

To prove this property, let us develop a reactance function $F(s)$ in a Taylor series around a point s located in the right-half s-plane in close vicinity of the imaginary axis, as shown in Fig. 10.7, sufficiently far, however, from a POLE.

$$F(s) = F(j\omega) + \frac{dF(j\omega)}{d(j\omega)}(s - j\omega) + \ldots$$

Since

$$s - j\omega = r\,e^{j\alpha}, \quad -\frac{\pi}{2} < \alpha < \frac{\pi}{2}$$

thus,

$$F(s) = F(j\omega) + \frac{dF(j\omega)}{d(j\omega)} r\,e^{j\alpha} + \ldots$$

For reactance functions $\mathrm{Re}\{F(j\omega)\} \equiv 0$, thus

$$\mathrm{Re}\{F(s)\} \approx \frac{dF(j\omega)}{d(j\omega)} r\cos(\alpha)$$

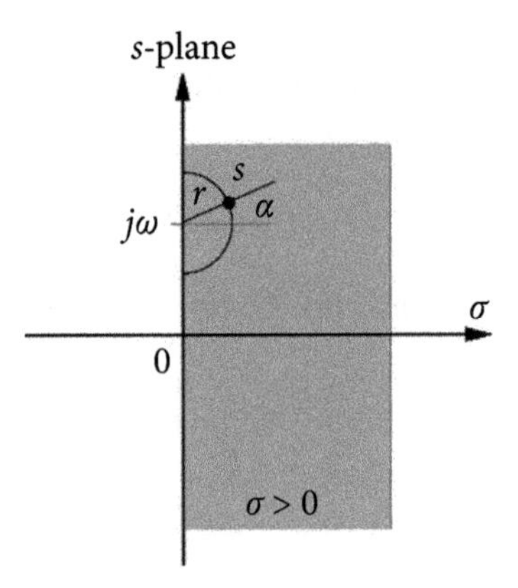

Figure 10.7 A point s in the right-half s-plane.

Since the cosine function in the range of the angle α variability is positive, the real part of $F(s)$ is positive on the condition that the derivative

$$\frac{d}{d(j\omega)}F(j\omega) > 0.$$

This conclusion applies directly to LC one-ports susceptance $B(\omega)$, since for the reactance function $F(j\omega) = Y(j\omega) = jB(\omega)$

$$\frac{d}{d(j\omega)}F(j\omega) = \frac{d}{d(j\omega)}jB(j\omega) = \frac{dB(\omega)}{d\omega} > 0$$

and to such one-ports reactance $X(\omega)$, since for $F(j\omega) = Z(j\omega) = jX(\omega)$

$$\frac{d}{d(j\omega)}F(j\omega) = \frac{d}{d(j\omega)}jX(\omega) = \frac{dX(\omega)}{d\omega} > 0.$$

It means that the susceptance $B(\omega)$ and reactance $X(\omega)$ of a reactance one-port can only increase with the frequency increase.

6. In a situation where the reactance function has multiple POLEs and ZEROs, they have to interlace each other. This is a conclusion from the previous one: the susceptance can increase between two ZEROs only if these ZEROs are separated by a POLE. The reactance can increase between two POLES only if these POLEs are separated by a ZERO. It is illustrated on the plot of susceptance variation versus frequency, shown in Fig. 10.8.

   At POLE frequencies, p_1, p_2, and p_3, meaning at frequencies where the susceptance $B(\omega)$ approaches infinity, the one-port is in the voltage (series) resonance. Its impedance has zero value, meaning the one-port behaves like a short-circuit. At ZERO frequencies, z_1 and z_2, meaning at frequencies where the susceptance $B(\omega)$ approaches zero, the one-port is in the current (parallel) resonance. Its impedance is infinite, meaning the one-port behaves as an open circuit. Because POLEs and ZEROs interlace each other, the voltage (series)

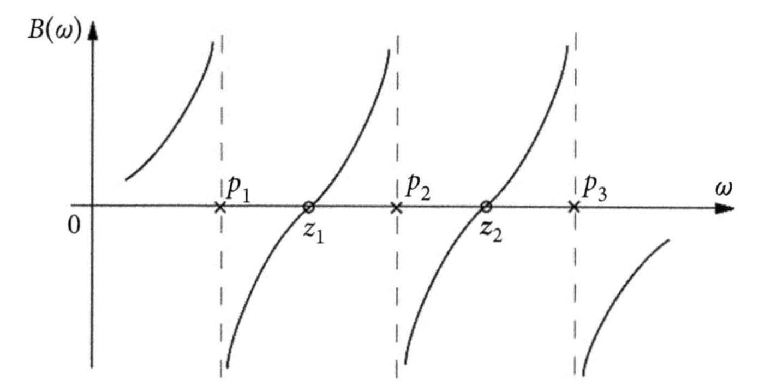

Figure 10.8 A plot of susceptance $B(\omega)$ variation versus frequency.

and current (parallel) resonances interlace each other too. The one-port has purely inductive or purely capacitive impedance between these resonances. When $B(\omega) > 0$, the one-port has a capacitive impedance, when $B(\omega) < 0$, the one-port has an inductive impedance.

7. Reactance functions are ratios of even and odd polynomials. When the odd polynomial is in the numerator, then as the variable s approaches zero, the function $F(s)$ approaches zero value, meaning it has ZERO at $s = 0$. When the odd polynomial is in the denominator, then as the variable s approaches zero then the function $F(s)$ approaches infinity, meaning the function has a POLE at $s = 0$, that is as s approaches zero,

$$F(s) \to \begin{cases} d_0\, s, & \text{ZERO at } s = 0 \\ d_0\, \dfrac{1}{s}, & \text{POLE at } s = 0 \end{cases}$$

The plots of susceptance $B(\omega)$ for these two situations is shown in Fig. 10.9.

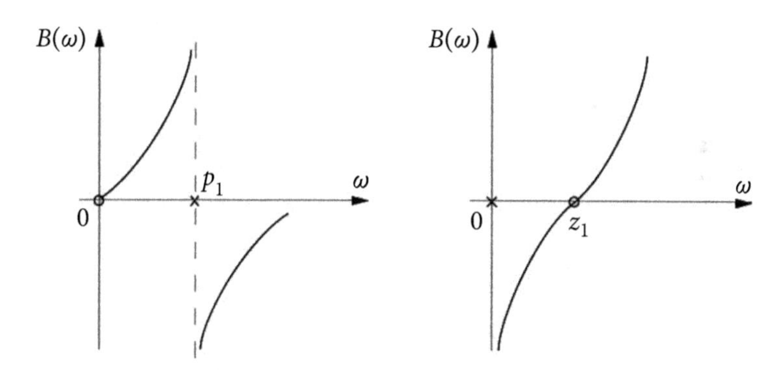

Figure 10.9 Two possible behaviors of susceptance $B(\omega)$ when ω approaches zero.

8. Degrees of the numerator and denominator polynomials have to differ by one. When the numerator polynomial is of the lower degree then as the variable s approaches infinity, the reactance function approaches zero, meaning it has a ZERO at infinity. When the numerator polynomial is of the higher degree,

then as the variable s approaches infinity, the reactance function approaches infinity, meaning it has a POLE at infinity:

$$F(s) \to \begin{cases} d_\infty s, & \text{POLE at } s \to \infty \\ d_\infty \dfrac{1}{s}, & \text{ZERO at } s \to \infty \end{cases}$$

Plots of $B(\omega)$ for these two situations are shown in Fig. 10.10.

Illustration 10.3 *The admittance*

$$Y(s) = \frac{a_5 s^5 + a_3 s^3 + a_1 s}{b_4 s^4 + b_2 s^2 + b_0}$$

has a ZERO at zero, a POLE at infinity, two ZEROs z_1 and z_2, and two POLEs p_1 and p_2, with

$$d_0 = a_1/b_0, \; d_\infty = a_5/b_4.$$

*****

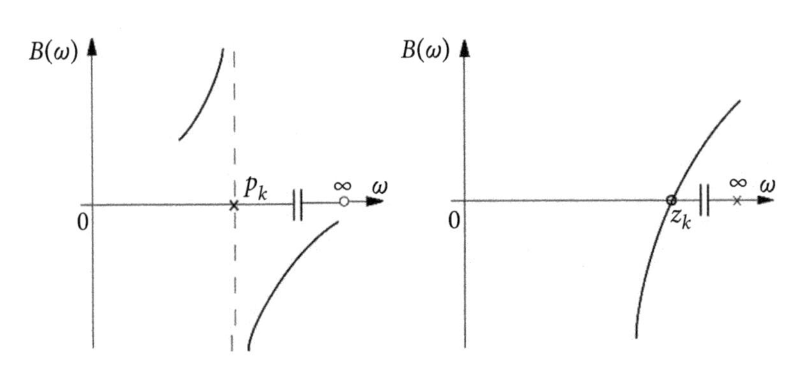

Figure 10.10 Two possible behaviors of susceptance $B(\omega)$ as ω approaches infinity.

The plot of susceptance $B(\omega)$ versus frequency ω is shown in Fig. 10.11.

Natural frequencies. When a reactance one-port is excited by the voltage Dirac pulse, $e(t) = \delta(t)$, as shown in Fig. 10.12a, then the current response, $i(t)$, of the one-port is equal to the Laplace inverted transform of its admittance, since

$$i(t) = \mathscr{L}^{-1}\{I(s)\} = \mathscr{L}^{-1}\{E(s)Y(s)\} = \mathscr{L}^{-1}\{Y(s)\} = \sum_r A_r \cos(p_r t) \tag{10.18}$$

which is a sum of oscillations with POLEs p_r frequencies. When the same one-port is excited by a current Dirac pulse, $j(t) = \delta(t)$, as shown in Fig. 10.12b, then the voltage response, $u(t)$, of the one-port is equal to the Laplace inverted transform of its impedance, since

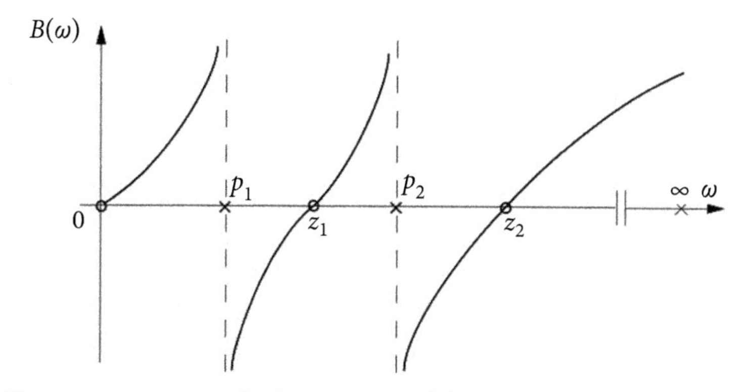

Figure 10.11 An example of susceptance $B(\omega)$ variation versus frequency ω.

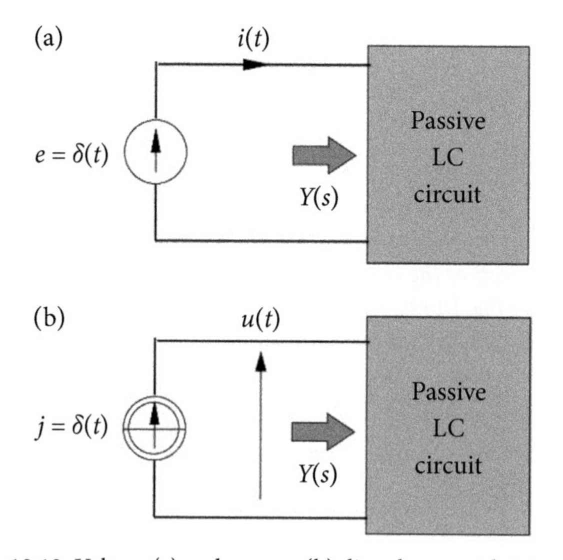

Figure 10.12 Voltage (a) and current (b) disturbance with Dirac pulse.

$$u(t) = \mathcal{L}^{-1}\{U(s)\} = \mathcal{L}^{-1}\{J(s)Z(s)\} = \mathcal{L}^{-1}\{Z(s)\} = \sum_k B_k \cos(z_k t). \tag{10.19}$$

Since the ideal voltage source is the source with zero impedance, while the current source is a source with an infinite impedance, then the frequencies of admittance POLEs p_r are frequencies of the current oscillations when the one-port is disturbed at short-circuited terminals, while frequencies of admittance ZEROs z_k are frequencies of the voltage oscillations at one-port open terminals. Therefore, POLEs and ZEROs other than that at $s = 0$ and at infinity, are referred to as *natural frequencies* of the reactance one-port. The number of natural frequencies, M, is equal to the number of POLEs and ZEROs, except those located at zero and infinity. They are not counted in this number.

10.5 Admittance of Shunt Reactance Compensator

When a reactance one-port is implemented as a shunt compensator, then such a compensator has to have specified susceptances for some supply voltage harmonics. When a load is to be compensated for harmonics of the order n from a set N, then the compensator susceptances $B_C(\omega)$ for $\omega = n\omega_1$ have to be equal to B_{Cn}. The general form of admittance $Y_C(s)$, specified by the number of POLEs and ZEROs, has to be found for that purpose. To find the number of POLEs and ZEROs, a constantly increasing susceptance $B_C(\omega)$, which for specified harmonic orders n assume values B_{Cn} has to be drafted. After that, POLEs and/or ZEROs at zero and infinity have to be added in such a way that $Y_C(s)$ could be a reactance function. This enables to express the compensator admittance in the form

$$Y_C(s) = f(s)\frac{(s^2+z_1^2)\ldots(s^2+z_n^2)}{(s^2+p_1^2)\ldots(s^2+p_m^2)}, \quad \text{with } f(s) = \begin{cases} As \\ A \\ \frac{A}{s} \end{cases} \quad \text{and } A > 0. \tag{10.20}$$

Illustration 10.4 *Let us assume that a reactance compensator has to compensate a load for the voltage fundamental harmonic and the third and the fifth order harmonics, which requires that compensator susceptances are* $B_{C1} = 2$ S, $B_{C3} = 1$ S, *and* $B_{C5} = -0.8$ S. *A draft of the compensator susceptance* $B_C(\omega)$ *which can have such values is shown in Fig. 10.13.*

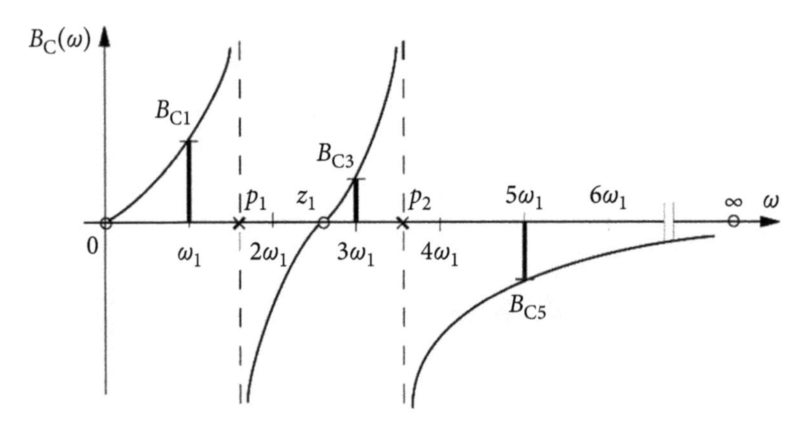

Figure 10.13 A draft of the susceptance with given values of B_{C1}, B_{C3}, and B_{C5}.

Admittance of such a compensator has to have one ZERO and two POLES with ZEROs at zero and infinity, meaning it has to have the form

$$Y_C(s) = As\frac{(s^2+z_1^2)}{(s^2+p_1^2)(s^2+p_2^2)}.$$

It is specified by four unknown parameters: A, z_1, p_1, *and* p_2. *There are, however, only three equations for their calculation, meaning* B_{C1}, B_{C3}, *and* B_{C5}. *Thus one of* *the unknown parameters has to be chosen at the designer's discretion with, of course, some limitations imposed by the plot of $B_C(\omega)$, as shown in Fig. 10.13. To simplify calculations, the fundamental frequency ω_1 is usually normalized by the assumption that $\omega_1 = 1$ rad/s.*

Let the ZERO z_1 be selected as a known parameter, for example $z_1 = 2\omega_1 = 2$. Thus, three equations for $s = j1$, $s = j3$, and $s = j5$ have to be solved for parameters A, p_1, p_2 calculation, namely

$$A(j\,1)\frac{\left(-1+2^2\right)}{\left(-1+p_1^2\right)\left(-1+p_2^2\right)} = jB_{C1}$$

$$A(j\,3)\frac{\left(-9+2^2\right)}{\left(-9+p_1^2\right)\left(-9+p_2^2\right)} = jB_{C3}$$

$$A(j\,5)\frac{\left(-25+2^2\right)}{\left(-25+p_1^2\right)\left(-25+p_2^2\right)} = jB_{C5}.$$

They can be simplified, with $p_1^2 \overset{\text{df}}{=} x$, $p_2^2 \overset{\text{df}}{=} y$, to the form

$$A\frac{3}{(x-1)(y-1)} = 2$$

$$3A\frac{-5}{(x-9)(y-9)} = 1$$

$$5A\frac{-21}{(x-25)(y-25)} = -0.8.$$

This set of equations has a solution: $A = 2.162$, $p_1 = 1.125$, $p_2 = 3.632$. Thus, the admittance of the compensator has the form

$$Y_C(s) = 2.162s\frac{(s^2+4)}{(s^2+1.266)(s^2+13.192)}.$$

*****

10.6 Foster Synthesis Procedures

One of the main features that differentiates circuit synthesis from circuit analysis is the possibility of the existence of equivalent solutions to the same synthesis problem. In particular, one-ports of different structures with different parameters can have the same admittance or impedance.

There are four basic procedures for developing the reactance one-port structure when its admittance is known. These are two *Foster procedures* and two *Cauer procedures*.

The one-port admittance $Y(s)$ can be developed in elementary fractions as follows:

$$Y(s) = f(s)\frac{(s^2+z_1^2)...(s^2+z_n^2)}{(s^2+p_1^2)...(s^2+p_m^2)} = a_0\frac{1}{s} + a_\infty s + \sum_{r=1}^{m}\frac{a_r s}{s^2+p_r^2} \tag{10.21}$$

with

$$a_0 = \lim_{s\to 0}\{sY(s)\}$$

$$a_\infty = \lim_{s\to\infty}\left\{\frac{1}{s}Y(s)\right\}$$

$$a_r = \lim_{s^2\to -p_r^2}\left\{\frac{s^2+p_r^2}{s}Y(s)\right\}.$$

The fractions in this decomposition stand for admittances of LC branches connected in parallel, as shown in Fig. 10.14. Such a procedure is referred to as the *Foster first procedure.*

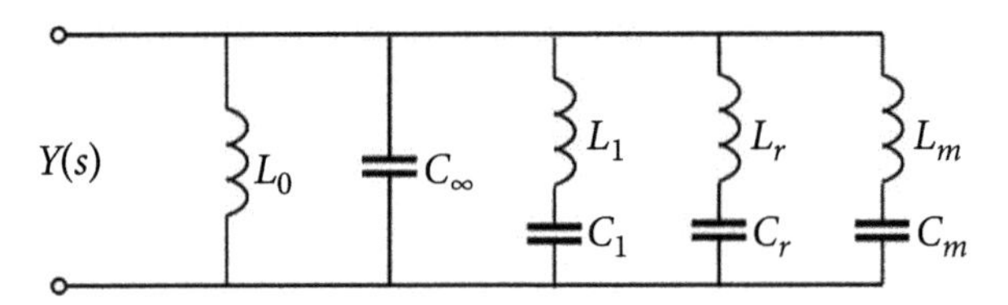

Figure 10.14 A reactance one-port with Foster first structure.

In this structure

$$L_0 = \frac{1}{a_0},\quad C_\infty = a_\infty,\quad L_r = \frac{1}{a_r},\quad C_r = \frac{a_r}{p_r^2}.$$

Illustration 10.5 *The admittance of the compensator found in Illustration 10.4 can be decomposed into elementary fractions as follows:*

$$Y_C(s) = 2.162\,s\frac{(s^2+4)}{(s^2+1.266)(s^2+13.192)} = a_0\frac{1}{s} + a_\infty s + \frac{a_1 s}{s^2+1.266} + \frac{a_2 s}{s^2+13.192}$$

with

$$a_0 = \lim_{s\to 0}\{sY_C(s)\} = 0,\ a_\infty = \lim_{s\to\infty}\{\frac{1}{s}Y_C(s)\} = 0$$

$$a_1 = \lim_{s^2\to -1.266}\left\{\frac{s^2+1.266}{s}Y_C(s)\right\} = 2.162\frac{(-1.266+4)}{(-1.266+13.192)} = 0.495$$

$$a_2 = \lim_{s^2\to -13.192}\left\{\frac{s^2+13.192}{s}Y_C(s)\right\} = 2.162\frac{(-13.192+4)}{(-13.192+1.266)} = 1.667.$$

Thus, the compensator has the structure shown in Fig. 10.15. Observe that inductance $L_0 = 1/a_0$ is infinite, meaning it is simply an open branch.

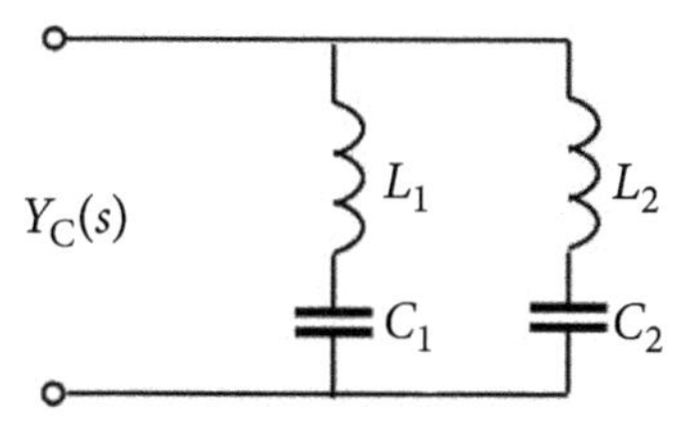

Figure 10.15 A compensator with Foster first structure.

The remaining parameters are

$$L_1 = \frac{1}{a_1} = \frac{1}{0.495} = 2.02\ \mathrm{H}$$

$$C_1 = \frac{a_1}{p_1^2} = \frac{0.495}{1.266} = 0.391\ \mathrm{F}$$

$$L_2 = \frac{1}{a_2} = \frac{1}{1.667} = 0.60\ \mathrm{H}$$

$$C_2 = \frac{a_2}{p_2^2} = \frac{1.667}{13.192} = 0.126\ \mathrm{F}.$$

*****

These values are surprisingly high. Observe, however, that the compensator admittance $Y_C(s)$ and parameters of the compensator were calculated, for the sake of computation simplification, under the assumption that the fundamental frequency of the supply voltage is $\omega_1 = 1$ rad/s. Such an assumption is referred to as *frequency normalization*. Moreover, often the level of admittance is normalized as well. It can be assumed, for example, that $B_{C1} = 1$ S. Thus, the parameters of the compensator shown in Fig. 10.15 are parameters of a normalized compensator. The admittance $Y_D(s)$ of a denormalized compensator is related to the admittance of the normalized compensator by the relation

$$Y_D(s) = kY\left(\frac{s}{\omega_1}\right).$$

Using the Foster decomposition, we can find the LC parameters of a denormalized compensator as follows:

$$Y_D(s) = kY\left(\frac{s}{\omega_1}\right) = k\left(a_0\frac{1}{s} + a_\infty\frac{s}{\omega_1} + \sum_{r=1}^{m}\frac{a_r\frac{s}{\omega_1}}{\left(\frac{s}{\omega_1}\right)^2 + p_r^2}\right) =$$

$$= (k\omega_1 a_0)\frac{1}{s} + \frac{ka_\infty}{\omega_1}a_\infty s + \sum_{r=1}^{m}\frac{(k\omega_1 a_r)\,s}{s^2 + (\omega_1 p_r)^2}.$$

Thus, denormalization requires that the circuit parameters are recalculated:

$$L_{D0} = \frac{1}{k\omega_1 a_0} = \frac{L_0}{k\omega_1}, \quad C_{D\infty} = \frac{ka_\infty}{\omega_1} = \frac{k}{\omega_1}C_\infty,$$
$$L_{Dr} = \frac{1}{k\omega_1 a_r} = \frac{L_r}{k\omega_1}, \quad C_{Dr} = \frac{k\omega_1 a_r}{(\omega_1 p_r)^2} = \frac{k}{\omega_1}C_r.$$

Observe that the compensator inductances and capacitances decline with the voltage angular frequency, ω_1. For example, for $f = 60$ Hz, meaning $\omega_1 = 377$ rad/s, and assuming that the admittance level is preserved, meaning $k = 1$, the parameters of the compensator shown in Fig. 10.15 are

$$L_{D1} = 5.36\ mH, C_{D1} = 0.846\ mF, L_{D2} = 1.59\ mH, C_{D2} = 0.334\ mF.$$

Instead of decomposing the admittance into elementary fractions, its inversion, meaning the impedance can be decomposed in such fractions, is

$$Z(s) = \frac{1}{Y(s)} = \frac{1}{f(s)}\frac{(s^2+p_1^2)...(s^2+p_m^2)}{(s^2+z_1^2)...(s^2+z_n^2)} = b_0\frac{1}{s} + b_\infty s + \sum_{r=1}^{n}\frac{b_r s}{s^2+z_r^2} \tag{10.23}$$

with

$$b_0 = \lim_{s\to 0}\left\{s\frac{1}{Y(s)}\right\}$$
$$b_\infty = \lim_{s\to\infty}\left\{\frac{1}{s}\frac{1}{Y(s)}\right\}$$
$$b_r = \lim_{s^2\to -z_r^2}\left\{\frac{s^2+z_r^2}{s}\frac{1}{Y(s)}\right\}.$$

The fractions in this decomposition stand for impedances of LC links connected in series, as shown in Fig. 10.16. Such a procedure is referred to as the *Foster second procedure.*

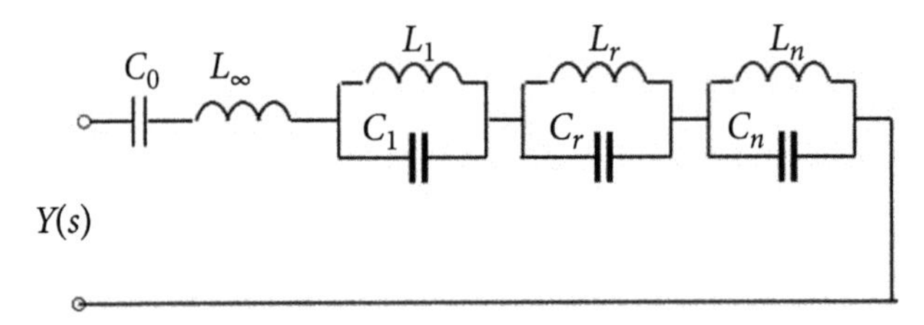

Figure 10.16 A reactance one-port with Foster second structure.

In this structure

$$C_0 = \frac{1}{b_0}, \quad L_\infty = b_\infty, \quad C_r = \frac{1}{b_r}, \quad L_r = \frac{b_r}{z_r^2}.$$

*****

Illustration 10.6 *The admittance of the compensator found in Illustration 10.4 can be decomposed according to the Foster second procedure into elementary fractions as follows:*

$$Z_C(s) = \frac{1}{Y_C(s)} = \frac{1}{2.162s}\frac{(s^2+1.266)(s^2+13.192)}{(s^2+4)} = b_0\frac{1}{s} + b_\infty s + \frac{b_1 s}{s^2+4}$$

with

$$b_0 = \lim_{s\to 0}\left\{s\frac{1}{Y_C(s)}\right\} = \frac{1}{2.162}\frac{(1.266)(13.192)}{4} = 1.93$$

$$b_\infty = \lim_{s\to\infty}\left\{\frac{1}{s}\frac{1}{Y(s)}\right\} = \lim_{s\to\infty}\left\{\frac{1}{2.162s^2}\frac{(s^2+1.266)(s^2+13.192)}{(s^2+4)}\right\} = 0.462$$
$$b_1 = \lim_{s^2\to -z_1^2}\left\{\frac{s^2+z_1^2}{s}\frac{1}{Y(s)}\right\} = \frac{(-4+1.266)(-4+13.192)}{2.162\,(-4)} = 2.90.$$

The compensator structure found in the Foster second procedure is shown in Fig. 10.17.

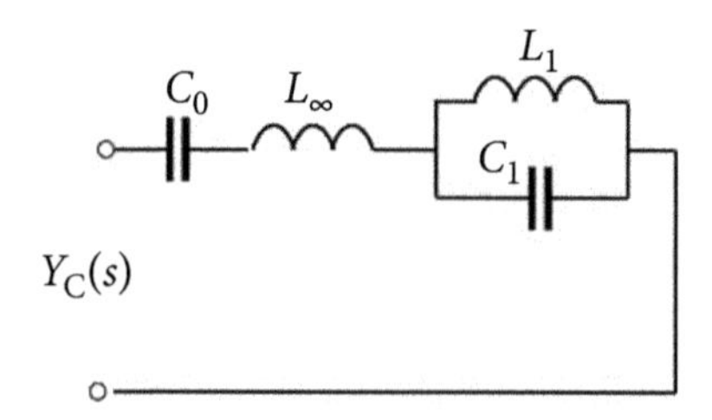

Figure 10.17 A compensator with Foster second structure.

Its parameters are equal to

$$C_0 = \frac{1}{b_0} = 0.518\ \text{F},\quad L_\infty = b_\infty = 0.462\ \text{H}$$

$$C_1 = \frac{1}{b_1} = 0.345\ \text{F},\quad L_1 = \frac{b_1}{z_1^2} = 0.725\ \text{H}.$$

*****

10.7 Cauer Synthesis Procedures

If the one-port admittance $Y(s)$ has a POLE at infinity, meaning the numerator polynomial is of a higher degree than the denominator, then a division of the numerator by denominator removes this POLE and consequently, the remainder has ZERO at infinity. Inversion of the remainder has the POLE at infinity which can be removed again by division of the remainder numerator by its denominator.

Sequential division of inverted remainders enables the presentation of the reactance admittance in a form of the following stair-like fraction

$$Y(s) = \frac{a_n s^n + a_{n-2} s^{n-2} + \dots}{b_m s^m + b_{m-2} s^{m-2} + \dots} = d_1 s + \cfrac{1}{d_2 s + \cfrac{1}{d_3 s + \cfrac{1}{d_4 s + \dots\dots}}}. \qquad (10.24)$$

This decomposition ends up in a finite number of steps with real, positive coefficients. Such a fraction specifies admittance of an LC one-port with a ladder structure, shown in Fig. 10.18 with parameters,

$$C_1 = d_1,\ L_2 = d_2,\ C_3 = d_3,\ L_4 = d_4,\ \dots..$$

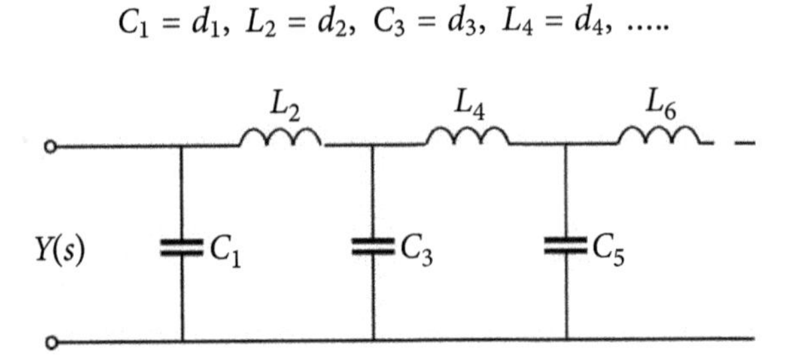

Figure 10.18 Reactance one-port with Cauer first structure.

When admittance $Y(s)$ does not have a POLE at infinity, then its inversion has such a POLE. Decomposition of admittance $Y(s)$ into a stars-like fraction begins in such a case from inversion of $Y(s)$. In effect, this fraction will not have the element $d_1 s$ and the branch with capacitance C_1.

Such a procedure is referred to as the *Cauer first procedure*. Observe, that in the case of the Foster first structure, branches were tuned to specific POLEs frequencies. In the case of the Foster second structure, links were tuned to specific ZEROs frequencies. The Cauer procedure results in a one-port with the ladder structure. There are no branches or links tuned to specific POLEs or ZEROs frequencies. All circuit LC elements contribute to natural frequencies.

Illustration 10.7 *The admittance of the compensator found in Illustration 10.4 can be rearranged as follows,*

$$Y_C(s) = 2.162\, s \frac{(s^2 + 4)}{(s^2 + 1.266)(s^2 + 13.192)} = \frac{2.162\, s^3 + 8.648\, s}{s^4 + 14.458\, s^2 + 16.701}$$

and decomposed according to the Cauer first procedure into a stair-like fraction

$$Y_C(s) = \frac{2.162s^3 + 8.648s}{s^4 + 14.458s^2 + 16.701} = \cfrac{1}{0.462s + \cfrac{1}{0.207s + \cfrac{1}{2.01s + \cfrac{1}{0.311}}}}$$

thus, because the coefficient $d_1 = 0$, the compensator does not have the first shunt capacitive branch with the capacitance C_1. Its structure is shown in Fig. 10.19. The parameters are:

$$L_2 = 0.462 \text{ H}, \quad C_3 = 0.207 \text{ F}, \quad L_4 = 2.01 \text{ H}, \quad C_5 = 0.311 \text{ F}.$$

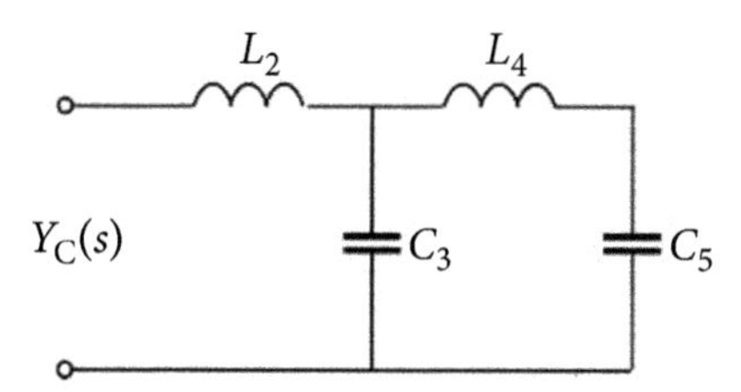

Figure 10.19 A compensator with Cauer first structure.

*****

Let us assume that the denominator of a reactance function $Y(s)$ is an odd polynomial. The polynomials can be rearranged in such a way that the power of s increases, namely

$$Y(s) = \frac{a_0 + a_2 s^2 + a_4 s^4 + \ldots.}{b_1 s + b_3 s^3 + a_5 s^5 + \ldots.}.$$

This means that the admittance has a POLE at zero. When the odd polynomial is in the numerator of $Y(s)$, then this polynomial is a denominator of the inversion, $1/Y(s)$, thus this inversion has a POLE at zero.

Division of the even polynomial of the numerator by the odd polynomial of the denominator removes the POLE at zero, meaning that the remainder has ZERO at zero. The inversion of the remainder has the POLE at zero again and this POLE can be removed again.

Continuation of divisions that remove the POLE at zero results in the expression of the function in a form of the following stair-like fraction:

$$Y(s) = \frac{a_0 + a_2 s^2 + a_4 s^4 + \ldots.}{b_1 s + b_3 s^3 + a_5 s^5 + \ldots.} = \frac{1}{d_1 s} + \cfrac{1}{\cfrac{1}{d_2 s} + \cfrac{1}{\cfrac{1}{d_3 s} + \cfrac{1}{d_4 s + \ldots\ldots}}}. \tag{10.25}$$

Such decomposition ends up in a finite number of steps with real, positive coefficients $d_1, d, d_3, \ldots, d_N$. Such a stair-like fraction is the impedance of a reactance one-port with the ladder structure shown in Fig. 10.20.

The parameters of such a one-port are

$$L_1 = d_1, \; C_2 = d_2, \; L_3 = d_3, \; C_4 = d_4, \ldots$$

When the odd polynomial is in the numerator, then the first term in this fraction, the inductor, does not have inductor L_1. Such a procedure is referred to as the *Cauer*

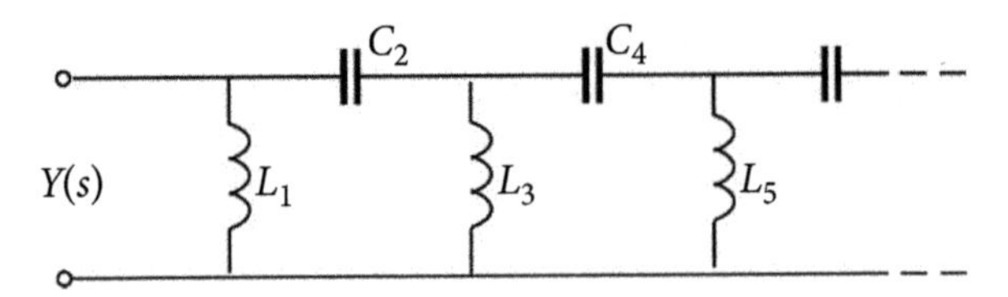

Figure 10.20 A reactance one-port with Cauer second structure.

second procedure. Similarly, as in the case of the Cauer first procedure, there are no branches or links tuned to specific POLEs or ZEROs frequencies. All circuit LC elements contribute to natural frequencies.

Illustration 10.8 *The admittance of the compensator found in Illustration 10.4 can be rearranged as follows,*

$$Y_C(s) = 2.162\, s \frac{(s^2+4)}{(s^2+1.266)(s^2+13.192)} = \frac{8.648\, s + 2.162\, s^3}{16.701 + 14.458\, s^2 + s^4}$$

and decomposed according to the Cauer second procedure into a stair-like fraction of the form

$$Y_C(s) = \frac{8.648\, s + 2.162\, s^3}{16.701 + 14.458\, s^2 + s^4} == \cfrac{1}{\cfrac{1}{0.519\, s} + \cfrac{1}{\cfrac{1}{1.19\, s} + \cfrac{1}{\cfrac{1}{0.128\, s} + \cfrac{1}{\cfrac{1}{0.757\, s}}}}}$$

The compensator structure is shown in Fig. 10.21. Its LC parameters are equal to

$$C_2 = 0.519\ \text{F},\ L_3 = 1.19\ \text{H},\ C_4 = 0.128\ \text{F},\ L_5 = 0.757\ \text{H}.$$

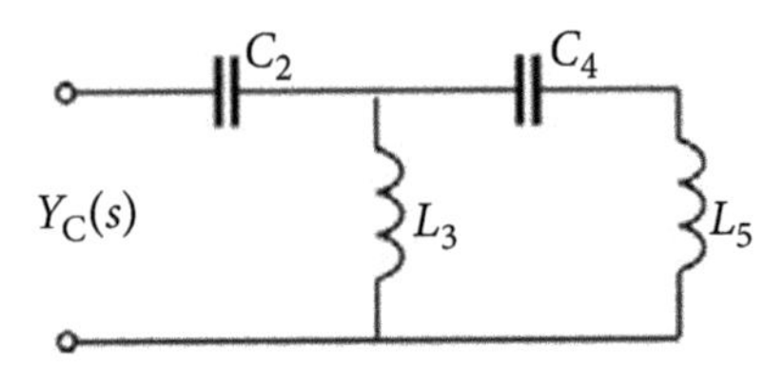

Figure 10.21 A compensator with Cauer second structure.

*****

Compensators obtained as a result of these four procedures are, of course, entirely equivalent as to their admittance $Y_C(s)$. They differ, however, in several features, such as the total inductance, capacitance, power rating of compensator components, grounding, capabilities of tuning, and so on. These are, moreover, not the only four structures. In the process of synthesis, procedures can be changed to different ones, which changes the compensator structure. The more advanced issues are not presented in this chapter, however. The reader is provided here with only some basic ideas on reactance compensator synthesis.

10.8 Harmonic Phase Shifter

Let us find a structure and parameters of a one-port which shifts the fundamental harmonic and third and fifth order current harmonics versus the supply voltage harmonics by 90^0, meaning a *harmonic phase shifter*, as shown in Fig. 10.22, without changing the contents of harmonics:

$$\frac{I_1}{U_1} = \frac{I_3}{U_3} = \frac{I_5}{U_5} \stackrel{\text{df}}{=} B.$$

At such conditions, the voltage and current are mutually orthogonal, that is, their scalar product $(u, i) = 0$ and their rms values are mutually proportional, $||i|| = B||u||$. One-ports with such properties are referred to [53] as *orthonormal one-ports.*

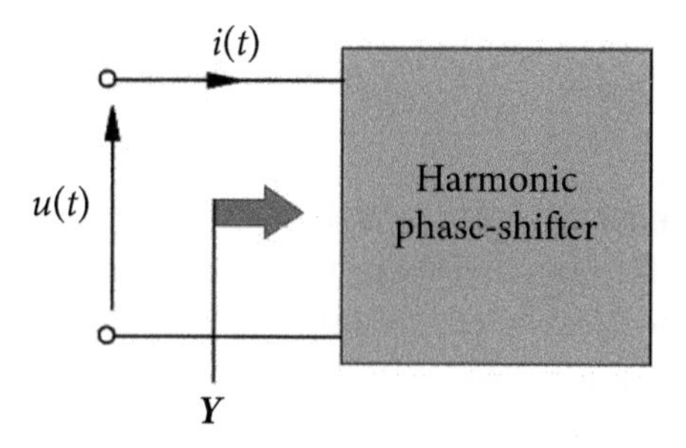

Figure 10.22 A harmonic phase-shifter.

In mathematical terms, the phase-shifter of the first, third, and fifth order harmonic, at the supply voltage

$$u(t) = \sqrt{2} \sum_{n=1,3,5} U_n \sin(n\omega_1 t - \beta_n)$$

should have the current

$$i(t) = \sqrt{2}B \sum_{n=1,3,5} U_n \sin(n\omega_1 t - \beta_n - 90^0).$$

Such a property concerning the phase shift has an ideal inductor. It changes the relative contents of harmonics, however, since

$$\frac{I_n}{U_n} = \frac{1}{n\,\omega_1 L}.$$

The admittance of one-port can be expressed as the ratio of Laplace Transforms of the supply voltage and the response current. Since the one-port is linear, its properties cannot depend on the voltage harmonics rms value U_n and their phase shift β_n. Therefore, its admittance is

$$Y(s) = \frac{\mathscr{L}\{i\}}{\mathscr{L}\{u\}} = \frac{\mathscr{L}\left\{B \sum\limits_{n=1,3,5} \cos n\,\omega_1 t\right\}}{\mathscr{L}\left\{\sum\limits_{n=1,3,5} \sin n\,\omega_1 t\right\}} = B\frac{\sum\limits_{n=1,3,5}\left(\frac{n\,\omega_1}{s^2+(n\,\omega_1)^2}\right)}{\sum\limits_{n=1,3,5}\left(\frac{s}{s^2+(n\,\omega_1)^2}\right)}. \tag{10.26}$$

It was proven in Ref. [53] that the admittance defined in such a way is a reactance function. To simplify its calculation and the one-port parameters, this admittance can be normalized by the assumption that $B = 1$ S and $\omega_1 = 1$ rad/s. At such assumption, the normalized admittance has the form:

$$Y(s) = \frac{9\,s^4 + 162\,s^2 + 345}{3\,s^5 + 70\,s^3 + 295\,s}.$$

Verification For the normalized fundamental frequency, $s = j\omega_1 = j1$,the admittance $Y(s)$ has the value

$$Y(j1) = \frac{9(j1)^4 + 162(j1)^2 + 345}{3(j1)^5 + 70(j1)^3 + 295(j1)} = -j1\text{ S}$$

thus the current fundamental harmonic is shifted versus the voltage fundamental by 90^0. Similarly, for the third and fifth order harmonics, $Y(j3) = Y(j5) = -j1$S.

Admittance $Y(s)$ when developed into a stair-like fraction

$$Y(s) = \frac{9\,s^4 + 162\,s^2 + 345}{3\,s^5 + 70\,s^3 + 295\,s} = \cfrac{1}{0.333s + \cfrac{1}{0.563s + \cfrac{1}{0.198s + \cfrac{1}{1.07s + \cfrac{1}{0.220s}}}}}$$

specifies the structure, shown in Fig. 10.23, of the harmonic phase shifter under design and its normalized parameters.

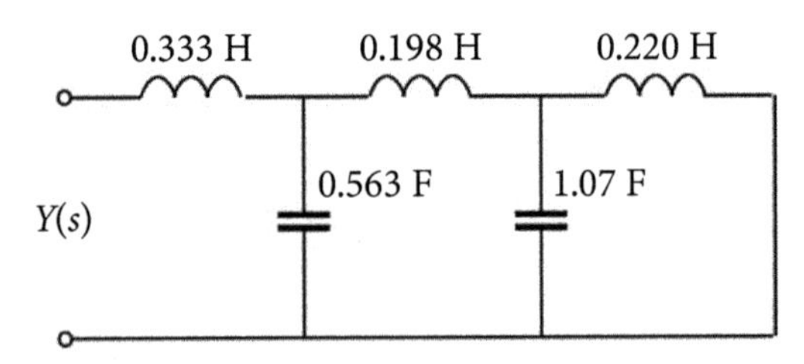

Figure 10.23 A structure and parameters of normalized harmonic phase-shifter for the fundamental, the 3rd, and 5th order harmonics.

*****

Summary

Synthesis of reactance one-ports that can be used as reactance compensators in electrical power systems is a well-developed area of electrical engineering. It is not taught usually in common courses on electrical circuits, however, so its potential for the compensators' development is not well recognized in power systems engineering.

Chapter 11
Capacitive Compensation

11.1 Capacitive Compensation at Sinusoidal Current

Capacitive compensators of reactive power Q in a form of capacitor banks, installed in distribution systems, are the most common devices for power factor improvement. Such capacitor banks are installed directly at terminals of customer loads with a low power factor at the cost of customers or along distribution lines or at power substations at the cost of energy providers. Such capacitive compensators improve the power factor by compensating the reactive current drawn by resistive-inductive (RL) loads, such as electrical motors, but also by fluorescent bulbs with magnetic ballasts.

The term "*power factor*" (PF) occurred at the end of the nineteenth century when the concept of electric powers was originally developed [2, 3, 4] and introduced to electrical engineering.

To deliver electric energy to a load, it has to be supplied with a voltage. It is specified by the rms value U. In response, the load draws the current of the rms value I. The energy delivered in some intervals to customer load is equal to the integral of the load active power P over that integral. To deliver this energy, the energy provider has to produce it and needs distribution equipment capable of operating at the voltage of the rms value U and conduct without overheating the current of the rms value I, thus the provider needs equipment capable of operating at the apparent power $S = U\,I$. The ratio of the active power P to the apparent power S

$$\frac{P}{S} \overset{\text{df}}{=} \lambda \tag{11.1}$$

specifies the effectiveness of this equipment utilization for energy delivery. It is called the "power factor". This definition applies to any structure and properties of the circuit, meaning to single-phase and three-phase circuits, and it is irrelevant whether the circuit is linear or not or whether the voltages and currents are sinusoidal or nonsinusoidal.

At a fixed value of the load active power P, the power factor λ affects the apparent power S of the distribution equipment, thus the cost of energy delivery and, consequently, the PF value affect the power system economy. Therefore, the power factor was the subject of studies [3, 4, 7] and measurements [6]. Various methods of power factor measurements are compiled in [167].

Powers and Compensation in Circuits with Nonsinusoidal Current. Leszek S. Czarnecki, Oxford University Press.
 DOI: 10.1093/oso/9780198879206.003.0011

The power factor λ must not be interpreted as the power efficiency of the system, which is defined as the ratio of the load and generator active powers. This efficiency for a simple generator load system is defined as

$$\eta \stackrel{\text{df}}{=} \frac{P}{P_{\text{G}}} \tag{11.2}$$

where P_{G} denotes the generator's active power. Observe that for a lossless generator and the supply line $P_{\text{G}} = P$, the efficiency $\eta = 1$, independently of the power factor λ value.

The power factor in single-phase circuits with sinusoidal voltages and currents (sv&c) is equal to

$$\lambda \stackrel{\text{df}}{=} \frac{P}{S} = \frac{UI\cos\varphi}{UI} = \cos\varphi, \tag{11.3}$$

meaning it is exclusively related to the phase shift φ between the load voltage and current. This conclusion does not apply, however, to three-phase circuits and circuits with nonsinusoidal voltages and currents (nv&c). It is not possible to explain the relation between the voltage and current in these situations in terms of the phase shift between them. Only definition (11.1), but not (11.3), is valid in such situations. The power factor can be lower than unity in three-phase circuits, as was shown in Chapter 7, even if the load is purely resistive but imbalanced. Thus, various phenomena in the load contribute to the power factor decline. The phase shift between the voltage and current is only one of them. The change of the load conductance with harmonic order, meaning the presence of the scattered current, the load-generated current harmonics, and the current asymmetry in three-phase circuits contribute to the power factor deterioration.

11.2 Detrimental Effects of Low PF

The low PF means that the supply current rms value is higher than that needed for energy delivery with the average rate equal to active power P. In other words, it means that the energy delivery at low PF requires a higher supply current rms value as compared to its delivery at high PF value.

The detrimental effects of the low power factor are directly related to the fact that at low PF the supply current rms value is higher than needed.

If the equivalent resistance of the supply source is R_{s}, then the active power loss ΔP_{s} in the source at the energy delivery is $\Delta P_{\text{s}} = R_{\text{s}}I^2$. The same is true with the power loss in the supply lines. This power loss increases fuel consumption in power plants and this increase is at the cost of the energy provider. The increase in the supply current rms value with a decline of the power factor causes an increase in the temperature of the distribution equipment. It can be overheated at a low power factor. Thus it reduces the equipment lifetime expectation and the distribution system supply capability. Consequently, equipment with a higher power rating is needed to supply loads with low PFs. In some situations, a new investment might be needed.

A low PF also increases the voltage drop on the supply source impedance. To explain the effect of low PF on the supply voltage, let us consider the equivalent circuit of the distribution system shown in Fig. 11.1, as seen at the load terminals.

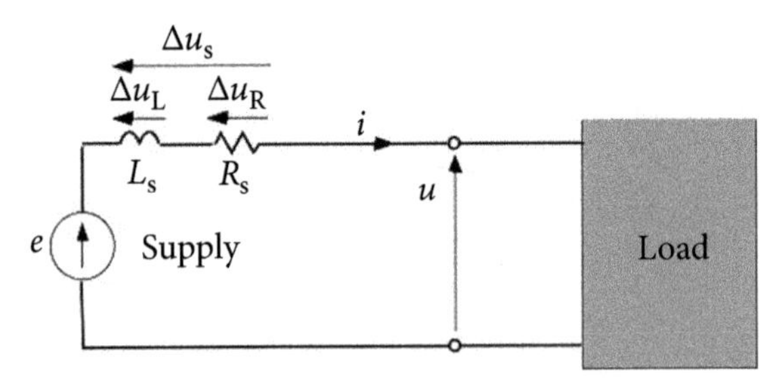

Figure 11.1 A single-phase load and equivalent parameters of the supply source.

It contains three components: distribution voltage e, equivalent resistance R_S, and inductance L_S. The resistance and reactance of the supply line and the closest transformer mainly contribute to equivalent resistance and inductance. The reactance X_s is usually a few times higher than the resistance R_S.

The load current causes the voltage drop on the supply impedance $\Delta \boldsymbol{U}_S$ equal to

$$\Delta U_s = E - U. \tag{11.4}$$

The magnitude of this drop is proportional to the current rms value, thus it increases with the PF decline, as shown in Fig. 11.2.

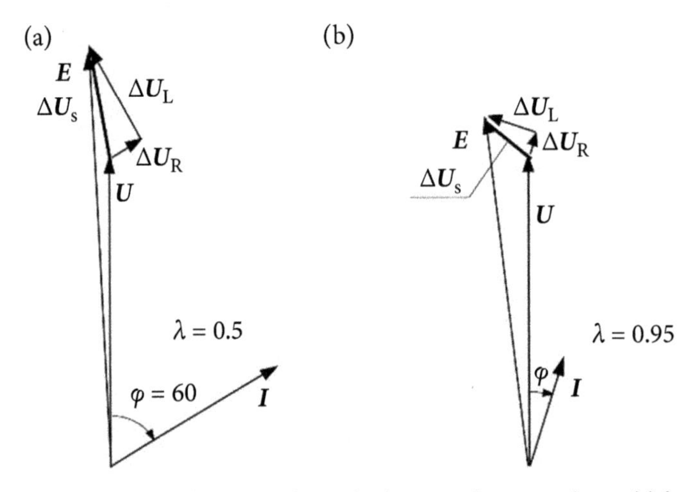

Figure 11.2 A diagram of the crms values of voltages at the power factor (a) $\lambda = 0.5$ and (b) at $\lambda = 0.95$.

The diagram of crms values in this figure is drawn for two loads that have the same active power but a different power factor λ. The voltage drop shown in Fig. 11.2 is exaggerated, however, because it should not be higher than 5% of the load voltage.

One should notice, however, that the *voltage loss*, $E - U$, that is, the difference of the rms values of the distribution and the load voltages, referred to as *voltage regulation* has a different meaning than the *voltage drop*, $\Delta \boldsymbol{U}_s = \boldsymbol{E} - \boldsymbol{U}$, that is, the difference of their crms values, since

$$\Delta U_r \overset{\text{df}}{=} E - U \neq |\boldsymbol{E} - \boldsymbol{U}| = Z_s I_s. \tag{11.4}$$

It is the change of the load voltage rms value from the idle state, when distribution system voltage rms value E can be measured, to its value at full loading U. In particular, the voltage drop, $Z_s\, I$, cannot be negative, while the voltage loss can have both negative and positive values, meaning that the rms value of the load voltage, u, can be higher than the distribution voltage, e. The voltage regulation is higher in circuits with a low PF, as is visible in the diagrams in Fig. 11.2.

When the distribution voltage rms value E is fixed, then the load with a low PF is supplied with a lower voltage U as compared to the load with a high PF. An increase in the distribution voltage might be needed to provide the rated value of the load voltage.

Thus, deterioration of the PF increases the cost of energy delivery. As it was shown, this extra cost has several components that depend on the PF in a sophisticated way. Let us compile these detrimental effects of a low power factor:

1. An increase of energy loss in distribution equipment, meaning an increase in the amount of energy that has to be produced and delivered to the system by generators.
2. An increase in temperature of the distribution system equipment, meaning a reduction of the life span of this equipment.
3. In a situation where the existing distribution equipment is not capable, due to overloading, of providing energy supply to new customers; new investments are needed.
4. An increase in the voltage drop on the distribution system impedance increases the load voltage variation with the change of the load power and consequently, voltage regulation in the distribution system could be needed.

The cost of these harmful effects is borne by the energy provider. Therefore, power utilities have various policies and methods of exerting pressure upon customers with low power factors aimed at PF improvement. These can take a form of a financial penalty for low PF or the energy is provided at a higher tariff. In a response to these policies, customers can retrofit low power factor energy receivers with higher PFs or install a compensator for its improvement. Let us explain these effects in more detail.

11.3 Power Factor Improvement with Capacitive Compensators

The power factor at the supply terminals can be improved by retrofitting the load equipment with equipment that has a higher power factor. For example, it is sufficient for that to fit the motor power to the motor's mechanical load power, otherwise a compensator of the reactive power is needed for PF improvement.

An over-excited synchronous machine can serve as such a compensator in three-phase circuits. In common single-phase distribution systems with resistive-inductive loads, capacitor banks, connected as shown in Fig. 11.3, are usually used for reactive power compensation.

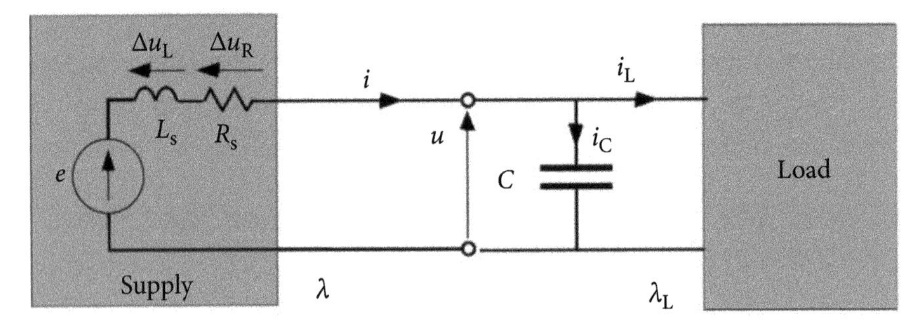

Figure 11.3 A single-phase circuit with compensating capacitor.

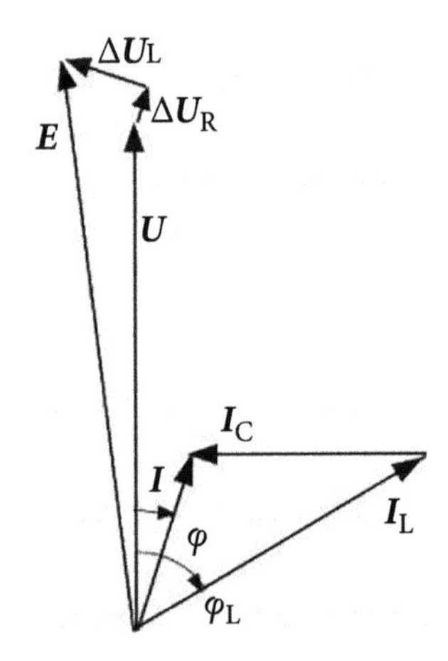

Figure 11.4 A diagram of voltages and currents crms values.

If the load current is inductive, the capacitor modifies the supply current crms value as shown in Fig. 11.4.

To improve the power factor from $\lambda_L = \cos\varphi_L$ to the value $\lambda = \cos\varphi$, the capacitor should change the reactive current of the supply by the value

$$I_C = \omega_1 CU = I_L \sin \varphi_L - I_L \sin \varphi. \tag{11.5}$$

If the load has the active power

$$P = UI_L\lambda_L$$

then the compensator should have the capacitance

$$C = \frac{P}{\omega_1 U^2 \lambda_L}[\sin(\mathrm{acos}\lambda_L) - \sin(\mathrm{acos}\lambda)]. \tag{11.6}$$

In particular, to improve the PF to unity, the capacitance should be equal to

$$C \stackrel{\mathrm{df}}{=} C_0 = \frac{P}{\omega_1 U^2 \lambda_L}\sin(\mathrm{acos}\lambda_L) = \frac{P}{\omega_1 U^2}\sqrt{\frac{1}{\lambda_L^2} - 1} = \frac{Q}{\omega_1 U^2}. \tag{11.7}$$

The benefits of a higher power factor are obtained at the cost of installation of the compensator and its operation. It can be assumed that the cost of the compensator is approximately proportional to the capacitor power ratings and, at a fixed supply voltage, to the capacitance C. At the same time reduction of the supply current rms value declines as the current phase shift approaches zero, thus economic benefits decline as the PF approaches unity.

A trade-off between these benefits and the compensator cost is, therefore, needed and consequently, compensators have usually a capacitance C lower than C_0. The choice of the capacitance value should be based on the evaluation of economical profits from the improved power factor and the cost of the compensator. The total cost is minimum at the PF referred to as an *economic power factor*. It is usually in the range of 0.90–0.95.

The dependence of the active power loss and the voltage loss in the distribution system on the compensator capacitance C is illustrated in the following example.

Illustration 11.1 *A single-phase load has the following parameters. At the load voltage rms value* $U = 230$V, *the load active power is* $P = 30$ kW *and its power factor,* $\lambda_L = 0.5$. *The short circuit power of the distribution system is equal to* $S_{sc} = 1$ MVA *and its reactance to resistance ratio at the fundamental frequency is* $\omega_1 L_s/R_s = 3$. *Assume* *that $f = 60$ Hz. For such parameters, the load current rms value is equal to*

$$I_{\mathrm{L}} = \frac{P}{U\lambda_{\mathrm{L}}} = 260.9 \text{ A}.$$

To calculate the loss of the active power in the supply, the value of its resistance R_s has to be known. Since the short-circuit power, S_{sc}, of distribution systems is not known usually with a high degree of accuracy, the resistance R_s can be evaluated, assuming that the distribution voltage, E, is approximately equal to the load voltage rms value, U, namely

$$S_{sc} \stackrel{\mathrm{df}}{=} EI_{sc} = \frac{E^2}{Z_s} \approx \frac{U^2}{Z_s}$$

where I_{sc} is the rms value of short-circuit current and Z_s is the magnitude of the supply source impedance. Hence

$$Z_s = R_s\sqrt{1 + \left(\frac{\omega_1 L_s}{R_s}\right)^2} \approx \frac{U^2}{S_{sc}} = 0.053\ \Omega,$$

$$R_s = 0.017\ \Omega, \quad X_s = 0.050\ \Omega.$$

Consequently, the active power loss in the supply is equal to

$$\Delta P_s = R_s I_L{}^2 = 1138 \text{ W}.$$

If the load voltage has the phase angle as shown in Fig. 11.4, that is, $\mathbf{U} = 230\ e^{j\pi/2}$ V, then the crms value of the distribution voltage, $\mathbf{E}$, of not compensated load is equal to

$$\mathbf{E} = \mathbf{U} + \mathbf{Z}_s\mathbf{I}_{\mathrm{L}} = j230 + (0.017 + j\,0.050) \times 260.9e^{j(\pi/2-\varphi)} = 243.5e^{j90.63^{\circ}} \text{ V}$$

thus the load voltage, U, is lower than the distribution voltage rms value E, by 13.5 V.

Assuming that the supply voltage radial frequency is $\omega_1 = 2\pi\,60 = 377$ rad/s, formula (11.7) results in the capacitance needed for total compensation of the reactive power, namely

$$C_0 = \frac{P}{\omega_1\, U^2 \lambda_{\mathrm{L}}} \sin\varphi_{\mathrm{L}} = \frac{30 \times 10^3}{377 \times 230^2 \times 0.5} \sin 60^{\circ} = 2.61 \text{ mF}.$$

The capacitor reduces the supply current rms value to

$$I = I_a = I_{\mathrm{L}} \cos\varphi_{\mathrm{L}} = 260.9 \cos 60^{\circ} = 130.45 \text{ A}.$$

and this current is in phase with the load voltage u. The active power loss is reduced after compensation to

$$\Delta P_s = R_s I^2 = 289 \text{ W}.$$

To keep the load voltage rms value U = 230 V, the distribution voltage crms value has to be equal to

$$\mathbf{E} = \mathbf{U} + \mathbf{Z}_s \mathbf{I} = j230 + (0.017 + j\,0.050) \times 130.45 e^{j\pi/2} = 232.3\, e^{j91.6^\circ} \text{ V},$$

thus at the compensated load the voltage loss in the supply is equal to only 2.3 V.

Changes in the power factor, the active power loss, and the voltage loss in the supply with the change of the compensating capacitance C are shown in Fig. 11.5.

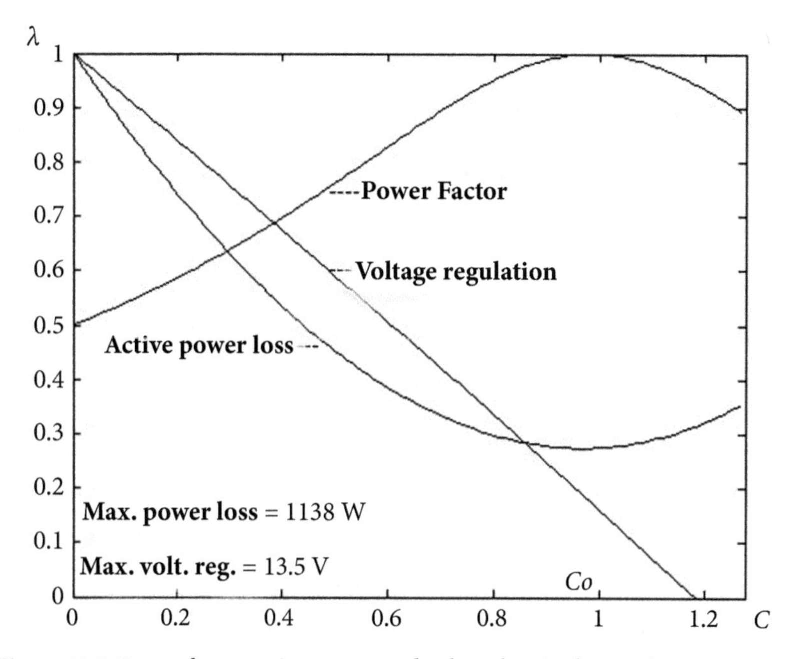

Figure 11.5 Power factor, active power, and voltage loss in the supply source versus capacitance C.

*****

Fig. 11.5 shows that the reduction in the active power loss ΔP_s declines when the compensating capacitance approaches C_0 value. On the other side, the voltage loss in the supply declines with the compensating capacitance at the same rate. For over-compensated loads, meaning when the capacitance is higher than C_0, this voltage loss could be even negative. In the situation considered in Illustration 11.1, the load voltage rms value U becomes higher than the distribution voltage rms value E when $C > 1.2\, C_0$.

11.4 Capacitive Compensation in the Presence of Harmonics

When the distribution voltage and the load current are sinusoidal, capacitor banks are very effective devices for the power factor improvement. This changes when the distribution voltage and/or the load current are distorted by harmonics. Analysis of the power theory developed by Shepherd and Zakikhani [39] and by Kusters and More [59], presented in Chapter 20, has shown that a lack of comprehension of power phenomena at reactive compensation in the presence of harmonics was the main cause of the failure of these two approaches to power theory development. This chapter is, therefore, mainly devoted to reactance compensation in single-phase circuits with nonsinusoidal supply voltage. Compensation of the reactive power in three-phase circuits can usually be combined with circuit balancing. Such compensation is discussed in Chapters 14 and 15. Before that, the concept of the power factor λ is discussed and the negative consequences of its low value are compiled.

The distribution voltage in Illustration 11.1 was assumed to be sinusoidal. Since the compensator's capacitive reactance declines with the harmonic order, while the source inductive reactance increases with this order, resonance can occur in the circuit for a harmonic frequency. The situation with a resonance between the compensator and the distribution system inductance for the fifth order is shown in the following illustration.

Illustration 11.2 *Let us assume that the same load as in Illustration 11.1 is compensated to the unity PF. To simplify calculations, let us normalize the fundamental frequency to* $\omega_1 = 1$ rad/s. *At such a frequency the compensating capacitance is* $C_0 = 0.982$ F. *Let us assume that the load is supplied with a voltage, which is distorted by the fifth order harmonic of* 1.5% *of the fundamental voltage value.*

The load conductance and susceptance for the fundamental harmonic are equal to

$$G_{L1} = \frac{P}{U^2} = 0.567 \text{ S}, \quad B_{L1} = -\frac{Q}{U^2} = -\frac{P\tan(\mathrm{acos}\lambda_L)}{U^2} = -0.982 \text{ S},$$

hence the load resistance and reactance for the fundamental have the values

$$R_{L1} = \frac{G_{L1}}{G_{L1}^2 + B_{L1}^2} = 0.441\ \Omega, \quad X_{L1} = -\frac{B_{L1}}{G_{L1}^2 + B_{L1}^2} = 0.763\ \Omega.$$

Thus for the fundamental harmonic, the load has impedances as shown in Fig. 11.6.

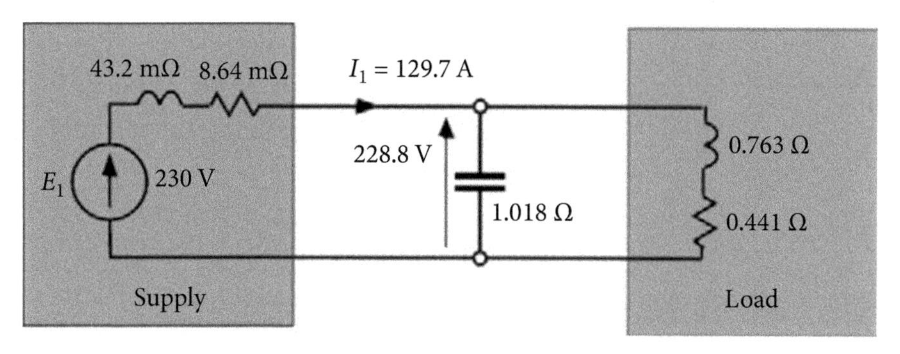

Figure 11.6 A circuit impedances for the fundamental harmonic.

Equivalent parameters of the distribution system are

$$R_{s1} = 8.64 \text{ m}\Omega, \; X_{s1} = 43.2 \text{ m}\Omega.$$

The impedance for the fundamental frequency of the compensated load as seen from the internal voltage of the distribution system is equal to

$$\mathbf{Z}_1 = \mathbf{Z}_{s1} + \mathbf{Z}_{LC1} = (0.00864 + j\,0.0432) + 1.764 = 1.772 + j\,0.0432 = 1.773\, e^{j1.4^\circ}\ \Omega.$$

Hence the rms value of the fundamental harmonic of the supply current is equal to

$$I_1 = \frac{E_1}{Z_1} = 129.7 \text{ A}.$$

Let us analyze this circuit for the fifth order harmonic. Each inductive reactance of the circuit has a value five times higher for that harmonic, while the reactance of the capacitor is five times lower. Consequently, the circuit has parameters for the fifth order harmonics as shown in Fig. 11.7.

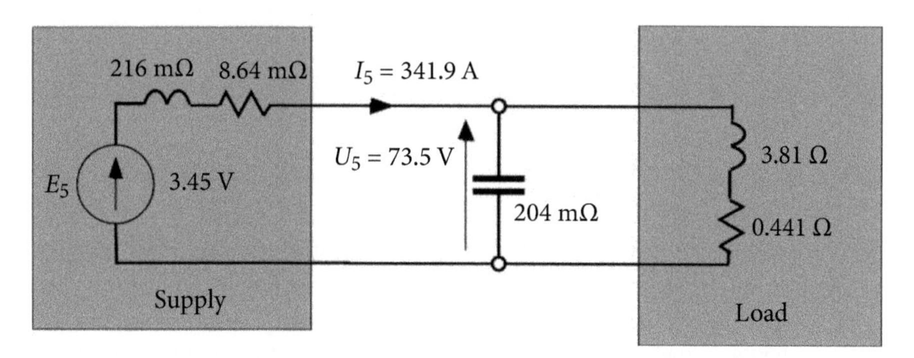

Figure 11.7 A circuit impedances for the 5th order harmonic.

The admittance of the load with a reactance compensator for the fifth order harmonic is

$$Y_{Lc5} = \frac{1}{-j0.204} + \frac{1}{0.441 + j3.81} = 0.030 + j4.65 \text{ S}$$

and the total impedance as seen from the distribution system

$$Z_5 = Z_{s5} + Z_{Lc5} = (0.00864 + j0.216) + \frac{1}{0.030 + j4.65} =$$
$$= (0.00864 + j0.216) + (0.00137 - j0.215)\ \Omega.$$

Observe that the distribution system reactance for the fifth order harmonic is almost equal to the negative value of the compensated load reactance for that frequency. In effect, they almost cancel each other out. Thus

$$Z_5 = Z_{s5} + Z_{Lc5} = 0.010 + j0.0011 = 0.01\, e^{j6.6°}\ \Omega$$

Consequently, the crms value of the fifth order the supply current harmonic is equal to

$$I_5 = \frac{E_5}{Z_5} = 341.9\, e^{j6.6°} \text{ A.}$$

The rms value of the supply current, which for the fundamental amounts to 129.7 A, increases in the presence of the fifth order harmonic of the rms value $E_5 = 1.5\%\ E_1$ *to*

$$||i|| = \sqrt{I_1^2 + I_5^2} = \sqrt{129.7^2 + 341.9^2} = 365.7 \text{ A.}$$

This increase of the supply current rms value caused by the fifth order harmonic is accompanied by the deterioration of the power factor from $\lambda = 1$, *to* $\lambda = 0.34$. *At the same time, the crms value of the fifth order voltage harmonic at the load terminals increases to*

$$U_5 = Z_{Lc5} I_5 = (0.00137 - j0.215) \times 341.9\, e^{j6.6°} = 73.5\, e^{-j83°} \text{ V.}$$

*****

This illustration demonstrates that even an insignificant distortion of the distribution voltage e of the order of 1.5%, can cause a drastic increase in distortion of the supply current and the load voltage, as well as the power factor λ deterioration. The resonance between the capacitance of the compensator and the inductance of the distribution system is responsible for this distortion since a *harmonic amplification* occurs at such a resonance.

11.5 Harmonic Amplification

In a response to the distribution voltage harmonic e_n the load voltage u contains the harmonic $u_n(e)$. The symbol $u_n(e)$ emphasizes that this harmonic is an effect of the distribution voltage e. Harmonic amplification concerning the voltage means that the rms value of the load voltage harmonic $u_n(e)$ is higher than the rms value of the distribution voltage harmonic e_n.

In a response to the load-generated harmonic j_n the supply current contains the harmonic $i_n(j)$. The symbol $i_n(j)$ emphasizes that this harmonic is an effect of the load-generated harmonic. Harmonic amplification concerning the current means that the rms value of the supply current harmonic $i_n(j)$ is higher than the rms value of harmonic j_n.

Customer single-phase harmonic-generating loads (HGLs) are usually composed of purely resistive loads, and RL loads. Both of them can be sources of harmonic current j.

An equivalent circuit with HGLs is shown in Fig. 11.8a.

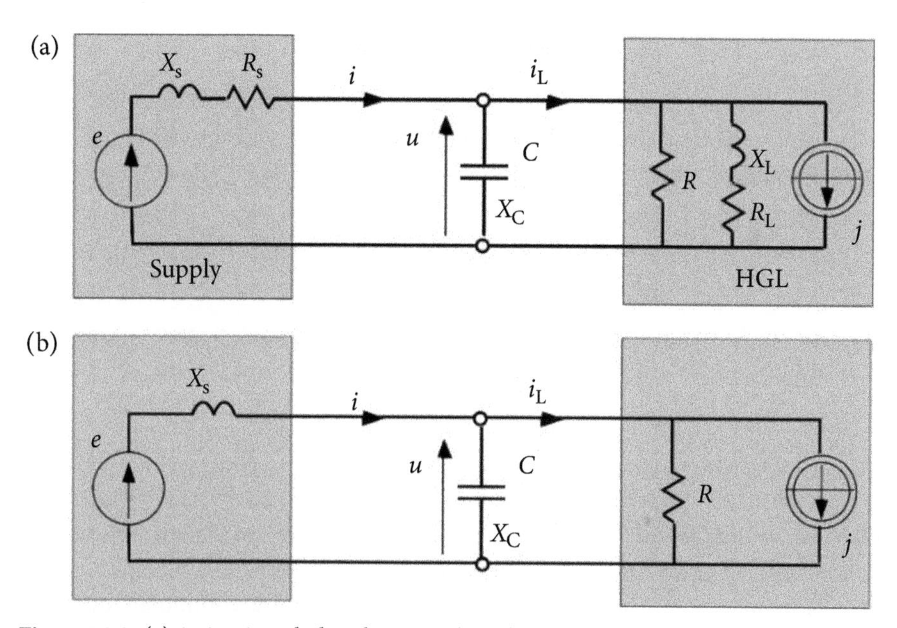

Figure 11.8 (a) A circuit with distribution voltage harmonics and load-generated current harmonics and (b) an equivalent circuit for higher orders harmonics.

It is assumed in this equivalent circuit that an HGL is composed of three branches, namely, purely resistive, R, RL, and an ideal current source j.

The current source j in this circuit represents current harmonics that occur in the load at the standard sinusoidal supply voltage. The current j changes, of course, with deviations of the supply voltage from its standard value. Since these deviations are not usually high, these changes will be neglected in the following analysis. This assumption enables us to regard the circuit as linear and use the superposition principle for its analysis. According to this principle, the supply current and the load voltage are considered as sums of independent effects of the distribution voltage e and generated current j, namely

$$i = i(e) + i(j), \qquad u = u(e) + u(j). \tag{11.8}$$

Symbols $i(e)$ and $u(e)$ denote components of the supply current and the load voltage which are effects of the distribution voltage e, while symbols $i(j)$ and $u(j)$ denote components of the supply current and the load voltage which are effects of the load-generated current j.

Because harmonic amplification in circuits with a capacitive compensator occurs above the fundamental frequency, let us analyze the properties of the system with an equivalent circuit shown in Fig. 12.8b in the range of high order harmonic frequency.

The load reactance, X_{Ln}, for harmonic order $n >> 1$, is usually much higher than the magnitude of the compensator reactance, X_{Cn}. Thus the branch X_{Ln} R_{Ln} can be removed from the equivalent circuit. Also, the distribution system reactance, X_{sn}, is usually much higher than its resistance, R_{sn}, thus it too can be neglected. Consequently, the circuit as shown in Fig. 11.8a can be approximated in a frequency range sufficiently high above the fundamental frequency by the circuit shown in Fig. 11.8b.

The voltage $u(e)$ has the Fourier transform $U(j\omega)$, which can be calculated using a nodal equation

$$U(j\omega) = \frac{E(j\omega)}{j\omega C + G + \frac{1}{j\omega L_s}} \frac{1}{j\omega L_s} = \frac{1}{1 - \omega^2 L_s C + j\omega L_s G} E(j\omega) \stackrel{\text{df}}{=} A(j\omega)\, E(j\omega). \tag{11.9}$$

Thus the dependence of the load voltage on the distribution voltage is specified by transmittance

$$A(j\omega) \stackrel{\text{df}}{=} \frac{U(j\omega)}{E(j\omega)} = \frac{1}{1 - \omega^2 L_s C + j\omega L_s G} = \frac{1}{1 - (\omega/\omega_r)^2 + j\omega L_s G}. \tag{11.10}$$

Transmittance $A(j\omega)$ will be referred to as a *load voltage-to-distribution voltage transmittance*. The transmittance as defined by formula (11.10) is a function of continuous frequency ω. The frequency of harmonics is not a continuous quantity, however, but it can assume only discrete values $\omega = n\omega_1 = \omega_n$. Thus harmonic amplification is specified by discrete values of the magnitude of transmittance $|A(j\omega_n)|$, equal to

$$A_n \stackrel{\text{df}}{=} |A(j\omega_n)| = \frac{|U(j\omega_n)|}{|E(j\omega_n)|} \stackrel{\text{df}}{=} \frac{U_n}{E_n}. \tag{11.11}$$

If for harmonic frequency $\omega_n = n\omega_1$ the magnitude of transmittance $|A(j\omega_n)|$ is higher than 1, then the load voltage harmonic u_n of such a frequency is amplified versus the distribution voltage harmonic e_n.

The frequency ω_r at which the sum of admittances the capacitor C and the distribution inductance L_s is equal to zero, that is

$$j\omega_r C + \frac{1}{j\omega_r L_s} = 0, \quad \text{or} \quad \frac{1}{\omega_r C} = \omega_r L_s$$

is the frequency of the series (voltage) resonance. Hence, this frequency is equal to

$$\omega_r = \frac{1}{\sqrt{L_s C}}. \tag{11.12}$$

The inductance L_s is a parameter of the distribution system, while capacitance C depends on the load properties and the designer of the compensator, since

$$C = k\, C_0 = k \frac{Q_1}{\omega_1\, U^2} \tag{11.13}$$

where coefficient k can be selected at the designer's discretion. At full compensation of the reactive power, $k = 1$; for $k < 1$, the load is under-compensated; for $k > 1$, the load is over-compensated.

The equivalent inductance L_s of the distribution system usually is not a well-known parameter. Its value can be obtained from the circuit modeling. It can be also calculated knowing the value of the short-circuit power S_{sc} at the bus where the compensator is installed and the expected reactance-to-resistance ratio. The obtained value is valid, however, for the fundamental frequency, while this inductance is to some degree dependent on the frequency. Therefore, the following analysis provides the only approximated value of the inductance L_s and the resonant frequency ω_r.

The short circuit power at the bus where the compensator is installed is equal to

$$S_{sc} = \frac{E^2}{\sqrt{R_{s1}^2 + \omega_1^2\, L_{s1}^2}}$$

where R_{s1} and L_{s1} denote the resistance and inductance of the distribution system at the fundamental frequency. Assuming that

$$\frac{\omega_1\, L_{s1}}{R_{s1}} \stackrel{\text{df}}{=} \xi$$

inductance L_{s1} can be calculated from the formula

$$L_{s1} = \frac{E^2}{\omega_1\, S_{sc}} \frac{\xi}{\sqrt{\xi^2 + 1}}. \tag{11.14}$$

When this inductance in the range of expected resonance is not known, it can be assumed that $L_s = L_{s1}$. Instead of resonant frequency ωS_r, it is more convenient to use a *relative resonant frequency*, Ω_r, defined as

$$\Omega_r \stackrel{\text{df}}{=} \frac{\omega_r}{\omega_1}.$$

This relative frequency, taking into account the formula (11.13), can be expressed as

$$\Omega_r = \frac{1}{\omega_1\sqrt{L_s C}} = \frac{1}{\sqrt{k}}\sqrt[4]{\left(1+\frac{1}{\xi^2}\right)}\sqrt{\frac{S_{sc}}{Q_1}}\left(\frac{U}{E}\right). \tag{11.15}$$

The short circuit and reactive powers are known usually with accuracy not higher than a few percent, therefore it is not possible to calculate the relative resonant frequency with high accuracy. Since, moreover, $\xi >> 1$ and $U \approx E$, the approximate value of this frequency is

$$\Omega_r \approx \frac{1}{\sqrt{k}}\sqrt{\frac{S_{sc}}{Q_1}}. \tag{11.16}$$

Thus, three major factors specify the relative resonant frequency of the circuit with a capacitive compensator. These are the short circuit power of the distribution system, reactive power of the load, and the compensator designer's decision. This last factor is strongly limited if the capacitor should compensate the system to an economical PF.

The magnitude of the transmittance $A(j\omega)$ at the resonant frequency is equal to

$$|A(j\omega_r)| = \frac{1}{\omega_r L_s G} = R\sqrt{\frac{C}{L_s}}.$$

It declines with the resistance of purely resistive loads, meaning with their active power, denoted as P_R. Taking into account the formula (12.7), this magnitude can be expressed as

$$|A(j\omega_r)| \approx \sqrt{k}\left(1+\frac{P_{RL}}{P_R}\right)\sqrt{\frac{S_{sc}}{P}\tan\varphi_L} \tag{11.17}$$

where P_{RL} denotes the active power of RL loads and P_R denotes the active power of purely resistive loads. When there are no purely resistive loads, meaning $P_R = 0$, this formula is, of course, not valid. In such a case the transmittance at the resonant frequency, $A(j\omega_r)$, is confined by the resistance R_s of the distribution system.

Illustration 11.3 *The plot of the magnitude of the voltage transmittance $A(j\omega)$ for two different ratios of the active power of resistive-to-RL loads, $P_{RL}/P_R = 10$ and $P_{RL}/P_R = 1$, is shown in Figs. 11.9 and 11.10. The transmittance was calculated for*

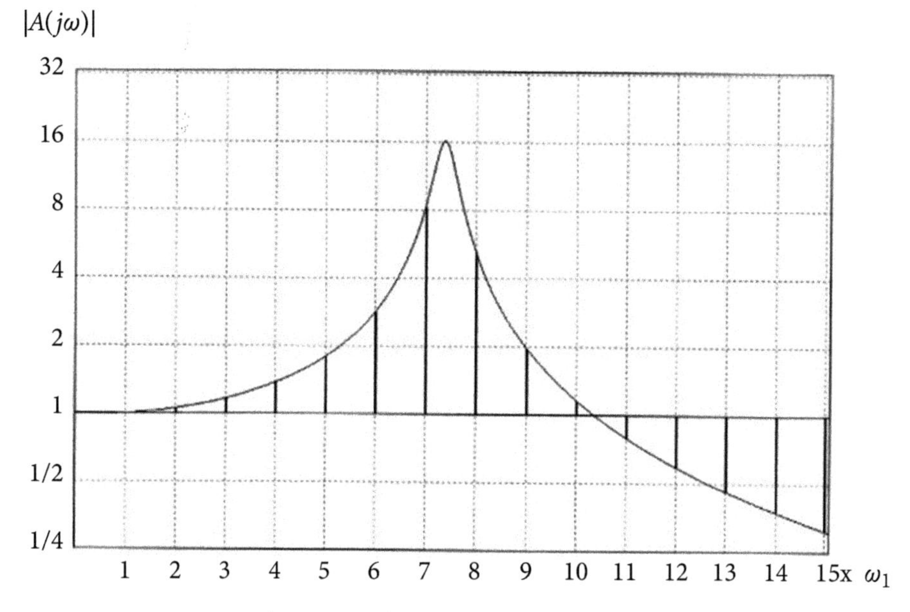

Figure 11.9 The magnitude of transmittance $A(j\omega)$ for $P_{RL}/P_R = 10$.

a distribution system with the short circuit power equal to $S_{SC} = 50P$, $\xi = 3$ *and* $\lambda_L = 0.71$.

Because transmittance $A(j\omega)$ *changes in a wide range, the plots are drawn on a logarithmic scale. The plots show the effect of resistive loads on the maximum value of the magnitude of transmittance* $A(j\omega_r)$*. At* $P_{RL}/P_R = 10$ *this magnitude is in the order of 16, while at* $P_{RL}/P_R = 1$*, it is reduced below 5.*

A comparison of plots in Figs 11.9 and 11.10 shows that purely resistive loads in circuits with a capacitive compensator, attenuate resonances and contribute to a reduction of the harmonic distortion.

*****

Assuming that the conductance $G = 1/R$ in the equivalent circuit of the circuit is zero, then the magnitude of transmittance $A(j\omega)$ is higher than 1 when

$$-1 < 1 - \omega^2 L_s C < 1,$$

meaning in the range of frequency

$$0 < \omega < \frac{\sqrt{2}}{\sqrt{L_s C}} = \sqrt{2}\,\omega_r.$$

All harmonics of the load voltage, u_n, in this range of frequency are amplified with versus the distribution voltage harmonics, e_n. Consequently, the load voltage can be more distorted than the distribution voltage. The range of frequency where harmonics are amplified will be referred to as a frequency range of *harmonic amplification*. Observe that the upper frequency of the range of harmonic amplification is

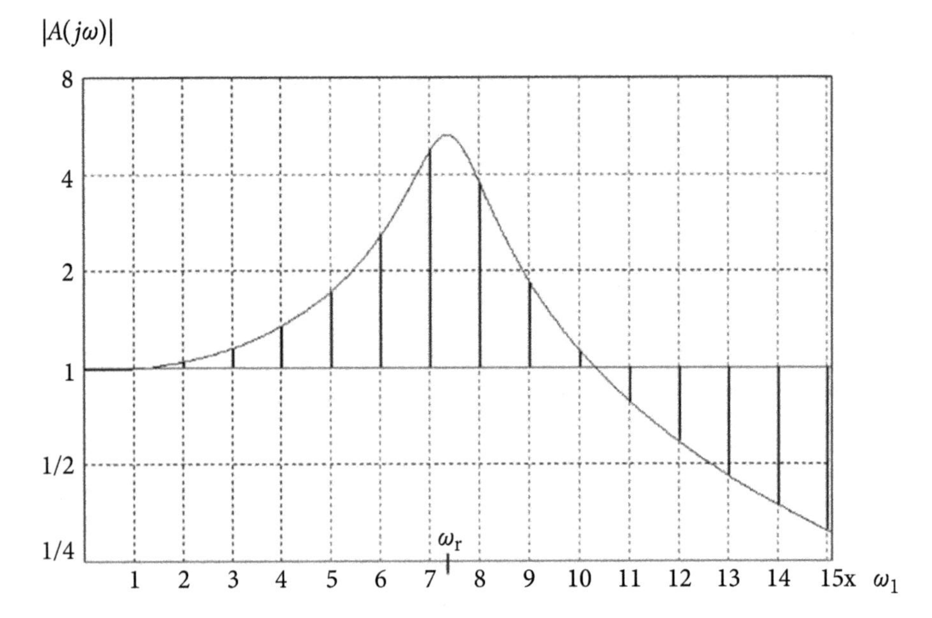

Figure 11.10 The magnitude of transmittance $A(j\omega)$ for P_{RL}/P_{R}= 1.

by a root of 2 higher than the resonant frequency. This is true, however, only when the circuit is idealized by the assumption that there are no purely resistive loads, that is, $G = 0$. Otherwise, this is the only approximation. Nonetheless, the accuracy of this approximation can be quite satisfactory. Observe, that the range of harmonic amplification in Figs 11.9 and 11.10, meaning the frequency, where the plot of $|A(j\omega)|$ crosses unity value, does not change visibly with the change of ratio P_R/P. At resonant frequency $\omega_r \approx 7.3\ \omega_1$, all voltage harmonics up to the tenth order are amplified. For

$$\omega < \sqrt{2}\,\omega_r$$

there is a range of *harmonic attenuation*. In the situation, as specified in Illustration 11.3, all the load voltage harmonics of the order higher than tenth order are attenuated.

11.6 Amplification of the Load-generated Current Harmonics

The capacitor as observed from the load is connected in parallel with the distribution system inductance and consequently, a *current resonance* has to occur in the circuit. The frequency properties of the circuit response to the load-generated harmonics can be specified in terms of the Fourier transform of the supply current, $I(j\omega)$. It is equal to

$$I(j\omega) = \frac{J(j\omega)}{j\omega C + G + \frac{1}{j\omega L_s}} \frac{1}{j\omega L_s} = \frac{1}{1 - \omega^2 L_s C + j\omega L_s G} J(j\omega). \tag{11.18}$$

The ratio of Fourier transforms of the supply current i(t) and the load-generated current $j(t)$ will be referred to as the *supply current-to-load-generated current transmittance*,

$$B(j\omega) \stackrel{\text{df}}{=} \frac{I(j\omega)}{J(j\omega)} = \frac{1}{1 - (\omega/\omega_r)^2 + j\omega L_s G}. \tag{11.19}$$

Comparison of this transmittance with formula (11.10) for the load voltage-to-distribution voltage transmittance $A(j\omega)$ shows that these two transmittances, although they have entirely different meanings, are numerically identical, that is,

$$B(j\omega) = \frac{I(j\omega)}{J(j\omega)} = A(j\omega) = \frac{U(j\omega)}{E(j\omega)}.$$

This means that the load-generated current harmonics are amplified in the supply current identically as the distribution voltage harmonics are amplified in the load voltage.

The values of transmittance $B(j\omega)$ for harmonic frequencies are

$$B_n \stackrel{\text{df}}{=} |B(j\omega_n)| = \frac{|I(j\omega_n)|}{|J(j\omega_n)|} \stackrel{\text{df}}{=} \frac{I_n}{J_n} = \frac{U_n}{E_n} = A_n. \tag{11.20}$$

Illustration 11.4 *A single-phase load shown in Fig. 11.11 when supplied with the voltage of the rms value* $U_1 = 240$ V *has the active power* $P = 20$ kW *and the PF* $\lambda_L = 0.71$. *The active power of a resistive load is* $P_R = 10$ kW, *while the rest of this power is the power of an RL load. The load generates a current harmonic of the seventh order of rms value* $J_7 = 1\%\ I_1$ *and the distribution voltage, e, contains the fifth order harmonic of rms value* $E_5 = 1\%\ E_1$. *The short-circuit power of the source is* $S_{sc} = 1.0$ MVA *with the reactance-to-resistance ratio equal to* $\xi = X_s/R_s = 3$.

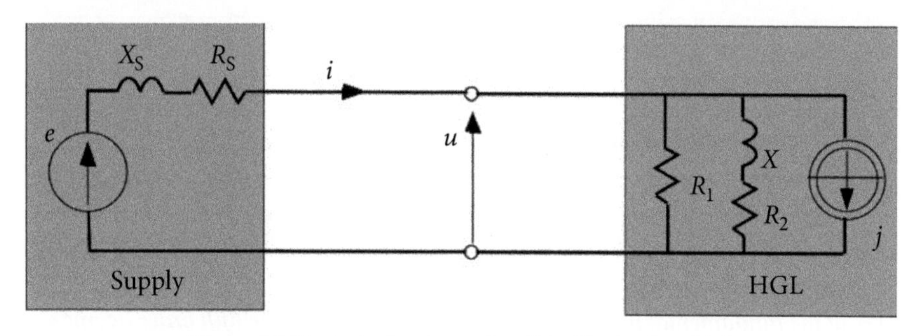

Figure 11.11 A single-phase circuit in Illustration 11.4.

Let us calculate the contents of the fifth and seventh order harmonic in the supply current and the load voltage after the load is compensated for the fundamental harmonic to unity power factor. The circuit parameters for the fundamental harmonic and the rms values of voltages and currents before compensation are shown in Fig. 11.12.

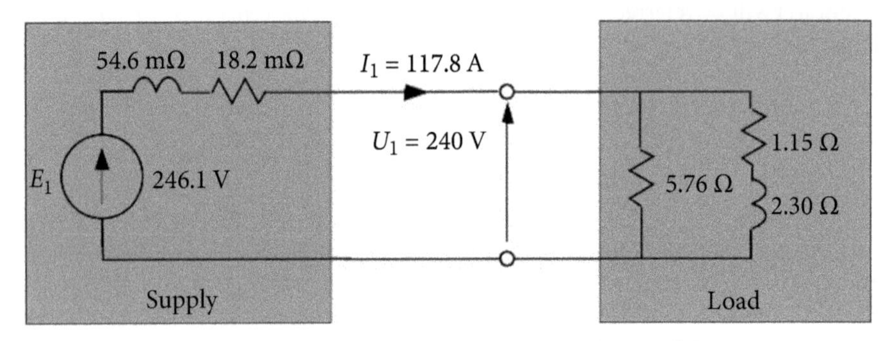

Figure 11.12 Parameters of the circuit for the fundamental frequency.

The active power dissipated in the supply source is $\Delta P_s = R_s I_1^{\,2} = 252$ W, *and the voltage loss in the source is* $\Delta U = 6.1$ V.

The load susceptance for the fundamental frequency is $B_1 = -\ 0.347$ S, *thus the shunt capacitance needed for the entire compensation of the load reactive power Q, at the normalized frequency* $\omega_1 = 1$ rad/s, *is equal to* $C_0 = 0.347$ F. *The rms value of voltages and currents after compensation are shown in Fig. 11.13.*

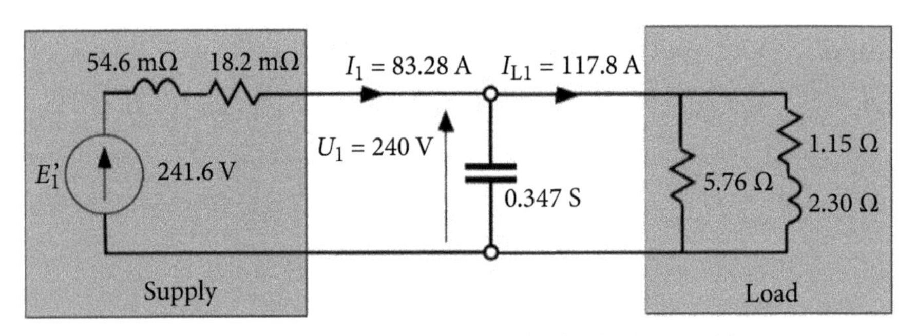

Figure 11.13 Results of the circuit analysis for the fundamental frequency.

Observe that the distribution voltage rms value E_1 *was recalculated such that the load voltage preserves its rms value before compensation,* $U_1 = 240$ V. *Thus compensation of the reactive power of the fundamental harmonic reduces the voltage loss in the supply source from* $\Delta U = 6.1$ V *to* 1.6 V *and the active power loss to* $\Delta P = 126$ W.

Assuming that the fifth order harmonic in the distribution voltage e amounts to 1% of the fundamental harmonic, that is, $E_5 = 2.4$ V, *then the fifth order harmonic of the load voltage* U_5, *and the supply current harmonic* I_5 *has the rms values shown in Fig. 11.14.*

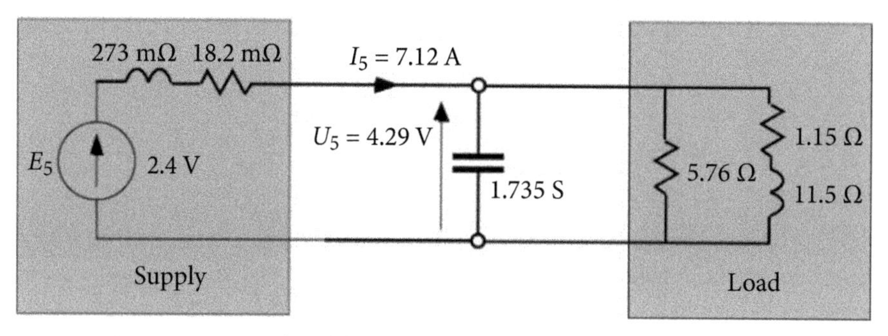

Figure 11.14 Results of the circuit analysis for the 5th order harmonic.

Amplification of the distribution voltage fifth order harmonic is specified by transmittance $A_5 = U_5/E_5 = 4.29/2.4 = 1.8$. *Thus, at* 1% *of the fifth harmonic in the distribution voltage, the load voltage contains* 1.8% *of that harmonic.*

Assuming that the seventh order harmonic in the load-generated current j *amounts to* 1% *of the load current fundamental harmonic* I_1, *i.e.,* $J_7 = 1.18\text{A}$, *the seventh order harmonic in the load voltage and the supply current has rms values* U_7 *and* I_7 *shown in Fig. 11.15.*

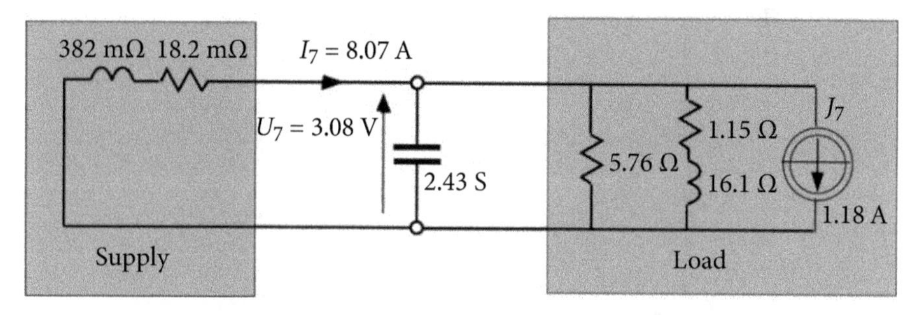

Figure 11.15 Results of the circuit analysis for the 7th order harmonic.

Amplification of the load-generated seventh order current harmonic is specified by transmittance $B_7 = I_7/J_7 = 8.07/1.18 = 6.8$. *At the same time, the compensator reduces the supply current fundamental harmonic by* λ, *thus the relative contents of the seventh order harmonic in the supply current is*

$$\frac{I_7}{I_1} \times 100 = \frac{I_7}{I_1\,\lambda} \times 100 = \frac{8.07}{117.8 \times 0.707} \times 100 = 9.7\%.$$

*****

11.7 Admittance as Seen from the Distribution System

Distribution voltage harmonics affect not only the load voltage but also the supply current. The Fourier transform of the supply current caused by the distribution voltage $i(e)$ is

$$I(j\omega) = \frac{E(j\omega)}{j\omega C + G + \frac{1}{j\omega L_s}} \frac{1}{j\omega L_s}(G + j\omega C) = Y_x(j\omega)E(j\omega) \tag{11.21}$$

where $Y_x(j\omega)$ is admittance as seen from the internal voltage e source of the distribution system, as shown in Fig. 11.16.

$$Y_x(j\omega) \stackrel{\text{df}}{=} \frac{I(j\omega)}{E(j\omega)} = \frac{G + j\omega C}{1 - (\omega/\omega_r)^2 + j\omega L_s G}. \tag{11.22}$$

Admittance $Y_x(j\omega)$, unlike transmittances $A(j\omega)$ and $B(j\omega)$, is not a relative and a dimensionless function. It changes with the level of the voltage rms value and the load power. Therefore, the frequency properties of this admittance are better specified by the ratio of this admittance to its value at the fundamental harmonic $|Y_x(j\omega_1)|$, namely

$$\Psi_x(j\omega) \stackrel{\text{df}}{=} \frac{Y_x(j\omega)}{|Y_x(j\omega_1)|}. \tag{11.23}$$

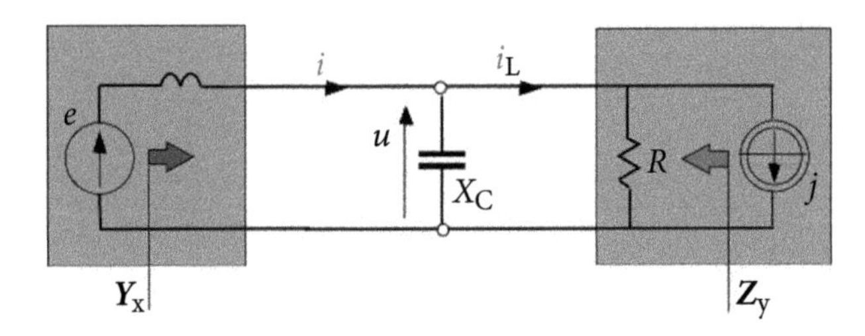

Figure 11.16 An equivalent circuit for harmonics.

Discrete values of this relative admittance are equal to

$$\Psi_{xn} \stackrel{\text{df}}{=} \frac{|Y_x(jn\omega_1)|}{|Y_x(j\omega_1)|} = \frac{|Y_{xn}|}{|Y_{x1}|}. \tag{11.24}$$

The relative admittance $\psi_x(j\omega)$ provides clear information on the expected contents of harmonics in the supply current due to the distribution voltage harmonics. A plot of the magnitude of the relative admittance $\psi_x(j\omega)$ as seen from the internal voltage e source of the distribution system for the system considered in Illustration 11.4 is shown in Fig. 11.17. This plot demonstrates how the circuit with a capacitive compensator is sensitive to distribution voltage harmonics. For example, the admittance as seen from the distribution system for the fifth order harmonic, Y_{x5}, is more than eight times higher than this admittance for the fundamental. Consequently, 1% of the fifth order harmonic in the distribution system voltage results in more than 8% contents of that harmonic in the supply current. Even worse is with harmonics of the frequency that are closer to the resonance frequency. In the situation illustrated in Fig. 11.17, the highest admittance is for the seventh order harmonic.

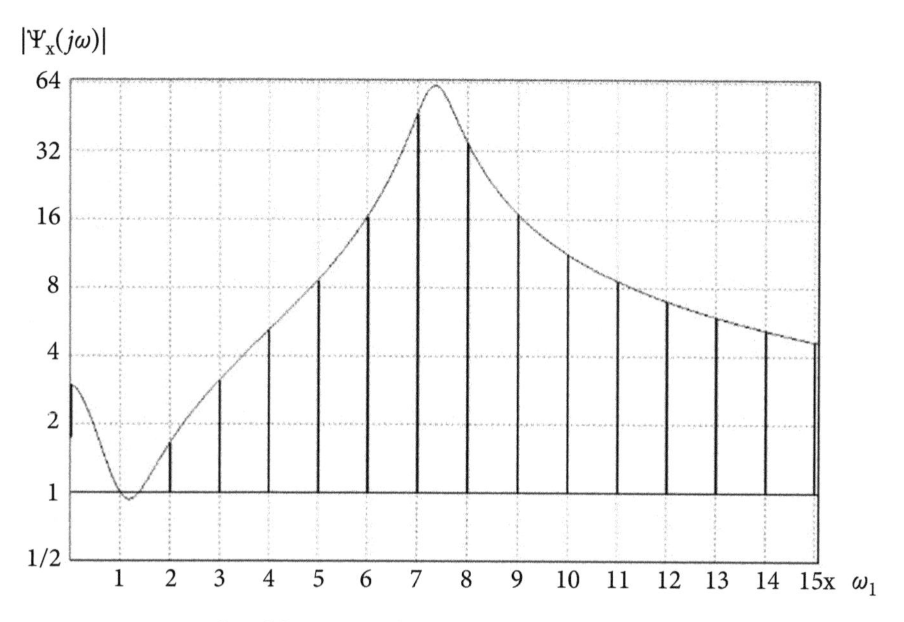

Figure 11.17 A plot of the magnitude of relative admittance $\psi_x(j\omega)$ for $P_{RL}/P_R = 1$.

Observe that all relative admittances for harmonics, ψ_{xn}, are substantially higher than unity. It applies even to harmonics in the range of attenuation, which for the circuit considered in illustration, is above tenth order harmonic. Thus, the loads with capacitive compensation are very sensitive to distortion in the distribution voltage.

11.8 Impedance as Seen from the Load-generated Current Source

In a response to the load-generated current harmonics, voltage $u(j)$ occurs at the load terminals. Its Fourier transform is equal to

$$U(j\omega) = \frac{J(j\omega)}{j\omega C + G + \frac{1}{j\omega L_s}} = \frac{j\omega L_s}{1 - \omega^2 L_s C + j\omega L_s G} J(j\omega) \stackrel{\text{df}}{=} Z_y(j\omega)\, J(j\omega). \qquad (11.25)$$

Symbol $Z_y(j\omega)$ denotes an impedance as seen from the load-generated current source, j, shown in Fig. 11.11. It is defined as

$$Z_y(j\omega) \stackrel{\text{df}}{=} \frac{U(j\omega)}{J(j\omega)} = \frac{j\omega L_s}{1 - \omega^2 L_s C + j\omega L_s G}. \tag{11.26}$$

The magnitude of this impedance for harmonic frequencies is equal to

$$Z_{y\,n} \stackrel{\text{df}}{=} |Z_y(j\omega_n)| = \frac{U_{s\,n}}{J_n}.$$

The voltage response to load-generated current harmonics can be particularly high at the resonant frequency of the circuit when the inductive admittance of the distribution system is equal to the capacitive admittance of the capacitor. This is a parallel (current) resonance and its frequency is the same as the frequency of the series (voltage) resonance, ω_r, for the distribution voltage harmonics.

Similarly, as the admittance $Y_x(j\omega)$ impedance $Z_y(j\omega)$ is not a dimensionless function, it depends on the level of the distribution voltage. The voltage response to the load-generated current harmonics is better specified, therefore, in terms of relative impedance, where the impedance of the load for the fundamental harmonic is the reference, since

$$\frac{|U(j\omega)|}{U_1} = \frac{|Z_y(j\omega)|}{Z_{L1}} \frac{|J(j\omega)|}{I_1} \stackrel{\text{df}}{=} \zeta_y(j\omega) \frac{|J(j\omega)|}{I_1}. \tag{11.27}$$

The relative contents of the nth order harmonic in the load voltage caused by the load-generated harmonic J_n is equal to

$$\frac{U_n}{U_1} = \frac{Z_{yn}\,J_n}{Z_{L1}\,I_1} = \zeta_{yn} \left(\frac{J_n}{I_1}\right) \tag{11.28}$$

where Z_{L1} is the load impedance for the fundamental frequency and

$$\zeta_{yn} \stackrel{\text{df}}{=} \frac{Z_{yn}}{Z_{L1}} = \frac{|Z_y(j\omega_n)|}{Z_{L1}} \tag{11.29}$$

is the relative impedance of the load and the distribution system as seen from the generated current source, j.

The plot of the relative impedance as seen from the source of the generated current j for the same circuit as considered in Illustration 11.3 is shown in Fig. 11.18.

This plot demonstrates that the load voltage is not sensitive to the load-generated harmonics. Only in the closest vicinity of the resonant frequency can the harmonic contents in the load voltage u be higher than these contents in the load-generated current j. In the situation considered, at $P_{RL}/P_R = 10$, $\zeta_{y7} = 1.56$. It means that 1% of the fifth and seventh order current harmonics generated by the load results in 1.56% contents of the seventh order harmonic, but only 0.25% of the fifth order harmonic in the load voltage.

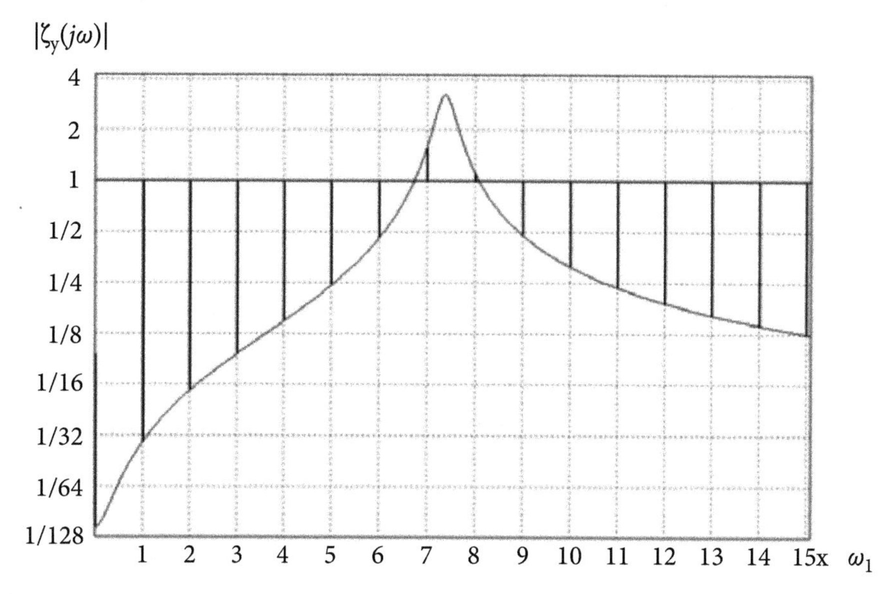

Figure 11.18 A plot of the magnitude of relative impedance, $\zeta_y(j\omega)$ for $P_{RL}/P_R = 10$.

11.9 Compensator Caused Harmonic Distortion

Transmittances A_n, B_n, relative admittance ψ_{xn}, and impedance ζ_{yn}, enable us to calculate the contents of individual voltage and current harmonics at the load terminals when the rms value of the distribution voltage and the load-generated current harmonics are known. A combined distortion, caused by all harmonics, is specified by *harmonic distortion*, δ, known also under the name of *total harmonic distortion* (THD). The THD, as used in the literature on the subject, does not usually differentiate distortion caused by the distribution voltage harmonics e_n from that caused by the load-generated harmonics, j_n. The acronym "THD" makes such differentiation confusing, but we must stick with it or it must be replaced by a much longer acronym. Therefore harmonic distortion will be described in this book in terms of harmonic distortion coefficient rather than in terms of THD.

The supply current i of any load, compensated or not compensated, can be decomposed into the fundamental harmonic, i_1, and a distorting current, i_d. The distorting current can be next decomposed into a distorted current caused by distribution voltage harmonics, $i_d(e)$, and a current caused by the generated current harmonics, $i_d(j)$, namely

$$i = i_1 + i_d = i_1 + i_d(e) + i_d(j) \tag{11.30}$$

The current $i_d(e)$ is composed of harmonics that occur in the supply current because of the distribution voltage e harmonics. The supply current distortion caused by these harmonics can be defined as the ratio of the rms value of distorting current

$||i_{\rm d}(e)||$ and the rms value of this current fundamental harmonic, I_1, that is,

$$\delta_{\rm i}(e) \stackrel{\rm df}{=} \frac{||i_{\rm d}(e)||}{I_1} = \frac{\sqrt{\sum_{n=2}^{\infty} I_n^2(e)}}{I_1} = \frac{\sqrt{\sum_{n=2}^{\infty} Y_{{\rm x}n}^2 E_n^2}}{I_1}. \tag{11.31}$$

Since $I_1 = Y_{\rm x1}E_1$ and $E_1 \sim U_1$, this formula can be expressed as

$$\delta_{\rm i}(e) = \sqrt{\sum_{n=2}^{\infty}\left(\frac{Y_{{\rm x}n}}{Y_{\rm x1}}\frac{E_n}{E_1}\right)^2} = \sqrt{\sum_{n=2}^{\infty}\left(\Psi_{{\rm x}n}\frac{E_n}{E_1}\right)^2} \tag{11.32}$$

meaning in terms of the relative admittance of the load, $\psi_{{\rm x}n}$, as seen from the distribution system. The supply current distortion caused by harmonics generated by the load is equal to

$$\delta_{\rm i}(j) \stackrel{\rm df}{=} \frac{||i_{\rm d}(j)||}{I_1} = \frac{\sqrt{\sum_{n=2}^{\infty} I_n^2(j)}}{I_1} = \sqrt{\sum_{n=2}^{\infty}\left(B_n\frac{J_n}{I_1}\right)^2}. \tag{11.33}$$

The distorting current caused by distribution voltage harmonics $i_{\rm d}(e)$ and those caused by current harmonics generated by the load $i_{\rm d}(j)$ may have different orders, and in such a case they are mutually orthogonal. They may have the same order, but their phases are usually mutually random. It can be assumed, therefore, that their rms values satisfy the relation

$$||i_{\rm d}||^2 = ||i_{\rm d}(e)||^2 + ||i_{\rm d}(j)||^2 \tag{11.34}$$

and consequently, distortion of the supply current can be calculated as

$$\delta_{\rm i} \stackrel{\rm df}{=} \frac{||i_{\rm d}||}{I_1} = \sqrt{\delta_{\rm i}^2(e) + \delta_{\rm i}^2(j)}. \tag{11.35}$$

Illustration 11.5 *A load at the supply voltage $U = 240$ V has the active power $P = 50$ kW, the ratio $P_{\rm RL}/P_{\rm R} = 1$ and power factor $\lambda_{\rm L} = 0.5$. It is supplied from a distribution system with the short-circuit power $S_{\rm sc} = 40P$ and reactance-to-resistance ratio, $\zeta = 1$. Let us calculate the current distortion if the distribution voltage contains harmonics of the relative value $E_3 = 1.5\%E_1$, $E_5 = 0.5\%E_1$ and $E_7 = 0.4\%E_1$, while the load generates harmonics of the relative value $J_3 = 5\%I_1$, $J_5 = 4\%I_1$ and $J_7 = 3\%I_1$. For harmonics of concern, the relative admittances $\psi_{{\rm x}n}$ of the circuit are $\psi_{\rm x3} = 6.0$, $\psi_{\rm x5} = 25.0$, $\psi_{\rm x7} = 22.6$ and current-to-current transmittances are $B_3 = 1.3$, $B_5 = 3.0$,*

$B_7 = 1.9$. For such data,

$$\delta_{\mathrm{i}}(e) = \sqrt{\sum_{n=2}^{\infty}\left(\Psi_{\mathrm{x}n}\frac{E_n}{E_1}\right)^2} = \sqrt{(6.0\times 1.5)^2 + (25.0\times 0.5)^2 + (22.6\times 0.4)^2} = 17.6\%$$

$$\delta_{\mathrm{i}}(j) = \sqrt{\sum_{n=2}^{\infty}\left(B_n\frac{J_n}{I_1}\right)^2} = \sqrt{(1.3\times 5)^2 + (3.0\times 4)^2 + (1.9\times 3)^2} = 14.6\%.$$

Distortion of the load-generated current is equal to

$$\delta_{\mathrm{j}} \stackrel{\mathrm{df}}{=} \frac{||j||}{I_1} = \sqrt{\sum_{n=2}^{\infty}\left(\frac{J_n}{I_1}\right)^2} = \sqrt{5^2+4^2+3^2} = 7.1\%$$

thus the distortion of the supply current is higher than the distortion of the load-generated current. Total distortion of the supply current is

$$\delta_{\mathrm{i}} = \sqrt{\delta_{\mathrm{i}}^2(e) + \delta_{\mathrm{i}}^2(j)} = \sqrt{17.6^2 + 14.6^2} = 22.8\%.$$

*****

The same approach can be applied to calculating the load voltage distortion. The load voltage, u, can be decomposed into the fundamental harmonic, u_1, and distorting voltage, u_{d}. This distorting voltage can be next decomposed into a voltage component that occurs because of the distribution voltage harmonics, $u_{\mathrm{d}}(e)$, and a component that occurs because of current harmonics generated in the load, $u_{\mathrm{d}}(j)$, namely

$$u = u_1 + u_{\mathrm{d}} = u_1 + u_{\mathrm{d}}(e) + u_{\mathrm{d}}(j).$$

Harmonic distortion caused by distribution voltage harmonics is a ratio of the rms value of distorting voltage $||u_{\mathrm{d}}(e)||$ and the load voltage fundamental harmonic, U_1, namely

$$\delta_{\mathrm{u}}(e) \stackrel{\mathrm{df}}{=} \frac{||u_{\mathrm{d}}(e)||}{U_1} = \frac{\sqrt{\sum_{n=2}^{\infty} U_n^2(e)}}{U_1} = \frac{\sqrt{\sum_{n=2}^{\infty} A_n^2 E_n^2}}{U_1}.$$

Since U_1 is approximately equal to E_1, hence

$$\delta_{\mathrm{u}}(e) \stackrel{\mathrm{df}}{=} \frac{||u_{\mathrm{d}}(e)||}{U_1} \approx \sqrt{\sum_{n=2}^{\infty}\left(A_n\frac{E_n}{E_1}\right)^2}. \tag{11.36}$$

Harmonic distortion caused by the load-generated current harmonics is a ratio of the rms value of distorting voltage $||u_{\mathrm{d}}(j)||$ and the load voltage fundamental harmonic, U_1, namely

$$\delta_u(j) \stackrel{\text{df}}{=} \frac{||u_{\rm d}(j)||}{U_1} = \frac{\sqrt{\sum\limits_{n=2}^{\infty} U_n^2(j)}}{U_1} = \frac{\sqrt{\sum\limits_{n=2}^{\infty} Z_{{\rm y}n}^2 J_n^2}}{Z_{\rm L1} I_1} = \sqrt{\sum_{n=2}^{\infty}\left(\zeta_{{\rm y}n}\frac{J_n}{I_1}\right)^2}. \tag{11.37}$$

Similarly, as it was with the distorting component $i_{\rm d}$ of the supply current, the rms value, the distorting component $u_{\rm d}$ of the load voltage can be calculated as

$$||u_{\rm d}||^2 = ||u_{\rm d}(e)||^2 + ||u_{\rm d}(j)||^2$$

and consequently, distortion of the load voltage can be calculated as

$$\delta_{\rm u} \stackrel{\text{df}}{=} \frac{||u_{\rm d}||}{U_1} = \sqrt{\delta_{\rm u}^2(e) + \delta_{\rm u}^2(j)}. \tag{11.38}$$

Illustration 11.6 *Relative impedances $\zeta_{{\rm y}n}$ as seen from the current source j for situation in Illustration 11.5 are equal to $\zeta_{{\rm y}3} = 0.14$, $\zeta_{{\rm y}5} = 0.53$, $\zeta_{{\rm y}7} = 0.47$, while the load-to-distribution voltage transmittances are $A_3 = 1.3$, $A_5 = 3.0$, $A_7 = 1.9$. Consequently*

$$\delta_{\rm u}(e) = \sqrt{\sum_{n=2}^{\infty}\left(A_n\frac{E_n}{E_1}\right)^2} = \sqrt{(1.3 \times 1.5)^2 + (3.0 \times 0.5)^2 + (1.9 \times 0.4)^2} = 2.6\%.$$

Observe that distortion of the distribution voltage is equal to

$$\delta_{\rm e} \stackrel{\text{df}}{=} \frac{||e_{\rm h}||}{E_1} = \sqrt{\sum_{n=2}^{\infty}\left(\frac{E_n}{E_1}\right)^2} = \sqrt{1.5^2 + 0.5^2 + 0.4^2} = 1.6\%,$$

thus the load voltage distortion caused by distortion voltage harmonics $\delta_v(e)$ is higher than the distribution voltage distortion. Additionally, the load voltage is distorted by the load-generated harmonics, which caused distortion

$$\delta_u(j) = \sqrt{\sum_{n=2}^{\infty}\left(\zeta_{{\rm y}n}\frac{J_n}{I_1}\right)^2} = \sqrt{(0.14 \times 5)^2 + (0.53 \times 4)^2 + (0.47 \times 3)^2} = 2.7\%$$

and the load voltage total distortion is equal to

$$\delta_{\rm u} = \sqrt{\delta_{\rm u}^2(e) + \delta_{\rm u}^2(j)} = \sqrt{2.6^2 + 2.7^2} = 3.7\%.$$

*****

11.10 Power Factor Components

The phenomenon of harmonic amplification in circuits with a capacitive compensator can increase the rms value of some harmonics of the supply current and the load voltage. Thus, this phenomenon increases the apparent power S and consequently, it reduces the power factor, λ.

If the contribution of the voltage and current harmonics to the active power P is neglected, the reduction of the PF can be expressed in terms of the voltage and current distortion as follows. The power factor is approximately equal to

$$\lambda_{\mathrm{L}} \stackrel{\mathrm{df}}{=} \frac{P}{S} = \frac{P}{\|u\|\,\|i_{\mathrm{L}}\|} \approx \frac{P_1}{\|u\|\,\|i_{\mathrm{L}}\|} = \frac{U_1 I_{\mathrm{L}1} \cos\varphi_{\mathrm{L}1}}{\|u\|\,\|i_{\mathrm{L}}\|}.$$

Since $\cos\varphi_{\mathrm{L}1} = \lambda_{\mathrm{L}1}$, meaning it is the power factor of the load at sinusoidal voltages and currents, and

$$\frac{\|u\|}{U_1} = \frac{\sqrt{U_1^2 + \|u_{\mathrm{d}}\|^2}}{U_1} = \sqrt{1+\delta_{\mathrm{u}}^2}, \quad \frac{\|i_{\mathrm{L}}\|}{I_{\mathrm{L}1}} = \frac{\sqrt{I_{\mathrm{L}1}^2 + \|i_{\mathrm{d}}\|^2}}{I_{\mathrm{L}1}} = \sqrt{1+\delta_{\mathrm{i}}^2} \tag{11.39}$$

we obtain

$$\lambda_{\mathrm{L}} \approx \frac{1}{\sqrt{1+\delta_{\mathrm{u}}^2}\sqrt{1+\delta_{\mathrm{i}}^2}}\,\lambda_{\mathrm{L}1} \tag{11.40}$$

where $\lambda_{\mathrm{L}1}$ is the power factor of the fundamental harmonic. Since it is equal to the $\cos\varphi_{\mathrm{L}1}$, that is, it is specified by the phase shift between the voltage and current fundamental harmonic, it is often referred to as a *displacement power factor*. According to formula (11.40), the voltage and current distortions, δ_{u} and δi, reduce the power factor below the value of the displacement power factor. Taking into account relations (11.36) and (11.39), the power factor at a distorted voltage and current can be presented in the form

$$\lambda_{\mathrm{L}} = \frac{1}{\sqrt{1+\|\delta_{\mathrm{u}}(e)\|^2+\|\delta_{\mathrm{u}}(j)\|^2}}\,\frac{1}{\sqrt{1+\|\delta_{\mathrm{i}}(e)\|^2+\|\delta_{\mathrm{i}}(j)\|^2}}\,\lambda_{\mathrm{L}1}. \tag{11.41}$$

It shows the effect of distribution voltage and load-generated current harmonics on the PF degradation. Observe, however, that the load voltage distortion is usually confined by standards to no more than 5%. At $\delta_{\mathrm{u}} = 0.05$, the term

$$\frac{1}{\sqrt{1+\delta_{\mathrm{u}}{}^2}} = 0.999,$$

thus it is negligible. Therefore, in most cases, only the supply current distortion contributes to the power factor decline, namely

$$\lambda_{\mathrm{L}} \approx \frac{1}{\sqrt{1+\|\delta_{\mathrm{i}}\|^2}}\,\lambda_{\mathrm{L}1} = \frac{1}{\sqrt{1+\|\delta_{\mathrm{i}}(e)\|^2+\|\delta_{\mathrm{i}}(j)\|^2}}\,\lambda_{\mathrm{L}1}. \tag{11.42}$$

Unfortunately, improvement of the displacement power factor using a capacitive compensator usually increases the supply current distortion. In effect, such a capacitor may reduce the PF instead of improving it.

Illustration 11.7 *A load at the voltage U = 240 V has the active power P = 10 kW and PF λ_L = 0.5. It is supplied from 60 Hz source with short-circuit power S_{sc} = 400 kVA, and reactance-to-resistance ratio ξ = 3. The distribution voltage is distorted with harmonics of the third, fifth, seventh, and eleventh order of relative value, E_3 = 0.5%, E_5 = 3%, E_7 = 1.5%, and E_{11} = 0.5% of the fundamental voltage rms value, E_1.*

When the distribution voltage is not distorted, then a shunt capacitor of capacitance C_0 = 0.80 mF improves the power factor λ to unity. It changes with the capacitance C value as shown in Fig. 11.19.

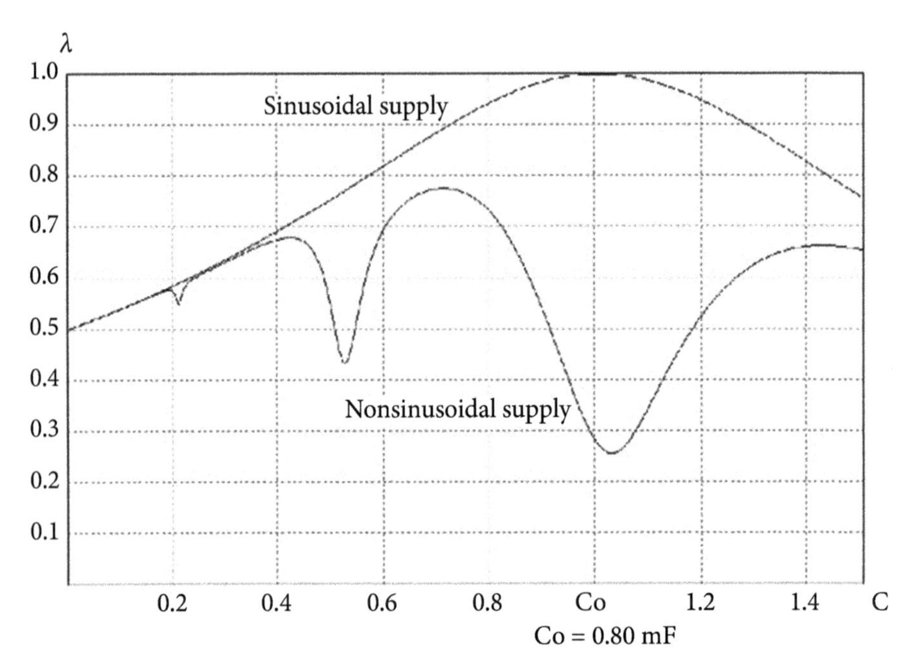

Figure 11.19 A plot of the power factor versus capacitance C.

In the presence of the distribution voltage distortion as specified above, this PF declines below λ = 0.3. The resonance for the fifth order harmonic is mainly responsible for this decline. The dips in the λ plot versus compensating capacitance C shown in Fig. 11.19, counting from the left side, are caused by resonances for the eleventh, seventh, and fifth order harmonics, respectively. They increase these harmonic contents in the supply current.

Due to resonances of the capacitor with the distribution system inductance, such compensators can improve the PF effectively only at the lack of harmonics or if resonances do not occur in the range of frequency where harmonics occur. It requires usually that the short-circuit power of the distribution system is a few hundred times higher than the load active power, meaning in very strong systems. This is illustrated in Fig. 11.20, which shows the change of magnitudes of transmittances $A(\omega)$ and

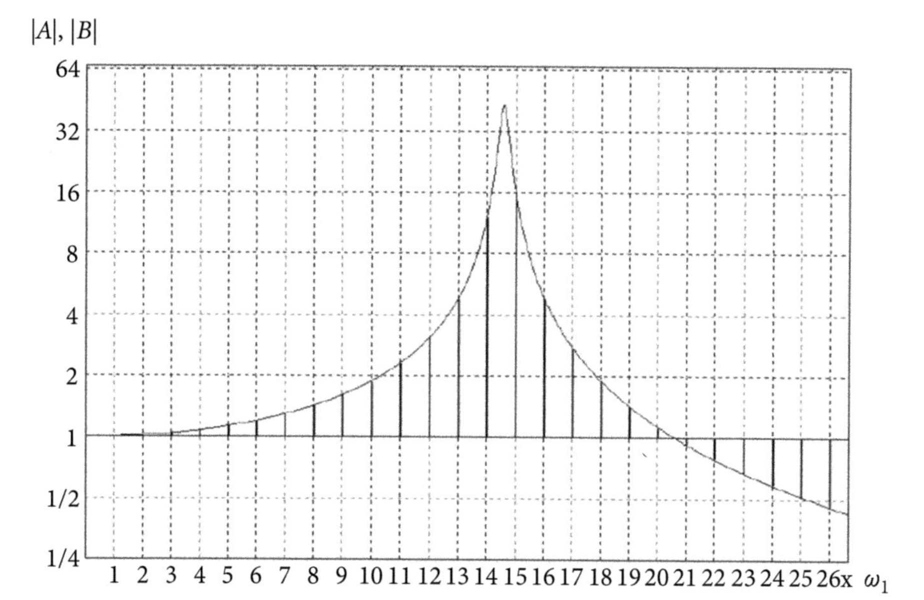

Figure 11.20 A plot of the magnitude of transmittances A and B versus frequency for $S_{sc}/Q = 200$.

$B(\omega)$ in a system with the short-circuit power-to-compensated reactive power ratio, $S_{sc}/Q = 200$, at $\zeta = 3$ and $P_{RL}/P_R = 0.1$.

The resonance in such a situation occurs between the fourteenth and fifteenth order harmonic. These harmonics are usually negligible, while the thirteenth order harmonic in common situations is well below 1% of the fundamental.

*****

11.11 Critical Capacitances and Resonant Frequency Control

Capacitances for which the plot of power factor λ versus capacitance C in the presence of the nth order harmonic has a dip, that is, a local minimum, are referred to as *critical capacitances* for that harmonic and denoted by C_{cn}. These are capacitances for which resonance occurs for that order harmonic, meaning,

$$\frac{1}{\sqrt{L_s\, C_{c\,n}}} = n\omega_1. \tag{11.43}$$

Thus, critical capacitances for $X_s >> R_s$, are approximately equal to

$$C_{cn} = \frac{1}{n^2}\frac{1}{\omega_1^2 L_s} \approx \frac{1}{n^2}\frac{S_{sc}}{\omega_1\, U^2} = \frac{1}{n^2}\frac{S_{sc}}{Q}\, C_0. \tag{11.44}$$

To avoid resonances for harmonics and, consequently, the current distortion increase and the power factor decline, the capacitance of the compensator should be selected as far as possible from critical values. Since the dominating harmonics in power systems are usually harmonics of the order $n = 3, 5, 7,\ldots$, critical capacitances should, first of all, be specified for these order harmonics.

Illustration 11.8 *A compensator of a load with reactive power* Q = 100 kVAr *installed at a bus with the short-circuit power* S_{sc} = 2.0 MVA *has critical capacitances*

$$C_{c3} = 2.2\ C_0, \quad C_{c5} = 0.8\ C_0, \quad C_{c7} = 0.41\ C_0.$$

The capacitance C_{c3}, *as much higher than* C_0, *is unlikely, of course, to occur in the circuit. Due to economic reasons, this capacitance is usually lower than* C_0. *Also, a protection system can disconnect some sections of the capacitor bank. Observe, however, that reduction of the compensator capacitance from* C_0 *to a value in the vicinity of* C_{c5}, *in the presence of the fifth order harmonic, can drastically reduce the power factor.*

*****

Since it is not possible to build distribution systems without an inductance and this inductance is beyond customer control, the resonance of the capacitive compensator with the distribution system in a presence of harmonics cannot be avoided. In effect, as shown in the numerical illustrations above, the compensator performance can be very sensitive to the distribution voltage distortion and the load-generated current harmonics.

Selection of the compensator capacitance C in such a way that its value is in the middle between critical values can reduce the supply current distortion and improve the power factor in some situations. It is enough to observe the plot of the PF shown in *Fig. 11.19* to conclude, that this improvement would not be high. At the same time, the load would be under-, or over-compensated. It is also worth observing that the compensation of reactive power Q is more affected by the change of the capacitance C than the resonant frequency Ω_r. Indeed, if $C = kC_0$, then the compensated reactive power $Q' = kQ$, while the relative resonant frequency changes to

$$\Omega_r' = \frac{1}{\sqrt{k}}\Omega_r \tag{11.45}$$

meaning it changes with the root of the coefficient k value.

Much more effective resonant frequency control and, consequently, PF improvement, as compared to pure capacitive compensator, can be provided by an LC compensator, that is, a capacitive compensator with an additional inductor.

An inductor in the compensator structure changes the frequency properties of the whole circuit. It can be connected in three ways as shown in *Fig. 11.21*.

When it is connected to the supply line as shown in *Fig. 11.21(a)*, it reduces the resonant frequency to

$$\Omega_r' = \frac{1}{\omega_1} \frac{1}{\sqrt{(L_s + L)C}} = \frac{1}{\sqrt{\left(1 + \frac{L}{L_s}\right)}} \Omega_r. \tag{11.46}$$

However, such an additional line inductor increases the voltage drop and consequently the dependence of the load voltage on the load power. Therefore, this structure is rather not used for improving the compensator performance. Only when there is resonance at a frequency very close to the fifth order harmonic, then a line inductor can move it down, towards the fourth order harmonic, usually of negligible level. It would not be appropriate to do the same when the resonance is in the vicinity of the seventh order harmonic. The line inductor, reducing the amplification of this harmonic, would increase the amplification of the harmonic of the fifth order.

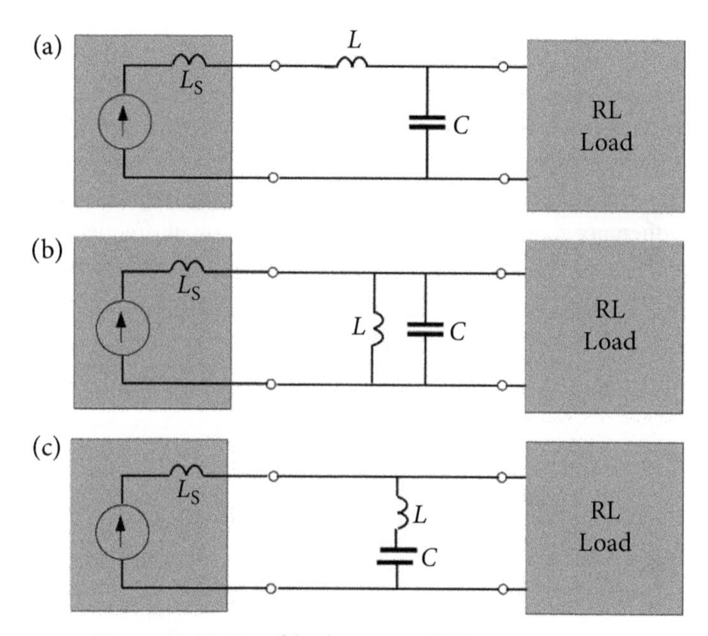

Figure 11.21 Possible structures of LC compensators.

When the inductor is connected in parallel to the capacitor, as shown in Fig. 11.21(b), it reduces the equivalent inductance as seen from the capacitor terminals, thus it increases the resonant frequency. This change is not high, however, and consequently the control of the resonant frequency is strongly limited, since the shunt inductance L has to be much higher than L_S. Otherwise, the supply source would be almost short-circuited by the shunt inductor.

This structure is sometimes used, however, not for the resonant frequency control but in controllable compensators of reactive power, where the level of reactive power compensation is controlled by a change of the equivalent inductance of the branch with a thyristor switch, connected as shown in Fig. 11.22(a).

The current rms value in the branch with the inductor is controlled by changing the thyristor firing angle α and consequently, the equivalent inductance L of that branch is controlled between L_s and zero value. Such a branch with an inductor and thyristor switch is referred to as athyristor switched inductor(TSI). The TSI enables control of the compensator susceptance B in the range

$$B_{\max} = \omega_1 C, \quad \text{for } \alpha = 180°,$$

$$B_{\min} = \omega_1 C - \frac{1}{\omega_1 L_c}, \quad \text{for } \alpha = 0$$

so that the circuit shown in *Fig. 11.22(a)* provides a thyristor-controlled susceptance (TCS), which has to be regarded as a harmonic current generating device. Its equivalent circuit is shown in *Fig. 11.22(b)*. It enables adaptive compensation of the reactive power. Unfortunately, the TSI is a source of current harmonics. Consequently, compensation of the reactive power with TCS is usually accompanied by strong harmonic distortion in the compensated circuit. The best performance of the compensator can be obtained using the compensator structure shown in *Fig. 11.21(c)*. The inductor connected in series with the capacitor enables control of the resonant frequency according to formula *(11.46)*, but without affecting the load voltage. Therefore, inductance value L can be much higher than distribution system inductance L_s, thus, a wide control of the resonant frequency is possible. It is the only structure that enables efficient control of the compensator resonant frequency. The term capacitive-inductive or LC compensator will be applied in this book to compensators only of this structure.

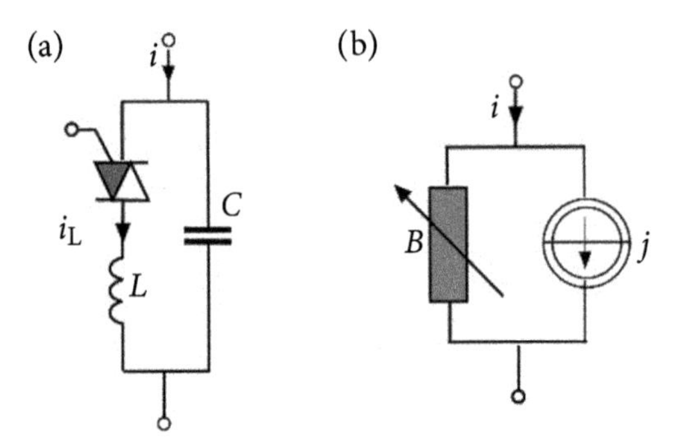

Figure 11.22 (a) A capacitive compensator with thyristor switched inductor (TSI) and (b) its equivalent circuit.

A TSI enables adaptive compensation of the reactive power. Unfortunately, the TSI is a source of current harmonics. Consequently, compensation of the reactive power with TCS is usually accompanied by strong harmonic distortion in the compensated circuit. The best performance of the compensator can be obtained using the compensator structure shown in *Fig. 11.21(c)*. The inductor connected in series with the capacitor enables control of the resonant frequency according to formula *(11.46)*, but without affecting the load voltage. Therefore, inductance value L can be much higher than distribution system inductance L_s, thus a wide control of the resonant frequency is possible. It is the only structure that enables efficient control of the compensator resonant frequency. The term **capacitive-inductive**, *or* **LC compensator**, will be applied in this book to compensators only of this structure.

Summary

This chapter demonstrates that the capacitive compensation in the presence of the distribution voltage and/or the load-generated current harmonics can be ineffective and can even cause degradation of the power factor and an increase in the voltage and current distortion.

Chapter 12
Resonant Harmonic Filters

12.1 Principle of Operation

Resonant harmonic filters (RHFs) are reactance devices installed in distribution systems for reducing harmonics, mainly current harmonics originated in nonlinear and/or periodically switched time-variant loads, such as power electronics devices, in particular three-phase rectifiers or AC/DC controlled converters. At the same time, RHFs provide a capacitive reactive power to the system and consequently, they improve the power factor at the bus where such filters are installed.

RHFs are built of capacitors and inductors and they are connected in parallel to harmonic-generating loads (HGLs). They are designed in such a way that they provide a low impedance path for selected harmonics, thus protecting the distribution system against the injection of these current harmonics into these systems. Consequently, RHFs should reduce distortion of both the supply current and the supply voltage.

Unlike filters used in communications systems, where these are low-power devices for processing information carried by electrical signals, RHFs are high-power devices of the rated power in the range from kilovolt amperes (kVA) to tens of megavolt amperes (MVA). Because of the power level, RHFs are not built as active devices but as passive devices. This is the major difference as compared to filters in communication systems, where filters are very often built as active devices, meaning with amplifiers, which require a power supply.

Originally, RHFs were developed for reducing current distortion in systems with high-power HGLs, usually high-power three-phase rectifiers, which generate harmonics mainly of the fifth, seventh, eleventh thirteenth, seventeenth... $6k \pm 1$ ...order. These days they are often used for reducing current harmonics generated by high-power three-phase AC/DC converters.

Such filters are built usually as three-phase devices, meaning they are indeed three single-phase filters, as symmetrical as possible, connected to each supply line. Sometimes, with an increase in the level of load-generated current harmonics, an existing capacitor bank for power factor improvement is decomposed into sections, equipped with inductors, and appropriately tuned so that it is converted into a resonant harmonic filter.

In the beginning, such filters were used in systems with low distortion of the distribution voltage and were connected at terminals of only individual loads. Currently, this situation has changed, because HGLs are much more common than they were before, and consequently, the distribution voltage is distorted more frequently.

Powers and Compensation in Circuits with Nonsinusoidal Current. Leszek S. Czarnecki, Oxford University Press.
 DOI: 10.1093/oso/9780198879206.003.0012

An RHF can reduce current harmonics originated in the load but can increase current harmonics that occur because of the distribution voltage harmonics. This changes the operating conditions of RHFs, reducing their effectiveness and making filter design much more complex. In some cases they are not even able to fulfill their objective of reducing harmonic distortion at the load terminals. More sophisticated devices for reducing harmonic distortion are needed in such situations, in particular *switching compensators*, discussed in Chapters 16 and 17, and *harmonics blocking compensators*, discussed in Section 9.8, can be used for that.

Resonant harmonic filters are composed of a few tuned resonant LC branches connected in parallel to the harmonics generating loads, as shown in Fig. 12.1. A line branch with impedance $\mathbf{Z}_A$ could sometimes be added for improving the filter effectiveness, but this weakens the supply; that is, the load voltage becomes more dependent on the load current in such a case.

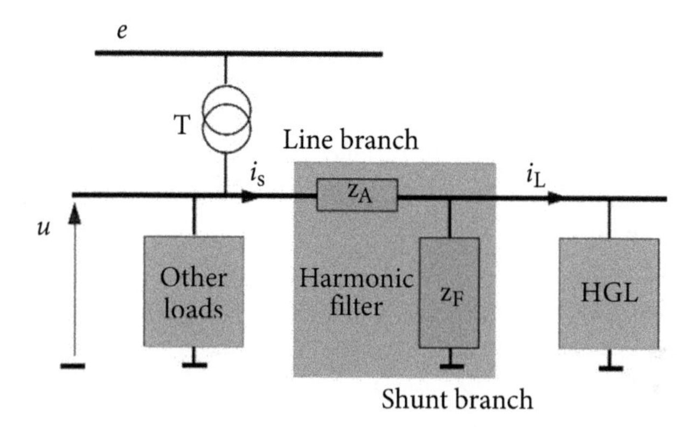

Figure 12.1 A supply bus with a harmonic filter and a line branch.

Moreover, the line branch could cause a loss of active power. Therefore, a connection of a series branch could be justified only if a shunt RHF alone is not capable of reducing the waveform distortion to the required level. Consequently, Resonant harmonic filters usually have only a shunt branch. The structure of the bus with an RHF can be simplified to the form shown in Fig. 12.2.

All loads supplied from the bus can be combined to a single equivalent HGL.

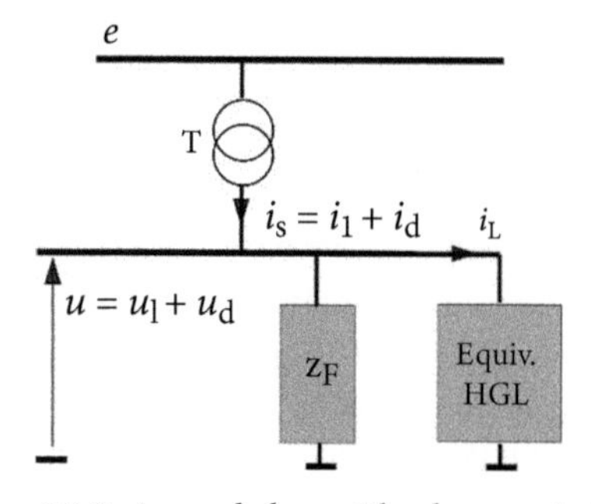

Figure 12.2 A supply bus with a harmonic filter.

Before properties and the filter design are discussed, a measure of the filter's effectiveness in reducing distortion of the bus voltage and the supply current is needed. Such a measure can be based on the change of the total harmonic distortion in the effect of the effect the installation of a filter makes. Harmonics of the bus voltage and the supply current are affected by a filter differently. Therefore, separate indicators for the effectiveness of reduction of the supply current distortion and the bus voltage distortion are needed.

In general, harmonic filters operate in such a way that they modify frequency characteristics at the bus where they are installed to provide a low impedance path for current harmonics generated in the loads supplied from that bus. Depending on this modification, harmonic filters can be classified as (i) *notch filters* and (ii) *high-pass filters*. High-pass filters provide a low impedance path for all load-generated current harmonics of the order higher than some selected frequency. Notch filters provide a low impedance path for a few distinctive harmonics and therefore RHFs can be classified as notch filters.

The RHF is composed of a few parallel LC branches as shown in Fig. 12.3(a). Branches are tuned to specific frequencies in the vicinity of the frequency of harmonics that should be reduced in the supply current i. The filter for the fundamental frequency is equivalent to a capacitor and its capacitance is used for reducing the inductive reactive power Q of the load.

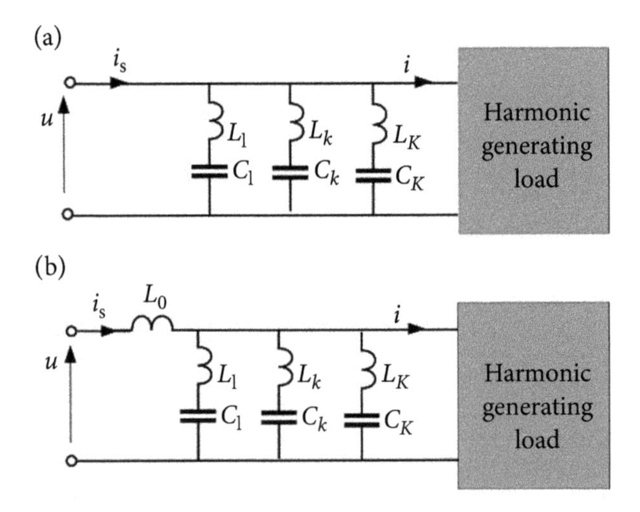

Figure 12.3 (a) A common notch filter and (b) a fixed POLE filter.

Notch and high-pass filters usually have only a shunt branch, in other words, they have the structure shown in Fig. 12.3. Sometimes, however, they are built with a line branch, meaning connected as shown in Fig. 12.1. In particular, a notch filter with a line inductor L_0, connected as shown in Fig. 12.3(b), enables control of the location of filter harmful resonances with the distribution system inductance. These resonances are POLES of the circuit transmittance and therefore, such a notch filter with series inductor L_0 is referred to [161] as a *fixed POLE filter*. To be accurate, the name "*fixed*

POLES filter" is associated with a method of design of a filter with a line inductor, the method that starts from the selection of POLEs frequency, so that POLEs are fixed at the designer's discretion.

Independently of the harmonic filter structure and properties, the filter creates a current divider shown in Fig. 12.4, meaning the load current i_L is divided by the filter into two currents: the current which flows from the distribution system i, and the filter current i_F. The shunt branch of the filter, of impedance $\mathbf{Z}_F$, is one leg of it. Another leg is formed by the distribution system, of impedance $\mathbf{Z}_s$, connected in series with, if it exists, the line branch of the filter of impedance $\mathbf{Z}_A$.

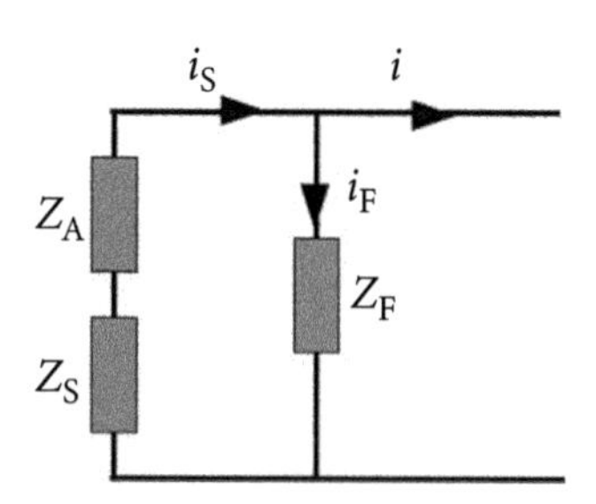

Figure 12.4 A current divider which is created by a harmonic filter.

The effectiveness of the filter in the reduction of current harmonics depends on the ratio of impedances of these two legs for harmonic frequencies. It means that the frequency properties of the distribution system impedance, $\mathbf{Z}_s$, are a major factor that affects the filter performance. This dependence of the harmonic filter performance on the distribution system impedance is particularly high when the filter has only the shunt branch, that is, $\mathbf{Z}_A = 0$. The lack of this line branch is a very common practice in distribution systems.

Inductors built for filters should be as linear as possible, therefore the magnetic core has to be assembled with an air gap, which reduces inductor nonlinearity. Otherwise, a phenomenon of ferroresonance might occur and consequently, the LC branch becomes a source of harmonics. Inductors for very high-power filters can be built without a magnetic core. Such inductors are referred to as *air-core inductors*, which are, naturally, linear.

Designing resonant harmonic filters is only apparently simple. This is because the filter reduces the load-originated current harmonics, which were taken into account in the filter design, but at the same time it increases the distribution system-originated current harmonics in the supply current. Moreover, usually, there are more load-originated current harmonics than the number of resonant branches to reduce them. In the lack of such branches these harmonics, referred to as *minor harmonics* can be amplified [138] in the supply current. Over the years, some methods of filter design and various recommendations based on engineering intuition were developed. Unfortunately, these recommendations often do not provide a solid fundamental for designing filters with satisfactory efficiency in reducing harmonic distortion.

12.2 Traditional Design of RHFs

Resonant harmonic filters for reducing the fifth, seventh, eleventh, and thirteenth order load-originated current harmonics, of the structure shown in Fig. 12.5, are designed traditionally by calculating the capacitance C_k and inductance L_k, $k = 5, 7, 11, 13$, in such a way that the branch has a resonance at the frequency specified as a tuning frequency

$$t_k = \frac{1}{\sqrt{L_k C_k}} \tag{12.1}$$

equal to or in the vicinity of frequency $\omega_k = k\omega_1$.

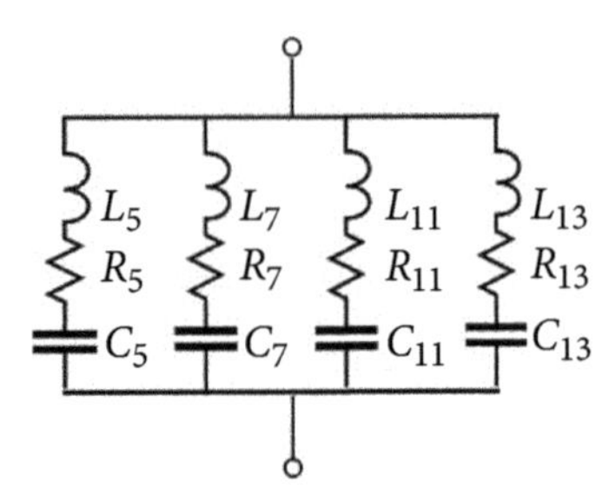

Figure 12.5 A four-branch RHF structure.

The reactive power of the fundamental harmonic, Q_k, compensated by such a branch is

$$Q_k = a_k Q = B_{1k}\, U_1^2 = \frac{\omega_1\, C_k}{1 - \omega_1^2\, L_k C_k} U_1^2 \tag{12.2}$$

where Q is the load reactive power per phase, a_k is the coefficient of the reactive power allocation to the branch $L_k C_k$ and B_{1k} is the branch susceptance for the fundamental harmonic. Combining formulae (12.1) and (12.2) we obtain

$$C_k = \left[1 - \left(\frac{\omega_1}{t_k}\right)^2\right] \frac{a_k\, Q}{\omega_1\, U_1^2}. \tag{12.3}$$

$$L_k = \frac{1}{t_k^2 C_k}. \tag{12.4}$$

The resistance R_k depends on the inductors' q-factor, q, defined as

$$q \stackrel{\text{df}}{=} \frac{\omega_1 L_k}{R_k}. \tag{12.5}$$

According to Ref. [156], for high voltage applications where air-core inductors are used, the q-factors of $50 < q < 150$ are typical, while for low voltage applications, iron-core inductors are needed with $10 < q < 50$.

Opinions concerning the reactive power allocation to particular branches are divided. According to [45], this allocation is irrelevant to the filter properties. Consequently, it could be assumed that each branch compensates the same reactive power, that is, allocation coefficients have the same value. Such filters will be referred to as *Type A filters* in this book. However, there are also other practices or recommendations. The reactive power allocation for a two-branch filter of the fifth and seventh order harmonics assumed in [141] is in the proportion of $Q_5/Q_7 = 2{:}1$, while in Ref. [136], this proportion is $Q_5/Q_7 = 8{:}3$. According to Ref. [142], the reactive power allocation should be "proportional to total harmonic current each filter will carry". Filters designed according to this particular recommendation will be referred to as *type B filters.*

In the presence of distribution voltage harmonics, the filter branches are tuned traditionally to a frequency below the harmonic frequency. It increases the branch reactance at the harmonic frequency and keeps it inductive, even if the capacitance of the capacitor bank declines in time. However, there are substantial differences in opinions on how much the branches should be detuned. Reference [141] assumes that filters are detuned by 5% below harmonic frequencies, while Ref. [62] suggests that detuning should be in the range of 3–10% below these frequencies. Indeed, detuning assumed in Ref. [162] amounts to 8% for all branches, that is, the relative detuning is the same for all branches. According to Ref. [136], the branches are detuned by 18 Hz, that is, the absolute detuning is the same. Branches are tuned to frequencies 4.7 ω_1 and 6.7 ω_1, respectively. It means that there is no clear recommendation concerning the filter detuning. Even the degree of detuning is not related to the level of these harmonics.

When a harmonic filter is under design, the attenuation of dominating, characteristic harmonics is the subject of the main concern. The AC/DC converters and other nonlinear loads supplied from the same bus also generate other non-characteristic harmonics. Their level is reported in numerous papers [62, 118, 141, 162, 175].

The traditional approach to filter design essentially neglects the presence of non-characteristic harmonics in the load current and the distribution voltage harmonics in the filter design process, considering them as a kind of minor [138] harmonics. Tuning the filter branches to a frequency below harmonic frequencies is a common countermeasure to the degrading effect of distribution voltage harmonics. Unfortunately, with all these recommendations, the filter performance might be still mediocre. More advanced methods of filter design based on a deeper analysis of the causes of this performance deterioration are needed.

12.3 Frequency Properties of RHFs

The equivalent circuit, per phase, of the filter, along with the load and distribution system, are shown in Fig. 12.6.

Observe that this equivalent circuit is drawn on the assumption that all loads are resistive-reactance loads. Resistances $R_1, \ldots R_K$ in this circuit do not represent separate

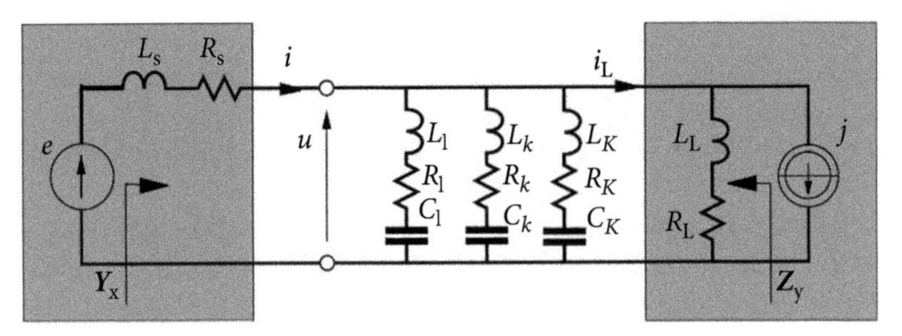

Figure 12.6 An equivalent circuit of the filter, the load, and a distribution system.

elements of the filter but rather equivalent resistances of the filter inductors, meaning they represent the active power losses both in inductor winding and in its magnetic core, due to eddy current and hysteresis. The current source j represents current harmonics generated in the load due to its nonlinearity or periodic time variance.

The equivalent circuit of the filter, the load, and the distribution system in the range of harmonic frequency, $\omega >> \omega_1$, could be substantially simplified. Distribution system reactance X_s for the fundamental frequency ω_1 is usually higher than this system resistance R_s approximately three to eight times. The load reactance X_L for the fundamental frequency is usually of the order of the load resistance R_L. The reactance X_s increases with the harmonic order, thus in the frequency range of $\omega >> \omega_1$, the resistance R_s, as much lower than ωL_s, can be neglected. The case is similar to the load resistance R_L. If, moreover, inductors are considered ideal, then the equivalent circuit in the range of harmonic frequency range could be simplified to the form shown in Fig. 12.7.

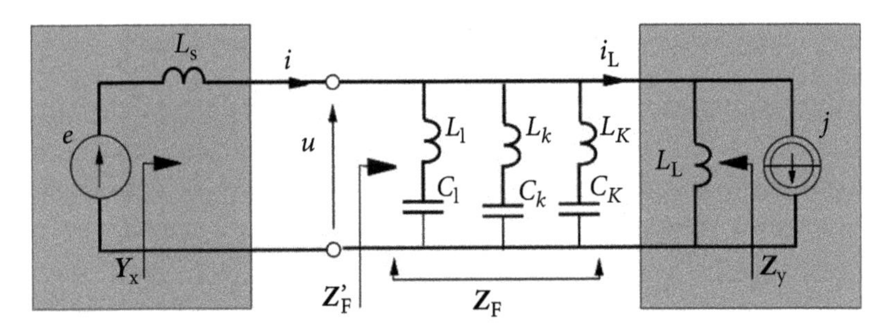

Figure 12.7 An equivalent circuit in the range of harmonic frequencies.

It is a reactance circuit with two excitation sources, e and j. The impedance of the filter and the load are equal to

$$Z'_F(j\omega) = \frac{Z_F(j\omega)Z_L(j\omega)}{Z_F(j\omega) + Z_L(j\omega)} = jX'_F(\omega) \tag{12.6}$$

and it is a *reactance function*, that is, its real part is equal to zero. The reactance has zero values at tuning frequencies, denoted by t_k, equal to

$$t_k \stackrel{\text{df}}{=} \frac{1}{\sqrt{L_k C_k}}$$

and these values are entirely under the designer's control. The reactance can be changed to a required value by changing the inductance or capacitance of an individual branch. These are frequencies at which the load is short-circuited by the filter. The plot of this reactance for a four-branch filter is shown in Fig. 12.8. At frequencies denoted by z_1, z_2, z_3, and z_4 there is a current (parallel) resonance in the filter and all parameters of the filter and the load (L_L) contribute to these frequencies. As was proven in Section 10.4, the derivative of any reactance functions versus frequency has to be positive:

$$\frac{d}{d\omega} X'_F(\omega) > 0.$$

Therefore the reactance $X'_F(\omega)$ has to change with frequency, as shown in Fig. 12.8.

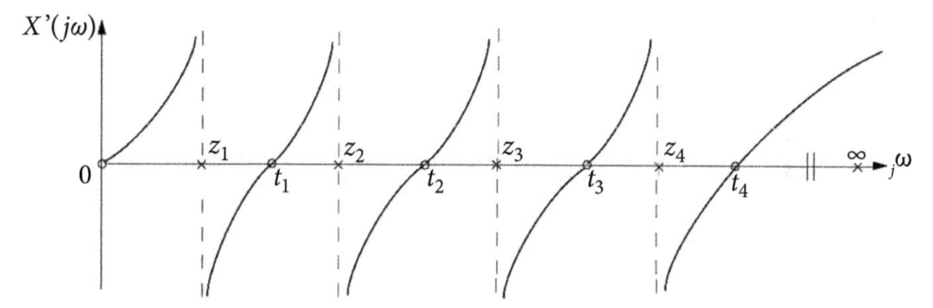

Figure 12.8 A plot of reactance $X_F'(j\omega)$ for four-branch RHF.

The tuning frequencies, where reactance has zero value, have to be separated by frequencies where it approaches infinity.

As observed from the distribution system internal voltage $e(t)$, the filter is connected in series with the system internal inductance L_s. Therefore, impedance as seen from this voltage source is

$$Z_x(j\omega) = j\omega L_s + jX'_F(\omega).$$

As an effect of the source inductance, this impedance $Z_x(j\omega)$ has zeros at cross-sections of the plot of reactance $X_F'(j\omega)$ with the line ωL_s, as shown in Fig. 12.9.

The admittance as seen from the internal voltage $e(t)$ of the distribution system

$$Y_x(j\omega) = \frac{1}{j\omega L_s + jX'_F(\omega)} = jB_x(j\omega)$$

approaches an infinite value at impedance $Z_x(j\omega)$ zeros. Its plot is shown in Fig. 12.10.

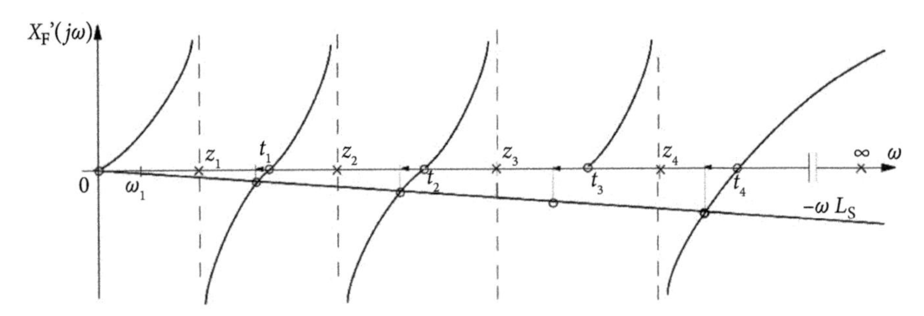

Figure 12.9 A plot of reactance $X_F'(j\omega)$, and the internal reactance of the supply source.

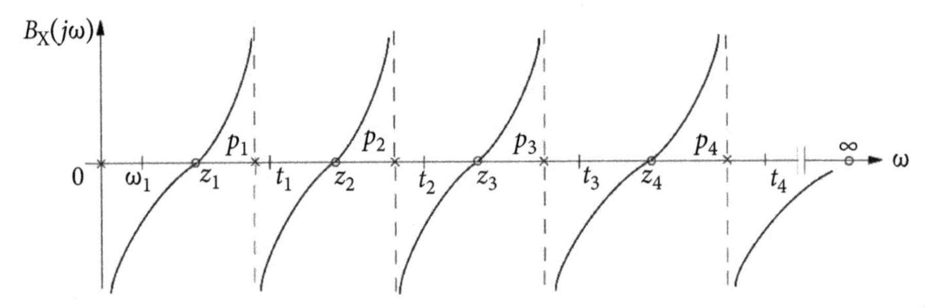

Figure 12.10 A plot of reactance $X_F'(j\omega)$.

Observe that the impedance of L_kC_k branches below tuning frequencies t_k is capacitive and consequently, a series resonance of these branches with the inductance L_s of the supply source has to occur. This is visible on the plot shown in Fig. 12.10 where below each tuning frequency t_k, susceptance B_x at points p_k approaches an infinite value, meaning a short-circuit is seen from the internal voltage $e(t)$ of the distribution system. These very high susceptances are, of course, confined by resistances of the filter branches and the supply. This resistance is mainly the resistance of the filter inductors, dependent on their q-factor.

Harmonic generating loads are nonlinear devices and consequently, the superposition principle cannot be used for the analysis of circuits with such loads. Such nonlinear devices are not operated in the wide range of the supply voltage, however, but at almost fixed voltage with an accuracy of only a few percent. At such a fixed voltage the load-generated distorting current $j(t)$ is also fixed. Therefore, it can be assumed that around the working point, specified by the voltage and current fundamental harmonic, the load is linear, so the superposition principle can be applied for the circuit analysis.

The load voltage $u(t)$ and the supply current $i(t)$ in a system with RHF can be expressed in terms of four transmittances, specified in terms of the Fourier transform of voltages and currents, that is they are functions of frequency, namely

$$A(j\omega) \stackrel{\text{df}}{=} \left.\frac{U(j\omega)}{E(j\omega)}\right|_{j(t)\equiv 0} = \frac{Z'_F(j\omega)}{Z_s(j\omega) + Z'_F(j\omega)} \tag{12.7}$$

$$B(j\omega) \overset{\text{df}}{=} \left.\frac{I(j\omega)}{J(j\omega)}\right|_{e(t)\equiv 0} = \frac{Z'_{\text{F}}(j\omega)}{Z_{\text{s}}(j\omega)+Z'_{\text{F}}(j\omega)}. \tag{12.8}$$

Observe that transmittances $A(j\omega)$ and $B(j\omega)$, physically different, are mathematically identical.

$$Y_{\text{x}}(j\omega) \overset{\text{df}}{=} \left.\frac{I(j\omega)}{E(j\omega)}\right|_{j(t)\equiv 0} = \frac{1}{Z_{\text{s}}(j\omega)+Z'_{\text{F}}(j\omega)} \tag{12.9}$$

$$Z_{\text{y}}(j\omega) \overset{\text{df}}{=} \left.\frac{U(j\omega)}{J(j\omega)}\right|_{e(t)\equiv 0} = \frac{Z_{\text{s}}(j\omega)Z'_{\text{F}}(j\omega)}{Z_{\text{s}}(j\omega)+Z'_{\text{F}}(j\omega)} \tag{12.10}$$

These transmittances can be easily found by modeling using, for example, PSpcie program.

A frequency scan at $E(j\omega)$ = 1V and $J(j\omega)$ = 0 provides directly transmittances $A(j\omega) = U(j\omega)$ and $Y_{\text{x}}(j\omega) = I(j\omega)$. A similar scan at $J(j\omega)$ = 1A and $E(j\omega) = 0$ provides directly transmittances $B(j\omega) = I(j\omega)$ and $Z_{\text{y}}(j\omega) = U(j\omega)$.

These transmittances can be approximated in the range of frequency $\omega >> \omega_1$ by

$$A(j\omega) = B(j\omega) = \frac{jX'_{\text{F}}(j\omega)}{j\omega L_{\text{s}} + jX'_{\text{F}}(j\omega)} \tag{12.11}$$

$$Y_{\text{x}}(j\omega) = \frac{1}{j\omega L_{\text{s}} + jX'_{\text{F}}(j\omega)} \tag{12.12}$$

$$Z_{\text{y}}(j\omega) = -\frac{\omega L_{\text{s}}\, X'_{\text{F}}(j\omega)}{j\omega L_{\text{s}} + jX'_{\text{F}}(j\omega)}. \tag{12.13}$$

All of them have the same denominator

$$D(j\omega) = j\omega L_{\text{s}} + jX'_{\text{F}}(j\omega) = Z_{\text{x}}(j\omega) \tag{12.14}$$

therefore ZEROs of the function $D(j\omega)$ specify frequencies at which all transmittances specified by formulae (12.11) to (12.13) approach infinity. These are POLEs of these transmittances. POLEs are frequencies of the voltage (series) resonances of the filter with the internal equivalent inductance L_{s} of the supply system; ZEROs are frequencies of the current (parallel) resonances of the filter with inductance L_{s}. Observe that POLEs p_k and ZEROs z_k are different from tuning frequencies t_{k}.

Transmittances $A(j\omega)$, $B(j\omega)$, $Y_{\text{x}}(j\omega)$, and $Z_{\text{y}}(j\omega)$ are functions of a continuous variable, meaning functions of frequency ω, while the frequency of harmonics is a discrete variable, equal to $\omega_n = n\omega_1$. Therefore relations between the load voltage and supply current harmonics, u_n and i_n, and sources of these harmonics, distribution voltage harmonics e_n, and the load-generated current harmonics j_n are specified by the magnitude of these transmittances at harmonic frequencies $n\omega_1$, namely

$$A_n \overset{\text{df}}{=} \left.\frac{U_n}{E_n}\right|_{j(t)\equiv 0} = |A(jn\omega_1)| \tag{12.15}$$

$$B_n \overset{\text{df}}{=} \frac{I_n}{J_n}\bigg|_{e(t)\equiv 0} = |B(jn\omega_1)| \tag{12.16}$$

$$Y_{xn} \overset{\text{df}}{=} \frac{I_n}{E_n}\bigg|_{j(t)\equiv 0} = |Y_x(jn\omega_1)| \tag{12.17}$$

$$Z_{yn} \overset{\text{df}}{=} \frac{U_n}{J_n}\bigg|_{e(t)\equiv 0} = |Z_y(jn\omega_1)|. \tag{12.18}$$

Illustration 12.1 *Let us design an RHF of the fifth, seventh, eleventh, and thirteenth order harmonics for a load of the active power per phase* $P = 150$ kW *and the power factor for the fundamental frequency* $\lambda_{L1} = 0.71$. *The load is supplied with the voltage of the rms value* $U_1 = 480$ V, *from a distribution system bus of the short-circuit power per phase* $S_{sc} = 30\,P$ *and the reactance-to-resistance ratio* $\xi = 5$. *Let us assume that the filter is designed as a type A filter with branches detuned from harmonic frequencies by 5%.*

At such detuning from harmonic frequency, resonant frequencies of the filter branches are $t_5 = 4.75\omega_1$, $t_7 = 6.65\omega_1$, $t_{11} = 10.45\omega_1$, $t_{13} = 12.35\omega_1$. *Since the reactive power of the fundamental harmonic* $Q_1 \approx Q = 150$ *kvar and for the* type A *filter the allocation coefficients are equal to* $a_5 = a_7 = a_{11} = a_{13} = 0.25$, *then, assuming that the fundamental frequency is normalized to* $\omega_1 = 1$ rad/s, *formulae* (13.1–2) *result in the filter parameters:*

$$C_5 = 0.154 \text{ F}, \quad L_5 = 0.287 \text{ H}, \quad C_7 = 0.158 \text{ F}, \quad L_7 = 0.143 \text{ H},$$
$$C_{11} = 0.160 \text{ F}, \quad L_{11} = 0.0572 \text{ H}, \quad C_{13} = 0.161 \text{ F}, \quad L_{13} = 0.0408 \text{ H},$$

Assuming that inductors' q-factor, $\omega_1 L_k/R_k = 50$, *the magnitude of transmittances* $A(j\omega)$ *and* $B(j\omega)$ *changes as shown in Fig. 12.11. Transmittances' values at ZEROs and POLEs change in a very wide range. Therefore, their plot in Fig. 12.11 was drawn on a logarithmic scale.*

*****

Minima of magnitudes $|A(j\omega)|$ and $|B(j\omega)|$ are at ZEROs z_k of these transmittances. The maxima, at POLEs p_k, are at frequencies located below ZEROs z_k and are shifted down from harmonic frequencies ω_k, $k = 5, 7, 11, 13$, but at the cost of an increase in the value of transmittances, A_k and B_k, for these harmonics. It is worth comparing these transmittances for a filter with not detuned branches. They are shown in Fig. 12.12.

This comparison shows that the effect of detuning on transmittances for other harmonics other than the 5th, 7th, 11th, and 13th orders, is mixed. Some are reduced, some are elevated. However, detuning increases, of course, transmittances for the fifth, seventh, eleventh, and thirteenth order harmonics. The plot of the magnitude of the relative admittance $Y_x(j\omega)$, meaning related to its value for the fundamental harmonic,

$$\Psi_x(j\omega) = \frac{Y_x(j\omega)}{Y_x(j\omega_1)},$$

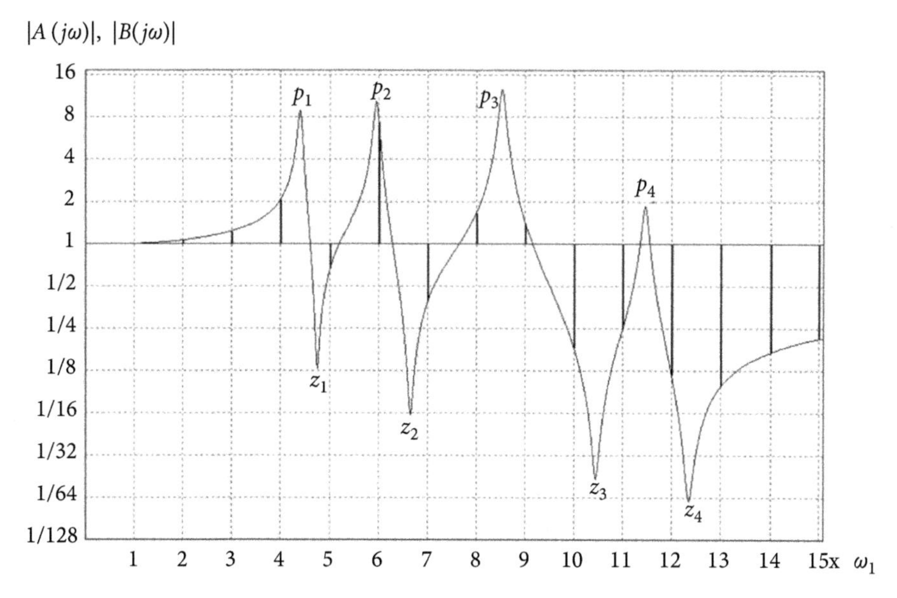

Figure 12.11 The magnitude of transmittances $A(j\omega)$ and $B(j\omega)$ for *the type A* filter detuned by 5%.

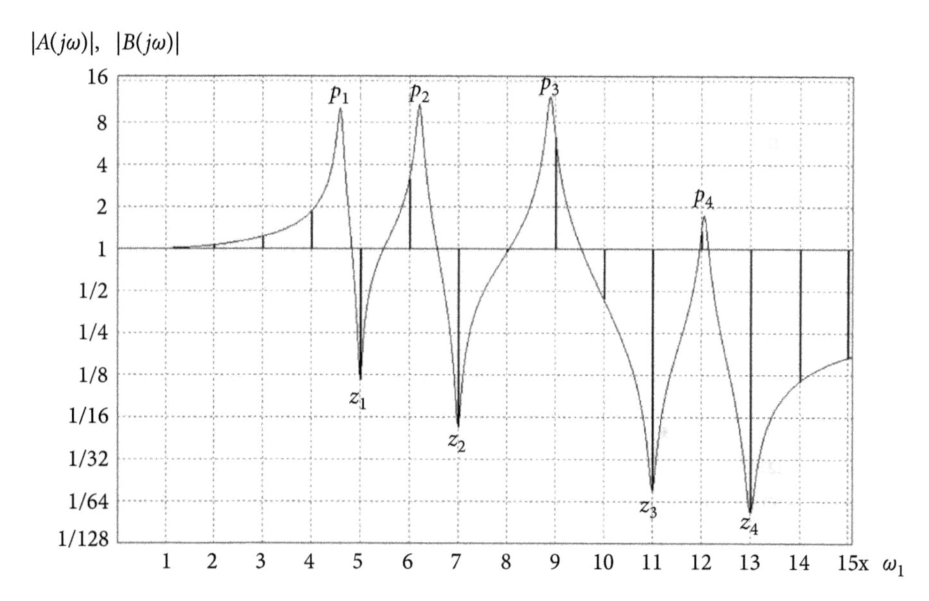

Figure 12.12 The magnitude of transmittances $A(j\omega)$ and $B(j\omega)$ for not detuned *the type A* filter.

for a not detuned, type A filter, is shown in Fig. 12.13. This plot demonstrates why the filter is so sensitive to the distribution system voltage harmonics. At such values of admittances for harmonics, even very low harmonic contents can cause very high distortion of the supply current, damaging the filter effectiveness.

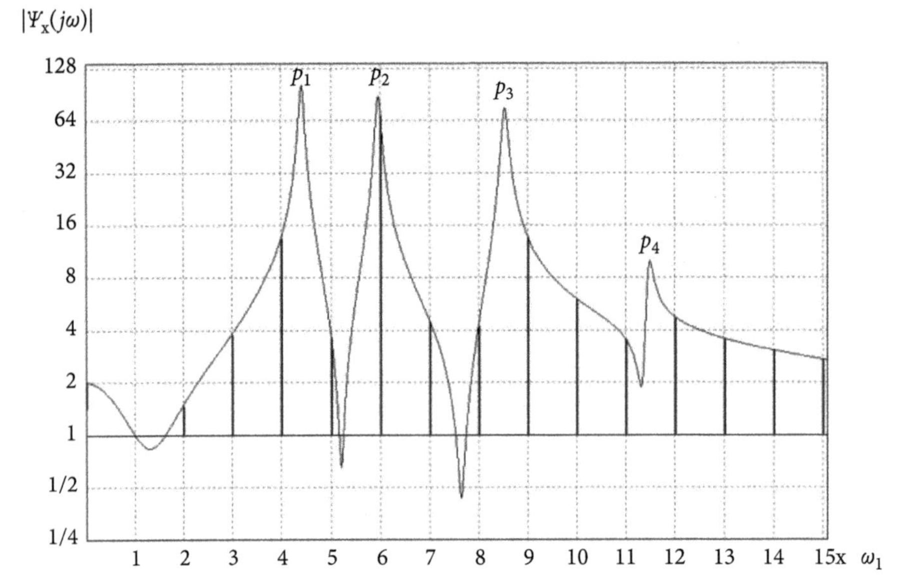

Figure 12.13 The magnitude of relative admittance $\psi_x(j\omega)$ for *type A* not detuned filter.

Detuning reduces the admittance of the filter for harmonics of the fifth, seventh, eleventh, and thirteenth order, which usually dominate the supply voltage in three-phase systems. This admittance for almost all harmonics is still more than four times higher than for the fundamental harmonic. Just the frequency features of this admittance are the main cause of a drastic decline in the filter effectiveness with the increase of harmonic distortion in the distribution voltage.

The load-generated harmonics $j_n(t)$ cause the load voltage $u(t)$ distortion. The effect of this generated current upon the voltage is specified by impedance $Z_y(j\omega)$, as seen from the current source j. The relative value of this impedance, meaning referenced to the load impedance for the fundamental harmonic,

$$\zeta_y(j\omega) = \frac{Z_y(j\omega)}{Z_{F'}\,(j\omega_1)}$$

for a detuned, the type A filter is shown in Fig. 12.14. This impedance value for almost all harmonics is well below unity, thus the voltage response to the load-generated harmonics is in common conditions, rather moderate. Nonetheless, if the load-generated current harmonic of the 5th order has the rms value, for example, *20%* of the fundamental, then at the impedance ζ_{y5} equal to, approximately, *1/10*, then the load voltage harmonic rms value U_5 would be of the order of *2%*, of the voltage fundamental rms value. Observing the plot in Fig. 12.14, we can conclude that detuning resonant branches from harmonic frequencies elevates the load voltage response to the load-generated current harmonics of the 5th, 7th, 11th, and 13th order.

The filter in Illustration 12.1 was designed according to the recommendation [45] that individual branches of the filter should compensate the same reactive power,

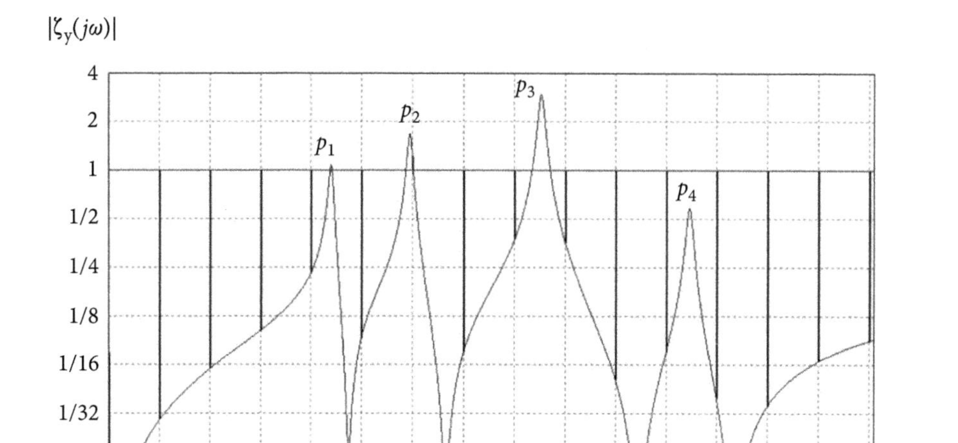

Figure 12.14 The magnitude of relative impedance $\zeta_y(j\omega)$ for the *type A* filter detuned by 5%.

meaning, as a type A filter. Other recommendations, such as, for example, presented in [142], say that the reactive power compensated by a branch that reduces a particular harmonic should be proportional to the contents of that harmonic in the load-generated current. A filter designed according to this recommendation is referred to as a *type B* filter. Coefficients a_k of the reactive power allocation to branch k for the *type B* filters can be calculated as follows. The reactive power is compensated entirely if

$$\sum_{k \in N} a_k = 1,$$

thus, the allocation coefficients should be equal to

$$a_k = \frac{J_k}{\sum\limits_{k \in N} J_k}.$$

Illustration 12.2 *Let us repeat the design of a resonant harmonic filter for the same conditions as in Illustration 12.1, assuming that the load generates current harmonics of the fifth, seventh, eleventh, and thirteenth order of the relative rms value* $J_5/I_1 = 16\%$, $J_7/I_1 = 13\%$, $J_{11}/I_1 = 7\%$, *and* $J_{13}/I_1 = 5\%$. *For such harmonics, the allocation coefficients should be equal to* $a_5 = 0.39$, $a_7 = 0.32$, $a_{11} = 0.17$, $a_{13} = 0.12$. *The plot of magnitude transmittances* $|A(j\omega)|$ *and* $|B(j\omega)|$ *is shown in Fig. 12.15.*

When this plot is compared with that in Fig. 12.11 for the type A filter, we can observe that transmittances for the fifth and seventh order harmonics of the type B filter have lower values than those for the type A filter. Although these transmittances for the eleventh and thirteenth order harmonics remain unchanged, the type B filter in the assumed situation is more efficient in the reduction of the fifth and the seventh order harmonics than the type A filter. Observe, however, that transmittances for the third, fourth, and the ninth order harmonics of the type B filter have higher values than those of the type A filter. Thus, the filter performance is more sensitive to other harmonics than of the fifth, seventh, eleventh, and thirteenth order.

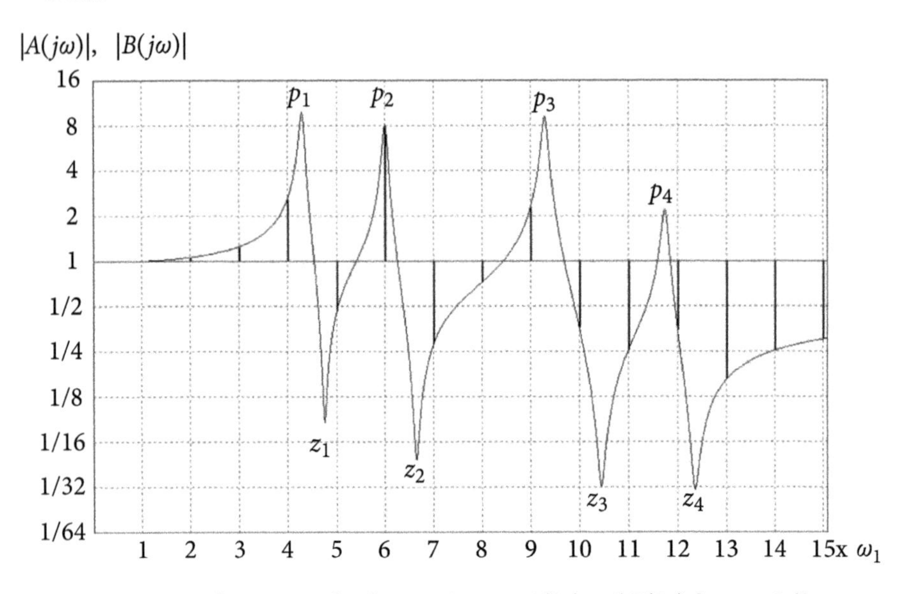

Figure 12.15 The magnitude of transmittances $A(j\omega)$ and $B(j\omega)$ for *type B* filter detuned by 5%.

The plot of the magnitude of the relative admittance $\psi_x(j\omega)$ for this filter is shown in Fig. 12.16.

When this admittance is compared with that in Fig. 12.12 for the type A filter, an increase in the values of ψ_{x5} and ψ_{x7} can be observed. It means that the filter is more sensitive to the fifth and seventh order harmonics in the distribution voltage than the type A filter. Because of an increase of A_n and B_n transmittances, it is also more sensitive to the third, fourth, and the ninth order harmonics in the distribution voltage and the load current.

*****

These two illustrations show that allocation of the reactive power to resonant branches in a degree proportional to the harmonic contents has mixed effects. This is because such general recommendations on the reactive power allocation and detuning do not provide a clue about how POLEs of the filter will be affected. Their

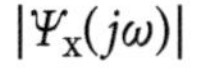

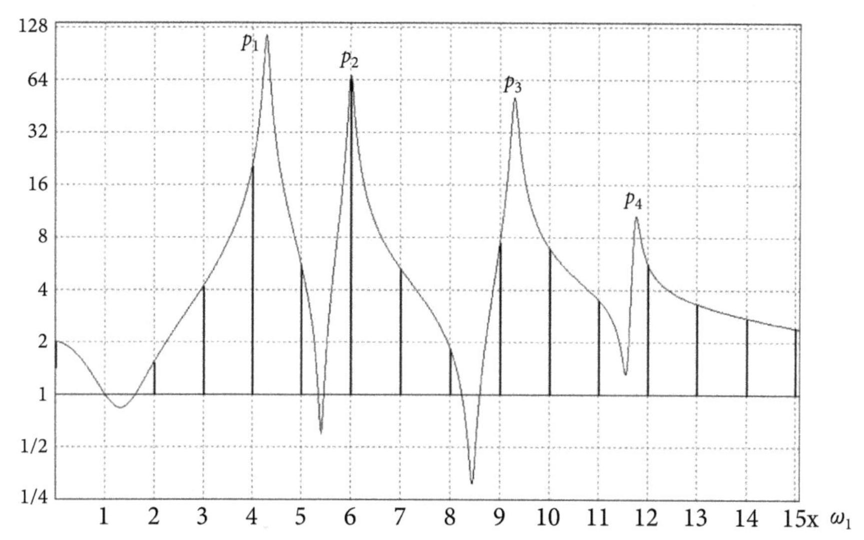

Figure 12.16 The magnitude of admittance $\psi_x(j\omega)$ for the *type B* filter detuned by 5%.

locations, meaning resonant frequencies of the filter with the distribution system reactance, have crucial importance for the filter performance. Unfortunately, independently of particular recommendations, the POLEs location does not depend on the single branch parameters, but it is specified by all of them and by the distribution system equivalent inductance. Only tuning frequencies and reactive power allocation coefficients are specified in conventional procedures of the resonant filter design. It is possible, however, to design a filter, at the cost of an additional line inductor, with not only tuning frequencies but also POLEs located at the designer's discretion. This means that it is possible to control frequencies at which all transmittances have maxima. Such a filter is referred to as a *fixed POLEs filter*.

12.4 Fixed POLEs Filter Design

A resonant harmonic filter with K branches is composed of $2K$ reactance elements. Their values are calculated in the conventional process of filter design having selected K tuning frequencies t_k and K reactive power allocation coefficients a_k. Instead of these allocation coefficients, the POLEs location p_k can be selected by a designer, thus moving off maxima of transmittances from harmonic frequencies.

The design procedure will be explained here for a four-branch filter, assuming that the frequency of the voltage is normalized to $\omega_1 = 1$ rad/s. The filter and the load in the range of harmonic frequencies are regarded as a reactance one-port, shown in Fig. 12.17.

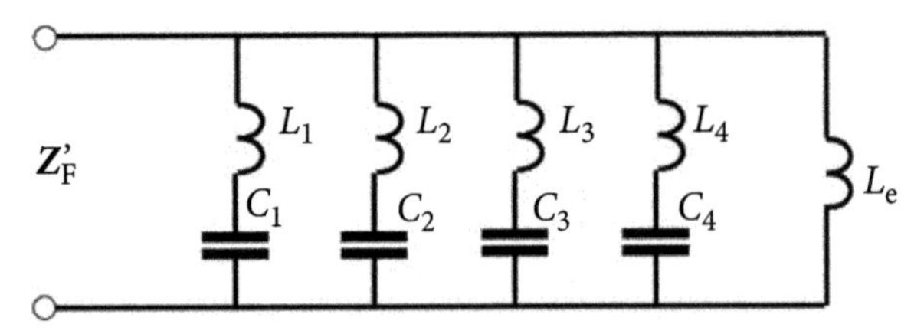

Figure 12.17 An equivalent circuit of the filter and the load at harmonic frequencies.

The impedance of this one-port has a ZERO at $\omega = 0$, a POLE when the frequency approaches infinity, and four ZEROs at tuning frequencies t_k of the filter. Observe, moreover, that at the entire compensation of the reactive power of the fundamental frequency, the one-port has to have a POLE at $\omega_1 = 1$ rad/s, since the reactive current of the fundamental frequency of a compensated load has to be equal to zero. Therefore, according to Chapter 10, the impedance of the one-port is a function of the complex variable $s = \sigma + j\omega$, of the form

$$Z'_{\rm F}(s) = s\frac{(s^2+t_1^2)(s^2+t_2^2)(s^2+t_3^2)(s^2+t_4^2)}{(s^2+1)(b_3s^6+b_3s^4+b_1s^2+b_0)}. \tag{12.19}$$

where b_k coefficients have to be real, positive numbers. It can be checked, that it has a ZERO for $s = 0$, and POLEs at infinity and $s = j1$.

As the complex frequency s approaches zero, the impedance of the one-port approaches the impedance of the inductor $L_{\rm e}$, that is,

$$Z'_{\rm F}(s) \underset{s\to 0}{\to} s\frac{t_1^2\, t_2^2\, t_3^2\, t_4^2}{b_0} = sL_{\rm e}. \tag{12.20}$$

Since the equivalent inductance of the load is known for a filter designer while the ZEROs t_k can be selected at his discretion, coefficient b_0 can be calculated as

$$b_0 = \frac{(t_1\ t_2\ t_3\ t_4)^2}{L_{\rm e}}. \tag{12.21}$$

The system POLEs are complex frequencies, $s = s_k$, where the impedance, as seen from the distribution system, is equal to zero, meaning

$$s_kL_{\rm s} + Z'_{\rm F}(s_k) = 0. \tag{12.22}$$

Since there has to be a POLE below each ZERO, this equation has to be fulfilled for four different POLEs, while the impedance specified by formula (12.19) has only three unknown parameters, b_3, b_2, and b_1. Such a system of equations cannot be solved. It can be solved only on the condition that one additional unknown is added to the set of equations. A line inductor of inductance $L_{\rm L}$, connected as shown in Fig. 12.18, could be such an unknown.

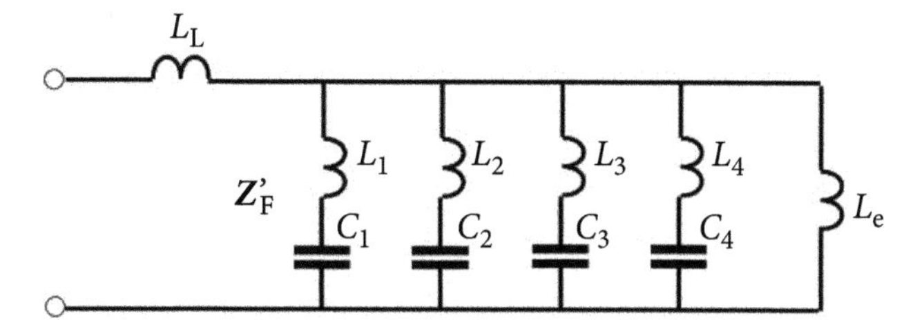

Figure 12.18 A four-branch filter with a line inductor.

It changes formula (13.22) to the form

$$s_k(L_s+L_L)+Z'_F(s_k) = 0. \tag{12.23}$$

Unfortunately, the line inductor affects the load voltage. It not only reduces the load voltage as compared to the distribution voltage but it makes the load voltage more dependent on the load power. From the load's perspective, it is simply supplied from a bus with reduced short-circuit power.

POLEs of reactance functions have to be located on the imaginary axis of the complex plane s, meaning $s_k = \pm jp_k$, thus

$$s_k^2 = -p_k^2.$$

Therefore, Eqn. (12.23) can be rewritten in the form

$$p_k(L_s + L_L) + p_k\frac{(t_1^2 - p_k^2)(t_2^2 - p_k^2)(t_3^2 - p_k^2)(t_4^2 - p_k^2)}{(1 - p_k^2)(-b_3p_k^6 + b_2p_k^4 - b_1 p_k^2 + b_0)} = 0. \tag{12.24}$$

There are four unknown parameters in this equation, namely the line inductance L_L and three parameters b_3, b_2, and b_1, while POLEs p_k can be selected at the filter designer's discretion.

Fixed POLEs filter design involves the search for the solution of a set of four equations. To avoid the necessity of handling very large numbers, it is important to normalize the system not only versus frequency, meaning assuming that the radial fundamental frequency is equal to $\omega_1 = 1$ rad/s, but also versus the impedance level. Such normalization is based on the assumption that the fundamental harmonic of the load voltage rms value is $U_N = 1$V and the load current fundamental harmonic rms value is $I_N = 1$ A, as shown in Fig. 12.19.

At such normalization, the load active power of the fundamental is equal to $P_N = \lambda_{L1}$, where λ_{L1} denotes the power factor for the fundamental frequency.

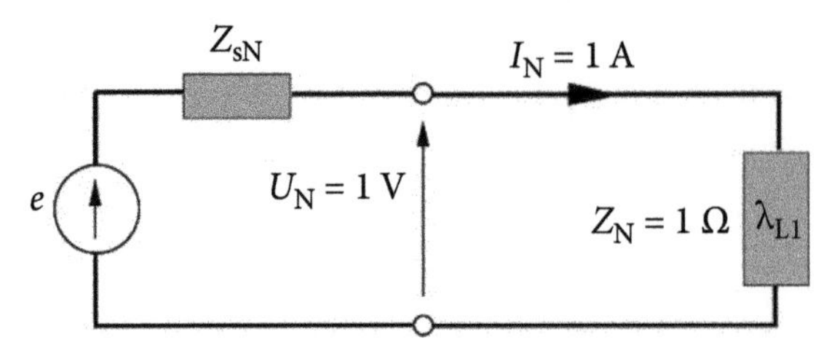

Figure 12.19 A normalized circuit.

When at the voltage rms value U, the load has the active power P, then the impedance of the normalized value Z_N can be denormalized with relation

$$Z = \frac{U^2\,\lambda_{L1}}{P} Z_N, \tag{12.25}$$

while the admittance can be denormalized with the formula

$$Y = \frac{P}{U^2\,\lambda_{L1}} Y_N. \tag{12.26}$$

Illustration 12.3 *Let us design a fixed POLEs filter for load which at voltage $U = 480$ V has the active power $P = 50$ kW and power factor $\lambda_{L1} = 0.71$. The load is supplied from a distribution bus with $S_{sc} = 50P$. The filter should be tuned to the fifth, seventh, eleventh, and thirteenth order harmonics and have POLEs at 3.7 rad/s, 5.5 rad/, 9.5 rad/s, and 12.5 rad/s.*

All calculations will be done for the system with the load impedance normalized to 1Ω, but to reduce symbols' complexity, the index "N" will be neglected.

The equivalent inductance of the load for the fundamental frequency L_e of the load can be calculated as follows:

$$\omega_1 L_e = Z \sin\varphi_{L1} = Z\sqrt{1-\lambda_{L1}^2} = 1\sqrt{1-0.71^2} = 0.71\ \Omega$$

hence L_e = 0.71 H. To calculate the equivalent inductance of the normalized distribution system, observe that

$$S_{sc} = \frac{E^2}{Z_s} \approx \frac{U^2}{Z_s}$$

thus neglecting the distribution system resistance,

$$\omega_1 L_s \approx \frac{U^2}{S_{sc}} = \frac{1}{50\,P} = \frac{1}{50\,\lambda_{L1}} = \frac{1}{50 \times 0.71} = 0.0282\ \Omega$$

and hence $L_s = 0.0282$ H. *The coefficient* b_0 *of the impedance is equal to*

$$b_0 = \frac{(z_1\, z_2\, z_3\, z_4)^2}{L_e} = \frac{(5 \times 7 \times 11 \times 13)^2}{0.71} = 17.7 \times 10^6.$$

With such parameters, Eqn. (13.24) for the POLE $p_1 = 3.7$ rad/s *has the form*

$$3.7(0.0282 + L_L) + 3.7\frac{(25 - 13.69)(49 - 13.69)(121 - 13.69)(169 - 13.69)}{(1 - 13.69)(-2566b_3 + 187.4\, b_2 - 13.69b_1 + 17.7 \times 10^6)} = 0$$

which can be simplified to

$$0.0282 + L_L + \frac{204.4}{b_3 - 0.073b_2 + 0.0053b_1 - 6905} = 0,$$

and next to

$$b_3 - 0.073b_2 + 0.0053b_1 - 6905 = -\frac{204.4}{0.0282 + L_L}.$$

If we denote

$$\frac{1}{0.0282 + L_L} \overset{\mathrm{df}}{=} x$$

then this equation can be simplified to a linear equation of the form

$$b_3 - 0.073b_2 + 0.0053b_1 + 204.4\,x = 6.905 \times 10^3.$$

Similarly, for the remaining POLEs,

$$b_3 - 0.0331\, b_2 + 0.0011\, b_1 - 1.53\, x = 640,$$

$$b_3 - 0.0111\, b_2 + 1.23 \times 10^{-4} b_1 + 0.0993\, x = 24,$$

$$b_3 - 0.0064\, b_2 + 4.1 \times 10^{-5}\, b_1 - 10.7 \times 10^{-3}\, x = 4.64.$$

This set of equations has the solution

$$b_3 = 47.2; \quad b_2 = 12.7 \times 10^3; \quad b_1 = 9.44 \times 10^5; \quad x = 13.4$$

and consequently, the filter impedance is equal to

$$Z'_F(s) = s\frac{(s^2 + 25)(s^2 + 49)(s^2 + 121)(s^2 + 169)}{(s^2 + 1)(47.2\, s^6 + 12.7 \times 10^3 s^4 + 9.44 \times 10^5 s^2 + 17.7 \times 10^6)}$$

while the line inductance should have the value

$$L_L = \frac{1}{x} - 0.0282 = \frac{1}{13.4} - 0.0282 = 0.0465\,\text{H}.$$

Before the filter parameters are calculated, it would be reasonable to verify whether or not the found impedance $Z_F'(s)$ satisfies Eqn. (13.24) at least in one POLEs, for example, $s = jp_2 = j5.5$ rad/s. Since

$$jp_2\,(L_s + L_L) + Z'_F\,(jp_2) = j\,5.5[(0.0282 + 0.0465)$$
$$+ \frac{(25 - 5.5^2)(49 - 5.5^2)(121 - 5.5^2)(169 - 5.5^2)}{(1 - 5.5^2)(-47.2 \times 5.5^6 + 12.7 \times 10^3 \times 5.5^4 + 9.44 \times 10^5 \times 5.5^2 + 17.7 \times 10^6)}$$
$$= j\,5.5 \times 2 \times 10^{-15} \approx 0$$

thus frequency p_2 is indeed the POLE of the system.

The obtained filter admittance

$$Y'_F\,(s) = \frac{(47.2\,s^6 + 12.7 \times 10^3 s^4 + 9.44 \times 10^5 s^2 + 17.7 \times 10^6)}{s(s^2 + 25)(s^2 + 49)(s^2 + 121)(s^2 + 169)(s^2 + 1)}$$

can be decomposed into fractions

$$Y'_F(s) = \frac{1}{sL_0} + \frac{1}{L_1}\frac{s}{s^2 + 25} + \frac{1}{L_2}\frac{s}{s^2 + 49} + \frac{1}{L_3}\frac{s}{s^2 + 121} + \frac{1}{L_4}\frac{s}{s^2 + 169}$$

with

$$\frac{1}{L_0} = \lim_{s \to 0}\,\{sY'_F(s)\} = \frac{17.7 \times 10^6}{25 \times 49 \times 121 \times 169} = \frac{b_0}{(z_1 z_2 z_3 z_4)^2} = \frac{1}{L_e}$$

$$\frac{1}{L_1} = \lim_{s \to -25}\left\{\frac{s^2 + 25}{s}Y'_F(s)\right\} =$$
$$= \frac{(1 - 25)(-47.2 \times 25^3 + 12.7 \times 10^3 \times 25^2 - 9.44 \times 10^5 \times 25 + 17.7 \times 10^6)}{-25(49 - 25)(121 - 25)(169 - 25)} = \frac{1}{0.269\text{ H}}.$$

Similarly,

$$\frac{1}{L_2} = \lim_{s \to -49}\left\{\frac{s^2 + 49}{s}Y'_F(s)\right\} = \frac{1}{0.0572\text{ H}}$$

$$\frac{1}{L_3} = \lim_{s \to -121}\left\{\frac{s^2 + 49}{s}Y'_F(s)\right\} = \frac{1}{0.0628\text{ H}}$$

$$\frac{1}{L_3} = \lim_{s \to -169}\left\{\frac{s^2 + 169}{s}Y'_F(s)\right\} = \frac{1}{0.107\text{ H}}.$$

To verify the correctness of the inductance calculation, let us verify whether the system has a POLE at p_2 frequency. Since

$$jp_2\,(L_s + L_L) + \frac{1}{Y'_F\,(jp_2)} = j\,5.5(0.0282 + 0.0465) +$$

$$+ \frac{1}{\dfrac{1}{j\,5.5\,L_0} + \dfrac{1}{L_1}\dfrac{j\,5.5}{25 - 5.5^2} + \dfrac{1}{L_2}\dfrac{j\,5.5}{49 - 5.5^2} + \dfrac{1}{L_3}\dfrac{j\,5.5}{121 - 5.5^2} + \dfrac{1}{L_4}\dfrac{j\,5.5}{169 - 5.5^2}} =$$

$$= -j\,1.01 \times 10^{-14} \approx 0$$

thus the system at inductances as calculated has a POLE, thus inductances of the filter were calculated correctly. Hence its capacitances, calculated from the formula

$$C_k = \frac{1}{t_k^2\,L_k},$$

are equal to

$$C_1 = 0.149\text{ F},\ C_2 = 0.356\text{ F},\ C_3 = 0.132\text{ F},\ C_4 = 0.055\text{ F}$$

The plot of the magnitude of the filter transmittances $A(j\omega)$ and $B(j\omega)$ is shown in Fig. 12.20.

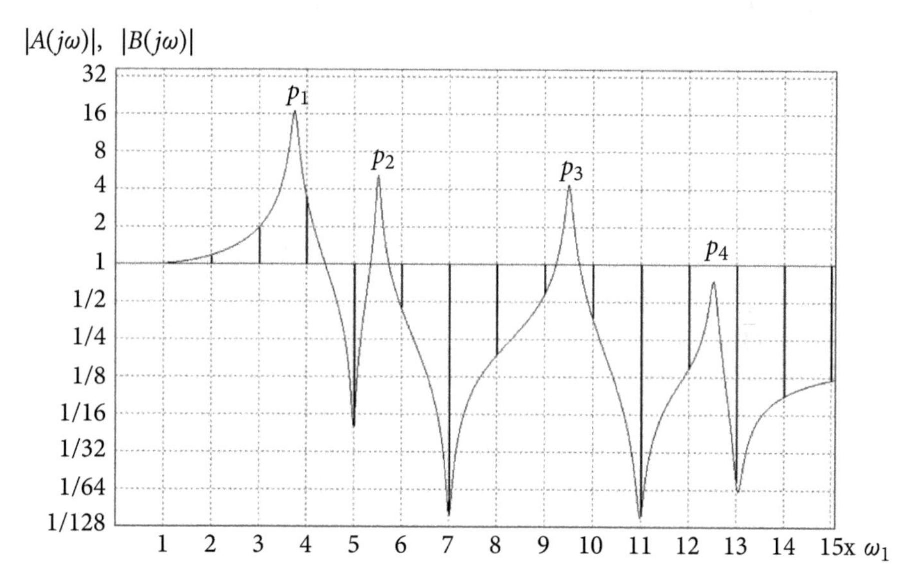

Figure 12.20 The magnitude of transmittances $A(j\omega)$ and $B(j\omega)$ for a fixed POLE filter.

It shows that only the third and fourth order harmonics of the distribution voltage and the load-generated current are amplified. All other harmonics are attenuated. The POLE p_1 in this illustration was selected at the frequency that was shifted *towards the fourth order harmonic since this harmonic is usually much lower than the third order harmonic.*

The plot of the magnitude of relative admittance as seen from the distribution is shown in Fig. 12.21.

*****

When it is compared with the plot in Fig. 12.21, then some reduction of admittances for higher-order harmonics can be observed, but admittances for the third and fourth order are higher.

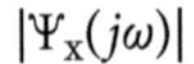

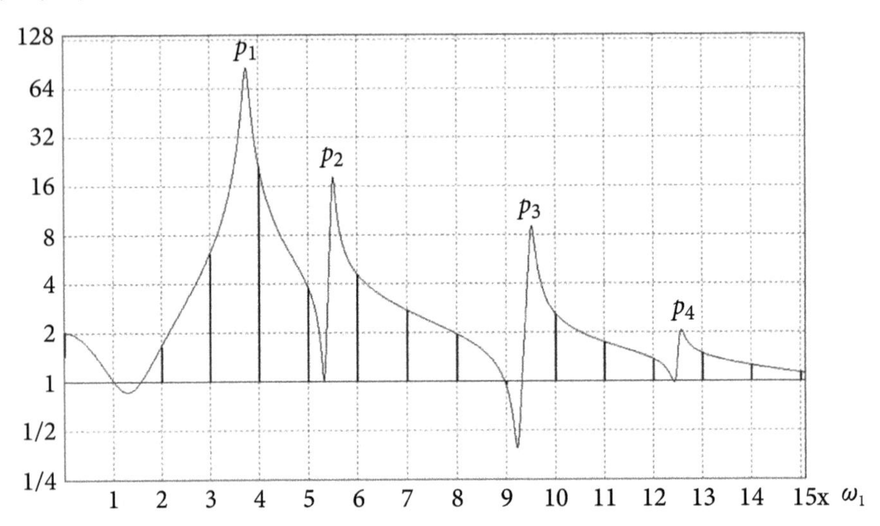

Figure 12.21 The magnitude of relative admittance $\psi(j\omega)$ for a fixed POLE filter.

Thus even if the POLEs location is set at the designer's discretion, the filter performance, though improved, remains sensitive to the distribution voltage harmonics.

The improvement of the filter performance is obtained, moreover, at the cost of the line inductor. There is not only some additional power loss in this inductor's resistance but first of all, it reduces the short-circuit power at the load terminal, meaning the voltage becomes more dependent on the load power variation. This reduction in the short-circuit power can be acceptable, however, for loads with fixed or low variation of power, since some loss of the load voltage can be eliminated by a higher tap setting of the supply transformer.

12.5 Filter Effectiveness

Traditional methods of RHFs design, as well as the design of fixed POLEs filters, are based to a large degree on engineering intuition and a set of recommended practices. Unfortunately, RHFs operate now in conditions where not only the load current but also supply voltage are distorted and harmonics of all orders are present in the supply voltage and the load current. The complexity of filter properties, conditions of their operation, and the effect of the distribution system parameters on their performance do not allow for the design of efficient filters based on only intuition. Optimization methods are necessary for that.

Optimization methods require first of all some measures of the filter efficiency, which can serve as optimization goals. Such measures can be defined as follows.

Resonant harmonic filters are installed to protect distribution systems against current harmonics generated in HGLs of rms value J_n, but at the same time, the internal voltage $e(t)$ of the system can contain harmonics of rms value E_n. These harmonics cause distortion of the supply current and the load voltage.

Let us decompose the supply current and the supply voltage of a harmonic generating load without any filter, as shown in Fig. 12.20, into the fundamental and distorting components

$$i_0(t) = i_{10}(t) + i_{d0}(t)$$

$$u_0(t) = u_{10}(t) + u_{d0}(t).$$

Index "0" emphasizes the lack of the filter.

Harmonic distortion of the supply current before an RHF is installed can be expressed as the ratio of the rms values of the distorting component and the fundamental, namely as

$$\delta_{i0} \stackrel{\text{df}}{=} \frac{||i_{d0}||}{||i_{10}||} = \frac{||i_{d0}||}{I_{10}}. \tag{12.27}$$

The supply current can be distorted because of the internal voltage $e(t)$ distortion and because of the load-generated current $j(t)$. Therefore it can be decomposed into components dependent on $e(t)$ and $j(t)$, namely

$$i_{d0} = i_{d0}(e) + i_{d0}(j). \tag{12.28}$$

The rms value of these components is equal to

$$||i_{d0}(e)|| = \sqrt{\sum_{n\in N} (Y_{xn}E_n)^2}$$

$$||i_{d0}(j)|| = \sqrt{\sum_{n\in N} (B_nJ_n)^2}$$

where transmittances Y_{xn} and B_n, as defined by formulae (12.17) and (12.16), are specified for the system without RHF, that is, in the equivalent circuit as shown in Fig. 12.22.

The current components in decomposition (12.28) can contain harmonics of the same order n. One could conclude, therefore, that they are not mutually orthogonal.

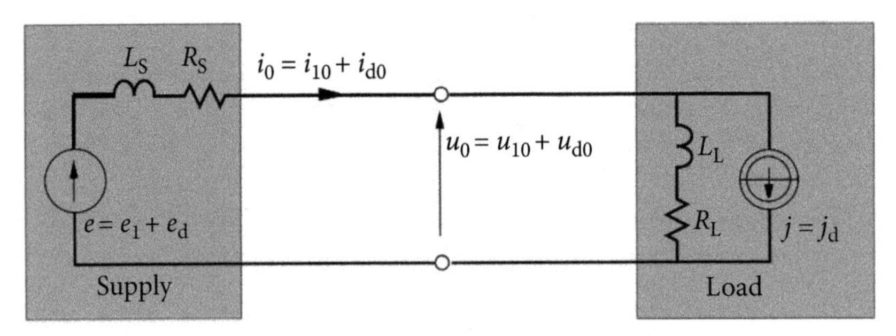

Figure 12.22 An equivalent circuit of a supply source and an HGL.

There is not any mutual relationship between voltage $e(t)$ and current $j(t)$ harmonics, however. Therefore, they can be regarded as mutually random. Random processes are mutually orthogonal, however. Therefore, the rms value of the total distorting current $i_{d0}(t)$ can be calculated as

$$||i_{d0}|| = \sqrt{||i_{d0}(e)||^2 + ||i_{d0}(j)||^2}. \tag{12.29}$$

Similarly, the load voltage distortion in the system without RHF can be defined as

$$\delta_{u0} \stackrel{df}{=} \frac{||u_{d0}||}{||u_{10}||} = \frac{||u_{d0}||}{U_{10}}. \tag{12.30}$$

The supply voltage can be distorted because of the internal voltage $e(t)$ distortion and because of the load-generated current $j(t)$. Therefore, it can be decomposed into components dependent on $e(t)$ and $j(t)$, namely

$$u_{d0} = u_{d0}(e) + u_{d0}(j). \tag{12.31}$$

The rms value of these components is equal to

$$||u_{d0}(e)|| = \sqrt{\sum_{n \in N} (A_n E_n)^2}$$

$$||u_{d0}(j)|| = \sqrt{\sum_{n \in N} (Z_{yn} J_n)^2}$$

where transmittances A_n and Z_{yn}, as defined by formulae (12.15) and (12.18), are specified for the system without RHF, that is, in the equivalent circuit as shown in Fig. 12.22. The rms value of the total distorting voltage $u_{d0}(t)$ can be calculated as

$$||u_{d0}|| = \sqrt{||u_{d0}(e)||^2 + ||u_{d0}(j)||^2}. \tag{12.32}$$

The most common application of RHFs is the reduction of harmonic distortion caused by three-phase rectifiers and AC/DC controlled converters. Such devices generate current harmonics of the order fifth, seventh, eleventh, and thirteenth ..., in general, of the order $n = 6k \pm 1$. They cause the supply current and the voltage distortion and this distortion is dependent on the short-circuit to active power, S_{sc}/P, ratio, as well as the reactance to resistance, X_s/R_s, ratio of the distribution system impedance.

Illustration 12.4 *Let us assume that a three-phase AC/DC converter operating at power factor for the fundamental $\lambda_{L1} = 0.8$, generates current harmonics of the relative rms value*

$$J_5/I_1 = 16\%, \quad J_7/I_1 = 11.4\%, \quad J_{11}/I_1 = 7.3\%, \quad J_{13}/I_1 = 6.1\%$$

The reactance to resistance ratio of the supply source $X_s/R_s = 5$. The values of the voltage and current distortion, specified in percent, for a few different values of the S_{sc}/P ratio, are tabulated in Table 12.1.

Table 12.1 The supply voltage and current distortion coefficients δ_{u0} and δ_{i0}.

S_{sc}/P	-	20	25	30	35	40	45	50
δ_{u0}	%	**8.4**	**6.9**	**5.9**	**5.1**	**4.5**	**4.1**	**3.7**
δ_{i0}	%	**19.8**	**20.2**	**20.4**	**20.6**	**20.8**	**20.9**	**21.0**

Table 12.1 shows that distortion of the supply voltage, especially at a lower level of the short-circuit power, needs that some measure for its reduction. There are two options: (i) increase in the short-circuit power of the supply, which means retrofitting the supply transformer with a transformer with higher power ratings; or (ii) installation of a harmonic filter. In particular, it could be an RHF.

Installation of a RHF, as shown in Fig. 12.23, changes the supply voltage and current to

$$i(t) = i_1(t) + i_d(t)$$

$$u(t) = u_1(t) + u_d(t)$$

and distortion coefficients to

$$\delta_i \stackrel{df}{=} \frac{||i_d||}{I_1}, \quad \delta_u \stackrel{df}{=} \frac{||u_d||}{U_1}. \tag{12.33}$$

The filter's effectiveness in the reduction of distortion of the supply current can be defined as

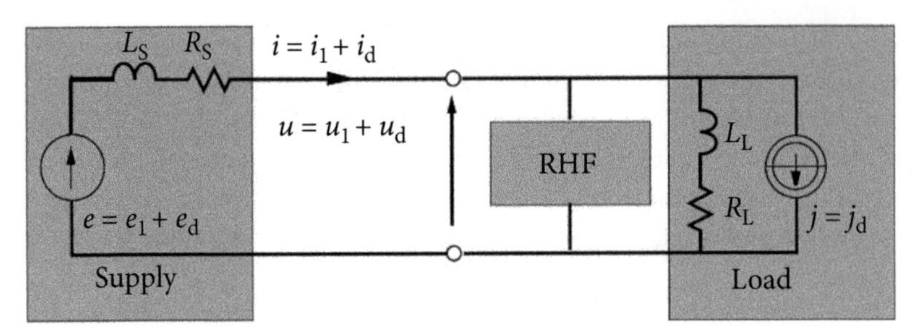

Figure 12.23 An equivalent circuit of a supply source, HGL with RHF.

$$\varepsilon_i \stackrel{\text{df}}{=} 1 - \frac{\delta_i}{\delta_{i0}}, \quad \varepsilon_u \stackrel{\text{df}}{=} 1 - \frac{\delta_u}{\delta_{u0}} \tag{12.34}$$

Observe that when a filter increases the total harmonic distortion, δ_I, then the filter effectiveness is defined in such a way becomes negative. These two measures of filter effectiveness, ε_I and ε_u, provide a measure for a comparison of filter effectiveness.

Illustration 12.5 *A resonant harmonic filter designed according to recommendations presented in [45], meaning a type A filter, and according to recommendation presented in [142], meaning a type B filter, for harmonics order fifth, seventh, eleventh, and thirteenth, detuned 12 Hz below harmonic frequencies $n\omega_1$, and built of reactors with q-factor equal to $q = 40$, has efficiency, depending on the S_{sc}/P ratio, tabulated in Table 12.2.*

Allocation coefficients for the type B filter were: $a_5 = 0.39$, $a_7 = 0.28$, $a_{11} = 0.18$, $a_{13} = 0.15$.

Effectiveness of both filters are tabulated for comparison in Table 12.3.

Table 12.2 Harmonic distortion of the supply voltage and current in a system with RHF.

				Type A				
S_{sc}/P	-	20	25	30	35	40	45	50
δ_u	%	2.0	1.8	1.7	1.6	1.5	1.4	1.3
δ_i	%	7.8	9.0	9.7	10.8	11.5	12.2	12.8
				Type B				
S_{sc}/P	-	20	25	30	35	40	45	50
δ_u	%	1.3	1.2	1.2	1.1	1.1	1.0	1.0
δ_i	%	5.0	6.0	6.8	7.5	8.2	8.9	9.5

One could observe that the type B filter is more effective than the type A filter. Both filters are more efficient at low short-circuit powers of the supply source than at high short-circuit powers. Moreover, they are more efficient in reducing the supply voltage distortion than the distortion of the supply current.

Table 12.3 Upper limit of the effectiveness of type A and type B filters.

Filter	S_{sc}/P	20	25	30	35	40	45
A	ε_u	**0.76**	**0.74**	**0.71**	**0.69**	**0.67**	**0.66**
	ε_i	**0.61**	**0.55**	**0.52**	**0.48**	**0.45**	**0.42**
B	ε_u	**0.85**	**0.83**	**0.80**	**0.78**	**0.76**	**0.74**
	ε_i	**0.75**	**0.70**	**0.67**	**0.64**	**0.61**	**0.57**

Filters' effectiveness compiled in Table 12.3 is regarded as an upper limit of this effectiveness because they were calculated at the assumption that filters operate at sinusoidal supply voltage and no harmonics are generated by the converter other than only the characteristic ones.

The effect of non-characteristic load-originated current harmonics on filter effectiveness was presented in Table 12.4. It was assumed that apart from characteristic harmonics there are also remaining ones, from the second to twelfth order of the same value. It was assumed that distortion caused by non-characteristic harmonics is 1%, which means that each of them has the rms value $J_n = 0.35\%$ of I_1.

Table 12.4 The effectiveness of type A and type B filters at 1% of the load current distortion by non-characteristic current harmonics.

Filter	S_{sc}/P	20	25	30	35	40	45
A	ε_u	**0.75**	**0.72**	**0.69**	**0.63**	**0.64**	**0.61**
	ε_i	**0.56**	**0.54**	**0.49**	**0.39**	**0.42**	**0.39**
B	ε_u	**0.57**	**0.65**	**0.71**	**0.67**	**0.73**	**0.73**
	ε_i	**0.51**	**0.53**	**0.61**	**0.61**	**0.59**	**0.56**

These data show that the effectiveness of the type B filter at low short-circuit power is more sensitive to non-characteristic harmonics than the type A filter, but for higher power it is quite the opposite.

Data on the filters' effectiveness compiled in Table 12.4 were obtained under the assumption that the internal voltage e(t) of the distribution system is sinusoidal. To conclude how the distortion of this voltage affects this efficiency, let us assume that apart from distortion caused by non-characteristic current harmonics $\delta_j = 1\%$, *also the internal voltage is distorted. Let us assume that this distortion is* $\delta_e = 2.5\%$. *Let us also assume, following the IEEE Standard 519, that 25% of this distortion is due to the presence of even order harmonics. It is assumed moreover, that the voltage harmonics rms value declines with the order, proportionally to* $1/n$.

The relative rms values of the voltage harmonics, calculated at such assumptions, are compiled in Table 12.5.

Table 12.5 The distribution voltage harmonics at $\delta_e = 2.5\%$.

Odd:	E_3	E_5	E_7	E_9	E_{11}	E_{13}
%	**1.80**	**1.10**	**0.77**	**0.60**	**0.50**	**0.42**
Even:	E_2	E_4	E_6	E_8	E_{10}	E_{12}
%	**0.50**	**0.32**	**0.17**	**0.12**	**0.10**	**0.07**

The effectiveness of filters of type A and type B, installed in a system with such voltage distortion are compiled in Table 12.6.

Table 12.6 The effectiveness of type A and type B filters at internal voltage distortion $\delta_e =$ 2.5% and 1% of the current distortion by non-characteristic harmonics.

Filter	S_{sc}/P	20	25	30	35	40	45
A	ε_u	**0.46**	**0.51**	**0.37**	**0.16**	**0.16**	**−0.32**
	ε_i	**0.11**	**0.33**	**0.23**	**−0.10**	**0.02**	**−0.47**
B	ε_u	**−0.26**	**0.20**	**0.36**	**0.37**	**0.33**	**0.32**
	ε_i	**−0.50**	**−0.03**	**0.29**	**0.36**	**0.36**	**0.34**

Data compiled in Table 12.6 show that even at a moderate level of voltage distortion the efficiency of RHFs, designed according to recommended practices, can substantially decline. At some level of the short-circuit to the active power ratio, this efficiency can be negative. The RHF can increase the voltage and/or the current distortion, instead of reducing it.

12.6 Optimized RHFs

Common methods of resonant harmonic filter design are based on the designer's decision on tuning frequencies of the filter and the amount of reactive power compensated by individual branches of the filter. Information on the contents of harmonics is essentially beyond the design process. Having information on which harmonics dominate in the load current, the designer makes the only decision on which of them should be reduced by the filter. The effect of such decisions on other harmonics is essentially beyond the designer's control. Consequently, the filter effectiveness in harmonics reduction cannot be predicted. It is an outcome of the entire design process. It depends, moreover, on the harmonic spectra both of load-generated current, j, and the distribution voltage, e, and can be far away from the effectiveness possible to achieve.

It seems that the interaction of RHF frequency properties with the load and distribution system-originated harmonics is too convoluted to successfully design an efficient filter based only on engineering intuition or recommendations. Optimization methods are needed for that. There is a great deal of variety of optimization methods that might be used for the RHF design and their selection is out of the scope of this chapter. Only results obtained using one of them, without going into details, are presented as an illustration that optimization enables finding parameters of a much more efficient filter than those calculated based on recommended practices.

Calculation of the filter parameters with the highest possible effectiveness requires that an optimization procedure is used in the design process. The objective of the optimization is not unique, however. Resonant harmonic filters affect the load voltage distortion in a different way than the supply current distortion. Filters parameters needed to obtain maximum effectiveness in the reduction of the current distortion, $\varepsilon_{\rm i}$, are different from those needed to obtain maximum effectiveness in the reduction of the voltage distortion, $\varepsilon_{\rm u}$. Which of these two is more important depends on the situation. When there are a lot of voltage harmonics-sensitive equipment supplied from the bus where the filter is to be installed, the filter should be more effective in the reduction of voltage harmonics. In the absence of such equipment, of more importance is the reduction of the supply current distortion. In general, the filter should maximize the effectiveness defined as a linear form of both of them. Optimization requires that its goal is defined. Let minimization of a general distortion

$$\delta = W_{\rm c}\,\delta_{\rm i} + W_{\rm v}\,\delta_{\rm u} \tag{12.35}$$

which is a linear form of the supply current and the load voltage distortions, with weight coefficients $W_{\rm c}$ and $W_{\rm v}$, selected at a designer's discretion, is such a goal.

More detailed measures could be based on effectiveness in the reduction of particular sets of harmonics, for example, such as odd or even order harmonics. Reduction of low- or high-order harmonics could be in some cases also a subject of interest. The cost of the compensator could be included in the optimization goal as well.

The generalized distortion of the filter, at fixed values of the load and the source parameters, as well as the load-generated harmonics J_n and internal voltage harmonics E_n, depends on the selection of reactive power allocation coefficients a_k and tuning frequencies $t_{\rm k}$.

For a filter of the fifth, seventh, eleventh, and thirteenth order harmonics, this generalized distortion is a function of eight parameters, namely

$$\delta = f(a_5, \ldots a_{13}, t_5, \ldots t_{13}) \tag{12.36}$$

At the assumption that $W_{\rm c} = W_{\rm v} = 0.5$, an optimization procedure resulted in an RHF which for internal voltage distortion $\delta_e = 2.5\%$ and 1% of the current distortion by non-characteristic harmonics has efficiencies compiled in Table 12.7.

The comparison of the filter efficiencies compiled in Table 12.7 with those compiled in Table 12.6 for a filter operating in identical situations shows that RHFs have

Table 12.7 The optimized filter effectiveness at internal voltage distortion $\delta_e = 2.5\%$ and 1% of the current distortion by non-characteristic harmonics.

S_{sc}/P	-	20	25	30	35	40	45
ε_u	%	**0.67**	**0.62**	**0.58**	**0.53**	**0.49**	**0.44**
ε_i	%	**0.57**	**0.54**	**0.52**	**0.48**	**0.44**	**0.39**

real potential for effective reduction of harmonic distortion, on the condition that optimization methods are used in their design. It is important to observe that allocation coefficients and tuning frequencies obtained in the process of optimization have been known to be very distant from recommended values. They are compiled in Table 12.8.

Table 12.8 The allocation coefficients and tuning frequencies for an optimized filter at internal voltage distortion $\delta_e = 2.5\%$ and 1% of the current distortion by non-characteristic harmonics.

S_{sc}/P	20	25	30	35	40	45
a_5	**0.09**	**0.11**	**0.11**	**0.12**	**0.15**	**0.09**
a_7	**0.11**	**0.30**	**0.61**	**0.55**	**0.73**	**0.69**
a_{11}	**0.60**	**0.46**	**0.21**	**0.14**	**0.08**	**0.10**
a_{13}	**0.20**	**0.14**	**0.07**	**0.20**	**0.04**	**0.13**
t_5/ω_1	**5.0**	**4.99**	**4.99**	**4.99**	**4.99**	**4.99**
t_7/ω_1	**6.99**	**7.00**	**7.02**	**6.95**	**6.88**	**6.86**
t_{11}/ω_1	**11.3**	**10.8**	**10.9**	**10.9**	**11.0**	**11.0**
t_{13}/ω_1	**13.00**	**13.0**	**13.0**	**12.7**	**13.0**	**13.0**

This table that the best allocation coefficients and tuning frequencies depend on the S_{sc}/P ratio. There are no particular patterns as to their values. This means that RHFs cannot be built as a sort of universal device but they have to be designed individually for specified conditions.

Conditions that have to be specified include not only the short-circuit power to active power ratio, but also the rms values of the load-generated harmonics J_n and the rms values of the internal voltage harmonics E_n. These rms values have to be distinguished from the rms values of the current and voltage harmonics observed at the supply terminals I_n and U_n in the system without the filter. This means the system identification is needed to provide data for an RHF design. Examples of a method of electrical systems identification concerning the rms the internal voltage harmonics E_n values and the load-generated harmonics J_n are presented in Refs [145, 146].

Summary

Resonant harmonic filters, until a few decades ago important tools for reducing harmonic distortions in distribution systems, have now lost their effectiveness. This is caused by an increase in the distribution voltage distortion. Having the capability of reducing the load-originated harmonics and distortion, RHFs increase the distribution system-originated distortion.

Chapter 13
Reactance Compensation in Single-Phase Circuits

13.1 Introduction

The development of a method of compensation of the reactive power in the presence of supply voltage distortion was one of the most important practical goals in the power theory (PT) of circuits and systems. Apart from the Currents' Physical Components (CPC)-based PT, all approaches to its development failed to provide the fundamentals of reactive power compensation.

Single-phase circuits are usually low-power circuits so such circuits handle a small amount of electric energy, therefore there are no commonly sufficient economic motivations for their compensation. Such motivations exist for compensation in three-phase circuits. Compensation in the presence of harmonics has turned out to be a very difficult problem in electrical engineering. Its solution for three-phase circuits had to be preceded by the solution for single-phase circuits, therefore there has been lengthy scientific effort aimed at creating fundamentals of compensation for single-phase loads in the presence of supply voltage distortion. The closest were the results presented by Shepherd and Zakikhani [39] and Kusters and Moore [59]. The most recent results were even recommended by the International Electrotechnical Commission (ICE) [57]. Unfortunately, these results were proven, in [69], to be incorrect.

13.2 Fundamentals of the CPC-Based Reactance Compensation

Fundamentals of compensation result from the concept of the current decomposition into physical components. It reveals which current components are harmful and how they could be reduced.

According to the CPC-based power theory, at nonsinusoidal supply voltage

$$u = \sqrt{2}\text{Re}\sum_{n\in N} \boldsymbol{U}_n\, e^{jn\omega_1 t}$$

Powers and Compensation in Circuits with Nonsinusoidal Current. Leszek S. Czarnecki, Oxford University Press.
 DOI: 10.1093/oso/9780198879206.003.0013

the current of LTI loads of the admittance for the nth order harmonic $\boldsymbol{Y}_n = G + jB_n$, is composed, according to Section 6.2, of the active, scattered, and reactive currents

$$i = i_{\rm a} + i_{\rm s} + i_{\rm r}.$$

These current, according to Section 6.3, are mutual orthogonal so that the supply source is loaded with the current of the rms value

$$||i|| = \sqrt{||i_{\rm a}||^2 + ||i_{\rm s}||^2 + ||i_{\rm r}||^2}$$

but only the active current $i_{\rm a}$ contributes to useful energy transfer to the load. The scatter and the reactive currents do not contribute to such a transfer but only to an increase in the supply current rms value.

When a lossless reactance compensator of susceptance for the nth order harmonic $B_{{\rm C}n}$ is connected at the load terminals, then it changes the reactive component of the supply current to the value

$$i_{\rm r}' = \sqrt{2}{\rm Re} \sum_{n \in N} j(B_{{\rm C}n} + B_n)\, \boldsymbol{U}_n\, e^{jn\omega_1 t}.$$

If for each harmonic order $n \in N$

$$B_{{\rm C}n} + B_n = 0 \tag{13.1}$$

then the reactive component of the supply current is reduced to zero. This is the condition that has to be satisfied by a reactance compensator of the reactive current. Such a compensator does not affect any parameter that specifies the scattered current, according to Section 6.5. Consequently, the scattered current remains unchanged. Thus the reactance compensator cannot improve the power factor to a value higher than

$$\lambda_{\max} = \frac{||i_{\rm a}||}{\sqrt{||i_{\rm a}||^2 + ||i_{\rm s}||^2}} \tag{13.2}$$

Observe that the load with the compensator as seen from the supply source for each such voltage harmonic is purely resistive, thus no series resonance with the distribution system inductance can occur. In this respect, the CPC-based compensation, meaning compensation that satisfies condition (13.1), differs substantially from purely capacitive compensation.

The condition for complete compensation of the reactive current (13.1) is simple but knowledge of the circuit synthesis is needed for finding the compensator structure and its LC parameters. Therefore the fundamentals of the circuit synthesis concerning the reactance circuits were outlined in Chapter 10. The procedure of this synthesis is explained in the following illustration.

Illustration 13.1 *A resistive-inductive* (RL) *load with parameters shown in Fig. 13.1 is supplied with the voltage*

$$u = \sqrt{2}\mathrm{Re}\,\{\boldsymbol{U}_1\, e^{j\omega_1 t} + \boldsymbol{U}_5\, e^{j5\omega_1 t}\}$$

and frequency normalized to $\omega_1 = 1$ rad/s. *Let us find the structure and calculate the parameters of a reactance compensator for the entire compensation of the reactive current.*

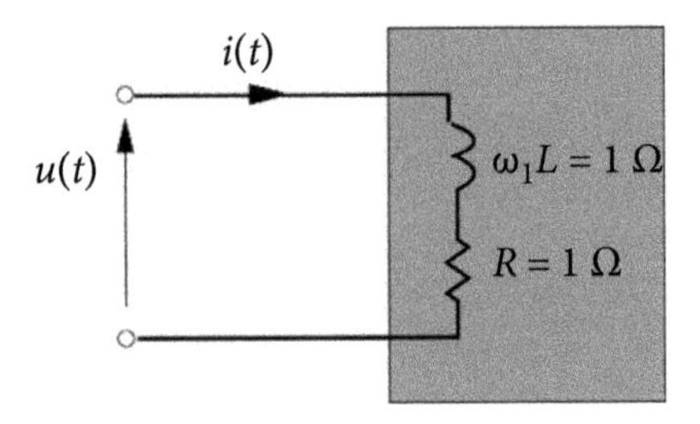

Figure 13.1 An example of RL load.

The load admittances for the fundamental and the fifth order harmonic are

$$\boldsymbol{Y}_1 = G_1 + jB_1 = \frac{1}{1+j2} = 0.20 - j0.40 = 0.45\, e^{-j63.4^0}\ \mathrm{S}$$

$$\boldsymbol{Y}_5 = G_5 + jB_5 = \frac{1}{1+j10} = 0.01 - j0.10 = 0.10\, e^{-j84.3^0}\ \mathrm{S}.$$

Thus the compensator should have for these harmonics the susceptance

$$B_{C1} = -B_1 = 0.4\ \mathrm{S}, \qquad B_{C5} = -B_5 = 0.1\ \mathrm{S}. \tag{13.3}$$

Because these two susceptances are of the same sign, they have to be separated by a POLE p_1 *and* ZERO z_1, *where the susceptance changes the sign. It would be beneficial, moreover, if the compensator had zero susceptance for high frequency. This would prevent the circuit against high-frequency noise. According to Chapter 10, the susceptance of reactance one-ports has to be increasing function of frequency, therefore it can change as shown in Fig. 13.2.*

Such variability requires that susceptance $B_C(\omega)$ *has the general form*

$$Y_C(s) = A\frac{s\left(s^2 + z_1^2\right)}{\left(s^2 + p_1^2\right)\left(s^2 + p_2^2\right)}$$

where coefficient A is a positive, real number. On the imaginary axis of the complex plane, where $s = j\omega$, *the reactance admittance changes as*

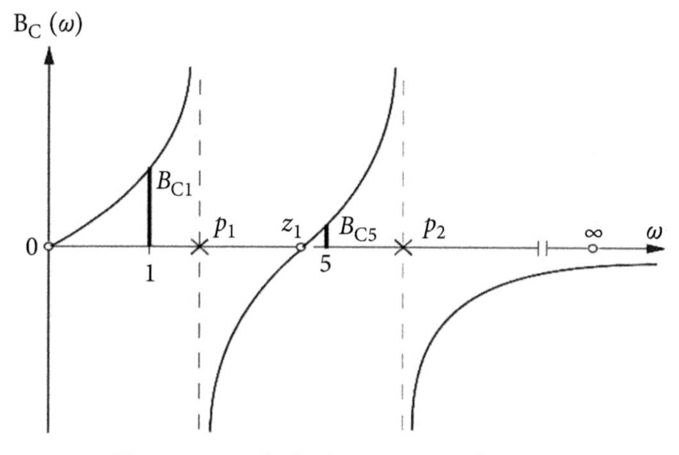

Figure 13.2 A draft of a susceptance function.

$$Y_C(j\omega) = jB_C(\omega) = jA\frac{\omega\left(z_1^2 - \omega^2\right)}{\left(p_1^2 - \omega^2\right)\left(p_2^2 - \omega^2\right)}.$$

It is specified by four parameters, $A, z_1, p_1,$ and p_2, while there are only two equations at points $\omega = 1$ rad/s and $\omega = 5$ rad/s, namely

$$A\frac{\left(z_1^2 - 1^2\right)}{\left(p_1^2 - 1^2\right)\left(p_2^2 - 1^2\right)} = 0.4, \quad A\frac{5\left(z_1^2 - 5^2\right)}{\left(p_1^2 - 5^2\right)\left(p_2^2 - 5^2\right)} = 0.1$$

for calculating these parameters. Therefore two of them can be selected at the compensator designer's discretion. These parameters affect the compensator properties and their selection has to satisfy some limitations which can be seen in Fig. 13.2, namely

$$A > 0, \qquad 1 < p_1 < z_1 < 5, \qquad 5 < p_2. \tag{13.4}$$

For example, we can assume that $A = 1$ and, to avoid the third order harmonic current in the supply in the case of this harmonic in the distribution voltage, it can be assumed that the compensator has a ZERO at $z_1 = 3$ rad/s. In such a case, the following equations have to be solved:

$$\frac{(9 - 1)}{\left(p_1^2 - 1\right)\left(p_2^2 - 1\right)} = 0.4, \quad \frac{5\,(9 - 25)}{\left(p_1^2 - 25\right)\left(p_2^2 - 25\right)} = 0.1.$$

The solution is

$$p_1 = 1.160 \text{ rad/s}, \; p_2 = 7.669 \text{ rad/s}$$

so that the compensator admittance has the form

$$Y_C(s) = \frac{s\,(s^2+9)}{(s^2+1.346)(s^2+58.820)}. \tag{13.5}$$

One should observe that this is not the only admittance that satisfies the compensator requirements. A different selection of two of four parameters would lead to a different susceptance function. Because there is an infinite number of options in selecting the coefficient A and ZERO z_1, that satisfy inequalities (13.4), there is an infinite number of admittances $Y_C(s)$ that satisfy condition (13.3).

There are several different structures of the reactance compensator that have the admittance $Y_C(s)$ of the form (13.5). They can be found using Foster and Cauer procedures explained in Chapter 10. For example, in the Cauer first procedure, the admittance $Y_C(s)$ can be expressed as a quotient of two polynomials

$$Y_C(s) = \frac{s\,(s^2+9)}{(s^2+1.346)(s^2+58.820)} = \frac{s^3+9\,s}{s^4+60.17\,s^2+79.17}$$

and by divisions of inverted remainders, it can be presented as a ladder fraction

$$Y_C(s) = \frac{s^3+9\,s}{s^4+60.17\,s^2+79.17} = \cfrac{1}{1.0\,s+\cfrac{1}{0.020\,s+\cfrac{1}{6.86\,s+\cfrac{1}{0.094\,s}}}}.$$

This is the admittance of the ladder one-port shown in Fig. 13.3.

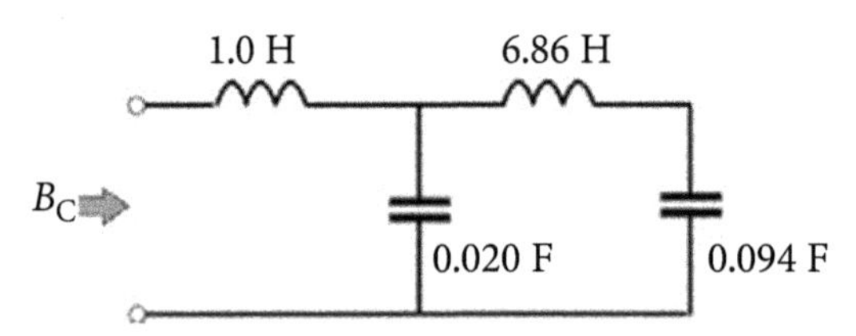

Figure 13.3 The reactance compensator developed according to the Cauer First Procedure.

According to the Foster first procedure, the compensator admittance can be decomposed into the following fractions

$$Y_C(s) = \frac{s\,(s^2+9)}{(s^2+1.346)(\,s^2+58.820)} = \frac{a_1 s}{s^2+1.346} + \frac{a_2 s}{s^2+58.820}$$

with

$$a_1 = \lim_{s^2 \to -1.346} \left\{ \frac{s^2 + 1.346}{s} Y_C(s) \right\} = \frac{-1.346 + 9}{-1.346 + 58.820} = 0.133$$

$$a_2 = \lim_{s^2 \to -58.820} \left\{ \frac{s^2 + 58.820}{s} Y_C(s) \right\} = \frac{-58.820 + 9}{-58.820 + 1.346} = 0.867.$$

Thus, the compensator has the structure shown in Fig. 13.4

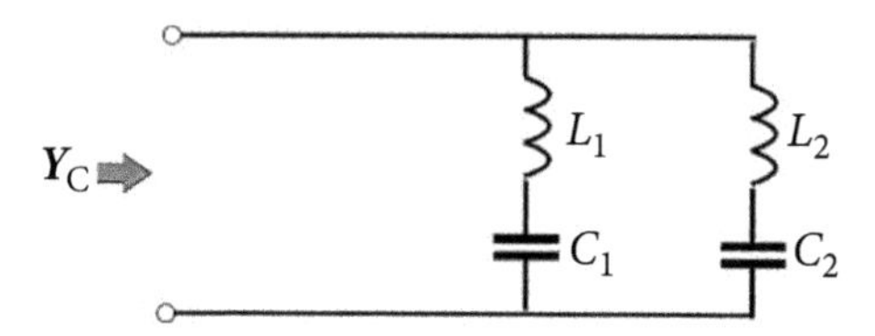

Figure 13.4 The reactance compensator developed according to the Foster First Procedure.

with parameters

$$L_1 = \frac{1}{a_1} = \frac{1}{0.133} = 7.52\text{H}, \; C_1 = \frac{a_1}{p_1^2} = \frac{0.133}{1.346} = 0.099\text{F},$$

$$L_2 = \frac{1}{a_2} = \frac{1}{0.867} = 0.60\text{H}, \; C_2 = \frac{a_2}{p_2^2} = \frac{0.867}{58.82} = 0.015\text{F}.$$

*****

The admittance of the compensated load is equal to

$$\boldsymbol{Y}_1{}' = jB_{C1} + \boldsymbol{Y}_1 = G_1 = 0.20 \text{ S}, \quad \boldsymbol{Y}_5{}' = jB_{C5} + \boldsymbol{Y}_5 = G_5 = 0.01 \text{ S}$$

thus it is purely resistive for the voltage harmonics. It is not equivalent to a resistor, however, because its resistance has a different value for the fundamental and the fifth order harmonic. Therefore the supply current after compensation contains a scattered current, thus the power factor after compensation remains lower than 1. Observe moreover that such a compensator compensates the reactive power only for distinctively specified harmonics. The compensated load is also not resistive for harmonics other than the fundamental and the fifth order. The compensator is, moreover, not simple. Therefore such compensation can be regarded as a theoretical solution rather than a practical one. Nonetheless, this kind of compensation provides a positive answer to a question asked for a long time in power theory: "Can the reactive current in circuits with nonsinusoidal supply voltage be compensated?" There is, however, another question asked sometimes in the power theory: "Can LTI loads supplied with a nonsinusoidal voltage be compensated to unity power factor?"

13.3 Compensator Complexity Reduction

Complete compensation of the reactive current by a reactance compensator in the presence of the supply voltage distortion requires that the compensator is composed of approximately two inductive-capacitive (LC) elements per voltage harmonic. Thus, with the increase of the number of harmonics, the idea of complete compensation of this current, although possible, has to be abandoned for the reactive current minimization with a compensator of reduced complexity.

When a purely capacitive compensator is excluded from considerations, for reasons explained in Chapter 11, the reactance compensator of the lowest complexity is a simple LC branch connected as shown in Fig. 13.5.

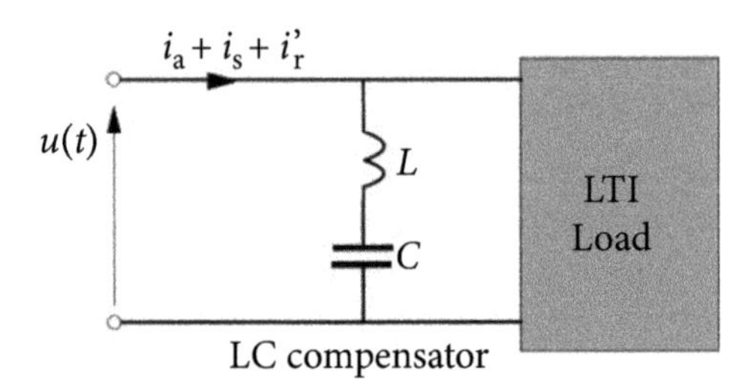

Figure 13.5 A TER compensator with an LTI load.

The series inductor should prevent the circuit against the voltage resonance of the compensator with the supply source internal inductance. Such a compensator composed of only two reactance devices connected as shown in Fig. 13.5 will be referred to as a two-element reactance (TER) compensator.

The susceptance of such a TER compensator for harmonic frequencies is

$$B_{Cn} = \frac{n\omega_1 C}{1 - n^2\omega_1^2 LC}.$$

It should have a capacitive impedance only for the fundamental harmonic to enable compensation of the reactive power of the load. This impedance for higher-order harmonics should be inductive. Since the susceptance for fundamental harmonic B_1 of common RL loads is negative, the susceptance of the compensator for the fundamental B_{C1} has to be positive. Thus the compensator LC parameters have to be selected such that

$$\omega_1^2 LC < 1. \tag{13.6}$$

To have the branch susceptance B_{Cn} for all harmonic orders $n > 1$ negative, the LC parameters have to be selected such that

$$n^2\omega_1^2 LC > 1. \tag{13.7}$$

Such a branch reduces the rms value of the reactive component of the supply current to

$$||i'_{\mathrm{r}}|| = \sqrt{\sum_{n\in N}\left(B_n + \frac{n\omega_1 C}{1 - n^2\omega_1^2 LC}\right)^2 U_n^2}.$$

If conditions (13.6) and (13.7) are satisfied, then the rms value $||i'_{\mathrm{r}}||$ of the reactive current is a monotonically declining function of L. It has no extremum. It has a minimum at a capacitance when its derivative versus the capacitance is zero, that is,

$$\frac{d}{dC}||i'_{\mathrm{r}}|| = 0.$$

If the inductance L of the TER compensator is selected such that a series resonance cannot occur for harmonic frequencies, then the change of the supply voltage with the change of the capacitance can be ignored. In such a case, at the assumption that $U_n = \text{const.}$, the condition for the minimum rms value results in optimum capacitance $C = C_{\mathrm{opt}}$, that satisfies the following equation

$$\sum_{n\in N}\frac{nB_n U_n^2}{\left(1 - n^2\omega_1^2 LC_{\mathrm{opt}}\right)^2} + \sum_{n\in N}\frac{n^2\omega_1 C_{\mathrm{opt}} U_n^2}{\left(1 - n^2\omega_1^2 LC_{\mathrm{opt}}\right)^3} = 0 \tag{13.8}$$

Unfortunately, the optimum capacitance C_{opt} in this equation is not in an explicit form, thus it cannot be solved versus this capacitance. A numerical procedure is needed to find it. In particular, it can be found as a limit of the sequence

$$C_1,\ C_2,\ \ldots\ldots\ldots\ldots C_i,\ C_{i+1},\ C_{i+2},\ \ldots\ldots$$

of capacitances calculated in an iterative formula

$$C_{i+1} = -\frac{\displaystyle\sum_{n\in N}\frac{nB_n U_n^2}{\left(1 - n^2\omega_1^2 LC_i\right)^2}}{\displaystyle\omega_1\sum_{n\in N}\frac{n^2 U_n^2}{\left(1 - n^2\omega_1^2 LC_i\right)^3}}. \tag{13.9}$$

A starting value, C_1, of this sequence can be the capacitance needed for the reactive power compensation at sinusoidal voltage. If C_0 is the capacitance of purely capacitive compensator needed for reactive power Q compensation, that is,

$$C_0 = \frac{Q_1}{\omega_1 U_1^2}$$

then, in the presence of inductance L, this value should be modified to

$$C_1 = \frac{C_0}{1 + \omega_1^2 L C_0}. \tag{13.10}$$

To avoid any resonance of the compensator for harmonic frequencies, inductance L should be selected such that the resonant frequency of the branch is below the frequency of the second harmonic:

$$\frac{1}{\sqrt{LC_1}} < 2\omega_1,$$

however after the optimum value C_{opt} of the capacitance is found, it should be verified that this condition is satisfied for the capacitance calculated. If needed, the inductance value should be adequately modified.

Such an approach secures the compensator against the resonances but can result in too high an inductance value, meaning too expansive a compensator. This inductance can be reduced if a less stringent requirement as to the resonant frequency can be accepted. In particular, in symmetrical three-phase, three-wire circuits the voltage harmonics below the third or even below the fifth order are usually negligible. Assuming that

$$\frac{1}{\sqrt{LC_1}} = \omega_{\mathrm{r}}$$

Eqn. (13.10) can be rearranged to

$$C_0 = \frac{C_1}{1 - \omega_1^2 L C_1} = \frac{C_1}{1 - \left(\frac{\omega_1}{\omega_{\mathrm{r}}}\right)^2}$$

and hence the starting capacitance C_1 for iterative calculation is equal to

$$C_1 = C_0 \left[1 - \left(\frac{\omega_1}{\omega_{\mathrm{r}}}\right)^2\right]. \tag{13.11}$$

Illustration 13.2 *At the supply voltage rms value* $U_1 = 230$ V *the load active power is equal to* $P = 50$ kW *at power factor* $\lambda_{\mathrm{L}} = 0.5$. *It is supplied from a bus with the short-circuit power* $S_{\mathrm{sc}} = 50\ P$ *and the reactance to resistance ratio* $X_{\mathrm{s}}/R_{\mathrm{s}} = 3$. *The distribution voltage contains harmonics of the order* $n = 3, 5, 7, 11$ *and relative rms* *values* $E_3 = 1\%$, $E_5 = 5\%$, $E_7 = 2\%$, *and* $E_{11} = 1\%$ *of the fundamental harmonic voltage.*

Let us assume that the fundamental frequency is normalized to $\omega_1 = 1$ rad/s. *The capacitance needed for entire compensating the reactive power of the fundamental harmonic is equal to* $C_0 = 0.98$ F. *Assuming that the resonance of the compensator is* $/_r = 2.5\ \omega_1$, *formula (13.11) results in the starting capacitance for iteration* $C_1 = 0.840\ C_0$. *With such staring capacitance, the iterative formula results in the sequence of capacitances which in only three iterations converges to* $C_{opt} = 0.843\ C_0$.

The plot of the power factor λ *versus capacitance C is shown in Fig. 13.6. One can observe that even the first step of iteration provides the value of the optimum capacitance with accuracy sufficient for practical applications.*

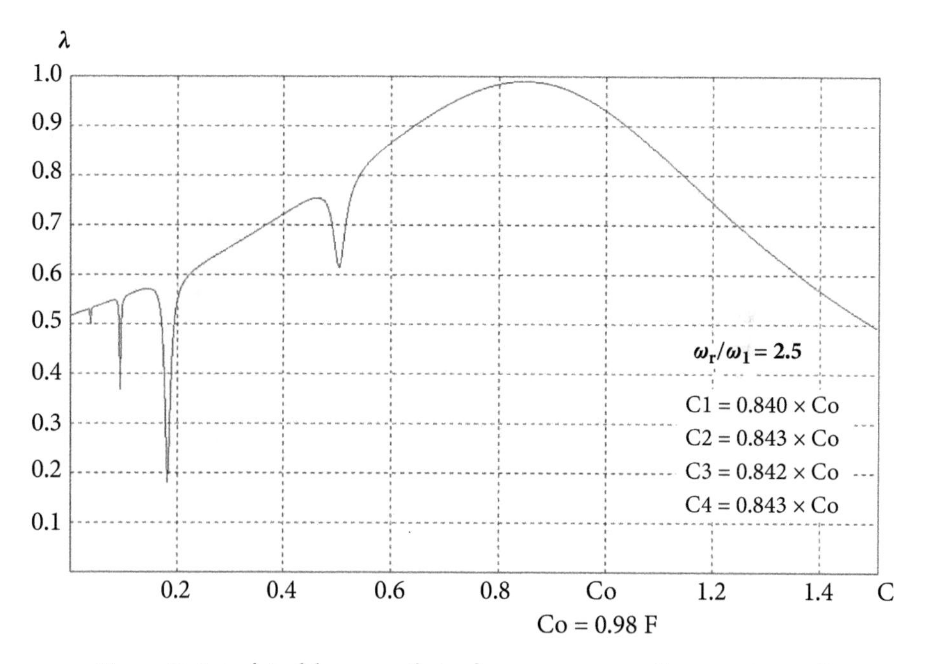

Figure 13.6 A plot of the power factor λ versus compensating capacitance C in Illustration 13.2.

The plot in Fig. 13.6 shows that the power factor declines drastically at a few values of the compensating capacitance, smaller than C_{opt}. *It is caused by the voltage resonance at harmonic frequencies, sequentially from the left, of the 11th, 7th, 5th, and the 3rd order of the compensator capacitance with the inductance as seen from the capacitor terminals.*

*****

13.4 Transmittances of the TER Compensator

The previous section enables the calculation of the value of the optimal capacitance C_{opt} of the TER compensator, however its frequency properties are not provided. The compensator designer should be aware of them.

Although such compensators can be regarded as single-branch resonant harmonic filters, the objective of their installation is different from that of harmonic filters. It should only compensate the reactive power of the load. The inductor is added to improve the compensator effectiveness in the presence of harmonics.

A simplified equivalent circuit of a circuit with the TER compensator in the range of harmonic frequencies is shown in Fig. 13.7. The load voltage to distribution voltage transmittance $A(j\omega)$ of the circuit with TER compensator is equal to

$$A(j\omega) \stackrel{\text{df}}{=} \frac{U(j\omega)}{E(j\omega)} = \frac{1 - \omega^2 LC}{1 - \omega^2 (L + L_s)C + j\omega L_s G(1 - \omega^2 LC)} = \frac{I(j\omega)}{J(j\omega)} \stackrel{\text{df}}{=} B(j\omega) \quad (13.12)$$

and it is equal to the supply current to generated current transmittance $B(j\omega)$.

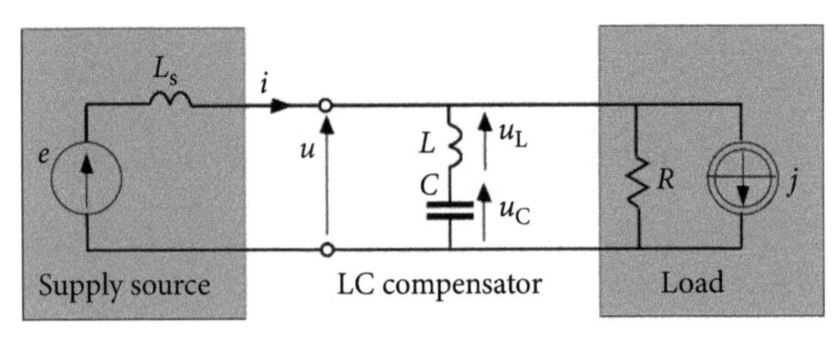

Figure 13.7 An equivalent circuit of a circuit with TER compensator in the range of harmonic frequencies.

It has zero value when $1-\omega^2 LC = 0$, that is, at the frequency

$$\omega_z = \frac{1}{\sqrt{LC}}.$$

The series resonance of the circuit occurs at frequency $\omega = \omega_r$ such that $1-\omega^2(L+L_s)C = 0$, meaning at the frequency

$$\omega_r = \frac{1}{\sqrt{(L_s + L)C}}.$$

Because the inductance as seen from the capacitor terminals, approximately equal to $L+L_s$, is higher than inductance L, thus, the series resonance of the circuit at frequency ω_r is below frequency ω_z. The transmittance at the resonant frequency is equal to

$$A(j\omega_r) = B(j\omega_r) = \frac{1}{j\omega L_s G} = \frac{R}{j\omega L_s}.$$

Illustration 13.3 *To illustrate the effect of the series inductor L on transmittance $A(j\omega)$, Fig. 13.8 presents the plot of the magnitude of this transmittance for the distribution circuit and the load parameters as assumed in Section 11.5, Illustration 11.3.*

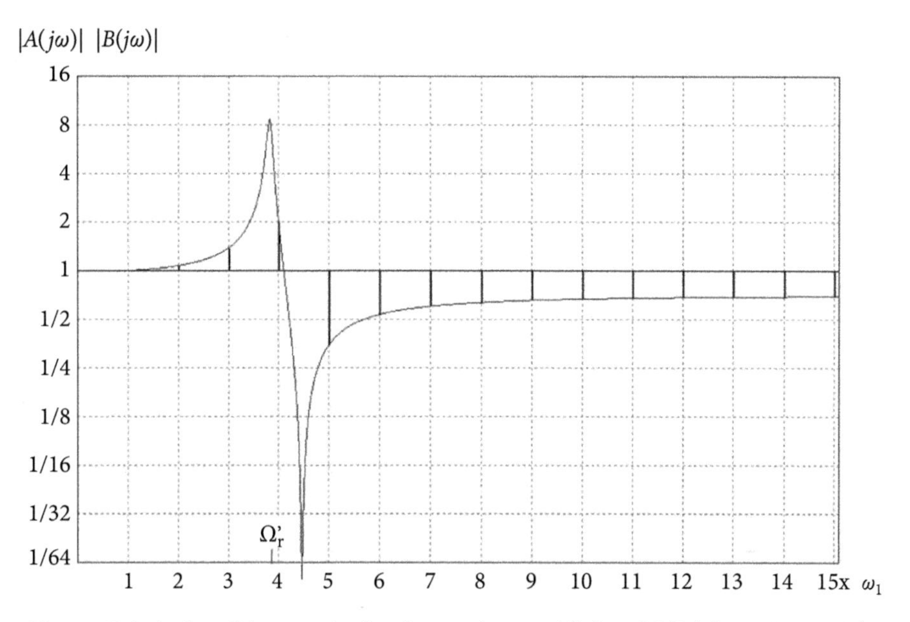

Figure 13.8 A plot of the magnitude of transmittance $A(j\omega)$ and $B(j\omega)$ for a circuit with TER compensator and $P_{RL}/P_R = 10$.

It was assumed in this illustration that the reactance of the series inductor for the fundamental frequency, $\omega_1 L = 5\%$ of the compensating capacitor reactance $1/\omega_1 C_0$ The inductor Q-factor was assumed to be equal to, $\omega_1 L/r_L = 50$, and active power ratio, explained in Section 11.5, equal to $P_{RL}/P_R = 10$.

The plot shows that the distribution voltage harmonics and the load-generated current harmonics are amplified almost in the whole range below the resonance frequency ω_r. When this plot is compared with that in Fig. 11.9 for a purely capacitive compensator, the substantial reduction of transmittances A_n and B_n can be observed. There is only some harmonic amplification in the closest vicinity of the resonant frequency, meaning for the fourth and the third order harmonics.

The admittance as seen from the distribution circuit is equal to

$$Y_x(j\omega) = \frac{I(j\omega)}{E(j\omega)} = \frac{G(1-\omega^2 LC) + j\omega C}{1-\omega^2(L+L_s)C + j\omega L_s G(1-\omega^2 LC)} \tag{13.13}$$

and it has a maximum at $\omega = \omega_z$, where it is confined by the system impedance, approximately equal to

$$Y_x(j\omega_z) = \frac{1}{j\omega_z L_s} = \frac{\omega_1}{\omega_z}\frac{1}{j\omega_1 L_s} \approx -j\frac{\omega_1}{\omega_z}\frac{1}{Z_{sc}} = -j\frac{\omega_1}{\omega_z}Y_{sc}.$$

Because this admittance is of the order of the short-circuit admittance of the distribution circuit, therefore in the presence of the distribution circuit harmonics of the frequency in a vicinity of the ω_z frequency, high distortion of the supply current can occur.

The plot of the magnitude of the relative admittance $\Psi_x(j\omega)$ for the circuit and compensator parameters as in Section 11.5, Illustration 11.3 for $P_{RL}/P_R = 1$, is shown in Fig. 13.9.

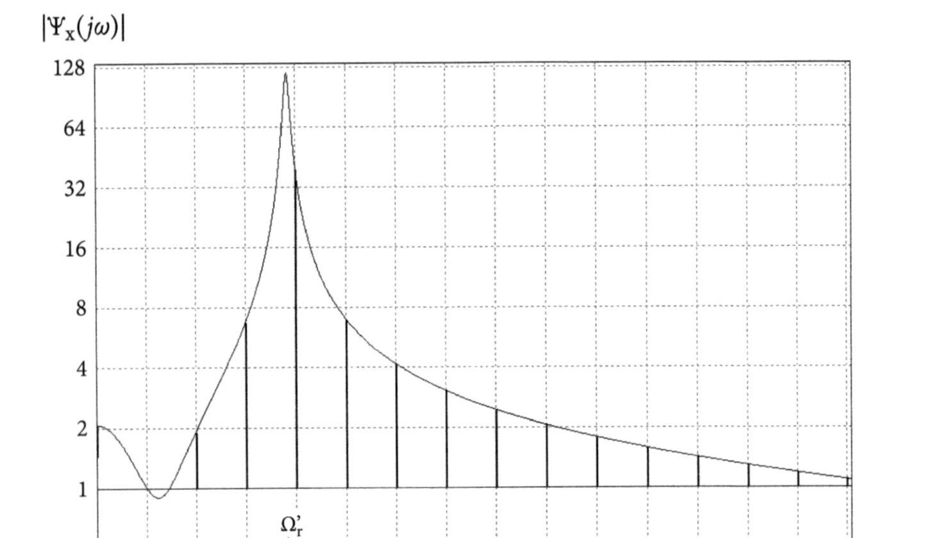

Figure 13.9 A plot of the magnitude of the relative admittance $\psi_x(j\omega)$ for $P_{RL}/P_R = 1$.

This plot when compared with its equivalent at a purely capacitive compensator, shown in Fig. 11.10, demonstrates that the series inductor substantially reduces the admittance as seen from the distribution circuit. Nonetheless, the compensator performance is still sensitive to the distribution voltage harmonics, especially of the frequency in the vicinity and below the resonant frequency.

The impedance as seen from the generated current source, j, is equal to

$$Z_y(j\omega) = \frac{U(j\omega)}{J(j\omega)} = \frac{j\omega L_s(1 - \omega^2 LC)}{1 - \omega^2(L + L_s)\,C + j\omega L_s G(1 - \omega^2 LC)}. \tag{13.14}$$

Its value is related to the load impedance for the fundamental frequency,

$$\zeta_{yn} = \frac{Z_{yn}}{Z_{L1}} = \frac{|Z_y(j\omega_n)|}{Z_{L1}}$$

and is shown for the circuit parameters as assumed previously, in Fig. 13.10.

This plot demonstrates that magnitudes of relative impedances ζ_n are well below one value, thus the load voltage in the presence of the LC compensator is not sensitive to the load-generated current harmonics.

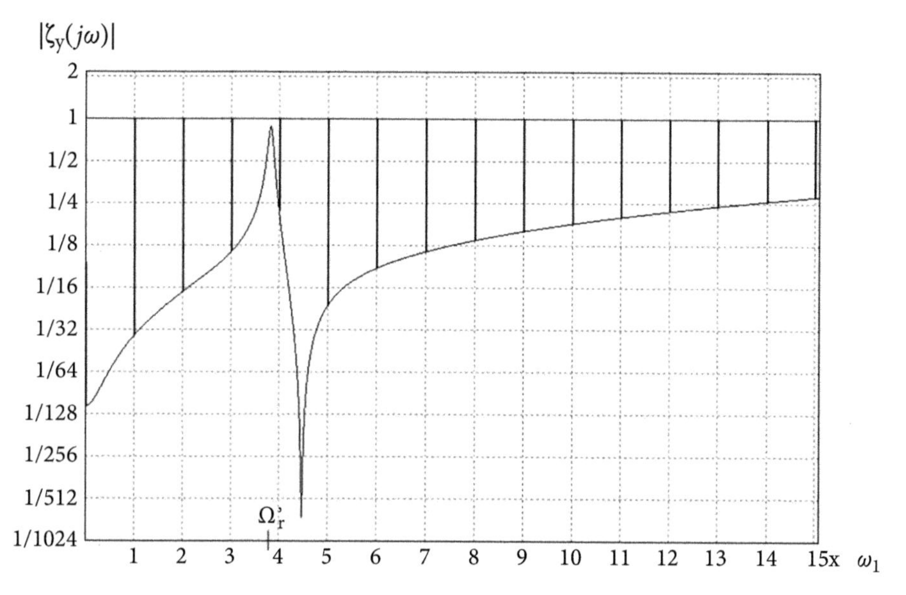

Figure 13.10 A plot of the magnitude of relative impedance as seen from the load-generated current source.

Comparing values of A_n, B_n, ψ_n, and ζ_n, it is easy to conclude that the high values of relative admittance, ψ_n, of the compensated load as seen from the distribution system, mean high sensitivity of the supply current to distribution voltage harmonics, which is the main issue at LC compensation. These values and consequently, the supply current sensitivity to distribution voltage harmonics, can be reduced by shifting the filter resonant frequency Ω_r' toward lower frequencies, which requires an increase in the series inductance L value.

*****

The reactive power of the fundamental harmonic, Q_1, compensated by a series LC branch is equal to

$$Q_1 = B_1\, U_1^2 = \frac{\omega_1 C}{1 - \omega_1^2 LC} U_1^2$$

thus the capacitance needed for the entire compensation of the reactive power is equal to

$$C \stackrel{\text{df}}{=} C_0' = \frac{Q_1}{\omega_1\left(U_1^2 + Q_1\omega_1 L\right)} = \frac{C_0}{1 + \omega_1^2 LC_0} = \frac{C_0}{1 + \vartheta} \tag{13.15}$$

where coefficient ϑ denotes the ratio of the series inductor L reactance to the reactance of the capacitor C_0,

$$\vartheta \stackrel{\text{df}}{=} \frac{X_L}{X_{C_0}} = \omega_1^2 L C_0.$$

Formula (13.15) shows that the capacitance $C_0{}'$ of the LC compensator needed for the entire compensation of the reactive power is lower than the capacitance C_0 of a purely capacitive compensator. Taking into account that the series inductor increases the reactive power of the fundamental harmonic which has to be compensated, this reduction in the capacitance value can be a surprising result. Observe, however, that the inductor reduces the impedance of the LC branch for the fundamental harmonic, thus the compensating current rms value and consequently, the capacitor voltage. It is because the power rating of the capacitor at LC compensation has to be higher than at purely capacitive compensation.

The capacitor voltage crms value is equal to

$$\boldsymbol{U}_{C1} = \frac{\boldsymbol{U}_1}{1 - \omega_1^2 L C'_0} = \left(1 + \omega_1^2 L C_0\right)\boldsymbol{U}_1 = \left(1 + \vartheta\right)\boldsymbol{U}_1,$$

thus the capacitor voltage rating has to be higher than the load voltage. The power rating of the capacitor for the entire compensation of the load reactive power of the fundamental harmonic has to be

$$S_C = \omega_1 C'_0\, U_1^2 = \omega_1 \frac{C_0}{1+\vartheta}\left[\left(1+\vartheta\right)U_1\right]^2 = \left(1+\vartheta\right) Q_1.$$

Since the series inductance is equal to

$$L = \frac{\vartheta}{\omega_1^2 C_0} = \vartheta \frac{U_1^2}{\omega_1 P \tan\varphi_L} \tag{13.16}$$

while the approximate value of the distribution system inductance is

$$L_s \approx \frac{U_1^2}{\omega_1\, S_{sc}},$$

thus the relative resonant frequency of the circuit and the load is entirely specified by the ratio of the inductor reactance to the reactance of the compensating capacitance, ϑ.

$$\Omega_r' \approx \frac{\Omega_r}{\sqrt{1 + \frac{S_{sc}}{P\tan\varphi_L}\vartheta}}.$$

This ratio, needed for moving the resonant frequency from Ω_r to Ω_r' is equal to

$$\vartheta = \left[\left(\frac{\Omega_r}{\Omega_r'}\right)^2 - 1\right]\frac{P}{S_{sc}}\tan\varphi_L. \tag{13.17}$$

Illustration 13.4 *The relative resonant frequency of the circuit with capacitive compensator and parameters $S_{SC} = 50P$ and $\lambda_L = 0.71$, is $\Omega_r = 7.4$. To shift this resonance below the frequency of the third order harmonic, to $\Omega_r' = 2.5$, the ratio of the inductor to capacitor reactance has to be $\vartheta = 0.155$. The plot of the magnitude of the relative admittance $Y_x(j\omega)$ for the circuit with the TER LC compensator is shown in Fig. 13.11.*

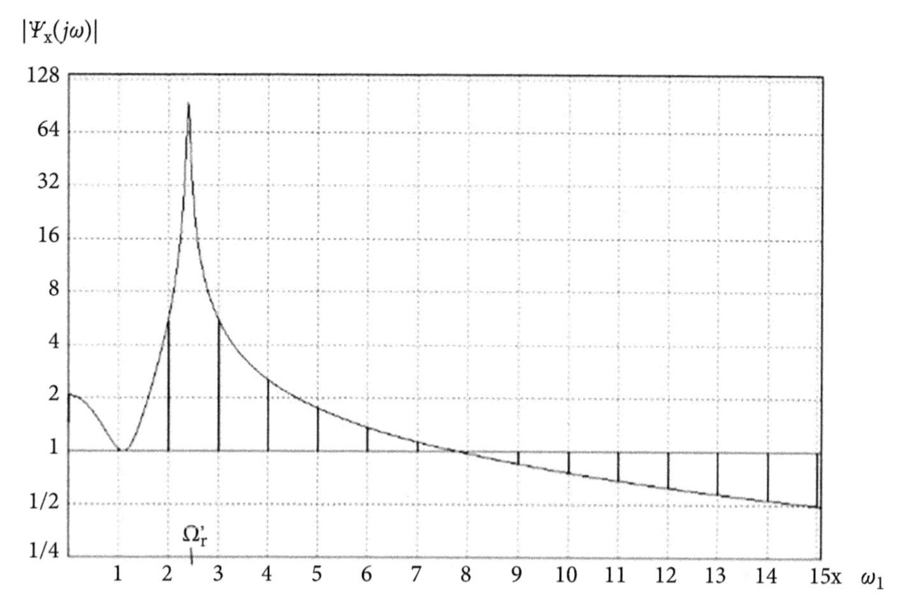

Figure 13.11 A plot of the magnitude of the relative admittance.

*****

13.5 A TER Compensator Control in Time Domain

The method of calculation of the parameters of a TER compensator presented in Section 13.3 and formula 13.9 for iterative calculation of the capacitance are based on the frequency domain approach. The supply voltage harmonic rms values U_n and the load susceptance B_n must be measured for that. Digital Signal Processing (DSP) system can calculate these values using the Discrete Fourier Transform (DFT), having a sufficient number of samples of the supply voltage and the load current in the whole period T of the voltage variability.

The optimum capacitance of a TER compensator can be found also in the time-domain, however, meaning, without DFT. The time-domain approach might be more convenient when the TER compensator should be used as an adaptive device for compensating loads with variable reactive power.

The supply current in a circuit with a TER compensator, shown in Fig. 13.12,

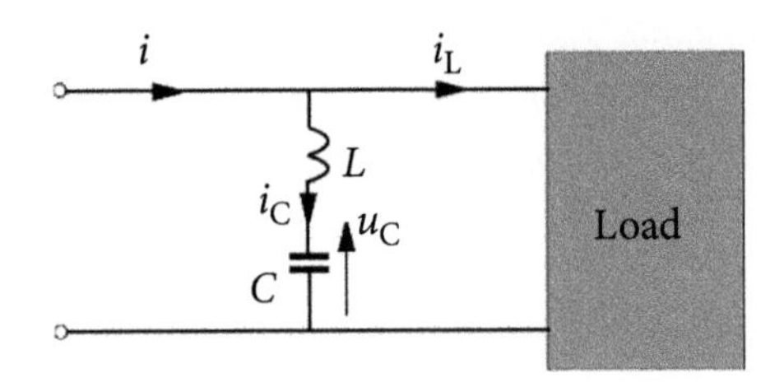

Figure 13.12 A circuit with a TER compensator.

can be expressed as

$$i(t) = i_L(t) + i_C(t) = i_L(t) + C\frac{d}{dt}u_C(t) = i_L(t) + Cu'_C(t).$$

where

$$u'_C(t) = \frac{d}{dt}u_C(t).$$

The square of the supply current rms value is

$$||i||^2 = ||i_L + Cu'_C||^2 = ||i_L||^2 + 2C(i_L, u'_C) + C^2||u'_C||^2$$

where

$$(i_L, u'_C) = \frac{1}{T}\int_0^T i_L(t)u'_C(t)\, dt$$

is the scalar product of the supply voltage and the capacitor's voltage derivative.

When the capacitance C of the compensator changes, then the rms value of the supply current changes as well. When it has a minimum value then its derivative, or derivative of its square, versus the capacitance C is equal to zero.

$$\frac{d}{dC}||i||^2 = 0$$

meaning

$$||i_L||\frac{d||i_L||}{dC} + (i_L, u'_C) + C\frac{d(i_L, u'_C)}{dC} + C||u'_C||^2 + C^2||u'_C||\frac{d||u'_C||}{dC} = 0.$$

This is the necessary condition for the supply current rms minimum value. It can be rearranged into the equation

$$a\,C^2 + bC + c = 0 \tag{13.18}$$

with coefficients

$$a = ||u'_\mathrm{C}||\frac{d||u'_\mathrm{C}||}{dC} \tag{13.19}$$

$$b = ||u'_\mathrm{C}||^2 + \frac{d(i_\mathrm{L}, u'_\mathrm{C})}{dC} \tag{13.20}$$

$$c = (i_\mathrm{L}, u'_\mathrm{C}) + ||i_\mathrm{L}||\frac{d||i_\mathrm{L}||}{dC}. \tag{13.21}$$

Thus, the optimal capacitance of the TER compensator can be calculated from Eqn. (13.18) with coefficients that can be provided by converters of the rms value of the load current $||i_\mathrm{L}||$ and derivative of the capacitor voltage $||u_\mathrm{C}'||$ and their scalar product $(i_\mathrm{L}, u_\mathrm{C}')$. These could be analog or digital converters. To measure derivatives of these values, the capacitance value C must be disturbed by ΔC, large enough to observe changes of $||i_\mathrm{L}||$, $||u_\mathrm{C}'||$, and $(i_\mathrm{L}, u_\mathrm{C}')$ values with sufficient accuracy. The change of this capacitance is easier when the TER compensator is built as an adaptive compensator with ON/OFF switched capacitor sections, as shown in Fig. 13.13.

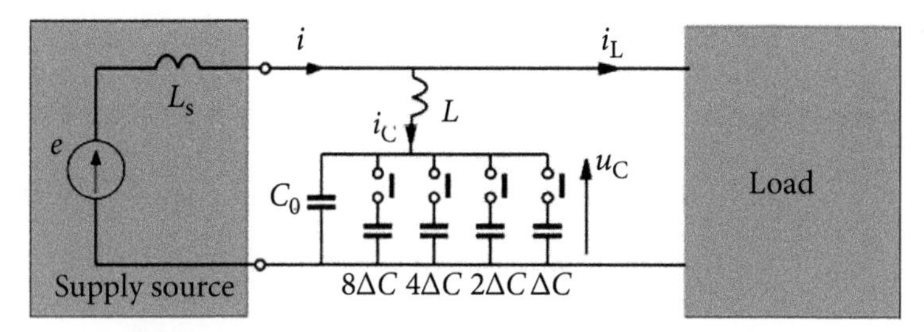

Figure 13.13 A circuit with an adaptive TER compensator.

A part of the capacitor bank of capacitance C_0 is separated as a permanent capacitance. Its value should prevent the circuit from the resonance of the compensator for the lowest order harmonic which occurs in the supply voltage. When h denotes the order of the lowest order harmonic in the supply voltage, and L_e is the inductance as seen from the capacitor terminals, approximately equal to

$$L_\mathrm{e} \approx L + L_\mathrm{s}$$

then the capacitance C_0 should be higher than the critical capacitance C_h at which a resonance for the harmonic of the h order can occur.

$$C_h = \frac{1}{n_h^2 \omega_1^2 L_\mathrm{e}}.$$

This is how the circuit is protected against any resonance.

The remaining part of the capacitor bank can be decomposed, for example, into four sections of the capacitance $8\Delta C$, $4\Delta C$, $2\Delta C$, and ΔC. They are connected by a switch which can be in one of sixteen states k. The ON/OFF positions of four switches of these sections are specified as the binary form s_k of these k states, namely

$$s_0 = 0000,\ s_1 = 0001,\ s_2 = 0010,\ s_3 = 0011,\ s_4 = 0100, \ldots s_{15} = 1111,$$

thus depending on the switch state k, the compensating capacitance can have sixteen different values, namely

$$C_k = C_0 + k\Delta C, \quad k = 0 \ldots 15$$

Let us assume that the compensator was switched from state 0 to state 1, meaning its capacitance was changed by ΔC, and the values

$$||i_{\rm L}||_0, ||i_{\rm L}||_1, ..||u'_{\rm C}||_0, ||u'_{\rm C}||_1, \ldots (i_{\rm L}, u'_{\rm C})_0, (i_{\rm L}, u'_{\rm C})_1,$$

were measured. Having these values, Eqn (13.18) can be approximated by

$$a_1 C^2 + b_1 C + c_1 = 0 \tag{13.22}$$

with coefficients

$$a_1 = ||u'_{\rm C}||_1 \frac{||u'_{\rm C}||_1 - ||u'_{\rm C}||_0}{\Delta C} \tag{13.23}$$

$$b_1 = ||u'_{\rm C}||_1^2 + \frac{(i_{\rm L}, u'_{\rm C})_1 - (i_{\rm L}, u'_{\rm C})_0}{\Delta C} \tag{13.24}$$

$$c_1 = (i_{\rm L}, u'_{\rm C})_1 + ||i_{\rm L}||_1 \frac{||i_{\rm L}||_1 - ||i_{\rm L}||_0}{\Delta C}. \tag{13.25}$$

The solution of (13.22) results in the capacitance C and the closest state k of the switch needed for this capacitance approximation. Now, the equation (13.18) can be approximated by

$$a_k C^2 + b_k C + c_k = 0 \tag{13.26}$$

with coefficients

$$a_k = ||u'_{\rm C}||_k \frac{||u'_{\rm C}||_k - ||u'_{\rm C}||_1}{C_k - C_1} \tag{13.27}$$

$$b_k = ||u'_{\rm C}||_k^2 + \frac{(i_{\rm L}, u'_{\rm C})_k - (i_{\rm L}, u'_{\rm C})_1}{C_k - C_1} \tag{13.28}$$

$$c_k = (i_{\rm L}, u'_{\rm C})_k + ||i_{\rm L}||_k \frac{||i_{\rm L}||_k - ||i_{\rm L}||_1}{C_k - C_1}. \tag{13.29}$$

When the solution of (13.26) differs from the solution of (13.22) by a value lower than ΔC, then the approximate value of the optimum capacitance was found. Otherwise, the iteration must be repeated.

Illustration 13.5 *An adaptive TER LC compensator is connected in the circuit shown in Fig. 13.14. The rms value of the supply voltage harmonics are equal to* $E_1 = 100$ V, $E_5 = 3.0$ V, $E_7 = 2.0$ V, $E_{11} = 1.0$ V, *the supply source inductance* $L_s = 0.15$ H, *and the supply source resistance* $R_s = 0.05\ \Omega$. *The voltage frequency is normalized to* $\omega_1 = 1$ rad/s. *The capacitor bank used for compensation has the capacitance* $C = 0.5$ F.

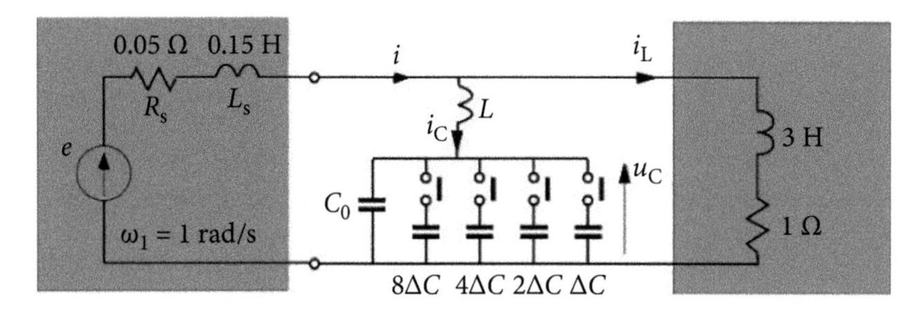

Figure 13.14 A circuit in Illustration 13.5.

The rms value of the supply current of the uncompensated load is $||i_L|| = 31.0$A *and the power factor of such a load* $\lambda = 0.31$.

The fixed parameters of the TER compensator, L and C_0, *can be selected at the designer's discretion in such a way that the series resonance of the compensator in the range of the supply voltage harmonics cannot occur. Let these be* $L = 0.4$ H *and* $C_0 = 0.14$ F. *Such a compensator reduces the supply current rms value to* $||i|| = 17.9$ A. *It improves the power factor to* $\lambda = 0.54$. *Since the inductance as seen from the capacitor terminals is approximately* $L_e = 0.55$ H, *the series resonance of the compensator can occur at the relative frequency approximately equal to* $\omega_r = 3.6$ rad/s, *meaning well below the lowest order harmonic, the fifth one of the supply voltage. The smallest section of the capacitor has the capacitance*

$$\Delta C = (0.5 - 0.14)/15 = 0.024 \text{ F}.$$

In the first step of the search for the optimum capacitance

$$C_1 = C_0 + \Delta C = 0.164 \text{ F, we obtain } \lambda = 0.62, \text{ and } ||i|| = 15.8 \text{ A}$$

and Eqn. (13.22) results in $C = C_2 = 0.28$ F. *Since* $(C_1 - C_0)/\Delta C = 5.8$, *the closest capacitance to this value is connected when the switch is in state* $s_2 = 6$.

In the second step of iteration, at capacitance

$$C_2 = C_0 + s_2\,\Delta C = 0.284\ \text{F},\ \text{we obtain}\ \lambda = 0.93,\ \text{and}\ ||i|| = 10.7\ \text{A}$$

and Eqn. (13.26) results in $C = C_3 = 0.255$ F. *Since* $(C_1 - C_0)/\Delta C = 4.8$, *the closest capacitance to this value is connected when the switch is in state* $s_3 = 5$.

In the third step of the iteration, at the capacitance

$$C_3 = C_0 + s_3\,\Delta C = 0.260\ \text{F},\ \text{we obtain}\ \lambda = 0.95,\ \text{and}\ ||i|| = 10.5\ \text{A}.$$

The next iterations do not change the switch state so they should be stopped with the switch in state $s_3 = 5$. *This procedure is illustrated in Fig. 13.15.*

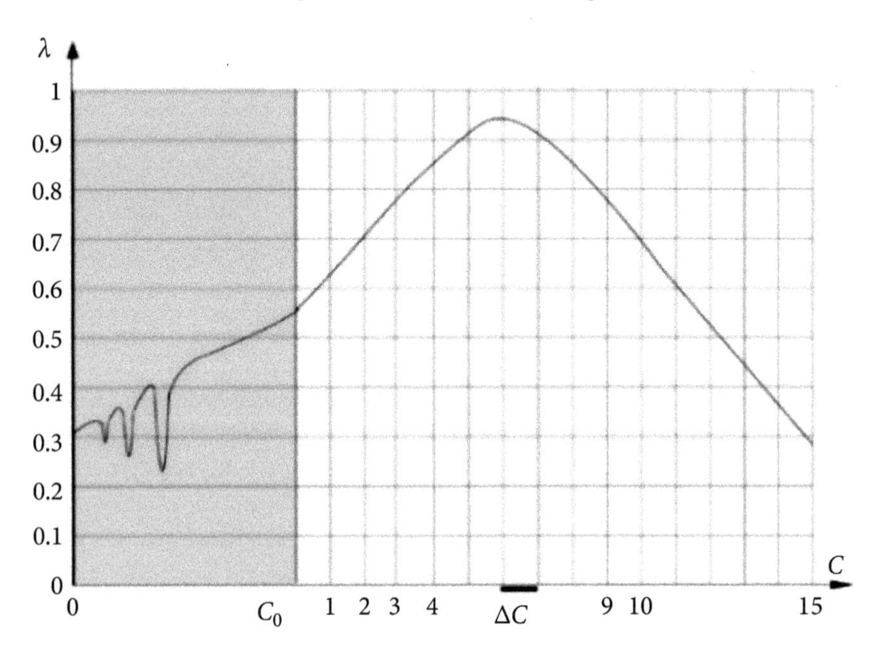

Figure 13.15 The dependence of the power factor on the capacitance in Illustration 13.5.

*****

13.6 Complete Reactance Compensation

In circuits with LTI loads and the sinusoidal supply voltage, the power factor can be improved by a reactance compensator to unity, that is, $\lambda = 1$. "Is it possible to compensate the load to unity power factor when the voltage is nonsinusoidal?" This question was asked for years in the electrical engineering community.

As was demonstrated in the previous sections, a compensator built in such a way that harmonic resonances cannot occur enables improvement of the power factor to almost unity value, so that the above question has theoretical rather than practical merits. Theoretical merits are no less important for knowledge than practical ones, however. Therefore, let us try to answer this question.

As was shown in Section 6.5, shunt reactance compensators are not capable of compensating the scattered current of the load. Such a load can be compensated to the unity power factor only on the condition that the load current does not contain any scattered component, which requires that the load conductance G_n for harmonic frequencies does not change with the order of the voltage harmonics. On the other side, as shown in Section 6.10, series reactance compensators are not capable of compensating the scattered voltage. A load can be compensated to the unity power factor by a series compensator only on the condition that the scattered voltage is equal to zero, which requires that the load resistance R_n for harmonic frequencies does not change with harmonic order.

Thus, the scatter current and the scattered voltage cannot be compensated because any shunt reactance compensator is not capable of changing the load conductance G_n, while any series reactance compensator is not capable of changing the load resistance R_n.

Observe, however, that even if any shunt reactance compensator cannot change the conductance G_n as seen from the supply source, it changes the resistance R_n.

Yet a parallel branch connected at the load terminals has susceptance $B_{\mathrm{p}n}$. It changes the resistance as seen from the supply source to

$$R_n = \mathrm{Re}\left\{\frac{1}{G_n + j\left(B_n + B_{\mathrm{p}n}\right)}\right\} = \frac{G_n}{G_n^2 + \left(B_n + B_{\mathrm{p}n}\right)^2}.$$

Susceptances $B_{\mathrm{p}n}$ can be selected to obtain a constant value of the resistance $R_n = R_{\mathrm{a}} = \mathrm{const}$.

It requires that equation

$$\frac{G_n}{G_n^2 + \left(B_n + B_{\mathrm{p}n}\right)^2} = R_{\mathrm{a}}$$

is solved versus the susceptance $B_{\mathrm{p}n}$. This solution has the form

$$B_{\mathrm{p}n} = -B_n \pm G_n\sqrt{\frac{G_{\mathrm{a}}}{G_n} - 1}. \tag{13.30}$$

The resistance R_{a} must be selected such that $G_{\mathrm{a}} \geq G_n$. This means that the shunt reactance branch converts the load to a load with a constant resistance R_n and such a converted load can be compensated to the unity power factor with a series reactance compensator. This means that compensation of LTI loads to unity power factor with a reactance compensator is possible, but the compensator has to be composed of two branches, shunt and series.

Illustration 13.6 *Let us consider the load shown in Fig. 13.16 supplied with the voltage*

$$u = \sqrt{2}\,\mathrm{Re}\,\{220\ e^{j\omega_1 t} + 20\ e^{j3\omega_1 t} + 10\ e^{j5\omega_1 t}\}\mathrm{V},\ \omega_1 = 1\ \mathrm{rad/s}.$$

The load has admittance for harmonics in the supply are equal to

$$Y_1 = G_1 + jB_1 = (0.838 - j0.208)\ \mathrm{S},$$
$$Y_3 = G_3 + jB_3 = (0.572 - j0.0207)\ \mathrm{S},$$
$$Y_5 = G_5 + jB_5 = (0.638 + j0.155)\ \mathrm{S}.$$

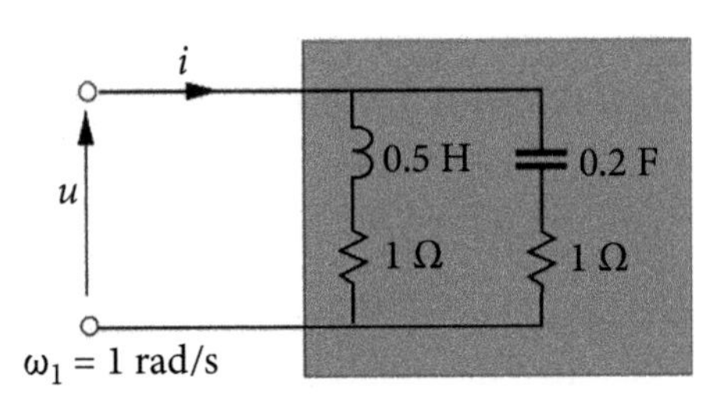

Figure 13.16 A load in illustration 13.6.

The highest conductance of G_1, G_2, and G_3 is the conductance G_1, so we can assume that $G_a = G_1 = 0.838$ S. It results from formula (13.30) that at such data the parallel branch of the compensator can have susceptances, respectively

$$B_{p1} = 0.208\ \mathrm{S},\ B_{p3} = -0.370\ \mathrm{S},\ B_{p5} = 0.202\ \mathrm{S}.$$

Such susceptance has a reactance one-port of admittance:

$$Y_p(s) = \frac{0.0377\,s^3 + 1.0628\,s}{s^2 + 5.762}.$$

The structure and parameters of such a one-port are shown in Fig. 13.17. The one-port converts the impedance of the load for harmonics to

$$Z_1' = 1.193\Omega$$
$$Z_3' = (1.193 + j0.813)\,\Omega,$$
$$Z_5' = (1.193 - j0.699)\,\Omega.$$

and, indeed, the resistance as seen by the supply source is the same for the 1st, 3rd, and 5th harmonic order. Thus the voltage of the modified load cannot contain the scattered component. Therefore, a series reactance compensator with reactance

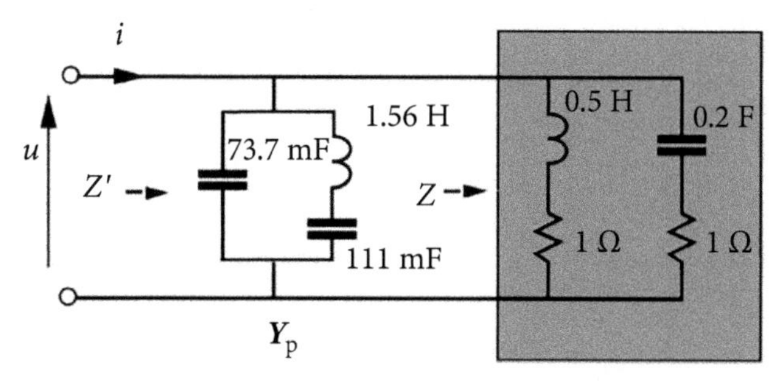

Figure 13.17 A load with a shunt reactance compensator.

$$X_{s1} = 0, \quad X_{s3} = -0.813\ \Omega, \quad X_{s5} = 0.669\ \Omega$$

can compensate such a load to unity power factor. Using synthesis procedures presented in Chapter 10, we can find that these reactances have the one-port of impedance

$$Z_s(s) = \frac{0.198\ s^4 + 2.622\ s^2 + 2.424}{s^3 + 6.891\ s}.$$

This result can be verified by calculating values $Z_s(j1)$, $Z_s(j3)$, and $Z_s(j5)$. The structure and parameters of the compensator are shown in Fig. 13.18.

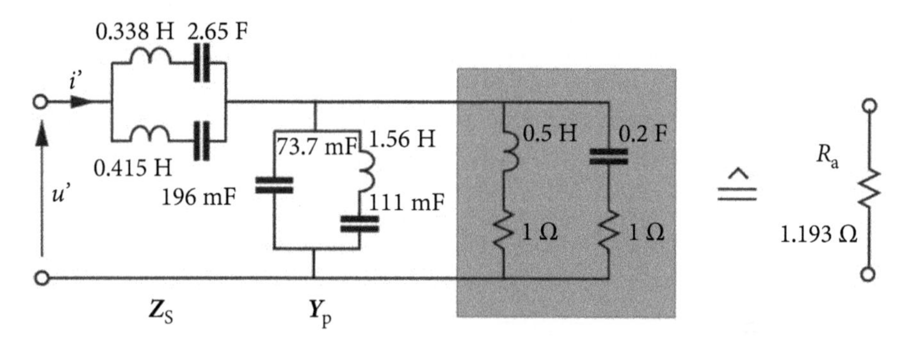

Figure 13.18 An LTI load with a reactance compensator that improves the power factor to the unity.

The compensated load for harmonics of the 1st, 3rd and 5th order is equivalent to a resistor of resistance $R_a = 1.193\ \Omega$, thus its power factor $\lambda = 1$.

The active power of the load at such compensation is not preserved, however. Its value before compensation is

$$P = \sum_{n=1,3,5} G_n U_n^2 = 40852\ \text{W}.$$

To preserve the load active power P after compensation, the rms value of the supply voltage

$$||u|| = \sqrt{U_1^2 + U_3^2 + U_5^2} = \sqrt{220^2 + 20^2 + 10^2} = 221.13 \text{ V}$$

should be changed by a transformer, as explained in Section 6.10, to

$$||u'|| = \sqrt{R_a P} = 220.76 \text{ V}.$$

This change of the supply voltage is well below variations acceptable by common standards, however, so the transformer is not needed.

*****

This illustration demonstrates that compensation of LTI loads by a reactance compensator to unity power factor in the presence of the supply voltage distortion is possible. Nonetheless, at such a compensator complexity, compensation to the unity power factor would not provide economic benefits.

Summary

Due to the usually low power of single-phase loads, such loads do not require compensation. Nonetheless, the development of methods of compensation of high-power three-phase loads requires that first of all, compensation at the level of single-phase circuits is well comprehended. Therefore the possibility of compensation in single-phase circuits in the presence of voltage and current distortion was investigated for decades. The solution was eventually found, as presented in this chapter, in the frame of the CPC-based power theory.

Chapter 14
Reactance Balancing Compensation in Three-Phase Three-Wire Circuits

14.1 Historical Background

Most electric energy is transmitted and utilized in three-phase circuits. Therefore, three-phase circuits are the area of main interest for investigations on the effectiveness of energy transfer improvement by compensation. Comprehension of compensation in a single-phase circuit is only a necessary step for the development of compensation methods in three-phase circuits.

When the load is balanced and is supplied with a sinusoidal and symmetrical voltage then it can be compensated, line by line, by a capacitor bank or a synchronous machine, as explained in Section 9.6. The load imbalance, harmonic distortion, and supply voltage asymmetry create new problems for methods of compensation in three-phase circuits.

The first balancing compensator for circuits with sinusoidal and symmetrical voltage was developed by Steinmetz and presented [7] in 1917. It is known as the Steinmetz circuit [247]. Investigations on reactance compensation and load balancing were continued with results reported in several papers, such as [136, 207, 311, 318, 319].

There were several different approaches to reactance balancing compensator design. It could be based, as initiated by Steinmetz [7], on searching for a circuit that will eliminate the oscillating component of the instantaneous power caused by the current asymmetry. Another approach was founded on the idea of elimination of the symmetrical components of the negative and the zero sequence from the supply current by reactance devices. Even optimization methods, which do not require very detailed knowledge of the power properties of electrical circuits, were used for that purpose [207, 136]. Optimization methods require a large number of computer calculations, however, and therefore are not well suited for adaptive compensators design. Adaptive compensation requires that the compensator parameters are specified by algebraic formulas that do not require intensive calculations.

Major progress in the methods of reactance balancing compensators design has occurred with the Currents' Physical Components (CPC)-based power theory development [96]. The CPC-based approach enables direct calculation of the compensator parameters using results of measurements of the voltage and current crms values at the load terminals.

Powers and Compensation in Circuits with Nonsinusoidal Current. Leszek S. Czarnecki, Oxford University Press.
 DOI: 10.1093/oso/9780198879206.003.0014

14.2 Compensation in Circuits with Sinusoidal Voltage

According to the CPC-based power theory, the supply current of linear, time-invariant (LTI) loads supplied with a symmetrical and sinusoidal voltage, as was developed in Section 7.4, is composed of the active, reactive, and unbalanced components, namely

$$\boldsymbol{i} = \boldsymbol{i}_a + \boldsymbol{i}_r + \boldsymbol{i}_u$$

Its three-phase rms value is equal to

$$||\boldsymbol{i}|| = \sqrt{||\boldsymbol{i}_a||^2 + ||\boldsymbol{i}_r||^2 + ||\boldsymbol{i}_u||^2}.$$

This value can be reduced to the active current three-phase rms value $||\boldsymbol{i}_a||$ on the condition that the reactive and the unbalanced currents are compensated.

A reactance compensator of the reactive and unbalanced currents referred to as a *reactance balancing compensator*, or as a *balancing compensator*, can have a structure as shown in Figure 14.1.

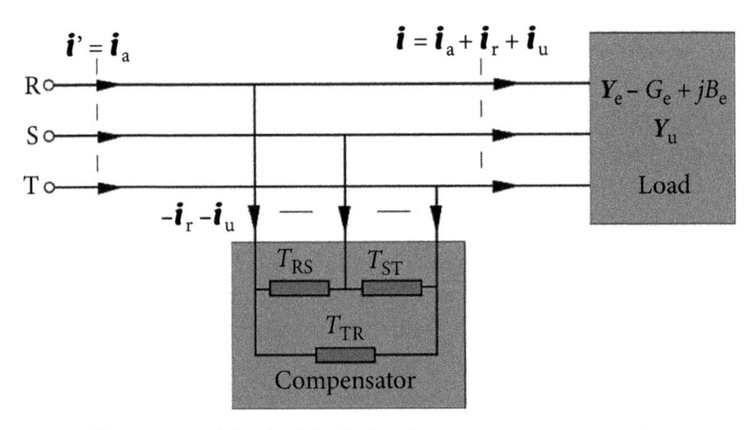

Figure 14.1 A load with a balancing reactance compensator.

It is built of three reactance elements, meaning capacitors and/or inductors, configured in Δ, though they can be configured in a Y structure as well. If the active power loss in the compensator is neglected, its active current is equal to zero. Such a compensator loads the supply source only with the reactive and unbalanced current.

The LTI load can be entirely characterized, according to the CPC approach, in terms of equivalent admittance $\boldsymbol{Y}_e$ and unbalanced admittance $\boldsymbol{Y}_u$.

To differentiate the compensator's line-to-line susceptance from such susceptances of the load, the compensator's branch susceptances will be denoted by T_{RS}, T_{ST}, and T_{TR}. Because it was assumed that the compensator is lossless, its line-to-line admittances are jT_{RS}, jT_{ST}, and jT_{TR}, respectively.

Such a compensator changes the reactive current of the supply source to the value

$$\boldsymbol{i}_{\mathrm{r}}' = \sqrt{2}\mathrm{Re}\,\{j[B_{\mathrm{e}} + (T_{\mathrm{ST}} + T_{\mathrm{TR}} + T_{\mathrm{RS}})]\,\mathbf{1}^{\mathrm{p}}\boldsymbol{U}\}\,e^{j\omega t}.$$

In particular, this current is reduced to zero when

$$B_{\mathrm{e}} + (T_{\mathrm{ST}} + T_{\mathrm{TR}} + T_{\mathrm{RS}}) = 0. \tag{14.1}$$

At the same time, the compensator changes the unbalanced current to

$$\boldsymbol{i}_{\mathrm{u}}' = \sqrt{2}\mathrm{Re}\,\{[\boldsymbol{Y}_{\mathrm{u}} - j(T_{\mathrm{ST}} + \alpha T_{\mathrm{TR}} + \alpha^* T_{\mathrm{RS}})]\,\mathbf{1}^{\mathrm{n}}\boldsymbol{U}\}\,e^{j\omega t}.$$

It is equal to zero when

$$\boldsymbol{Y}_{\mathrm{u}} - j(T_{\mathrm{ST}} + \alpha T_{\mathrm{TR}} + \alpha^* T_{\mathrm{RS}}) = 0. \tag{14.2}$$

The compensator parameters are specified by three susceptances T_{RS}, T_{ST}, and T_{TR}. Since condition (14.2) is expressed in terms of complex numbers, it must be satisfied both for the real and imaginary parts, thus

$$\mathrm{Re}\{\boldsymbol{Y}_{\mathrm{u}} - j(T_{\mathrm{ST}} + \alpha T_{\mathrm{TR}} + \alpha^* T_{\mathrm{RS}})\} = 0 \tag{14.2a}$$

$$\mathrm{Im}\{\boldsymbol{Y}_{\mathrm{u}} - j(T_{\mathrm{ST}} + \alpha T_{\mathrm{TR}} + \alpha^* T_{\mathrm{RS}})\} = 0, \tag{14.2b}$$

we have three independent equations for the compensator parameters calculation. Since

$$\alpha = 1\,e^{j120^\circ} = -\frac{1}{2} + j\frac{\sqrt{3}}{2}, \qquad \alpha^* = 1\,e^{-j120^\circ} = -\frac{1}{2} - j\frac{\sqrt{3}}{2}$$

Eqn. (14.2a) can be rearranged to the form

$$\mathrm{Re}\{\boldsymbol{Y}_{\mathrm{u}}\} + \frac{\sqrt{3}}{2}T_{\mathrm{TR}} - \frac{\sqrt{3}}{2}T_{\mathrm{RS}} = 0 \tag{14.3a}$$

while Eqn. (14.2b) to can be expressed as

$$\mathrm{Im}\{\boldsymbol{Y}_{\mathrm{u}}\} - T_{\mathrm{ST}} + \frac{1}{2}T_{\mathrm{TR}} + \frac{1}{2}T_{\mathrm{RS}} = 0. \tag{14.3b}$$

Eqns (14.1), (14.3a), and (14.3b) have solution

$$\begin{aligned} T_{\mathrm{RS}} &= (\sqrt{3}\mathrm{Re}\boldsymbol{Y}_{\mathrm{u}} - \mathrm{Im}\boldsymbol{Y}_{\mathrm{u}} - B_{\mathrm{e}})/3 \\ T_{\mathrm{ST}} &= (2\mathrm{Im}\boldsymbol{Y}_{\mathrm{u}} - B_{\mathrm{e}})/3 \\ T_{\mathrm{TR}} &= (-\sqrt{3}\mathrm{Re}\boldsymbol{Y}_{\mathrm{u}} - \mathrm{Im}\boldsymbol{Y}_{\mathrm{u}} - B_{\mathrm{e}})/3. \end{aligned} \tag{14.4}$$

meaning, line-to-line susceptances T_{XY} of a reactance compensator that reduces both the reactive $\boldsymbol{i}_{\mathrm{r}}$ and unbalanced $\boldsymbol{i}_{\mathrm{u}}$ currents to zero value.

When susceptance T_{XY}, calculated from formula (14.4), is a positive number, then a capacitor of capacitance

$$C_{XY} = \frac{T_{XY}}{\omega} \tag{14.5}$$

should be connected between X and Y terminals. When this susceptance is negative, then an inductor of inductance

$$L_{XY} = \frac{1}{\omega T_{XY}} \tag{14.6}$$

should be connected. Such a compensator improves the power factor λ to unity and removes the supply current asymmetry.

Illustration 14.1 *Let us calculate the parameters of a balancing compensator for the entire reduction of the reactive and unbalanced currents in the circuit shown in Figure 14.2 assuming that the transformer is ideal, of turn ratio 1:1.*

The circuit, as seen from the primary side of the transformer, is equivalent to a circuit with a load in a Δ structure, with only one line-to-line branch between lines R and S. Its admittance is equal to

$$Y_{RS} = \frac{1}{3 + j1} = 0.30 - j0.10 \text{ S}.$$

Thus the equivalent susceptance of the load is equal to $B_e = -0.10$ S, *while the unbalanced admittance is*

$$Y_u = \text{Re}Y_u + j\text{Im}Y_u = -\alpha^* Y_{RS} = 0.316\, e^{j41.6°} = 0.236 + j0.210 \text{ S}.$$

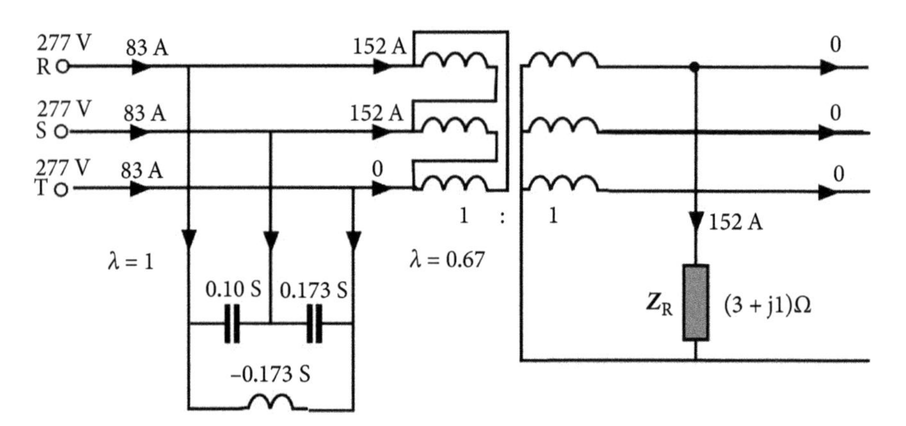

Figure 14.2 An example of an unbalanced load with a balancing compensator.

Hence, formula (14.4) results in the following branch susceptances of the compensator

$$T_{RS} = 0.10\ \text{S},\quad T_{ST} = 0.173\ \text{S},\quad T_{TR} = -0.173\ \text{S}.$$

Because susceptances T_{RS} and T_{ST} are positive, capacitors should be connected between lines RS and lines ST. The susceptance T_{TR} is negative, hence an inductor should be connected between lines TR. The structure and parameters of the balancing compensator are shown in Figure 14.2. The compensator reduces the reactive and unbalanced currents to zero so that it reduces the three-phase rms value of the supply current vector from value $||\boldsymbol{i}||$ = 215 A to $||\boldsymbol{i}'||$ = 144 A. Such a balancing compensator restores the supply current symmetry and improves the power factor from λ = 0.67 to unity.

*****

This illustration shows that the CPC-based approach to compensation provides a very simple and transparent method of calculation of the balancing compensator parameters. It confirms the applicability of this approach to reactance balancing compensator design.

14.3 Compensation in Circuits with Asymmetrical Sinusoidal Voltage

The method of compensator parameters calculation as presented in the previous section was developed under the assumption that the supply voltage is symmetrical. When this assumption is not satisfied, the developed method is no longer valid.

According to Section 7.16, formula (7.93), the current of unbalanced LTI load supplied from a source of sinusoidal but asymmetrical voltage is composed of four physical components

$$\boldsymbol{i} = \boldsymbol{i}_a + \boldsymbol{i}_r + \boldsymbol{i}_u^p + \boldsymbol{i}_u^n$$

which are mutually orthogonal, so that their three-phase rms values satisfy the relationship

$$||\boldsymbol{i}||^2 = ||\boldsymbol{i}_a||^2 + ||\boldsymbol{i}_r||^2 + ||\boldsymbol{i}_u^p||^2 + ||\boldsymbol{i}_u^n||^2.$$

The unbalanced current of the load is composed of the positive and negative sequence symmetrical components, namely

$$\boldsymbol{i}_u = \boldsymbol{i}_u^p + \boldsymbol{i}_u^n$$

Similarly, as in LTI circuits with symmetrical supply voltage, the reactive and unbalanced currents, $\boldsymbol{i}_r$ and $\boldsymbol{i}_u$, cause a decline in the power factor λ at the supply terminals. These currents can be reduced by a shunt reactance compensator. It can have a Δ configuration as shown in Figure 14.3. Let us assume that it is built of lossless reactance elements of susceptances T_{RS}, T_{ST}, and T_{TR}. The compensated load is specified, as was explained in Section 7.16, in terms of four admittances $\boldsymbol{Y}_b$, $\boldsymbol{Y}_d$, $\boldsymbol{Y}_u^p$, and $\boldsymbol{Y}_u^n$. These admittances of the compensator have additional index C.

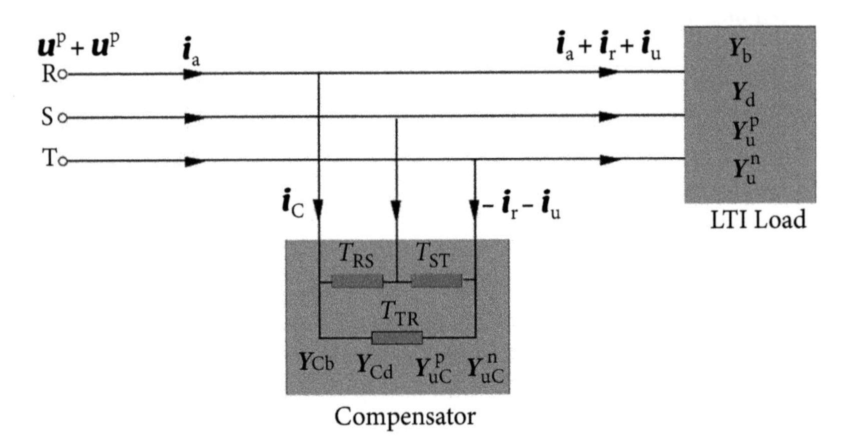

Figure 14.3 A three-phase LTI load with a reactance compensator.

The vector of crms values of the reactive current $\boldsymbol{i}_{Cr}$ of such a compensator is

$$\boldsymbol{I}_{Cr} = jB_{Cb}\,\boldsymbol{U}$$

where its balanced susceptance is

$$B_{Cb} = \frac{T_{RS}\,U_{RS}^2 + T_{ST}\,U_{ST}^2 + T_{TR}\,U_{TR}^2}{||\boldsymbol{u}||^2}. \tag{14.7}$$

Unbalanced admittances of such a compensator are

$$\boldsymbol{Y}_{uC}^p = -j(T_{ST} + \alpha T_{TR} + \alpha^* T_{RS}) \tag{14.8}$$

$$\boldsymbol{Y}_{uC}^n = -j(T_{ST} + \alpha^* T_{TR} + \alpha T_{RS}) \tag{14.9}$$

while the asymmetry-dependent unbalanced admittance

$$Y_{Cd} = j\frac{2a}{1+a^2}[T_{ST}\cos\psi + T_{TR}\cos(\psi - \frac{2\pi}{3}) + T_{RS}\cos(\psi + \frac{2\pi}{3})].$$

The compensator reduces the reactive current to zero on the condition that

$$B_{Cb} + B_b = 0$$

and it reduces the unbalanced current to zero on the condition that the sum of the unbalanced current of the load, specified by formula (7.85), and the compensator are equal to zero, that is,

$$(\boldsymbol{Y}_{\mathrm{Cd}} + \boldsymbol{Y}_{\mathrm{d}})\,\boldsymbol{U} + \mathbf{1}^{\mathrm{n}}(\boldsymbol{Y}^{\mathrm{p}}_{\mathrm{uC}} + \boldsymbol{Y}^{\mathrm{p}}_{\mathrm{u}})\,\boldsymbol{U}^{\mathrm{p}} + \mathbf{1}^{\mathrm{p}}(\boldsymbol{Y}^{\mathrm{n}}_{\mathrm{uC}} + \boldsymbol{Y}^{\mathrm{n}}_{\mathrm{u}})\boldsymbol{U}^{\mathrm{n}} = \mathbf{0}.$$

Coefficients at the voltage in this equation are identical for each line. Therefore, the unbalanced current is reduced to zero on the condition that for a single line, in particular for line R,

$$(\boldsymbol{Y}_{\mathrm{Cd}} + \boldsymbol{Y}_{\mathrm{d}})\,\boldsymbol{U}_{\mathrm{R}} + (\boldsymbol{Y}^{\mathrm{p}}_{\mathrm{uC}} + \boldsymbol{Y}^{\mathrm{p}}_{\mathrm{u}})\,\boldsymbol{U}^{\mathrm{p}} + (\boldsymbol{Y}^{\mathrm{n}}_{\mathrm{uC}} + \boldsymbol{Y}^{\mathrm{n}}_{\mathrm{u}})\boldsymbol{U}^{\mathrm{n}} = 0$$

or

$$(\boldsymbol{Y}_{\mathrm{Cd}} + \boldsymbol{Y}_{\mathrm{d}})(\boldsymbol{U}^{\mathrm{p}} + \boldsymbol{U}^{\mathrm{n}}) + (\boldsymbol{Y}^{\mathrm{p}}_{\mathrm{uC}} + \boldsymbol{Y}^{\mathrm{p}}_{\mathrm{u}})\,\boldsymbol{U}^{\mathrm{p}} + (\boldsymbol{Y}^{\mathrm{n}}_{\mathrm{uC}} + \boldsymbol{Y}^{\mathrm{n}}_{\mathrm{u}})\boldsymbol{U}^{\mathrm{n}} = 0. \tag{14.10}$$

Since

$$\frac{\boldsymbol{U}^{\mathrm{n}}}{\boldsymbol{U}^{\mathrm{p}}} = \boldsymbol{a} = a e^{j\psi}$$

Eqn (14.10) can be rearranged to

$$(\boldsymbol{Y}^{\mathrm{p}}_{\mathrm{C}} + \boldsymbol{Y}^{\mathrm{p}}_{\mathrm{u}}) + (\boldsymbol{Y}^{\mathrm{n}}_{\mathrm{uC}} + \boldsymbol{Y}^{\mathrm{n}}_{\mathrm{u}})a e^{j\psi} + (\boldsymbol{Y}_{\mathrm{Cd}} + \boldsymbol{Y}_{\mathrm{d}})(1 + a e^{j\psi}) = 0. \tag{14.11}$$

The asymmetry-dependent unbalanced admittance, specified in Section 7.16, for the load by formula (7.82), for the compensator can be written in the form

$$\boldsymbol{Y}_{\mathrm{Cd}} = (\boldsymbol{c}_1 T_{\mathrm{ST}} + \boldsymbol{c}_2 T_{\mathrm{TR}} + \boldsymbol{c}_3 T_{\mathrm{RS}}) \tag{14.12}$$

where

$$\begin{aligned}
\boldsymbol{c}_1 &\stackrel{\mathrm{df}}{=} j\frac{2a\cos\psi}{1 + a^2}\\
\boldsymbol{c}_2 &\stackrel{\mathrm{df}}{=} j\frac{2a\cos(\psi - 120^{\circ})}{1 + a^2}\\
\boldsymbol{c}_3 &\stackrel{\mathrm{df}}{=} j\frac{2a\cos(\psi - 240^{\circ})}{1 + a^2}.
\end{aligned} \tag{14.13}$$

With formula (14.13), Eqn. (14.12) can be rearranged to the form

$$\boldsymbol{F}_1\, T_{\mathrm{RS}} + \boldsymbol{F}_2\, T_{\mathrm{ST}} + \boldsymbol{F}_3\, T_{\mathrm{TR}} + \boldsymbol{F}_4 = 0 \tag{14.14}$$

where

$$\boldsymbol{F}_1 = \boldsymbol{c}_3(1 + a e^{j\psi}) - j(\boldsymbol{a}^* + \alpha\, a e^{j\psi})$$

$$\boldsymbol{F}_2 = \boldsymbol{c}_1(1 + ae^{j\psi}) - j(1 + ae^{j\psi})$$

$$\boldsymbol{F}_3 = \boldsymbol{c}_2(1 + ae^{j\psi}) - j(\alpha + \alpha^* ae^{j\psi})$$

$$\boldsymbol{F}_4 = (1 + ae^{j\psi})\boldsymbol{Y}_{\rm d} + \boldsymbol{A}^{\rm p} + (1 + ae^{j\psi})\boldsymbol{Y}^{\rm n}$$

Eqn (14.14) must be satisfied both for the real and the imaginary parts, so that

$$\begin{aligned} &T_{\rm RS}{\rm Re}\boldsymbol{F}_1 + T_{\rm ST}{\rm Re}\boldsymbol{F}_2 + T_{\rm TR}{\rm Re}\boldsymbol{F}_3 + {\rm Re}\boldsymbol{F}_4 = 0 \\ &T_{\rm RS}{\rm Im}\boldsymbol{F}_1 + T_{\rm ST}{\rm Im}\boldsymbol{F}_2 + T_{\rm TR}{\rm Im}\boldsymbol{F}_3 + {\rm Im}\boldsymbol{F}_4 = 0. \end{aligned} \tag{14.15}$$

When the condition (14.15) is combined with formula (14.7) the third equation is obtained

$$T_{\rm RS}\, U_{\rm RS}^2 + T_{\rm ST}\, U_{\rm ST}^2 + T_{\rm TR}\, U_{\rm TR}^2 = -B_{\rm b}||\boldsymbol{u}||^2 \tag{14.16}$$

Eqns (14.15) and (14.16) can be expressed in the following matrix form, referred to as a *compensator equation*:

$$\begin{bmatrix} U_{\rm RS}^2 & U_{\rm ST}^2 & U_{\rm TR}^2 \\ {\rm Re}\boldsymbol{F}_1 & {\rm Re}\boldsymbol{F}_2 & {\rm Re}\boldsymbol{F}_3 \\ {\rm Im}\boldsymbol{F}_1 & {\rm Im}\boldsymbol{F}_2 & {\rm Im}\boldsymbol{F}_3 \end{bmatrix} \begin{bmatrix} T_{\rm RS} \\ T_{\rm ST} \\ T_{\rm TR} \end{bmatrix} = \begin{bmatrix} -B_{\rm b}||\boldsymbol{u}||^2 \\ -{\rm Re}\boldsymbol{F}_4 \\ -{\rm Im}\boldsymbol{F}_4 \end{bmatrix} \tag{14.17}$$

Illustration 14.2 *Let us calculate the parameters of the compensator for the unbalanced load shown in Figure 14.4, assuming that it is supplied with strongly asymmetrical voltage.*

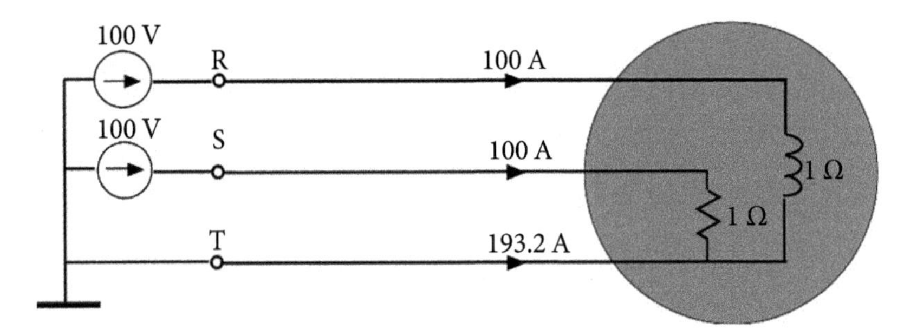

Figure 14.4 An example of a circuit with a very high load imbalance and very high supply voltage asymmetry.

The crms values of the positive and negative sequence symmetrical components in the circuit analyzed are equal to

$$\begin{bmatrix} \boldsymbol{U}^{\rm p} \\ \boldsymbol{U}^{\rm n} \end{bmatrix} = \frac{1}{3}\begin{bmatrix} 1, & \alpha, & \alpha^* \\ 1, & \alpha^*, & \alpha \end{bmatrix} \begin{bmatrix} 100 \\ 100e^{-j120^\circ} \\ 0 \end{bmatrix} = \begin{bmatrix} 66.66 \\ 33.33\, e^{j60^\circ} \end{bmatrix} \text{V}.$$

The three-phase rms values of the supply voltage symmetrical components are

$$||\mathbf{u}^{\mathrm{p}}|| = \sqrt{3}\, U^{\mathrm{p}} = \sqrt{3} \times 66.66 = 115.47 \text{ V}$$

$$||\mathbf{u}^{\mathrm{n}}|| = \sqrt{3}\, U^{\mathrm{n}} = \sqrt{3} \times 33.33 = 57.73 \text{ V}$$

and consequently, the three-phase rms value of the supply voltage is

$$||\mathbf{u}|| = \sqrt{||\mathbf{u}^{\mathrm{p}}||^2 + ||\mathbf{u}^{\mathrm{n}}||^2} = \sqrt{115.47^2 + 57.73^2} = 129.1 \text{ V}.$$

Since the active and reactive powers are equal to $P = 10$ kW *and* $Q = 10$ kVar, *the equivalent balanced admittance* Y_{b} *of the load in the circuit shown in Figure 14.4 is equal to*

$$Y_{\mathrm{b}} = G_{\mathrm{b}} + jB_{\mathrm{b}} = \frac{P - jQ}{||\mathbf{u}||^2} = 0.600 - j\,0.600 \text{ S}.$$

The complex coefficient of the supply voltage asymmetry is equal to

$$\boldsymbol{a} = a\, e^{j\psi} = \frac{\boldsymbol{U}^{\mathrm{n}}}{\boldsymbol{U}^{\mathrm{p}}} = \frac{33.33\, e^{j60°}}{66.66} = 0.50\, e^{j60°}.$$

The crms values of the supply voltage measured against the artificial zero are

$$\boldsymbol{U}_{\mathrm{R}} = \boldsymbol{U}^{\mathrm{p}} + \boldsymbol{U}^{\mathrm{n}} = 88.18\, e^{j19.1°} \text{ V},$$

$$\boldsymbol{U}_{\mathrm{S}} = \alpha^* \boldsymbol{U}^{\mathrm{p}} + \alpha \boldsymbol{U}^{\mathrm{n}} = 88.18\, e^{-j139.1°} \text{ V},$$

$$\boldsymbol{U}_{\mathrm{T}} = \alpha \boldsymbol{U}^{\mathrm{p}} + \alpha^* \boldsymbol{U}^{\mathrm{n}} = 33.33\, e^{j120.0°} \text{ V}$$

while the crms values of line-to-line voltages are equal to

$$\boldsymbol{U}_{\mathrm{RS}} = \boldsymbol{U}_{\mathrm{R}} - \boldsymbol{U}_{\mathrm{S}} = 173.21\, e^{j30°} \text{ V},$$

$$\boldsymbol{U}_{\mathrm{ST}} = \boldsymbol{U}_{\mathrm{S}} - \boldsymbol{U}_{\mathrm{T}} = 100\, e^{-j90°} \text{ V},$$

$$\boldsymbol{U}_{\mathrm{TR}} = \boldsymbol{U}_{\mathrm{T}} - \boldsymbol{U}_{\mathrm{R}} = 100\, e^{j180°} \text{ V}.$$

The load unbalanced admittances are equal to

$$\boldsymbol{Y}^{\mathrm{p}} = -(Y_{\mathrm{ST}} + \alpha Y_{\mathrm{TR}} + \alpha^* Y_{\mathrm{RS}}) = -[1 + \alpha(-j\,1)] = 1.932\, e^{-j165°} \text{ S},$$

$$\boldsymbol{Y}^{\mathrm{n}} = -(Y_{\mathrm{ST}} + \alpha^* Y_{\mathrm{TR}} + \alpha Y_{\mathrm{RS}}) = -[1 + \alpha^*(-j\,1)] = 0.518\, e^{-j105°} \text{ S},$$

$$\begin{aligned} \boldsymbol{Y}_{\mathrm{d}} &= \frac{2a}{1 + a^2}[Y_{\mathrm{ST}} \cos\psi + Y_{\mathrm{TR}} \cos(\psi - \frac{2\pi}{3}) + Y_{\mathrm{RS}} \cos(\psi + \frac{2\pi}{3})] = \\ &= \frac{2 \times 0.5}{1 + 0.5^2}[\cos(60°) - j\cos(60° - 120°)] = 0.566\, e^{-j45°} \text{ S}. \end{aligned}$$

With these values of the load admittances:

$$c_1 = j\frac{2a\cos\psi}{1+a^2} = j\frac{2\times 0.6\cos 60^\circ}{1+0.5^2} = j0.4,$$

$$c_2 = j\frac{2a\cos(\psi - 120^\circ)}{1+a^2} = j0.4,$$

$$c_3 = j\frac{2a\cos(\psi - 240^\circ)}{1+a^2} = -j0.8.$$

Hence

$$F_1 = c_3(1 + ae^{j\psi}) - j(\alpha^* + \alpha a e^{j\psi}) = 0.52,$$

$$F_2 = c_1(1 + ae^{j\psi}) - j(1 + ae^{j\psi}) = 0.26 - j\,0.75,$$

$$F_3 = c_2(1 + ae^{j\psi}) - j(\alpha + \alpha^* a e^{j\psi}) = 0.26 + j\,0.75,$$

$$F_4 = (1 + ae^{j\psi})Y_d + Y^p + (1 + ae^{j\psi})Y^n = -1.01 - j1.01.$$

With such coefficients, the compensator equation has the form

$$\begin{bmatrix} 3\times 10^4 & 10^4 & 10^4 \\ -0.52 & 0.26 & 0.26 \\ 0 & -0.75 & 0.75 \end{bmatrix} \begin{bmatrix} T_{RS} \\ T_{ST} \\ T_{TR} \end{bmatrix} = \begin{bmatrix} 10^4 \\ 1.01 \\ 1.01 \end{bmatrix}.$$

With regard to the compensator branch susceptances, this equation has the solution

$$T_{RS} = -0.58\ \text{S}, \quad T_{ST} = 0.69\ \text{S}, \quad T_{TR} = 2.04\ \text{S}$$

Assuming that the supply voltage frequency is $f = 50$ Hz, *thus* $\omega = 314$ rad/s, *then the compensator parameters are*

$$L_{RS} = -\frac{1}{\omega T_{RS}} = 5.49\ \text{mH}, \quad C_{ST} = \frac{T_{ST}}{\omega} = 2.20\ \text{mF}, \quad C_{TR} = \frac{T_{TR}}{\omega} = 6.50\ \text{mF}.$$

The circuit with the balancing compensator and compensation results is shown in Figure 14.5.

The reactance balancing compensator with such parameters compensates entirely the reactive and unbalanced currents, reducing the three-phase rms value of the supply current from $||\boldsymbol{i}|| = 239.4$ A *to* $||\boldsymbol{i}|| = 77.5$ A, *which increases the power factor from* $\lambda = 0.32$ *to* $\lambda = 1$. *The load with the compensator is equivalent to a balanced resistive load of conductance* G_b. *Observe that the supply current remains asymmetrical, however, because of the supply voltage asymmetry.*

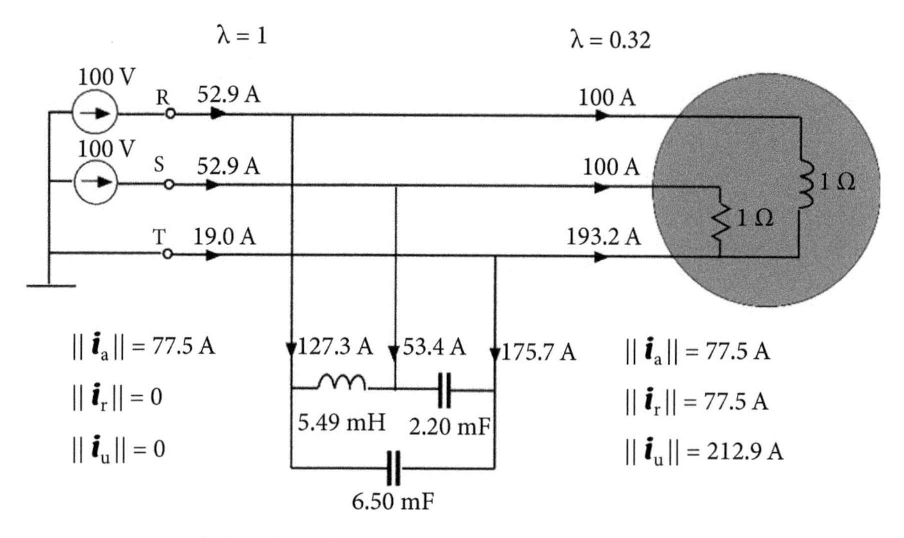

Figure 14.5 An unbalanced load with a reactance compensator and compensation results.

14.4 Compensation in Circuits with Nonsinusoidal Voltage

Compensation of the reactive current of balanced LTI loads in the presence of the supply voltage distortion does not differ from such compensation in single-phase circuits. It can be performed line by line, with single-phase reactance compensators, as discussed in Chapter 13. Only the circuit does not have to be compensated for the zero-sequence harmonics, meaning harmonics of the order $n = 3k$, mainly the third order harmonic, which cannot occur in the supply current of such balanced loads. It means that the compensation of such loads, for readers acquainted with Chapter 13, is trivial. When the load is unbalanced, it cannot be compensated line by line, however. More advanced methods are needed for that. Such methods can be developed using the CPC-based power theory of three-phase, three wire (3p3w) circuits with nonsinusoidal supply voltage, as presented in Section 7.9, and the methods of reactance compensation in single-phase circuits, as presented in Section 13.1.

Let the supply voltage $\boldsymbol{u}$ be nonsinusoidal but symmetrical and composed of harmonics of the order n from a set N, thus

$$\boldsymbol{u} = \sum_{n \in N} \boldsymbol{u}_n.$$

The supply current of LTI loads is composed only of harmonics $\boldsymbol{i}_n$ of the order from the same set N, that is,

$$\boldsymbol{i} = \sum_{n \in N} \boldsymbol{i}_n.$$

If the load is LTI, then the superposition principle can be applied for the supply current calculation, meaning the supply current harmonics $||\boldsymbol{i}_n||$ can be calculated harmonic by harmonic. The LTI load can be specified for an individual harmonic in terms of the equivalent admittance Y_{en} and the unbalanced admittance $\boldsymbol{Y}_{un}$, as shown in Figure 14.6.

The supply current harmonic $||\boldsymbol{i}_n||$ can be decomposed, according to Section 7.9, into three components

$$\boldsymbol{i}_{cn} = G_{en}\,\boldsymbol{u}_n$$

$$\boldsymbol{i}_{rn} = \sqrt{2}\mathrm{Re}\{jB_{en}\;\mathbf{1}_n U_n\, e^{jn\omega_1 t}\}$$

$$\boldsymbol{i}_{un} = \sqrt{2}\mathrm{Re}\{Y_{un}\;\mathbf{1}_n^* U_n\, e^{jn\omega_1 t}\}.$$

Let a lossless reactance compensator, with Δ structure and line-to-line susceptances for harmonic frequencies T_{RSn}, T_{STn}, and T_{TRn} be connected in parallel to the load, as shown in Figure 14.6.

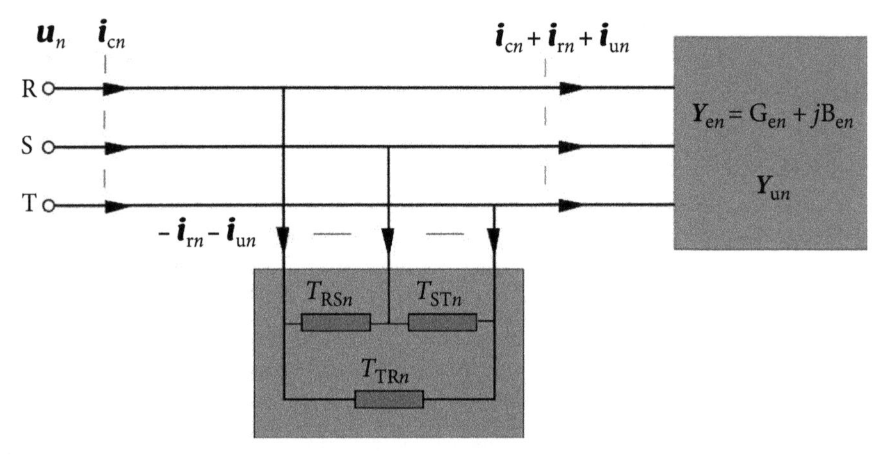

Figure 14.6 An LTI load with a balancing compensator and their equivalent parameters for the n-th order harmonic.

The equivalent susceptance of the compensator for the nth order harmonic is

$$B_{Cn} = T_{STn} + T_{TRn} + T_{RSn}$$

and its unbalanced admittance for that harmonic,

$$Y_{Cu\,n} = -j(T_{STn} + \beta T_{TRn} + \beta^* T_{RSn}).$$

The compensator reduces entirely the reactive current of the nth order harmonic, $\boldsymbol{i}_{rn}$, on the condition that

$$B_{en} + (T_{STn} + T_{TRn} + T_{RSn}) = 0 \tag{14.18}$$

and it reduces entirely the unbalanced current, $\boldsymbol{i}_{un}$, of the same harmonic order on the condition that

$$Y_{u\,n} - j(T_{STn} + \beta T_{TRn} + \beta^* T_{RSn}) = 0. \tag{14.19}$$

According to definition (7.23), the complex rotation coefficient β depends on the harmonic sequence index, specified by integer s. For harmonics of the positive symmetrical sequence $s = 1$, thus $\beta = \alpha$, while for harmonics of the negative sequence $s = -1$, thus $\beta = \alpha^*$. Consequently, Eqns (14.18) and (14.19) have the solution

$$\begin{aligned} T_{RSn} &= \frac{1}{3}(s\sqrt{3}\mathrm{Re}\, Y_{un} - \mathrm{Im}\, Y_{un} - B_{en}) \\ T_{STn} &= \frac{1}{3}(2\mathrm{Im}\, Y_{un} - B_{en}) \\ T_{TRn} &= -\frac{1}{3}(s\sqrt{3}\mathrm{Re}\, Y_{un} + \mathrm{Im}\, Y_{un} + B_{en}). \end{aligned} \tag{14.20}$$

with the compensator susceptances T_{RSn}, T_{STn}, and T_{TRn}, denoted generally by T_{XYn}, dependent on the harmonic sequence. If the compensator has these susceptances for each harmonic order n from the set of the supply voltage harmonics N, it compensates both the reactive and unbalanced currents entirely. Since susceptance T_{XYn} of branches of a reactance compensator, calculated with formulae (14.20), has to be a real, positive, or negative number, such a compensator can be always built. Procedures presented in Chapter 10 can be used for that.

The nth order harmonic of the supply current that remains after compensation is equal to

$$\boldsymbol{i}_{cn} = G_{en}\, \boldsymbol{u}_n.$$

This current can be decomposed as follows

$$\boldsymbol{i}_{cn} = G_{en}\, \boldsymbol{u}_n = G_{en}\, \boldsymbol{u}_n + (G_{en} - G_e)\, \boldsymbol{u}_n = \boldsymbol{i}_{an} + \boldsymbol{i}_{sn}$$

thus it is composed of harmonic components of the active and the scattered currents.

Illustration 14.3 *A source of a symmetrical three-phase voltage is loaded with a resistive single-phase load, as shown in Figure 14.7, with resistance $R = 3\Omega$. The rms value of the supply voltage fundamental harmonic is equal to $U_1 = 220$ V. The supply voltage contains the fifth order harmonic of the rms value equal to $U_5 = 5\,\%\, U_1$ and the seventh order harmonic with $U_7 = 3\,\%\, U_1$. The coupling transformer is ideal with a turn ratio 1:1.*

The circuit analysis provides the following results. The three-phase rms value of the supply current is $||\boldsymbol{i}|| = 179.6$ A, while the rms value of the active current is $||\boldsymbol{i}_a|| =$ 127.0 A. The source is not loaded, of course, with the reactive and the scattered currents, but only with the unbalanced current, $\boldsymbol{i}_u$, of the rms value $||\boldsymbol{i}_u|| = 127.0$ A. The power factor of such a load is $\lambda = 0.71$.

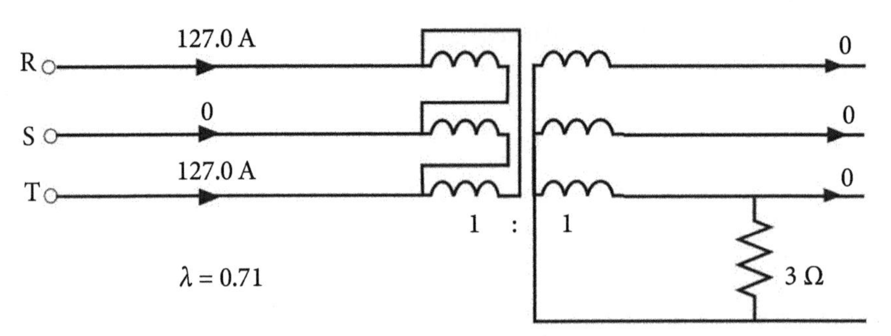

Figure 14.7 An example of a circuit with a resistive unbalanced load.

The fundamental and the seventh order harmonics are of positive order ($s = 1$), while the fifth order harmonic is of the negative order ($s = -1$). Since the equivalent circuit of the load has only one non-zero line-to-line admittance of the same value for each harmonic, $\mathbf{Y}_{\mathrm{TR}} = 1/R = 0.333$ S, thus the unbalanced admittances of the load for harmonics that are present in the supply voltage are equal to

$$\mathbf{Y}_{\mathrm{u}1} = \mathbf{Y}_{\mathrm{u}7} = 0.333\, e^{-j60°}\ \mathrm{S} \quad \text{and} \quad \mathbf{Y}_{\mathrm{u}5} = 0.333\, e^{j60°}\ \mathrm{S}.$$

The susceptance B_{en} of such a load is equal, of course, to zero.

Having the values of the unbalanced admittance and equivalent susceptance of the load for the supply voltage harmonics n = 1, 5, and 7, susceptances of the compensator branches can be calculated from formulae (14.20).

The susceptance of the RS branch should be equal to

$$T_{\mathrm{RS}1} = 0.1925\ \mathrm{S},\quad T_{\mathrm{RS}5} = -0.1925\ \mathrm{S},\quad T_{\mathrm{RS}7} = 0.1925\ \mathrm{S}$$

meaning it must be capacitive for the fundamental and the seventh order harmonic, while it should be inductive for the fifth order harmonic.

The susceptance of the ST branch should be equal to

$$T_{\mathrm{ST}1} = -0.1925\ \mathrm{S},\quad T_{\mathrm{ST}5} = 0.1925\ \mathrm{S},\quad T_{\mathrm{ST}7} = -0.1925\ \mathrm{S}$$

meaning it must be inductive for the fundamental and the seventh order harmonic, while it should be capacitive for the fifth order harmonic.

It results from formulae (14.20) that the susceptance of the TR branch should be equal to zero for each harmonic order, thus the TR branch is not needed. The structure and parameters of the compensator that satisfies these requirements are shown in Figure 14.8.

The compensator improves the power factor to unity and reduces entirely the supply current asymmetry. The load with the compensator is equivalent to a resistive balanced load, with the resistance from line-to-zero equal for each line to $R_{\mathrm{R}} = R_{\mathrm{S}} = R_{\mathrm{T}} = 3\ \Omega$. The synthesis of the compensator branches was based on Chapter 10, with the assumption that $\omega_1 = 1$ rad/s.

Parameters of the compensator at other frequencies can be found by denormalization. If a capacitor and inductor at the angular frequency of 1 rad/s are denoted *by C_N and L_N, respectively, then at other frequencies of the supply voltage, $C = C_N/\omega_1$ and $L = L_N\,\omega_1$. Observe also that the structure and parameters of the compensator as shown in Figure 14.8 are not the only ones. There are an infinite number of reactance one-ports that differ as to parameters and even structure, but can have the same required susceptances.*

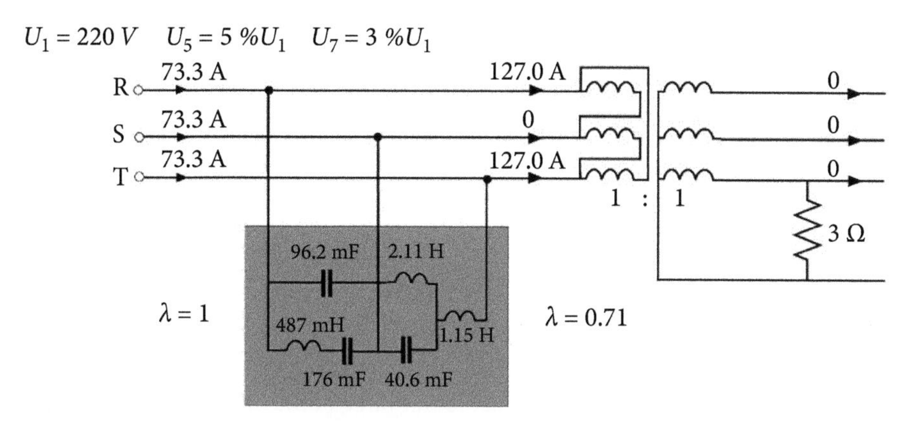

Figure 14.8 An example of a three-phase circuit with a reactance balancing compensator.

*****

This illustration demonstrates the possibility of a complete compensation of the unbalanced current at distorted supply voltage. At the same time, it shows the compensator complexity. In this illustration, there are two reactance elements for each supply voltage harmonic.

14.5 Reduction of the Compensator Complexity

A balancing compensator would have the lowest complexity if it has no more than one reactance element per line. If a capacitor is needed as such an element, however, then a resonance in the range of harmonic frequencies with the supply system inductance will occur. To avoid any resonance, branches of the compensator should have an inductive impedance in the range of harmonics. Thus only inductors or capacitors with a series inductor are acceptable as the balancing compensator branches. Structures of acceptable branches and plots of their susceptances are shown in Figure 14.9.

The susceptance of a compensator's branches with no more than two reactance elements branches will be denoted by D, and the vector or the branch currents will be denoted by $\boldsymbol{j}_D$, as shown in Figure 14.10. Such a compensator does not compensate the unbalanced and reactive currents but only modifies them. When the compensator is lossless, then it does not affect equivalent conductance for harmonics as seen from

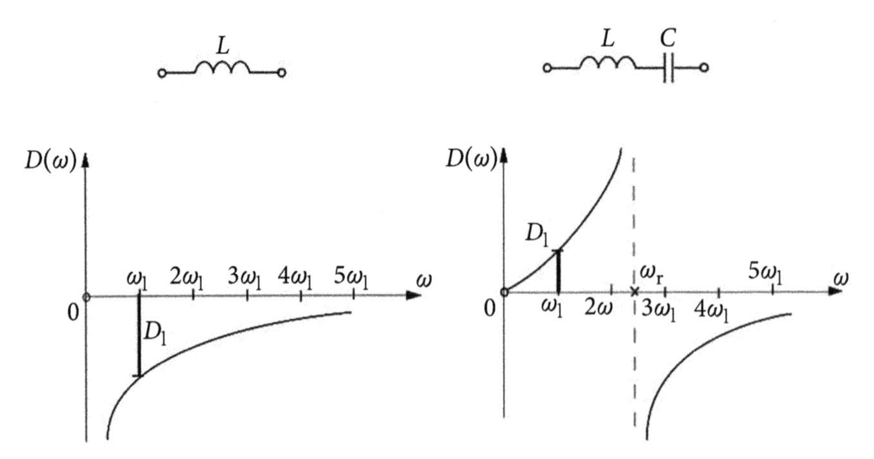

Figure 14.9 Acceptable branches of a balancing compensator and their susceptances.

the distribution system. It remains equal to the value of the load conductance G_{en}, and consequently, the current $\boldsymbol{i}_{cn}$ remains unchanged.

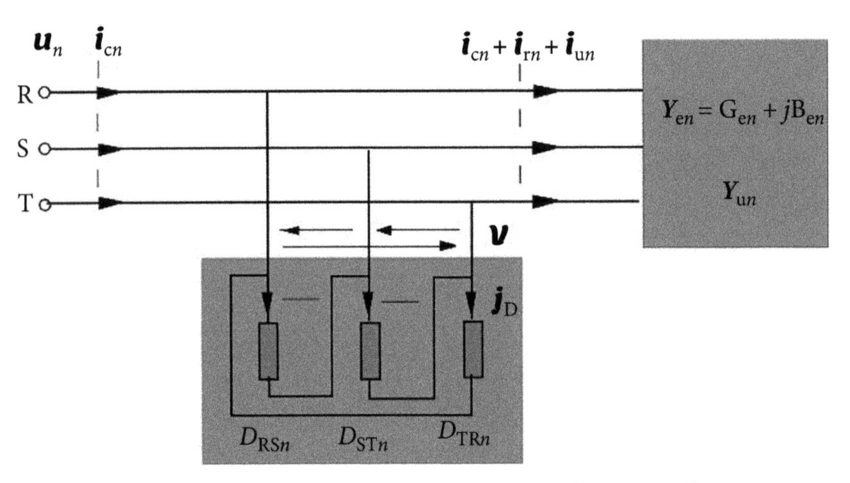

Figure 14.10 A load with a shunt balancing compensator with no more than two reactance elements branches.

Let us denote by $\boldsymbol{D}_{\text{n}}$ the matrix of branch susceptances of such a compensator

$$\boldsymbol{D}_n \stackrel{\Delta}{=} \begin{bmatrix} D_{\text{RS}n} & 0 & 0 \\ 0 & D_{\text{ST}n} & 0 \\ 0 & 0 & D_{\text{TR}n} \end{bmatrix}.$$

If symbol $\boldsymbol{V}_n$ denotes the vector of crms values of harmonics of the line-to-line voltages

$$\boldsymbol{V}_n \stackrel{\Delta}{=} \begin{bmatrix} V_{\mathrm{RS}n} \\ V_{\mathrm{ST}n} \\ V_{\mathrm{TR}n} \end{bmatrix},$$

then the vector of branch currents of such a compensator is equal to

$$\boldsymbol{j}_{\mathrm{D}} = \sqrt{2}\mathrm{Re}\sum_{n\in N} \boldsymbol{D}_n\, \boldsymbol{V}_n\, e^{jn\omega_1 t}. \tag{14.21}$$

Let us assume that the compensator for total compensation of the unbalanced and reactive currents, shown in Figure 14.11

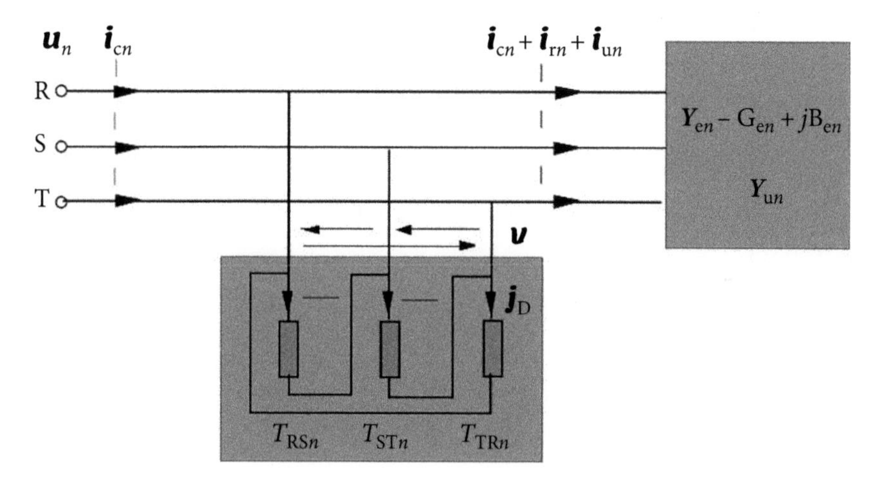

Figure 14.11 A load with a shunt balancing compensator for total compensation of the unbalanced and reactance currents.

has susceptances for the voltage harmonics $T_{\mathrm{RS}n}$, $T_{\mathrm{ST}n}$, and $T_{\mathrm{TR}n}$ arranged in the matrix

$$\boldsymbol{T}_n \stackrel{\Delta}{=} \begin{bmatrix} \boldsymbol{T}_{\mathrm{RS}n} & 0 & 0 \\ 0 & \boldsymbol{T}_{\mathrm{ST}n} & 0 \\ 0 & 0 & \boldsymbol{T}_{\mathrm{TR}n} \end{bmatrix}.$$

The vector of branch current of such a compensator, denoted by $\boldsymbol{j}_{\mathrm{T}}$, is equal to

$$\boldsymbol{j}_{\mathrm{T}} = \sqrt{2}\mathrm{Re}\sum_{n\in N} \boldsymbol{T}_n\, \boldsymbol{V}_n\, e^{jn\omega_1 t}. \tag{14.22}$$

Let us define the distance of vectors $\boldsymbol{j}_\mathrm{D}$ and $\boldsymbol{j}_\mathrm{r}$ as the three-phase rms value of their difference

$$d \stackrel{\mathrm{df}}{=} ||\boldsymbol{j}_\mathrm{T} - \boldsymbol{j}_\mathrm{D}|| \stackrel{\mathrm{df}}{=} ||\Delta\boldsymbol{j}|| \tag{14.23}$$

where the difference of these vectors can be expressed as

$$\Delta\boldsymbol{j} = \sqrt{2}\mathrm{Re}\sum_{n\in N}(\boldsymbol{T}_n - \boldsymbol{D}_n)\,\boldsymbol{V}_n\, e^{jn\omega_1 t} = \sqrt{2}\mathrm{Re}\sum_{n\in N}\Delta\boldsymbol{J}_n\, e^{jn\omega_1 t} \tag{14.24}$$

A balancing compensator with susceptances $D_{\mathrm{RS}n}$, $D_{\mathrm{ST}n}$, and $D_{\mathrm{TR}n}$ minimizes the unbalanced and reactive currents rms value on the condition that the distance d of vectors $\boldsymbol{j}_\mathrm{D}$ and $\boldsymbol{j}_\mathrm{T}$ is the lowest possible.

The square of this distance is equal to

$$d^2 = ||\Delta\boldsymbol{j}||^2 = \mathrm{Re}\sum_{n\in N}\Delta\boldsymbol{J}_n^\mathrm{T}\,\Delta\boldsymbol{J}_n^* = \sum_{n\in N}(\Delta J_{\mathrm{RS}n}^2 + \Delta\boldsymbol{J}_{\mathrm{ST}n}^2 + \Delta\boldsymbol{J}_{\mathrm{TR}n}^2) =$$
$$= \sum_{n\in N}[(T_{\mathrm{RS}n} - D_{\mathrm{RS}n})^2\, V_{\mathrm{RS}n}^2 + (T_{\mathrm{ST}n} - D_{\mathrm{ST}n})^2\, V_{\mathrm{ST}n}^2 + (T_{\mathrm{TR}n} - D_{\mathrm{TR}n})^2\, V_{\mathrm{TR}n}^2]. \tag{14.25}$$

It is equal to the sum of three, non-negative, mutually independent terms, of the general form

$$d_{\mathrm{XY}}^2 = \sum_{n\in N}(T_{\mathrm{XY}n} - D_{\mathrm{XY}n})^2 V_{\mathrm{XY}n}^2 = \sum_{n\in N} d_{\mathrm{XY}n}^2 \tag{14.26}$$

where indices XY denote terminals RS, ST, and TR, respectively. Thus the distance between vectors $\boldsymbol{j}_\mathrm{r}$ and $\boldsymbol{j}_\mathrm{D}$ is the shortest when each of the terms in the formula (14.25) has the lowest value.

The rms value of the voltage fundamental harmonic $V_{\mathrm{XY}1}$ is usually at least twenty times higher than the rms value of any voltage harmonic, $V_{\mathrm{XY}n}$. Thus if a resonance will not occur in any of the compensator branches in the vicinity of the voltage harmonic, then the difference of susceptances for the fundamental frequency $T_{\mathrm{XY}1} - D_{\mathrm{XY}1}$ is the dominating term in the distance of vectors $\boldsymbol{j}_\mathrm{T}$ and $\boldsymbol{j}_\mathrm{D}$.

Therefore compensator branches should be built in such a way that susceptances $T_{\mathrm{XY}1}$ and $D_{\mathrm{XY}1}$ should have the same sign. This means that if susceptance $T_{\mathrm{XY}1}$ is negative, the compensator should have an inductor between terminals X and Y since inductors have negative susceptance. Inductance L_{XY} of such a branch should minimize the term

$$d_{\mathrm{XY}}^2 = \sum_{n\in N}\left(T_{\mathrm{XY}n} + \frac{1}{n\omega_1 L_{\mathrm{XY}}}\right)^2 V_{\mathrm{XY}n}^2. \tag{14.27}$$

When susceptance $T_{\mathrm{XY}1}$ is positive, then the compensator should have a capacitor, but in series with an inductor between X and Y terminals. The inductance value should be selected in such a way that the brunch susceptance

$$D_{XYn} = \frac{n\omega_1 C_{XY}}{1 - n^2\omega_1^2 L_{XY} C_{XY}}$$

is capacitive for the fundamental, but inductive in the range of harmonic frequencies. It requires that the resonant frequency of the branch

$$\omega_r = \frac{1}{\sqrt{L_{XY} C_{XY}}}$$

is higher than the fundamental frequency. Both parameters should be selected in such a way that the term

$$d_{XY}^2 = \sum_{n\in N} \left(T_{XYn} - \frac{n\omega_1 C_{XY}}{1 - n^2\omega_1^2 L_{XY} C_{XY}} \right)^2 V_{XYn}^2. \tag{14.28}$$

The inductance that minimizes the distance d_{XY} for a purely inductive branch can be found from the condition

$$\frac{d}{d L_{XY}} \{d_{XY}^2\} = 0$$

applied to the formula (14.27). If terminals XY are neglected for simplicity's sake, we obtain the optimum inductance of the compensator branch, equal to

$$L_{opt} = -\frac{\sum_{n\in N} \frac{V_n^2}{n^2}}{\omega_1 \sum_{n\in N} \frac{T_n V_n^2}{n}}. \tag{14.29}$$

At the voltage distortion at a level of a few percent, the optimum capacitance can be approximated with high accuracy by the value $L_{opt} .= -1/\omega_1 T_1$.

Minimization of the distance d_{XY} for an LC branch, with the square specified with formula (14.28), results in the optimal capacitance C_{opt}, which has to satisfy the following relation

$$\sum_{n\in N} \frac{n T_n V_n^2}{(1 - n^2\omega_1^2 L\, C_{opt})^2} + \sum_{n\in N} \frac{n^2 \omega_1 C_{opt} V_n^2}{(1 - n^2\omega_1^2 L\, C_{opt})^3} = 0. \tag{14.30}$$

This relation is similar to formula (13.8) in Section 13.2 for optimum capacitance value at the reactive current rms value minimization with an LC compensator. It is an implicit formula for C_{opt}, which requires an iterative calculation of this capacitance as a limit of sequence C_1, C_2,... C_i, C_{i+1},...C_{opt}. where

$$C_{i+1} = \frac{\sum\limits_{n\in N} \frac{n\,T_n V_n^2}{\left(1 - n^2\omega_1^2 LC_i\right)^2}}{\omega_1 \sum\limits_{n\in N} \frac{n^2 V_n^2}{\left(1 - n^2\omega_1^2 LC_i\right)^3}}. \tag{14.31}$$

The inductance is a parameter in this calculation, chosen in such a way that the resonance in the LC branch is located between the fundamental frequency and the frequency of higher-order harmonics present in the supply voltage. For the load balancing and reactive current compensation at the fundamental frequency, a capacitance C_0 that satisfies the relation

$$T_1 - \frac{\omega_1 C_0}{1 - \omega_1^2 LC_0} = 0$$

is needed and this capacitance can be selected for starting iterative calculation of the capacitance optimum value. Denoting resonant frequency of the branch

$$\omega_r = \frac{1}{\sqrt{LC_0}}$$

the susceptance of the capacitor should be equal to

$$\omega_1 C_0 = [1 - \left(\frac{\omega_1}{\omega_r}\right)^2] T_1. \tag{14.32}$$

This is usually the susceptance of the optimal capacitance, that is, C_{opt} $C_{\text{opt}} \approx C_0$, meaning the iterative calculation does not change the needed capacitance very much.

Illustration 14.4 *Let us find parameters of a balancing reactance compensator of reduced complexity for the load as assumed in Illustration 14.3, and shown in Figure 14.7.*

Branch susceptances of the compensator for total compensation of the unbalanced and reactive currents at the fundamental frequency, as it was calculated in Illustration 14.3, are equal to

$$T_{RS1} = 0.1925 \text{ S}, \quad T_{ST1} = -0.1925 \text{ S}, \quad T_{TR1} = 0.$$

The susceptance T_{RS1} is positive, thus LC branch is needed between the R and S terminals. The susceptance T_{ST1} is negative, thus an inductor must be connected between terminals S and T. Assuming that the frequency is normalized to $\omega_1 = 1$

rad/s, *and the resonant frequency of the LC branch is* $\omega r = 3.5\omega_1$, *the parameters of the compensator are*

$$C_{\text{RSopt}} \approx C_{\text{RS0}} = [1 - \left(\frac{\omega_1}{\omega_r}\right)^2]\frac{T_{\text{RS1}}}{\omega_1} = 176.8 \text{ mF},$$

$$L_{\text{RS}} = \frac{1}{\omega_r^2\, C_{\text{RSopt}}} = 46.18 \text{ mH},$$

$$L_{\text{STopt}} \approx -\frac{1}{\omega_1\, T_{\text{ST1}}} = -\frac{1}{-0.1925} = 5.195 \text{ H}.$$

The compensator structure is shown in Figure 14.12.

Analysis of this circuit shows that some amount of the unbalanced current remains after compensation. Its three-phase rms value is

$$||\boldsymbol{i}'_{\text{u}}|| = 13.5 \text{ A}.$$

Moreover, the compensator injects some reactive current, originally not present because the load is purely resistive and the transformer ideal. Its three-phase rms value is

$$||\boldsymbol{i}'_{\text{r}}|| = 17.2 \text{ A}.$$

Such a compensator reduces the supply current three-phase rms value from $||\boldsymbol{i}|| =$ 179.6 A *to* $||\boldsymbol{i}'|| =$ 128.9 A, *improves the power factor to* $\lambda = 0.98$ *and restores to large degree the supply current symmetry. It does not enable total compensation but only minimization of the supply's current three-phase rms value. This result is obtained with a compensator composed of only three reactance elements.*

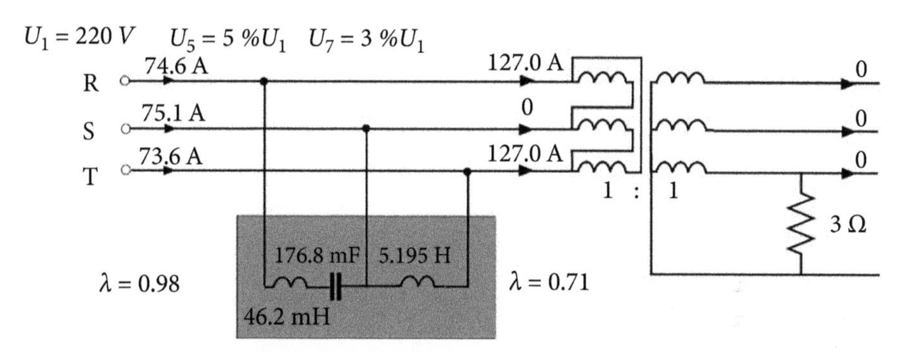

Figure 14.12 A balancing compensator for minimization of the supply current three-phase rms value.

*****

This approach to minimization of the supply current three-phase rms value and its asymmetry presented above and illustrated for a load supplied from an infinite bus can also be applied effectively in situations where the distribution system has a relatively high inductive impedance. This is because an inductive character of the impedance of the compensator branches in the range of harmonic frequencies eliminates the possibility of a resonance between the compensator and the distribution system. This is shown in the following illustration with relatively strong distortion of the supply voltage and relatively high inductive impedance.

Illustration 14.5 *An unbalanced load is supplied from a distribution system with symmetrical voltage rms value for the fundamental to E_1 = 120V. The supply voltage is distorted with the fifth and the seventh order harmonics with rms values 5V and 3V, respectively. The transformer is assumed to be ideal with a turn ratio 1:1.*

The circuit and results of its analysis are shown in Figure 14.13.

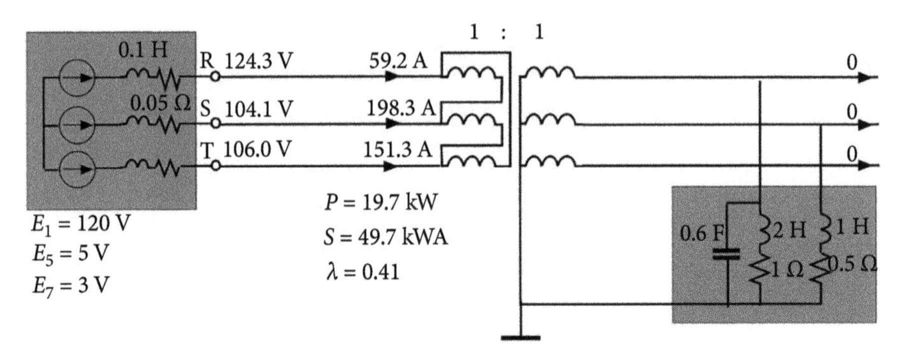

Figure 14.13 An example of a circuit with unbalanced load.

In response to asymmetrical line currents of the rms value $||i_R||$ = 59.2 A, $||i_S||$ = 198.3 A *and* $||i_T||$ = 151.3 A. *Due to line currents asymmetry, the supply voltage is asymmetrical. The rms value of line voltages are* $||u_R||$ = 124.3 V, $||u_S||$ = 104.1 V, *and* $||u_T||$ = 106.0 V. *The load power factor is equal to* λ = 0.40.

Load equivalent admittances for the voltage harmonics are equal to

$$Y_{e1} = 0.6 - j0.6 \text{ S}, \quad Y_{e5} = 0.03 + j2.7 \text{ S}, \quad Y_{e7} = 0.015 + j3.4 \text{ S},$$

and unbalanced admittances:

$$Y_{u1} = -0.473 + j1.07 \text{ S}, \quad Y_{u5} = 2.50 + j1.64 \text{ S}, \quad Y_{u7} = -3.60 + j2.21 \text{ S}$$

Susceptances of the balancing compensator branches calculated according to formula (14.20) are equal to

$$\begin{array}{lll} T_{RS1} = -0.431 \text{ S}, & T_{RS5} = -2.89 \text{ S}, & T_{RS7} = -4.13 \text{ S}, \\ T_{ST1} = 0.915 \text{ S}, & T_{ST5} = 0.192 \text{ S}, & T_{ST7} = 0.145 \text{ S}, \\ T_{TR1} = 0.115 \text{ S}, & T_{TR5} = -0.006 \text{ S}, & T_{TR7} = 0.003 \text{ S}, \end{array}$$

The susceptance for the fundamental frequency of the RS branch is negative, thus the balancing compensator should have an inductor in that branch of the inductance, at frequency normalized to $\omega_1 = 1$ rad/s, *approximately equal to*

$$L_{\rm RS} = -1/T_{\rm RS1} = 2.32 \text{ H}.$$

The susceptance for the fundamental frequency of the two remaining branches are positive thus these two branches should be built of inductor and capacitor in series.

Assuming that the resonant frequency $\omega_r = 2.2$ rad/s, *we obtain from formula (14.30) the following parameters*

$$\begin{array}{ll} L_{\rm ST} = 0.255 \text{ H}, & C_{\rm ST} = 0.742 \text{ F} \\ L_{\rm TR} = 2.02 \text{ H}, & C_{\rm TR} = 0.094 \text{ F} \end{array}.$$

The circuit with the compensator and the results of its analysis are shown in Figure 14.14.

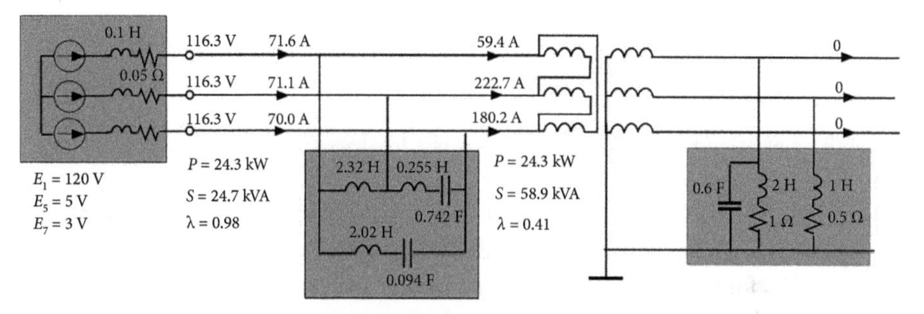

Figure 14.14 An example of a circuit with an unbalanced load and a balancing compensator.

Despite the limited complexity of the compensator, it effectively improves the power factor, reduces the current asymmetry, and restores the symmetry of the supply voltage. It is worth observing that it even increases the load active power from $P = 19.7$ kW *to* 24.3 kW. *It is because a reduction of the supply current rms value by the compensator increases the supply voltage rms value from* $||\boldsymbol{u}|| = 256.4$ V *before compensation to* $||\boldsymbol{u}|| = 292.6$ V.

*****

14.6 Compensation at Asymmetrical Supply Voltage and Nonsinusoidal Voltage and Currents (nv&c)

Since all physical components of the load current are sums of harmonics, reactance compensation can be analyzed by using a harmonic-by-harmonic approach. Therefore, for any harmonic of the $n^{\rm th}$ order, the load with a balancing reactance compensator can be presented as shown in Figure 14.15. As was shown in Section

7.17, LTI loads can be specified in terms of four admittances for harmonics: Y_{bn}, Y^{p}_{un}, Y^{n}_{un}, and Y_{dn}.

A reactance compensator connected as shown in Figure 14.15, changes the vector of the load line current harmonic from $\boldsymbol{i}_n$ to $\boldsymbol{i}'_n$.

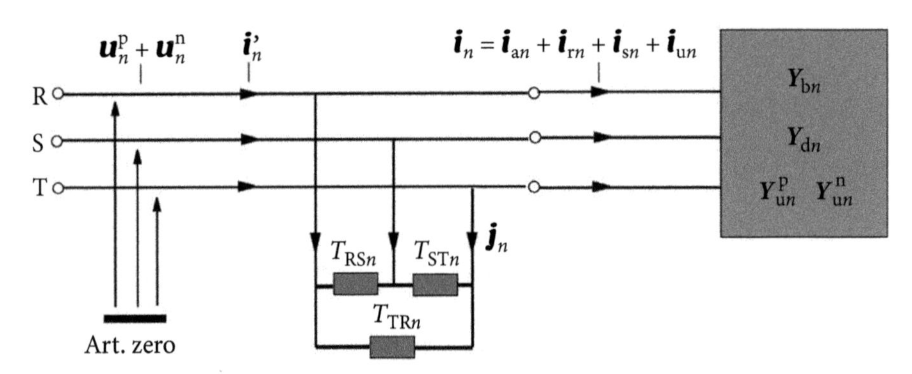

Figure 14.15 Equivalent admittances of the load and a compensator for the nth order harmonic.

It will be assumed that the compensator is a lossless device, and the load is supplied from an ideal voltage source, that is with zero internal impedance. At such assumptions, the compensator does not affect the voltage and the active power at the supply source terminals. Consequently, the balanced conductance G_b and conductances G_{bn} for harmonics remain unchanged, which means that the compensator cannot affect the scattered current $\boldsymbol{i}_s$. Thus in the presence of the scattered current, the unity power factor cannot be achieved by a reactance compensator. Only the reactance and unbalanced currents can be compensated.

Since the line-to-line admittances of the compensator for harmonic frequencies are

$$Y_{STn} = jT_{STn}, \quad Y_{TRn} = jT_{TRn}, \quad Y_{RSn} = jT_{RSn}$$

the compensator can be specified for harmonic frequencies in terms of the balanced susceptance

$$B_{Cb\,n} = \frac{T_{STn}U^2_{STn} + T_{TRn}U^2_{TRn} + T_{RSn}U^2_{RSn}}{||\boldsymbol{u}_n||^2} \tag{14.33}$$

and unbalanced admittances

$$Y^{p}_{Cu\,n} = -j(T_{STn} + \alpha T_{TRn} + \alpha^* T_{RSn}) \tag{14.34a}$$

$$Y^{n}_{Cu\,n} = -j(T_{STn} + \alpha^* T_{TRn} + \alpha T_{RSn}) \tag{14.34b}$$

$$Y_{Cdn} = jb_n[T_{STn}\cos\psi_n + T_{TRn}\cos(\psi_n - \frac{2\pi}{3}) + T_{RSn}\cos(\psi_n + \frac{2\pi}{3})]. \tag{14.35}$$

The compensator reduces entirely the reactive current harmonic of the nth order if its susceptance $B_{\text{Cb}n}$ satisfies the condition

$$B_{\text{Cb}n} + B_{\text{b}n} = 0. \tag{14.36}$$

Taking (7.108) into account, the nth order harmonic of the unbalanced current $\boldsymbol{i}_{\text{u}}$ in the reference line R of the load with the compensator is equal to

$$\boldsymbol{I}_{\text{Ru}n} = (\boldsymbol{Y}_{\text{Cd}n} + \boldsymbol{Y}_{\text{d}n})\boldsymbol{U}_{\text{R}n} + (\boldsymbol{Y}^{\text{p}}_{\text{Cu}n} + \boldsymbol{Y}^{\text{p}}_{\text{u}n})\boldsymbol{U}^{\text{p}}_{n} + (\boldsymbol{Y}^{\text{n}}_{\text{Cu}n} + \boldsymbol{Y}^{\text{n}}_{\text{u}n})\boldsymbol{U}^{\text{n}}_{n}.$$

This harmonic of the unbalanced current is compensated entirely on the condition that

$$(\boldsymbol{Y}_{\text{Cd}n} + \boldsymbol{Y}_{\text{d}n})\boldsymbol{U}_{\text{R}n} + (\boldsymbol{Y}^{\text{p}}_{\text{Cu}n} + \boldsymbol{Y}^{\text{p}}_{\text{u}n})\boldsymbol{U}^{\text{p}}_{n} + (\boldsymbol{Y}^{\text{n}}_{\text{Cu}n} + \boldsymbol{Y}^{\text{n}}_{\text{u}n})\boldsymbol{U}^{\text{n}}_{n} = 0. \tag{14.37}$$

It can be satisfied if condition (14.37) is satisfied, separately, for the real and the imaginary part. In such a way three equations, which are needed for calculating three susceptances $T_{\text{RS}n}$, $T_{\text{ST}n}$, and $T_{\text{TR}n}$ are obtained.

On the condition that $\boldsymbol{U}^{\text{p}}_{\text{n}} \neq 0$, formula (14.37) can be modified to

$$(\boldsymbol{Y}_{\text{Cd}n} + \boldsymbol{Y}_{\text{d}n})(1 + \boldsymbol{a}_n) + (\boldsymbol{Y}^{\text{p}}_{\text{Cu}n} + \boldsymbol{Y}^{\text{p}}_{\text{u}n}) + (\boldsymbol{Y}^{\text{n}}_{\text{Cu}n} + \boldsymbol{Y}^{\text{n}}_{\text{u}n})\boldsymbol{a}_n = 0. \tag{14.38}$$

with the complex coefficient $\boldsymbol{a}_n$, defined by (7.110). In this equation, the unbalanced admittances of the compensator are specified by formulae (14.34). Assuming that

$$\boldsymbol{c}_{1n} \stackrel{\text{df}}{=} jb_n \cos\psi_n, \quad \boldsymbol{c}_{2n} \stackrel{\text{df}}{=} jb_n \cos(\psi_n - \frac{2\pi}{3}), \quad \boldsymbol{c}_{3n} \stackrel{\text{df}}{=} jb_n \cos(\psi_n + \frac{2\pi}{3}) \tag{14.39}$$

the unbalanced admittance $\boldsymbol{Y}_{\text{Cd}n}$ can be expressed in the form

$$\boldsymbol{Y}_{\text{Cd}n} = \boldsymbol{c}_{1n} T_{\text{ST}n} + \boldsymbol{c}_{2n} T_{\text{TR}n} + \boldsymbol{c}_{3n} T_{\text{RS}n} \tag{14.40}$$

and Eqn. (14.38) can be rearranged to a linear form of the compensator susceptances

$$\boldsymbol{A}_n T_{\text{ST}n} + \boldsymbol{B}_n T_{\text{TR}n} + \boldsymbol{C}_n T_{\text{RS}n} + \boldsymbol{D}_n = 0. \tag{14.41}$$

When we denote

$$\boldsymbol{d}_n \stackrel{\text{df}}{=} \boldsymbol{a}_n + 1, \quad \boldsymbol{f}_n \stackrel{\text{df}}{=} \alpha^* + \alpha \boldsymbol{a}_n, \quad \boldsymbol{g}_n \stackrel{\text{df}}{=} \alpha + \alpha^* \boldsymbol{a}_n \tag{14.42}$$

then the coefficients in the form (14.41) are equal to

$$\boldsymbol{A}_n = (\boldsymbol{c}_{1n} - j1)\, \boldsymbol{d}_n, \quad \boldsymbol{B}_n = \boldsymbol{c}_{2n}\boldsymbol{d}_n - j\cdot \boldsymbol{g}_n. \tag{14.43}$$

$$\boldsymbol{C}_n = \boldsymbol{c}_{3n}\boldsymbol{d}_n - j\cdot \boldsymbol{f}_n, \quad \boldsymbol{D}_n = \boldsymbol{Y}^{\text{n}}_{\text{u}n}\boldsymbol{a}_n + \boldsymbol{Y}_{\text{d}n}\boldsymbol{d}_n + \boldsymbol{Y}^{\text{p}}_{\text{u}n}. \tag{14.44}$$

The condition (14.33) for the nth order harmonic can be rearranged to the form

$$U_{\text{ST}\,n}^2\, T_{\text{ST}n} + U_{\text{TR}\,n}^2\, T_{\text{TR}n} + U_{\text{RS}\,n}^2\, T_{\text{RS}n} = -B_{\text{b}n}||\boldsymbol{u}_n||^2. \tag{14.45}$$

The linear form (14.41) has to be satisfied for the real and the imaginary part of it, thus

$$\text{Re}\{\boldsymbol{A}_n\, T_{\text{ST}n} + \boldsymbol{B}_n\, T_{\text{TR}n} + \boldsymbol{C}_n\, T_{\text{RS}n} + \boldsymbol{D}_n\} = 0 \tag{14.46}$$

$$\text{Im}\{\boldsymbol{A}_n\, T_{\text{ST}n} + \boldsymbol{B}_n\, T_{\text{TR}n} + \boldsymbol{C}_n\, T_{\text{RS}n} + \boldsymbol{D}_n\} = 0. \tag{14.47}$$

Eqns (14.45–14.47) specify the compensator's line-to-line susceptances for the nth order harmonic. They can be written in the form

$$\begin{bmatrix} U_{\text{ST}n}^2, & U_{\text{TR}n}^2, & U_{\text{RS}n}^2 \\ \text{Re}\boldsymbol{A}_n, & \text{Re}\boldsymbol{B}_n, & \text{Re}\boldsymbol{C}_n \\ \text{Im}\boldsymbol{A}_n, & \text{Im}\boldsymbol{B}_n, & \text{Im}\boldsymbol{C}_n \end{bmatrix} \begin{bmatrix} T_{\text{ST}\,n} \\ T_{\text{TR}n} \\ T_{\text{RS}n} \end{bmatrix} = \begin{bmatrix} -B_{\text{b}n}||\boldsymbol{u}_n||^2 \\ -\text{Re}\boldsymbol{D}_n \\ -\text{Im}\boldsymbol{D}_n \end{bmatrix}. \tag{14.48}$$

Illustration 14.6 *For the circuit used in Illustration 7.8 in Section 7.15 for numerical verification with load parameters compiled in Table 7.1, the compensator Eqn. (14.48) results in susceptances of the compensator's branches compiled in Table 14.1.*

Table 14.1 Compensator's line-to-line susceptance for the voltage harmonics.

	n	1	3	5	7
$T_{\text{ST}\,n}$	S	−0.0567	0.0032	0.0093	−0.0023
$T_{\text{TR}n}$	S	−0.3530	−1.187	−2.282	−3.374
$T_{\text{TR}n}$	S	0.2519	−1.348	−2.327	−3.350

Having the susceptance of the compensator branches for the supply voltage harmonics, their structure and parameters can be found using the procedures described in Chapter 10.

It can be observed that despite a total compensation of the unbalanced and reactive currents, the vector of the supply current is not proportional to the load voltage vector. It is because the supply current still contains the scattered current $\boldsymbol{i}_{\text{s}}$ which is not proportional to the voltage vector. Only the active current $\boldsymbol{i}_{\text{a}}$ is proportional to that vector. The results of compensation are shown in Figure 14.16.

*****

Because the compensator in this illustration compensates the reactive and unbalanced current for the fundamental harmonic and three higher-order ones, approximately eight reactance devices are needed for each branch of the compensator. Consequently, such a reactance compensator, due to its complexity does not have practical but only theoretical value. Therefore the structure of the compensator's

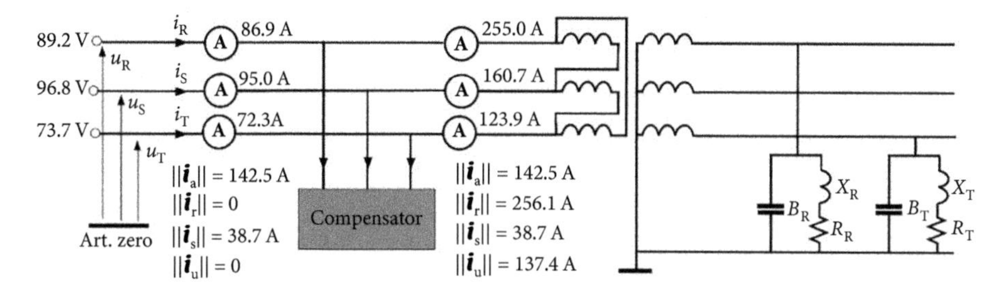

Figure 14.16 Results of compensation of the reactive and unbalanced currents.

branches is not shown in this illustration, and parameters are not calculated. The values of branch susceptance compiled in Table 14.1 could serve for the calculation of the parameters of a compensator of reduced complexity.

The reduction of the compensator complexity of a compensator designed in such a way that it takes the supply voltage asymmetry into account does not differ from this reduction, discussed in Section 14.5, when the supply voltage is symmetrical. The values of the susceptance for harmonic frequencies of the compensator branches provide the starting point for calculating the parameters of the compensator with reduced complexity.

The effect of compensation by a compensator of complexity reduced to the compensator with two reactance devices by the compensator branch is illustrated below on an example of a circuit more realistic than that in Illustration 14.6. The supply voltage only has asymmetry of the order of a few percent.

Illustration 14.7 *A three-phase voltage source such that $e_S = 0.97e_R$, $e_T = 0.97e_R$, and harmonic distortion such that $E_3 = E_5 = E_7 = 4\%$ of $E_1 = 100$ V, supplies from an ideal transformer, a single-phase LTI load, as shown in Figure 14.17. It was assumed in the illustration that the supply source is relatively weak, with the short circuit power S_{sc} only twenty times higher than the load active power, and the reactance-to-resistance ratio for the fundamental frequency $X_s/R_s = 5$.*

The results of the CPC-based analysis and the three-phase rms values of the supply current physical components are shown in Figure 14.17.

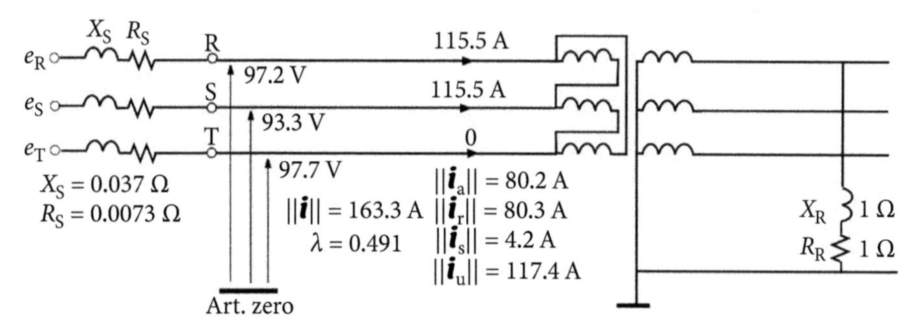

Figure 14.17 A circuit analyzed in numerical illustration 14.7.

The susceptance of the compensator's branches for harmonic frequencies needed for complete compensation of the reactive and unbalanced currents are compiled in Table 14.2.

Table 14.2 Compensator's line-to-line susceptance for harmonic frequencies.

	n	1	3	5	7
T_{STn}	S	0.297	520	−0.022	0.012
T_{TRn}	S	−0.286	0.683	0.022	−0.011
T_{TRn}	S	0.503	−0.384	−0.192	−0.142

The LC parameters of the compensator of reduced complexity, at the assumption that inductances L_k of the compensator branches were selected in such a way that the resonance in these branches is located below the frequency of the second order harmonic, that is, it was assumed that

$$\frac{1}{\sqrt{L_k C_{0k}}} = 1.5\omega_1.$$

The connection of the compensator to the circuit affects the voltage at the compensator terminals, in other words, the rms values U_{kn} in formulae (14.29) and (14.30). Consequently, their new values have to be found and the calculation of the compensator LC parameters has to be repeated. Thus their calculation is an iterative process. It results in

$$L_{ST} = 2.74 \text{ H}, \quad C_{ST} = 0.162 \text{ F}$$

$$L_{TR} = 3.50 \text{ H}$$

$$L_{ST} = 1.59 \text{ H}, \quad C_{ST} = 0.279 \text{ F}.$$

The results of minimization of the reactive and unbalanced currents by the compensator of reduced complexity are shown in Figure 14.18.

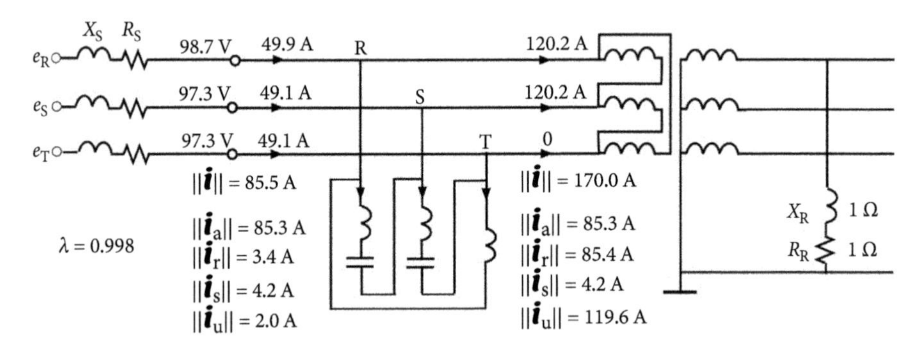

Figure 14.18 Results of minimization of the reactive and unbalanced currents.

These results demonstrate that at the cost of small, not compensated reactive and unbalanced currents, the complexity of the compensator can be substantially reduced. Despite substantial distortion of the supply voltage and its asymmetry, the reduced complexity compensator has enabled to improve power factor from $\lambda = 0.49$ *to* $\lambda = 0.998$*, thus practically to unity value.*

*****

14.7 Adaptive Balancing Compensation

Three-phase loads can be composed of three-phase devices, such as three-phase motors which by nature are balanced, and aggregates of single-phase devices that burden individual lines to a different degree, causing some level of imbalance. In traction systems, where motors are in movement, energy can be delivered only by one line. Thus, such traction systems are three-phase circuits with single-phase loads. The train engine is supplied sequentially from different lines which changes drastically the load imbalance. AC arc furnaces are other examples of high-power loads with a high level of imbalance, which occurs when one of three arcs is extinct.

A reactance balancing compensator must have an adaptive property for compensating such variable loads. This means that at least some parameters of the compensator must be adjusted to the load variations.

The capacitance of a capacitor bank can be controlled by ON/OFF switching the bank sections. Some precautions are needed, however, to avoid charging/discharging current pulses at the instant of time when two capacitor sections with different voltages are connected in parallel. For example, the instantaneous voltage sensors, installed at the capacitor bank sections, can enable switching ON only when the section voltage is the same.

The inductance of an inductor with a nonlinear ferromagnetic core can be controlled by saturating the core with a DC current, as shown in Figure 14.19(a). This current shifts the operating point of the inductor on the induction flux density B and the magnetic field intensity H characteristic, shown in Figure 14.19(b), of the core.

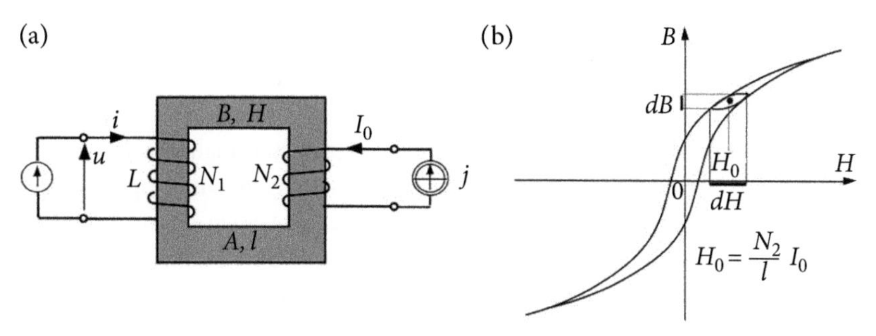

Figure 14.19 (a) An inductor with saturating winding and dc current source, and (b) magnetic core $B = f(H)$ characteristic.

The shift of the operation point changes the slope of $B = f(H)$ characteristic, thus the magnetic permeability in the operation point is approximately equal to

$$\mu\mu_0 = \frac{dB}{dH}$$

and consequently, the differential inductance of the inductor at the point of operation is

$$L = \mu\mu_0 \frac{A}{l} N_1^2.$$

Symbol A denotes the area of the core cross-section and l denotes the average length of the magnetic path in the core. Since the $B = f(H)$ characteristic at the operation point specified by the DC current I_0, other than zero, is not symmetrical, the inductor even at sinusoidal supply voltage is a source of current harmonics not only of odd but also even order.

The inductance of a transformer as seen from its primary side can be changed also by switching ON/OFF an inductor connected on its secondary side, as shown in Figure 14.20.

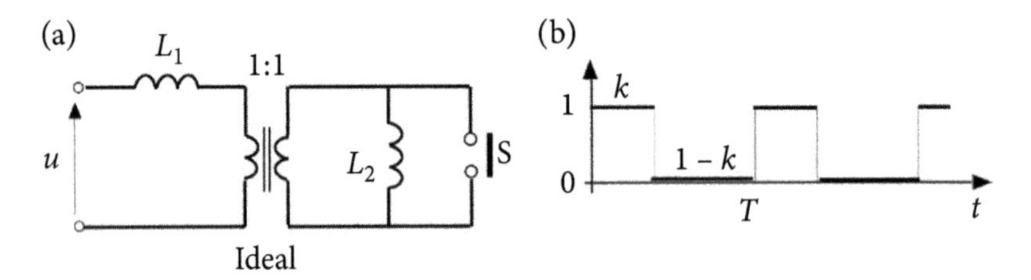

Figure 14.20 (a) An inductor with a periodic switch, and (b) the switch states.

If k denotes the duty factor of the switch, then the mean value of the inductance is approximately equal to

$$L_e \approx L_1 + (1 - k)L_2.$$

The accuracy of this approximation depends on the selected switching period T_s. It should be selected sufficiently longer than the transients caused by switching. Transients are attenuated by energy loss in the circuit. Their duration can be reduced only by an increase of this loss, thus by reduction of the inductor quality factor, however.

A thyristor ON/OFF switching of an inductor [45, 46], as shown in Figure 14.21(a), provides the most common method of inductance control. Such a branch is referred to as a *thyristor switched inductor* (TSI). There are two thyristors in such a TSI branch, one is fired at the angle α, and the second is fired at the angle $180^\circ + \alpha$. For simplicity's sake, however, diagrams of TSI branches are drawn in circuit schematics usually as shown in Figure 14.21(b).

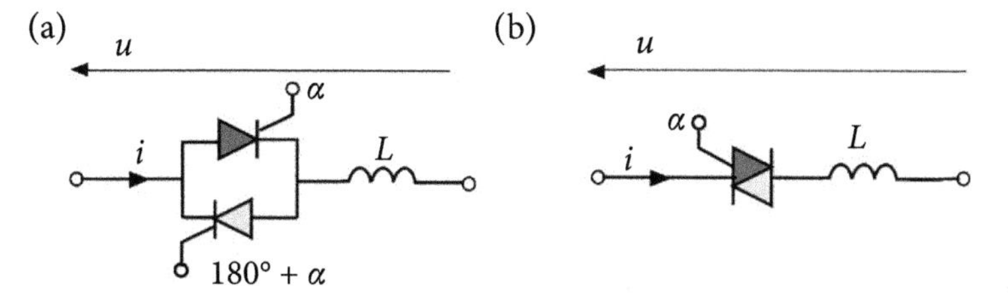

Figure 14.21 (a) A thyristor switched inductor (TSI) and (b) as it is drawn in schematics.

The current waveform in the branch TSI, assuming that the supply voltage is sinusoidal, is shown in Figure 14.22. Symbol i_0 in this figure denotes TSI current at $\alpha = 0$.

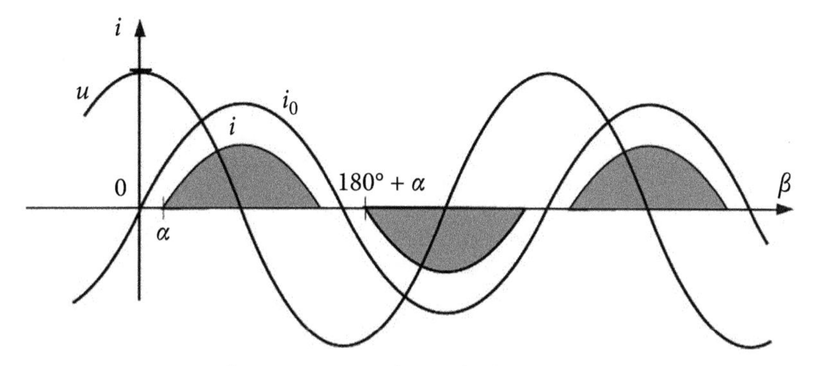

Figure. 14.22 The currents waveform of a thyristor switched inductor.

The ratio of the crms value of a sinusoidal supply voltage $\boldsymbol{U}_1$ at TSI terminals to the crms value of the branch current fundamental harmonic $\boldsymbol{I}_1$, namely

$$Z_1 = \frac{\boldsymbol{U}_1}{\boldsymbol{I}_1} = jX_1 = -j\frac{\pi}{\pi - 2\alpha - \sin 2\alpha}\omega_1 L$$

specifies the branch impedance for the fundamental harmonic, and the equivalent inductance of the TSI branch. It is equal to

$$L_e = \frac{\pi}{\pi - 2\alpha - \sin 2\alpha} = L_e(\alpha). \qquad (14.49)$$

It changes, from inductance L, at $\alpha = 0$, to infinity, at α approaching 90°. Thus, at sinusoidal supply voltage, a TSI can be regarded as a parallel connection, shown in Figure 14.23, of the equivalent inductance L_e and a current source of odd order harmonics j_h other than the fundamental one i_1.

A thyristor switched inductor, connected in parallel to a capacitor, was used [45] for the construction of controllable compensators of reactive power, shown in Figure 14.24.

At thyristors' non-zero firing angle, the current of a branch with TSI is distorted. The third order harmonic is the dominating one. When TSIs are used in a compensator of only the reactive current, the compensator operates as a balanced device.

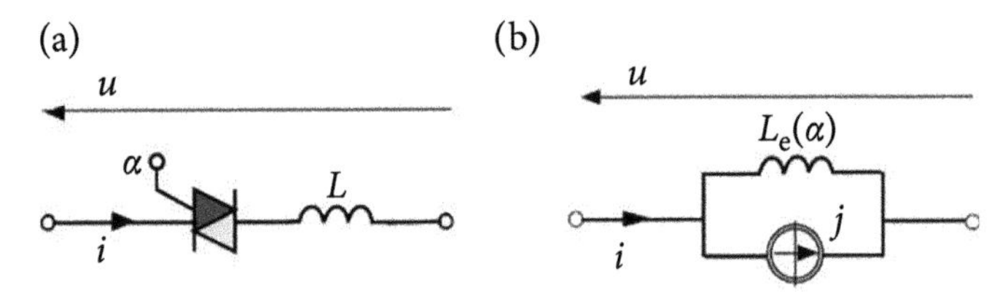

Figure 14.23 (a) A thyristor switched inductor (TSI), and (b) its equivalent circuit.

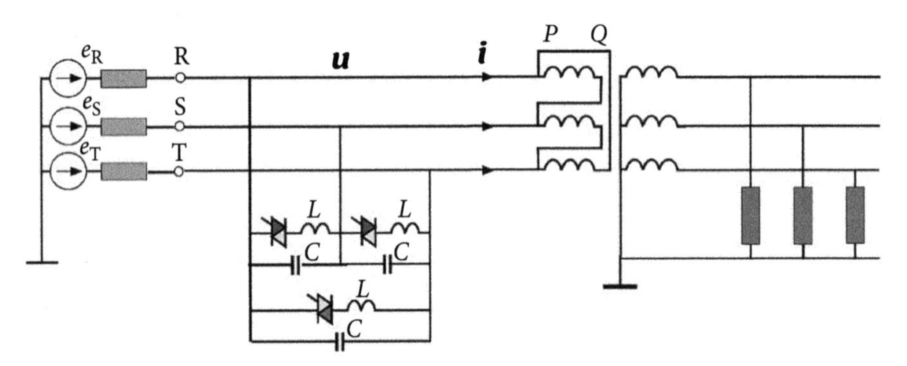

Figure 14.24 A controllable compensator of reactive power with TSI.

With regard to the voltage at the compensator branch, each TSI is fired at the same firing angle. Consequently, each line-to-line branch of the compensator generates the third order current harmonic of the same rms value and phase. Therefore, it flows in the loop formed by the capacitors, not leaving the compensator. This is no longer valid when TSIs are used in balancing compensators because such compensators are not balanced devices. The third order current harmonic generated in the compensator branches can have different rms values. Consequently, it can leave the compensator, along with other current harmonics generated in the TSI, and cause the supply current distortion.

The injection of the third order harmonic into the supply source can be reduced [82] by an LC filter connected at the TSI terminals, as shown in Figure 14.25, and tuned to the frequency of the third order harmonic.

The one-ports, composed of TSI and additional reactance elements in various configurations that enable shaping the value of the one-port susceptance are referred to [125] as *thyristor controlled susceptance* (TCS) one-ports or branches.

Observe that the susceptance of the filter of the third order harmonic, originated in the TSI branch, is for the fundamental harmonic capacitive, thus this filter can serve as the compensator of the fundamental harmonic reactive power Q_1. Therefore the compensator can be regarded as composed of two parts: a fixed-parameter compensator and an adaptive compensator with TSI inductor. This enables the reduction of the current ratings of thyristors used in the adaptive part.

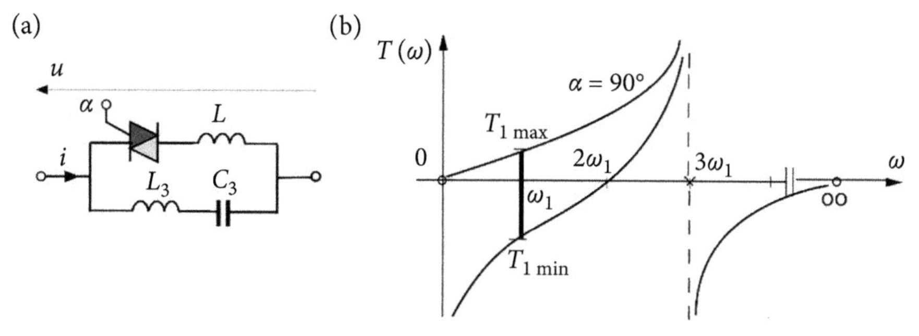

Figure 14.25 (a) A TCS branch with a filter of the 3rd order harmonic, and (b) plot of its susceptance $T(\omega)$.

The susceptance of such a branch changes with frequency as shown in Figure 14.25(b). This susceptance T_1 for the fundamental frequency is equal to

$$T_1 = \frac{9}{8}\omega_1 C_3 - \frac{1}{\omega_1 L_e(\alpha)} = T_1(\alpha).$$

With the thyristor firing angle changes from $\alpha = 0$ to $\alpha = 90^\circ$, this susceptance T_1 changes from a minimum to a maximum value

$$T_1(0) = T_{1\,\min} = \frac{9}{8}\omega_1 C_3 - \frac{1}{\omega_1 L}, \qquad T_1(90^\circ) = T_{1\,\max} = \frac{9}{8}\omega_1 C_3$$

The compensator designer should know the range of this susceptance change, meaning the values $T_{1\min}$ and $T_{1\max}$. Having them, the branch parameters can be calculated, namely

$$C_3 = \frac{8}{9}\frac{T_{1\,\max}}{\omega_1}, \qquad L = \frac{1}{\omega_1(T_{1\,\max} - T_{1\,\min})}$$

The effectiveness in the third order harmonic attenuation of the TCS branch shown in Figure 14.25(a) depends on the internal impedance of the source which supplies the branch. This effectiveness can be enhanced by a series inductor connected as shown in Figure 14.26(a).

The susceptance of such a branch changes with frequency as shown in Figure 14.26(b). This susceptance T_1 for the fundamental frequency is equal to

$$T_1 = \frac{9\,\omega_1^2 L_e(\alpha)\,C_3 - 8}{\omega_1 L_e(\alpha) - \omega_1 L_0\,[9\,\omega_1^2 L_e(\alpha)\,C_3 - 8]} = T_1(\alpha). \tag{14.50}$$

With the thyristor firing angle change from $\alpha = 0$ to $\alpha = 90^\circ$, this susceptance T_1 changes from a minimum to a maximum value

$$T_1(0) = T_{1\,\min} = \frac{9\omega_1^2 L C_3 - 8}{8\omega_1(L_0 + L) - 9\omega_1^3 L\,L_0 C_3}, \qquad T_1(90^\circ) = T_{1\,\max} = \frac{9\omega_1 C_3}{8 - 9\omega_1^2 L_0 C_3}. \tag{14.51}$$

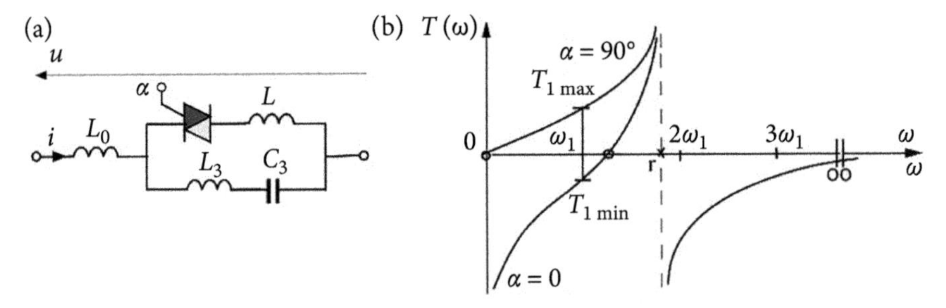

Figure 14.26 (a) A TSC branch with 3rd order harmonic filter and a series inductor, and (b) plot of its susceptance $T(\omega)$.

At some frequency, denoted by ω_r, a voltage resonance of the whole TSC branch occurs. Its susceptance approaches infinity. Since the equivalent inductance of the TSI branch changes with the firing angle, the frequency ω_r changes as well. Its maximum value is in the thyristor ON state. Its relative value, referenced to the fundamental frequency, is equal to

$$\frac{\omega_r}{\omega_1} = \Omega = \sqrt{8}\sqrt{\frac{L_3(L + L_0)}{L\,L_0}}. \tag{14.52}$$

To avoid resonance at the second order harmonic, which can be present in the supply voltage, the parameters of the TCS branch should be selected in such a way that the relative resonance frequency (14.52) is below 2.

The conditions (14.51) and (14.52), and the resonance frequency of the L_3C_3 branch, can be rearranged versus one of four parameters of the LSC branch. If inductance L_0 is such a parameter, then it must satisfy the equation

$$a_3L_0^3 + a_2L_0^2 + a_1L_0 + a_0 = 0. \tag{14.53}$$

When the frequency ω_1 is normalized to 1 rad/s, then coefficients of Eqn. (14.53) are

$$a_0 = 9 - \Omega^2 \tag{14.54}$$

$$a_1 = (27 - 11\Omega^2)T_{1\,max} \tag{14.55}$$

$$a_2 = \frac{2(9 - 5\Omega^2)T_{1\,max} + 9(1 - \Omega^2)T_{1\,min}}{T_{1\,max}} \tag{14.56}$$

$$a_3 = 9(1 - \Omega^2)T_{1\,min}\,T_{1\,max}^2. \tag{14.57}$$

thus they can be specified in terms of the minimum and maximum susceptance of the branch and relative resonance frequency.

Having solved Eqn. (14.53), the parameters of the TCS branch can be expressed in terms of the inductance L_0 as follows

$$L_3 = \frac{L_0\, T_{1\,\max} + 1}{8T_{1\,\max}} \tag{14.58}$$

$$C_3 = \frac{1}{9L_3} \tag{14.59}$$

$$L = \frac{T_{1\,\min} T_{1\,\max} L_0^{\;2} + (T_{1\,\min} + T_{1\,\max}) L_0 + 1}{T_{1\,\max} - T_{1\,\min}}. \tag{14.60}$$

These results show that the required range of susceptance change and the selected frequency of the resonance of the compensator which cannot be avoided enable the calculation of the adaptive compensator parameters. The firing angle of thyristors of the compensator branches cannot be found from the formula (14.49), however. This equation cannot be solved versus the firing angle α. A look-up table must be created instead. For specific firing angle α, formula (14.49) provides the equivalent inductance L_e of TSI. Having this inductance, formula (14.50) provides the susceptance T_1 of the TCS branch. This procedure, repeated from $\alpha = 0$ to $\alpha = 90^\circ$, results in a look-up table that enables the selection of the thyristor firing angle for the needed value of the compensator branch susceptance.

The reader should be aware that the presented method of adaptive balancing by a reactance compensator enables balancing only for the fundamental harmonic, however. The right selection of the compensator structure and parameters only prevent the circuit from an increase of the supply current distortion.

14.8 Adaptive Compensation of DC Currents of AC Arc Furnaces

Due to the level of power and randomness of the currents' waveform, AC arc furnaces present the most challenging problems for compensation. At the same time, a relatively low level of the power factor and low loading quality (LQ), require their improvement.

The compensation of the reactive and unbalanced currents of AC arc furnaces does not differ from the compensation of other high-power loads. The same applies to the reduction of harmonics. The only difference is that a DC component can occur in the AC furnace supply currents, while the method of compensation presented previously is not capable of reducing this component.

The molten iron and the graphite electrode in an AC arc furnace sequentially exchange, twice the period T, their role as the anode and the cathode for the furnace arc. Since they are not electrically symmetrical, the arc current's positive pulse can differ from this current's negative pulse. DC components can occur in the arc furnace supply currents due to this asymmetry. They can have the highest value when the arc in some directions is not fired. This situation was shown in Fig. 2.54.

There is not very much information in the published reports on the DC components in arc furnace currents, although some can be found [301, 317], because instrumentation transformers are usually used for the measurement of these currents and DC components are not visible on their secondary side. We can only infer their presence from the presence of the even order harmonics in these currents. They can occur only if these currents do not have the negative symmetry versus values shifted by half of the period T. At the lack of such symmetry also a DC component can occur in these currents.

The DC components in the furnace currents are harmful because of the extra loss of energy in the furnace transformer. Moreover, the DC components create a constant magnetic flux that shifts the working point on the hysteresis loop of the core towards its saturation. This contributes to extra distortion of the furnace supply current on both sides of the furnace transformer. In particular, the current on the primary side can contain not only the odd order but also the even order harmonics, including the DC component. The reduction of the DC components requires their measurement.

DC Current Sensor. Since the instrumentation transformers are useless for the DC currents measurement, sensors without any transformer coupling are required. These could, for example, be resistive sensors, of the structure shown in Figure 14.27.

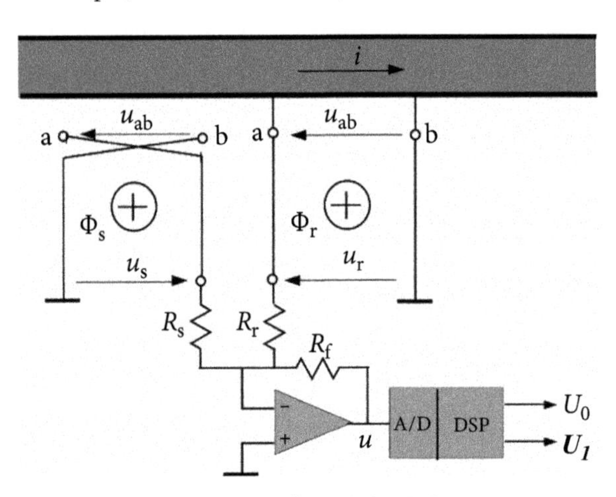

Figure 14.27 A structure of a resistive DC current sensor.

The voltage between two points, a and b, on the surface of the arc furnace supply conductor can be measured for that. The measurement circuit is coupled with the magnetic field of the supply conductor, however, so that the sensor provides the voltage

$$u_r(t) = u_{ab}(t) + \frac{d}{dt}\Phi_r(t)$$

with the induced component that can be much higher than the voltage drop between points a and b. This induced voltage can be compensated by another sensor that provides the voltage

$$u_s(t) = u_{ab}(t) - \frac{d}{dt}\Phi_s(t).$$

When these two sensors are geometrically identical as much as possible and superimposed one over another, so that

$$\Phi_s(t) \approx \Phi_r(t).$$

Then, assuming that $R_s = R_r$, the differential signal at the amplifier output is

$$u(t) = -\frac{R_f}{R_r}[u_r(t) + u_s(t)] = -2\frac{R_f}{R_r}u_{ab}(t) = -2\frac{R_f}{R_r}r\,i(t). \tag{14.61}$$

This voltage, sampled by an A/D converter, provides data for a discrete signals processing (DSP) system that calculates the mean value of this voltage U_0. It is proportional to the line current DC component but the coefficient proportionality, meaning the resistance r between points a and b is not a known parameter. Thus the value U_0 cannot be recalculated to the value I_0. A process of the calibration of such a sensor is needed.

A sensor with a common instrumentation current transformer can be used for that. Such a sensor enables measurement of the line current $i(t)$ so that the calculation of the rms value of the line current fundamental harmonic I_1. Having this value, the amplification coefficient $2R_f/R_r$ can be selected such that the DC sensor will provide the same value I_1.

Compensator location. The compensator can be located on the primary or the secondary side of the furnace transformer. It depends on the location of the arc stabilizing inductors. Usually, these inductors are located on the primary side, so the compensator and harmonic filters are located as shown in Figure 14.28.

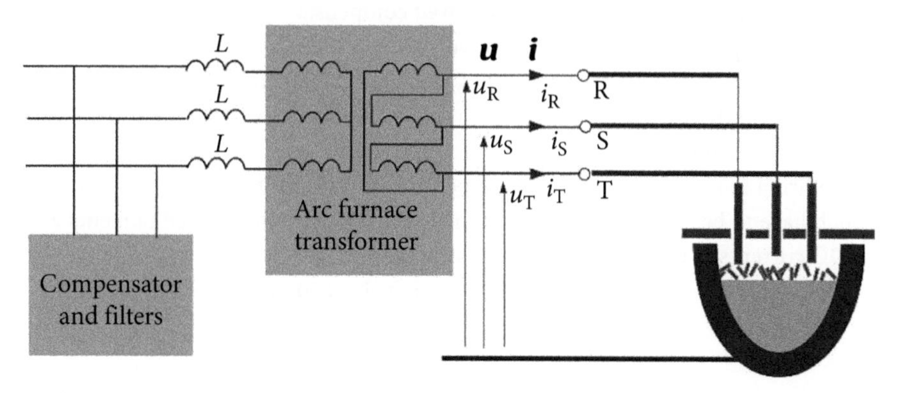

Figure 14.28 An arc furnace with inductors at the primary side of the furnace transformer.

Unfortunately, at such a location, the compensator does not reduce energy loss in the arc furnace transformer. It has to be connected to the secondary side of this transformer for that. This means, however, that inductors for the arc stabilization are connected also on the secondary side, as shown in Figure 14.29, which is not very common.

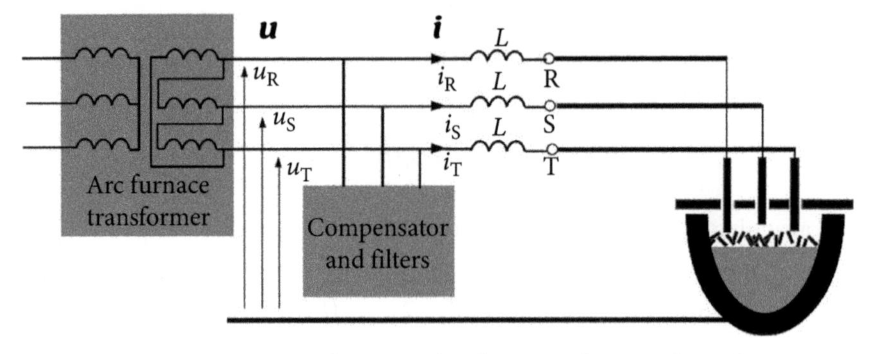

Figure 14.29 An arc furnace with inductors at the secondary side of the furnace transformer.

Only at such a location, meaning on the secondary side, can the compensator have the capability of reducing the transformer current components that do not contribute to the useful energy transmission to the furnace.

The components of the compensator that have fixed parameters, meaning the filters of the third harmonic can form a separate compensator which combined with the filter of other harmonics can be used for the compensation of the average value of the reactive power Q of the arc furnace. The remaining part of the compensator with TSI is shown in Figure 14.30.

When a TSI compensator is used as a compensator of the unbalanced and/or the reactive currents, thyristors in each branch are switched with a delay of 180°. No DC current is produced at such firing. When this condition is not satisfied then a DC current occurs in the compensator and it can be used for compensation of the DC currents produced by the arc furnace.

For that purpose, the control of a TSI-based compensator with thyristors switched asymmetrically in the positive and negative direction of the branch current, as shown in Figure 14.31, has to be analyzed. The adjective “asymmetrically” means that the firing angles of thyrsitors in particular branches of the compensator, denoted by α and β, are not mutually equal.

The firing angles of thyristors in the positive and the negative direction in each branch have to be selected such that at the same time the unbalanced currents and the DC currents of the arc furnace are compensated.

Since harmonics of the compensator's currents have to satisfy the relationship

$$j_{RSn} - j_{TRn} = i_{CRn}$$

$$j_{STn} - j_{RSn} = i_{CSn}$$

$$i_{CRn} + i_{CSn} + i_{CTn} = 0$$

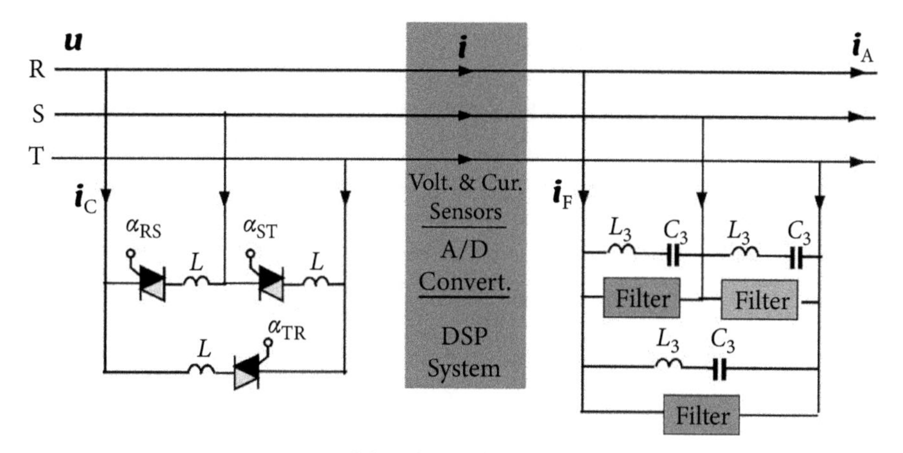

Figure 14.30 A structure of the adaptive balancing compensator and filters.

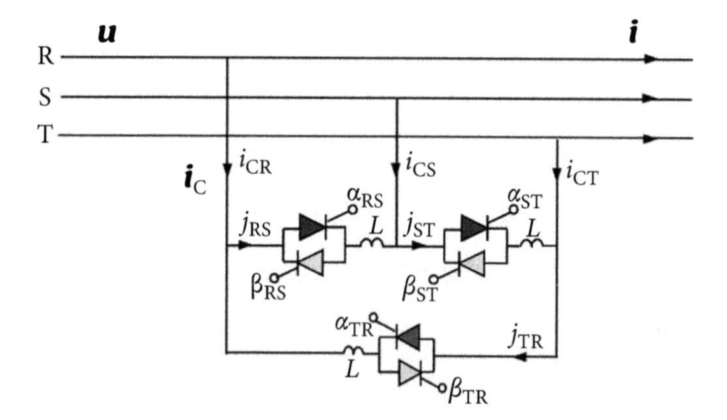

Figure 14.31 An adaptive balancing compensator with asymmetrically switched thyristors.

thus the crms values of the fundamental harmonic of TSI branches are

$$J_{RS1} = \frac{1}{3}(\boldsymbol{I}_{CS1} - \boldsymbol{I}_{CR1})$$

$$J_{ST1} = \frac{1}{3}(-2\boldsymbol{I}_{CS1} - \boldsymbol{I}_{CR1})$$

$$J_{TR1} = \frac{1}{3}(\boldsymbol{I}_{CS1} + 2\boldsymbol{I}_{CR1}) \tag{14.62}$$

while the DC component of these currents should have the mean values

$$J_{RS0} = \frac{1}{3}(I_{CS0} - I_{CR0})$$

$$J_{ST0} = \frac{1}{3}(-2\,I_{CS0} - I_{CR0})$$

$$J_{TR0} = \frac{1}{3}(I_{CS0} + 2\,I_{CR0}). \tag{14.63}$$

The DSP system of the compensator can provide the values on the right sides of formulae (14.62) and (14.63), so the needed values on the left sides can be calculated.

Thyristors' Firing Angles Selection. The crms value of the current fundamental harmonic and the mean value of the DC component in each branch of the compensator depend on the firing angles of thyristors connected in the positive and the negative direction of the branch's current. This dependence is identical for the TSI compensator's each branch, thus it is enough to find this dependence for only one of them, thus ignoring, as shown in Figure 14.32, indices RS, ST, and TR.

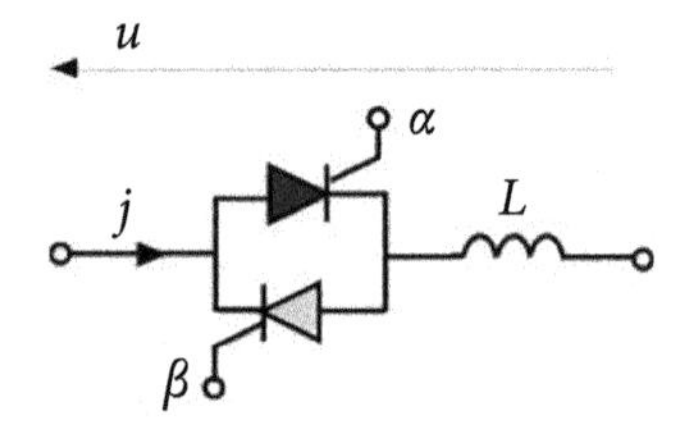

Figure 14.32 A TSI branch with a separate firing angles control.

The current of the branch is a sum of currents of thyristors connected in the branch current's positive and negative directions, shown in Figure 14.33, namely

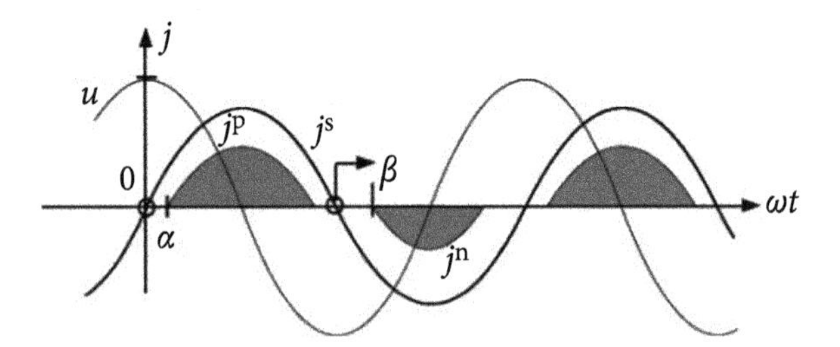

Figure 14.33 Current components in a TSI branch with a separate firing angles control.

$$j(t) = j^{\mathrm{p}}(t) + j^{\mathrm{n}}(t)$$

The symbol j^{s} in this figure denotes the steady-state current of the branch with both thyristors in the ON state.

The crms value of the fundamental harmonic of such a current is

$$J_1 = \frac{U_1}{2\pi\omega_1 L}[2(\pi - \alpha - \beta) + \sin 2\alpha + \sin 2\beta]e^{-j\frac{\pi}{2}} = f_1(\alpha, \beta) \tag{14.64}$$

and its mean value

$$J_0 = \frac{\sqrt{2}U_1}{\pi\omega_1 L}(\cos\alpha - \cos\beta) = f_0(\alpha, \beta). \tag{14.65}$$

For each needed value of the compensator branch current fundamental harmonic crms value $\boldsymbol{J}_1$ there is an infinite number of pairs of thyristors' firing angles (α, β) at which that value $\boldsymbol{J}_1$ can be obtained. When these angles can have only discrete values that differ by Δa, there is a finite number M of such pairs. For example, assuming that $\Delta a = 1^{\circ}$, then in the range of the firing angle change from 0 to 90°, there are no more than ninety such pairs, thus M is not higher than 90. The same is valid as to firing angles for producing by the compensator branch the needed mean value J_0 of the current DC component.

Having the formula (14.64), a look-up table for angles α, β changing in the range from 0° to 90° with step Δa, thus of the dimension (M x M), for the crms value of the branch current fundamental harmonic $\boldsymbol{J}_1$ can be created.

$$J_1^{\mathrm{d}} = \begin{bmatrix} J_1^{\mathrm{d}}(0,0) & \dots & J_1^{\mathrm{d}}(0,\beta) & \dots & J_1^{\mathrm{d}}(0,90) \\ \dots & & \dots & & \dots \\ J_1^{\mathrm{d}}(\alpha,0) & \dots & J_1^{\mathrm{d}}(\alpha,\beta) & \dots & J_1^{\mathrm{d}}(\alpha,90) \\ \dots & & \dots & & \dots \\ J_1^{\mathrm{d}}(90,0) & \dots & J_1^{\mathrm{d}}(90,\beta) & \dots & J_1^{\mathrm{d}}(90,90) \end{bmatrix}. \tag{14.66}$$

The upper index "d" means that such a value results from the formula (14.64). This is not the needed crms value of the branch's current fundamental harmonic.

A similar look-up table can be created for the mean value of the compensator branch current, namely

$$J_0^{\mathrm{d}} = \begin{bmatrix} J_0^{\mathrm{d}}(0,0) & \dots & J_0^{\mathrm{d}}(0,\beta) & \dots & J_0^{\mathrm{d}}(0,90) \\ \dots & & \dots & & \dots \\ J_0^{\mathrm{d}}(\alpha,0) & \dots & J_0^{\mathrm{d}}(\alpha,\beta) & \dots & J_0^{\mathrm{d}}(\alpha,90) \\ \dots & & \dots & & \dots \\ J_0^{\mathrm{d}}(90,0) & \dots & J_0^{\mathrm{d}}(90,\beta) & \dots & J_0^{\mathrm{d}}(90,90) \end{bmatrix}. \tag{14.67}$$

When simultaneously the compensator branch should provide the current that has the crms value of the fundamental harmonic equal to $\boldsymbol{J}_1$ and has a DC component of the mean value equal to J_0, then thyristors should be fired at angles α, and β for which the value of the form

$$|\boldsymbol{J}_1^{\mathrm{d}}(\alpha,\beta) - \boldsymbol{J}_1| + |J_0^{\mathrm{d}}(\alpha,\beta) - J_0| \tag{14.68}$$

is minimum. Having the look-up tables $\boldsymbol{J}_1^{\mathrm{d}}$ and J_0^{d} calculated and stored in computer memory, a search algorithm that calculates the form (14.68), can identify the needed pair of firing angles (α, β). At such firing angles, the TSI-based compensator will compensate the fundamental harmonic of the reactive and unbalanced currents as well as the DC currents of the furnace.

Illustration 14.8 *As the numerical illustration of the method discussed in this chapter, the same AC arc furnace is used as analyzed in [295]. It is a reference furnace which supplied from the transformer with the reactance-to-resistance ratio $X_s/R_s = 5$ with* *the secondary voltage rms value $U = 700$ V. The equivalent resistance of cables, electrodes, and the arc was assumed to be $R = 0.25\ \Omega$, the equivalent reactance $\omega_1 L = 1\Omega$. It was assumed that the arc ignites when the its voltage is higher than $U_0 = 300$ V. At such assumptions, the furnace operates [295] with a power factor $\lambda = 0.71$ and the apparent power of the furnace is $S = 0.49$ MVA. Such assumptions simplify modeling, while the results obtained for the reference furnace can be relatively easily recalculated for that of a real furnace, or at least provide information on mutual proportions of some major electrical quantities.*

Results of modeling the furnace with regard to supply voltages and currents rms values and powers in state s0 are shown in Figure 14.34.

When the arc in line S extinguishes (state s1), then voltages, currents, powers, and the power factor change to values shown in Figure 14.35.

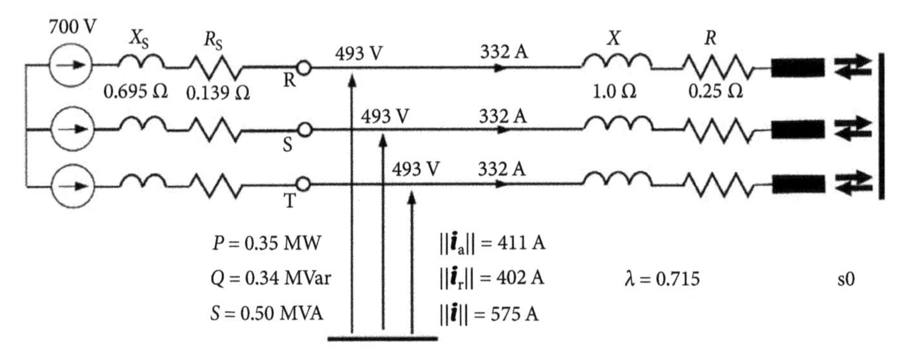

Figure 14.34 Equivalent parameters of the AC arc furnace and results of its analysis at symmetrical (state s0) operation.

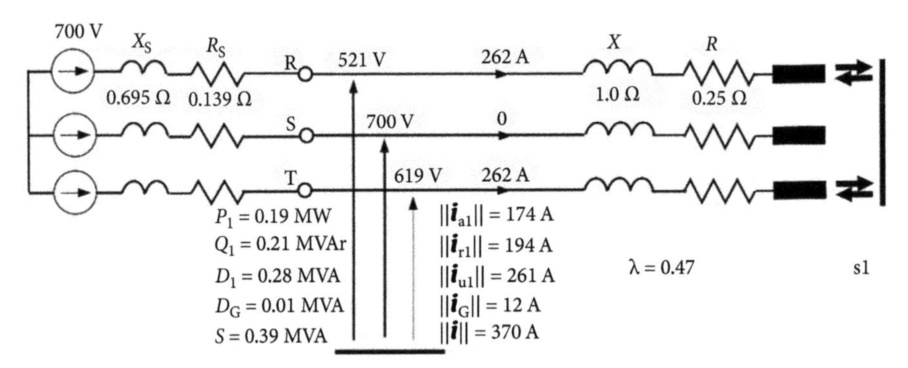

Figure 14.35 Results of the arc furnace analysis at two-arcs (state s1) operation.

These values at the unidirectional arc in line S (state s2) are shown in Figure 14.36. The waveforms of voltages and currents at the supply terminals at this state of operation are shown in Figure 14.37. It can be observed, as expected, that line currents and voltages have lost their negative symmetry versus their values shifted by half of the period. They have DC components and even order harmonics.

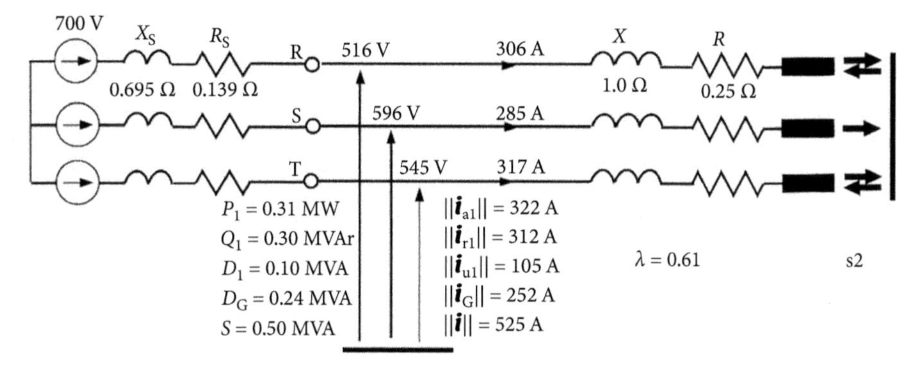

Figure 14.36 Results of the arc furnace analysis at the unidirectional arc (state s2) operation.

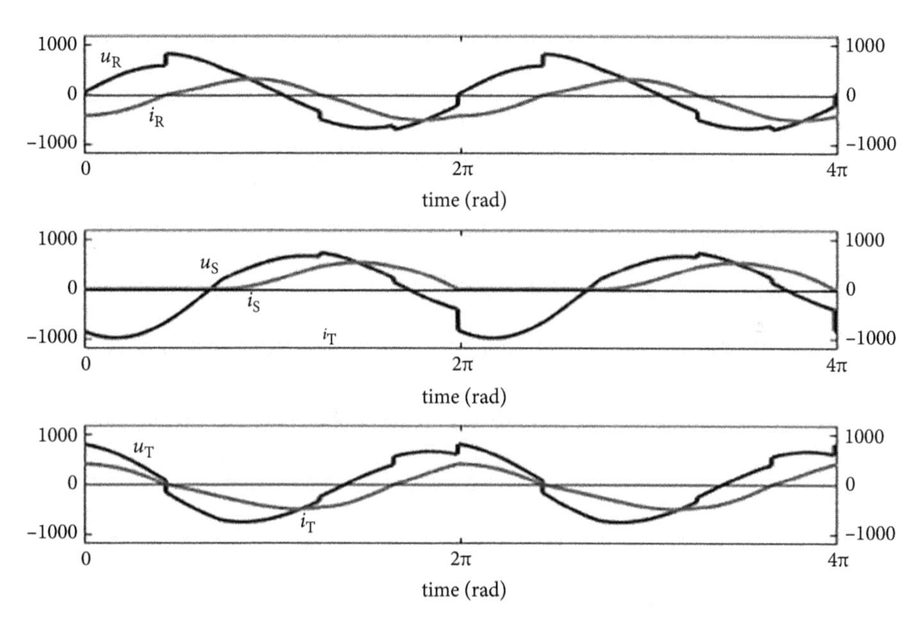

Figure 14.37 Waveforms of voltages and currents at the furnace terminals in the arc furnace state s2.

To reduce the current of the TSI balancing compensator, the reactive power along some furnace-originated current harmonics can be compensated by a fixed-parameter reactance compensator. Let us assume, following [299], that this compensator compensates reactive power Q_1 of the furnace in the state s0, along with the second and the third order current harmonics. When each branch of the filter compensates half of the reactive power Q_1 of the furnace, then the filter should have parameters shown in Figure 14.38.

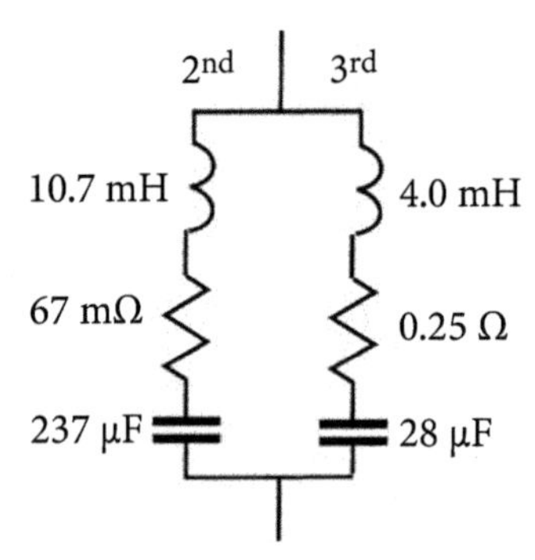

Figure 14.38 Parameters of one branch of the fixed-parameter compensator.

Having the reactive power as well as the second and the third order current harmonics compensated by the fixed-parameters compensator, the TSI-based adaptive compensator should compensate unbalanced and DC components of the furnace current. A search algorithm that operates according to formula (14.68) provides firing angles of the compensator thyristors such that these two components are eliminated from the furnace supply current, changing their waveforms to that shown in Figure 14.39.

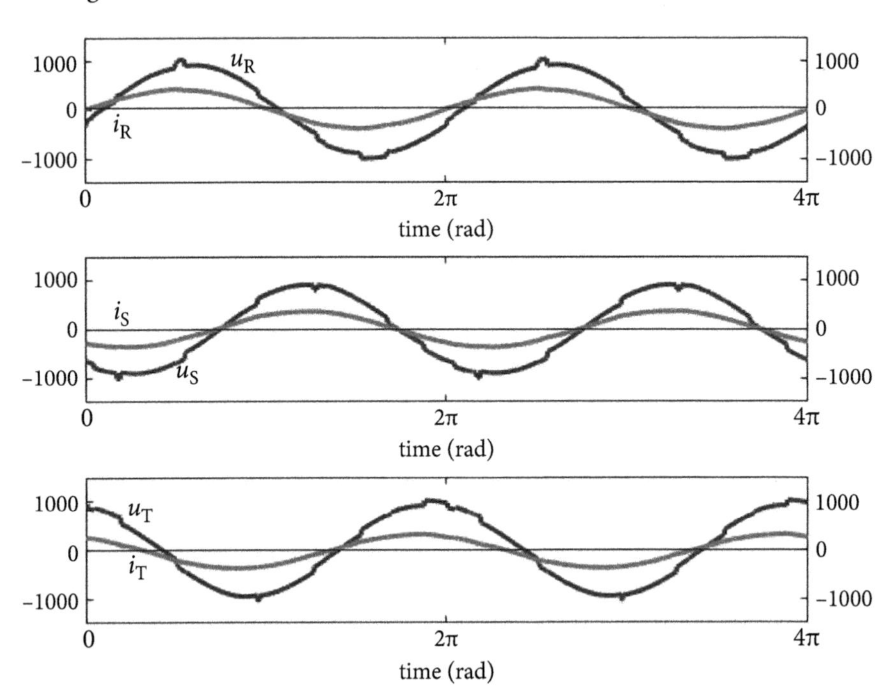

Figure 14.39 The waveforms of the furnace currents and supply voltages after the unbalanced and DC currents compensation.

The results of the adaptive compensation of the unbalanced and DC currents of the reference AC arc furnace are compiled in Table 14.3

Table 14.3 The results of compensation.

State:		s0		s1		s2	
Compens.		No	Yes	No	Yes	No	Yes
$\|\boldsymbol{i}_{\mathrm{a}}\|$	A	411	533	174	254	322	443
$\|\boldsymbol{i}_{\mathrm{r}}\|$	A	402	81	194	45	312	65
$\|\boldsymbol{i}_{\mathrm{u}}\|$	A	0	0.2	261	42	105	39
$\|\boldsymbol{i}_{\mathrm{G}}\|$	A	13	10	27	12	252	48
$\|\boldsymbol{i}\|$	A	575	539	370	262	525	452
λ		**0.71**	**0.99**	**0.47**	**0.97**	**0.61**	**0.98**

*****

Summary

The CPC-based approach to the PT provides fundamentals for the reactance compensation of reactive and unbalanced currents in the presence of the supply voltage harmonic distortion and asymmetry. It also provides fundamentals for adaptive compensation of these currents' fundamental harmonic and even adaptive compensation of their DC components.

Chapter 15
Reactance Balancing Compensation in Three-Phase Circuits with Neutral

15.1 Historical Background

Distribution electrical systems in residential subdivisions, commercial buildings, and manufacturing plants supply not only balanced three-phase equipment, mainly three-phase motors, but also aggregates of single-phase loads. Such distribution systems are supplied from transformers in Δ/Y configuration and are equipped with a neutral conductor, which forms a three-phase with neutral or three-phase four-wire (3p4w) circuit, on the secondary side of the supply transformer.

The neutral conductor increases the structural complexity of the circuit as compared to three-phase circuits, but this increase is not only caused by the higher number of supply conductors, four instead of three of them. The role of the neutral conductor is different from the role of the remaining three line conductors. It enables the supply of single-phase energy receivers. Therefore, three-phase circuits with neutral cannot be regarded as a simple generalization of three-phase, three-wire (3p3w) to 3p4w circuits. Moreover, as shown in Chapter 8, the analysis of 3p4w circuits is more complex than 3p3w ones. The first can be described in terms of only four quantities, two line currents, and two line voltages measured versus the artificial zero. Six quantities, three line currents, and three line voltages are needed for that in 3p4w circuits.

Apart from the increased analytical complexity, nonlinear or/and time-variable single-phase devices supplied in 3p4w circuits can generate harmonics of the order $n = 3k$, mainly the third order harmonic. These current harmonics result in the presence of such harmonics in line voltages. At the load imbalance, not only symmetrical components of voltages and current of the positive and negative sequence but also the zero sequence components can occur in the circuit.

The detrimental effects of a degraded power factor in such circuits, mainly energy loss, are located to a large degree in the supply transformer. They can be reduced by compensation on the secondary side of the transformer, thus in the 3p4w circuit supplied by Y windings of such a transformer.

As to goals, reactance compensation in a 3p4w circuit does not differ from compensation in three-phase circuits without the neutral. The neutral conductor changes the properties of the circuit and this affects the methods of compensation, however.

Powers and Compensation in Circuits with Nonsinusoidal Current. Leszek S. Czarnecki, Oxford University Press.
 DOI: 10.1093/oso/9780198879206.003.0015

Of all areas of compensation, such as compensation in single-phase circuits or 3p3w circuits, compensation in 3p4w circuits seems to be the most important because most electric energy is utilized in these kinds of circuits and systems. Moreover, compensation on the secondary side of the Δ/Y transformer, that is, in the 3p4w circuit, can increase the load voltage, and reduces its asymmetry and distortion to a higher degree as compared to compensation performed on the primary side of the transformer.

Because of the power level, compensation of the reactive and unbalanced currents in large manufacturing plants could be above the capability of switching compensators built of power transistors, however. Only reactance compensators can have sufficient power for that.

Despite its importance, there is a relatively low number of published research on reactance compensation in 3p4w circuits. This is mainly because power theory, being focused for decades on single-phase loads, does not provide theoretical fundamentals for such compensation. Although some results of methods of design of compensators 3p4w circuits were published [22, 28, 115, 258, 294, 307], the development of such methods is substantially retarded. A controversy [218, 284, 235, 308] regarding how to describe the power properties of four-wire circuits is the main reason for that. Because of the lack of such a description, only optimization methods could provide parameters of a reactance compensator. Optimization methods might not be appropriate for control of adaptive reactance compensators, when the speed of control is crucial, however. Formulae that would enable direct calculation of the compensator's inductances and capacitances from algebraic formulae are needed.

A method of reactance balancing compensator synthesis for four-wire circuits with the assumption that the supply voltage is sinusoidal was developed in [278]. At the same time, the CPC-based power theory was generalized in [282] to four-wire circuits with nonsinusoidal supply voltage. The results obtained in [278] and [282], covered in Chapter 8, provide a starting point for the development of a method of balancing four-wire circuits with nonsinusoidal supply voltage.

15.2 Partial Compensation at Sinusoidal Voltage and Currents (sv&c)

According to Chapter 8, formula (8.11), three currents, $\boldsymbol{i}_{\mathrm{r}}$, $\boldsymbol{i}_{\mathrm{u}}^{\mathrm{n}}$, and $\boldsymbol{i}_{\mathrm{u}}^{\mathrm{z}}$ contribute in 3p4w circuits with sinusoidal voltages and currents to the power factor degradation. These currents occur due to the load reactance and the load imbalance. Each of them can be compensated by a reactance compensator. Currents $\boldsymbol{i}_{\mathrm{r}}$, and $\boldsymbol{i}_{\mathrm{u}}^{\mathrm{n}}$ can be compensated by a compensator configured in a Δ structure, as discussed in Chapter 13, or configured in a Y structure. The unbalanced current of the zero sequence $\boldsymbol{i}_{\mathrm{u}}^{\mathrm{z}}$ cannot be compensated by any compensator configured in Δ, however. Therefore, let us assume that the compensator is configured in Y, as shown in Fig. 15.1.

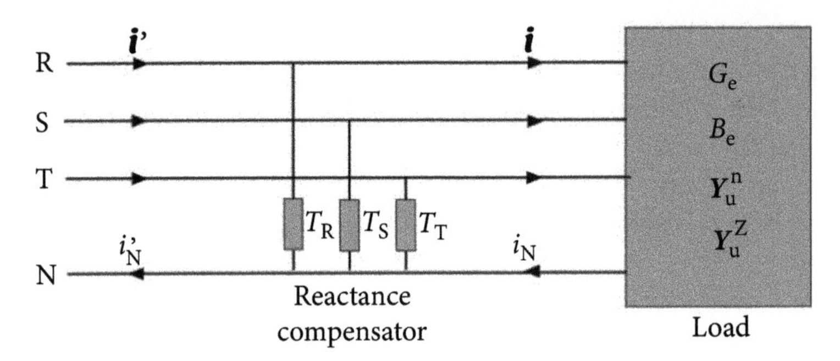

Figure 15.1 An LTI load with reactance compensator in a Y structure.

It is composed of three reactance devices, inductors, or capacitors, of susceptance T_R, T_S, and T_T. It is assumed here that these devices are lossless, meaning their conductance G is equal to zero.

According to Chapter 8, the equivalent susceptance of such a compensator is equal to

$$B_{eC} = \frac{1}{3}(T_R + T_S + T_T)$$

while unbalanced admittances of the negative and zero sequence

$$Y_{uC}^{n} = \frac{1}{3}j\,(T_R + \alpha T_S + \alpha^* T_T)$$

$$Y_{uC}^{z} = \frac{1}{3}j\,(T_R + \alpha^* T_S + \alpha T_T).$$

The reactive current of the load is compensated completely on the condition that

$$B_{eC} + B_e = 0. \tag{15.1}$$

The unbalanced current of the negative sequence of the load is compensated completely on the condition that

$$Y_{uC}^{n} + Y_{u}^{n} = 0 \tag{15.2}$$

while for the zero sequence of this current, the condition has the form

$$Y_{uC}^{z} + Y_{u}^{z} = 0. \tag{15.3}$$

The two last conditions must be satisfied for real and imaginary parts of the unbalanced admittance, thus each of them stands for two conditions. Consequently, compensation of the currents $\boldsymbol{i}_r$, $\boldsymbol{i}_u^n$, and $\boldsymbol{i}_u^z$ requires that three unknown susceptances, T_R, T_S, and T_T, of the compensator branches satisfy five equations. Such equations are mutually contradictory. Thus, compensation of these three currents by the compensator shown in Fig. 15.1 is not possible. Only one of two unbalanced currents, $\boldsymbol{i}_{\mathrm{u}}^{\mathrm{n}}$ or $\boldsymbol{i}_{\mathrm{u}}^{\mathrm{z}}$, along with the reactive current $\boldsymbol{i}_{\mathrm{r}}$ can be compensated. The compensator shown in Fig. 15.1 enables only partial compensation.

The decision on which of them should be compensated depends on the compensator designer. Let us suppose that the designer concludes that the unbalanced current of the negative sequence along with the reactive current should be compensated. Thus, the compensator susceptances must satisfy the following equations

$$\mathrm{Re}\left\{\frac{1}{3}j\,(T_{\mathrm{R}} + \alpha T_{\mathrm{S}} + \alpha^* T_{\mathrm{T}}) + \boldsymbol{Y}_{\mathrm{u}}^{\mathrm{n}}\right\} = 0$$

$$\mathrm{Im}\left\{\frac{1}{3}j\,(T_{\mathrm{R}} + \alpha T_{\mathrm{S}} + \alpha^* T_{\mathrm{T}}) + \boldsymbol{Y}_{\mathrm{u}}^{\mathrm{n}}\right\} = 0$$

$$\frac{1}{3}(T_{\mathrm{R}} + T_{\mathrm{S}} + T_{\mathrm{T}}) + B_{\mathrm{e}} = 0.$$

Their solution results in the compensator parameters

$$\begin{aligned} T_{\mathrm{R}} &= -2\,\mathrm{Im}\,\boldsymbol{Y}_{\mathrm{u}}^{\mathrm{n}} - B_{\mathrm{e}} \\ T_{\mathrm{S}} &= \sqrt{3}\mathrm{Re}\,\boldsymbol{Y}_{\mathrm{u}}^{\mathrm{n}} + \mathrm{Im}\,\boldsymbol{Y}_{\mathrm{u}}^{\mathrm{n}} - B_{\mathrm{e}}. \\ T_{\mathrm{T}} &= -\sqrt{3}\mathrm{Re}\,\boldsymbol{Y}_{\mathrm{u}}^{\mathrm{n}} + \mathrm{Im}\,\boldsymbol{Y}_{\mathrm{u}}^{\mathrm{n}} - B_{\mathrm{e}} \end{aligned} \tag{15.4}$$

A compensator with such susceptance of the compensator's branches satisfies Eqn. (15.2), thus its negative sequence unbalanced admittance$\boldsymbol{Y}_{\mathrm{uC}}^{\mathrm{n}}$ is equal to $-\boldsymbol{Y}_{\mathrm{u}}^{\mathrm{n}}$. Thus the compensator reduces the unbalanced current of the negative sequence $\boldsymbol{i}_{\mathrm{u}}^{\mathrm{n}}$, along with the reactive current $\boldsymbol{i}_{\mathrm{r}}$, completely. The compensator is an unbalanced device connected between the supply lines of the load and the neutral so that it can draw an unbalanced current of the zero sequence from the supply source. To calculate it, let us calculate the zero sequence admittance $\boldsymbol{Y}_{\mathrm{uC}}^{\mathrm{z}}$ of this compensator:

$$\begin{aligned} \boldsymbol{Y}_{\mathrm{uC}}^{\mathrm{z}} &= \frac{1}{3}j\,(T_{\mathrm{R}} + \alpha^* T_{\mathrm{S}} + \alpha T_{\mathrm{T}}) = \\ &= \frac{1}{3}j\left[(-2\,\mathrm{Im}\,\boldsymbol{Y}_{\mathrm{u}}^{\mathrm{n}} - B_{\mathrm{e}}) + \alpha^*\left(\sqrt{3}\mathrm{Re}\,\boldsymbol{Y}_{\mathrm{u}}^{\mathrm{n}} + \mathrm{Im}\,\boldsymbol{Y}_{\mathrm{u}}^{\mathrm{n}} - B_{\mathrm{e}}\right) + \right. \\ &\quad \left. + \,\alpha\left(-\sqrt{3}\mathrm{Re}\,\boldsymbol{Y}_{\mathrm{u}}^{\mathrm{n}} + \mathrm{Im}\,\boldsymbol{Y}_{\mathrm{u}}^{\mathrm{n}} - B_{\mathrm{e}}\right)\right] = \mathrm{Re}\,\boldsymbol{Y}_{\mathrm{u}}^{\mathrm{n}} - j\,\mathrm{Im}\,\boldsymbol{Y}_{\mathrm{u}}^{\mathrm{n}} = \boldsymbol{Y}_{\mathrm{u}}^{\mathrm{n}*}. \end{aligned} \tag{15.5}$$

Thus such a compensator reduces the negative sequence component $\boldsymbol{i}_{\mathrm{u}}^{\mathrm{n}}$ of the unbalanced current to zero but changes the zero sequence component $\boldsymbol{i}_{\mathrm{u}}^{\mathrm{z}}$ to

$$\boldsymbol{i}_{\mathrm{u}}^{\mathrm{z}} = \sqrt{2}\mathrm{Re}\left\{(\boldsymbol{Y}_{\mathrm{uC}}^{\mathrm{z}} + \boldsymbol{Y}_{\mathrm{u}}^{\mathrm{z}})\mathbf{1}^{\mathrm{z}}\boldsymbol{U}_{\mathrm{R}}\, e^{j\omega_1 t}\right\} = \sqrt{2}\mathrm{Re}\left\{(\boldsymbol{Y}_{\mathrm{u}}^{\mathrm{n}*} + \boldsymbol{Y}_{\mathrm{u}}^{\mathrm{z}})\mathbf{1}^{\mathrm{z}}\boldsymbol{U}_{\mathrm{R}} e^{j\omega_1 t}\right\} \tag{15.6}$$

which must have a higher rms value than that before compensation.

Illustration 15.1 *Let us calculate branch susceptances of a reactance compensator for compensation of the negative sequence component of the load current for the unbalanced load shown in Fig. 15.2. Calculate the rms values of the line currents before and after compensation, assuming that* $U_R = 120$ V *and* $\omega_1 = 1$ rad/s.

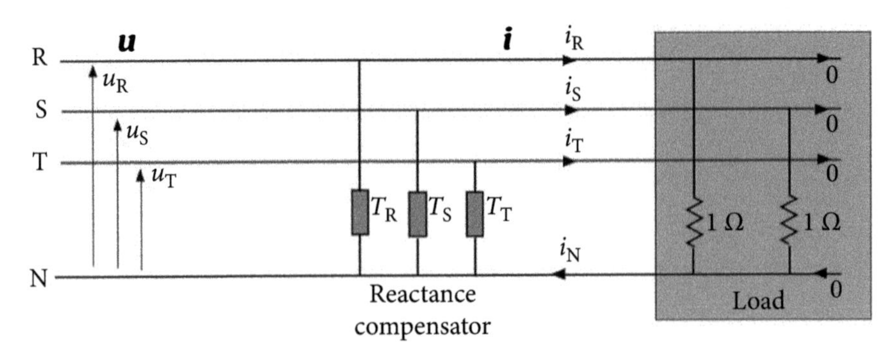

Figure 15.2 An example of a load with reactance compensator.

For the load in Fig. 15.2, $\boldsymbol{Y}_R = 1$ S, $\boldsymbol{Y}_S = 1$ S, $\boldsymbol{Y}_T = 0$, *thus* $B_e = 0$ *and*

$$\boldsymbol{Y}_u^n = \frac{1}{3}(\boldsymbol{Y}_R + \alpha \boldsymbol{Y}_S + \alpha^* \boldsymbol{Y}_T) = \frac{1}{3}(1+\alpha) = 0.333\, e^{j60^\circ} = 0.167 + j\,0.288 \text{ S}.$$

The compensator's susceptances are equal to

$$T_R = -2\,\text{Im}\,\boldsymbol{Y}_u^n - B_e = -0.577 \text{ S}$$

$$T_S = \sqrt{3}\,\text{Re}\,\boldsymbol{Y}_u^n + \text{Im}\,\boldsymbol{Y}_u^n - B_e = 0.577 \text{ S}$$

$$T_T = -\sqrt{3}\,\text{Re}\,\boldsymbol{Y}_u^n + \text{Im}\,\boldsymbol{Y}_u^n - B_e = 0$$

so that the compensator parameters are

$$L_R = -\frac{1}{\omega_1 T_R} = 1.73 \text{ H}, \qquad C_S = \frac{T_S}{\omega_1} = 0.577 \text{ F}.$$

The circuit with the compensator and compensation results are shown in Fig. 15.3. The compensator modifies the crms value of the supply currents to

$$\boldsymbol{I}'_R = 138.6\, e^{-j30^\circ} \text{ A}, \qquad \boldsymbol{I}'_S = 138.6\, e^{-j90^\circ} \text{ A},$$

and consequently, the negative sequence component of this current to

$$\boldsymbol{I}'^n = \frac{1}{3}(138.6\, e^{-j30^\circ} + \alpha^* \times 138.6\, e^{-j90^\circ}) = 0.$$

Unfortunately it increases the supply and the neutral currents. Despite the unbalanced current of the negative sequence reduction, the final effect of compensation is quite opposite to expectations.

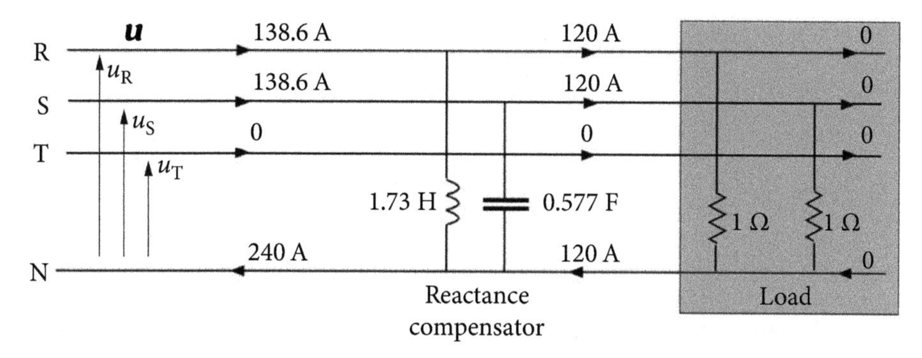

Figure 15.3 A load with compensator of the negative sequence component of the unbalanced current.

******

Let us assume that the compensator designer selected the zero sequence component $\boldsymbol{i}_{\mathrm{u}}^{\mathrm{z}}$ of the unbalanced current, along with the reactive current $\boldsymbol{i}_{\mathrm{r}}$ as the goal of compensation. In such a case the compensator branch susceptances must satisfy the equations

$$\mathrm{Re}\left\{\frac{1}{3}j\,(T_{\mathrm{R}}+\alpha^{*}T_{\mathrm{S}}+\alpha T_{\mathrm{T}})+\boldsymbol{Y}_{\mathrm{u}}^{\mathrm{z}}\right\}=0$$

$$\mathrm{Im}\left\{\frac{1}{3}j\,(T_{\mathrm{R}}+\alpha^{*}T_{\mathrm{S}}+\alpha T_{\mathrm{T}})+\boldsymbol{Y}_{\mathrm{u}}^{\mathrm{z}}\right\}=0$$

$$\frac{1}{3}(T_{\mathrm{R}}+T_{\mathrm{S}}+T_{\mathrm{T}})+B_{\mathrm{e}}=0$$

which have the solution

$$\begin{aligned}T_{\mathrm{R}}&=-2\,\mathrm{Im}\,\boldsymbol{Y}_{\mathrm{u}}^{\mathrm{z}}-B_{\mathrm{e}}\\ T_{\mathrm{S}}&=-\sqrt{3}\mathrm{Re}\,\boldsymbol{Y}_{\mathrm{u}}^{\mathrm{z}}+\mathrm{Im}\,\boldsymbol{Y}_{\mathrm{u}}^{\mathrm{z}}-B_{\mathrm{e}}.\\ T_{\mathrm{T}}&=\sqrt{3}\mathrm{Re}\,\boldsymbol{Y}_{\mathrm{u}}^{\mathrm{z}}+\mathrm{Im}\,\boldsymbol{Y}_{\mathrm{u}}^{\mathrm{z}}-B_{\mathrm{e}}\end{aligned}\tag{15.7}$$

A compensator with such branch susceptances satisfies Eqn. (15.3), thus its negative sequence unbalanced admittance $\boldsymbol{Y}_{\mathrm{uC}}^{\mathrm{z}}$ is equal to $-\boldsymbol{Y}_{\mathrm{u}}^{\mathrm{z}}$. Thus the compensator reduces the unbalanced current of the zero sequence $\boldsymbol{i}_{\mathrm{u}}^{\mathrm{z}}$, along with the reactive current $\boldsymbol{i}_{\mathrm{r}}$, completely. The compensator is an unbalanced device connected between the supply lines of the load and the neutral so that it can draw an unbalanced current of the negative sequence from the supply source. To calculate it, let us calculate the negative sequence admittance $\boldsymbol{Y}^{\rm n}_{\rm uC}$ of this compensator:

$$\boldsymbol{Y}^{\rm n}_{\rm uC} = \frac{1}{3}j\,(T_{\rm R} + \alpha T_{\rm S} + \alpha^* T_{\rm T}) = \frac{1}{3}j\Big[\,(-2\,{\rm Im}\,\boldsymbol{Y}^{\rm z}_{\rm u} - B_{\rm e}) +$$
$$+\,\alpha\left(-\sqrt{3}{\rm Re}\,\boldsymbol{Y}^{\rm z}_{\rm u} + {\rm Im}\,\boldsymbol{Y}^{\rm z}_{\rm u} - B_{\rm e}\right) +$$
$$+\,\alpha^*\left(\sqrt{3}{\rm Re}\,\boldsymbol{Y}^{\rm z}_{\rm u} + {\rm Im}\,\boldsymbol{Y}^{\rm z}_{\rm u} - B_{\rm e}\right)\Big] = {\rm Re}\,\boldsymbol{Y}^{\rm z}_{\rm u} - j\,{\rm Im}\,\boldsymbol{Y}^{\rm z}_{\rm u} = \boldsymbol{Y}^{\rm z^*}_{\rm u}. \tag{15.8}$$

Thus such a compensator reduces the zero sequence component $\boldsymbol{i}^{\rm z}_{\rm u}$ of the unbalanced current to zero but changes the negative sequence component $\boldsymbol{i}^{\rm n}_{\rm u}$ to

$$\boldsymbol{i}^{\rm n}_{\rm u} = \sqrt{2}{\rm Re}\{(\boldsymbol{Y}^{\rm n}_{\rm uC} + \boldsymbol{Y}^{\rm n}_{\rm u})\,\mathbf{1}^{\rm n}\boldsymbol{U}_{\rm R}\,e^{j\omega_1 t}\} = \sqrt{2}{\rm Re}\left\{\left(\boldsymbol{Y}^{\rm z^*}_{\rm u} + \boldsymbol{Y}^{\rm n}_{\rm u}\right)\mathbf{1}^{\rm n}\boldsymbol{U}_{\rm R}\,e^{j\omega_1 t}\right\}. \tag{15.9}$$

Illustration 15.2 *Let us calculate branch susceptances of a reactance compensator for compensation of the zero sequence component of the load current for the same load as in Illustration 15.1, assuming that the compensator should compensate the zero sequence component of the unbalanced current and the reactive current.*

Since the zero sequence unbalanced admittance is

$$\boldsymbol{Y}^{\rm z}_{\rm u} = \frac{1}{3}(\boldsymbol{Y}_{\rm R} + \alpha^*\boldsymbol{Y}_{\rm S} + \alpha\boldsymbol{Y}_{\rm T}) = \frac{1}{3}(1 + \alpha^* 1) = 0.333\,e^{-j60^\circ} = 0.167 - j\,0.288\ {\rm S}$$

the compensator's susceptances are equal to

$$T_{\rm R} = -2\,{\rm Im}\,\boldsymbol{Y}^{\rm z}_{\rm u} - B_{\rm e} = 0.567\ {\rm S}$$
$$T_{\rm S} = -\sqrt{3}{\rm Re}\,\boldsymbol{Y}^{\rm z}_{\rm u} + {\rm Im}\,\boldsymbol{Y}^{\rm z}_{\rm u} - B_{\rm e} = -0.567\ {\rm S}$$
$$T_{\rm T} = \sqrt{3}{\rm Re}\,\boldsymbol{Y}^{\rm z}_{\rm u} + {\rm Im}\,\boldsymbol{Y}^{\rm z}_{\rm u} - B_{\rm e} = 0,$$

thus it is composed of reactance devices of parameters

$$C_{\rm R} = \frac{T_{\rm R}}{\omega_1} = 0.577\ {\rm F}, \qquad L_{\rm S} = -\frac{1}{\omega_1 T_{\rm S}} = 1.73\ {\rm H}$$

The circuit with the compensator and compensation results are shown in Fig. 15.4. The compensator modifies the crms value of the supply currents to

$$\boldsymbol{I}'_{\rm R} = 138.6\,e^{j30^\circ}\ {\rm A}, \qquad \boldsymbol{I}'_{\rm S} = 138.6\,e^{-j150^\circ}\ {\rm A}.$$

Unfortunately it increases the supply current rms value.

*****

The reasoning presented above, supported by numerical illustrations, shows that the tempting idea of reducing the compensator complexity by compensation of only one of two components of the unbalanced current by a compensator in a Y configuration is not right. Such partial compensation does not reduce the supply current.

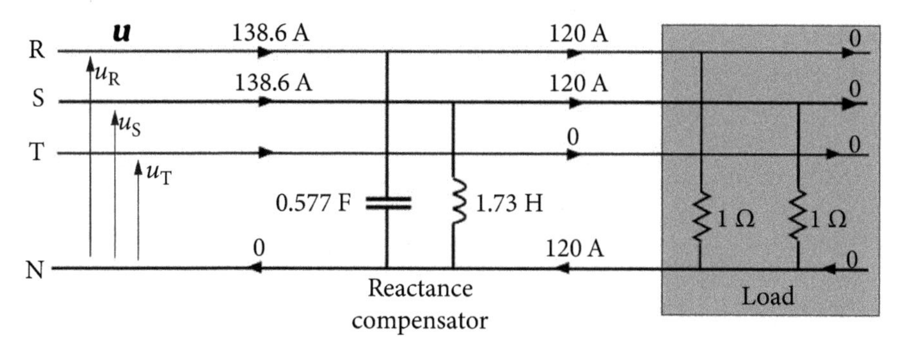

Figure 15.4 An unbalanced load with a reactance compensator of the zero sequence component of unbalanced current.

Partial compensation is possible, however, if instead of a compensator in a Y configuration a compensator in a Δ configuration, shown in Fig. 15.5, is used for compensation of the unbalanced current of the negative sequence.

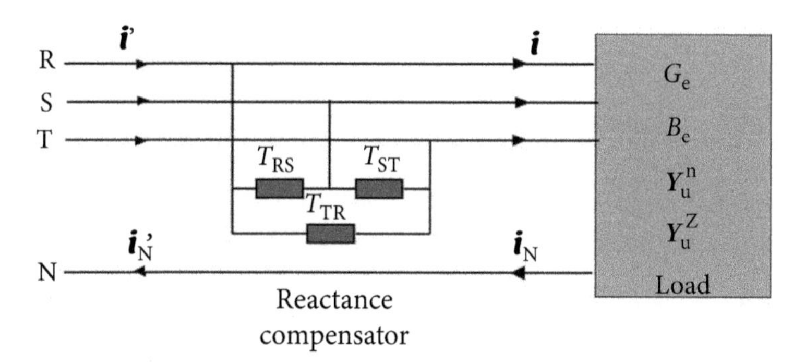

Figure 15.5 LTI load with reactance compensator in Δ structure.

According to Chapter 14, formula (14.4), line-to-line susceptances of such a compensator must satisfy the following equations

$$T_{RS} = \left(\sqrt{3}\mathrm{Re}\,\boldsymbol{Y}_u^n - \mathrm{Im}\,\boldsymbol{Y}_u^n - B_e\right)/3$$

$$T_{ST} = (2\,\mathrm{Im}\,\boldsymbol{Y}_u^n - B_e)/3$$

$$T_{TR} = \left(-\sqrt{3}\mathrm{Re}\,\boldsymbol{Y}_u^n - \mathrm{Im}\,\boldsymbol{Y}_u^n - B_e\right)/3.$$

Because such a compensator does not have branches connected to the neutral, it cannot affect the zero sequence unbalanced admittance and consequently the zero sequence unbalanced current. It only compensates the unbalanced current of the negative sequence.

Illustration 15.3 *Let us calculate susceptances of a compensator in Δ configuration for compensation of the negative sequence component of the load unbalanced current for the same load as in Illustration 15.1.*

The unbalanced admittance of the negative sequence, calculated in Illustration 15.1, is equal to

$$\boldsymbol{Y}_{\mathrm{u}}^{\mathrm{n}} = 0.333e^{j60^{\circ}} = 0.167 + j\,0.288\ \mathrm{S},$$

hence the susceptances

$$T_{\mathrm{RS}} = \left(\sqrt{3} \times 0.167 - 0.288\right)/3 = 0$$

$$T_{\mathrm{ST}} = (2 \times 0.288)/3 = 0.192\ \mathrm{S}$$

$$T_{\mathrm{TR}} = \left(-\sqrt{3} \times 0.167 - 0.288\right)/3 = -0.192\ \mathrm{S},$$

and the compensator LC parameters are equal to

$$C_{\mathrm{ST}} = \frac{T_{\mathrm{ST}}}{\omega_1} = 0.192\ \mathrm{F}, \qquad L_{\mathrm{TR}} = -\frac{1}{\omega_1\, T_{\mathrm{TR}}} = 5.21\ \mathrm{H}.$$

The results of compensation are shown in Fig. 15.6.

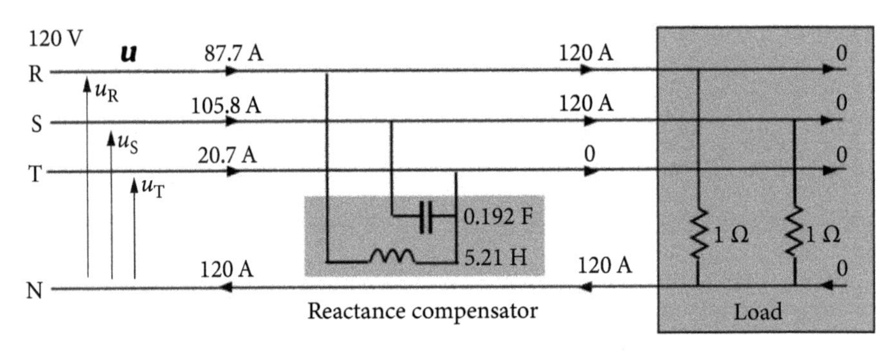

Figure 15.6 A load, compensator, and results of partial compensation.

*****

Thus, reactance partial compensation in 3p4w circuits with sinusoidal voltage is confined only to the negative sequence component of the unbalanced current and the reactive current. These currents can be completely reduced using compensators configured in Δ. Compensators configured in Y cannot be used for that.

15.3 Complete Compensation at sv&c

As was shown in the previous section, the unbalanced current of the zero sequence cannot be reduced by partial compensation, thus it is not possible to obtain the unity power factor by such compensation. Complete compensation is needed for that. The compensator can have the structure shown in Fig. 15.7, or that shown in Fig. 15.8.

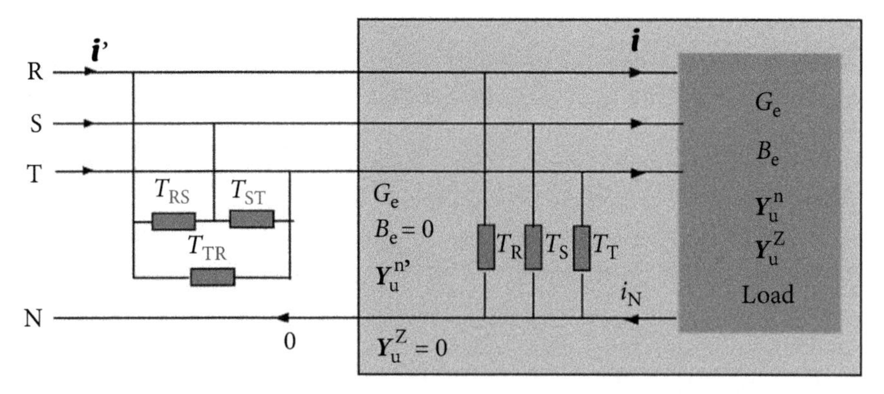

Figure 15.7 An LTI load with reactance compensators of Y and Δ structure.

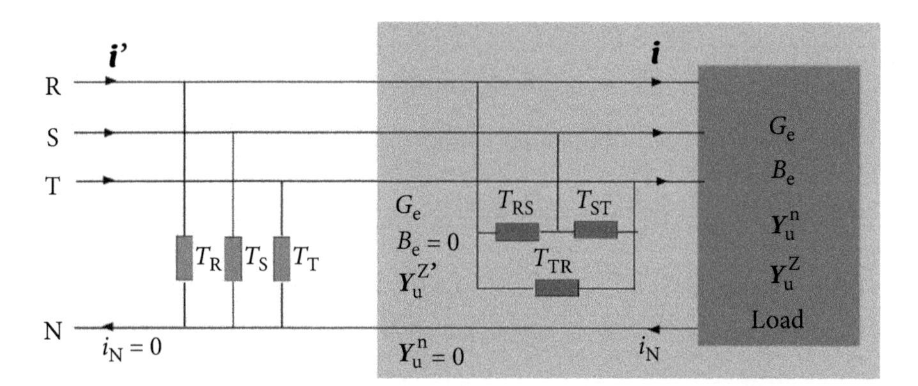

Figure 15.8 An LTI load with reactance compensators of Y and Δ structure.

In both cases, compensation is performed in two separate steps by sub-compensators configured as a Y or a Δ. In the first step, which stands for partial compensation, only one of two unbalanced currents is compensated.

When the compensator has the structure shown in Fig. 15.7, the unbalanced current of the zero sequence is compensated. When the compensator has the structure shown in Fig. 15.8, the unbalanced current of the negative sequence is compensated in the first step. The reactive current can be compensated in the first or the second step. According to Figs 14.7 and 14.8, this current is compensated in the first step.

After the first step of compensation is performed, the partially compensated load, enclosed in Figs 15.7 and 15.8 by the shadowed area, draws from the supply source only one unbalanced current. This remaining current can be completely compensated in the second step by a sub-compensator configured in Δ, as shown Fig. 15.7, or a sub-compensator configured in a Y, as shown in Fig. 15.8. Thus both unbalanced currents, along with the reactive current, can be completely compensated, and consequently the power factor can reach a unity value.

Illustration 15.4 *Consider the sub-compensator configured in Y, found in Illustration 15.2, as the result of the first step of compensation and complete the second step.*

The sub-compensator configured in a Y, shown in Fig. 15.4, compensated entirely the unbalanced current of the zero sequence $\boldsymbol{i}_{\mathrm{u}}^{\mathrm{z}}$ *and has changed the unbalanced current of the negative sequence* $\boldsymbol{i}_{\mathrm{u}}^{\mathrm{n}}$*, according to formula (15.9). It means, that the unbalanced admittance of the negative sequence of the partially compensated load is equal to*

$$\boldsymbol{Y}_{\mathrm{u}}^{\mathrm{n}\,\prime} = \boldsymbol{Y}_{\mathrm{u}}^{\mathrm{z}^{*}} + \boldsymbol{Y}_{\mathrm{u}}^{\mathrm{n}} = (0.167 - j\,0.288)^{*} + 0.167 + j\,0.288 = 0.334 + j\,0.576\ \mathrm{S},$$

thus susceptance of the compensator branches is equal to

$$T_{\mathrm{RS}} = \left(\sqrt{3}\mathrm{Re}\,\boldsymbol{Y}_{\mathrm{u}}^{\mathrm{n}\,\prime} - \mathrm{Im}\boldsymbol{Y}_{\mathrm{u}}^{\mathrm{n}\,\prime}\right)/3 = 0$$

$$T_{\mathrm{ST}} = (2\,\mathrm{Im}\,\boldsymbol{Y}_{\mathrm{u}}^{\mathrm{n}\,\prime})/3 = 0.384\ \mathrm{S}$$

$$T_{\mathrm{TR}} = \left(-\sqrt{3}\mathrm{Re}\,\boldsymbol{Y}_{\mathrm{u}}^{\mathrm{n}\,\prime} - \mathrm{Im}\boldsymbol{Y}_{\mathrm{u}}^{\mathrm{n}\,\prime}\right)/3 = -0.384\ \mathrm{S}.$$

The compensator should be composed of reactance devices with parameters

$$C_{\mathrm{ST}} = \frac{T_{\mathrm{ST}}}{\omega_1} = 0.384\ \mathrm{F}, \qquad L_{\mathrm{TR}} = -\frac{1}{\omega_1\,T_{\mathrm{TR}}} = 2.60\ \mathrm{H}.$$

The results of compensation are shown in Fig. 15.9.

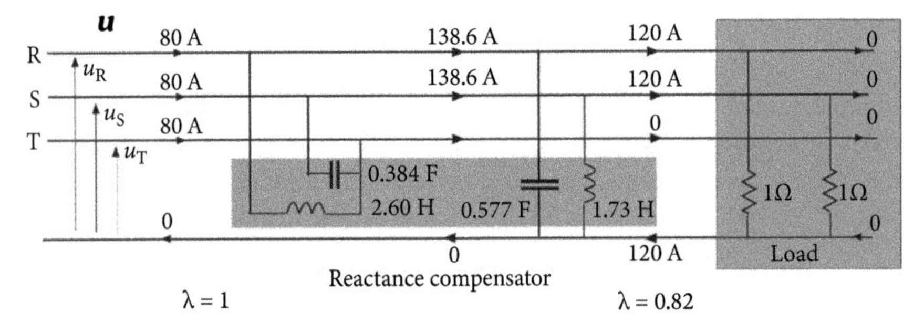

Figure 15.9 An unbalanced load, balancing compensator, and results of compensation

Three-phase systems that supply electric locomotives stand for a borderline case of 3p4w unbalanced circuits. Locomotives are supplied using a pantograph from a single overhead line, but the circuit as a whole is built usually as a three-phase circuit with rails that serve as the neutral conductor. The line that supplies the locomotive can be switched with another line. The switching points decompose the supply line into sectors, as shown in Fig. 15.10. When more than only one locomotive is supplied in the system, and at the same moment they are in different supply sectors, then the load imbalance is naturally reduced.

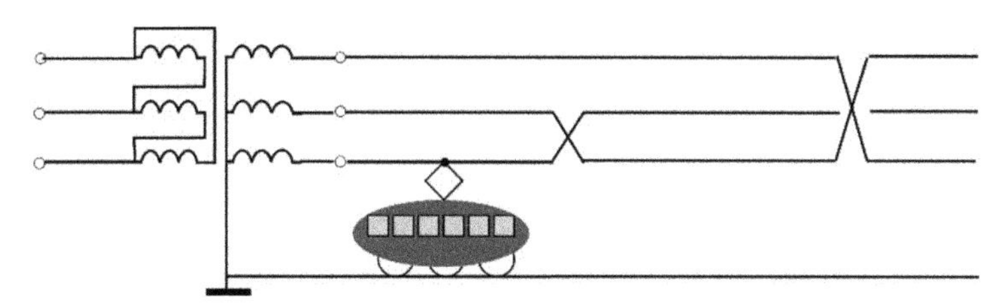

Figure 15.10 A supply system of AC electrical locomotives and supply sectors.

A variety of drives are used in electrical locomotives but independently of a specific drive, it burdens the supply system with active and reactive powers. They change in a wide range during locomotive acceleration, running at a constant speed, deceleration, changes in the terrain slopes, or even wind changes.

From the supply perspective, electrical locomotives stand for low-quality loads with a low power factor. This can be improved by compensation. Due to the change in the supply sectors and power variability, compensation should have adaptive properties. The first step for such an adaptive compensation is compensation under the assumption that the load parameters are fixed.

Illustration 15.5 *An electrical locomotive supplied with the line-to-neutral voltage rms value* $U = 3000$ V *operates with the active power* $P = 5$MW *and reactive power* $Q = 5$MVAr. *Let us calculate the parameters of a reactance balancing compensator of the structure shown in Fig. 15.7, assuming that the locomotive is supplied by T line.*

The locomotive line T admittance is equal to

$$Y_{\rm T} = G_{\rm T} + jB_{\rm T} = \frac{P - jQ}{U_{\rm T}^2} = \frac{3\times10^6 - j3\times10^6}{(3\times10^3)^2} = 0.333 - j0.333 = 0.333\sqrt{2}\,e^{-j45^\circ}\ \text{S}.$$

The equivalent admittance

$$Y_{\rm e} = G_{\rm e} + jB_{\rm e} = \frac{1}{3}(Y_{\rm R} + Y_{\rm S} + Y_{\rm T}) = \frac{1}{3}(0.3333 - j0.3333)\ \text{S} = 0.1111 - j0.1111\ \text{S}.$$

The unbalanced admittance of the zero sequence

$$\boldsymbol{Y}_{\rm u}^{\rm z} = \frac{1}{3}(\boldsymbol{Y}_{\rm R} + \alpha^*\boldsymbol{Y}_{\rm S} + \alpha\boldsymbol{Y}_{\rm T}) = \frac{1}{3}\alpha\boldsymbol{Y}_{\rm T} = \frac{1}{3}1\,e^{j120^\circ}\,0.3333\sqrt{2}\,e^{-j45^\circ}$$
$$= 0.0407 + j0.1517\ \text{S}.$$

The unbalanced admittance of the negative sequence

$$\boldsymbol{Y}_{\mathrm{u}}^{\mathrm{n}} = \frac{1}{3}(\boldsymbol{Y}_{\mathrm{R}} + \alpha\boldsymbol{Y}_{\mathrm{S}} + \alpha^{*}\boldsymbol{Y}_{\mathrm{T}}) = \frac{1}{3}\alpha^{*}\boldsymbol{Y}_{\mathrm{T}} = \frac{1}{3}1\, e^{-j120^{\circ}}\, 0.3333\sqrt{2}\, e^{-j45^{\circ}}$$
$$= -0.1517 - j\,0.0407 \text{ S}.$$

The three-phase rms values of the Currents' Physical Components (CPC) of the locomotive are

$$||\boldsymbol{i}_{\mathrm{a}}|| = G_{\mathrm{e}}||\boldsymbol{u}|| = 0.1111 \times 3000\sqrt{3} = 576.8 \text{ A}$$
$$||\boldsymbol{i}_{\mathrm{r}}|| = |B_{\mathrm{e}}|\ ||\boldsymbol{u}|| = 0.1111 \times 3000\sqrt{3} = 576.8 \text{ A}$$
$$||\boldsymbol{i}_{\mathrm{u}}^{\mathrm{z}}|| = Y_{\mathrm{u}}^{\mathrm{z}}||\boldsymbol{u}|| = 0.1111\sqrt{2} \times 3000\sqrt{3} = 815.7 \text{ A}$$
$$||\boldsymbol{i}_{\mathrm{u}}^{\mathrm{n}}|| = Y_{\mathrm{u}}^{\mathrm{n}}||\boldsymbol{u}|| = 0.1111\sqrt{2} \times 3000\sqrt{3} = 815.7 \text{ A}$$

so that its power factor

$$\lambda = \frac{P}{S} = \frac{||\boldsymbol{i}_{\mathrm{a}}||}{||\boldsymbol{i}||} = \frac{||\boldsymbol{i}_{\mathrm{a}}||}{\sqrt{||\boldsymbol{i}_{\mathrm{a}}||^2 + ||\boldsymbol{i}_{\mathrm{r}}||^2 + ||\boldsymbol{i}_{\mathrm{u}}^{\mathrm{z}}||^2 + ||\boldsymbol{i}_{\mathrm{u}}^{\mathrm{n}}||^2}} = 0.41.$$

Susceptances of the configured in Y sub-compensator of the zero sequence unbalanced current $\boldsymbol{i}_{\mathrm{u}}^{\mathrm{z}}$ and the reactive current $\boldsymbol{i}_r$ are equal to

$$T_{\mathrm{R}} = -2\,\mathrm{Im}\,\boldsymbol{Y}_{\mathrm{u}}^{\mathrm{n}} - B_{\mathrm{e}} = -0.1923 \text{ S}$$
$$T_{\mathrm{S}} = \sqrt{3}\mathrm{Re}\,\boldsymbol{Y}_{\mathrm{u}}^{\mathrm{n}} + \mathrm{Im}\,\boldsymbol{Y}_{\mathrm{u}}^{\mathrm{n}} - B_{\mathrm{e}} = 0.1923 \text{ S}$$
$$T_{\mathrm{T}} = -\sqrt{3}\mathrm{Re}\,\boldsymbol{Y}_{\mathrm{u}}^{\mathrm{n}} + \mathrm{Im}\,\boldsymbol{Y}_{\mathrm{u}}^{\mathrm{n}} - B_{\mathrm{e}} = 0.3333 \text{ S}$$

so that the Y sub-compensator parameters are

$$L_{\mathrm{R}} = -\frac{1}{\omega_1 T_{\mathrm{R}}} = 5.200 \text{ H}, \qquad C_{\mathrm{S}} = \frac{T_{\mathrm{S}}}{\omega_1} = 0.1923 \text{ F}, \qquad C_{\mathrm{S}} = \frac{T_{\mathrm{S}}}{\omega_1} = 0.3333 \text{ F}.$$

The unbalance admittance of the negative sequence as seen from the supply source is

$$\boldsymbol{Y}_{\mathrm{u}}^{\mathrm{n}'} = \boldsymbol{Y}_{\mathrm{u}}^{\mathrm{z}^*} + \boldsymbol{Y}_{\mathrm{u}}^{\mathrm{n}} = 0.1571\, e^{-j75^{\circ}} + 0.1571\, e^{-j120^{\circ}} = -0.1111 - j\,0.1924 \text{ S}$$

thus susceptance of the compensator branches is equal to

$$T_{\mathrm{RS}} = \left(\sqrt{3}\mathrm{Re}\,\boldsymbol{Y}_{\mathrm{u}}^{\mathrm{n}'} - \mathrm{Im}\,\boldsymbol{Y}_{\mathrm{u}}^{\mathrm{n}'}\right)/3 = 0$$
$$T_{\mathrm{ST}} = (2\,\mathrm{Im}\,\boldsymbol{Y}_{\mathrm{u}}^{\mathrm{n}'})/3 = -0.1283 \text{ S}$$
$$T_{\mathrm{TR}} = \left(-\sqrt{3}\mathrm{Re}\,\boldsymbol{Y}_{\mathrm{u}}^{\mathrm{n}'} - \mathrm{Im}\,\boldsymbol{Y}_{\mathrm{u}}^{\mathrm{n}'}\right)/3 = 0.1283 \text{ S}.$$

The compensator can be composed of reactance devices with parameters

$$L_{\mathrm{ST}} = -\frac{1}{\omega_1\, T_{\mathrm{ST}}} = 7.79 \text{ H}, \qquad C_{\mathrm{TR}} = \frac{T_{\mathrm{TR}}}{\omega_1} = 0.1283 \text{ F}.$$

The results of compensation are shown in Fig. 15.11.

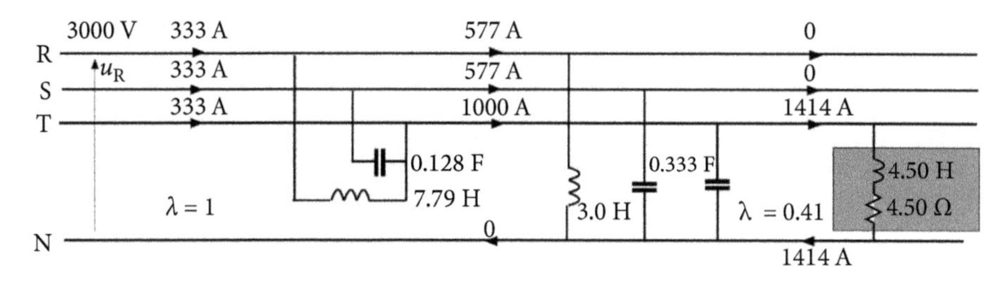

Figure 15.11 A reactance balancing compensator of a locomotive and results of compensation.

*****

15.4 Compensation at Nonsinusoidal Voltages and Currents (nv&c)

As was shown in Chapter 8, the current of LTI load in 3p4w circuits with nonsinusoidal supply voltage is composed of six physical components

$$\boldsymbol{i} = \boldsymbol{i}_{\mathrm{a}} + \boldsymbol{i}_{\mathrm{s}} + \boldsymbol{i}_{\mathrm{r}} + \boldsymbol{i}_{\mathrm{u}}^{\mathrm{p}} + \boldsymbol{i}_{\mathrm{u}}^{\mathrm{n}} + \boldsymbol{i}_{\mathrm{u}}^{\mathrm{z}}$$

with only one current component, the active current $\boldsymbol{i}_{\mathrm{a}}$, which contributes to permanent energy transfer to the load. The remaining ones are harmful because they contribute to the increase of the supply current three-phase rms value $||\boldsymbol{i}||$. The power factor of such a load is

$$\lambda = \frac{P}{S} = \frac{||\boldsymbol{i}_{\mathrm{a}}||}{||\boldsymbol{i}||} = \frac{||\boldsymbol{i}_{\mathrm{a}}||}{\sqrt{||\boldsymbol{i}_{\mathrm{a}}||^2 + ||\boldsymbol{i}_{\mathrm{s}}||^2 + ||\boldsymbol{i}_{\mathrm{r}}||^2 + ||\boldsymbol{i}_{\mathrm{u}}^{\mathrm{p}}||^2 + ||\boldsymbol{i}_{\mathrm{u}}^{\mathrm{n}}||^2 + ||\boldsymbol{i}_{\mathrm{u}}^{\mathrm{z}}||^2}}.$$

Apart from the scattered current $\boldsymbol{i}_{\mathrm{s}}$, which cannot be compensated by any shunt reactance compensator, four remaining harmful supply current components can be reduced by reactance compensators.

As explained earlier in this chapter, reactance compensation of the unbalanced and reactive currents in 3p4w circuits with sinusoidal supply voltage requires that a compensator is built of two sub-compensators. One of them should be connected between the supply lines and the neutral conductor in Y configuration, while the other should be connected between supply lines and has a Δ configuration, as shown in Fig. 15.12. The same applies to individual harmonics of the positive and negative sequence order, but not to the zero sequence harmonics. The sub-compensator configured in Δ cannot compensate the current harmonics that are in particular supply lines in the same phase, meaning the zero sequence component of the unbalanced current. The sub-compensator for compensation of such current components should be connected between the supply lines and the neutral conductor.

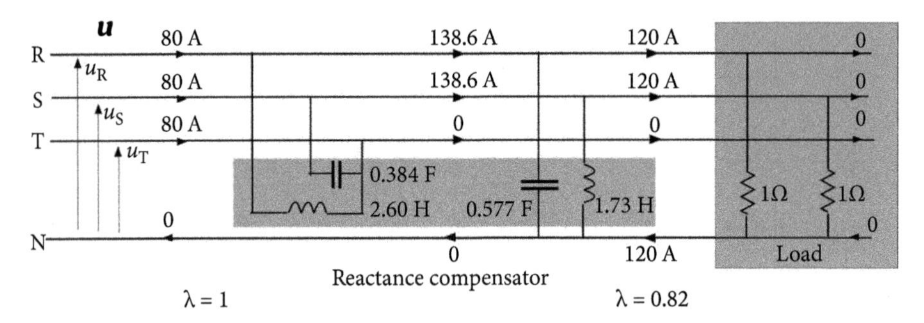

Figure 15.12 A compensator structure and susceptances for the nth order harmonic.

Compensator susceptances T_{Rn}, T_{Sn}, T_{Tn} and T_{RSn}, T_{STn}, T_{TRn} for the fundamental harmonic and other harmonics of the positive sequence order $n = 1,4,7,\ldots 3k+1$ can be calculated similarly as in the case of purely sinusoidal supply voltage, with Eqns (15.4) and (14.4). Only the load admittances for the fundamental frequency have to be replaced by these admittances for harmonic frequencies. Assuming that along with the zero sequence component of the unbalanced current, $\boldsymbol{i}^{z}_{un}$, the reactive current $\boldsymbol{i}_{rn}$ is also compensated by a Y-configured sub-compensator, the susceptances, T_{Rn}, T_{Sn}, and T_{Tn}, are equal to

$$\begin{aligned} T_{Rn} &= -2\,\mathrm{Im}\,\boldsymbol{Y}^{z}_{un} - B_{en} \\ T_{Sn} &= \sqrt{3}\mathrm{Re}\,\boldsymbol{Y}^{z}_{un} + \mathrm{Im}\,\boldsymbol{Y}^{z}_{un} - B_{en} \\ T_{Tn} &= -\sqrt{3}\mathrm{Re}\,\boldsymbol{Y}^{z}_{un} + \mathrm{Im}\,\boldsymbol{Y}^{z}_{un} - B_{en}. \end{aligned} \tag{15.10}$$

When compensation at sinusoidal voltage was discussed, it was shown in formula (15.8) that the Y sub-compensator that compensates the zero sequence component, $\boldsymbol{i}^{z}_{un}$ of the unbalanced current has the negative sequence unbalanced admittance equal to $\boldsymbol{Y}^{n}_{uCn} = \boldsymbol{Y}^{z*}_{un}$. The same applies to all harmonics of the order of the positive sequence. Therefore, assuming that

$$\boldsymbol{Y}^{n\prime}_{un} = \boldsymbol{Y}^{n}_{uCn} + \boldsymbol{Y}^{n}_{un} = \boldsymbol{Y}^{z*}_{un} + \boldsymbol{Y}^{n}_{un} \tag{15.11}$$

susceptances of the Δ-configured sub-compensator, T_{RSn}, T_{STn}, and T_{TRn} should have the values

$$\begin{aligned} T_{RSn} &= \left(\sqrt{3}\mathrm{Re}\,\boldsymbol{Y}^{n'}_{un} - \mathrm{Im}\,\boldsymbol{Y}^{n'}_{un}\right)/3 \\ T_{STn} &= (2\,\mathrm{Im}\,\boldsymbol{Y}^{n'}_{un})/3 \\ T_{TRn} &= -\left(\sqrt{3}\mathrm{Re}\,\boldsymbol{Y}^{n'}_{un} + \mathrm{Im}\,\boldsymbol{Y}^{n'}_{un}\right)/3. \end{aligned} \tag{15.12}$$

A compensator with such susceptances reduces to zero the rms value of the zero sequence component of the unbalanced current, $||\boldsymbol{i}^{z}_{un}||$, and the negative sequence component of this current, $||\boldsymbol{i}^{n}_{un}||$, along with the rms value of reactive current, $||\boldsymbol{i}_{rn}||$, for the positive sequence harmonics.

The presence of harmonics of negative sequence order $n = 2, 5, ..., 3k-1$ in the supply voltage causes the unbalanced current $\boldsymbol{i}_{un}$ can contain zero and positive sequence components $\boldsymbol{i}^{z}_{un}$ and $\boldsymbol{i}^{p}_{un}$.

The zero sequence component of the unbalanced current, $\boldsymbol{i}^{z}_{un}$, can be compensated completely, along with the reactive current $\boldsymbol{i}_{rn}$, by a Y-configured sub-compensator. The formulae for calculating branch susceptance of this sub-compensator are similar to formulae (15.10), only the real part of the zero sequence unbalanced admittance $\boldsymbol{Y}^{z}_{un}$ enters these formulae with the opposite sign, namely

$$\begin{aligned} T_{Rn} &= -2\,\mathrm{Im}\,\boldsymbol{Y}^{z}_{un} - B_{en} \\ T_{Sn} &= -\sqrt{3}\mathrm{Re}\,\boldsymbol{Y}^{z}_{un} + \mathrm{Im}\,\boldsymbol{Y}^{z}_{un} - B_{en} \\ T_{Tn} &= \sqrt{3}\mathrm{Re}\,\boldsymbol{Y}^{z}_{un} + \mathrm{Im}\,\boldsymbol{Y}^{z}_{un} - B_{en}. \end{aligned} \tag{15.13}$$

The Y-configured sub-compensator with such susceptances reduces the zero sequence component of the unbalanced current rms value, $||\boldsymbol{i}^{z}_{un}||$, along with the rms value of reactive current, $||\boldsymbol{i}_{rn}||$, for the harmonics of negative sequence order to zero. The positive sequence unbalanced admittance of such a compensator is

$$\begin{aligned} \boldsymbol{Y}^{p}_{uCn} &= \frac{1}{3}\,j\,(T_{Rn} + a^{*}\,T_{Sn} + \alpha T_{Tn}) = \\ &= \frac{1}{3}\,j\,[(-2\,\mathrm{Im}\,\boldsymbol{Y}^{z}_{un} - B_{en}) + \alpha^{*}(\sqrt{3}\mathrm{Re}\,\boldsymbol{Y}^{z}_{un} + \mathrm{Im}\,\boldsymbol{Y}^{z}_{un} - B_{en}) + \alpha(-\sqrt{3}\mathrm{Re}\,\boldsymbol{Y}^{z}_{un} \\ &\quad + \mathrm{Im}\,\boldsymbol{Y}^{z}_{un} - B_{en})] = \\ &= \mathrm{Re}\,\boldsymbol{Y}^{z}_{un} - j\,\mathrm{Im}\,\boldsymbol{Y}^{z}_{un} = \boldsymbol{Y}^{z^{*}}_{un}. \end{aligned} \tag{15.14}$$

Therefore, the positive sequence unbalanced admittance of the load with the Y-configured sub-compensator changes to

$$\boldsymbol{Y}^{p'}_{un} = \boldsymbol{Y}^{p}_{uCn} + \boldsymbol{Y}^{p}_{un} = \boldsymbol{Y}^{z^{*}}_{un} + \boldsymbol{Y}^{p}_{un}. \tag{15.15}$$

Compensation of the unbalanced current caused by negative order harmonics with a Δ configured compensator was discussed in Chapter 14. Its susceptances can be calculated with formulae (14.20). They can be applied now for calculating susceptances of the Δ sub-compensator, namely

$$\begin{aligned} T_{\mathrm{RS}n} &= -\left(\sqrt{3}\,\mathrm{Re}\,\boldsymbol{Y}^{\mathrm{p'}}_{\mathrm{u}n} + \mathrm{Im}\,\boldsymbol{Y}^{\mathrm{p'}}_{\mathrm{u}n}\right)/3 \\ T_{\mathrm{ST}n} &= (2\,\mathrm{Im}\,\boldsymbol{Y}^{\mathrm{p'}}_{\mathrm{u}n})/3 \\ T_{\mathrm{TR}n} &= \left(\sqrt{3}\,\mathrm{Re}\,\boldsymbol{Y}^{\mathrm{p'}}_{\mathrm{u}n} - \mathrm{Im}\,\boldsymbol{Y}^{\mathrm{p'}}_{\mathrm{u}n}\right)/3. \end{aligned} \tag{15.16}$$

A compensator with such susceptances reduces to zero the three-phase rms value of the zero sequence component of the unbalanced current, $||\boldsymbol{i}^{\mathrm{z}}_{\mathrm{u}n}||$, and the positive sequence component of this current, $||\boldsymbol{i}^{\mathrm{p}}_{\mathrm{u}n}||$, along with the three-phase rms value of the reactive current $||\boldsymbol{i}_{\mathrm{r}n}||$, for negative sequence harmonics.

When the voltage applied to an unbalanced load with a neutral conductor, contains harmonics of the zero sequence order, n = 3, 6, 9,..., $3k$, the unbalanced current can contain, according to formula (8.45), positive and negative sequence components, $\boldsymbol{i}^{\mathrm{p}}_{\mathrm{u}n}$, and $\boldsymbol{i}^{\mathrm{n}}_{\mathrm{u}n}$, as well as the reactive current component, $\boldsymbol{i}_{\mathrm{r}n}$. One of the components of the unbalanced current can be compensated by a Y-configured sub-compensator. Let us assume that this is the negative-sequence component, $\boldsymbol{i}^{\mathrm{n}}_{\mathrm{u}n}$. It can be compensated, along with the reactive current, by a Y-configured sub-compensator with branch susceptances

$$\begin{aligned} T_{\mathrm{R}n} &= -2\,\mathrm{Im}\,\boldsymbol{Y}^{\mathrm{n}}_{\mathrm{u}n} - B_{\mathrm{e}n} \\ T_{\mathrm{S}n} &= -\sqrt{3}\,\mathrm{Re}\,\boldsymbol{Y}^{\mathrm{n}}_{\mathrm{u}n} + \mathrm{Im}\,\boldsymbol{Y}^{\mathrm{n}}_{\mathrm{u}n} - B_{\mathrm{e}n} \\ T_{\mathrm{T}n} &= \sqrt{3}\,\mathrm{Re}\,\boldsymbol{Y}^{\mathrm{n}}_{\mathrm{u}n} + \mathrm{Im}\,\boldsymbol{Y}^{\mathrm{n}}_{\mathrm{u}n} - B_{\mathrm{e}n}. \end{aligned} \tag{15.17}$$

The positive sequence unbalanced admittance of such a sub-compensator for harmonics of the zero sequence is

$$\begin{aligned} \boldsymbol{Y}^{\mathrm{p}}_{\mathrm{uC}n} &= \frac{1}{3}j\,(T_{\mathrm{R}n} + \alpha T_{\mathrm{S}n} + \alpha^*\, T_{\mathrm{T}n}) = \\ &= \frac{1}{3}j\,[(-2\,\mathrm{Im}\,\boldsymbol{Y}^{\mathrm{n}}_{\mathrm{u}n} - B_{\mathrm{e}n}) + \alpha(-\sqrt{3}\,\mathrm{Re}\,\boldsymbol{Y}^{\mathrm{n}}_{\mathrm{u}n} + \mathrm{Im}\,\boldsymbol{Y}^{\mathrm{n}}_{\mathrm{u}n} - B_{\mathrm{e}n}) + \alpha^*(\sqrt{3}\,\mathrm{Re}\,\boldsymbol{Y}^{\mathrm{n}}_{\mathrm{u}n} \\ &\quad + \mathrm{Im}\,\boldsymbol{Y}^{\mathrm{n}}_{\mathrm{u}n} - B_{\mathrm{e}n})] = \\ &= \mathrm{Re}\,\boldsymbol{Y}^{\mathrm{n}}_{\mathrm{u}n} - j\,\mathrm{Im}\,\boldsymbol{Y}^{\mathrm{n}}_{\mathrm{u}n} = \boldsymbol{Y}^{\mathrm{n}^*}_{\mathrm{u}n}. \end{aligned} \tag{15.18}$$

Therefore, the positive sequence unbalanced admittance of the load with the Y-configured sub-compensator changes to

$$\boldsymbol{Y}^{\mathrm{p'}}_{\mathrm{u}n} = \boldsymbol{Y}^{\mathrm{p}}_{\mathrm{uC}n} + \boldsymbol{Y}^{\mathrm{p}}_{\mathrm{u}n} = \boldsymbol{Y}^{\mathrm{n}^*}_{\mathrm{u}n} + \boldsymbol{Y}^{\mathrm{p}}_{\mathrm{u}n}. \tag{15.19}$$

Unfortunately, this remaining component of the unbalanced current cannot be compensated, as it was earlier explained, by a Δ-configured sub-compensator.

The zero sequence harmonics that might be taken into account is usually the third order harmonic. Therefore, the Y-configured sub-compensator can reduce the three-phase rms value of the unbalanced current of the third order harmonic only if

$$|Y_{u3}^{n^*} + Y_{u3}^{p}| < |Y_{u3}^{p}|$$

but this depends only on the load. In general, this inequality does not have to be fulfilled. It means that the unbalanced current caused by the voltage harmonics of the zero sequence order cannot be compensated by a reactance compensator. Only the reactive current caused by such harmonics can be compensated.

Thus, similarly, as it is in single-phase circuits with LTI loads and nonsinusoidal supply voltage, the unity power factor in 3p4w circuits cannot be obtained at reactance compensation. It is not only the scattered current $\boldsymbol{i}_s$ that cannot be compensated. In the presence of the zero sequence harmonics in the supply voltage, one component of the zero sequence component of the unbalanced current cannot be compensated either. Usually, this is the current $\boldsymbol{i}^z{}_{u3}$. Therefore, the maximum power factor which can be achieved by reactance compensation cannot be higher than

$$\lambda_{max} = \frac{||\boldsymbol{i}_a||}{||\boldsymbol{i}||_{min}} = \frac{||\boldsymbol{i}_a||}{\sqrt{||\boldsymbol{i}_a||^2 + ||\boldsymbol{i}_s||^2 + ||\boldsymbol{i}_{u3}^z||^2}}. \tag{15.20}$$

Illustration 15.6 *Let us assume that the load shown in Fig. 15.13, with $\omega_1 L = R = 0.5\ \Omega$, is supplied with a symmetrical voltage of the fundamental harmonic rms value $U_1 = 240$ V, distorted by the third, fifth, and seventh order harmonics of relative rms value $U_3 = 2\%U_1$, $U_5 = 3\%U_1$ and $U_7 = 1.5\%U_1$.*

The rms values of line-to-neutral voltage harmonics and the load admittance for harmonic frequencies are compiled in Table 15.1.

Table 15.1 The rms values of the supply voltage harmonics and the load admittance.

n	U_n [V]	$Y_{Tn} = G_{Tn} + jB_{Tn}$ [S]
1	240	$1.000 - j1.000$
3	4.8	$0.200 - j0.600$
5	7.2	$0.0769 - j0.385$
7	3.6	$0.0400 - j0.280$

The results of the circuit analysis and the three-phase rms values of the Current's Physical Components are shown in Fig. 15.13.

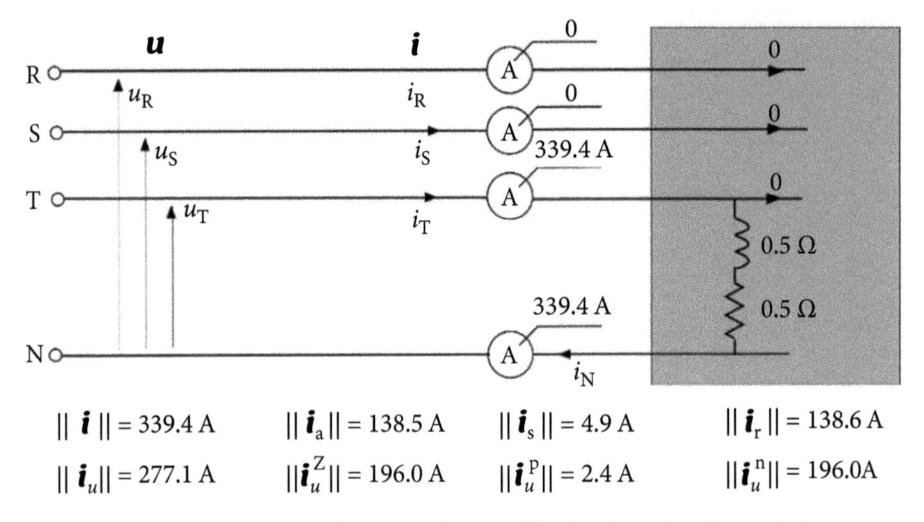

Figure 15.13 An example of an unbalanced load and results of its analysis.

The active power of the load is

$$P = \sum_{n \in N} G_{Tn} U_{Tn}^2 = 57.609 \text{ kW}.$$

The supply voltage three rms value

$$||\boldsymbol{u}|| = \sqrt{\sum_{n \in N} ||\boldsymbol{u}_n||^2} = \sqrt{3 \sum_{n \in N} U_n^2} = 416.01\text{V}.$$

Thus the equivalent conductance of the load is

$$G_e = \frac{P}{||\boldsymbol{u}||^2} = 0.3329 \text{ S}$$

and the three-phase rms value of the active current

$$||\boldsymbol{i}_a|| = G_e ||\boldsymbol{u}|| = \frac{P}{||\boldsymbol{u}||} = 138.5 \text{ A}.$$

The values of equivalent conductance G_{en}, susceptance B_{en}, calculated according to (8.34) and (8.38), while the magnitude of unbalanced admittances $\mathbf{Y}_{un}^{p}$, $\mathbf{Y}_{un}^{n}$, and $\mathbf{Y}_{un}^{z}$ for harmonic frequencies, calculated according to (8.47–8.49), are compiled in Table 15.2.

The three-phase rms value of the scattered current is

$$||\boldsymbol{i}_s|| = \sqrt{3 \sum_{n \in N} [(G_{en} - G_e)\, U_n]^2} = 4.9 \text{ A}$$

Table 15.2 Equivalent parameters of the load for harmonic frequencies.

n	G_{en}	B_{en}	Y^{z}_{un}	Y^{p}_{un}	Y^{n}_{un}
-	mS	mS	mS	mS	mS
1	333	−333	471	0	471
3	66.6	−200	0	211	211
5	25.6	−128	131	131	0
7	13.3	−93.3	94.3	0	94.3

and the reactive current

$$||\boldsymbol{i}_{r}|| = \sqrt{3\sum_{n\in N}(B_{en}U_{n})^{2}} = 138.6\,\text{A}.$$

The three-phase rms value of the unbalanced current components are

$$||\boldsymbol{i}^{z}_{u}|| = \sqrt{3\sum_{n\in N}(Y^{z}_{un}U_{n})^{2}} = 196.0\,\text{A}$$

$$||\boldsymbol{i}^{p}_{u}|| = \sqrt{3\sum_{n\in N}(Y^{p}_{un}U_{n})^{2}} = 2.4\,\text{A}$$

$$||\boldsymbol{i}^{n}_{u}|| = \sqrt{3\sum_{n\in N}(Y^{n}_{un}U_{n})^{2}} = 196.0\,\text{A}$$

and consequently, the three-phase rms value of the unbalanced current is

$$||\boldsymbol{i}_{u}|| = \sqrt{||\boldsymbol{i}^{p}_{u}||^{2} + ||\boldsymbol{i}^{n}_{u}||^{2} + ||\boldsymbol{i}^{z}_{u}||^{2}} = 277.1\,\text{A}.$$

The three-phase rms value of the load current, calculated as the root of the sum of squares of rms values of the line currents, is

$$||\boldsymbol{i}|| = \sqrt{||i_{R}||^{2} + ||i_{S}||^{2} + ||i_{T}||^{2}} = ||i_{T}|| = \sqrt{\sum_{n\in N}(Y_{Tn}U_{n})^{2}} = 339.4\,\text{A}.$$

This value can be used for verification of the decomposition of the load current into physical components since the root of the sum of the squares of their three-phase rms values should result in the same value of $||\boldsymbol{i}||$. *Indeed*

$$||\boldsymbol{i}|| = \sqrt{||\boldsymbol{i}_{a}||^{2} + ||\boldsymbol{i}_{s}||^{2} + ||\boldsymbol{i}_{r}||^{2} + ||\boldsymbol{i}_{u}||^{2}} = 339.4\,\text{A}$$

which confirms the numerical correctness of calculations of three-phase rms values of the CPC. The results of the load analysis are compiled in Fig. 15.14. The power factor of the load is $\lambda = P/S = 0.408$.

Table 15.3 Unbalanced admittances of the load.

n	Y^{z}_{un}[mS]	Y^{p}_{un}[mS]	Y^{n}_{un}[mS]
1	$122 + j455$	0	$-455 - j122$
3	0	$-206 + j42.3$	$140 + j158$
5	$-124 + j41.9$	$58.2 + j68.3$	0
7	$74.2 + j58.2$	0	$-87.5 + j35.1$

Unbalanced admittances of the zero, positive, and negative sequence for particular supply voltage harmonics, which are needed for balancing compensator design, calculated from formulae (8.47–8.49), have values compiled in Table 15.3.

The susceptance of the ideal compensator branches can be calculated from formulae (15.10–15.17) has the values compiled in Table 15.4.

A compensator with such susceptance of its branches affects the supply current three-phase rms values as shown in Fig. 15.14.

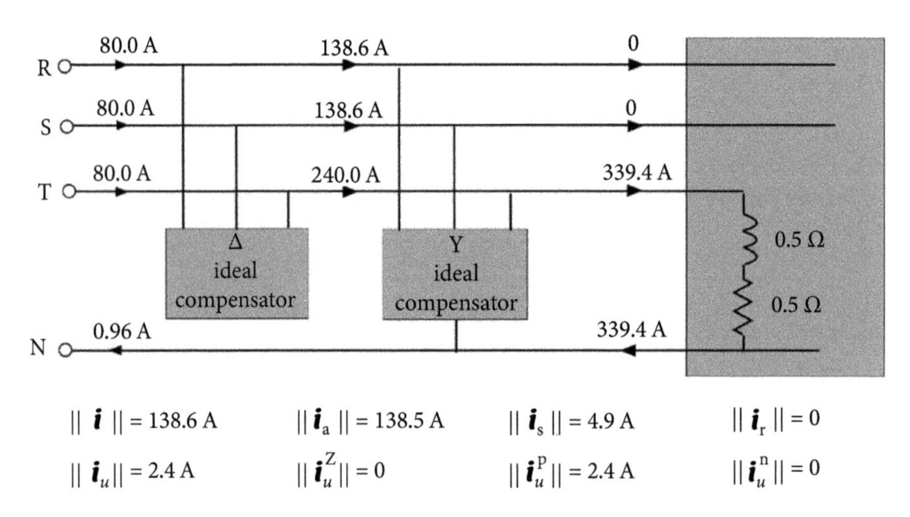

Figure 15.14 Results of an ideal compensation.

It does not reduce the scattered current and the unbalanced current of the positive sequence, but it improves the power factor to $\lambda = 0.999$.

As explained above, the Δ-structure compensator can compensate the positive sequence component of the unbalanced current or the negative sequence component, but not both. Since the former component is much lower than the latter, it is reasonable to select the negative sequence component as a goal for compensation, so that the positive sequence component has to remain in the supply lines. The current in

Table 15.4 The susceptance of the ideal compensator branches.

n	T_{Rn}	T_{Sn}	T_{Tn}	T_{RSn}	T_{STn}	T_{TRn}
	mS	mS	mS	mS	mS	mS
1	−577	577	1000	0	−385	385
3	115	−115	600	0	0	0
5	44.4	−44.4	385	0	29.6	−29.6
7	−23.1	23.1	280	0	−15.4	15.5

the neutral conductor (0.96A) is composed of the third order harmonic of the active current.

The reactance compensator, generally composed of six branches, is composed of only five branches, because the Δ-configured sub-compensator compensates only one but not two current components. Each of these branches has to have four values T_{Xn} *of the susceptance as specified in Table 15.4. Such branches can be synthesized according to the information in Chapter 10, unfortunately, approximately six to eight reactance elements are needed to build each of these five branches. Thus such compensation only has theoretical rather than practical merits. Reduction of the compensator complexity is necessary.*

15.5 Reduction of the Compensator Complexity

Compensator complexity can be reduced if the goal of complete compensation of the unbalanced and reactive currents is abandoned for the only reduction of these currents rms values by a compensator built of a lower number of reactance devices. It should minimize the three-phase rms value of the supply current.

The simplest branches of a reactance compensator are purely inductive or capacitive. Since a purely capacitive branch in the compensator could cause a resonance for harmonic frequencies with an inductive impedance of the supply source, such capacitive branches are not acceptable. Therefore, to avoid such resonances the compensator should be built exclusively of L or LC branches shown in Fig. 15.15.

Figure 15.15 Acceptable branches of a reduced complexity compensator.

To avoid confusion, the susceptance of a compensator of the reduced complexity is denoted by D_n. This susceptance, depending on the branch structure, has for harmonic frequencies the value

$$D_n = -\frac{1}{n\omega_1 L} \qquad \text{or} \qquad D_n = \frac{n\omega_1 C}{1 - n^2\omega_1^2 LC}.$$

A compensator for the unbalanced and reactive currents rms value minimization can have the same structure as the ideal compensator, only its vector of branch currents will change, as shown in Fig. 15.16, from $\boldsymbol{j}_{\mathrm{T}}$ to $\boldsymbol{j}_{\mathrm{D}}$.

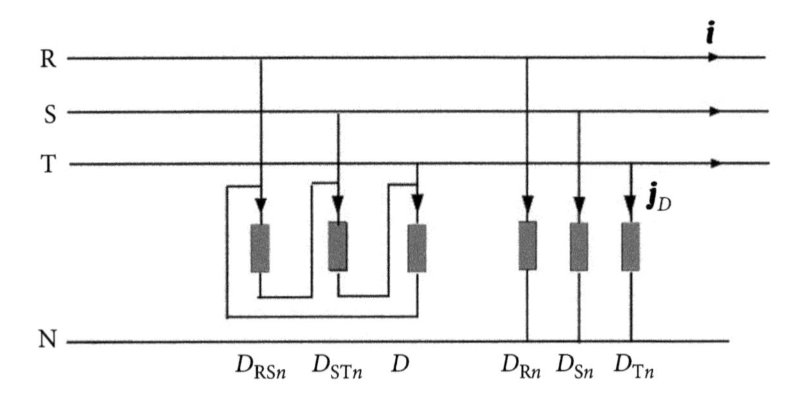

Figure 15.16 A compensator with branches of reduced complexity and D_n susceptance.

Let us denote the vector of crms values of the voltage harmonics on the compensator branches by

$$\boldsymbol{U}_n = [U_{\mathrm{RS}n},\ U_{\mathrm{ST}n},\ U_{\mathrm{TR}n},\ U_{\mathrm{R}n},\ U_{\mathrm{S}n},\ U_{\mathrm{T}n}]^{\mathrm{T}}.$$

The vector of currents of the ideal compensator branches can be presented in the form

$$\boldsymbol{j}_{\mathrm{T}} = \sqrt{2}\mathrm{Re}\sum_{n\in N} j\,\boldsymbol{T}_n\,\boldsymbol{U}_n\,e^{jn\omega_1 t}$$

where

$$\boldsymbol{T}_n = \begin{bmatrix} T_{\mathrm{RS}n} & 0 & 0 & 0 & 0 & 0 \\ 0 & T_{\mathrm{ST}n} & 0 & 0 & 0 & 0 \\ 0 & 0 & T_{\mathrm{TR}n} & 0 & 0 & 0 \\ 0 & 0 & 0 & T_{\mathrm{R}n} & 0 & 0 \\ 0 & 0 & 0 & 0 & T_{\mathrm{S}n} & 0 \\ 0 & 0 & 0 & 0 & 0 & T_{\mathrm{T}n} \end{bmatrix}.$$

For the compensator with branches shown in Fig. 15.15, this vector changes to the form

$$\boldsymbol{j}_{\mathrm{D}} = \sqrt{2}\mathrm{Re}\sum_{n\in N} j\,\boldsymbol{D}_n\,\boldsymbol{U}_n\,e^{jn\omega_1 t}$$

where

$$\boldsymbol{D}_n = \begin{bmatrix} D_{\mathrm{RS}n} & 0 & 0 & 0 & 0 & 0 \\ 0 & D_{\mathrm{ST}n} & 0 & 0 & 0 & 0 \\ 0 & 0 & D_{\mathrm{TR}n} & 0 & 0 & 0 \\ 0 & 0 & 0 & D_{\mathrm{R}n} & 0 & 0 \\ 0 & 0 & 0 & 0 & D_{\mathrm{S}n} & 0 \\ 0 & 0 & 0 & 0 & 0 & D_{\mathrm{T}n} \end{bmatrix}.$$

Vectors $\boldsymbol{j}_{\mathrm{T}}$ and j_{D} can be regarded as elements in a six-dimensional space. If the matrix of branch susceptance is denoted by $\boldsymbol{B}$, then the length of the vector $\boldsymbol{j}_{\mathrm{B}}$ is defined as

$$||\boldsymbol{j}_{\mathrm{B}}|| = \sqrt{\frac{1}{T}\int_0^T \boldsymbol{j}_{\mathrm{B}}^{\mathrm{T}}(t)\boldsymbol{j}_{\mathrm{B}}(t)\mathrm{d}t} = \mathrm{Re}\sum_{n\in N}(j\,\mathrm{B}_n\,\boldsymbol{U}_n)^* = \mathrm{Re}\sum_{n\in N}\mathrm{B}_n^2|\boldsymbol{U}_n|^2.$$

In this formula

$$|\boldsymbol{U}_n|^2 = [U_{\mathrm{RS}n}^2,\ U_{\mathrm{ST}n}^2,\ U_{\mathrm{TR}n}^2,\ U_{\mathrm{R}n}^2,\ U_{\mathrm{S}n}^2,\ U_{\mathrm{T}n}^2]^{\mathrm{T}}.$$

The distance between vectors $\boldsymbol{j}_{\mathrm{T}}$ and $\boldsymbol{j}_{\mathrm{D}}$ in this space can be defined as

$$d = ||\boldsymbol{j}_{\mathrm{T}} - \boldsymbol{j}_{\mathrm{D}}||.$$

The *LC* parameters of a compensator which is to minimize the three-phase rms value of the supply current should be selected such that this distance is minimum.

The difference between the branch currents vectors $\boldsymbol{j}_{\mathrm{T}}$ and $\boldsymbol{j}_{\mathrm{D}}$ can be expressed as

$$\boldsymbol{j}_{\mathrm{T}} - \boldsymbol{j}_{\mathrm{D}} = \sqrt{2}\mathrm{Re}\sum_{n\in N} j[\boldsymbol{T}_n - \boldsymbol{D}_n]\,\boldsymbol{U}_n\,e^{jn\omega_1 t} = \sqrt{2}\mathrm{Re}\sum_{n\in N}\Delta\boldsymbol{J}_n e^{jn\omega_1 t}$$

where

$$\Delta\boldsymbol{J}_n = j\begin{bmatrix} (T_{\mathrm{RS}n} - D_{\mathrm{RS}n})\boldsymbol{U}_{\mathrm{RS}n} \\ (T_{\mathrm{ST}n} - D_{\mathrm{ST}n})\boldsymbol{U}_{\mathrm{ST}n} \\ (T_{\mathrm{TR}n} - D_{\mathrm{TR}n})\boldsymbol{U}_{\mathrm{TR}n} \\ (T_{\mathrm{R}n} - D_{\mathrm{R}n})\boldsymbol{U}_{\mathrm{R}n} \\ (T_{\mathrm{S}n} - D_{\mathrm{S}n})\boldsymbol{U}_{\mathrm{S}n} \\ (T_{\mathrm{T}n} - D_{\mathrm{T}n})\boldsymbol{U}_{\mathrm{T}n} \end{bmatrix}$$

and consequently the distance d of these vectors is

$$d = ||\boldsymbol{j}_{\mathrm{T}} - \boldsymbol{j}_{\mathrm{D}}|| = \sqrt{\sum_{n\in N}\Delta\boldsymbol{J}_n^{\mathrm{T}}\Delta\boldsymbol{J}_n^*} = \sqrt{\sum_{n\in N}\sum_{k\in K}(T_{kn} - D_{kn})^2\,U_{kn}^2} \tag{15.21}$$

where k is an index from the set

$$K = \{\mathrm{RS}, \mathrm{ST}, \mathrm{TR}, \mathrm{R}, \mathrm{S}, \mathrm{T}\}$$

The sequence of summation over harmonic orders n and the branch index k in the last formula can be switched, so that the distance d can be expressed as

$$d = \sqrt{\sum_{k\in K}\sum_{n\in N} (T_{kn} - D_{kn})^2\, U_{nk}^2} = \sqrt{\sum_{k\in K} d_k^2}.$$

Since the terms under the root in this formula are non-negative numbers so mutual cancellation is not possible, the distance d is minimum on the condition that all distances d_k for individual branches

$$d_k = \sqrt{\sum_{n\in N} (T_{kn} - D_{kn})^2\, U_{kn}^2} \tag{15.22}$$

are minimum, that is, for each branch of the compensator

$$\sum_{n\in N} (T_{kn} - D_{kn})^2\, U_{kn}^2 = \text{Min.} \tag{15.23}$$

Susceptances T_{kn} in this formula are known from formulae (15.10–15.17). Susceptances D_{kn} dependent on LC parameters of the compensator branches, shown in Fig. 15.15, should minimize (15.22).

The supply voltage rms value of the fundamental harmonic U_{k1} in common distribution systems is much higher than the rms value of other harmonics. Because of that, the component of (15.23) with U_{k1} is the dominating one. Therefore the term $(T_{k1} - D_{k1})$ should be as close to zero as possible. Thus, when $T_{k1} < 0$, the compensator branch should be chosen such that $D_{k1} < 0$, that is, the inductive branch. When $T_{k1} > 0$, the branch should be chosen such that $D_{k1} > 0$, that is, the LC branch. Consequently, for purely inductive branches, the inductance L_k should be chosen such that

$$\sum_{n\in N}\left(T_{kn} + \frac{1}{n\omega_1 L_k}\right)^2 U_{kn}^2 = Min. \tag{15.24}$$

For LC branches, the inductance L_k and capacitance C_k should be chosen such that

$$\sum_{n\in N}\left(T_{kn} - \frac{n\omega_1 C_k}{1 - n^2\omega_1^2 L_k C_k}\right)^2 U_{kn}^2 = \text{Min.} \tag{15.25}$$

Condition (15.23) is satisfied when the derivative of (15.24) versus L_k is zero:

$$\frac{d}{dL_k}\left\{\sum_{n\in N}\left(T_{kn} + \frac{1}{n\omega_1 L_k}\right)^2 U_{kn}^2\right\} = 0$$

and this condition results in the optimum value of the branch inductance

$$L_{k,\mathrm{opt}} = -\frac{1}{\omega_1}\frac{\sum_{n\in N}\frac{1}{n^2}U_{kn}^2}{\sum_{n\in N} T_{kn}\frac{1}{n}U_{kn}^2}. \tag{15.26}$$

The form on the left side of (15.25) is a function of two variables, the inductance L_k and capacitance C_k. Considering that inductance L_k is a continuously declining function, meaning it does not have a minimum for any finite value of L_k, it can be selected at a designer's discretion. When it is selected, the capacitance C_k can be calculated such that (15.25) is the minimum. When it is minimum, the derivative of (15.25) versus C_k is zero, that is,

$$\frac{d}{dC_k}\left\{\sum_{n\in N}\left(T_{kn} - \frac{n\omega_1 C_k}{1 - n^2\omega_1^2 L_k C_k}\right)^2 U_{kn}^2\right\} = 0.$$

It results in the equation

$$\sum_{n\in N}\frac{T_{kn}\, nU_{kn}^2}{1 - n^2\omega_1^2 L_k C_k} - \sum_{n\in N}\frac{n^2\omega_1 C_k U_{kn}^2}{(1 - n^2\omega_1^2 L_k C_k)^2} = 0$$

which cannot be solved directly versus the optimum value of the capacitance C_k. Numerical methods are needed for that. In particular, it can be solved in an iterative process with the formula

$$C_{k,s+1} = \frac{\sum_{n\in N}\frac{T_{kn}\, nU_{kn}^2}{1 - n^2\omega_1^2 L_k C_{k,s}}}{\omega_1\sum_{n\in N}\frac{n^2 U_{kn}^2}{(1 - n^2\omega_1^2 L_k C_{k,s})^2}} \tag{15.27}$$

which results in a sequence of capacitances

$$C_{k,0},\ C_{k,1},\ C_{k,2}, \ldots\ldots C_{k,s},\ C_{k,s+1},$$

convergent to the optimum capacitance $C_{k,\mathrm{opt}}$ which minimizes (15.25). The capacitance for which the left side of (15.25) is zero for the fundamental harmonic can be selected as the first term $C_{k,0}$ of this sequence.

Illustration 15.7 *Having susceptances $T_{X\underline{n}}$ of the ideal compensator of the load analyzed in Illustration 15.6, which are compiled in Table 15.4, calculate LC parameters of the compensator of the reduced complexity.*

Since the susceptance of the branches ST and R is negative, they can be purely inductive with the inductance calculated from (15.26). The susceptance of the remaining branches is positive, these branches should be built as series LC branches.

Table 15.5 LC parameters of a reduced complexity compensator.

	Line:	R	S	T	RS	ST	TR
L	mH	1730	770	444	0	2600	1155
C	mF	0	399	691	0	0	266

Assuming that the resonance of that branch is at frequency ω_r = 2.5 rad/s, the iterative formula (15.27) results in the LC parameters compiled in Table 15.5.

The results of compensation are shown in Fig. 15.17. The power factor is improved by the compensator of the reduced complexity to $\lambda = 0.994$.

The presented approach enables a reduction in the number of reactance devices needed for the compensator construction from approximately forty to only eight, without any n effectiveness.

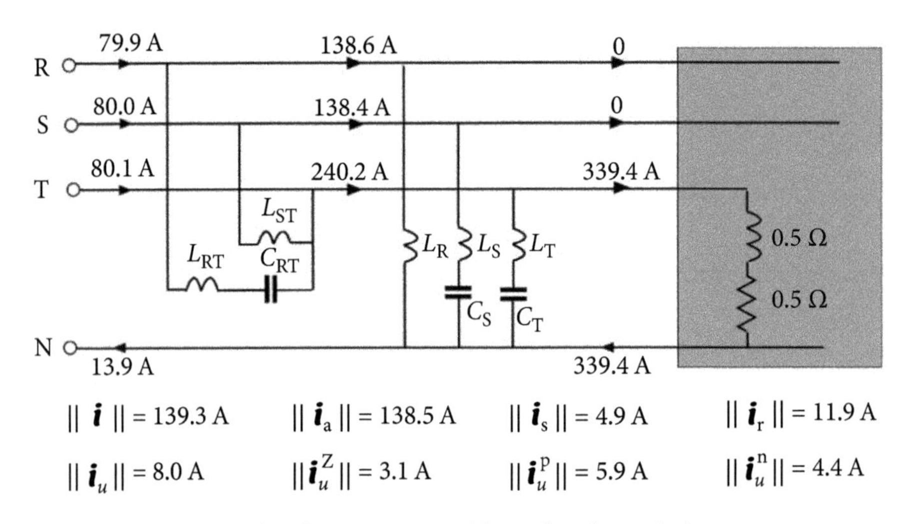

Figure 15.17 Results of compensation with a reduced complexity compensator.

*****

Summary

The chapter demonstrates that the reactive and unbalanced currents in four-wire circuits can be compensated at sinusoidal and nonsinusoidal supply voltage, with high effectiveness, by a reactance compensator of relatively low complexity. The presented method does not have any limitations as to the load power and can be used even in the highest power manufacturing plants, which usually have three-phase distribution systems with a neutral conductor.

Chapter 16
Switching Compensators

16.1 Introduction

Reactance compensators can compensate the load only at a few harmonic frequencies. When they are built as adaptive devices then compensation is confined only to the fundamental harmonic. They cannot compensate fast varying components and adaptability is not their natural property. These deficiencies, at the cost of lower power ratings, can be covered by switching compensators (SCs).

Like reactance compensators or filters, shunt switching compensators are used to modify the supply current of the source, according to some compensation objectives. Despite this, switching compensators are devices that operate at fundamentally different principles compared to reactance compensators or filters.

Although SCs can also be used as series compensators, we are confined in this chapter to switching compensators connected only in parallel to the load, meaning to shunt switching compensators (SSCs).

There is a fundamental difference between reactance compensators and switching compensators. The current at the input terminals of a reactance compensator occurs as its response to the voltage at these terminals and this response depends on the compensator admittance. The same is with resonant harmonic filters. The current at the input terminals of a switching compensator occurs as a current replica of the compensator reference signal.

There is major confusion in the power engineering community concerning switching compensators. There are different names for these devices. They are called commonly "active power filters", "active harmonic filters", or "power conditioners". Unfortunately, the switching compensator is not an active but a passive device. "*Active*" are devices that deliver electric energy to a circuit, but SC consumes this energy, thus it is a "*passive*" device. It does not "*filter*" harmful components of the supply current but *compensates* them by injecting these current components, with the opposite sign, into the supply source. This injection is not confined, moreover, only to "*harmonics*". These could also be the reactive and unbalanced currents and even non-periodic transients. Also, the operation of a switching compensator has nothing in common with "*power conditioning*".

16.2 Operation Principle

The switching compensator's basic structure is shown in Fig. 16.1. It is composed of a three-phase inverter, built of power transistors, an input inductive filter of L

Powers and Compensation in Circuits with Nonsinusoidal Current. Leszek S. Czarnecki, Oxford University Press.
 DOI: 10.1093/oso/9780198879206.003.0016

inductance, capacitive energy storage of C capacitance, and a few digital systems needed for the compensator control.

It reproduces a *reference signal*, worked out by a Digital Signal Processing (DSP) system, according to specified compensation objectives. The DSP system obtains rough data on the load voltage and current $\boldsymbol{u}$ and $\boldsymbol{i}$, from a Data Acquisition (DA) system, equipped with voltage and current sensors and analog-to-digital (A/D) converters. It conveys information on the load to the DSP system in the form of sequences of digitized samples $\boldsymbol{u}(k)$ and $\boldsymbol{i}(k)$ of voltages and currents. The DSP system performs an analysis of the load power properties and reveals the load current components that should be compensated. Based on this analysis, it generates the reference signal for the compensator, $\boldsymbol{j}_{\text{ref}}$, which should be reproduced as the compensator current $\boldsymbol{j}$.

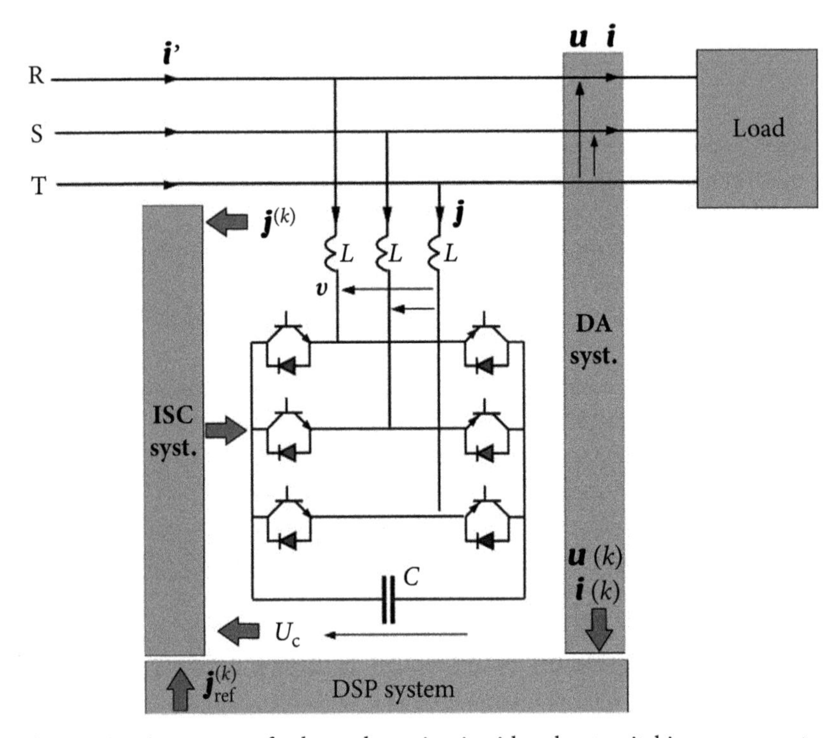

Figure 16.1 A structure of a three-phase circuit with a shunt switching compensator.

The reference signal is used for generating switching ON-OFF signals for the inverter transistors by Inverter Switches Control (ISC) applied to transistor driving circuits. The ISC system should also control DC voltage U_{C} needed for the inverter operation. A comparison of the reference signal $\boldsymbol{j}_{\text{ref}}$ with the compensator current $\boldsymbol{j}$ provides the feedback for the compensator control. Microcontrollers which can be programmed to fulfill tasks of the ISC are available on the market.

The supply voltage does not affect the terminal current $\boldsymbol{j}$ of the SSC directly. This voltage affects the compensator current only indirectly, by affecting the reference signal. Therefore, SSCs can be regarded as controlled current sources. Thus they do not change the frequency properties of the circuit and do not cause resonant conditions, which could be so crucial when a circuit is compensated by a reactive compensator or filter.

Pulse Width Modulated (PWM) inverters. The output current of switching compensators $\boldsymbol{j}$ is shaped by a Pulse Width Modulated (PWM) inverter, shown in Fig. 16.1, operated usually at a switching frequency of several kHz. A PWM inverter is built for three-phase, three-wire switching compensators, usually of six fully controlled semiconductor power switches, such as Insulated Gate Bipolar Transistors (IGBT) or Metal Oxide Semiconductor Field Effect Transistors (MOSFET). To provide DC voltage U_C for the inverter operation, an SSC also needs an energy storage device, usually in the form of a capacitor C. The inverter shown in Fig. 16.1 is sometimes referred to as a Two-Level Voltage Source Inverter (VSI). For higher voltage implementations a Three-Level VSI can be used. Its structure is shown in Fig. 16.2.

Such an inverter, built of similar transistors as that in Fig. 16.1, can have higher power ratings and lower switching noise.

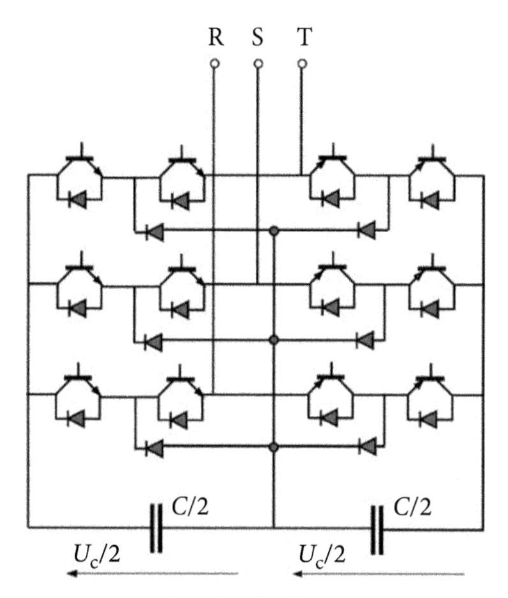

Figure 16.2 A three-level voltage source inverter.

The supply voltage $\boldsymbol{u}$ in the compensated distribution system, shown in Fig. 16.1, is sinusoidal or distorted by some harmonics, meaning this voltage is a set of continuous quantities. At the same time, the switching compensator can produce only a stair-like voltage $\boldsymbol{v}$ of the instantaneous value dependent on the ON-OFF state of six inverter switches that connect the capacitor DC voltage U_C to the inverter output terminals. The ON states of switches in the same branches, which would short circuit the capacitor C, have to be excluded from the possible states count.

Combinations of ON-OFF states of six switches, create no more than eight possible states, denoted as S_0, S_1, S_2, S_3, S_4, S_5, S_6, and S_7, and consequently, no more than eight different values of the inverter output voltage $\boldsymbol{v}_s$, denoted as $\boldsymbol{v}_0$, $\boldsymbol{v}_1$, $\boldsymbol{v}_2$, $\boldsymbol{v}_3$, $\boldsymbol{v}_4$, $\boldsymbol{v}_5$, $\boldsymbol{v}_6$, and $\boldsymbol{v}_7$. In the case of a compensator with Three-Level VSI, the stair-like output voltage $\boldsymbol{v}$ is shaped by the state of twelve switches. Therefore to reduce distortion of the compensator current $\boldsymbol{j}$, the compensator cannot be connected to the distribution system directly, but by line inductors L which serve as a low-pass filter.

The needed value of the compensator terminals' current

$$\boldsymbol{j} \stackrel{\text{df}}{=} [j_R, j_S, j_T]^T$$

is calculated by the DSP system to satisfy the compensation objectives, set at the compensator designer's discretion.

Inverter DC Voltage Control. Because PWM inverters are built of power transistors, they need a DC voltage or a DC current source for their operation. Inverters shown in Figs 16.1 and 16.2, with DC voltage provided by a capacitor, are referred to as voltage source PWM inverters and just such inverters are commonly used as SSCs.

When the inverter is not controlled, meaning all transistors are in the OFF state, the inverter operates as a three-phase rectifier. The capacitor C is charged to the maximum value of the line-to-line supply voltage. Thus, the capacitor voltage is approximately equal to

$$U_C \approx \sqrt{2}\sqrt{3}\, U_R = \sqrt{6}\, U_R.$$

This voltage is not sufficient for the compensator operation and has to be elevated. Energy stored in a capacitor is

$$W_e = \frac{1}{2} C U_C^2$$

thus the capacitor voltage U_C can be controlled by controlling the amount of energy stored in the capacitor electric field. At fixed capacitance C, its voltage is

$$U_C = \sqrt{\frac{2\, W_e}{C}}. \tag{16.1}$$

This energy has to be delivered to the capacitor by a component of the compensator terminal current $\boldsymbol{j}(t)$, which is in phase with the supply voltage $\boldsymbol{u}(t)$. This is an active component of the compensator current, denoted by $\boldsymbol{j}_a(t)$. It has to be present in the compensator current because of the energy loss in the inverter, inductors, and even in the capacitor. If the active power associated with this loss in the compensator is denoted by ΔP_C, then the compensator has to draw the active current

$$\boldsymbol{j}_a(t) = \frac{\Delta P_C}{||\boldsymbol{u}||^2}\boldsymbol{u}(t) = G_{Ce}\,\boldsymbol{u}(t) \tag{16.2}$$

where G_{Ce} is an equivalent conductance of the compensator. This component has to be added to the compensator reference signal which is reproduced as the compensator terminal current.

The equivalent conductance G_{Ce} is not known by the compensator designer, who uses it as a coefficient k for calculating the active current $\boldsymbol{j}_a(t)$. If this coefficient is lower than the equivalent conductance G_{Ce}, then the energy in the compensator is dissipated at the cost of energy stored in the capacitor, and consequently, its voltage U_C declines. When this coefficient is higher than G_{Ce}, then energy conveyed by the active current $\boldsymbol{j}_a(t)$ is not dissipated entirely in the compensator but stored in the capacitor C. Consequently its voltage increases.

The capacitor voltage U_C is kept at a needed level U_{ref} by the control which usually includes proportional and integrating components, meaning it operates according to the following formula:

$$\boldsymbol{j}_a(t) = [k + \int (U_{ref} - U_C)\,dt]\,\boldsymbol{u}(t). \tag{16.3}$$

This current has to be added to the compensator reference current $\boldsymbol{j}_{ref}(t)$ in the ISC system.

Compensator Input Equation. The DSP system provides numerical information on the needed terminal current I think, $\boldsymbol{j}$ by a *reference signal*, $\boldsymbol{j}_{ref}$, in a form of digital code for the compensator control. The compensator should be controlled by the ISC system in such a way that

$$\boldsymbol{j} = \boldsymbol{j}_{ref}$$

as closely as possible. It requires that the output voltage of the compensator be controlled in such a way that it is equal to

$$v_R(t) = u_R(t) - [R j_R(t) + L\frac{d}{dt}j_R(t)]$$
$$v_S(t) = u_S(t) - [R j_S(t) + L\frac{d}{dt}j_S(t)]$$
$$v_T(t) = u_T(t) - [R j_T(t) + L\frac{d}{dt}j_T(t)] \tag{16.4}$$

where R denotes the inductor resistance. It is assumed here that these three inductors have the same parameters, R and L. These three differential equations can be rewritten in the matrix form as

$$\boldsymbol{v}(t) = \boldsymbol{u}(t) - (R + L\frac{d}{dt})\boldsymbol{j}(t). \tag{16.5}$$

The DSP system does not handle continuous values of the supply voltage $\boldsymbol{u}(t)$ and the reference signal, $\boldsymbol{j}_{\text{ref}}(t)$, but their discrete values at instants of time $t_k = kT_s$, namely

$$\boldsymbol{u}(k) \stackrel{\text{df}}{=} \boldsymbol{u}(kT_s)$$

$$\boldsymbol{j}_{\text{ref}}(k) \stackrel{\text{df}}{=} \boldsymbol{j}_{\text{ref}}(kT_s)$$

where T_s denotes the sampling period of the DA system. It is specified by sampling frequency f_s, that is, $T_s = 1/f_s$, selected at the compensator designer's discretion. It could be in the range of a few to several kHz.

Therefore, the compensator output voltage can be calculated only in discrete instants of time

$$\boldsymbol{v}(k) = \boldsymbol{u}(k) - \left[R\boldsymbol{j}(k) + L\frac{\Delta\boldsymbol{j}(k)}{T_s}\right] \tag{16.6}$$

where $\Delta\,\boldsymbol{j}(k)$ denotes the increase of the compensator current $\boldsymbol{j}$ in the sampling interval T_s.

The current which should be injected by the compensator to the distribution system, at the instant kT_s, calculated by a DSP system, $\boldsymbol{j}(k)$ is known, similarly as it is known this current increase, $\Delta\,\boldsymbol{j}(k)$. Then, if the resistive-inductive (RL) parameters of the filter are known, the voltage at the compensator output at the same instant kT_s, $\boldsymbol{v}(k)$ can be calculated.

Vectors $\boldsymbol{u}(k)$ and $\boldsymbol{j}(k)$ have discrete values, with the number of discrete levels dependent on the resolution of the A/D converter used in the DA system. For example, for an A/D converter with a ten-bit resolution, this number is equal to 2^{10}.

16.3 Clarke Vector

Clarke Transform. Handling the matrix three-phase equation of the form (16.5) can be simplified by transforming it to a single-phase equation using the Clarke Transform, which enables the presentation of three-phase voltages $u_R(t)$, $u_S(t)$, and $u_T(t)$ in orthogonal α and β coordinates.

Phase quantities such as $x_R(t)$, $x_S(t)$, and $x_T(t)$ can be interpreted as a presentation of a quantity $\boldsymbol{x}(t)$ in phase coordinates R, S, and T. To emphasize this interpretation the symbol of vector $\boldsymbol{x}(t)$ can be written in the form

$$\begin{bmatrix} x_R(t) \\ x_S(t) \\ x_T(t) \end{bmatrix} \stackrel{\text{df}}{=} \boldsymbol{x}_{RST}(t) \stackrel{\text{df}}{=} \boldsymbol{x}_{RST}.$$

The same vector $\boldsymbol{x}(t)$ can be specified in orthogonal 0, α, and β coordinates with a Clarke Transform, which has the form

$$\boldsymbol{x}_{\alpha\beta 0}(t) \stackrel{\text{df}}{=} \boldsymbol{x}_{\alpha\beta 0} \stackrel{\text{df}}{=} \begin{bmatrix} x_0(t) \\ x_\alpha(t) \\ x_\beta(t) \end{bmatrix} = k \begin{bmatrix} \frac{1}{\sqrt{2}} & \frac{1}{\sqrt{2}} & \frac{1}{\sqrt{2}} \\ 1 & -\frac{1}{2} & -\frac{1}{2} \\ 0 & \frac{\sqrt{3}}{2} & -\frac{\sqrt{3}}{2} \end{bmatrix} \begin{bmatrix} x_R(t) \\ x_S(t) \\ x_T(t) \end{bmatrix}.$$

For quantities such that

$$x_R(t) + x_S(t) + x_T(t) \equiv 0$$

$x_0(t) \equiv 0$ and the Clarke Transform can be written in a simplified form

$$\boldsymbol{x}_{\alpha\beta}(t) \stackrel{\text{df}}{=} \boldsymbol{x}_{\alpha\beta} \stackrel{\text{df}}{=} \begin{bmatrix} x_\alpha(t) \\ x_\beta(t) \end{bmatrix} = k \begin{bmatrix} 1 & -\frac{1}{2} & -\frac{1}{2} \\ 0 & \frac{\sqrt{3}}{2} & -\frac{\sqrt{3}}{2} \end{bmatrix} \begin{bmatrix} x_R(t) \\ x_S(t) \\ x_T(t) \end{bmatrix}.$$

Coefficient k in this Transform has to be selected in such a way that the Transform preserves the norm or the length of the transformed quantity $\boldsymbol{x}(t)$.

The norm of a periodic quantity or rms value is denoted in this book by $||\boldsymbol{x}||$. In the case of the Clarke Transform, it has to preserve its instantaneous norm. To distinguish these two norms, the latest will be denoted

$$\langle\langle \boldsymbol{x} \rangle\rangle = \sqrt{\boldsymbol{x}^T \boldsymbol{x}}.$$

When this norm is expressed in phase coordinates, then

$$\langle\langle \boldsymbol{x}_{RST} \rangle\rangle^2 = x_R^2 + x_S^2 + x_T^2 = x_R^2 + x_S^2 + (-x_R - x_S)^2 = 2(x_R^2 + x_R x_S + x_S^2)$$

while when it is expressed in α and β coordinates

$$\langle\langle \boldsymbol{x}_{\alpha\beta} \rangle\rangle^2 = x_\alpha^2 + x_\beta^2 = k^2 \,[(\frac{3}{2} x_R)^2 + (\frac{\sqrt{3}}{2} x_R + \sqrt{3}\, x_S)^2] = 3k^2\,(x_R^2 + x_R x_S + x_S^2).$$

These norms are mutually equal, that is,

$$\langle\langle \boldsymbol{x}_{\alpha\beta} \rangle\rangle = \langle\langle \boldsymbol{x}_{RST} \rangle\rangle$$

if $k = \sqrt{\frac{2}{3}}$. Hence

$$\boldsymbol{x}_{\alpha\beta} = \begin{bmatrix} x_\alpha(t) \\ x_\beta(t) \end{bmatrix} = \sqrt{\frac{2}{3}} \begin{bmatrix} 1 & -\frac{1}{2} & -\frac{1}{2} \\ 0 & \frac{\sqrt{3}}{2} & -\frac{\sqrt{3}}{2} \end{bmatrix} \begin{bmatrix} x_R(t) \\ x_S(t) \\ x_T(t) \end{bmatrix} = \begin{bmatrix} \sqrt{\frac{2}{3}} & -\frac{1}{\sqrt{6}} & -\frac{1}{\sqrt{6}} \\ 0 & \frac{1}{\sqrt{2}} & -\frac{1}{\sqrt{2}} \end{bmatrix} \begin{bmatrix} x_R(t) \\ x_S(t) \\ x_T(t) \end{bmatrix}.$$

Since $x_T(t) = -x_R(t) - x_S(t)$, the Clarke Transform can be simplified to the form

$$\boldsymbol{x}_{\alpha\beta}(t) = \begin{bmatrix} x_\alpha(t) \\ x_\beta(t) \end{bmatrix} = \begin{bmatrix} \sqrt{\frac{3}{2}}, & 0 \\ \frac{1}{\sqrt{2}}, & \sqrt{2} \end{bmatrix} \begin{bmatrix} x_R(t) \\ x_S(t) \end{bmatrix} \stackrel{\text{df}}{=} \mathbf{C} \begin{bmatrix} x_R(t) \\ x_S(t) \end{bmatrix} = \mathbf{C}\, \mathcal{X}_{RS}(t) \tag{16.7}$$

where $\mathcal{X}_{RS}(t)$ denotes a reduced three-phase vector. To simplify notation, the quantity in α and β coordinates will be denoted in this chapter by

$$\boldsymbol{x}_{\alpha\beta}(t) = \boldsymbol{x}_{\alpha\beta} \stackrel{\text{df}}{=} \boldsymbol{x}^{C}.$$

The three-phase quantity $\boldsymbol{x}(t)$, expressed in α and β coordinates by $x_\alpha(t)$ and $x_\beta(t)$, can be considered as the real and imaginary parts of a complex quantity defined as

$$\mathbf{X}^{C}(t) \stackrel{\text{df}}{=} x_\alpha(t) + jx_\beta(t) = [1, j]\, \boldsymbol{x}^{C}(t). \tag{16.8}$$

This is referred to as a *Clarke Vector*, referred sometimes also as a *Space Vector*. The justification for this last term is not clear, however.

The Clarke Vector can be visualized on the complex plane as shown in Fig. 16.3. It rotates with the angular frequency, ω_1.

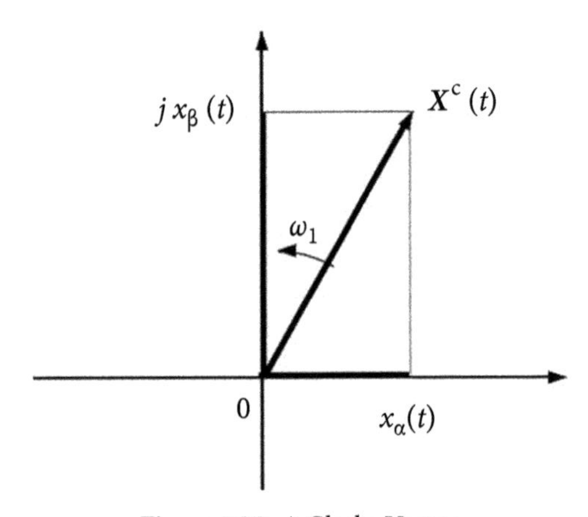

Figure 16.3 A Clarke Vector.

Illustration 16.1 *Calculate the Clarke Vector for the three-phase symmetrical voltage*

$$u_R = 277\sqrt{2}\cos(\omega_1 t - 30°)\ \mathrm{V}$$
$$u_S = 277\sqrt{2}\cos(\omega_1 t - 150°)\ \mathrm{V}$$
$$u_T = 277\sqrt{2}\cos(\omega_1 t + 90°)\ \mathrm{V}.$$

According to (16.8)

$$\boldsymbol{u}^C(t) = \mathbf{C}\begin{bmatrix} u_R(t) \\ u_S(t) \end{bmatrix} = 277\sqrt{2}\begin{bmatrix} \sqrt{3/2}, & 0 \\ 1/\sqrt{2}, & \sqrt{2} \end{bmatrix}\begin{bmatrix} \cos(\omega_1 t - 30°) \\ \cos(\omega_1 t - 150°) \end{bmatrix}$$
$$= 277\sqrt{3}\begin{bmatrix} \cos(\omega_1 t - 30°) \\ \sin(\omega_1 t - 30°) \end{bmatrix}\mathrm{V}.$$

Thus, the Clarke Vector of the voltage is equal to

$$\boldsymbol{U}^C(t) = u_\alpha(t) + ju_\beta(t) = 277\sqrt{3}[\cos(\omega_1 t - 30°) + j\sin(\omega_1 t - 30°)] = 277\sqrt{3}\, e^{j(\omega_1 t - 30°)}.$$

The Clarke Vector of symmetrical sinusoidal voltage $\boldsymbol{u}$(t) rotates on the complex plane with the angular frequency ω_1, as it is illustrated in *Fig. 16.4.* Its trajectory is a circle of the radius $r = U^C = 277\sqrt{3}$ V, meaning the rms value of the Clarke Vector $\boldsymbol{U}^C$ is equal of the line-to-line voltage rms value U_{xy} and the phase angle of the Clarke Vector $\boldsymbol{U}^C$ is equal to the phase angle of the voltage u_R(t), that is, $\alpha = -30°$.

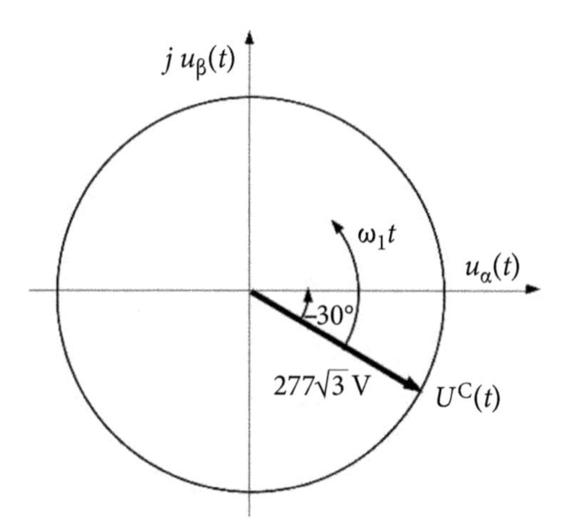

Figure 16.4 The Clarke Vector of a symmetrical sinusoidal voltage.

This illustration shows that the Clarke Vector conveys all information on the original three-phase quantity. Indeed, this original three-phase quantity can be reconstructed with the reverse Clarke Transform

$$\begin{bmatrix} x_{\mathrm{R}}(t) \\ x_{\mathrm{S}}(t) \end{bmatrix} = \begin{bmatrix} \sqrt{2/3}, & 0 \\ -1/\sqrt{6}, & 1/\sqrt{2} \end{bmatrix} \begin{bmatrix} x_{\alpha}(t) \\ x_{\beta}(t) \end{bmatrix} = \mathbf{C}^{-1} \begin{bmatrix} x_{\alpha}(t) \\ x_{\beta}(t) \end{bmatrix}. \tag{16.9}$$

The control of switching compensators is usually described in terms of the compensator three line currents $j_{\mathrm{R}}(t)$, $j_{\mathrm{S}}(t)$, and $j_{\mathrm{T}}(t)$, and three line-to-ground voltages $v_{\mathrm{R}}(t)$, $v_{\mathrm{S}}(t)$, and $v_{\mathrm{T}}(t)$. Since one current and one voltage in a three-wire circuit are dependent on the remaining two, such a description in terms of six quantities hides two intrinsic dependencies. Thus, the control can be described more explicitly when only two compensator currents and two voltages are used.

The switching compensator's Eqn. (16.5) for phase quantities can be transformed into the equation for the Clarke Vectors

$$\boldsymbol{V}^{\mathrm{C}}(t) = \boldsymbol{U}^{\mathrm{C}}(t) - (R + L\frac{d}{dt})\boldsymbol{J}^{\mathrm{C}}(t). \tag{16.10}$$

Time delay, needed for signal processing, is neglected for simplification in this equation. Voltage $\boldsymbol{U}^{\mathrm{C}}(k)$ in this equation is calculated by a DSP system, based on the supply voltage samples. Current $\boldsymbol{J}^{\mathrm{C}}(k)$ should reproduce the reference signal calculated by the DSP system according to the compensator designer's decision on which components of the load current $\boldsymbol{i}(t)$ should be compensated. Voltage $\boldsymbol{V}^{\mathrm{C}}(k)$ should be provided by the switching compensator.

16.4 Inverter Switching Modes

The PWM inverter enables control of the compensator line current by controlling the voltage on the inverter output terminals.

Since the sum of the compensator currents has to be equal to zero at each instant of time, only two compensator line currents, say $j_{\mathrm{R}}(t)$ and $j_{\mathrm{S}}(t)$, have to be controlled. The third current cannot be other than $j_{\mathrm{T}}(t) = -[j_{\mathrm{R}}(t) + j_{\mathrm{S}}(t)]$. This is fulfilled independently of whether there is an inductor in the T line or not, as shown in Fig. 16.5. Although it is commonly used, it is not needed for normal compensator operation. To control these two line currents of the compensator, two line-to-line output voltages of the compensator have to be controlled. These could be voltages, referenced to the same common point, say voltages v_{RT} and v_{ST}.

The Clarke Transform in the form presented by formula (16.7) is specified for line-to-zero quantities. Its form for phase-to-phase quantities is therefore needed.

Since

$$\begin{aligned} v_{\mathrm{RT}} &= v_{\mathrm{R}} - v_{\mathrm{T}} = v_{\mathrm{R}} - (-v_{\mathrm{R}} - v_{\mathrm{S}}) = 2v_{\mathrm{R}} + v_{\mathrm{S}} \\ v_{\mathrm{ST}} &= v_{\mathrm{S}} - v_{\mathrm{T}} = v_{\mathrm{S}} - (-v_{\mathrm{R}} - v_{\mathrm{S}}) = 2v_{\mathrm{S}} + v_{\mathrm{R}} \end{aligned} \tag{16.11}$$

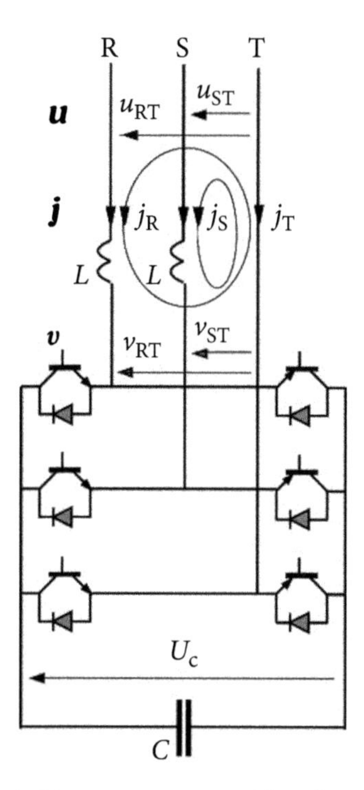

Figure 16.5 A switching compensator with only two line inductors.

line-to-zero voltages v_R and v_S can be expressed in terms of line-to-line voltages v_{RT} and v_{ST}, namely

$$\begin{aligned} v_R &= \frac{1}{3}(2v_{RT} - v_{ST}) \\ v_S &= \frac{1}{3}(2v_{ST} - v_{RT}) \end{aligned} \tag{16.12}$$

hence the voltage $\boldsymbol{v}$ in Clarke coordinates,

$$\boldsymbol{v}^C(t) = \begin{bmatrix} v_\alpha(t) \\ v_\beta(t) \end{bmatrix} = \mathbf{C} \begin{bmatrix} v_R \\ v_S \end{bmatrix} = \frac{1}{3} \begin{bmatrix} \sqrt{3/2}, & 0 \\ 1/\sqrt{2}, & \sqrt{2} \end{bmatrix} \begin{bmatrix} 2\,v_{RT} - v_{ST} \\ 2\,v_{ST} - v_{RT} \end{bmatrix}$$

can be rearranged to the form

$$\boldsymbol{v}^C(t) = \begin{bmatrix} v_\alpha(t) \\ v_\beta(t) \end{bmatrix} = \begin{bmatrix} \sqrt{2/3}, & -1/\sqrt{6} \\ 0, & 1/\sqrt{2} \end{bmatrix} \begin{bmatrix} v_{RT}(t) \\ v_{ST}(t) \end{bmatrix} \stackrel{\text{df}}{=} \mathbf{D} \begin{bmatrix} v_{RT}(t) \\ v_{ST}(t) \end{bmatrix}. \tag{16.13}$$

The inverter can be in one of eight states which provide one of eight fixed values $\boldsymbol{v}_s$ of the output voltage. Therefore, the needed value of the vector $\boldsymbol{v}(k)$

can be obtained only as a linear form of inverter voltage in different states, $\boldsymbol{U}_m$ and $\boldsymbol{U}_n$, namely

$$\boldsymbol{U}(k) = a_m \boldsymbol{U}_m + a_n \boldsymbol{U}_n \tag{16.14}$$

where a_m and a_n denote relative intervals of time, τ_m and τ_n, when the inverter is in state S_m and S_n, namely

$$a_m \stackrel{\mathrm{df}}{=} \frac{\tau_m}{T_s}, \quad a_n \stackrel{\mathrm{df}}{=} \frac{\tau_n}{T_s}. \tag{16.15}$$

The width, in time, of ON and OFF states of the inverter's switches has to be calculated in each sampling interval T_s to obtain the required value of the inverter output voltage $\boldsymbol{U}_k$. Inverters with the output voltage controlled by a change of the width of ON-OFF states of switches are referred to as PWM inverters.

A two-level inverter is built of six power transistor switches with a diode that provides a current path for current of the opposite direction that can flow through the transistor. A pair of such switches with symbols of components is shown in Fig. 16.6.

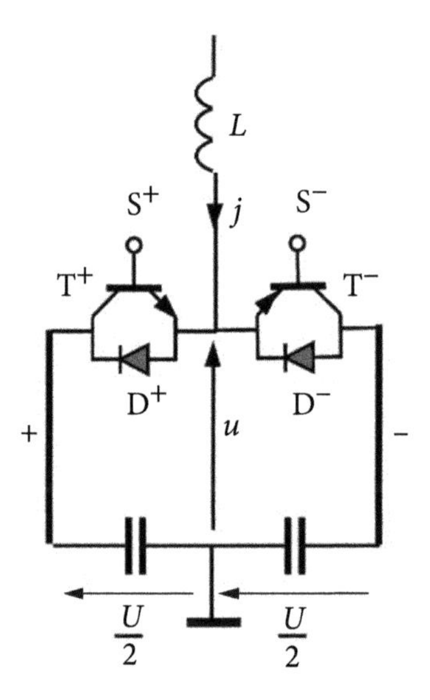

Figure 16.6 A pair of switches.

Control of the inverter requires that there is full control over ON-OFF switching. Transistors enable such full control, but only for one direction of the current, marked by the emitter arrow. Diodes, however, are not controlled devices. One might doubt whether the switches shown in Fig. 16.6 provide full ON-OFF control of a bidirectional current in the compensator output lines. Therefore, a detailed analysis of such a switch would be desirable.

Symbols T^+ and T^-, as well as D+ and D^-, denote transistors and diodes connected to the positive (+) and negative (−) DC bars of the inverter. Symbols S^+ and S^- denote switching signals of transistors, with logical values 1 (transistor in ON state) and 0 (transistor in OFF state). To avoid a short circuit of the capacitor, the switches have to be controlled such that both of them cannot be in the ON state at the same time, meaning the logic product ⊗ of control signals has to be equal to zero

$$S^+ \otimes S^- \equiv 0.$$

The change of the value of switching signals, S^+ and S^-, and the change of direction of the output current *i*, creates six different combinations of the switch conditions, thus it operates in one of six different modes. These combinations are shown in Fig. 16.7.

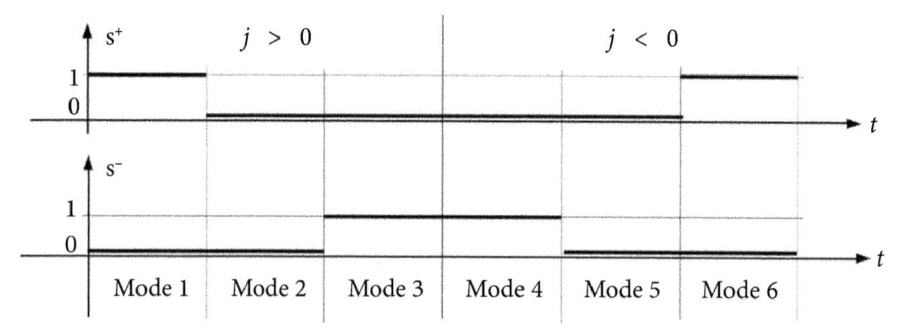

Figure 16.7 Combinations of the switching signal values and the sign of output current.

Observe that to avoid situations that, due to transients, both switches are in the ON state, there have to be intervals of time, where both switching signals have zero value, thus Mode 2 and Mode 5 are needed. The operation of the switch in these modes is shown in Fig. 16.8.

The purpose of the switch is to connect, according to the value of the switching signal, the output line of the inverter to the positive or negative DC bus.

Current paths in Fig. 16.8 show that at positive output current *j*, the output line is not connected to the negative bus before switching signal S^- is equal to 1, meaning in Mode 2. When this current is negative, the output line is not connected to the positive DC bus before S^+ is equal to 1, meaning in Mode 5. Thus, in Modes 2 and 5 the output line is not connected to the DC buses according to the switching signal values. Therefore, these two modes, necessary for avoiding switching hazards, should be as short as possible. Sequential switching of the inverter switches changes the state of the inverter and the voltage at the node $u(t)$ as shown in Fig. 16.9.

The logical state of transistors that connect terminals R, S, and T to the positive DC voltage bus specifies the *state of the inverter* in a binary or a decimal number. According to Fig. 16.10 with sequential switching of only one switch, the inverter progresses through the sequence: State 4 ≫ State 6 ≫ State 2 ≫ State 3 ≫ State 1 ≫ state 5 ≫ State 4... .

To produce zero voltage at the inverter output terminals, all switches connected to the positive DC bus have to be in ON or OFF states, as shown in Fig. 16.11. This

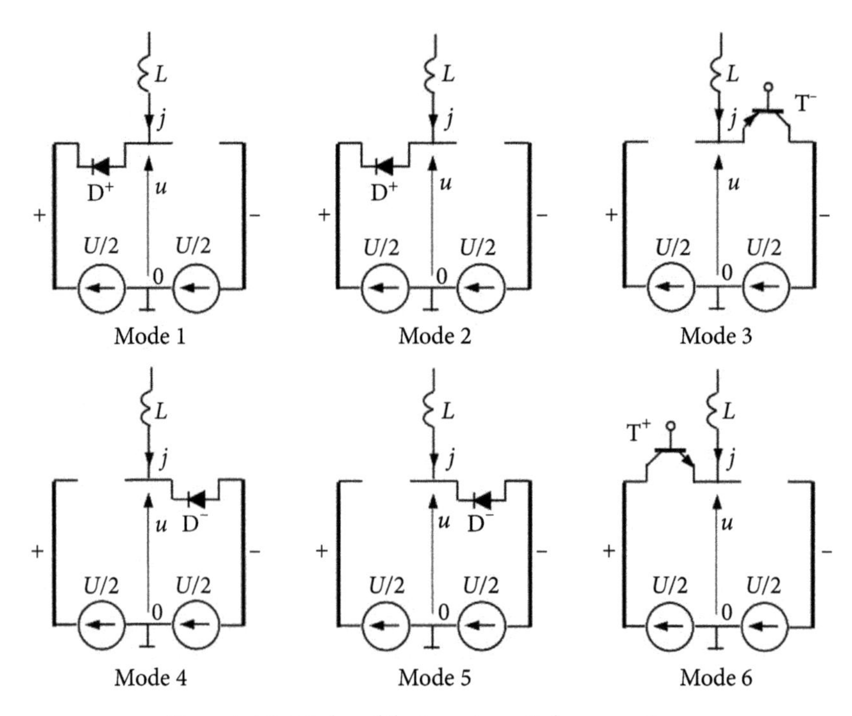

Figure 16.8 Modes of the inverter switches operations.

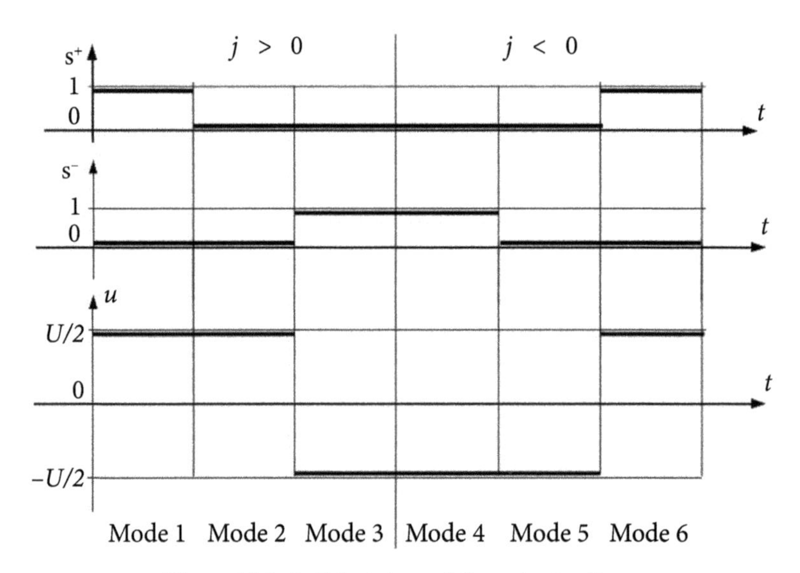

Figure 16.9 Switch states and the output voltage.

creates two additional states of the inverter. Because the inverter output lines are short circuited in these two states, they can be referred to as *short-circuit states*. One of these states can be reached with only one transistor switched ON at any instant of the switching sequence.

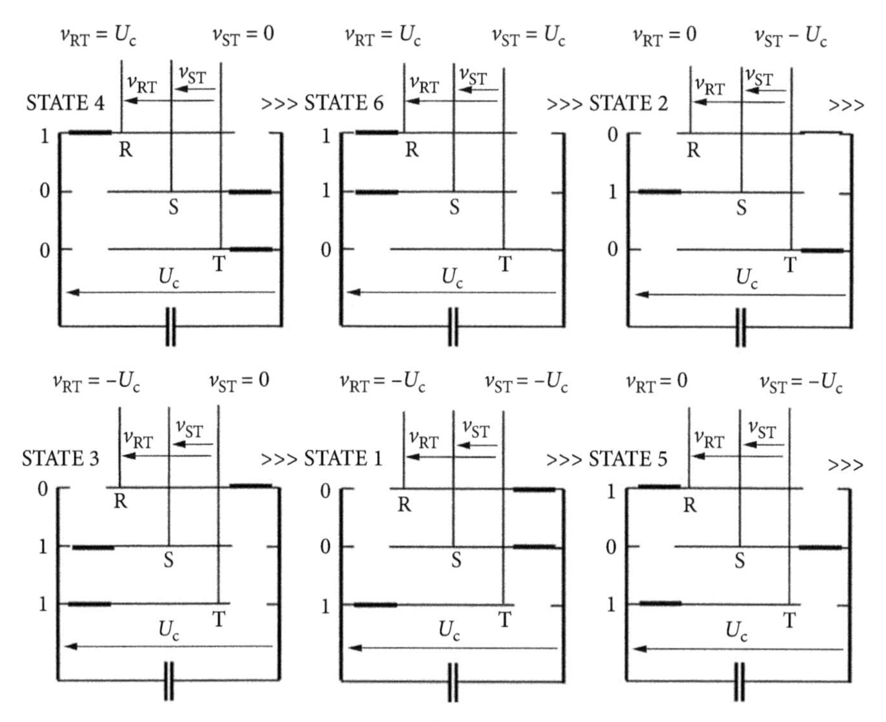

Figure 16.10 The inverter states.

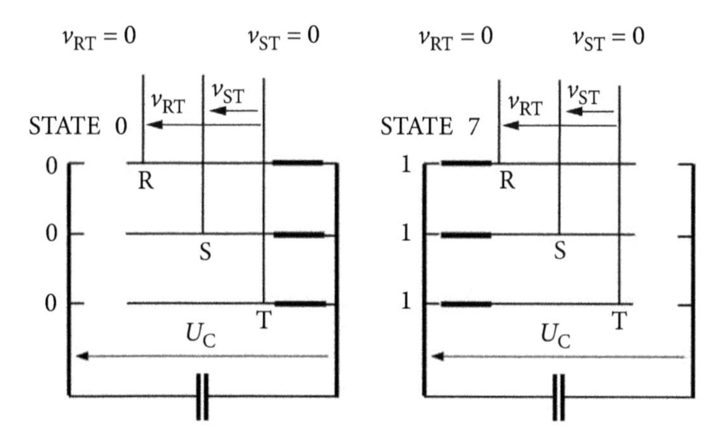

Figure 16.11 Short circuit states.

The inverter output voltage $\boldsymbol{V}$ can have only seven values $\boldsymbol{V}_s$, where index s denotes the state decimal number, $\boldsymbol{V}_0$, $\boldsymbol{V}_1$, $\boldsymbol{V}_2$, $\boldsymbol{V}_3$, $\boldsymbol{V}_4$, $\boldsymbol{V}_5$, $\boldsymbol{V}_6$. The voltage in state 7 is equal to that in state 0, that is, $\boldsymbol{V}_7 = \boldsymbol{V}_0 = \boldsymbol{0}$. The entries of vectors $\boldsymbol{V}_s$ in the line-to-line voltage form, meaning values of voltages v_{RT}, and v_{ST}, are compiled in Figs 16.10 and 16.11, respectively.

Clarke Vector of Inverter Voltage. Since the output voltage of the inverter can have only seven discrete values, the Clarke Vector of this voltage, denoted generally by $\boldsymbol{V}^{\mathrm{C}}(t)$, can have only seven constant values

$$\boldsymbol{V}_s^{\mathrm{C}} \stackrel{\mathrm{df}}{=} V_s^{\mathrm{C}}\, e^{j\phi_s} = [1,j]\, \boldsymbol{V}_s^{\mathrm{C}} = [1,j]\ \mathbf{D}\begin{bmatrix} v_{\mathrm{RT}} \\ v_{\mathrm{ST}} \end{bmatrix}_s = [1,j]\begin{bmatrix} \sqrt{2/3}, & -1/\sqrt{6} \\ 0, & 1/\sqrt{2} \end{bmatrix}\begin{bmatrix} v_{\mathrm{RT}} \\ v_{\mathrm{ST}} \end{bmatrix}_s. \tag{16.16}$$

With this general expression, for sequential states of the inverter, we obtain

$$\boldsymbol{V}_1^{\mathrm{C}} = [1,j]\begin{bmatrix} \sqrt{2/3}, & -1/\sqrt{6} \\ 0, & 1/\sqrt{2} \end{bmatrix}\begin{bmatrix} -U_{\mathrm{c}} \\ -U_{\mathrm{c}} \end{bmatrix} = \sqrt{\frac{2}{3}}\, U_{\mathrm{c}}\, e^{-j120^\circ}$$

$$\boldsymbol{V}_2^{\mathrm{C}} = [1,j]\begin{bmatrix} \sqrt{2/3}, & -1/\sqrt{6} \\ 0, & 1/\sqrt{2} \end{bmatrix}\begin{bmatrix} 0 \\ U_{\mathrm{c}} \end{bmatrix} = \sqrt{\frac{2}{3}}\, U_{\mathrm{c}}\, e^{j120^\circ}$$

$$\boldsymbol{V}_3^{\mathrm{C}} = [1,j]\begin{bmatrix} \sqrt{2/3}, & -1/\sqrt{6} \\ 0, & 1/\sqrt{2} \end{bmatrix}\begin{bmatrix} -U_{\mathrm{c}} \\ 0 \end{bmatrix} = \sqrt{\frac{2}{3}}\, U_{\mathrm{c}}\, e^{j180^\circ}$$

$$\boldsymbol{V}_4^{\mathrm{C}} = [1,j]\begin{bmatrix} \sqrt{2/3}, & -1/\sqrt{6} \\ 0, & 1/\sqrt{2} \end{bmatrix}\begin{bmatrix} U_{\mathrm{c}} \\ 0 \end{bmatrix} = \sqrt{\frac{2}{3}}\, U_{\mathrm{c}}$$

$$\boldsymbol{V}_5^{\mathrm{C}} = [1,j]\begin{bmatrix} \sqrt{2/3}, & -1/\sqrt{6} \\ 0, & 1/\sqrt{2} \end{bmatrix}\begin{bmatrix} 0 \\ -U_{\mathrm{c}} \end{bmatrix} = \sqrt{\frac{2}{3}}\, U_{\mathrm{c}}\, e^{-j60^\circ}$$

$$\boldsymbol{V}_6^{\mathrm{C}} = [1,j]\begin{bmatrix} \sqrt{2/3}, & -1/\sqrt{6} \\ 0, & 1/\sqrt{2} \end{bmatrix}\begin{bmatrix} U_{\mathrm{c}} \\ U_{\mathrm{c}} \end{bmatrix} = \sqrt{\frac{2}{3}}\, U_{\mathrm{c}}\, e^{j60^\circ}.$$

Thus, Clarke Vectors of the inverter output voltage $\boldsymbol{v}$ in particular states s have the same module

$$V_s^{\mathrm{C}} \stackrel{\mathrm{df}}{=} V^{\mathrm{C}} \stackrel{\mathrm{df}}{=} \sqrt{\frac{2}{3}}\, U_{\mathrm{C}} = \text{const.}$$

with different angles ϕ_s which is an integer multiple of 60^0.

Since the length of the Clarke Vector cannot be lower than the minimum value specified by formula (16.1), the capacitor voltage cannot be lower than

$$U_{\mathrm{C}} \geq \sqrt{\frac{3}{2}}\, V_{\min}^{\mathrm{C}} = \sqrt{\frac{3}{2}}\left\{ U^{\mathrm{C}} + \omega_1 L\ [(B_{1\mathrm{e}} + A)U^{\mathrm{C}} + \sum_{n\in N} n\, I_n^{\mathrm{C}} \right\}. \tag{16.17}$$

The remaining two Clarke Vectors $\boldsymbol{V}_0^{\mathrm{C}} = \boldsymbol{V}_7^{\mathrm{C}} = \mathbf{0}$. Clarke Vectors of the inverter's possible values of the output voltage $\boldsymbol{v}$ are shown in Fig. 16.12.

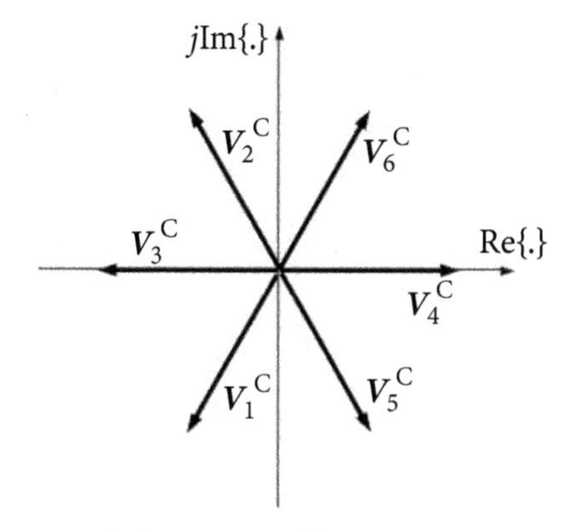

Figure 16.12 Clarke Vectors of the inverter output voltage $\boldsymbol{v}$.

These values change with the angle ϕ_s sequential increases from 0 to 360^0 with only one switch change from the OFF to the ON state between adjacent sectors.

Clarke Vectors $\boldsymbol{V}_s^C$ of the inverter output voltage in particular states s divide the complex plane into six polar zones, Z_0, Z_1, Z_2, Z_3, Z_4, and Z_5, with the same polar angle of 60^0, as shown in Fig. 16.13.

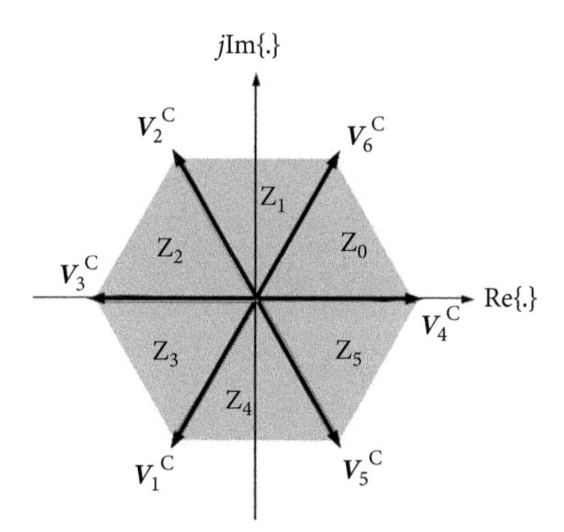

Figure 16.13 The zones of the complex plane of inverter voltage Clarke Vectors.

The zones are confined by Clarke Vectors of inverter states. For example, zone Z_0 is confined by vectors $\boldsymbol{V}_4^C$ and $\boldsymbol{V}_6^C$. Zone Z_4 is confined by vectors $\boldsymbol{V}_1^C$, and $\boldsymbol{V}_5^C$, and so on. The vector with the lower angle ϕ_s will be referred to as the *right border* of the zone and denoted with index "n", while the vector with higher ϕ_s will be referred to as the *left border* of the zone and denoted with index "m". For example, for zone Z_5, the borders are: $\boldsymbol{V}_n^C = \boldsymbol{V}_5^C$ and $\boldsymbol{V}_m^C = \boldsymbol{V}_4^C$.

16.5 Inverter Switching Control

The compensator control is not continuous, but discrete. The reference signal is calculated for sequential instants, kT_s. All calculations are performed on data sampled at the instant $(k-1)T_s$, applied at instant kT_s, and the effect can be observed at the instant $(k+1)T_s$. Taking into account that there are usually several hundred samples per single period T, let us neglect here, for simplification, the delay between data acquisition and control effects, that is, let us assume that all happen at the same instant of time kT_s.

The needed Clarke Vector of the inverter output voltage at instant kT_s is

$$\boldsymbol{V}^{\mathrm{C}}(k) \stackrel{\mathrm{df}}{=} V^{\mathrm{C}}(k)e^{j\Theta(k)} = \boldsymbol{U}^{\mathrm{C}}(k) - (R + L\frac{\Delta}{T_s})\boldsymbol{J}_{\mathrm{r}}^{\mathrm{C}}(k) \tag{16.18}$$

where

$$\boldsymbol{U}^{\mathrm{C}}(k) \stackrel{\mathrm{df}}{=} u_{\alpha}(k) + i\,u_{\beta}(k) = [1,\ i]\,\boldsymbol{u}^{\mathrm{C}}(k) = [1,\ i]\ \mathbf{D}\begin{bmatrix} u_{\mathrm{RT}}(k) \\ u_{\mathrm{ST}}(k) \end{bmatrix}$$

$$\boldsymbol{J}_{\mathrm{r}}^{\mathrm{C}}(k) \stackrel{\mathrm{df}}{=} j_{\mathrm{r}\alpha}(k) + i\,j_{\mathrm{r}\beta}(k) = [1,\ i]\boldsymbol{j}_{\mathrm{r}}^{\mathrm{C}}(k) = [1,\ i]\ \mathbf{D}\begin{bmatrix} j_{\mathrm{rR}}(k) \\ j_{\mathrm{rS}}(k) \end{bmatrix}.$$

The symbol $i = \sqrt{-1}$ was used to avoid confusion in the last formulae instead of j, which also denotes the compensator line current. This vector is located in one of six zones $Z_0, \ldots Z_5$, as illustrated in Fig. 16.14.

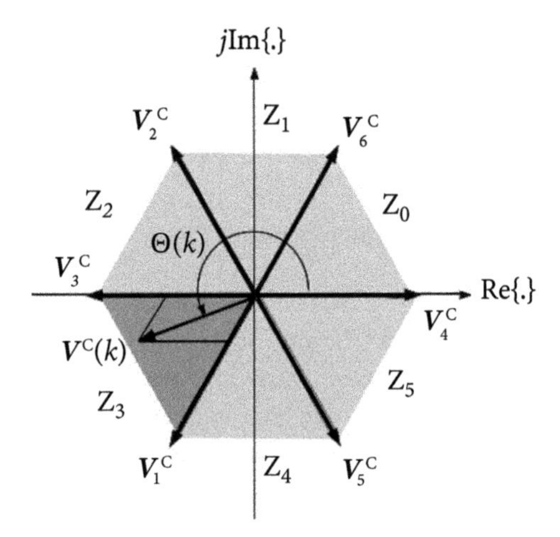

Figure 16.14 An example of Clarke Vector of an inverter voltage at instant kT_s, located in zone Z_3.

The zone number has to be identified because the needed voltage can be obtained only as a linear form of voltages that confines that zone, namely

$$V^{C}(k) = a_m V^{C}_{m} + a_n V^{C}_{n}. \tag{16.19}$$

Coefficients of this form, a_{m} and a_n, are duty factors of switches, meaning the ratio of time τ_m in state m and time τ_n in state n to the sampling period, T_{s}, namely

$$\tau_m = a_m T_{\mathrm{s}}, \quad \tau_n = a_n T_{\mathrm{s}}.$$

Since usually $\tau_m + \tau_n < T_{\mathrm{s}}$, for the remaining part of the sampling period T_{s}

$$\tau_0 = T_{\mathrm{s}} - (\tau_m + \tau_n) \tag{16.20}$$

the inverter should not contribute to the output voltage, meaning it should be in a short-circuit state, in other words, in state 0 or state 7. The duty factor of the inverter in the short-circuit state is

$$a_0 \stackrel{\mathrm{df}}{=} \frac{\tau_0}{T_{\mathrm{s}}} = 1 - (a_m + a_n). \tag{16.21}$$

Since zones Z_0, Z_1, ...Z_5 are confined by different vectors, the calculation of duty factors a_m and a_n in these zones would result in a different formula. This could be avoided if the Clarke Vector of the inverter voltage needed at instant kT_{s}, originally located in zone Z_{s}, specified by the formula

$$s(k) = int\left\{\frac{\Theta(k)}{60}\right\} \tag{16.22}$$

is rotated to zone Z_0 confined by vectors

$$A \stackrel{\mathrm{df}}{=} V^{C}_{4}, \quad Ae^{j60^\circ} \triangleq V^{C}_{6}$$

shown in Fig. 16.15.

After duty factors a_m and a_n are calculated for zone Z_0, they specify the linear form of the Clarke Vector in the original zone Z_{s}. The Clark Vector of the inverter voltage rotated to zone Z_0 has the value

$$\boldsymbol{B} \stackrel{\mathrm{df}}{=} Be^{j\phi} = V^{C}(k)e^{j[\Theta(k)] \bmod (60)} \tag{16.23}$$

that is, $B \triangleq V^{C}(k)$ and $\phi \triangleq [\Theta(k)] \bmod(60)$. These three vectors have to satisfy the relationship

$$Be^{j\phi} = a_n Ae^{j60^\circ} + a_m A$$

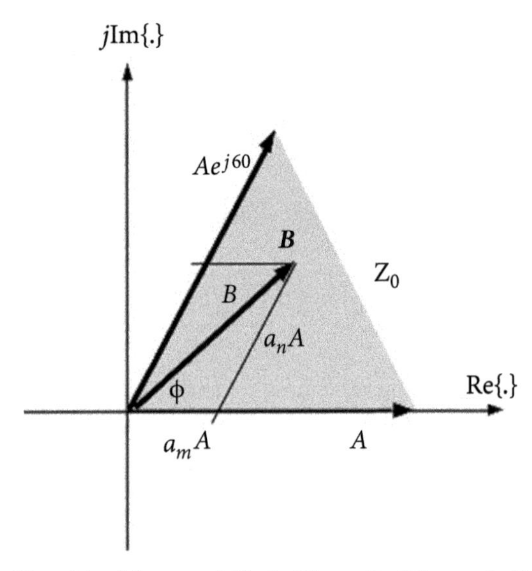

Figure 16.15 Zone Z_0 with rotated Clarke Vector ***B*** of the needed inverter voltage.

which can be presented in the rectangular form

$$B\cos\phi + jB\sin\phi = a_nA\cos 60° + j\,a_nA\sin 60° + a_mA$$

$$B\cos\phi + jB\sin\phi = a_nA\frac{1}{2} + j\,a_nA\frac{\sqrt{3}}{2} + a_mA.$$

From equations for the real and the imaginary part, formulae for duty factors can be found:

$$a_n = \frac{2}{\sqrt{3}}\frac{B}{A}\sin\phi, \tag{16.24}$$

$$a_m = \frac{B}{A}\cos\phi - \frac{a_n}{2} = \frac{B}{A}(\cos\phi - \frac{1}{\sqrt{3}}\sin\phi). \tag{16.25}$$

Indeed, let us assume that $\phi = 0$, then $a_n = 0$, while $a_m = B/A$, meaning the inverter required output voltage is obtained entirely by the inverter in state 4. Now, let us assume that $\phi = 60^0$, then $a_n = B/A$, while $a_m = 0$, meaning the inverter output voltage is obtained entirely by the inverter in state 5. When $\phi = 30^0$ then

$$a_n = a_m = \frac{1}{\sqrt{3}}\frac{B}{A}$$

meaning the inverter should be in states 4 and 5 for the same interval of time.

Illustration 16.2 *The Clarke Vector of the inverter voltage $\boldsymbol{u}(k)$ at some instant of time should be equal to $\mathbf{V}^{\mathrm{C}}(k) = 250\ e^{j200\mathrm{deg}}$ [V]. Which state of the inverter can provide this voltage and how long it should stay in these states if the capacitor voltage is $U_{\mathrm{C}} = 500$ [V] and the switching frequency $f_s = 18$ [kHz].*

The Clarke Vector at this time instant is in the sector

$$s(k) = int\left\{\frac{\Theta(k)}{60^\circ}\right\} = int\left\{\frac{200^\circ}{60^\circ}\right\} = 3$$

confined by vectors $\mathbf{V}_3^{\mathrm{C}}$ and $\mathbf{V}_1^{\mathrm{C}}$. States 3 and 1, which provide voltages $\boldsymbol{u}_3$ and $\boldsymbol{u}_1$ as well as state 0 are shown in Fig. 16.16.

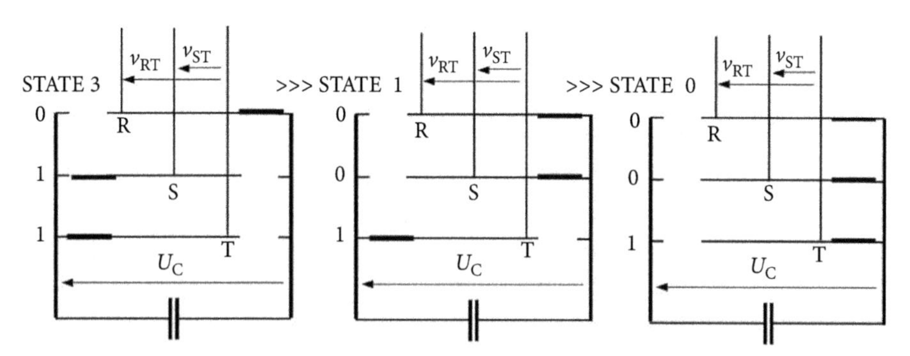

Figure 16.16 Inverter states in Illustration 16.1.

Vector $\mathbf{V}^{\mathrm{C}}(k)$ rotated to sector zero is equal to

$$\boldsymbol{B} \stackrel{\mathrm{df}}{=} Be^{j\phi} = V^{\mathrm{C}}(k)e^{j[\Theta(k)]\ \mathrm{mod}\ (60)} = 250\ e^{j20^\circ}\,[\mathrm{V}]$$

while

$$A = \sqrt{\frac{2}{3}}\,U_{\mathrm{C}} = \sqrt{\frac{2}{3}}500 = 408\ [\mathrm{V}].$$

Duty factors a_{m} and a_{n} are equal to

$$a_n = \frac{2}{\sqrt{3}}\frac{B}{A}\sin\phi = \frac{2}{\sqrt{3}}\frac{250}{408}\sin(20^\circ) = 0.24$$

$$a_m = \frac{B}{A}(\cos\phi - \frac{1}{\sqrt{3}}\sin\phi) = \frac{250}{408}(\cos 20^\circ - \frac{1}{\sqrt{3}}\sin 20^\circ) = 0.45.$$

The duty factor in the zero state is

$$a_0 = 1 - (a_m + a_n) = 1 - (0.45 + 0.24) = 0.31.$$

Since the sampling period $T_s = 1/f_s = 1/18 \times 10^3 = 55$ [μs], thus the inverter should be in state 3 over time interval

$$\tau_m = a_m T_s = 0.45 \times 55 = 24.7 \ [\mu s]$$

in state 1 over the interval

$$\tau_n = a_n T_s = 0.24 \times 55 = 13.3 \ [\mu s]$$

and in state 0 over the interval

$$\tau_0 = a_0 T_s = 0.31 \times 55 = 17.0 \ [\mu s].$$

*****

It was previously assumed for simplicity's sake that data acquisition, signal processing, and compensator control are in the same instant of time, kT_s. To be more accurate, the DA system provides data on the load voltages and current at the instant $(k-1)T_s$. and gating signals calculation is completed before instant kT_s when they are applied to switching transistors and the effects of the control upon the compensator current $\boldsymbol{j}$ can be observed at the instant $(k+1)T_s$. This sequence is illustrated in Fig. 16.17.

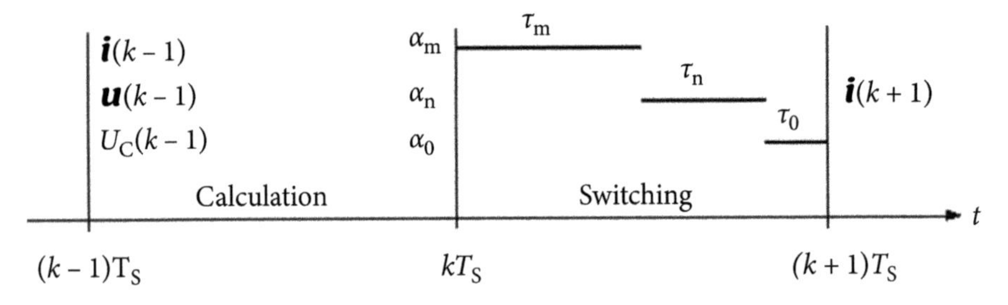

Figure 16.17 Control timing.

After data on the load voltages and currents at the instant $(k-1)T_s$ are provided for the DSP system, the Clarke Vector of the inverter voltage $\boldsymbol{V}(k-1)$

$$V^{C}_{ref}(k-1) = \boldsymbol{U}^{C}(k-1) - \left[R\boldsymbol{J}^{C}_{ref}(k-1) + L\frac{\Delta \boldsymbol{J}^{C}_{ref}(k-1)}{T_s} \right] \tag{16.26}$$

and switches duty factors a_m, a_n, and a_0 can be calculated. Having these duty factors, timing signals can be applied by the ISC system to the inverter switches after instant kT_s. The inverter voltage will affect the compensator current $\boldsymbol{j}$ at the instant $(k+1)T_s$.

The control of a switching compensator as presented above can be summarized as follows. The DSP system calculates the reference signal $\boldsymbol{j}_{\text{ref}}$ and the Clarke Vector of the inverter voltage $\boldsymbol{v}$ needed to reproduce the reference signal as the compensator terminal current $\boldsymbol{j}$. The ISC system generates gating signals for the inverter switches to produce the calculated voltage as the voltage $\boldsymbol{v}$ at the inverter terminals.

Parameters RL of the line inductors used for the inverter voltage $\boldsymbol{v}$ calculation are known with limited accuracy, however. Also, switches are not ideal. In effect, the compensator current $\boldsymbol{j}$ can differ from the reference signal $\boldsymbol{j}_{\text{ref}}$. The results of compensation remain unknown, however. To reduce the difference between the reference signal and the current produced, this current has to be measured, and compared with the reference signal and the difference observed has to be used for the compensator control. Thus, feedback is needed for the compensator operation.

Closed-loop control. The Clark Vector of this current can be compared with the reference current and their difference

$$\boldsymbol{J}^{\text{C}}(k+1) - \boldsymbol{J}^{\text{C}}_{\text{ref}}(k+1) = \varepsilon\boldsymbol{J}^{\text{C}}(k+1), \tag{16.27}$$

can be found. When this control error is not taken into account at the instant $(k+1)T_s$, then

$$\boldsymbol{V}^{\text{C}}_{\text{ref}}(k+1) = \boldsymbol{U}^{\text{C}}(k+1) - [R\boldsymbol{J}^{\text{C}}_{\text{ref}}(k+1) + L\frac{\Delta\boldsymbol{J}^{\text{C}}_{\text{ref}}(k+1)}{T_s}] \tag{16.28}$$

To obtain the correct value of the compensator current, the inverter voltage should have the Clarke Vector

$$\begin{aligned}\boldsymbol{V}^{\text{C}}(k+1) &= \boldsymbol{U}^{\text{C}}(k+1) - [R\boldsymbol{J}^{\text{C}}(k+1) + L\frac{\Delta\boldsymbol{J}^{\text{C}}(k+1)}{T_s}] = \\ &= \boldsymbol{U}^{\text{C}}(k+1) - \{R[\boldsymbol{J}^{\text{C}}(k+1) + \varepsilon\boldsymbol{J}^{\text{C}}(k+1) + L\frac{\Delta\,[\boldsymbol{J}^{\text{C}}(k+1) + \varepsilon\boldsymbol{J}^{\text{C}}(k+1)]}{T_s}]\} = \\ &= \boldsymbol{V}^{\text{C}}_{\text{ref}}(k+1) - R\,\varepsilon\boldsymbol{J}^{\text{C}}(k+1) - L\frac{\Delta\,\varepsilon\boldsymbol{J}^{\text{C}}(k+1)}{T_s}.\end{aligned}$$

If the last term on the right side, as small, is neglected, then the Clarke Vector of the inverter voltage should be corrected to

$$\boldsymbol{V}^{\text{C}}(k+1) = \boldsymbol{V}^{\text{C}}_{\text{ref}}(k+1) - R\,\varepsilon\boldsymbol{J}^{\text{C}}(k+1). \tag{16.29}$$

The last term in (16.29) introduces the feedback to the compensator control and this feedback may cause the compensator instability. Therefore, the feedback should be controlled by a coefficient A selected as smaller than 1, that is, the Clarke Vector of the inverter voltage should be calculated rather as

$$\boldsymbol{V}^{\mathrm{C}}(k+1) = \boldsymbol{V}^{\mathrm{C}}_{\mathrm{ref}}(k+1) - A\,R\,\varepsilon \boldsymbol{J}^{\mathrm{C}}(k+1). \tag{16.30}$$

If instability occurs, the value of the A coefficient should be reduced.

16.6 Energy Flow and Storage

An SSC shapes its current $\boldsymbol{j}$ by controlling switches of a PWM inverter which requires some DC voltage or a DC current for its operation. Usually, this is a DC voltage on a capacitor connected as shown in Fig. 16.18 of capacitance C. This capacitor is supplied through the inverter from the grid being compensated. Energy has to be stored in the capacitor to provide the inverter's DC voltage. If W denotes the stored energy then the capacitor voltage is equal to

$$u_{\mathrm{C}} = \sqrt{\frac{2\,W}{C}}.$$

The reference signal for the compensator control should be proportional to unwanted components of the load current, that is to the reactive, unbalanced, and harmonic currents. When the compensator current does not contain any active component, then the energy is not delivered to the compensator permanently. It can only oscillate, but with zero mean value. Currents cannot flow in the compensator and the filter without energy dissipation, however. At zero active current the energy is dissipated at the cost of energy stored in the capacitor. Consequently, the capacitor voltage U_{C} declines. To keep this voltage, the compensator current has to contain an active component, $\boldsymbol{j}_{\mathrm{a}}$. Thus, the compensator has to be controlled in such a way that its current contains an active component. Its rms value $||\boldsymbol{j}_{\mathrm{a}}||$ should be proportional to the active power loss, ΔP.

A flow of energy between the grid and the capacitor, as well as energy loss in the compensator and the filter, affect the capacitor voltage. Therefore, the energy flow and its storage are important issues for the compensator operation.

Let us analyze energy flow for a simplified sinusoidal situation, assuming that the switching noise is neglected. The compensator current $\boldsymbol{j}$ in such a situation can be composed of the active, reactive, and unbalanced components, namely

$$\boldsymbol{j} = \boldsymbol{j}_{\mathrm{a}} + \boldsymbol{j}_{\mathrm{r}} + \boldsymbol{j}_{\mathrm{u}}$$

and consequently, the rate of energy flow to the compensator from the supply is

$$p(t) \stackrel{\mathrm{df}}{=} \frac{dW(t)}{dt} = \boldsymbol{u}^{\mathrm{T}}\boldsymbol{i} = \boldsymbol{u}^{\mathrm{T}}(\boldsymbol{i}_{\mathrm{a}} + \boldsymbol{i}_{\mathrm{r}} + \boldsymbol{i}_{\mathrm{u}}) \stackrel{\mathrm{df}}{=} p_{\mathrm{a}}(t) + p_{\mathrm{r}}(t) + p_{\mathrm{u}}(t). \tag{16.31}$$

The instantaneous power associated with the active current

$$p_{\rm a} \stackrel{\rm df}{=} \boldsymbol{u}^{\rm T}\boldsymbol{j}_{\rm a} = \text{const.} = \Delta P$$

is equal to the power loss in the compensator and the input filter. The instantaneous power associated with the reactive current

$$p_{\rm r} \stackrel{\rm df}{=} \boldsymbol{u}^{\rm T}\boldsymbol{j}_{\rm r} \equiv 0$$

meaning that there is no energy flow associated with compensation of the load reactive current at any instant of time. The instantaneous power associated with the unbalanced current

$$p_{\rm u} \stackrel{\rm df}{=} \boldsymbol{u}^{\rm T}\boldsymbol{j}_{\rm u} = -3AU^2\cos(2\omega t + \psi) = -D\cos(2\omega t + \psi)$$

oscillates with the amplitude of the compensated load unbalanced power D. Consequently

$$p = \frac{dW}{dt} = \boldsymbol{u}^{\rm T}\boldsymbol{j} = p_{\rm a} + p_{\rm r} + p_{\rm u} = \Delta P - D\cos(2\omega t + \psi). \tag{16.32}$$

thus the energy at the input terminals of the compensator can oscillate only because of the compensating current asymmetry.

Resistance of the inverter switches, inductors, and the capacitor causes a loss of energy in the compensator and filter. We can assume, however, that there are line resistors, connected as shown in Fig. 17.18, of an equivalent resistance $R_{\rm e}$ that causes the same loss of energy, while the compensator and the filter inductors are lossless.

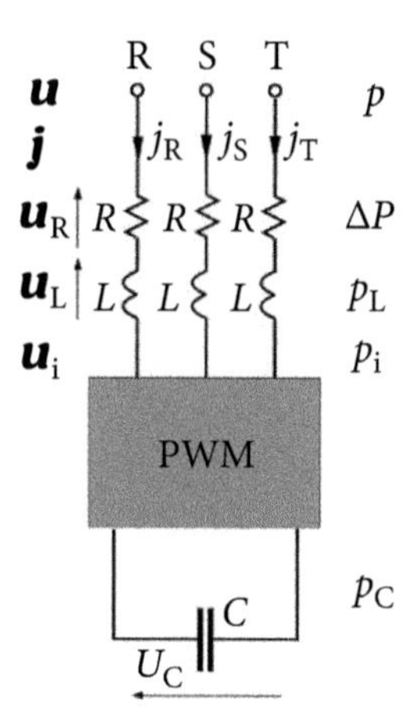

Figure 16.18 A switching compensator with an equivalent resistance.

The instantaneous power at terminals R, S, and T can be expressed as

$$p = \boldsymbol{u}^{\mathrm{T}}\boldsymbol{j} = (\boldsymbol{u}_{\mathrm{i}} + \boldsymbol{u}_{\mathrm{L}} + \boldsymbol{u}_{\mathrm{R_e}})^{\mathrm{T}}\boldsymbol{j} = p_{\mathrm{i}} + p_{\mathrm{L}} + p_{\mathrm{R_e}}.$$

The PWM inverter has no capability of energy storage, thus $p_{\mathrm{i}}(t) \equiv p_{\mathrm{C}}(t)$, and hence, the instantaneous power at the capacitor terminals can be calculated as

$$p_{\mathrm{C}} = p_{\mathrm{i}} = p - p_{\mathrm{R_e}} - p_{\mathrm{L}}$$

where

$$p_{Re} \stackrel{\mathrm{df}}{=} \boldsymbol{u}_{Re}^{\mathrm{T}}\boldsymbol{j} = R_{\mathrm{e}}\langle\langle\boldsymbol{j}\rangle\rangle^2 = R_{\mathrm{e}}(j_{\mathrm{R}}^2 + j_{\mathrm{S}}^2 + j_{\mathrm{T}}^2)$$

and $\langle\langle\boldsymbol{j}\rangle\rangle$ denotes the instantaneous norm or length of the three-phase vector $\boldsymbol{j}$, defined as

$$\langle\langle\boldsymbol{j}\rangle\rangle \stackrel{\mathrm{df}}{=} \sqrt{\boldsymbol{j}^{\mathrm{T}}\boldsymbol{j}}. \tag{16.33}$$

The instantaneous power of the inductor is

$$p_{\mathrm{L}} \stackrel{\mathrm{df}}{=} \boldsymbol{u}_{\mathrm{L}}{}^{\mathrm{T}}\boldsymbol{j} = \frac{d}{dt}W_L = \frac{d}{dt}\{\frac{1}{2}L(j_{\mathrm{R}}^2 + j_{\mathrm{S}}^2 + j_{\mathrm{T}}^2)\} = \frac{1}{2}L\frac{d}{dt}\langle\langle\boldsymbol{j}\rangle\rangle^2.$$

Thus the instantaneous power of the capacitor p_{C} at a specified instantaneous power p at R, S, and T terminals depends on the variability of the square of the instantaneous norm of the compensator current vector

$$\langle\langle\boldsymbol{j}\rangle\rangle^2 = j_{\mathrm{R}}^2 + j_{\mathrm{S}}^2 + j_{\mathrm{T}}^2. \tag{16.34}$$

When the compensator current is symmetrical, meaning it is composed of only the active and reactive currents and consequently the current is symmetrical, with

$$j_{\mathrm{Rb}} = j_{\mathrm{Ra}} + j_{\mathrm{Rr}} = \sqrt{2}J_{\mathrm{a}}\cos\omega t + \sqrt{2}J_{\mathrm{r}}\sin\omega t = \sqrt{2}J_{\mathrm{b}}\cos(\omega t - \varphi),$$

then

$$\begin{aligned}\langle\langle\boldsymbol{j}\rangle\rangle^2 &= j_{\mathrm{Rb}}^2 + j_{\mathrm{Sb}}^2 + j_{\mathrm{Tb}}^2 = 2J_{\mathrm{b}}^2[\cos^2(\omega t - \varphi) + \cos^2(\omega t - 120° - \varphi) \\ &\quad + \cos^2(\omega t + 120° - \varphi)] = \\ &= 3J_{\mathrm{b}}^2 + J_{\mathrm{b}}^2[\cos 2(\omega t - \varphi) + \cos 2(\omega t - 120° - \varphi) \\ &\quad + \cos 2(\omega t + 120° - \varphi)] = 3J_{\mathrm{b}}^2.\end{aligned}$$

When the compensator current, apart from the active and reactive components, contains an unbalanced current, which is symmetrical, but of negative sequence, then

$$\begin{aligned}\langle\langle \boldsymbol{j} \rangle\rangle^2 &= j_R^2 + j_S^2 + j_T^2 = (j_{Rb} + j_{Ru})^2 + (j_{Sb} + j_{Su})^2 + (j_{Tb} + j_{Tu})^2 = \\ &= 3J_b^2 + 3J_u^2 + 2(j_{Rb} \cdot j_{Ru} + j_{Sb} \cdot j_{Su} + j_{Tb} \cdot j_{Tu}) = \\ &= 3J_b^2 + 3J_u^2 + 4J_b J_u[\cos(\omega t - \varphi)\cos(\omega t + \psi) + \\ &\qquad + \cos(\omega t - 120° - \varphi)\cos(\omega t + 120° + \psi) + \\ &\qquad + \cos(\omega t + 120° - \varphi)\cos(\omega t - 120° + \psi)] = \\ &= 3[J_b^2 + J_u^2 + 2J_b J_u \cos(2\omega t - \varphi + \psi)]. \qquad (16.35)\end{aligned}$$

It is worth observing that the instantaneous length $\langle\langle \boldsymbol{j} \rangle\rangle$ of the vector has an oscillating component only if both rms values J_b and J_u are not equal to zero. When one of them is zero, this length is constant.

Knowing the variability of the instantaneous norm $\langle\langle \boldsymbol{j} \rangle\rangle$ of the compensator current vector $\boldsymbol{j}$, let us calculate the instantaneous power p_C of the capacitor,

$$p_C = p - p_{R_e} - p_L = \Delta P - D\cos(2\omega t + \psi) - R_e\langle\langle \boldsymbol{j} \rangle\rangle^2 - \frac{1}{2}L\frac{d}{dt}\langle\langle \boldsymbol{j} \rangle\rangle^2.$$

The energy loss in the compensator is not specified by the norm $\langle\langle \boldsymbol{j} \rangle\rangle$ of the compensator current, but by its rms value, $||\boldsymbol{j}||$, equal to

$$||\boldsymbol{j}|| = \sqrt{\frac{1}{T}\int_0^T (j_R^2 + j_S^2 + j_T^2)dt} = \sqrt{\frac{1}{T}\int_0^T \langle\langle \boldsymbol{j} \rangle\rangle^2 dt} = \sqrt{3}\sqrt{J_b^2 + J_u^2}.$$

Since the energy loss in the compensator and the filter is confined to energy dissipation on the equivalent resistance R_e, while the remaining circuit is lossless

$$\Delta P - 3R_e(J_b^2 + J_u^2) = 0$$

thus, the instantaneous power of the capacitor $p_C = p_C(t)$ does not have any DC component. It is an oscillating quantity, equal to

$$p_C = -D\cos(2\omega t + \psi) + 6\sqrt{R_e^2 + \omega^2 L^2}\, J_b J_u \cos(2\omega t - \varphi - \alpha + \psi)$$

that is, it is composed of two oscillating components of the same frequency. This result means that the filter increases the amplitude of energy oscillation at the capacitor terminals as compared to this oscillation on terminals R, S, and T.

The filter reactance is usually selected such that $\omega L = X \gg R_e$. Moreover, the active component of the compensator current, caused by energy dissipation in the compensator and the filter, is rather substantially lower than the compensating reactive current, meaning that

$$\sqrt{R_e^2 + \omega^2 L^2} \approx \omega L, \quad J_b \approx J_r, \quad \varphi \approx 90°.$$

Consequently

$$p_C = -(D + 6\omega L J_r J_u)\cos(2\omega t + \psi) = -A\cos(2\omega t + \psi)$$

and therefore, taking into account that

$$J_r = ||\boldsymbol{j}_r||/\sqrt{3}, \qquad J_u = ||\boldsymbol{j}_u||/\sqrt{3},$$

the amplitude of the instantaneous power oscillations on the capacitor terminals is approximately equal to

$$A = D + 2X||\boldsymbol{j}_r||\,||\boldsymbol{j}_u|| = D(1 + 2\frac{|Q|}{||\boldsymbol{u}||^2}X). \tag{16.36}$$

The obtained formula relates the rate of energy flow at the capacitor terminals to the unbalanced and reactive powers of the compensated load,

$$D = ||\boldsymbol{j}_u||\,||\boldsymbol{u}||, \quad Q = \pm||\boldsymbol{j}_r||\,||\boldsymbol{u}||,$$

and the selected reactance X of the filter.

Observe that the phase of the instantaneous power p_C of the capacitor is irrelevant for the energy storage and can be neglected so its variability and the variability of the energy stored in the capacitor can be presented as shown in Fig. 16.19.

The instantaneous power, p_a, associated with the compensator active current, $\boldsymbol{j}_a$, at a sinusoidal supply voltage, has a constant value. Thus, this current affects only the mean value of the energy stored in the capacitor, but not the oscillating component of this energy. Consequently, only the mean value of the capacitor voltage depends on the active current, but the voltage ripples do not. Assuming that the difference between the maximum and minimum value of the capacitor voltage is

$$(u_C)_{max} - (u_C)_{min} = \Delta u_C,$$

then its maximum voltage is

$$(u_C)_{max} = U_{C0}\left(1 + \frac{1}{2}\frac{\Delta u_C}{U_{C0}}\right) = \sqrt{\frac{2(W_0 + \Delta W)}{C}}$$
$$= \sqrt{\frac{2W_0}{C}\left(1 + \frac{\Delta W}{W_0}\right)} \approx \sqrt{\frac{2W_0}{C}}\left(1 + \frac{1}{2}\frac{\Delta W}{W_0}\right),$$

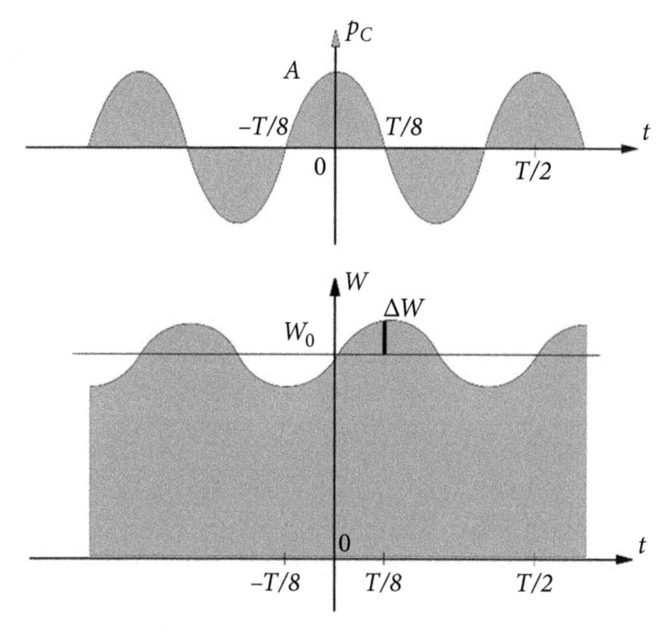

Figure 16.19 The variability of instantaneous power and energy stored in the capacitor.

where

$$U_{C0} = \sqrt{\frac{2W_0}{C}}$$

denotes the mean value of the capacitor voltage, while

$$\frac{\Delta u_C}{U_{C0}} \approx \frac{\Delta W}{W_0}$$

is the relative value of the voltage ripples. Due to the instantaneous power of the capacitor variability, the stored energy changes by

$$2\Delta W = \int_{-T/8}^{T/8} p_C\, dt = \int_{-T/8}^{T/8} A \cos 2\omega t\, dt = \frac{A}{\omega}$$

hence the capacitor voltage ripples are approximately equal to

$$\Delta u_C \approx U_{C0}\frac{\Delta W}{W_0} = \frac{1}{\omega\, C U_{C0}} A = D(1 + 2\frac{|Q|}{||\boldsymbol{u}||^2} X)\frac{1}{\omega\, C U_{C0}}. \tag{16.37}$$

This result shows that the capacitor voltage has the lowest ripples when the compensator compensates only the unbalanced power D of the load. Compensation of the load reactive power Q increases these ripples, but this increase can be controlled by the selection of the filter reactance X or the capacitance C. The mean value of the capacitance voltage U_{C0} is selected with regard to the PWM inverter operation. Observe that

$$\frac{|Q|}{||\boldsymbol{u}||^2} = |B_e|$$

thus this ratio is specified by the equivalent susceptance of the load. Hence, the voltage ripples increase at a compensation of reactive power Q depends on the product of the load susceptance and the selected filter reactance, $|B_e|\,X$.

Illustration 16.3 *A SSC operated at frequency $f = 50$ Hz, is applied for compensation of an LTI load that at line-to-line voltage rms value $U = 380$ V has the unbalanced power $D = 150$ kVA. The PWM inverter should be supplied with the voltage $U_{C0} = 1.2\,U$, with ripples maximum value not higher than $\Delta u_C = 10$ V. Calculate the capacitance C for a situation, when apart from the unbalanced power, the compensator should also compensate the reactive power $Q = 150$ kVAr, assuming that $|B_e|\,X = 0.5$.*

The condition concerning the ripples maximum value is satisfied at the load reactive power $Q = 0$, according to formula (16.37), when the capacitance is higher than

$$C > \frac{D}{\omega\, U_{C0}\, \Delta\, u_C} = C_0 = \frac{150 \times 10^3}{314 \times (1.2 \times 380) \times 10} = 105 \text{ mF}.$$

When the same compensator compensates the reactive power, then

$$C > \frac{D(1 + 2\,|B_e|\,X)}{\omega\, U_{C0}\, \Delta\, u_C} = (1 + 1)C_0 = 210 \text{ mF}.$$

*****

16.7 Switching Noise

A PWM inverter does not deliver a continuous but a stair-like voltage, which only approximates the needed continuous voltage $\boldsymbol{v}$. Therefore to increase the accuracy of this approximation, there is a tendency to increase the switching frequency of the inverter as much as possible. It can be in the order of a few or several kHz. It depends mainly on the required switching power of power transistors. Since the energy in the transistor switch is dissipated mainly not when the switch is in the ON or the OFF state but during the intervals when it switches between these two states, the increase in switching frequency increases energy loss and consequently it reduces the switching power of transistor switches. It also contributes to the switching noise. How this noise occurs is explained below.

Due to a stray capacitance between each turn of the inductor wire, inductors have some capacitance. This stray capacitance between individual turns, shown in Fig. 16.20a, can be represented by one equivalent capacitance of C value, connected as shown in Fig. 16.20b.

Due to this capacitance, a capacitive current occurs in the supply line of the compensator

$$j_{\mathrm{HF}} = C_{\mathrm{e}}\frac{d}{dt}(u - v). \tag{16.38}$$

As compared to the inverter phase voltage v, controlled with switches operated at the frequency of a few or several kHz, the distribution line voltage u is a slowly varying quantity, hence

$$j_{\mathrm{HF}} \approx -C_{\mathrm{e}}\frac{d}{dt}v. \tag{16.39}$$

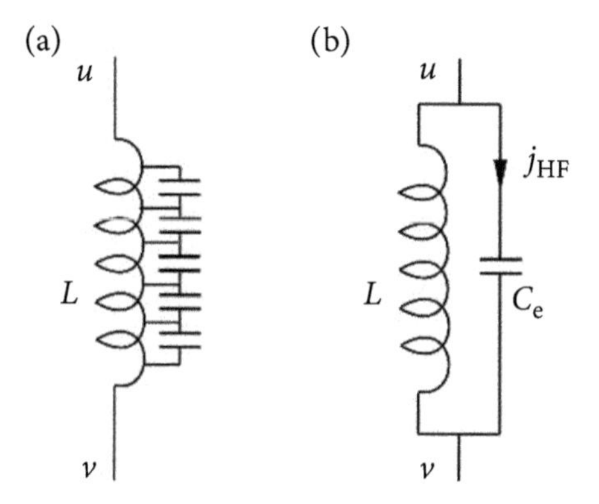

Figure 16.20 (a) An inductor with marked stray capacitances and (b) its equivalent circuit with a single capacitance.

Line-to-line voltages v of the inverter can have only three values, U_{C}, $-U_{\mathrm{C}}$, or 0, with some intermediate values in transients, as shown in Fig. 16.21a. These transients are approximated in this figure with straight lines, but of different duration for switching OFF-ON and ON-OFF. In response to the inverter voltage changes, current pulses of random duration, as shown in Fig. 16.21b, occur in the compensator current. The duration of these pulses is random because the duty factors a_n and a_m have random values. These pulses form a *switching noise* of the compensator.

The switching noise and consequently switching compensators can be a source of electromagnetic interference (EMI). It can spread out directly in the distribution system, disturbing other devices supplied from that system, and by the magnetic field created by the compensator current. Since the switching noise contains high-frequency components, this interference can be very arduous. In particular, video and computer-like devices can be disturbed, therefore the level of switching noise produced by switching compensators can be a matter of concern.

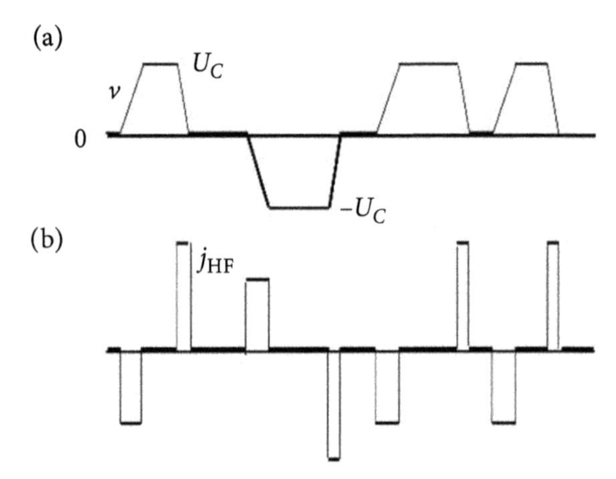

Figure 16.21 (a) A line-to-line inverter voltage and (b) capacitive current.

Compensators should be built in such a way that this noise is as low as possible. When its level is still not acceptable, EMI filters are needed.

Switching noise occurs in the effect of fast changes of the inverter output voltage in transients between ON and OFF states. It is reduced with the reduction of the voltage change rate, meaning when switching is slower. Unfortunately, it increases energy dissipated in the transistors and consequently it reduces their switching power. The switching noise can be reduced by the reduction of the equivalent capacitance C_e of line inductors, therefore, these inductors should be built in such a way that this capacitance is as low as possible.

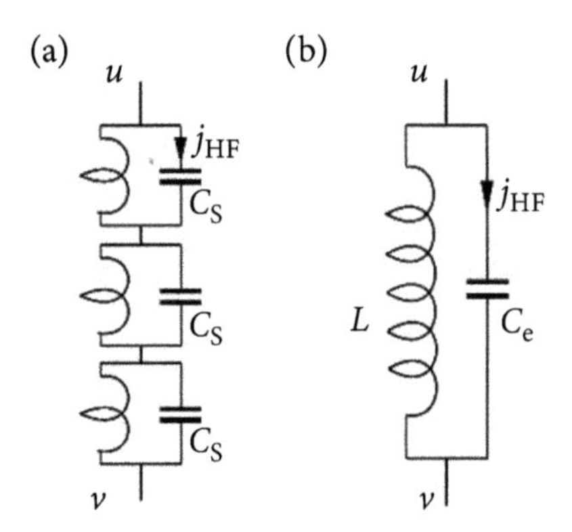

Figure 16.22 (a) An inductor decomposed into segments and (b) its an equivalent circuit.

The stray capacitance declines with an increase in the distance between wires. Also, the decomposition of the inductor into a few series segments, as shown in Fig. 16.22a, separated by some distance, substantially reduces this capacitance.

16.8 Switching Compensator Control in Terms of CPC

Shunt switching compensators are commonly controlled using the Instantaneous Reactive Power (IRP) p-q Theory, developed by Akagi, Kanazawa, and Nabae [72] in 1983, and there is a lot of papers on such control.

Unfortunately, as was demonstrated in papers [201, 242, 245, 269] and discussed in Chapter 21, the IRP p-q Theory misinterprets power phenomena in three-phase circuits. Namely, it claims that the instantaneous power p after compensation should be constant and the compensator control signal should be shaped in such a way that this power indeed is constant. This is not the right goal of compensation, however. In the presence of the supply voltage distortion, the instantaneous power p of even an ideal resistive, balanced load, that is, the load with unity power factor, is not constant. The same is true when the supply voltage, even sinusoidal, is no longer symmetrical. This major misinterpretation of power properties, suggested by the IRP p-q Theory, means that when this theory is used for generating the reference signal for the control of a switching compensator, which operates at distorted or/and asymmetrical supply voltage, this signal is erroneous. Instead of improving, this compensator can degrade the power factor.

As demonstrated in Section 7.18, the harmonic generating loads in three-phase circuits supplied with nonsinusoidal voltage can be decomposed into the Currents' Physical Components (CPC)

$$\boldsymbol{i}(t) = \boldsymbol{i}_{\mathrm{Ca}}(t) + \boldsymbol{i}_{\mathrm{Cs}}(t) + \boldsymbol{i}_{\mathrm{Cr}}(t) + \boldsymbol{i}_{\mathrm{Cu}}(t) + \boldsymbol{i}_{\mathrm{G}}(t).$$

In this decomposition, only the active current $\boldsymbol{i}_{\mathrm{Ca}}(t)$ transfers energy to the load permanently. The remaining ones are harmful. All of them can be compensated by a switching compensator on the condition that the reference signal for the compensator control is

$$\boldsymbol{i}_{\mathrm{ref}}(t) = \boldsymbol{i}_{\mathrm{Cs}}(t) + \boldsymbol{i}_{\mathrm{Cr}}(t) + \boldsymbol{i}_{\mathrm{Cu}}(t) + \boldsymbol{i}_{\mathrm{G}}(t). \tag{16.40}$$

This decomposition of the reference signal has only cognitive merit, however. It says which component in the supply current can be compensated. It is not needed for generating this signal. It can be calculated without the load current decomposition into CPCs, namely, it is equal to

$$\boldsymbol{i}_{\mathrm{ref}}(t) = \boldsymbol{i}(t) - \boldsymbol{i}_{\mathrm{Ca}}(t) \tag{16.41}$$

thus it is enough to calculate the active current $\boldsymbol{i}_{\mathrm{Ca}}(t)$ for that.

We can move to a more advanced goal of compensation. As explained in Section 7.19, in circuits with unbalanced, harmonics generating loads only a part of the active power stands for the useful power. The working active power P_{w} is the only useful component of this power. It is associated with the working active current $\boldsymbol{i}_{\mathrm{w}}(t)$. The reduction of the supply current $\boldsymbol{i}(t)$ to $\boldsymbol{i}_{\mathrm{w}}(t)$ can be regarded as the ultimate objective of compensation. The procedure of such compensation is shown in the following illustration.

Illustration 16.4 *The load in the circuit shown in Fig. 16.23 is composed of an AC/DC converter operated at the firing angle $\alpha = 45^0$ and an unbalanced RL load. The internal voltage $\boldsymbol{e}$ of the supply source is sinusoidal and symmetrical, with the line-to-neutral voltage rms value $U = 230$ V, The supply short-circuit power of the supply source $S_{\mathrm{sc}} = 1.5$ MVA, and $X_s/R_s = 3$.*

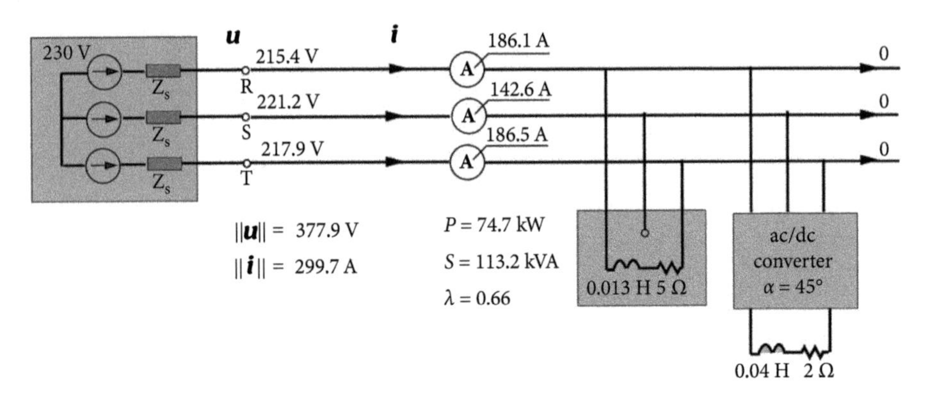

Figure 16.23 A three-phase circuit with unbalanced HGL.

The circuit modeling, with the assumption that the AC/DC converter is lossless, results in the rms values of the supply voltages and currents shown in Fig. 16.23. The waveforms of the voltage $u_{\mathrm{R}}(t)$ and line currents $i_{\mathrm{R}}(t)$ and $i_{\mathrm{S}}(t)$ are shown in Fig. 16.24.

The crms values of the line-to-ground voltages u_{R} and u_{S} fundamental harmonic are

$$\boldsymbol{U}_{\mathrm{R1}} = 210.6\, e^{-j91.6^{\circ}}\ \mathrm{V}$$

$$\boldsymbol{U}_{\mathrm{S1}} = 216.1 e^{-j211.7^{\circ}}\ \mathrm{V}$$

so that the crms value of the working voltage, using Formula (3.84) and (3.85), is

$$\boldsymbol{U}_{\mathrm{w}} = U_{\mathrm{w}}\, e^{j\alpha_{\mathrm{w}}} \stackrel{\mathrm{df}}{=} \boldsymbol{U}_1^{\mathrm{p}} = \frac{1}{3}\,[1,\ \alpha,\ \alpha^*] \begin{bmatrix} \boldsymbol{U}_{\mathrm{R1}} \\ \boldsymbol{U}_{\mathrm{S1}} \\ \boldsymbol{U}_{\mathrm{T1}} \end{bmatrix} = \frac{1}{\sqrt{3}} \left[1 e^{j30^{\circ}},\ 1 e^{j90^{\circ}}\right] \begin{bmatrix} \boldsymbol{U}_{\mathrm{R1}} \\ \boldsymbol{U}_{\mathrm{S1}} \end{bmatrix} =$$

$$= \frac{1}{\sqrt{3}} \left[1 e^{j30^{\circ}},\ 1 e^{j90^{\circ}}\right] \begin{bmatrix} 210.6\, e^{-j91.6^{\circ}} \\ 216.1 e^{-j211.8^{\circ}} \end{bmatrix} = 213.2 e^{-j92.1^{\circ}}\ \mathrm{V}.$$

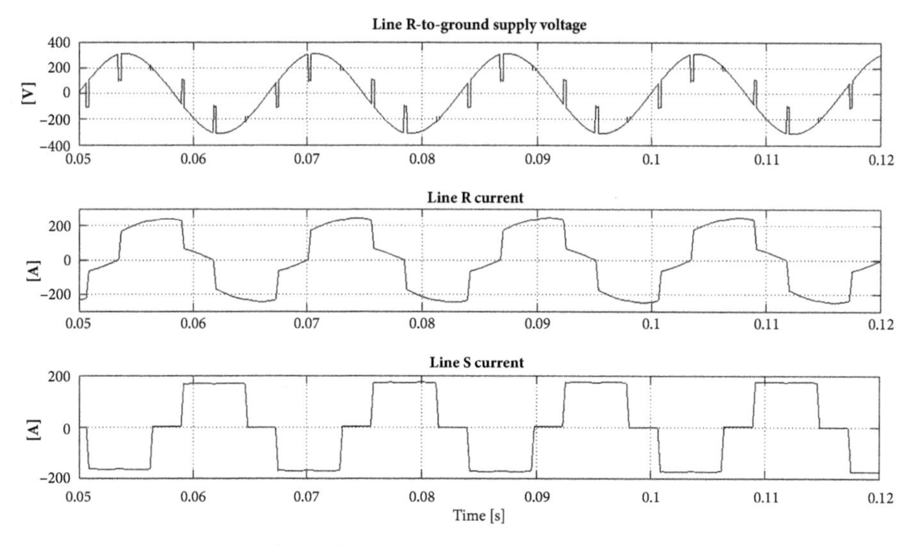

Figure 16.24 The load voltage u_R, line currents i_R, and i_S waveforms.

Similarly, the crms values of the line currents i_R *and* i_S *fundamental harmonic are equal to*

$$\boldsymbol{I}_{R1} = 182.2\, e^{-j145.3^0}\ \text{A}$$

$$\boldsymbol{I}_{S1} = 133.3\, e^{-j257.1^0}\ \text{A}$$

so that the symmetrical component of the positive sequence of the load current fundamental harmonic is

$$\boldsymbol{I}_1^{\text{p}} = \frac{1}{\sqrt{3}}\left[1e^{j30^\circ},\ 1e^{j90^\circ}\right]\begin{bmatrix} I_{R1} \\ I_{S1} \end{bmatrix} = \frac{1}{\sqrt{3}}\left[1e^{j30^\circ},\ 1e^{j90^\circ}\right]\begin{bmatrix} 182.2\, e^{-j145.3^\circ} \\ 133.3\, e^{-j257.1^\circ} \end{bmatrix}$$

$$= 166.3 e^{-j137.2^\circ}\ \text{A}.$$

The working conductance of the load, defined by (7.140), is equal to

$$G_{\text{W}} = \frac{P_1^{\text{P}}}{||\boldsymbol{u}_1^{\text{P}}||^2} = \text{Re}\left\{\frac{\boldsymbol{U}_1^{\text{P}}\boldsymbol{I}_1^{\text{P}^*}}{(U_1^{\text{P}})^2}\right\} = \frac{I_1^{\text{P}}\cos\varphi_1^{\text{P}}}{U_1^{\text{P}}} = \frac{166.3}{213.2}\cos(45.1^\circ) = 0.550\ \text{S}.$$

The vector of the working voltage has the waveform

$$\boldsymbol{u}_{\text{w}}(t) = \begin{bmatrix} u_{\text{wR}}(t) \\ u_{\text{wS}}(t) \\ u_{\text{wT}}(t) \end{bmatrix} = \sqrt{2}\text{Re}\left\{\begin{bmatrix} 1 \\ \alpha^* \\ \alpha \end{bmatrix} \boldsymbol{U}_{\text{w}}\, e^{j\omega_1 t}\right\} = 213.2\sqrt{2}\begin{bmatrix} \sin(\omega_1 t - 2.1^\circ) \\ \sin(\omega_1 t - 122.1^\circ) \\ \sin(\omega_1 t - 242.1^\circ) \end{bmatrix}\ \text{V}.$$

The vector of the working current waveform is

$$\boldsymbol{i}_{\mathrm{w}}(t) = \begin{bmatrix} i_{\mathrm{wR}}(t) \\ i_{\mathrm{wS}}(t) \\ i_{\mathrm{wT}}(t) \end{bmatrix} = G_{\mathrm{w}}\,\boldsymbol{u}_{\mathrm{w}}(t) = 117.3\sqrt{2}\begin{bmatrix} \sin(\omega_1 t - 2.1^\circ) \\ \sin(\omega_1 t - 122.1^\circ) \\ \sin(\omega_1 t - 242.1^\circ) \end{bmatrix}\ \mathrm{A}.$$

Reduction of the supply current to this working current is the objective of compensation in this illustration. To reach this objective, the reference signal should have the form

$$\boldsymbol{i}_{\mathrm{ref}}(t) = \boldsymbol{i}(t) - \boldsymbol{i}_{\mathrm{w}}(t) = \boldsymbol{i}(t) - G_{\mathrm{w}}\,\boldsymbol{u}_{\mathrm{w}}(t).$$

This formula only presents the concept of control. Indeed, the reference signal has to be worked out by a DSP system, which receives information on the load voltages and currents from AD sampling converters in discrete instants of time $t_k = kT/K$, where T is the supply voltage period and K is the number of samples per period. So that the DSP has to calculate the value of the reference signal,

$$\boldsymbol{i}_{\mathrm{ref}}(t_k) = \boldsymbol{i}_{\mathrm{ref}}(k) = \boldsymbol{i}(k) - \boldsymbol{i}_{\mathrm{w}}(k) = \boldsymbol{i}(k) - G_{\mathrm{w}}\,\boldsymbol{u}_{\mathrm{w}}(k).$$

This means that the DSP system has to calculate the working conductance G_{w}

$$G_{\mathrm{w}} = \frac{P_1^{\mathrm{p}}}{\|\boldsymbol{u}_1^{\mathrm{p}}\|^2} = \mathrm{Re}\frac{\boldsymbol{U}_1^{\mathrm{p}}\boldsymbol{I}_1^{\mathrm{p}*}}{(U_1^{\mathrm{p}})^2} = \frac{I_1^{\mathrm{p}}}{U_1^{\mathrm{p}}}\cos\varphi_1^{\mathrm{p}}$$

and the instantaneous value of the working voltage at instant t_k for that.

The crms value of the fundamental harmonic of a periodic sequence of $x_m = x(m)$ of discrete samples at the instant t_k can be calculated with the Discrete Fourier Transform (DFT)

$$\boldsymbol{X}_1(k) = \frac{\sqrt{2}}{K}\sum_{m=k-K+1}^{m=k} x_m\, e^{-j\frac{2\pi}{K}m}. \tag{16.42}$$

When the complex values of the exponential function

$$\exp(m) = \frac{\sqrt{2}}{K}e^{-j\frac{2\pi}{K}m} = \frac{\sqrt{2}}{K}\cos\left(\frac{2\pi}{K}m\right) - j\frac{\sqrt{2}}{K}\sin\left(\frac{2\pi}{K}m\right)$$

are stored in a look-up table, then 2K multiplications are needed for calculating the crms value of the fundamental harmonic. This number can be reduced by a recursive calculation. The samples taken in period T have to be stored for that in a buffer of the

DFT system for such recursive calculations. Having these samples stored, the formula (16.42) can be rearranged to a recursive form

$$\mathbf{X}_1(k) = \frac{\sqrt{2}}{K}\sum_{m=k-K+1}^{m=k} x_m\, e^{-j\frac{2\pi}{K}m} = \frac{\sqrt{2}}{K}\sum_{m=k-K}^{m=k-1} x_m\, e^{-j\frac{2\pi}{K}m} + (x_k - x_{k-K})\frac{\sqrt{2}}{K}e^{-j\frac{2\pi}{K}k} =$$

$$= \mathbf{X}_1(k-1) + (x_k - x_{k-K})\frac{\sqrt{2}}{K}e^{-j\frac{2\pi}{K}k} = \mathbf{X}_1(k-1) + (x_k - x_{k-K})\mathbf{W}_1(k). \tag{16.43}$$

If the real and imaginary parts of the complex coefficient

$$\mathbf{W}_1(k) = \frac{\sqrt{2}}{K}e^{-j\frac{2\pi}{K}k} = \frac{\sqrt{2}}{K}\cos\left(\frac{2\pi}{K}k\right) - j\frac{\sqrt{2}}{K}\sin\left(\frac{2\pi}{K}k\right) \tag{16.44}$$

are stored in a look-up table, then only two multiplications of real numbers are needed for updating $\mathbf{X}_1(k-1)$ *to* $\mathbf{X}_1(k)$. *Moreover, when the sequence* x_m *is periodic then for each k*

$$x_k = x_{k-K}$$

and consequently, $\mathbf{X}_1(k) = \mathbf{X}_1(k-1)$. *The sequential crms values have to be updated only if the periodicity of voltages and currents in the compensated circuit is disturbed. Recursive calculations can accumulate the rounding error, however. Therefore, to reduce this error, random rounding of sequential recursive results in needed.*

The number of samples per period K should be sufficiently high to avoid aliasing error. It should be two times higher than the highest order harmonic. It was assumed in this illustration that harmonics of the order higher than $n = 15$ *can be ignored so* $K = 32$. *Thus, to calculate the crms values* $\mathbf{U}_{R1}$, $\mathbf{U}_{S1}$, $\mathbf{I}_{R1}$, *and* $\mathbf{I}_{S1}$ *in a recursive procedure, four buffers for thirty-two samples of* u_{Rk}, u_{Sk}, i_{Rk}, *and* i_{Sk} *in a binary format, and the look-up table for thirty-two complex values of* $\mathbf{W}_1(\mathrm{k})$ *are needed. Having these, the DSP system provides all crms values* $\mathbf{U}_{R1}$, $\mathbf{U}_{S1}$, $\mathbf{I}_{R1}$, *and* $\mathbf{I}_{S1}$ *with only eight multiplications. Having stored values*

$$\frac{1}{\sqrt{3}}e^{j30^\circ} \text{ and } \frac{1}{\sqrt{3}}e^{j90^\circ}$$

the crms values of the positive sequence symmetrical components of the fundamental harmonic of the load voltages and currents, $\mathbf{U}_1^{\mathrm{p}}$ *and* $\mathbf{I}_1^{\mathrm{p}}$, *can be calculated with only six multiplications. The instantaneous values of the working current at instants* t_k *are equal to*

$$\boldsymbol{i}_{\mathrm{w}}(k) = G_{\mathrm{w}}\, \boldsymbol{u}_{\mathrm{w}}(k) = G_{\mathrm{w}}\sqrt{2}\mathrm{Re}\left\{\left[\begin{array}{c} 1 \\ \alpha^* \\ \alpha \end{array}\right] U_1^{\mathrm{p}}\, e^{-j\frac{2\pi}{K}k}\right\}. \tag{16.45}$$

Having the complex values of the terms

$$G_{\mathrm{W}}\sqrt{2}\,e^{-j\frac{2\pi}{K}k}, \qquad \alpha^* G_{\mathrm{W}}\sqrt{2}\,e^{-j\frac{2\pi}{K}k}, \qquad \alpha G_{\mathrm{W}}\sqrt{2}\,e^{-j\frac{2\pi}{K}k}$$

stored in a look-up table, only six multiplications of real numbers are needed for the calculation of the instantaneous values of the working current $\boldsymbol{i}_{\mathrm{w}}(k)$*. Thus, the presented algorithm enables the calculation of the reference signal for the switching compensator control with a relatively low computational burden. The results of compensation are shown in Fig. 16.25.*

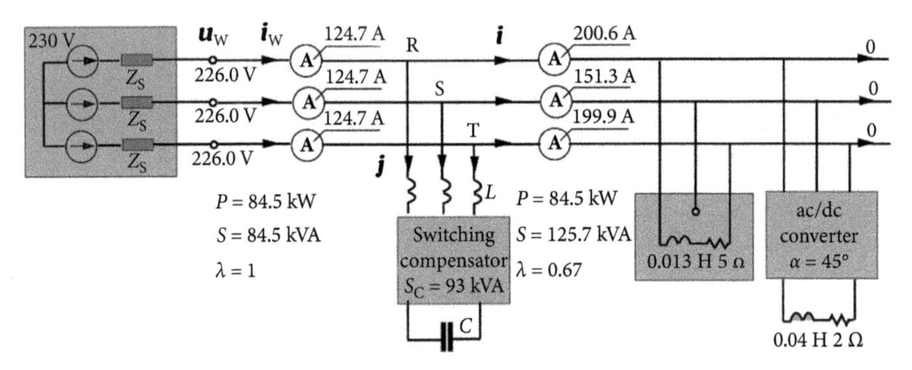

Figure 16.25 The results of compensation.

The compensator reduces to zero all harmful components of the supply currents: the reactive, reflected, and the load-generated harmonic current. Only the working current remains. Observe, that the load active power after compensation, P = 84.5 kW, is higher than before compensation, shown in Fig. 16.23, where it is equal to P = 74.7 kW. It is because the reduction of the supply current rms value by the compensator increases the voltage provided to the load. The waveforms of the compensator currents as well the supply voltage and current of the compensated load are shown in Fig. 16.26.

****

In the illustration above, it was assumed that the supply source voltage was sinusoidal and symmetrical. When these conditions are not satisfied, the reference signal should be generated according to (16.41). It means that the DSP system has to calculate the values of the active current $\boldsymbol{i}_{\mathrm{Ca}}(t)$, defined by (7.115), at instances $t = t_k$ is equal to

$$\boldsymbol{i}_{\mathrm{Ca}}(k) = G_{\mathrm{Cb}}\, \boldsymbol{u}_{\mathrm{C}}(k). \tag{16.46}$$

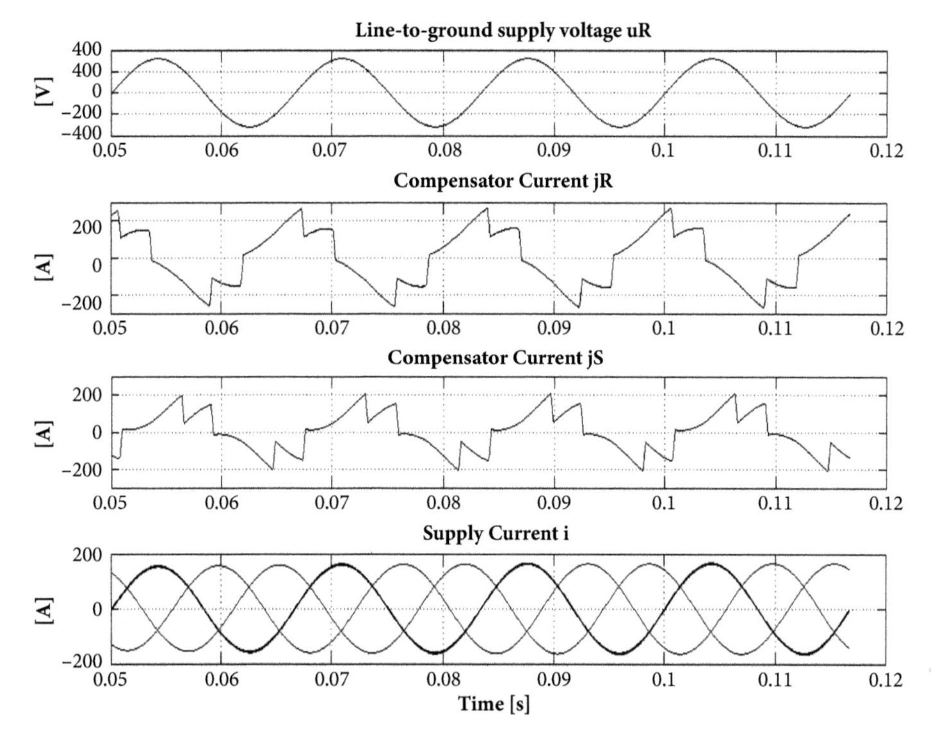

Figure 16.26 Waveforms of the supply voltage u_{R}, compensator currents j_{R}, and j_{S}, as well as the supply currents after compensation in Illustration 16.4.

The conductance G_{Cb}, defined by (7.116), is constant or, when the periodicity of voltages and current is disturbed, a slowly varying parameter, so that its value can be updated once a period.

Identification of the set N_{C} of harmonics requires that the Fourier analysis of the load voltages and currents is performed, and the sign of the harmonic active power is found, as explained in Section 7.12.

The crms value of the nth order harmonic of a periodic sequence of $x_m = x(m)$ of discrete samples at the instant t_k can be calculated with the DFT

$$\boldsymbol{X}_n(k) = \frac{\sqrt{2}}{K} \sum_{m=k-K+1}^{m=k} x_m\, e^{-jn\frac{2\pi}{K}m}. \tag{16.47}$$

When the complex values of the exponential function

$$\exp(nm) = \frac{\sqrt{2}}{K} e^{-jn\frac{2\pi}{K}m} = \frac{\sqrt{2}}{K}\cos\left(n\frac{2\pi}{K}m\right) - j\frac{\sqrt{2}}{K}\sin\left(n\frac{2\pi}{K}m\right)$$

are stored in a look-up table, then $2K$ multiplications are needed for calculating the crms value of a single harmonic. This number can be reduced by a recursive calculation. The samples taken in period T have to be stored for that in a buffer of the DFT system for such recursive calculations. Having these samples stored, the formula (16.47) can be rearranged to a recursive form

$$\boldsymbol{X}_n(k) = \frac{\sqrt{2}}{K}\sum_{m=k-K+1}^{m=k} x_m e^{-jn\frac{2\pi}{K}m} = \frac{\sqrt{2}}{K}\sum_{n=k-K}^{n=k-1} x_m e^{-jn\frac{2\pi}{K}m} + (x - x_{k-K})\frac{\sqrt{2}}{K}e^{-jn\frac{2\pi}{K}k} =$$

$$= \boldsymbol{X}_n(k-1) + (x_k - x_{k-K})\frac{\sqrt{2}}{K}e^{-jn\frac{2\pi}{K}(k)} = \boldsymbol{X}_n(k) + (x_k - x_{k-K})\boldsymbol{W}_n(k). \tag{16.48}$$

If the real and imaginary parts of the complex coefficient

$$\boldsymbol{W}_n(k) = \frac{\sqrt{2}}{K}e^{-jn\frac{2\pi}{K}k} = \frac{\sqrt{2}}{K}\cos\left(n\frac{2\pi}{K}k\right) - j\frac{\sqrt{2}}{K}\sin\left(n\frac{2\pi}{K}k\right) \tag{16.49}$$

are stored in a look-up table, then only two multiplications of real numbers are needed for updating the crms value of the nth order harmonic from $\boldsymbol{X}_n(k-1)$ to $\boldsymbol{X}_n(k)$. Recursive calculations can accumulate the rounding error, however. Therefore, to reduce this error, random rounding of sequential recursive results in needed.

The crms values of harmonics $\boldsymbol{X}_n$ have been averaged over the period T in their definition, so that as long as the voltages and currents remain periodic, meaning the samples taken at the period T distance, do not change, that is,

$$x_k = x_{k-K}$$

the crms values of harmonics do not change as well. Thus, from (16.49) comes the result that

$$\boldsymbol{X}_n(k-1) = \boldsymbol{X}_n(k) = \boldsymbol{X}_n$$

are constant.

The set N_{C} of the harmonic orders n is composed of these orders for which harmonic active power P_n

$$P_n = U_{\mathrm{R}n}I_{\mathrm{R}n}\cos\varphi_{\mathrm{R}n} + U_{\mathrm{S}n}I_{\mathrm{S}n}\cos\varphi_{\mathrm{S}n} + U_{\mathrm{T}n}I_{\mathrm{T}n}\cos\varphi_{\mathrm{T}n} \tag{16.50}$$

is positive. Having values of the cosine function stored in a look-up table, three multiplications are needed for calculating this power. Along with eight multiplications needed for calculating the crms values $\boldsymbol{U}_{\mathrm{R}n}$, $\boldsymbol{U}_{\mathrm{S}n}$, $\boldsymbol{I}_{\mathrm{R}n}$, and $\boldsymbol{I}_{\mathrm{S}n}$, eleven multiplications of real numbers are needed for calculating the active power of each harmonic in the supply voltage in three-phase, three-wire circuits, and hence

$$P_C = \sum_{n \in N_C} P_n. \tag{16.51}$$

The square of the three-phase rms value of the voltage $\boldsymbol{u}_C$ is

$$||\boldsymbol{u}_C||^2 = \sum_{n \in N_C} (U_{Rn}^2 + U_{Sn}^2 + U_{Tn}^2) \tag{16.52}$$

so the conductance G_{Cb}, defined by (7.116), can be calculated. According to (16.46), the active current $\boldsymbol{i}_{Ca}(t)$ at instant t_k is

$$\boldsymbol{i}_{Ca}(k) = G_{Cb}\,\boldsymbol{u}_C(k) = G_{Cb}\sqrt{2}\mathrm{Re}\left\{\sum_{n \in N_C} \begin{bmatrix} U_{Rn} \\ U_{Sn} \\ U_{Tn} \end{bmatrix} e^{-jn\frac{2\pi}{K}k}\right\}. \tag{16.53}$$

Having the complex values

$$G_{Cb}\sqrt{2}\,e^{-jn\frac{2\pi}{K}k} = G_{Cb}\sqrt{2}\,\cos\left(n\frac{2\pi}{K}k\right) - j\,G_{Cb}\sqrt{2}\,\sin\left(n\frac{2\pi}{K}k\right)$$

stored in a look-up table, only six multiplications of real numbers are needed for the calculation of each harmonic component of the active current $\boldsymbol{i}_{Ca}(t)$.

Illustration 16.5 *Let us evaluate the number of multiplications needed for generating the reference signal according to the CPC control, which has to be performed after each sampling at the frequency of* $f_s = 15$ *kHz, and evaluate the time interval available for that. The control algorithm should take into account the possible presence of the fifth, seventh, eleventh, and thirteenth order harmonics.*

The number of needed multiplications after each sampling for calculation of four crms values, formula (16.48), for five harmonics $4 \times 2 \times 5 = 40$ *multiplications. Moreover, the instantaneous value of the active current* $\boldsymbol{i}_{Cb}(k)$ *should be calculated with the formula (7.116), which requires* $3 \times 2 \times 5 = 30$ *multiplications, thus 70 multiplications have to be performed. The time interval available for that is* $\Delta t = 1/f_s = 1/15 \times 10^3 = 66.7$ *μs.*

Additionally, for every period T, the conductance value G_{Cb} *should be updated. It requires that the active power* P_C *with formula (16.50) and (16.51) is calculated, which requires, for five harmonics,* $6 \times 5 = 30$ *multiplications, and the power of the three-phase rms value* $||\boldsymbol{u}_C||^2$*, with formula (16.52), has to be calculated, which requires* $6 \times 5 = 30$ *multiplications, thus all together 60 multiplications per a period T.*

******

This illustration shows that generation of the reference signal according to CPC-based power theory for the switching compensator control is not very demanding as to the number of required calculations.

Summary

Switching compensators provide a simple tool for reducing the reactive and unbalanced currents and the load-originated current harmonics. They have, moreover, natural adaptive properties. Their power is limited to medium power, however, and they can create a high-frequency noise. Such compensators can be controlled using the CPC-based approach to PT with not a very demanding computational burden.

Chapter 17
Hybrid Compensators

17.1 Introduction

As compared to adaptive reactance compensators, switching compensators have relatively low power ratings. This is due to the much lower switching power of power transistors compared to thyristors. It is roughly ten times lower. At the same time, reactance compensators do not have the capabilities of compensation of rapidly varying harmful components of the supply current. Compensators discussed in this book are built as devices connected in parallel to loads for improving the loading quality (LQ). To improve the supply quality (SQ), a compensator has to be connected in series to the load.

Compensators composed of two compensators with different principles of operation or compensation goals are referred to as *hybrid compensators*. However, without more details, the term "hybrid" does not specify the compensator.

17.2 Low-Frequency/High-Frequency Hybrid Compensators

Switching power of transistors which is necessary for switching compensator construction declines with the switching frequency, and this power limits the power ratings of such compensators.

The harmful components of the supply current which are to be reduced by the compensator have different frequencies. The reactive and/or unbalanced currents, which usually dominate the supply current, are mainly of the fundamental frequency. Harmonic components of this current are usually much lower than the reactive and/or unbalanced currents.

The switching frequency needed for shaping the compensating current of the fundamental frequency is much lower than that needed for shaping the high-frequency harmonic compensating currents. Therefore, it is reasonable to compensate the reactive and unbalanced current by an SC of higher power but a low switching frequency and compensate the higher order current harmonics by an SC compensator with a high switching frequency but a reduced power. Such a strategy can enhance the SC compensation capability to higher power loads.

Control of a hybrid compensator composed of a high frequency (HF) sub-compensator and a low frequency (LF) sub-compensator, requires two reference

Powers and Compensation in Circuits with Nonsinusoidal Current. Leszek S. Czarnecki, Oxford University Press.
 DOI: 10.1093/oso/9780198879206.003.0017

signals. The discrete signals processing (DSP) system which generates these signals has to be capable of separating the HF harmful components of the supply current from the components of the fundamental frequency. Such algorithms can be developed only with power theories formulated in the frequency domain. Such a capability has the Currents' Physical Components (CPC)-based power theory. Power theories formulated in the time domain cannot be used for that.

The LF sub-compensator, meaning higher power but lower switching frequency, could be a source of substantial switching noise. This noise, along with the load-generated current harmonics, can be reduced by the HF sub-compensator when these two sub-compensators are connected in the configuration shown in Fig. 17.1.

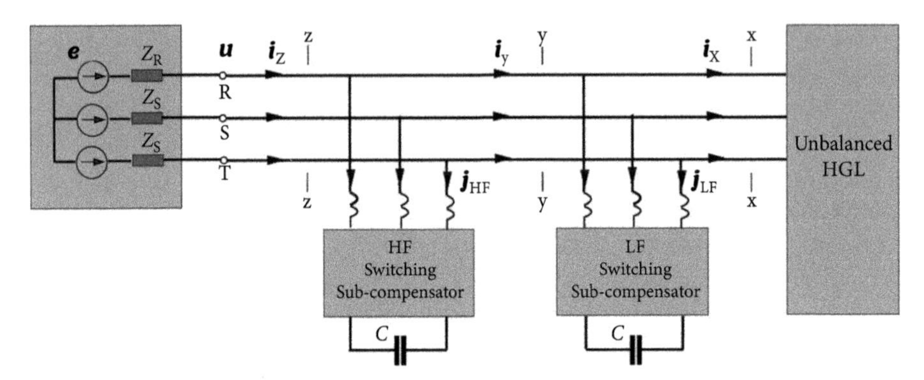

Figure 17.1 Configuration of a hybrid LF/HF compensator.

The LF sub-compensator should reduce the harmful components of the fundamental harmonic in the supply current. When the supply voltage is sinusoidal and symmetrical, then the reduction of the fundamental harmonic $\boldsymbol{i}_1(t)$ to the working current $\boldsymbol{i}_{\rm w}(t)$ could be the compensation objective for the LF sub-compensator. The reference signal for its control should have the form

$$\boldsymbol{i}_{\rm ref}^{\rm LF}(t) = \boldsymbol{i}_1(t) - \boldsymbol{i}_{\rm w}(t). \tag{17.1}$$

At such a reference signal, the LF sub-compensator compensates only the reactive and unbalanced currents of the fundamental harmonic. It does not compensate the load-originated harmonic current $\boldsymbol{i}_{\rm G}$. Instead, the LF sub-compensator is the source of a switching noise current $\boldsymbol{i}_{\rm sn}$ of relatively high magnitude and low frequency. This low frequency makes the noise obstinate for filtering. Thus, the current in the cross-section y–y is

$$\boldsymbol{i}_{\rm y}(t) = \boldsymbol{i}_{\rm yw}(t) + \boldsymbol{i}_{\rm G}(t) + \boldsymbol{i}_{\rm sn}(t) \tag{17.2}$$

Observe that the working current in this cross-section has changed. It is because the LF sub-compensator is a passive device so it consumes some amount of energy.

The switching noise current $\boldsymbol{i}_{\rm sn}$ is not periodic but a random current, so that it is orthogonal to periodic ones, thus the three-phase rms value of the current in the cross-section y–y satisfies the relationship

$$||\boldsymbol{i}_y||^2 = ||\boldsymbol{i}_{yw}||^2 + ||\boldsymbol{i}_G||^2 + ||\boldsymbol{i}_{sn}||^2. \tag{17.3}$$

A reader should be only aware that the orthogonality for non-periodic quantities is defined in a way different from that for periodic ones. Consequently, there is some level of approximation in the last relationship.

Reduction of the supply current fundamental harmonic to the active current $\boldsymbol{i}_{Cb}(t)$ could be the objective of compensation of the LF sub-compensator at nonsinusoidal supply voltage, so that the reference signal should have the form

$$\boldsymbol{i}_{ref}^{LF}(t) = \boldsymbol{i}_1(t) - \boldsymbol{i}_{Cb}(t) \tag{17.4}$$

At such a control, the harmful components of the fundamental frequency will be eliminated from the supply current. Only the HF current components originated in the load and currents caused by the transistors' switching in the LF sub-compensator but not effectively filtered $\boldsymbol{i}_{sn}$ can remain in the supply current. Since the LF sub-compensator consumes some energy, it can change both the working current $\boldsymbol{i}_w(t)$ and the active current $\boldsymbol{i}_{Cb}(t)$. Therefore, the reference signal for the HF sub-compensator should have the form

$$\boldsymbol{i}_{ref}^{HF}(t) = \boldsymbol{i}_y(t) - \boldsymbol{i}_{yw}(t) \tag{17.5}$$

or

$$\boldsymbol{i}_{ref}^{HF}(t) = \boldsymbol{i}_y(t) - \boldsymbol{i}_{yCb}(t). \tag{17.6}$$

Details of such signals calculation at instants t_k, by DSP systems, with only small modifications, are discussed in Section 16.8.

The sub-compensators can have, depending on the designer's discretion, separate data acquisition (DA) and DSP systems, as shown in Fig. 17.2, or these systems are common as shown in Fig. 17.3. In the first case, both sub-compensators are controlled in an open feedback loop so that instability of controlled sub-compensators cannot occur.

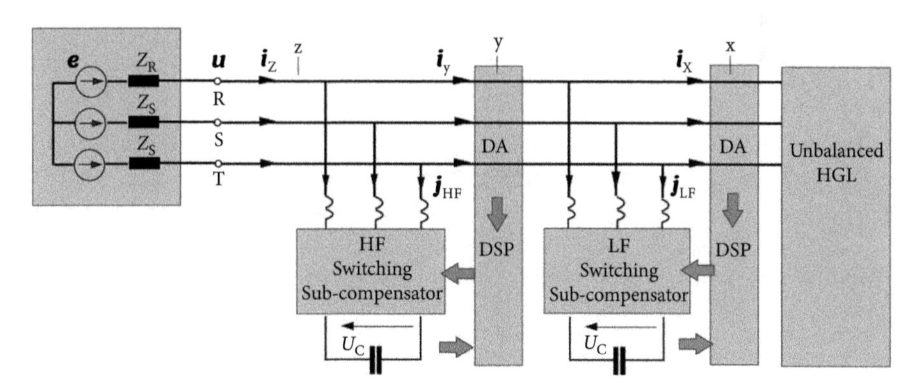

Figure 17.2 A hybrid LF/HF compensator with separate DA and DSP systems.

When DA and DSP systems are located as shown in Fig. 17.3, then an LF sub-compensator is controlled in a closed feedback loop: measurements are performed on the results of compensation by LF sub-compensator. Instability of the LF sub-compensator can occur at such a control.

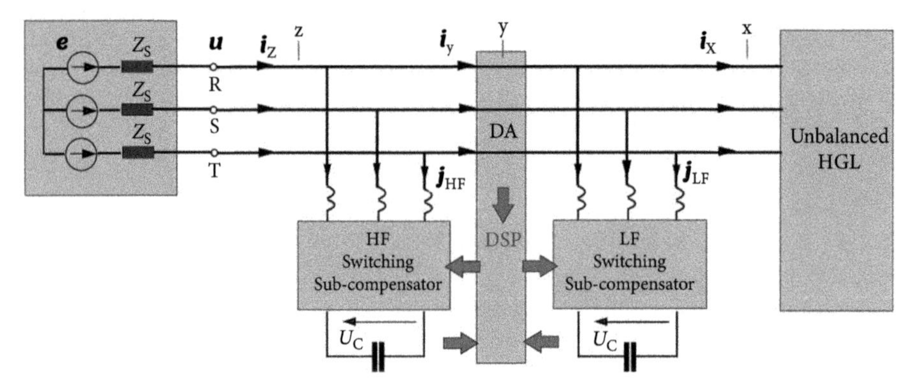

Figure 17.3 A hybrid LF/HF compensator with a common DA and DSP system.

Illustration 17.1 *The load, shown in Fig. 16.23, with line current waveforms shown in Fig. 16.24, which was the subject of compensation by a single SC in Illustration 16.4, is compensated by a hybrid LF/HF compensator with separate DA and DSP systems. The reference signals were generated according to formulae (17.1) and (17.3). The results of compensation are shown in Fig. 17.4.*

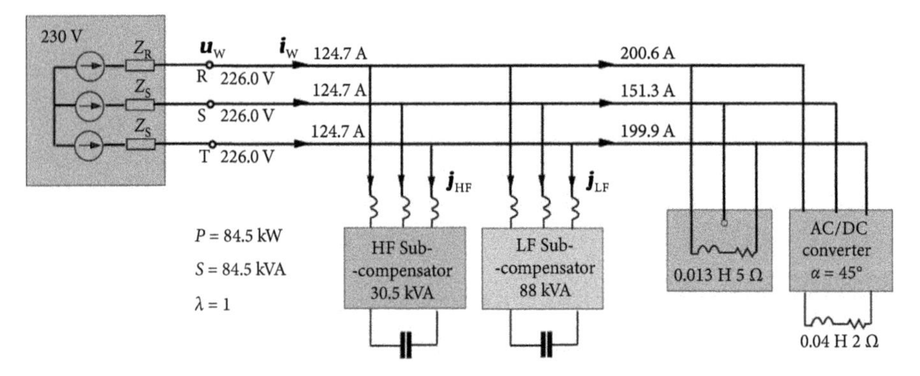

Figure 17.4 Results of compensation of the load are shown in Fig. 16.23, by hybrid LF/HF compensator.

The load voltage, the compensating currents of the LF and HF sub-compensators, and the supply current after compensation are shown in Fig. 17.5.

******

Observe that replacement of a sole SC by a hybrid LF/HF SC compensator substantially reduces the currents which are switched by the SC transistors at high

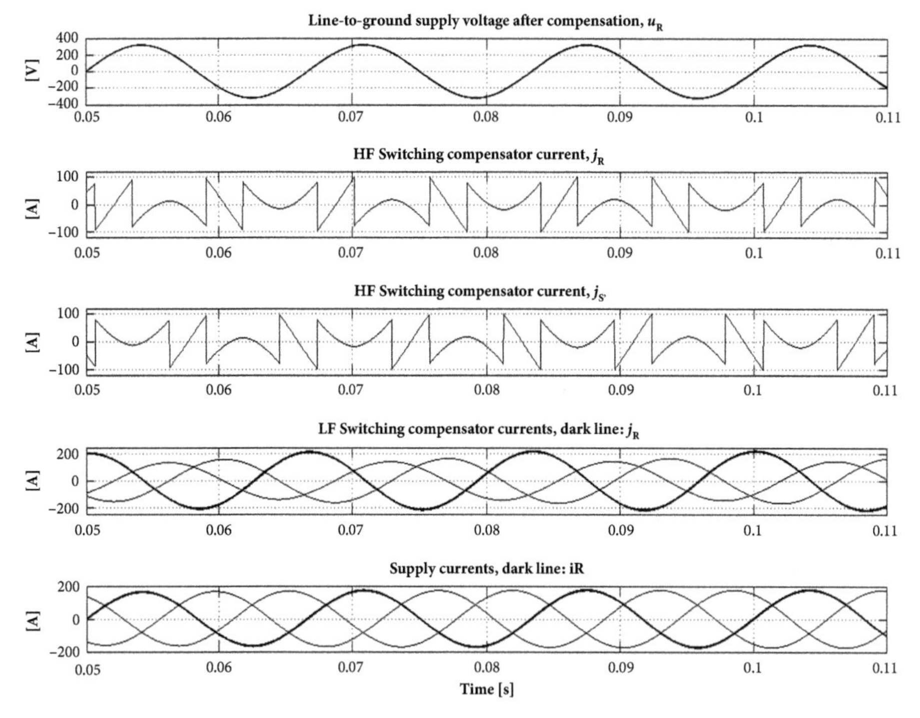

Figure 17.5 Waveforms of the LF and HF sub-compensators currents, as well as the supply voltage and current waveforms after hybrid LF/HF compensation.

frequency, meaning the high-frequency noise. Thus, such a hybrid LF/HF SC compensator has a lower level of electromagnetic interference (EMI) as compared to a sole switching compensator. Each sub-compensators has lower power ratings than the sole SC, which increases compensation capabilities, but this enhancement in the situation as considered in Illustration 17.1 was relatively small. A major enhancement can be obtained by the replacement of the LF switching sub-compensator by a reactance compensator.

17.3 Reactance/HF Switching Hybrid Compensators

When the load power is of the order of tens or hundreds of megavolt amperes (MVA) then such loads can be compensated exclusively by reactance compensators. Such compensators cannot compensate the load-originated harmonic currents, however, and this creates a space for using HF switching compensators for doing that, thus for a Reactance/HF Switching Hybrid Compensator. The harmful currents' major component of the fundamental frequency is compensated by a reactance sub-compensator, while the load-originated current harmonics, which have a much lower level than the fundamental, are compensated by HF switching compensator of the reduced power ratings. A structure of such a Reactance/HF Switching Hybrid Compensator is shown in Fig. 17.6.

When the load power changes in a wide range, the reactance sub-compensator can be built as an adaptive device with thyristor-controlled susceptance (TCS) branches, as discussed in Section 15.6. When the load power does not change in a wide range then the lack of complete compensation can be economically acceptable, however, so that the reactance sub-compensator can be built as a fixed parameter device.

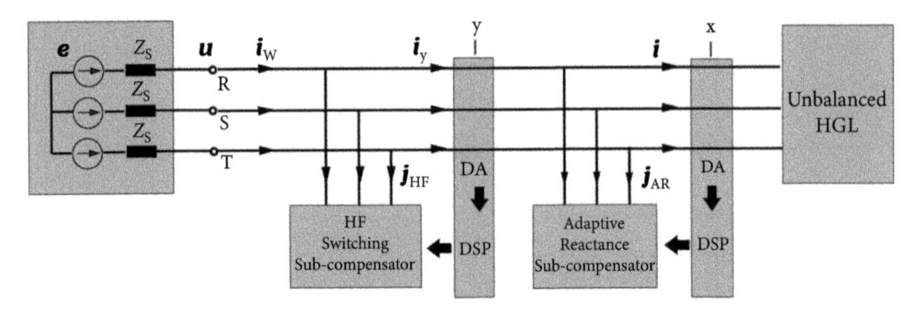

Fig. 17.6 Structure of a hybrid compensator.

Sub-compensators could have separate DA and DSP systems, as is shown in Fig. 17.6, or combined. Since the voltage and current instrumentation transformers are expensive components of the compensator, a decision on common DA and DSP systems can reduce the compensator cost. When the reactance sub-compensator is built as a fixed parameters device, the control circuit for it is not needed.

Since the current harmonics are to be compensated by the HF sub-compensator, the reactance compensator should compensate the load only at the fundamental frequency. It simplifies its design. Only the crms values of two lines voltages and two lines currents of the fundamental harmonic $\boldsymbol{U}_{R1}$, $\boldsymbol{U}_{S1}$, $\boldsymbol{I}_{R1}$, and $\boldsymbol{I}_{S1}$ are needed for that. When the reactance compensator is to be a fixed parameter device, then a single measurement of these values provides sufficient data for calculating the compensator parameters, as was explained in Section 14.2.

When the reactance sub-compensator is to be an adaptive device built of branches with thyristor-controlled susceptance, discussed in Section 15.6, then the presence of the HF sub-compensator allows for some reduction of expectation as to harmonics attenuation by such branches. The series inductor can be ignored, so that the TCS branch could have the structure as shown in Fig. 14.25(a).

The adaptive reactance sub-compensator has to be equipped, of course, with the DA and DSP systems. The firing angle of thyristors of TCS branches cannot be changed more often than three times in the single period T of the supply voltage, so that there is plenty of time for calculation of crms values $\boldsymbol{U}_{R1}$, $\boldsymbol{U}_{S1}$, $\boldsymbol{I}_{R1}$, and $\boldsymbol{I}_{S1}$ needed for sub-compensator control, especially if the DFT, which specifies these values, is calculated using the iterative formula (16.43).

The line current in cross-section y–y, denoted by $\boldsymbol{i}_{\mathrm{y}}$, is reduced as compared to the load current $\boldsymbol{i}$ by the reactive and unbalanced currents of the fundamental frequency $\boldsymbol{i}_{\mathrm{r1}}$ and $\boldsymbol{i}_{\mathrm{u1}}$ but enlarged by current harmonics generated by thyristors switching in TCS branches. The energy loss in these branches increases also the working current $\boldsymbol{i}_{\mathrm{w}}$. When the reactance sub-compensator is a fixed parameters device, then the current $\boldsymbol{i}_{\mathrm{y}}$ can also contain some reactive and unbalanced currents but of substantially reduced value. All harmful components of the supply current are compensated by the HF sub-compensator if the reference signal for its control is equal to

$$\boldsymbol{i}_{\mathrm{ref}}^{\mathrm{HF}}(t) = \boldsymbol{i}_{\mathrm{y}}(t) - \boldsymbol{i}_{\mathrm{yw}}(t) \tag{17.7}$$

It means that the DSP system, having data provided by the DA system, should calculate, according to formula (16.45), the values of the working current at sequential instants t_k, meaning the instantaneous values $\boldsymbol{i}_{\mathrm{yw}}(t)$.

17.4 Hybrid Compensators of Ultra-High Power Loads

The highest power loads, such as manufacturing plants, refineries, or metallurgic plants, do not have three-phase three-wire (3p3w) distribution systems but three-phase circuits with a neutral conductor, meaning three-phase four-wire (3p4w) distribution systems. Such circuits are supplied from Δ/Y transformers with a grounded neutral.

The level of power, the supply quality, and the configuration of such loads impose some conditions upon their compensation. Such ultra-high power loads are often sole loads with dedicated high power supply lines. Occasionally, even with dedicated power plants for their supply. Because of that such loads could be immune to disturbances from other loads. Consequently, the supply quality concerning the supply voltage distortion and asymmetry of such ultra-high power plants could be very high.

To confine the load voltage rms variation with the change of the load power to usually no more than 5%, the supply transformers in electrical distribution systems should have power ratings at least twenty times higher than the load power. Transformers in common distribution systems have relative power ratings much higher than that. It is difficult, however, to satisfy this expectation due to the availability or cost of such transformers, when the load power reaches hundreds or millions of MVA. Therefore, the supply transformers for ultra-high power loads could have relatively low power and consequently relatively high impedance. The voltage drop by the load-originated current harmonics and unbalanced current on such transformer's impedance could therefore be relatively high and be the main source of the SQ degradation in the distribution system inside of such plants. Additionally, a substantial amount of energy could be lost in the transformer that supplies such an ultra-high power load.

Therefore, to improve the SQ inside such a plant, reduce the energy loss, and even to reduce the demand for the main transformer power ratings, a compensator located as close as possible to secondary terminals of the supply transformer is needed. Only a reactance/HF switching hybrid compensator can be used for that.

As was discussed in Chapter 15, reactance compensators in three-phase circuits with neutral, meaning in 3p4w circuits, have to be composed of two sub-compensators, one configured in a grounded Y, and the second in Δ. It compensates reactive, zero, and the negative-sequence symmetrical components of the unbalanced components of the supply current fundamental harmonic. The load and the reactance compensator-originated harmonics are compensated by the HF switching compensator. The structure of the circuit and the compensator are shown in Fig. 17.7.

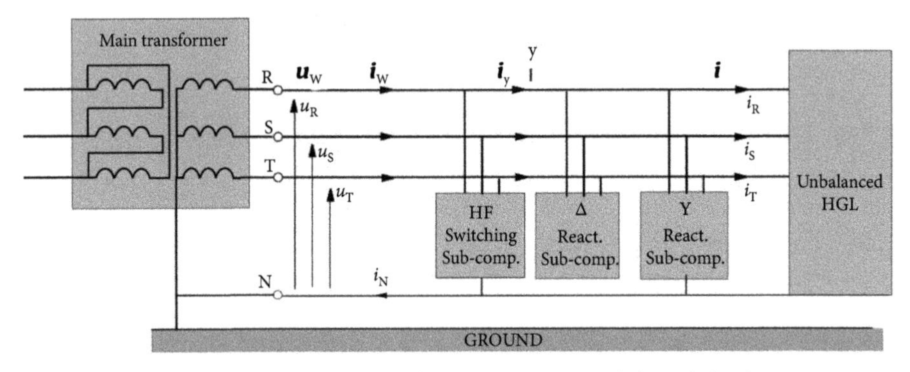

Fig. 17.7 Structure of a 3p4w circuit with a reactance/HF switching hybrid compensator.

The supply voltage distortion and asymmetry as observed at the main transformer secondary side in the considered circuits are largely the results of the voltage drop on the transformer by the load-originated current harmonics and the unbalanced current. This drop will decline with compensation, therefore it is reasonable to assume at the compensator design that the distribution voltage, due to reasons explained before, is sinusoidal and symmetrical. Therefore, the parameters of the reactance sub-compensator can be calculated, as demonstrated in Section 15.3 for circuits with a sinusoidal supply voltage. Having these parameters calculated, the reactance sub-compensator can operate as a fixed-parameters sub-compensator, or it can be converted, as shown in Section 15.6, into an adaptive sub-compensator. In the first situation, DA and DSP systems are not needed. In the second situation, the reactance sub-compensator has to be equipped with the DA and DSP systems which would be able to update the thyristors' firing angle. This angle for each of six TSC branches of the reactance Δ/Y sub-compensator can be updated no more often than once a period T.

The reactance sub-compensator reduces the line current in cross-section y by the reactive current, the zero sequence and negative sequence symmetrical components of the unbalanced current, but increases it by a harmonic current $\boldsymbol{i}_T(t)$ generated in

TSC branches by thyristors switching, and some active current due to energy loss in these branches, $\boldsymbol{i}_{\mathrm{Ta}}(t)$, namely

$$\boldsymbol{i}_{\mathrm{y}}(t) = \boldsymbol{i}(t) - [\boldsymbol{i}_{\mathrm{r}}(t) + \boldsymbol{i}_{\mathrm{u}}^{\mathrm{z}}(t) + \boldsymbol{i}_{\mathrm{u}}^{\mathrm{n}}(t)] + \boldsymbol{i}_{\mathrm{T}}(t) + \boldsymbol{i}_{\mathrm{Ta}}(t) \tag{17.8}$$

This current can be reduced to the working component if the reference signal of the HF switching sub-compensator is equal to

$$\boldsymbol{i}_{\mathrm{ref}}^{\mathrm{HF}}(t) = \boldsymbol{i}_{\mathrm{y}}(t) - \boldsymbol{i}_{\mathrm{yw}}(t).$$

Its value at instants t_k can be calculated as explained in Section 16.8.

Such a hybrid reactance/HF switching compensator is capable of restoring sinusoidal and symmetrical supply voltage in the plant, thus improving the supply quality and reducing energy loss at its delivery.

17.5 Compensation of Highly Variable Loads

The load power is usually a slowly varying quantity. Sometimes it is constant or it changes slowly in minutes or even hours. There are machines with continuously changing power in a wide range, however. Rock crushers or mills are examples of these kinds of machines. The power of a rock crusher changes with the random size and/or hardness of the crushed rock. Very high-power squeezers in metallurgic plants are other examples of loads with high power variability. These can be referred to as *Fast Variable Loads* (FVLs).

High variability of the load power can affect the voltage in the distribution system, causing disturbance of other loads. Thus, such loads degrade the SQ of the distribution system. It also causes high variability of the supply current rms value, meaning such loads have low loading quality (LQ). This variability increases the energy loss at its delivery. It is minimum when energy is delivered at the supply current constant rms value and grows up with the rms value variations increase. To improve the LQ and SQ compensation of such loads can be required.

The reactive and unbalanced currents of such a load can be compensated by an adaptive reactance compensator even at high variability of the rms value of these currents. With an increase of power variability, only the accuracy of such a reactance compensation can decline. The question is whether the variability of the active component of the supply current of highly variable load can be reduced or not by a compensator. It certainly cannot be a reactance compensator. It cannot change the active current. A switching compensator is needed for that.

From the power perspective, the main difference of highly variable loads from the loads discussed previously is the variability of the active power P. An example of such variability is shown in Fig. 17.8.

Compensation is always associated with some flow of energy between the distribution system and the compensator. A much higher capability of energy storage by the compensator is needed at a compensation of FVLs.

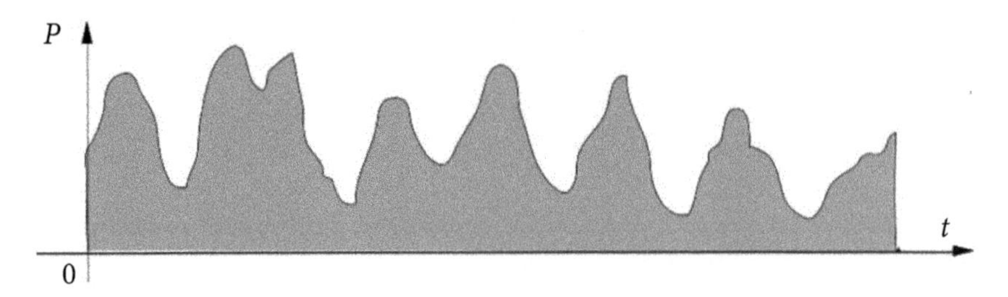

Figure 17.8 An example of active power variation of an FVL.

When an SC is used for reactive current compensation, energy storage is not needed because, as shown in Section 7.7, the presence of the reactive current does not cause any oscillations of energy. Such oscillations occur at a compensation of the unbalanced current so that the compensator should be able of storing energy transferred to the compensator. Some oscillations of energy occur also at a compensation of the load-originated current harmonics, but the capacitor of the SC compensator usually provides a sufficient capability for energy storage, especially that the average value, over period T, of energy transferred by the unbalanced current and current harmonics is zero. It is not the same with the active current, however. The SC can generate this current for reducing its rms value variability in the supply lines only if a substantial amount of energy can be stored and released by the compensator.

There are several electrical devices dedicated to a large amount of energy storage/release operations. To use them in compensators of highly variable loads, such devices should be able to switch between storage and release modes in a time interval of the order of a single period of the voltage variation T, or no more than only a few of such periods. A flywheel energy storage/recovery device, of the structure shown in Fig. 17.9, can be used for that.

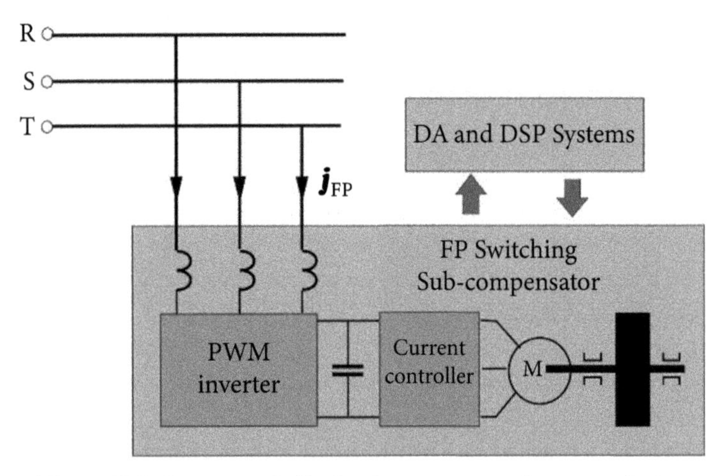

Figure 17.9 Structure of a flywheel energy storage/recovery device.

The energy is stored in such a device is stored as kinetic energy of a disk that rotates at high speed. The disk is driven by a synchronous three-phase machine, usually with a permanent magnet, which operates in motoring/ generation mode controlled by a switching mode inverter.

The idea of the active current rms value variability reduction is simple. When the load active power is below some median value, the electric energy supplied from the distribution circuit is stored as kinetic energy of the flywheel. The synchronous machine operates in the motoring mode and speeds up the flywheel. When this power is higher than the median value, the load is supplied with energy recovered from this flywheel. It drives the synchronous machine which operates as a generator. The source operates as if it is burdened with the median active power as an effect of this kind of compensation.

Voltages and currents in circuits with highly variable loads could be non-periodic but can be regarded as semi-periodic quantities as discussed in Chapter 4.

By analogy to the running active power in a single-phase circuit, defined as

$$\tilde{P} = \frac{1}{T}\int_{t-T}^{t} u(t)i(t)dt$$

the running active power in three-phase circuits

$$\tilde{P} = \frac{1}{T}\int_{t-T}^{t} \mathbf{u}^{\mathrm{T}}(t)\mathbf{i}(t)\, dt$$

can be defined. A *median active power* is the average running active power calculated over an interval T_{a} larger than the period T, namely

$$\overline{\overline{P}} = \frac{1}{T_{\mathrm{a}}}\int_{t-T_{\mathrm{a}}}^{t} \tilde{P}(t)\, dt \tag{17.9}$$

Observe that a double bar was used to denote the medium active power. The difference between these two active powers

$$\tilde{P}_{\Delta} = \tilde{P} - \overline{\overline{P}} \tag{17.10}$$

is illustrated in Fig. 17.10.

The supply current of highly variable loads contains, apart from other current physical components, a *fluctuating component of the active current*,

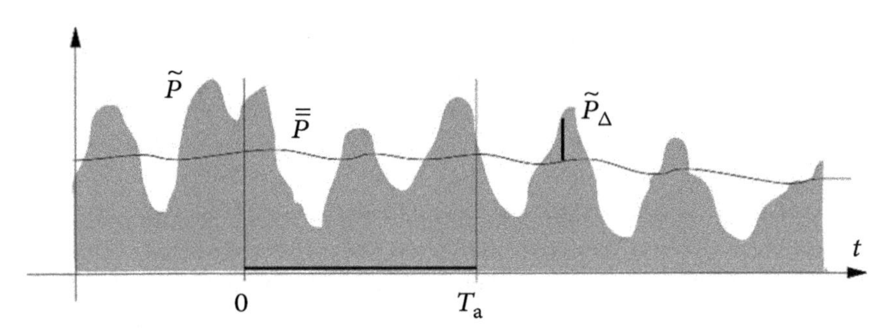

Figure 17.10 An example of active running power variation and its deviation from the median active power.

$$\boldsymbol{i}_{\rm Fa}(t) \stackrel{\rm df}{=} \boldsymbol{i}_{\rm a}(t) - \boldsymbol{i}_{\rm Ma}(t) \tag{17.11}$$

where

$$\boldsymbol{i}_{\rm Ma}(t) \stackrel{\rm df}{=} G_{\rm M}\,\boldsymbol{u}(t), \quad G_{\rm M} = \frac{\bar{\bar{P}}}{||\boldsymbol{u}||^2} \tag{17.12}$$

is the active current of a load equivalent to the original load but with medium active power.

The fluctuating component $\boldsymbol{i}_{\rm Fa}$ can be compensated, along with the reactive and unbalanced currents of the fundamental frequency by a *fluctuating power* (FP) *switching sub compensator* with the reference signal

$$\boldsymbol{i}_{\rm ref}^{\rm FP}(t) = \boldsymbol{i}_1(t) - \boldsymbol{i}_{\rm Ma}(t). \tag{17.13}$$

Such a compensator should of course have sufficient capability for energy storage. It should be able to store energy with the power P_s at least equal to $(\tilde{P}_\Delta)_{\rm max}$.

The current harmonics generated in the load and the FP sub-compensator can be compensated by an HF switching sub-compensator with the reference signal

$$\boldsymbol{i}_{\rm ref}^{\rm HF}(t) = \boldsymbol{i}_{\rm y}(t) - \boldsymbol{i}_{\rm yMa}(t)$$

calculated by a DSP system based on measurements performed by a DA system in cross-section y.

Illustration 17.2 *The algorithm is applied for an illustration to a hybrid switching compensator of a three-phase highly variable load, composed of a six-pulse AC/DC converter which provides controlled DC voltage to a motor drive, and a single-phase RL load, as shown in Fig. 17.11. The line-to-ground distribution voltage rms value is* $U_{\rm R} = 230$ V. *The short circuit power of the distribution system* $S_{\rm sc} = 1.5$ MVA *and* *the distribution system reactance to resistance ratio for the fundamental frequency is $X_s/R_s = 3$.*

The state of the circuit cannot be specified, of course, in terms of powers, the rms values of line currents, and voltages, since they change in time. Records of their variability are needed for that.

Results of the circuit modeling are shown in Fig. 17.12. The figure shows the variability of line R-to-ground voltage, line R and S currents, and running active power. The variation of the median active power at the assumption that the median is calculated for twelve periods of the supply voltage and shown in the bottom plot.

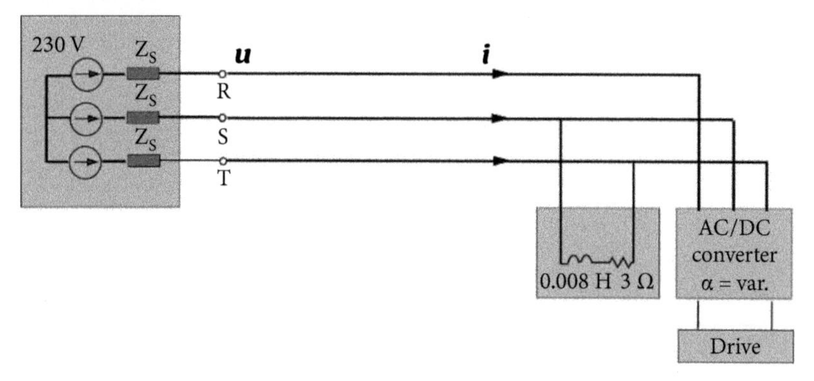

Figure 17.11 An example of a circuit with a highly variable load.

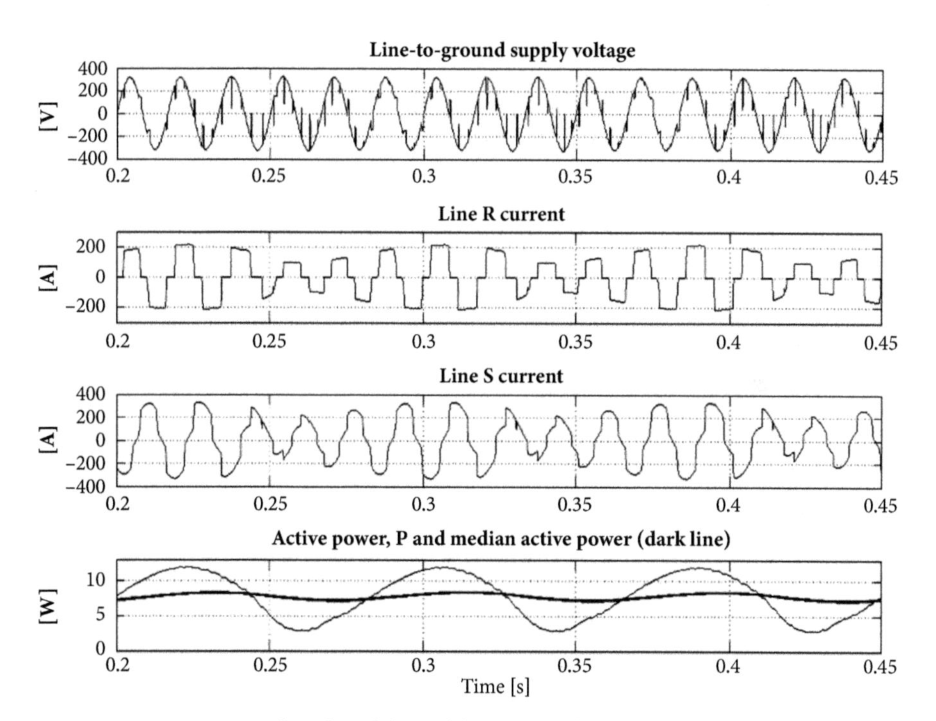

Figure 17.12 Results of modeling of the circuit which is shown in Fig. 17.11.

A circuit with the hybrid compensator, composed of the FP switching sub-compensator and the HF switching compensator, is shown in Fig. 17.13.

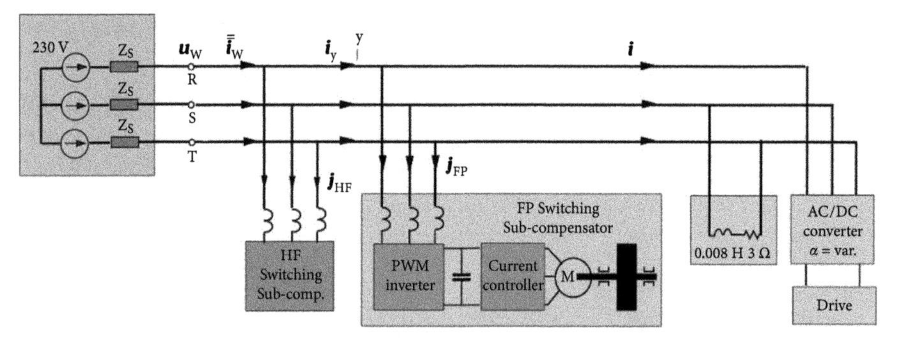

Figure 17.13 A circuit with a hybrid switching compensator.

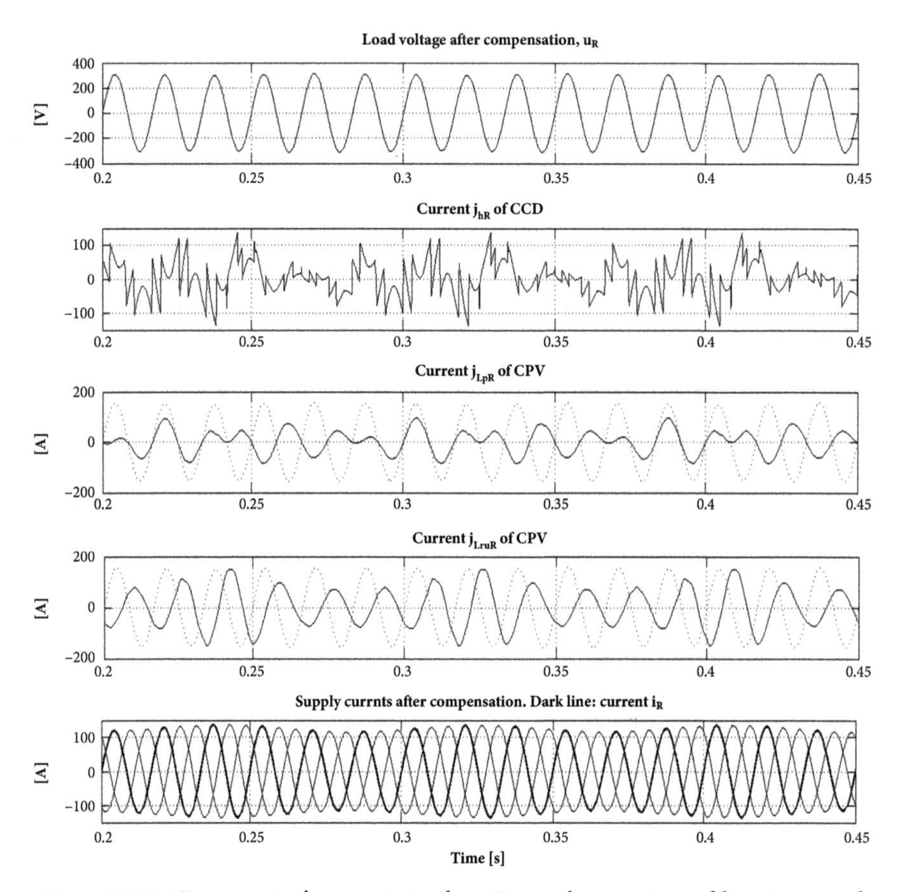

Figure 17.14 Compensator's currents in phase R, supply currents, and line-to-ground voltage after compensation and the supply current.

Reference signals for the compensator of the current distortion (CCD) and for the compensator of the power variation (CPV), at the assumption that reference signals are reproduced ideally, as well as results of compensation in respect to the supply currents and the load voltage, are shown in Fig. 17.14. For clarity's sake, the compensator currents for only line R are shown in this figure.

Some variability of the current amplitude in the bottom figure is visible. It remains because the median value of the load active power was calculated for only twelve periods of the supply voltage.

******

Summary

Hybrid compensation can substantially extend the possibility of compensation of electrical loads. When a switching compensator is combined in a hybrid structure with a reactance compensator, then the power of compensated loads could be several times higher than at simple switching compensation. Since the cost of thyristors and associated electronic drivers could be the main component of the compensator's cost, hybrid compensation provides a path for this cost reduction. At the same time, switching compensators enable compensation of the load features that cannot be handled by simple reactance compensators.

PART C
CONTROVERSIES AND DISPUTES

Introduction

There are several expectations, both of a cognitive and a practical nature, concerning the power theory of electrical circuits. Cognitive, because as humans we attempt to understand phenomena in the world we live in: in this case, the phenomena that accompany energy flow in electrical circuits. Practical, because power theory should provide fundamentals for solutions to various technological and economic issues which occur when energy is being delivered to customers.

Power properties of electrical circuits are studied and described in terms of various power theories, formulated from different perspectives, with different goals of a cognitive or practical nature. Thus what are the right goals of such theories? There is essentially no ground to answer such a question.

There is one power theory that was not challenged as erroneous, however. It is the power theory of single-phase circuits with sinusoidal voltages and currents. Power properties of such circuits are specified in terms of three powers: the active, reactive, and apparent, P, Q, and S powers which satisfy the relation

$$S^2 = P^2 + Q^2.$$

This theory is commonly accepted in the electrical engineering community. Let us summarize the general features of the power theory of single-phase circuits with sinusoidal voltages and currents.

1) It provides a physical interpretation of all power-related phenomena in the circuit.
2) All powers can be expressed in terms of equivalent parameters of the load.
3) The powers specify the technical parameters of the circuit components.
4) It provides fundamentals for measurements needed for the power system operation and financial accounts.
5) It specifies the power factor and relates it to the load parameters.
6) It provides data needed for compensator design.

The first features, (1) and (2), have cognitive merits. The remaining ones are of practical importance. Feature (3) enables circuit design. Feature (4) enables the system supervision and energy accounts. Feature (5) enables the power factor improvement by modification of the load parameters. The last, Feature (6), enables this power factor improvement by compensation.

This power theory describes, in power terms, a very simple circuit, however. Present-day power systems are much more complex and consequently they need much more sophisticated power theories. Nonetheless, it can serve as a reference for what we can expect from a power theory of circuits with nonsinusoidal and asymmetrical voltages and currents. One could expect that the power theory of circuits under such conditions would have the Features (1) to (6) as enumerated above, meaning that it would provide equally solid fundamentals on the power properties of such circuits, comparable with the power theory of single-phase circuits with sinusoidal voltages and currents.

There was a remarkable effort of innumerable scientists aimed at the development of such a power theory. Observing the number of scientists involved, some have devoted their whole professional life to the problem, and more than a century-long research exists on it. We can conclude that the development of the power theory of electrical circuits with nonsinusoidal voltages and currents was one of the most difficult problems of electrical engineering.

Originally, this was an academic rather than a practical problem. There were not many sources of voltage and current distortion in electrical systems, so the voltages and currents were usually sinusoidal with a high level of accuracy. This has changed with the development of power electronics, computers, video, and computer-like equipment, as well as electronically driven home appliances, and fluorescent- or LED-based lighting in homes and commercial buildings. All such loads draw nonsinusoidal currents, distorting the distribution voltage. In effect, electrical engineering has lost the capability of describing such loads in power terms. The need to develop the power theory of electrical circuits has become very urgent. In response to this need, a few recurring meetings were established for selecting power definitions and methods of compensation in circuits with nonsinusoidal voltages and currents, namely:

IEEE Working Group on Power Definitions in Systems Under Nonsinusoidal Conditions, chaired by A. Emanuel.

International Workshop on Power Definitions and Measurement Under Nonsinusoidal Conditions, bi-annual meetings, run by Milano Polytechnic, Italy, chaired by A. Ferrero.

International School on Nonsinusoidal Currents and Compensation (ISNCC), bi-annual meetings, run by Zielona Gora University, Poland, chaired by L.S. Czarnecki.

Most of the power theories developed as a result of all these efforts do not provide Features (1) to (6) as effects power theory has on single-phase circuits with sinusoidal voltage and current. But despite this, many results that were obtained have gained wide recognition and some have even been adopted as national and international standards. As a result, they have influenced studies on powers and compensation for decades, in the wrong way. The power theory suggested in 1927 by Budeanu is the most convincing example of this. It was disseminated by standards, books, papers, and academic teaching for more than sixty years until it was challenged [87] in 1987 as being erroneous to its very roots. All attempts aimed at the development of methods of compensation based on that theory have failed but in the meantime. it derailed studies on compensation for decades.

Most of the power theories developed satisfy only some of the expectations of the electrical engineering community. Moreover, in the long process of studies on the power properties of electrical circuits, several misconceptions, wrong power definitions, and incorrect physical interpretations have occurred and have been disseminated in electrical engineering over several decades of academic teaching. Most of the present-day misconceptions concerning power phenomena, definitions, and physical interpretations in circuits with nonsinusoidal voltages and currents were inherited from previous generations of electrical engineers. They are deeply rooted, and it is not easy now to get rid of such misinterpretations. Therefore, before the power phenomena are explained in the right way and the methods of compensation are developed, some common misconceptions ought to be discussed and cleared.

It is interesting to observe that all power theories are mathematically correct. Any mathematical error disqualifies the concept on the spot. Most of them, apart from the Currents' Physical Components (CPC)-based power theory, have failed, even using the right mathematical tools, to recognize correctly the physical phenomena that determine the effectiveness of the energy transfer between the supply source and the load. Comprehension of the physics behind energy transfer has occurred to be much more challenging than mathematics.

Any power theory has to identify and describe the power-related phenomena in electrical circuits correctly, however. This is not only important for cognitive merits but also very practical ones. Without the right understanding of these phenomena, effective compensation might not be possible.

This Part will investigate how particular power theories handle the power-related physical phenomena in electrical circuits.

Chapter 18
Budeanu's Power Theory Misconceptions

18.1 Misconceptions Related to Budeanu's Reactive Power

The reactive power in single-phase circuits with nonsinusoidal voltages and currents, introduced [14] in 1927, by C.I. Budeanu, Professor at the Bucharest University, Romania, should be defined as

$$Q = Q_{\mathrm{B}} \stackrel{\mathrm{df}}{=} \sum_{n=1}^{\infty} U_n I_n \sin \varphi_n. \tag{18.1}$$

It was interpreted as a result and a sort of measure of energy oscillations between the supply source and the load. This interpretation was challenged as incorrect in [87], published in 1987, based on the following reasoning.

In single-phase circuits with a sinusoidal supply voltage and current

$$u(t) = \sqrt{2}U \cos \omega_1 t, \quad i(t) = \sqrt{2}I \cos\left(\omega_1 t - \varphi\right)$$

the rate of energy flow between the supply source and the load is

$$p(t) \stackrel{\mathrm{df}}{=} \frac{dW}{dt} = u(t)i(t) = 2\,UI \cos \omega_1 t \cos\left(\omega_1 t - \varphi\right) = P(1 + \cos 2\omega_1 t) + Q \sin 2\omega_1 t.$$

Thus, the reactive power Q is indeed equal to the amplitude of the oscillating component of the instantaneous power $p(t)$. It occurs only when such oscillations do exist and is a sort of measure of them.

The same conclusions apply to a single harmonic of the voltage and current of the nth order

$$u_n(t) = \sqrt{2}\, U_n \cos n\omega_1 t, \quad i_n(t) = \sqrt{2}\, I_n \cos(n\omega_1 t - \varphi_n).$$

The rate of energy flow associated with such a harmonic is

$$p_n(t) = u_n(t)\; i_n(t) = P_n(1 + \cos 2n\omega_1 t) + Q_n \sin 2n\omega_1 t$$

Powers and Compensation in Circuits with Nonsinusoidal Current. Leszek S. Czarnecki, Oxford University Press.
 DOI: 10.1093/oso/9780198879206.003.0018

where

$$Q_n = U_n I_n \sin \varphi_n$$

is the reactive power of the nth order harmonic, thus it is equal to the amplitude of energy oscillations at the frequency $2n\omega_1$. This conclusion does not, however, apply to the reactive Budeanu's reactive power which is defined as the sum of harmonic reactive powers Q_n, namely

$$Q_{\rm B} = \sum_{n=1}^{\infty} U_n I_n \sin \varphi_n = \sum_{n=1}^{\infty} Q_n$$

because these powers, depending on harmonic phase shift φ_n, can be positive or negative and consequently, they can cancel mutually. In particular, having non-zero values, their sum, meaning the reactive power $Q_{\rm B}$, could be zero. Thus, the reactive power as defined by Budeanu is not associated with energy oscillations between the supply source and the load. Moreover, the sum of amplitudes of oscillations of different frequencies does not describe any physical phenomenon in electrical circuits. Even if the sum of these amplitudes Q_n is equal to zero, energy oscillations can exist.

Illustration 18.1 *Let us consider a circuit shown in Fig. 18.1*

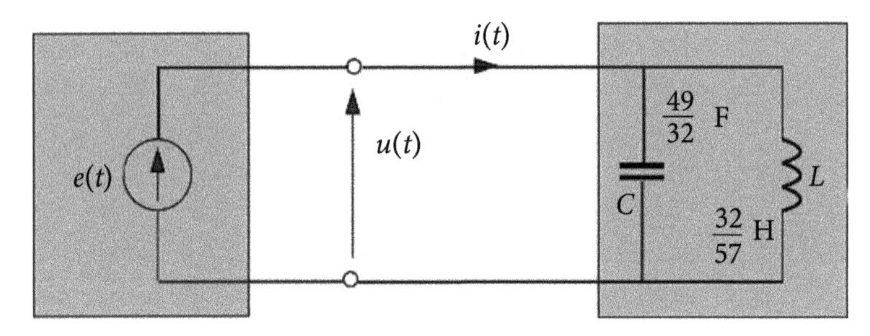

Figure 18.1 An example of a circuit with zero Budeanu's reactive power $Q_{\rm B}$.

with the supply voltage

$$u(t) = \sqrt{2}(100 \sin \omega_1 t + 25 \sin 3\omega_1 t) \text{ V}, \quad \omega_1 = 1 \text{ rad/s}.$$

The load admittance for the supply voltage harmonics of the order $n = 1$ and $n = 3$ is equal to

$$Y_1 = -j1/4 \text{ S}, \quad Y_3 = j4 \text{ S}$$

so that the load current is

$$i(t) = \sqrt{2}[25 \sin(\omega_1 t - \pi/2) + 100 \sin(3\omega_1 t + \pi/2)]\text{A}.$$

The reactive power according to Budeanu's definition is

$$Q_B = U_1 I_1 \sin\varphi_1 + U_3 I_3 \sin\varphi_3 = 100 \times 25 \sin 90° + 25 \times 100 \sin(-90°) = 0.$$

Despite the zero reactive power Q_B, *energy in the considered circuit oscillates between the supply source and the load. It is because the instantaneous power* $p(t) = u(t)\, i(t)$ *changes as shown in Fig. 18.2.*

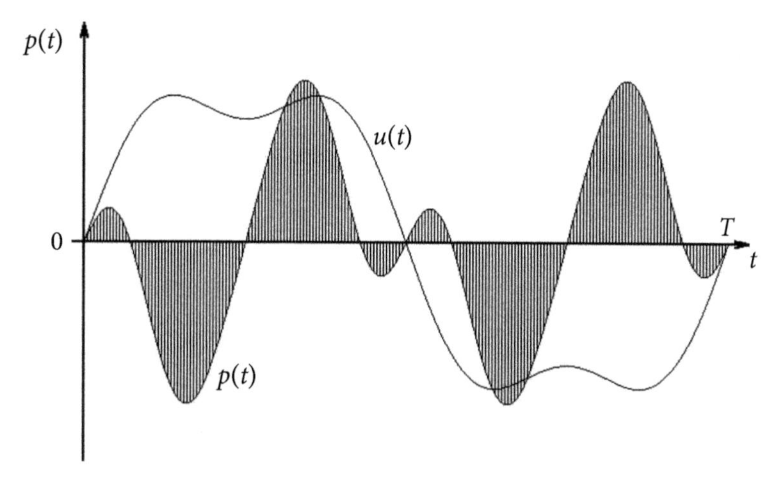

Figure 18.2 The rate of energy flow $p(t)$ in the circuit shown in Fig. 16.1 with zero Budeanu's reactive power Q_B.

******

Budeanu defined the reactive power in the frequency domain, meaning using harmonics. It was also defined [65] in the time-domain using the Hilbert Transform, which for quantity $x(t)$ is specified as

$$\mathscr{H}\{x(t)\} = \frac{1}{\pi}\mathrm{PV}\int_{-\infty}^{\infty} \frac{x(\tau)}{\tau - t} d\tau \tag{18.2}$$

where PV denotes the principal value of the integral. With this Transform, Budeanu's reactive power can be defined as

$$Q_B = \frac{1}{T}\int_0^T u(t)\mathscr{H}\{i(t)\}\, dt \tag{18.3}$$

This definition opened the way [33, 35] for developing methods of Budeanu's reactive power measurement [48], [50], [64], and [79].

18.2 Budeanu's Reactive Power and Power Balance Principle

The reactive power as defined by Budeanu satisfies the Power Balance Principle (PBP). It means, that if the circuit is energetically isolated, then the sum of reactive powers Q_B of all branches of the circuit is equal to zero. This property is used in some debates on Budeanu's reactive power as an argument for its physical meaning.

This conclusion is not right, however. The PBP for Budeanu's reactive power can be developed only from the Tellegen Theorem [25], which describes mathematical but not physical properties of electrical circuits. It cannot be developed from the Energy Conservation Principle (ECP) which provides the fundamentals for the active PBP.

According to Tellegen Theorem, voltages and currents of two different but the same topology circuits composed of K branches, shown in Fig. 18.3,

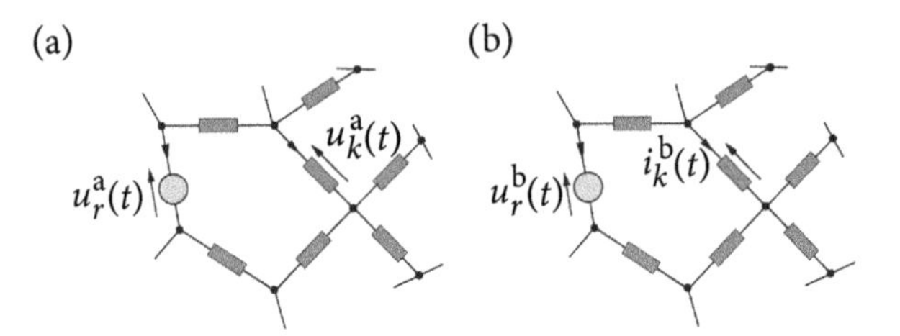

Figure 18.3 Two circuits with the same topology.

satisfy the relationship

$$\sum_{k=1}^{K} u_k^a(t) i_k^b(t) \equiv 0. \tag{18.4}$$

When these circuits are identical and all source voltages and currents in circuit (b) are Hilbert transforms of source voltages and currents in circuit (a), then all branch currents in circuit (b) are Hilbert transforms of branch currents in the circuit (a), namely

$$i_k^b(t) \equiv \mathscr{H}\{i_k^a(t)\}$$

hence

$$\sum_{k=1}^{K} u_k^a(t) i_k^b(t) \equiv \sum_{k=1}^{K} u_k^a(t) \mathscr{H}\{i_k^a(t)\} \equiv \sum_{k=1}^{K} u_k(t) \mathscr{H}\{i_k(t)\} \equiv 0.$$

The mean value over period T of this sum

$$\frac{1}{T}\int_0^T \sum_{k=1}^{K} u_k(t)\mathscr{H}\{i_k(t)\}dt = \sum_{k=1}^{K} \frac{1}{T}\int_0^T u_k(t)\mathscr{H}\{i_k(t)\}dt = \sum_{k=1}^{K} Q_k = 0$$

results in the PBP for Budeanu's reactive power. This is a conclusion only from mathematical but not physical properties of electrical circuits, however.

Unfortunately the lack of any physical meaning of the reactive power defined by Budeanu was not recognized early enough. This definition of reactive power was regarded as correct for more than sixty years, it was endorsed by the IEEE Standard Dictionary of Electrical and Electronics Terms [154] in 1997 and even included in the IEEE Standard 1459 of Power Terms [172] in 2000.

18.3 Misconceptions Related to Budeanu's Distortion Power

In his paper on electric powers at nonsinusoidal voltages and currents, Budeanu introduced to power theory the concept of distortion power, defined as

$$D \stackrel{\text{df}}{=} \sqrt{S^2 - P^2 - Q_B^2} \tag{18.5}$$

It was interpreted by Budeanu as a power caused by the mutual distortion of the load voltage and current. Definition (18.5) does not relate the distortion power D to the load parameters and the supply voltage, however. Let us now find this relation.

Definition of the distortion power D can be rearranged as follows:

$$D = \sqrt{S^2 - P^2 - Q_B^2} =$$
$$= \sqrt{\sum_{n \in N_0} U_n^2 \sum_{n \in N_0} I_n^2 - \left(\sum_{n \in N_0} U_n I_n \cos \varphi_n \right)^2 - \left(\sum_{n \in N} U_n I_n \sin \varphi_n \right)^2}.$$

To simplify this expression, let us assume that the voltage has only two harmonics of the order $n = 1$ and $n = 2$. In such a case

$$D^2 = (U_1^2 + U_2^2)(I_1^2 + I_2^2) - (U_1 I_1 \cos \varphi_1 + U_2 I_2 \cos \varphi_2)^2 - (U_1 I_1 \sin \varphi_1 + U_2 I_2 \sin \varphi_2)^2 =$$
$$= U_1^2 U_2^2 [Y_1^2 + Y_2^2 - 2 Y_1 Y_2 \cos(\varphi_1 - \varphi_2)].$$

Let us observe that

$$|\boldsymbol{Y}_1 - \boldsymbol{Y}_2|^2 = (\boldsymbol{Y}_1 - \boldsymbol{Y}_2)(\boldsymbol{Y}_1 - \boldsymbol{Y}_2)^* = Y_1^2 + Y_2^2 - 2\text{Re}\{\boldsymbol{Y}_1 \boldsymbol{Y}_2^*\} = Y_1^2 + Y_2^2 - 2 Y_1 Y_2 \cos(\varphi_1 - \varphi_2)$$

and hence

$$D^2 = U_1^2 U_2^2 |\boldsymbol{Y}_1 - \boldsymbol{Y}_2|^2.$$

When the supply voltage has harmonics of the order from the set N_0, then the last formula can be generalized to the form

$$D^2 = \frac{1}{2}\sum_{r\in N_0}\sum_{s\in N_0} A_{rs}$$

with

$$A_{rs} = U_r^2 I_s^2 + U_s^2 I_r^2 - 2U_r U_s I_r I_s \cos(\varphi_r - \varphi_s) = U_r^2 U_s^2 |\boldsymbol{Y}_r - \boldsymbol{Y}_s|^2 \geq 0$$

while $\boldsymbol{Y}_r$ and $\boldsymbol{Y}_s$ are the load admittances for harmonic frequencies $r\omega_1$ and $s\omega_1$. Thus the distortion power can be expressed in terms of the rms value of the supply voltage harmonics and the load admittance for harmonic frequencies

$$D = \sqrt{\frac{1}{2}\sum_{r\in N_0}\sum_{s\in N_0} U_r^2\, U_s^2 |\boldsymbol{Y}_r - \boldsymbol{Y}_s|^2}. \tag{18.6}$$

Since terms A_{rs} are non-negative, distortion power D can be equal to zero on the condition that for harmonics with non-zero rms values U_r and U_s, all terms $A_{rs} = 0$, which requires that for each $r, s \in N_0$,

$$\boldsymbol{Y}_r = \boldsymbol{Y}_s.$$

Thus the distortion power of the load is equal to zero if and only if for harmonics of the supply voltage the load impedance $\boldsymbol{Y}_n$ has the same value,

$$\boldsymbol{Y}_n = Y_n\, e^{j\,\varphi_n} = \text{const.} \tag{18.7}$$

This is *the necessary condition* for zero distortion power D. This is not the condition for the lack of mutual distortion of the load current versus the supply voltage, however.

The load current is not distorted versus the supply voltage, but only shifted, if it can be expressed as

$$i(t) = Y u(t - \tau)$$

where τ is the current delay versus the supply voltage $u(t)$. Assuming that the voltage is periodic and its harmonics have crms values $\boldsymbol{U}_n$, then from the shifting property results that the current harmonics have the crms value

$$\boldsymbol{I}_n = Y\boldsymbol{U}_n\, e^{-jn\omega_1\tau}.$$

Thus the load current is not distorted versus the supply voltage on the condition that

$$Y_n = Ye^{-jn\omega_1\tau} = Ye^{-jn\varphi_1} \tag{18.8}$$

meaning on the condition that the magnitude of the load admittance does not depend on the order of the voltage harmonics, while the argument φ_n changes linearly with this order. This is the necessary condition for the lack of mutual distortion between the load voltage and current. Conditions (18.7) and (18.8) can be satisfied simultaneously for nonsinusoidal voltages and currents only if the phase shift angle between the voltage and current harmonics $\varphi_n = 0$, meaning for only purely resistive loads. Distortion power D of such loads is zero, however. Otherwise, when condition (18.7) is satisfied, thus, distortion power $D = 0$, the condition (18.8) is not satisfied, thus the load current is distorted versus the supply voltage. When the load current is not distorted versus the supply voltage, thus the condition (18.8) is satisfied, then the condition (18.7) is not satisfied, thus distortion power D cannot be equal to zero. Consequently, Budeanu's interpretation of the distorted power is entirely erroneous.

Illustration 18.2 *The circuit shown in Fig. 18.4, supplied with the voltage*

$$u(t) = \sqrt{2}\,(100 \sin \omega_1 t + 50 \sin 3\omega_1 t)\ \text{V}, \quad \omega_1 = 1\ \text{rad/s}$$

has for the supply voltage harmonics the admittance

$$Y_1 = Y_3 = 1\, e^{j\frac{\pi}{2}}\ \text{S}.$$

The load satisfies the condition (18.7), thus its distortion power D *is zero. The load current is equal to*

$$i(t) = \sqrt{2}\left[100 \sin\left(\omega_1 t + \frac{\pi}{2}\right) + 50 \sin\left(3\, \omega_1 t + \frac{\pi}{2}\right)\right]\ \text{A}.$$

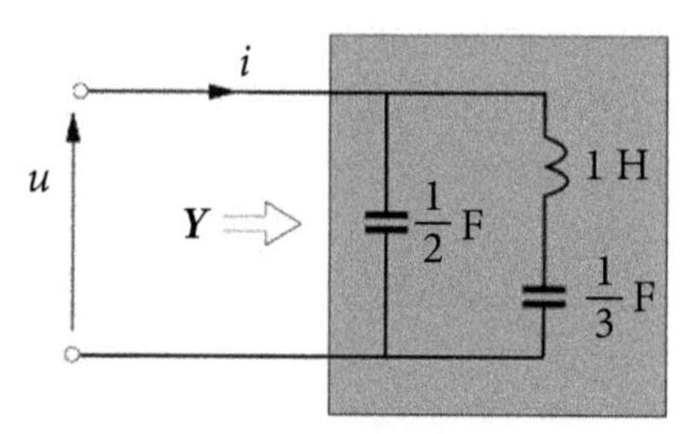

Figure 18.4 A load with zero distortion D power but with the current distorted versus the supply voltage.

and its waveform, along with the load voltage waveform, are shown in Fig. 18.5.

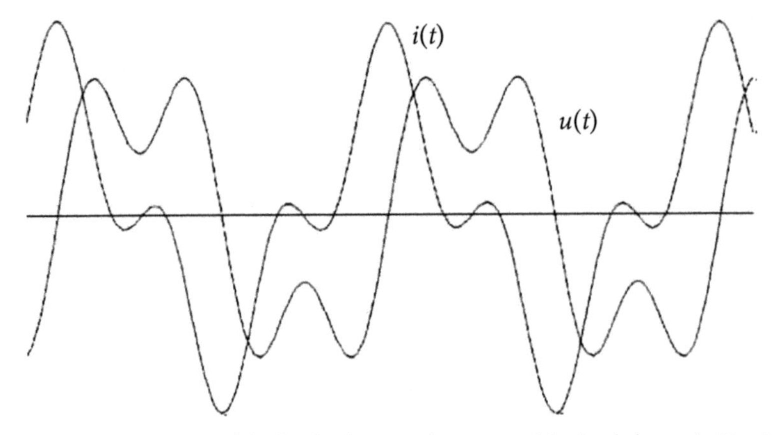

Figure 18.5 Waveforms of the load voltage and current of the load shown in Fig. 16.4, with zero distortion power D.

Despite the distortion power being equal to zero, this current is distorted versus the voltage. The load admittances satisfy, however, the condition (5.19) for zero distortion power, D.

Illustration 18.3 *The load shown in Fig. 18.6 is supplied with a nonsinusoidal voltage*

$$u(t) = \sqrt{2}\,(100 \sin \omega_1 t + 50 \sin 3\omega_1 t)\mathrm{V}, \quad \omega_1 = 1 \text{ rad/s}$$

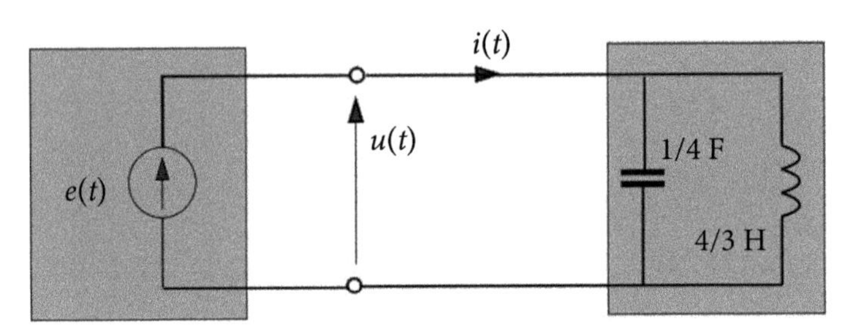

Figure 18.6 An example of a circuit with non-zero distortion power D but without any mutual voltage and current distortion.

Its admittance for the supply voltage harmonics is

$$Y_1 = \frac{1}{2} e^{-j\pi/2}\ \mathrm{S}, \quad Y_3 = \frac{1}{2} e^{-j3\pi/2} = \frac{1}{2} e^{j\pi/2}\ \mathrm{S}.$$

The load admittance for harmonic frequencies does not satisfy the condition (18.7), thus the load distortion power D *cannot be zero. It is equal to*

$$\mathrm{D} = \frac{2}{3} 10\,\sqrt{2}\,\mathrm{kVA}.$$

The load current in this circuit is

$$i(t) = \frac{2}{3}100\sqrt{2}\left[\sin\left(\omega_1 t - \frac{\pi}{2}\right) + \sin\left(3\omega_1 t + \frac{\pi}{2}\right)\right] =$$
$$= \frac{2}{3}100\sqrt{2}\left[\sin\omega_1\left(t - \frac{T}{4}\right) + \sin\, 3\omega_1\left(t - \frac{T}{4}\right)\right] = \frac{2}{3}u\left(t - \frac{T}{4}\right).$$

This is shifted versus the supply voltage by the quarter of the period T. *Its waveform, along with the load voltage waveform, are shown in Fig. 18.7.*

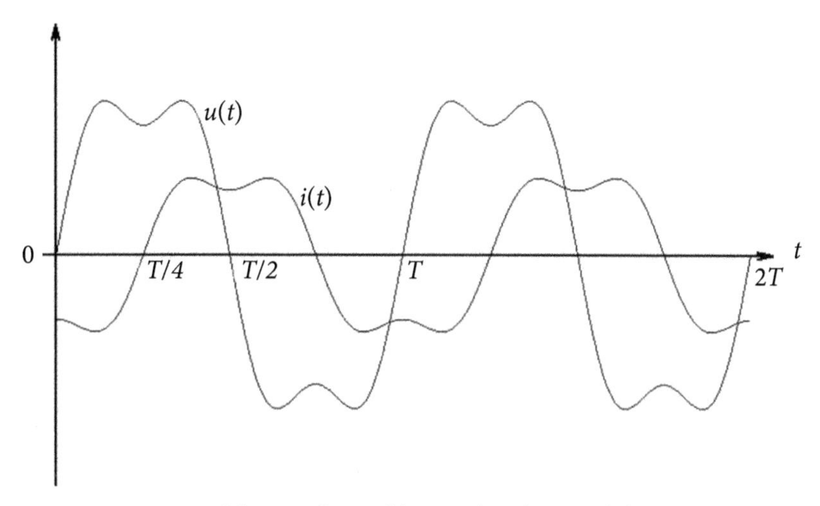

Figure 18.7 The waveform of the supply voltage and the current of the load shown in Fig. 18.6.

******

This illustration demonstrates that distortion power D has nothing in common with the current distortion. As demonstrated, it has even properties that contradict its name. It has occurred in electrical engineering only as a result of the erroneous definition of reactive power.

18.4 Usefulness Budeanu's PT for Compensation

The reactive power Q in circuits with sinusoidal voltages and currents is a quantity of major practical importance. It affects the power factor. The knowledge of this power value makes designing compensators possible. The capacitance needed for compensation of the load to the unity power factor is equal to

$$C = \frac{Q}{\omega_1 U^2}.$$

Therefore there were expectations that the reactive power as defined by Budeanu would provide fundamentals for compensation at nonsinusoidal supply voltage. Unfortunately, all attempts aimed at developing methods of power factor improvement using Budeanu's definition of reactive power have failed. The reason for this failure is relatively simple.

The supply current of the load at sinusoidal supply voltage can be expressed in the form

$$i(t) = \sqrt{2}\, I \cos\left(\omega_1 t - \varphi\right) = \sqrt{2}\frac{P}{U} \cos \omega_1 t + \sqrt{2}\frac{Q}{U} \sin \omega_1 t$$

thus its rms value is

$$||i|| = \sqrt{\left(\frac{P}{U}\right)^2 + \left(\frac{Q}{U}\right)^2}. \tag{18.9}$$

This rms value is minimum, at unchanged active power P and the load voltage rms value U, only if the reactive power Q is equal to zero. It can be reduced to zero by a capacitor or a synchronous machine and this improves the power factor to unity.

When the load current is distorted but periodic, it can be expressed as a sum of harmonics

$$i(t) = \sum_{n=0}^{N} i_n(t).$$

Its nth order harmonic can be expressed in the form

$$i_n(t) = \sqrt{2}\, I_n \cos(n\omega_1 t = \varphi_n) = \sqrt{2}\,\frac{P_n}{U_n} \cos\; n\omega_1 t + \sqrt{2}\,\frac{Q_n}{U_n} \sin\; n\omega_1 t.$$

Thus the rms value of this harmonic is

$$||i_n|| = \sqrt{\left(\frac{P_n}{U_n}\right)^2 + \left(\frac{Q_n}{U_n}\right)^2}.$$

Since harmonics are mutually orthogonal, the current rms value is equal to

$$||i|| = \sqrt{\sum_{n=0}^{N} ||i_n||^2} = \sqrt{\sum_{n=0}^{N}\left(\frac{P_n}{U_n}\right)^2 + \sum_{n=0}^{N}\left(\frac{Q_n}{U_n}\right)^2}. \tag{18.10}$$

This expression shows that the current rms value does not depend on the sum of the harmonic reactive powers Q_n

$$\sum_{n=1}^{\infty} Q_n = Q_{\text{B}}$$

meaning, on the Budeanu's reactive power, but the term

$$\sum_{n=1}^{N}\left(\frac{Q_n}{U_n}\right)^2 \tag{18.11}$$

which has nothing in common with this power. The supply current rms value is minimum not when Budeanu reactive power is equal to zero, but when the term (18.11) is equal to zero. Since this is a sum of positive terms, it can be zero only if each reactive power of individual harmonics Q_n is equal to zero. Budeanu's reactive power Q_B would be equal to zero in such a case, but its zero value *is only a necessary but not sufficient condition* for the supply current rms minimum value. Therefore, reduction of the reactive power Q_B is not sufficient for reducing the supply current rms value, $||i||$, meaning, for the power factor improvement.

Illustration 18.4 *Let us consider a load shown in Fig. 18.8,*

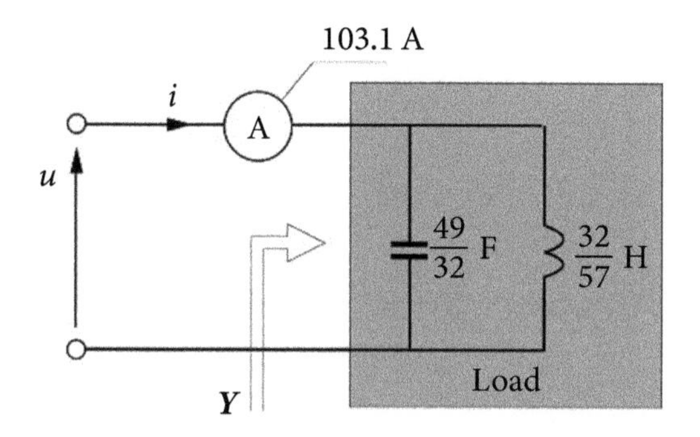

Figure 18.8 An example of a load.

supplied with the voltage

$$u(t) = \sqrt{2}\,(100 \sin \omega_1 t + 25 \sin 3\omega_1 t)\text{V}, \quad \omega_1 = 1 \text{ rad/s}.$$

The load admittance for the fundamental and the third order harmonics are, respectively, equal to $\mathbf{Y}_1 = -j1/4$ *S and* $\mathbf{Y}_3 = j4$ *S and consequently, the load current is equal to*

$$i(t) = \sqrt{2}\,[25 \ \sin\,(\omega_1 t - 90^0) + 100 \ \sin\,(3\omega_1 t + 90^0)] \text{ A}.$$

The supply current rms value, shown by an ammeter, is

$$||i|| = \sqrt{I_1^2 + I_3^2} = \sqrt{100^2 + 25^2} = 103.1 \text{ A}.$$

The load can be compensated by a shunt reactance compensator such that its admittance for the fundamental and the third order harmonic compensate the load admittance, meaning they are of the opposite sign, namely

$$Y_{C1} = j1/4\,\text{S},\;\; Y_{C3} = -j4\,\text{S}.$$

The structure and parameters of a reactance compensator that has such admittances are shown in Fig. 18.9. It compensates the reactive power of individual harmonics entirely and reduces the supply current rms to zero value. Budeanu's reactive power of the load is zero and, of course, remains zero after compensation. Thus, the reduction of the supply current rms value has nothing in common with Budeanu's reactive power reduction.

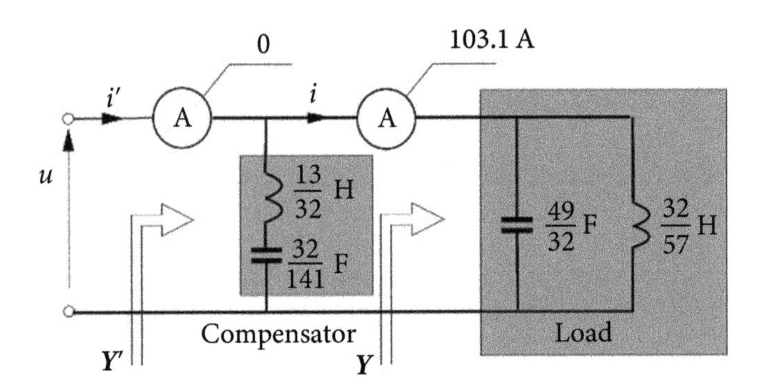

Figure 18.9 A load with a reactance Compensator designed in the frame of the CPC.

******

Summary

Budeanu's power theory stands for the first phase of power theory's development. By its end, Milic, [36], Nowomiejski [65], and Fisher [67] provided very advanced and sophisticated mathematical forms for Budeanu's power theory. Unfortunately, despite it Budeanu's power theory misinterprets power-related physical phenomena in electrical circuits entirely. In addition, it does not provide fundamentals for power factor improvement, even in single-phase circuits with linear, time-invariant loads. Because definitions of the reactive and distortion powers Q_B and D, were endorsed by the IEEE Standard Dictionary of Electrical and Electronics Terms, Budeanu's power theory has exerted a negative impact on the studies on power properties and compensation in circuits with nonsinusoidal voltages and currents.

Chapter 19
Deficiencies of Fryze's Power Theory

19.1 Active and Reactive Currents Interpretations

S. Fryze, a Professor at the Lwow Polytechnic, Poland, and an Honorary Member of the Polish Physical Society strongly emphasized the need for understanding the physical phenomena behind power theory (PT). As a student, the author of this book attended Fryze's lectures on electrical circuits, which were always accompanied by physical demonstrations and experiments.

The concept of the load current $i(t)$ instead of the load apparent power S decomposition, as introduced [17, 18] by Fryze, was his major contribution to PT development. It was because the current flow is a sort of physical phenomenon, while the apparent power S is only a conventional quantity. The same can be said regarding the concept of the current components orthogonality, because individual components of the load current contribute to energy loss at its delivery independently of each other on the condition that these components are mutually orthogonal.

The concept of the active current $i_{\mathrm{a}}(t)$ as introduced by Fryze and interpreted by him as the minimal current needed for the load supply at nonsinusoidal voltage $u(t)$ with the active power P was also a very important contribution to PT, but Fryze's interpretation of this current is only partially true. It is true only in circuits with LTI, meaning the loads that are not the sources of the load current distortion. In circuits with current harmonics generating loads, energy at harmonic frequencies can flow from the load to the supply source. The active current $i_{\mathrm{a}}(t)$ is not, as shown in Section 6.11, the minimum current needed to supply the HGL with the active power P, however. Such a load has to be supplied with a current higher than the active current.

According to Fryze, the load current decomposition into an active $i_{\mathrm{a}}(t)$ and a reactive $i_{\mathrm{Fr}}(t)$ component was a decomposition only into a useful and a useless component. The reactive current $i_{\mathrm{Fr}}(t)$ was not endowed with interpretation other than that. This is not the physical interpretation, however. It was not explained what physical phenomena in the circuit are responsible for the presence of the reactive current defined according to Fryze, even in circuits with LTI loads. The same is with circuits with HGLs. When Fryze's decomposition of the load current is compared with the Currents' Physical Components-based decomposition, we can conclude that the physical nature of the reactive current $i_{\mathrm{Fr}}(t)$ is complex and dependent on the load properties. In circuits with LTI loads, the reactive current $i_{\mathrm{Fr}}(t)$ also includes the

Powers and Compensation in Circuits with Nonsinusoidal Current. Leszek S. Czarnecki, Oxford University Press.
 DOI: 10.1093/oso/9780198879206.003.0019

scattered current. In circuits with HGLs, moreover, it includes the load-generated harmonic current. Thus, the PT of electrical circuits with nonsinusoidal voltage, as developed by Fryze, has provided only a very limited interpretation of physical phenomena in such circuits.

19.2 Reactance Compensation

Power theory as suggested by Fryze was confined only to the load current and the apparent power S decomposition at nonsinusoidal supply voltage. Compensation, similarly to Illovici and Budeanu, was beyond the scope of Fryze's considerations. Much later it was discovered that Fryze's power theory creates fundamentals neither for reactance compensator design [83, 153] nor for switching compensator control. These arguments are presented below.

Compensation of the reactive power was not the subject of Fryze's PT but power factor improvement is one of the main practical objectives of PT development. Therefore we can ask how much information Fryze's PT provides about the possibility of power factor improvement.

The power equation of single-phase loads with nonsinusoidal supply voltage according to Fryze's PT has the form

$$S^2 = P^2 + Q_F^2$$

with the reactive power

$$Q_F \stackrel{\text{df}}{=} ||u||\ ||i_{Fr}||.$$

The power factor of the load can be expressed in terms of powers

$$\lambda = \frac{P}{S} = \frac{P}{\sqrt{P^2 + Q_F^2}}$$

In a single-phase circuit with a sinusoidal supply voltage and current the value of the reactive power, along with the active power, enables not only the calculation of the power factor λ but also a calculation of the compensating capacitance that improves this factor to unity. The reactive power Q_F does not have these qualities, however. This is demonstrated with the following illustration.

Illustration 19.1 *Let us consider the load shown in Fig. 19.1, supplied with the voltage*

$$u(t) = 100\sqrt{2}(\sin\omega_1 t + \sin 3\,\omega_1 t)\,\text{V}, \quad \omega_1 = 1\text{rad/s}.$$

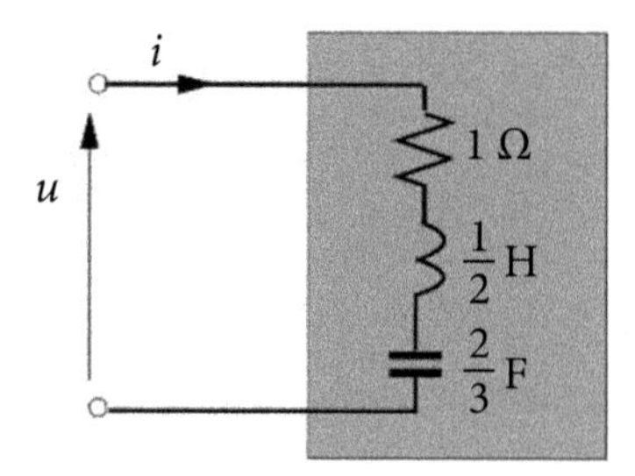

Figure 19.1 An example of a load.

The load admittance for the fundamental and the third order harmonics are

$$\boldsymbol{Y}_1 = 0.5 + j0.5 = 0.5\sqrt{2}\, e^{j45^0}\ \text{S}, \qquad \boldsymbol{Y}_3 = 0.5 - j0.5 = 0.5\sqrt{2}\, e^{-j45^0}\ \text{S},$$

Thus the rms value of the fundamental and the third order harmonic of the supply current are equal to

$$I_1 = Y_1\, U_1 = 70.1\ \text{A}, \quad I_3 = Y_3\, U_3 = 70.1\ \text{A}.$$

Thus the supply source apparent power amounts to

$$S = ||u||\, ||i|| = \sqrt{U_1^2 + U_3^2} \times \sqrt{I_1^2 + I_3^2} = \sqrt{100^2 + 100^2} \times \sqrt{70.1^2 + 70.1^2} =$$
$$= 140.2 \times 100 = 14.0\ \text{kVA}$$

while the load active power is

$$P = \sum_{n=1,3} G_n\, U_n^2 = 0.5 \times 100^2 + 0.5 \times 100^2 = 10\ \text{kW}.$$

The load reactive power calculated according to Fryze's definition is

$$Q_{\text{F}} = \sqrt{S^2 - P^2} = \sqrt{14.0^2 - 10^2} = 10\ \text{kVAr}.$$

The power factor of the load $\lambda = P/S = 0.71$. *The power factor of such a load can be improved with a shunt reactance compensator that for the fundamental and the third order harmonics has admittance*

$$\boldsymbol{Y}_{\text{C}1} = -j0.5\ \text{S}, \qquad \boldsymbol{Y}_{\text{C}3} = j0.5\ \text{S}.$$

Such admittance has a compensator shown in Fig. 19.2.

Such a compensator reduces the admittance as seen by the supply source to

$$\boldsymbol{Y}_1' = 0.5\ \text{S}, \qquad \boldsymbol{Y}_3' = 0.5\ \text{S},$$

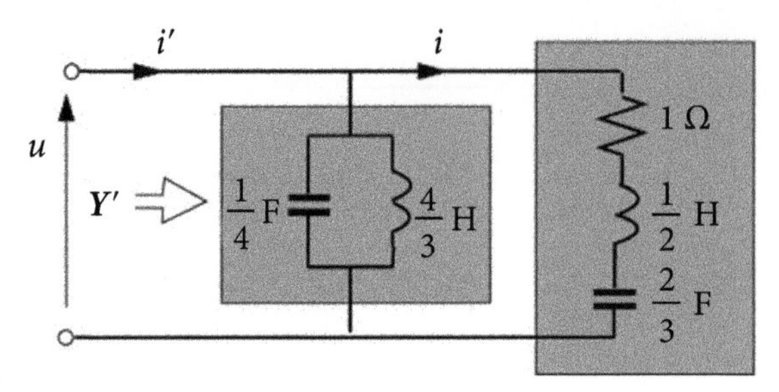

Figure 19.2 A load with a shunt reactance compensator.

thus the rms value of the supply current harmonics after compensation are

$$I'_1 = Y'_1\, U_1 = 50\,\text{A}, \quad I'_3 = Y'_3\, U_3 = 50\,\text{A}$$

and its rms value is

$$||i'|| = \sqrt{(I'_1)^2 + (I'_3)^2} = \sqrt{50^2 + 50^2} = 70.1\,\text{A}.$$

Thus, the compensator reduces the apparent power to

$$S' = ||u||\ ||i'|| = 140.2 \times 70.1 = 10\,\text{kVA}.$$

This improves the power factor of the supply source to $\lambda = 1$. It should be noted that parameters of such a compensator cannot be calculated using Fryze's PT, however. Instead, they were calculated using the CPC-based PT.

Let the same voltage as assumed above supply a slightly different load, shown in Fig. 19.3.

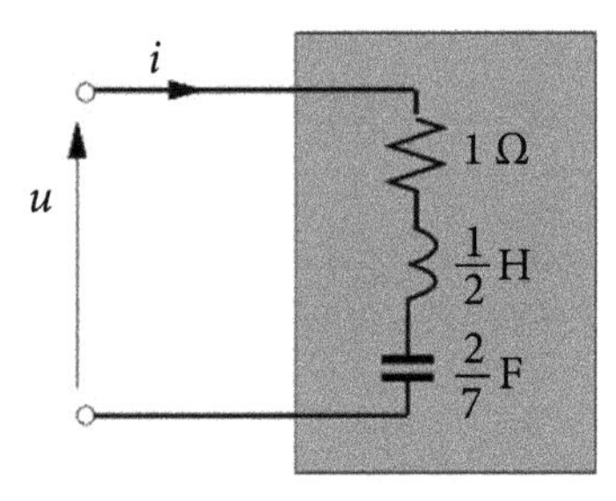

Figure 19.3 An example of a load.

The load admittance for the fundamental and the third order harmonics are

$$Y_1 = 0.1 + j0.3 = 0.316e^{j72^0}\ \text{S}, \qquad Y_3 = 0.9 - j0.3 = 0.95\ e^{-j18.4^0}\ \text{S},$$

thus the rms value of the fundamental and the third order harmonic of the supply current are equal to

$$I_1 = Y_1\, U_1 = 31.6\,\text{A}, \qquad I_3 = Y_3\, U_3 = 95\,\text{A}.$$

The supply current rms value of the load is

$$||i|| = \sqrt{I_1^2 + I_3^2} = 100\,\text{A}.$$

Thus the apparent power S of the supply source, similar to the circuit in Illustration 19.1, is

$$S = ||\,u||\;||i\,|| = 140.2 \times 100 = 14.0\,\text{kVA}$$

and similarly, the load active power is

$$P = \sum_{n=1,3} G_n\, U_n^2 = 0.1 \times 100^2 + 0.9 \times 100^2 = 10\,\text{kW}.$$

Thus these two loads do not differ concerning the active, reactive, and apparent powers, and consequently as to the power factor, namely for both of them

$$S = 14.0\,\text{kVA}, \qquad Q_{\text{F}} = 10\,\text{kVAr}, \qquad \lambda = 0.71$$

meaning these two loads are identical in terms of Fryze's PT. The shunt reactance compensator that compensates the load susceptance should have admittance for harmonic frequencies equal to

$$Y_{\text{C1}} = -j0.3\ \text{S}, \qquad Y_{\text{C3}} = j0.3\ \text{S}.$$

The structure and parameters of such a compensator are shown in Fig. 19.4.

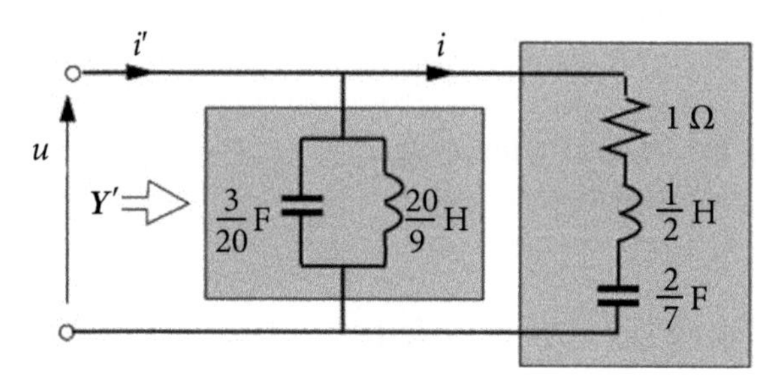

Figure 19.4 Load with a shunt reactance compensator.

Such a compensator changes the admittance as seen from the supply source to

$$Y_1 = 0.5\,\text{S}, \qquad Y_3 = 0.5\,\text{S},$$

thus the rms value of the supply current harmonics after compensation are

$$I_1' = Y_1'\,U_1 = 10\,\text{A}, \qquad I_3' = Y_3'\,U_3 = 90\,\text{A}$$

and the rms value of the supply current

$$||i'|| = \sqrt{(I'_1)^2 + (I'_3)^2} = \sqrt{10^2 + 90^2} = 90.5\,\text{A}.$$

The compensator reduces the apparent power to

$$S' = ||u||\,||i'|| = 140.2 \times 90.5 = 12.7\text{kVA}$$

which is higher than the load active power, however. Thus some amount of the reactive power

$$Q_F = \sqrt{S^2 - P^2} = \sqrt{12.7^2 - 10^2} = 8\text{kVAr}$$

remains after compensation and consequently, the power factor is improved only to $\lambda = 0.78$ *but not to unity as the circuit shown in Fig. 19.2. Observe, moreover, that the load admittance for the fundamental and the third order harmonics after compensation is a real number, thus the supply current harmonics are in phase with voltage harmonics. Further reactance compensation of such a load is not possible, however. Despite non-zero reactive power* Q_F*, the reactance of the compensated load is no longer reactive but resistive. The compensator has reduced this power to the scattered power* D_s*.*

******

Fryze's PT does not provide any explanation for the question that can be asked in connection with the obtained result: *Why does the power factor of the compensated load not achieve unity in the situation where the supply current harmonics are in phase with the supply voltage harmonics?* This is because Fryze's reactive power Q_F remains a composite quantity. A part of it, the scattered power Q_s, cannot be compensated by any shunt reactance compensator.

19.3 Switching Compensation

Although Fryze's PT does not provide fundamentals for reactance compensator design, there are views [226] that it can be used in an algorithm for a switching compensator (SC) control. A single-phase idea of such a compensator is shown in Fig. 19.5.

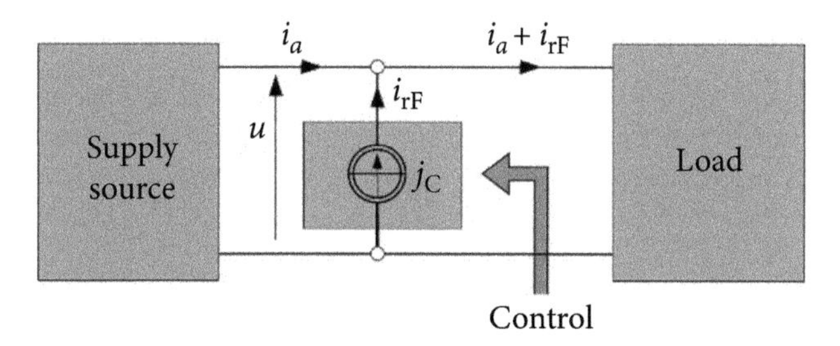

Figure 19.5 A circuit with a switching compensator in the form of a controlled current source.

The compensator injects a current generated by a controlled current source into the distribution system. The waveform of the injected current is shaped with transistors switched at high frequency in such a way that its waveform follows the reference signal. This current can compensate any undesirable component of the supply current.

Such a compensator should reduce the useless component of the supply current, and Fryze's PT interprets the reactive current $i_{Fr}(t)$ just in such a way so the reference signal for the compensator should be equal to

$$i_{ref}(t) = i_{Fr}(t) = i(t) - i_a(t).$$

Apparently, the rationale behind such control is reasonable. It is [99] erroneous, however, when such compensation is analyzed in detail. This is demonstrated with the following illustration.

Illustration 19.2 *Let's consider a circuit shown in Fig. 19.6*

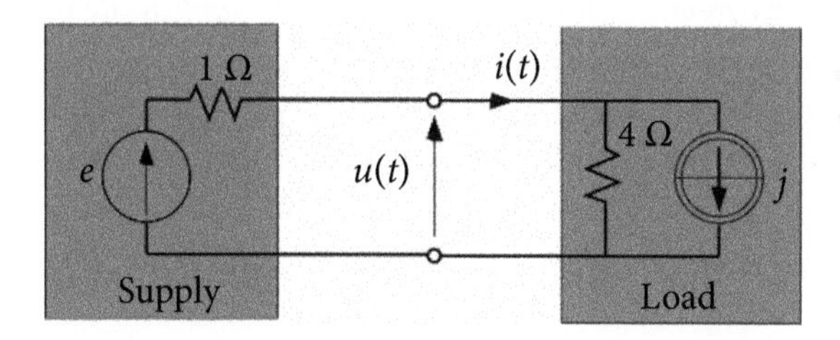

Figure 19.6 An example of a circuit with a harmonic generating load.

with sinusoidal supply voltage

$$e = 100\sqrt{2}\sin\omega_1 t \text{ V}$$

and a load that generates the third order current harmonic

$$j = 50\sqrt{2}\,\sin 3\omega_1 t \text{ A.}$$

This circuit simulates a common situation, where a supply source with sinusoidal voltage delivers energy to a harmonic-generating load. Only parameters of the circuit in this illustration were selected to emphasize some phenomena.

The voltage and current at the load terminals are equal to, respectively,

$$u = 80\sqrt{2}\sin\omega_1 t - 40\sqrt{2}\sin 3\omega_1 t \text{ V}$$

$$i = 20\sqrt{2}\sin\omega_1 t + 40\sqrt{2}\sin 3\omega_1 t \text{ A,}$$

thus the active power measured at the load terminals is

$$P = \frac{1}{T}\int_0^T u\,i\,dt = \sum_{n=1,3} U_n I_n \cos\varphi_n = 1600 - 1600 = 0$$

and consequently there is no active component in the supply current, since

$$i_{\rm a}(t) = \frac{P}{||u||^2}u(t) \equiv 0.$$

The supply current contains in the sense of Fryze's PT only the reactive current

$$i_{\rm rF}(t) = i(t) = 20\sqrt{2}\sin\omega_1 t + 40\sqrt{2}\sin 3\,\omega_1 t = i_1(t) + i_3(t).$$

Compensation of this current of course stops any delivery of energy to the load, which does not make much sense. A compensator should be able to detect the presence of the third order harmonic in the supply current and compensate it, but this current harmonic is not equal to Fryze's reactive current $i_{Fr}(t)$*.*

Thus, Fryze's PT only apparently provides fundamentals for control of switching compensators. In the situation considered, the compensator should not affect the energy delivery at a fundamental frequency. The power factor in the regarded circuit was reduced by the phenomenon of a current harmonic generation by an HGL

19.4 Fryze's Power Theory and Harmonics

Being very sensitive to physical phenomena in electrical circuits, Fryze rightly concluded that harmonics are not physical but only mathematical entities. Section 1.6 presents details of this reasoning with a circuit called sometimes a "Fryze's circuit".

Therefore Fryze suggested that harmonics should not be used for the description of energy flow in electrical circuits.

At the same time, however, Fryze did not recognize that as only mathematical entities, harmonics and Fourier series provide powerful tools for the analysis of electrical circuits with nonsinusoidal voltages and currents. Without harmonics, it could be very difficult to express a periodic quantity in an analytical form. Therefore, rejection of the concept of harmonics was the major deficiency of Fryze's PT.

Moreover, without harmonics, Fryze's PT was not capable of creating fundamentals for reactance compensation. Reactance compensators can be synthesized, based on the frequency properties of reactance one-ports, only in the frequency domain. Methods of synthesis of reactance one-ports in the time domain are not yet developed, thus one of the most important methods of compensation was not available in terms of that power theory. Moreover, we cannot design resonant harmonic filters which make the reduction of harmonic distortion in several situations possible.

Moreover, Fryze was aware that Budeanu's concepts of the reactive and distortion powers were erroneous. His concern was confined, however, only to harmonics used in the definition of the reactive power Q_B. Not using the concept of harmonics, he failed to demonstrate that there is no physical phenomenon in electrical circuits that is specified by reactive power Q_B, as was done in Section 18.1. The same holds for Budeanu's distortion power D. He failed to demonstrate that it was not associated, as was proved in Section 18.2, with the mutual distortion of the load voltage and current. Further, having a strong physical orientation toward electrical circuits and power theory, the rejection of harmonic tools did not allow him to use this tool for proving that the reactive and distortion powers Q_F and D_s do not describe any physical phenomenon. Just the opposite: in his last, posthumously published article [81] he demonstrated that these two powers can be defined without using the concept of harmonics. Somehow this has strengthened the opinion that Budeanu's reactive and distortion powers could have some physical meaning. At the same time, Sections 18.1 and 18.3 show that when harmonics are used for the circuit analysis, the proofs that these two powers are not associated with any physical phenomena in electrical circuits are simple to derive.

Rejecting the concept of harmonics, Fryze has confined the cognitive capabilities of the power theory he developed.

Summary

Fryze's approach has introduced to power theory a few very important and permanent concepts. First was the concept of the active current, and the load current rather than the load apparent power decomposition, as well as the concept of current components orthogonality. Equally important was his conclusion that the power factor declines only apparently because of energy oscillations.

Nonetheless, Fryze's opinion that power properties of electrical circuits should be specified only in the time-domain, has negatively affected the power theory development. All attempts of developing it in that domain eventually failed.

It seems, that despite some deficiencies, Fryze was the scientist who contributed to the development of the power theory much more than anyone else, at least in the early stage of its development.

Chapter 20
Deficiencies of the Kusters and Moore's PT

20.1 Interpretation of Currents in the Kusters and Moore's PT

From a perspective of compensation, the power theory (PT) suggested [59] in 1980 by Kusters and Moore (K&M), covered briefly in Section 5.9, was a major breakthrough. It has enabled the calculation and measurement [63] of the optimal compensating capacitance of resistive-inductive (RL) loads supplied with a nonsinusoidal voltage. Kusters and Moore's approach has resulted in the same optimal capacitance as that proposed by Shepherd and Zakikhani but its value can be obtained entirely in the time domain, meaning without any use of the concept of harmonics. It was regarded as the major progress in methods of compensation.

According to Kusters and Moore, the optimal compensating capacitance in the presence of voltage distortion is equal to

$$C_{\mathrm{opt}} = -\frac{Q_C}{||u||\ ||\dot{u}||} \tag{20.1}$$

where $\dot{u}$ denotes the supply voltage derivative. This result was supported by Page [60] and Fodor [68], recognized by the International Electrotechnical Commission [57], and recommended for use in electrical engineering for calculating compensating capacitance in the presence of the supply voltage distortion.

The load current decomposition as suggested by the K&M power theory

$$i(t) = i_{\mathrm{a}}(t) + i_{\mathrm{qC}}(t) + i_{\mathrm{qCr}}(t) \tag{20.2}$$

is oriented only at a capacitive compensation. The current $i_{\mathrm{a}}(t)$ is the active current as defined by Fryze. The current component $i_{\mathrm{qC}}(t)$ is the maximum current that can be compensated by a shunt capacitor. The current component $i_{\mathrm{qCr}}(t)$ is a residual reactive current which, according to K&M PT, is a current that cannot be compensated by a reactance compensator. This interpretation of the current $i_{\mathrm{qCr}}(t)$ is wrong, however. When the load is linear and time invariant (LTI), the current $i_{\mathrm{qCr}}(t)$ can contain a scattered current $i_{\mathrm{s}}(t)$ as defined in the Currents' Physical Components (CPC)-based PT. It cannot be compensated by any shunt reactance compensator. The remaining component of the residual capacitive current, $\Delta i(t) = i_{\mathrm{qCr}}(t) - i_{\mathrm{s}}(t)$, is

Powers and Compensation in Circuits with Nonsinusoidal Current. Leszek S. Czarnecki, Oxford University Press.

DOI: 10.1093/oso/9780198879206.003.0020

still a sort of reactive current that can be compensated, as demonstrated in [78], by a reactance compensator. This is because the sum $i_{qC}(t) + \Delta i(t)$ has to be equal to the reactive current $i_r(t)$ as defined in CPC-based PT, that is,

$$i_{qC}(t) + \Delta if(t) = i_r(t) \tag{20.3}$$

Thus, Kusters and Moore's PT misinterprets the power properties of electrical circuits and the possibility of their compensation.

20.2 Kusters and Moore's PT and Capacitive Compensation

The correctness of the method of the optimal capacitance calculation, as suggested by the K&M PT, was challenged by the author of this book in Refs [69] and [78].

The K&M approach is confined entirely to capacitive compensation alone. The supply sources in distribution systems usually have inductive impedance, consequently compensating capacitors cause amplification of the supply voltage harmonics and the load-generated current harmonics, as was discussed in detail in Sections 11.3–11.6. Thus, capacitive compensation degrades the supply quality (SQ) in the compensated distribution system. Therefore, the power factor improvement in the presence of the supply voltage distortion and/or harmonics generating loads should not be founded on capacitive compensation. More advanced compensators are needed for that. Moreover, because the supply system impedance is inductive and it increases with the harmonic order while compensating branch is capacitive and its impedance declines with this order, the connection of this branch to the distribution system can change the voltage on the compensator, meaning the optimal capacitance calculated from the K&M formula (20.1) could have an incorrect value. Its calculation should be repeated with actual values of quantities in that formula. A change of the capacitance value changes the voltage on the compensator again. The process has to be repeated. It can be convergent or not. Even if the process is convergent, it does not mean that it converges to the optimum value of the compensating capacitance. It is demonstrated by the following illustration.

Illustration 20.1 *The RL load of the series structure, shown in Fig. 20.1(a), when supplied with a sinusoidal voltage of the rms value* $U = 230$ V *has the active power* $P = 35$ kW *and the power factor* $\lambda = 0.5$.

Parameters of the equivalent parallel circuit, shown in Fig. 30.1(b), for the fundamental frequency, of the load are

$$G_1 = \frac{P_1}{U_1^2} = \frac{35 \times 10^3}{230^2} = 0.661\text{S}, \quad B_1 = -\frac{Q_1}{U_1^2} = -G_1 \text{tg}\,\varphi = -1.146\ \text{S}.$$

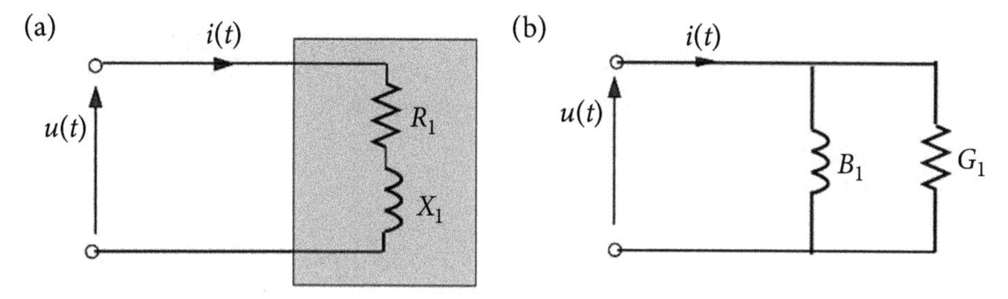

Figure 20.1 An example of the load (a) and its equivalent parallel circuit (b).

Parameters of the series RL load, shown in Fig. 20.1(a), are

$$R_1 = \frac{G_1}{G_1^2 + B_1^2} = \frac{0.661}{0.661^2 + 1.146^2} = 0.378\ \Omega, \quad X_1 = -\frac{B_1}{G_1^2 + B_1^2} = 0.654\ \Omega.$$

This load is supplied as shown in Fig. 20.2 from the voltage source with the short-circuit power $S_{sc} = 1\text{MVA}$ *and the reactance-to-resistance ratio for the fundamental frequency,* $X_{s1}/R_{s1} = 5$*, and the voltage harmonics of the relative value:*

$$E_1 = 230\,\text{V}, \quad E_5 = 3.0\%\,E_1, \qquad E_7 = 1.5\%\,E_1, \quad E_{11} = 0.5\%\,E_1.$$

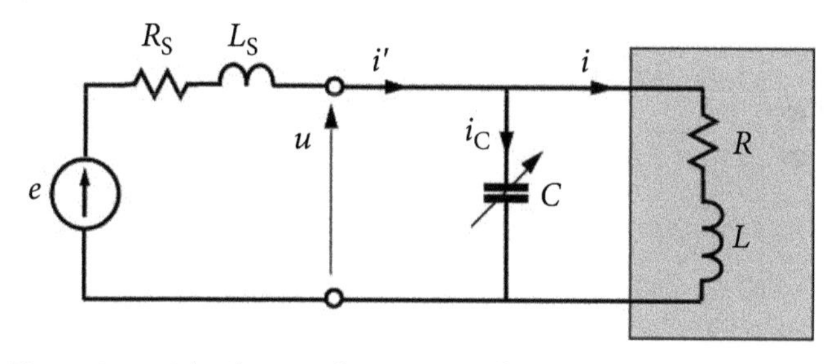

Figure 20.2 A load, a supply source, and a compensating capacitor.

The equivalent parameters of the supply source for the fundamental frequency are equal to

$$R_{s1} = 0.0103\,\Omega, \quad X_{s1} = 0.052\,\Omega.$$

The power factor λ *of the load is to be improved by a capacitor. The change of the power factor with capacitance C, at sinusoidal supply voltage and at its distortion as assumed in this illustration, is shown in Fig. 20.3. It was assumed in the modeling that the fundamental frequency is normalized to* $\omega_1 = 1$ *rad/s.*

The plots in Fig. 20.3 were found by modeling. When the supply voltage is sinusoidal, then the optimal capacitance C_O *can be calculated using the known formula for such a capacitance*

$$\omega_1 C_o = -B_1 = 1.146\,\text{S}.$$

When the supply voltage is nonsinusoidal then the resonance of the compensating capacitance with the supply source inductance L_s can occur, causing the sharp decline of the power factor, even at distortion that is very common in distribution systems.

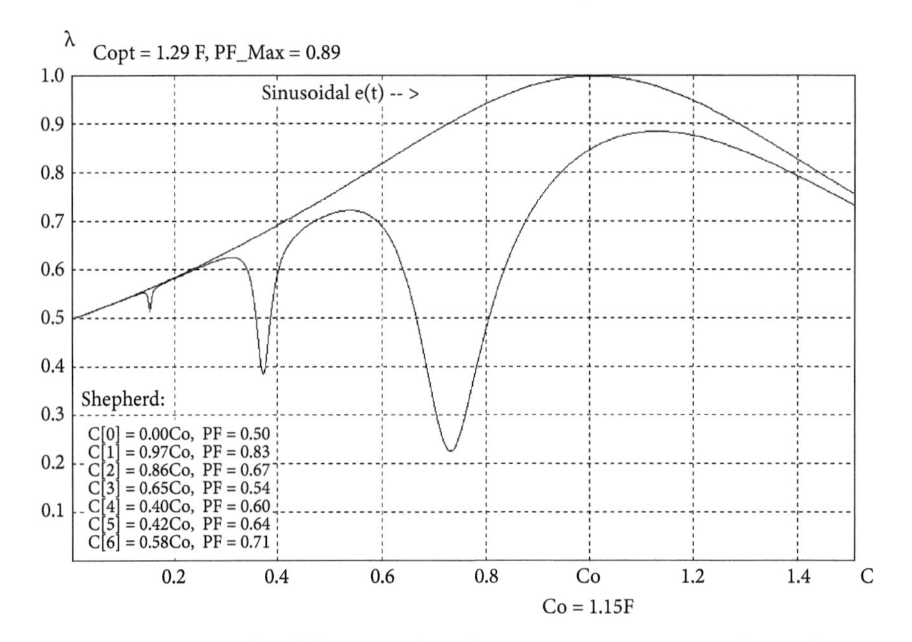

Figure 20.3 A plot of the power factor λ versus compensator capacitance C at sinusoidal and at distorted voltage.

******

The optimal capacitance at nonsinusoidal supply voltage was found in this illustration by modeling, This requires that parameters of the load and the supply source are available, however. The knowledge of only the voltage and current at the load terminals is not sufficient for such modeling.

The K&M PT was to provide a method of the optimal capacitance calculation, but in situations that are common in distribution systems, such as assumed in this illustration, the method can fail. Let us show this in detail.

Since the capacitive reactive power, as shown in Section 5.10 is

$$Q_C \stackrel{\text{df}}{=} ||u||\ ||i_{qC}||\text{sgn}(\dot{u}, i)$$

while for RL loads the function: $\text{sgn}(\dot{u},i) = -1$, hence, the formula (20.1) for the optimal capacitance can be simplified to

$$C_{\text{opt}} = \frac{||i_{qC}||}{||\dot{u}||}. \tag{20.4}$$

Calculation of this capacitance does not require that the rms value of the supply voltage and current harmonics are known, but when they are known, as in this illustration, they can be used for this capacitance calculation, according to formula (20.4). Formula (20.4) has to be expressed in terms of harmonics for that.

The capacitive reactive current is defined in the K&M power theory as

$$i_{qC}(t) \stackrel{\text{df}}{=} \frac{(\dot{u}, i)}{||\dot{u}||^2}\dot{u}(t).$$

Thus, at the supply voltage

$$u \stackrel{\text{df}}{=} U_0 + \sqrt{2}\,\text{Re} \sum_{n\in N} \boldsymbol{U}_n\, e^{jn\omega_1 t}$$

its derivative is

$$\dot{u} = \sqrt{2}\,\text{Re}\{\omega_1 \sum_{n\in N} jn\, \boldsymbol{U}_n\, e^{jn\omega_1 t}\}$$

and its rms value

$$||\dot{u}|| = \omega_1 \sqrt{\sum_{n\in N} n^2\, U_n^2}.$$

Since

$$i = I_0 + \sqrt{2}\,\text{Re} \sum_{n\in N} \boldsymbol{I}_n\, e^{jn\omega_1 t} = G_0\, I_0 + \sqrt{2}\text{Re} \sum_{n\in N} \left(G_n + jB_n\right)\boldsymbol{U}_n e^{jn\omega_1 t}$$

the scalar product

$$(\dot{u}, i) = \text{Re} \sum_{n\in N} jn\omega_1 \boldsymbol{U}_n \boldsymbol{I}_n^* = \omega_1 \sum_{n\in N} nB_n\, U_n^2.$$

The rms value of the capacitive reactive current is

$$||i_{qC}|| = \frac{|(\dot{u}, i)|}{||\dot{u}||}$$

hence

$$C_{\text{opt}} = \frac{||i_{qC}||}{||\dot{u}||} = \frac{|(\dot{u}, i)|}{||\dot{u}||^2} = \frac{|\sum_{n\in N} nB_n\, U_n^2|}{\omega_1 \sum_{n\in N} n^2\, U_n^2}. \tag{20.5}$$

This is the same formula as suggested by Shepherd and Zakikhani's PT, discussed in Section 5.6.

The load susceptance for the supply voltage harmonics is equal to

$$B_1 = -1.146\text{ S},\quad B_5 = -0.301\text{ S},\qquad B_7 = -0.217\text{ S},\quad B_{11} = -0.138\text{ S}.$$

The rms value of the harmonics at the load terminals without a capacitor is equal to

$$U_1 = 215.7\text{ V},\quad U_5 = 6.4\text{ V},\quad U_7 = 3.2\text{ V},\quad U_{11} = 0.32\text{ V}.$$

The capacitance calculated from formula (20.5) is equal to $C = 1.11$ F. and will be denoted, as the first calculated value by C[1]. Such a capacitor improves the power factor to $\lambda = 0.83$. It is not, however, the maximum value of the power factor. The plot of the power factor λ versus compensating capacitance C, shown in Fig. 20.3, shows that this maximum value is $\lambda_{max} = 0.89$ at optimum capacitance $C_{opt} = 1.29$ F.

When a capacitor of the previously calculated capacitance C[1] = 1.11 F is connected to the circuit, it changes the rms value of the voltage harmonics at the load terminals to

$$U_1 = 227.9\text{ V},\quad U_5 = 19.0\text{ V},\quad U_7 = 2.0\text{ V},\quad U_{11} = 0.2\text{ V}.$$

Since the capacitor affects the rms values U_n of the load voltage harmonics, one could suggest that an iterative calculation might provide the valid value of the compensator capacitance. Such an iterative approach requires that the capacitance value is updated and calculations are repeated, resulting in a sequence of capacitances, $C[n]$. Such sequence should converge to C_{opt}.

The capacitance calculated at modified voltage harmonics is C[2] = 0.98 F. The power factor at such capacitance is $\lambda = 0.67$ and it changes the rms values of the load voltage harmonics to

$$U_1 = 226.4\text{ V},\quad U_5 = 34.2\text{ V},\quad U_7 = 2.4\text{ V},\quad U_{11} = 0.2\text{ V}$$

and the capacitance recalculated at such voltage harmonics is C[3] = 0.75 F. Thus, the process of the capacitance calculation should be again repeated.

Unfortunately, the process of sequential calculation of the optimal capacitance is not convergent. One could observe that the process of the optimal capacitance calculation is not convergent for a common load at the supply voltage distortion common in distribution systems. Thus, the K&M formula (20.1) for the optimum capacitance calculation can be useless in common industrial conditions.

It is because the K&M formula (20.1) was developed under the assumption that the load voltage does not depend on the compensating capacitor. This is true only at an infinite short-circuit power of the supply source, meaning for sources with zero internal impedance. In common industrial situations, such as those in the last illustration, where the supply source impedance is strongly inductive, the capacitor affects the load voltage harmonics to such a degree that this formula provides an erroneous value of the compensating capacitance or the process of this capacitance calculation is not convergent at all.

Summary

The method developed by Kusters and Moore applies only to a capacitive compensation which, however, should be used in a situation of the supply voltage distortion because such compensation can cause an amplification of this distortion. Moreover, it can result in an adverse effect on the power factor than expected.

Chapter 21
Misinterpretations of the Instantaneous Reactive Power p-q Theory

21.1 Could Three-Phase Loads be Identified Instantaneously?

According to the authors of the Instantaneous Reactive Power (IRP) p-q Theory, presented in [72] in 1983, its development was a response to "the demand to instantaneously compensate the reactive power". They asserted in Ref. [111] that this concept has been developed to enable instantaneous compensation of fast varying loads.

The adverb "*instantaneous*" in the name of this theory and definitions of p and q powers in terms of the instantaneous value of voltages and currents, suggests the possibility of instantaneous identification and compensation of the reactive power of three-phase loads. This suggestion was supported by some authors who claimed [186, 209] that using IRP p-q Theory as a control algorithm makes instantaneous compensation possible.

Indeed, power properties of three-phase three-wire electrical loads are specified in terms of the IRP p-q Theory, (see Section 5.13) by only two powers, namely the instantaneous active power

$$p \stackrel{\text{df}}{=} u_\alpha i_\alpha + u_\beta i_\beta \tag{21.1}$$

and the instantaneous reactive power

$$q \stackrel{\text{df}}{=} u_\alpha i_\beta - u_\beta i_\alpha \tag{21.2}$$

These powers are defined in the IRP p-q Theory in terms of three-phase voltages and currents transformed to α and β coordinates by the Clarke Transform. For quantities that satisfy the identity

$$x_\text{R}(t) + x_\text{S}(t) + x_\text{T}(t) \equiv 0$$

it converts the phase quantities from lines R, S, and T into α, β coordinates, with the Transform

$$\begin{bmatrix} x_\alpha \\ x_\beta \end{bmatrix} \stackrel{\text{df}}{=} \begin{bmatrix} \sqrt{3/2}, & 0 \\ 1/\sqrt{2}, & \sqrt{2} \end{bmatrix} \begin{bmatrix} x_\text{R} \\ x_\text{S} \end{bmatrix} \stackrel{\text{df}}{=} \mathbf{C} \begin{bmatrix} x_\text{R} \\ x_\text{S} \end{bmatrix}. \tag{21.3}$$

Powers and Compensation in Circuits with Nonsinusoidal Current. Leszek S. Czarnecki, Oxford University Press.
 DOI: 10.1093/oso/9780198879206.003.0021

These two instantaneous powers p and q can be measured/calculated almost instantaneously. Only a few multiplications and additions are needed for that, which means these quantities can provide, according to [72], data for instantaneous compensation. Does it mean, however, that having instantaneous values of these two powers, then the power properties of the load are identified instantaneously? The answer to this question is important because without instantaneous identification of these properties, instantaneous compensation is not possible. To verify the claim that the power properties of the load can be identified instantaneously, let us analyze two circuits in the following illustrations.

Illustration 21.1 *Let us assume that a resistive load, connected as shown in Figure 21.1, is supplied from a source of sinusoidal, symmetrical, positive-sequence voltage, such that*

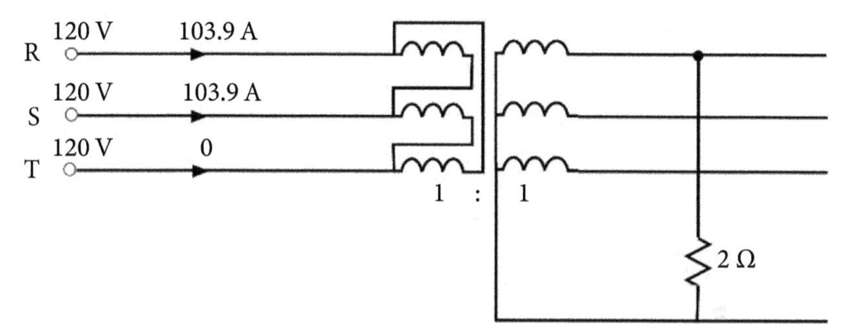

Figure 21.1 An example of the circuit with a resistive unbalanced load.

$$u_{\mathrm{R}} = \sqrt{2}\, U \cos\omega_1 t, \quad U = 120\ V.$$

At such assumptions, the supply voltage in α and β coordinates is equal to

$$\begin{bmatrix} u_\alpha \\ u_\beta \end{bmatrix} = \mathbf{C} \begin{bmatrix} \sqrt{2}\, U \cos \omega_1 t \\ \sqrt{2}\, U \cos(\omega_1 t - 120^{\circ}) \end{bmatrix} = \begin{bmatrix} \sqrt{3}\, U \cos \omega_1 t \\ \sqrt{3}\, U \sin \omega_1 t \end{bmatrix}$$

Since the line currents are equal to

$$i_{\mathrm{R}} = \sqrt{2}\, I \cos(\omega_1 t + 30^{\circ}) = -i_S, \quad I = 103.9\,\mathrm{A}, \quad i_T = 0$$

the supply current in the α and β coordinates is

$$\begin{bmatrix} i_\alpha \\ i_\beta \end{bmatrix} = \mathbf{C} \begin{bmatrix} i_{\mathrm{R}} \\ -i_{\mathrm{R}} \end{bmatrix} = \begin{bmatrix} \sqrt{3}\, I \cos(\omega_1 t + 30^{\circ}) \\ -I \cos(\omega_1 t + 30^{\circ}) \end{bmatrix}$$

and consequently, the instantaneous active power of such a load is

$$p = u_\alpha i_\alpha + u_\beta i_\beta = \sqrt{3}UI\,[1 + \cos 2(\omega_1 t + 30^\mathrm{o})] \tag{21.4}$$

and the instantaneous reactive power is

$$q = u_\alpha i_\beta - u_\beta i_\alpha = -\sqrt{3}\ UI \sin 2(\omega_1 t + 30^\mathrm{o}). \tag{21.5}$$

Let us observe that at

$$\omega_1 t + 30^\mathrm{o} = 45^\mathrm{o}$$

the instantaneous powers satisfy the relationship

$$p = -q. \tag{21.6}$$

Now, let us calculate the same powers in a purely reactive circuit shown in Figure 21.2, supplied as previously.

The line currents in such a circuit are equal to

$$i_\mathrm{R} = \sqrt{2}\,I \cos(\omega_1 t - 60^\mathrm{o}), \quad i_S = -i_\mathrm{R}, \quad i_T = 0, \quad I = 103.92\mathrm{A}$$

thus, the line currents in α and β coordinates are equal to

$$\begin{bmatrix} i_\alpha \\ i_\beta \end{bmatrix} = \mathbf{C} \begin{bmatrix} \sqrt{2}\,I\cos(\omega_1 t - 60^\mathrm{o}) \\ -\sqrt{2}\,I\cos(\omega_1 t - 60^\mathrm{o}) \end{bmatrix} = \begin{bmatrix} \sqrt{3}\,I\cos(\omega_1 t - 60^\mathrm{o}) \\ -I\cos(\omega_1 t - 60^\mathrm{o}) \end{bmatrix}.$$

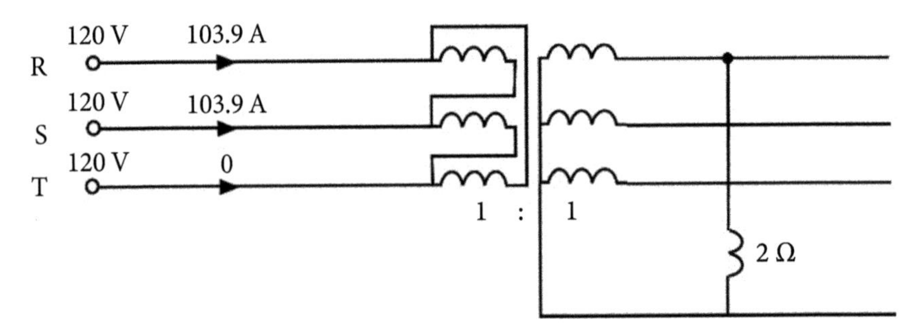

Figure 21.2 An example of a three-phase circuit with a purely reactive load.

Consequently, the instantaneous real and imaginary powers are

$$p = u_\alpha i_\alpha + u_\beta i_\beta = \sqrt{3}\, UI\cos(2\omega_1 t - 30^\circ). \tag{21.7}$$

$$q = u_\alpha i_\beta - u_\beta i_\alpha = -\sqrt{3}\, UI\,[1 + \sin(2\omega_1 t - 30^\circ)]. \tag{21.8}$$

Let us observe that at

$$2\omega_1 t = 30^\circ$$

the instantaneous powers satisfy the relationship

$$p = -q. \tag{21.9}$$

Thus, according to (21.6) and (21.9), two entirely different loads can have identical pairs of instantaneous powers p and q. It means that knowing such a pair, the loads shown in Figs 21.1 and Figure 21.2 cannot be distinguished from each other. Loads cannot be identified with powers p and q instantaneously.

***

One could observe, moreover, that the instantaneous reactive power q can occur in a purely resistive load, that is one with zero reactive power Q, as it is in the circuit shown in Figure 21.1. Also, the instantaneous active power p can occur in a purely reactive load, that is with zero active power P, as it is in the circuit shown in Figure 21.2. Thus, the names "instantaneous active power" and "instantaneous reactive power" do not fit the traditional and well-established adjectives commonly used in electrical engineering: "active" and "reactive". This is even more visible in the circuit shown in Figure 21.3, with parameters of the load selected such that both the active and reactive powers, P and Q, are zero.

Illustration 21.2 *Let us assume that a load, connected as shown in Figure 21.3, is supplied as in the circuit shown in Figure 21.1.*

The line currents in such a circuit are equal to

$$i_R = \sqrt{2}\, I\cos(\omega_1 t - 60^\circ), \quad i_S = \sqrt{2}\, I\cos(\omega_1 t + 60^\circ), \quad I = 103.92\ \text{A}$$

thus the line currents in α and β coordinates are equal to

$$\begin{bmatrix} i_\alpha \\ i_\beta \end{bmatrix} = \mathbf{C}\begin{bmatrix} \sqrt{2}\, I\cos(\omega_1 t - 60^\circ) \\ \sqrt{2}\, I\cos(\omega_1 t + 60^\circ) \end{bmatrix} = \begin{bmatrix} \sqrt{3}\, I\cos(\omega_1 t - 60^\circ) \\ -\sqrt{3}\, I\sin(\omega_1 t - 60^\circ) \end{bmatrix}.$$

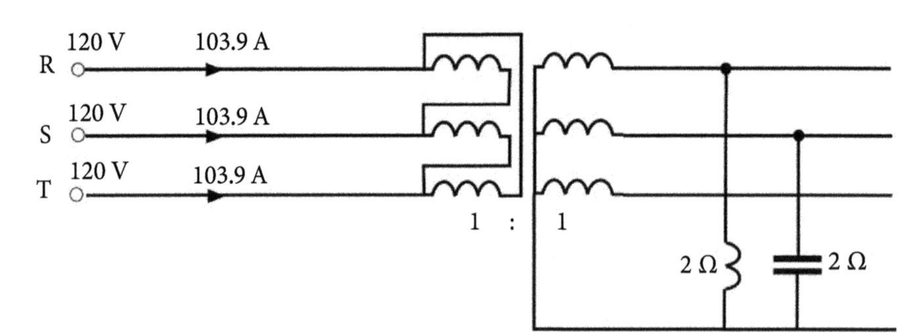

Figure 21.3 An example of a three-phase circuit with zero active and reactive powers but with instantaneous active and reactive powers.

Hence the instantaneous active and reactive powers are equal to

$$p = u_\alpha i_\alpha + u_\beta i_\beta = 3UI\cos(2\omega_1 t - 60^\circ) \tag{21.10}$$

$$q = u_\alpha i_\beta - u_\beta i_\alpha = -3UI\sin(2\omega_1 t - 60^\circ). \tag{21.11}$$

Thus the instantaneous active and reactive powers occur in the circuit despite of zero active and reactive powers. Also, when

$$\cos(2\omega_1 t - 60^\circ) = \sin(2\omega_1 t - 60^\circ)$$

then

$$p = -q.$$

*****

It means that a pair of instantaneous powers is observed which is identical to that in the circuit with a purely resistive load shown in Figure 21.1 and the circuit with a purely reactive load shown in Figure 21.2. Moreover, the adjectives "active" and "reactive" with regard to the instantaneous powers p and q have nothing in common with the resistivity or reactivity of the load. Using them can lead to a misinterpretation of the power properties of electrical circuits and the possibility of their compensation. One could observe, moreover, that the active instantaneous power is nothing else than the instantaneous power $p(t)$, defined as the rate of the energy flow from the supply source to the load.

21.2 Instantaneous Powers and Load Identification

As shown in the previous section, three-phase loads cannot be identified instantaneously in terms of the instantaneous active and reactive powers, p and q. Even a purely resistive load cannot be distinguished from a purely reactive one. Would such identification be possible, however, having these two powers recorded over the whole period T?

The answer is no. To show this, let us express the instantaneous powers p and q in terms of line voltages and currents. The instantaneous active and reactive powers p and q are defined in terms of the IRP p-q Theory using the load line voltages and currents transformed to the α and β coordinates. Clark's Transform is a linear operation, however, thus it has to preserve the power properties of the load. Unfortunately, it hides the physical meaning of these electrical powers. Therefore, to grasp this meaning it is better to return to line voltages and currents expressed in three-phase coordinates, R, S, and T.

The instantaneous active power is nothing else than the instantaneous power $p(t)$ and in three-phase, three-wire circuits it can be expressed as

$$p(t) = \boldsymbol{u}(t)^{\mathrm{T}}\boldsymbol{i}(t) = u_{\mathrm{RT}}(t)i_{\mathrm{R}}(t) + u_{\mathrm{ST}}(t)i_{\mathrm{S}}(t). \tag{21.12}$$

Voltages and currents in α and β coordinates are related to these quantities in phase coordinates by transform (21.3), thus, the instantaneous reactive power q can be rearranged to the form

$$\begin{aligned} q &= u_{\alpha}i_{\beta} - u_{\beta}i_{\alpha} = \\ &= \sqrt{\tfrac{3}{2}}\, u_{\mathrm{R}}\left(\tfrac{1}{\sqrt{2}}i_{\mathrm{R}} + \sqrt{2}\,i_{\mathrm{S}}\right) - \left(\tfrac{1}{\sqrt{2}}u_{\mathrm{R}} + \sqrt{2}\,u_{\mathrm{S}}\right)\sqrt{\tfrac{3}{2}}\, i_{\mathrm{R}} = \sqrt{3}\,(u_{\mathrm{R}}i_{\mathrm{S}} - u_{\mathrm{S}}\,i_{\mathrm{R}}). \end{aligned} \tag{21.13}$$

Let us assume that the supply voltage is symmetrical and sinusoidal, thus $\boldsymbol{u}(t) = \boldsymbol{u}_1(t)$, while the current is symmetrical, but distorted by harmonics generated in the load:

$$\boldsymbol{i}(t) = \sum_{n\in N} \boldsymbol{i}_n(t).$$

At such assumptions, the formula for the instantaneous active power p can be expressed as follows

$$p(t) = \boldsymbol{u}^{\mathrm{T}}(t)\boldsymbol{i}(t) = \boldsymbol{u}_1^{\mathrm{T}}(t)\sum_{n\in N}\boldsymbol{i}_n(t) = \sum_{n\in N}\boldsymbol{u}_1^{\mathrm{T}}(t)\boldsymbol{i}_n(t) = \sum_{n\in N} p_n(t) \tag{21.14}$$

where

$$p_n(t) = p_n = u_{\mathrm{RT}1}i_{\mathrm{R}n} + u_{\mathrm{ST}1}i_{\mathrm{S}n}. \tag{21.15}$$

Observe, that notation in the last formula was simplified by neglecting the symbol of time t. The same will apply to the formula below.

The instantaneous reactive power q can be also expressed as the sum of this power for individual harmonics, namely

$$q(t) = q = u_\alpha i_\beta - u_\beta i_\alpha = \sqrt{3}\,(u_\mathrm{R} i_\mathrm{S} - u_\mathrm{S} i_\mathrm{R}) = \sqrt{3}\left(u_{\mathrm{R}1} \sum_{n\in N} i_{\mathrm{S}n} - u_{\mathrm{S}1} \sum_{n\in N} i_{\mathrm{R}n}\right) =$$

$$= \sqrt{3}\left(\sum_{n\in N} u_{\mathrm{R}1} i_{\mathrm{S}n} - \sum_{n\in N} u_{\mathrm{S}1} i_{\mathrm{R}n}\right) = \sum_{n\in N} q_n = \sum_{n\in N} q_n(t) \quad (21.16)$$

where

$$q_n(t) = q_n = \sqrt{3}(u_{\mathrm{R}1} i_{\mathrm{S}n} - u_{\mathrm{S}1} i_{\mathrm{R}n}). \quad (21.17)$$

Illustration 21.3 *Let us calculate the instantaneous power for a balanced resistive load shown in Fig. 21.4, which generates a current harmonic of the fifth order.*

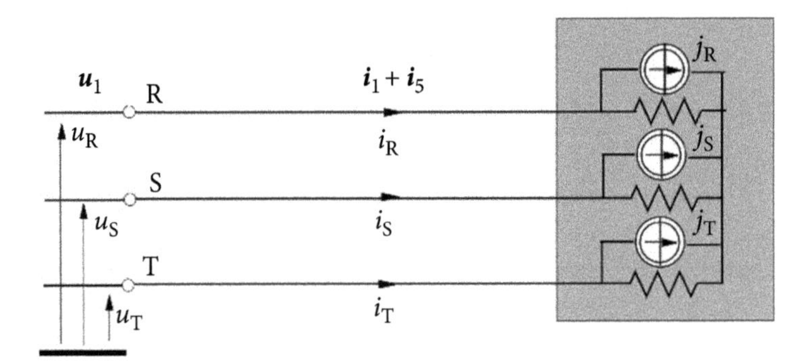

Figure 21.4 A resistive balanced load that generates the 5th order current harmonic.

Let us assume that the supply voltage and the current at terminal R *are*

$$u_\mathrm{R} = u_{\mathrm{R}1} = \sqrt{2}\, U_1 \cos \omega_1 t,\; i_\mathrm{R} = \sqrt{2}\, I_1 \cos \omega_1 t + \sqrt{2}\, I_5 \cos (5\omega_1 t + \alpha_5).$$

At such assumptions, line-to-line voltages have waveforms

$$u_{\mathrm{RT}1} = \sqrt{6}\, U_1 \cos (\omega_1 t - 30^\mathrm{o})$$
$$u_{\mathrm{ST}1} = \sqrt{6}\, U_1 \cos (\omega_1 t - 90^\mathrm{o}).$$

Since the fifth order current harmonic is a negative sequence harmonic, then

$$i_\mathrm{S} = \sqrt{2}\, [I_1 \cos (\omega_1 t - 120^\mathrm{o}) + I_5 \cos (5\omega_1 t + \alpha_5 + 120^\mathrm{o})].$$

At such conditions, the instantaneous active power is equal to

$$p = u_{\mathrm{RT}}\, i_{\mathrm{R}} + u_{\mathrm{ST}}\, i_{\mathrm{S}} = 2\sqrt{3}\,U_1 \cos(\omega_1 t - 30^{\mathrm{o}})[I_1 \cos\omega_1 t + I_5 \cos(5\omega_1 t + \alpha_5)] +$$
$$+ \sqrt{3}\,U_1 \cos(\omega_1 t - 90^{\mathrm{o}})[I_1 \cos(\omega_1 t - 120^{\mathrm{o}}) +$$
$$+ I_5 \cos(5\omega_1 t + \alpha_5 + 120^{\mathrm{o}})] =$$
$$= 3U_1 I_1 + 3U_1 I_5 \cos(6\omega_1 t + \alpha_5) =$$
$$= p_1 + p_5 = \bar{p} + \tilde{p}. \tag{2.18}$$

The instantaneous active power is decomposed in formula (21.18) directly into the constant and oscillating components. These two components were obtained explicitly only because waveforms of voltages and currents as well as rms values U_1, I_1, and I_5 were assumed to be known in this illustration. The IRP p-q Theory is formulated in the time domain, so that it does not need Fourier analysis of voltages and currents. Without this analysis, the components of the instantaneous active power p, which was calculated by formula. (21.18), meaning the constant and the oscillating components, $\bar{p}$ and $\tilde{p}$, *are not known explicitly, however. A Digital Signal Processing (DSP) system provides only sequential values of p at instants t_k. A low-pass or a high-pass filter is needed for the decomposition of this power into the constant and oscillating components. The result of filtering for the situation as discussed has the form*

$$\bar{p} = A,\ \tilde{p} = B\cos(6\omega_1 t + \alpha_5) \tag{21.19}$$

but the filter does not provide any information on factors, shown in formula (21.18) that contribute to A and B values.

The instantaneous reactive power of such a load, according to formula (21.16) is

$$q = q_1 + q_5.$$

The load in this illustration for the fundamental harmonic is resistive and balanced, so that

$$q_1(t) \equiv 0$$

while according to formula (21.17)

$$q_5 = \sqrt{3}\,(u_{\mathrm{R1}} i_{\mathrm{S5}} - u_{\mathrm{S1}} i_{\mathrm{R5}}) = 2\sqrt{3}\,U_1\{\cos\omega_1 t[I_5 \cos(5\omega_1 t + \alpha_5 + 120^{\mathrm{o}})] -$$
$$- \cos(\omega_1 t - 120^{\mathrm{o}})[I_5 \cos(5\omega_1 t + \alpha_5)]\} =$$
$$= 3U_1 I_5 \sin(6\omega_1 t + \alpha_5 + \pi). \tag{20.20}$$

Thus, at the output of the filter, we obtain

$$\bar{q} = 0, \qquad \tilde{q} = B \sin(6\omega_1 t + \beta) \tag{21.21}$$

but the filter does not provide any information on factors, shown in formula (21.20) that contribute to the amplitude B value.

To check how the order n of the load-originated current harmonic affects the instantaneous powers, let us assume that all other parameters of the circuit in Figure 21.5 are kept unchanged, but instead of the fifth order harmonic, the load shown in that figure generates the seventh order current harmonic, that is,

$$u_R = u_{R1} = \sqrt{2}\, U_1 \cos\omega_1 t, \; i_R = \sqrt{2}\, I_1 \cos\omega_1 t + \sqrt{2}\, I_7 \cos(7\omega_1 t + \alpha_7).$$

Since the seventh order current harmonic is the positive sequence, then the current in the line S is

$$i_S = \sqrt{2}\,[I_1 \cos(\omega_1 t - 120^o) + I_7 \cos(7\omega_1 t + \alpha_7 - 120^o)].$$

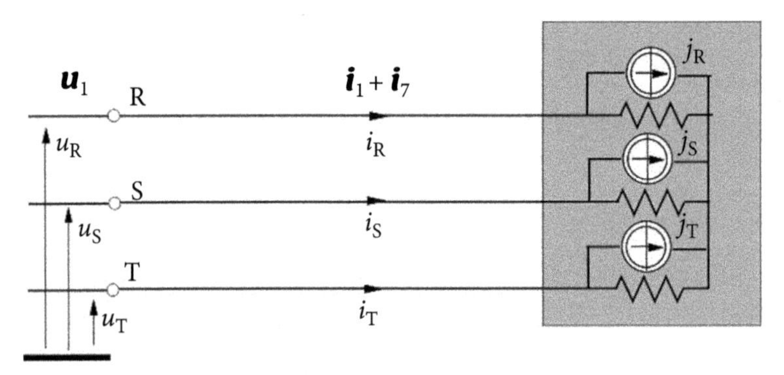

Figure 21.5 A resistive balanced load that generates the 7th order current harmonic.

The instantaneous active power is

$$\begin{aligned} p = u_{RT} i_R + u_{ST} i_S &= 2\sqrt{3}\, U_1 \cos(\omega_1 t - 30^o)[I_1 \cos\omega_1 t + I_7 \cos(7\omega_1 t + \alpha_7)] + \\ &+ \sqrt{3}\, U_1 \cos(\omega_1 t - 90^o)[I_1 \cos(\omega_1 t - 120^o) + I_7 \cos(7\omega_1 t + \alpha_7 - 120^o)] = \\ &= 3U_1 I_1 + 3U_1 I_7 \cos(6\omega_1 t + \alpha_7) = \\ &= p_1 + p_7 = \bar{p} + \tilde{p}. \end{aligned} \tag{21.22}$$

Thus, the change in the order of the load-generated current harmonic from n = 5 to n = 7 does not affect the instantaneous active power. After filtering the DSP output signal, we obtain, as previously

$$\bar{p} = A, \; \tilde{p} = B\cos(6\omega_1 t + \alpha_7) \tag{21.23}$$

thus only the initial phase angle can be different, meaning equal to α_5 or α_7.

The instantaneous reactive power associated with the presence of the seventh order load-originated current harmonic is

$$q_7 = \sqrt{3}\,(u_{R1}i_{S7} - u_{S1}i_{R7}) = 2\sqrt{3}\,U_1\{\cos\omega_1 t[I_7\cos(7\omega_1 t + \alpha_7 - 120^{\circ})]-$$
$$-\cos(\omega_1 t - 120^{\circ})[I_7\cos(7\omega_1 t + \alpha_7)]\} =$$
$$= 3U_1I_7\sin(6\omega_1 t + \alpha_7) \tag{21.24}$$

thus, only the phase of this power is affected, but not its frequency. A low-pass filter provides the values

$$\bar{q} = 0, \qquad \tilde{q} = B\sin(6\omega_1 t + \beta). \tag{21.25}$$

It means that the loads shown in Figs. 21.4 and 21.5 cannot be distinguished in terms of the instantaneous powers, even if their values over the whole period are recorded. In terms of p and q powers, these two loads are identical. It creates a challenge for the interpretation of the instantaneous powers. This challenge is demonstrated in the following illustration.

Illustration 21.4 *Now, let us assume that the load, shown in Figure 21.6, generates both the fifth and the seventh order current harmonics.*

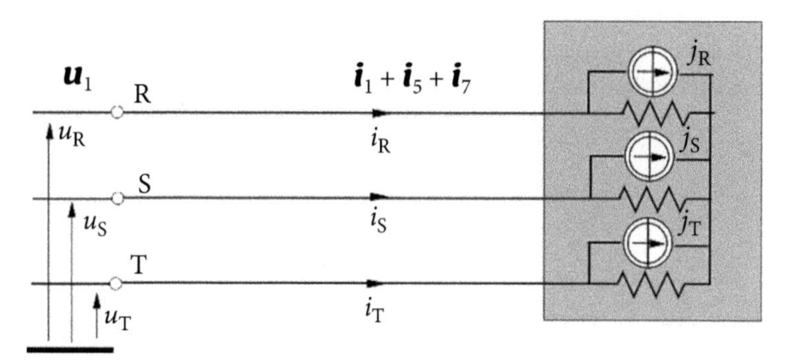

Figure 21.6 A resistive balanced load that generates the 5th and 7th order current harmonics.

The oscillating component of the instantaneous active power for such a load is

$$\tilde{p} = p_5 + p_7 = 3U_1[I_5\cos(6\omega_1 t + \alpha_5) + I_7\cos(6\omega_1 t + \alpha_7)] \tag{21.26}$$

while the instantaneous reactive power is

$$\tilde{q} = q_5 + q_7 = 3U_1[I_5\sin(6\omega_1 t + \alpha_5 + \pi) + I_7\sin(6\omega_1 t + \alpha_7)]. \tag{21.27}$$

Formulae (21.26) and (21.27) show that oscillating components of the instantaneous active and reactive powers are not associated with particular power properties of *the load, but only with rms values of harmonics, I_5, I_7, and their phases α_5, α_7. In particular, if $I_7 = I_5$ and $\alpha_7 = \alpha_5$ then*

$$\tilde{q} \equiv 0, \quad \tilde{p} = 2p_5 = 6\, U_1 I_5 \cos(6\omega_1 t + \alpha_5) \tag{21.28}$$

while if $\alpha_7 = \alpha_5 + \pi$, then

$$\tilde{p} \equiv 0, \quad \tilde{q} = 2q_5 = -6\, U_1 I_5 \sin(6\omega_1 t + \alpha_5). \tag{21.29}$$

*****

Observe that the phase angle α of the load current harmonics does not affect the power properties of the system, but it affects the instantaneous powers p and q. Thus, these powers are not associated with any particular power properties of the load.

21.3 IRP p-q Theory Compensation Objective Misconceptions

The IRP p-q Theory was developed [72] mainly to provide control fundamentals for switching compensators, referred to as "active power filters". According to IRP p-q Theory, each load is specified by the instantaneous active power, which has a constant and an oscillating part, namely

$$p = \bar{p} + \tilde{p}$$

and the instantaneous reactive power q. A switching compensator can be controlled in such a way that it burdens the distribution system with specified instantaneous active and reactive powers, p_C and q_C. According to the IRP p-q Theory, the switching compensator should be controlled in such a way that after compensation the instantaneous reactive power at the supply terminals is zero and the instantaneous active power is constant. It means that the compensator should burden the distribution system with the negative value of the oscillating component of the instantaneous active power and the negative value of the instantaneous reactive power, as is shown in Figure 21.7.

This conclusion of the IRP p-q Theory as to the compensation goal was challenged in papers [242] and [245], where it was shown that it is not correct.

It was assumed in these papers that an ideal, meaning purely resistive and balanced load, the load with unity power factor, thus not requiring any compensation, shown in Figure 21.8, is compensated with a switching compensator controlled according to the IRP p-q Theory.

Let us assume that the supply voltage is distorted by only one harmonic of the fifth order. It can be presented in the form

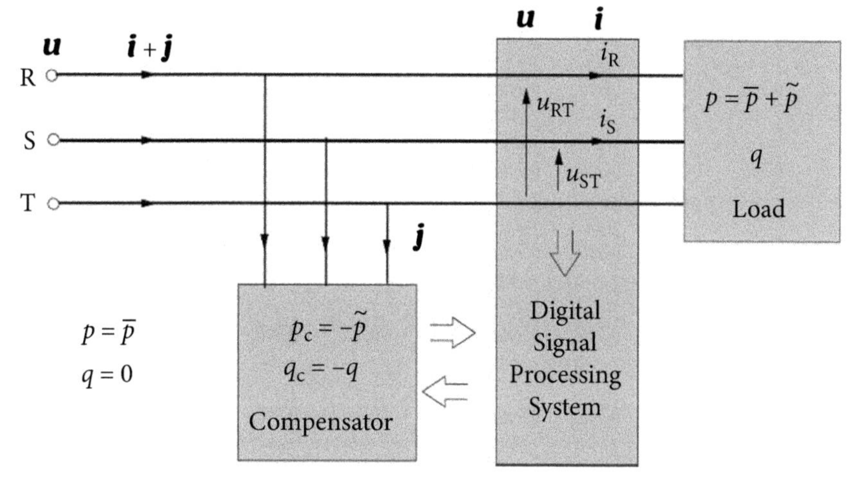

Figure 21.7 Control of a switching compensator according to IRP p-q Theory.

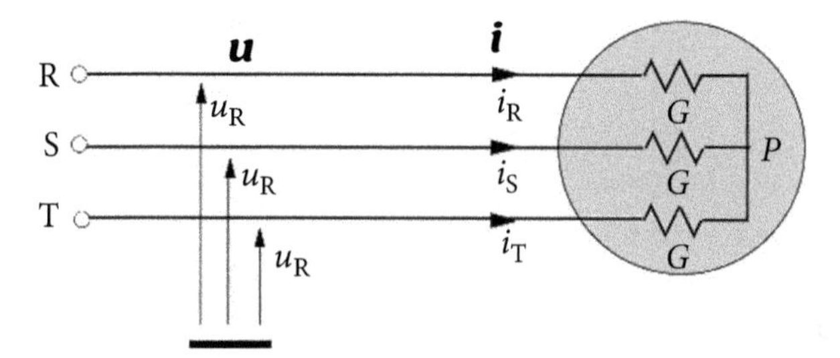

Figure 21.8 An ideal purely resistive and balanced load.

$$\boldsymbol{u}(t) = \boldsymbol{u} = \begin{bmatrix} u_R \\ u_S \\ u_T \end{bmatrix} = \begin{bmatrix} u_{R1} \\ u_{S1} \\ u_{T1} \end{bmatrix} + \begin{bmatrix} u_{R5} \\ u_{S5} \\ u_{T5} \end{bmatrix} = \boldsymbol{u}_1 + \boldsymbol{u}_5.$$

Let for the sake of simplicity assume that

$$u_{R1} = \sqrt{2}\, U_1 \cos \omega_1 t,\; u_{R5} = \sqrt{2}\, U_5 \cos 5\omega_1 t.$$

The load current at such a supply contains the fifth order harmonic and can be presented in the form

$$\boldsymbol{i}(t) = \boldsymbol{i} = \begin{bmatrix} i_R \\ i_S \\ i_T \end{bmatrix} = \begin{bmatrix} i_{R1} \\ i_{S1} \\ i_{T1} \end{bmatrix} + \begin{bmatrix} i_{R5} \\ i_{S5} \\ i_{T5} \end{bmatrix} = \boldsymbol{i}_1 + \boldsymbol{i}_5$$

with

$$i_{R1} = \sqrt{2}\, GU_1 \cos \omega_1 t,\; i_{R5} = \sqrt{2}\, GU_5 \cos 5\omega_1 t.$$

At such a voltage, the instantaneous power $p(t)$ is equal to

$$p(t) = \frac{d}{dt}W(t) = \boldsymbol{u}^{\mathrm{T}}(t)\boldsymbol{i}(t) = [\boldsymbol{u}_1 + \boldsymbol{u}_5]^{\mathrm{T}}[\boldsymbol{i}_1 + \boldsymbol{i}_5].$$

It can be rearranged to the form

$$p(t) = [\boldsymbol{u}_1 + \boldsymbol{u}_5]^{\mathrm{T}}[\boldsymbol{i}_1 + \boldsymbol{i}_5] = \boldsymbol{u}_1^{\mathrm{T}}\boldsymbol{i}_1 + \boldsymbol{u}_5^{\mathrm{T}}\boldsymbol{i}_5 + \boldsymbol{u}_5^{\mathrm{T}}\boldsymbol{i}_1 + \boldsymbol{u}_1^{\mathrm{T}}\boldsymbol{i}_5 =$$
$$= P_1 + P_5 + 6\ G\ U_1\ U_5 \cos 6\omega_1 t = \overline{p} + \tilde{p} \qquad (21.30)$$

where P_1 and P_5 denote the active power of the first and the fifth order harmonics. This formula shows that, despite the unity power factor, the instantaneous power $p(t)$ of such a load has the oscillating component

$$\tilde{p} = 6\ G\ U_1 U_5 \cos 6\omega_1 t. \qquad (21.31)$$

The presence of the oscillating component of the instantaneous active power does not degrade the power factor of the load. In its presence, the power factor can be equal to 1. Therefore, the conclusion of the IRP p-q Theory that it should be compensated is erroneous. If one were to follow this conclusion, however, and connect a switching compensator, as shown in Figure 21.9.

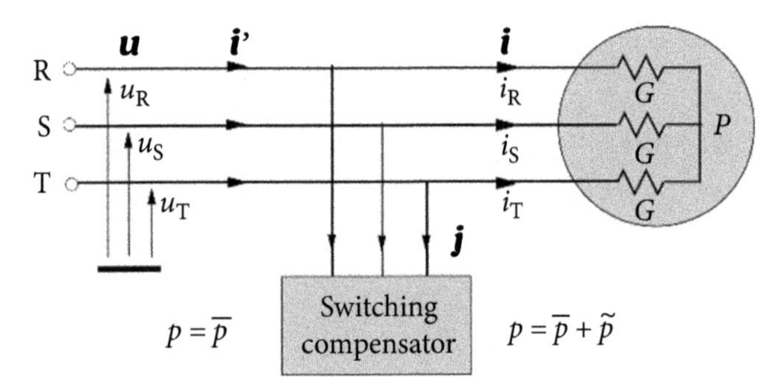

Figure 21.9 A balanced purely resistive load with a switching compensator.

then the compensator controlled according to the IRP p-q Theory would inject, as shown in paper [242], the following current

$$j(t) = \begin{bmatrix} j_{\mathrm{R}}(t) \\ j_{\mathrm{S}}(t) \end{bmatrix} = \frac{-2\sqrt{2}\ GU_1U_5 \cos(6\omega_1 t)}{U_1^2 + U_5^2 + 2U_1U_5\cos(6\omega_1 t)} \begin{bmatrix} u_{\mathrm{R}}(t) \\ u_{\mathrm{S}}(t) \end{bmatrix}$$

into the load supply lines. This is a nonsinusoidal current, thus it would increase the supply current distortion and reduce the power factor.

Similar conclusions as to the IRP p-q Theory compensation goal can be drawn when the supply voltage of an ideal load, as in Figure 21.8, is sinusoidal but asymmetrical.

To show this, let us assume that the voltages of the positive and negative sequences at terminal R are equal, respectively, to

$$u_{\mathrm{R}}^{\mathrm{p}} = \sqrt{2}\,U^{\mathrm{p}} \cos \omega_1 t, \qquad u_{\mathrm{R}}^{\mathrm{n}} = \sqrt{2}\,U^{\mathrm{n}} \cos \omega_1 t.$$

Thus, the supply voltage can be presented in the form

$$\boldsymbol{u} = \sqrt{2}U^{\mathrm{p}} \begin{bmatrix} \cos \omega_1 t \\ \cos(\omega_1 t - 2\pi/3) \\ \cos(\omega_1 t + 2\pi/3) \end{bmatrix} + \sqrt{2}\,U^{\mathrm{n}} \begin{bmatrix} \cos \omega_1 t \\ \cos(\omega_1 t + 2\pi/3) \\ \cos(\omega_1 t - 2\pi/3) \end{bmatrix} = \boldsymbol{u}^{\mathrm{p}} + \boldsymbol{u}^{\mathrm{n}}.$$

At such a voltage, the load current is also asymmetrical, namely

$$\boldsymbol{i} = G\boldsymbol{u} = G(\boldsymbol{u}^{\mathrm{p}} + \boldsymbol{u}^{\mathrm{n}}) = \boldsymbol{i}^{\mathrm{p}} + \boldsymbol{i}^{\mathrm{n}}.$$

The instantaneous power $p(t)$ of such a load is equal to

$$p(t) = \frac{d}{dt}W(t) = \boldsymbol{u}^{\mathrm{T}}(t)\boldsymbol{i}(t) = [\boldsymbol{u}^{\mathrm{p}} + \boldsymbol{u}^{\mathrm{n}}]^{\mathrm{T}}[\boldsymbol{i}^{\mathrm{p}} + \boldsymbol{i}^{\mathrm{n}}]$$

and it can be presented in the form

$$p(t) = [\boldsymbol{u}^{\mathrm{p}} + \boldsymbol{u}^{\mathrm{n}}]^{\mathrm{T}}[\boldsymbol{i}^{\mathrm{p}} + \boldsymbol{i}^{\mathrm{n}}] = \boldsymbol{u}^{\mathrm{pT}}\boldsymbol{i}^{\mathrm{p}} + \boldsymbol{u}^{\mathrm{nT}}\boldsymbol{i}^{\mathrm{n}} + \boldsymbol{u}^{\mathrm{pT}}\boldsymbol{i}^{\mathrm{n}} + \boldsymbol{u}^{\mathrm{nT}}\boldsymbol{i}^{\mathrm{p}} =$$

$$= P^{\mathrm{p}} + P^{\mathrm{n}} + 6\,G\,U^{\mathrm{p}}\,U^{\mathrm{n}} \cos 2\omega_1 t = \bar{p} + \tilde{p} \qquad (21.32)$$

where P^{p} and P^{n} are the active powers of the positive and the negative sequence symmetrical components of the voltages and currents. This formula shows that at asymmetrical supply voltage the instantaneous power has the oscillating component,

$$\tilde{p} = 6GU^{\mathrm{p}}U^{\mathrm{n}} \cos 2\omega_1 t$$

but the load has a power factor equal to one. Consequently, compensation is not needed. However, when it would be installed to compensate, according to IRP p-q Theory, this oscillating component of the instantaneous active power p, then this compensator would inject a distorted current into supply lines. For example, as was shown in detail in paper [245], the compensator's R line current should have the waveform

$$j_{\mathrm{R}} = \frac{-2\sqrt{2}\,G\,(U^{\mathrm{p}} + U^{\mathrm{n}})U^{\mathrm{p}}\,U^{\mathrm{n}} \cos \omega_1 t \cos 2\omega_1 t}{U^{\mathrm{p}2} + U^{\mathrm{n}2} + 2U^{\mathrm{p}}U^{\mathrm{n}} \cos 2\omega_1 t}.$$

It would distort an originally sinusoidal supply current and increase its three-phase rms value, thus such a compensator would reduce the power factor. It confirms once again that the IRP p-q Theory compensation goal is erroneous. It is correct only when the three-phase load is supplied from a source of a sinusoidal and symmetrical voltage.

Summary

The Instantaneous Reactive Power p-q Theory misinterprets the power properties of electrical circuits. It erroneously suggests that the instantaneous identification of electrical properties of three-phase loads and hence the instantaneous compensation, are possible. Moreover, it erroneously suggests that the compensation of the oscillating component of the active instantaneous power should be the objective for the compensation aimed at the power factor improvement.

Chapter 22
Conservative Power Theory Misconceptions

22.1 Misinterpretation of the "Reactive Energy"

The Conservative Power Theory (CPT) describes power properties and fundamentals of compensation in three-phase circuits [236, 255, 272], but some major doubts as to the correctness of CPT occur even when it is applied to single-phase circuits.

To be valid for a three-phase circuit, power theory (PT) has to be credible first when used for describing power properties and compensation in single-phase circuits. Therefore, the discussion on the CPT in this chapter is confined to only single-phase circuits.

The central quantity in the CPT is the "reactive energy" defined in [236] as

$$W \overset{\text{df}}{=} (\widehat{u}, i) \overset{\text{df}}{=} \frac{1}{T}\int_{0}^{T} \widehat{u}(t)i(t)dt \tag{22.1}$$

where the symbol $\widehat{u}$ denotes an unbiased voltage integral. It is used to define the reactive current in this theory. One can observe, however, that when a capacitor voltage is sinusoidal, namely

$$u(t) = \sqrt{2}\, U \sin \omega_1 t,$$

then capacitor current and the unbiased voltage integral are

$$i(t) = \sqrt{2}\, \omega C\, U \cos \omega_1 t, \quad \widehat{u}(t) = -\sqrt{2}\frac{1}{\omega} U \cos \omega_1 t$$

hence

$$W = (\widehat{u}, i) = \frac{1}{T}\int_{0}^{T} \widehat{u}(t)i(t)dt = -CU^2 < 0,$$

thus the quantity W is negative, so that it cannot be interpreted as energy. The physical meaning of this "reactive energy" W in Ref. [236] is not specified, however.

Powers and Compensation in Circuits with Nonsinusoidal Current. Leszek S. Czarnecki, Oxford University Press.
 DOI: 10.1093/oso/9780198879206.003.0022

Let us calculate the "reactive energy" W for a pure reactance load of admittance for the nth order harmonic equal to

$$Y_n = jB_n$$

supplied with nonsinusoidal voltage

$$u(t) = \sum_{n\in N} u_n(t) = \sqrt{2}\sum_{n\in N} U_n \cos n\omega_1 t.$$

The current of such a purely reactive load can be expressed in the form

$$i(t) = \sum_{n\in N} i_n(t) = -\sqrt{2}\sum_{n\in N} B_n U_n \sin n\omega_1 t. \tag{22.2}$$

The unbiased integral of such a voltage is

$$\hat{u}(t) = \sum_{n\in N} \hat{u}_n(t) = \sqrt{2}\sum_{n\in N} \frac{U_n}{n\,\omega_1} \sin n\omega_1 t. \tag{22.3}$$

The "reactive energy" W of purely reactance load at such a voltage is

$$W = (\hat{u}, i) = \sum_{n\in N} (\hat{u}_n, i_n) = \sum_{n\in N} W_n = -\sum_{n\in N} \frac{B_n}{n\omega_1} U_n^2. \tag{22.4}$$

Thus this "energy" has a similarity to Budeanu's reactive power

$$Q_{\mathrm{B}} = \sum_{n\in N} U_n I_n \sin\varphi_n = \sum_{n\in N} Q_n = \sum_{n\in N} B_n U_n^2 \tag{22.5}$$

only with the load susceptance B_n, recalculated to $-B_n/(n\omega_1)$. Since the susceptance B_n can be positive or negative, then as in the case of Budeanu's reactive power, individual terms of the "reactive energy" W, specified by formula (22.4), can cancel each other out.

22.2 "Reactive Energy" and Energy Conservation Principle

The "reactive energy" W satisfies the Power Balance Property (PBP). It means that in any circuit confined by a sphere with zero energy transfer across it, and composed of K branches, as shown in Fig. 22.1

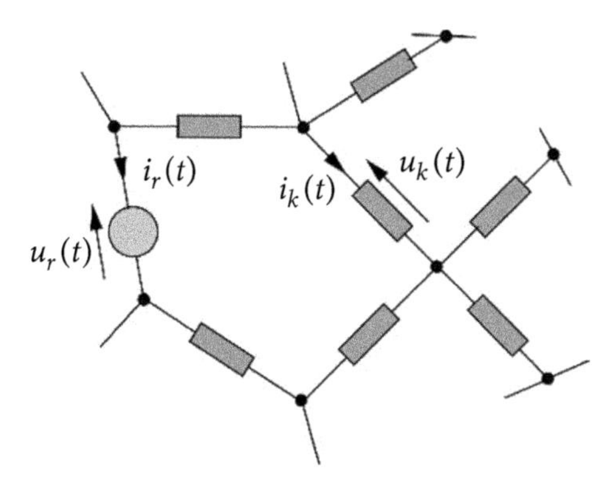

Figure 22.1 A circuit with K branches.

$$\frac{1}{T}\int_0^T \sum_{k=1}^{K} \hat{u}_k(t) i_k(t) dt = \sum_{k=1}^{K} W_k = 0 \tag{22.6}$$

the sum of "reactive energies" of individual branches W_k is equal to zero.

This is an important property. The power theory developed by Tenti and coworkers [236] was called a "Conservative Power Theory" to emphasize the importance of the "reactive energy" W for the power properties of electrical circuits and to stress its physical sense.

Unfortunately, unlike as it is with the active power P, the PBP for the "reactive energy" W is not the conclusion from the Energy Conservation Principle (ECP), which is a physical principle, but from the Tellegen Theorem [25], which is only a mathematical but not a physical property of electrical circuits.

The PBP for the "reactive energy" W can be developed from the Tellegen Theorem as follows. (This Theorem and its application for developing the PBP were explained in Section 18.2.)

Having two circuits of identical topology, as shown in Fig. 18.3, according to the Tellegen Theorem, the voltage and current of their branches have to satisfy the identity

$$\sum_{k=1}^{K} u_k^{\mathrm{a}}(t) i_k^{\mathrm{b}}(t) \equiv 0.$$

When the voltage and current of the sources in the circuit as shown in Fig. 18.3(a) are selected in such a way that they are equal to the unbiased integral of source voltages and currents in the circuit shown in Fig. 18.3(b), then assuming that

$$u_k^{\mathrm{a}}(t) \equiv \hat{u}_k(t), \qquad i_k^{\mathrm{b}}(t) \equiv i_k(t) \tag{22.7}$$

then

$$\sum_{k=1}^{K} u_k^{\mathrm{a}}(t) i_k^{\mathrm{b}}(t) = \sum_{k=1}^{K} \widehat{u}_k(t) i_k(t) \equiv 0$$

and hence

$$\frac{1}{T}\int_0^T \sum_{k=1}^{K} \widehat{u}_k(t) i_k(t)\, dt = \sum_{k=1}^{K} (\widehat{u}_k, i_k) = \sum_{k=1}^{K} W_k = 0. \tag{22.8}$$

It was not demonstrated that such a property can be developed from the ECP, however.

As was shown in Section 18.2, the reactive power Q_{B} as defined by Budeanu satisfies the PBP developed from the Tellegen Theorem but not from the Energy Conservation Principle. Consequently, Budeanu's power theory is no less "Conservative" than Tenti's Conservative PT.

22.3 "Reactive Energy" and Stored Energy

The physical meaning of the "reactive energy" W as defined by [236] is not clear, however. It is, as claimed, the average energy stored in reactive elements of the circuit. This raises some questions. Let us consider a parallel inductive-capacitive (LC) load, shown in Fig. 22.2 with the supply voltage

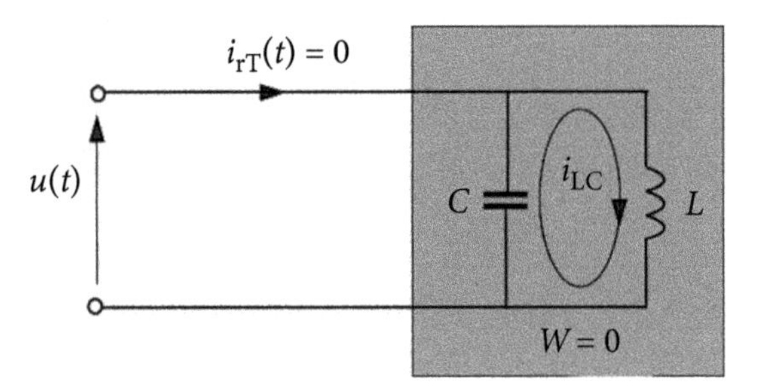

Figure 22.2 A current in an *LC* loop.

$$u(t) = \sqrt{2}\, U \cos \omega_1 t.$$

Energy stored in the electric field of the capacitor is

$$E_{\mathrm{e}}(t) = \frac{1}{2} C\, u^2(t) = C\, U^2 \cos^2 \omega_1 t$$

and energy stored in the magnetic field of the inductor

$$E_{\mathrm{m}}(t) = \frac{1}{2} L\, i^2(t) = \frac{U^2}{\omega_1^2 L} \sin^2 \omega_1 t.$$

Thus energy stored in such a load is

$$E(t) = E_{\mathrm{m}}(t) + E_{\mathrm{e}}(t) = \frac{U^2}{\omega_1^2 L} \sin^2 \omega_1 t + C\, U^2 \cos^2 \omega_1 t, \tag{22.9}$$

however, this stored energy is not equal to the "reactive energy"

$$W = W_L + W_C = \frac{U^2}{\omega_1^2 L} - U^2 C. \tag{22.10}$$

Thus, the "reactive energy" W is not the energy $E(t)$ stored in the magnetic and electric fields of reactance elements of a circuit. This is particularly visible at the resonance when the relationship $1/\omega_1 L = \omega_1 C$ is satisfied. In such a case, the energy stored in such LC load is

$$E(t) = \frac{U^2}{\omega_1^2 L} = \text{Const.}$$

while the "reactive energy" $W = 0$. Thus it cannot be regarded as energy or averaged energy stored in such a load. Energy $E(t)$ stored in electric and magnetic fields causes a current in the LC loop, as shown in Fig. 22.2. Its presence in that loop cannot be explained in terms of the "reactive energy" W.

Another doubt as to the physical interpretation of the "reactive energy" W raises the possibility of its presence in circuits without any capability of energy storage. To show this, let us consider a purely resistive load with a TRIAC, shown in Fig. 22.3.

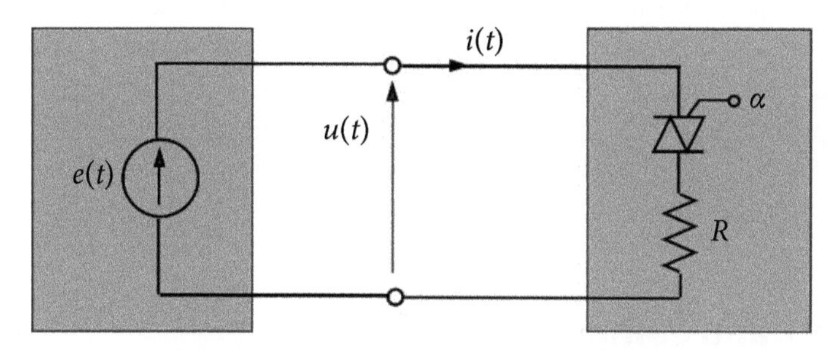

Figure 22.3 A resistive load with a TRIAC.

At the sinusoidal supply voltage

$$u(t) = \sqrt{2}\, U \sin \omega_1 t$$

the load current has the waveform as shown in Fig. 22.4.

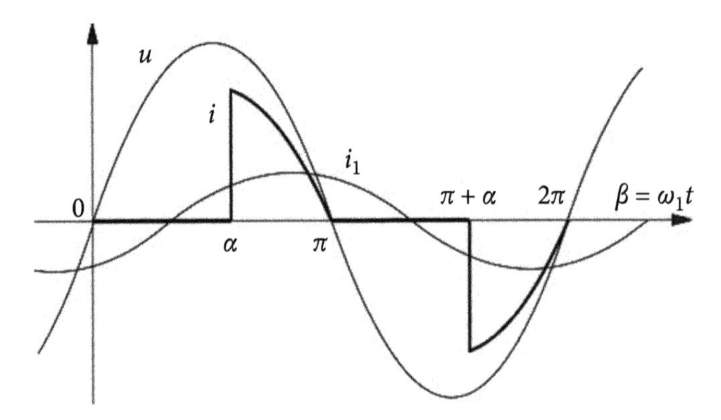

Figure 22.4 Waveforms of the voltage, current, and its fundamental harmonic of a TRIAC.

The current fundamental harmonic of the load current

$$i_1(t) = \sqrt{2}\, I_1 \sin(\omega_1 t - \varphi_1)$$

is shifted versus the voltage, as shown in Fig. 22.4. The supply voltage unbiased integral is

$$\widehat{u}(t) = \sqrt{2}\,\frac{U}{\omega}\cos\omega_1 t$$

so that the "reactive energy" W is equal to

$$W = (\widehat{u}, i) = \sum_{n=1}^{\infty} (\widehat{u}_n, i_n) = (\widehat{u}, i_1) = \frac{UI_1}{\omega_1}\sin\varphi_1. \tag{22.11}$$

Thus loads without any capability of energy storage could have the "reactive energy" W. This conforms to the previous conclusion that the "reactive energy" W cannot be associated with the phenomenon of energy storage. Consequently, the physical meaning of "reactive energy" is not clear. In any case, whatever this interpretation, it should not depend on the load properties. It should be the same for linear and nonlinear loads and the same for reactive and resistive ones.

22.4 CPT and Compensation

The reactive current is defined in the CPT as

$$i_{\mathrm{rT}}(t) \stackrel{\mathrm{df}}{=} \frac{W}{||\widehat{u}||^2}\widehat{u}(t). \tag{22.12}$$

According to the CPT, the reactive current $i_{rT}(t)$ of RL loads can be compensated by a capacitor of the capacitance

$$C = \frac{W}{||u||^2} \tag{22.13}$$

connected at the load terminals. Thus, the CPT enables the calculation of the compensating capacitance of RL loads.

This means that considering compensation, the CPT only reached the result obtained earlier by Kusters and Moore [59]. Unfortunately, both approaches share the same deficiencies, discussed in detail in Section 20.2. The compensating capacitor causes amplification of the supply voltage harmonics and the load-generated current harmonics. Therefore, such a capacitive compensation in the presence of harmonics should be avoided. Moreover, since the supply sources have usually inductive impedance, then connecting the capacitor at the supply source terminals changes the supply voltage, the "reactive energy" W, and consequently, the connected capacitor has wrong capacitance C. Its calculation has to be repeated. At the recalculated capacitance, the supply voltage, and the "reactive energy" W change again, however. As was shown in Section 20.2, in situations common in distribution systems, the iterative calculation of the capacitance may not converge. Unfortunately, even if the supply source has a negligible impedance and calculation of the compensating capacitance, according to formula (22.13), is possible, the CPT does not provide fundamentals for the power factor λ improvement. This is shown in the following illustration.

Illustration 22.1 *Let us calculate the compensating capacitance, according to the CPT, for the load shown in Fig. 18.6 at the supply voltage as assumed in Illustration 18.3.*

$$u(t) = \sqrt{2}\,(100 \sin \omega_1 t + 30 \sin 3\omega_1 t)\,\text{V}, \quad ||u|| = 104.4\,\text{V}, \; \omega_1 = 1\,\text{rad/s}.$$

The admittances of such a load, shown in Fig. 22.5, for the voltage harmonics are

$$Y_1 = jB_1 = -j1/2\ \text{S}, \qquad Y_3 = jB_3 = j1/2\ \text{S}.$$

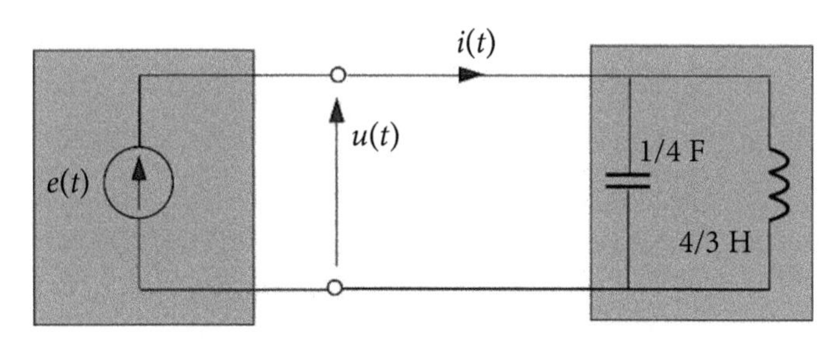

Figure 22.5 A circuit with a reactive load.

The "reactive energy" W of such load is equal to

$$W = -\sum_{n\in\{1,3\}} B_n \frac{U_n^2}{n\omega_1} = 4.85\,\text{kJ}.$$

Since

$$||\hat{u}|| = \sqrt{\sum_{n\in\{1,3\}} \left(\frac{U_n}{n\omega_1}\right)^2} = \sqrt{\left(\frac{U_1}{\omega_1}\right)^2 + \left(\frac{U_3}{3\,\omega_1}\right)^2} = 100.5\,\text{Vt}$$

the rms value of the reactive current $i_{\text{rT}}(t)$ *is*

$$||i_{\text{rT}}|| = ||\frac{W}{||\hat{u}||^2}\,\hat{u}(t)|| = \frac{|W|}{||\hat{u}||} = \frac{4850}{100.50} = 48.26\,\text{A}.$$

The load current rms value is

$$||i|| = \sqrt{\sum_{n\in\{1,3\}} (Y_n\, U_n)^2} = \sqrt{(0.5\times 100)^2 + (0.5\times 30)^2} = 52.2\,\text{A}.$$

Since the active component $i_{\text{a}}(t)$ *does not exist in the load current, the rms value of the void current, as defined in the CPT*

$$i_{\text{v}}(t) \stackrel{\text{df}}{=} i(t) - i_{\text{a}}(t) - i_{\text{rT}}(t) \tag{22.14}$$

is equal to

$$||i_{\text{v}}|| = \sqrt{||i||^2 - ||i_{\text{rT}}||^2} = \sqrt{52.2^2 - 48.26^2} = 19.9\,\text{A}$$

so that the distortion power of the load is equal to

$$D_{\text{T}} = ||i_{\text{v}}||\;||u|| = 19.90\times 104.40 = 2.08\,\text{kVA}.$$

This circuit was analyzed in Illustration 18.3 and it was shown there that the load current is not distorted versus the supply voltage but only shifted, as was shown in Fig. 18.7. Thus the CPT conclusion that there is a distortion power in this circuit is erroneous.

A shunt capacitor of the capacitance

$$C = \frac{W}{||u||^2} = \frac{4850}{104.4^2} = 0.445\ \text{F}$$

compensates the reactive current $i_{rT}(t)$ entirely since the "reactive energy" W after compensation is zero. The supply current of the compensated load is

$$i'(t) = \sqrt{2}\,\text{Re}\{(j\omega_1 C + \boldsymbol{Y}_1)\,\boldsymbol{U}_1 e^{j\omega_1 t} + (j3\omega_1 C + \boldsymbol{Y}_3)\,\boldsymbol{U}_3 e^{j3\omega_1 t}\} = \sqrt{2}\,\text{Re}\{\boldsymbol{I}_1' e^{j\omega_1 t} + \boldsymbol{I}_3' e^{j3\omega_1 t}\}$$

with the rms values of the current harmonics

$$I_1' = |(\omega_1 C + B_1)|U_1 = |(0.445 - 0.5)| \times 100 = 5.5\ \text{A}$$

$$I_3' = |(3\omega_1 C + B_3)|U_3 = |(3 \times 0.445 + 0.5)| \times 30 = 55.0\ \text{A}$$

so that the rms value of the supply current after compensation is

$$||i'|| = \sqrt{I_1'^2 + I_3'^2} = 55.3\ \text{A}.$$

This illustration confirms the opinion that compensation of the "reactive energy" W and consequently, the reactive current $i_{rT}(t)$, does not contribute to a reduction of the supply current. The effects of the CPT-based capacitive compensation are shown in Fig. 22.6.

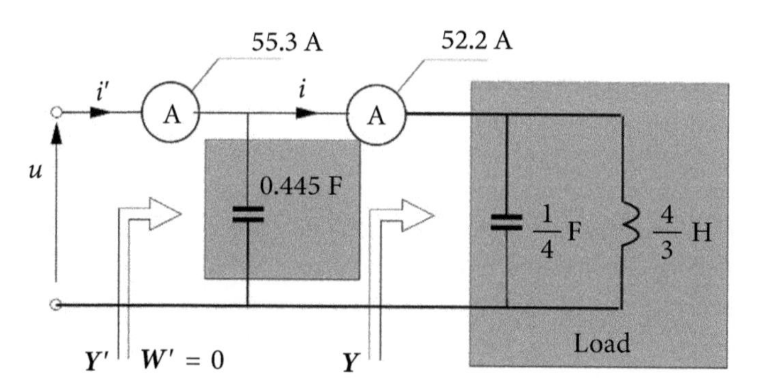

Figure 22.6 Effects of compensation of the "reactive energy" W by a capacitor.

Thus, the CPT conclusion that compensation should be founded on the "reactive energy" W reduction is erroneous. At the same time, the CPT, being formulated in the time domain, does not provide any frequency domain data on the load properties needed for a reactance compensator synthesis. Having such data, and using a CPC-based approach to reactance compensation, one could conclude that complete compensation of the load shown in Fig. 22.5, can be achieved by a shunt compensator of impedance

$$Z_C(j\omega) = j\omega L + \frac{1}{j\omega C}, \quad L = 1\,\text{H}, \quad C = \frac{1}{3}\,\text{F}.$$

Indeed, such a compensator reduces the supply current as this is shown in Fig. 22.7.

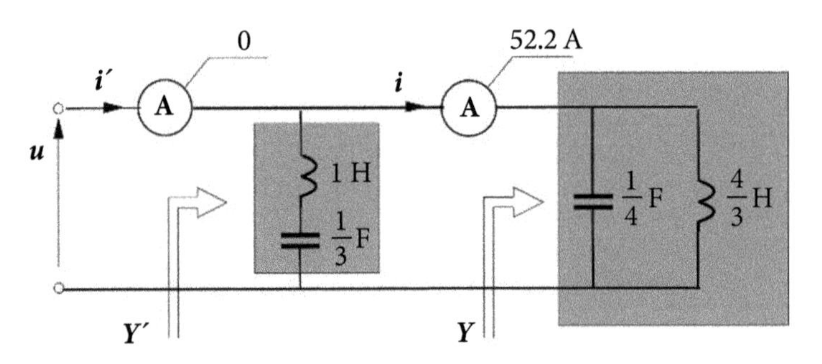

Figure 22.7 Effects of the CPC-based compensation.

*****

Summary

The CPT theory does not interpret the power properties of single-phase circuits correctly. The "reactive energy" W is not energy and is not associated with the energy storage in the load. The reactive current in the CPT does not have any physical meaning. Capacitive compensation of this current affects the void current rms value. The distortion power in the CPT is not associated with the voltage and current mutual distortion. These conclusions have a strong analogy to conclusions drawn on Budeanu's PT. The "reactive energy" W satisfies the Power Balance Principle similarly to Budeanu's reactive power Q_B, and both are "conservative", not having any physical meaning. The concept of distortion power in both theories is misleading; both theories moreover cannot be used for compensation.

Chapter 23
Meta-Theory of Electric Power

23.1 Meaning of the Meta-Theory of Electric Power

The power theory of electrical circuits is a part of electrical engineering (EE) which belongs to the empirical sciences. Knowledge in the empirical sciences takes the form of a set of opinions, statements, or theorems drawn from observations with conclusions supported by logic and mathematics. All are hypotheses that cannot be proven, but they can be falsified. According to Karl Popper [249], taking as evidence a population of 1,000 white swans will not prove the hypothesis that all swans are white, while the presence of a single black swan will falsify this hypothesis. Such "black swans" are used several times in this book, mainly in Part C, in discussions on various EE opinions circulating as to the power properties of electrical circuits and compensation.

Hypotheses formulated by any power theory can be falsified by measurement or by mathematical and logical analysis. Unfortunately, the falsification based on measurements in real circuits can be challenged as unreliable because of limited knowledge of the circuit parameters and the accuracy of measurement. Hypothesis falsification based on the analysis of an idealized model of the circuit can be much more reliable, therefore the falsification of various hypotheses and opinions on the power properties of electrical circuits in this book was based on the mathematical analysis of idealized models of these circuits.

The approach to verification and falsification, selection of tools for the development of power theory, or its objectives are not the subject of power theory itself but the subject of a superior structure of knowledge. These issues are the subjects of a knowledge structure which is pitched at a higher level than individual power theories, something that can be referred to as a *meta-theory of electric power*.

The meta-theory of electric power does not deal with individual conclusions of particular concepts of power theory concerning power properties of electrical circuits and compensation but rather with general capabilities of such concepts. The meta-theory of electrical powers provides some general fundamentals for the comparison of various tools and approaches to power theory development.

23.2 What Is Power Theory and Its Objectives?

The term "power theory" was used for the first time by Fryze [17] in 1931. He applied the term to his explanation of the difference between the apparent and the

Powers and Compensation in Circuits with Nonsinusoidal Current. Leszek S. Czarnecki, Oxford University Press.
 DOI: 10.1093/oso/9780198879206.003.0023

active power of single-phase loads. This explanation is referred to as Fryze's power theory. The same is true of Budeanu's power theory. Originally, an explanation of the difference between the apparent and the active powers in power terms was the only objective of these theories. Later this objective was enlarged by the search for a method of this difference in power reduction using a capacitor, as was the case of the Shepherd and Zakikhani and Kusters and Moore's theories. Finally, more sophisticated methods of compensation, such as reactance, switching, and hybrid compensations were included in power theory.

Some authors endowed the theories they developed with their names or were referenced in literature by the names of original contributors. This was the case of Shepherd and Zakikhani's power theory; Kusters and Moore's power theory; the Instantaneous Reactive Power p-q Theory by Nabae and Akagi; the Buchholz-Fryze-Depenbrock Method developed by Depenbrock; the Currents' Physical Components-based power theory by Czarnecki; and the Conservative Power Theory by Tenti. All these have provided conclusions specific to these theories, and some of these conclusions overlapped with the conclusions reached in other theories; some were right and some were erroneous.

We can release power theories from being attached to specific authors or having particular names if we consider power theory to be the compound knowledge about power properties of electrical circuits accumulated by all the individual power theories developed and posited by specific individuals. Instead, the power theory can be regarded [135] as a sort of database, composed of true statements on the power properties of electrical circuits and methods of compensation. Thus, the term "power theory" can have two meanings, one associated with specific individuals and theories, the other more generic. Each new conclusion on power properties and compensation can contribute to this "general" power theory, as long as it is not challenged as being erroneous. Even recognized misconceptions can contribute to this power theory.

23.3 Domains of Power Theory

From the very beginning of power theory development, there was a major controversy as to the "domain" in which it should be formulated. One school of its development suggested that the power properties of electrical circuits should be described in terms of their frequency properties, in the *frequency domain*. Although used by Illovici, Budeanu was the main propagator of this domain. This approach has resulted in us using the concept of harmonics. Largely due to Fryze's objections, this approach was challenged by the concept of the *time domain*. The time domain was regarded as being closer to physical phenomena in the circuit and consequently superior to the frequency domain, based on the concept of harmonics, which are only mathematical not physical entities.

Measurement technology also contributed to the inferiority of the frequency domain in practical applications. For a long time, harmonics were measured only by analog filters. Such filters can measure only their resonant harmonic filter (rms) value but not the phase. Consequently, practical implementations of results obtained in the frequency domain were strongly limited. More than forty years were needed for the construction of the first meter that could measure the reactive power as defined by Budeanu in the frequency domain. This situation has changed with the development of analog-to-digital (A/D) converters and digital signal processing (DSP) systems. The rms and phase of the voltage and current harmonics are easily measured these days. Only a few lines of code are needed now for digital measurement of the reactive power and other quantities defined in the frequency domain. Such measurements can be performed almost in real time, or with only a negligible delay.

In this situation, most power theories were developed in the time domain. Only Budeanu's, Shepherd and Zakikhani's, as well as Czarnecki's theories were developed in the frequency domain. Fryze, Kusters and Moore, Depenbrock, Nabae and Akagi, as well as Tenti developed their power theories in the time domain.

Since the time domain ignores frequency properties of electrical circuits, all theories founded on that domain were not able to explain and describe energy-flow effects of the supply voltage distortion, as well as such effects caused by the load current distortion, due to load nonlinearity and/or periodic switching. Such theories were not able to reveal the fact that the change of the load conductance with the harmonic order increases the load apparent power, thus it reduces the power factor. It was because the term "conductance for the harmonic order" did not exist in the time domain's vocabulary. Thus, time domain power theories were not able to identify the scattered current or even identify it within the frame of CPC-based PT; these theories were not able to explain and describe it in mathematical terms. Moreover, time domain approaches were not able to identify a very important and common phenomenon of the generation of currents harmonics by nonlinear or periodically switched load and the flow of energy from harmonic-generating load (HGL) to the supply source.

Power properties of linear three-phase loads with a neutral conductor, when supplied from a source of nonsinusoidal voltage, are specified in terms of six currents. Only one of these currents, the active current, was identified in the time domain. The remaining five were identified in the frequency domain.

Moreover, the time-domain approach to power theory development has excluded any possibility of developing methods of reactance compensation other than purely capacitive ones. Data on the frequency properties of the load are necessary for such a reactance compensator synthesis.

Nonetheless, some ideas crucial to power theory have emerged from the time domain approach. These were the load current instead of the power decomposition, the concept of the active current, and the current components' orthogonality. These days, these concepts are commonly used in the frequency domain approach to power theory.

Apart from Nabae and Akagi's IRP p-q power theory, all theories founded on the time-domain approach are based on averaging the power-related quantities over the period T of the supply voltage. In opposition to these theories, Nabae and Akagi's power theory attempted to specify the power properties of electrical circuits instantaneously. Two powers, defined as instantaneous ones, p and q, were to provide information on the instantaneous power properties of the loads. Unfortunately, instantaneous identification of power properties of electrical loads is not possible. Such impossibility is visible even in elementary single-phase circuits. To demonstrate it, let us try to answer the following question: knowing instantaneous values at a given instant t_k of the voltage and current of a load enclosed in the black box, shown in Fig. 23.1, then what is in the box?

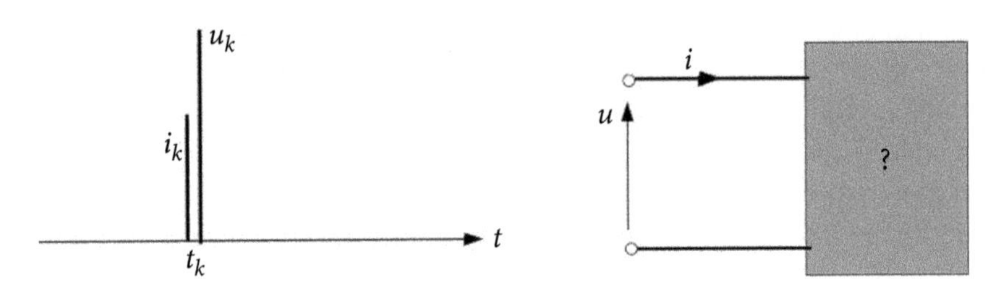

Figure 23.1 A black box with an unknown load, and a pair of instantaneous values of the voltage and current at its terminals.

The answer is quite evident. It is not possible to conclude what is in the box. This answer can be supported by a few possibilities, shown in Fig. 23.2.

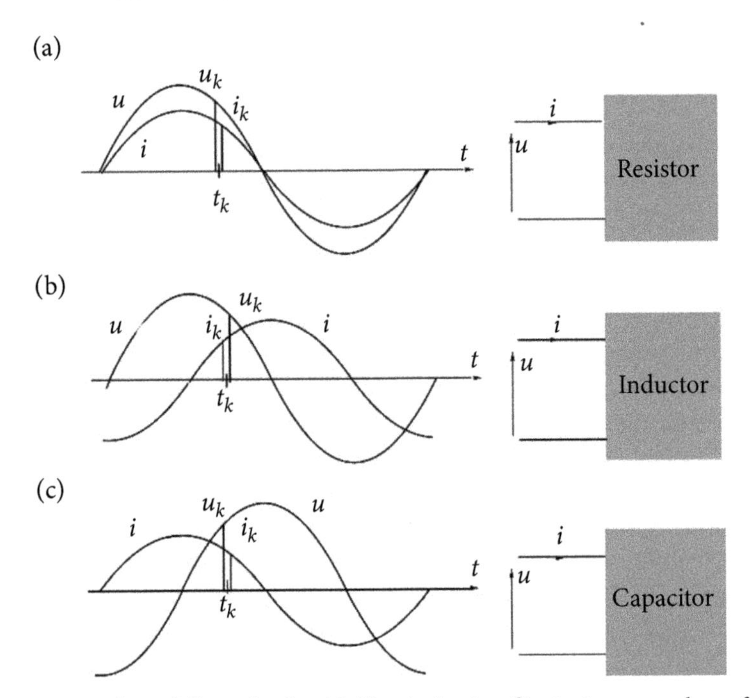

Figure 23.2 Three different loads with identical pairs of instantaneous values of the voltage and current at their terminals.

Thus, if instantaneous identification of power properties of electrical loads is not possible in single-phase circuits with sinusoidal supply voltage, it cannot be possible, as shown in Section 21.1, in three-phase circuits, which are the subjects of Nabae and Akagi's IRP p-q Theory.

Despite this obvious conclusion, one could speculate that by observing instantaneous values of the voltage and current at the load terminal over the whole period T, we might conclude that we will be able to identify the power properties of such a load. Unfortunately, this is not true. Apart from the IRP p-q Theory, all power theories are based on values averaged over the period. Even the complex rms value of harmonics cannot be calculated without averaging. Thus, observation of the load over a given period is not sufficient to identify its power properties.

The frequency domain approach to PT development has proven to be much more effective than the time domain one. It brought a trap with it, however. Harmonics are regarded by some engineers and scientists in the EE community as physical entities, while they are only mathematical objects. Consequently, energy oscillations that occur in certain results of the energy flow analysis using Fourier series are regarded as if they exist physically. In 1931, Fryze presented a circuit, shown in Fig. 23.3, and the reasoning which demonstrated that the interpretation of energy flow in that circuit using the frequency domain and considering energy oscillations at harmonic frequencies as the existing ones, could lead to nonsense.

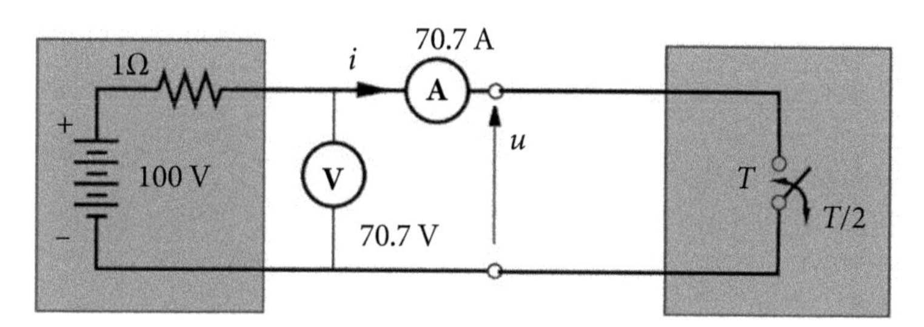

Figure 23.3 A circuit with a periodic switch.

The load current and voltage waveforms in this circuit are shown in Fig. 23.4.

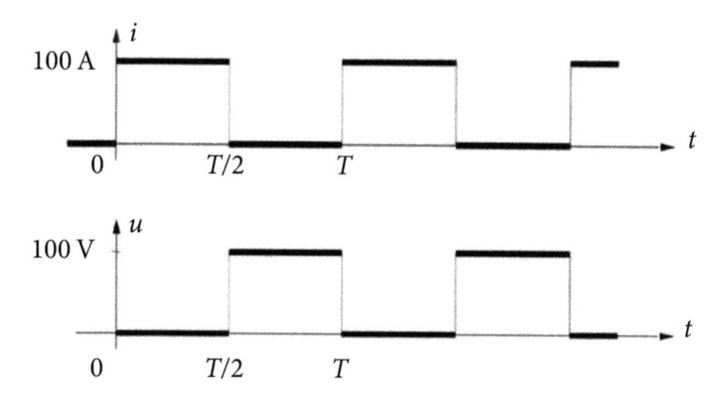

Figure 23.4 The voltage and current waveform at the periodic switch terminals.

They can be expressed using the Fourier series as

$$u(t) = U_0 + \sqrt{2}\sum_{n=1}^{\infty} U_n \cos(n\omega_1 t + \alpha_n) = \sum_{n=0}^{\infty} u_n$$

$$i(t) = I_0 + \sqrt{2}\sum_{n=1}^{\infty} I_n \cos(n\omega_1 t + \beta_n) = \sum_{n=0}^{\infty} i_n.$$

The rate of the energy flow from the supply source to the load, expressed in the frequency domain, is equal to

$$p(t) = \frac{dW}{dt} = u(t)i(t) = \sum_{r=0}^{\infty} u_r(t) \sum_{s=0}^{\infty} i_s(t) = P_0 + \sum_{n=1}^{\infty} A_n \cos(n\omega_1 t + \phi_n) = \sum_{n=0}^{\infty} p_n(t)$$

meaning it is an infinite sum of oscillating components $p_n(t)$. A calculation of this infinite sum is needed for obtaining the instantaneous power $p(t)$ at each instant of time. At the same time, in the time domain it is directly observable that apart from the points of the voltage and current discontinuities, there is no flow of energy in this circuit, since in all other points

$$p(t) = u(t)i(t) \equiv 0.$$

Energy cannot flow through the open and closed switch, but the state of the switch is not visible in the frequency domain.

The frequency domain approach to power theory requires Fourier analysis, which involves more computations than the time domain. This is not a major obstacle at the present time, taking into account the processing powers of computers and microprocessors. Nonetheless, at an adaptive compensation of fast varying loads, reduction of the computational burden can have some advantages. This can be obtained only when the fundamental harmonic is separated from the voltage and current, while all other harmonics are treated together. Separation of the fundamental harmonic means taking a frequency domain approach, while leaving the remaining part of the voltage and current not decomposed into harmonics means taking a time domain approach. This frequency domain approach to the fundamental harmonic, while the time domain approach to the distorting component of the voltage and current, can be referred to as a *hybrid domain* approach.

It could be justified, as was observed by Depenbrock [55], by the fact that the fundamental harmonic has special importance for energy transmission. The energy which is useful for customers is mainly transferred by the fundamental harmonic, while others to a large degree disturb transmission and the load performance.

This hybrid domain approach does not have any cognitive merit since the power-related phenomena that are associated with harmonics cannot be identified in such a domain. It does not have any meaning, however, for compensation since harmonics should and can be compensated irrespective of the cause they have occurred in the circuit.

Summary

The objective of power theory development, the evaluation of the mathematical tools used for this, and the results obtained are not the subject of power theories under development but the subject of the meta-theory of electric power. Because electrical engineering belongs to the empirical sciences, all results obtained in the development of power theories take the form of a hypothesis that cannot be proven. Positive verifications of these theories can only make the hypothesis more and more probable. A single negative result falsifies them, however.

Chapter 24
Miscellaneous Issues

24.1 Has the Reactive Power Q Any Physical Meaning?

The author of this book has had several opportunities to run seminars and conferences on powers for engineers, scientists, and Ph.D. students involved in power engineering in various countries. He has observed that a great majority of participants in perhaps two-thirds of these professional meetings used to associate the reactive power and degradation of the power factor with energy oscillation between the supply source and the load. The remaining participants were usually confused on the subject. These opinions and confusions used to occur even if the voltages and currents in the circuit were regarded as sinusoidal. Moreover, it seems that most engineers and scientists in the electrical engineering community hold the very strong opinion that the reactive power Q is a physical quantity, meaning that there is a physical phenomenon in electrical circuits specified by this power. This opinion is bolstered by university teaching and numerous books on power system engineering, for example [51] or [177]. Indeed, the meaning of the reactive power Q in courses on electrical circuits is usually explained for single-phase linear circuits with sinusoidal supply voltage, as shown in Section 1.2, as the amplitude of the bidirectional component of the instantaneous power, as illustrated in Fig. 1.3, meaning with energy oscillations between the supply source and the load.

Even if the existence of the energy oscillations in three-phase lines which connect the synchronous generators, bus stations, and customer loads does not have any practical importance, this is important for cognitive reasons. As electrical engineers, we should know whether this energy oscillates in these lines or not. We should know whether the reactive power describes an existing physical phenomenon or not.

It was shown in Section 1.4 that in three-phase circuits the reactive power can exist, however, without any energy oscillations between the supply source and a reactive load.

Some opponents of this opinion argue that such oscillations do exist in individual lines of the supply but they are canceled mutually and consequently, only apparently are these oscillations not visible. Indeed, the voltage-current product, say, for line R, in a symmetrical three-phase circuit with sinusoidal supply voltage is equal to

$$u_{\mathrm{R}}(t)i_{\mathrm{R}}(t) = \frac{P}{3}(1 + \cos 2\omega_1 t) + \frac{Q}{3}\sin 2\omega_1 t \tag{24.1}$$

Powers and Compensation in Circuits with Nonsinusoidal Current. Leszek S. Czarnecki, Oxford University Press.
 DOI: 10.1093/oso/9780198879206.003.0024

thus it contains the oscillating component. Does this product stand for the instantaneous power of line R, however?

The instantaneous power $p(t)$ is the rate of energy flow to a space in the circuit, confined in such a way that electric energy to this space can be delivered only by the circuit terminals, as is shown in Fig. 24.1.

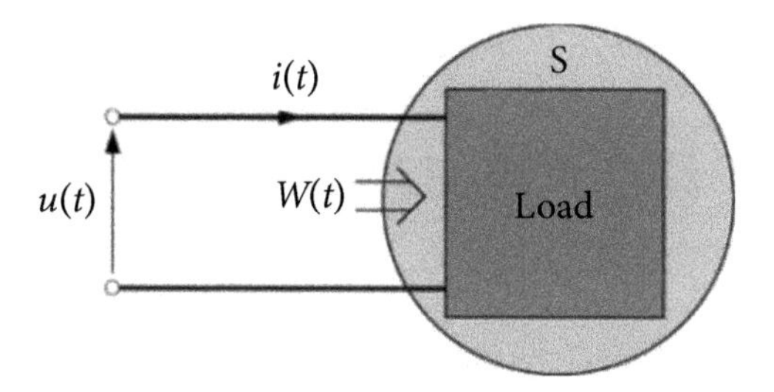

Figure 24.1 A single-phase load isolated as to electric energy flow by space S.

Thus, the instantaneous power for such a load is

$$p(t) = \frac{dW}{dt} = u(t)i(t). \tag{24.2}$$

According to this definition, the instantaneous power $p(t)$ of three-phase loads, confined by the space S, shown in Fig. 24.2, can be expressed as

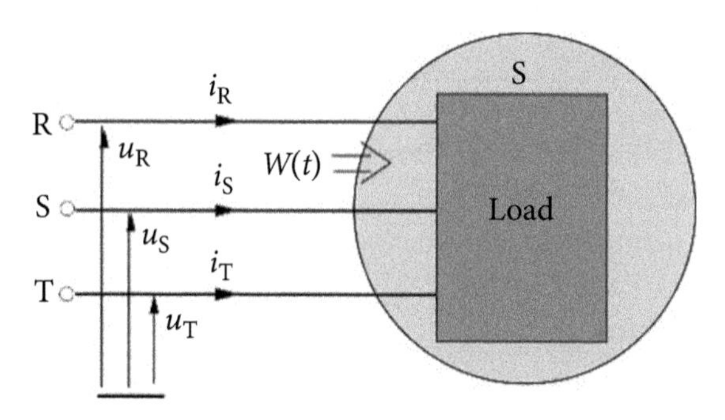

Figure 24.2 A three-phase load isolated as to electric energy flow by space S.

$$p(t) = \frac{dW}{dt} = u_R(t)i_R(t) + u_S(t)i_S(t) + u_T(t)i_T(t). \tag{24.3}$$

Unfortunately, it is not possible in three-phase circuits to find a space that would enclose only one supply terminal. An illustration of the attempt to isolate one line by such a space is shown in Fig. 24.3.

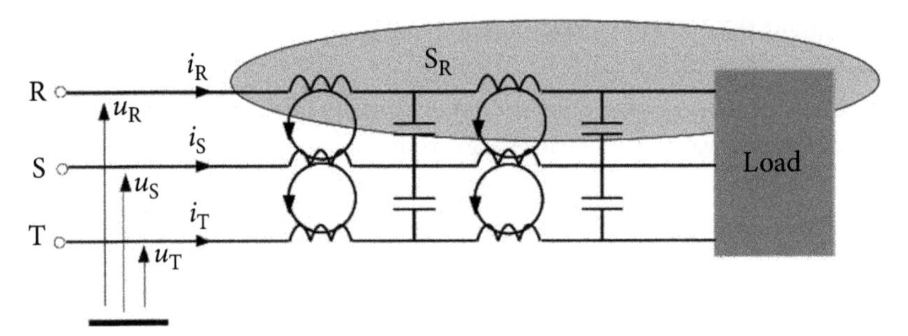

Figure 24.3 A three-phase load with an example of the attempt of isolating only one supply line.

Without such a space, the voltage-current product for a single line, for example line R, cannot be regarded as the instantaneous power, namely

$$u_R(t)i_R(t) \neq p_R(t).$$

Moreover, the node voltages in electrical circuits are relative quantities of the value dependent on the reference node, which could be grounded. If, for example, terminal R is grounded, assuming of course that this is the only grounded point in the circuit, as t is shown in Fig. 24.4,

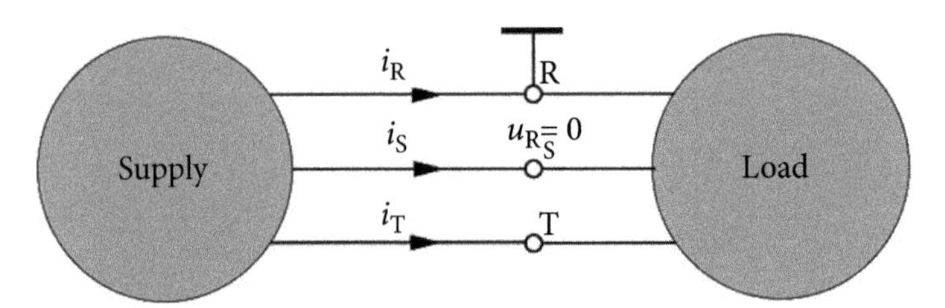

Figure 24.4 A three-phase load with grounded terminal R.

then, the voltage-current product for line R

$$u_R(t)i_R(t) \equiv 0.$$

This grounding, so that the zero value of the voltage-current product for line R, does not affect the energy flow in the circuit and the instantaneous power of the load, as defined by formula (24.3). It means that any oscillations in single lines of three-phase circuits, that hypothetically cancel mutually, do not exist. The reactive power Q in such circuits cannot be associated with energy oscillations.

There are also opinions in the electrical engineering community that the reactive power Q occurs in the electrical circuits because of their capability, in the presence of reactance elements, of energy storage in electric and magnetic fields of such elements. Indeed, despite the lack of energy oscillations in the circuit shown in Fig. 1.7, the energy is stored in the load. Apparently, this is a convincing opinion. Unfortunately, as shown in Section 1.2, the reactive power can occur in a purely resistive circuit, that is, without any capability of energy storage.

It seems that the meaning or physical interpretation of the reactive power should not depend on the circuit features. It should be the same for single- or multi-phase circuits with linear or nonlinear, time-invariant, or switched loads. If we agree with such an opinion, the presence of reactive power in a circuit cannot be associated with energy oscillations or with energy storage. Thus, is there any physical phenomenon that could be specified by the reactive power?

The reactive power Q occurs because of the phase shift between the supply voltage and the load current. In single-phase circuits with sinusoidal voltage and current, it is equal to

$$Q = UI \sin \varphi.$$

It does not specify the phase-shift phenomenon, however. The value of the reactive power depends also on the supply voltage and the load current rms values which are not related to the phase-shift between the voltage and current. We can only say that the reactive power is *associated* with the phenomenon of the phase shift.

The reactive power is sometimes defined with the formula

$$Q \stackrel{\mathrm{df}}{=} \frac{1}{T} \int_{0}^{T} u(t)\, i\left(t - \frac{T}{4}\right) dt$$

in other words, as a mean value of a quantity that resembles the instantaneous power $p(t)$. The similarity of this definition to the definition of the active power P might suggest that the reactive power Q is also a physical quantity. This is an incorrect conclusion, however. This formula defines the reactive power Q throughout a current shifted against the voltage, that is, by the quantity $i(t{-}T/4)$. Such a quantity does not exist in the circuit at the instant t when the value of the voltage $u(t)$ is specified. The quantity specified by formula (24.4) is only a mathematical and not a physical entity. This formula does not describe any physical phenomenon.

Thus, we must close this section with the conclusion that there is no physical phenomenon in electrical circuits that could be specified by the reactive power Q. It is only associated with the phase shift between the load current and the supply voltage. It is not associated, however, either with energy oscillations or energy storage.

24.2 Comments to the German Standard DIN 40110

Opinions that the reactive power is associated with a bidirectional flow of energy between the supply source and the load were bolstered in 1972 by the German Standard DIN 40110 [37], by decomposition of the instantaneous power $p(t)$ of an

LTI load into its positive and negative components of the mean values P^{p} and P^{n}. Assuming that

$$u(t) = \sqrt{2}\,U \sin \omega_1 t, \quad i(t) = \sqrt{2}\,I \sin(\omega_1 t - \varphi),$$

this rationale and decomposition are illustrated in Fig. 24.5.

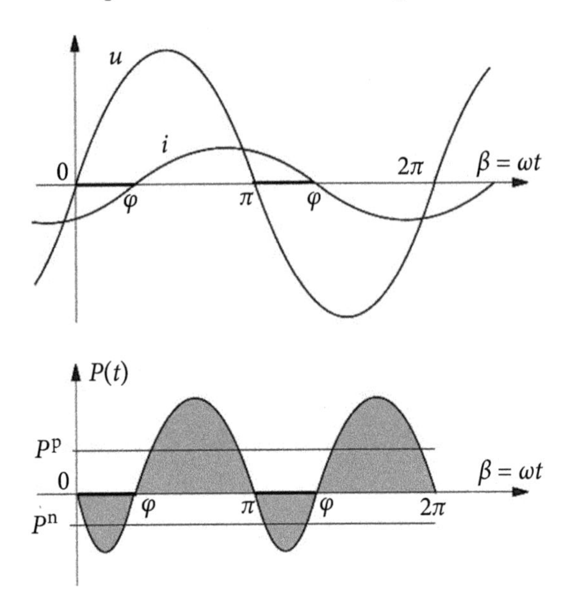

Figure 24.5 Instantaneous power waveform and its intervals when its flow to the supply source.

Indeed,

$$P = \frac{1}{2\pi} \int_{0}^{2\pi} u(\beta) i(\beta) d\beta = P^{n} + P^{p} \tag{24.5}$$

where

$$P^{n} = 2\frac{UI}{\pi} \int_{0}^{\varphi} \sin\beta \sin(\beta - \varphi) d\beta = \frac{UI}{\pi}[\varphi \cos\varphi - \sin\varphi] = \frac{\varphi}{\pi} P - \frac{1}{\pi} Q \tag{24.6}$$

$$P^{p} = 2\frac{UI}{\pi} \int_{\varphi}^{\pi} \sin\beta \sin(\beta - \varphi) d\beta = \frac{UI}{\pi}[(\pi - \varphi)\cos\varphi + \sin\varphi] = \frac{\pi - \varphi}{\pi} P + \frac{1}{\pi} Q. \tag{24.7}$$

The last two formulae show that the reactive power Q is an intricate function of the positive and negative mean values of powers P^{p} and P^{n}. It does not introduce anything to the physical interpretation of the reactive power.

24.3 Can Energy Rotate around Three-Phase Supply Lines?

The Instantaneous Reactive Power (IRP) p-q Theory, developed in [72], is based on two power quantities. One of them is the instantaneous active power, which is identical to the instantaneous power $p(t)$, and it was only its name inside the IRP p-q Theory that was changed. Another power quantity in the IRP p-q Theory is the instantaneous imaginary or reactive power, q. Because the IRP p-q Theory is one of the major power theories, the question of the physical meaning of the q power is quite natural.

In paper [72], which introduced this theory, the physical meaning of the instantaneous reactive power q was not provided. Eventually, Akagi, Watanabe, and Aredes concluded in their book [226], published in 2007, that "the imaginary power q is proportional to quantity of energy that is being exchanged between the phases of the system". Fig. 24.6 summarizes the above explanations about the real and imaginary powers. Fig. 24.6, copied along with the captions from Ref. [226], is shown below.

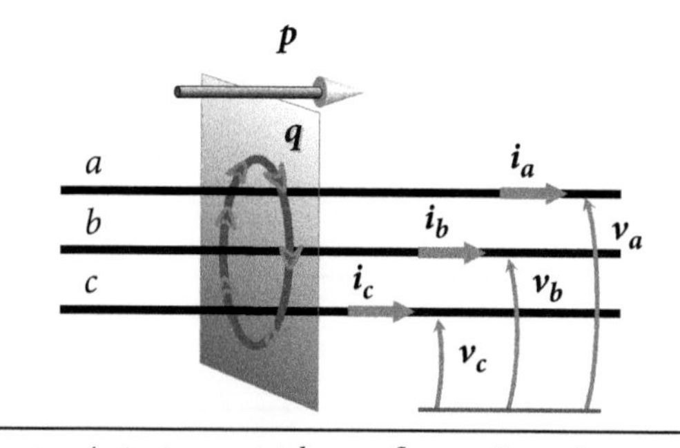

p : *instantaneous total energy flow per time unit;*
q : *energy exchanged between the phases without transferring energy.*

Figure 24.6 The physical meaning of the instantaneous real and imaginary powers according to Ref. [72].

The reader may observe, however, that the presented explanation of the meaning of the imaginary power q does not fit Fig. 24.6 because in the text below the figure states "q: energy exchanged between phases", while in that figure this energy q rotates around the supply line. Because this is not clear, we should verify, as was done in [269], if either of these two flows of energy is possible. Moreover, quantities p and q are not energies. They have the dimension of power.

The energy in electromagnetic fields flows in the direction of the Poynting Vector (PV),

$$\vec{P} = \vec{E} \times \vec{H} \tag{24.8}$$

where $\vec{E}$ denotes the electric field intensity, while $\vec{H}$ denotes the magnetic field intensity, which at any point of the space is perpendicular to the plane that is spanned on vectors of the electric and magnetic field intensities at that point. It is perpendicular to each of them, as shown in Fig. 24.7.

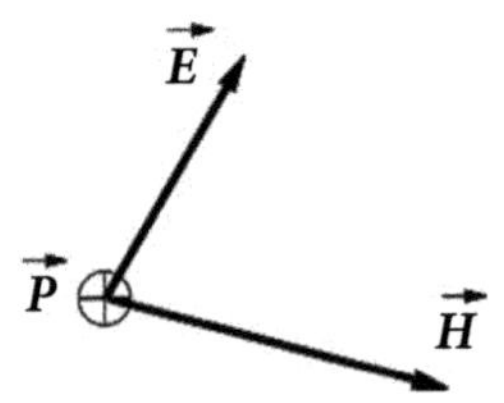

Figure 24.7 The electric and magnetic field intensities and the Poynting Vector.

Let us check now whether the situation shown in Fig. 24.5 is possible or not, in other words *can the energy rotate around the supply line?* Let us assume that the supply line, composed of conductors a, b, and c, is flat. If the energy rotates around such a line, then the PV at point x, in the conductors plane, should be perpendicular to that plane, as shown in Fig. 24.8. It is not possible, however, because the magnetic field intensity created by the line currents at point x is perpendicular to that plane, while these two vectors have to be perpendicular to each other. Thus, the energy cannot rotate around the supply line.

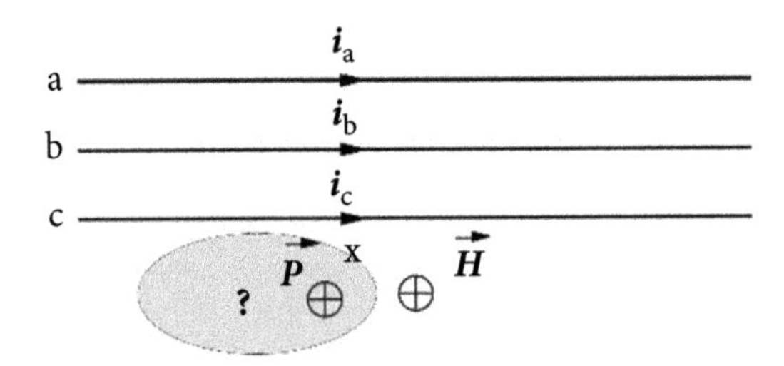

Figure 24.8 The magnetic field intensity at a point x in the conductors' plane.

Let us verify now whether "energy is being exchanged between phases" or not. If the energy flows between phases, then the PV should be perpendicular to conductors, meaning it should be oriented as shown in Fig. 24.9. Its sign for this discussion does not matter.

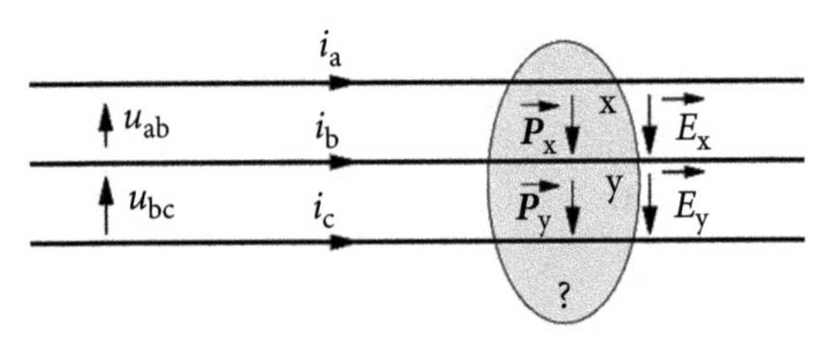

Figure 24.9 The electric field intensity between resistive-less conductors

When conductors are ideal, meaning their resistance can be neglected, then the electric field intensity has to be perpendicular to the conductor surface. Thus, the PV cannot be perpendicular to conductors, as shown in Fig. 24.9, because it has to be perpendicular to the electric field intensity.

Only when the line conductors have resistance can the electric field intensity have a component along the conductor surface and the PV a component toward conductors. In such a case, some amount of energy flows to conductors, where it is dissipated as heat. This dissipated energy has nothing in common, however, with "energy exchange between phases".

Thus, this reasoning based on a very fundamental principle of electromagnetic fields, meaning on the PV concept, demonstrates that the physical interpretation of the instantaneous imaginary power q, as suggested in Ref. [226], is erroneous.

24.4 Poynting Vector and Power Theory

The PV is a fundamental concept of electromagnetic field theory concerning energy flow. It is used extensively in such problems as energy radiation by antennas and other radiating high-frequency devices with distributed parameters. Therefore, in recent discussions [166, 180] on definitions and interpretations of powers in circuits with nonsinusoidal voltages and currents, in discussions on power theory development, the PV is increasingly often referred to as the very basis of this theory. There are even suggestions that the PV should be applied as a basis for power quality evaluation.

The PV is interpreted as the surface density of the rate of energy flow. Its direction is perpendicular to the surface specified by vectors of the electric and magnetic field intensities.

Suggestions that the PV concept should be considered as fundamental to the power theory stem from the fact that the flux of this vector to volume V through its surface S is equal to the rate of energy W flow to this volume, that is, to the instantaneous power, $p\,(t) = dW/dt$, of all devices confined by the surface S. Indeed, let us assume that electric energy is delivered to a load exclusively by its three-phase, three-wire supply lines, with the supply terminals denoted by R, S, and T, line-to-ground voltages u_R, u_S, and u_T and line currents i_R, i_S, and i_T, as shown in Fig. 24.10.

The flux of the PV through surface S, that crosses the supply lines only once, in such a case is equal to

$$\oiint_{\mathrm{S}} \vec{\boldsymbol{P}} \bullet d\vec{\boldsymbol{S}} = u_\mathrm{R} i_\mathrm{R} + u_\mathrm{S} i_\mathrm{S} + u_\mathrm{T} i_\mathrm{T} = p(t).$$

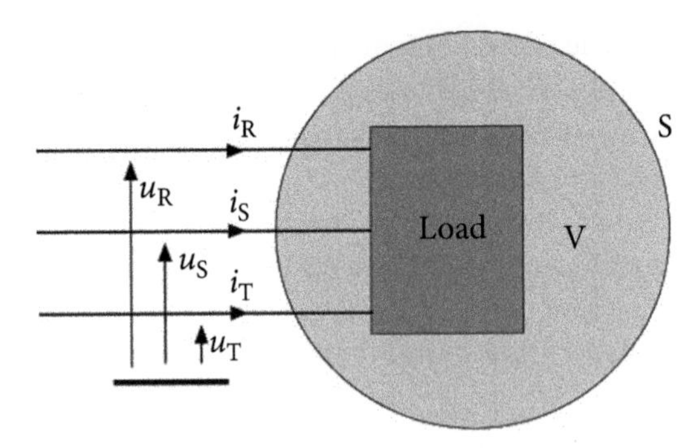

Figure 24.10 A three-phase load confined. by surface S.

The direct relation of the PV to Maxwell equations, fundamental equations of electrical engineering, and energy flow, its storage, and dissipation in electromagnetic fields make the claims that power theory should be founded on the PV very appealing.

Unfortunately, as was pointed out in [217], PV does not provide any information on power properties and the power-related phenomena in electrical circuits other than on instantaneous and active powers alone. Power theory is founded, however, not only on the phenomenon of energy transfer, specified in terms of the instantaneous and the active powers, but also on the conventional quantities such as the apparent power, and on powers that are only associated with physical phenomena, such as reactive, scattered, or unbalanced powers and on the load-generated harmonic power. The PV cannot be used for their identification and calculation. Consequently, power theory cannot be founded on the concept of the PV.

According to the scientists who initiated the power theory development, such as Steinmetz, Budeanu, Fryze, Depenbrock, and numerous others, its subject is the explanation of the difference between the apparent power S and the active power P at the load terminals. Observe, moreover, that the apparent power is not a physical but a conventional quantity, so it cannot be calculated with the PV. The same applies to the rms values of the load voltages and currents.

24.5 Geometric Algebra in Power Theory

The main mathematical tool used for the power theory development in the frequency domain and the hybrid domain is the algebra of complex numbers (ACN). This algebra has proven to be a very efficient and reliable mathematical instrument in power theory development. It has enabled solving all main issues of power theory regarding the description and interpretation of power-related phenomena in single- and three-phase circuits with linear and nonlinear loads and with periodically switched devices. It has enabled the development of methods of compensation by reactance, switching, and hybrid compensators in such circuits.

ACN has been challenged as incorrect [261], however, with a suggestion that it should be replaced by geometric algebra (GA).

A search for new mathematical tools is generally beneficial because such tools, when found, can provide a new perspective and results. Unfortunately, the authors of [261] did not demonstrate that GA, as applied to power theory provides such new results, even when applied to single-phase circuits with linear time-invariant loads. As it was shown in the paper [309], GA provides results obtained forty years earlier with ACN and is well behind the present state of power theory development of single-phase circuits, however. GA-based results do not provide either any new interpretations of power phenomena in electrical circuits or methods of power factor improvement. Even worse, the results known long ago were obtained at the cost of using much more sophisticated mathematical tools. No new property of electrical circuits was revealed, and no new method of compensator design was developed using GA.

Authors who suggest that instead of ACN, GA should be used for such investigations also suggest a replacement of the very fundamental power quantity, namely, the "apparent power" S by "a multivector". It is very unlikely that power systems engineers would be able to accept this mysterious term instead of the well-known and existing one, in use now for more than a century. One could also observe that GA is not usually taught in basic mathematics courses. ACN is taught in such courses.

Summary

This chapter shows that there are still several debatable and controversial issues related to the power properties of electrical circuits. A few of them were discussed. The history of power theory development has taught us that a many new issues may yet arise.

Literature

The book is written in such a way that extra literature is not needed. Titles of books and journal papers compiled below reflect recognition of the scientists who established the fundamentals of our present knowledge on the power properties of electrical circuits, such as Fourier, Steinmetz, Fortesque, and innumerable others.

The titles compiled in this list do not represent the entire body of literature on the subjects discussed in this book, however. The number of such papers and books could be in the order of a few thousand. The list largely comprises only those books and papers written in English, with some exceptions written by such great contributors to the power theory development as Steinmetz or Buchholz in German or Fryze in Polish.

This book presents the power theory from the perspective of CPC-based power theory and this is reflected in this literature selection. Throughout the book, only references comprising original research that have established schools of thought on the subject matter have been included, not followers of those schools.

Most of the items in this list are not referenced in the book. They can be used for more advanced studies and can be easily selected because most of the titles are self-explanatory.

[1] Fourier, J.B.J. *Theorie Analytique de la Chaleur* (Analytical Theory of Heat), Paris 1822, translation by A. Freeman (1878), Cambridge University Press, Cambridge, 2009.

[2] Steinmetz, Ch.P. "Is a phase-shift in the current of an electric arc?" (in German), *Elektrotechnische Zeitschrift*, Heft 42, 567–568, 1892.

[3] Houston E.J., Kennelly A.E. "On the causes producing phase difference in AC", *Electrical World*, Vol. 6, June 1895, 150–158, 1895.

[4] Heyland, A. "The practical significance of the power factor", *The Electrical Review*, Vol. 2, 45–50, 1896.

[5] MacGahan, P. "Power factor meter", USA Patent No. 695 913, 1898.

[6] Ganz, A.F. "The physical meaning of power factor", *Journal of the Franklin Institute*, Vol. 11, 429–435. 1906.

[7] Steinmetz, Ch.P. *Theory and Calculation of Electrical Apparatus*. McGraw-Hill Book Comp., New York, 1917.

[8] Fortesque, C.L. "Methods of symmetrical coordinates applied to the solution of polyphase networks", *Trans. AIEE*, Vol. 37, No. 3, 217–250, 1918.

[9] Lyon, W.V. "Reactive power and unbalanced circuits", *Electr. World*, Vol. 75, No. 25, 1417–1420, 1920.

[10] A.I.E.E. Committee, "Apparent power in three-phase systems", *Trans. AIEE*, Vol. 39, 1450–1455, 1920.

[11] Buchholz, F. "Die Drehstrom-Scheinleistung bei ungleichmabiger Belastung der drei Zweige", *Licht und Kraft*, Vol. 2, 9–11, 1922.

[12] Kriger, L. "Der einfluss des Hg-Gleichrichters auf den Leistungsfaktor", *ETZ*, Vol. 41, 286–290, 1923.

[13] Illovici, M.A. "Definition et mesure de la puissance et de l'energie reactives", *Bull. Soc. Franc. Electriciens*, Vol. 5, 931–956, 1925.

[14] Budeanu, C.I. *Puissances reactives et fictives*. Institut Romain de l'Energie, Bucharest, Publication. N4, 1927.

[15] Weber, E. "Die elektrische Leistung in allgemeinen Wechselstromkreis", *ETZ*, Vol. 47, 1547–1557, 1929.

[16] Brune, O. "Synthesis of finite two-terminal network", *J Math Phys.*, Vol. 10, 191–236, 1931.
[17] Fryze, S. "Active, reactive and apparent powers in systems with distorted waveform" (in Polish) *Przeglad Elektrotechniczny*, Z. 7, 193–203, 1931; Z. 8, 225–234, Z. 22, 673–676, 1932.
[18] Fryze, S. "Wirk-, Blind-, und Scheinleistung in Elektrischen Stromkreisen mit nicht-sinusoidalen Verlauf von Strom und Spannung", *ETZ*, No. 25, 193–202, 1932.
[19] Besicovitch, A.S. *Almost Periodic Functions*. Cambridge University Press, Cambridge, Vol. XIII, 180–185, 1932.
[20] Hommel, G. "Leistungsbegriffe und Leistungsfaktor bei ein- und mehrphasigen Wechselstromen", *Arch. fur Elektrotechnik*, Vol. 28, 326, 1934.
[21] Curtis, H.L., Silsbee, F.B. "Definitions of power and related quantities", *Trans. AIEE*, Vol. 54, 394–404, 1935.
[22] Quade, W. "Zusammensetzung der Wirk-, Blind-, und Scheinleistung bei Wechselstromen beliebiger Kurvenform und neue Leistungsdefinition fur unsymetrische Mehr-phasensysteme belibieger Kurvenform", *ETZ*, Vol. 58, 1312–1322, 1937.
[23] Emde, F. "Scheinleistung eines nicht sinusformigen Drehstromes", Elektrotechn, und Maschinen, Vol. 55, 557– 567, 1937.
[24] Rosenzweig, I. "Symbolic, multidimensional vector calculations as a method of polyphase system analysis" (in Polish), *Czasopismo Techniczne*, Tom LVI, 6–11, 1939.
[25] Tellegen, B.D.H. "A general network theorem with applications", *Philips Research Reports* (Philips Research Laboratories), Vol. 7, 259–269, 1952.
[26] Cauer, W. *Synthesis of Linear Communication Networks*. McGraw-Hill, New York, 1958.
[27] Balabanian, N. *Network Synthesis*. Prentice-Hall, Englewood Cliffs, 1958.
[28] Nedelcu, V.N. "Die enheitliche Leistungstheorie der unsymmetrischen mehrwelligen Mehr-phasensysteme", *ETZ-A*, No. 5, 153–157, 1963.
[29] Depenbrock, M. "Blind- und Scheinlaistung in einphasing gespeisten Netzwerken", *ETZ-A*. Year 85, Heft. 13, 385–391, 1964.
[30] Nowomiejski, Z., Cichowska, Z. "Unbalanced three-phase systems" (in Polish), *Scientific Letters of Gliwice University of Technology, Elektryka*, No. 17, 25–76, 1964.
[31] Fack, H. "Scheinleistung in Drehstromsystemen", *ETZ*, H.1, 55, 1967.
[32] Nowomiejski, Z. "Analyse elektrischen Kreise mit periodischen nicht sinusodalformigen Vorgangen", *Wissenschaftliche Zeitschrift der Elektrotechnik*, Vol. 8, 244–254, 1967.
[33] Czarnecki, L.S. "Theoretical problems of the realizability of the Hilbert transformation", *Acta Imeko*, 277–395, 1967.
[34] Antoniu, S.I., Leon, M. "Linear electronic model for the determination of active and reactive powers in nonsinusoidal state", *Acta Imeko*, 362–369, 1967.
[35] Czarnecki, L.S. "Synthesis of circuits that realize the Hilbert Transform", Ph.D. Dissertation, *Gliwice University of Technology*, Poland, 1969.
[36] Milic, M. "Integral representation of powers in periodic non-sinusoidal steady state and the concept of generalized powers", *IEEE Trans. on Education*, 107–109, 1970.
[37] German Standards DIN 40110, Wechselstromgrossen, 303–310, 1972.
[38] Czarnecki, L.S. "Reactive power meter for systems with nonsinusoidal voltages and currents" (in Polish), *Scientific Letters of Gliwice Univ. of Technology, Elektryka*, No. 36, 61–69, and Polish Patent No. 75834, 1972.
[39] Shepherd, W., Zakikhani, P. "Suggested definition of reactive power for nonsinusoidal systems", *Proc. IEE*, Vol. 119, No. 9, 1361–1362, 1972.
[40] Sharon, D. "Reactive power definitions and power factor improvement in nonlinear systems", *Proc. IEE*, Vol. 120, No. 6, 704–706, 1973.
[41] Czarnecki, L.S. "Reactive power meter for nonsinusoidal systems", Polish Patent No. 75 834, 1973.

[42] Fack, H. *Blindleistung Grundblindleistung und Verzerrungsleistung fur stationaren Verlauf von Spannungen und Strom.* Publication PTB-E-1, PTB-Bericht, 1–71, 1974.
[43] Czarnecki, L.S. "Reactive power meter for nonsinusoidal systems", Polish Patent No. 85 524, 1975.
[44] Gyugyi, L., Strycula, E. "Active ac power filters", *IEEE Trans. on Ind. Appl.*, Vol. IA-12, No. 3, 529–535, 1976.
[45] Steeper, D.A., Stratford, R.P. "Reactive power compensation and harmonic suppression for industrial power systems using thyristor converters", *IEEE Trans. on Ind. Appl.*, Vol. IA-12, No. 3, 232–254, 1976.
[46] Gyugyi, L., Otto, R.A., Putman, T.H. "Principles and applications of stationary thyristor-controlled shunt compensators", *IEEE Trans. on Power App. and Systems*, 1935–1945, 1976.
[47] Depenbrock, M. "Kompensation schnell verandliche Blindstrome", *ETZ-A*, Bd. 98, 408–411, 1977.
[48] Sawicki J. "The measurement of reactive power $\sum$ UIsinφ", *Acta Imeko*, Vol. II, 23–31, 1977.
[49] Lopez, R.A., Asquerino, J.C.M., Rodrigez-Izguierdo, G. "Reactive power meter for nonsinusoidal systems", *IEEE Trans. on Instr. Meas.*, Vol. IM-26, No. 3, 256–260, 1977.
[50] Czarnecki, L.S. "Reactive power meter for nonsinusoidal systems", Polish Patent No. 85 524, 1977.
[51] Elgerd, O.I. *Basic Electric Power Engineering.* Addison-Wesley, Menlo Park, 1977.
[52] Grandpierr, M. Trannoy, B. "A stationary power device to rebalance and compensate reactive power in three-phase network", *Proc. of 1977 IAS Annual Conference*, 127–135, 1977.
[53] Czarnecki, L.S. "One-ports with orthonormal properties", *Int. Journal on Circuit Theory and Appl.*, Vol. 6, 65–73, 1978.
[54] Shepherd, W., Zand, P. *Energy Flow and Power Factor in Nonsinusoidal Circuits.* Cambridge University Press, Cambridge, 1978.
[55] Depenbrock, M, *Wirk, und Blindleistungen periodischer Strome in Ein- und Mehrphasensystemen mit periodischen Spannungen beliebiger Kurvenform.* ETG-Fachtagung, 6, Berlin & Offenbach, VDE-VERLAG, 17–62, 1979.
[56] Harms, G. "Blindleistung im Energieeinsatz", *Elektrotechnik*, Vol. 61, H. 23, 1979.
[57] International Electrotechnical Commission, Techn. Committee No. 25, Working Group 7, Report: "*Reactive Power and Distortion Power*", Doc. No. 25, 113 (Secr.), 1979.
[58] Klinger, G. "L-C Kompensation und Symmetrierung fur Mehrphasensysteme mit beliebigen Spannungsverlauf", *ETZ-A*, H. 2, 57–61, 1979.
[59] Kusters, N.L., Moore, W.J.M. "On the definition of reactive power under nonsinusoidal conditions", *IEEE Trans. Pow. Appl. Syst.*, Vol. PAS-99, 1845–1854, 1980.
[60] Page, C. "Reactive power in nonsinusoidal systems", *IEEE Trans. on Instr. Meas.*, Vol. IM-29, No. 4, 420–423, 1980.
[61] Harms, G. "Blindleistund im Energieeinsatz", *Elektrotechnika*, Vol. 62, H. 6, 18–22, 1980.
[62] Gonzales, D.A., McCall, J.C. "Design of filters to reduce harmonic distortion in industrial power systems", *Proc. of IEEE Ann. Meeting*, Toronto, Canada, 361–365, 1980.
[63] Czarnecki, L.S. "Converter of the optimal capacitance to DC voltage", *Electronics Letters*, Vol. 17, No. 12, 427–428, and Polish Patent No. 134 452, 1981.
[64] Czarnecki, L.S. "Measurement principle of a reactive power meter for nonsinusoidal systems", *IEEE Trans. Instr. Meas.*, Vol. 6, No. 30, 209–212, 1981.
[65] Nowomiejski, Z. "Generalized theory of electric power", *Archiv fur Elekt.* Vol. 63, 177–182, 1981.
[66] Czarnecki, L.S. "Minimization of distortion power of nonsinusoidal source applied to linear loads", *Proc. IEE, Part C*, Vol. 128, No. 4, 208–210, 1981.

[67] Fisher, H.D. "Bemerkungen zu Leistungsbegriffen bei Stromen und Spannungen mit Oberschwingungen", *Archiv fur Elektrotechnik*, Vol. 64, 289–295, 1982.

[68] Fodor, G., Tevan, T. "Power and compensation in networks in periodic state", *Archiv fur Elektrotechnik*, Vol. 65, 27–40, 1982.

[69] Czarnecki, L.S. "Additional discussion to 'Reactive power under nonsinusoidal conditions'", *IEEE Trans. on Power and Systems*, Vol. PAS-102, No. 4, 1023–1024, 1983.

[70] Tschappu, F. "Considerations on the definition and measurement of apparent power S with the presence of harmonics in the network", *Landis & Gyr Review*, Vol 32, 9–17, 1983.

[71] Czarnecki, L.S. "Orthogonal components of a linear load supplied from a source of distorted voltage" (in Polish), *Scientific Letters of Gliwice Univ. of Techn., Elektryka*, No. 86, 1983.

[72] Akagi, H., Kanazawa, Y., Nabae, A. "Generalized theory of the instantaneous reactive power in three-phase circuits", *Proc. of IPEC*, Tokyo 1993, 1375–1380, 1983.

[73] Czarnecki, L.S. "An orthogonal decomposition of the current of nonsinusoidal voltage source applied to nonlinear loads", *Int. Journal on Circuit Theory and Appl.*, Vol. 11, 235–239, 1983.

[74] Czarnecki, L.S. "Measurement of the individual harmonic reactive power in nonsinusoidal systems", *IEEE Trans. Instr. Meas.*, Vol. IM-32, 383–385, 1983.

[75] Czarnecki, L.S. "Considerations on the reactive power in nonsinusoidal situations", *IEEE Trans. Instr. Measurement*, Vol. IM-34, 399–404, 1984.

[76] Czarnecki, L.S. "Measurement principle of the reactive current RMS value and the load susceptances for harmonic frequencies meter", *IEEE Trans. IM.*, Vol. IM-34, 14–17, 1984.

[77] Akagi, H., Kanazawa, Y., Nabae, A. "Instantaneous reactive power compensators comprising switching devices without energy storage components", *IEEE Trans. on IA.*, Vol. IA-20, 625–630, 1984.

[78] Czarnecki, L.S. "Comments on reactive powers as defined by Kusters and Moore for nonsinusoidal systems", *Rozprawy Elektrotechniczne*, Tom XXX, Z. 3–4, 1089–1099, 1984.

[79] Czarnecki, L.S. "Circuits that realize Hilbert transformation and their application for reactive power meters in nonsinusoidal system", *Arch. Elektrot.*, Tom XXXII, Z. 3/4, 415–475, 1984.

[80] Czarnecki, L.S. "A method of reactive power meter for nonsinusoidal systems construction based on a single-band modulation", *Gliwice Univ. of Techn., Elektryka*, No. 88, 31–39, 1984.

[81] Fryze, S. "Theoretical and physical fundamentals for the active, reactive and apparent power definitions in multi-phase systems with distorted voltages and currents" (in Polish), *Scientific Letters of Gliwice Univ. of Techn., Elektryka*, No. 100, 29–46, posthumous publication, 1985.

[82] Haque, S. E., Malik, N.H., Shepherd, W. "Operation of fixed capacitor-thyristor controlled reactor (FC-TCR) power compensators", *IEEE Trans.*, PAS-104, No. 6, 1385–1390, 1985.

[83] Czarnecki, L.S. "Power theories of periodic nonsinusoidal systems", *Rozprawy Elektrotechniczne*, Vol. 31, z. 3–4, 659–685, 1985.

[84] Akagi, H., Nabae, A., Atoh S. "Control strategy of active power filters using multiple voltage source PWM converters", *IEEE Trans. Ind. Appl.*, Vol. IA-22, 460–465, 1986.

[85] Richards, G.G., Tan, O.T., Czarnecki, L.S. "Comments on 'Considerations on the reactive power in nonsinusoidal situations'", *IEEE Trans. Instr. Meas.*, Vol. IM-37, No. 4, 365–366, 1986.

[86] Koch, K. "Bestimmung von Groβen in Mehr-Leitr Systemen", *ETZ Archiv*, Vol. 8, No. 10, 313–318, 1986.

[87] Czarnecki, L.S. "What is wrong with the Budeanu concept of reactive and distortion powers and why it should be abandoned", *IEEE Trans. IM*, Vol. IM-36, No. 3, 834–837, 1987.
[88] Hanzelka, Z. "Use of static compensators in symmetrization and compensation of asymmetrical electric loads", *Elektrotechnika*, Vol. 6, No. 4, 341–351, 1987.
[89] Czarnecki, L.S. "Minimization of reactive power in nonsinusoidal situation", *IEEE Trans. Instr. Meas.*, Vol. IM-36, No. 1, 18–22, 1987.
[90] Depenbrock, M. "The FBD-method, a generalized applicable tool for analyzing power relations", *IEEE Trans. on Pow. Del.*, Vol. 8, No. 2, 381–387, 1987.
[91] Czarnecki, L.S. "Converter of the optimal compensating capacitance in a nonsinusoidal system into a DC voltage", Polish Patent No. 134 452, 1987.
[92] Willems, J.L. "Power factor correction for distorted bus voltages", *Electr. Mach. and Power Syst.*, Vol. 13, 207–218, 1987.
[93] Czarnecki, L.S. "Orthogonal decomposition of the current in a three-phase non-linear asymmetrical circuit with nonsinusoidal voltage", *IEEE Trans. on IM*, Vol. IM-37, No. 1, 30–34, 1988.
[94] Czarnecki, L.S., Lasicz, A. "Active, reactive, and scattered currents in circuits with non-periodic voltage of a finite energy", *IEEE Trans. on IM*, Vol. IM-37, No. 3, 398–402, 1988.
[95] Kloss, A. *Harmonics* (in German). Technische Akademie Wuppertal, VDE-Verlag, Berlin-Offenbach, 1989.
[96] Czarnecki, L.S. "Reactive and unbalanced currents compensation in three-phase circuits under nonsinusoidal conditions", *IEEE Trans. Instr. Meas.*, Vol. IM-38, No. 3, 754–459, 1989.
[97] Czarnecki, L.S. "Comments on "Measurement and compensation of fictitious power under nonsinusoidal voltage and current conditions", *IEEE Trans. IM*, Vol. IM-38, No. 3, 839–841, 1989.
[98] Johnson, D.E., Johnson, L.R., Hilburn, J.L., et al. *Electric Circuit Analysis*. Prentice-Hall, Englewood Cliffs, 1989.
[99] Czarnecki, L.S., Swietlicki, T. "Powers in nonsinusoidal networks, their analysis, interpretation and measurement", *IEEE Trans. on IM*, Vol. IM-39, No. 2, 340–345, 1990.
[100] Czarnecki, L.S. "Time-domain approach to the reactive current minimization in nonsinusoidal situations", *IEEE Trans. Instr. Meas.*, Vol. IM-39, No. 5, 698–703, June 1990.
[101] Emanuel, A.E. "Powers in nonsinusoidal situations. A review of definitions and physical meanings", *IEEE Trans. on Pow. Del.*, Vol. 5, No. 3, 1377–1389, 1990.
[102] Czarnecki, L.S. "Discussion to 'Powers in nonsinusoidal situations—a review of definitions and physical meaning'", *IEEE Trans. Power Deliv.*, Vol. 5, No. 3, 1377–1378, 1990.
[103] Emanuel, A.E., Czarnecki, L.S. "Correspondence on 'Power components in systems with sinusoidal and nonsinusoidal voltages'", *Proc. IEE, Part. B*, Vol. 137, No. 3, 194–196, 1990.
[104] Czarnecki, L.S. "Comments on 'A new control philosophy for power electronic converters as fictitious power compensators'", *IEEE Trans. on Pow. Elect.*, Vol. 5, No. 4, 503–504, 1990.
[105] Czarnecki, L.S. "Scattered and reactive current, voltage, and power in circuits with nonsinusoidal waveforms and their compensation", *IEEE Trans. IM*, Vol. 40, No. 3, 563–567, 1991.
[106] Sun, S.Q., Kiyokawa, H. "Decomposition of Czarnecki's reactive current and reactive power", *Proc. IEE, Part B*, Vol. 138, No. 3, 125–128, 1991.
[107] Czarnecki, L.S. "Distortion power in systems with nonsinusoidal voltage", *Proc. IEE, Part B*, Vol. 139, No. 3, 276–280, 1992.

[108] Czarnecki, L.S. “Two algorithms of the fundamental harmonic complex RMS value calculation”, *Archiv fur Elektrotechnik*, Vol. 75, 163–168, 1992.

[109] Czarnecki, L.S. “Minimization of unbalanced and reactive currents in three-phase asymmetrical circuits with nonsinusoidal voltage”, *Proc. IEE, Part B*, Vol. 139, No. 4, 347–354, 1992.

[110] IEEE Std 519-1992: “IEEE Recommended Practices and Requirements for Harmonic Control in Electrical Power Systems”, 1992.

[111] Akagi, H., Nabae, A. “The p-q theory in three-phase systems under nonsinusoidal conditions”, *Europ. Trans. Elec. Pow.*, Vol. 3, No. 1, 27–31, 1993.

[112] Czarnecki, L.S. “Physical reasons of current RMS value increase in power systems with nonsinusoidal voltage”, *IEEE Trans. Power Deliv.*, Vol. 8, No. 1, 437–447, 1993.

[113] Depenbrock, M. “The FBD-method, a generalized applicable tool for analyzing power relations”, *IEEE Trans. on Power Del.*, Vol. 8, No. 2, 381–387, 1993.

[114] Czarnecki, L.S. “Current and power equations at a bi-directional flow of harmonic active power in circuits with rotating machines”, *Europ. Trans. Elec. Pow.*, Vol. 3, No. 1, 45–52, 1993.

[115] Lee, S.Y., Wu, C.J. “On-line reactive power compensation schemes for unbalanced three-phase four-wire distribution systems”, *IEEE Trans. on Power Del.*, Vol. 8, No. 4, 1235–1239, 1993.

[116] Makram, E., Varadan, S. “Analysis of reactive power and power factor correction in the presence of harmonics and distortion”, *Electric Power Systems Research*, Vol. 26, 211–218, 1993.

[117] Czarnecki, L.S. “Power factor improvement of three-phase unbalanced loads with nonsinusoidal voltage”, *Europ. Trans. Elec. Pow.*, Vol. 3, No. 1, 67–74, 1993.

[118] Cameron, M.M. “Trends in power factor correction with harmonic filtering”, *IEEE Trans. on IA.*, Vol. IA-29, No. 1, 60–65, 1993.

[119] Krogeric, A.F., Reszewic, KK. Trejmanic, E.P., et al. *Powers of Alternating Currents.* Latvia Academy of Science, Riga, 1993.

[120] Watanabe, E.H., Stephan, M., Aredes, M. “New concept of instantaneous active and reactive powers in electrical systems”, *IEEE Trans. on Pow. Del.*, Vol. 8, No. 2, 697–703, 1993.

[121] Czarnecki, L.S. “Supply and loading quality improvement in sinusoidal power systems with unbalanced loads supplied with asymmetrical voltage”, *Arch. fur Elektrot.*, Vol. 77, 169–177, 1994.

[122] Rossetto, P., Tenti, P. “Evaluation of instantaneous power terms in multi-phase systems: techniques and application to power-conditioning equipment”, *Europ. Trans. Elec. Pow.*, Vol. 4, No. 6, 469–475, 1994.

[123] Czarnecki, L.S. “Comments on ‘Apparent power—a misleading quantity in nonsinusoidal power theory: Are all non-sinusoidal power theories doomed to fail’”, *Europ. Trans. Elec. Pow.*, Vol. 4, No. 5, 427–431, 1994.

[124] Superti-Furga, G. “Searching for generalization of the reactive power—a proposal”, *Europ. Trans. Elec. Pow.*, Vol. 4, No. 5, 411–420, 1994.

[125] Czarnecki, L.S., Hsu, M.S. “Thyristor controlled susceptances for balancing compensators operated under nonsinusoidal conditions”, *Proc. IEE, Part B*, Vol. 141, No. 4, 177–185, 1994.

[126] Czarnecki, L.S., Tan, O.T. “Evaluation and reduction of harmonic distortion caused by solid-state voltage controllers of induction motors”, *IEEE Trans. on Energy Conversion*, Vol. 9, No. 3, 528–534, 1994.

[127] Czarnecki, L.S. “Comments on ‘Apparent and reactive powers in three-phase systems: in search of a physical meaning and better resolution’”, *Europ. Trans. Elec. Pow.*, Vol. 4, No. 5, 421–424, 1994.

[128] Czarnecki, L.S. "Dynamic, power quality oriented approach to theory and compensation of asymmetrical systems under nonsinusoidal conditions", *Europ. Trans. Elec. Pow.*, Vol. 5, 347–358, 1994.

[129] Depenbrock, M., Marshal, D.A., van Wyk, J.D. "Formulating requirements for universally applicable power theory as control algorithm in power compensators", *Europ. Trans. Elec. Pow.*, Vol. 4, No. 6, 445–456, 1994.

[130] Czarnecki, L.S. "Misinterpretations of some power properties of electric circuits", *IEEE Trans. on Pow. Del.*, Vol. 9, No. 4, 1760–1770, 1994.

[131] Czarnecki, L.S. "A combined time-domain and frequency domain approach to hybrid compensation in unbalanced nonsinusoidal systems", *Europ. Trans. Elec. Pow.*, Vol. 4, No. 6, 477–484, 1994.

[132] Staudt, V. "Differences between compensation of instantaneous non-active power and total non-active power", *Proc. of Int. Conf. on Harmonics in Pow. Syst.*, ICHPS, Bologna, 381–387, 1994.

[133] Czarnecki, L.S., Hsu, M.S. "Thyristor controlled susceptances for balancing compensators operated under nonsinusoidal conditions", *Proc. IEE, Part B*, Vol. 141, No. 4, 177–185, 1994.

[134] Depenbrock, M., Skudelny, H.-Ch. "Dynamic compensation of non-active power using the FBD method—basic properties demonstrated by benchmark examples", *Europ. Trans. Elec. Pow.*, Vol. 4, No. 5, 381–388, 1994.

[135] Czarnecki, L.S. "Power theory of electric circuits: a common database of power related phenomena and properties", *Europ. Trans. Elec. Pow.*, Vol. 4, No. 6, 491–498, 1994.

[136] Almonte, R.L., Ashley, A.W. "Harmonics at the utility industrial interface: a real world example", *IEEE Trans. on Ind. Appl.*, Vol. 31, No. 6, 1419–1426, 1995.

[137] Czarnecki, L.S. "Power related phenomena in three-phase unbalanced systems", *IEEE Trans. on Pow. Del.*, Vol. 10, No. 3, 1168–1176, 1995.

[138] Czarnecki, L.S. "Effect of minor harmonics on the performance of resonant harmonic filters in distribution systems", *Proc. IEE, Part B*, Vol. 144, No. 5, 349–356, 1995.

[139] Czarnecki, L.S., Hsu, M.S., Chen, G. "Adaptive balancing compensator", *IEEE Trans. on Pow. Del.*, Vol. 10, No. 3, 1663–1669, 1995.

[140] Czarnecki, L.S. "Equivalent circuits of unbalanced loads supplied with symmetrical and asymmetrical voltage and their identification", *Archiv fur Elektrotechnik*, Vol. 78, 165–168, 1995.

[141] Peeran, S.M., Cascadden, C.W.P. "Application, design, and specification of harmonic filters for variable frequency drives", *IEEE Trans. on IA*, Vol. 31, No. 4, 841–847, 1995.

[142] Bonner, J.A., et al. "Selecting ratings for capacitors and reactors in applications involving multiple single-tuned filters", *IEEE Trans. on Pow. Del.*, Vol. 10, No. 1, 547–555, 1995.

[143] Czarnecki, L.S. "Comments on active power flow and energy accounts in electrical systems with nonsinusoidal voltage and asymmetry", *IEEE Trans., PD.*, Vol. 11, No. 3, 1244–1250, 1996.

[144] Salmeron, P., Montano, J.C. "Instantaneous power components in poly-phase systems under nonsinusoidal conditions", *IEE Proc. on Science, Meas. Techn.*, Vol. 143, No. 2, 239–297, 1996.

[145] Czarnecki, L.S., Staroszczyk, Z. "Dynamic on-line measurement of equivalent parameters of three-phase systems for harmonic frequencies", *Europ. Trans. on Electric Pow.*, Vol. 6, No. 5, 329–335, 1996.

[146] Czarnecki, L.S., Staroszczyk, Z. "On-line measurement of equivalent parameters for harmonic frequencies of a power distribution system and load", *IEEE Trans. on Instr. Meas.*, Vol. 45, No. 2, 467–472, 1996.

[147] Czarnecki, L.S., Chen, G., Staroszczyk, Z. "Application of running quantities for control of adaptive hybrid compensator", *Europ. Trans. Elec. Pow.*, Vol. 6, No. 3, 337–344, 1996.

[148] Czarnecki, L.S. "Power theory of electrical circuits with quasi-periodic waveforms of voltages and currents", *Europ. Trans. Elec. Pow.*, Vol. 6, No. 5, 321–328, 1996.

[149] Czarnecki, L.S. "Comments on active power flow and energy accounts in electrical systems with nonsinusoidal voltage and asymmetry", *IEEE Trans. on PD*, Vol. 11, No. 3, 1244–1250, 1996.

[150] Peng, F.Z., Lai, J-S. "Generalized instantaneous reactive power theory for three-phase power systems", *IEEE Trans. IM.*, Vol. IM-45, No. 1, 293–297, 1996.

[151] Sharon, D. "Power factor definition and power transfer quality in nonsinusoidal systems", *IEEE Trans. on Instrum. and Measur.*, Vol. 45, No. 3, 728–733, 1996.

[152] Czarnecki, L.S. "Application of running quantities for a control of an adaptive hybrid compensator", *Europ. Trans. Elec. Pow.*, Vol. 6, No. 5, 337–344, 1996.

[153] Czarnecki, L.S. "Budeanu and Fryze: Two frameworks for interpreting power properties of circuits with nonsinusoidal voltages and currents", *Archiv fur Elektr.*, Vol. 81, No. 2, 5–15, 1997.

[154] IEEE, *The New Standard Dictionary of Electrical and Electronics Terms*. IEEE, London, 1997.

[155] Czarnecki, L.S. "Effect of minor harmonics on the performance of resonant harmonic filters in distribution systems", *Proc. IEE, Part B*, Vol. 144, No. 5, 349–356, 1997.

[156] Phipps, J.K. "A transfer function approach to harmonic filter design", *IEEE Industry Appl. Magazine*, 68–82, 1997.

[157] Czarnecki, L.S. "Minimization od reactive power under nonsinusoidal conditions", *IEEE Trans. Instr. Meas.*, Vol. 36, No. 1, 18–22, 1997.

[158] Czarnecki, L.S. "Powers and compensation in circuits with periodic voltages and currents. Part 1. Budeanu's power theory: 60 years of illusions", *Journal on Electrical Power Quality and Utilization*, Vol. III, No. 1, 37–43, 1997.

[159] Czarnecki, L.S. "Powers and compensation in circuits with periodic voltages and currents. Part 2. Outline of the history of power theory development", *Journal on Electrical Power Quality and Utilization*, Vol. III, No. 2, 37–46, 1997.

[160] Peng, F.Z., Ott Jr, G.W., Adams, D.J. "Harmonic and reactive power compensation based on the generalized instantaneous reactive power theory for three-phase four-wire systems", *IEEE Trans. on Power Electronics*, Vol. 13, No. 6, 1174–1181, 1998.

[161] Czarnecki, L.S. "Common and fixed-poles resonant harmonic filters", *Europ. Trans. Elec. Power., ETEP*, Vol. 8, No. 5, 345–351, 1998.

[162] Wu, C.J., Chiang, S.-S., Jen, C.-J. "Investigation and mitigation of harmonic amplification problems caused by single-tuned filters", *IEEE Trans. on Pow. Del.*, Vol. 13, No. 3, 800–806, 1998.

[163] le Roux, W., van Wyk, J.D. "Correspondence and difference between the FBD and Czarnecki current decomposition methods for linear loads", *Europ. Trans. Elec. Pow.*, Vol. 8, No. 5, 329–336, 1998.

[164] Hsu, S.M., Czarnecki, L.S. "Harmonic blocking compensator: frequency properties", *Europ. Trans. Elec. Pow.*, Vol. 8, No. 5, 37–46, 1998.

[165] Czarnecki, L.S. "Energy flow and power phenomena in electrical circuits: illusions and reality", *Archiv fur Elektrotechnik*, Vol. 82, No. 4, 10–15, 1999.

[166] Cekareski, Z., Emanuel, A.E. "On the physical meaning of nonactive powers in three-phase systems", *Power Engineering Review, IEEE*, Vol. 19, No. 7, 46–47, 1999.

[167] Czarnecki, L.S. "Power factor measurement" in (editors' names to be inserted here), *Wiley Encyclopedia of Electrical and Electronics Engineering* (Vol. 16, 668–679). John Wiley & Sons, (location of publisher to be inserted here), John G. Webster, 1999.

[168] Huang, S-J., Wu, J-C. "A control algorithm for three-phase three-wired active power filters under nonideal mains voltages", *IEEE Trans. on Power Electr.*, Vol. 14, No. 4, 753–760, 1999.

[169] Morendaz-Eguilaz, J.M., Peracaula, J. "Understanding AC power using the generalized instantaneous reactive power theory: considerations for instrumentation of three-phase electronic converters", *Proc. of the IEEE Int. Symp., Ind. Electr.* Vol. 99, No. 3, 1273–1277, 1999.

[170] Aller, J.M., Bueno, A., Restrepo, J.A. "Advantages of the instantaneous reactive power definitions in three-phase system measurement", *IEEE Power Eng. Review*, Vol. 19, No. 6, 54–56, 1999.

[171] Lee, C.-Y. "Effects of unbalanced voltage on the operation performance of three-phase induction motors", *IEEE Trans. on Energy Conversion*, Vol. 14, No. 2, 202–208, 1999.

[172] IEEE Trial use standard for the measurement of electric power quantities under sinusoidal, nonsinusoidal, balanced, and unbalanced conditions, *IEEE STD. 1459*, 2000.

[173] Oriega, L.C., De Oliviera, M.C., Barros Neto, J.B. "Load compensation in four-wire electrical power systems", *Proc. of Int. Conf. on Power System Techn*, Vol. 3, 1975–1580, 2000.

[174] Czarnecki, L.S. "Harmonics and power phenomena" in (editor names to be inserted here), *Wiley Encyclopedia of Electrical and Electronics Engineering* (Suppl. 1, 195–218). John Wiley & Sons, John G. Webster, (publisher location), 2000.

[175] Chou, C.-J. at al. "Optimal planning of large passive harmonic filters set at high voltage level", *IEEE Trans. on Pow. Syst.*, Vol. 15, No. 1, 433–441, 2000.

[176] Ghassemi, F. "A new apparent power and power factor with non-sinusoidal waveforms", *IEE Proc. Gen. Transm. and Distr.*, Vol. 147, No. 6. 146–152, 2000.

[177] Carlson, A.B. *Circuits.* Brooks/Cole, Thomson Learning, Pacific Grove, 2000.

[178] Mishra, M.K., Joshi, A., Ghosh, A. "Unified shunt compensator algorithm based on generalized instantaneous reactive power theory", *IEEE Trans. on Gen. Trans. and Distr.*, Vol. 148, No. 6, 583–589, 2001.

[179] Rawa, H. "Energy and power in electrical systems" (in Polish), *Przegląd Elektrotechniczny*, Vol. LXXVII, No. 5, 141–147, 2001.

[180] Cekareski, Z., Emanuel, A.E. "Poynting Vector and the power quality of electric energy", *Europ. Trans. Elec. Pow.*, Vol. 11, No. 6, 375–382, 2001.

[181] Wang, Y.J. "Analysis of effects of three-phase voltage unbalance on induction motors with emphasis on the angle of the complex unbalanced factor", *IEEE Trans. on Energy Conversion*, Vol. 16, No. 3, 270–275, 2001.

[182] Czarnecki, L.S., Chen, G. "Compensation of semi-periodic currents", *Europ. Trans. Elec. Pow.*, Vol. 12, No. 1, 33–39, 2002.

[183] Rens, A.P.J., Swart, P.H. "Investigating the validity of the Czarnecki three-phase power definitions", *Proc. of IEEE Africon*, 815–821, 2002. 0-7803-7570-X/02

[184] Peng, F.Z., Tolbert, L.M., Qian, Z. "Definitions and compensation of non-active current in power systems", 33rd IEEE Annual Electronics Specialists Conf., Cat. No. 02CH377289, Cairns, Australia, 1779–1784, 2002.

[185] Czarnecki, L.S. "Circuits with semi-periodic currents: Main features and power properties", *Europ. Trans. Elec. Pow.*, Vol. 12, No. 1, 41–46, 2002.

[186] Montano, J.C., Salmeron, P. "Strategies of instantaneous compensation for three-phase four-wire circuits", *IEEE Trans. Pow. Del.*, Vol. 17, 1079–1084, 2002.

[187] Kim, K., Blaabjerg, F., Bak-Jensen, B. "Instantaneous power compensation in three-phase system using p-q-r theory", *IEEE Trans. on Power Electr.*, Vol. 17, No. 5, 701–710, 2002.

[188] Paretto L., Sasdelli R., Tinarelli R., Rossi, C. "Measurement on electrical power systems under bi-tone conditions by using the virtual time-domain approach", *Europ. Trans. Elec. Pow.*, ETAP, Vol. 12, No. 1, 5–9, 2002.

[189] Depenbrock, M., Staudt, V., Wrede, H. "A theoretical investigation of original and modified instantaneous power theory applied to four-wire systems", *IEEE Trans. on Ind. Appl.*, Vol. 39, No. 4, 1160–1168, 2003.

[190] Czarnecki, L.S. "Powers in three-phase circuits and their misinterpretations by the Instantaneous Reactive Power p-q Theory", *Przegląd Elektrotechniczny*, Vol. 79, No. 10, 658–603, 2003.

[191] Tenti, P., Mattavelli, P. "A time-domain approach to power term definitions under non-sinusoidal conditions", *Proc. of the Six Int. Workshop on Power Definitions and Measurement under Nonsinusoidal Conditions*, Milan, 2003.

[192] Gajić, Z. *Linear Dynamic Systems and Signals.* Prentice Hall, Englewood Cliffs, 2003.

[193] Hsu, S.M., Czarnecki, L.S. "Adaptive blocking compensator", *IEEE Trans. on Pow. Del.*, Vol. 18, No. 3, 895–902, 2003.

[194] Levi-Ari, H., Stanković, A.M. "Hilbert space techniques for modeling and compensation of reactive power in energy systems", *IEEE Trans. on CAS-I*, Vol. 50, No. 4, 540–556, 2003.

[195] Czarnecki, L.S. "Comparison of the instantaneous reactive power, p-q, theory with the theory of current's physical components", *Archiv fur Elektrotechnik*, Vol. 85, No. 1, 21–28, 2004.

[196] Czarnecki, L.S. "Currents' Physical Components and powers in systems with pulsing flow of energy", *Przeglad Elektrotechniczny*, Vol. LXXX, No. 6, 560–569, 2004.

[197] Aziz, M.M.A., El-Zahab, E.E., Ibrahim, A.M. "LC compensator for power factor correction of nonlinear loads", *IEEE Trans on Pow. Del.*, Vol. 19, No, 1, 331–335, 2004.

[198] Czarnecki, L.S., Mendrela, E.A., Ginn, H.L. "Decomposition of pulsed load current for hybrid compensation", *L'Energia Elettrica*, Vol. 81, 150–156, 2004.

[199] Balci, M.E., Hocaoglu, M.H. "Comparison of power definitions for reactive power compensation in nonsinusoidal conditions", *Proc. of Int. Conf on Harmonics and Quality of Power (ICHQP)*, Lake Placid, 519–524, 2004.

[200] Czarnecki, L.S., Ginn, H.L. "Effects of damping on the performance of resonant harmonic filters", *IEEE Trans. on Pow. Del.*, Vol. 19, No. 3, 846–853, 2004.

[201] Czarnecki, L.S. "On some misinterpretations of the Instantaneous Reactive Power p-q Theory", *IEEE Trans. on Power Electronics*, Vol. 19, No. 3, 828–836, 2004.

[202] Jeon, S.-J. "Discussion of "Instantaneous reactive power p-q theory and power properties of three-phase systems"", *IEEE Trans. on Power Del.*, Vol. 23, No. 3, 1694–1696, 2004.

[203] Willems, J.L. "Reflections on apparent power and power factor in nonsinusoidal polyphase systems", *IEEE Trans. on Power Del.*, Vol. 19, No. 3, 835–840, 2004.

[204] Czarnecki, L.S. "Consideration on the concept of the Poynting Vector contribution to power theory development", *L'Energia Elettrica*, Vol. 81, 64–74, 2004.

[205] Lathi, B.P. *Linear Systems and Signals.* Oxford University Press, Oxford, 2005.

[206] Czarnecki, L.S. *Powers in Circuits with Nonsinusoidal Voltages and Currents* (in Polish), Warsaw University of Technology Publication Office, Warsaw, 2005.

[207] Mayer. D., Kropik, P. "New approach to symmetrization of three-phase networks", *Int. Journ. of Electrical Engineering*, Vol. 56, No. 5–6, 156–161, 2005.

[208] Czarnecki, L.S., Ginn, H.L. "The effect of the design method on the efficiency of resonant harmonic filters", *IEEE Trans. on Pow. Del.*, Vol. 20, No. 1, 286–291, 2005.

[209] Montano, C.J., Salmeron P., Thomas P. "Analysis of power loss for instantaneous compensation of three-phase four-wire systems", *IEEE Trans. on Pow. Electr.*, Vol. 20, No. 4, 901–907, 2005.

[210] Czarnecki, L.S. "Currents' Physical Components (CPC) in circuits with nonsinusoidal voltages and currents. Part 1: Single-phase linear circuits", *Electrical Power Quality and Utilization Journal*, Vol. XI, No. 2, 3–14, 2005.

[211] Sommariva, A.M. "Power analysis of one-port under periodic multi-sinusoidal linear operations", *IEEE Trans. on Circuits and Systems—I*, Vol. 53, No. 9, 2068–2074, 2006.

[212] Czarnecki, L.S. "Currents' Physical Components (CPC) in circuits with nonsinusoidal voltages and currents. Part 2: Three-phase linear circuits", *Electrical Power Quality and Utilization Journal*, Vol. X, No. 1, 1–10, 2006.

[213] Ginn, H.L., Czarnecki, L.S. "Optimization of resonant harmonic filters", *IEEE Trans. on Pow. Del.*, Vol. 21, No. 3, 1445–1451, 2006.

[214] Czarnecki, L.S. "Energy oscillation and non-active powers in terms of the CPC power theory and the Poynting Vector", *Przeglad Elektrotechniczny*, Vol. 82, No. 6, 1–7, 2006.

[215] Sommariva, A.M. "Power analysis of one-ports under periodic multi-sinusoidal linear operation", *IEEE Trans. on Circuits and Systems I*, Vol. 53, No. 9, 2068–2074, 2006.

[216] Czarnecki, L.S. "Comments on 'The Instantaneous Reactive Power p-q Theory—Power theory or a useful algorithm of switching compensator control'", *Przeglad Elektrotechniczny*, Vol. 82, No. 7–8, 164–167, 2006.

[217] Czarnecki, L.S. "Could power properties of three-phase systems be described in terms of the Poynting Vector?", *IEEE Trans. on Pow. Del.*, Vol. 21, No. 1, 339–344, 2006.

[218] de Leon, F., Cohen, J. "Discussion of 'Generalized theory of instantaneous reactive quantity for multiphase power systems'", *IEEE Trans. on Pow. Del.*, Vol. 21, No. 1, 540–541, 2006.

[219] Czarnecki, L.S. "Instantaneous Reactive Power p-q Theory and power properties of three-phase systems", *IEEE Trans. on Pow. Del.*, Vol. 21, No. 1, 362–367, 2006.

[220] Tenti, P., Tedeschi, E., Mattavelli, P. "Compensation techniques based on reactive power conservation", *Proc. of the 7th Intern. Workshop on Power Definition and Measurements under Nonsinusoidal Conditions*, Cagliari (Italy), 2006.

[221] Ginn, H.L., Czarnecki, L.S. "An optimization-based method for selection of resonant harmonic filter parameters", *IEEE Trans. on Pow. Del.*, Vol. 21, No, 3, 1445–1451, 2006.

[222] De Leon, F., Cohen, J. "A practical approach to power factor definitions: Transmission losses, reactive power compensation, and machine utilization", 1–7, 2006. DOI: 1-4244-0493-2/06/$20.00 C 1EEE

[223] Czarnecki, L.S. "Physical interpretation of the reactive power in terms of the CPC power theory", *Electrical Power Quality and Utilization Journal*, Vol. XIII, No. 1, 89–95, 2007.

[224] Czarnecki, L.S. "Powers of asymmetrically supplied loads in terms of the CPC power theory", *Electrical Power Quality and Utilization Journal*, Vol. XIII, No. 1, 97–104, 2007.

[225] Morsi, W.G., El-Hawary, M.E. "Defining power components in nonsinusoidal unbalanced poly-phase systems: The issues", *IEEE Trans. on Power Del.*, Vol. 22, No. 4, 2428–2437, 2007.

[226] Akagi, H., Watanabe, E.H., Aredes M. *Instantaneous Power Theory and Applications to Power Conditioning.* IEEE Press, Piscataway, 2007.

[227] Czarnecki, L.S. "Quasi-instantaneous generation of reference signals for hybrid compensator control", *Electrical Power Quality and Utilization Journal*, Vol. l. XIII, No. 2, 33–38, 2007.

[228] Spath, H. "A general purpose definition of active current and non-active power based on German Standard DIN 40110", *Electrical Engineering*, Vol. 89, No. 3, 167–175, 2007.

[229] Firlit, A. "Current's physical components theory and p-q power theory in the control of the three-phase shunt active power filter", *Electrical Power Quality and Utilisation Journal*, Vol. XIII, No. 1, 59–66, 2007.

[230] Superti-Furga, G., Todeschini, G. "Discussion on instantaneous p-q strategies for control of active filters", *IEEE Trans. on Pow. Electronics*, Vol. 23, No. 4, 1945–1955, 2008.

[231] Sekara, T.B., Mikulovic, J.C., Djuriscic, Z.R. "Optimal reactive compensators in power systems under asymmetrical and nonsinusoidal conditions", *IEEE Trans. on Pow. Del.*, Vol. 23, No. 2, 974–984, 2008.

[232] Czarnecki, L.S. "Currents' Physical Components (CPC) concept: A fundamental for power theory", *Przeglad Elektrotechniczny*, Vol. 84, 28–37, 2008.

[233] Ginn, H.L., Chen, G.D. "Flexible active compensator control for variable compensation objectives", *IEEE Trans. on Power Electronics*, Vol. 23, No. 6, 2931–2941, 2008.

[234] Bitoleanu, A., Popescu, M., Dobriceanu, M. "Current decomposition methods based on p-q and CPC theories for active filtering reasons", *WSEAS Trans. on Circuits and Systems*, Vol. 7, No. 10, 869–878, 2008.

[235] Staudt, V. "Fryze-Buchholz-Depenbrock: A time-domain power theory", *Przeglad Elektrotechn*, Vol. 84, 12–11, 2008.

[236] Tedeschi, E., Tenti, P. "Cooperative design and control of distributed harmonic and reactive compensators", *Przegląd Elektrotechniczny*, Vol. 85, 23–27, 2008.

[237] Ginn, H., Chen, G. "Switching compensator control strategy based on CPC Power Theory", *Przegląd Elektrotechniczny*, Vol. 85, 38–47, 2008.

[238] Watanabe, E., Akagi, H., Aredes, M. "Instantaneous p-q Power Theory for compensating nonsinusoidal systems", *Przegląd Elektrotechniczny*, Vol. 85, 167–169, 2008.

[239] Czarnecki, L.S., Pearce, S.E. "Compensation objectives and CPC–based generation of reference signals for shunt switching compensator control", *IET Pow. Electr.*, Vol. 2, No. 1, 33–41, 2009.

[240] Monteiro, L.F.C., Afonso, J.L., Pinto, J.G., et al. "Comparison algorithms based on the *p-q* and CPC theories for switching compensators in micro-grids", *Proc. of COBEP 09 Power Electr. Conf.*, Brazil, 32–40, 2009. DOI: 9781-4244-3370-4/09/$25.00-0/09 C IEEE.

[241] Chen, G., Jiang, Y., Zhou, H. "Practical issues of recursive DFT in active power filter based on CPC Power Theory", Asia-Pacific Power and Energy Conf., Wuhan, China, 2009. DOI: 10.1109/appeec.2009.4918621.

[242] Czarnecki, L.S. "Effect of supply voltage harmonics on IRP-based switching compensator control", *IEEE Trans. on Power Electronics*, Vol. 24, No. 2, 2009. 483–488, 2009.

[243] Czarnecki, L.S. "Comments on the paper: 'Instantaneous p-q Theory for compensating nonsinusoidal systems'", *Przegląd Elektrotechniczny*, Vol. 85, 167–169, 2009.

[244] Herrera, R.S., Salmeron, P. "Present point of view about the instantaneous reactive power theory", *IET Power Electronics*, Vol. 2, 484–495, 2009.

[245] Czarnecki, L.S. "Effect of supply voltage asymmetry on IRP p-q-based switching compensator control", *IET Proc. on Power Electronics*, 2009.

[246] Mauri, G., Moneta, D., Bettoni, C. "Energy conservation and smart grids: new challenges for multimetering infrastructures", IEEE Power Tech. Conference, Bucharest, 1–5, 2009.

[247] Sainz, L., Caro, M., Caro, E. "Analytical study of series resonance in power systems with the Steinmetz circuit", *IEEE Trans. on Power Del.*, Vol. 24, No. 4, 2090–2099, 2009.

[248] Czarnecki, L.S. "Comments on the paper: 'About the possibility of developing a uniform approach to the reactive power definition in time-domain'", *Przegląd Elektrotechn.*, Vol. 85, No. 6, 164–166, 2009.

[249] Popper, K. *The Logic of Scientific Discovery*. Routledge Classic, London, 2010.

[250] Czarnecki, L.S., Pearce, S.E. "CPC-based comparison of compensation goals in systems with nonsinusoidal voltages and currents", *Przegląd Elektrotechniczny*, Vol. 86, No. 6, 22–29, 2010.

[251] Mikulovic, J.C., Sekara, T.B. "A new formulation of apparent power for nonsinusoidal unbalanced polyphase systems", *Proc. of Intern. School on Nonsinusoidal Currents and Compensation (ISNCC)*, Łagów, Poland, 180–185, 2010.

[252] Mohammadnezhah, H., Ftuhi-Firazbad, M. "Impact of penalty-reward mechanism on the performance of electric distribution systems and regulator budget", *IET Gen. Trans. Distr.*, Vol. 4, No. 7, 770–779, 2010.

[253] De Leon, F., Cohen, J. "AC power theory from Poynting Theorem; Accurate identification of power components in nonlinear circuits", *IEEE Trans. on Pow. Del.*, Vol. 25, No. 4, 2104–2112, 2010.

[254] Arendse, C., Atkinson-Hope G. "Design of Steinmetz symmetrizer and application in unbalanced network", *Proc. of UPEC 2010 Conference*, Cardiff, Wales, 2010.

[255] Tenti P., Parades H.K.M., Marafao F.P., et al. "Accountability in smart microgrids based on Conservative Power Theory", *IEEE Trans. on Instr. Meas.*, Vol. 60, No. 9, 3059–3069, 2011.

[256] Jeon, S-J., Willems, J.L. "Reactive power compensation in multi-line systems under sinusoidal unbalanced conditions", *Int. Journal on Circuit Theory and Applications*, Vol. 39, 211–224, 2011.

[257] Czarnecki, L.S. "Power theories and meta-theory of powers in electrical circuits", *Przegląd Elektrotechniczny*, Vol. 87, No. 8, 197–200, 2011.

[258] Orts-Grau, S., Gimeno-Sales, F.J., Sugei-chiet, S. "Selective shunt active power compensator applied in four-wire electrical systems based on IEEE Std. 1459", *IEEE Trans. on PD.*, Vol. 23. No. 4, 2563–2574, 2011.

[259] Ginn, H.L. "CPC-based converter control for systems with non-ideal supply voltage", *Przeglad Elektrotechniczny*, Vol. 87, No. 1, 8–13, 2011.

[260] Czarnecki, L.S. "Is the active power useful? For what we should pay?" (in Polish), *Automatyka, Elektryka, Zakłócenia*, (Control, Electrical Eng., Disturbances), Vol. 5, No. 5, 100–108, 2011.

[261] Castro-Nunez, M. Castro-Puche, R. "The IEEE Standard 1459, the CPC Power Theory, and geometric algebra in circuits with nonsinusoidal sources and linear loads", *IEEE Trans. on Circuits and Syst.*, Vol. 59, No. 12, 2980–2990, 2012.

[262] Czarnecki L.S. "Working, reflected and detrimental active powers", *IET on Generation, Transmission and Distribution*, Vol. 6, No. 3, 233–239, 1912.

[263] Guo, J., Xiao, X., Shun, T. "Discussion on instantaneous reactive power theory and currents' physical components theory", *Proc. of 15th Int. Conf. on Harmonics and Quality of Power (ICHQP)*, Hong Kong, 427–431, 2012.

[264] Kukacka, L., Kraus, J., Bubla, V., et al. "CPC and IEEE power theory–application for offline waveform data analysis", 22nd Int. Conf. on Electricity Distribution, Stockholm, Paper No. 434, 2013.

[265] Masoudipour, I., Samet, H. "Comparison of various reactive power definitions in non-sinusoidal networks with the practical data of electrical arc furnaces", 22nd Int. Conf. on Electricity Distribution, Stockholm, 2013. DOI: 10.1049/cp.2013.1144.

[266] Czarnecki, L.S. "Meta-theory of electric powers and present state of power theory of circuits with periodic voltages and currents", *Przegląd Elektrotechniczny*, Vol. 89, No. 6, 26–31, 2013.

[267] Mostafa, G. "In quest of experimental support for the increased real and reactive power in the Czarnecki Power Model under nonsinusoidal waveforms", 2nd Int. Conf. on Advances in Electrical Engineering (ICAEE), 2013. DOI: 10.1109/ICAEE.2013.6750335.

[268] Grunbaum, R., Ekstrom, P., Hellstrom, A-A. "Powerful reactive power compensation of a very large electric arc furnace", Int. Conf. on Power Eng., Energy and Electr. Drives, Istanbul, Turkey, 2013. DOI: 10.1109/PowerEng.2013.6635619.

[269] Czarnecki, L.S. "Constraints of the Instantaneous Reactive Power p-q Theory", *IET Power Electronics*, Vol. 7, No. 9, 2201–2208, 2014.

[270] Jeltsema, D., van der Wounde, J. "Currents' Physical Components (CPC) in the time domain: single-phase systems", European Control Conference (CC), Strasburg, 2014.

[271] Czarnecki, L.S., Toups, T.N. "Working and reflected active powers of harmonics generating single-phase loads", *Przegląd Elektrotechniczny*, Vol. 90, No. 10, 7–10, 2014.

[272] Tenti P., Costabeber A., Mattavelli P., et al. "Load characterization and revenue metering under non-sinusoidal and asymmetrical operation", *IEEE Trans. on IM.*, 2014.

[273] Czarnecki, L.S., Haley, P.H. "Currents' Physical Components (CPC) in four-wire systems with nonsinusoidal symmetrical voltage", *Przegląd Elektrotechniczny*, Vol. 91, No. 6, 48–55, 2015.

[274] Czarnecki, L.S., Bhattarai, P.D. "Currents' Physical Components (CPC) in three-phase systems with asymmetrical voltage", *Przegląd Elektrotechniczny*, Vol. 91, No. 6, 40–47, 2015.

[275] Czarnecki, L.S., Bhattarai, P.D. "Reactive compensation of LTI loads in three-wire systems at asymmetrical voltage" *Przegląd Elektrotechniczny*, Vol. 91, No. 12, 7–11, 2015.

[276] Czarnecki, L.S. "Currents' Physical Components (CPC) in systems with semi-periodic voltages and currents", *Przegląd Elektrotechniczny*, Vol. 91, No. 6, 25–31, 2015.

[277] Calamero, N. "Defining the unique signatures of loads using the Currents' Physical Components theory and Z-Transform", *IEEE Trans. on Ind. Informatics*, Vol. 11, No. 1, 155–165, 2015.

[278] Czarnecki, L.S., Haley, P.H. "Unbalanced power in four-wire systems and its reactive compensation", *IEEE Trans. on Pow. Del.*, Vol. 30, No. 1, 53–63, 2015.

[279] Czarnecki, L.S., Toups, T.N. "Working and reflected active powers of three-phase loads", *Przegląd Elektrotechniczny*, Vol. 91, No. 11, 149–153, 2015.

[280] Martell, F., Izaguirre, A.R., Macias, M.E. "CPC Power Theory for analysis arc furnaces", *Przeglad Elektrotechniczny*, Vol. 92, No. 6, 138–142, 2016.

[281] Li., W., Rahmani, B., Liu, G. "Expansion of Current Physical Components (CPC) to three-phase four-wire systems under non-ideal waveforms by AUPQS", *Proc. of Sixth Int. Conf., on Instr. & Measurement, Computer, Communication and Control*, Harbin, China, 99–103, 2016.

[282] Czarnecki, L.S., Haley, P.H. "Power properties of four-wire systems with nonsinusoidal symmetrical voltage", *IEEE Trans. on Pow. Del.*, Vol. 31, No. 2, 513–521, 2016.

[283] Kukacka, L., Kraus, J., Zissis, G. "Review of AC power theories under stationary and non-stationary, clean and distorted conditions", *IET on Generation, Transmission and Distribution*, Vol. 10, No. 1, 221–231, 2016.

[284] Mikulovic, J., Sekava, T., Skrbio, B. "Currents' Physical Components (CPC) power theory for three-phase four-wire systems", Mediterranean Conf. on Power Generation Transmission Distribution, Belgrade, 2016. DOI: 101049/cp.2016.1061.

[285] Czarnecki, L.S. "Comparison of the Conservative Power Theory (CPT) with Budeanu's power theory", *Annals of the University of Craiova*, No. 40, 1–8, 2016.

[286] Wang, L.M., Lam, C.S., Wong, M.C. "Selective compensation of distortion, unbalanced and reactive power of Thyristor-controlled LC coupling hybrid active power filter (TCLC-HAPF)", *IEEE Trans. on Pow. Electronics*, 32(12), 9065–9077, 2017.

[287] Czarnecki, L.S., Bhattarai, P.D. "A method of calculation of LC parameters of balancing compensators for AC arc furnaces", *IEEE Trans. on PD.*, Vol. 32, No. 2, 688–695 2017.

[288] Al-BayAty, H., Ambroze, M., Ahmed, M.Z. "New effective power terms and right-angled triangle (RAT) power theory", *Int. Jour. Electrical Power and Energy Syst.*, Vol. 88, 133–140, 2017.

[289] Bhattarai, P.D., Czarnecki, L.S. "Currents' Physical Components (CPC) of the supply current of unbalanced LTI loads at asymmetrical and nonsinusoidal voltage", *Przegląd Elektrotechniczny*, Vol. 93, No. 9, 30–35, 2017.

[290] Bucci, G., Ciancetta, F., Ometto, A. "Survey about classical and innovative definitions of the power quantities under nonsinusoidal conditions", *Int. Journal of Electrical Power and Energy Systems*, Vol. 18, No. 3, 13–23, 2017.

[291] Czarnecki, L.S. "Critical comments to the Standard DIN 401100. Alleged effect of energy oscillation on the power factor and critical verification of a physical meaning of reactive power", *Electrical Eng., Control, and Disturbances*, http://www.elektro-innowacje.pl, 2017.

[292] Pana, A., Molnar Matei F., Baloi, A. "A smart solution for a smart grid: unbalanced reactive power compensation", Electric Vehicles Int. Conf. and Show, EV2017, Bucuresti, 2017.

[293] Czarnecki, L.S. "CPC-founded clarification of decline of the effectiveness of the energy transfer in power systems", *Annals University of Craiova*, Vol. 41, No. 1, 1–8, 2017.

[294] Mikulovic, J., Skrbic, B., Durisic, Z. "Power definitions for polyphase systems based on Fortescue's symmetrical components", *Int. Journal of Electrical Power and Energy Systems*, Vol. 98, 455–462, 2018.

[295] Czarnecki, L.S., Almousa, M., Gadiraju, V.M. "Why the electric arc nonlinearity improves the power factor of AC arc furnaces?", IEEE Intern. Conf. on Environment and Electrical Engineering (EEEIC 2018), Palermo, Session N2-TS4 576, 2018.

[296] Zajkowski, K., Rusica, I., Palkova, Z. "The use of CPC theory for energy description of two nonlinear receivers", MATEC Web Conf., Vol. 178, 2018. DOI 10.1051/matecconf/20181178090008.

[297] Czarnecki, L.S., Gadiraju, V.M., Shindi, A. "Effectiveness of harmonic filters of AC arc furnaces at an uneasy mode of operation", IEEE Intern. Conf. on Environment and Electrical Engineering (EEEIC 2018), Palermo, Session N2-TS4 517, 2018.

[298] Das, S., Chatteerjee, D., Goswami, S.K. "Are reactive power compensation scheme for unbalanced four-wire system using virtual Y-TCR model", *IEEE Trans in Ind. Electronics*, Vol. 65, No. 4, 3210–3219, 2018.

[299] Czarnecki, L.S., Ezeonwumelu, I.L. "Considerations on a direct balancing of ultra-high power AC arc furnaces in an uneasy state", IEEE Intern. Conf. on Environment and Electrical Engineering (EEEIC 2018) Palermo, Session N2 TS4 519, 2018.

[300] Grasso, F., Luchetta, A., Manetti, S. "Improvement of power flow analysis based on Currents' Physical Components (CPC) theory", IEEE Int. Symp. Circ. Syst. (ISCAS), Florence, 2018, DOI: 10.1109/ISCAS.20188351223.

[301] Martell Chavez, F., Macias Garcia, M.E., Izaguirre Alegria, A. "Electrical efficiency of arc furnaces considering the load generated currents defined by the CPC power theory, IEEE Intern. Conf. on Environment and Electrical Engineering (EEEIC 2018), Palermo, Session N2 TS4 607, 2018.

[302] Czarnecki, L.S. "CPC–based reactive balancing of linear loads in four-wire supply systems with nonsinusoidal voltage", *Przegląd Elektrotechniczny*, Vol. 95, No. 9, 1–8, 2019.

[303] Czarnecki, L.S. "Currents' Physical Components (CPC)-based power theory. A review, Part I: Power properties of electrical circuits and systems", *Przegląd Elektrotechniczny*, Vol. 95, No. 10, 1–11, 2019.

[304] Czarnecki, L.S., Bhattarai, P.D. "CPC–based reactive compensation of linear loads supplied with asymmetrical nonsinusoidal voltage", *Annels University of Craiova*, Vol. 42, No. 10, 1–9, 2019.

[305] Skarbic, B., Mikulovic, J., Sekara, T. "Extension of the CPC power theory to four-wire power systems with non-sinusoidal and unbalanced voltages", *Int. Journal of Electrical Power and Energy Systems*, Vol. 105, 341–350, 2019.

[306] Czarnecki, L.S. "On physical phenomena that determine the effectiveness of energy transfer in electrical systems", *Automatyka, Elektryka, Zakłócenia*, 2019. <http://www.epismo–aez.pl>.

[307] Jeon, S.-J. "Passive-component-based reactive power compensation in a non-sinusoidal multi-line system", *Electrical Engineering*, Vol. 102, 1567–1577, 2020.

[308] Malengret, M., Gaunt, C.T. "Active currents, power factor, and apparent power for practical power delivery systems", *IEEE Access*, Vol. 8, 133095–133113, 2020.

[309] Czarnecki, L.S., Almousa, M. "What is wrong with the paper 'The IEEE Standard 1459, the CPC power theory and geometric algebra in circuits with nonsinusoidal sources and linear loads?'", *Przegląd Elektrotechniczny*, Vol. 96, No. 7, 1–7, 2020.

[310] Czarnecki, L.S. "Currents' Physical Components (CPC)–based power theory. A review, Part II: Filters and reactive, switching, and hybrid compensators", *Przegląd Elektrotechniczny*, Vol. 96, No. 4, 1–10, 2020.

[311] Soljan, Z., Hodynski, G., Zajkowski, M. "Balancing reactive compensation at three-phase four-wire systems with a sinusoidal and asymmetrical voltage source", *Bulletin of Polish Academy of Sciences—Technical Sciences*, Vol. 68, No. 1, 71–79, 2020.

[312] Sapurov, M., Bleizgys, V., Macaitis, V. "Asymmetric compensation of reactive power using Thyristor-controlled reactors", *Symmetry—Basel*, Vol. 12, No. 6, 2020.

[313] Skopec, A., Stec, C. "Theoretical background for compensation methods of reactive power of non-sinusoidal currents in the frequency and time domain. Energetic interpretation of the Wroclaw mathematical identity", *Przegląd Elektrotechniczny*, Vol. 96, No. 2, 1–12, 2020.

[314] Bielecka, A., Wojciechowski, D. "Compensation of supply current harmonics, reactive power, and unbalance load current balance in the closed-loop control of a shunt active power filter", *Scientific Journal of the Maritime Szczecin Univ.*, Vol. 61, No. 133, 9–16, 2020.

[315] Czarnecki, L.S., Gadiraju, V. "A study on direct reduction of harmonics of ultra-high power AC arc furnaces in an uneasy mode of operation", Int. Conf. Environment and Electrical Engineering IEEE-EEEIC, Madrid, June 2020.

[316] Coelho, R.D., Brito, N.S.D., Lima, E.M. "Effects of current decomposition on power calculation in nonsinusoidal conditions", *Electrical Engineering*, 102(4), 2325–2339, 2020.

[317] Martel-Chavez, F., Marcias-Garcia, M., Izaguirre-Alegria, A. "Performance parameters of electric arc furnaces based on the Currents' Physical Components power theory", *IEEE Trans. on Industry Applications*, Vol. 56, No. 6, 6076–6082, 2020.

[318] Soljan, Z., Hodynski, G., Zajkowski, M. "CPC-based currents' description in three-phase four-wire systems with asymmetrical nonsinusoidal waveforms", *Bulletin of Polish Academy of Sciences—Technical Sciences*, Vol. 68, No. 5, 1127–1134, 2020.

[319] Pana, A., Baloi. A., Molnar-Matei, F. "New method of the susceptance of a balancing capacitive compensator for a three-phase four-wire distribution network", *American Journal of Electrical Power and Energy Systems*, Vol. 115, 1–16, 2020.

[320] Zajkowski, K. "Two-stage reactive compensation in three-phase four-wire systems at nonsinusoidal periodic waveforms" in *Electric Power Systems Research*. V. 184, Elsevier, 2020. DOI: 10.1016/j.epsr.

[321] Czarnecki, L.S., Almousa, M. "Conversion of fixed-parameters compensator in four-wire system with nonsinusoidal voltage into adaptive compensator", *Przegląd Elektrotechniczny*, Vol. 97, No. 7, 1–6, 2021.

[322] Soljan, Z., Holdynski, G., Zajkowski, M. "CPC-based minimizing of balancing compensators in four-wire nonsinusoidal asymmetrical systems", *Energies*, Vol. 14, No. 7, 2021. DOI: 10.3390/en14071815.

[323] Czarnecki, L.S., Almousa, M. "Adaptive balancing by reactive compensators of three-phase linear loads supplied by nonsinusoidal voltage from four-wire lines", *American Journal of Electrical Power and Energy Systems*, Vol. 10, No. 3, 32–42, 2021.

[324] Czarnecki, L.S. "Do energy oscillations degrade energy transfer in electrical systems?", *IEEE Transactions on Industry Applications*, IEEE, Trans. IA, Vol. 57, No. 2, 1314–1324, 2021.

[325] Haase, H. "Definition of reactive power as a mean value", *Elektrotechnik und Informationstechnik*, Vol. 136, No. 6, 438–441, 2021.

[326] Petroianu, A.I. *Bridging Circuits and Fields. Fundamental Questions in Power Theory*. CRC Press, Boca Raton, 2022.

[327] Czarnecki, L.S. Almousa, M. "Adaptive compensation of the DC current component of AC arc furnace", *Przegląd Elektrotechniczny*, Vol. 99, No. 2, 14–20, 2023. DOI: 10.15199/48.2023.02.02.

Index